The sun is just a star.

Nuclear reactions make energy.

Earth · · · · · · · Moon
(not to scale)

First hominids

One-inch line

Goal line

Ten thousand years ago, on the 0.0026 inch line, humans begin building cities and modern civilization begins.

50
40
30
20
10

Formation of the sun and planets from a cloud of interstellar gas and dust

Life begins in Earth's oceans.

Cambrian explosion 540 million years ago: Life in Earth's oceans becomes complex.

Life first emerges onto the land.

Age of Dinosaurs

TODAY

40
30
20
10

Over billions of years, generation after generation of stars have lived and died, cooking the hydrogen and helium of the big bang into the atoms of which you are made. Study the last inch of the time line to see the rise of human ancestors and the origin of civilization. Only in the last flicker of a moment on the time line have astronomers begun to understand the story.

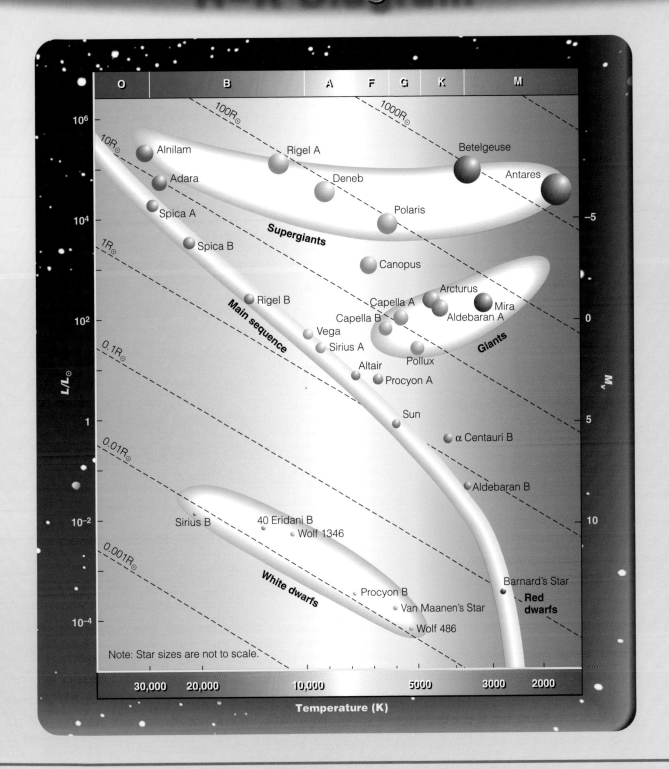

The H–R diagram is the key to understanding stars, their birth, their long lives, and their eventual deaths. Luminosity ($L/L_\odot$) refers to the total amount of energy that a star emits in terms of the sun's luminosity, and the temperature refers to the temperature of its surface. Together, the temperature and luminosity of a star locate it on the H–R diagram and tell astronomers its radius, its family relationships with other stars, and a great deal about its history and fate.

The terrestrial or Earthlike planets lie very close to the sun, and their orbits are hardly visible in a diagram that includes the outer planets.

Mercury, Venus, Earth and its moon, and Mars are small worlds made of rock and metal with little or no atmospheric gases.

The outer worlds of our solar system orbit far from the sun. Jupiter, Saturn, Uranus, and Neptune are Jovian or Jupiter-like planets much bigger than Earth. They contain large amounts of low-density gases.

Pluto is a tiny, icy world orbiting far from the sun, and some astronomers argue that it isn't really a planet at all.

This book is designed to use arrows to alert you to important concepts in diagrams and graphs. Some arrows point things out, but others represent motion, force, or even the flow of light. Look at arrows in the book carefully and use this Flash Reference card to catch all of the arrow clues.

The Terrestrial Worlds

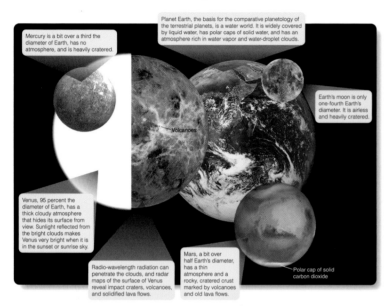

Mercury is a bit over a third the diameter of Earth, has no atmosphere, and is heavily cratered.

Planet Earth, the basis for the comparative planetology of the terrestrial planets, is a water world. It is widely covered by liquid water, has polar caps of solid water, and has an atmosphere rich in water vapor and water-droplet clouds.

Earth's moon is only one-fourth Earth's diameter. It is airless and heavily cratered.

Volcanoes

Venus, 95 percent the diameter of Earth, has a thick cloudy atmosphere that hides its surface from view. Sunlight reflected from the bright clouds makes Venus very bright when it is in the sunset or sunrise sky.

Radio-wavelength radiation can penetrate the clouds, and radar maps of the surface of Venus reveal impact craters, volcanoes, and solidified lava flows.

Mars, a bit over half Earth's diameter, has a thin atmosphere and a rocky, cratered crust marked by volcanoes and old lava flows.

Polar cap of solid carbon dioxide

Planetary Orbits

Venus
Sun
1 AU
Mercury
Earth

Enlarged to show relative size
Earth
Sun

Pluto in January 1979
Area of Figure 1-6
Mars
Jupiter
Saturn
Uranus
Neptune
Pluto
Pluto in March 1999

The Outer Worlds

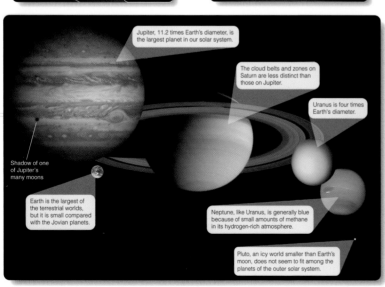

Jupiter, 11.2 times Earth's diameter, is the largest planet in our solar system.

The cloud belts and zones on Saturn are less distinct than those on Jupiter.

Uranus is four times Earth's diameter.

Shadow of one of Jupiter's many moons

Earth is the largest of the terrestrial worlds, but it is small compared with the Jovian planets.

Neptune, like Uranus, is generally blue because of small amounts of methane in its hydrogen-rich atmosphere.

Pluto, an icy world smaller than Earth's moon, does not seem to fit among the planets of the outer solar system.

Point at things:

You are here

Force:

Process flow:

Measurement:

Direction:

Radio waves, infrared, photons:

Motion:

Rotation 2-D   *Rotation 3-D*   *Linear*

Light flow:
Updated arrow style

Focal length

• See page 478 for the terrestrial planets. See pages 6 and 7 for the two orbital diagrams. See page 555 for the outer worlds.

What are we?
How do we know?

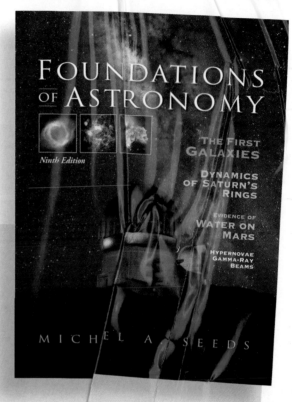

Using astronomy to understand science— using science to understand who we are

Foundations of Astronomy, **Ninth Edition**, not only shows students their place in the universe—it helps them answer the ultimate question of human existence: *What are we?* While great minds throughout history have tackled this very question, best-selling author Mike Seeds leads students on their *own* exploration of what we are and how we know. With this fascinating and engaging approach, Seeds weaves the student experience into the very fabric of astronomy, giving readers a conceptual understanding of their place in the big picture.

Seeds uses the interplay of evidence and theory to help students understand what we know about the universe and to show them how science works as a way to understand nature. He avoids dry recitation of the facts and continually illustrates how students can comprehend their place in the universe by grasping a small set of physical laws. *Foundations of Astronomy* incorporates mathematics, but the book's arguments do not depend on mathematical reasoning, keeping even math-averse students engaged.

Build the perfect course solution to fit your needs

Because every course is as unique as its instructor— and its students—you need a program tailored to your distinct needs. *Foundations of Astronomy* is supported by a range of teaching and learning solutions that allow you to craft the best course for your approach and workflow needs.

The text provides the foundation—our powerful technology tools are the building blocks that help you customize *your* course our specifications:

Enhanced WebAssign market-leading homework management system allows you to transform the way you track your student progress . . . *see page 8 for details.*

 AceAstronomy™, an online tutorial and assessment resource that is seamlessly the text through *Active Figures* . . . *see the inside back or details.*

Virtual Astronomy Laboratories, an interactive way for students to explore and discover the universe from their own computers . . . *inside back cover for details.*

Planetarium e, including **TheSky**™ Student Edition, St ™, and **RedShift**™ College Edition . . . 7 *for more information.*

Join ecture aids, such as **JoinIn**™ on urningPoint® for audience response systems, and the powerful **Multimedia** structor's Resources CD-ROM . . . *see* page 8 formation.

THOMSON

™

BROOKS/COLE

PREVIEW

1

What are we?

Students and professors find great clarity and accessibility in Seeds's presentation of astronomy. His engaging writing style and powerful use of visuals reveal our place in the universe and help students experience the excitement of discovery.

▲ This edition's **all-new art program** presents this visual science in a consistent and striking new style. The book's renowned concept art spreads are now numbered and emphasized in the text narrative to make these two-page visual summaries easier to analyze and synthesize. Enhanced spreads combine several ideas over two pages, providing an opportunity for students to process information visually.

Planetarium software help students understand their place in the universe

Three different planetarium software options are available on request, allowing you even more flexibility designing your course. Choose from **TheSky™ Student Edition** or **Starry Night™** (both at no additional cost) or **RedShift™ College Edition** (an additional cost). *Exploring TheSky* at the end of each chapter provide interesting and relevant exercises for use with the planetarium software package of your choosing. See page 7 for more information about these programs.

■ **Active Figure 12-9**

When a star runs out of hydrogen at its center, the core of helium contracts to a small size, becomes very hot, and begins nuclear fusion in a shell (blue). The outer layers of the star expand and cool. The red giant star shown here has an average density much lower than the air at Earth's surface. Here $M_\odot$ stands for the mass of the sun, and $R_\odot$ stands for the radius of the sun. (Illustration design by the author)

Ace★Astronomy™ Log into AceAstronomy and select this chapter to see Active Figure "Inside Stars." Compare the interior of stars with that of the sun.

■ **Figure 12-10**

Stars like the sun suffer a helium flash, but more massive stars begin helium fusion without a helium flash. Stars less than 0.4 times the mass of the sun ... sequence is

Many of the text's figures are brought to life through *Active Figures*, found on **AceAstronomy** and on the **Instructor's Multimedia Manager CD-ROM**.

▲ The H–R diagrams have been **completely redesigned** to help students understand the main idea at a glance. Plus, a new tear-out reference card in the front of the book serves as a quick and portable reference for the key H–R diagram, comparative planetology, and the new arrow styles.

■ **Figure 21-12**

Mercury's orbital motion and rotation are related. It orbits the sun in 88 days and rotates on its axis in two-thirds of that time. One full day on Mercury from noon to noon takes two full orbits.

▶ Compelling guided discovery figures lead students step-by-step through several frames of art to demonstrate processes that occur over time. Seeds provides brief guiding comments as though he himself were explaining each frame to the student.

Mercury is a bit over one-third the diameter of Earth, but its high density must mean it has a large iron core. The amount of heat it retains is unknown. (NASA)

Celestial Profile 4: Mercury

Motion:

Average distance from the sun	0.387 AU (5.79 × 10⁷ km)
Eccentricity of orbit	0.2056
Maximum distance from the sun	0.467 AU (6.97 × 10⁷ km)
Minimum distance from the sun	0.306 AU (4.59 × 10⁷ km)
Inclination of orbit to ecliptic	7°0′16″
Average orbital velocity	47.9 km/s
Orbital period	87.969 days (0.24085 y)

NEW *Celestial Profiles* (formerly *Data Files*) use Earth as a basis of comparison to help students understand the characteristics of other planets. Each profile features relevant and fun facts about the planet being discussed.

Mercury lies very close to the sun and completes an orbit in only 88 days. For this reason, the ancients named the planet after Mercury, the fleet-footed messenger of the gods. The name is also applied to the element mercury, which is also known as quicksilver because it is a heavy, quickly flowing silvery liquid at room temperatures.

can be explained by Newton's laws but at just the rate predicted by Einstein's theory of general relativity. The orbital motion of Mercury is taken as strong confirmation of the curvature of space-time as predicted by general relativity.

VISUALLY COMPELLING

3

How do we know?

Seeds makes the scientific process accessible by showing students that science is simply a logical attempt to understand nature. He includes a wealth of features to help students grasp this logic.

Building Scientific Arguments

How does inflation theory solve the flatness problem?
This question requires a carefully constructed argument. The flatness problem can be stated as a question: Why is the universe so nearly flat? After all, the density of matter in the universe could be anything from zero to infinite, but the observed density of baryonic and dark matter adds up to about 27 percent of the critical density that would produce a flat universe. Furthermore, the density must have been astonishingly close to the critical density when the universe was very young. Otherwise it would not be so close now. The inflationary theory solves this problem by proposing that the universe underwent a moment of rapid inflation when it was a tiny fraction of a second old. That inflation drove the universe toward flatness just as blowing up a balloon makes a spot on the balloon nearly flat.

Understanding theory in cosmology is critically important, but science depends on evidence. Now build an argument based on evidence. **What evidence can you cite that the expansion of the universe is accelerating?**

■   ■   ■

Figure 18-26
This graph shows how commonly irregularities of different sizes occur in the cosmic microwave background radiation. Irregularities of about 1 degree in diameter are most common. Models of the universe that are open or closed are ruled out. The data fit a flat model of the universe very well. Crosses on data points show the uncertainty in the measurements.

acceleration, so over 95 percent of the universe is not understood. Hearing this, another astronomer suggested a better phrase was "cosmology in crisis." Certainly there are further mysteries to be explored, but cosmologists are growing more confident that they can describe the overall properties of the universe.

Connections: As you climbed the cosmology pyramid, you negotiated a lot of steps. Most were easy and all were logical. They have carried you up to some sweeping insights, and you have traced the origin of the universe, the origin of the elements, the birth of galaxies, and the birth and death of stars. Now it is time for you to add one final class of objects—planets. How do Earth and other planets fit into this grand scheme of origins? That is the subject of the following chapters.

◀ *Building Scientific Arguments* questions at the end of each section contain a carefully designed question to help students learn how to construct scientific arguments. A short answer follows each question to emphasize how scientists construct logical arguments from observations, evidence, theories, and natural laws. A further question then gives students a chance to construct their own arguments on a related issue.

Building Scientific Arguments

What evidence can you cite that life is at least possible on other worlds?
A good scientic argument is based on evidence, but it also includes careful analysis. The evidence is limited almost entirely to Earth, but it is promising. Fossils show that life originated in Earth's oceans almost 4 billion years ago, and biologists have proposed relatively simple chemical processes that could have formed these first reproducing molecules. Fossils show that life developed from a very slow beginning into more and more complex creatures that filled the oceans. The pace of evolution quickened dramatically about half a billion years ago, the beginning of the Cambrian period, when life began taking on complex forms; later, when life emerged onto the surface of the land, it evolved rapidly to produce the tremendous diversity of today. Human intelligence has been a very recent development; it is only a few million years old.

If this process occurred naturally on Earth, then it seems reasonable that it could have occurred on other worlds as well. This conclusion hinges mostly on a reasonable belief that the laws of nature apply on other worlds as they do on Earth. Now expand your argument: **What conditions should you expect of other worlds that host life?**

■   ■   ■

Window on Science | 18-2

Why Scientists Speak Carefully: Words Lead Thoughts

There are certain words you should never say, not even as a joke. You should never call your friend "fool," for example, because you might begin to think of your friend as a fool. Words lead thoughts. It works in advertising and politics, and it can work in science, too. Using words carelessly can lead you to think carelessly.

In science, there are certain ways to say things, not because scientists are sticklers for good grammar, but because if they say things wrongly they may begin to think things wrongly. For example, a biologist would never let you say a beehive knows it must store food. "No, don't say it that way," the biologist would object. "The hive is just a collection of individual creatures, and it can't 'know' anything.

Even the individual bees don't really 'know' things. The instinctive behavior of individual bees causes food to accumulate in the hive. That's the way to say it."

All scientists are careful of language because careless words can mislead. You would never refer to the center of the universe, for example, but you must also be careful not to say things like "galaxies flying away from the big bang." Those words imply a center to the expansion of the universe, and you know the expansion must be centerless. Rather, you should say "galaxies flying away from each other."

Whatever science you study, notice the customary ways of using words. It is not just a matter of convention. It is a matter of careful thought.

You should not say that a bee "knows" how to make honey. How you speak guides how you think.

▲ *Window on Science* boxes draw on examples from everyday life to illustrate the logical processes that scientists use to learn about nature, building on the "how do we know?" theme of the text. Boxes include discussions such as how scientists use statistical evidence, how they reason by analogy, how they speak, and how they build a scientific model in place of scientific hypothesis.

10. How can astronomers detect neutrinos from the sun?
11. How can neutrino oscillation explain the solar neutrino problem?
12. What evidence can you give that sunspots are magnetic?
13. How does the Babcock model explain the sunspot cycle?
14. What does the spectrum of a prominence reveal? What does its shape reveal?
15. How can solar flares affect Earth?
16. The upper two images here show two solar phenomena. What are they, and how are they related? How do they differ?
17. The lower image was recorded in the extreme ultraviolet by the SOHO spacecraft. Explain the features you see.

(Images courtesy Daniel Good and NOAO)

(NASA/SOHO)

NEW New "learning to look" questions at the end of each chapter ask students to look at a photo and analyze what they see. Astronomy is a visual science, and these new questions teach students to identify critical concepts visually.

How does it fit into the big picture?

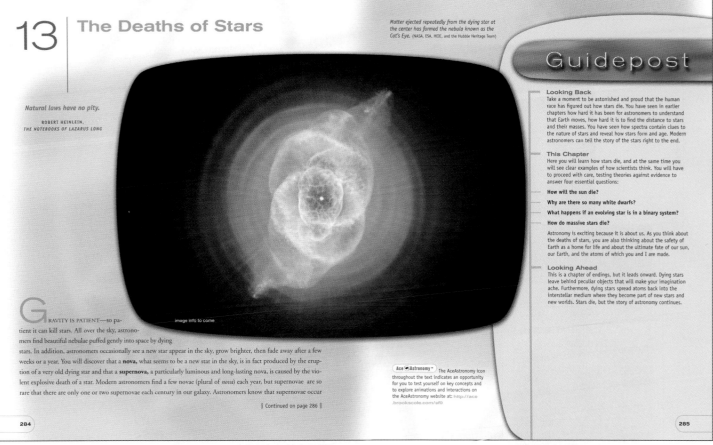

13 | The Deaths of Stars

Matter ejected repeatedly from the dying star at the center has formed the nebula known as the Cat's Eye. (NASA, ESA, HEIC, and the Hubble Heritage Team)

Natural laws have no pity.

ROBERT HEINLEIN,
THE NOTEBOOKS OF LAZARUS LONG

GRAVITY IS PATIENT—so patient it can kill stars. All over the sky, astronomers find beautiful nebulae puffed gently into space by dying stars. In addition, astronomers occasionally see a new star appear in the sky, grow brighter, then fade away after a few weeks or a year. You will discover that a **nova**, what seems to be a new star in the sky, is in fact produced by the eruption of a very old dying star and that a **supernova**, a particularly luminous and long-lasting nova, is caused by the violent explosive death of a star. Modern astronomers find a few novae (plural of *nova*) each year, but supernovae are so rare that there are only one or two supernovae each century in our galaxy. Astronomers know that supernovae occur

‖ Continued on page 286 ‖

image info to come

284

Guidepost

Looking Back
Take a moment to be astonished and proud that the human race has figured out how stars die. You have seen in earlier chapters how hard it has been for astronomers to understand that Earth moves, how hard it is to find the distance to stars and their masses. You have seen how spectra contain clues to the nature of stars and reveal how stars form and age. Modern astronomers can tell the story of the stars right to the end.

This Chapter
Here you will learn how stars die, and at the same time you will see clear examples of how scientists think. You will have to proceed with care, testing theories against evidence to answer four essential questions:

How will the sun die?

Why are there so many white dwarfs?

What happens if an evolving star is in a binary system?

How do massive stars die?

Astronomy is exciting because it is about us. As you think about the deaths of stars, you are also thinking about the safety of Earth as a home for life and about the ultimate fate of our sun, our Earth, and the atoms of which you and I are made.

Looking Ahead
This is a chapter of endings, but it leads onward. Dying stars leave behind peculiar objects that will make your imagination ache. Furthermore, dying stars spread atoms back into the interstellar medium where they become part of new stars and new worlds. Stars die, but the story of astronomy continues.

Ace ◆ Astronomy™ The AceAstronomy icon throughout the text indicates an opportunity for you to test yourself on key concepts and to explore animations and interactions on the AceAstronomy website at: http://ace .brookscole.com/sf9

285

▲ This edition emphasizes an understanding of the **"big picture"** using the scientific process. Each chapter-opening *Guidepost* connects previous and future chapters to the material the students are about to read, highlighting the essential questions within that chapter. Plus, *Connections* sections appear at the end of each chapter section to further illustrate how one concept seamlessly connects to the next.

▶ The essential questions act as the organizing elements in the chapter *Summaries* to encourage students to gain a deeper understanding of the material rather than just memorize a list of facts.

Study and Review Tools

Summary

13-1 | Lower-Main-Sequence Stars

How will the sun die?
■ Red dwarfs less massive than about 0.4 solar mass are completely mixed and will have very little hydrogen left when they die. They cannot ignite a hydrogen-fusion shell, so they cannot become giant stars. They will remain on the main sequence for many times the present age of the universe.

■ Medium-mass stars between about 0.4 and 4 solar masses, including the sun, become cool giants and fuse helium but cannot fuse carbon.

■ Medium-mass stars lose mass into space while they are giant stars and then collapse to become white dwarfs.

■ If the collapsed star becomes hot enough, it can ionize the gas it has ejected, first as a slow wind and then as a fast wind to turn on a planetary nebula.

Why are there so many white dwarfs?
■ White dwarfs are the cores of sunlike stars that have collapsed until their gas becomes degenerate. They have no nuclear fuels and slowly cool off; they will eventually become black dwarfs.

■ A white dwarf cannot support its own weight by degenerate pressure if its mass is greater than the Chandrasekhar limit of 1.4 solar masses.

■ Medium-mass stars up to about 8 solar masses can probably lose enough mass to die as white dwarfs. This explains the large number of white dwarfs found in space.

■ A type Ia supernova can occur when mass transferred onto a white dwarf pushes it over the Chandrasekhar limit and it collapses suddenly, fusing all of its carbon at once.

■ A type Ib supernova occurs when a massive star loses its outer layers of hydrogen in a binary system. Later its iron core can collapse, and it can explode as a type Ib supernova.

■ The spectra of type II supernovae contain hydrogen lines, but the spectra of type Ia and type Ib supernovae do not.

■ Supernova remnants are the expanding shells of gas ejected by supernova explosions. They are filled with hot gas. The Crab Nebula is a supernova remnant formed by the supernova of AD 1054.

■ The supernova of 1987 is believed to have been produced by a peculiar type II supernova explosion. Neutrinos were detected and are evidence that the iron core collapsed and presumably formed a neutron star. The nebula the supernova produced is still developing.

New Terms

nova (p. 284)	Roche surface (p. 293)
supernova (p. 284)	Lagrangian point (p. 293)
thermal pulse (p. 287)	accretion disk (p. 294)
planetary nebula (p. 288)	supernova (type I) (p. 299)
compact object (p. 289)	supernova (type II) (p. 299)
black dwarf (p. 289)	carbon deflagration (p. 300)
Chandrasekhar limit (p. 289)	synchrotron radiation (p. 301)

5

What have we discovered?

This ninth edition is fully updated with new images, new discoveries, and new understandings that take students to the very frontier of human knowledge.

Figure 23-18

The Huygens probe descended by parachute through the thick atmosphere of Titan and photographed the surface. From an altitude of 8 km (26,000 ft), the surface showed clear signs that a liquid, almost certainly liquid methane, had drained over the surface and into the lowlands. Once Huygens landed on the surface, it radioed back photos showing a level plain and chunks of ice rounded by a moving liquid. (ESA/NASA/JPL/University of Arizona)

Visual-wavelength images

Drainage channels were cut by flowing liquid methane.

Possible methane fog

Icy grapefruit-size rocks on Titan are bathed in orange light from its hazy atmosphere.

◀ Seeds shares with students the latest evidence of a new world beyond Pluto, new evidence of dark energy and the acceleration of the universe, methane rain on Titan and ice volcanoes on Enceladus, evidence of ancient water on Mars, and a look inside supernova explosions.

Visual-wavelength image   Visual-wavelength image

Host galaxy

Visual-wavelength image

Figure 14-26

Alerted by gamma-ray detectors on satellites, observers used one of the VLT 8.2-meter telescopes on a mountaintop in Chile to image the location of a gamma-ray burst only hours after the burst. The image at top left shows that fading glow of the eruption. The image at top right, recorded 13 years before, reveals no trace of an object at the location of the gamma-ray burst. The Hubble Space Telescope image at bottom was recorded a year later and reveals a very faint galaxy at the location of the gamma-ray burst. (ESO and NASA)

▶ Chapter 18, "Cosmology in the 21st Century," starts with a major section that leads students from their simple expectations about the universe to an insight into the nature of the big bang. True to the book's theme, the very first sentence helps students place themselves in the universe: *"Look at your thumb. The matter in your thumb was present in the fiery beginning of the universe."*

This chapter's second major section builds further, leading students into sophisticated ideas of space-time curvature. The third major section focuses on the newest theories and observations and brings students' understandings into the 21st century.

axies and quasars, with redshifts a bit over 10. In contrast, the radiation from the hot gas of the big bang has a redshift of about 1100. That means the light emitted by the big bang gases arrives at Earth as infrared radiation and short radio waves. You can't see it with your eyes, but it should be detectable with infrared and radio telescopes. Unlike the Gettysburg Address, the big bang isn't over, and Earth's astronomers should be able to see it happening if they can detect the radiation it emitted. That amazing discovery is the subject of the next section.

The Cosmic Background Radiation

If radiation is now arriving from the big bang, then it should be detectable. The story of that discovery begins in the mid-1960s

when two Bell Laboratories physicists, Arno Penzias and Robert Wilson, were using a horn antenna to measure the radio brightness of the sky (■ Figure 18-8). Their measurements showed a peculiar noise in the system, which they at first attributed to droppings from pigeons living inside the antenna. Perhaps they would have enjoyed cleaning the antenna more if they had known they would win the 1978 Nobel Prize for physics for the discovery they were about to make.

When the antenna was cleaned, they measured the brightness of the sky at radio wavelengths and again found the low-level radio noise. The pigeons were innocent, but what was causing the signal?

The explanation for the noise goes back to 1948, when George Gamow predicted that the early big bang would be very hot and

Microwave radiation from the sky enters the horn and is focused into the instrument room.

In 1965, Arno Penzias (right) and Robert Wilson first detected the background radiation with an unused horn antenna.

The horn could be rotated about two axes to scan the entire sky.

Launched in 1989, the COBE satellite showed the background radiation followed the black body curve.

$T = 2.725 \pm 0.002$ K

Intensity

Wavelength (cm)

0.05   0.1   0.5   1

All-sky COBE map of tiny variations in the background radiation.

Figure 18-8

When the cosmic microwave background radiation was first detected in 1963, technology did not allow measurements at many wavelengths. Not until infrared detectors could be put in orbit was it conclusively shown that the background radiation, as predicted by theory, followed the black body curve. (Photo, AT&T Archives; COBE map, NASA)

Build your perfect course solution with these innovative technology options

Foundations of Astronomy is supported by a range of teaching and learning solutions that allow you to craft the best course for your approach and workflow needs. The text provides the ideal foundation—our powerful technology tools are the building blocks that help you customize *your* course to fit *your* specifications.

Starry Night™

Starry Night™ is an incredibly powerful planetarium software package, yet makes learning about the night sky accessible to everyone. Containing data on 1 million stars as well as 110 deep-space objects such as galaxies, star clusters, and nebulae, it allows the user to view the sky from an exact location on Earth, or from elsewhere within 20,000 light-years of our solar system. The user also has the option to travel through 200,000 years, from the remote past to the distant future. The user even has the option to create personal QuickTime™ movies of the experience. Available packaged on request with the textbook at no additional charge; contact your Thomson Brooks/Cole representative for ordering information.

TheSky™ Student Edition

The student edition of this popular software turns students' computers into powerful personal planetariums. Loaded with data on 118,000 stars and 13,000 deep-sky objects with images, it allows students to view the universe from any location on Earth, and at any point in time from 4,000 years ago to 8,000 years in the future. Students will see the sky in motion, view constellations, see moon phases, search the database for objects, display both lunar and solar eclipses, print star charts, and more. Available packaged on request with the textbook at no additional charge; contact your Thomson representative for ordering information.

TheSky™ Workbook (with CD-ROM)

This workbook provides students with examples, exercises, and in-depth explanations of astronomical topics found in **TheSky™ Student Edition** planetarium software. Topics include the motions of Earth and other planetary objects, the naming conventions and locations of astronomical bodies, lunar phases, and more.

RedShift™ College Edition

RedShift™ College Edition offers a stunning combination of text, graphics, movies, narration, music, illustrations, multiple-choice and matching exercises, applets, animations, and more. **RedShift™ College Edition** is an expanded version of the award-winning, internationally renowned **RedShift™** series. With a host of features such as *Nightly News*, and the astronomical event predictor *Sky Diary*, **RedShift™ College Edition** is the most comprehensive planetarium software available. Supplemental **RedShift™** workbook also available; contact your Thomson representative for packaging information.

ALSO AVAILABLE:

The Night Sky Planisphere

Latitude 20-30: 0-534-40746-3
Latitude 30-40: 0-534-40747-1
Latitude 40+: 0-534-40745-5

The Night Sky planisphere from the David Chandler Company is an ideal portable reference for observational exercises. Each planisphere is printed with dark stars on a light background for easy nighttime readability. The constellations are drawn simply, emphasizing the brighter stars. Available packaged on request with the textbook at no additional charge; contact your Thomson representative for ordering information.

Concept Mastery and Application
Turn to the inside back cover to learn more about these resources

AceAstronomy™

Empower your students and help them to take charge of their learning with **AceAstronomy™**, the first assessment-centered learning tool for astronomy! Seamlessly tied to the text through *Active Figures* (interactive animations of art from the text), this powerful resource helps students gauge their unique study needs and then gives them a *Personalized Learning Plan* that focuses their study time on the concepts they most need to master.

Virtual Astronomy Laboratories
0-534-49087-5

Virtual Astronomy Labs are an online, interactive way for students to learn. Twenty hands-on exercises complement the topics in the text. Many of the text chapters conclude with references to these labs in the *Media Cluster* section.

Visit **http://ace.brookscole.com/sf9** to learn more about our technology solutions, including interactive live demonstrations of **AceAstronomy** and **Virtual Astronomy Labs**.

More ways to craft your perfect course

Multimedia Manager with Instructor's Resources: A Microsoft® PowerPoint® Tool
0-495-01587-3

This easy-to-use multimedia lecture tool allows you to quickly assemble art, photos, and multimedia with notes to create fluid lectures. The **Multimedia Manager** includes every image from *Foundations of Astronomy,* **Ninth Edition**, as well as a collection of *Active Figure* animations and additional simulations. Preformatted Microsoft® PowerPoint® presentations have been assembled for each text chapter to facilitate generation of lecture presentations. The simple interface makes it easy for you to incorporate graphics, digital video, animations, and audio clips into your lectures.

JoinIn™ on TurningPoint®
for Audience Response Systems
0-495-10742-5

Transform your classroom and assess your students' progress with instant in-class quizzes and polls! Thomson Brooks/Cole is now pleased to offer you book-specific **JoinIn™** content for Response Systems specifically tailored to *Foundations of Astronomy,* **Ninth Edition**. Our exclusive agreement to offer **TurningPoint®** software lets you pose book-specific questions and display students' answers seamlessly within the Microsoft® PowerPoint® slides of your own lecture, in conjunction with the "clicker" hardware of your choice. Enhance how your students interact with you, your lecture, and each other. For college and university adopters only. Contact your Thomson representative for more information about **JoinIn on TurningPoint** and our exclusive infrared or radio frequency hardware solutions.

Enhanced WebAssign
The reliable and easy-to-use functionality of the market-leading **WebAssign** system combines with the superior content of this text to create the perfect homework management solution. With **Enhanced WebAssign**, you can assign selected end-of-chapter problems and collect student scores in an online gradebook. Special problems ask students to analyze photos and graphs, manipulate interactive *Active Figures*, or assess their conceptual knowledge. Contact your Thomson representative for more information about **Enhanced WebAssign** for *Foundations of Astronomy*.

Online Instructor's Manual with Test Bank
0-495-01592-X

This resource manual, written by Gene Byrd from the University of Alabama, contains additional lectures and key instructor resources including suggested course syllabi; listings of main concepts; chapter outlines and notes; classroom demonstrations; answers to review, end-of-chapter, and critical inquiry questions; recommended readings; and a complete test bank with answers. Available exclusively for download from the password-protected Instructor's Website.

ExamView® Computerized Testing
0-495-01588-1

Create, deliver, and customize tests and study guides (both print and online) in minutes with this easy-to-use assessment and tutorial system. **ExamView®** offers both a *Quick Test Wizard* and an *Online Test Wizard* that guide you step-by-step through the process of creating tests. You can build tests of up to 250 questions using up to 12 question types. Using the complete word processing capabilities of ExamView, you can enter an unlimited number of new questions or edit questions from our test bank.

Due to contractual reasons, certain ancillaries are available only in higher education or U.S. domestic markets. Minimum purchases may apply to receive the ancillaries at no charge. For more information, please contact your local Thomson representative.

RESOURCES

Foundations of Astronomy

About the Author

Mike Seeds is Professor of Astronomy at Franklin and Marshall College, where he has taught astronomy since 1970. His research interests have focused on peculiar variable stars and the automation of astronomical telescopes. He has been serving since 1987 as Principal Astronomer in charge of the Phoenix 10, the first fully robotic telescope, located in southern Arizona. In 1989, he received the Christian R. and Mary F. Lindback Award for Distinguished Teaching. In addition to teaching, writing, and research, Mike has published educational tools for use in computer-smart classrooms. His interest in the history of astronomy led him to offer upper-level courses "Archaeoastronomy" and "Changing Concepts of the Universe." He has also published educational software for preliterate toddlers. Mike is Senior Consultant in the creation of the telecourse *Astronomy: Observations and Theories*.

He is the author of *Astronomy: The Solar System and Beyond,* Fourth Edition (2005), and *Horizons: Exploring the Universe,* Ninth Edition (2005), published by Brooks/Cole.

About the Cover

Located atop a 13,800-foot-high volcano in Hawaii, the Gemini North telescope is 8.1 m (26.6 feet) in diameter. Here it is shown in front of the Milky Way in Sagittarius. The center of our galaxy is visible as glowing clouds of billions of stars partially obscured by clouds of dust that lie in the plane of our disk-shaped galaxy. (Photo: Kris Koenig, 2005)

9

NINTH EDITION

Foundations of Astronomy

Michael A. Seeds

Joseph R. Grundy Observatory
Franklin and Marshall College

THOMSON

BROOKS/COLE

Australia • Canada • Mexico • Singapore • Spain
United Kingdom • United States

For Emery and Helen Seeds

THOMSON

BROOKS/COLE

Foundations of Astronomy, **Ninth Edition**

Michael A. Seeds

Astronomy Acquisitions Editor: *Chris Hall*
Development Editor: *Alyssa White*
Assistant Editor: *Brandi Kirksey*
Editorial Assistant: *Jessica Jacobs*
Editorial Intern: *David Marks*
Technology Project Manager: *Samuel Subity*
Publisher/Executive Editor: *David Harris*
Marketing Manager: *Mark Santee*
Marketing Assistant: *Michele Colella*
Managing Marketing Communications Manager: *Bryan Vann*
Project Manager, Editorial Production: *Teri Hyde*
Executive Art Director: *Vernon Boes*
Print and Media Buyer: *Karen Hunt*
Permissions Editor: *Stephanie Lee, Bob Kauser*
Production Service: *Heckman & Pinette*
Text Designer: *Linda Beaupré/Stone House Art*

Art Editor: *Lisa Torri*
Photo Researcher: *Mike Seeds and Kathleen Olson*
Copy Editor: *Margaret Pinette*
Illustrator: *Precision Graphics*
Cover Designer: *Irene Morris*
Cover Image: *Milky Way Over Gemini, North Observatory: Kris Koenig; Left: Helix Nebula, Hubble Images: NASA, ESA, C. R. O'Dell (Vanderbilt University), M. Melxner and P. I. McCullough (STScI); Middle: Large Magellanic Cloud N11, Hubble Images: NASA, C. Aguilera, S. Points, and C. Smith (CTIO), and Z. Levay (STScI); Right: Supernova Remnant Menagerie: NASA, ESA, HEIC, and The Hubble Heritage Team (STScI/ AURA) with acknowledgment to Y.-H. Chu and R. M. Williams.*
Cover Printer: *Transcontinental-Interglobe*
Compositor: *Thompson Type*
Printer: *Transcontinental-Interglobe*

Printed in Canada
1 2 3 4 5 6 7 10 09 08 07 06

Library of Congress Control Number: 2005936361

ISBN 0-495-01578-4

Thomson Higher Education
10 Davis Drive
Belmont, CA 94002-3098
USA

For more information about our products, contact us at:
Thomson Learning Academic Resource Center
1-800-423-0563

For permission to use material from this text or product, submit a request online at **http://www.thomsonrights.com.**
Any additional questions about permissions can be submitted by e-mail to **thomsonrights@thomson.com.**

Brief Contents

Contents

Part 1: Exploring the Sky

Windows on Science

Concept Art Portfolios

Part 2: The Stars

Windows on Science

Concept Art Portfolios

Part 4: The Solar System

Windows on Science

Concept Art Portfolios

A Note to the Student

From Mike Seeds

Hi,

I'm really glad you are taking an astronomy course. You are going to see some amazing things, from the icy rings of Saturn to monster black holes. Our universe is so beautiful, it is sad to think that not everyone gets to take an astronomy course.

Two Goals

You will learn a lot of new stuff in this course, but there are two things I hope you find especially satisfying. This astronomy course will help you answer two important questions:

- What are we?
- How do we know?

By "What are we?" I mean, "Where do we fit into the history of the universe?" The atoms you are made of had their first birthday in the big bang when the universe began, but those atoms have been cooked and remade inside stars, and now they are inside you. Where will they be in a billion years? Astronomy is the only course on campus than can tell you that story, and it is a story that everyone should know.

By "How do we know?" I mean, "How does science work?" How can anyone know there was a big bang? In today's world, you need to think carefully about the things so-called experts say. You should demand explanations. Scientists have a special way of knowing based on evidence. Scientific knowledge isn't just opinion or policy or marketing or public relations. It is humanity's best understanding of nature. To understand the world around you, you should understand how science works.

These two questions are the focus of astronomy, and they are just for you. You need to know the answers to these questions so you can appreciate how wonderful the universe is and how special you are.

Favorite Stars

Along the way, I want to introduce you to eight of my favorite stars. Maybe you will add a few favorites of your own, but these are eight stars I enjoy looking at in the sky. When I see Betelgeuse in the winter sky, for example, I see more than just a bright red star. In my imagination, I see a star over 800 times larger than the sun—an aging star that will soon die, perhaps in a violent explosion. Here's my list of favorite stars:

Sirius	**Brightest star in the sky**	**Winter**
Betelgeuse	**Bright red star in Orion**	**Winter**
Rigel	**Bright blue star in Orion**	**Winter**
Aldebaran	**Red eye of Taurus the Bull**	**Winter**
Polaris	**The North Star**	**Year round**
Vega	**Bright star overhead**	**Summer**
Spica	**Bright southern star**	**Summer**
Alpha Centauri	**Nearest star to the sun**	**Spring, far south**

You can see Polaris year round, but Sirius, Betelgeuse, Rigel, and Aldebaran are in the winter sky. Spica is in the summer sky, and Vega is visible evenings in the late summer or fall. Alpha Centauri is a special star, but you will have to travel at least as far south as southern Florida to get a glimpse of it above the southern horizon. Use the charts here along with the star charts in the back of this book to locate these stars.

Every one of these stars has its own secret, and when you find out why they are my favorites, you'll enjoy finding them in the sky just as I do.

Expect to Be Astonished

One reason astronomy is exciting is that astronomers discover new things every day.

Astronomers expect to be astonished. You can share in the excitement because I've worked hard to include the newest images, the newest discoveries, and the newest insights that will take you, in an introductory course, to the frontier of human knowledge. You'll see new evidence of ancient oceans and lakes on Mars and a vast ocean on Jupiter's moon Europa. You'll visit the moon Titan, where it rains liquid methane, and Pluto, where most gases freeze solid. You'll see stars die, and you'll share the struggle to understand new evidence that the expansion of the universe is speeding up. Huge telescopes in space and on remote mountaintops provide a daily dose of excitement that goes far beyond sensationalism. These new discoveries in astronomy are exciting because they are about us. They tell us more and more about what we are.

As you read this book, notice that it is not organized as lists of facts for you to memorize. That could make even astronomy boring. Rather this book is organized to show you how scientists use evidence and theory to create logical arguments that show how nature works. Look at the list of special features that follows this note. Those features were carefully designed to help you understand astronomy as evidence and theory. Once you see science as logical arguments, you hold the key to the universe.

Do Not Be Humble

As a teacher, my quest is simple. I want you to understand your place in the universe—not just your location in space but your location in the unfolding history of the physical universe. Not only do I want you to know where you are and what you are in the universe, but I want you to understand how scientists know. By the end of this book, I want you to know that the universe is very big, but that it is described by a small set of rules and that we humans have found a way to figure out the rules—a method called *science*.

Do not be humble. Astronomy tells us that the universe is vast and powerful, but it also tells us that we are astonishing creatures. We humans are the parts of the universe that think. You are a small creature, but remember that it is your human brain that is capable of understanding the depth and beauty of the cosmos.

To appreciate your role in this beautiful universe, you must learn more than just the facts of astronomy. You must understand what we are and how we know. Every page of this book reflects that ideal.

Mike Seeds
mike.seeds@fandm.com

Key Content and Pedagogical Changes to the Ninth Edition

- Chapter 8, "The Sun," has been thoroughly rewritten and reorganized. The new presentation is designed to help you see the sun as a physical object in which energy is generated in its core, flows outward through the interior, and generates the activity we see on the sun's surface.

- The Concept Art Portfolios were redesigned to aid you in navigating through these key visual summaries. The flow through the material has been clarified by adding numerical markers and arrows to better guide you to a stronger conceptual understanding.

- The Guideposts and Summaries were redesigned to make them a more effective study tool. The chapter-opening Guideposts contain a short set of Essential Questions focused on the key ideas in each chapter. The questions are answered in the narrative and then are presented as bulleted insights in the Summary.

- All figures have been updated, redesigned, and carefully annotated to help you pick out the critical details related to the discussions. In particular, the H–R diagrams have been fully revitalized to help you better understand the nature of stars.

- A new type of question has been added to the end-of-chapter material. "Learning to Look" questions give you a chance to critically examine images to test your understanding of fundamental concepts.

- Of course, I have updated the newest exciting discoveries in astronomy, including images from the surface of Titan, gamma ray bursts, volcanism on icy moons, the Kuiper belt and the status of Pluto, dark energy, the search for water on Mars, supernova explosions, extrasolar planets, and much more.

Special Features

- **Special two-page Concept Art Portfolios** provide an opportunity for you to create your own understanding and share in the satisfaction that scientists feel as they uncover the secrets of nature.

- **Guided discovery figures** illustrate important ideas visually and guide you to understand relationships and contrasts interactively.

- **Windows on Science** provide a parallel commentary on how all of science works. For example, the Windows point out where astronomers use statistical evidence, where they are reasoning by analogy, and where they are building a scientific model rather than a scientific hypothesis.

- **Focus on Fundamentals** will help you understand five concepts from physics that are critical to understanding modern astronomy.

- **Guideposts** on the opening page of each chapter help you see the organization of the book. Cast as a series of Essential Questions that are answered within the text narrative, each Guidepost further connects the chapter with preceding and following chapters to provide an overall organizational guide.

- **Building Scientific Arguments** at the end of each text section are carefully designed questions to help you review and synthesize concepts from the section. A short answer follows to show how scientists construct scientific arguments from observations, evidence, theories, and natural laws that lead to a conclusion. A further question then gives you a chance to construct your own argument on a related issue.

- **End-of-Chapter Review Questions** are designed to help you review and test your understanding of the material.

- **End-of-Chapter Discussion Questions** go beyond the text and invite you to think critically and creatively about scientific questions. You can think about these questions yourself or discuss them in class.
- **Media Clusters** list all of the Web resources you can call on to illustrate, review, and expand the material in each chapter.
- **Critical Inquiries for the Web** challenge you to use the World Wide Web to explore further and to think creatively and analytically about astronomy.
- **Exploring *TheSky*** are experiments you can perform with the software *TheSky* or, by making slight modifications, with any planetarium software. These experiments will allow you to see the sky and celestial bodies in new ways on your own computer screen.

As additional aids, notice that most versions of this book also include access to the following electronic enhancements:

- **Virtual Astronomy Laboratories.** This set of online lab exercises covers 20 of the most important concepts of introductory astronomy in an interactive online environment. You may be asked to run simulations, observe animations, perform calculations, or collect data to help you master these fundamental concepts. The labs cover topics from helioseismology to dark matter and allow you to submit your results electronically to your instructor or print them out to hand in. The Media Cluster at the end of each chapter in this textbook notes which labs correlate to that chapter.
- **AceAstronomy.** Take charge of your learning with the first assessment-centered student learning tool for astronomy. AceAstronomy uses a series of diagnostic tests to provide you with a personalized learning plan, so you can begin to maximize your study time with a host of interactive tutorials and quizzes that help you focus on what you need to learn to master astronomy.

Acknowledgments

I started writing astronomy textbooks in 1973, and over the years I have had the guidance of a great many people who care about astronomy and teaching.

I would like to thank all of the students and teachers who have responded so enthusiastically to *Foundations of Astronomy*. Their comments and suggestions have been very helpful in shaping this book.

I would especially like to thank the reviewers whose careful analysis and thoughtful suggestions have been invaluable in completing this new edition.

Many observatories, research institutes, laboratories, and individual astronomers have supplied figures and diagrams for this edition. Their names are given on the page with the illustrations they provided, and I would like to thank them specifically for their generosity.

Writing about every branch of astronomy is a daunting task, and I could not do it without helpful contributions from experts in various fields. I especially want to think Dana Backman, George Jacoby, Victoria Kaspi, and William Keel for their helpful guidance on technical issues.

Certain unique diagrams in Chapters 11, 12, 13, and 14 are based on figures I designed for my article "Stellar Evolution," which appeared in *Astronomy*, February 1979.

I am happy to acknowledge the use of images and data from a number of important programs. In preparing materials for this book I used NASA's Sky View facility located at NASA Goddard Space Flight Center. I have used atlas images and mosaics obtained as part of the Two Micron All Sky Survey (2MASS), a joint project of the University of Massachusetts and the Infrared Processing and Analysis Center/California Institute of Technology, funded by the National Aeronautics and Space Administration and the National Science Foundation. A number of solar images are used by the courtesy of the SOHO consortium, a project of international cooperation between ESA and NASA.

It is always a pleasure to work with the Thomson Learning team at Brooks/Cole and Wadsworth. Special thanks go to all of the people who have contributed to this project, including Teri Hyde, Kathleen Olson, Sam Subity, Lisa Torri, Vernon Boes, Mark Santee, and Brandi Kirksey. I have enjoyed working on production with Margaret Pinette and Bill Heckman of Heckman & Pinette, and I appreciate their understanding and goodwill. I would especially like to thank my editor Chris Hall and my development editor Alyssa White for their help and guidance on this project.

Most of all, I would like to thank my wife, Janet, and my daughter, Kate, for putting up with "the books." They know all too well that textbooks are made of time.

Mike Seeds

Manuscript Reviewers

Steve Desch, Arizona State University
Michael Frey, California State University, Long Beach
Marc Gagné, West Chester University
Paul Hintzen, California State University, Long Beach
Don McCarthy, University of Arizona
Scott Miller, Penn State University
Sharon L. Montgomery, Clarion University of Pennsylvania
Richard Scarborough, University of Nebraska, Lincoln

1 | The Scale of the Cosmos

*The longest journey
begins with a single step.*

CONFUCIUS

Visual-wavelength image

Y OU ARE ABOUT to embark on a
voyage out to the end of the universe. Marco
Polo journeyed east, Columbus west, but you will travel
outward away from your home on Earth, out past the moon, sun, and other planets, past the stars you see in the
evening sky, and past billions more that can be seen only with the aid of the largest telescopes. You will journey
through great whirlpools of stars to the most distant galaxies visible from Earth—and then you will continue on,
carried only by experience and imagination looking for the structure of the universe itself.   Imagination is your
key to discovery; it will be your scientific time machine transporting you into the past and into the future. Go back
to watch the birth of the universe, the formation of the first stars, and the origin of the sun and Earth. Then rush

| Continued on page 4 |

Guidepost

Getting Started

You are already an expert on astronomy. You have enjoyed sunsets and moonrises. You have admired the stars and may know a few constellations. You have probably read about Mars and eclipses, and you may have read about distant galaxies. That is more than most Earthlings know about astronomy. But you have a right to know more.

You are a planetwalker, and you should understand what it means to live on a planet that whirls around a star drifting through a universe of stars and galaxies. You owe it to yourself to know where you are. That is the first step to knowing what you are.

This Chapter

Here you will take a quick trip, a cosmic zoom, from objects you know up to the largest things in the universe. That quick survey will answer four essential questions:

Where is Earth in relation to the sun, the planets, and the galaxies?

How do astronomers talk about these huge astronomical distances?

Which objects are big, and which are small?

Are there other worlds like Earth?

This chapter will give you a sense of the scale of the universe that will carry you to new discoveries.

Looking Ahead

It is easy to learn a few facts, but it is the relationships among facts that are important. This chapter will give you the sense of scale that you need to understand where you are in the universe. The remaining chapters of this book will fill in the details, cite evidence and theories, and illustrate the wonderful intricacy and beauty of the universe. It is an exciting journey, and it begins here.

into the future to see what will happen when the sun dies and Earth withers.

Although you will discover a beginning to the universe, you will not find an edge in space. No matter how far you voyage, you will not run into a wall or limit beyond which you cannot go. Rather you will discover evidence that our universe may be infinite, that it may extend in all directions without limit. Such vastness may dwarf our earthly dimensions but not our human curiosity and imagination.

Astronomy is more than the study of stars and planets. It is the study of the universe in which we humans exist. You and I live on a small planet circling a small sun drifting through the universe, but astronomy can take us beyond these boundaries and help us not only see where we are in the universe but understand what we are. You have a right to know these things. Perhaps you have a duty to know them.

Do not be humble. Although astronomical sizes and distance may dwarf you, remember that you are an intelligent creature, and you are capable of understanding your universe. It is, after all, yours.

Astronomy will introduce you to sizes, distances, and times far beyond your usual experience on Earth. Your task in this chapter is to grasp the meaning of these unfamiliar sizes, distances, and times. Believe it or not, the solution lies in a single word, *scale.*

In this chapter, you will compare objects of different sizes to grasp the scale of the universe.

Let's begin with something familiar. ■ Figure 1-1 shows a region about 52 feet across occupied by a human being, a sidewalk, and a few trees—all objects whose size you can understand. Each successive picture in this chapter will show you a region of the universe that is 100 times wider than the preceding picture. That is, each step will widen your field of view by a factor of 100.

In ■ Figure 1-2 your field of view widens by a factor of 100, and you can see an area 1 mile in diameter. The arrow points to the scene shown in the preceding photograph. People, trees, and sidewalks have vanished, but now you can see a college campus and the surrounding streets and houses. The dimensions of houses and streets are familiar. This is the world you know, and you can relate such objects to the scale of your body.

You started your adventure using feet and miles, but you should use the metric system of units. Not only is it used by all scientists around the world, but it makes calculations much easier. If you are not already familiar with the metric system, or if you need a review, study Appendix A before reading on.

■ **Figure 1-2**

(USGS)

■ **Figure 1-1**

(Michael A. Seeds)

range of "colors," from X rays to radio waves, to reveal sights invisible to unaided human eyes.

At the next step in your journey, you will see our entire planet (■ Figure 1-4), which is 12,756 km in diameter. Earth rotates on its axis once a day, exposing half of its surface to daylight at any particular moment. The photo shows most of the daylight side of the planet. The blurriness at the extreme right is the sunset line. The rotation of Earth carries you eastward, and as you cross the sunset line into darkness, you say the sun has set. It is the rotation of the planet that causes the cycle of day and night.

Earth's interior is made mostly of iron and nickel, and its crust is mostly silicate rocks. Only a thin layer of water makes up

■ Figure 1-3
(NASA infrared photograph)

■ Figure 1-4
(NASA)

The photo in Figure 1-2 is 1 mile in diameter. A mile equals 1.609 kilometers, so you can see in the photo that a kilometer is a bit over two-thirds of a mile—a short walk across a neighborhood.

The view in ■ Figure 1-3 spans 160 kilometers. In this infrared photo, the green foliage shows up as various shades of red. The college campus is now invisible, and the patches of gray are small cities, with the suburbs of Philadelphia visible at the lower right. At this scale, you see the natural features of Earth's surface. The Allegheny Mountains of southern Pennsylvania cross the image in the upper left, and the Susquehanna River flows southeast into Chesapeake Bay. What look like white bumps are a few puffs of clouds.

These features are a reminder that you live on the surface of a changing planet. Forces in Earth's crust pushed the mountain ranges up into parallel folds, like a rug wrinkled on a polished floor. The clouds tell you that Earth's atmosphere is rich in water, which falls as rain and erodes the mountains, washing material down the rivers and into the sea. Mountains and valleys are only temporary features on Earth; they are constantly changing. As you explore the universe, you will come to see that it, like Earth's surface, is always evolving.

Take a closer look at Figure 1-3 and notice the red color. This is an infrared photograph in which healthy green leaves and crops show up as red. Human eyes are sensitive to only a narrow range of colors. As you explore the universe, you will learn to use a wide

the oceans, and the atmosphere is only a few hundred kilometers deep. On the scale of this photograph, the depth of the atmosphere on which life depends is less than the thickness of a piece of thread.

Enlarge your field of view by a factor of 100, and you will see a region 1,600,000 km wide (■ Figure 1-5). Earth is the small blue dot in the center, and the moon, whose diameter is only one-fourth that of Earth, is an even smaller dot along its orbit 380,000 km from Earth.

These numbers are so large that it is inconvenient to write them out. Astronomy is the science of big numbers, and you will use numbers much larger than these to discuss the universe. Rather than writing out these numbers as in the previous paragraph, it is convenient to write them in **scientific notation.** This is nothing more than a simple way to write numbers without writing lots of zeros. In scientific notation, you would write 380,000 as 3.8×10^5. If you are not familiar with scientific notation, read the section on powers of 10 notation in the Appendix. The universe is too big to discuss without using scientific notation.

When you once again enlarge your field of view by a factor of 100 (■ Figure 1-6), Earth, the moon, and the moon's orbit all lie in the small red box at lower left. But now you can see the sun and two other planets that are part of our solar system. Our **solar system** consists of the sun, its family of planets, and some smaller bodies such as moons and comets.

Like Earth, Venus and Mercury are **planets,** small, non-luminous bodies that shine by reflected light. Venus is about the size of Earth, and Mercury is a bit larger than Earth's moon. On this diagram, they are both too small to be seen as anything but tiny dots. The sun is a **star,** a self-luminous ball of hot gas that generates its own energy. The sun is 109 times larger in diameter than Earth (inset), but it too is nothing more than a dot in this diagram.

■ **Figure 1-6**

(NOAO)

■ **Figure 1-5**

(NASA)

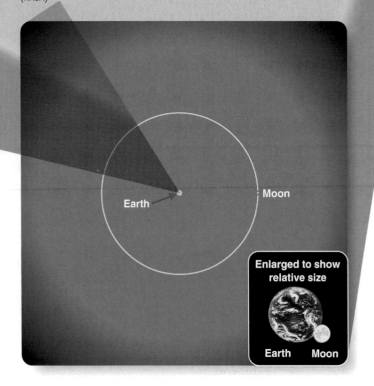

This diagram has a diameter of 1.6×10^8 km. One way astronomers deal with large numbers is to define new units. The average distance from Earth to the sun is a unit of distance called the **astronomical unit (AU),** a distance of 1.5×10^{11} m. Using this unit, you can say that the average distance from Venus to the sun is about 0.7 AU. The average distance from Mercury to the sun is about 0.39 AU.

The orbits of the planets are not perfect circles, and this is particularly apparent for Mercury. Its orbit carries it as close to the sun as 0.307 AU and as far away as 0.467 AU. You can see this variation in the distance from Mercury to the sun in Figure 1-6. Earth's orbit is more circular, and its distance from the sun varies by only a few percent.

Your first field of view was only 52 feet (about 16 m) in width. After only six steps of enlarging by a factor of 100, you can now see the entire solar system (■ Figure 1-7). Your field of view is 1 trillion (10^{12}) times wider than in your first view. The details of the preceding figure are now lost in the red square at the center of this diagram. You see only the brighter, more widely separated objects as you enlarge your view. The sun, Mercury, Venus, and Earth lie so close together that you cannot separate them at this scale. Mars, the next outward planet, lies only 1.5 AU from the sun. In contrast, Jupiter, Saturn, Uranus, Neptune, and Pluto are so far from the sun that they are easy to place in this diagram. These are cold worlds far from the sun's warmth. Light from the sun reaches Earth in only 8 minutes, but it takes over 4 hours to reach Neptune. Notice that Pluto's orbit is so elliptical that Pluto can come closer to the sun than Neptune does, as Pluto did between 1979 and 1999.

When you again enlarge your field of view by a factor of 100, the solar system vanishes (■ Figure 1-8). The sun is only a point of

In ■ Figure 1-9, your field of view has expanded to a diameter a bit over 1 million AU. The sun is at the center, and you can see a few of the nearest stars. These stars are so distant that it is not reasonable to give their distances in astronomical units. To express distances so large, astronomers define a new unit of distance, the light-year. One **light-year (ly)** is the distance that light travels in one year, roughly 10^{13} km or 63,000 AU. The diameter of your field of view is 17 ly. The nearest star to the sun, Alpha Centauri, is 4.2 ly from Earth. In other words, light from Alpha Centauri takes 4.2 years to reach Earth.

Do you remember that Alpha Centauri is one of our Favorite Stars? (See p. xi for a list of Favorite Stars.) It is in the southern sky, so it is invisible from all but the southernmost parts of the United States where it occasionally peeks above the southern horizon. If you ever have the chance, you should locate the sun's nearest companion in space. In a later chapter, you will discover that Alpha Centauri is actually three stars orbiting around each other. Our Favorite Stars have interesting secrets.

Although stars are roughly the same size as the sun, they are so far away that you cannot see them as anything but points of light. Even with the largest telescopes on Earth, you would see only points of light when you looked at stars, and any planets that might circle those stars are much too small and too faint to be visible. Using indirect methods, astronomers have found nearly 200 planets orbiting other stars.

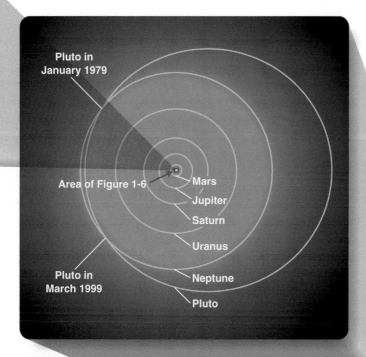

■ **Figure 1-7**

light, and all the planets and their orbits are now crowded into the small red square at the center. The planets are too small and reflect too little light to be visible so near the brilliance of the sun.

Nor are any stars visible except for the sun. The sun is a fairly typical star, and it seems to be located in a fairly average neighborhood in the universe. Although there are many billions of stars like the sun, none is close enough to be visible in this diagram, which shows an area only 11,000 AU in diameter. The stars are typically separated by distances about 10 times larger than the diameter of this diagram. You will see stars in the next field of view, but, except for the sun at the center, this diagram is empty.

It is difficult to grasp the isolation of the stars. If the sun were represented by a golf ball in New York City, the nearest star would be another golf ball in Chicago. Except for the widely scattered stars and a few atoms of gas drifting between the stars, the universe is nearly empty.

■ **Figure 1-8**

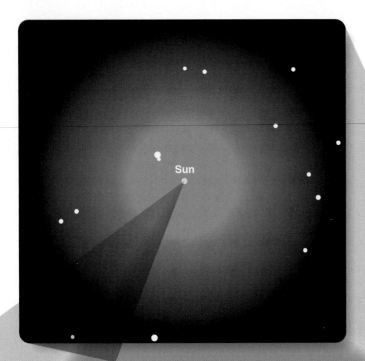

Milky Way, and our galaxy is called the **Milky Way Galaxy.** Of course, no one can journey far enough into space to look back and photograph our home galaxy, so the photo in Figure 1-11 shows a galaxy similar to our own. Our sun would be invisible in such a photo, but if you could see it, you would find it in the disk of the galaxy about two-thirds of the way out from the center.

Our galaxy, like many others, has graceful **spiral arms** winding outward through the disk. You will discover that stars are born in great clouds of gas and dust as they pass through the spiral arms.

■ Figure 1-10

This box ■ represents the relative size of the previous frame. (NOAO)

In Figure 1-9, the sizes of the dots represent not the sizes of the stars but their brightness. This is the custom in astronomical diagrams, and it is also how star images are recorded on photographs. Bright stars make larger spots on a photograph than faint stars. The size of a star image in a photograph tells you not how big the star is but only how bright it looks.

In ■ Figure 1-10, you expand your field of view by another factor of 100, and the sun and its neighboring stars vanish into the background of thousands of other stars. The field of view is now 1700 ly in diameter. Of course, no one has ever journeyed thousands of light-years from Earth to look back and photograph the solar neighborhood, so this is a representative photograph of the sky. The sun is a relatively faint star that would not be easily located in a photo at this scale.

What you do not see in this photograph is critically important. You do not see the thin gas that fills the spaces between the stars. Although those clouds of gas are thinner than the best vacuum on Earth, it is those clouds that give birth to new stars. Our sun formed from such a cloud about 5 billion years ago. You will see evidence of star formation in your next field of view.

If you expand your field of view by a factor of 100, you see our galaxy (■ Figure 1-11). A **galaxy** is a great cloud of stars, gas, and dust bound together by the combined gravity of all the matter. Galaxies range from 1500 to over 300,000 ly in diameter and can contain over 100 billion stars. In the night sky, you see our galaxy as a great, cloudy wheel of stars ringing the sky as the

Ours is a fairly large galaxy, roughly 75,000 ly in diameter. Only a century ago astronomers thought it was the entire universe—an island universe of stars in an otherwise empty vastness. Now they know that our galaxy is not unique. Indeed ours is only one of many billions of galaxies scattered throughout the universe.

As you expand your field of view by another factor of 100, our galaxy appears as a tiny luminous speck surrounded by other specks (■ Figure 1-12). This diagram includes a region 17 million ly in diameter, and each of the dots represents a galaxy. Notice that our galaxy is part of a cluster of a few dozen galaxies.

Galaxies are commonly grouped together in such clusters. Some of these galaxies have beautiful spiral patterns like our own galaxy, but others do not. Some are strangely distorted. One of the mysteries of modern astronomy is what produces these differences among the galaxies. Astronomers have found some important clues and have developed a fascinating theory.

If you again expand your field of view, you see that the clusters of galaxies are connected in a vast network (■ Figure 1-13). Clusters are grouped into superclusters—clusters of clusters— and the superclusters are linked to form long filaments and walls outlining voids that seem nearly empty of galaxies. These appear to be the largest structures in the universe. Were you to expand your field of view another time, you would probably see a uniform fog of filaments and voids. When you puzzle over the origin of these structures, you are at the frontier of human knowledge.

■ Figure 1-13

This box ■ represents the relative size of the previous frame. (Detail from galaxy map from M. Seldner, B. L. Siebers, E. J. Groth, and P. J. E. Peebles, *Astronomical Journal 82* [1977])

■ Figure 1-11

(© Anglo-Australian Telescope Board)

Milky Way Galaxy

■ Figure 1-12

The first hurdle in studying astronomy is keeping a proper sense of scale. Remember that each of the billions of galaxies contains billions of stars. Later you will see clear evidence that many of those stars have families of planets like our solar system, and on some of those planets liquid-water oceans and protective atmospheres may have sheltered the spark of life. It is possible that at least a few other planets are inhabited by intelligent creatures who share our curiosity and our wonder at the scale of the cosmos.

Summary

Your goal in this chapter is to preview the scale of astronomical objects. To do so, you journeyed outward from a familiar campus scene by expanding your field of view by factors of 100. Only 12 such steps took you to the largest structures in the universe.

Where is Earth in relation to the sun, the planets, and the galaxies?

■ You live on planet Earth, which orbits our star, the sun, once a year. As Earth rotates once a day, you see the sun rise and set.

■ The other planets in our solar system, Mercury, Venus, Mars, Jupiter, Saturn, Uranus, Neptune, and Pluto, orbit the sun in ellipses that are nearly circular.

■ The sun is just one out of the billions of stars that fill our home galaxy, the Milky Way spiral galaxy.

■ Our galaxy is just one of billions of galaxies that fill the universe in great clusters, clouds, filaments, and walls.

How do astronomers talk about these huge astronomical distances?

■ Astronomers use the metric system because it simplifies calculations and scientific notation for very large or very small numbers.

■ The astronomical unit (AU) is the average distance from Earth to the sun. Mars, for example, orbits 1.5 AU from the sun. The light-year (ly) is the distance light can travel in one year. The nearest star is 4.2 ly from the sun.

Which objects are big, and which are small?

■ The moon is only a fourth the diameter of Earth, but the sun is 109 times larger in diameter than Earth—a typical size for a star.

■ The solar system includes the sun at the center and all of the planets that orbit around it.

■ Galaxies contain many billions of stars. Our galaxy is about 75,000 ly in diameter and contains over 100 billion stars.

■ The largest things in the universe are the vast filaments and walls containing many clusters of galaxies.

Are there other worlds like Earth?

■ Many stars seem to have planets, but such small, distant worlds are difficult to detect. Fewer than 200 or so have been found so far, but planets seem to be common, so you can probably trust that there are other worlds like Earth.

New Terms

scientific notation (p. 6)

solar system (p. 6)

planet (p. 6)

star (p. 6)

astronomical unit (AU) (p. 6)

light-year (ly) (p. 7)

galaxy (p. 8)

Milky Way (p. 8)

Milky Way Galaxy (p. 8)

spiral arm (p. 8)

Review Questions

Ace ◯ Astronomy™ Assess your understanding of this chapter's topics with additional quizzing and animations at http://ace .brookscole.com/sf9

1. What is the largest dimension you have personal knowledge of? Have you run a mile? Hiked 10 miles? Run a marathon?

2. In Figure 1-4, the division between daylight and darkness is at the right on the globe of Earth. How do you know this is the sunset line and not the sunrise line?

3. What is the difference between our solar system, our galaxy, and the universe?

4. Look at Figure 1-6. How can you tell that Mercury follows an elliptical orbit? Can you detect the elliptical shape of any other orbits in this figure or the next?

5. Which is the outermost planet in our solar system? Why does that change?

6. Why are light-years more convenient than miles, kilometers, or astronomical units for measuring certain distances?

7. Why is it difficult to detect planets orbiting other stars?

8. What does the size of the star image in a photograph tell you?

9. What is the difference between the Milky Way and the Milky Way Galaxy?

10. What are the largest known structures in the universe?

11. Of the objects listed here, which would be contained inside the object shown in the photograph at the right? Which would contain the object in the photo? Stars, planets, galaxy clusters, filaments, spiral arms

(Bill Schoening/ NOAO/AURA/NSF)

Problems

1. The diameter of Earth is 7928 miles. What is its diameter in inches? In yards? If the diameter of Earth is expressed as 12,756 km, what is its diameter in meters? In centimeters?

2. If a mile equals 1.609 km and the moon is 2160 miles in diameter, what is its diameter in kilometers?

3. One astronomical unit is about 1.5×10^8 km. Explain why this is the same as 150×10^6 km.

4. Venus orbits 0.7 AU from the sun. What is that distance in kilometers?

5. Light from the sun takes 8 minutes to reach Earth. How long does it take to reach Mars?

6. The sun is almost 400 times farther from Earth than is the moon. How long does light from the moon take to reach Earth?

7. If the speed of light is 3×10^5 km/s, how many kilometers are in a light-year? How many meters?

8. How long does it take light to cross the diameter of our Milky Way Galaxy?

9. The nearest galaxy to our own is about 2 million light-years away. How many meters is that?

10. How many galaxies like our own would it take laid edge to edge to reach the nearest galaxy? (*Hint:* See Problem 9.)

Media Cluster

Critical Inquries for the Web

1. Locate photographs of Earth taken from space. What do cities look like? Can you see highways? Is the presence of our civilization detectable from space?

2. Locate photographs of nearby galaxies and compare them with photos of very distant galaxies. What kind of detail is invisible for distant galaxies?

3. One of the biggest clusters of galaxies is the Virgo cluster. Find out how many and what kind of galaxies are in the cluster. Is it nearby or far away?

Exploring *TheSky*

1. Locate and center one example of each of three different types of objects:

   a. A planet, such as Saturn. Find its rising and setting time. Such objects have distances measured in astronomical units (AU).
      How to proceed: Decide on the object you want to locate. Then find and center the object by pressing the **Find** button on the **Object Toolbar.** The second method is to press the **F** key. The third is to click **Edit,** then **Find.** Once you have the **Object Information** window, press the **center** button.

   b. A star. All stars in *TheSky* belong to our Milky Way Galaxy. Give the star's name, its magnitude, and its distance in light-years.
      How to proceed: Click on any star, which brings up an **Object Information** window containing a variety of information about the star.

   c. A galaxy; give its name and/or its designation.
      How to proceed: To show galaxies, click on the **Galaxies** button in the **Object Toolbar,** then click on any galaxy. Distances to galaxies are on the order of millions and billions of light years.

2. Look at the solar system from beyond Pluto by clicking on **View** and then on **3D Solar System Mode.** Tip the solar system edge-on and then face-on. Zoom in to see the inner planets. Under **Tools,** set the **Time Skip Increment** to 1 day and then go **forward in time** to watch the planets move.

3. Identify some of the brightest constellations located along the Milky Way. (*Hint:* See **View, Reference Lines.**)

 Go to the Brooks/Cole Astronomy Resource Center (**http://astronomy.brookscole.com**) for critical thinking exercises, articles, and additional readings from InfoTrac College Edition, Brooks/Cole's online student library.

2 | The Sky

Visual-wavelength image

THE NIGHT SKY is the rest of the
universe as seen from our planet. When you look
up at the stars, you look out through a layer of air only a few
hundred kilometers deep. Beyond that, space is nearly empty, with stars scattered light-years apart. Here you will begin
your search for the natural laws that govern the universe by trying to understand what the universe looks like. ▌ As
you read this chapter, keep in mind that you live on a planet. The stars are scattered into the void all around you, most
very distant and some closer. Our planet rotates on its axis once a day and revolves around the sun in its orbit once a year,
and that makes the sky appear to revolve around you in a daily and yearly cycle. Not only does the sun rise and set, but
so also do the stars. The entire sky seems to whirl around you because our planet whirls through the universe.

▌ Continued on page 14 ▌

Guidepost

Looking Back

Astronomy is about us. As we learn about astronomy, we learn about ourselves. We search for an answer to the question, "What are we?" The cosmic zoom in the previous chapter showed you Earth in relation to other objects in the universe. Even in that quick preview, as you learned about stars and galaxies, you were also learning about yourself.

This Chapter

Now it is time to return to Earth and look closely at the sky. You have noticed stars and may know a few constellations and star names, but now you need to refine your terms and answer a few essential questions:

How do astronomers refer to stars?

How can you compare the brightness of the stars?

How does the sky move as Earth moves?

What causes the seasons?

How do astronomical cycles affect Earth's climate?

Answering these questions will tell you a great deal about yourself and your home on planet Earth.

Looking Ahead

In the next chapter, you will learn more about the beautiful cycles of the sky, and in this chapter and the next it is important to remember that the motions you see in the heavens are produced in part by the motions of Earth. Chapter 4 will show you how hard it was for humanity to understand that we live on a moving planet. In fact, you will discover that modern science was born when people tried to understand the motions in the sky.

Ace★Astronomy™ The AceAstronomy icon throughout the text indicates an opportunity for you to test yourself on key concepts and to explore animations and interactions on the AceAstronomy website at: http://ace .brookscole.com/sf9

ON A DARK NIGHT far from city lights, you can see a few thousand stars in the sky. As you begin your study of the sky, the first step is to organize what you see by naming groups of stars and individual stars and by specifying the brightness of individual stars. That will make the sky familiar territory, and you will be ready to explore further.

Constellations

All around the world, ancient cultures celebrated heroes, gods, and mythical beasts by naming groups of stars—**constellations** (■ Figure 2-1). You should not be surprised that the star patterns do not look like the creatures they represent any more than Columbus, Ohio, looks like Christopher Columbus. The constellations "celebrate" the most important mythical figures in each culture. The constellations named within Western culture originated in Mesopotamia over 5000 years ago, with other constellations added by Babylonian, Egyptian, and Greek astronomers during the classical age. Of these ancient constellations, 48 are still in use today.

Different cultures grouped stars and named constellations differently. The constellation you probably know as Orion was

■ **Figure 2-1**

The constellations are an ancient heritage handed down for thousands of years as celebrations of great heroes and mythical creatures. Here Sagittarius and Scorpius hang above the southern horizon.

known as Al Jabbar, the giant, to the ancient Syrians, as the White Tiger to the Chinese, and as Prajapati in the form of a stag in ancient India. The Pawnee Indians knew the constellation Scorpius as two groupings. The long tail of the scorpion was the Snake, and the two bright stars at the tip of the scorpion's tail were the Two Swimming Ducks.

Many ancient cultures around the world, including the Greeks, northern Asians, and Native Americans, associated the stars of the Big Dipper with a bear. The concept of the celestial bear may have crossed the land bridge into North America with the first Americans over 10,000 years ago. Some of the constellations you see in the sky may be among the oldest surviving traces of human culture.

To the ancients, a constellation was a loose grouping of stars. Many of the fainter stars were not included in any constellation, and regions of the southern sky, not being visible to the ancient astronomers of northern latitudes, were not organized into constellations. Constellation boundaries, when they were defined at all, were only approximate (■ Figure 2-2a), so a star like Alpheratz could be thought of as part of Pegasus or part of Andromeda. In recent centuries, astronomers have added 40 modern constellations to fill gaps, and in 1928 the International Astronomical Union established 88 official constellations with clearly defined boundaries (Figure 2-2b). Consequently, a constellation now represents not a group of stars but an area of the sky, and any star within the boundaries of the area belongs to the constellation. Because the entire sky is covered by constellations, every star is a member of one and only one constellation.

In addition to the 88 official constellations, the sky contains a number of less formally defined groupings called **asterisms.** The Big Dipper, for example, is a well-known asterism that is part of the constellation Ursa Major (the Great Bear). Another asterism is the Great Square of Pegasus (Figure 2-2b), which includes three stars from Pegasus and one from Andromeda. The star charts at the end of this book will introduce you to the brighter constellations and asterisms.

Although you refer to constellations and asterisms by name, most are made up of stars that are not physically associated with one another. Some stars may be many times farther away than others and moving through space in different directions. The only thing they have in common is that they lie in approximately the same direction from Earth (■ Figure 2-3).

The Names of the Stars

In addition to naming groups of stars, ancient astronomers named the brighter stars,

Goat), and Aldebaran (the Follower of the Pleiades) are beautiful additions to the mythology of the sky.

Giving the stars individual names is not very helpful, because you can see thousands of stars, and these names do not help you locate the star in the sky. In which constellation is Antares, for example? In 1603, Bavarian lawyer Johann Bayer published an atlas of the sky called *Uranometria* in which he assigned lowercase Greek letters to the brighter stars of each constellation in approximate order of brightness. Astronomers have used those Greek letters ever since. (See the Appendix table with the Greek alphabet.) In this way, the brightest star is usually designated α (alpha), the second-brightest β (beta), and so on (■ Figure 2-4). To identify a star in this way, give the Greek letter followed by the possessive form of the constellation name, such as α Scorpii (sometimes written alpha Scorpii) for Antares. That designation reveals that Antares is in the constellation Scorpius and that it is probably the brightest star in the constellation.

■ Figure 2-2

(a) In antiquity, constellation boundaries were poorly defined, as shown on this map by the curving dotted lines that separate Pegasus from Andromeda. (From Duncan Bradford, *Wonders of the Heavens*, Boston: John B. Russell, 1837) (b) Modern constellation boundaries are precisely defined by international agreement.

and modern astronomers still use many of those names. The constellation names come from Greek versions translated into Latin—the language of science from the fall of Rome to the 19th century—but most star names come from ancient Arabic, though much altered by the passing centuries. The name of Betelgeuse, the bright red star in Orion, for example, comes from the Arabic *yad al jawza*, meaning "armpit of Jawza [Orion]." Names such as Sirius (the Scorched One), Capella (the Little She

■ Figure 2-3

The stars you see in the Big Dipper—the brighter stars of the constellation Ursa Major, the Great Bear—are not at the same distance from Earth. You see the stars in a group in the sky because they lie in the same general direction as seen from Earth. The size of the star dots in the star chart represents the apparent brightness of the stars.

The brighter stars in a constellation are usually given Greek letters in order of decreasing brightness.

α Orionis is also known as Betelgeuse.

Orion

β Orionis is also known as Rigel.

In Orion β is brighter than α, and κ is brighter than η. Fainter stars do not have Greek letters or names, but if they are located inside the constellation boundaries, they are part of the constellation.

■ Figure 2-4

Stars in a constellation can be identified by Greek letters and by names derived from Arabic. The spikes on the star images in the photograph were produced by the optics in the camera. (William Hartmann)

This method of identifying a star's brightness, however, is only approximate. To be precise, you need an accurate way of referring to the brightness of stars, and for that you must consult one of the first great astronomers.

The Brightness of Stars

Astronomers measure the brightness of stars using the **magnitude scale,** a system that first appeared in the writings of the ancient astronomer Claudius Ptolemy about 140 AD. The system may have originated earlier than Ptolemy, and most astronomers attribute it to the Greek astronomer Hipparchus (160–127 BC) (■ Figure 2-5).

The ancient astronomers divided the stars into six classes. The brightest were called first-class stars and those that were fainter, second-class. The scale continued downward to sixth-class stars, the faintest visible to the human eye. Thus, the larger the magnitude number, the fainter the star. This makes sense if you think of the bright stars as first-class stars and the faintest stars visible as sixth-class stars.

Hipparchus is believed to have compiled the first star catalog, and he may have used the magnitude system in that catalog.

Almost 300 years later Ptolemy used the magnitude system in his own catalog, and successive generations of astronomers have continued to use the system. Later astronomers had to revise the ancient magnitude system to include very faint and very bright stars. Telescopes reveal many stars fainter than those the eye can detect, so the magnitude scale was extended to include these faint stars. The Hubble Space Telescope, currently the most sensitive astronomical telescope, can detect stars as faint as 30th magnitude. In contrast, the brightest stars are actually brighter than the first brightness class in the ancient magnitude system. Consequently, modern astronomers extended the magnitude system into negative numbers to account for brighter objects. For instance, Favorite Star Vega (alpha Lyrae) is almost zero magnitude at 0.04, and Sirius, the brightest star in the sky, has a magnitude of -1.42. You can even place the sun on this scale at -26.5 and the moon at -12.5 (■ Figure 2-6).

These numbers are known as **apparent visual magnitudes (m_v),** and they describe how the stars look to human eyes observing from Earth. Although some stars emit large amounts of infrared or ultraviolet light, human eyes can't see it, and it is not included in the apparent visual magnitude. The subscript "v" reminds you that visual magnitudes include only the light human eyes can see. Another problem is the distance to the stars. Very

■ Figure 2-5

Hipparchus (2nd century BC) was the first great observational astronomer. He is believed to have compiled the first star catalog and may have invented the magnitude scale. He is honored here on a Greek stamp that also shows one of his observing instruments.

The scale of apparent visual magnitudes extends into negative numbers to represent the brightest objects and to positive numbers larger than 6 to represent objects fainter than the human eye can see.

distant stars look fainter, and nearby stars look brighter. Apparent visual magnitude ignores the effect of distance and tells you only how bright the star looks as seen from Earth.

Magnitude and Intensity

Nearly every star catalog from today back to the time of the ancients uses the magnitude scale of stellar brightness. However, brightness is subjective, depending on such things as the physiology of the eye and the psychology of perception. To be accurate in measurements and calculations astronomers use the more precise term *intensity*—a measure of the light energy from a star that hits 1 square meter in 1 second. That means astronomers need a way to convert between magnitudes and intensities.

The human eye senses the brightness of objects by comparing the ratios of their intensities. If two stars have intensities I_A and I_B, they can be compared by writing the ratio of their intensities, I_A/I_B. Hipparchus's magnitude system is based on a constant intensity ratio of about 2.5 for each magnitude. That is, if two stars differ by 1 magnitude, then they have an intensity ratio of about 2.5. If they differ by 2 magnitudes, their ratio is 2.5 × 2.5, and so on.

In the 19th century, astronomers began measuring starlight, and they realized they needed to define a mathematical magnitude system that was precise; but for convenience they wanted a system that agreed at least roughly with that of Hipparchus. The stars that ancient astronomers classified as first and sixth differ by 5 magnitudes and have an intensity ratio of almost exactly 100, so the modern system of magnitudes specifies that a magnitude difference of 5 magnitudes corresponds to an intensity ratio of exactly 100. That means that 1 magnitude corresponds to an intensity ratio of 2.512, the fifth root of 100. That is, $100 = (2.512)^5$.

By giving the ancient magnitude scale this precise definition, astronomers can compare the light from two stars. The light from a first-magnitude star is 2.512 times more intense than that from a second-magnitude star. The light from a third-magnitude star is $(2.512)^2$ more intense than the light from a fifth-magnitude star. In general, the intensity ratio equals 2.512 raised to the power of the magnitude difference. That is:

$$\frac{I_A}{I_B} = (2.512)^{(m_B - m_A)}$$

For instance, if two stars differ by 6.32 magnitudes, you can calculate the ratio of their intensities as $2.512^{6.32}$. Your pocket calculator tells you that the ratio is 337.

When you know the intensity ratio and want to find the magnitude difference, you might like to rearrange the equation above and write it as

$$m_A - m_B = 2.5 \, \text{Log} \left(\frac{I_B}{I_A} \right)$$

For example, if you knew that the light from Sirius is 24.2 times more intense than light from Polaris, you could find the magnitude difference easily. It is just 2.5 Log (24.2). Your pocket calculator tells you that the logarithm of 24.2 is 1.38, so the magnitude difference is 2.5 × 1.38, which equals 3.4 magnitudes. Sirius is 3.4 magnitudes brighter than Polaris.

This modern magnitude system has some big advantages. It compresses a tremendous range of intensity into a small range of magnitudes (■ Table 2-1) and makes it possible for modern astronomers to compare their measurements with all the recorded

■ **Table 2-1 I Magnitude and Intensity**

Magnitude Difference	Intensity Ratio
0	1
1	2.5
2	6.3
3	16
4	40
5	100
6	250
7	630
8	1600
9	4000
10	10,000
⋮	⋮
15	1,000,000
20	100,000,000
25	10,000,000,000
⋮	⋮

Scientific Arguments: The Structure of Science

An argument can be a shouting match, but another definition is "a discourse intended to persuade." Scientists construct arguments as part of the business of science not because they want to persuade others they are right, but because they want to test their ideas. For example, a team of biologists might construct a scientific argument to show that ants communicate by smell. If they do their best and cite all the evidence and all the theories and the argument seems convincing, they gain confidence that they do understand how ants "talk" to each other. If the scientists discover a hole in their argument, they know they have more work to do.

A scientific argument is a logical presentation of evidence and theory with interpretations and explanations that help you understand some aspect of nature. Scientists would be free to include any evidence or theory that helps persuade, but they must observe one fundamental rule of science: They must be totally honest—they must include all of the evidence and

all of the theories. The purpose of a scientific argument is to test understanding, not to win votes or sell soap. Dishonesty in a scientific argument is self-deluding, and scientists consider such dishonesty disgraceful.

In the chapters ahead, you will meet lots of scientific arguments about sophisticated theories and detailed evidence involving everything from the formation of Earth to the fate of the universe, but you can begin using scientific arguments now as a way to review your reading. If you can organize what you know about naming stars, for example, and explain why most star names come from Arabic but most constellation names are Latin, and why Greek letters give you clues to the brightness of stars, you will be building a scientific argument that will help you line up the facts you know and convert them to real understanding.

Scientists think in scientific arguments not to convince others, but to test their own understanding. (Roger Ressmeyer/Corbis)

Each section of this book ends with a review tool called "Building Scientific Arguments." Use these as a model to organize your thoughts as you review each concept in a section, and they will help you understand nature the way scientists do.

---

measurements of the past, right back to the first star catalogs over 2000 years ago.

It's time for you to review the preceding section on constellations, star names, and magnitudes. Before you begin, read **Window on Science 2-1** and notice how scientists organize information. How you review is critical when you study a science. Memorizing facts won't help you much, but organizing your understanding into scientific arguments will line the facts up in a meaningful way. The review tool that follows, "Building Scientific Arguments," will help you review an important concept from the section, but you should use the same process to review each concept from the beginning of the section to the end.

Building Scientific Arguments

Nonastronomers sometimes complain that the magnitude scale is awkward. Why would they think it is awkward, and how did it get that way?
Two things might make the magnitude scale seem awkward. First, it is backward; the bigger the magnitude number, the fainter the star. Of course, that arose because ancient astronomers were not measuring the brightness of stars but rather classifying them, and first-class stars would be brighter than

second-class stars. The second awkward feature of the magnitude scale is its mathematical relation to intensity. If two stars differ by one magnitude, one is about 2.5 times brighter than the other. But if they differ by two magnitudes, one is 2.5 × 2.5 times brighter. This mathematical relationship arises because of the way human eyes perceive brightness as ratios of intensity.

Now build your own scientific argument to analyze the following question: **If the magnitude scale is so awkward, why do you suppose astronomers have used it for over two millennia?**

■   ■   ■

Connections: In this section, you have found a way to identify stars by name and by brightness. Now it is time to look at the sky as a whole and notice its motion.

2-2 The Sky and Its Motion

THE SKY ABOVE seems to be a great blue dome in the daytime and a sparkling ceiling at night. Learning to look at the sky requires that you begin thousands of years ago.

Frameworks for Thinking about Nature: Scientific Models

In everyday language, you use the word *model* in various ways—fashion model, model airplane, model student—but scientists use the word in a very specific way. A **scientific model** is a carefully devised mental conception of how something works, a framework that helps scientists think about some aspect of nature. For example, astronomers use the celestial sphere as a way to think about the motions of the sky, sun, moon, and stars.

Although a scientific model is a mental conception, it can take many forms. Some models are quite abstract—the psychologist's model of how the human mind processes visual information into images, for instance. But other models are so specific that they can be expressed as a set of mathematical equations. For example, an astronomer might use a set of equations to describe in detail how gas falls into a black hole. You could refer to such a calculation as a model. Of course, you could use metal and plastic to build a celestial globe, but the thing you built wouldn't really be the model any more than the equations

are a model. A scientific model is a mental conception, an idea that helps you think about nature.

On the other hand, a model is not meant to be a statement of truth. The celestial sphere is not real; you know the stars are scattered through space at various distances, but you can imagine a celestial sphere and use it to help think about the sky. A scientific model does not have to be true to be useful. Chemists, for example, think about the atoms in molecules by visualizing them as balls joined together by rods. This model of a molecule is not fully correct, but it is a helpful way to think about molecules; it gives chemists a framework within which to organize their ideas.

Because scientific models are not meant to be totally true, you must always remember the assumptions on which they are based. If you begin to think a model is true, it can be misleading instead of helpful. The celestial sphere, for instance, can help you think about the sky, but you must remember that it is only

The ancient celestial sphere is a useful model of the sky.

a mental crutch. The universe is much larger and much more interesting than this ancient scientific model of the heavens.

The Celestial Sphere

Ancient astronomers believed the sky was a great sphere surrounding Earth with the stars stuck on the inside like thumbtacks in the ceiling. Modern astronomers know that the stars are scattered through space at different distances, but it is still convenient to think of the sky as a great celestial sphere.

As you study **The Sky around Us** on pages 20–21, notice three important points:

1 The sky appears to rotate westward around Earth each day, but that is a consequence of the eastward rotation of Earth. That produces day and night.

2 Astronomers measure distance across the sky as angles and express them as degrees, minutes, and seconds.

3 What you can see of the sky depends on where you are on Earth. If you lived in Australia, you would see many constellations and asterisms invisible from North America, but you would never see the Big Dipper. Remember our Favorite Star Alpha Centauri? It is in the southern sky and isn't visible from most of the United States. You could just glimpse it above the southern horizon if you were in Miami, but you could see it easily from Australia.

The celestial sphere is an example of a scientific model, a common feature of scientific thought (**Window on Science 2-2**). Notice that a scientific model does not have to be true to be useful. Many scientific models are discussed in the chapters that follow, and you will discover that the most useful models are often quite obviously flawed descriptions of the true facts.

In addition to the daily motion of the sky, Earth's rotation adds a second motion to the sky that can be detected only over centuries.

Precession

Over 2000 years ago, Hipparchus compared a few of his star positions with those made nearly two centuries before and realized that the celestial poles and equator were slowly moving across the sky. Later astronomers understood that this motion is caused by Earth's toplike motion.

If you have ever played with a gyroscope or top, you have seen how the spinning mass resists any change in the direction of its axis of rotation. The more massive the top and the more rapidly it spins, the more difficult it is to change the direction of its axis of rotation. But you probably recall that the axis of even the most

The Sky around Us

1 The eastward rotation of Earth causes the sun, moon, and stars to move westward in the sky as if the **celestial sphere** were rotating westward around Earth. From any location on Earth you see only half of the celestial sphere, the half above the **horizon**. The **zenith** marks the top of the sky above your head, and the **nadir** marks the bottom of the sky directly under your feet. The drawing at right shows the view for an observer in North America. An observer in South America would have a dramatically different horizon, zenith, and nadir.

The apparent pivot points are the **north celestial pole** and the **south celestial pole** located directly above Earth's north and south poles. Halfway between the celestial poles lies the **celestial equator**. Earth's rotation defines the directions you use every day. The **north point** and **south point** are the points on the horizon closest to the celestial poles. The **east point** and the **west point** lie halfway between the north and south points. The celestial equator always touches the horizon at the east and west points.

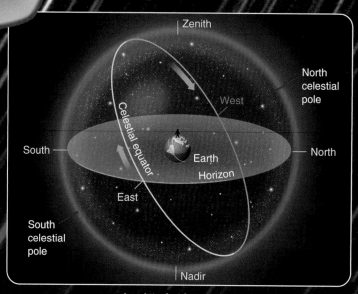

Ace ✺ Astronomy™

Log into AceAstronomy and select this chapter to see Active Figure "Celestial Sphere." Notice how each location on Earth has its unique horizon.

AURA/NOAO/NSF

Looking north

Looking east

Looking south

1a This time exposure of about 30 minutes shows stars as streaks, called star trails, rising behind an observatory dome. The camera was facing northeast to take this photo. The motion you see in the sky depends on which direction you look, as shown at right. Looking north, you see the star Polaris, the North Star, located near the north celestial pole. As the sky appears to rotate westward, Polaris hardly moves, but other stars circle the celestial pole. Looking south from a location in North America, you can see stars circling the south celestial pole, which is invisible below the southern horizon.

Ace ✺ Astronomy™

Log into AceAstronomy and select this chapter to see Active Figure "Rotation of the Sky." Look in different directions and compare the motions of the stars.

Astronomers measure distance across the sky as angles.

Angular distance

North celestial pole
Zenith
Latitude 90°

North celestial pole
Zenith
W
S — — N
Latitude 60°   E

2 Astronomers might say, "The star was only 2 degrees from the moon." Of course, the stars are much farther away than the moon, but when you think of the celestial sphere, you can measure distances *on the sky* as **angular distances** in degrees, minutes of arc, and seconds of arc. A **minute of arc** is 1/60th of a degree, and a **second of arc** is 1/60th of a minute of arc. Then the **angular diameter** of an object is the angular distance from one edge to the other. The sun and moon are each about half a degree in diameter, and the bowl of the Big Dipper is about 10° wide.

North celestial pole
Zenith
W
S — — N
Latitude 30°   E

3 What you see in the sky depends on your latitude as shown at right. Imagine that you begin a journey in the ice and snow at Earth's North Pole with the north celestial pole directly overhead. As you walk southward, the celestial pole moves toward the horizon, and you can see further into the southern sky. The angular distance from the horizon to the north celestial pole always equals your latitude (L)—the basis for celestial navigation. As you cross Earth's equator, the celestial equator would pass through your zenith, and the north celestial pole would sink below your northern horizon.

North celestial pole
Zenith
W
S — — N
Latitude 0°   E

A few circumpolar constellations

Cassiopeia

Cepheus

Perseus

Rotation of sky

Rotation of sky

Polaris

Ursa Minor

Ursa Major

South celestial pole
Zenith
W
S — — N
Latitude −30°   E

3a **Circumpolar constellations** are those that never rise or set. From mid-northern latitudes, as shown at left, you see a number of familiar constellations circling Polaris and never dipping below the horizon. As the sky rotates, the pointer stars at the front of the Big Dipper always point toward Polaris. Circumpolar constellations near the south celestial pole never rise as seen from mid-northern latitudes. From a high latitude such as Norway, you would have more circumpolar constellations, and from Quito, Ecuador, located on Earth's equator, you would have no circumpolar constellations at all.

rapidly spinning top sweeps around in a conical motion. The weight of the top tends to make it tip over, and this combines with its rapid rotation to make its axis sweep around in a conical motion called **precession** (■ Figure 2-7a).

Earth spins like a giant top, but it does not spin upright in its orbit; it is inclined 23.5° from vertical. Earth's axis of rotation happens to be pointed toward a spot near the star Polaris, the north star, and Earth's large mass and rapid rotation keep its axis fixed in space. The axis would not wander at all if Earth were a perfect sphere. However, Earth, because of its rotation, has a slight bulge around its middle, and the gravity of the sun and moon pull on this bulge, tending to twist Earth upright in its orbit. The combination of these forces and Earth's rotation causes Earth's axis to precess in a conical motion, taking about 26,000 years for one cycle (Figure 2-7b).

Because the celestial poles and equator are defined by Earth's rotational axis, precession moves these reference marks on the sky. You would notice no change at all from night to night or year to year, but precise measurements reveal the precessional motion of the celestial poles and equator.

Over centuries, precession has dramatic effects. Egyptian records show that 4800 years ago the north celestial pole was pointed near the star Thuban (α Draconis).

The pole is now approaching Polaris and will be closest to it in about 2100. In about 12,000 years, the pole will have moved to within 5° of Vega (α Lyrae). Figure 2-7c shows the path followed by the north celestial pole. You will discover in later chapters that precession is common among rotating celestial bodies.

As you study astronomy, notice the special terms used to describe such things as precession and the celestial sphere. You need to know those terms, but science is about understanding nature and not about naming its parts (**Window on Science 2-3**). Science is more than just vocabulary.

Building Scientific Arguments

Does everyone see the same circumpolar constellations? Here you must use your imagination and build your argument with great care. You can use the celestial sphere as a convenient model of the sky. A circumpolar constellation is one that does not set or rise. Which constellations are circumpolar depends on your latitude. If you live on Earth's equator, you see all the constellations rising and setting, and there are no circumpolar constellations at all. If you live at Earth's North Pole, all the constellations north of the celestial equator never set, and all the constellations south of the celestial equator never rise. In that case, every constellation is circumpolar. At intermediate latitudes, the circumpolar regions are caps on the sky whose angular radius equals the latitude of the observer. If you live in Iceland, the caps are very large, and if you live in Egypt, near the equator, the caps are much smaller.

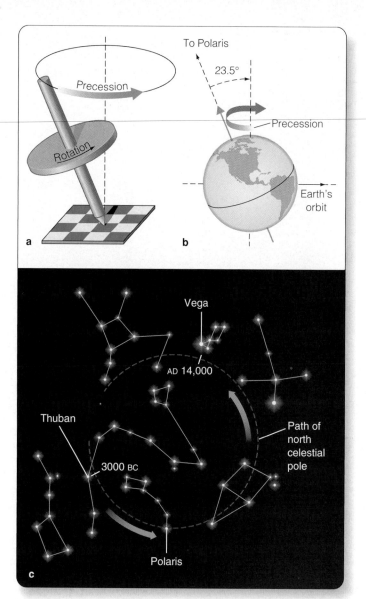

■ **Figure 2-7**

Precession. (a) A spinning top precesses in a conical motion around the perpendicular to the floor because its weight tends to make it fall over. (b) Earth precesses around the perpendicular to its orbit because the gravity of the sun and moon tend to twist it upright. (c) Precession causes the north celestial pole to drift among the stars, completing a circle in 26,000 years.

Locate Ursa Major and Orion on the star charts at the end of this book. For people in Canada, Ursa Major is circumpolar, but people in Mexico see most of this constellation slip below the horizon. From much of the United States, some of the stars of Ursa Major set, and some do not. In contrast, Orion rises and sets as seen from nearly everywhere on Earth. Explorers at Earth's poles, however, never see Orion rise or set.

Now use the argument you have just built. **How would you improve the definition of a circumpolar constellation to clarify the status of Ursa Major? Would your definition help in the case of Orion?**

Understanding versus Naming: The True Goal of Science

One of the fascinations of science is that it can reveal things you normally do not sense, such as a slow drift in the direction of Earth's axis. Furthermore, science can take your imagination into realms beyond your experience, from the inside of an atom to the inside of a star. Because these experiences are so unfamiliar, you need a vocabulary of technical terms just to talk about them, and that leads to a common confusion between naming things and understanding things.

The first step in understanding something is naming the parts, but it is only the beginning. The scientist's true goal is not naming but understanding. Consequently, biologists studying plants go beyond memorizing the names of the flowers' colors. They search for evidence to test their hypothesis that specific plants have developed specific flower colors to attract spe-cific kinds of insects for efficient pollination. Naming the colors is only the beginning.

Some people complain about jargon in science, but the terminology is unfamiliar only because it describes phenomena you don't notice in everyday life. Scientists don't use this specialized vocabulary because they want to be pretentious or confusing. They use proper terminology so they can discuss nature with precision.

You must know the proper words, but real understanding comes from being able to use that vocabulary to discuss the way nature works. You must be able to tell the stories that scientists call theories and hypotheses. You must be able to use technical terms precisely to cite the evidence that makes us think those stories are true. That goes far beyond merely memorizing terminology; that is real understanding.

Naming the parts is only the first step in under-standing how something works. (Royalty-Free/Corbis)

In folklore, naming a thing gives us power over it—recall the story of Rumpelstiltskin. In science, naming things is only the first step toward understanding.

■ ■ ■

Connections: The motions of the celestial sphere may seem to have little to do with your life, but at the end of this chapter you will see how precession may be one of the causes of the ice ages. Before you can think about ice ages, however, you must consider Earth's orbital motion around the sun and the resulting apparent motion of the sun. That affects you directly through the seasons.

(2-3) The Cycles of the Sun

PERHAPS THE MOST OBVIOUS CYCLES in the sky are those of the sun, but those apparent motions are produced by Earth's rotation and revolution. **Rotation** is the turning of a body about an axis through its center. Thus Earth rotates on its axis once a day. **Revolution** is the motion of a body around a point located outside the body, and Earth revolves around the sun once a year.

The cycle of day and night is caused by the rotation of Earth on its axis. In ■ Figure 2-8 you can see that four people in different places on Earth have different times of day, and that Earth's rotation carries you across the daylight side and then across the night side of Earth, producing the cycle of day and night.

Earth's rotation takes a day and causes day and night. Earth's revolution around the sun takes a year and produces the cycle of the seasons. To understand that motion, you must imagine that you can make the sun fainter.

The Annual Motion of the Sun

Even in the daytime, the sky is filled with stars, but the glare of sunlight fills Earth's atmosphere with scattered light, and you can see only the brilliant sun during the daytime. If the sun were fainter and you could see the stars, you would notice that day by day the sun was moving slowly eastward against the background of stars. That motion is caused by Earth's motion along its orbit as it revolves around the sun.

In ■ Figure 2-9, you can see how the sun appears in front of the stars of the constellation Sagittarius on January 1 and how Earth's motion along its orbit causes the sun to appear to move eastward against the constellations. By March 1 the sun appears to be in Aquarius.

■ Figure 2-8

Looking down on Earth from above the North Pole shows how the time of day or night depends on your location on Earth.

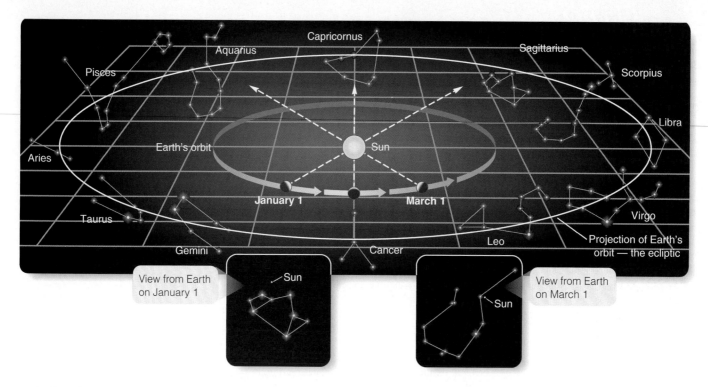

Earth's motion around the sun makes the sun appear to move against the background of the stars. Earth's circular orbit is thus projected on the sky as the circular path of the sun, the ecliptic. If you could see the stars in the daytime, you would notice the sun crossing in front of the distant constellations as Earth moves along its orbit.

Ace◯Astronomy™ Log into AceAstronomy and select this chapter to see the Active Figure called "Constellations in Different Seasons." Notice how the constellations in the morning sky move into the evening sky as the year passes.

Of course, it is not quite correct to say that the sun is "in Aquarius." The sun is only 93 million miles away, and the stars of Aquarius are at least a million times farther away. But in March of each year, you can see the sun against the background of the stars in Aquarius, and thus you could say, "The sun is in Aquarius."

If you continue watching the sun against the background of stars throughout the year, you could plot its path on a star chart. After one full year, you would see the sun begin to retrace its path as it continued its annual cycle of motion around the sky. This line, the apparent path of the sun in its yearly motion around the sky, is called the **ecliptic.** Another way to define the ecliptic is to say it is the projection of Earth's orbit on the sky. If the sky were a great screen illuminated by the sun at the center, then the shadow cast by Earth's orbit would be the ecliptic. Yet a third way to define the ecliptic is to refer to it as the plane of Earth's orbit. These three definitions of the ecliptic are equivalent, and it is worth considering them all because the ecliptic is one of the most important reference lines on the sky.

Earth circles the sun in 365.25 days, and consequently the sun appears to circle the sky in the same period. That means the sun, traveling 360° around the ecliptic in 365.25 days, travels about 1° eastward each day, about twice its angular diameter. You don't notice this motion because you can't see the stars in the daytime, but the motion of the sun has an important consequence that you do notice—the seasons.

The Seasons

The seasons are caused by a simple fact: Earth's axis of rotation is tipped 23.5° from the perpendicular to its orbit (■ Figure 2-10). As you study **The Cycle of the Seasons** on pages 26–27, notice two important principles:

❶ The seasons are not caused by any variation in the distance from Earth to the sun. Earth's orbit is nearly circular, so it is always about the same distance from the sun.

❷ The seasons are caused by the changes in solar energy that Earth's northern and southern hemispheres receive at different times of the year. Because of circulation patterns in Earth's atmosphere, the northern and southern hemispheres are mostly isolated from each other and exchange little heat.

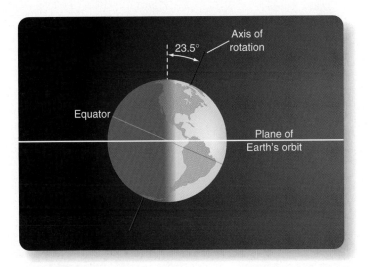

■ Figure 2-10

Earth's axis is inclined 23.5° from the perpendicular to the plane of its orbit. As it moves along its orbit, its axis of rotation remains fixed, pointing at the North Star.

When one hemisphere receives more solar energy than the other, it grows rapidly warmer.

In ancient times, the cycle of the seasons and the solstices and equinoxes were celebrated with rituals and festivals. Shakespeare's play *A Midsummer Night's Dream* describes the enchantment of the summer solstice night. (In Shakespeare's time, the equinoxes and solstices were taken to mark the midpoint of the seasons.) Many North American Indians marked the summer solstice with ceremonies and dances. Early church officials placed Christmas day in late December to coincide with an earlier pagan celebration of the winter solstice.

Building Scientific Arguments

If Earth had a significantly elliptical orbit, how would its seasons be different?

Sometimes, as you review your understanding, it is helpful to build a scientific argument with one factor slightly exaggerated. Suppose Earth had an elliptical orbit so that perihelion occurred in July and aphelion in January. At perihelion, Earth would be closer to the sun, and the entire surface of Earth would be a bit warmer. If that happened in July, it would be summer in the northern hemisphere and winter in the southern hemisphere, and both would be warmer than they are now. It could be a dreadfully hot summer in Canada, and southern Argentina could have a mild winter. Six months later, at aphelion, Earth would be a bit farther from the sun, and if that occurred in January, winter in northern latitudes could be frigid. Argentina, in the southern hemisphere, could be experiencing an unusually cool summer.

Of course, this doesn't happen. Earth's orbit is nearly circular, and the seasons are caused not by a variation in the distance of Earth from the sun but by the inclination of Earth in its orbit.

Nevertheless, Earth's orbit is slightly elliptical. Earth passes perihelion about January 4 and aphelion about July 4. Although Earth's oceans tend to store heat and reduce the importance of this effect, this very slight variation in distance does affect the seasons. Now use your scientific argument to analyze the seasons. **Does the elliptical shape of Earth's orbit make your winters warmer or cooler?**

■   ■   ■

Connections: The sun is not the only body that moves along the ecliptic. The moon and planets maintain constant traffic along the sun's highway.

 Log into AceAstronomy and select this chapter to see Astronomy Exercise "Sunrise through the Seasons" to watch the sun moving.

 Log into AceAstronomy and select this chapter to see Astronomy Exercise "The Seasons" and see how the sun's altitude changes.

(2-4) The Motion of the Planets

THE ECLIPTIC IS IMPORTANT because it is the path the sun follows around the sky, and that motion gives rise to the seasons. In the next chapter, you will consider the moon's cycle around the ecliptic, but here you should make note of the motion of the planets. Not only are most of the planets visible to the naked eye, but their motion is the basis for one of the longest-living superstitions in human history—that our fate is written in the stars.

The Moving Planets

Most of the planets of our solar system are visible to the unaided eye, though they produce no light of their own. They shine by reflected sunlight. Mercury, Venus, Mars, Jupiter, and Saturn are all visible to the naked eye, but Uranus is usually too faint to be seen (magnitude 5.6 at its brightest), and Neptune is never bright enough. Pluto is even fainter, and you would need a large telescope to find it. Although Uranus, Neptune, and Pluto are too faint to see, their motions are the same as those of the other planets.

All the planets of our solar system move in nearly circular orbits around the sun. If you were looking down on the solar system from the north celestial pole, you would see the planets moving in the same counterclockwise direction around their orbits (Chapter 1). The farther they are from the sun, the more slowly the planets move.

1 You can use the celestial sphere to help you think about the seasons. The celestial equator is the projection of Earth's equator on the sky, and the ecliptic is the projection of Earth's orbit on the sky. Because Earth is tipped in its orbit, the ecliptic and equator are inclined to each other by 23.5° as shown at right. As the sun moves eastward around the sky, it spends half the year in the southern half of the sky and half of the year in the northern half. That causes the seasons.

The sun crosses the celestial equator going northward at the point called the **vernal equinox**. The sun is at its farthest north at the point called the **summer solstice**. It crosses the celestial equator going southward at the **autumnal equinox** and reaches its most southern point at the **winter solstice**.

North celestial pole

Celestial equator

Autumnal equinox

Winter solstice

Ecliptic

Summer solstice

23.5°

Vernal equinox

South celestial pole

1a The seasons are defined by the dates when the sun crosses these four points, as shown in the table at the right. *Equinox* comes from the word for "equal"; the day of an equinox has equal amounts of daylight and darkness. *Solstice* comes from the words meaning "sun" and "stationary." *Vernal* comes from the word for "green." The "green" equinox marks the beginning of spring.

Event	Date*	Season
Vernal equinox	March 20	Spring begins
Summer solstice	June 22	Summer begins
Autumnal equinox	September 22	Autumn begins
Winter solstice	December 22	Winter begins

* Give or take a day due to leap year and other factors.

Ace Astronomy™

Log into AceAstronomy and select this chapter to see Active Figure "Seasons" and watch Earth orbiting the sun.

1b On the day of the summer solstice in late June, Earth's northern hemisphere is inclined toward the sun, and sunlight shines almost straight down at northern latitudes. At southern latitudes, sunlight strikes the ground at an angle and spreads out. North America has warm weather, and South America has cool weather.

Earth's axis of rotation points toward Polaris, and, like a top, the spinning Earth holds its axis fixed as it orbits the sun. On one side of the sun, Earth's northern hemisphere leans toward the sun; on the other side of its orbit, it leans away. However, the direction of the axis of rotation does not change.

23.5°

To Polaris

40° N latitude

Equator

40° S latitude

Sunlight nearly direct on northern latitudes

To sun →

Sunlight spread out on southern latitudes

Earth at summer solstice

Summer solstice light

1c Light striking the ground at a steep angle spreads out less than light striking the ground at a shallow angle. Light from the summer-solstice sun strikes northern latitudes from nearly overhead and is concentrated.

Winter solstice light

Light from the winter-solstice sun strikes northern latitudes at a much steeper angle and spreads out. The same amount of energy is spread over a larger area, so the ground receives less energy from the winter sun.

2 The two causes of the seasons are shown at right for someone in the northern hemisphere. First, the noon summer sun is higher in the sky and the winter sun is lower, as shown by the longer winter shadows. Thus winter sunlight is more spread out. Second, the summer sun rises in the northeast and sets in the northwest, spending more than 12 hours in the sky. The winter sun rises in the southeast and sets in the southwest, spending less than 12 hours in the sky. Both of these effects mean that northern latitudes receive more energy from the summer sun, and summer days are warmer than winter days.

At summer solstice

At winter solstice

Ace☉Astronomy™

Log into AceAstronomy and select this chapter to see Active Figure "Path of the Sun" and see this figure from the inside.

Sunlight spread out on northern latitudes →

← To sun

Sunlight nearly direct → on southern latitudes

23.5°   To Polaris

40° N latitude

Equator

40° S latitude

1d On the day of the winter solstice in late December, Earth's northern hemisphere is inclined away from the sun, and sunlight strikes the ground at an angle and spreads out. At southern latitudes, sunlight shines almost straight down and does not spread out. North America has cool weather and South America has warm weather.

Earth's orbit is only very slightly elliptical. About January 4, Earth is at **perihelion**, its closest point to the sun, when it is only 1.7 percent closer than average. About July 4, Earth is at **aphelion**, its most distant point from the sun, when it is only 1.7 percent farther than average. This small variation does not significantly affect the seasons.

When you look for planets in the sky, you will always find them near the ecliptic, because their orbits lie in nearly the same plane as the orbit of Earth. As they orbit the sun, they appear to move generally eastward along the ecliptic. (In Chapter 4, you will discover an exception to this general eastward motion.) In fact, the word *planet* comes from a Greek word meaning "wanderer." Mars moves completely around the ecliptic in slightly less than two years, but Saturn, being farther from the sun, takes nearly 30 years.

As seen from Earth, Venus and Mercury can never move far from the sun because their orbits are inside Earth's orbit. They sometimes appear near the western horizon just after sunset or near the eastern horizon just before sunrise. Venus is easier to locate because its larger orbit carries it higher above the horizon than Mercury (■ Figure 2-11). Mercury's orbit is so small that it can never get farther than about 28° from the sun. Consequently, it is usually hard to see against the sun's glare and is often

■ Figure 2-11

Mercury and Venus follow orbits that keep them near the sun, and they are visible only soon after sunset or before sunrise. Venus takes 584 days to move from morning sky to evening sky and back again, but Mercury zips around in only 116 days.

hidden in the clouds and haze near the horizon. At certain times when it is farthest from the sun, however, Mercury shines brightly and can be located near the horizon in the evening or morning sky. (See the Appendix for the best times to observe Venus and Mercury.)

By tradition, any planet visible in the evening sky is called an **evening star,** although planets are not stars. Any planet visible in the sky shortly before sunrise is called a **morning star.** Perhaps the most beautiful is Venus, which can become as bright as minus fourth magnitude. As Venus moves around its orbit, it can dominate the western sky each evening for about half a year, but eventually its orbit carries it back toward the sun, and it is lost in the haze near the horizon. In a few weeks it reappears in the dawn sky as a brilliant morning star.

Astrology

Seen from Earth, the planets move gradually eastward along the ecliptic, but they don't follow the ecliptic exactly. Also, each travels at its own pace and seems to speed up and slow down at various times. To the ancients, this complex motion reflected the moods of the sky gods, and astrology was born.

Ancient astrologers defined a **zodiac,** a band 18° wide centered on the ecliptic, as the highway the planets follow. They divided this band into 12 segments named for the constellations along the ecliptic—the signs of the zodiac. A **horoscope** shows the location of the sun, moon, and planets among the zodiacal signs with respect to the horizon at the moment of a person's birth as seen from that longitude and latitude. Even if astrology worked, the generalized horoscopes published in newspapers and tabloids can't have been calculated accurately for individual readers.

Astrology buffs argue that a person's personality, life history, and fate are revealed in his or her horoscope, but the evidence contradicts this belief. Astrology has been tested many times over the centuries, and it just doesn't work. Believers, however, don't give up on it. Thus, astrology is a superstition that depends on blind belief and not a science that depends on evidence (**Window on Science 2-4**).

One reason astronomers find astrology irritating is that it has no link to the physical world. For example, precession has moved the constellations so that they no longer match the zodiacal signs. Whatever sign you were "born under," the sun was probably in the previous zodiacal constellation. In fact, if you were born on or between November 30 and December 17, the sun was passing through a corner of the nonzodiacal constellation Ophiuchus, and you have no official zodiacal sign.* Furthermore, astronomers like to point out, there is no mechanism by which

---

* The author of this book was born on December 14 and thus has no astrological sign. An astronomer friend claims that the author must therefore have no personality.

Astrology and Pseudoscience

Astronomers have a low opinion of astrology, not so much because it is groundless but because it pretends to be a science. It is a pseudoscience, from the prefix *pseudo*, meaning false. There are many examples of pseudoscience, and it is illuminating to consider the difference between pseudoscience and science.

A pseudoscience is a set of beliefs that appear to be based on scientific ideas but that fail to obey the most basic rules of science. For example, some years ago a fad arose in which people placed objects under pyramids made of paper, plastic, wire, and so on. The claim was that the pyramidal shape would focus cosmic forces on anything inside and so preserve fruit, sharpen razor blades, and do other miraculous things. Many books promoted this idea, but simple experiments showed that any shape would protect a piece of fruit from airborne spores and allow it to dry without rotting. Likewise, any shape would allow oxidation to improve the cutting edge of a razor blade. In short, experimental evidence contradicted the claim. Nevertheless, supporters of the theory declined to abandon or revise their claims. Thus, the fad of pyramid power was a pseudoscience.

One characteristic of a pseudoscience is that it appeals to our needs and desires. Thus, some pseudoscientific claims are self-fulfilling. For example, some people bought pyramidal tents to put over their beds and thus improve their rest. While there is no logical mechanism by which such a tent could affect a sleeper, because people wanted and expected the claim to be true they slept more soundly. Many pseudoscientific claims involve medical cures, ranging from magnetic bracelets and crystals to focus spiritual power to astonishingly expensive and illegal treatments for cancer. Logic is a stranger to pseudoscience, but human fears and needs are not.

Astrology is a pseudoscience. Over the centuries, astrology has been tested repeatedly, and no correlation has been found. But it survives, and its supporters disregard any evidence that it doesn't work. Like all pseudosciences, astrology is not open to revision in the face of contradictory evidence. Furthermore, astrology fulfills our human need to believe that there is order and meaning to our lives. It may comfort us to believe that our sweetheart has rejected us because of the motions of the planets rather than to admit that we behaved

Astrology may be the oldest pseudoscience.

badly on our last date. Comfort aside, astrology is a poor basis for life decisions.

Human nature and human needs probably ensure that pseudoscientific beliefs will continue to plague us, like emotional viruses propagating from person to person. But if we recognize them for what they are we can more easily guide our lives by rational principles and not by giving credit for our successes and blame for our failings to the stars.

---

the planets could influence us. The gravitational influence of a doctor who is delivering a baby is many times more powerful than the gravitational influence of the planets.

The arguments in the preceding paragraph actually miss the point. Astrology is not related to the physical world at all. It does not matter what constellation the sun occupies, because astrology divides the zodiac into equal mathematical sections, sometimes called houses, and it does not matter to the believer in astrology that the constellations don't match. Furthermore, the physical mechanism is beside the point for the true believer. Astrology is not so much an astronomical superstition as it is a mathematical superstition.

Astrology makes sense only when you think of the world as the ancients did. They believed in multiple sky gods whose moods altered events on Earth. They believed in supernatural influences between natural events and human events. The ancients did not understand natural forces such as gravity as the mechanisms that cause things to happen, and thus they did not believe in cause and effect as modern people do. If a house burned, they might conclude that it caught fire because it was cursed and not because someone was careless with an oil lamp.

Modern science left astrology behind centuries ago, but astrology survives as a fascinating part of human history—an early attempt to understand the meaning of the sky.

Building Scientific Arguments

Planets like Mars can sometimes be seen rising in the east as the sun sets. Why is the same not true for Mercury or Venus? In this case, the argument hinges on simple geometry. Mars has an orbit outside the orbit of Earth, and that means it can reach a location opposite the sun in the sky. As the sun sets in the west, Mars rises in the east. But Mercury and Venus follow orbits that are smaller than the orbit of Earth, so they can never reach a point opposite the sun in the sky.

As they follow their orbits around the sun, Earthlings see them gradually swing out on the east side of the sun, reach a maximum distance from the sun, swing back toward the sun, move out to the west side of the sun, reach a maximum distance west, and then move back toward the sun again.

Because they never get very far from the sun in the sky, they can never be seen rising in the east as the sun sets.

The planetary motions in the sky are produced by the orbital motions of the planets around the sun. Build a geometrical argument to analyze the following question: **What would you see if the planets followed orbits that were all in exactly the same plane as the orbit of Earth?**

■ ■ ■

Connections: Modern astronomers have a low tolerance for astrological superstition, yet the motions of the heavenly bodies do affect human lives. The motion of the sun produces the seasons, and, as you will discover in Chapter 5, the moon governs the tides. In addition, small changes in Earth's orbit may partially control global climate.

(2-5) Astronomical Influences on Earth's Climate

WEATHER IS WHAT HAPPENS TODAY; climate is the average of what happens over decades. Earth has gone through past episodes, called ice ages, when the worldwide climate was cooler and dryer and thick layers of ice covered northern latitudes.

The earliest known ice age occurred about 570 million years ago and the next about 280 million years ago. The most recent ice age began only about 3 million years ago and is still going on. You are living in one of the periodic episodes when the glaciers melt and Earth grows slightly warmer. The current warm period began about 20,000 years ago.

Ice ages seem to occur with a period of roughly 250 million years, and cycles of glaciation within ice ages occur with a period of about 40,000 years. These cyclic changes have an astronomical origin.

The Hypothesis

Sometimes a theory or hypothesis is proposed long before scientists can find the critical evidence to test it. That happened in 1920 when Yugoslavian meteorologist Milutin Milankovitch proposed what became known as the **Milankovitch hypothesis**—that changes in the shape of Earth's orbit, in precession, and in inclination affect Earth's climate and trigger ice ages. Let's examine each of these three motions in turn.

First, astronomers know that the elliptical shape of Earth's orbit varies slightly over a period of about 100,000 years. At present, Earth's orbit carries it 1.7 percent closer than average to the sun during northern-hemisphere winters and 1.7 percent farther away in northern-hemisphere summers. This makes the northern climate very slightly less extreme, and that is critical—most of the landmass where ice can accumulate is in the northern hemisphere. If Earth's orbit became more elliptical, northern summers might be too cool to melt all of the snow and ice from the previous winter. That would make glaciers grow larger.

A second factor is also at work. Precession causes Earth's axis to sweep around a cone with a period of about 26,000 years, and that changes the location of the seasons around Earth's orbit. Northern summers now occur when Earth is 1.7 percent farther from the sun, but in 13,000 years northern summers will occur on the other side of Earth's orbit where Earth is slightly closer to the sun. Northern summers will be warmer, which could melt all of the previous winter's snow and ice and prevent the growth of glaciers.

The third factor is the inclination of Earth's equator to its orbit. Currently at 23.5°, this angle varies from 22° to 24° with a period of roughly 41,000 years. When the inclination is greater, seasons are more severe.

In 1920, Milankovitch proposed that these three factors cycled against each other to produce complex periodic variations in Earth's climate and the advance and retreat of glaciers (■ Figure 2-12a). But no evidence was available to test the theory in 1920, and scientists treated it with skepticism. Many thought it was laughable.

The Evidence

By the middle 1970s, Earth scientists could collect the data that Milankovitch needed. Oceanographers could drill deep into the seafloor and collect samples, and geologists could determine the age of the samples from the natural radioactive atoms they contained. From all this, scientists constructed a history of ocean temperatures that convincingly matched the predictions of the Milankovitch hypothesis (Figure 2-12b).

The evidence seemed very strong, and, by the 1980s, the Milankovitch hypothesis was widely discussed as the leading hypothesis. But science follows a mostly unstated set of rules that holds that a hypothesis must be tested over and over against all available evidence (**Window on Science 2-5**). In 1988, scientists discovered contradictory evidence.

A water-filled crack in Nevada called Devil's Hole contains deposits of the mineral calcite. Diving with scuba gear, scientists drilled out samples of the calcite and analyzed the oxygen atoms found there. For 500,000 years, layers of calcite have built up in Devil's Hole, recording in their oxygen atoms the temperature of the atmosphere when rain fell there. Finding the ages of the mineral samples was difficult, but the results seemed to show that the previous ice age ended thousands of years too early to have been caused by Earth's motions.

These contradictory findings are irritating because we naturally prefer certainty, but such circumstances are common in science. The disagreement between ocean floor samples and Devil's Hole samples triggered a scramble to understand the problem. Were the ages of one or the other set of samples wrong? Were the ancient temperatures wrong? Or were scientists misunderstanding the significance of the evidence?

In 1997, a new study of the ages of the samples confirmed that those from the ocean floor are correctly dated. This seems to

The Foundation of Science: Evidence

Science is based on evidence. Every theory and conclusion must be supported by evidence obtained from experiments or from observation. If a theory is supported by many pieces of evidence but is clearly contradicted by a single experiment or observation, scientists quickly abandon it. For a theory to be true, there can be no contradictory evidence.

Of course, scientists argue about the significance of particular evidence and often disagree on the interpretation of evidence. Some observations may seem significant at first glance, but upon closer examination you may find that the procedure was flawed and so

the piece of evidence is not important. Or you might conclude that the observational fact is correct but is being misinterpreted. The observation may not mean what it seems. Some of the most famous disagreements in science, such as those surrounding Galileo and Darwin, have arisen over the interpretation of well-established factual evidence.

Furthermore, scientists are not allowed to be selective in considering evidence. A lawyer in court can call a certain witness and intentionally fail to ask a critical question that would reveal evidence harmful to the lawyer's case. But a scientist may not ignore any known evidence. The difference in their methods is

revealing. The lawyer is attempting to prove only one side of the case and rightly may ignore contradictory evidence. The scientist, however, is searching for the truth and so must test any theory against all available evidence. In a sense, the scientist, in dealing with evidence, must act as both the prosecution and the defense.

As you read about any science, look for the evidence in the form of measurements or observations. Every theory or conclusion should have supporting evidence. If you can find and understand the evidence, the science will make sense. All scientists, from astronomers to zoologists, demand evidence. You should, too.

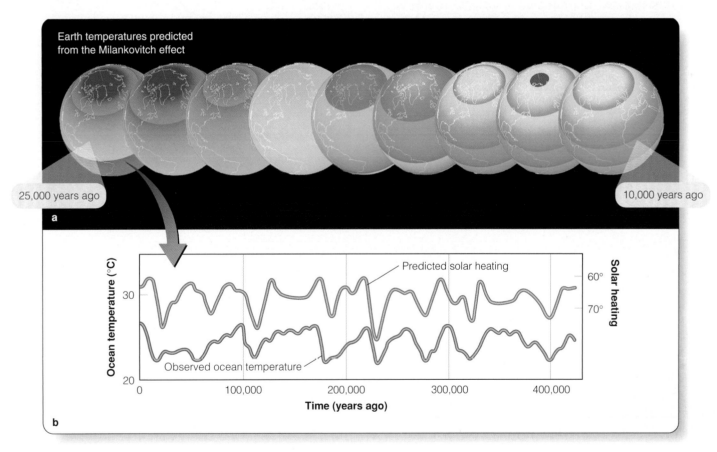

Figure 2-12

(a) Mathematical models of the Milankovich effect can be used to predict temperatures on Earth over time. In these Earth globes, cool temperatures are represented by violet and blue and warm temperatures by yellow and red. These globes show the warming that occurred beginning 25,000 years ago, which ended the last ice age. (Courtesy Arizona State University, Computer Science and Geography Departments) (b) Over the last 400,000 years, changes in ocean temperatures measured from fossils found in sediment layers from the seabed match calculated changes in solar heating. (Adapted from Cesare Emiliani) give scientists renewed confidence in the Milankovitch hypothesis. But the same study found that the ages of the Devil's Hole samples are also correct. Evidently the temperatures at Devil's Hole record local climate changes in the region that became the southwestern United States. The ocean floor samples record global climate changes, and they fit well with the Milankovitch hypothesis. In this way, the Milankovitch hypothesis, though widely accepted today, is still being tested as scientists try to understand the world we live on.

Building Scientific Arguments

How do precession and the shape of Earth's orbit interact to affect Earth's climate?

Here, as in an earlier section of this chapter, exaggeration is a useful analytical tool in your argument. If you exaggerate the variation in the shape of Earth's orbit, you can see dramatically the influence of precession. At present, Earth reaches perihelion during winter in the northern hemisphere and aphelion during summer. The variation in distance is only about 1.7 percent, and that difference doesn't cause much change in the severity of the seasons. But if Earth's orbit were much more elliptical, then winter in the northern hemisphere would be much warmer, and summer would be much cooler.

Now you can see the importance of precession. As Earth's axis precesses, it points gradually in different directions, and the seasons occur at different places in Earth's orbit. In 13,000 years, northern winter will occur at aphelion, and, if Earth's orbit were highly elliptical, northern winter would be terrible. Similarly, summer would occur at perihelion, and the heat would be awful. Such extremes might deposit large amounts of ice in the winter but then melt it away in the hot summer, thus preventing the accumulation of glaciers.

Continue this analysis by exaggeration in your own scientific argument. **What effect would precession have if Earth's orbit were more circular?**

■   ■   ■

Connections: This chapter is about the sky, but it has told you a great deal about Earth. You have discovered that Earth's orbital motion produces the apparent motion of the sun along the ecliptic, causes the seasons, and may even cause the cycle of the ice ages. As is often the case, astronomy helps you understand what you are and where you are. In this case, your study of the sky has helped you understand what it means to live on a planet. However, this discussion has omitted one of the most dramatic objects in the sky—the moon. That omission will be repaired in the next chapter.

Summary

2-1 | The Stars

How do astronomers refer to stars?

■ Astronomers divide the sky into 88 constellations. Although the constellations originated in Greek and Middle Eastern mythology, the names are Latin. Even the modern constellations, added to fill in the spaces between the ancient figures, have Latin names.

■ The names of stars usually come from ancient Arabic, although modern astronomers often refer to a star by its constellation and a Greek letter assigned according to its brightness within the constellation.

How can you compare the brightness of the stars?

■ The magnitude system is the astronomer's brightness scale. First-magnitude stars are brighter than second-magnitude stars, which are brighter than third-magnitude stars, and so on. The magnitude you see when you look at a star in the sky is its apparent visual magnitude, which does not take into account its distance from Earth.

■ Apparent visual magnitude, m_v, includes only the light that human eyes can see.

2-2 | The Sky and Its Motion

How does the sky move as Earth moves?

■ The celestial sphere is a model of the sky, carrying the celestial objects around Earth. Because Earth rotates eastward, the celestial sphere appears to rotate westward on its axis. The northern and southern celestial poles are the pivots on which the sky appears to rotate.

■ The celestial equator, an imaginary line around the sky above Earth's equator, divides the sky in half.

■ Astronomers often refer to angles "on" the sky as if the stars, sun, moon, and planets were equivalent to spots painted on a plaster ceiling. These angular distances are unrelated to the true distance between the objects in light-years.

■ What you see of the celestial sphere depends on your latitude. Much of the southern hemisphere of the sky is not visible from northern latitudes. To see that part of the sky, you would have to travel southward over Earth's surface.

■ The angular distance from the horizon to the north celestial pole always equals your latitude. This is the basis for celestial navigation.

■ The gravitational forces of the moon and sun act on the spinning Earth and cause it to precess like a top. Earth's axis of rotation sweeps around in a conical motion with a period of 26,000 years, and consequently the celestial poles and celestial equator move slowly against the background of the stars.

2-3 | The Cycles of the Sun

What causes the seasons?

■ Because Earth orbits the sun, the sun appears to move eastward along the ecliptic through the constellations. It circles the entire zodiac in a year.

■ Because the ecliptic is tipped 23.5° to the celestial equator, the sun spends half the year in the northern celestial hemisphere and half in the southern celestial hemisphere.

■ In the summer, the sun is above the horizon longer and shines more directly down on the ground. Both effects cause warmer weather.

2-4 | The Motion of the Planets

■ In addition to the sun, the visible planets also move along the ecliptic, and their positions give rise to the ancient superstition called astrology.

■ Mercury and Venus follow orbits inside Earth's orbit and never move far from the sun. They are visible in the east before dawn or in the west after sunset.

2-5 | Astronomical Influences on Earth's Climate

How do astronomical cycles affect Earth's climate?

■ Changes in the shape of Earth's orbit, in its precession, and in its axial tilt can alter the planet's heat balance and seem to be at least partly responsible for the ice ages and glacial periods.

New Terms

constellation (p. 14)	second of arc (p. 21)
asterism (p. 14)	angular diameter (p. 21)
magnitude scale (p. 16)	circumpolar constellation (p. 21)
apparent visual magnitude (m_v) (p. 16)	precession (p. 22)
scientific model (p. 19)	rotation (p. 23)
celestial sphere (p. 20)	revolution (p. 23)
horizon (p. 20)	ecliptic (p. 24)
zenith (p. 20)	vernal equinox (p. 26)
nadir (p. 20)	summer solstice (p. 26)
north celestial pole (p. 20)	autumnal equinox (p. 26)
south celestial pole (p. 20)	winter solstice (p. 26)
celestial equator (p. 20)	perihelion (p. 27)
north point (p. 20)	aphelion (p. 27)
south point (p. 20)	evening star (p. 28)
east point (p. 20)	morning star (p. 28)
west point (p. 20)	zodiac (p. 28)
angular distance (p. 21)	horoscope (p. 28)
minute of arc (p. 21)	Milankovitch hypothesis (p. 30)

Review Questions

Ace⊘Astronomy™ Assess your understanding of this chapter's topics with additional quizzing and animations at **http://ace .brookscole.com/sf9**

1. Why are most modern constellations composed of faint stars or located in the southern sky?

2. What does a star's Greek-letter designation tell you that its ancient Arabic name does not?

3. From your knowledge of star names and constellations, which of the following stars in each group is the brighter and which is the fainter? Explain your answers. (a) α Ursae Majoris; ε Ursae Majoris (b) ε Scorpii; α Pegasus (c) α Telescopium; α Orionis

4. Give two reasons why the magnitude scale might be confusing.

5. Why do modern astronomers continue to use the celestial sphere when they know that stars are not all at the same distance?

6. Define the celestial poles and the celestial equator.

7. From what locations on Earth is the north celestial pole not visible? the south celestial pole? the celestial equator?

8. If Earth did not turn on its axis, could you still define an ecliptic? Why or why not?

9. Give two reasons why winter days are colder than summer days.

10. How do the seasons in Earth's southern hemisphere differ from those in the northern hemisphere?

11. Why should the eccentricity of Earth's orbit make winter in Earth's northern hemisphere different from winter in the southern hemisphere?

12. The stamp at right shows the constellation Orion. Explain why this looks odd to residents of the northern hemisphere.

Discussion Questions

1. Have you thought of the sky as a ceiling? as a dome overhead? as a sphere around Earth? as a limitless void?

2. How would the seasons be different if Earth were inclined 90° instead of 23.5°? 0° instead of 23.5°?

Problems

1. If one star is 6.3 times brighter than another star, how many magnitudes brighter is it?

2. If one star is 40 times brighter than another star, how many magnitudes brighter is it?

3. If two stars differ by 7 magnitudes, what is their intensity ratio?

4. If two stars differ by 8.6 magnitudes, what is their intensity ratio?

5. If star A is third magnitude and star B is fifth magnitude, which is brighter and by what factor?

6. If star A is magnitude 4 and star B is magnitude 9.6, which is brighter and by what factor?

7. By what factor is the sun brighter than the full moon? (*Hint:* See Figure 2-6.)

8. What is the angular distance from the north celestial pole to the summer solstice? to the winter solstice?

9. As seen from your latitude, what is the angle between the north celestial pole and the northern horizon? between the southern horizon and the noon sun at the summer solstice?

Media Cluster

To access the resources in the Media Cluster, log into AceAstronomy at **http://ace.brookscole .com/sf9** and select Chapter 2.

ACTIVE FIGURES

The Celestial Sphere
Observers on Earth see the sky as a celestial sphere. This animation lets you select your vantage point and visualize that location's half-dome of the sky.

Rotation of the Sky
This animation shows how the stars (including the sun) appear to rotate in the sky every 24 hours, as seen by northern hemisphere observers.

Constellations in Different Seasons
This animation lets you observe how the constellations change with the seasons as Earth moves through its orbit of the sun.

Constellations from Different Latitudes
Use this animation to observe how a change in latitude provides a different view of the constellations.

Seasons
Visit different cities around the world and watch the seasons pass as Earth orbits the sun.

Path of the Sun
Use a shadow stick to observe the altitude of the sun as it crosses the sky in different seasons of the year.

ASTRONOMY EXERCISES

Sunrise through the Seasons
In this animation, you can observe how the position of the sunrise changes throughout the year.

The Seasons
Earlier in this chapter, you saw how the angle of the sun in the sky brings about warmer days in the summer and colder days in the winter for the northern hemisphere. In this animation, you can select the latitude and hemisphere to compare the angle of the sun for various points around the world.

Critical Inquiries for the Web

1. Nearly all cultures have populated the sky with gods, heroes, animals, and objects. What can you learn on the Web about non-Western constellations?

2. Who was Orion? How is he related to the scorpion in the sky?

3. What holidays, rituals, special foods, and beliefs are associated with the winter solstice?

4. What can you find out about Milutin Milankovitch? What is the latest news about the Milankovitch hypothesis?

Exploring *TheSky*

1. As discussed in this chapter, Earth's rotation about its own axis gives the impression that the whole sky rotates around the north celestial pole in a period of one day. This apparent motion of the celestial sphere is difficult to notice because it happens so slowly. However, *TheSky* makes it possible to simulate this motion at a pace that is easy to observe by using a feature called **Time Skip.** Observe and describe the apparent motion of the sky as you see it looking north, east, south, and west.

How to proceed: Set the **Time Skip Increment** (a dropdown menu on the **Time Skip Toolbar**) to 1 minute, and click on the **Go Forward** button to begin the simulation. View the sky from the four cardinal directions, due north, south, east, and west. (You'll find **Time Skip** under the **Tools** menu as well.)

2. In which constellation was the sun located on the date of your birth? (*Hint:* Click **Data** and then **Site Information** to set the location, date, and time of your birth. Turn on constellation figures, constellation boundaries, and labels. Then find the sun.)

3. Set your location to Earth's North Pole and the date to the summer solstice. Turn on the ecliptic and then step forward in time through a day to see what happens to the sun. Repeat for the autumnal equinox and the winter solstice.

4. Repeat Problem 3 above for a location on Earth's equator.

Go to the Brooks/Cole Astronomy Resource Center (**http:// astronomy.brookscole.com**) for critical thinking exercises, articles, and additional readings from InfoTrac College Edition, Brooks/Cole's online student library

3 | Cycles of the Moon

Even a man who is

pure in heart

And says his prayers by night

May become a wolf when

the wolfbane blooms

And the moon shines

full and bright.

PROVERB FROM OLD WOLFMAN MOVIES

Visual-wavelength image

D
ID ANYONE EVER WARN YOU,
"Don't stare at the moon—you'll go crazy"?*
For centuries, the superstitious have associated the moon
with insanity. The word *lunatic* comes from a time when even doctors thought that the insane were "moonstruck." A
"mooncalf" is someone who has been crazy since birth, and the word is probably related to the belief that moonlight
can harm unborn children. Everyone "knows" that people act less rationally when the moon is bright, but everyone is

‖ Continued on page 38 ‖

* When I was very small, my grandmother told me if I gazed at the moon, I might go
crazy. But it was too beautiful, and I ignored her warning. I secretly watched the moon
from my window, became fascinated by the sky, and grew up to be an astronomer.

Guidepost

Looking Back

In the preceding chapter, you watched Earth rotating on its axis, which makes the sun rise and set. You watched Earth revolving along its orbit around the sun, which produces the cycle of the seasons. These two cycles so completely dominate your life on Earth that you hardly notice them. But there are other cycles in the sky, and now you are ready to study one of the most dramatic and beautiful actors on the celestial stage.

This Chapter

The moon is the brightest object in the night sky, and it moves rapidly against the background of stars, changing its shape and occasionally producing strange events called eclipses. You may feel you know the moon well, but you will probably find some surprises as you answer four essential questions about Earth's satellite:

Why does the moon go through phases?

What causes a lunar eclipse?

What causes a solar eclipse?

How can eclipses be predicted?

This chapter illustrates two powerful ways to analyze certain kinds of problems. Many problems in astronomy depend on seeing how light and shadow move through space in three dimensions. Phases and eclipses are good case studies. Also, if a process repeats, you can analyze it by searching for cycles. The seemingly complex motions of the sun and moon become simple and elegant when you see them as cycles.

Looking Ahead

Once you have a 21st-century understanding of your world and its motion, you will be ready to read the next chapter, where you will see how Renaissance astronomers analyzed what they saw in the sky and came to a revolutionary conclusion—that we live on a planet.

Ace⊙Astronomy™ The AceAstronomy icon throughout the text indicates an opportunity for you to test yourself on key concepts and to explore animations and interactions on the AceAstronomy website at: http://ace .brookscole.com/sf9 wrong. Careful studies of hospital, school, and police records show that there is no real correlation between the moon and erratic behavior.

The moon is so bright and its cycles through the sky are so dramatic we *expect* it to influence us, and many people are disappointed to learn the moon does not influence us. Some simply refuse to believe the evidence.

The cycles of the moon are indeed dramatic, and the moon is a rich source of myths and traditions. Many different cultures around the world explain eclipses of the sun as invisible monsters devouring the sun (■ Figure 3-1). A Chinese story tells of two astronomers, Hsi and Ho, who were too drunk to predict the solar eclipse of October 22, 2137 BC, or perhaps failed to conduct the proper ceremonies to scare away the dragon snacking on the sun. When the emperor recovered from the terror of the eclipse, he had the two astronomers beheaded. The cycles of the moon provide a wealth of beauty in the sky and an important part of the cultural beliefs of Earth's peoples.

This chapter discusses the lunar cycles of phases and eclipses. Studying these events will help you understand what you see in the night sky. They will also introduce you to some of the most basic concepts in astronomy.

You will discover how light and shadow move through space, how the size and distance of an object affect what you see, and how cyclic events can be analyzed by searching for their patterns.

Most of all, this chapter helps you to begin thinking of your home as a world in space.

3-1 The Changeable Moon

STARTING THIS EVENING, look for the moon in the sky. If the night is cloudy or the moon is in the wrong part of its orbit, you may not see it, but keep trying on successive evenings, and within a week or two you will see the moon. Then watch for the moon on following evenings, and you will see it following its orbit around Earth and cycling through its phases as it has done for billions of years.

The Motion of the Moon

If you watch the moon night after night, you will notice two things about its motion. First, you will see it moving eastward against the background of stars; second, you will notice that the markings on its face don't change. These two observations will help you understand the motion of the moon and the origin of the moon's phases.

The moon moves rapidly among the constellations. If you watch the moon for just an hour, you can see it move eastward against the background of stars by slightly more than its angular diameter. In the previous chapter, you discovered that the moon

■ **Figure 3-1**

(a) A 12th-century Mayan symbol believed to represent a solar eclipse. The black-and-white sun symbol hangs from a rectangular sky symbol, and a voracious serpent approaches from below. (b) This Chinese representation of a solar eclipse shows a dragon flying in front of the sun. (From the collection of Yerkes Observatory) (c) This wall carving from the ruins of a temple at Vijayanagara in southern India symbolizes a solar eclipse as two snakes approaching the disk of the sun. (T. Scott Smith) is about 0.5° in angular diameter, so it moves eastward a bit more than 0.5° per hour. In 24 hours, it moves 13°. Each night when you look at the moon, you see it about 13° eastward of its location the night before. This eastward movement is the result of the motion of the moon along its orbit around Earth.

The Cycle of Phases

The changing shape of the moon as it orbits Earth is one of the most easily observed phenomena in astronomy. Everyone has noticed the full moon rising dramatically or a thin crescent moon hanging in the evening sky. Study **The Phases of the Moon** on pages 40–41 and notice three important points:

1 First, the moon always keeps the same side facing Earth, and you never see the far side of the moon. "The man in the moon" is produced by the familiar features on the moon's near side.

2 Second, the changing shape of the moon as it passes through its cycle of phases is produced by sunlight illuminating different parts of the side of the moon you can see.

3 The third thing to notice is the difference between the orbital period of the moon around Earth and the length of the lunar phase cycle. That difference is a good illustration of how your view from Earth is produced by the combined motions of Earth and other heavenly bodies such as the sun and moon.

You can make a moon-phase dial from the middle diagram on page 40 by covering the lower half of the moon's orbit with a sheet of paper, aligning the edge of the paper to pass through the word "Full" at the left and the word "New" at the right. Push a pin through the edge of the paper at Earth's North Pole to make a pivot, and under the word "Full" write on the paper "Eastern Horizon." Under the word "New" write "Western Horizon." The paper now represents the horizon you see facing south.

You can set your moon-phase dial for a given time by rotating the diagram behind the horizon paper. Set the dial to sunset by turning the diagram until the human figure labeled "sunset" is standing at the top of the Earth globe; the dial shows, for example, that the full moon at sunset would be at the eastern horizon.

We Earthlings always see the same side of the moon looking down on us, but the changing shadows make the man in the moon shift his moods as the moon cycles through its phases (■ Figure 3-2). Watch for the moon and enjoy the cycle of phases. It is one of the perks of living on this planet.

Ace Astronomy™ Log into AceAstronomy and select this chapter to see Astronomy Exercise "Phases of the Moon." You can take control of the moon's phases.

Ace Astronomy™ Log into AceAstronomy and select this chapter to see Astronomy Exercise "Moon Calendar." Find the phases of the moon for any month.

Building Scientific Arguments

Why is the moon sometimes visible in the daytime?
Lots of people are surprised when they notice the moon in the daytime sky, but you aren't surprised because you can explain it with a simple scientific argument involving the geometry of the moon's motion. The full moon rises at sunset and sets at sunrise, so it is visible in the night sky but never in the daytime sky. At other phases, it is possible to see the moon in the daytime. For example, when the moon is a waxing gibbous moon, it rises a few hours before sunset. If you look in the right spot, you can see it in the late afternoon in the southeast sky. It looks pale and washed out because of sunlight in Earth's atmosphere, but it is quite visible once you notice it. You can also locate the waning gibbous moon in the morning sky. It sets an hour or two after sunrise, so you would look for it in the southwestern sky in the morning.

| Waxing crescent | First quarter | Waxing gibbous | Full moon |

■ Figure 3-2

In this sequence of the waxing moon, you see the same face of the moon, the same mountains, craters, and plains, but the changing direction of sunlight produces the lunar phases. (UCO/Lick Observatory)

The Phases of the Moon

1 As the moon orbits Earth, it rotates to keep the same side facing Earth as shown at right. Consequently you always see the same features on the moon, and you never see the far side of the moon. A mountain on the moon that points at Earth will always point at Earth as the moon revolves and rotates.

(Not to scale)

Ace ◯ Astronomy™

Log into AceAstronomy and select this chapter to see Active Figure "Lunar Phases" and take control of this diagram.

First quarter

Waxing gibbous

Waxing crescent

Sunset

North Pole

Midnight — Noon

Earth's rotation

Sunrise

Full

New

← Sunlight

Waning gibbous

Waning crescent

Third quarter

2 As seen at left, sunlight always illuminates half of the moon. Because you see different amounts of this sunlit side, you see the moon cycle through phases. At the phase called "new moon," sunlight illuminates the far side of the moon, and the side you see is in darkness. At new moon you see no moon at all. At full moon, the side you see is fully lit, and the far side is in darkness. How much you see depends on where the moon is in its orbit.

Notice that there is no such thing as the "dark side of the moon." All parts of the moon experience day and night in a month-long cycle.

In the diagram at the left, you see that the new moon is close to the sun in the sky, and the full moon is opposite the sun. The time of day depends on the observer's location on Earth.

2a The first two weeks of the cycle of the moon are shown below by its position at sunset on 14 successive evenings. As the moon grows fatter from new to full, it is said to wax.

The first quarter moon is one week through its 4-week cycle.

Gibbous comes from the Latin word for humpbacked.

Waxing gibbous

Waxing crescent

The full moon is two weeks through its 4-week cycle.

New moon is invisible near the sun

THE SKY AT SUNSET

Full moon rises at sunset

East

South

West

3 The moon orbits eastward around Earth in 27.32 days, its **sidereal period.** This is how long the moon takes to circle the sky once and return to the same position among the stars.

A complete cycle of lunar phases takes 29.53 days, the moon's **synodic period.** (Synodic comes from the Greek words for "together" and "path.")

To see why the synodic period is longer than the sidereal period, study the star charts at the right.

Although you think of the lunar cycle as being about 4 weeks long, it is actually 1.53 days longer than 4 weeks. The calendar divides the year into 30-day periods called months (literally "moonths") in recognition of the 29.53 day synodic cycle of the moon.

New moon

Ecliptic

Sun

New moon

Sagittarius

Scorpius

The sun and moon are near each other at new moon.

One sidereal period after new moon

Ecliptic

Sun

Sagittarius

Moon

Scorpius

One sidereal period after new moon, the moon has returned to the same place among the stars, but the sun has moved on along the ecliptic.

One synodic period after new moon

Sun

Ecliptic

New moon

Sagittarius

Scorpius

One synodic period after new moon, the moon has caught up with the sun and is again at new moon.

You can use the diagram on the opposite page to determine when the moon rises and sets at different phases.

TIMES OF MOONRISE AND MOONSET

Phase	Moonrise	Moonset
New	Dawn	Sunset
First quarter	Noon	Midnight
Full	Sunset	Dawn
Third quarter	Midnight	Noon

2b The last two weeks of the cycle of the moon are shown below by its position at sunrise on 14 successive mornings. As the moon shrinks from full to now, it is said to wane.

The first quarter moon is 3 weeks through its 4-week cycle.

Waning crescent

Waning gibbous

New moon is invisible near the sun

THE SKY AT SUNRISE

Full moon sets at sunrise

East

South

West

A simple scientific argument analyzing the motion of the moon can explain a lot about what you see and what you don't see. **Why is it extremely difficult to see the crescent moon in the daytime?**

■　　■　　■

Connections: The phases of the moon don't affect you directly—you don't act crazier than usual at full moon. The phases are a decoration in the sky, but occasionally, the moon provides a more dramatic treat.

(3-2) Lunar Eclipses

A **LUNAR ECLIPSE** occurs at full moon when the moon moves through Earth's shadow. Because the moon shines only by reflected sunlight, it gradually darkens as it enters the shadow.

Earth's Shadow

Earth's shadow consists of two parts. The **umbra** is the region of total shadow. If you were floating in space in the umbra of Earth's shadow, you would see no portion of the sun; it would be completely hidden behind Earth. However, if you moved into the **penumbra,** you would be in partial shadow and would see part of the sun peeking around Earth's edge. In the penumbra, the sunlight is dimmed but not extinguished.

You can construct a model of Earth's shadow by pressing a map tack into the eraser of a pencil and holding the tack between a lightbulb a few feet away and a white cardboard screen (■ Figure 3-3). The lightbulb represents the sun, and the map tack represents Earth. If you hold the screen close to the tack, you will see that the umbra is nearly as large as the tack and that the penum-

bra is only slightly larger. However, if you move the screen away from the tack, the umbra shrinks and the penumbra expands. Beyond a certain point, the shadow has no dark core at all, indicating that the screen is beyond the end of the umbra.

The umbra of Earth's shadow is about 1.4 million km (860,000 mi) long and points directly away from the sun. A giant screen placed in the shadow at the average distance of the moon would reveal a dark umbra about 9000 km (5700 mi) in diameter, and the faint outer edges of the penumbra would mark a circle about 16,000 km (10,000 mi) in diameter. For comparison, the moon's diameter is only 3476 km (2160 mi). Consequently, when the moon's orbit carries it through the umbra, it has plenty of room to become completely immersed in shadow.

Total Lunar Eclipses

A lunar eclipse occurs when the moon passes through Earth's shadow and grows dark. If the moon passes through the umbra and no part of the moon remains outside the umbra in the partial sunlight of the penumbra, the eclipse is a **total lunar eclipse.**

■ Figure 3-4 illustrates the stages of a total lunar eclipse as a three-dimensional diagram (**Window on Science 3-1**) showing Earth, its shadows, and the path of the moon. As the moon begins to enter the penumbra, it is only slightly dimmed, and a casual observer may not notice anything odd. After an hour, the moon is deeper in the penumbra and dimmer; and, once it begins to enter the umbra, you see a dark bite on the edge of the lunar disk. The moon travels its own diameter in an hour, so it takes about an hour to enter the umbra completely and become totally eclipsed.

Even when the moon is totally eclipsed, it does not disappear completely. Sunlight, bent by Earth's atmosphere, leaks into the umbra and bathes the moon in a faint glow. Because blue

■ Figure 3-3

The shadows cast by a map tack resemble the shadows of Earth and the moon. The umbra is the region of total shadow; the penumbra is the region of partial shadow.

3-D Relationships

Much of science is explained in diagrams; one particular type of diagram can be confusing. When artists draw three-dimensional diagrams on flat sheets of paper, they use lots of artistic clues that have been discovered over the centuries. Perspective, shading, color, and shadows help you see the three-dimensional figure if the drawing is familiar—a house, for example. But when a drawing shows something that is unfamiliar, as is often the case in science, the artist's clues don't work as well. When you look at drawings of a molecule, a nerve cell, layers of rock under the ocean, or Earth and its shadows, you must pay special attention to the three-dimensional nature of the figure.

When you see a three-dimensional diagram in any science book, it helps to decide what the point of view is. For example, if the diagram were a photograph, where was the camera? In the drawing in Figure 3-7, the camera would have to have been in space looking back at Earth and the moon. Of course, astronauts have never been that far from Earth, but you can make such a voyage in your imagination, and that helps you understand the geometry of eclipses. Other three-dimensional diagrams have other points of view, so always be sure to first determine the point of view when you look at any three-dimensional diagram.

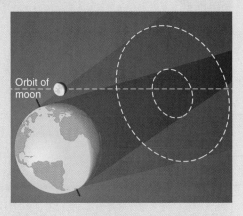

If this were a photograph, where would the photographer have been located?

light is scattered more easily than red light, it is red light that penetrates through Earth's atmosphere to illuminate the moon in a coppery glow. If you were on the moon during totality and looked back at Earth, you would not see any part of the sun because it would be entirely hidden behind Earth. However, you would see Earth's atmosphere illuminated from behind by the sun in a spectacular sunset completely ringing Earth. It is the red glow from this sunset that gives the totally eclipsed moon its reddish color.

How dim the totally eclipsed moon becomes depends on a number of things. If Earth's atmosphere is especially cloudy in those regions that must bend light into the umbra, the moon will be darker than usual. An unusual amount of dust in Earth's atmosphere (from volcanic eruptions, for instance) also causes a dark eclipse. Also, total lunar eclipses tend to be darkest when the moon's orbit carries it through the center of the umbra (■ Figure 3-5).

As the moon moves through Earth's umbral shadow, you can see that the shadow is circular. From this the Greek philosopher Aristotle (384–322 BC) concluded that Earth had to be a sphere, because only a sphere could cast a shadow that was always circular.

Depending on the geometry of the eclipse, the moon can take as long as 1 hour 40 minutes to cross the umbra and another hour to emerge into the penumbra. Still another hour passes as it emerges into full sunlight. A total eclipse of the moon, including the penumbral stage, can take almost six hours from start to finish.

■ Figure 3-4

During a total lunar eclipse, the moon passes through Earth's shadow, as shown in this multiple-exposure photograph. A longer exposure was used to record the moon while it was totally eclipsed. The moon's path appears curved in the photo because of photographic effects. (© 1982 Dr. Jack B. Marling)

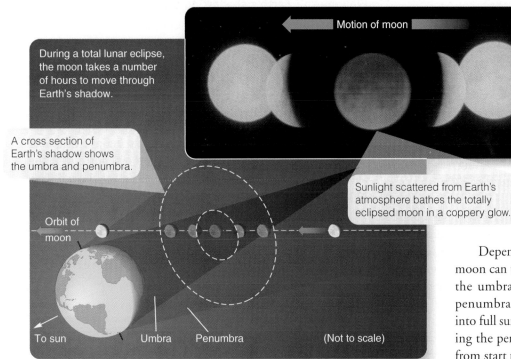

Motion of moon

During a total lunar eclipse, the moon takes a number of hours to move through Earth's shadow.

A cross section of Earth's shadow shows the umbra and penumbra.

Sunlight scattered from Earth's atmosphere bathes the totally eclipsed moon in a coppery glow.

Orbit of moon

To sun Umbra Penumbra (Not to scale)

■ Figure 3-5

During a total lunar eclipse, the moon turns coppery red. In this photo, the moon is darkest toward the lower right, the direction toward the center of the umbra. The edge of the moon at upper left is brighter because it is near the edge of the umbra. (Celestron International)

Partial and Penumbral Lunar Eclipses

Not all eclipses of the moon are total. Because the moon's orbit is inclined a bit over 5° to the plane of Earth's orbit, the moon might not pass through the center of the umbra.

If the moon's orbit carries the full moon too far north or south of the umbra, the moon may only partially enter the umbra. The resulting **partial lunar eclipse** is usually not as dramatic as a total lunar eclipse. Because part of the moon remains outside the umbra, it receives some direct sunlight and looks bright in contrast with the dark part of the moon inside the umbra. Unless the moon almost completely enters the umbra, the glare from the illuminated part of the moon drowns out the fainter red glow inside the umbra. Partial lunar eclipses are interesting because part of the full moon is darkened, but they are not as beautiful as a total lunar eclipse.

If the orbit of the moon carries the moon far enough north or south of the umbra, the moon may only pass through the penumbra and never reach the umbra. Such **penumbral eclipses** are not dramatic at all. In the partial shadow of the penumbra, the moon is only partially dimmed. Most people glancing at a penumbral eclipse would not notice any difference from a full moon.

Total, partial, or penumbral, lunar eclipses are interesting events in the night sky and are not difficult to observe. When the moon passes through Earth's shadow, the eclipse is visible from anywhere on Earth's dark side. One or two lunar eclipses occur in most years. Consult ■ Table 3-1 to find the next lunar eclipse visible in your part of the world.

■ Table 3-1 | Total and Partial Eclipses of the Moon, 2006 to 2013

Year	Time* of Mideclipse (GMT)	Length of Totality (Hr:Min)	Length of Eclipse[†] (Hr:Min)
2006 Sept. 7	18:52	Partial	1:30
2007 Mar. 3	23:22	1:14	3:40
2007 Aug. 28	10:38	1:30	3:32
2008 Feb. 21	3:27	0:50	3:24
2008 Aug. 16	21:11	Partial	3:08
2009 Dec. 31	19:24	Partial	1:00
2010 June 26	11:40	Partial	2:42
2010 Dec. 21	8:18	1:12	3:28
2011 June 15	20:13	1:40	3:38
2011 Dec. 10	14:33	0:50	3:32
2012 June 4	11:03	Partial	2:08

*Times are Greenwich Mean Time. Subtract 5 hours for Eastern Standard Time, 6 hours for Central Standard Time, 7 hours for Mountain Standard Time, and 8 hours for Pacific Standard Time. From your time zone, lunar eclipses that occur between sunset and sunrise will be visible, and those at midnight will be best placed.

[†]Does not include penumbral phase.

Building Scientific Arguments

Why doesn't Earth's shadow on the moon look red during a partial lunar eclipse?

Once again, you need to build an argument based on geometry, but now you must consider the direction of sunlight. During a partial lunar eclipse, part of the moon protrudes from Earth's umbral shadow into sunlight. This part of the moon is very bright compared to the fainter red light inside Earth's shadow, and the glare of the reflected sunlight makes it difficult to see the red glow. If a partial eclipse is almost total, so that only a small sliver of moon extends out of the shadow into sunlight, you can sometimes detect the red glow in the shadow.

Of course, this red glow does not happen for every planet–moon combination in the universe. Adapt your argument for a new situation. **Would a moon orbiting a planet that had no atmosphere glow red during a total eclipse? Why or why not?**

■   ■   ■

Connections: Lunar eclipses are slow and stately. For drama and excitement, there is nothing like a solar eclipse, as you will discover in the next section.

3-3 Solar Eclipses

A SOLAR ECLIPSE OCCURS when the moon moves between Earth and the sun. If the moon covers the disk of the sun completely, the eclipse is a **total solar eclipse.** If the moon covers only part of the sun, the eclipse is a **partial solar eclipse.** During a particular solar eclipse, people in one place on Earth may see a total eclipse, while people only a few hundred kilometers away see a partial eclipse.

These spectacular sights are possible because we on Earth are very lucky. Our moon has the same angular diameter as our sun, so it can cover the sun almost exactly. That lucky coincidence allows you to see total solar eclipses.

The Angular Diameter of the Sun and Moon

You learned about the angular diameter of an object in Chapter 2; now you need to think carefully about how the size and distance of an object like the moon determine its angular diameter. This is the key to understanding solar eclipses.

Linear diameter is simply the distance between an object's opposite sides. You use linear diameter when you order a 16-inch pizza—the pizza is 16 inches in diameter. The linear diameter of the moon is 3476 km. The angular diameter of an object is the angle formed by lines extending toward you from opposite sides of the object and meeting at your eye (■ Figure 3-6). Clearly, the farther away an object is, the smaller its angular diameter.

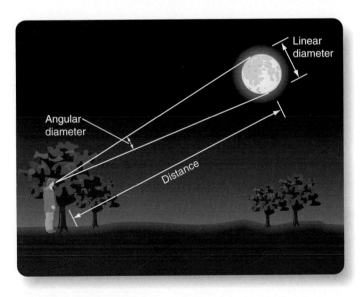

■ Active Figure 3-6

The angular diameter of an object is related to its linear diameter and also to its distance.

Ace ⟲ Astronomy™ Log into AceAstronomy and select this chapter to see the Active Figure called "Small-Angle Formula." Notice that both distance and linear diameter affect an object's angular diameter.

The **small-angle formula** gives you a way to figure out the angular diameter of any object, whether it is a pizza, the moon, or a galaxy. In the small-angle formula, you should always express angular diameter in seconds of arc* and always use the same units for distance and linear diameter:

$$\frac{\text{angular diameter}}{206{,}265''} = \frac{\text{linear diameter}}{\text{distance}}$$

You can use this formula to find any one of these three quantities if you know the other two; here you are interested in finding the angular diameter of the moon.

The moon has a linear diameter of 3476 km and a distance from Earth of about 384,000 km. What is its angular diameter? The moon's linear diameter and distance are both given in the same units, kilometers, so you can put them directly into the small-angle formula:

$$\frac{\text{angular diameter}}{206{,}265''} = \frac{3476 \text{ km}}{384{,}000 \text{ km}}$$

To solve for angular diameter, you can multiply both sides by 206,265 and find that the angular diameter is 1870 seconds of arc. If you divide by 60, you get 31 minutes of arc or, dividing by 60 again, about 0.5°. The moon's orbit is slightly elliptical, so it can sometimes look a bit larger or smaller, but its angular diameter is always close to 0.5°.

You can repeat this calculation for the angular diameter of the sun. The sun is 1.39×10^6 km in linear diameter and 1.50×10^8 km from Earth. If you put these numbers into the small-angle formula, you will discover that the sun has an angular diameter of 1900 seconds of arc, which is 32 minutes of arc or about 0.5°. Earth's orbit is slightly elliptical, and consequently the sun can sometimes look slightly larger or smaller, but it, like the moon, is always close to 0.5° in angular diameter.

By fantastic good luck, you live on a planet with a moon that is almost exactly the same angular diameter as your sun. When the moon passes in front of the sun, it is almost exactly the right size to block the brilliant surface of the sun. Then you see the most exciting sight in astronomy—a total solar eclipse. There are few other worlds where this can happen, because the angular diameters of the sun and a moon rarely match so closely. To see this beautiful sight, all you have to do is arrange to be in the moon's shadow when the moon crosses in front of the sun. That means you have to chase the moon's shadow wherever it sweeps over Earth's surface.

The Moon's Shadow

Like Earth's shadow, the moon's shadow consists of a central umbra of total shadow and a penumbra of partial shadow. What you see when the moon crosses in front of the sun depends on where you

---

* The number 206,265″ is the number of seconds of arc in a radian. When you divide by 206,265″, you convert the angle from seconds of arc to radians.

are in the moon's shadow. The moon's umbral shadow produces a spot of darkness roughly 269 km (167 mi) in diameter on Earth's surface (■ Figure 3-7). (The exact size of the umbral shadow depends on the location of the moon in its elliptical orbit and the angle at which the shadow strikes Earth.) If you are in this spot of total shadow, you see a total solar eclipse. If you are just outside the umbral shadow but in the penumbra, you see part of the sun peeking around the moon, and the eclipse is partial. Of course, if you are outside the penumbra, you see no eclipse at all.

Because of the orbital motion of the moon, its shadow sweeps across Earth at speeds of at least 1700 km/h (1060 mph). To be sure of seeing a total solar eclipse, you must select an appropriate eclipse (■ Table 3-2), plan far in advance, travel to the right place on Earth, and place yourself in the **path of totality,** the path swept out by the umbral spot.

Total Solar Eclipses

A total solar eclipse begins when you first see the edge of the moon encroaching on the sun's disk. This is the moment when the edge of the penumbra sweeps over your location.

During the partial phase, part of the sun remains visible, and it is hazardous to look at the eclipse without protection. Dense filters and exposed film do not necessarily provide protection, because some filters do not block the invisible heat radiation (infrared) that can burn the retina of your eyes. This has led officials to

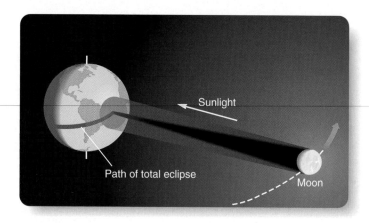

■ **Figure 3-7**

Observers in the path of totality see a total solar eclipse when the umbral shadow sweeps over them. Those in the penumbra see a partial eclipse.

warn the public not to look at solar eclipses and has even frightened some people into locking themselves and their children into windowless rooms during eclipses. In fact, the sun is a bit less dangerous than usual during an eclipse because part of the brilliant surface is covered by the moon. But an eclipse is dangerous in that it can tempt people to look at the sun directly and burn their eyes.

The safest and simplest way to observe the partial phases of a solar eclipse is to use pinhole projection. Poke a small pinhole

■ Table 3-2	Total and Annular Eclipses of the Sun, 2006 to 2016**			
Date	Total/Annular (T/A)	Time of Mideclipse* (GMT)	Maximum Length of Total or Annular Phase (Min:Sec)	Area of Visibility
2006 Mar. 29	T	10^h	4:07	Atlantic, Africa, Turkey
2006 Sept. 22	A	12^h	7:09	N.E. of S. America, Atlantic
2008 Feb. 7	A	4^h	2:14	S. Pacific, Antarctica
2008 Aug. 1	T	10^h	2:28	Canada, Arctic, Siberia
2009 Jan. 26	A	8^h	7:56	S. Atlantic, Indian Ocean
2009 July 22	T	3^h	6:40	Asia, Pacific
2010 Jan. 15	A	7^h	11:10	Africa, Indian Ocean
2010 July 11	T	20^h	5:20	Pacific, S. America
2012 May 20	A	23^h	5:46	Japan, N. Pacific, W. U.S.
2012 Nov. 13	T	22^h	4:02	Australia, S. Pacific
2013 May 10	A	0^h	6:04	Australia, Pacific
2013 Nov. 3	AT	13^h	1:40	Atlantic, Africa
2015 March 20	T	10^h	2:47	N. Atlantic, Arctic
2016 March 9	T	2^h	4:10	Borneo, Pacific
2016 Sept. 1	A	9^h	3:06	Atlantic, Africa, Indian Oc.

The next major total solar eclipse visible from the United States will occur on August 21, 2017.

*Times are Greenwich Mean Time. Subtract 5 hours for Eastern Standard Time, 6 hours for Central Standard Time, 7 hours for Mountain Standard Time, and 8 hours for Pacific Standard Time.

^{h}hours.

**There are no total or annular eclipses of the sun during 2014.

■ Figure 3-8

A safe way to view the partial phases of a solar eclipse. Use a pinhole in a card to project an image of the sun on a second card. The greater the distance between the cards, the larger (and fainter) the image will be.

in a sheet of cardboard. Hold the sheet with the hole in sunlight and allow light to pass through the hole to a second sheet of cardboard (■ Figure 3-8). On a day when there is no eclipse, the result is a small, round spot of light that is an image of the sun. During the partial phases of a solar eclipse, the image shows the dark silhouette of the moon obscuring part of the sun. These pinhole images of the partially eclipsed sun can also be seen in the shadows under trees as sunlight peeks through the tiny openings between the leaves and branches. This can produce an eerie effect just before totality as the remaining sliver of sun produces thin crescents of light on the ground under trees.

Throughout the partial phases of a solar eclipse, the moon gradually covers the bright disk of the sun (■ Figure 3-9). Totality begins as the last sliver of the sun's bright surface disappears behind the moon. This is the moment when the edge of the umbra sweeps over your location. So long as any of the sun is visible, the countryside is bright; but, as the last of the sun disappears, dark falls in a few seconds. Automatic streetlights come on, car drivers switch on their headlights, and birds go to roost. The darkness of totality depends on a number of factors, including the weather at the observing site, but it is usually dark enough to make it difficult to read the settings on cameras.

The totally eclipsed sun is a spectacular sight. With the moon covering the bright disk of the sun, called the **photosphere,*** you

_____

* The photosphere, corona, chromosphere, and prominences will be discussed in detail in Chapter 8. Here the terms are used as the names of features you see during a total solar eclipse.

A Total Solar Eclipse

The moon moving from the right just begins to cross in front of the sun.

The disk of the moon gradually covers the disk of the sun.

Sunlight begins to dim as more of the sun's disk is covered.

During totality, pink prominences are often visible.

A longer-exposure photograph during totality shows the fainter corona.

■ Figure 3-9

This sequence of photos shows the first half of a total solar eclipse. (Daniel Good)

can see the sun's faint outer atmosphere, the **corona,** glowing with a pale, white light so faint you can safely look at it directly. This corona is made of low-density, hot gas, which is given a wispy appearance by the solar magnetic field, as shown in the last frame of Figure 3-9. Also visible just above the photosphere is a thin

(a) During a total solar eclipse, the moon covers the photosphere, and the white corona and pink prominences are visible. Note the streamers in the corona caused by the sun's magnetic field. (Daniel Good) (b) The diamond ring effect can sometimes occur momentarily at the beginning or end of totality if a small segment of the photosphere peeks out through a valley at the edge of the lunar disk. (National Optical Astronomy Observatory)

layer of bright gas called the **chromosphere.** The chromosphere is often marked by eruptions on the solar surface called **prominences** (■ Figure 3-10a), which glow with a clear, pink color due to the high temperature of the gases involved. The small-angle formula reveals that a large prominence is about 3.5 times the diameter of Earth.

Totality cannot last longer than 7.5 minutes under any circumstances, and the average is only 2 to 3 minutes. Totality ends when the sun's bright surface reappears at the trailing edge of the moon. This corresponds to the moment when the trailing edge of the moon's umbra sweeps over the observer.

Just as totality begins or ends, a small part of the photosphere can peek out from behind the moon through a valley at the edge of the lunar disk. Although it is intensely bright, such a small part of the photosphere does not completely drown out the fainter corona, which forms a silvery ring of light with the brilliant spot of photosphere gleaming like a diamond (Figure 3-10b). This **diamond ring effect** is one of the most spectacular of astronomical sights, but it is not visible during every solar eclipse. Its occurrence depends on the exact orientation and motion of the moon.

Once totality is over, daylight returns quickly, and the corona and chromosphere vanish. Not too many years ago astronomers traveled great distances to forbidding places to get their instruments into the path of totality and study the faint outer corona visible only during the few minutes of a total solar eclipse. Now many of those observations can be made every day by solar telescopes in space, but eclipse fans still journey to exotic places for the thrill of seeing a total solar eclipse.

Not every solar eclipse is total. Sometimes when the moon crosses in front of the sun, it is too small to fully cover the sun. That can happen because the orbit of the moon is elliptical and its distance from Earth varies. When the moon is at **perigee,** its closest point to Earth, the moon is almost 12 percent larger in angular diameter than when it is at **apogee,** its most distant point

from Earth. Because Earth's orbit around the sun is very slightly elliptical, the sun's angular diameter can vary by a total of about 3.4 percent. If a solar eclipse occurs when the moon's angular diameter is too small, you would see an **annular eclipse,** a solar eclipse in which a ring (or annulus) of the photosphere is visible around the disk of the moon (■ Figure 3-11). With a portion of the photosphere visible, the eclipse never becomes total; it never quite gets dark; and you can't see the prominences, chromosphere, and corona. An annular eclipse swept across the United States on May 10, 1994.

Building Scientific Arguments

If you were on Earth watching a total solar eclipse, what would astronauts on the moon see when they looked at Earth?
Building this argument requires that you change your point of view and imagine seeing the geometry from a new direction. Astronauts on the moon could see Earth only if they were on the side that faces Earth. Because solar eclipses always happen at new moon, the near side of the moon would be in darkness, and the far side of the moon would be in full sunlight. The astronauts would be standing in darkness, and they would be looking at the fully illuminated side of Earth. They would see a "full Earth." The moon's shadow would be crossing Earth; and, if the astronauts looked closely, they might be able to see the spot of darkness where the moon's umbral shadow touched Earth. It would take hours for the shadow to cross Earth.

Standing on the moon and watching the moon's umbral shadow sweep across Earth would be a cold, tedious assignment. Perhaps you can imagine a more interesting assignment for the astronauts. **What would astronauts on the moon see while people on Earth were seeing a total lunar eclipse?**

■　　　■　　　■

The angular diameters of the moon and sun vary slightly because the orbits of the moon and Earth are slightly eliptical.

If the moon is too far from Earth during a solar eclipse, the umbra does not reach Earth's surface.

Sunlight

Path of annular eclipse

Moon

Annular eclipse of 1994

Disk of sun

Disk of moon centered in front of the sun

■ **Figure 3-11**

An annular eclipse occurs when the moon is far enough from Earth that its umbral shadow does not reach Earth's surface. From Earth, you see an annular eclipse because the moon's angular diameter is smaller than the angular diameter of the sun. In the photograph of the annular eclipse of 1994, the dark disk of the moon is almost exactly centered on the bright disk of the sun. (Moon: UCO/Lick Observatory; Sun: Daniel Good)

Connections: Eclipses of the sun and moon are often dramatic and mysterious. Ancient astronomers studied them and found ways to predict the coming of an eclipse. In the section that follows, you will see how eclipses can be predicted from a basic understanding of the motion of the sun and the moon.

3-4 Predicting Eclipses

To make exact eclipse predictions, you would need to calculate the precise motions of the sun and moon, and that requires a computer and proper software. Such software is available for desktop computers, but it isn't necessary if you are satisfied with making less exact predictions. In fact, many primitive peoples, such as the builders of Stonehenge and the ancient Maya, are believed to have made eclipse predictions.

You should think about eclipse prediction for three reasons. First, it is an important part of the history of science. Second, it illustrates how apparently complex phenomena can be analyzed in terms of cycles. Third, eclipse prediction will exercise your mental muscles and force you to see Earth, the moon, and the sun as objects moving through space.

Conditions for an Eclipse

You can predict eclipses by understanding the conditions that make them possible. As you begin to think about these conditions, be sure you are aware of your point of view. (See Window on Science 3-1.) Later you will change your point of view, but to begin you must imagine that you can look up into the sky from your home on Earth and see the sun moving along the ecliptic and the moon moving along its orbit.

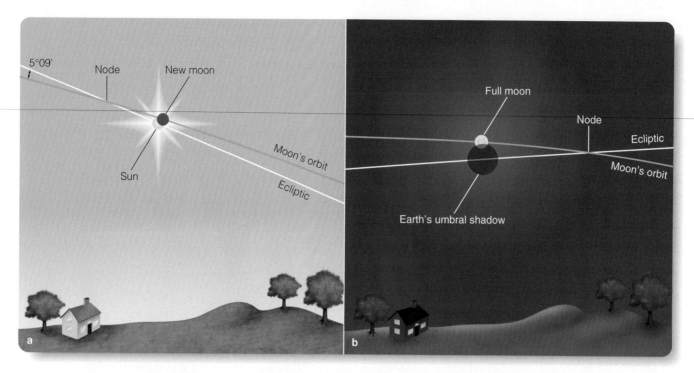

■ Figure 3-12

Eclipses can occur only near the nodes of the moon's orbit. (a) A solar eclipse occurs when the moon meets the sun near a node. (b) A lunar eclipse occurs when the sun and moon are near opposite nodes. Partial eclipses are shown here for clarity.

The orbit of the moon is tipped 5°8′43″ to the plane of Earth's orbit, so you see the moon follow a path tipped by that angle to the ecliptic. Each month, the moon crosses the ecliptic at two points called **nodes.** It crosses at one node going southward, and two weeks later it crosses at the other node going northward.

Eclipses can only occur when the sun is near one of the nodes of the moon's orbit. A solar eclipse happens at new moon if the moon passes in front of the sun. Most new moons pass too far north or too far south of the sun to cause an eclipse. Only when the sun is near a node in the moon's orbit can the moon cross in front of the sun, as shown in ■ Figure 3-12a. A lunar eclipse doesn't happen at every full moon because most full moons pass too far north or too far south of the ecliptic and miss Earth's shadow. The moon can enter Earth's shadow only when the shadow is near a node in the moon's orbit, and that means the sun must be near the other node. This is shown in Figure 3-12b.

So there are two conditions for an eclipse: The sun must be crossing a node, and the moon must be crossing either the same node (solar eclipse) or the other node (lunar eclipse). That means, of course, that solar eclipses can occur only when the moon is new, and lunar eclipses can occur only when the moon is full.

Now you know the secret of predicting eclipses. An eclipse can only occur during a period called an **eclipse season** during which the sun is close to a node in the moon's orbit. For solar eclipses, an eclipse season is about 32 days long. Any new moon

during this period will produce a solar eclipse. For lunar eclipses, the eclipse season is a bit shorter, about 22 days. Any full moon in this period will encounter Earth's shadow and be eclipsed.

This makes eclipse prediction easy. All you have to do is keep track of where the moon crosses the ecliptic (where the nodes of its orbit are). When the sun is near one of these nodes, you can predict that the nearest new moon will cause a solar eclipse and the nearest full moon will cause a lunar eclipse. This system works fairly well, and ancient astronomers such as the Maya may have used such a system. You could be a very successful ancient Mayan astronomer with what you know about eclipse seasons, but you can do even better if you change your point of view.

The View from Space

Change your point of view and imagine that you are looking at the orbits of Earth and the moon from a point far away in space. You would see the moon's orbit as a small disk tipped at an angle to the larger disk of Earth's orbit. As Earth orbits the sun, the moon's orbit remains fixed in direction. The nodes of the moon's orbit are the points where it passes through the plane of Earth's orbit; an eclipse season occurs each time the line connecting these nodes, the **line of nodes,** points toward the sun. Study ■ Figure 3-13 and notice that the line of nodes does not point at the sun in the example at lower left, and no eclipses are possible. At lower right, the line of nodes points toward the sun, and the shadows produce eclipses.

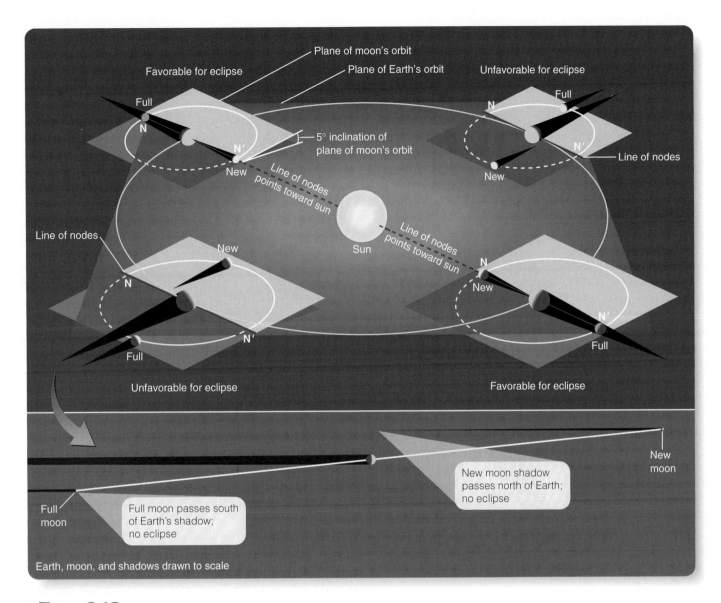

Figure 3-13

The moon's orbit is tipped about 5° to Earth's orbit. The nodes N and N' are the points where the moon passes through the plane of Earth's orbit. If the line of nodes does not point at the sun, the long narrow shadows miss, and there are no eclipses at new moon and full moon. At those parts of Earth's orbit where the line of nodes points toward the sun, eclipses are possible at new moon and full moon.

The shadows of Earth and moon, seen from space, are very long and thin, as shown in the lower part of Figure 3-13. It is easy for them to miss their mark at new moon or full moon and fail to produce an eclipse. Only during an eclipse season, when the line of nodes points toward the sun, do the long, skinny shadows produce eclipses.

If you watched for years from your point of view in space, you would see the orbit of the moon precess like a hubcap spinning on the ground. This precession is caused mostly by the gravitational influence of the sun, and it makes the line of nodes rotate once every 18.6 years. People back on Earth see the nodes slipping westward along the ecliptic 19.4° per year, and the sun takes only 346.62 days (an **eclipse year**) to return to a node. This means

that, according to the calendar, the eclipse seasons begin about 19 days earlier every year (■ Figure 3-14).

The cyclic pattern of eclipses shown in Figure 3-14 should give you another clue how to predict eclipses. Eclipses follow a pattern, and if you were an ancient astronomer and understood the pattern, you could predict eclipses without ever knowing what the moon was or how an orbit works. All you need know is the cycle.

The Saros Cycle

Ancient astronomers could predict eclipses in a crude way using the eclipse seasons, but they could have been much more accurate if they recognized that eclipses occur following certain patterns.

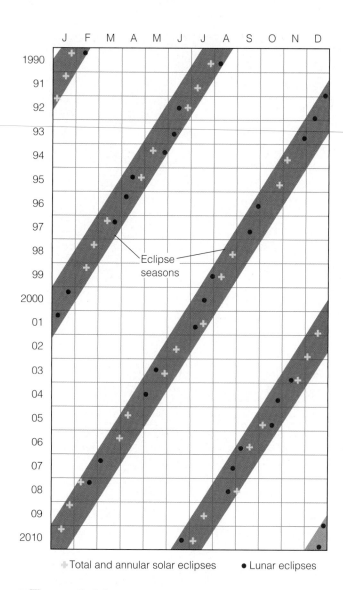

J F M A M J J A S O N D

1990
91
92
93
94
95
96
97
98
99
2000
01
02
03
04
05
06
07
08
09
2010

Eclipse seasons

✛ Total and annular solar eclipses   ● Lunar eclipses

■ Figure 3-14

A calendar of eclipse seasons. Each year the eclipse seasons begin about 19 days earlier. Any new moon or full moon that occurs during an eclipse season results in an eclipse. Only total and annular eclipses are shown here.

The most important of these is the **saros cycle** (sometimes referred to simply as the saros). After one saros cycle of 18 years $11\frac{1}{3}$ days, the pattern of eclipses repeats. In fact, *saros* comes from a Greek word that means "repetition."

The eclipses repeat because, after one saros cycle, the moon and the nodes of its orbit return to the same place with respect to the sun. One saros contains 6585.321 days, which is equal to 223 lunar months. Therefore, after one saros cycle the moon is back to the same phase it had when the cycle began. But one saros is also equal to 19 eclipse years. After one saros cycle, the sun has returned to the same place it occupied with respect to the nodes of the moon's orbit when the cycle began. If an eclipse occurs on a given day, then 18 years $11\frac{1}{3}$ days later the sun, the moon, and the nodes of the moon's orbit return to nearly the same relationship, and the eclipse occurs all over again.

■ Figure 3-15

The saros cycle at work. The total solar eclipse of March 7, 1970, recurred after 18 years $11\frac{1}{3}$ days over the Pacific Ocean. After another interval of 18 years $11\frac{1}{3}$ days, the same eclipse was visible from Asia and Africa. After a similar interval, the eclipse will again be visible from the United States.

Although the eclipse repeats almost exactly, it is not visible from the same place on Earth. The saros cycle is one-third of a day longer than 18 years 11 days. When the eclipse happens again, Earth will have rotated one-third of a turn farther east, and the eclipse will occur one-third of the way westward around Earth (■ Figure 3-15). That means that after three saros cycles—a period of 54 years 1 month—the same eclipse occurs in the same part of Earth.

One of the most famous predictors of eclipses was the Greek philospher Thales of Miletus (about 640–546 BC), who supposedly learned of the saros cycle from the Chaldeans, who had discovered it. No one knows which eclipse Thales predicted, but some scholars suspect the eclipse of May 28, 585 BC. In any case, the eclipse occurred at the height of a battle between the Lydians and the Medes, and the mysterious darkness in midafternoon so startled the two factions that they concluded a truce.

In fact, many historians doubt that Thales actually predicted the eclipse. It would have been very difficult to gather enough information about past eclipses; total solar eclipses are rare as seen from any one place. If you stay in one city, you will see a total solar eclipse about once in 360 years. Also, 585 BC is very early for the Greeks to have known of the saros cycle. The important point is not that Thales did it, but that he could have done it. If he had had records of past eclipses of the sun visible from the area, he could have discovered that they tended to recur with a period of 54 years 1 month (three saros cycles). Indeed, he could have predicted the eclipse without ever understanding what the sun and moon were or how they moved.

Building Scientific Arguments

Why can't two successive full moons be totally eclipsed? Most people suppose that eclipses occur at random or in some pattern so complex you need a big computer to make predictions. You know the geometry is fairly simple, so you can build a simple argument to analyze this question. Remember that a total lunar eclipse occurs when the moon passes through Earth's shadow. Earth's shadow always points toward the ecliptic exactly opposite the sun, and most of the time the moon will pass north or south of Earth's shadow, and there will be no eclipse. A lunar eclipse can happen only when the sun is near one node and the moon crosses Earth's shadow at the other node.

Now you can apply what you know about the moon's phases. An eclipse season for a total lunar eclipse is only 22 days long, but the moon takes 29.5 days to go from one full moon to the next. If one full moon is totally eclipsed, the next full moon 29.5 days later will occur long after the end of the eclipse season and there will be no eclipse.

Now use your knowledge of the cycles of the sun and moon to revise your argument. **Why can the sun be eclipsed by two successive new moons?**

■   ■   ■

Connections: Predicting eclipses isn't very hard, and many ancient peoples were probably familiar enough with the cycles of the sun and moon to know when eclipses were likely. Some people call those first predictions early science, but science involves understanding nature, not just predicting events. In the next chapter, you will see how modern science was born when astronomers tried to understand the cycles they saw in the sky.

Study and Review Tools

Summary

3-1 | The Changeable Moon

Why does the moon go through phases?
- The moon orbits eastward around Earth once a month.
- The moon rotates on its axis as it orbits around Earth and keeps the same side facing Earth throughout the month.
- Because you see the moon by reflected sunlight, its shape appears to change as it orbits Earth and sunlight illuminates different amounts of the side facing Earth.
- The lunar phases wax from new moon to first quarter to full moon and wane from full moon to third quarter to new moon.
- A complete cycle of lunar phases takes 29.53 days.

3-2 | Lunar Eclipses

What causes a lunar eclipse?
- If the full moon passes through Earth's shadow, sunlight is cut off, and the moon darkens in a lunar eclipse. If the moon only grazes the shadow, the eclipse is partial or penumbral and not total.
- The totally eclipsed moon looks copper-red because of sunlight scattered through Earth's atmosphere.

3-3 | Solar Eclipses

What causes a solar eclipse?
- A total solar eclipse occurs if the new moon passes directly between the sun and Earth and the moon's shadow sweeps over Earth's surface. Observers inside the path of totality see a total eclipse, and those just outside the path of totality see a partial eclipse if the penumbra sweeps over their location.
- During a total eclipse of the sun, the bright photosphere of the sun is covered, and the fainter corona, chromosphere, and prominences become visible.
- If during a solar eclipse the moon is in the farther part of its orbit, near apogee rather than perigee, its angular diameter is not large enough to cover the entire photosphere, and you see only an annular eclipse.

3-4 | Predicting Eclipses

How can eclipses be predicted?
- Solar eclipses must occur at new moon, and lunar eclipses must occur at full moon.
- The moon's orbit crosses the ecliptic at two locations called nodes, and eclipses can occur only when the sun is crossing a node. During these periods, called eclipse seasons, a new moon will cause a solar eclipse, and a full moon can cause a lunar eclipse. Knowing when the eclipse seasons occur would allow you to guess which new moons and full moons could cause eclipses.
- Eclipses follow a pattern called the saros cycle. After one saros of 18 years $11\frac{1}{3}$ days, the pattern of eclipses repeats. Some ancient astronomers knew of the saros cycle and used it to predict eclipses.

New Terms

sidereal period (p. 41)	penumbra (p. 42)
synodic period (p. 41)	total eclipse (lunar or solar) (pp. 42, 45)
lunar eclipse (p. 42)	
umbra (p. 42)	

partial eclipse (lunar or solar)
(pp. 44, 45)

penumbral eclipse (p. 44)

small-angle formula (p. 45)

path of totality (p. 46)

photosphere (p. 47)

corona (p. 47)

chromosphere (p. 48)

prominences (p. 48)

diamond ring effect (p. 48)

perigee (p. 48)

apogee (p. 48)

annular eclipse (p. 48)

nodes (p. 50)

eclipse season (p. 50)

line of nodes (p. 50)

eclipse year (p. 51)

saros cycle (p. 52)

Review Questions

 Assess your understanding of this chapter's topics with additional quizzing and animations at http://ace .brookscole.com/sf9

1. Which lunar phases would be visible in the sky at dawn? at midnight?

2. If you looked back at Earth from the moon, what phase would Earth have when the moon was full? new? a first-quarter moon? a waxing crescent?

3. If a planet has a moon, must that moon go through the same phases that Earth's moon displays?

4. Could a solar-powered spacecraft generate any electricity while passing through Earth's umbral shadow? through Earth's penumbral shadow?

5. Draw the umbral and penumbral shadows onto the diagram in the middle of page 40. Explain why lunar eclipses can occur only at full moon and solar eclipses can occur only at new moon.

6. How did lunar eclipses lead Aristotle to conclude that Earth was a sphere?

7. Why isn't the corona visible during partial or annular solar eclipses?

8. Why can't the moon be eclipsed when it is halfway between the nodes of its orbit?

9. Why aren't solar eclipses separated by one saros cycle visible from the same location on Earth?

10. How could Thales of Miletus have predicted the date of a solar eclipse without observing the location of the moon in the sky?

11. The stamp at right shows a crescent moon. Explain why the moon could never look this way.

12. The photo at right shows the annular eclipse of May 30, 1984. How is it different from the annular eclipse shown in Figure 3-11?

Discussion Questions

1. How would eclipses be different if the moon's orbit were not tipped with respect to the plane of Earth's orbit?

2. Are there other planets in our solar system from whose surface you could see a lunar eclipse? a total solar eclipse?

3. Can you detect the saros cycle in Figure 3-14?

Problems

1. Identify the phases of the moon if on March 21 the moon is located at the point on the ecliptic called (a) the vernal equinox, (b) the autumnal equinox, (c) the summer solstice, (d) the winter solstice.

2. Identify the phases of the moon if at sunset the moon is (a) near the eastern horizon, (b) high in the southern sky, (c) in the southeastern sky, (d) in the southwestern sky.

3. About how many days must elapse between first-quarter moon and third-quarter moon?

4. If on March 1 the full moon is near the star Spica, when will the moon next be full? When will it next be near Spica?

5. How many times larger than the moon is the diameter of Earth's umbral shadow at the moon's distance? (Hint: See the photo in Figure 3-4.)

6. Use the small-angle formula to calculate the angular diameter of Earth as seen from the moon.

7. During solar eclipses, large solar prominences are often seen extending 5 minutes of arc from the edge of the sun's disk. How far is this in kilometers? in Earth diameters?

8. If a solar eclipse occurs on October 3: (a) Why can't there be a lunar eclipse on October 13? (b) Why can't there be a solar eclipse on December 28?

9. A total eclipse of the sun was visible from Canada on July 10, 1972. When did this eclipse occur next? From what part of Earth was it total?

10. When will the eclipse described in Problem 9 next be total as seen from Canada?

11. When will the eclipse seasons occur during the current year? What eclipse(s) will occur?

Ace ☾ Astronomy™ To access the resources in the Media Cluster, log into AceAstronomy at **http://ace.brookscole .com/sf9** and select Chapter 3.

Small-Angle Formula
The small-angle formula animation illustrates how the size and distance of an object affect its angular size and how well this simple animation works.

Lunar Phases
Take control of the position of the moon in its orbit and notice how its phases change as seen from Earth.

Phases of the Moon
Sunlight always illuminates half of the moon. Because you see different amounts of this sunlit side, you see the moon cycle through phases. Explore this relationship further with this interactive exercise.

Moon Calendar
This exercise lets you select any month or year in the past, present, or future and see the moon phases for that month.

Critical Inquiries for the Web

1. Search the web for myths about the moon. What common themes do different cultures ascribe to the moon and its cycle of phases?

2 Search for web pages showing photos and observations of a recent total solar eclipse. Can you find web pages that show cultural responses to the eclipse such as celebrations or religious ceremonies?

3. Most people see more total lunar eclipses in a lifetime than total solar eclipses. Why is this so? Compare the regions of visibility for a number of past and upcoming eclipses and determine which future events will be visible from your area.

Exploring *TheSky*

1. How many days must elapse between new moons? Use the **Moon Phase Calendar** under the **Tools** menu to locate new moons.

2. Where would you have to go to see a total solar eclipse in the next year or two? Use the **Eclipse Finder** under the **Tools** menu to select a total solar eclipse. Check **Show Path of Totality** to see a map of Earth.

3. View the total solar eclipse you located in activity 2. In the **Eclipse Finder**, click **View** to see the sun and moon from your present location. Then use **Site Information** under the **Data** menu to change your location to a spot in the path of totality. Set the **Time Step** to 5 minutes and click the **arrows** to move the moon across the sun.

4. Locate and observe an annular eclipse.

5. Locate and observe a total lunar eclipse.

Go to the Brooks/Cole Astronomy Resource Center **(http:// astronomy.brookscole.com)** for critical thinking exercises, articles, and additional readings from InfoTrac College Edition, Brooks/Cole's online student library.

4 | The Origin of Modern Astronomy

How you would burst out laughing, my dear Kepler, if you would hear what the greatest philosopher of the Gymnasium told the Grand Duke about me . . .

FROM A LETTER BY GALILEO GALILEI

GALILEO LAUGHED AT JOKES just as you do. We tend to forget that famous scientists in history laughed and joked, felt satisfaction, and experienced frustration day by day. The astronomers you are about to meet loved their work, and they must have laughed often and taken great satisfaction when their observations went well. But there were moments of frustration too. The history of astronomy is the story of great scientific advances made by astronomers working in many lands, but it is also the story of people who loved the sky and felt joy in their work. ❚ As you read, notice how two subplots twist through the story. The first plot involves astronomers trying to understand the place of the Earth. Is it the center of the universe? Does it move? This is not an idle question of philosophy; it led Galileo before the Inquisition.

❚ Continued on page 58 ❚

Astronomers like Galileo Galilei and Johannes Kepler struggled to understand the place of the Earth and the motion of the planets.

Guidepost

Looking Back

The previous chapters gave you a modern view of the heavens. You are now familiar with what you see in the sky, and you also can imagine how the Earth, moon, and sun move through space. Now you are ready to understand one of the most sweeping revolutions in human thought: the realization that we live on a planet.

This Chapter

All cultures have tried to explain their place under the sky, and the ancient Greek philosophers developed an elaborate theory. By the 16th century, however, many astronomers were uncomfortable with the ancient theory that Earth sat unmoving at the center of a spherical universe. In this chapter, you will discover how an astronomer named Copernicus changed the ancient theory, how Galileo Galilei changed the rules of the debate, and how astronomers changed the way they thought about motion in the heavens. Here you will find answers to four essential questions:

How did the ancients describe the place of the Earth?

How did Copernicus change the place of the Earth?

Why was Galileo condemned by the Inquisition?

How did Copernican astronomers solve the puzzle of planetary motion?

This chapter is not just about the history of astronomy. As they struggled to understand Earth and the heavens, the astronomers of the Renaissance invented a new way of understanding nature—a way of thinking that is now called science.

Looking Ahead

This is a book of science, and every chapter that follows will use the methods that were invented when Copernicus tried to repair the ancient theory that Earth was the center of the universe. One of the greatest adventures in that story began when an Englishman wondered why objects fall. That key to the universe is the subject of the next chapter.

The second plot, the puzzle of planetary motion, reveals astronomers observing the sky and trying to understand how the sun, moon, and planets move. That puzzle led eventually to the discovery of gravity, understanding orbital motion, and putting astronauts in space.

This story goes far beyond the birth of astronomy. The quest for the place of the Earth and the puzzle of planetary motion became, in the 16th and 17th centuries, the center of a storm of controversy over the best way to understand our world and ourselves. That conflict is now known as the scientific revolution. As you read about the birth of astronomy, notice that you are also reading about the birth of modern science.

4-1 The Roots of Astronomy

ASTRONOMY HAS ITS ORIGIN in that most noble of all human traits, curiosity. Just as modern children ask their parents what the stars are and why the moon changes, so did ancient peoples ask themselves those same questions. The answers, often couched in mythical or religious terms, reveal a great reverence for the order of the heavens.

Archaeoastronomy

Most of the history of astronomy is lost forever. You can't go to a library or search the Internet to find out what the first astronomers thought about their world because they left no written record. The study of the astronomy of ancient peoples has been called **archaeoastronomy**, and, in spite of the fog of eons gone, it tells you one thing: trying to understand the heavens is part of human nature.

Perhaps the best-known example of archaeoastronomy is also a huge tourist attraction. Stonehenge, standing on Salisbury Plain in southern England, was built in stages from about 3000 BC to about 1800 BC, a period extending from the late Stone Age into the Bronze Age. Though the public is most familiar with the massive stones of Stonehenge, those were added late in its history. In its first stages, Stonehenge consisted of a circular ditch slightly larger in diameter than the length of a football field, with a concentric bank just inside the ditch and a long avenue leading away toward the northeast. A massive stone, the Heelstone, stood then, as it does now, outside the ditch in the opening of the avenue.

As early as AD 1740, the English scholar W. Stukely suggested that the avenue pointed toward the rising sun at the summer solstice, but few accepted the idea. More recently, astronomers have recognized significant astronomical alignments at Stonehenge. Seen from the center of the monument, the summer-solstice sun rises behind the Heelstone. Other sight lines point toward the most northerly and most southerly risings of the moon (■ Figure 4-1).

The significance of these alignments has been debated. Some have claimed that the Stone Age people who built Stonehenge

■ **Figure 4-1** ▶

The central horseshoe of upright stones is only the most obvious part of Stonehenge. The best-known astronomical alignment at Stonehenge is the summer solstice sun rising over the Heelstone. Although a number of astronomical alignments have been found at Stonehenge, experts debate their significance. (Photo: Jamie Backman)

were using it as a device to predict lunar eclipses. After studying eclipse prediction in the previous chapter, you understand that predicting eclipses is easier than most people assume. You could use Stonehenge to predict eclipses, but was that the intention of the people who built it? Some experts doubt that eclipse prediction was the main use of the monument. The truth may never be known. The builders of Stonehenge had no written language and left no records of their intentions. Nevertheless, the presence of solar and lunar alignments at Stonehenge and at many other Stone Age monuments dotting England and continental Europe shows that so-called primitive peoples were paying detailed attention to the sky. The roots of astronomy lie not in sophisticated science and logic but in human curiosity and wonder.

Astronomical alignments in sacred structures are common all around the world. For example, many tombs are oriented toward the rising sun, and Newgrange, a 5000-year-old passage-grave in Ireland (■ Figure 4-2), faces southeast so that, at dawn on the day of the winter solstice, light from the rising sun shines into its long passageway and illuminates the central chamber. No one today knows what the alignment meant to the builders of Newgrange, and many experts doubt that it was actually intended to be a tomb. Whatever its original purpose, Newgrange is clearly a sacred site linked by its alignment to the order and power of the sky. Astronomical alignments in sacred structures are not just an ancient custom. Many cathedrals are carefully oriented to face east.

Building astronomical alignments into structures gives them meaning by connecting them with the heavens. Navajo Indians of the American southwest, for example, have for centuries built their hogans with the door facing east so they can greet the rising sun each morning with prayers and offerings.

Some alignments may have served calendrical purposes. The 2000-year-old Temple of Isis in Dendera, Egypt, was build to align with the rising point of the bright star Sirius. The first appearance of this star in the dawn twilight marked the flooding of the Nile, so it was an important calendar indicator. The symbolism goes further. According to Egyptian mythology, the goddess Isis was associated with the star Sirius, and her husband, Osirus, was linked to the constellation you know as Orion and also to the Nile, the source of Egypt's agricultural fertility.

An American site in New Mexico is known as the Sun Dagger, but there is no surviving mythology to help you understand it. At noon on the day of the summer solstice, a narrow dagger of sunlight shines across the center of a spiral carved on a cliff face high above the desert floor (■ Figure 4-3). The purpose of the

Sunrise on the morning of the summer solstice

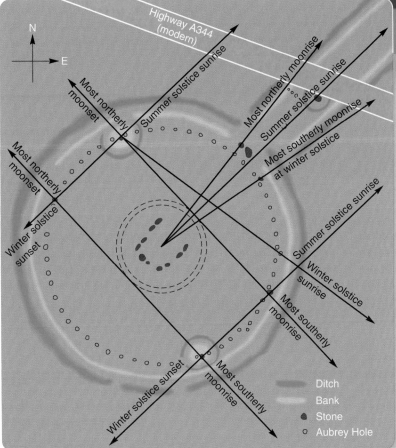

Ditch
Bank
Stone
Aubrey Hole

■ Figure 4-2

Newgrange was built on a small hill in Ireland about 3200 BC. A long passage-way extends from the entryway back to the center of the mound, and sunlight shines down the passageway into the central chamber at dawn on the day of the winter solstice. Other passage graves have similar alignments, but their purpose is unknown. (bottom, Newgrange: Benelux Press/Index Stock Imagery)

Sun Dagger is open to debate, but similar examples have been found throughout the American southwest. It may have been more a symbolic and ceremonial marker than a precise calendrical indicator. In any case, it is just one of the many astronomical alignments that ancient people built into their structures to link themselves with the sky.

Some scholars are looking not at temples but at small artifacts from thousands of years ago. Scratches on certain bone and stone implements seem to follow a pattern and may be an attempt to keep a record of the phases of the moon (■ Figure 4-4). Some scientists contend that humanity's first attempts at writing were stimulated by a desire to record and predict lunar phases.

Archaeoastronomy is uncovering the earliest roots of astronomy and simultaneously revealing some of the first human efforts at systematic inquiry. The most important lesson of archaeoastronomy is that humans don't have to be technologically sophisticated to admire and study the universe.

One thing about archaeoastronomy is especially sad. Although archaeoastronomy can show how ancient people observed the sky, their thoughts about their universe are in many cases lost. Many had no written language. In other cases, the written record has been lost. Dozens, perhaps hundreds, of beautiful Mayan manuscripts, for instance, were burned by Spanish missionaries who believed that the books were the work of Satan. Only four of these books have survived, and all four contain astronomical references. One contains sophisticated tables that allowed the Maya to predict the motion of Venus and eclipses of the moon. No one will ever know what was burnt.

The fate of the Mayan books illustrates one reason why histories of astronomy usually begin with the Greeks. Some of their writing has survived, and you can discover what they thought about the shape and motion of the heavens.

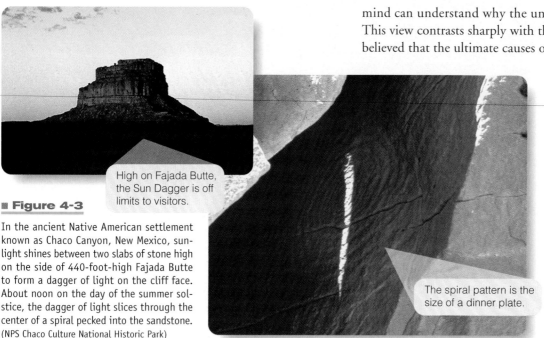

High on Fajada Butte, the Sun Dagger is off limits to visitors.

The spiral pattern is the size of a dinner plate.

■ Figure 4-3

In the ancient Native American settlement known as Chaco Canyon, New Mexico, sunlight shines between two slabs of stone high on the side of 440-foot-high Fajada Butte to form a dagger of light on the cliff face. About noon on the day of the summer solstice, the dagger of light slices through the center of a spiral pecked into the sandstone. (NPS Chaco Culture National Historic Park)

The Astronomy of Greece

Greek astronomy was based on the astronomy of Babylon and Egypt, but these astronomies were heavily influenced by religion and astrology. The astronomers of that time studied the motions of the heavens as a way of worship and divination. The Greek astronomers studied astronomy in an entirely new way—they tried to understand the universe.

This new attitude toward the heavens, a truly scientific attitude, was made possible by two early Greek philosophers. Thales of Miletus (c. 624–547 BC) lived and worked in what is now Turkey. He taught that the universe is rational and that the human

■ Figure 4-4

A fragment of a 27,000-year-old mammoth tusk found at Gontzi in Ukraine contains scribe marks on its edge, simplified in this drawing. These markings have been interpreted as a record of four cycles of lunar phases. Although controversial, such finds suggest that some of the first human attempts at recording events in written form were stimulated by astronomical phenomena.

mind can understand why the universe works the way it does. This view contrasts sharply with those of earlier cultures, which believed that the ultimate causes of things are mysteries beyond human understanding. To Thales and his followers, the mysteries of the universe are mysteries because they are unknown, not because they are unknowable.

The second philosopher who made the new scientific attitude possible was Pythagoras (c. 570–500 BC). He and his students noticed that many things in nature seem to be governed by geometrical or mathematical relations. Musical notes, for example, are related in a regular way to the lengths of plucked strings. This led Pythagoras to propose that all nature was underlain by musical principles, by which he meant mathematics. One result of this philosophy was the later belief that the harmony of the celestial movements produced actual music, the music of the spheres. But, at a deeper level, the teachings of Pythagoras made Greek astronomers look at the universe in a new way. Thales said that the universe could be understood, and Pythagoras said that the underlying rules were mathematical.

In trying to understand the universe, Greek astronomers did something that Babylonian astronomers had never done—they tried to construct descriptions based on geometrical forms. Anaximander (c. 611–546 BC) described a universe made up of wheels filled with fire: The sun and moon are holes in the wheels through which the flames can be seen. Philolaus (fifth century BC) argued that Earth moves in a circular path around a central fire (not the sun), which is always hidden behind a counterearth located between the fire and Earth. This was the first theory to suppose that Earth is in motion.

Plato (428–347 BC) was not an astronomer, but his teachings influenced astronomy for 2000 years. Plato argued that the reality humans see is only a distorted shadow of a perfect, ideal form. If human observations are distorted, then observation can be misleading, and the best path to truth is through pure thought on the ideal forms that underlie nature.

Plato also argued that the most perfect geometrical form was the sphere, and therefore, he said, the perfect heavens must be made up of spheres rotating at constant rates carrying objects around in circles. Consequently, later astronomers tried to describe the motions of the heavens by imagining multiple, rotating spheres. This became known as the principle of **uniform circular motion.**

■ Figure 4-5

The spheres of Eudoxus explain the motions in the heavens by means of nested spheres rotating about various axes at different rates. Earth is located at the center.

■ Figure 4-6

Aristotle, honored on this Greek stamp, wrote on such a wide variety of subjects and with such deep insight that he became the great authority on all matters of learning. His opinions on the nature of Earth and the sky were widely accepted for almost two millennia.

Pythagoras had taught that Earth is a sphere and that the other heavenly bodies were divine, perfect spheres moving in perfect circles. Eudoxus of Cnidus (409–356 BC), a student of Plato, combined a system of 27 nested spheres rotating at different rates about different axes at constant rates to produce a mathematical description of the motions of the universe (■ Figure 4-5).

At the time of the Greek philosophers, it was common to refer to systems such as that of Eudoxus as descriptions of the world, where the word *world* included not only Earth but all of the heavenly spheres. The reality of these spheres was open to debate. Some thought of the spheres as nothing more than mathematical ideas that described motion in the world model, while others began to think of the spheres as real objects made of perfect celestial material. Aristotle, for example, seems to have thought of the spheres as real.

Aristotle and the Nature of Earth

Aristotle (384–322 BC), one of Plato's students, taught and wrote on philosophy, history, politics, ethics, poetry, drama, and so on (■ Figure 4-6). Because of his sensitivity and insight, he became the great authority of antiquity, and astronomers for almost 2000 years cited him as their authority in adopting the Greek model of the universe.

Much of what Aristotle wrote about scientific subjects was wrong, but he was not a scientist. He was trying to understand the universe by creating a system of formal philosophy. Modern science depends on evidence and hypothesis, but scientific methods had not been invented at the time of Aristotle. Rather, he attempted to use the most basic observations combined with first principles (ideas that he believed were obviously true) to understand the world. The perfection of the heavens was, for Aristotle, a first principle.

Aristotle believed that the universe was divided into two parts—Earth, corrupt and changeable; and the heavens, perfect and unchanging. Like most of his predecessors, he believed that Earth was the center of the universe, so his model is called a **geocentric universe.** The heavens surrounded Earth, and he devised 55 crystalline spheres turning at different rates and at different angles to carry the sun, moon, and planets across the sky. The lowest sphere, that of the moon, marked the boundary between the changeable imperfect region of Earth and the unchanging perfection of the celestial realm above the moon.

Because he believed Earth to be immobile, Aristotle had to make these spheres whirl westward around Earth to produce day and night and move more slowly with respect to one another to produce the motions of the sun, moon, and planets against the background of the stars. Because his model was geocentric, he taught that Earth could be the only center of motion. All of his whirling spheres had to be centered on Earth. Like most other Greek philosophers, Aristotle viewed the universe as a perfect heavenly machine that was not many times larger than Earth itself.

About a century after Aristotle, the Alexandrian philosopher Aristarchus proposed a theory that Earth rotated on its axis and revolved around the sun. This theory is, of course, correct, but most of the writings of Aristarchus were lost, and his theory was not well known. Later astronomers rejected any suggestion that

Earth could move because it conflicted with the teachings of the great philosopher Aristotle.

Aristotle had taught that Earth had to be a sphere because it always casts a round shadow during lunar eclipses, but he could only estimate its size. About 200 BC, Eratosthenes, working in the great library in Alexandria, found a way to calculate Earth's radius. He learned from travelers that the city of Syene (Aswan) in southern Egypt contained a well into which sunlight shone vertically on the day of the summer solstice. This told him that the sun was at the zenith at Syene; but, on that same day in Alexandria, he noted that the sun was $\frac{1}{50}$ of the circumference of the sky (about 7°) south of the zenith.

Because sunlight comes from such a great distance, its rays arrive at Earth traveling almost parallel. That allowed Eratosthenes to use simple geometry to find that the distance from Alexandria to Syene was $\frac{1}{50}$ of Earth's circumference (■ Figure 4-7).

To find Earth's circumference, Eratosthenes had to know the distance from Alexandria to Syene. Travelers told him it took 50 days to cover the distance, and he knew that a camel can travel about 100 stadia per day. That meant the total distance was about 5000 stadia. If 5000 stadia is $\frac{1}{50}$ of Earth's circumference, then Earth must be 250,000 stadia around, and, dividing by 2π, Eratosthenes found Earth's radius to be 40,000 stadia.

How accurate was Eratosthenes? The stadium (singular of *stadia*) had different lengths in ancient times. If you assume 6 stadia to the kilometer, then Eratosthenes's result was too big by only 4 percent. If he used the Olympic stadium, his result was 14 percent too big. In any case, this was a much better measurement of Earth's radius than Aristotle's estimate, which was much too small, about 40 percent of the true radius.

You might think this is just a disagreement between two ancient philosophers, but it had an interesting consequence. At the time of Christopher Columbus, all educated people knew the world was round and not flat, but they weren't sure how big it was. Columbus, like many others, adopted Aristotle's diameter for Earth, so he thought Earth was small enough that he could sail west and reach Japan and the Spice Islands of the East Indies. If he had accepted Eratosthenes' diameter, Columbus would never have risked the voyage. As it turned out, he and his crew were lucky that North America was in the way. If there had been open ocean all the way, they would have starved to death long before they reached Japan.

Aristotle was a philosopher and, following Plato's recommendation, made few direct observations. Eratosthenes made some measurements to find the diameter of Earth, but the next person you need to meet was a real astronomer who observed the sky in detail. Little is known about him, but he is clearly the first great observer. Hipparchus (see Figure 2-5) lived during the second century BC, about two centuries after Aristotle. He is usually credited with the invention of trigonometry, he compiled the first star catalog, and he discovered precession (Chapter 2). Instead of describing the motion of the sun and moon using nested spheres, as most Greek philosophers did, Hipparchus proposed that the sun and moon traveled around circles with Earth near, but not at, their centers. These off-center circles are now known as **eccentrics.** Hipparchus recognized that he could produce the same motion by having the sun, for instance, travel around a small circle that followed a larger circle around Earth. The compounded circular motion that he devised became the key element in the masterpiece of the last great astronomer of classical times, Ptolemy.

The Ptolemaic Universe

Claudius Ptolemaeus was one of the great astronomer-mathematicians of antiquity. His nationality and birth date are unknown, but he lived and worked in the Greek settlement at

On the day of the summer solstice, sunlight fell to the bottom of a well at Syene, but the sun was about $\frac{1}{50}$ of a circle (7°) south of the zenith at Alexandria. From this, Eratosthenes was able to calculate Earth's radius.

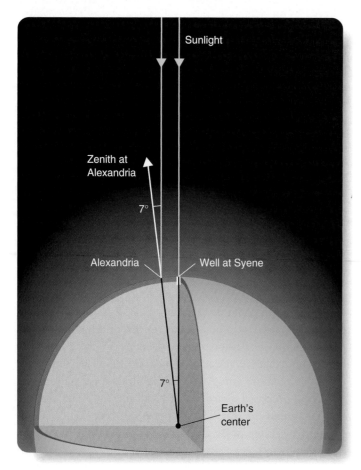

Alexandria in what is now Egypt about AD 140. He ensured the survival of Aristotle's universe by transforming it into a sophisticated mathematical model.

Study **The Ancient Universe** on pages 64–65 and notice three important ideas that show how first principles influenced early descriptions of the universe and its motions:

❶ Ancient philosophers and astronomers accepted as first principles that Earth was located at the center of the universe and that the heavens moved in uniform circular motion.

❷ Second, notice how the observed motion of the planets, the evidence, did not fit the theory very well. The retrograde loops the planets made were very difficult to explain using geocentrism and uniform circular motion.

❸ Finally, notice how Ptolemy attempted to explain the motion of the planets by devising a small circle rotating along the edge of a larger circle, which enclosed a slightly off-center Earth. He even allowed the speed of the planets to vary slightly as they circled Earth. Ptolemy lived roughly five centuries after Aristotle, and although Ptolemy believed in the Aristotelian universe, he was interested in a different problem—the motion of the planets. He was a brilliant mathematician, and he was mainly interested in creating a mathematical description of the motions he saw in the heavens. For him, first principles took second place to mathematical precision.

Aristotle's universe, as embodied in the mathematics of Ptolemy, dominated ancient astronomy, but it was wrong. The planets don't follow circles at uniform speeds. At first the Ptolemaic system predicted the positions of the planets well; but, as centuries passed, errors accumulated. If your watch gains only one second a year, it will keep time well for many years, but the error gradually accumulates. After a century your watch will be 100 seconds fast. So, too, did the errors in the Ptolemaic system gradually accumulate as the centuries passed. Islamic and later European astronomers tried to update the system, computing new constants and adjusting epicycles. In the middle of the 13th century,

a team of astronomers supported by King Alfonso X of Castile studied the *Almagest* for 10 years. Although they did not revise the theory very much, they simplified the calculation of the positions of the planets using the Ptolemaic system and published the result as *The Alfonsine Tables*, the last great attempt to make the Ptolemaic system of practical use.

Ace✷Astronomy™ Log into AceAstronomy and select this chapter to see Astronomy Exercise "Parallax I." Notice that the parallax angle depends on the length of the baseline.

Building Scientific Arguments

How did the astronomy of Hipparchus and Ptolemy violate the principles of the early Greek philosophers Plato and Aristotle?

Today, scientific arguments depend on evidence and theory; but, in classical times, they started from first principles. Hipparchus and Ptolemy lived very late in the history of classical astronomy, and they concentrated more on the mathematical problems and less on philosophical principles. They replaced the perfect spheres of Plato with nested circles in the form of epicycles and deferents. Earth was moved slightly away from the center of the deferent, so their models of the universe were not exactly geocentric, and the epicycles moved uniformly only as seen from the equant. The celestial motions were no longer precisely uniform, and the principles of geocentrism and uniform circular motion were weakened.

The work of Hipparchus and Ptolemy led eventually to a new understanding of the heavens, but first astronomers had to abandon uniform circular motion. Construct a scientific argument in the classical style based on first principles to answer the question: **Why did Plato argue for uniform circular motion?**

■   ■   ■

Connections: Ptolemy was the last of the great classical astronomers, and his work dominated astronomical thought for almost 1500 years. The collapse of the Ptolemaic model and the rise of a new model of the universe make an exciting story, but it is not just the story of an astronomical idea. It includes the invention of science as a new way of knowing and understanding what we are and where we are.

(4-2) The Copernican Revolution

NICOLAUS COPERNICUS (■ Figure 4-8) triggered an Earthshaking revision in human thought by proposing that the universe is not centered on Earth but rather on the sun. Such a model is called a **heliocentric universe.** That idea eventually brought

■ Figure 4-8

Copernicus proposed that the sun and not Earth was the center of the universe. Notice the heliocentric model on this stamp issued in 1973 to commemorate the 500th anniversary of his birth.

The Ancient Universe

1 For 2000 years, the minds of astronomers were shackled by a pair of ideas. The Greek philosopher Plato argued that the heavens were perfect. Because the only perfect geometrical shape is a circle and the only perfect motion is uniform motion, Plato concluded that all motion in the heavens must be made up of combinations of circles turning at uniform rates. This was called **uniform circular motion.**

Plato's student Aristotle argued that Earth was imperfect and lay at the center of the universe. That is, he argued for a **geocentric universe.** He devised a model universe with 55 spheres turning at different rates and at different angles to carry the seven known planets (the moon, Mercury, Venus, the sun, Mars, Jupiter, and Saturn) across the sky.

Aristotle was known as the greatest philosopher in the ancient world, and his authority, lasting for 2000 years, chained the minds of astronomers and forced them to expect the universe to be geocentric and to move in uniform circular motion. See model at right by Peter Apian.

From *Cosmographica* by Peter Apian (1539).

Seen by left eye

Seen by right eye

1a Ancient astronomers believed that Earth did not move because they saw no **parallax,** the apparent motion of an object because of the motion of the observer. To demonstrate parallax, close one eye and cover a distant object with your thumb held at arm's length. Switch eyes, and your thumb appears to shift position as shown at left. If Earth moves, ancient astronomers reasoned, you should see the sky from different locations at different times of the year, and you should see parallax distorting the shapes of the constellations. They saw no parallax, so they concluded Earth could not move. Actually, the parallax of the stars is too small to see with the unaided eye.

2 Planetary motion was a big problem for ancient astronomers. In fact, the word *planet* comes from the Greek word for "wanderer," referring to the eastward motion of the planets against the background of the fixed stars. The planets did not, however, move at a constant rate, and they could occasionally stop and move westward for a few months before resuming their eastward motion. This backward motion is called **retrograde motion.**

Jan. 29, 2008

East

Nov. 15, 2007

Position of Mars at 5-day intervals

Ecliptic

Dec. 12, 2005

Oct. 3, 2005

Taurus

West

Betelgeuse

Every 2.14 years, Mars passes through a retrograde loop. Two successive loops are shown here. Each loop occurs further east along the ecliptic and has its own shape.

Orion

2a Simple uniform circular motion centered on Earth could not explain retrograde motion, so ancient astronomers combined uniformly rotating circles much like gears in a machine to try to reproduce the motion of the planets.

Rigel

3   Uniformly rotating circles were key elements of ancient astronomy. Claudius Ptolemy created a mathematical model of the Aristotelian universe in which the planet followed a small circle called the **epicycle** that slid around a larger circle called the **deferent.** By adjusting the size and rate of rotation of the circles, he could approximate the retrograde motion of a planet. See illustration at right.

To adjust the speed of the planet, Ptolemy supposed that Earth was slightly off center and that the center of the epicycle moved such that it appeared to move at a constant rate as seen from the point called the **equant.**

To further adjust his model, Ptolemy added small epicycles riding on top of larger epicycles (not shown here), producing a highly complex model.

3a   Ptolemy's great book *Mathematical Syntaxis* (c. AD 140) contained the details of his model. Islamic astronomers preserved and studied the book through the Middle Ages, and they called it *Al Magisti* (The Greatest). When the book was found and translated from Arabic to Latin in the 12th century, it became known as *Almagest*.

3b   The Ptolemaic model of the universe shown below was geocentric and based on uniform circular motion. Note that Mercury and Venus were treated differently from the rest of the planets. The centers of the epicycles of Mercury and Venus had to remain on the Earth–Sun line as the sun circled Earth through the year.

Equants and smaller epicycles are not shown here. Some versions contained nearly 100 epicycles as generations of astronomers tried to fine-tune the model to better reproduce the motion of the planets.

Notice that this modern illustration shows rings around Saturn and sunlight illuminating the globes of the planets, features that could not be known before the invention of the telescope.

Ace ◯ Astronomy™   Go to AceAstronomy and click Active Figures to see "Epicycles." Notice how the counter-clockwise rotation of the epicycle produces retrograde motion.

Galileo before the Inquisition and changed forever how we think of our world and ourselves. This Copernican Revolution was much more than an upheaval in astronomy; it marks the birth of modern science.

Copernicus the Revolutionary

When Copernicus proposed that the universe was heliocentric, he risked controversy. According to Aristotle, the most perfect region was the starry sphere, and the most imperfect was Earth's center. In this way the classical geocentric universe matched the commonly held Christian geometry of heaven and hell, and any-one who criticized the geometry of the Aristotelian universe challenged Christian belief and thus risked a charge of heresy.

Throughout his life, Copernicus was associated with the Church. His uncle, by whom he was raised and educated, was an important bishop in Poland; and, after studying canon law and medicine in some of the finest universities in Europe, Copernicus became a canon of the Church at the unusually young age of 24. He was secretary and personal physician to his powerful uncle for 15 years. When his uncle died, Copernicus moved to quarters adjoining the cathedral in Frauenburg, where he was a canon. No doubt his long association with the Church added to his reluctance to publish controversial ideas.

■ **Figure 4-9**

(a) The Copernican universe as reproduced in *De Revolutionibus*. Earth and all the known planets revolve in separate circular orbits about the sun (Sol) at the center. The outermost sphere carries the immobile stars of the celestial sphere. Notice the orbit of the moon around Earth (Terra). (Yerkes Observatory) (b) The model was elegant not only in its arrangement of the planets but also in their motions. Orbital speed (blue arrows) decreased from Mercury, the fastest, to Saturn, the slowest. Compare the elegance of this model with the complexity of the Ptolemaic model on page 65.

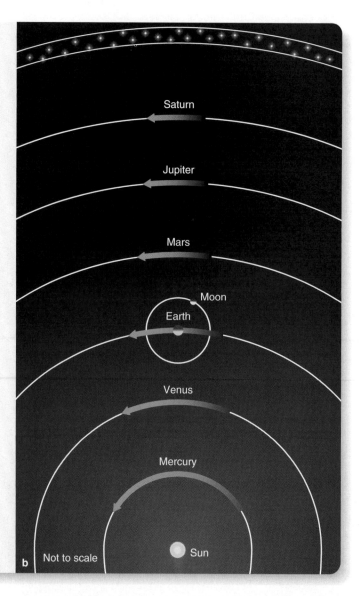

He first wrote about a heliocentric universe sometime before 1514 in a short pamphlet that he distributed in handwritten form and in some cases anonymously. Over the years, Copernicus worked on his book *De Revolutionibus Orbium Coelestium (On the Revolutions of the Celestial Spheres)* (■ Figure 4-9a), but he hesitated to publish it even though other astronomers knew of his theories.

This was a time of rebellion—Martin Luther was speaking harshly about fundamental Church teachings, and others, scholars and scoundrels, questioned the authority of the Church. Even astronomy could stir argument; moving Earth from its central place was a controversial and perhaps heretical idea. Copernicus was concerned about how his ideas would be received, but in 1540 he allowed the visiting astronomer Joachim Rheticus (1514–1576) to publish an account of the Copernican universe in Rheticus's book *Prima Narratio (First Narrative)*. In 1542, Copernicus sent the manuscript for *De Revolutionibus* off to be printed. He died in the spring of 1543 before the printing was completed.

The Copernican Model

You should notice the difference between the Copernican hypothesis and the Copernican model. The Copernican hypothesis was that the universe was heliocentric, and that was correct. The Copernican model was the arrangement and motion of the planets that Copernicus devised to explain the motions in the sky. The model won gradual acceptance, but, as you will discover, it was flawed.

The most important idea in *De Revolutionibus* was the placement of the sun at the center of the universe. That single innovation had an astonishing consequence—the retrograde motion of the planets was immediately explained in a straightforward way without the large epicycles that Ptolemy used. In the Copernican system, Earth moves faster along its orbit than the planets that lie further from the sun. Consequently, Earth periodically overtakes and passes these planets, and they appear to slow and fall behind (■ Figure 4-10). Because the planetary orbits do not lie in precisely the same plane, a planet does not resume its eastward motion in precisely the same path it followed earlier. Consequently, it describes a loop whose shape depends on the angle between the orbital planes.

Copernicus could explain retrograde motion without epicycles, and that was impressive. But he could not do away with epicycles completely. Copernicus, a classical astronomer, had tremendous respect for the old concept of uniform circular motion. In fact, he objected strongly to Ptolemy's use of the equant. It seemed arbitrary to Copernicus, a direct violation of the elegance of Aristotle's philosophy of the heavens. Copernicus called equants "monstrous" in that they violated uniform circular motion. In his model, Copernicus returned to a strong belief in uniform circular motion. Although he did not need epicycles to explain retrograde motion, Copernicus discovered that the sun, moon, and planets suffered small variations in their motions—variations that

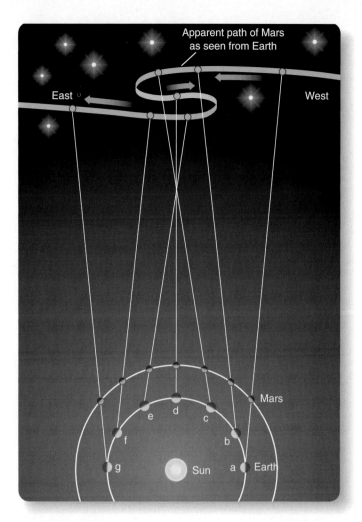

■ Figure 4-10

The Copernican explanation of retrograde motion: As Earth overtakes Mars (a–c), Mars appears to slow its eastward motion. As Earth passes Mars (d), Mars appears to move westward. As Earth draws ahead of Mars (e–g), Mars resumes its eastward motion against the background stars. Compare with the illustration of retrograde motion on page 64. The positions of Earth and Mars are shown at equal intervals of one month.

he could not explain with the concept of uniform circular motion centered on the sun. Today, we recognize those variations as typical of objects following elliptical orbits, but Copernicus held firmly to uniform circular motion, so he had to use small epicycles to produce these minor variations in the motions of the sun, moon, and planets.

Because Copernicus imposed uniform circular motion on his model, it could not accurately predict the motions of the planets. *The Prutenic Tables* (1551) were based on the Copernican model, and they were not significantly more accurate than *The Alfonsine Tables* (1251), which were based on Ptolemy's model. Both could be in error by as much as 2°, four times the angular diameter of the full moon.

The Copernican *model* was inaccurate, but the Copernican *hypothesis* that the universe is heliocentric was correct. There are probably a number of reasons why the hypothesis gradually won

Creating New Ways to See Nature: Scientific Revolutions

The Copernican Revolution is often cited as the perfect example of a scientific revolution. Over a few decades, astronomers abandoned a way of thinking about the universe that was almost 2000 years old and adopted a new set of ideas and assumptions. The American philosopher of science Thomas Kuhn has referred to a commonly accepted set of scientific ideas and assumptions as a scientific **paradigm.** The pre-Copernican astronomers had a geocentric paradigm that included uniform circular motion and the perfection of the heavens. That paradigm survived for many centuries until a new generation of astronomers was able to overthrow the old paradigm and establish a new paradigm that included heliocentrism and Earth's motion.

A scientific paradigm is powerful because it shapes your perceptions. It determines what you judge to be important questions and what you judge to be significant evidence. Consequently, it is often difficult to recognize how your paradigms limit what you can understand.

For example, the geocentric paradigm contained problems that seem obvious today, but because astronomers before Copernicus lived and worked inside that paradigm, they had difficulty seeing the problems. Overthrowing an outdated paradigm is not easy, because you must learn to see nature in an entirely new way. Galileo and Kepler saw nature from a new paradigm that would have been almost incomprehensible to astronomers of earlier centuries.

You can find examples of scientific revolutions in many fields, including biology, geology, genetics, and psychology. These scientific revolutions have been difficult and controversial because they have involved the overthrow of accepted paradigms. But that is why scientific revolutions are exciting. They provide not just a new idea or a new theory, but an entirely new insight into how nature works— a new way of seeing the world.

The ancients believed the stars were attached to a starry sphere. (NOAO and Nigel Sharp)

---

acceptance in spite of the inaccuracy of the model's predictions. The most important factor may be the elegance of the idea. Placing the sun at the center of the universe produced a symmetry among the motions of the planets that was pleasing to the eye and to the intellect (Figure 4-9b). All of the planets moved in the same direction at speeds that were simply related to their distance from the sun. In the Ptolemaic model, Venus and Mercury were treated differently from the other planets. In the Copernican model, all planets were treated the same. In this way, the model may have won support not for its accuracy, but for its elegance.

The most astonishing consequence of the Copernican hypothesis was not what it said about the sun, but what it said about Earth. By placing the sun at the center, Copernicus made Earth move along a path around the sun just as the other planets did, and that made Earth a planet. By revealing that Earth is a planet, Copernicus revolutionized humanity's view of our place in the universe and triggered a controversy that would eventually bring Galileo before the Inquisition.

Although astronomers throughout Europe read and admired *De Revolutionibus,* they did not usually accept the Copernican hypothesis. The mathematics was elegant, and the astronomical observations and calculations were of tremendous value. Yet few astronomers believed, at first, that the sun actually was the center of the planetary system and that Earth moved. How the Copernican hypothesis became gradually recognized as correct has been named the Copernican Revolution because it involved not just the adoption of a new idea but a total revolution in the way astronomers thought about the place of the Earth (**Window on Science 4-1**).

Galileo the Defender

Most people know two facts about Galileo, and both facts are wrong. You should begin by getting those facts right: Galileo did not invent the telescope, and he was not condemned by the Inquisition for believing that Earth moved around the sun. Then why is Galileo so important that in 1979, almost 400 years after his trial, the Vatican reopened his case? As you read about Galileo, you will discover that his trial concerned not just the place of the Earth but a new way of finding truth, a new way of knowing about the world, a method known today as science.

Galileo Galilei (■ Figure 4-11) was born in Pisa, a city in what is now Italy, in 1564, and he studied medicine at the university there. His true love, however, was mathematics; and, although he had to leave school early because of financial difficulties, he returned only four years later as a professor of mathematics. Three years later, he became professor of mathematics at the university at Padua. He remained there for 18 years.

During this time, Galileo seems to have adopted the Copernican model, although he admitted in a 1597 letter to the Ger-

Galileo, remembered as the great defender of Copernicanism, also made important discoveries in
the physics of motion. He is honored here on an
Italian 2000-lira note. The reverse side shows one
of his telescopes at lower right and a modern observatory above it.

man astronomer Kepler that he did not
support Copernicanism publicly because
of the criticism such a declaration would
bring. It was the telescope that drove
Galileo to publicly defend the heliocentric
model.

Galileo did not invent the telescope.
It was apparently invented around 1608
by lens makers in Holland. Galileo, hearing descriptions in the fall of 1609, was
able to build telescopes in his workshop.
Galileo was not the first person to look at
the sky through a telescope, but he was
the first person to observe the sky system-
atically and apply his observations to the
theoretical problem of the day—the place
of the Earth.

What Galileo saw through his telescopes was so amazing he rushed a small
book into print. *Sidereus Nuncius (The Sidereal Messenger)* reported three major discoveries. First, the moon was not perfect.

It had mountains and valleys on its surface, and Galileo used the
shadows to calculate the height of the mountains. The Ptolemaic
model held that the moon was perfect, but Galileo showed that
it was not only imperfect, it was a world like Earth.

The second discovery reported in the book was that the Milky
Way was made up of a myriad of stars too faint to see with the
unaided eye. While intriguing, this discovery could not match
the third discovery. Galileo's telescope revealed four new "planets"
circling Jupiter, planets known today as the Galilean moons of
Jupiter (■ Figure 4-12).

The moons of Jupiter supported the Copernican model over
the Ptolemaic model. Critics of Copernicus had said Earth could
not move because the moon would be left behind. But Jupiter
moved yet kept its satellites, so Galileo's discovery suggested that
Earth, too, could move and keep its moon. Also, the Ptolemaic
model included the Aristotelian belief that all heavenly motion
was centered on Earth at the center of the universe. Galileo showed
that Jupiter's moons revolve around Jupiter, so there could be other
centers of motion.

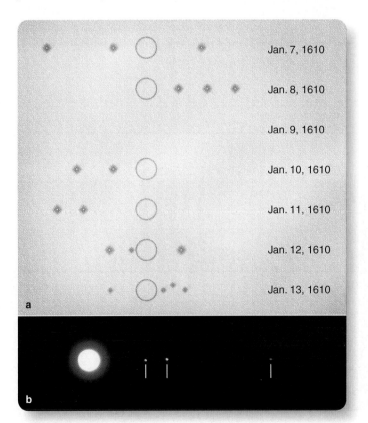

Jan. 7, 1610

Jan. 8, 1610

Jan. 9, 1610

Jan. 10, 1610

Jan. 11, 1610

Jan. 12, 1610

Jan. 13, 1610

a

b

■ **Figure 4-12**

(a) On the night of January 7, 1610, Galileo saw three small "stars" near the
bright disk of Jupiter and sketched them in his notebook. On subsequent
nights (excepting January 9, which was cloudy), he saw that the stars were
actually four moons orbiting Jupiter. (b) This photo taken through a modern
telescope shows the overexposed disk of Jupiter and three of the four Galilean moons. (Grundy Observatory)

Sometime after *Sidereus Nuncius* was published, Galileo made another discovery that made Jupiter's moons even stronger evidence supporting the Copernican universe. Eventually, he was able to measure the orbital periods of the four moons, and he found that the innermost moon moved fastest and that the moons further from Jupiter moved more slowly. In this way, Jupiter's moons made up a harmonious system ruled by Jupiter, just as the planets in the Copernican universe were a harmonious system ruled by the sun (Figure 4-9b). The similarity isn't proof, but Galileo saw it as an argument that the solar system was sun centered and not Earth centered.

Also soon after *Sidereus Nuncius* was published, Galileo made two additional discoveries. When he observed the sun, he discovered sunspots, raising the suspicion that the sun was less than perfect. Further, by noting the movement of the spots, he concluded that the sun was a sphere and that it rotated on its axis. When he observed Venus, Galileo saw that it was going through phases like those of the moon. In the Ptolemaic model, Venus moves around an epicycle centered on a line between Earth and the sun. In that epicycle, it would always be seen as a crescent (■ Figure 4-13a). But Galileo saw Venus go through a complete set of phases, proving that it did indeed revolve around the sun (Figure 4-13b).

Sidereus Nuncius was very popular and made Galileo famous. He became chief mathematician and philosopher to the Grand Duke of Tuscany in Florence. In 1611, Galileo visited Rome and was treated with great respect. He had long, friendly discussions with the powerful Cardinal Barberini, a man with a deep interest in the new discoveries being made. But Galileo also made enemies. Personally, Galileo was outspoken, forceful, and sometimes tactless. He enjoyed debate, but most of all he enjoyed being right. In lectures, debates, and letters he offended important people who questioned his telescopic discoveries.

By 1616, Galileo was the center of a storm of controversy. Some critics said he was wrong, and others said he was lying. Some refused to look through a telescope lest it mislead them, and others looked and claimed to see nothing (hardly surprising given the awkwardness of those first telescopes) (■ Figure 4-14). Pope Paul V decided to end the disruption, so when Galileo visited Rome in 1616 Cardinal Bellarmine interviewed him privately and ordered him to cease debate. There is some controversy today about the nature of Galileo's instructions, but he did not pursue astronomy for some years after the interview. Books relevant to Copernicanism were banned, including *De Revolutionibus,* although owners were allowed to keep their books if they changed certain phrases to make it clear that the central place of the sun was only a theory and not a fact.

In 1621 Pope Paul V died, and his successor, Pope Gregory XV, died in 1623. The next pope was Galileo's friend Cardinal Barberini, who took the name Urban VIII. Galileo rushed to Rome hoping to have the prohibition of 1616 lifted; and, although the new pope did not revoke the orders, he did encourage Galileo. When he got back home, Galileo began to write his great defense of the Copernican model, finally completing it on December 24, 1629. After some delay, the book was approved by both the local censor in Florence and the head censor of the Vatican in Rome. It was printed in February 1632.

Called *Dialogo Dei Due Massimi Sistemi (Dialogue Concerning the Two Chief World Systems),* it confronts the ancient astronomy of Aristotle and Ptolemy with the Copernican model and with telescopic observations as evidence. Galileo wrote the book as a debate among three friends. Salviati, a swift-tongued defender of Copernicus, dominates the book; Sagredo is intelligent but largely uninformed. Simplicio is the dismal defender of Ptolemy. In fact, he does not seem very bright.

■ **Figure 4-13**

(a) If Venus moved in an epicycle centered on the Earth–sun line (see page 65), it would always appear as a crescent. (b) Galileo's telescope showed that Venus goes through a full set of phases, proving that it must orbit the sun.

■ Figure 4-14

Galileo's telescope made him famous, and he demonstrated his telescope and discussed his observations with powerful people. Some thought the telescope was the work of the devil and would deceive anyone who looked. In any case, Galileo's discoveries produced intense, and in some cases, angry debate. (Yerkes Observatory)

The publication of *Dialogo* created a sensation, and it was sold out by August 1632, when the Inquisition ordered sales stopped. The book was a clear defense of Copernicus, and, either intentionally or unintentionally, Galileo exposed the pope's authority to ridicule. Urban VIII was fond of arguing that, as God was omnipotent, He could construct the universe in any form while making it appear to humans to have a different form, and thus its true nature could not be deduced by mere observation. Galileo placed the pope's argument in the mouth of Simplicio, and Galileo's enemies showed the passage to the pope as an example of Galileo's disrespect. The pope thereupon ordered Galileo to face the Inquisition.

The Trial of Galileo

The trial of Galileo was one of the turning points in the history of science and human learning, but historians still argue about what happened and why. The trial involved the highest religious principles and the lowest of behind-the-scenes political maneuvering. One thing you can be sure of: The trial changed the way humanity thought about the world and marked the beginning of modern science as a way to understand nature.

Galileo was interrogated by the Inquisition four times and was threatened with torture. He must have thought often of Giordano Bruno, tried, condemned, and burned at the stake in Rome in 1600. One of Bruno's offenses had been Copernicanism. But the trial did not center on Galileo's belief in Copernicanism. After all, *Dialogo* had been approved by two censors. Rather, the trial

centered on the instructions given Galileo in 1616. From his file in the Vatican, his accusers produced a record of the meeting between Galileo and Cardinal Bellarmine that included the statement that Galileo was "not to hold, teach, or defend in any way" the principles of Copernicus. Some historians believe that this document, which was signed neither by Galileo nor by Bellarmine nor by a legal secretary, was a forgery. Others suspect it may be a draft that was never used. In any case, it is possible that Galileo's true instructions were much less restrictive. But Bellarmine was dead and could not testify at Galileo's trial.

The Inquisition condemned him not for heresy but for disobeying the orders given him in 1616. On June 22, 1633, at the age of 70, kneeling before the Inquisition, Galileo read a recantation admitting his errors. Tradition has it that as he rose he whispered, *"E pur si muove"* ("Still it moves"), referring to Earth. Although he was sentenced to life imprisonment, he was actually confined at his villa for the next 10 years, perhaps through the intervention of the pope. He died there on January 8, 1642, 99 years after the death of Copernicus.

Galileo was not condemned for heresy, nor was the Inquisition interested when he tried to defend Copernicanism. He was tried and condemned on a charge you might call a technicality. Then why is his trial so important that historians have studied it for almost four centuries? Why have some of the world's greatest authors, including Bertolt Brecht, written about Galileo's trial? Why in 1979 did Pope John Paul II create a commission to reexamine the case against Galileo?

To understand the trial, you must recognize that it was the result of a conflict between two ways of understanding our universe. Plato had argued that observation was deceptive, and that the only way to find truth was through pure thought. Since the Middle Ages, scholars had taught that the only path to true understanding was through religious faith. St. Augustine (AD 354–430) wrote *"Credo ut intelligame,"* which can be translated as, "Believe in order to understand." But Galileo and other scientists of the Renaissance used their own observations to try to understand the universe; and, when their observations contradicted Scripture, they assumed their observations of reality were correct and that Scripture was not being correctly understood (■ Figure 4-15). Galileo paraphrased Cardinal Baronius in saying, "The Bible tells us how to go to heaven, not how the heavens go."

Galileo's discoveries produced intense, and in some cases angry, debate. Various passages of Scripture seemed to contradict observation. For example, Joshua is said to have commanded the sun to stand still, not Earth to stop rotating (Joshua 10:12–13). In response to such passages, Galileo argued that you should "read the book of nature"—that is, you should observe the universe with your own eyes.

The trial of Galileo was not about the place of the Earth. It was not about Copernicanism. It wasn't really about the instructions Galileo received in 1616. It was about the birth of modern science as a rational way to understand our universe. The

■ Figure 4-15

Although he did not invent it, Galileo will always be associated with the telescope because it was the source of the observational evidence he used to try to understand the universe. By depending on evidence instead of first principles, Galileo led the way to the invention of modern science as a way to know about the natural world.

commission appointed by John Paul II in 1979, reporting its conclusions in October 1992, said of Galileo's inquisitors, "This subjective error of judgment, so clear to us today, led them to a disciplinary measure from which Galileo 'had much to suffer.'" Galileo was not found innocent in 1992 so much as the Inquisition was forgiven for having charged him in the first place.

The gradual change from reliance on personal faith to reliance on scientific observation came to a climax with the trial of Galileo. Since that time, scientists have increasingly reserved Scripture and religious faith for ethical guidance and personal comfort and have depended on systematic observation to describe the physical world. The final verdict of 1992 was an attempt to bring some balance to this conflict. In his remarks on the decision, Pope John Paul II said, "A tragic mutual incomprehension has been interpreted as the reflection of a fundamental opposition between science and faith. . . . this sad misunderstanding now belongs in the past." Galileo's trial is over, but it continues to echo through history as part of humanity's struggle to understand the place of the Earth in our universe.

How were Galileo's observations of the moons of Jupiter evidence against the Ptolemaic model?

Modern scientific arguments depend critically on evidence, and that started with Galileo. He presented his arguments in the form of evidence and conclusions, and the moons of Jupiter were key evidence. Ptolemaic astronomers argued that Earth could not move or it would lose its moon, but even in the Ptolemaic universe Jupiter moved, and Galileo's telescope showed that it kept its moons. Evidently, Earth could move and not leave its moon behind. Furthermore, moons circling Jupiter did not fit the classical belief that all motion was centered on Earth. Obviously there could be other centers of motion. Finally, the orbital periods of the moons were related to their distance from Jupiter, just as the orbital periods of the planets were, in the Copernican system, related to their distance from the sun. This similarity suggested that the sun rules its harmonious family of planets just as Jupiter rules its harmonious family of moons.

Of all of Galileo's telescopic observations, the moons of Jupiter caused the most debate. But there was more. Use the evidence to build an argument to answer the following: **How did craters on the moon and the phases of Venus weigh against the Ptolemaic model?**

■   ■   ■

Connections: Galileo faced the Inquisition because of a conflict between two ways of knowing and understanding the world. The Church taught faith and understanding through revelation, but the scientists of the age were inventing a new way to understand nature that relied on evidence, a way of knowing now called science. In astronomy, evidence means observations, so it is time to turn your attention from the philosophical problem of the place of the Earth to the observational problem of the motion of the planets.

(4-3) The Puzzle of Planetary Motion

WHILE GALILEO WAS TEACHING mathematics at Pisa and Padua, two other astronomers were working in northern Europe, beyond the sway of the Inquisition. Although they, too, struggled to understand the place of the Earth, they approached the problem differently. They used the most accurate observations of the positions of the planets to try to discover the rules that govern planetary motion. When that puzzle was finally solved, astronomers understood why the Ptolemaic model of the universe does not work well and how to make the Copernican universe a precise predictor of planetary motion.

Tycho the Observer

The great observational astronomer of this story is a Danish nobleman, Tycho Brahe, born December 14, 1546, only three years after the publication of *De Revolutionibus* (■ Figure 4-16). As a son of a powerful noble family, he was well known for his vanity and lordly manners.

Tycho's college days were eventful. He was officially studying law with the expectation that he would enter Danish politics, but he made it clear to his family that his real interest was astronomy and mathematics. It was also during his college days that he fought a duel and received a wound that disfigured his nose. For the rest of his life, he wore false noses made of gold and silver and stuck on with wax. The disfigurement probably did little to improve his disposition.

Tycho's first astronomical observations were made while he was a student. In 1563, Jupiter and Saturn passed very near each other in the sky, nearly merging into a single point on the night of August 24. Tycho found that the *Alfonsine Tables* were a full month in error and that the *Prutenic Tables* were in error by a number of days. These discrepancies dismayed Tycho and sparked his interest in the motions of the planets.

In 1572, a brilliant "new star" (now called Tycho's supernova) appeared in the sky. Such changes in the sky puzzled classically trained astronomers. Aristotle had argued that the starry sphere was perfect and unchanging, and therefore such new stars had to lie closer to Earth than the moon. Layers below the moon were thought to be less perfect and thus more changeable than the starry sphere.

Tycho carefully measured the position of the star time after time through the night and found that it displayed no parallax. To understand the significance of this observation, you must note that Tycho, like most astronomers of his time, believed Earth was fixed at the center of the universe. Consequently, he believed the heavens rotated westward around Earth once a day. The new star, according to classical astronomy, represented a change in the heavens and therefore had to lie below the sphere of the moon. In that case, Tycho reasoned, the new star would appear slightly too far east when it was in the eastern sky and slightly too far west later in the night when it was carried into in the western sky (■ Figure 4-17). That is an example of parallax. Tycho could detect no parallax in the position of the new star, so he concluded that it must lie above the sphere of the moon and was probably on the starry sphere itself. That was an astonishing discovery because it contradicted Aristotle's belief that the starry sphere was perfect and unchanging.

No one before Tycho could have made this discovery because no one had ever measured the positions of stars so accurately. Tycho had great confidence in the precision of his measurements; so, when he failed to detect parallax for the new star, he knew it was important evidence against the Ptolemaic theory. He announced his discovery in a small book, *De Stella Nova (The New Star),* published in 1573.

The book attracted the attention of astronomers throughout Europe, and soon Tycho was famous. Although Tycho had finished at the university, he was unemployed, idling at the family home and embarrassing his powerful family. His fame allowed his family to introduce him to the court of the Danish King Frederik II, and there the king offered Tycho funds to build an observatory on the island of Hveen just off the Danish coast. Tycho also received a steady source of income as landlord of a coastal

■ **Figure 4-16**

Tycho Brahe (1546–1601) was, during his lifetime, the most famous astronomer in the world. Proud of his noble rank, he wears the elephant medal awarded him by the king of Denmark. His artificial nose is suggested in this engraving.

■ **Figure 4-17**

If an object lay lower than the celestial sphere, reasoned Tycho Brahe, then it should be seen east of its average position as it was rising and west of its average position when it was setting. Because he did not detect this daily parallax, he concluded the new star of 1572 had to lie on the celestial sphere.

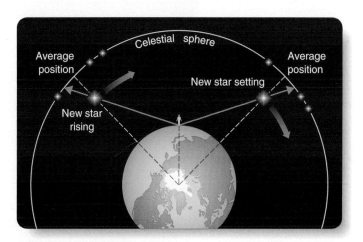

district from which he collected rents. (He was not a popular landlord.) On Hveen, Tycho constructed a luxurious home with six towers specially equipped for astronomy and populated it with servants, assistants, and a dwarf to act as jester. Soon Hveen was an international center of astronomical study.

Tycho Brahe's Legacy

Tycho Brahe made no lasting contribution to astronomical theory. Because he could measure no parallax for the stars, he concluded that Earth had to be stationary, and that led him to reject the Copernican hypothesis. However, he also rejected the Ptolemaic model because of its inaccurate predictions. Instead, he devised a complex model in which Earth was the immobile center of the universe around which the sun and moon moved. The other planets circled the sun. You might find this arrangement familiar; it is really the Copernican model with Earth held stationary and the sun allowed to move around Earth. In this way, Tycho preserved the central, immobile Earth that most astronomers believed was described in scripture (■ Figure 4-18a). Although Tycho's model, the Tychonic Universe, was very popular at first, the Copernican model replaced it within a century.

The true value of Tycho's work was observational. Because he was able to devise new and better instruments, he was able to make highly accurate observations of the positions of the stars, sun, and planets. Tycho had no telescopes—they were not invented until the next century—so his observations were made by the naked eye peering along sights. All of his instruments were designed to measure angles in the sky. For example, his quadrant could measure angles up to 90° above the horizon (Figure 4-18b). By designing and building large instruments with great care, he

was able to measure angles to high precision. He measured the positions of 777 stars to better than 4 minutes of arc and regularly measured the positions of the sun, moon, and planets during the 20 years he stayed on Hveen.

Unhappily for Tycho, King Frederik II died in 1588, and his young son took the throne. Suddenly Tycho's temper, vanity, and noble presumptions threw him out of favor. In 1596, taking most of his instruments and books of observations, he went to Prague, the capital of Bohemia, and became imperial mathematician to the Holy Roman Emperor Rudolph II. His goal was to revise *The Alfonsine Tables* and publish the revision as a monument to his new patron. It would be called *The Rudolphine Tables.*

Tycho did not intend to base *The Rudolphine Tables* on the Ptolemaic system but rather on his own Tychonic system, proving once and for all the validity of his hypothesis. To assist him, he hired a few mathematicians and astronomers, including one Johannes Kepler. Then in November 1601, Tycho collapsed at a nobleman's home. Before he died, 11 days later, he asked Rudolph II to make Kepler imperial mathematician. Thus the newcomer, Kepler, became Tycho's replacement (though at one-sixth Tycho's salary).

Kepler the Analyst

No one could have been more different from Tycho Brahe than Johannes Kepler (■ Figure 4-19). He was born December 27, 1571, to a poor family in a region now included in southwestern Germany. His father was unreliable and shiftless, principally employed as a mercenary soldier fighting for whoever paid enough. He finally failed to return from a military expedition, either because he was killed or because he found other circum-

■ Figure 4-18

(a) Tycho Brahe's model of the universe held that Earth was fixed at the center of the starry sphere. The moon and sun circled Earth, while the planets circled the sun. (b) Much of Tycho's success was due to his skill in designing large, accurate instruments. In this engraving of his mural quadrant, the figure of Tycho, his dog, and the background scene are painted on the wall within the arc of the quadrant. The observer (Tycho himself) at the extreme right peers through a sight out the loophole in the wall at the upper left to measure an object's angular distance above the horizon. (Granger Collection, New York)

Sun · Venus · Saturn · Jupiter · Mars · Moon · Mercury · Earth

Tycho Brahe's universe was geocentric.

stances more to his liking. Kepler's mother was apparently an unpleasant and unpopular woman. She was accused of witchcraft in her later years, and Kepler had to defend her in a trial that dragged on for three years. She was finally acquitted but died the following year.

Kepler was the oldest of six children, and his childhood was no doubt unhappy. The family was not only poor but suffered from an absentee father whom Kepler described as "vicious, inflexible, quarrelsome, and doomed to a bad end." In addition, Kepler was never healthy, even as a child, so it is surprising that he did well in school, eventually winning promotion to a Latin school and finally a scholarship to the university at Tübingen, where he studied to become a Lutheran pastor.

During his last year of study, Kepler accepted a job in Graz teaching mathematics and astronomy. Evidently, he was not a good teacher. He had few students his first year and none at all his second. His superiors put him to work teaching a few introductory courses and preparing an annual almanac that contained astronomical, astrological, and weather predictions. Through good luck, in 1595 some of his weather predictions were fulfilled, and he gained a reputation as an astrologer and seer; even in later life he earned money by publishing almanacs.

While still a college student, Kepler had become a believer in the Copernican theory, and at Graz he used his extensive spare time to study astronomy. By 1596, the same year Tycho left Hveen, Kepler was ready to solve the mystery of the universe. That year he published a book called *The Forerunner of Dissertations on the Universe, Containing the Mystery of the Universe.* Like nearly all scientific works of the time, the book was in Latin, and it is now known as *Mysterium Cosmographicum.*

The book begins with a long appreciation of Copernicanism and then goes on to speculate on the spacing of the planetary orbits. Kepler, as a Copernican, knew of six planets circling the sun—Mercury, Venus, Earth, Mars, Jupiter, and Saturn. According to his model, the spheres containing the six planets were separated from one another, and their relative sizes were fixed by the five regular solids*—the cube, tetrahedron, dodecahedron, icosahedron, and octahedron (Figure 4-19). Kepler gave astrological, numerological, and musical arguments for his theory and so followed the tradition of Pythagoras in believing that the order of the universe is underlain by musical (meaning mathematical) principles.

In the second half of the book, Kepler tried to fit the five solids to the planetary circles of the Copernican theory, a complex problem involving three-dimensional trigonometry and geometry. He sent copies to Tycho and to Galileo, and neither seemed very impressed with his theory, but his mathematical abilities shown brightly in his calculations.

Life was unsettled for Kepler because of the persecution of Protestants in the region, so when Tycho invited him to Prague in 1600, he went readily, eager to work with the famous astronomer. Tycho's sudden death in 1601 left Kepler in a position to use the observations from Hveen to analyze the motions of the planets and complete *The Rudolphine Tables.* Tycho's family, recognizing that Kepler was a Copernican and guessing that he would not follow the Tychonic system in completing *The Rudolphine Tables,* sued to recover the instruments and books of observations. The legal wrangle went on for years. The family did recover the instruments Tycho had brought to Prague; Kepler seems to have had little interest in them, perhaps because

■ **Figure 4-19**

Johannes Kepler (1571–1630) was Tycho Brahe's sucessor. This diagram, based on one drawn by Kepler, shows how he believed the sizes of the celestial spheres carrying the outer three planets are determined by spacers (blue) consisting of the regular solids. Inside the sphere of Mars, the remaining regular solids (not shown here) separate the spheres of the Earth, Venus, and Mercury. The sun lay at the very center of this Copernican universe based on geometrical spacers.

---

*A regular solid is a three-dimensional body each of whose faces is the same. For example, a cube is a regular solid each of whose faces is a square.

of his poor eyesight. However, Kepler had the books of observations, and he kept them.

Whether Kepler had any legal right to Tycho's records is debatable, but he put them to good use. He began by studying the motion of Mars, trying to deduce from the observations how the planet moves. By 1606, he had solved the puzzle of planetary motion. The orbit of Mars is an ellipse and not a circle, he said, and with that he abandoned the 2000-year-old belief in the circular motion of the planets. But he discovered that the mystery was even more complex: The planets do not move at a uniform speed along their elliptical orbits. Kepler's analysis showed that they move faster when closer to the sun and slower when farther away. With those two brilliant discoveries, Kepler abandoned both circular motion and uniform motion.

Kepler published his results in 1609 in a book called *Astronomia Nova (The New Astronomy)*. Like Copernicus's book, *Astronomia Nova* did not become an instant best seller. It is written in Latin for other scientists and is highly mathematical. In some ways, the book is surprisingly advanced. For instance, Kepler discusses the force that holds the planets in their orbits, a question that Isaac Newton considered later in his recognition of gravity as a natural force.

Despite the abdication of Kepler's patron Rudolph II in 1611, Kepler continued his astronomical studies. He wrote about a supernova that had appeared in 1604 (now known as Kepler's supernova) and about comets, and he wrote a textbook about Copernican astronomy. In 1619, he published *Harmonice Mundi (The Harmony of the World),* in which he returned to the cosmic mysteries of *Mysterium Cosmographicum.* The main thing of note in *Harmonice Mundi* is his discovery that the radii of the planetary orbits are related to the planets' orbital periods. That and his two previous discoveries are now recognized as Kepler's three laws of planetary motion.

Kepler's Three Laws of Planetary Motion

Although Kepler dabbled in the philosophical arguments of the day, he was a mathematician, and his triumph was the solution of the problem of the motion of the planets. The key to his solution was the ellipse.

An **ellipse** is a figure drawn around two points called the foci in such a way that the distance from one focus to any point on the ellipse and back to the other focus equals a constant. You can easily draw ellipses with two thumbtacks and a loop of string. Press the thumbtacks into a board, loop the string about the tacks, and place a pencil in the loop. If you keep the string taut as you move the pencil, it traces out an ellipse (■ Figure 4-20a).

The geometry of an ellipse is described by two simple numbers. The **semimajor axis, *a,*** is half of the longest diameter. The **eccentricity** of an ellipse, *e,* is the distance from either focus to the center of the ellipse divided by the semimajor axis. If you want to draw a circle with the string and tacks as shown in Figure 4-20a, you would move the two thumbtacks together, which shows that a circle is really just an ellipse with eccentricity equal to zero. As you move the thumbtacks farther apart, the ellipse becomes flatter, and the eccentricity moves closer to 1.

Kepler used ellipses to describe the motion of the planets in three fundamental rules that have been tested and confirmed so many times that astronomers now refer to them as natural laws **Window on Science 4-2**). They are commonly called Kepler's laws of planetary motion (■ Table 4-1).

Kepler's first law states that the orbits of the planets around the sun are ellipses with the sun at one focus. Thanks to the precision of Tycho's observations and the sophistication of Kepler's mathematics, Kepler was able to recognize the elliptical shape of the orbits, even though they are nearly circular. Of the planets known to Kepler, Mercury has the most elliptical orbit, but even it deviates only slightly from a circle (■ Figure 4-21).

Kepler's second law states that a line from the planet to the sun sweeps over equal areas in equal intervals of time. This means

Keep the string taut, and the pencil point will follow an ellipse.

String

Focus   Focus

The sun is at one focus, but the other focus is empty.

a

■ Figure 4-20

The geometry of elliptical orbits: Drawing an ellipse with two tacks and a loop of string is easy. The semimajor axis, *a,* is half of the longest diameter. The sun lies at one of the foci of the elliptical orbit of a planet.

Hypothesis, Theory, and Law: Levels of Confidence

Even scientists misuse the words *hypothesis*, *theory*, and *law*. You must try to distinguish these terms from one another because they are key elements in science.

A **hypothesis** is a single assertion or conjecture that must be tested. It could be true or false. "All Texans love chili" is a hypothesis. To know whether it is true or false, you need to test it against reality by making observations or performing experiments. Copernicus asserted that the universe was heliocentric; his assertion was a hypothesis subject to testing.

In Chapter 2, you saw that a model (Window on Science 2-1) is a description of some natural phenomenon; it can't be right or wrong. A model is not a conjecture of truth but merely a convenient way to think about a natural phenomenon. Consequently, a model such as the celestial sphere cannot be a hypothesis. Copernicus used his hypothesis to build a model, but they are not the same thing.

A **theory** is a system of rules and principles that can be applied to a wide variety of circumstances. A theory may have begun as one or more hypotheses, but it has been tested, expanded, and generalized. Many textbooks refer to the "Copernican theory," but some historians argue that it was not complete and had not been tested enough to be a theory. It is probably better to call it the Copernican hypothesis.

A **natural law** is a theory that has been refined, tested, and confirmed so often scientists have great confidence in it. Natural laws are the most fundamental principles of scientific knowledge. Kepler's laws are good examples.

Confidence is the key to understanding these terms. Scientists have more confidence in a theory than in a hypothesis and great confidence in a natural law. Nevertheless, scientists are not always consistent about these words. For example, Einstein's theory of relativity is much more accurate than Newton's laws of gravity and motion, but tradition dictates that no one refers to Einstein's laws of relativity and Newton's theories of gravity and motion. Darwin's theory of evolution has been tested many times, and scientists have great confidence in it, but no one refers to Darwin's law of evolution. These distinctions are subtle and sometimes depend more on custom than on levels of confidence.

A fossil of a 500-million-year-old trilobite: Darwin's theory of evolution has been tested many times and is widely accepted, but by custom it is called a theory and not a law. (From the collection of John Coolidge III)

■ Table 4-1 | Kepler's Laws of Planetary Motion

I.   The orbits of the planets are ellipses with the sun at one focus.

II. A line from a planet to the sun sweeps over equal areas in equal intervals of time.

III. A planet's orbital period squared is proportional to its average distance from the sun cubed:

$$P_{yr}^2 = a_{AU}^3$$

that when the planet is closer to the sun and the line connecting it to the sun is shorter, the planet moves more rapidly to sweep over the same area that is swept over when the planet is farther from the sun. So the planet in Figure 4-21 (inset) would move from point A to point B in one month, sweeping over the area shown. But when the planet is farther from the sun, one month's motion would be shorter, from A' to B'.

The time that a planet takes to travel around the sun once is its orbital period, P, and its average distance from the sun equals the semimajor axis of its orbit, a. Kepler's third law says these two quantities are related: The orbital period squared is proportional to the semimajor axis cubed. If you measure P in years and a in astronomical units, you can summarize the third law as:

$$P_{yr}^2 = a_{AU}^3$$

The subscripts are reminders that you must measure the period in years (yr) and the semimajor axis in astronomical units (AU).

You can use Kepler's third law to make simple calculations. For example, Jupiter's average distance from the sun is 5.20 AU. What is its orbital period? If a equals 5.20, then a^3 equals 140.6. The orbital period must be the square root of 140.6, which equals about 11.8 years.

You should note that Kepler's three laws are empirical. That is, they describe a phenomenon without explaining why it occurs. Kepler derived them from Tycho's extensive observations, not from any fundamental assumption or theory. In fact, Kepler never knew

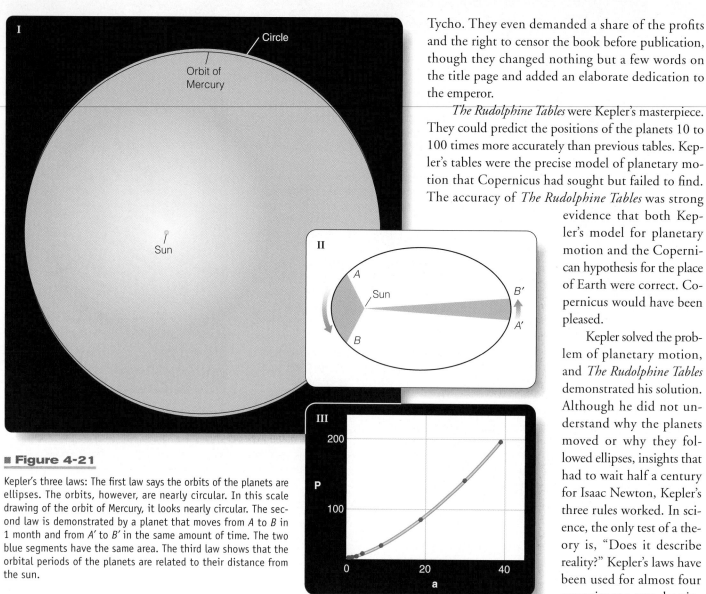

■ Figure 4-21

Kepler's three laws: The first law says the orbits of the planets are ellipses. The orbits, however, are nearly circular. In this scale drawing of the orbit of Mercury, it looks nearly circular. The second law is demonstrated by a planet that moves from *A* to *B* in 1 month and from *A'* to *B'* in the same amount of time. The two blue segments have the same area. The third law shows that the orbital periods of the planets are related to their distance from the sun.

Tycho. They even demanded a share of the profits and the right to censor the book before publication, though they changed nothing but a few words on the title page and added an elaborate dedication to the emperor.

The Rudolphine Tables were Kepler's masterpiece. They could predict the positions of the planets 10 to 100 times more accurately than previous tables. Kepler's tables were the precise model of planetary motion that Copernicus had sought but failed to find. The accuracy of *The Rudolphine Tables* was strong evidence that both Kepler's model for planetary motion and the Copernican hypothesis for the place of Earth were correct. Copernicus would have been pleased.

Kepler solved the problem of planetary motion, and *The Rudolphine Tables* demonstrated his solution. Although he did not understand why the planets moved or why they followed ellipses, insights that had to wait half a century for Isaac Newton, Kepler's three rules worked. In science, the only test of a theory is, "Does it describe reality?" Kepler's laws have been used for almost four centuries as a true description of orbital motion.

what held the planets in their orbits or why they continued to move around the sun. His books are a fascinating blend of careful observation, mathematical analysis, and mystical theory.

The Rudolphine Tables

In spite of Kepler's recurrent involvement with astrology and numerology, he continued to work on *The Rudolphine Tables*. At last, in 1627, they were ready, and he financed their printing himself, dedicating them to the memory of Tycho Brahe. In fact, Tycho's name appears in larger type on the title page than Kepler's own. This is especially surprising when you recall that the tables were based on the heliocentric model of Copernicus and the elliptical orbits of Kepler and not on the Tychonic system. The reason for Kepler's evident deference was Tycho's family, still powerful and still intent on protecting the memory of

Building Scientific Arguments

What were the main differences among *The Alfonsine Tables*, *The Prutenic Tables*, and *The Rudolphine Tables*?
Each of these tables was an expression of a different theory, and comparing theory with evidence is the heart of scientific arguments. All three of these tables predicted the motions of the sun, moon, and planets, but only *The Rudolphine Tables* proved accurate. *The Alfonsine Tables*, produced in Toledo around AD 1250, were based on the Ptolemaic model, so they were geocentric and used uniform circular motion; consequently, they were not very accurate. *The Prutenic Tables*, published in 1551, were based on the Copernican model, and so were heliocentric. But because *The Prutenic Tables* included the classical principle of uniform circular motion, they were no more accurate than *The Alfonsine Tables*. Kepler's *Rudol-*

phine Tables were Copernican in that they were heliocentric, but they included Kepler's three laws of planetary motion and consequently could predict the positions of the planets 10 to 100 times more accurately than previous tables.

One of the main reasons for the success of the Copernican hypothesis was not that it was accurate but that it was elegant. Compare the motions of the planets and the explanation of retrograde motion in the Copernican model with those in the Ptolemaic and Tychonic models. **How was the Copernican model more elegant?**

■ ■ ■

Connections: Kepler died on November 15, 1630. During his life he had been an astrologer, a mystic, a numerologist, and a seer, but he became one of the world's great astronomers. His work not only overthrew the concept of uniform circular motion that had shackled the minds of astronomers for over 2000 years but also opened the way to the empirical study of planetary motion, a study that lies at the heart of modern astronomy.

4-4 Modern Astronomy

YOU CAN DATE THE ORIGIN of modern astronomy from the 99 years between the deaths of Copernicus and Galileo (1543 to 1642); it was an age of transition. That period marked the change from the Ptolemaic model of the universe to the Copernican model with the attendant controversy over the place of the Earth. But that same period also marked a transition in the nature of astronomy in particular and science in general, a transition illustrated in the resolution of the puzzle of planetary motion. The puzzle was not solved by philosophical arguments about the perfection of the heavens or by debate over the meaning of scripture. It was solved by precise observation and careful computation, techniques that are the foundation of modern science.

The discoveries of Kepler and Galileo found acceptance in the 1600s because the world was in transition. Astronomy was not the only thing changing during this period. The Renaissance is commonly taken to be the period between 1300 and 1600, and these 99 years of astronomical history lie at the culmination of the reawakening of learning in all fields (■ Figure 4-22). Ships

■ Figure 4-22

The 99 years between the death of Copernicus in 1543 and the death of Galileo in 1642 marked the transition from the ancient astronomy of Ptolemy and Aristotle to the revolutionary theory of Copernicus, and, simultaneously, the invention of science as a way of understanding the world.

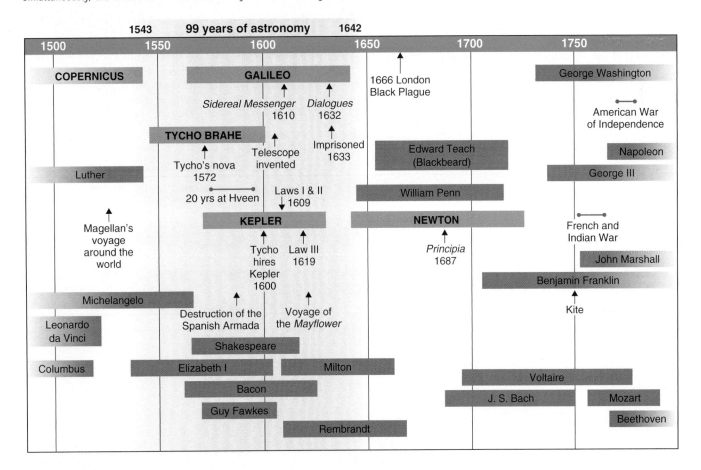

were sailing to new lands and encountering new cultures. The world was open to new ideas and new observations. Martin Luther remade religion, and other philosophers and scholars reformed their areas of human knowledge. Had Copernicus not published his hypothesis, someone else would have suggested that the universe is heliocentric. History was ready to shed the Ptolemaic system.

In addition, this period marks the beginning of the modern scientific method. Beginning with Copernicus, scientists such as Tycho, Kepler, and Galileo depended more and more on evidence, observation, and measurement. This, too, is coupled to the Renaissance and its advances in metalworking and lens making. Before the story told in this chapter began, no astronomer had looked through a telescope, because one could not be made. By 1642, not only telescopes but also other sensitive measuring instruments had transformed science into something new and precise. As you can imagine, scientists were excited by these discoveries, and they founded scientific societies that increased the exchange of observations and hypotheses and stimulated more and better work. The most important advance, however, was the application of mathematics to scientific questions. Kepler's work demonstrated the power of mathematical analysis; and, as the quality of these numerical tools improved, the progress of science accelerated. This story of the birth of modern astronomy is actually the story of the birth of modern science as well.

Building Scientific Arguments

Why was it so hard for astronomers to abandon the Ptolemaic model?

The central position of Earth and uniform circular motion were part of a paradigm—a set of ideas that people accepted as obvious. Because they all worked within that paradigm, they could not see its faults as well as you can today. It did not occur to them to test obvious ideas. Also, astronomers before the time of the Copernican Revolution were not accustomed to thinking in scientific arguments based on theory and evidence. Observations that did not fit the paradigm were not taken as seriously as they would be today. Only after the time of Galileo did scholars begin to think and reason like modern scientists. Scientific thinking is so common today, you take it for granted. When you read about new music players in a magazine like *Consumer Reports,* hear about blood tests at a trial, or try two different kinds of toothpaste to see which you like best, you are thinking scientifically and depending on evidence. Perhaps you are surprised that something as obvious as scientific thinking had to be invented.

There was yet another reason why astronomers found it hard to let go of the Ptolemaic model. The central place of the Earth had serious theological meaning. **Why did moving Earth away from the center make some astronomers hesitate to adopt the Copernican theory?**

■   ■   ■

Connections: You have read that Kepler solved the puzzle of planetary motion, but his solution seems incomplete. Why do the planets move as they do? The story will continue in the next chapter with the life of Isaac Newton. You will see how his accomplishments had their origins in the work of Galileo and led to Einstein's deeper understanding of gravity.

Summary

4-1 | The Roots of Astronomy

How did the ancients describe the place of the Earth?

■ Archaeoastronomy is the study of the astronomy of ancient peoples.

■ Many cultures around the world observed the sky and marked important alignments. Structures such as Stonehenge, Newgrange, the Sun Dagger, and Egyptian temples have astronomical alignments.

■ In most cases, ancient cultures, having no written language, left no detailed records of their astronomical beliefs.

■ Greek astronomy, derived in part from Babylon and Egypt, is better known because written documents have survived.

■ Classical philosophers accepted as a first principle that Earth was the unmoving center of the universe. Another first principle was that the heavens were perfect, so philosophers such as Plato argued that, because the sphere was the most perfect geometrical form, the heavens must be made up of spheres in uniform rotation. This led to the belief in uniform circular motion.

■ The geocentric universe became part of the teachings of the great philosopher Aristotle, who argued that the sun, moon, and stars were carried around Earth on rotating crystalline spheres.

■ Aristotle's estimate for the size of Earth was only about one-third of its true size. Eratosthenes used the well at Syene to measure the diameter of Earth and got an accurate estimate.

■ About 140 AD, Ptolemy gave mathematical form to Aristotle's model in the *Almagest*. Ptolemy preserved the principles of geocentrism and uniform circular motion, but he added epicycles, deferents, and equants to better predict the motions of the planets.

4-2 | The Copernican Revolution

How did Copernicus change the place of the Earth?

■ Copernicus devised a heliocentric model. He preserved the principle of uniform circular motion, but he argued that Earth rotates on its axis and revolves around the sun once a year. His theory was controversial because it contradicted Church teaching.

■ Copernicus published his theory in his book *De Revolutionibus* in 1543, the same year he died.

■ Because Copernicus kept uniform circular motion as part of his theory, his model did not predict the motions of the plants well, but it did offer a simple explanation of retrograde motion without using big epicycles.

■ One reason the Copernican model won converts was that it was more elegant. Venus and Mercury were treated the same as all the other planets, and the velocity of each planet was related to its distance from the sun.

Why was Galileo condemned by the Inquisition?

■ Galileo used the newly invented telescope to observe the heavens, and he recognized the significance of what he saw there. His discoveries of the phases of Venus, the satellites of Jupiter, the mountains of the moon, and other phenomena helped undermine the Ptolemaic universe.

■ Galileo based his analysis on observational evidence. In 1633, he was condemned before the Inquisition for refusing to halt his defense of Copernicanism.

4-3 | The Puzzle of Planetary Motion

How did Copernican astronomers solve the puzzle of planetary motion?

■ Tycho Brahe developed his own model in which the sun and moon circled Earth and the planets circled the sun.

■ Tycho's great contribution was to compile detailed observations over a period of 20 years, observations that were later used by Kepler.

■ Kepler inherited Tycho's books of observations in 1601 and used them to uncover three laws of planetary motion. He found that the planets follow ellipses with the sun at one focus, that they move faster when near the sun, and that a planet's orbital period squared is proportional to its orbital radius cubed.

■ Kepler's final book, *The Rudolphine Tables* (1627), combined heliocentrism with elliptical orbits and predicted the positions of the planets well.

4-4 | Modern Astronomy

■ The 99 years from the death of Copernicus to the death of Galileo marked the birth of modern science. From that time on, science depended on evidence to test theories and relied on the analytic methods first demonstrated by Kepler.

New Terms

archaeoastronomy (p. 58)

uniform circular motion (p. 60)

geocentric universe (p. 61)

eccentric (p. 62)

heliocentric universe (p. 63)

parallax (p. 64)

retrograde motion (p. 64)

epicycle (p. 65)

deferent (p. 65)

equant (p. 65)

paradigm (p. 68)

ellipse (p. 76)

semimajor axis, *a* (p. 76)

eccentricity, *e* (p. 76)

hypothesis (p. 77)

theory (p. 77)

natural law (p. 77)

Review Questions

Ace ◎ Astronomy™ Assess your understanding of this chapter's topics with additional quizzing and animations at http://ace .brookscole.com/sf9

1. What evidence is there that early human cultures observed astronomical phenomena?

2. Why did Plato propose that all heavenly motion was uniform and circular?

3. How do the epicycles of Mercury and Venus differ from those of Mars, Jupiter, and Saturn?

4. Why did Copernicus have to keep small epicycles in his model?

5. Explain how each of Galileo's telescopic discoveries contradicted the Ptolemaic theory.

6. Galileo was condemned, but Kepler, also a Copernican, was not. Why not?

7. When Tycho observed the new star of 1572, he could detect no parallax. Why did that result undermine belief in the Ptolemaic system?

8. Does Tycho's model of the universe explain the phases of Venus that Galileo observed? Why or why not?

9. How do the first two of Kepler's three laws overthrow one of the basic beliefs of classical astronomy?

10. What is the difference between a hypothesis, a theory, and a law?

11. How did *The Alfonsine Tables, The Prutenic Tables,* and *The Rudolphine Tables* differ?

12. What three astronomical objects are represented here? What are the two rings?

13. Whose observatory is shown here? Why are there no telescopes?

Discussion Questions

1. Historian of science Thomas Kuhn has said that *De Revolutionibus* was a revolution-making book but not a revolutionary book. How was it an old-fashioned, classical book?

2. Why might Tycho Brahe have hesitated to hire Kepler? Why do you suppose he appointed Kepler his scientific heir?

3. How does the modern controversy over creationism and evolution reflect two ways of knowing about the physical world?

Problems

1. Draw and label a diagram of the eastern horizon from northeast to southeast and label the rising point of the sun at the solstices and equinoxes. (See page 27 and Figure 4-1).

2. If you lived on Mars, which planets would exhibit retrograde motion? Which would never be visible as crescent phases?

3. Galileo's telescope showed him that Venus has a large angular diameter (61 seconds of arc) when it is a crescent and a small angular diameter (10 seconds of arc) when it is nearly full. Use the small-angle formula to find the ratio of its maximum distance to its minimum distance. Is this ratio compatible with the Ptolemaic universe shown on page 65?

4. Galileo's telescopes were not of high quality by modern standards. He was able to see the moons of Jupiter, but he never reported seeing features on Mars. Use the small-angle formula to find the angular diameter of Mars when it is closest to Earth. How does that compare with the maximum angular diameter of Jupiter?

5. If a planet has an average distance from the sun of 4 AU, what is its orbital period?

6. If a space probe is sent into an orbit around the sun that brings it as close as 0.5 AU and as far away as 5.5 AU, what will be its orbital period?

7. Pluto orbits the sun with a period of 247.7 years. What is its average distance from the sun?

Ace Astronomy™ To access the resources in the Media Cluster, log into AceAstronomy at **http://ace.brookscole .com/sf9** and select Chapter 4.

Eratosthenes' Experiment
Instead of light in a well, watch the shadow of a vertical stick from three different places on Earth. From that, you could calculate the size of Earth.

Epicycles
Watch Mars move against the background of stars as its epicycle carries it closest to Earth. You can see the classical explanation of retrograde motion.

Eratosthenes' Calculations
Reproduce Eratosthenes' experiment and find the size of Earth. Then take the experiment to Earth's moon and to Jupiter.

Parallax I
As you learned in this chapter, parallax is the apparent motion of an object because of the motion of the observer. In this animation, see how a spaceship appears to move against a background of stars due to Earth's motion.

Critical Inquiries for the Web

1. The trial of Galileo is an important event in the history of science. You now know, and the Roman Catholic Church now recognizes, that Galileo's view was correct, but what were the arguments on both sides of the issue as it was unfolding? Research the Internet for documents that chronicle the trial, Galileo's observations and publications, and the position of the Church. Use this information to outline the case for and against Galileo in the context of the times in which the trial occurred.

2. Take a "virtual tour" of Stonehenge by browsing the Web for information related to this megalithic site and the possibility that it was used for astronomical purposes. (Be careful to use legitimate scientific and historical websites in your survey.) Summarize the various astronomical alignments evident in the layout of the site.

Exploring *TheSky*

1. Go to Stonehenge in southern England and watch the sunrise on the morning of the summer solstice. Where does it rise on the morning of the winter solstice?

2. Observe Mars going through its retrograde motion. (*Hint:* Use **Reference Lines** under the **View** menu to turn on the ecliptic. Be sure you are in **Free Rotation** under the **Orientation** menu. Locate Mars and use the time skip arrows to watch it move.)

3. Compare the size of the retrograde loops made by Mars, Jupiter, and Saturn.

4. Can you recognize the effects of Kepler's second law in the orbital motion of any of the planets? (*Hint:* Use **3D Solar System Mode** under the **View** menu.)

5. Can you recognize the effects of Kepler's third law in the orbital motion of the planets?

Go to the Brooks/Cole Astronomy Resource Center (**http:// astronomy.brookscole.com**) for critical thinking exercises, articles, and additional readings from InfoTrac College Edition, Brooks/Cole's online student library.

5 | Newton, Einstein, and Gravity

Nature and Nature's laws
lay hid in night:
God said, "Let Newton be!"
and all was light.

ALEXANDER POPE

ISN'T IT WEIRD THAT Isaac Newton is said to have "discovered" gravity in the late 17th century—as if people didn't have gravity before that, as if they floated around holding onto tree branches? Newton's accomplishment is more impressive than just a discovery. Everyone experienced gravity without noticing it. Newton realized that a force had to exist that made things fall, and that changed the way people thought about nature (■ Figure 5-1). ▌ Isaac Newton was born in Woolsthorpe, England, on December 25, 1642, and on January 4, 1643. This was not a biological anomaly but a calendrical quirk. Most of Europe, following the lead of the Catholic countries, had adopted the Gregorian calendar, but Protestant England continued to use the Julian calendar. So December 25 in England was January 4 in Europe. If you take the

▌ Continued on page 86 ▌

The motion of the planets was not well understood until Newton discovered gravity. Much later, Einstein found an even better way to understand gravity.

Guidepost

Looking Back

In the last chapter, you saw how Renaissance astronomers struggled to understand the place of the Earth and the motion of the planets. Although Johannes Kepler seemed to solve the puzzle of planetary motion, his three laws only *described* how the planets moved. He never really understood why the planets moved in elliptical orbits.

This Chapter

Now you are ready to finish the adventure. In this chapter you will see how Galileo tried to understand motion and how Newton discovered a wonderful way to understand gravity. But then you will see how Einstein found an even better explanation for motion and gravity. Here you will find answers to five essential questions:

What happens when an object falls?

How did Newton discover gravity?

How does gravity explain orbital motion?

How does gravity explain the tides?

How did Einstein refine our understanding of motion and gravity?

Looking Ahead

Gravity rules. Every object in the universe attracts every other object. The moon orbiting Earth, matter falling into black holes, and the overall structure of the universe are dominated by gravity. The rest of this book will tell the story of matter and gravity, and you will see how modern astronomers have carried on the discoveries made by Galileo, Newton, and Einstein. The universe is a swirling waltz of matter dancing to the music of gravity, and you are going along for the ride.

Ace ◖Astronomy™ The AceAstronomy icon throughout the text indicates an opportunity for you to test yourself on key concepts and to explore animations and interactions on the AceAstronomy website at: http://ace .brookscole.com/sf9

■ Figure 5-1

Space stations and astronauts, as well as planets, moons, stars, and galaxies, follow paths called orbits that are described by three simple laws of motion and a theory of gravity first understood by Isaac Newton (1642–1727). Newtonian physics is adequate to send astronauts to the moon and analyze the rotation of the largest galaxies. (NASA/JSC)

English date, then Newton was born in the same year that Galileo Galilei died.

Newton went on to become one of the greatest scientists who ever lived, but even he admitted the debt he owed to those who had studied nature before him. He said, "If I have seen farther than other men, it is because I stood on the shoulders of giants." One of those giants was Galileo. Although Galileo is remembered as the defender of Copernicanism, he was also a talented scientist who studied the motions of falling bodies.

For over two centuries following the publication of Newton's works, astronomers used Newton's laws to describe the universe. Then, early in the 20th century, Albert Einstein proposed a new way to describe gravity. The new theory did not replace Newton's laws but rather showed that they were only approximately correct and could be seriously in error under special circumstances. Einstein's theories further extend the scientific understanding of the nature of gravity. Just as Newton had stood on the shoulders of Galileo, Einstein stood on the shoulders of Newton.

5-1 Galileo and Newton

JOHANNES KEPLER DISCOVERED three laws of planetary motion, but he never understood why the planets move along their orbits. At one place in his writings, he wonders if they are pulled along by magnetic forces emanating from the rotating sun. At another place, he considers and dismisses the idea that the planets are pushed along their orbits by angels.

Newton refined Kepler's model of planetary motion but did not perfect it. In science, a model is an intellectual conception of how nature works (Window on Science 2-1). No model is perfect. Kepler's model was better than Aristotle's, but Newton improved Kepler's model by expanding it into a general theory of motion and gravity. In fact, most scientists now refer to Newton's *law* of gravity. Whether it is called a model, a theory, or a law, it is not perfect. Newton never understood what gravity was. It was as mysterious as an angel pushing the moon inward toward Earth instead of forward along the moon's orbit.

To understand science, you must understand the nature of scientific descriptions. The scientist studies nature by either creating new theories or refining old theories. Yet a theory can never be perfect, because it can never represent the universe in all its intricacies. Instead, a theory must be a limited approximation of a single phenomenon, such as orbital motion. It is fitting that Newton's discoveries all began with Kepler's fellow Copernican, Galileo.

Galileo and Motion

Even before Galileo built his first telescope, he had begun studying the motion of freely moving bodies (■ Figure 5-2). After the Inquisition condemned and imprisoned him in 1633, he continued his study of motion. He seems to have realized that he would have to understand motion before he could truly understand the Copernican system. That he was eventually able to formulate principles that later led Newton to the laws of motion and the theory of gravity is a tribute to Galileo's ability to set aside the authority of the ancients and think for himself.

The authority of the age was Aristotle, whose ideas on motion were hopelessly confused. Aristotle said that the world is made up of four classical elements: earth, water, air, and fire, each located in its proper place. The proper place for the element earth was the center of the universe, and the proper place for water was just above earth. Air and then fire formed higher layers and above them was the realm of the planets and stars. (You can see the four layers of the classical elements in the diagram at the top of page 64.) If displaced, the four elements were believed to have a natural tendency to move toward their proper place in the cosmos. Things made up mostly of air or fire—smoke, for instance—tend to move upward. Things composed mostly of earth and water—wood, rock, flesh, bone, and so on—tend to move downward toward the proper place of earth and water. According to Aristotle, objects fall downward because they are moving toward their proper place.*

Aristotle called these motions **natural motions** to distinguish them from **violent motions** produced, for instance, when you push on an object and make it move other than toward its proper place. According to Aristotle, such motions stop as soon as the force is removed. To explain how an arrow could continue to move upward even after it had left the bowstring, he said currents in the air around the arrow carried it forward even though the bowstring was no longer pushing it.

These ideas about natural and violent motion, and the necessity of a force to preserve motion, were still accepted theory in Galileo's time. In fact, in 1590, when Galileo was 26, he wrote a short work called *De Motu (On Motion)* that deals with the proper places of objects and their natural motions.

---

* This is one reason why Aristotle had to have a geocentric universe. If Earth's center had not also been the center of the cosmos, his explanation of gravity would not have worked.

■ **Figure 5-2**

Although Galileo is often associated with the telescope, as on this Italian stamp, he also made systematic studies of the motion of falling bodies and made discoveries that led to the law of inertia.

In Galileo's time and for the two preceding millennia, scholars had commonly tried to resolve problems of science by referring to authority. To analyze the flight of a cannonball, for instance, they would turn to the writings of Aristotle and other classical philosophers and try to deduce what those philosophers would have said on the subject. This generated a great deal of discussion but little real progress. Galileo broke with this tradition and conducted his own experiments.

He began by studying the motions of falling bodies, but he quickly discovered that the velocities were so great and the times so short that he could not measure them accurately. Consequently, he began using polished bronze balls rolling down gently sloping inclines. In that instance, the velocity is lower and the time longer. Using an ingenious water clock, he was able to measure the time the balls took to roll given distances down the incline, and he correctly recognized that these times are proportional to the times taken by falling bodies.

He found that falling bodies do not fall at constant rates, as Aristotle had said, but are accelerated. That is, they move faster with each passing second. Near Earth's surface, a falling object will have a velocity of 9.8 m/s (32 ft/s) at the end of 1 second, 19.6 m/s (64 ft/s) after 2 seconds, 29.4 m/s (96 ft/s) after 3 seconds, and so on. Each passing second adds 9.8 m/s (32 ft/s) to the object's velocity (■ Figure 5-3). In modern terms, this is called the **acceleration of gravity** at Earth's surface.

Galileo also discovered that the acceleration does not depend on the weight of the object. This, too, is contrary to the teachings of Aristotle, who believed that heavy objects, containing more earth and water, fall with higher velocity. Galileo found that the acceleration of a falling body is the same whether it is heavy or light. According to some accounts, he demonstrated this by dropping balls of iron and wood from the top of the Leaning Tower of Pisa to show that they would fall together and hit the ground at the same time (■ Figure 5-4a). In fact, he probably didn't perform this experiment. It would not have been conclusive anyway because of air resistance. More than 300 years later, Apollo 15 astronaut David Scott, standing on the airless moon, demonstrated

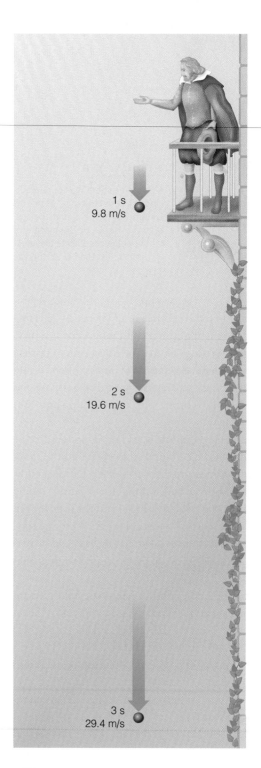

1 s
9.8 m/s

2 s
19.6 m/s

3 s
29.4 m/s

■ **Figure 5-3**

Galileo found that a falling object is accelerated down-ward. Each second, its velocity increases by 9.8 m/s (32 ft/s).

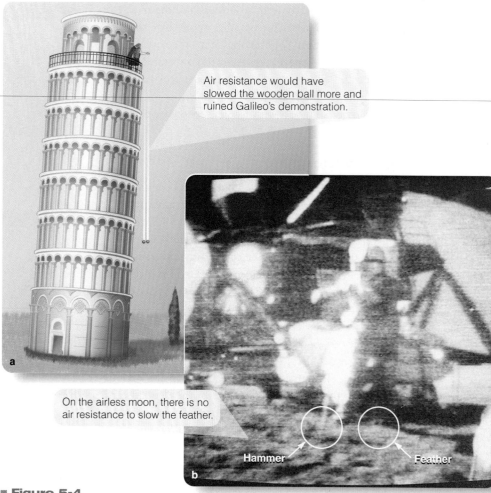

Air resistance would have slowed the wooden ball more and ruined Galileo's demonstration.

On the airless moon, there is no air resistance to slow the feather.

Hammer   Feather

b

■ **Figure 5-4**

(a) According to tradition, Galileo demonstrated that the acceleration of a falling body is indepen-dent of its weight by dropping balls of iron and wood from the Leaning Tower of Pisa. In fact, air resistance would have confused the result. (b) In a historic television broadcast from the moon on August 2, 1971, David Scott dropped a hammer and a feather at the same instant. They fell with the same acceleration and hit the surface together. (NASA)

an object rolling down an incline is accelerated and that an ob-ject rolling up the same incline is decelerated. If the incline were perfectly horizontal and frictionless, he reasoned, there could be no acceleration or deceleration to change the object's velocity, and, in the absence of friction, the object would continue to move for-ever. In his own words, "any velocity once imparted to a moving body will be rigidly maintained as long as the external causes of acceleration or retardation are removed."

This is contrary to Aristotle's belief that motion can continue only if a force is present to maintain it. In fact, Galileo's statement is a perfectly valid summary of the law of inertia, which became Newton's first law of motion.

Galileo published his work on motion in 1638, two years after he had become entirely blind and only four years before his death. The book was called *Mathematical Discourses and Demon-strations Concerning Two New Sciences, Relating to Mechanics and to Local Motion.* It is known today as *Two New Sciences.*

Galileo's discovery by dropping a feather and a steel geologist's hammer. They hit the lunar surface at the same time (Figure 5-4b).

Having described natural motion, Galileo turned his atten-tion to violent motion—that is, motion directed other than to-ward an object's proper place in the cosmos. He pointed out that

The book is a brilliant achievement for a number of reasons. To understand motion, Galileo had to abandon the authority of the ancients, devise his own experiments, and draw his own conclusions. In a sense, this was the first example of experimental science. But Galileo also had to generalize his experiments to discover how nature worked. Though his apparatus was finite and plagued by friction, he was able to imagine an infinite, frictionless plane on which a body moves at constant velocity. In his workshop, the law of inertia was obscure; but, in his imagination, it was clear and precise.

Ace✸Astronomy™   Log into AceAstronomy and select this chapter to see Astronomy Exercise "Falling Bodies." Try one of the most famous experiments Galileo never did.

Newton and the Laws Of Motion

Newton's three laws of motion (■ Table 5-1) are critical to understanding gravity and orbital motion. They apply to any moving object, from an automobile driving along a highway to galaxies colliding with each other.

The first law is really a restatement of Galileo's law of inertia. An object continues at rest or in uniform motion in a straight line unless acted upon by some force. Astronauts drifting in space will travel at constant rates in straight lines forever if no forces act on them (■ Figure 5-5a).

Newton's first law also explains why a projectile continues to move after all forces have been removed—for instance, why an arrow continues to move after leaving the bowstring. The object continues to move because it has momentum. You can think of an object's **momentum** as a measure of its amount of motion.

Momentum equals velocity times mass. Obviously velocity is important. A low-velocity object such as a paper clip tossed across a room has little momentum, and you could easily catch it in your hand. But the same paper clip fired at the speed of a rifle bullet would have tremendous momentum, and you would not dare try to catch it. Momentum also depends on the mass of an object (**Focus on Fundamentals 1**). To see how, imag-

■ **Table 5-1 I Newton's Three Laws of Motion**

I. A body continues at rest or in uniform motion in a straight line unless acted upon by some force.

II. The acceleration of a body is inversely proportional to its mass, directly proportional to the force, and in the same direction as the force.

III. To every action, there is an equal and opposite reaction.

ine that, instead of tossing a paper clip, someone tosses you a bowling ball. A bowling ball contains much more mass than a paper clip and therefore has much greater momentum at the same velocity.

Newton's first law explains the consequences of the conservation of momentum. When Newton said that momentum is conserved, he meant that it remains constant until something acts to change it. A moving object has a given amount of momentum. To change that momentum, you must exert some force on the object to change either the speed or the direction. Newton's first law and the concept of momentum came from the work of Galileo.

Newton's second law of motion discusses forces, and Galileo did not talk about forces. Galileo spoke instead of accelerations. Newton saw that an acceleration is the result of a force acting on a mass (Figure 5-5b). Newton's second law is commonly written as:

$$F = ma$$

As always, you must carefully define terms when you look at an equation. An **acceleration** is a change in velocity, and a **velocity** is a directed rate of motion. Rate of motion means, of course, a speed, but the word *directed* has a special meaning. Speed itself does not have any direction associated with it, but velocity does. If you drive a car in a circle at 55 mph, your speed is constant, but your velocity is changing because your direction of motion is changing. So an object experiences an acceleration if its speed changes or if its direction of motion changes. Every

■ **Figure 5-5**

Newton's three laws of motion.

Mass

One of the most fundamental parameters in science is **mass,** a measure of the amount of matter in an object. A bowling ball, for example, contains a large amount of mass, but a child's rubber ball contains less matter than the bowling ball, and it is said to be less massive.

Mass is not the same as weight. Your weight is the force that Earth's gravity exerts on the mass of your body. Because gravity pulls you downward, you press against the bathroom scale, and you can measure your weight. Floating in space, you would have no weight at all; a bathroom scale would be useless. But your body would still contain the same amount of matter, so you would still have mass.

Sports analogies illustrate the importance of mass in dramatic ways. A bowling ball, for example, must be massive in order to have a large effect on the pins it strikes. Imagine trying to knock down all the pins with a bowling ball no more massive than a balloon. Even in space, where the bowling ball would be weightless, a low-mass bowling ball would have little effect on the pins. On the other hand, runners want track shoes that have low mass and thus are easy to move. Imagine trying to run a 100-meter dash wearing track shoes that were as massive as bowling balls. They would be very hard to move, and it would be difficult to accelerate away from the starting block. The shot put takes muscle because the shot is massive,

not because it is heavy. Imagine throwing the shot in space where it would have no weight. It would still be massive, and it would take great effort to start it moving.

Mass is a unique measure of the amount of material in an object. Using the metric system (Appendix A), mass is measured in kilograms.

Mass is not the same as weight.

100 kg

automobile has three accelerators—the gas pedal, the brake pedal, and the steering wheel. All three change the car's velocity.

In a way, the second law is just common sense; you experience it every day. The acceleration of a body is proportional to the force applied to it. This is reasonable. If you push gently against a grocery cart, you expect a small acceleration. The second law of motion also says that the acceleration depends on the mass of the body. This, too, is reasonable. If your grocery cart were filled with bricks and you pushed it gently, you would expect very little result. If it were full of inflated balloons, however, it would move easily in response to a gentle push. Finally, the second law says that the resulting acceleration is in the direction of the force. This is also what you would expect. If you push on a cart that is not moving, you expect it to begin moving in the direction you push.

The second law of motion is important because it establishes a precise relationship between cause and effect (**Window on Science 5-1**). Objects do not just move. They accelerate due to the action of a force. Moving objects do not just stop. They decelerate due to a force. Also, moving objects don't just change direction for no reason. Any change in direction is a change in velocity and requires the presence of a force. Aristotle said that objects move because they have a tendency to move. Newton said that objects move due to a specific cause, a force.

Newton's third law of motion specifies that for every action there is an equal and opposite reaction. In other words, forces must occur in pairs directed in opposite directions. For example, if you stand on a skateboard and jump forward, the skateboard will shoot away backward. As you jump, your feet exert a force

against the skateboard, which accelerates it toward the rear. But forces must occur in pairs, so the skateboard must exert an equal but opposite force on your feet that accelerates your body forward (Figure 5-5c).

Mutual Gravitation

The three laws of motion led Newton to consider the force that causes objects to fall. The first and second laws tell you that falling bodies accelerate downward because some force must be pulling downward on them. Newton wondered what that force could be.

Newton was also aware that some force has to act on the moon. The moon follows a curved path around Earth, and motion along a curved path is accelerated motion. The second law says that an acceleration requires a force, so a force must be making the moon follow that curved path.

Newton wondered if the force that holds the moon in its orbit could be the same force that causes apples to fall—gravity. He was aware that gravity extends at least as high as the tops of mountains, but he did not know if it could extend all the way to the moon. He believed that it could, but he thought it would be weaker at greater distances, and he guessed that its strength would decrease as the square of the distance increased.

This relationship, the **inverse square law,** was familiar to Newton from his work on optics, where it applied to the intensity of light. A screen set up 1 meter from a candle flame receives a certain amount of energy on each square meter. However, if that screen is moved to a distance of 2 meters, the light that originally illuminated one square meter must cover four square me-

The Unstated Assumption of Science: Cause and Effect

One of the most often used and least often stated principles of science is cause and effect, and you could argue that Newton's second law of motion was the first clear statement of the principle.

Ancient philosophers such as Aristotle argued that objects moved because of tendencies. They said that earth and water, and objects made mostly of earth and water, had a natural tendency to move toward the center of the universe. This natural motion had no cause but was inherent in the nature of the objects. But Newton's second law says $F = ma$. If an object (of mass m) changes its motion (a in the equation), then it must be acted on by a force

(F in the equation). Any effect (a) must be the result of a cause (F).

The principle of cause and effect goes far beyond motion. The principle of cause and effect gives scientists confidence that every effect has a cause. Hearing loss in certain laboratory rats, color changes in certain chemical dyes, and explosions on certain stars are all effects that must have causes. All of science is focused on understanding the causes of the effects you see around you. If the universe were not rational, then you could never expect to discover causes. Newton's second law of motion was arguably the first clear statement that the behavior of the universe depends rationally on causes.

Cause and effect: Why did this star explode in 1992? There must have been a cause. (ESA/STScI and NASA)

ters (■ Figure 5-6). Consequently, the intensity of the light is inversely proportional to the square of the distance to the screen.

Newton made two assumptions that enabled him to predict the strength of Earth's gravity at the distance of the moon. He assumed that the strength of gravity follows the inverse square law and that the critical distance is not the distance from Earth's surface but the distance from Earth's center. Because the moon is about

■ Figure 5-6

As light radiates away from a source, it is spread thinner and becomes less intense. Here the light falling on one square meter on the inner sphere must fall on four square meters on a sphere twice as big. This shows how the intensity of light is inversely proportional to the square of the distance.

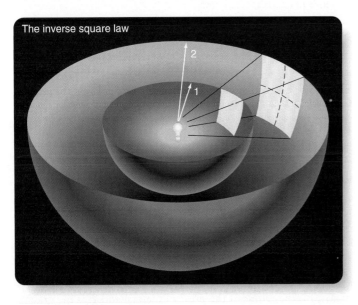

The inverse square law

60 Earth radii away, Earth's gravity at the distance of the moon should be about 60^2 times less than at Earth's surface. Instead of being 9.8 m/s² at Earth's surface, it should be about 0.0027 m/s² at the distance of the moon.

Now, Newton wondered, could this acceleration keep the moon in orbit? He knew the moon's distance and its orbital period, so he could calculate the actual acceleration needed to keep it in its curved path. The answer is 0.0027 m/s². To the accuracy of Newton's data for Earth's radius, it was exactly what his assumptions predicted. The moon is held in its orbit by gravity, and gravity obeys the inverse square law.

Newton's third law says that forces always occur in pairs, and this leads to the conclusion that gravity is mutual. If Earth pulls on the moon, then the moon must pull on Earth. Gravitation is a general property of the universe. The sun, the planets, and all their moons must also attract each other by mutual gravitation. In fact, every particle of mass in the universe must attract every other particle, and Newtonian gravity is often called universal mutual gravitation.

Clearly the force of gravity depends on mass. Your body is made of matter, and you have your own personal gravitational field. But your gravity is weak and does not attract personal satellites orbiting around you. Larger masses have stronger gravity. From an analysis of the third law of motion, Newton realized that the mass that resists acceleration in the first law must be the same as the mass associated with gravity. Then when two bodies attract one another and accelerate toward each other, the two masses must be equally involved. Newton performed precise experiments with pendulums and confirmed this equivalence between the mass that resists acceleration and the mass that causes gravity.

From this, combined with the inverse square law, he was able to write the famous formula for the gravitational force between two masses, M and m:

$$F = -\frac{GMm}{r^2}$$

The constant G is the gravitational constant, and r is the distance between the masses. The negative sign means that the force is attractive, pulling the masses together and making r decrease. In plain English, Newton's law of gravitation says: The force of gravity between two masses M and m is proportional to the product of the masses and inversely proportional to the square of the distance between them.

Newton's description of gravity was a difficult idea for physicists of his time to accept because it is an example of action at a distance. Earth and moon exert forces on each other although there is no physical connection between them. Modern scientists resolve this problem by referring to gravity as a **field.** Earth's presence produces a gravitational field directed toward Earth's center. The strength of the field decreases according to the inverse square law. Any particle of mass in that field experiences a force that depends on the mass of the particle and the strength of the field at the particle's location. The resulting force is directed toward the center of the field.

The field is an elegant way to describe gravity, but it does not say what gravity is. Later in this chapter, when you learn about Einstein's theory of curved space-time, you will get a better idea of what gravity really is.

Building Scientific Arguments

What do the words *universal* and *mutual* mean when you say "universal mutual gravitation"?
Newton argued that the force that makes an apple accelerate downward is the same as the force that accelerates the moon and holds it in its orbit. You can learn more by thinking about Newton's third law of motion, which says that forces always occur in pairs. If Earth attracts the moon, then the moon must attract Earth. That is, gravitation is *mutual* between any two objects.

Furthermore, if Earth's gravity attracts the apple and the moon, then it must attract the sun, and the third law says that the sun must attract Earth. But if the sun attracts Earth, then it must also attract the other planets and even distant stars, which, in turn, must attract the sun and each other. Step by step, Newton's third law of motion leads to the conclusion that gravitation must apply to all masses in the universe. That is, gravitation must be *universal*.

Aristotle explained gravity in a totally different way. Use Aristotle's argument that accounted for a falling apple: **Could that explanation account for a hammer falling on the surface of the moon?**

5-2 Orbital Motion and Tides

ORBITAL MOTION AND TIDES are two different kinds of gravitational phenomena. As you think about the orbital motion of the moon and planets, you need to think about how gravity pulls on an object. When you think about tides, you must think about how gravity pulls on different parts of an object. Analyzing these two kinds of phenomena will give you a deeper insight into how gravity works.

Orbits

To understand how an object can orbit another object, you must see orbital motion as Newton did. Objects in orbit are falling. You can explore Newton's insight by analyzing the motion of objects orbiting Earth.

Study **Orbiting Earth** on pages 94–95 and notice three important ideas:

1 An object orbiting Earth is actually falling (being accelerated) toward Earth's center. The object continuously misses Earth because of its orbital velocity.

2 Also, notice that objects orbiting each other actually revolve around their center of mass.

3 Finally, notice the difference between closed orbits and open orbits. If you want to leave Earth never to return, you must give your spaceship a high enough velocity so it will follow an open orbit.

Ace Astronomy™ Log into AceAstronomy and select this chapter to see Astronomy Exercise "Orbital Motion." Experiment with an object in orbit.

Orbital Velocity

If you were about to ride a rocket into orbit, you would have a critical question. "How fast must I go to stay in orbit?" An object's circular velocity is the lateral velocity the object must have to remain in a circular orbit. If you assume that the mass of your spaceship is small compared with the mass of the object you expect to orbit, Earth in this case, then the circular velocity is:

$$V_c = \sqrt{\frac{GM}{r}}$$

In this formula, M is the mass of the central body in kilograms, r is the radius of the orbit in meters, and G is the gravitational constant, 6.67×10^{-11} m³/s²kg. This formula is all you need to calculate how fast an object must travel to stay in a circular orbit.

For example, how fast does the moon travel in its orbit? Earth's mass is 5.98×10^{24} kg, and the radius of the moon's orbit is 3.84×10^8 m. Then the moon's velocity is:

$$V_c = \sqrt{\frac{6.67 \times 10^{-11} \times 5.98 \times 10^{24}}{3.84 \times 10^8}} = \sqrt{\frac{39.9 \times 10^{13}}{3.84 \times 10^8}}$$

$$= \sqrt{1.04 \times 10^6} = 1020 \text{ m/s} = 1.02 \text{ km/s}$$

This calculation shows that the moon travels 1.02 km along its orbit each second. That is the circular velocity at the distance of the moon.

A satellite just above Earth's atmosphere is only about 200 km above Earth's surface, or 6578 km from Earth's center, so Earth's gravity is much stronger, and the satellite must travel much faster to stay in a circular orbit. You can use the formula above to find that the circular velocity just above Earth's atmosphere is about 7790 m/s, or 7.79 km/s. This is about 17,400 miles per hour and shows why putting satellites into Earth orbit takes such large rockets. Not only must the rocket lift the satellite above Earth's atmosphere, but the rocket must tip over and accelerate the satellite to circular velocity.

Calculating Escape Velocity

If you launch a rocket upward, it will consume its fuel in a few moments and reach its maximum speed. From that point on, it will coast upward. How fast must a rocket travel to coast away from Earth and escape? Of course, no matter how far it travels, it can never escape from Earth's gravity. The effects of Earth's gravity extend to infinity. It is possible, however, for a rocket to travel so fast initially that gravity can never slow it to a stop. That means it could leave Earth.

The escape velocity is the velocity required to escape from the surface of an astronomical body. Here you are interested in escaping from Earth or a planet; later chapters will consider the escape velocity from stars, galaxies, and even a black hole.

The escape velocity, V_e, is given by a simple formula:

$$V_e = \sqrt{\frac{2GM}{r}}$$

Here G is the gravitational constant 6.67×10^{-11} m³/s²kg, M is the mass of the astronomical body in kilograms, and r is its radius in meters. (Notice that this formula is very similar to the formula for circular velocity.)

You can find the escape velocity from Earth by looking up its mass, 5.98×10^{24} kg, and its radius, 6.38×10^6 m. Then the escape velocity is:

$$V_e = \sqrt{\frac{2 \times 6.67 \times 10^{-11} \times 5.98 \times 10^{24}}{6.38 \times 10^6}} = \sqrt{\frac{7.98 \times 10^{14}}{6.38 \times 10^6}}$$

$$= \sqrt{1.25 \times 10^8} = 11,200 \text{ m/s} = 11.2 \text{ km/s}$$

This is equal to about 25,000 miles per hour.

Notice that the formula says that the escape velocity from a body depends on both its mass and radius. A massive body might have a low escape velocity if it has a very large radius. You will meet such objects in the discussion of giant stars. On the other hand, a rather low-mass body could have a very large escape velocity if it had a very small radius, a condition you will meet in the discussion of black holes.

Circular velocity and escape velocity are two aspects of Newton's laws of gravity and motion. Once Newton understood gravity and motion, he could do what Kepler had failed to do—he could explain why the planets obey Kepler's laws of planetary motion.

Ace Astronomy™ Log into AceAstronomy and select this chapter to see Astronomy Exercise "Escape Velocity." Try to escape from a planet.

Kepler's Laws Reexamined

Now that you understand Newton's laws, gravity, and orbital motion, you can understand Kepler's laws of planetary motion in a new way.

Kepler's first law says that the orbits of the planets are ellipses with the sun at one focus. Kepler wondered why the planets keep moving along these orbits, and now you know the answer. They move because there is nothing to slow them down. Newton's first law says that a body in motion stays in motion unless acted on by some force. The gravity of the sun accelerates the planets inward toward the sun and holds them in their orbits, but it doesn't pull backward on the planets, so they don't slow to a stop. With no friction, they must continue to move.

The orbits of the planets are ellipses because gravity follows the inverse square law. In one of his most famous problems, Newton proved that if a planet moves in a closed orbit under the influence of an attractive force that follows the inverse square law, then the planet must follow an elliptical path.

Kepler's second law says that a planet moves faster when it is near the sun and slower when it is farther away. Once again, Newton's discoveries explain why. Earlier you saw that a body moving on a frictionless surface will continue to move in a straight line until it is acted on by some force; that is, the object has momentum. In a similar way, an object set rotating on a frictionless surface will continue rotating until something acts to speed it up or slow it down. Such an object has **angular momentum,** a measure of the rotation of the body about some point. A planet circling the sun has a given amount of angular momentum; and, with no outside influences to alter its motion, it must conserve its angular momentum. That is, its angular momentum must remain constant.

Mathematically, a planet's angular momentum is the product of its mass, velocity, and distance from the sun. This explains why a planet must speed up as it comes closer to the sun along an elliptical orbit. Its angular momentum is conserved, so as its distance from the sun decreases, its velocity must increase. In the

Orbiting Earth

1 You can understand orbital motion by thinking of a cannonball falling around Earth in a circular path. Imagine a cannon on a high mountain aimed horizontally as shown at right. A little gunpowder gives the cannonball a low velocity, and it doesn't travel very far before falling to Earth. More gunpowder gives the cannonball a higher velocity, and it travels farther. With enough gunpowder, the cannonball travels so fast it never strikes the ground. Earth's gravity pulls it toward Earth's center, but Earth's surface curves away from it at the same rate it falls. It is in orbit. The velocity needed to stay in a circular orbit is called the **circular velocity**. Just above Earth's atmosphere, circular velocity is 7790 m/s or about 17,400 miles per hour, and the orbital period is about 90 minutes.

A satellite above Earth's atmosphere feels no friction and will fall around Earth indefinitely.

North Pole

Earth satellites eventually fall back to Earth if they orbit too low and experience friction with the upper atmosphere.

Ace⊙Astronomy™

Go to AceAstronomy and click Active Figures to see "Newton's Cannon" and fire your own version of Newton's cannon.

1a A **geosynchronous satellite** orbits eastward with the rotation of Earth and remains above a fixed spot — ideal for communications and weather satellites.

A Geosynchronous Satellite

At a distance of 42,250 km (26,260 miles) from Earth's center, a satellite orbits with a period of 24 hours.

The satellite orbits eastward, and Earth rotates eastward under the moving satellite.

The satellite remains fixed above a spot on Earth's equator.

Ace⊙Astronomy™

Go to AceAstronomy and click Active Figures to see "Geosynchronous Orbit" and place your own satellite into geosynchronous orbit.

1b According to Newton's first law of motion, the moon should follow a straight line and leave Earth forever. Because it follows a curve, Newton knew that some force must continuously accelerate it toward Earth — gravity. Each second the moon moves 1020 m (3350 ft) eastward and falls about 1.6 mm (about 1/16 inch) toward Earth. The combination of these motions produces the moon's curved orbit. The moon is falling.

Straight line motion of the moon

Motion toward Earth

Curved path of moon's orbit

Earth

1c Astronauts in orbit around Earth feel weightless, but they are not "beyond Earth's gravity," to use a term from old science fiction movies. Like the moon, the astronauts are accelerated toward Earth by Earth's gravity, but they travel fast enough along their orbits that they continually "miss the Earth." They are literally falling around Earth. Inside or outside a spacecraft, astronauts feel weightless because they and their spacecraft are falling at the same rate. Rather than saying they are weightless, you should more accurately say they are in free fall.

NASA

2 To be precise you should not say that an object orbits Earth. Rather the two objects orbit each other. Gravitation is mutual, and if Earth pulls on the moon, the moon pulls on Earth. The two bodies revolve around their common **center of mass,** the balance point of the system.

2a Two bodies of different mass balance at the center of mass, which is located closer to the more massive object. As the two objects orbit each other, they revolve around their common center of mass as shown at right. The center of mass of the Earth–moon system lies only 4708 km (2926 miles) from the center of Earth — inside the Earth. As the moon orbits the center of mass on one side, the Earth swings around the center of mass on the opposite side.

Center of mass

3 **Closed orbits** return the orbiting object to its starting point. The moon and artificial satellites orbit Earth in closed orbits. Below, the cannonball could follow an elliptical or a circular closed orbit. If the cannonball travels as fast as **escape velocity,** the velocity needed to leave a body, it will enter an open orbit. An **open orbit** does not return the cannonball to Earth. It will escape.

Ace ☯ Astronomy™ Go to AceAstronomy and click Active Figures to see "Center of Mass." Change the mass ratio to move the center of mass.

Hyberbola

Parabola

A cannonball with a velocity greater than escape velocity will follow a hyperbola and escape from Earth.

A cannonball with escape velocity will follow a parabola and escape.

North Pole

Ellipse

Circle

Ellipse

3a As described by Kepler's Second Law, an object in an elliptical orbit has its lowest velocity when it is farthest from Earth (apogee), and its highest velocity when it is closest to Earth (perigee). Perigee must be above Earth's atmosphere, or friction will rob the satellite of energy and it will eventually fall back to Earth.

Energy

One of the most fundamental ideas in science is **energy.** Physicists define energy as the ability to do work, but you might paraphrase that definition as the ability to produce a change. A moving body has energy called **kinetic energy.** A planet moving along its orbit, a cement truck rolling down the highway, and a golf ball sailing down the fairway all have the ability to produce a change. Imagine colliding with any of these objects!

Energy need not be represented by motion. Sunlight falling on a green plant, on photographic film, or on unprotected skin can produce chemical changes, and thus light is a form of energy. Batteries and gasoline are examples of chemical energy, and uranium fuel rods contain nuclear energy. A tank of hot water contains thermal energy.

Potential energy is the energy an object has because of its position in a gravitational field. A bowling ball on a shelf above your desk has potential energy. It is only potential, however, and doesn't produce any changes until the bowling ball descends onto your desk. Obviously the higher the shelf, the more potential energy the ball has.

Much of science is the study of how energy flows from one place to another place producing changes. A biologist might study the way a nerve cell transmits energy along its length to a muscle, while a geologist might study how energy flows as heat from Earth's interior and deforms Earth's surface. In such processes, you see energy being transformed from one state to another. Sunlight (energy) is absorbed by ocean plants and stored as sugars and starches (energy). When the plant dies, it and other ocean life are buried and become oil (energy), which gets pumped to the surface and burned in automobile engines to produce motion (energy).

Aristotle believed that all change originated in the motion of the starry sphere and flowed down to Earth. Modern science has found a more sophisticated description of the continual change you see around you. In a way, science is the study the way energy flows through the world and produces change. Energy is the pulse of the natural world.

Using the metric system (Appendix A), you express energy in **joules** (abbreviated **J**). One joule is about as much energy as that given up when an apple falls from a table to the floor.

Energy is the ability to cause change.

same way, the planet's velocity must decrease as its distance from the sun increases.

This conservation of angular momentum is actually a common human experience. Skaters spinning slowly can draw their arms and legs closer to their axis of rotation and, through conservation of angular momentum, spin faster (■ Figure 5-7). To slow their rotation, they again extend their arms. Similarly, divers can spin rapidly in the tuck position and slow their rotation by stretching into the extended position.

Do you like this explanation of the second law? You may not be very comfortable with it if you are not fully familiar with angular momentum. Scientists often use a well-understood idea like the conservation of angular momentum as a stepping-stone to understand a new concept, but you should hop back one stepping-stone and look at the second law again. Imagine you are in an elliptical orbit around the sun. As you round the most distant part of the ellipse, aphelion, you begin to move back closer to the sun, and the sun's gravity pulls you slightly forward in your orbit. You pick up speed as you fall closer to the sun, so, of course, you go faster as you approach the sun. As you round the closest point to the sun, perihelion, you begin to move away from the sun, and the sun's gravity pulls slightly backward on you, slowing you down as you climb away from the sun. So the second law and the conservation of angular momentum make sense when you analyze them in terms of forces and motions.

Kepler's third law is also explained by a conservation law, but in this case it is the law of conservation of energy (**Focus on Fundamentals 2**). A planet orbiting the sun has a spe-

■ Figure 5-7

Skaters demonstrate conservation of angular momentum when they spin faster by drawing their arms and legs closer to their axis of rotation.

cific amount of energy that depends only on its average distance from the sun. That energy can be divided between energy of motion and energy stored in the gravitational attraction between the planet and the sun. The energy of motion depends on how fast the planet moves, and the stored energy depends on the size of its orbit. The relation between these two kinds of energy is fixed by Newton's laws. That means there has to be a fixed relationship between the rate at which a planet moves around its orbit and the size of the orbit—between its orbital period P and the orbit's semi-major axis a. This is just Kepler's third law.

Newton's Version of Kepler's Third Law

The equation for circular velocity is actually a version of Kepler's third law, as you can prove with three lines of simple algebra. The result is one of the most useful formulas in astronomy.

The equation for circular velocity, as you have seen, is:

$$V_c = \sqrt{\frac{GM}{r}}$$

The orbital velocity of a planet is simply the circumference of its orbit divided by the orbital period:

$$V = \frac{2\pi r}{P}$$

If you substitute this for V in the equation for circular velocity and solve for P^2, you will get:

$$P^2 = \frac{4\pi^2}{GM} r^3$$

Here M is just the total mass of the system in kilograms. For a planet orbiting the sun, you can use the mass of the sun for M, because the mass of the planet is negligible compared to the mass of the sun. (In a later chapter, you will apply this formula to two stars orbiting each other, and then the mass M will be the sum of the two masses.) For a circular orbit, r equals the semi-major axis a, so this formula is a general version of Kepler's third law, $P^2 = a^3$. In Kepler's version, you used the units AU and years, but in Newton's version of the formula, you should use units of meters, seconds, and kilograms. G, of course, is the gravitational constant.

This is a powerful formula. Astronomers use it to find the masses of bodies by observing orbital motion. If, for example, you observed a moon orbiting a planet and you could measure the size of the moon's orbit, r, and the orbital period, P, you could use this formula to solve for M, the total mass of the planet plus the moon. There is no other way to find masses in astronomy, and, in later chapters, you will see this formula used to find the masses of stars, galaxies, and planets.

This discussion is a good illustration of the power of Newton's work. By carefully defining motion and gravity and by giving them mathematical expression, Newton was able to derive new truths, among them Newton's version of Kepler's third law.

His work transformed the mysterious wanderings of the planets into understandable motions that follow simple rules. In fact, his discovery of gravity explained something else that had mystified philosophers for millennia—the ebb and flow of the oceans.

Tides and Tidal Forces

Newton understood that gravity is mutual—Earth attracts the moon, and the moon attracts Earth—and that means the moon's gravity can explain the ocean tides. But Newton also realized that gravitation is universal, and that means there is much more to tides than just Earth's oceans.

Tides are caused by small differences in gravitational forces. For example, Earth's gravity attracts your body downward with a force equal to your weight. The moon is less massive than Earth and more distant, so the moon attracts your body with a force equal to roughly 0.0003 percent of your weight. You don't notice that tiny force, but Earth's oceans respond dramatically.

The side of Earth facing the moon is about 4000 miles closer to the moon than is the center of Earth. Consequently, the moon's gravity, tiny though it is at the distance of Earth, is just a bit stronger when it acts on the near side of Earth than on the center. It pulls on the oceans on the near side of Earth a bit more strongly than on Earth's center, and the oceans respond by flowing into a bulge of water on the side of Earth facing the moon. There is also a bulge on the side away from the moon, which develops because the moon pulls more strongly on Earth's center than on the far side. Thus the moon pulls Earth away from the far-side oceans, which flow into a bulge on the far side as shown at the top of ■ Figure 5-8.

You might wonder: If Earth and moon accelerate toward each other, why don't they smash together? In fact, they would smash together in just a couple weeks, except that they are moving sideways, and they keep missing. That is, they are orbiting around their common center of mass. The ocean tides are caused by the accelerations Earth and the oceans feel as they move around that center of mass.

The moon's tidal forces are not confined to water. The rocky bulk of Earth also responds to these tidal forces. Although you don't notice, Earth flexes, and the mountains and plains rise and fall by a few centimeters in response to the moon's gravitational acceleration.

You can see dramatic evidence of this effect if you watch the ocean shore for a few hours. Though Earth rotates on its axis, the tidal bulges remain fixed with respect to the moon. As the turning Earth carries you and your beach into a tidal bulge, the ocean water deepens, and the tide crawls up the sand. The tide does not so much "come in" as you are carried into the tidal bulge. Later, when Earth's rotation carries you out of the bulge, the ocean becomes shallower, and the tide falls. Because there are two bulges on opposite sides of Earth, the tides rise and fall twice a day on an ideal coast.

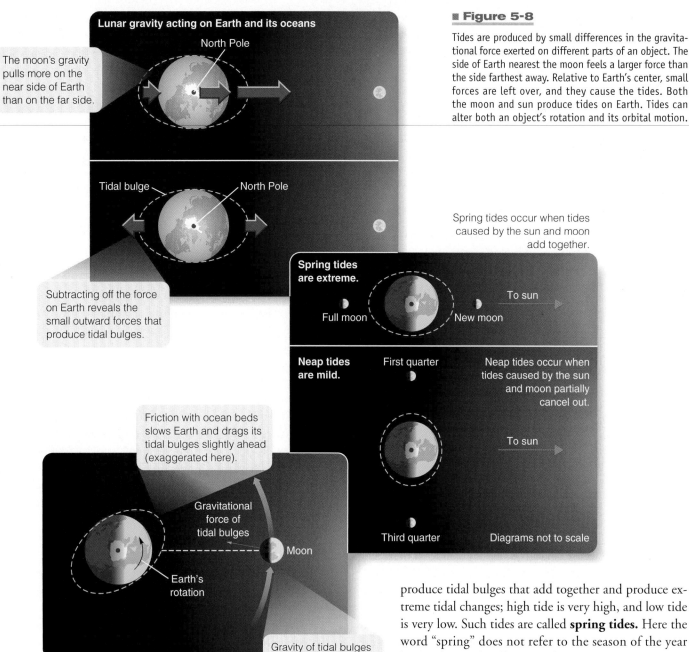

■ Figure 5-8

Tides are produced by small differences in the gravitational force exerted on different parts of an object. The side of Earth nearest the moon feels a larger force than the side farthest away. Relative to Earth's center, small forces are left over, and they cause the tides. Both the moon and sun produce tides on Earth. Tides can alter both an object's rotation and its orbital motion.

Lunar gravity acting on Earth and its oceans

North Pole

The moon's gravity pulls more on the near side of Earth than on the far side.

Tidal bulge

North Pole

Subtracting off the force on Earth reveals the small outward forces that produce tidal bulges.

Spring tides occur when tides caused by the sun and moon add together.

Spring tides are extreme.

Full moon

New moon

To sun

Neap tides are mild.

First quarter

Neap tides occur when tides caused by the sun and moon partially cancel out.

To sun

Third quarter

Diagrams not to scale

Friction with ocean beds slows Earth and drags its tidal bulges slightly ahead (exaggerated here).

Gravitational force of tidal bulges

Moon

Earth's rotation

Gravity of tidal bulges pulls the moon forward and alters its orbit.

In reality, the tidal cycle at any given location can be quite complex because of the latitude of the site, shape of the shore, winds, and so on. Tides in the Bay of Fundy (New Brunswick, Canada), for example, occur twice a day and can exceed 40 feet. In contrast, the northern coast of the Gulf of Mexico has only one tidal cycle a day of roughly 1 foot.

Gravity is universal, so the sun, too, produces tides on Earth. The sun is roughly 27 million times more massive than the moon, but it lies almost 400 times farther from Earth. Consequently, tides on Earth caused by the sun are less than half those caused by the moon. At new moon and at full moon, the moon and sun produce tidal bulges that add together and produce extreme tidal changes; high tide is very high, and low tide is very low. Such tides are called **spring tides.** Here the word "spring" does not refer to the season of the year but to the rapid welling up of water. Spring tides occur twice a month, at new and full moon. At first- and third-quarter moons, the sun and moon pull at right angles to each other, and the sun's tides cancel out some of the moon's tides. These less-extreme tides are called **neap tides,** and they do not rise very high or fall very low. The word *neap* comes from an obscure Old English word, *nep,* that seems to have meant "lacking power to advance." Spring tides and neap tides are illustrated in Figure 5-8.

Galileo tried to understand tides, but not until Newton described gravity could astronomers analyze tidal forces and recognize their surprising effects. For example, the friction of the ocean waters with the ocean beds slows Earth's rotation and makes the length of a day grow by 0.0023 seconds per century. Fossils of an-

Testing a Theory by Prediction

When you read about any science, you should notice that scientific theories face in two directions. They look back into the past and explain phenomena previously observed. For example, Newton's laws of motion and gravity explained how the planets moved. But theories also face forward in that they make predictions about what you should find as you explore further. In this way, Newton's laws allowed astronomers to calculate the orbits of comets, predict their return, and eventually understand their origin.

Scientific predictions are important in two ways. First, if a theory leads to a prediction and scientists later discover the prediction was true, the theory is confirmed, and scientists gain confidence that it is a true description of nature. But predictions are important in science in a second way. Using an existing theory to make a prediction may lead to an unexplored avenue of knowledge. For example, the first theories of genetics made predictions that confirmed the genetic theory of inheritance, but those predictions also created a new understanding of how living creatures evolve.

As you read about any scientific theory, think about both what it can explain and what it can predict.

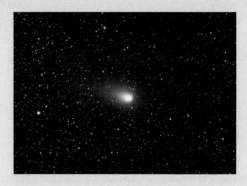

The motion of comets can be predicted using Newton's laws. (Nigel Sharp/NOAO/AURA/NSF)

cient corals confirm that only 900 million years ago Earth's day was 18 hours long. In addition, Earth's gravitation exerts tidal forces on the moon, and although there are no bodies of water on the moon, friction within the flexing rock has slowed the moon's rotation to the point that it now keeps the same face toward Earth.

Tidal forces can also affect orbital motion. Friction with the ocean beds drags the tidal bulges eastward out of a direct Earth–moon line. These tidal bulges contain a large amount of mass, and their gravitational field pulls the moon forward in its orbit, as shown at the bottom of Figure 5-8. As a result, the moon's orbit is growing larger by about 4 cm a year, an effect that astronomers can measure by bouncing laser beams off reflectors left on the lunar surface by the Apollo astronauts.

Newton's gravitation is much more than just the force that makes apples fall. In later chapters, you will see how tides can pull gas away from stars, rip galaxies apart, and melt the interiors of small moons orbiting near massive planets. Tidal forces produce some of the most surprising and dramatic processes in the universe.

Astronomy after Newton

Newton published his work in July 1687 in a book called *Philosophiae Naturalis Principia Mathematica* (*Mathematical Principles of Natural Philosophy*), now known simply as *Principia* (■ Figure 5-9). It is one of the most important books ever written. *Principia* (pronounced *Prin KIP ee uh*) changed astronomy, changed science, and changed the way people think about nature.

Principia changed astronomy and ushered in a new age. No longer did astronomers appeal to the whim of the gods to explain things in the heavens. No longer did they speculate on why the planets wander across the sky. *Principia* says that the motions of the heavenly bodies are governed by simple, universal rules that describe the motions of everything from planets to falling apples. Suddenly the universe was understandable in simple terms.

Newton's laws of motion and gravity made it possible for astronomers to calculate the orbits of planets and moons. Not only could they explain how the heavenly bodies move, they could predict future motions (**Window on Science 5-2**). This subject, known as gravitational astronomy, dominated

■ Figure 5-9

Newton, working from the discoveries of Galileo and Kepler, derived three laws of motion and the principle of mutual gravitation. He and some of his discoveries are honored on this English pound note. Notice the diagram of orbital motion in the background and the open copy of *Principia* in Newton's hands.

astronomy for almost 200 years and is still important. It included the calculation of the orbits of comets and asteroids and the theoretical prediction of the existence of two planets, Neptune and Pluto.

Principia also changed science in general. The works of Copernicus and Kepler had been mathematical, but no book before had so clearly demonstrated the power of mathematics as a language of precision. Newton's arguments in *Principia* were so powerful an illustration of the quantitative study of nature that scientists around the world adopted mathematics as their most powerful tool.

Also, *Principia* changed the way people thought about nature. Newton showed that the rules that govern the universe are simple. Particles move according to three rules of motion and attract each other with a force called gravity. These motions are predictable, and that makes the universe a vast machine based on a few simple rules. It is complex only in that it contains a vast number of particles. In Newton's view, if he knew the location and motion of every particle in the universe, he could, in principle, derive the past and future of the universe in every detail. This mechanical determinism has been undermined by modern quantum mechanics, but it dominated science for more than two centuries during which scientists thought of nature as a beautiful clockwork that would be perfectly predictable if they knew how all the gears meshed.

Most of all, Newton's work broke the last bonds between science and formal philosophy. Newton did not speculate on the good or evil of gravity. He did not debate its meaning. Not more than a hundred years before, scientists would have argued over the "reality" of gravity. Newton didn't care for these debates. He wrote, "It is enough that gravity exists and suffices to explain the phenomena of the heavens."

Building Scientific Arguments

How do Newton's laws of motion explain the orbital motion of the moon?

Natural laws are powerful because they explain so much. The key is often to build an argument step by step. If Earth and the moon did not attract each other, the moon would move in a straight line in accord with Newton's first law of motion and vanish into space in a few months. Instead, gravity pulls the moon toward Earth's center, and the moon accelerates toward Earth. This acceleration is just enough to pull the moon away from its straight-line motion and cause it to follow a curve around Earth. In fact, it is correct to say that the moon is falling, but because of its lateral motion it continuously misses Earth.

Every orbiting object is falling toward the center of its orbit but is moving laterally fast enough to compensate for the inward motion, and it follows a curved orbit. That is an elegant argument, but it raises a question: **How can astronauts float inside spacecraft in a "weightless" state? Why might "free fall" be a more accurate term?**

■　　　■　　　■

Connections: Newton's laws of motion and gravitation are critical in astronomy not only because they describe orbital motion and tides but also because they describe nearly all interactions between astronomical bodies. Newton made other discoveries, and you will meet those in later chapters. Now you are ready to jump forward a bit more than two centuries to see how Einstein described gravity in a new and powerful way.

5-3 Einstein and Relativity

IN THE EARLY YEARS of the last century, Albert Einstein (1879–1955) (■ Figure 5-10) began thinking about how motion and gravity interact. He soon gained international fame by showing that Newton's laws of motion and gravity were only partially correct. The revised theory became known as the theory of relativity. As you will see, there are really two theories of relativity.

Special Relativity

Einstein began by thinking about how moving observers see events around them. His analysis led him to the first postulate of relativity, also known as the principle of relativity:

> **First postulate** (the principle of relativity): Observers can never detect their *uniform* motion except relative to other objects.

You may have experienced the first postulate while sitting on a train in a station. You suddenly notice that the train on the next track has begun to creep out of the station. However, after several moments you realize that it is your own train that is moving and that the other train is still motionless on its track.

■ Figure 5-10

Einstein has become a symbol of the brilliant scientist, but his fame began when he was a young man and thought deeply about the nature of motion. That led him to revolutionary insights into the meaning of space and time and a new understanding of gravity.

Consider another example. Suppose you are floating in a spaceship in interstellar space and another spaceship comes coasting by (■ Figure 5-11a). You might conclude that it is moving and you are not, but someone in the other ship might be equally sure that you are moving and it is not. Of course, you could just look out a window and compare the motion of your spaceship with a nearby star, but that just expands the problem. Which is moving, your spaceship or the star? The principle of relativity says that there is no experiment you can perform to decide which ship is moving and which is not. This means that there is no such thing as absolute rest—all motion is relative.

Because neither you nor the people in the other spaceship could perform any experiment to detect your absolute motion through space, the laws of physics must have the same form in both spaceships. Otherwise, experiments would give different results in the two ships, and you could decide who was moving. So, a more general way of stating the first postulate refers to these laws of physics:

> **First postulate** (alternate version): The laws of physics are the same for all observers, no matter what their motion, so long as they are not *accelerated*.

The words *uniform* and *accelerated* are important. If either spaceship were to fire its rockets, then its velocity would change. The crew of that ship would know it because they would feel the acceleration pressing them into their couches. Accelerated motion, therefore, is different—the pilots of the spaceships can always tell which ship is accelerating and which is not. The postulates of relativity discussed here apply only to observers in *uniform* motion, which means *unaccelerated* motion. That is why the theory is called **special relativity.**

The first postulate led to the conclusion that the speed of light must be constant for all observers. No matter how you are moving, your

measurement of the speed of light has to give the same result (Figure 5-11b). This became the second postulate of special relativity:

> **Second postulate:** The velocity of light is constant and will be the same for all observers independent of their motion relative to the light source.

■ **Figure 5-11**

(a) The principle of relativity says that observers can never detect their uniform motion, except relative to other objects. Thus, neither of these travelers can decide who is moving and who is not. (b) If the velocity of light depended on the motion of the observer through space, then these travelers could perform measurements inside their spaceships to discover who was moving. If the principle of relativity is correct, then the velocity of light must be a constant for all observers.

Remember, this is required by the first postulate; if the velocity of light were not constant, then the pilots of the spaceships could decide who was moving.

Once Einstein had accepted the basic postulates of relativity, he was led to some startling discoveries. Newton's laws of motion and gravity worked well as long as distances were small and velocities were low. But when you begin to think of very large distances or very high velocities, Newton's laws are no longer adequate to describe what happens. Instead, you must use relativistic physics. For example, special relativity shows that the observed mass of a moving particle depends on its velocity. The higher the velocity, the greater the mass of the particle. This is not significant at low velocities, but it becomes very important as the velocity approaches the velocity of light. Such increases in mass are observed whenever physicists accelerate atomic particles to high velocities (■ Figure 5-12).

This discovery led to yet another insight. The relativistic equations that describe the energy of a moving particle predict that the energy of a motionless particle is not zero. Rather, its energy at rest is $m_0 c^2$. This is of course the famous equation:

$$E = m_0 c^2$$

The c is the speed of light, and the m_0 is the mass of the particle when it is at rest. This simple formula suggests that mass and energy are related, and you will see in later chapters how nature can convert one into the other inside stars.

For example, suppose that you convert 1 kg of matter into energy. You must express the velocity of light as 3×10^8 m/s, and your result is 9×10^{16} joules (J) (approximately equal to a 20-megaton nuclear bomb). (Recall that a joule is a unit of energy roughly equivalent to the energy given up when an apple falls from a table to the floor.) This simple calculation shows that the energy equivalent of even a small mass is very large.

■ Figure 5-12

The observed mass of moving electrons depends on their velocity. As the ratio of their velocity to the velocity of light, v/c, gets larger, the mass of the electrons in terms of their mass at rest, m/m_0, increases. Such relativistic effects are quite evident in particle accelerators, which accelerate atomic particles to very high velocities.

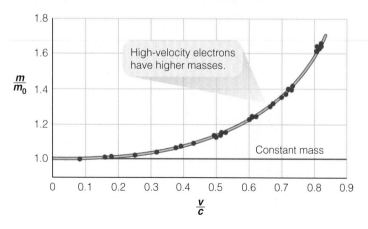

Other relativistic effects include the slowing of moving clocks and the shrinkage of lengths measured in the direction of motion. A detailed discussion of the major consequences of the special theory of relativity is beyond the scope of this book. Instead, let's consider Einstein's second advance, the general theory.

The General Theory of Relativity

In 1916, Einstein published a more general version of the theory of relativity that dealt with accelerated as well as uniform motion. This **general theory of relativity** contained a new description of gravity.

Einstein began by thinking about observers in accelerated motion. Imagine an observer sitting in a windowless spaceship. Such an observer cannot distinguish between the force of gravity and the inertial forces produced by the acceleration of the spaceship (■ Figure 5-13). This led Einstein to conclude that gravity and acceleration are related, a conclusion now known as the equivalence principle:

> **Equivalence principle:** Observers cannot distinguish locally between inertial forces due to acceleration and uniform gravitational forces due to the presence of a massive body.

This should not surprise you. Earlier in this chapter you read that Newton concluded that the mass that resists acceleration is the same as the mass that exerts gravitational forces. He even performed an elegant experiment with pendulums to test the equivalence of the mass related to motion and the mass related to gravity.

The importance of the general theory of relativity lies in its description of gravity. Einstein concluded that gravity, inertia, and acceleration are all associated with the way space is related to time. This relation is often referred to as curvature, and a one-line description of general relativity explains a gravitational field as a curved region of space-time:

> **Gravity according to general relativity:** Mass tells space-time how to curve, and the curvature of space-time (gravity) tells mass how to accelerate.

So you feel gravity because Earth's mass causes a curvature of space-time. The mass of your body responds to that curvature by accelerating toward Earth's center. According to general relativity, all masses cause curvature, and the larger the mass, the more severe the curvature. That's gravity.

Confirmation of the Curvature of Space-Time

Einstein's general theory of relativity has been confirmed by a number of experiments, but two are worth mentioning here because they were among the first tests of the theory. One involves Mercury's orbit, and the other involves eclipses of the sun.

Johannes Kepler understood that the orbit of Mercury is elliptical, but only since 1859 have astronomers known that the

I feel gravity. I must be on the surface of a planet.

I feel gravity. I must be on the surface of a planet.

■ **Figure 5-13**

(a) An observer in a closed spaceship on the surface of a planet feels gravity. (b) In space, with the rockets smoothly firing and accelerating the spaceship, the observer feels inertial forces that are equivalent to gravitational forces.

long axis of the orbit sweeps around the sun in a motion called precession (■ Figure 5-14). The total observed precession is 5600.73 seconds of arc per century (as seen from Earth), which equals about 1.5° per century. This precession is produced by the gravitation of Venus, Earth, and the other planets. However, when astronomers used Newton's description of gravity, they calculated that the precession should amount to only 5557.62 seconds of arc per century. So Mercury's orbit is advancing 43.11 seconds of arc per century faster than Newton's law predicts.

This is a tiny effect. Each time Mercury returns to perihelion, its closest point to the sun, it is about 29 km (18 miles) past the position predicted by Newton's laws. This is such a small distance compared with the planet's diameter of 4850 km that it could never have been detected had it not been cumulative. Each orbit, Mercury gains 29 km, and in a century it gains over 12,000 km— more than twice its own diameter. This tiny effect, called the advance of perihelion of Mercury's orbit, accumulated into a serious discrepancy in the Newtonian description of the universe.

The advance of perihelion of Mercury's orbit was one of the first problems to which Einstein applied the principles of general relativity. First he calculated how much the sun's mass curves space-time in the region of Mercury's orbit, and then he calculated how Mercury moves through the space-time. The theory predicted that the curved space-time should cause Mercury's orbit to advance by 43.03 seconds of arc per century, well within the observational accuracy of the excess (Figure 5-14b).

Einstein was elated with this result, and he would be even happier with modern studies that have shown that Mercury, Venus, Earth, and even Icarus, an asteroid that comes close to the sun, have orbits observed to be slipping forward due to the curvature of space-time near the sun (■ Table 5-2). This same effect has been detected in pairs of stars that orbit each other.

A second test of the curvature of space-time was directly related to the motion of light through the curved space-time near the sun. The equations of general relativity predicted that light would be deflected by curved space-time, just as a rolling golf

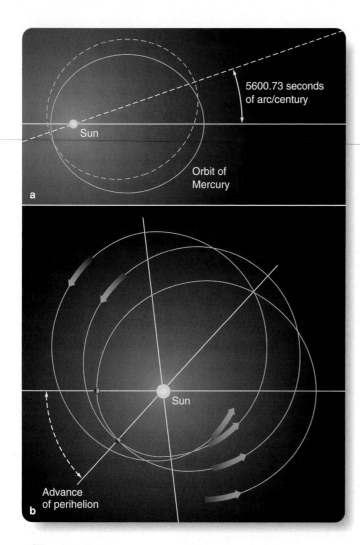

Figure 5-14

(a) Mercury's orbit precesses 5600.73 seconds of arc per century—43.11 seconds of arc per century faster than predicted by Newton's laws. (b) Even when you ignore the influences of the other planets, Mercury's orbit is not a perfect ellipse. Curved space-time near the sun distorts the orbit from an ellipse into a rosette. The advance of Mercury's perihelion is exaggerated about a million times in this figure.

ball is deflected by undulations in a putting green. Einstein predicted that starlight grazing the sun's surface would be deflected by 1.75 seconds of arc (■ Figure 5-15). Starlight passing near the sun is normally lost in the sun's glare, but during a total solar eclipse stars beyond the sun could be seen. As soon as Einstein published his theory, astronomers rushed to observe such stars and test the curvature of space-time.

The first solar eclipse following Einstein's announcement in 1916 was June 8, 1918. It was cloudy at some observing sites, and results from other sites were inconclusive. The next occurred on May 29, 1919, only months after the end of World War I, and was visible from Africa and South America. British teams went to both Brazil and Príncipe, an island off the coast of Africa. Months before the eclipse, they photographed that part of the sky where the sun would be located during the eclipse and mea-

■ Table 5-2 | Precession in Excess of Newtonian Physics

PLANET	Observed Excess Precession (Sec of arc per century)	Relativistic Prediction (Sec of arc per century)
Mercury	43.11 ± 0.45	43.03
Venus	8.4 ± 0.48	8.6
Earth	5.0 ± 1.2	3.8
Icarus	9.8 ± 0.8	10.3

sured the positions of the stars on the photographic plates. Then, during the eclipse, they photographed the same star field with the eclipsed sun located in the middle. After measuring the plates, they found slight changes in the positions of the stars. During the eclipse, the positions of the stars on the plates were shifted outward, away from the sun (■ Figure 5-16). If a star had been located at the edge of the solar disk, it would have been shifted outward by about 1.8 seconds of arc. This represents good agreement with the theory's prediction.

This test has been repeated at many total solar eclipses since 1919, with similar results. The most accurate results were obtained in 1973 when a Texas–Princeton team measured a deflection of 1.66 ± 0.18 seconds of arc—good agreement with Einstein's theory.

The general theory of relativity is critically important in modern astronomy. You will meet it again in the discussion of black holes, distant galaxies, and the big bang universe. The theory revolutionized modern physics by providing a theory of gravity based on the geometry of curved space-time. Thus,

■ Figure 5-15

Like a depression in a putting green, the curved space-time near the sun deflects light from distant stars and makes them appear to lie slightly farther from the sun than their true positions.

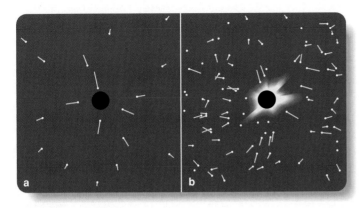

Figure 5-16

(a) Schematic drawing of the deflection of starlight by the sun's gravity. Dots show the true positions of the stars as photographed months before the eclipse. Lines point toward the positions of the stars during the eclipse. (b) Actual data from the eclipse of 1922. Random uncertainties of observation cause some scatter in the data, but in general the stars appear to move away from the sun by 1.77 seconds of arc at the edge of the sun's disk. The deflection of stars is magnified by a factor of 2300 in both (a) and (b).

Galileo's inertia and Newton's mutual gravitation are shown to be not just descriptive rules but fundamental properties of space and time.

Building Scientific Arguments

What does the equivalence principle tell you?
The equivalence principle says that there is no observation you can make inside a closed spaceship to distinguish between uniform acceleration and gravitation. Of course, you could open a window and look outside, but then you would no longer be in a closed spaceship. As long as you make no outside observations, you can't tell whether your spaceship is firing its rockets and accelerating through space or resting on the surface of a planet where gravity gives you weight.

Einstein took the equivalence principle to mean that gravity and acceleration through space-time are somehow related. The general theory of relativity gives that relationship mathematical form and shows that gravity is really a distortion in space-time that physicists refer to as curvature. Consequently, you can say "mass tells space-time how to curve, and space-time tells mass how to move." The equivalence principle led Einstein to an explanation for gravity.

Einstein began his work by thinking carefully about common things such as what you feel when you are moving uniformly or accelerating. This led him to deep insights now called postulates. Special relativity sprang from two postulates. **Why does the second postulate have to be true if the first postulate is true?**

■   ■   ■

Connections: Your study of the origin of astronomy began with the builders of Stonehenge and reaches the modern day with Einstein's general theory of relativity. Now that you have seen where astronomy came from, you are ready to see how it helps you understand your place the universe. Your first question could be "How do astronomers get information?" The answer involves the astronomer's most basic tool, the telescope, and that is the subject of the next chapter.

Study and Review Tools

Summary

5-1 | Galileo and Newton

What happens when an object falls?
- Galileo found that a falling object is accelerated; that is, it falls faster and faster with each passing second. The rate at which it accelerates, termed the acceleration of gravity, is 9.8 m/s² (32 ft/s²) at Earth's surface and does not depend on the weight of the object.
- According to tradition, Galileo demonstrated this by dropping balls of iron and wood from the Leaning Tower of Pisa to show that they would fall together.
- Galileo stated the law of inertia. In the absence of friction, a moving body on a horizontal plane will continue moving forever.

How did Newton discover gravity?
- Newton adopted Galileo's law of inertia as the first of three laws of motion. The second law says that a change in motion, an acceleration, must be caused by a force, and the third law says that forces occur in pairs acting in opposite directions.
- Newton realized that the curved path of the moon meant that it was being accelerated away from a straight-line path, and that required the presence of a force—gravity.
- From his mathematical analysis, Newton was able to show that the force of gravity between two masses is proportional to the product of their masses and inversely proportional to the square of the distance between them.

5-2 | Orbital Motion and Tides

How does gravity explain orbital motion?
- An object in space near Earth would move along a straight line and quickly leave Earth were it not for Earth's gravity accelerating the object toward Earth's center and forcing it to follow a curved path, an orbit. Objects in orbit around Earth are falling (being accelerated) toward Earth's center.

- If there is no friction, the object will fall around its orbit forever.

- Newton's laws explain Kepler's three laws of planetary motion. The planets follow elliptical orbits because gravity follows the inverse square law. The planets move faster when closer to the sun and slower when farther away because they conserve angular momentum. A planet's orbital period squared is proportional to its orbital radius cubed because the moving planet conserves energy.

- An object in a closed orbit follows an elliptical path. A circle is just a special ellipse of zero eccentricity.

- If a body's velocity equals or exceeds the escape velocity, V_e, it will follow a parabola or hyperbola. These orbits are termed "open" because the object never returns to its starting place.

How does gravity explain the tides?

- Tides are caused by differences in the force of gravity acting on different parts of a body.

- Tides on Earth occur because the moon's gravity pulls more strongly on the near side of Earth than on the center of Earth. A tidal bulge occurs on the far side because the moon's gravity is slightly weaker there than at the center of Earth.

- Tides produced by the moon combine with tides produced by the sun to cause extreme tides (called spring tides) at new and full moons. The moon and sun work against each other to produce less-extreme tides (neap tides) at quarter moons.

- Friction from tides can slow the rotation of a rotating world, and the gravitational pull of tidal bulges can make orbits change slowly.

5-3 | Einstein and Relativity

How did Einstein refine the understanding of motion and gravity?

- Einstein published two theories that extended Newton's laws of motion and gravity, the special theory of relativity and the general theory of relativity.

- The special theory of relativity says that uniform (unaccelerated) motion is relative. Observers cannot detect their uniform motion through space except relative to outside objects. This is known as the first postulate.

- This leads to the second postulate: the speed of light is a constant for all observers.

- A consequence of special relativity is that mass and energy are related.

- The general theory of relativity says that a gravitational field is a curvature of space-time caused by the presence of a mass. For example, Earth's mass curves space-time, and the mass of your body responds to that curvature by accelerating toward Earth's center.

- The curvature of space-time was confirmed by the slow advance in perihelion of the orbit of Mercury and by the deflection of starlight observed during a 1919 total solar eclipse.

New Terms

natural motion (p. 87)	angular momentum (p. 93)
violent motion (p. 87)	circular velocity (p. 94)
acceleration of gravity (p. 87)	geosynchronous satellite (p. 94)
momentum (p. 89)	center of mass (p. 95)
acceleration (p. 89)	closed orbit (p. 95)
velocity (p. 89)	escape velocity (p. 95)
mass (p. 90)	open orbit (p. 95)
inverse square law (p. 90)	energy (p. 96)
field (p. 92)	kinetic energy (p. 96)

potential energy (p. 96)	neap tide (p. 98)
joule (J) (p. 96)	special relativity (p. 101)
spring tide (p. 98)	general theory of relativity (p. 102)

Review Questions

Ace⊙Astronomy™ Assess your understanding of this chapter's topics with additional quizzing and animations at **http://ace .brookscole.com/sf9**

1. Why wouldn't Aristotle's explanation of gravity work if Earth was not the center of the universe?

2. According to the principles of Aristotle, what part of the motion of a baseball pitched across the home plate is natural motion? What part is violent motion?

3. If you drop a feather and a steel hammer at the same moment, they should hit the ground at the same instant. Why doesn't this work on Earth, and why does it work on the moon?

4. What is the difference between mass and weight? between speed and velocity?

5. Why did Newton conclude that some force had to pull the moon toward Earth?

6. Why did Newton conclude that gravity has to be mutual and universal?

7. How does the concept of a field explain action at a distance? Name another kind of field also associated with action at a distance.

8. Why can't a spacecraft go "beyond Earth's gravity"?

9. What is the center of mass of the Earth–moon system? Where is it?

10. How do planets orbiting the sun and skaters doing a spin conserve angular momentum?

11. Why is the period of an open orbit undefined?

12. How does the first postulate of special relativity imply the second?

13. When you ride a fast elevator upward, you feel slightly heavier as the trip begins and slightly lighter as the trip ends. How is this phenomenon related to the equivalence principle?

14. From your knowledge of general relativity, would you expect radio waves from distant galaxies to be deflected as they pass near the sun? Why or why not?

15. Why can the object shown at the right be bolted in place and used 24 hours a day without adjustment?

16. Why is it a little bit misleading to say that this astronaut is weightless?

Larry Mulvehill/The Image Works

(NASA/JSC)

Discussion Questions

1. How did Galileo idealize his inclines to conclude that an object in motion stays in motion until it is acted on by some force?

2. Give an example from everyday life to illustrate each of Newton's laws.

3. People who lived before Newton may not have believed in cause and effect as strongly as you do. How do you suppose they saw their daily lives?

Problems

1. Compared with the strength of Earth's gravity at its surface, how much weaker is gravity at a distance of 10 Earth radii from Earth's center? at 20 Earth radii?

2. Compare the force of lunar gravity on the surface of the moon with the force of Earth's gravity at Earth's surface.

3. If a small lead ball falls from a high tower on Earth, what will be its velocity after 2 seconds? after 4 seconds?

4. What is the circular velocity of an Earth satellite 1000 km above Earth's surface? (*Hint:* Earth's radius is 6380 km.)

5. What is the circular velocity of an Earth satellite 36,000 km above Earth's surface? What is its orbital period? (*Hint:* Earth's radius is 6380 km.)

6. What is the orbital period of an imaginary satellite orbiting just above Earth's surface? Ignore friction with the atmosphere.

7. Repeat the previous problem for Mercury, Venus, the moon, and Mars.

8. Describe the orbit followed by the slowest cannonball on page 94 on the assumption that the cannonball could pass freely through Earth. (Newton got this problem wrong the first time he tried to solve it.)

9. If you visited an asteroid 30 km in radius with a mass of 4×10^{17} kg, what would be the circular velocity at its surface? A major league fastball travels about 90 mph. Could a good pitcher throw a baseball into orbit around the asteroid?

10. What is the orbital period of a satellite orbiting just above the surface of the asteroid in Problem 9?

11. What would be the escape velocity at the surface of the asteroid in Problem 9? Could a major league pitcher throw a baseball off of the asteroid?

Media Cluster

Ace Astronomy™ To access the resources in the Media Cluster, log into AceAstronomy at **http://ace.brookscole.com/sf9** and select Chapter 5.

Geosynchronous Orbit
Can you put your satellite into geosynchronous orbit so that it can communicate with a ground antenna 24 hours a day? Can you move it from one antenna to the next?

Center of Mass
Adjust the masses of the stars in a binary star system and see how the orbits change. Do the sizes of the orbits correctly reflect the ratio of the masses?

Falling Bodies
According to tradition, Galileo demonstrated that the acceleration of a falling body is independent of its weight by dropping various objects from the Leaning Tower of Pisa. You can repeat his experiment with this interactive exercise.

Escape Velocity
Escape velocity is the initial velocity an object needs to escape from the surface of a celestial body. See if you can determine the correct escape velocity for the rocket in this animation.

Orbital Motion
Explore how the orbit of a hypothetical planet is affected by changing the mass of the sun, distance to the sun, and eccentricity of orbit in this animation.

Critical Inquiries for the Web

1. Einstein's general theory of relativity predicts the curvature of space-time, but here on Earth you have little opportunity to observe such effects. Find an astronomical situation in which space-time curvature is evident from observations and describe the effect of the curvature on what astronomers see when they view these objects.

2. Communications satellites are obvious uses of the geosynchronous orbit, but can you think of other uses for such orbits? Find an Internet site

that uses or displays information gleaned from geosynchronous orbit that provides a useful service.

  Go to the Brooks/Cole Astronomy Resource Center (**http://astronomy.brookscole.com**) for critical thinking exercises, articles, and additional readings from InfoTrac College Edition, Brooks/Cole's online student library.

6 | Light and Telescopes

The strongest thing that's

given us to see with's

A telescope. Someone in

every town

Seems to me owes it to the

town to keep one.

ROBERT FROST, "THE STAR-SPLITTER,"
*NEW HAMPSHIRE: A POEM WITH NOTES
AND GRACE NOTES* (NEW YORK: HENRY
HOLT AND CO., 1923), pp. 27–30.

D O YOU ENJOY BEING outdoors on a clear, starry night? Astronomers appreciate the beauty of the night like everyone else, but they have an additional insight. They know that starlight is going to waste. Every night, light from the stars falls on trees, oceans, roofs, and parking lots, and it is all wasted. To an astronomer, nothing is so precious as starlight. It is the only link to the sky, and the astronomer's quest is to gather as much starlight as possible and extract from it the secrets of the stars. ▌ The telescope is the emblematic tool of the astronomer because its purpose is to gather and concentrate light for analysis. Nearly all of the interesting objects in the sky are faint sources of light, so modern astronomers are driven to build the largest possible telescopes to gather the maximum amount of light (▪ Figure 6-1).

▌ Continued on page 110 ▌

Guidepost

Looking Back

In the early chapters of this book, you looked at the sky the way ancient astronomers did, with the unaided eye. In the last chapter, you got a glimpse through Galileo's telescope, and it revealed astonishing things about Jupiter, the moon, and Venus. Now it is time to examine the tools of the modern astronomer.

This Chapter

You should begin by noting that these tools gather and focus light and its related forms of radiation. For that reason, the first of the essential questions is about mysterious light:

What is light?

How do telescopes work, and how are they limited?

What kind of instruments do astronomers use to record and analyze light?

Why do astronomers use radio telescopes?

Why must some telescopes go into space?

As you answer these questions, you will discover how important it is to understand light in all its forms.

Looking Ahead

Astronomy is almost entirely an observational science. Astronomers cannot visit distant galaxies and far-off worlds, so they must observe using astronomical telescopes. Twenty chapters remain, and every one will discuss information gathered by telescopes.

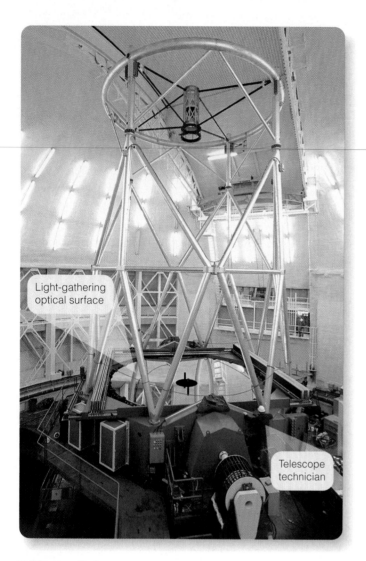

Light-gathering optical surface

Telescope technician

■ **Figure 6-1**

Astronomical telescopes are often very large to gather large amounts of starlight. The Northern Gemini Telescope stands over 19 m (60 ft) high when pointed straight up. Its main mirror is 8.1 m (26.5 ft) in diameter—larger than some classrooms. The dome of this telescope is shown at the left of the photo on the opening page of this chapter. (NOAO/AURA/NSF)

If you wish to gather visible light, a normal telescope will do; but, as you will soon see, visible light is only one kind of radiation from the stars. Astronomers can extract information from other forms of radiation by using specialized telescopes. Radio telescopes, for example, provide an entirely different view of the sky. Some of these specialized telescopes can be used from Earth's surface, but some must go into orbit above Earth's atmosphere. Telescopes that observe X rays, for instance, must be placed in space.

As you study the sophisticated telescopes and instruments that modern astronomers use, keep in mind Robert Frost's suggestion: In every town, someone should keep a telescope. Astronomy is more than technology and scientific analysis. It tells us what we are, and every town should have a telescope to keep us looking upward.

JUST AS A BOOK on bread baking might begin with a discussion of flour, this chapter on telescopes begins with a discussion of light—not just visible light, but the entire range of radiation from the sky.

Light as a Wave and a Particle

If you have admired the colors in a soap bubble, you have seen light behave as a wave. But when light enters the light meter on your camera, it behaves as a particle. How it behaves depends on how you observe it—it is both wave and particle.

You experience waves whenever you hear sound. Sound waves are a mechanical disturbance that travels through the air from source to ear. Sound requires a medium; so, on the moon, where there is no air, there can be no sound. In contrast, light is made up of electric and magnetic fields that can travel through empty space. Unlike sound, light does not require a medium, and so it can travel through a perfect vacuum. There is no sound on the moon, but there is plenty of sunlight.

Light is merely one form of radiation, called **electromagnetic radiation** because it is associated with changing electric and magnetic fields that travel through space and transfer energy from one place to another. When light enters your eye, the fluctuating electric and magnetic fields carry energy that stimulates nerve endings, and you see light.

The oscillating electric and magnetic fields that constitute electromagnetic radiation move through space at about 300,000 km/s (186,000 mi/s). This speed is commonly referred to as the speed of light, c, but it is in fact the speed of all such radiation in a vacuum.

It may seem odd to use the word *radiation* when you speak of light. The word can be used to refer to high-energy particles emitted from radioactive atoms, and you have learned to be a little bit concerned when you see the word *radiation*. But it really refers to anything that spreads outward from a source. Light radiates from a source, so you can correctly refer to light as a form of radiation.

Electromagnetic radiation is a wave phenomenon; that is, it is associated with a periodically repeating disturbance, or wave. You are familiar with waves in water. If you disturb a quiet pool of water, waves spread across the surface. Imagine that you use a meter stick to measure the distance between the successive peaks of a wave. This distance is the **wavelength,** usually represented by the Greek letter lambda (λ). If you were measuring ripples in a pond, you might find that the wavelength is a few centimeters, whereas the wavelength of ocean waves might be a hundred meters or more. There is no restriction on the wavelength of electromagnetic radiation. Wavelengths can range from smaller than the diameter of an atom to larger than that of Earth.

Because all electromagnetic radiation travels at the speed of light, wavelength is related to **frequency,** the number of cycles that pass in one second. Short-wavelength radiation has a high frequency; long-wavelength radiation has a low frequency. To understand this, imagine watching an electromagnetic wave race past while you count its peaks (■ Figure 6-2). If the wavelength is short, you will count many peaks in one second; if the wavelength is long, you will count few peaks per second. The dials on radios are marked in frequency, but they could just as easily be marked in wavelength. The relation between wavelength and frequency is a simple one:

$$\lambda = \frac{c}{f}$$

That is, the wavelength equals the speed of light *c* divided by the frequency *f*. Notice that the larger (higher) the frequency, the smaller (shorter) the wavelength. In most cases, astronomers use wavelength rather than frequency.

Radio waves can have wavelengths from a few millimeters for microwaves to kilometers. In contrast, the wavelength of light is so short that you will need more convenient units. This book uses **nanometers (nm)** because this unit is consistent with the International System of units. One nanometer is 10^{-9} meter, and visible light has wavelengths that range from about 400 nm to about 700 nm. Another unit that astronomers commonly use, and a unit that you will see in many references on astronomy, is the **Angstrom (Å).** One Angstrom is 10^{-10} meter, and visible light has wavelengths between 4000 Å and 7000 Å.

You may find radio astronomers describing wavelengths in centimeters or millimeters, and infrared astronomers often refer to wavelengths in micrometers (or microns). One micrometer (μm) is 10^{-6} meter. Whatever unit is used to describe the wavelength, you must keep in mind that all electromagnetic radiation is the same phenomenon.

■ Figure 6-2

All electromagnetic waves travel at the speed of light. The wavelength is the distance between successive peaks. The frequency of the wave is the number of peaks that pass you in one second.

What exactly is electromagnetic radiation? Is it a particle or a wave? Throughout his life, Newton believed that light was made up of particles, but modern physicists now recognize that light can behave as both particle and wave. The modern model of light is more complete than Newton's, and it refers to "a particle of light" as a **photon.** You can recognize its dual nature by thinking of it as a bundle of waves.

The amount of energy a photon carries depends on its wavelength. The shorter the wavelength, the more energy the photon carries; the longer the wavelength, the less energy it contains. This is easy to remember because short wavelengths have high frequencies, and you would naturally expect rapid fluctuations to be more energetic. A simple formula expresses the relationship between energy and wavelength:

$$E = \frac{hc}{\lambda}$$

Here *h* is Planck's constant (6.6262×10^{-34} joule s), *c* is the speed of light (3×10^{8} m/s), and λ is the wavelength in meters. A photon of visible light carries a very small amount of energy, but a photon with a very short wavelength can carry much more.

The Electromagnetic Spectrum

A spectrum is an array of electromagnetic radiation in order of wavelength. You are most familiar with the spectrum of visible light, which you see in rainbows. The colors of the spectrum differ in wavelength, with red having the longest wavelength and violet the shortest. The visible spectrum is shown at the top of ■ Figure 6-3.

The average wavelength of visible light is about 0.00005 cm. You could put 50 light waves end to end across the thickness of a sheet of household plastic wrap. Measured in nanometers, the wavelength of visible light ranges from about 400 to 700 nm. Just as you sense the wavelength of sound as pitch, you sense the wavelength of light as color. Light near the short-wavelength end of the visible spectrum (400 nm) looks violet to your eyes, and light near the long-wavelength end (700 nm) looks red.

Figure 6-3 shows how the visible spectrum makes up only a small part of the entire electromagnetic spectrum. Beyond the red end of the visible spectrum lies **infrared radiation,** where wavelengths range from 700 nm to about 0.1 cm. Your eyes are not sensitive to this radiation, but your skin senses it as heat. A "heat lamp" is just a bulb that gives off principally infrared radiation.

Beyond the infrared part of the electromagnetic spectrum lie radio waves. Microwaves have wavelengths of a millimeter to a few centimeters and are used for radar and long-distance telephone communication. Longer wavelengths are used for UHF and VHF television transmissions. FM, military, governmental, and ham radio signals have wavelengths up to a few meters, and AM radio waves can have wavelengths of kilometers.

The distinction between the wavelength ranges is not sharp. Long-wavelength infrared radiation and the shortest microwave

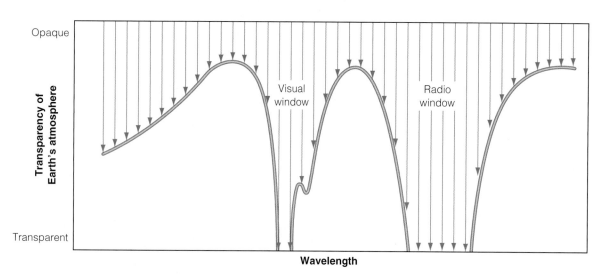

▪ Figure 6-3

The spectrum of visible light, extending from red to violet, is only part of the electromagnetic spectrum. Most radiation is absorbed in Earth's atmosphere, and only radiation in the visual window and the radio window can reach Earth's surface.

radio waves are the same. Similarly, there is no clear division between the short-wavelength infrared and the long-wavelength part of the visible spectrum. It is all electromagnetic radiation.

Look once again at the electromagnetic spectrum in Figure 6-3 and notice that electromagnetic waves shorter than violet are called **ultraviolet.** Electromagnetic waves even shorter are called X rays, and the shortest are gamma rays. Again, the boundaries between these wavelength ranges are not clearly defined.

X rays and gamma rays can be dangerous, and even ultraviolet photons have enough energy to do harm. Small doses of ultraviolet produce a suntan and larger doses sunburn and skin cancers. Contrast this to the lower-energy infrared photons. Individually they have too little energy to affect skin pigment, a fact that explains why you can't get a tan from a heat lamp. Only by concentrating many low-energy photons in a small area, as in a microwave oven, can you transfer significant amounts of energy.

Astronomers are interested in electromagnetic radiation because it carries clues to the nature of stars, planets, and other celes-

tial objects. Earth's atmosphere is opaque to most electromagnetic radiation, as shown by the graph at the bottom of Figure 6-3. Gamma rays, X rays, and some radio waves are absorbed high in Earth's atmosphere, and a layer of ozone (O_3) at an altitude of about 30 km absorbs ultraviolet radiation. Water vapor in the lower atmosphere absorbs the longer wavelength infrared radiation. Only visible light, some shorter wavelength infrared, and some radio waves reach Earth's surface through two wavelength regions called **atmospheric windows.** Obviously, if you wish to study the sky from Earth's surface, you must look out through one of these windows.

Ace⊙Astronomy™ Log into AceAstronomy and select this chapter to see Astronomy Exercise "The Electromagnetic Spectrum." Explore different wavelength regions.

Building Scientific Arguments

What could you see if your eyes were sensitive only to X rays?

As you build this scientific argument, you must imagine a totally new situation. That is sometimes a powerful tool in the critical analysis of an idea. In this case, you might at first expect to be able to see through walls, but remember that your eyes detect only light that already exists. There are almost no X rays bouncing around at Earth's surface, so if you had X-ray eyes, you would be in the dark and would be unable to see anything. Even when you looked up at the sky, you would see nothing because Earth's atmosphere is not transparent to X rays. If Superman can see through walls, it is not because his eyes can detect X rays.

But now imagine a slightly different situation and modify your argument. **Would you be in the dark if your eyes were sensitive only to radio wavelengths?**

■   ■   ■

Connections: Now that you know something about electromagnetic radiation, you are ready to meet the tools astronomers use to gather and analyze that radiation.

6-2 Optical Telescopes

ASTRONOMERS BUILD OPTICAL TELESCOPES to gather light and focus it into sharp images. This requires sophisticated optical and mechanical designs, and it leads astronomers to build gigantic telescopes on the tops of high mountains. As you begin, you will need to learn some technical terms that describe telescopes, but remember your real goal. You want to understand how different kinds of telescopes work and why some are better than others.

Two Kinds of Telescopes

Astronomical telescopes focus light into an image in one of two ways, as shown in ■ Figure 6-4. A lens bends (refracts) the light as it passes through the glass and brings it to a focus to form a small

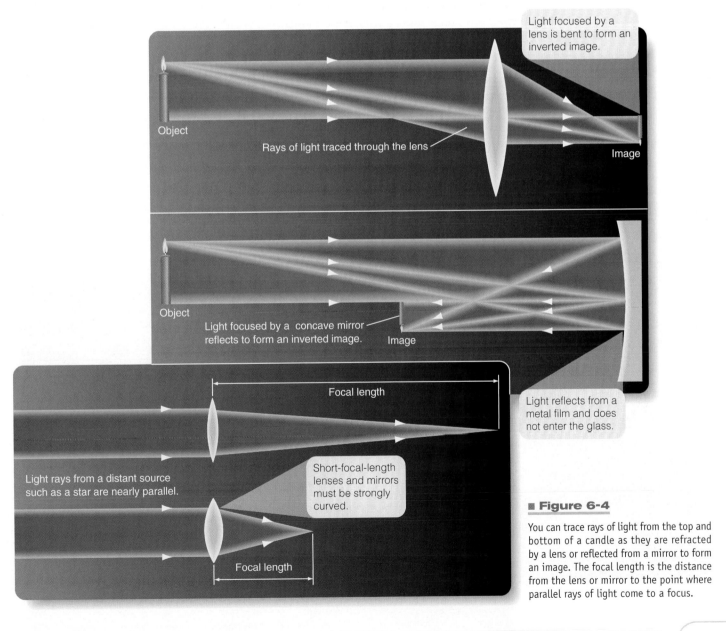

Light focused by a lens is bent to form an inverted image.

Object

Rays of light traced through the lens

Image

Object

Light focused by a concave mirror reflects to form an inverted image.

Image

Light reflects from a metal film and does not enter the glass.

Focal length

Light rays from a distant source such as a star are nearly parallel.

Short-focal-length lenses and mirrors must be strongly curved.

Focal length

■ **Figure 6-4**

You can trace rays of light from the top and bottom of a candle as they are refracted by a lens or reflected from a mirror to form an image. The focal length is the distance from the lens or mirror to the point where parallel rays of light come to a focus.

inverted image. A mirror—a concave piece of glass with a reflective surface—forms an image by reflecting the light. In either case, the **focal length** is the distance from the lens or mirror to the image formed of a distant light source, such as a star. Short-focal-length lenses and mirrors must be strongly curved, and long-focal-length lenses and mirrors are less strongly curved. Grinding the proper shape on a lens or mirror is a delicate, time-consuming, and expensive process.

Because there are two ways to focus light, there are two kinds of astronomical telescopes. **Refracting telescopes** use a large lens to gather and focus the light, whereas **reflecting telescopes** use a concave mirror. The advantages of the reflecting telescope have made it the preferred design for modern observatories.

The main lens in a refracting telescope is called the **primary lens,** and the main mirror in a reflecting telescope is called the **primary mirror.** These are also called the **objective lens** and **mirror.** Both kinds of telescopes form a very small, inverted image that is difficult to observe directly, so astronomers use a small lens

■ Active Figure 6-5

(a) A refracting telescope uses a primary lens to focus starlight into an image that is magnified by a lens called an eyepiece. The primary lens has a long focal length, and the eyepiece has a short focal length. (b) A reflecting telescope uses a primary mirror to focus the light by reflection. A small secondary mirror reflects the starlight back down through a hole in the middle of the primary mirror to the eyepiece.

Ace◖Astronomy™ Log into AceAstronomy and select this chapter to see the Active Figure "Refractors and Reflectors." Watch light pass through the optics.

called the **eyepiece** to magnify the image and make it convenient to view (■ Figure 6-5).

Refracting telescopes suffer from a serious optical distortion that limits their usefulness. When light is refracted through glass, shorter wavelengths bend more than longer wavelengths, and blue light, having shorter wavelengths, comes to a focus closer to the lens than does red light (■ Figure 6-6a). If you focus the eyepiece on the blue image, the red light is out of focus, and you see a red blur around the image. If you focus on the red image, the blue light blurs. The color separation is called **chromatic aberration.** Telescope designers can grind a telescope lens of two components made of different kinds of glass and so bring two different wavelengths to the same focus (Figure 6-6b). This does improve the image, but these **achromatic lenses** are not totally free of chromatic aberration, because other wavelengths still blur. Telescopes made with such lenses were popular until the end of the 19th century.

The primary lens of a refracting telescope is very expensive to make because it must be achromatic, and the glass must be pure and flawless because the light passes through the lens. The four surfaces must be ground precisely, and the lens can be supported only along its edge. The largest refracting telescope in the world was completed in 1897 at Yerkes Observatory in Wisconsin. Its lens is 1 m (40 in.) in diameter and weighs half a ton. Larger refracting telescopes are prohibitively expensive.

Reflecting telescopes are much less expensive because the light reflects from the front surface of the mirror. Consequently

■ Figure 6-6

(a) A normal lens suffers from chromatic aberration because short wavelengths bend more than long wavelengths. (b) An achromatic lens, made in two pieces of two different kinds of glass, can bring any two colors to the same focus, but other colors remain slightly out of focus.

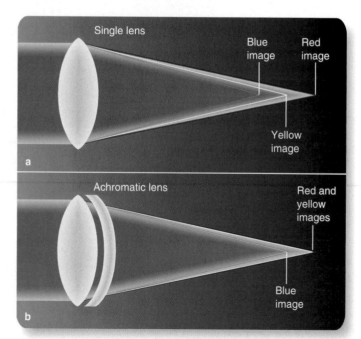

only the front surface need be ground to precise shape. Also, the glass of the mirror need not be perfectly transparent, and the mirror can be supported over its back surface to reduce sagging. Most important, reflecting telescopes do not suffer from chromatic aberration because the light is reflected from the metallic film on the front surface of the mirror and never enters the glass. For these reasons, every large astronomical telescope built since the beginning of the 20th century has been a reflecting telescope.

Ace✸Astronomy™ Log into AceAstronomy and select this chapter to see Astronomy Exercise "Lenses: Focal Length." Create different lenses and observe their focal lengths.

Ace✸Astronomy™ Log into AceAstronomy and select this chapter to see Astronomy Exercise "Telescopes: Objective Lens and Eyepiece." See how the two lenses interact to produce an image.

The Powers of a Telescope

A telescope can aid your eyes in only three ways—the three powers of a telescope. They make images brighter, more detailed, and larger.

Most interesting celestial objects are faint sources of light, so you need a telescope that can gather large amounts of light to produce a bright image. **Light-gathering power** refers to the ability of a telescope to collect light. Catching light in a telescope is like catching rain in a bucket—the bigger the bucket, the more rain it catches (■ Figure 6-7). Light-gathering power is proportional to the area of the telescope objective. A lens or mirror with a large area gathers a large amount of light. The area of a circular lens or mirror of diameter D is just πr^2 or, written in terms of the diameter, the area is $\pi(D/2)^2$. To compare the relative light-gathering powers (LGP) of two telescopes A and B, you can calculate the ratio of the areas of their objectives, which reduces to the ratio of their diameters (D) squared.

$$\frac{LGP_A}{LGP_B} = \left(\frac{D_A}{D_B}\right)^2$$

For example, suppose you compared a telescope 24 cm in diameter with a telescope 4 cm in diameter. The ratio of the diameters is 24/4, or 6, but the larger telescope does not gather 6 times as much light. Light-gathering power increases as the ratio of diameters squared, so it gathers 36 times more light than the smaller telescope. This example shows the importance of diameter in astronomical telescopes. Even a small increase in diameter produces a large increase in light-gathering power and allows astronomers to study much fainter objects.

The second power, **resolving power,** refers to the ability of the telescope to reveal fine detail. Because light acts as a wave, it produces a small **diffraction fringe** around every point of light in the image, and you cannot see any detail smaller than the fringe (■ Figure 6-8). Astronomers can't eliminate diffraction fringes,

■ **Figure 6-7**

Gathering light is like catching rain in a bucket. A large-diameter telescope gathers more light and has a brighter image than a smaller telescope of the same focal length.

but the larger a telescope is in diameter, the smaller the diffraction fringes are. That means the larger the telescope, the better its resolving power.

As you will see later, resolving power is worse for longer wavelengths, but if you consider only optical telescopes, you can estimate the resolving power by calculating the angular distance between two stars that are just barely visible through the telescope as two separate images. Astronomers say the two images are "resolved," meaning they are separated from each other. The resolving power, α, in seconds of arc, equals 11.6 divided by the diameter of the telescope in centimeters:

$$\alpha = \frac{11.6}{D}$$

For example, the resolving power of a 25-cm telescope is 11.6 divided by 25, or 0.46 second of arc. No matter how perfect the telescope optics, this is the smallest detail you can see through that telescope.

In addition to resolving power, two other factors—lens quality and atmospheric conditions—limit the detail you can see through a telescope. A telescope must contain high-quality optics to achieve its full potential resolving power. Even a large telescope reveals little detail if its optics are marred with imperfections. Also, when you look through a telescope, you are looking up through miles of turbulent air in Earth's atmosphere, which makes the image dance and blur, a condition called **seeing.** On a

Resolving Power and the Resolution of a Measurement

Have you ever seen a movie in which the hero magnifies a newspaper photo and reads some tiny detail? It isn't really possible, because newspaper photos are made up of tiny dots of ink, and no detail smaller than a single dot will be visible no matter how much you magnify the photo. In fact, all images are made up of elements of some sort, and that means there is a limit to the amount of detail you can see in an image. In an astronomical image, the resolution is often set by seeing. It is foolish to attempt to see a detail in the image that is smaller than the resolution.

This limitation is true of all measurements in science. A zoologist might be trying to measure the length of a live snake, or a sociologist might be trying to measure the attitudes of people toward drunk driving, but both face limits to the resolution of their measurements. The zoologist might specify that the snake was 43.28932 cm long, and the sociologist might say that 98.2491 percent of people oppose drunk driving, but a critic might point out that it isn't possible to make these measurements that accurately. The resolution of the techniques does not justify the accuracy implied by all of those digits.

Science is based on measurement, and whenever you make a measurement you should ask yourself how accurate that measurement can be. The accuracy of the measurement is limited by the resolution of the measurement technique, just as the amount of detail in a photograph is limited by its resolution.

A high-resolution image of Mars reveals details such as mountains, craters, and the southern polar cap. (NASA)

night when the atmosphere is unsteady and the images are blurred, the seeing is bad (■ Figure 6-9). Even under good seeing conditions, the detail visible through a large telescope is limited, not by its diffraction fringes, but by the air through which the telescope must look. A telescope performs better on a high mountaintop where the air is thin and steady, but even there Earth's atmosphere limits the detail the best telescopes can reveal to about 0.5 second of arc.

This limitation on the amount of information in an image is related to the limitation on the accuracy of a measurement. All measurements have some built-in uncertainty (**Window on Science 6-1**), and scientists must learn to work within those limitations.

The third and least important power of a telescope is **magnifying power,** the ability to make the image bigger. Because the amount of detail you can see is limited by the seeing conditions and the resolving power, very high magnification does not necessarily show more detail. Also, you can change the magnification by changing the eyepiece, but you cannot alter the telescope's light-gathering power or resolving power without changing the diameter of the objective lens or mirror, and that would be so expensive you might as well build a whole new telescope.

■ Active Figure 6-8

(a) Stars are so far away that their images are points, but the wave nature of light surrounds each star image with diffraction fringes (much magnified in this computer model). (b) Two stars close to each other have overlapping diffraction fringes and become impossible to detect separately. (Computer model by M. A. Seeds)

Ace◯Astronomy™ Log into AceAstronomy and select this chapter to see Active Figure "Resolution and Telescopes." You can control telescope diameter and watch resolution change.

a   b

Visual-wavelength image

You can calculate the magnification of a telescope by dividing the focal length of the objective by the focal length of the eyepiece:

$$M = \frac{F_\text{o}}{F_\text{e}}$$

For example, if a telescope has an objective with a focal length of 80 cm and you use an eyepiece whose focal length is 0.5 cm, the magnification is 80/0.5, or 160 times.

Notice that the two most important powers of the telescope, light-gathering power and resolving power, depend on the diameter of the telescope. This explains why astronomers refer to telescopes by diameter and not by magnification. Astronomers will refer to a telescope as an 8-meter telescope or a 10-meter telescope, but they would never identify a telescope as a 200-power telescope.

The search for light-gathering power and high resolution explains why nearly all major observatories are located far from big cities and usually on high mountains. Astronomers avoid cities because **light pollution,** the brightening of the night sky by light scattered from artificial outdoor lighting, can make it impossible to see faint objects (■ Figure 6-10). In fact, many residents of cities are unfamiliar with the beauty of the night sky because they can

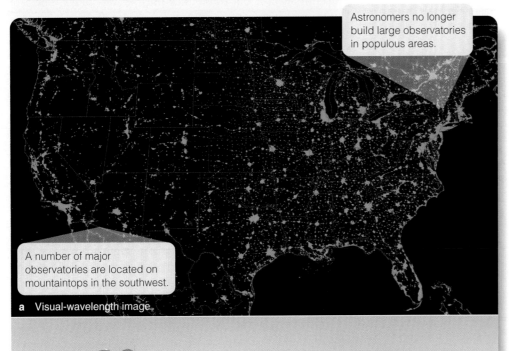

Astronomers no longer build large observatories in populous areas.

A number of major observatories are located on mountaintops in the southwest.

a Visual-wavelength image

Paranal Observatory
Altitude: 2635 m (8660 ft)
Location: Atacama desert of northern Chile
Nearest city: Antofagasta 120 km (75 mi)

b

■ **Figure 6-10**

(a) This satellite view of the continental United States at night shows the light pollution and energy waste produced by outdoor lighting. Observatories cannot be located near large cities. (NOAA) (b) The domes of four giant telescopes are visible at upper left at Paranal Observatory, built by the European Southern Observatory. The Atacama Desert is believed to be the driest place on Earth. (ESO) see only the brightest stars. Nevertheless, nature's own light pollution, the moon, is so bright it drowns out fainter objects, and astronomers are often unable to observe on the nights near full moon when faint objects cannot be observed even with the largest telescopes on high mountains.

Astronomers prefer to place their telescopes on carefully selected high mountains. The air there is thin and more transparent. The air is very dry at high altitudes and is more transparent to infrared radiation. Most important, astronomers select mountains where the air flows smoothly and is not turbulent. This produces the best seeing. Building an observatory on top of a high mountain far from civilization is difficult and expensive, as you can imagine from the photo in Figure 6-10, but the dark sky and steady seeing make it worth the effort.

Ace⟲Astronomy™ Log into AceAstronomy and select this chapter to see Astronomy Exercise "Telescopes and Resolution I." What affects resolution most in the design of a telescope?

Ace⟲Astronomy™ Log into AceAstronomy and select this chapter to see Astronomy Exercise "Telescopes and Resolution II." How does wavelength affect the resolution of a telescope?

Ace⟲Astronomy™ Log into AceAstronomy and select this chapter to see Astronomy Exercise "Particulate, Heat, and Light Pollution." What happens when the air near a city is polluted by dust, heat, or light?

Buying a Telescope

Thinking about how to shop for a new telescope will not only help you if you decide to buy one but will also illustrate some important points about astronomical telescopes.

Assuming you have a fixed budget, you should buy the highest-quality optics and the largest-diameter telescope you can afford. Of the two things that limit what you see, optical quality is under your control. You can't make the atmosphere less turbulent, but you should buy good optics. If you buy a telescope from a toy store and it has plastic lenses, you shouldn't expect to see very much. Also, you want to maximize the light-gathering power of your telescope, so you want to purchase the largest-diameter telescope you can afford. Given a fixed budget, that means you should buy a reflecting telescope rather than a refracting telescope. Not only will you get more diameter per dollar, but your telescope will not suffer from chromatic aberration.

You can safely ignore magnification. Department stores and camera stores may advertise telescopes by quoting their magnification, but it is not an important number. What you can see is fixed by light-gathering power, optical quality, and Earth's atmosphere. Besides, you can change the magnification by changing eyepieces.

Other things being equal, you should choose a telescope with a solid mounting that will hold the telescope steady and allow it to point at objects easily. Computer-controlled pointing systems are available for a price on many small telescopes. A good telescope on a poor mounting is almost useless.

You might be buying a telescope to put in your backyard, but you must think about the same issues astronomers consider when they design giant telescopes to go on mountaintops. Designing new, giant telescopes has led astronomers to solve some traditional problems in new ways, as you will see in the next section.

New-Generation Telescopes

For most of the 20th century, astronomers faced a serious limitation on the size of astronomical telescopes. Traditional telescope mirrors were made thick to avoid sagging that would distort the reflecting surface, but those thick mirrors were heavy. The 5-m (200-in.) mirror on Mount Palomar weighs 14.5 tons. These traditional telescopes were big, heavy, and expensive.

Modern astronomers have solved these problems in a number of ways. Study **Modern Astronomical Telescopes** on pages 120–121 and notice four important advances in telescope design made possible by high-speed computers:

❶ Traditional telescopes use large, solid, heavy mirrors to focus starlight.

❷ Astronomers can now build simpler, lighter-weight telescope mountings and depend on computers to move the telescope and follow the westward motion of the stars as Earth rotates.

❸ Notice that computer control of the shape of telescope mirrors allows the use of thin, lightweight mirrors—either "floppy" mirrors or segmented mirrors. Lowering the weight of the mirror lowers the weight of the rest of the telescope and makes it stronger and less expensive. Also, thin mirrors cool faster at nightfall and produce better images.

❹ Also notice the way astronomers use high-speed computers to reduce seeing distortion caused by Earth's atmosphere. Only a few decades ago, many astronomers argued that it wasn't worth building more large telescopes on Earth's surface because of the limitations set by seeing. Now a number of new giant telescopes have been built, and more are in development that can partially overcome the seeing problem.

Did you notice that astronomical telescopes must be aligned with the north celestial pole? Polaris, the North Star, is one of our favorite stars in the list at the beginning of this book. It marks the location of the north celestial pole. Equatorial mountings have an axis that points toward Polaris, and alt-azimuth telescopes are run by computers, which align their motion with Polaris. Even telescopes in the southern hemisphere, where the north celestial pole lies below the horizon, must tip their hats toward Polaris. That's one reason Polaris deserves to be one of your favorite stars; whenever you notice Polaris in the night sky, think of all the

Astronomers have begun building multiple telescopes to solve different problems. Two Gemini telescopes have been built, one in Hawaii and one in Chile, to observe the entire sky. The Large Binocular Telescope (LBT) carries two 8.4-m mirrors that combine their light. The four telescopes of the VLT are housed in separate domes, but they can combine to work as a very large telescope. (Gemini: NOAO/AURA/NSF; LBT: Large Binocular Telescope Project and European Industrial Engineer; VLT: ESO)

The LBT has the light-gathering power of a 11.8-m telescope and the resolving power of a 22.8-m telescope.

The mirrors in the VLT telescopes are 8.2 m in diameter.

The two Gemini telescopes have 8.1-m mirrors with active optics.

astronomical telescopes in backyards and observatories all over the world that bow toward Polaris.

High-speed computers have allowed astronomers to build new, giant telescopes as shown in ■ Figure 6-11. An international collaboration of astronomers have built the Gemini telescopes with 8.1-m thin mirrors. One is located in the northern hemisphere and one in the southern hemisphere to cover the entire sky. The European Southern Observatory has built the Very Large Telescope (VLT) high in the remote Andes Mountains of north-ern Chile. The VLT consists of four telescopes with computer-controlled mirrors 8.2 m in diameter and only 17.5 cm (6.9 in.) thick. The four telescopes can work singly or can combine their light to work as one large telescope. Italian and American astronomers are building the Large Binocular Telescope, which carries a pair of 8.4-m mirrors on a single mounting.

High-speed computers have improved astronomical telescopes in another way that might surprise you. Computer control and data handling have made possible huge surveys of the sky in which millions of objects are observed. The Sloan Digital Sky Survey, for example, is mapping the sky, measuring the position and brightness of 100 million stars and galaxies at a number of wavelengths. The Two-Micron All Sky Survey (2MASS) has mapped the entire sky at three wavelengths in the infrared. Other surveys are being made at many other wavelengths. Every night large telescopes scan the sky, and billions of bytes of data are compiled automatically in immense sky atlases. Astronomers will study those data banks for decades to come.

The days when astronomers worked beside their telescopes through long, dark, cold nights are nearly gone. The complexity and sophistication of telescopes require a battery of computers, and almost all research telescopes are run from control rooms that astronomers call "warm rooms." Astronomers don't need to be kept warm, but computers demand comfortable working conditions (■ Figure 6-12).

■ Figure 6-12

In the control room of the 4-meter telescope atop Kitt Peak National Observatory, the telescope operator at left manages the operation and safety of the telescope. The astronomer at right operates the instruments, records data, and makes decisions on the observing program. Astronomers work through the night controlling the computers that control the telescope and its instruments. (NOAO/AURA/NSF)

Modern Astronomical Telescopes

1 The traditional telescopes described on this page are limited by complexity, weight, and Earth's atmosphere. Modern solutions are shown on the opposite page.

In larger telescopes the light can be focused to a **prime focus** position high in the telescope tube as shown at the right. Although it is a good place to image faint objects, the prime focus is inconvenient for large instruments. A **secondary mirror** can reflect the light through a hole in the primary mirror to a **Cassegrain focus.** This focal arrangement may be the most common form of astronomical telescope.

Secondary mirror

With the secondary mirror removed, the light converges at the prime focus. In large telescopes, astronomers can ride inside the prime-focus cage, although most observations are now made by instruments connected to computers in a separate control room.

Traditional mirrors are thick to prevent the optical surface from sagging and distorting the image as the telescope is moved around the sky. Large mirrors can weigh many tons and are expensive to make and difficult to support. Also, they cool slowly at nightfall. Expansion and contraction in the cooling mirror causes distortion in the images.

The Cassegrain focus is convenient and has room for large instruments.

1a Smaller telescopes are often found with a **Newtonian focus,** the arrangement that Isaac Newton used in his first reflecting telescope. The Newtonian focus is inconvenient for large telescopes as shown at right.

Newtonian focus

Thin correcting lens

Schmidt-Cassegrain telescope

1b Many small telescopes such as the one on your left use a **Schmidt-Cassegrain focus.** A thin correcting plate improves the image but is too slightly curved to introduce serious chromatic aberration.

1c Shown below, the 4-meter Mayall Telescope at Kitt Peak National Observatory in Arizona can be used at either the prime focus or the Cassegrain focus. Note the human figure at lower right.

Prime focus cage

Secondary mirror

Primary mirror (inside)

Cassegrain focus

Astronomer

Equatorial mounting

To north celestial pole

Westward rotation about polar axis follows stars.

Polar axis

To north celestial pole

North Pole

Eastward rotation of Earth

Alt-azimuth mounting

Computer control of motion about both axes follows stars.

To north celestial pole

North Pole

Eastward rotation of Earth

2 Telescope mountings must contain a **sidereal drive** to move smoothly westward and counter the eastward rotation of Earth. The traditional **equatorial mounting** (far left) has a **polar axis** parallel to Earth's axis, but the modern **alt-azimuth mounting** (near left) moves like a cannon — up and down and left to right. Such mountings are simpler to build but need computer control to follow the stars.

3 Unlike traditional thick mirrors, thin mirrors, sometimes called floppy mirrors as shown at right, weigh less and require less massive support structures. Also, they cool rapidly at nightfall and there is less distortion from uneven expansion and contraction.

Floppy mirror

Computer-controlled thrusters   Support structure

3a Grinding a large mirror may remove tons of glass and take months, but new techniques speed the process. Some large mirrors are cast in a rotating oven that causes the molten glass to flow to form a concave upper surface. Grinding and polishing such a preformed mirror is much less time consuming.

3b Mirrors made of segments are economical because the segments can be made separately. The resulting mirror weighs less and cools rapidly. See image at right.

Segmented mirror

Computer-controlled thrusters   Support structure

3c Both floppy mirrors and segmented mirrors sag under their own weight. Their optical shape must be controlled by computer-driven thrusters under the mirror in what is called **active optics.**

3d As shown below, the two largest telescopes in the world, the Keck I and Keck II telescopes in Hawaii, contain segmented mirrors 10 m in diameter.

W. M. Keck Observatory

The thrusters are located behind the mirror segments in this photo of the Keck I mirror. The technician is sitting in the front of the light baffle over the Cassegrain hole in the center of the mirror.

Edge of mirror

Adaptive optics in telescopes

Adaptive optics off

Object appears to be a single star.

Adaptive optics on

Object revealed as a pair of stars.

1 second of arc

Paul Kalas

4 **Adaptive optics** uses high-speed computers to monitor the image distortion caused by Earth's atmosphere and adjust the optics many times a second to compensate. This can reduce the blurring due to seeing and dramatically improve image quality in Earth based telescopes.

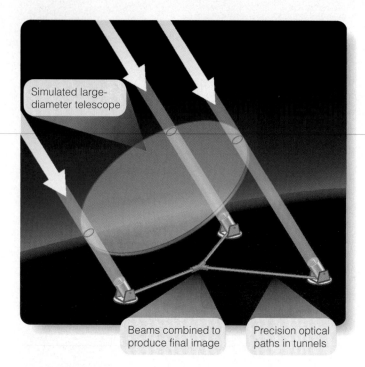

Simulated large-diameter telescope

Beams combined to produce final image

Precision optical paths in tunnels

■ **Figure 6-13**

In an astronomical interferometer, smaller telescopes can combine their light through specially designed optical tunnels to simulate a larger telescope with a resolution set by the separation of the smaller telescopes.

Interferometry

One of the reasons astronomers build big telescopes is to increase resolving power, and astronomers have been able to achieve very high resolution by connecting multiple telescopes together to work as if they were a single telescope. This method of synthesizing a larger telescope is known as **interferometry** (■ Figure 6-13).

To work as an interferometer, the separate telescopes must combine their light through a network of mirrors, and the path that each light beam travels must be controlled so that it does not vary more than some small fraction of the wavelength. Turbulence in Earth's atmosphere constantly distorts the light, and high-speed computers must continuously adjust the light paths. Recall that the wavelength of light is very short, roughly 0.0005 mm, so building optical interferometers is one of the most difficult technical problems that astronomers face. Infrared- and radio-wavelength interferometers are slightly easier to build because the wavelengths are longer. In fact, as you will discover later in this chapter, the first astronomical interferometers worked at radio wavelengths.

The VLT shown in Figure 6-11 consists of four 8.2-m telescopes that can operate separately, but they can be linked together through underground tunnels with three 1.8-m telescopes on the same mountaintop. The resulting optical interferometer provides the resolution of a telescope 200 meters in diameter. Other telescopes can work as interferometers. The two Keck 10-m telescopes can be used as an interferometer. The CHARA array on Mt. Wilson combines six 1-meter telescopes to create the equivalent of a telescope one-fifth of a mile in diameter. The Large Binocular Telescope shown in Figure 6-11 can be used as an interferometer.

Although turbulence in Earth's atmosphere can be partially averaged out in an interferometer, plans are being made to put interferometers in space. The Space Interferometry Mission, for example, will work at visual wavelengths and study everything from the cores of erupting galaxies to planets orbiting nearby stars.

Building Scientific Arguments

Why do astronomers build observatories at the tops of mountains?

To build this argument you need to think about the powers of a telescope. Astronomers have joked that the hardest part of building a new observatory is constructing the road to the top of the mountain. It certainly isn't easy to build a large, delicate telescope at the top of a high mountain, but it is worth the effort. A telescope on top of a high mountain is above the thickest part of Earth's atmosphere. There is less air to dim the light, and there is less water vapor to absorb infrared radiation. Even more important, the thin air on a mountaintop causes less disturbance to the image, and consequently the seeing is better. A large telescope on Earth's surface has a resolving power much better than the distortion caused by Earth's atmosphere. So it is limited by seeing, not by its own diffraction. It really is worth the trouble to build telescopes atop high mountains.

Astronomers not only build telescopes on mountaintops, they also build gigantic telescopes many meters in diameter. Revise your argument to focus on telescope design. **What are the problems and advantages in building such giant telescopes?**

■   ■   ■

Connections: Astronomers sometimes refer to a telescope that produces distorted images as a "light bucket." In a sense, all astronomical telescopes are light buckets because the light they focus into images reveals very little until it is recorded and analyzed by special instruments attached to the telescopes.

(6-3) Special Instruments

LOOKING THROUGH A TELESCOPE doesn't tell you much. A star looks like a point of light. A planet looks like a little disk. A galaxy looks like a hazy patch. To use an astronomical telescope to learn about the universe, you must be able to analyze the light the telescope gathers. Special instruments attached to the telescope make that possible.

Galaxy NGC 891 in true color. It is edge-on and contains thick dust clouds.

Visual-wavelength image

In this image, color shows brightness. White and red are brightest, and yellow and green are dimmer.

Visual image in false color

In these negative images of NGC 891, the sky is white and the stars are black.

Visual-wavelength negative images

Imaging Systems

The original imaging device in astronomy was the photographic plate. It could record faint objects in long time exposures and could be stored for later analysis. But photographic plates have been almost entirely replaced in astronomy by electronic imaging systems.

Most modern astronomers use **charge-coupled devices (CCDs)** to record images. A CCD is a specialized computer chip containing roughly a million microscopic light detectors arranged in an array about the size of a postage stamp. These devices can be used like small photographic plates, but they have dramatic advantages. They can detect both bright and faint objects in a single exposure, are much more sensitive than photographic plates, and can be read directly into computer memory for later analysis. Although CCDs for astronomy are extremely sensitive and therefore expensive, less sophisticated CCDs are used in video and digital cameras.

The image from a CCD is stored as numbers in computer memory, so it is easy to manipulate the image to bring out details that would not otherwise be visible. For example, astronomical images are often reproduced as negatives with the sky white and the stars dark. This makes the faint parts of the image easier to see (■ Figure 6-14). Astronomers also manipulate images to produce **false-color images** in which the colors represent different levels of intensity and are not related to the true colors of the object. You can see an example in Figure 6-14. In fact, false-color images are common in many fields (**Window on Science 6-2**).

Measurements of intensity and color were made in the past using a photometer, a highly sensitive light meter attached to a

■ Figure 6-14

Astronomical images can be manipulated in many ways to bring out details. The photo of the galaxy at upper left is dark, and the details of the dust clouds in the disk of the galaxy do not show well. The two negative images of the galaxy have been produced to show the dust clouds more clearly. (C. Hawk, B. Savage, N. A. Sharp NOAO/WIYN/NSF) The image at upper right shows two interacting galaxies known as Arp 273. The visual-wavelength image has been given false color according to brightness. (NOAO/WIYN/NSF)

telescope. Today, however, most such measurements are made directly on CCD images. Because the CCD image is easily digitized, brightness and color can be measured more easily and more accurately than on photographic plates.

The Spectrograph

To analyze light in detail, astronomers need to spread the light out according to wavelength to form a spectrum, a task performed by a **spectrograph.** You can understand how this works if you imagine reproducing an experiment performed by Isaac Newton in 1666. Newton bored a small hole in the window shutter of his bedroom to admit a thin beam of sunlight. When he placed a prism in the beam, it spread the light into a beautiful spectrum splashed across his bedroom wall. From this Newton concluded that white light was made of a mixture of all the colors.

Newton didn't think in terms of wavelength, but you can use that modern concept to see that the light passing through the prism is bent at an angle that depends on the wavelength. Violet (short wavelength) bends most, and red (long wavelength) least. In this way, the white light entering the prism is spread into a spectrum (■ Figure 6-15). You could build a spectrograph with a prism to spread the light and a lens to guide the light into a camera.

False-Color Images and Reality

Astronomical images are usually recorded in digital form directly into computer memory, so astronomers often add false color to images to reveal things otherwise hard to see. They might exaggerate colors already present, but more often they add colors to represent levels of brightness. Of course, the colors used are entirely arbitrary. One astronomer might like shades of blue, and another might use all of the colors ranging from violet to red. To understand the image, you need to know the meaning of the colors—that white and red are bright areas, for instance, and that yellow and green are dimmer.

Astronomers also use false color to create images recorded at wavelengths not visible to the human eye. Radio, infrared, ultraviolet, and X-ray images are often displayed as false-color images. You might think of them as maps in which the colors show the relative intensity of the radiation in different parts of the image. Again the choice of colors is up to the astronomer analyzing the image.

False-color images are common in astronomy, but they are also used in other sciences. Doctors often analyze medical X-ray images, CAT scans, and so on by converting the images to false color. Biologists use false color to analyze microscopic photographs, and geologists use false color to study photographs of Earth recorded by a satellite at various wavelengths.

Even a simple photograph of a galaxy or a glowing cloud of gas in space is a kind of false-color image. The object is much too faint to see with the human eye, and even if you flew in a spaceship to within a short distance, it would still be too faint to see with your eyes. The photograph you see is a time exposure in which the light was allowed to accumulate over an extended period ranging from seconds to hours. That makes the galaxy or gas cloud visible. Even a black-and-white photograph of a glowing cloud of gas shows an aspect of the universe you can never see with unaided eyes. In that sense, all astronomical images are false-

A false-color X-ray image of the expanding cloud of gas left behind by the explosion of a massive star. (NASA/CXC/Rutgers/J. Hughes)

color images. Their virtue is not that they are colorful but that they are meaningful.

Nearly all modern spectrographs use a grating in place of a prism. A **grating** is a piece of glass with thousands of microscopic parallel grooves scribed onto its surface. Different wavelengths of light reflect from the grating at slightly different angles, so white light is spread into a spectrum. You have probably noticed this effect when you look at the closely spaced lines etched onto a compact disk; as you move the disk about, different colors flash across its surface. You could build a modern spectrograph by using a high-quality grating to spread the light into a spectrum and a CCD camera to record the spectrum.

The spectrum of an astronomical object can contain hundreds of spectral lines produced by the atoms in the object. Because astronomers must identify these lines and measure the wavelengths, they use a **comparison spectrum** as a calibration of their spectrograph. Special bulbs built into the spectrograph produce bright lines given off by such atoms as thorium and argon or neon. The wavelengths of these spectral lines have been measured to high precision in the laboratory, so astronomers can use spectra of these light sources as guides to measure wavelengths and identify spectral lines in the spectrum of a star, galaxy, or planet.

Because astronomers understand how light interacts with matter, a spectrum carries a tremendous amount of information (as you will see in the next chapter), and that makes a spectrograph the astronomer's most powerful instrument. An astronomer recently remarked, "We don't know anything about an object till we get a spectrum," and that is only a slight exaggeration.

Building Scientific Arguments

What is the difference between light going through a lens and light passing through a prism?

When you think about natural processes, it is often helpful to compare similar things and scientific arguments often make such comparisons. A few simple rules explain most natural events, so the similarities are often revealing. A refracting telescope producing chromatic aberration and a prism dispersing light into a spectrum are two examples of the same thing, but one is bad and one is good. When light passes through the curved surfaces of a lens, different wavelengths are bent by slightly different amounts, and the different colors of light come to focus at different focal lengths. This produces the color fringes in an image called chromatic aberration, and that's bad. But the surfaces of a prism are made to be precisely flat, so all of the light enters the prism at the same angle, and any given wavelength is bent by the same amount. Consequently, white light is dispersed into a spectrum. You could call the dispersion of light by a prism "controlled chromatic aberration," and that's good.

Now you can build your own argument comparing similar things. CCDs have been very good for astronomy, and they have almost completely replaced photographic plates. **How are CCD chips similar to photographic plates, and how are they better?**

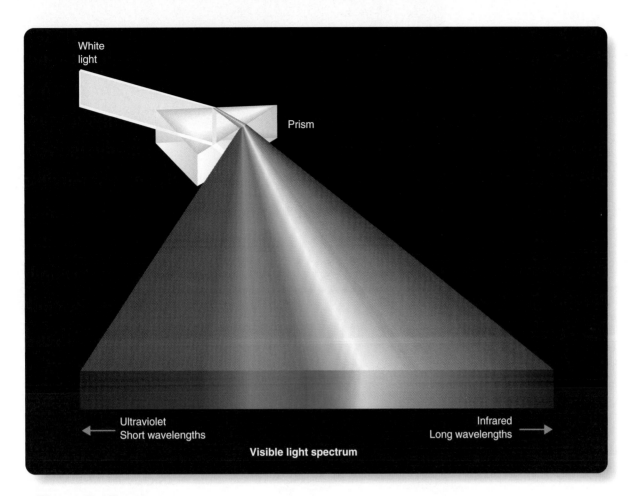

White light

Prism

Ultraviolet
Short wavelengths

Infrared
Long wavelengths

Visible light spectrum

■ **Figure 6-15**

A prism bends light by an angle that depends on the wavelength of the light. Short wavelengths bend most and long wavelengths least. Thus, white light passing through a prism is spread into a spectrum.

■ ■ ■

Connections: So far, this discussion has been limited to visual wavelengths. Now it is time for you to explore the rest of the electromagnetic spectrum.

6-4 Radio Telescopes

ALL THE TELESCOPES and instruments you have discussed so far look out through the visible light window in Earth's atmosphere, but there is another window running from a wavelength of 1 cm to about 1 m (see Figure 6-3). By building the proper kinds of instruments, astronomers can study the universe through this radio window.

Operation of a Radio Telescope

A radio telescope usually consists of four parts: a dish reflector, an antenna, an amplifier, and a recorder (■ Figure 6-16). The components, working together, make it possible for astronomers to detect radio radiation from celestial objects.

The dish reflector of a radio telescope, like the mirror of a reflecting telescope, collects and focuses radiation. Because radio waves are much longer than light waves, the dish need not be as smooth as a mirror; wire mesh will reflect all but the shortest wavelength radio waves. But don't be surprised if you see a photo of a radio telescope that doesn't have a dish shape. There are a number of ways to collect radio energy from the sky. Nevertheless, dish-shaped reflectors are very common at radio observatories.

Though a radio telescope's dish may be many meters in diameter, the antenna may be as small as your hand. Like the antenna on a TV set, its only function is to absorb the radio energy and direct it along a cable to an amplifier. After amplification, the signal goes to some kind of recording instrument. Most radio observatories record data directly into computer memory.

However it is recorded, an observation with a radio telescope measures the amount of radio energy coming from a specific point on the sky, and that causes two problems. For one thing, the intensity at one spot doesn't tell you much, so the radio telescope must be scanned over an object, a cloud of gas, for example, to produce a map of the radio intensity at different points. The

■ **Figure 6-16**

In most radio telescopes, a dish reflector concentrates the radio signal on the antenna. The signal is then amplified and recorded. For all but the shortest radio waves, wire mesh is an adequate reflector (photo). (Courtesy Seth Shostak/SETI Institute)

Antenna

Cable

Dish reflector

Amplifier

Computer

■ **Figure 6-17**

(a) A contour map of a baseball stadium shows regions of similar admission prices. The most expensive seats are those behind home plate. (b) A false-color-image radio map of Tycho's supernova remnant, the expanding shell of gas produced by the explosion of a star in 1572. The radio contour map has been color-coded to show intensity. (Courtesy NRAO)

Seat prices in a baseball stadium
Red most expensive
Violet least expensive

a

Radio energy map
Red strongest
Violet weakest

b

second problem is that humans can't see radio waves, so astronomers draw maps in which contours mark areas of similar radio intensity. You could compare such a map to a seating diagram for a baseball stadium in which the contours mark areas in which the seats have the same price (■ Figure 6-17a). Contour maps are very common in radio astronomy and are often reproduced using false colors (Figure 6-17b)

Limitations of a Radio Telescope

A radio astronomer works under three handicaps: poor resolution, low intensity, and interference. You remember that the resolving power of an optical telescope depends on the diameter of the objective lens or mirror. It also depends on the wavelength of the radiation. At very long wavelengths, like those of radio waves, images become fuzzy because the diffraction fringes are very large. As with an optical telescope, there is no way to improve the resolving power without building a bigger telescope. Consequently, radio telescopes must be quite large.

Even so, the resolving power of a radio telescope is not good. A dish 30 m in diameter receiving radiation with a wavelength of 21 cm has a resolving power of about 0.5°. Such a radio telescope would be unable to reveal any details in the sky smaller than the moon. Fortunately, radio astronomers can combine two or more radio telescopes to form a **radio interferometer** capable of much higher resolution. For example, the Very Large Array (VLA) consists of 27 dish antennas spread in a Y-shape across the New Mexico desert (■ Figure 6-18). In combination, they have the resolving power of a radio telescope 36 km (22 mi) in diameter. The VLA can resolve details smaller than 1 second of arc. Eight new dish antennas being added across New Mexico will give the VLA 10 times better resolving power. Another large radio interferometer, the Very Long Baseline Array (VLBA), consists of matched radio dishes spread from Hawaii to the Virgin Islands and has an effective diameter almost as large as Earth.

The second handicap radio astronomers face is the low intensity of the radio signals. You saw earlier that the energy of a photon depends on its wavelength. Photons of radio energy have such long wavelengths that their individual energies are quite low. In order to get strong signals focused on the antenna, the radio astronomer must build large collecting dishes.

■ **Figure 6-18**

The Very Large Array uses 27 radio dishes, which can be moved to different positions along a Y-shaped set of tracks across the New Mexico desert. They are shown here in the most compact arrangement. Signals from the dishes are combined to create very-high-resolution radio maps of celestial objects. (NRAO/AUI/NSF)

The largest fully steerable radio telescope in the world is at the National Radio Astronomy Observatory in Green Bank, West Virginia (■ Figure 6-19a). The telescope has a reflecting surface 100 meters in diameter, big enough to hold an entire football field, and can be pointed anywhere in the sky. Its surface consists of 2004 computer-controlled panels that adjust to maintain the shape of the reflecting surface.

The largest radio dish in the world is 300 m (1000 ft) in diameter. So large a dish can't be supported in the usual way, so it is built into a mountain valley in Arecibo, Puerto Rico.

■ **Figure 6-19**

(a) The largest steerable radio telescope in the world is the GBT located in Green Bank, West Virginia. With a diameter of 100 m, it stands higher than the Statue of Liberty. (Mike Bailey: NRAO/AUII) (b) The 300-m (1000-ft) radio telescope in Arecibo, Puerto Rico, hangs from cables over a mountain valley. The Arecibo Observatory is part of the National Astronomy and Ionosphere Foundation operated by Cornell University and the National Science Foundation. (David Parker/SPL/Photo Researchers, Inc.)

The reflecting dish is a thin metallic surface supported above the valley floor by cables attached near the rim, and the antenna hangs above the dish on cables from three towers built on three mountain peaks that surround the valley (Figure 6-19b).

Although this telescope can look only overhead, the operators can change its aim slightly by moving the antenna and by waiting for Earth's rotation to point the telescope in the proper direction. This may sound clumsy, but the telescope's ability to detect weak radio sources, together with its good resolution, makes it one of the most important radio observatories in the world.

The third handicap the radio astronomer faces is interference. A radio telescope is an extremely sensitive radio receiver listening to radio signals thousands of times weaker than artificial radio and TV transmissions. Such weak signals are easily drowned out by interference. Sources of such interference include everything from poorly designed transmitters in Earth satellites to automobiles with faulty ignition systems. To avoid this kind of interference, radio astronomers locate their telescopes as far from civilization as possible. Hidden deep in mountain valleys, they are able to listen to the sky protected from human-made radio noise.

Advantages of a Radio Telescope

Building large radio telescopes in isolated locations is expensive, but three factors make it all worthwhile. First, and most important, a radio telescope can reveal clouds of cool hydrogen in space. Because 90 percent of the atoms in the universe are hydrogen, that is important information. Large clouds of cool hydrogen are completely invisible to normal telescopes because they produce no visible light of their own and reflect too little to be detected on photographs. However, cool hydrogen emits a radio signal at the specific wavelength of 21 cm. (You will see how the hydrogen produces this radiation in the discussion of the gas clouds in space in Chapter 10.) These hydrogen clouds are important because, for one thing, they are the places where stars are born. The only way astronomers can detect these clouds of gas is with a radio telescope that receives the 21-cm radiation, so that is one reason that radio telescopes are important.

The second reason is related to dust in space. Because radio signals have relatively long wavelengths, they can penetrate the vast clouds of dust that obscure astronomers' view at visual wavelengths. Light waves are short, and they interact with tiny dust grains floating in space; as a result, the light is scattered and never gets through the dust to reach optical telescopes on Earth. However, radio signals from far across the galaxy pass unhindered through the dust, giving radio astronomers an unobscured view.

Finally, radio telescopes are important because they can detect objects that are more luminous at radio wavelengths than at visible wavelengths. This includes everything from the coldest clouds of gas to the hottest stars. Some of the most distant objects in the universe, for instance, are detectable only at radio wavelengths.

Why do optical astronomers build big telescopes, while radio astronomers build groups of widely separated smaller telescopes?
Once again you can learn a lot by building a scientific argument based on comparison. Optical astronomers build large telescopes to maximize light-gathering power, but the problem for radio telescopes is resolving power. Because radio waves are so much longer than light waves, a single radio telescope can't see details in the sky much smaller than the moon. By linking radio telescopes miles apart, radio astronomers build a radio interferometer that can simulate a radio telescope miles in diameter and thus increase the resolving power.

The difference between the wavelengths of light and radio waves makes a big difference in building the best telescopes. Keep that difference in mind as you build a new argument: **Why don't radio astronomers want to build their telescopes on mountaintops as optical astronomers do?**

■   ■   ■

Connections: Earth's atmosphere causes trouble for astronomers in two ways. It distorts images, and it absorbs radiation at many wavelengths. The only way to avoid these limitations completely is to send telescopes above the atmosphere, into space.

(6-5) Astronomy from Space

YOU HAVE LEARNED about the observations that ground-based telescopes can make through the two atmospheric windows in the visible and radio parts of the electromagnetic spectrum. Most of the rest of the electromagnetic radiation—infrared, ultraviolet, X ray, and gamma ray—never reaches Earth's surface. To observe at these wavelengths, telescopes must fly above the atmosphere in high-flying aircraft, rockets, balloons, and satellites. The only exceptions are observations that can be made in the near-infrared and the near-ultraviolet.

The Ends of the Visual Spectrum

Astronomers can observe in the near-infrared just beyond the red end of the visible spectrum. You can't see this light, but some of it leaks through the atmosphere in narrow, partially open atmospheric windows scattered from 1200 nm to about 40,000 nm. Infrared astronomers usually measure wavelength in micrometers (10^{-6} meters), so they refer to this wavelength range as 1.2 to 40 micrometers (or microns for short). In this range, much of the radiation is absorbed by water vapor, but carbon dioxide and oxygen molecules also absorb infrared. As you saw earlier in this chapter, it is an advantage to place telescopes on mountaintops

Infrared astronomers can often observe with the dome lights on. Their instruments are not usually sensitive to visible light.

SOFIA will fly at roughly 12 km (over 40,000 ft) to get above most of Earth's atmosphere.

Adding liquid nitrogen to the camera on a telescope is a familiar task for astronomers.

■ **Figure 6-20**

Comet Hale–Bopp hangs in the sky over the 3-meter NASA Infrared Telescope Facility (IRAF) atop Mauna Kea. The air at high altitudes is so dry that it is transparent to shorter infrared photons. SOFIA will fly so high it will be able to observe infrared wavelengths that cannot be observed from mountaintops. Most astronomical CCD cameras must be cooled to low temperatures, and this is especially true for infrared cameras. (Hale–Bopp over observatory: William Keel/IRAF; SOFIA: NASA; Camera: Kris Koenig/Coast Learning Systems)

where the air is thin and dry. For example, a number of important infrared telescopes observe from the 4150-m (13,600-ft) summit of Mauna Kea in Hawaii. At this altitude, the telescopes are above much of the water vapor in Earth's atmosphere (■ Figure 6-20).

The far-infrared range, which includes wavelengths longer than 40 micrometers, carries clues to the nature of comets, planets, forming stars, and other cool objects, but these wavelengths are absorbed high in Earth's atmosphere—much higher than mountaintops. Infrared telescopes have flown to high altitudes under balloons and in airplanes. NASA is now building the Stratospheric Observatory for Infrared Astronomy (SOFIA), a Boeing 747 that will carry a 2.5-m telescope, control systems, and a team of technicians and astronomers to the fringes of the atmosphere. Once at that altitude, they can open a door above the telescope and make extended infrared observations as the plane flies a precisely calculated path. You can see the door in the photo in Figure 6-20. Some infrared wavelengths are too strongly absorbed by the atmosphere to be observed from even the highest-flying aircraft. The only way to make those infrared observations is to put the telescope in space above the atmosphere.

Wherever infrared telescopes are based, they have one thing in common—cooling. If a telescope observes at far-infrared wavelengths, then it must be cooled to low temperature. Infrared radiation is emitted by heated objects, and if the telescope is warm it will emit many times more infrared radiation than that coming from a distant object. Imagine trying to look for rabbits at night through binoculars that are themselves glowing. In a tele-

scope observing in the near-infrared, only the detector, the element on which the infrared radiation is focused, must be cooled, usually with liquid nitrogen, as shown in Figure 6-20. To observe in the far-infrared, however, the entire telescope must be cooled.

At the short-wavelength end of the spectrum, astronomers can observe in the near-ultraviolet. Your eyes don't detect this radiation, but it can be recorded by photographic plates and CCDs. Wavelengths shorter than about 290 nm, the far-ultraviolet, are completely absorbed by the ozone layer extending from 20 km to about 40 km above Earth's surface. No mountaintop is that high, and no airplane can fly to such an altitude. To observe in the far-ultraviolet or beyond at X-ray or gamma-ray wavelengths, telescopes must be in space above the atmosphere.

Telescopes in Space

To observe far beyond the ends of the visible spectrum, astronomical telescopes must go above Earth's atmosphere into space. This is very expensive and difficult, but it is the only way to study some processes. Stars, for example, are born inside clouds of dust and gas, and visible wavelengths cannot escape from these dust clouds. Only observations in the infrared can reveal the secrets of star formation. Black holes are small and hard to detect, but matter falling into a black hole emits X rays. Telescopes in space can explore such processes as these that are invisible from within Earth's atmosphere.

One of the most successful space telescopes was the International Ultraviolet Explorer (IUE), launched in 1978. It carried a

1 The Hubble Space Telescope was carried into orbit by the Space Shuttle in 1990. The telescope contains a 2.4-m (96-in.) mirror and can observe from the near-infrared to the near-ultraviolet.

Orbiting above Earth's blurring atmosphere, Hubble is limited only by diffraction in its optics. It can detect details 10 times smaller than Earth-based telescopes.

The telescope, as big as a large bus, has been visited twice by astronauts, who repaired equipment and installed new instruments. Named after Edwin Hubble, the astronomer who discovered the expansion of the universe, the telescope has been tremendously productive observing everything from the weather on Mars to the most distant galaxies visible in the universe.

Hubble image of Mars and its polar cap.

Visual

Hubble image of a nebula around an aging star.

Visual

Hubble image of a dust-filled galaxy.

Visual

2 The Compton Gamma Ray Observatory at left was in orbit from 1991 to 2000. It made observations of very-high-energy photons, helping astronomers understand such violently active objects as neutron stars and black holes.

2a The Chandra X-Ray Observatory was placed in orbit 1/3 of the way to the moon in 1999. It is nearly 14 m (45 ft) long and carries highly precise mirrors 1.2 m (47 in.) in diameter. X rays would penetrate into regular mirrors, so Chandra's mirrors are designed as cylinders polished on the inside so that X rays just graze the surface and are focused onto detectors. The telescope was named after the late Indian-American Nobel laureate Subrahmanyan Chandrasekhar, who was a pioneer in many branches of theoretical astronomy.

Chandra can detect X-ray emitting objects 50 times fainter and resolve details 10 times smaller than any previous X-ray telescope.

Chandra X-ray images

Two galaxies collide and trigger the birth of stars.

Saturn emits X-rays from near its equator.

Very hot gas is trapped in a cluster of galaxies.

Spitzer infrared image

A comet glows in the infrared.

2b The Spitzer Space Telescope (at left) observes in the infrared. It was launched in 2003 and named in honor of the late astronomer Lyman Spitzer, a leader in space astronomy. The telescope is cooled to –273°C (–459°F) so it cannot orbit the warm Earth. Instead it is in an orbit around the sun and will drift slowly away from Earth during its lifetime. Protected from sunlight by a heat screen, it can observe a wide range of astronomical objects.

Dust warmed by hot young stars glows in the disk of this spiral galaxy.

Spitzer infrared image

Heat screen

Spitzer infrared image

An infrared image penetrates a dusty nebula to reveal newborn stars just beginning to shine.

3 The James Webb Space Telescope (at right) is planned as the next great observatory in space. Named after the early director of NASA who oversaw the planning of the Apollo moon landings, the telescope will carry a 6.5-m (256-in.) segmented mirror made of the metal beryllium. It will observe without a telescope tube from behind a multilayered sunscreen. Launch is planned for 2011.

Segmented mirror

Sunscreen

telescope only 45 cm (18 in.) in diameter and was expected to last only a year or two, but it became the little telescope that could. It made many observations and exciting discoveries until it finally failed in 1996.

Many space telescopes are small satellites designed to make specific observations for a short period, but some are large general-purpose telescopes. Over two decades ago, astronomers developed a plan to place a series of great observatories in space. Those space telescopes have revolutionized human understanding of what we are and where we are in the universe. Study **The Great Observatories in Space** on pages 130–131 and notice three points:

1 Not only can a telescope in space observe at a wide range of wavelengths, but it is above the atmospheric blurring called seeing. The Hubble Space Telescope observes mostly at visual wavelengths and has the advantage of sharp images undistorted by seeing.

2 Notice how telescopes must be specialized for their wavelength range. The Compton Gamma Ray Observatory had special detectors, the Chandra X-Ray Observatory must have cylindrical mirrors, and the Spitzer infrared observatory must have cooled optics.

3 The Hubble Space Telescope has been maintained by visits from astronauts, but such visits are expensive, and the future of Hubble is in doubt. Astronauts cannot reach the Chandra and Spitzer telescopes, and the Compton Observatory was removed from orbit in 2000. Space observatories have limited lifetimes, and astronomers are already planning the next great observatory in space. The new James Webb Space Telescope will not be available for many years.

These great observatories in space are controlled from research centers on Earth and are open to proposals from any astronomer with a good idea; but competition is fierce, and only the most worthy projects win approval.

Cosmic Rays

All of the radiation you have read about in this chapter has been electromagnetic radiation. **Cosmic rays,** however, are not really rays; they are subatomic particles traveling at tremendous velocities that strike Earth's atmosphere from space. Almost no cosmic rays reach the ground, but they do smash gas atoms in the upper atmosphere, and fragments of those atoms shower down on you day and night over your entire life. These secondary cosmic rays are passing through you as you read this sentence.

Some cosmic-ray research can be done from high mountains or high-flying aircraft; but, to study cosmic rays in detail, detectors must go into space. A number of cosmic-ray detectors have been carried into orbit, but this area of astronomical research is just beginning to bear fruit.

Astronomers can't be sure what produces cosmic rays. Because they are atomic particles with electric charges, they are deflected by the magnetic fields spread through our galaxy, and that means you can't tell where they are coming from. The space between the stars is a glowing fog of cosmic rays. Some lower-energy cosmic rays come from the sun, but observations show that at least some cosmic rays are produced by the violent explosions of dying stars.

At present, cosmic rays largely remain an exciting mystery. You will meet them again in future chapters.

Building Scientific Arguments

Why can infrared astronomers observe from high mountain-tops, while X-ray astronomers must observe from space? Once again, you can analyze this question by building a scientific argument based on comparison. Infrared radiation is absorbed by water vapor in Earth's atmosphere. If you built an infrared telescope on top of a high mountain, you would be above most of the water vapor in the atmosphere, and you could collect some infrared radiation from the stars. The longer-wavelength infrared radiation is absorbed much higher in the atmosphere, so you couldn't observe it from our mountaintop. Similarly, X rays are absorbed in the uppermost layers of the atmosphere, and you would not be able to find any mountain high enough to get an X-ray telescope above those absorbing layers. To observe the stars at X-ray wavelengths, you would need to put your telescope in space, above Earth's atmosphere.

You can see why X-ray and far-infrared telescopes must observe from space. Now build another argument based on comparison: **Why must the Hubble Space Telescope be in space when it observes in the visual-wavelength range?**

■   ■   ■

Connections: The tools of the astronomer are designed to gather radiation from the sky and extract information. Perhaps no tool is as important as the spectrograph because no form of observation is as loaded with information as a spectrum. In the next chapter, you will see how astronomers can harvest the information in a star's spectrum.

Summary

6-1 | Radiation: Information from Space

What is light?

■ Light is the visible form of electromagnetic radiation, an electric and magnetic disturbance that transports energy at the speed of light. The electromagnetic spectrum includes gamma rays, X rays, ultraviolet radiation, visible light, infrared radiation, and radio waves.

■ You can think of a particle of light, a photon, as a bundle of waves that acts sometimes as a particle and sometimes as a wave.

■ The energy a photon carries depends on its wavelength. The wavelength of visible light, usually measured in nanometers (10^{-9} m), ranges from 400 nm to 700 nm. Infrared and radio photons have longer wavelengths and carry less energy. Ultraviolet, X-ray, and gamma-ray photons have shorter wavelengths and carry more energy.

6-2 | Optical Telescopes

How do telescopes work, and how are they limited?

■ Astronomers use telescopes to gather light, resolve fine detail, and magnify the image. The first two of these three powers of the telescope depend on the telescope's diameter. Consequently, astronomers strive to build telescopes with large diameters.

■ A refracting telescope uses a lens to bend the light and focus it into an image. Because of chromatic aberration, refracting telescopes cannot bring all colors to the same focus, resulting in color fringes around the images. An achromatic lens partially corrects for this, but such lenses are expensive and cannot be made much larger than about 1 m in diameter.

■ Reflecting telescopes use a mirror to focus the light and are less expensive than refracting telescopes of the same diameter. Also, reflecting telescopes do not suffer from chromatic aberration. Most recently built large telescopes are reflectors.

■ Astronomers build observatories on high mountains for two reasons. Turbulence in Earth's atmosphere blurs the image of an astronomical telescope, a phenomenon that astronomers refer to as seeing. Atop a mountain, the air is steady, and the seeing is better; the air on a mountaintop is also thin and dry and is more transparent, especially in the infrared.

■ Sometimes astronomical telescopes can be linked together to form an interferometer, which has a resolution equivalent to that of a telescope as large in diameter as the separation between the telescopes.

6-3 | Special Instruments

What kind of instruments do astronomers use to record and analyze light?

■ For many decades astronomers used photographic plates to record images at the telescope, but modern electronic systems such as CCD cameras have replaced photographic plates in most applications.

■ Spectrographs using prisms or a grating spread starlight out according to wavelength to form a spectrum revealing hundreds of spectral lines produced by atoms in the object being studied.

6-4 | Radio Telescopes

Why do astronomers use radio telescopes?

■ Astronomers use radio telescopes for three reasons: They can detect cool hydrogen in space; they can see through dust clouds that block visible light; and they can detect certain objects invisible at other wavelengths.

■ Most radio telescopes contain a dish reflector, an antenna, an amplifier, and a data recorder. Such a telescope can record the intensity of the radio energy coming from a spot on the sky. Scans of small regions are used to produce radio maps.

■ Because of the long wavelength, radio telescopes have very poor resolution, and astronomers often link separate radio telescopes together to form a radio interferometer capable of resolving much finer detail.

6-5 | Astronomy from Space

Why must some telescopes go into space?

■ Earth's atmosphere is transparent in two wavelength ranges called atmospheric windows, the visual window and the radio window. At other wavelengths, the atmosphere absorbs radiation. To observe at other wavelengths, telescopes must go into space.

■ Earth's atmosphere distorts and blurs images. Telescopes in orbit are above this seeing distortion and are limited only by diffraction in their optics.

■ Cosmic rays are not electromagnetic radiation; they are subatomic particles such as electrons and protons traveling at nearly the speed of light. They can best be studied from above Earth's atmosphere.

New Terms

electromagnetic radiation (p. 110)

wavelength (p. 110)

frequency (p. 111)

nanometer (nm) (p. 111)

Angstrom (Å) (p. 111)

photon (p. 111)

infrared radiation (p. 111)

ultraviolet radiation (p. 112)

atmospheric window (p. 112)

focal length (p. 114)

refracting telescope (p. 114)

reflecting telescope (p. 114)

primary lens, mirror (p. 114)

objective lens, mirror (p. 114)

eyepiece (p. 114)

chromatic aberration (p. 114)

achromatic lens (p. 114)

light-gathering power (p. 115)

resolving power (p. 115)

diffraction fringe (p. 115)

seeing (p. 115)

magnifying power (p. 116)

light pollution (p. 117)

prime focus (p. 120)

secondary mirror (p. 120)

Cassegrain focus (p. 120)

Newtonian focus (p. 120)

Schmidt–Cassegrain focus (p. 120)

sidereal drive (p. 121)

equatorial mounting (p. 121)

polar axis (p. 121)

alt-azimuth mounting (p. 121)

active optics (p. 121)

adaptive optics (p. 121)

interferometry (p. 122)

charge-coupled device (CCD) (p. 123)

false-color image (p. 123)

spectrograph (p. 123)

grating (p. 124)

comparison spectrum (p. 124)

radio interferometer (p. 127)

cosmic ray (p. 132)

Review Questions

Assess your understanding of this chapter's topics with additional quizzing and animations at http://ace.brookscole.com/sf9

1. Why would you not plot sound waves in the electromagnetic spectrum?

2. If you had limited funds to build a large telescope, which type would you choose, a refractor or a reflector? Why?

3. Why do nocturnal animals usually have large pupils in their eyes? How is that related to astronomical telescopes?

4. Why do optical astronomers sometimes put their telescopes at the tops of mountains, while radio astronomers sometimes put their telescopes in deep valleys?

5. Optical and radio astronomers both try to build large telescopes but for different reasons. How do these goals differ?

6. What are the advantages of making a telescope mirror thin? What problems does this cause?

7. Small telescopes are often advertised as "200 power" or "magnifies 200 times." As someone knowledgeable about astronomical telescopes, how would you improve such advertisements?

8. Not long ago an astronomer said, "Some people think I should give up photographic plates." Why might she change to something else?

9. What purpose do the colors in a false-color image or false-color radio map serve?

10. How is chromatic aberration related to a prism spectrograph?

11. Why would radio astronomers build identical radio telescopes in many different places around the world?

12. Why do radio telescopes have poor resolving power?

13. Why must telescopes observing in the far-infrared be cooled to low temperatures?

14. What might you detect with an X-ray telescope that you could not detect with an infrared telescope?

15. The moon has no atmosphere at all. What advantages would you have if you built an observatory on the lunar surface?

16. The two images at the right show a star before and after an adaptive optics system was switched on. What causes the distortion in the first image, and how does adaptive optics correct the image?

(ESO)

17. The X-ray image at right shows the remains of an exploded star. Explain why images recorded by telescopes in space are often displayed in false color rather than in the "colors" received by the telescope.

(NASA/CXC/PSU/ S. Park)

Discussion Questions

1. Why does the wavelength response of the human eye match so well the visual window of Earth's atmosphere?

2. Most people like beautiful sunsets with brightly glowing clouds, bright moonlit nights, and twinkling stars. Most astronomers don't. Why?

Problems

1. The thickness of the plastic in plastic bags is about 0.001 mm. How many wavelengths of red light is this?

2. What is the wavelength of radio waves transmitted by a radio station with a frequency of 100 million cycles per second?

3. Compare the light-gathering powers of one of the Keck telescopes and a 0.5-m telescope.

4. How does the light-gathering power of one of the Keck telescopes compare with that of the human eye? (*Hint:* Assume that the pupil of your eye can open to about 0.8 cm.)

5. What is the resolving power of a 25-cm telescope? What do two stars 1.5 seconds of arc apart look like through this telescope?

6. Most of Galileo's telescopes were only about 2 cm in diameter. Should he have been able to resolve the two stars mentioned in Problem 5?

7. How does the resolving power of the 5-m telescope compare with that of the Hubble Space Telescope? Why does the HST outperform the 5-m telescope?

8. If you build a telescope with a focal length of 1.3 m, what focal length should the eyepiece have to give a magnification of 100 times?

9. Astronauts observing from a space station need a telescope with a light-gathering power 15,000 times that of the human eye, capable of resolving detail as small as 0.1 second of arc, and having a magnifying power of 250. Design a telescope to meet their needs. Could you test your design by observing stars from Earth?

10. A spy satellite orbiting 400 km above Earth is supposedly capable of counting individual people in a crowd. Roughly what minimum-diameter telescope must the satellite carry? (*Hint:* Use the small-angle formula.)

To access the resources in the Media Cluster, log into AceAstronomy at http://ace.brookscole .com/sf9 and select Chapter 6.

ACTIVE FIGURES

Refractors and Reflectors
Compare the two basic types of telescope designs and the paths light takes through each with this animation.

Resolution and Telescopes
This animation allows you to contrast the high resolution of larger-aperture telescopes with the lower resolution of smaller ones.

ASTRONOMY EXERCISES

Lenses: Focal Length

Astronomers use lenses to focus starlight into an image. You can change the shape and material of the lens in this animation to study how these parameters affect the focal length.

Telescopes: Objective Lens and Eyepiece

Use this animation to study how the objective lens and eyepiece work together to form an image in a telescope. You can change parameters for the diameter of the lens and the focal lengths of the objective lens and eyepiece.

Telescopes and Resolution I

In this simulation, you can vary the focal length of the objective lens, the focal length of the eyepiece lens, and the diameter of the objective lens. Try them all and see which factor (or factors) determines how sharp the image is.

Telescopes and Resolution II

Get a feel for how wavelength affects the resolution of a telescope by comparing the lens diameter needed to observe craters on the moon with a visible-light telescope versus other types such as radio and microwave telescopes.

Particulate, Heat, and Light Pollution

These three animations let you study how environmental variables can affect viewing quality. How does this fit in with what you learned in this chapter about the locations of major astronomical telescopes?

The Electromagnetic Spectrum

In this exercise, study the properties of radiation, such as wavelength and sources of various parts of the electromagnetic spectrum.

VIRTUAL ASTRONOMY LABS

Lab 4: Solar Wind and Cosmic Rays

This lab begins with an overview of the properties of the sun's atmosphere and how energetic particles escape and travel through the solar system. The lab ends with a discussion of cosmic rays.

Critical Inquiries for the Web

1. Research in chemistry, physics, and biology is supported in part by industry. Because astronomy has few industrial applications, it is not well supported by industry. Visit websites for major observatories and find out who pays their bills.

2. Visit websites for the major observatories in space, such as Hubble Space Telescope and Chandra, and check on their latest status. What kinds of observations are they making, and what kinds of discoveries are they making?

3. Astronomers are leaders in efforts to reduce electric power wasted on outdoor lighting because it causes light pollution. Find websites on light pollution and see if your town is making any effort to preserve the beauty of the night sky and simultaneously save electrical power.

Exploring *TheSky*

1. Astronomical telescopes using equatorial mountings must be aligned precisely with the north celestial pole. Locate Polaris and determine how far it is from the north celestial pole. (*Hint:* Use **Reference Lines** under the **View** menu and check **Grid** under Equatorial. Be sure the spacing is set to auto/fine. Then locate the Little Dipper and zoom in on Polaris.)

Go to the Brooks/Cole Astronomy Resource Center (**http://astronomy.brookscole.com**) for critical thinking exercises, articles, and additional readings from InfoTrac College Edition, Brooks/Cole's online student library.

7 | Starlight and Atoms

Awake! for Morning in the

Bowl of Night

Has flung the Stone that puts

the Stars to Flight:

And Lo! the Hunter of the

East has caught

The Sultan's Turret in a

Noose of Light.

THE RUBÁIYÁT OF OMAR KHAYYÁM,
TRANS. EDWARD FITZGERALD

Visual-wavelength image

THE UNIVERSE IS FILLED with fabulously beautiful clouds of glowing gas illuminated by brilliant stars, but it is all hopelessly beyond reach. No laboratory jar on Earth holds a sample labeled "star stuff," and no space probe has ever visited the inside of a star. The stars are far away, and the only information you can obtain about them comes hidden in starlight (■ Figure 7-1). Whatever you want to know about stars you must catch in a noose of light.   ▌   Earthbound humans knew almost nothing about stars until the early 19th century, when the Munich optician Joseph von Fraunhofer studied the solar spectrum and found it interrupted by some 600 dark lines. As scientists realized that the lines were related to the various atoms in the sun and found that stellar spectra had similar patterns of lines, the door to an understanding of stars finally opened.

▌ Continued on page 138 ▌

Clouds of glowing gas illuminated by hot, bright stars lie thousands of light years across space, but clues hidden in starlight tell a story of star birth and star death. (ESO)

Guidepost

Looking Back
In the last chapter you read how hard astronomers work to gather light from the stars using giant telescopes on mountaintops and in space. You also read how spectrographs can spread light out into spectra. Now you are ready to see what all the fuss is about.

This Chapter
This chapter explains how light interacts with matter and how astronomers must understand that interaction to understand stars. Here you will find answers to five essential questions:

How do stars produce light?

What is an atom?

How do atoms interact with light?

What kind of spectra do you see when you look at celestial objects?

What can you learn from a star's spectrum?

This chapter marks a change in the way you will look at nature. Up to this point, you have been thinking about what you can see with your eyes alone or aided by telescopes. In this chapter, you begin using modern astrophysics, the application of physics to study the sky. Now you can search out secrets of the stars that lie beyond what you can see.

Looking Ahead
The analysis of spectra is a powerful tool, and in the next chapter you will use that tool to study the sun. In the chapters that follow, you will study other suns—the stars.

Ace ☉ Astronomy™ The AceAstronomy icon throughout the text indicates an opportunity for you to test yourself on key concepts and to explore animations and interactions on the AceAstronomy website at: http://ace .brookscole.com/sf9

Visual-wavelength image

■ Figure 7-1

What's going on here? The sky is filled with beautiful and mysterious objects that lie far beyond your reach—in the case of the nebula NGC 6751, about 6500 ly beyond your reach. The only way to understand such objects is by analyzing their light. Such an analysis reveals that this object is a dying star surrounded by the expanding shell of gas it ejected a few thousand years ago. You will learn more about this phenomenon in Chapter 13. (NASA Hubble Heritage Team/STScI/AURA)

In this chapter, you will go through that door by considering how stars produce light and how atoms interact with light to produce spectral lines. The first step is to consider the hydrogen atom because it is the most common atom in the universe as well as the simplest. Other atoms are larger and more complicated, but in many ways their properties resemble those of hydrogen.

Once you understand how an atom's structure can interact with light to produce spectral lines, you will recognize certain patterns in stellar spectra. By classifying the spectra according to these patterns, you can arrange the stars in a sequence according to temperature. One of the most important pieces of information revealed in a star's spectrum is its temperature.

But, properly analyzed, a stellar spectrum can tell you much more. The spectrum contains information about the chemical composition of the star and the star's motion relative to Earth.

 Starlight

IF YOU LOOK at the stars in the constellation Orion, you will notice that they are not all the same color (see Figure 2-4). One of our Favorite Stars, Betelgeuse, in the upper left corner, is quite red; another Favorite Star, Rigel, in the lower right corner, is blue.

These differences in color arise from the way the stars produce light, and they provide an important clue to the temperatures of stars.

Temperature and Heat

Temperature is one of the defining characteristics of a star. That is, if you know a star's temperature and a few other properties, such as size, you can understand the star. But if you don't know a star's temperature, you can know almost nothing about it.

A gas is made up of particles—atoms and molecules—that are in constant motion, colliding with one another millions of times a second. The **temperature** of a gas is a measure of the average kinetic energy of the particles. Recall that kinetic energy is energy of motion, so if a gas is hot, the particles must be moving very rapidly; if the gas is cool, the particles are moving more slowly. The motion of the particles represents stored energy, and you feel that energy as heat. Read **Focus on Fundamentals 3** and notice the distinction between the temperature, the stored energy, and the heat you feel.

Astronomers express the temperature of stars on the **Kelvin temperature scale.** The Kelvin scale sets its zero point at **absolute zero,** the temperature at which the particles of a gas have no remaining kinetic energy that could be extracted as heat. Absolute zero is $-273.2°C$, which is $-459.7°F$. (See Appendix A.) The temperatures of the stars range from less than 2000 K to 40,000 K or more. To see how astronomers can estimate the temperatures of the stars from starlight, you must understand how stars produce light.

The Origin of Starlight

The starlight you see comes from gases in the outer surface layers of the star—the photosphere. The gases deep inside the star also emit light, but it is absorbed before it can escape, and the low-density gas above the photosphere is too thin to emit significant amounts of light. Starlight comes from the photosphere of a star, that layer of gases dense enough to emit significant amounts of light but thin enough to allow the light to escape. (Recall that you met the photosphere of the sun in Chapter 3.) So when astronomers refer to a star as "hot" they are referring to the temperature of its visible surface—its photosphere.

To see how the photosphere of a star can produce light, you must consider two things, how photons are produced and how the temperature of a material is related to motion among its atoms.

First, a photon can be produced by a changing electric field. An **electron** is a negatively charged subatomic particle, and if you disturb the motion of an electron, the sudden change in the electric field around it can produce an electromagnetic wave. For example, if you run a comb through your hair while standing near a radio tuned to an AM station, you can produce popping noises on the radio. The moving comb disturbs the electrons in both the comb and your hair, building static electricity. The sudden sparks of static electricity produce electromagnetic waves that the radio

Temperature, Heat, and Thermal Energy

One of the most common misconceptions in science involves temperature. People often say "temperature" when they really mean "heat," and sometimes they say "heat" when they mean something entirely different. This is a fundamental idea, so you need to understand the differences.

When something is hot, the particles in the object, be they atoms or molecules, are moving rapidly. Temperature is a measure of the average motion of the particles. (Mathematically, temperature is proportional to the square of the average velocity.) In a hot object, the particles race around at higher speeds than in a cool object. If you have your temperature taken, it will probably be 98.6°F, an indication that the atoms and molecules in your body are moving about at a normal pace. If you measure the temperature of a month-old baby, the thermometer should register the same temperature, showing that the atoms and molecules in the baby's body are moving at the same average velocity as the atoms and molecules in your body.

The energy of the moving particles in a body is called **thermal energy.** You have much more mass than the baby, so you must contain more thermal energy even though you have the same temperature. The thermal energy in your body and in the baby's body have the same intensity (temperature) but different amounts. People often confuse temperature and thermal energy, so you must be careful to distinguish between them. Temperature is an intensity, and thermal energy is an amount.

Many people say "heat" when they should say thermal energy. Heat is the thermal energy that moves from a hot object to a cool object. If two objects have the same temperature, you and the infant for example, there is no transfer of thermal energy and no heat. This is a fine distinction, but when you hear someone say "heat," check to see if he or she doesn't really mean thermal energy.

What's the difference between temperature and heat?

You may have burned yourself on cheese pizza, but you probably haven't burned yourself on green beans. At the same temperature, cheese holds more thermal energy than green beans. It isn't the temperature that burns your tongue, but the flow of thermal energy, and that's heat.

PRESSURE | MASS | ENERGY | **TEMPERATURE AND HEAT** | DENSITY

---

picks up as pops and crackles. This illustrates an important principle: Any change in the motion of an electron can generate an electromagnetic wave.

Second, recall that temperature is related to the motion among the particles in a material. In the gases of a hot star, the atoms move faster, on average, than the atoms in a cool star.

Now put these two ideas together, and you can understand why a hot object glows. The hotter an object is, the more violent the motion among its particles. The agitated particles collide with electrons, and when electrons are accelerated, part of the energy is carried away as electromagnetic radiation in the form of a photon. The radiation emitted by a heated object is called **black body radiation,** a name that refers to the way a perfect emitter of radiation would behave. A perfect emitter would also be a perfect absorber and, at room temperature, would look black. You will often find the term *black body radiation* referring to objects that glow brightly.

Black body radiation is quite common. In fact, it is responsible for the light emitted by an incandescent lightbulb. Electricity flowing through the wire filament of the lightbulb heats the wire to high temperature, and it glows. You can also recognize the light emitted by a heated horseshoe in the blacksmith's forge as black body radiation. Many objects in astronomy, including stars, emit radiation approximately as if they were black bodies.

Hot objects emit black body radiation, but so do cold objects. Ice cubes are cold, but their temperature is higher than absolute zero, so they contain some thermal energy and must emit some black body radiation. The coldest gas drifting in space has a temperature only a few degrees above absolute zero, but it too emits black body radiation.

Two important features of black body radiation will help you understand what you see when you look at a star. First, the hotter an object is, the more black body radiation it emits. Hot objects emit more radiation because their agitated particles travel faster and collide more often. So, of course, you expect a glowing coal from a fire to emit more total energy than an ice cube of the same size.

The second feature is the relationship between the temperature of the object and the wavelengths of the photons it emits. The wavelength of a photon emitted when a particle collides with an electron depends on the violence of the collision. Only a violent collision can produce a short-wavelength (high-energy) photon. Because extremely violent collisions don't occur very often, short-wavelength photons are rare. Similarly, most collisions are not extremely gentle, so long-wavelength (low-energy) photons are also rare. Consequently, black body radiation is made up of photons with a distribution of wavelengths, and very short and very long wavelengths are rare. The **wavelength of maximum** intensity (λ_{max}), the wavelength at which the object emits the most radiation, occurs at some intermediate wavelength.

■ Figure 7-2 shows the intensity of radiation versus wavelength for three objects of different temperatures. The curves are high in the middle and low at either end, which tells you that these objects emit most intensely at intermediate wavelengths. The total area under each curve is proportional to the total energy emitted, and you can see that the hotter object emits more total energy than the cooler objects. Look closely at the curves, and you will notice that the wavelength of maximum intensity depends on temperature. The hotter the object is, the shorter the wavelength

■ Figure 7-2

Black body radiation from three bodies at different temperatures demonstrates that a hot body radiates more total energy and that the wavelength of maximum intensity is shorter for hotter objects. The hotter object here will look blue to your eyes, while the cooler object will look red.

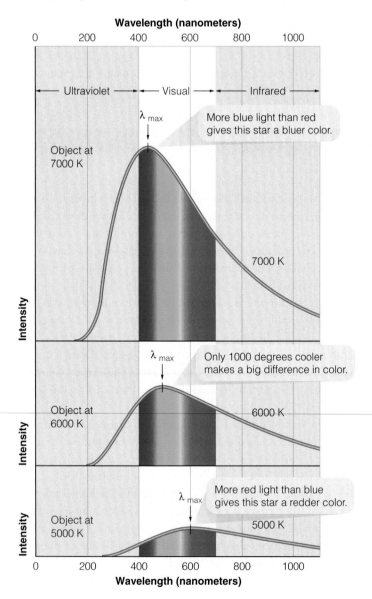

of maximum intensity. Notice in this figure how temperature determines the color of a glowing black body. The hotter object emits more blue light than red and thus looks blue, and the cooler object emits more red light than blue and consequently looks red. Now you can understand why two of our Favorite Stars, Betelgeuse and Rigel, have such different colors. Betelgeuse is cool and looks red, but Rigel is hot and looks blue

Notice that cool objects may emit little visible radiation but are still producing black body radiation. For example, the human body has a temperature of 310 K and emits black body radiation mostly in the infrared part of the spectrum. Infrared security cameras can detect burglars by the radiation they emit, and mosquitoes can track you down in total darkness by homing in on your infrared radiation. Although humans emit lots of infrared radiation, you rarely emit higher energy photons; and you almost never emit X-ray or gamma-ray photons. Your wavelength of maximum intensity lies in the infrared part of the spectrum.

Two Radiation Laws

The two features of black body radiation that you have just considered can be given precise mathematical form, and they have proven so dependable, they are known as laws. One law is related to energy and one to color.

As you saw in the previous section, a hot object emits more black body radiation than a cool object. That is, it emits more energy. Recall from Chapter 5 that energy is expressed in units called joules (J); 1 joule is about the energy of an apple falling from a table to the floor. The total radiation given off by 1 square meter of the object in joules per second equals a constant number, represented by σ, times the temperature raised to the fourth power.* This relationship is called the Stefan–Boltzmann law:

$$E = \sigma T^4 \; (\text{J/s/m}^2)$$

How does this help you understand stars? Suppose a star the same size as the sun had a surface temperature that was twice as hot as the sun's surface. Then each square meter of that star would radiate not twice as much energy but 2^4, or 16, times as much energy. From this law you can see that a small difference in temperature can produce a very large difference in the amount of energy emitted.

The second radiation law is related to the color of stars. In the previous section, you saw that hot stars look blue and cool stars look red. Wien's law tells you that the wavelength at which a star radiates the most energy, its wavelength of maximum intensity (λ_{max}), depends only on the star's temperature:

$$\lambda_{max} = \frac{3,000,000}{T}$$

---

*For the sake of completeness, you can note that the constant σ equals $5.67 \times 10^{-8} \; \text{J/m}^2\text{s degree}^4$.

That is, the wavelength of maximum radiation in nanometers equals 3 million divided by the temperature on the Kelvin scale.

This is a powerful tool in astronomy, because it means you can relate the temperature of a star and its wavelength of maximum intensity. For example, you might find a star that has a surface temperature of 3000 K. Then its wavelength of maximum intensity would be 3,000,000/3000, or 1000 nm—in the near-infrared. Later you will meet objects much hotter than most stars; such objects radiate most of their energy at very short wavelengths. The hottest stars, for instance, radiate most of their energy in the ultraviolet.

Now you can understand how astronomers can estimate the temperature of a star's surface from its color. The astronomers use the same technique that doctors and nurses use to measure a patient's body temperature. Medical personnel use a small device that detects the infrared radiation emerging from the patient's ear. You might suspect the device depends on the Stephan–Boltzmann law and measures the intensity of the infrared radiation. A person with a fever will emit more energy than a healthy person. However, a healthy person with a large ear canal would emit more energy than a person with a small ear canal, so measuring intensity wouldn't be accurate. The device actually depends on Wien's law in that it measures the "color" of the infrared radiation. A patient with a fever will emit at a slightly shorter wavelength of maximum intensity, and the infrared radiation emerging from that person's ear will be a tiny bit "bluer" than normal. Remember that the wavelength of maximum intensity depends on the temperature, so even a small fever produces changes in the color of the infrared radiation. Astronomers use the same technique to find the temperatures of the stars; they observe the color of the starlight.

Ace✸Astronomy™ Log into AceAstronomy and select this chapter to see Astronomy Exercise "Black Body." Change the temperature and watch the black body curve change.

Ace✸Astronomy™ Log into AceAstronomy and select this chapter to see Astronomy Exercise "Stefan–Boltzmann Law." Watch the brightness of a star change as you adjust its temperature.

Building Scientific Arguments

Why does the wavelength of maximum intensity depend on temperature?

Remember as you build any scientific argument that it must proceed logically, step by step. No gaps are allowed. In this case, you can begin with the temperature of an object. The hotter the object is, the more rapidly its particles move. That means that the typical collision between a particle and an electron will be more violent in a hotter body, and more violent collisions will accelerate the electrons more violently and, on average, produce higher-energy, shorter-wavelength photons. So the wavelength of maximum intensity depends on the temperature of the body. That's Wien's law. It all makes sense when you think about it step by step.

Now change the argument to answer a different question: **Why does the Stefan–Bolzmann law work?**

■   ■   ■

Connections: You have been thinking about atoms in a general way, but now it is time to get specific. What is an atom, and how can it interact with light?

(7-2) Atoms

THE ATOMS IN THE SURFACE LAYERS of stars leave their marks on the light the stars emit. By understanding what atoms are and how they interact with light, you can decode the spectra of the stars.

A Model Atom

To think about atoms and how they can interact with light, you need a working model of an atom. In Chapter 2, you used a working model of the sky, the celestial sphere. You identified and named the important parts and described how they were located and how they interacted. So you should begin your study of atoms by creating a model of an atom.

Your model atom contains a positively charged **nucleus** at the center; this nucleus consists of two kinds of particles. **Protons** carry a positive electrical charge, and **neutrons** have no charge. That means the nucleus has a net positive charge.

The nucleus in this model atom is surrounded by a whirling cloud of orbiting electrons, low-mass particles with a negative charge. In a normal atom, the number of electrons equals the number of protons, and the positive and negative charges balance to produce a neutral atom. Because protons and neutrons each have a mass 1836 times greater than that of an electron, most of the mass of the atom lies in the nucleus. The hydrogen atom is the simplest of all atoms. The nucleus is a single proton orbited by a single electron, with a total mass of only 1.67×10^{-27} kg, about a trillionth of a trillionth of a gram.

An atom is mostly empty space. To see this, imagine constructing a simple scale model. The nucleus of a hydrogen atom is a proton with a diameter of about 0.0000016 nm, or 1.6×10^{-15} m. If you multiply this by one trillion (10^{12}), you can represent the nucleus of your model atom with a grape seed, which is about 0.16 cm in diameter. The region of a hydrogen atom containing the whirling electron has a diameter of about 0.4 nm, or 4×10^{-10} m. Multiplying by a trillion magnifies the diameter to about 400 m, or about 4.5 football fields laid end to end (■ Figure 7-3). When you imagine a grape seed in the midst of a sphere

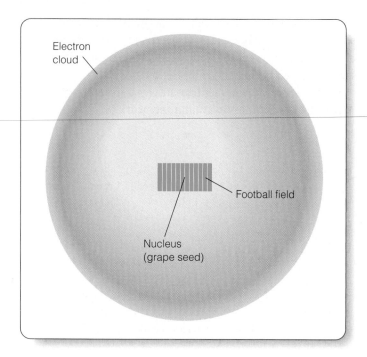

■ Figure 7-3

Magnifying a hydrogen atom by 10^{12} makes the nucleus the size of a grape seed and the diameter of the electron cloud about 4.5 times longer than a football field. The electron itself is still too small to see.

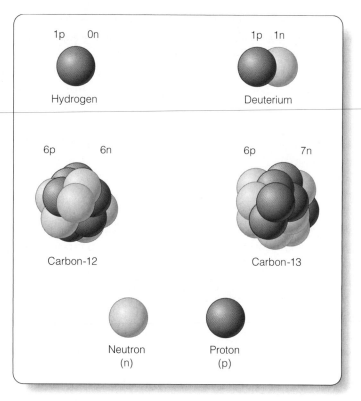

■ Figure 7-4

Some common isotopes. A rare isotope of hydrogen, deuterium, contains a proton and a neutron in its nucleus. Two isotopes of carbon are carbon-12 and carbon-13.

4.5 football fields in diameter, you can see that an atom is mostly empty space.

Different Kinds of Atoms

There are over a hundred kinds of atoms, called chemical elements. Which element an atom is depends only on the number of protons in the nucleus. For example, carbon has six protons in its nucleus. An atom with one more proton than this is nitrogen, and an atom with one proton fewer is boron.

Although the number of protons in an atom of an element is fixed, you could change the number of neutrons in an atom's nucleus without changing the atom significantly. For instance, if you added a neutron to a carbon nucleus, you would still have carbon, but it would be slightly heavier than normal carbon. Atoms that have the same number of protons but a different number of neutrons are **isotopes.** Carbon has two stable isotopes. One form contains six protons and six neutrons for a total of 12 particles and is thus called carbon-12. Carbon-13 has six protons and seven neutrons in its nucleus (■ Figure 7-4).

Protons and neutrons are bound tightly into the nucleus, but the electrons are held loosely in the electron cloud. Running a comb through your hair creates a static charge by removing a few electrons from their atoms. This process is called **ionization,** and the atom that has lost one or more electrons is an **ion.** A carbon atom is neutral if it has six electrons to balance the positive charge of the six protons in its nucleus. If you ionize the atom by remov-

ing one or more electrons, the atom is left with a net positive charge. Under some circumstances, an atom may capture one or more extra electrons, giving it more negative charges than positive. Such a negatively charged atom is also considered an ion.

Atoms that collide may form bonds with each other by exchanging or sharing electrons. Two or more atoms bonded together form a **molecule.** Atoms do collide in stars, but the high temperatures cause violent collisions that are unfavorable for chemical bonding. Only in the coolest stars are the collisions gentle enough to permit the formation of chemical bonds. You will see later that the presence of molecules such as titanium oxide (TiO) in a star is a clue that the star is very cool. In later chapters, you will see that molecules can form in cool gas clouds in space and in the atmospheres of planets.

Electron Shells

So far you have been thinking of the cloud of the whirling electrons in a general way, but now it is time to be more specific as to how the electrons behave within the cloud.

The electrons are bound to the atom by the attraction between their negative charge and the positive charge on the nucleus. This attraction is known as the **Coulomb force,** after the French physicist Charles-Augustin de Coulomb (1736–1806). To ionize

Quantum Mechanics: The World of the Very Small

Quantum mechanics is the set of rules that describe how atoms and subatomic particles behave. When you think about large objects such as stars, planets, aircraft carriers, and hummingbirds, you don't have to think about quantum mechanics, but on the atomic scale, particles behave in ways that seem unfamiliar.

One of the principles of quantum mechanics specifies that you cannot know simultaneously the exact location and motion of a particle. This is why physicists refer to the electrons in an atom as if they were a cloud of negative charge surrounding the nucleus. Because you can't know the position and motion of the electron, you can't really describe it as a small particle following an orbit. You can use that image as a model to help your imagination, but the reality is much more interesting, and describing the electrons as a charge cloud provides a better and more sophisticated model of an atom.

This raises some serious questions about reality. Is an electron really a particle at all? Quantum mechanics can describe particles as waves and waves as particles. If you can't know simultaneously the position and motion of a specific particle, how can you know how it will react to a collision with a photon or another particle? The answer is that you can't know, and that seems to violate the principle of cause and effect (Window on Science 5-1).

Needless to say, you can't expect to explore these discrepancies here. Scientists and philosophers of science continue to struggle with the meaning of reality on the quantum-mechanical level. Here you should note that the reality you see on the scale of stars and hummingbirds is only part of nature. You have constructed a model to help think about nature on the scale of atoms, but the truth is much more interesting and much more exciting than anything you see on larger scales. Although you will use models of atoms to study stars, there is still much to learn about the atoms themselves.

---

an atom, you need a certain amount of energy to pull an electron away from the nucleus. This energy is the electron's **binding energy,** the energy that holds it to the atom.

An electron may orbit the nucleus at various distances. If the orbit is small, the electron is close to the nucleus, and a large amount of energy is needed to pull it away. Consequently, its binding energy is large. An electron orbiting farther from the nucleus is held more loosely, and less energy will pull it away. That means it has less binding energy. The size of an electron's orbit is related to the energy that binds it to the atom.

Nature permits atoms only certain amounts (quanta) of binding energy, and the laws that describe how atoms behave are called the laws of **quantum mechanics** (**Window on Science 7-1**). Much of this discussion of atoms is based on the laws of quantum mechanics.

Because atoms can have only certain amounts of binding energy, your model atom can have orbits of only certain sizes, called **permitted orbits.** These are like steps in a staircase: You can stand on the number-one step or the number-two step, but not on the number-one-and-one-quarter step. The electron can occupy any permitted orbit but not orbits in between.

The arrangement of permitted orbits depends primarily on the charge of the nucleus, which in turn depends on the number of protons. Consequently, each kind of element has its own pattern of permitted orbits (■ Figure 7-5). Isotopes of the same ele-

Hydrogen nuclei have one positive charge; the electron orbits are not tightly bound.

Only the innermost orbits are shown.

Boron nuclei have 5 positive charges; the electron orbits are more tightly bound.

Hydrogen · · · Helium · · · Boron

■ **Figure 7-5**

The electron in an atom may occupy only certain, permitted orbits. Because different elements have different charges on their nuclei, the elements have different, unique patterns of permitted orbits.

ments have nearly the same pattern because they have the same number of protons. However, ionized atoms have orbital patterns that differ from their un-ionized forms. Thus the arrangement of permitted orbits differs for every kind of atom and ion.

How many hydrogen atoms would it take to cross the head of a pin?

This is not a frivolous question. In answering it, you will discover how small atoms really are, and you will see how powerful physics and mathematics can be as a way to understand nature. Many scientific arguments are convincing because they have the precision of mathematics. To begin, assume that the head of a pin is about 1 mm in diameter. That is 0.001 m. The size of a hydrogen atom is represented by the diameter of the electron cloud, roughly 0.4 nm. Because 1 nm equals 10^{-9} m, you can multiply and discover that 0.4 nm equals 4×10^{-10} m. To find out how many atoms would stretch 0.001 m, you can divide the diameter of the pinhead by the diameter of an atom. That is, divide 0.001 m by 4×10^{-10} m, and you get 2.5×10^6. It would take 2.5 million hydrogen atoms lined up side by side to cross the head of a pin.

Now you can see how tiny an atom is and also how powerful a bit of physics and mathematics can be. It reveals a view of nature beyond the capability of your eyes. Now build an argument using another bit of arithmetic: **How many hydrogen atoms would you need to add up to the mass of a paper clip (1 g)?**

■ ■ ■

Connections: Astronomers are experts on electron orbits because the electrons in those orbits can interact with light. Such interactions fill starlight with clues to the secrets of the stars.

(7-3) The Interaction of Light and Matter

IF LIGHT DID NOT INTERACT with matter, you would not be able to see these words. In fact, you would not exist, because, among other problems, photosynthesis would be impossible, and there would be no grass, wheat, bread, beef, cheeseburgers, or any other kind of food. The interaction of light and matter makes your life possible, and it also makes it possible for you to understand the universe.

You should begin your study of light and matter by considering the hydrogen atom. As you read earlier, hydrogen is both simple and common. Roughly 90 percent of all atoms in the universe are hydrogen.

The Excitation of Atoms

Each orbit in an atom represents a specific amount of binding energy, so physicists commonly refer to the orbits as **energy levels.** Using this terminology, you can say that an electron in its small-est and most tightly bound orbit is in its lowest permitted energy level. You could move the electron from one energy level to another by supplying enough energy to make up the difference between the two energy levels. It would be like moving a flowerpot from a low shelf to a high shelf; the greater the distance between the shelves, the more energy you would need to raise the pot. The amount of energy needed to move the electron is the energy difference between the two energy levels.

If you move the electron from a low energy level to a higher energy level, you can call the atom an **excited atom.** That is, you have added energy to the atom in moving its electron. If the electron falls back to the lower energy level, that energy is released.

An atom can become excited by collision. If two atoms collide, one or both may have electrons knocked into a higher energy level. This happens very commonly in hot gas, where the atoms move rapidly and collide often.

Another way an atom can get the energy that moves an electron to a higher energy level is to absorb a photon. Only a photon with exactly the right amount of energy can move the electron from one level to another. If the photon has too much or too little energy, the atom cannot absorb it. Because the energy of a photon depends on its wavelength, only photons of certain wavelengths can be absorbed by a given kind of atom. ■ Figure 7-6 shows the lowest four energy levels of the hydrogen atom along with three photons the atom could absorb. The longest-wavelength photon has only enough energy to excite the electron to the second energy level, but the shorter-wavelength photons can excite the electron to higher levels. A photon with too much or too little energy cannot be absorbed. Because the hydrogen atom has many more energy levels than shown in Figure 7-6, it can absorb photons of many different wavelengths.

■ **Figure 7-6**

A hydrogen atom can absorb only those photons that move the atom's electron to one of the higher-energy orbits. Here three different photons are shown along with the change they would produce if they were absorbed.

Atoms, like humans, cannot exist in an excited state forever. The excited atom is unstable and must eventually (usually within 10^{-6} to 10^{-9} seconds) give up the energy it has absorbed and return its electron to the lowest energy level. Because the electrons eventually tumble down to this bottom level, physicists call it the **ground state.**

When the electron drops from a higher to a lower energy level, it moves from a loosely bound level to one more tightly bound. The atom then has a surplus of energy—the energy difference between the levels—that it can emit as a photon. Study the sequence of events in ■ Figure 7-7 to see how an atom can absorb and emit photons. Because each type of atom or ion has its unique set of energy levels, each type absorbs and emits photons with a unique set of wavelengths. As a result, you can identify the elements in a gas by studying the characteristic wavelengths of light absorbed or emitted.

The process of excitation and emission is a common sight in urban areas at night. A neon sign glows when atoms of neon gas in the glass tube are excited by electricity flowing through the tube. As the electrons in the electric current flow through the gas, they collide with the neon atoms and excite them. As you have seen, immediately after an atom is excited, its electron drops back to a lower energy level, emitting the surplus energy as a photon of a certain wavelength. The photons emitted by excited neon produce a reddish-orange glow. Signs of other colors, erroneously called "neon," contain other gases or mixtures of gases instead of pure neon.

The Formation of a Spectrum

The spectrum of a star is formed as light passes outward through the gases near its surface. Study **Atomic Spectra** on pages 146–147. Notice three important properties of spectra:

1 There are three kinds of spectra described by three simple rules. When you see one of these types of spectra, you can recognize the kind of matter that emitted the light.

2 Notice that the wavelengths of the photons that are absorbed or emitted are determined by the atomic energy levels in the atoms. The emitted photons coming from a hot cloud of hydrogen gas have the same wavelengths as the photons absorbed by hydrogen atoms in the gases of a star. Although the hydrogen atom produces many spectral lines from the ultraviolet to the infrared, only three are visible to human eyes.

3 Most modern astronomy books display spectra as graphs of intensity versus wavelength. Be sure you see the connection between dark absorption lines and dips in the graphed spectrum.

Spectra are filled to bursting with information about the sources of the light; but, to extract that information, astronomers must be experts on the interaction of light and matter. Electrons moving among the orbits within atoms can reveal the secrets of the stars.

Ace◐Astronomy™   Log into AceAstronomy and select this chapter to see Astronomy Exercise "Emission and Absorption Spectra." This exercise will give you an inside look at atoms as they interact with photons.

Building Scientific Arguments

What spectrum would you see if you observed molten iron? To analyze this situation, you can begin your argument with the well-understood Kirchhoff's laws. Molten iron is a dense liquid, and the atoms and molecules collide so often they emit all wavelengths, so you would see a continuous spectrum. That is Kirchhoff's first law. But in order to see the molten iron, you would have to look through the hot vapors rising from it. Photons on their way to your spectrograph would pass through these gases, and atoms in the gases would absorb certain wavelengths. So what you would really see would be the continuous spectrum of the molten iron with weak absorption lines caused by the gases above the iron. This is what Kirchhoff's third law describes.

Now expand this argument. Suppose you had a very sensitive spectrograph that could look at the hot gases above the molten iron from the side so as to avoid looking directly at the molten iron. **What kind of spectrum would the hot gases emit?**

■   ■   ■

Connections: Whatever kind of spectrum astronomers look at, the most common spectral lines are the Balmer lines of hydrogen, the only hydrogen lines that can be studied from Earth's surface. In the next section, you will see how the Balmer lines work like a thermometer to measure a star's temperature.

■ **Figure 7-7**

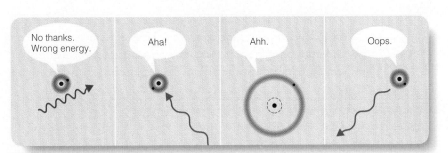

An atom can absorb a photon only if the photon has the correct amount of energy. The excited atom is unstable and within a fraction of a second returns to a lower energy level, reradiating the photon in a random direction.

Atomic Spectra

1 To understand how to analyze a spectrum, begin with a simple incandescent lightbulb. The hot filament emits black body radiation, which forms a **continuous spectrum.**

An **absorption spectrum** results when radiation passes through a cool gas. In this case you can imagine that the lightbulb is surrounded by a cool cloud of gas. Atoms in the gas absorb photons of certain wavelengths, which are missing from the spectrum, and you see their positions as dark **absorption lines.** Such spectra are sometimes called **dark-line spectra.**

An **emission spectrum** is produced by photons emitted by an excited gas. You could see **emission lines** by turning your telescope aside so that photons from the bright bulb did not enter the telescope. The photons you would see would be those emitted by the excited atoms near the bulb. Such spectra are also called **bright-line spectra.**

Telescope Spectrograph

Continuous spectrum

Gas atoms

Absorption spectrum

Emission spectrum

1a The spectrum of a star is an absorption spectrum. The denser layers of the photosphere emit black body radiation. Gases in the atmosphere of the star absorb their specific wavelengths and form dark absorption lines in the spectrum.

Absorption spectrum

GUSTAV ROBERT KIRCHHOFF
1824
1887

$\Sigma I_v = 0$
$\Sigma U_v = 0$

30

DEUTSCHE BUNDESPOST BERLIN

1b In 1859, long before scientists understood atoms and energy levels, the German scientist Gustav Kirchhoff formulated three rules, now known as **Kirchhoff's laws**, that describe the three types of spectra.

KIRCHHOFF'S LAWS

Law I: The Continuous Spectrum

A solid, liquid, or dense gas excited to emit light will radiate at all wavelengths and thus produce a continuous spectrum.

Law II: The Emission Spectrum

A low-density gas excited to emit light will do so at specific wavelengths and thus produce an emission spectrum.

Law III: The Absorption Spectrum

If light comprising a continuous spectrum passes through a cool, low-density gas, the result will be an absorption spectrum.

954.6 nm
1005.0 nm
1093.8 nm
1281.8 nm
1875.1 nm

388.9 nm
397.0 nm
410.2 nm
434.0 nm
486.1 nm
656.3 nm

93.8 nm
95.0 nm
97.2 nm
102.6 nm
121.5 nm

Paschen series
(IR)

H_ζ

Balmer series
(Visible-UV)

H_β
H_α

Lyman
series
(UV)

Nucleus

2 The electron orbits in the hydrogen atom are shown here as energy levels. When an electron makes a **transition** from one orbit to another, it changes the energy stored in the atom. In this diagram, arrows pointed inward represent transitions that result in the emission of a photon. If the arrows pointed outward, they would represent transitions that result from the absorption of a photon. Long arrows represent large amounts of energy and correspondingly short-wavelength photons.

2a Transitions in the hydrogen atom can be grouped into series—the **Lyman series**, **Balmer series**, **Paschen series**, and the like. Transitions and the resulting spectral lines are identified by Greek letters. Only the first few transitions in the first three series are shown at left.

2b In this drawing (right) of the hydrogen spectrum, emission lines in the infrared and ultraviolet are shown as gray. Only the first three lines of the Balmer series are visible to human eyes.

2c Excited clouds of gas in space emit light at all of the Balmer wavelengths, but you see only the red, blue, and violet photons blending to create the pink color typical of ionized hydrogen.

Visual-wavelength image

AURA/NOAO/NSF

The shorter-wavelength lines in each series blend together.

3 Modern astronomers rarely work with spectra as bands of light. Spectra are usually recorded digitally, so it is easy to represent them as graphs of intensity versus wavelength. Here the artwork above the graph suggests the appearance of a stellar spectrum. The graph below reveals details not otherwise visible and allows comparison of relative intensities. Notice that dark absorption lines in the spectrum appear as dips in the curve of intensity.

Intensity

H_γ
H_β
H_α

500   600   700
Wavelength (nm)

2000 nm
1500 nm
1000 nm
500 nm
100 nm

Infrared
Visible
Ultraviolet

Paschen lines
Balmer lines
Lyman lines

H_α
H_β
H_γ

7-4 Stellar Spectra

IN LATER CHAPTERS, you will use spectra to study galaxies and planets, but here you can begin by studying the spectra of stars. Such spectra are the easiest to understand, and the nature of stars is central to the study of all celestial objects.

The Balmer Thermometer

You can use the Balmer absorption lines as a thermometer to find the temperatures of stars. From the discussion of black body radiation, you know how to estimate temperature from color—red stars are cool, and blue stars are hot. (Remember that stellar spectra tell you about the surface of the star; that's where the light comes from.) You can estimate temperature from color, but the Balmer lines give you much greater accuracy.

The Balmer thermometer works because the Balmer absorption lines are produced only by atoms whose electrons are in the second energy level. If the star is cool, there are few violent collisions between atoms to excite the electrons, and most atoms have their electrons in the ground state. If most electrons are in the ground state, they can't absorb photons in the Balmer series. As a result, you should expect to find weak Balmer absorption lines in the spectra of cool stars.

In the surface layers of stars hotter than about 20,000 K, on the other hand, there are many violent collisions between atoms, exciting electrons to high energy levels or knocking the electron clear out of some atoms; that is, some atoms are ionized. Consequently, few hydrogen atoms have their electron in the second orbit to form Balmer absorption lines, and you should expect hot stars, like cool stars, to have weak Balmer absorption lines.

At an intermediate temperature, roughly 10,000 K, the collisions are just right to excite large numbers of electrons into the second energy level. With many atoms excited to the second level, the gas absorbs Balmer wavelength photons strongly and produces strong Balmer lines.

To summarize, the strength of the Balmer lines depends on the temperature of the star's surface layers. Both hot and cool stars have weak Balmer lines, but medium-temperature stars have strong Balmer lines.

Theoretical calculations can predict just how strong the Balmer lines should be for stars of various temperatures. Such calculations are the key to finding temperatures from stellar spectra. The curve in ■ Figure 7-8a shows the strength of the Balmer lines for various stellar temperatures. You could use this as a temperature indicator, except that the curve gives two answers. A star with Balmer lines of a certain strength might have either of two temperatures, one high and one low. How do you know which is the right answer? You must examine other spectral lines to choose the correct temperature.

You have seen how the strength of the Balmer lines depends on temperature. Temperature has a similar effect on the spectral lines of other elements, but the temperature at which the lines reach maximum strength differs for each element (Figure 7-8b).

Hydrogen Balmer lines are strongest for medium-temperature stars.

Lines of ionized calcium are strongest at lower temperatures than the hydrogen Balmer lines.

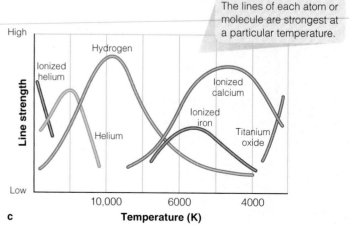

The lines of each atom or molecule are strongest at a particular temperature.

■ Figure 7-8

The strength of spectral lines can tell you the temperature of a star. (a) Balmer hydrogen lines alone are not enough because they give two answers. Balmer lines of a certain strength could be produced by a hotter star or a cooler star. (b) Adding another atom to the diagram helps, and (c) adding many atoms and molecules to the diagram creates a precise tool to find the temperatures of stars.

If you add these elements to your graph, you get a powerful tool for finding the stars' temperatures (Figure 7-8c).

How do you use this tool? You determine a star's temperature by comparing the strengths of its spectral lines with your graph. For instance, if you recorded the spectrum of a star and found medium-strength Balmer lines and strong helium lines, you could conclude it had a temperature of about 20,000 K. But if the star had weak hydrogen lines and strong lines of ionized iron, you would assign it a temperature of about 5800 K, similar to that of the sun.

The spectra of stars cooler than about 3000 K contain dark bands produced by molecules such as titanium oxide (TiO). Because of their structure, molecules can absorb photons at many wavelengths, producing numerous, closely spaced spectral lines that blend together to form bands. These molecular bands appear only in the spectra of the coolest stars because, as mentioned before, molecules in cool stars are not subject to the violent collisions that would break up molecules in hotter stars. Consequently, the presence of dark bands in a star's spectrum indicates that the star is very cool.

From stellar spectra, astronomers have found that the hottest stars have surface temperatures above 40,000 K and the coolest about 2000 K. Compare these with the surface temperature of the sun, which is about 5800 K.

Spectral Classification

You have seen that the strengths of spectral lines depend on the surface temperature of the star. From this you can conclude that all stars of a given temperature should have similar spectra. If you learn to recognize the pattern of spectral lines produced by a 6000 K star, for instance, you need not use Figure 7-8c every time you see that kind of spectrum. You can save time by classifying stellar spectra rather than analyzing each one individually.

The first widely used classification system was devised by astronomers at Harvard during the 1890s and 1900s. One of them, Annie J. Cannon, personally inspected and classified the spectra of over 250,000 stars. The spectra were first classified in groups labeled A through Q, but some groups were later dropped, merged with others, or reordered. The final classification includes the seven major **spectral classes,** or **types,** still used today: O, B, A, F, G, K, M.*

This sequence of spectral types, called the **spectral sequence,** is important because it is a temperature sequence. The O stars are the hottest, and the temperature decreases down to the M stars, the coolest. For maximum precision, astronomers divide each spectral class into 10 subclasses. For example, spectral class A consists of the subclasses A0, A1, A2, . . . A8, A9. Next comes F0, F1, F2, and so on. This finer division gives a star's temperature to an accuracy within about 5 percent. The sun, for example, is not just a G star, but a G2 star, with a temperature of 5800 K.

Astronomers classify a star by the lines and bands in its spectrum, as shown in ■ Table 7-1. For example, if it has weak Balmer lines and lines of ionized helium, it must be an O star. This table is based on the same information shown in Figure 7-8c.

■ Figure 7-9 shows 13 stellar spectra ranging from the hottest at the top to the coolest at the bottom. Notice how special features change gradually from hot to cool stars. Although these spectra are attractive, astronomers rarely work with spectra as color images. Rather, they display spectra as graphs of intensity versus wavelength with dark absorption lines as dips in the graph (■ Figure 7-10). Such graphs show more detail than photographs. Notice also that the overall curves are similar to black body curves. The wavelength of maximum intensity is in the infrared for the coolest stars and in the ultraviolet for the hottest stars.

Compare Figures 7-9 and 7-10 and notice how the strength of spectral lines depends on temperature. Note that the Balmer lines are strongest in A stars, where the temperature is moderate but still high enough to excite the electrons in hydrogen atoms

---

*Generations of astronomy students have remembered the spectral sequence using the mnemonic "Oh, Be A Fine Girl (Guy), Kiss Me." More recent suggestions from students include, "Oh Boy, An F Grade Kills Me," and "Only Bad Astronomers Forget Generally Known Mnemonics."

■ Table 7-1 I Spectral Classes

Spectral Class	Approximate Temperature (K)	Hydrogen Balmer Lines	Other Spectral Features	Naked-Eye Example
O	40,000	Weak	Ionized helium	Meissa (O8)
B	20,000	Medium	Neutral helium	Achernar (B3)
A	10,000	Strong	Ionized calcium weak	Sirius (A1)
F	7,500	Medium	Ionized calcium weak	Canopus (F0)
G	5,500	Weak	Ionized calcium medium	Sun (G2)
K	4,500	Very weak	Ionized calcium strong	Arcturus (K2)
M	3,000	Very weak	TiO strong	Betelgeuse (M2)

■ Figure 7-9

These spectra show stars from hot O stars at the top to cool M stars at the bottom. The Balmer lines of hydrogen are strongest about A0, but the two closely spaced lines of sodium in the yellow are strongest for very cool stars. Helium lines appear only in the spectra of the hottest stars. Notice that the helium line visible in the top spectrum has nearly but not exactly the same wavelength as the sodium lines visible in cooler stars. Bands produced by the molecule titanium oxide are strong in the spectra of the coolest stars. (AURA/NOAO/NSF)

to the second energy level, where they can absorb Balmer wavelength photons. In the hotter stars (O and B), the Balmer lines are weak because the higher temperature excites the electrons to energy levels above the second or ionizes the atoms. The Balmer lines in cooler stars (F through M) are also weak but for a different reason. The lower temperature cannot excite many electrons to the second energy level, so few hydrogen atoms are capable of absorbing Balmer wavelength photons.

The spectral lines of other atoms also change from class to class. Helium is visible only in the spectra of the hottest classes, and titanium oxide bands only in the coolest. Two lines of ionized calcium increase in strength from A to K and then decrease from K to M. Because the strength of these spectral lines depends on temperature, it requires only a few moments to study a star's spectrum and determine its temperature.

Now you can learn something new about our Favorite Stars. Sirius, brilliant in the winter sky, is an A1 star; and Vega, bright overhead in the summer sky, is an A0 star. They have nearly the same temperature and color and strong Balmer lines in their spectra. The bright red star in Orion is Betelgeuse, a cool M2 star, but blue-white Rigel is a hot B8 star. Polaris, the North Star, is an F8 star a bit hotter than our sun, and Alpha Centauri, the closest star to the sun, seems to be a G2 star just like the sun.

The study of spectral types is a century old, but astronomers continue to discover new types of stars. The **L dwarfs,** found in 1998, are cooler and fainter than M stars. The L dwarfs are

■ Figure 7-10 ▶

Modern digital spectra are often represented by graphs of intensity versus wavelength with dark absorption lines appearing as sharp dips in the curves. The hottest stars are at the top and the coolest at the bottom. Hydrogen Balmer lines are strongest at about A0, while lines of ionized calcium (CaII) are strong in K stars. Titanium oxide (TiO) bands are strongest in the coolest stars. Compare these spectra with Figures 7-8c and 7-9. (Courtesy NOAO, G. Jacoby, D. Hunter, and C. Christian) clearly a different type of star. The spectra of M stars contain bands produced by metal oxides such as titanium oxide, but L dwarf spectra contain bands produced by molecules such as iron hydride (FeH) (■ Figure 7-11). The **T dwarfs,** discovered in 2000, are even cooler and fainter than L dwarfs. Their spectra show absorption by methane (CH_4) and water vapor. The development of giant telescopes and highly sensitive infrared cameras and spectrographs is allowing astronomers to find and study these coolest of stars.

The Composition of the Stars

It seems as though it should be easy to find the composition of the sun and stars just by looking at their spectra, but it turns out to be a difficult task that wasn't well understood until the 1920s.

The story of how astronomers first discovered the composition of the stars is worth telling, not only because it is the story of an important American astronomer who never got proper credit, but also because the story illustrates the temperature dependence of spectral features. The story begins in England.

As a child in England, Cecilia Payne (1900–1979) excelled in classics, languages, mathematics, and literature, but her first love was astronomy. After finishing Newnham College in Cambridge, she left England, sensing that there were no opportunities in England for a woman of science. In 1922, Payne arrived at Harvard, where she eventually earned her Ph.D., although her degree was awarded by Radcliffe because Harvard did not then admit women.

In her thesis, Payne attempted to relate the strength of the absorption lines in stellar spectra to the physical conditions in the atmospheres of the stars. This was not easy because, as you have seen in this chapter, a given spectral line can be weak because the atom is rare or because the temperature is too high or too low for that atom to be able to absorb efficiently. If you see sodium lines in a star's spectrum, you can be sure that the star contains sodium atoms, but if you see no sodium lines, you must consider the possibility that the star is too hot or too cool for sodium to produce spectral lines.

Payne's problem was to untangle these two factors and find the true temperatures of the stars and the true abundance of the atoms in their atmospheres. Recent advances in atomic physics gave her the theoretical tools she needed. About the time Payne left Newnham College, Indian physicist Meghnad Saha published his work on the ionization of atoms. Drawing from such theoretical work, Payne was able to show

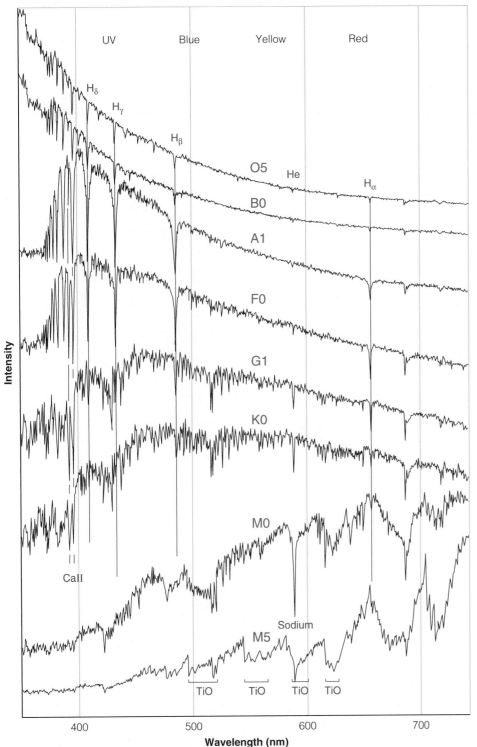

These six infrared spectra show the dramatic differences between L dwarfs and T dwarfs. Spectra of M stars show titanium oxide bands (TiO), but L and T dwarfs are so cool that TiO molecules do not form. Other molecules such as iron hydride (FeH), water (H_2O), and methane (CH_4) can form in these very cool stars. (Adapted from Thomas R. Geballe, Gemini Observatory, from a graph that originally appeared in *Sky and Telescope Magazine,* February 2005, p. 37.)

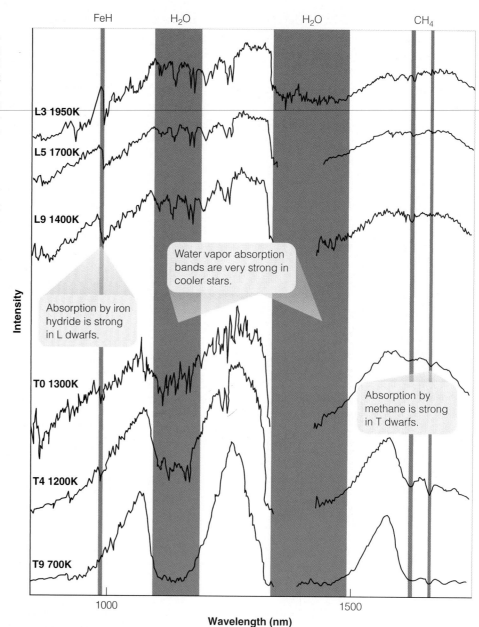

that over 90 percent of the atoms in stars (including the sun) were hydrogen and most of the rest helium (■ Table 7-2). The heavier atoms like calcium, sodium, and iron seem more abundant only because they are better at absorbing photons at the temperatures of stars.

At the time, astronomers found it hard to believe Payne's abundances of hydrogen and helium. They especially found the abundance of helium unacceptable. After all, hydrogen lines are at least visible in most stellar spectra, but helium lines are almost invisible in the spectra of all but the hottest stars. Rather, nearly all astronomers assumed that the stars had roughly the same composition as Earth's surface; that is, they believed that the stars were composed mainly of heavier atoms such as carbon, silicon, iron, aluminum, and so on. Even the most eminent astronomers dismissed Payne's result as illusory. Faced with this pressure and realizing the limited opportunities available to women in science in the 1920s, Payne could not press her discovery.

■ **Table 7-2 | The Most Abundant Elements in the Sun**

Element	Percentage by Number of Atoms	Percentage by Mass
Hydrogen	91.0	70.9
Helium	8.9	27.4
Carbon	0.03	0.3
Nitrogen	0.008	0.1
Oxygen	0.07	0.8
Neon	0.01	0.2
Magnesium	0.003	0.06
Silicon	0.003	0.07
Sulfur	0.002	0.04
Iron	0.003	0.1

It was 1929 before astronomers generally understood the importance of temperature in measurements of composition derived from stellar spectra. At that point, astronomers recognized that stars are mostly hydrogen and helium, but Payne received no credit.

Payne worked for many years as a staff astronomer at the Harvard College Observatory with no formal position on the faculty. She married Russian astronomer Sergei Gaposchkin in 1934 and was afterward known as Cecilia Payne-Gaposchkin. In 1956, when Harvard accepted women to its faculty, she was appointed a full professor and chair of the Harvard astronomy department.

Cecilia Payne-Gaposchkin's work on the chemical composition of the stars illustrates the importance of fully understanding the interaction between light and matter. Only a detailed understanding of the physics could lead her to the correct composition. As you turn your attention to other information that can be derived from stellar spectra, you will again discover the importance of understanding light.

Ace Astronomy™ Log into AceAstronomy and select this chapter to see Astronomy Exercise "Stellar Atomic Absorption Lines." You can change the chemical composition of a star and watch its spectrum change.

The Doppler Effect

Surprisingly, one of the pieces of information hidden in a spectrum is the velocity of the light source. Astronomers can measure the wavelengths of lines in a star's spectrum and find the velocity of the star. The **Doppler effect** is the apparent change in the wavelength of radiation caused by the motion of the source.

You can detect the Doppler shift in sound. Sound is not electromagnetic radiation, of course; it is a mechanical wave transmitted through air, but because it is a wave, it is subject to the Doppler effect. Sounds with long wavelengths have low pitches, and sounds with short wavelengths have higher pitches. When a car is approaching you on the highway, you hear its sound waves with a slightly shorter wavelength, and they have a higher pitch. As the car passes you, the pitch drops, and you hear slightly longer wavelengths. That's the Doppler effect.

If a star is moving toward Earth, the lines in its spectrum will be shifted slightly toward shorter wavelengths. That is, they are shifted toward the blue end of the spectrum—a **blueshift.** If a star is moving away from Earth, the lines are shifted slightly toward the red end of the spectrum—a **redshift.** These Doppler shifts are small and don't change the color of the star. But it is easy to detect these changes in wavelength in a star's spectrum.

■ Figure 7-12a shows the Doppler effect in two spectra of the star Arcturus. The lines in the top spectrum are slightly blueshifted because the spectrum was recorded when Earth, following its orbit, was moving toward Arcturus. The lines in the bottom spectrum are redshifted because it was recorded six months later, when Earth was moving away from Arcturus. The greater the shift, the greater the velocity, so astronomers can measure velocities this way.

You may be quite familiar with the Doppler effect. Meteorologists use Doppler radar to measure the velocities of clouds and weather systems. A large redshift close to a large blueshift is a warning of a possible tornado. Or perhaps you have had your own velocity measured by police radar as you drive down the highway. Radar can even be used in sports to measure the speed of a baseball or tennis ball. It's just redshifts and blueshifts.

The Doppler effect tells you how rapidly the distance between you and the source of light is increasing or decreasing. It does not matter whether you are moving or the star is moving. Only the relative velocity is important. Also, the Doppler shift is sensitive only to the part of the velocity directed away from you or toward you. This part of the velocity is called the **radial velocity** (V_r), as shown in ■ Figure 7-13.

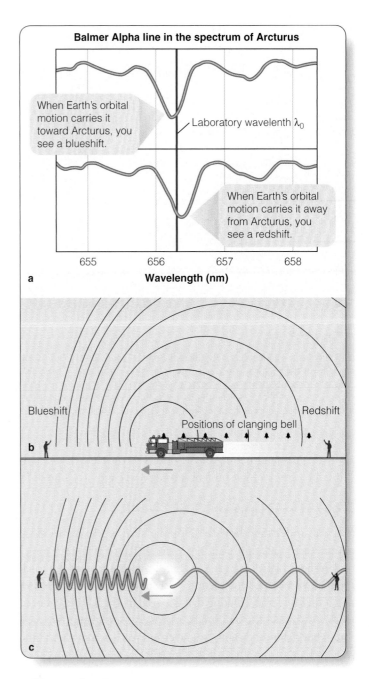

■ Figure 7-12

The Doppler effect. (a) Earth's orbital motion causes small Doppler sifts in a star's spectrum. (b) The clanging bell on a moving fire truck produces sounds that move outward (black circles). An observer ahead of the truck hears the clangs closer together, while an observer behind the truck hears them farther apart. (c) A moving source of light emits waves that move outward (black circles). An observer in front of the light source observes a shorter wavelength (a blueshift), and an observer behind the light source observes a longer wavelength (a redshift).

The Doppler effect occurs for any kind of radiation, not just light. Radio astronomers, for instance, observe the Doppler effect when they measure the wavelength of radio waves coming from objects moving toward or away from Earth. Whatever wavelength range you consider—radio, X rays, gamma rays—you can

■ Figure 7-13

(a) Police radar can measure only the radial part of your velocity (V_r) as you drive down the highway, not your true velocity along the pavement (V). That is why police using radar never park far from the highway. (b) From Earth, astronomers can use the Doppler effect to measure the radial velocity (V_r) of a star, but they cannot measure its true velocity, V, through space.

will observe a longer wavelength. This is shown in Figure 7-12c, where the peaks of the waves appear compressed in front of the moving light source and stretched out behind the source.

Of course, how great the change in wavelength is depends on the velocity. Just as a slowly moving car has a smaller Doppler effect in sound than does a high-speed airplane, a slowly moving star has a smaller Doppler effect in light than does a high-speed star. Consequently, you can measure the velocity by measuring the amount by which the spectral lines are shifted in wavelength. A large Doppler shift means a high radial velocity.

Calculating the Doppler Velocity

It is easy to calculate the radial velocity of an object from its Doppler shift. The formula is a simple ratio relating the radial velocity V_r divided by the speed of light c to the change in wavelength, λ, divided by the unshifted wavelength, λ_0:

$$\frac{V_r}{c} = \frac{\Delta\lambda}{\lambda_0}$$

For example, suppose you observed a line in a star's spectrum with a wavelength of 600.1 nm. Laboratory measurements show that the line should have a wavelength of 600 nm. That is, its unshifted wavelength is 600 nm. What is the star's radial velocity? First note that the change in wavelength is 0.1 nm:

$$\frac{V_r}{c} = \frac{0.1}{600} = 0.000167$$

Multiplying by the speed of light, 3×10^5 km/s, gives the radial velocity, 50 km/s. Because the wavelength is shifted to the red (lengthened), the star must be receding.

Now that you understand the Doppler shift you can understand a final illustration of the information hidden in stellar spectra. Even the shapes of the spectral lines can reveal secrets about the stars.

Ace ◆ Astronomy™ Log into AceAstronomy and select this chapter to see Astronomy Exercise "Doppler Shift." Compare the Doppler shift for a source of sound with that for a source of light.

The Shapes of Spectral Lines

When astronomers refer to the shape of a spectral line, they mean the variation of intensity across the line. An absorption line, for instance, is darkest in the center and brighter to each side. Two examples are shown in ■ Figure 7-14.

The exact shape of a line can reveal a great deal about a star, but the most important characteristic is the width of the line. Spectral lines are not perfectly narrow; if they were, they would be undetectable. They have a natural width because nature allows an atom some leeway in the energy it may absorb or emit. In the absence of all other effects, spectral lines have a natural width of about 0.001 to 0.00001 nm—very narrow indeed.

The natural widths of spectral lines are not important in most branches of astronomy because other effects smear out the

still refer to a lengthening of the observed wavelength as a redshift and a shortening of the observed wavelength as a blueshift.

How the Doppler Shift Works

You can understand how the Doppler shift works by thinking about a fire truck approaching with a bell clanging once a second. When the bell clangs, the sound travels ahead of the truck to reach your ears. One second later, the bell clangs again, but not at the same place. During that one second, the fire truck moved closer to you, so the bell is closer at its second clang. Now the sound has a shorter distance to travel and reaches your ears a little sooner than it would have if the fire truck were not approaching. The third time the bell clangs, it is even closer. By timing the bell, you could observe that the clangs are slightly less than 1 second apart, all because the fire truck is approaching. If the fire truck were moving away from you, you would hear the clangs sounding more than one second apart, because each successive clang of the bell would occur farther away.

Figure 7-12b shows a fire truck moving toward one observer and away from another observer. The position of the bell at each clang is shown by the small black bells with the sound spreading outward as black circles. You can see that the clangs are squeezed together ahead of the fire truck and stretched apart behind.

Now you can substitute a source of light waves for the clanging of the bell. If the source is approaching, then each time the source emits the peak of a wave it will be slightly closer to you, and you will observe a shorter wavelength. If it is moving away, you

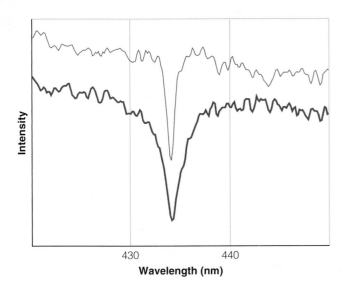

■ Figure 7-14

Here you see two dark absorption lines magnified from the spectra of two A1 stars. The upper line is quite narrow, but the bottom line is much broader. Because the two stars have the same spectral type, you can trust that they must have the same temperature. The stars differ not in temperature but in gas density. The star with the narrow spectral lines has a very low-density atmosphere. Precise observations of the shapes of spectral lines can reveal a great deal about stars. (Courtesy NOAO, G. Jacoby, D. Hunter, and C. Christian)

lines and make them much broader. For example, if a star spins rapidly, the Doppler effect will broaden the spectral lines. As the star rotates, one side will recede from Earth, and the other side will approach Earth. Light from the receding side will be redshifted, and light from the approaching side will be blueshifted, so any spectral lines will be broadened. Astronomers can measure a star's rotation rate from the width of its spectral lines.

Another important process is called **Doppler broadening.** To consider this process, imagine that you photograph the spectrum of a jar full of hydrogen atoms (■ Figure 7-15). Because the gas has some thermal energy (it is not at absolute zero), the

gas atoms are in motion. Some will be coming toward your spectrograph, and some will be receding. Most, of course, will not be traveling very fast, but some will be moving very quickly. The photons emitted by the atoms approaching you will have slightly shorter wavelengths because of the Doppler effect, and photons emitted by atoms receding from you will have slightly longer wavelengths. Thus, the Doppler shifts due to the motions of the individual atoms will smear the spectral line out and make it broader. This summary describes the Doppler broadening of an emission line, but the effect is the same for absorption lines.

The extent of Doppler broadening depends on the temperature of the gas. If the gas is cold, the atoms travel at low velocities, and the Doppler shifts are small (Figure 7-15a). If the gas is hot, however, the atoms travel faster, Doppler shifts are larger, and the lines will be wider (Figure 7-15b). Sometimes astronomers will estimate the temperature of a cloud of gas in space by looking at the widths of its spectral lines.

Another form of broadening, **collisional broadening,** is caused by collisions between atoms, and consequently it depends on the density of the gas (**Focus on Fundamentals 4**). Densities in astronomy cover an enormous range, from one atom per cubic centimeter in space to millions of tons of atoms per cubic centimeter inside dead stars. Clearly, you would expect such densities to affect the way atoms collide with one another and how they absorb and emit photons.

Collisional broadening spreads out spectral lines when the atoms absorb or emit photons while they are colliding with other

■ Figure 7-15

Doppler broadening. The atoms of a gas are in constant motion. Photons emitted by atoms moving toward the observer will have slightly shorter wavelengths, and those emitted by atoms moving away will have slightly longer wavelengths. This broadens the spectral line. If the gas is cool (a), the atoms do not move very fast, the Doppler shifts are small, and the line is narrow. If the gas is hot (b), the atoms move faster, the Doppler shifts are larger, and the line is broader.

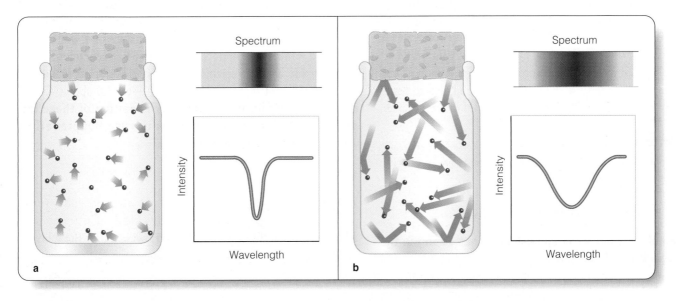

Density

You are about as dense as an average star. What does that mean? As you study astronomy, you will use the term *density* often, so you should be sure to understand this fundamental concept. **Density** is a measure of the amount of matter in a given volume. Density is expressed as mass per volume, such as grams per cubic centimeter. The density of water, for example, is about 1 g/cm³, and you are almost as dense as water.

To get a feel for density, imagine holding a brick in one hand and a similar-sized block of Styrofoam in the other hand. You can easily tell that the brick contains more matter than the Styrofoam block, even though both are the same size. The brick weighs more than the Styrofoam, but it isn't really the weight that you should consider. Rather, you should think about the mass of the two objects. In space, where they have no weight, the brick and the Styrofoam would still have mass, and you could tell just by moving them around that the brick contains more mass than the Styrofoam. For example, imagine tapping each object gently

against your ear. The massive brick would be easy to distinguish from the low-mass Styrofoam block, even in weightlessness.

When you think of density, you divide mass by volume, and your mind makes that comparison at an instinctive level. You sense the density of an object just by handling it. Gift shops sometimes sell imitation rocks made of Styrofoam as humorous gifts. "Rocks" made of Styrofoam seem odd when you han-

A brick would be dense even in space where it had no weight.

dle them because your brain expects rocks to be dense.

Density is a fundamental idea in science because it is a general property of materials. Metals tend to be dense; lead, for example, has a density of about 7 g/cm³. Rock, in contrast, has a density of 3 to 4 g/cm³. Water and ice have densities of about 1 g/cm³. If you knew that a small moon orbiting Saturn had a density of 1.5 g/cm³, you could immediately draw some conclusions about what kinds of materials the little moon might be made of—ice and a little rock, but not much metal. The density of an object is a basic clue to its composition.

Astronomical bodies can have dramatically different densities. The gas in a nebula can have a very low density, but the same kind of gas in a star can have a much higher density. The sun, for example, has an average density of about 1 g/cm³, about the same as your body. As you study astronomical objects, pay special attention to their densities. Density is fundamental.

PRESSURE | MASS | ENERGY | TEMPERATURE AND HEAT | **DENSITY**

---

atoms, ions, or electrons. The collisions disturb the energy levels in the atoms, making it possible for the atoms to absorb a slightly wider range of wavelengths. Because of this, the spectral lines are wider. Because atoms in a dense gas collide more often than atoms in a low-density gas, collisional broadening depends on the density of the gas. Temperature is also an important factor. Atoms in a hot gas travel faster and collide more often and more violently than atoms in a cool gas. The two spectral lines in Figure 7-14 illustrate the effect of density.

Once again, the physics of the interaction of light and matter provides a tool to understand starlight. In later chapters, you will see how astronomers use the widths of spectral lines to better understand clouds of gas in space, stars, and even distant galaxies.

(Ace ⓢ Astronomy™) Log into AceAstronomy and select this chapter to see Astronomy Exercise "Stellar Rotation." See an interesting application of the Doppler shift.

Building Scientific Arguments

Why are helium lines weak and calcium lines strong in the visible spectrum of the sun?

To analyze this problem, you must recall that the ability of an atom or ion to absorb light depends on temperature. Helium is quite abundant in the sun, but the surface of the sun is too cool to excite helium atoms and enable them to easily absorb visible-wavelength photons. On the other hand, calcium is easily ionized, and calcium atoms that have lost an electron are very good absorbers of photons at temperatures like those at the sun's surface. Consequently, calcium lines are strong in the solar spectrum even though calcium ions are rare, and helium lines are weak even though helium atoms are common.

When you create a scientific argument, you must include all of the important factors. In this case, both composition and temperature are important. **But why is neither of these factors important when astronomers use the Doppler effect to measure a star's radial velocity?**

■　　■　　■

Connections: The spectra of the stars are filled with clues about the gas emitting the light. Much of astronomy is based on unraveling these clues. You will begin in the following chapter by studying the nearest star—the sun.

Summary

7-1 | Starlight

How do stars produce light?

- Stars emit light from the hot gases at their surfaces. Light from deeper layers does not escape.

- Temperature is a measure of the average kinetic energy of the particles in a gas. In a hot gas, the particles move fast: in a cool gas, they move slowly.

- Thermal energy is the total kinetic energy in the particles of a material. Heat is the flow of thermal energy from hot regions to cool regions.

- In a hot material, the rapidly moving atoms or molecules collide with electrons, and when electrons are accelerated, part of the energy is emitted as electromagnetic radiation. This is known as black body radiation.

- The hotter a black body is, the more it radiates and the shorter is its wavelength of maximum intensity, λ_{max}. This allows astronomers to estimate the temperatures of stars from their colors.

7-2 | Atoms

What is an atom?

- An atom consists of a nucleus surrounded by a cloud of electrons. The nucleus is made up of positively charged protons and uncharged neutrons.

- The number of protons in an atom determines which element it is. Atoms of the same element (that is, having the same number of protons) with different numbers of neutrons are called isotopes.

- A neutral atom is surrounded by a number of negatively charged electrons equal to the number of protons in the nucleus. An atom that has lost or gained an electron is called an ion.

- The electrons in an atom may occupy various permitted orbits around the nucleus but not orbits in between.

7-3 | The Interaction of Light and Matter

How do atoms interact with light?

- The size of an electron's orbit depends on the energy stored in the motion of the electron.

- An electron may be excited to a higher orbit during a collision between atoms, or it may move from one orbit to another by absorbing or emitting a photon of the proper energy.

- Kirchhoff's laws describe the formation of spectra. A hot solid, liquid or dense gas will emit black body radiation, producing a continuous spectrum. An excited low-density gas will produce an emission (bright-line) spectrum. Light passing through a cool, low-density gas will form an absorption (dark-line) spectrum.

7-4 | Stellar Spectra

What kind of spectra do you see when you look at celestial objects?

- Because orbits of only certain energies are permitted in an atom, photons of only certain wavelengths can be absorbed or emitted. Each kind of atom has its own characteristic set of spectral lines. The hydrogen atom has the Lyman series of lines in the ultraviolet, the Balmer series in the visible, and the Paschen series (and others) in the infrared.

- If light passes through a low-density gas on its way to your telescope, the gas can absorb photons of certain wavelengths, and you see dark lines in the spectrum at those positions. Such a spectrum is called an absorption spectrum.

- If you look at a low-density gas that is excited to emit photons, you see bright lines in the spectrum. Such a spectrum is called an emission spectrum.

What can you learn from a star's spectrum?

- Nearly all stellar spectra are absorption spectra, and the hydrogen lines seen there are Balmer lines. In cool stars, the Balmer lines are weak because atoms are not excited out of the ground state. In hot stars, the Balmer lines are weak because atoms are excited to higher orbits or are ionized. Only at medium temperatures are the Balmer lines strong.

- The strength of spectral lines in a star's spectrum can tell you its temperature. In its simplest form, this amounts to classifying the star's spectrum in the spectral sequence: O, B, A, F, G, K, M.

- Long after the spectral sequence was created, astronomers found the L and T stars at temperatures even cooler than the M stars.

- A spectrum can reveal the chemical composition of the stars. The presence of spectral lines of a certain element shows that that element must be present in the star. But astronomers must proceed with care. Lines of a certain element may be weak or absent if the star is too hot or too cool.

- The wavelengths of spectral lines provide clues to the motions of the stars. When a star is approaching, you observe slightly shorter wavelengths (a blueshift); when it is receding, you observe slightly longer wavelengths (a redshift). This Doppler effect reveals a star's radial velocity, that part of its velocity directed toward or away from Earth.

- The shape of a spectral line can reveal the rotation of a star because one side of the star approaches Earth while the other side recedes. The Doppler shift broadens the star's spectral lines.

- Doppler broadening can make spectral lines wider because the atoms in a hot gas move very rapidly and at any moment some are receding and some are approaching.

- Collisional broadening occurs in a dense gas where the atoms collide often. This widens spectral lines.

New Terms

temperature (p. 138)	quantum mechanics (p. 143)
Kelvin temperature scale (p. 138)	permitted orbit (p. 143)
absolute zero (p. 138)	energy level (p. 144)
electron (p. 138)	excited atom (p. 144)
thermal energy (p. 139)	ground state (p. 145)
black body radiation (p. 139)	continuous spectrum (p. 146)
wavelength of maximum intensity (λ_{max}) (p. 139)	absorption spectrum (dark-line spectrum) (p. 146)
nucleus (p. 141)	absorption line (p. 146)
proton (p. 141)	emission spectrum (bright-line spectrum) (p. 146)
neutron (p. 141)	
isotope (p. 142)	emission line (p. 146)
ionization (p. 142)	Kirchhoff's laws (p. 146)
ion (p. 142)	transition (p. 147)
molecule (p. 142)	Lyman series (p. 147)
Coulomb force (p. 142)	Balmer series (p. 147)
binding energy (p. 143)	Paschen series (p. 147)

spectral class or type (p. 149)

spectral sequence (p. 149)

L dwarf (p. 150)

T dwarf (p. 151)

Doppler effect (p. 153)

blueshift (p. 153)

redshift (p. 153)

radial velocity (V_r) (p. 153)

Doppler broadening (p. 155)

collisional broadening (p. 155)

density (p. 156)

13. The nebula shown at right is mostly low-density hydrogen excited to emit photons. What kind of spectrum would you expect this nebula to produce?

14. If the nebula in the image here crosses in front of the star and the nebula and star have different radial velocities, what might the spectrum of the star look like?

(T. Rector, University of Alaska, and WIYN/NURO/AURA/NSF)

Review Questions

Ace◐Astronomy™ Assess your understanding of this chapter's topics with additional quizzing and animations at **http://ace .brookscole.com/sf9**

1. Why can a good blacksmith judge the temperature of a piece of heated iron by its color?

2. Why do hot stars look bluer than cool stars?

3. Use black body radiation to explain why you could hold a cigarette between your lips and light it with a match, but you couldn't hold the cigarette in your lips while you light it by sticking the tip into a bonfire. The match and bonfire have about the same temperature. (This is just one of the hazards of smoking.)

4. What is the difference between a neutral atom, an ion, and an excited atom?

5. How do the energy levels in an atom determine which wavelength photons it can absorb or emit?

6. What kind of spectrum would you expect to record if you observed molten lava? In practice, to view molten lava you must look through gases boiling out of the lava. What kind of spectrum might you see in that case?

7. Why do the strengths of the Balmer lines depend on the temperature of the star?

8. Why does a stellar spectrum tell you about the surface layers but not the deeper layers of the star?

9. Explain the similarities among Figure 7-8, Figure 7-9, Figure 7-10, and Table 7-1.

10. Why would you not expect to see TiO bands in the spectra of hot stars?

11. Imagine that you observed a star's spectrum and found that the lines of a certain element were not present. Would you be safe in concluding that the star did not contain this element? Why or why not?

12. If a star moves exactly perpendicular to a line connecting it to Earth, will the Doppler effect change its spectrum? Why or why not?

13. Why would you expect a star that rotates very rapidly to have broad spectral lines? What else could broaden the lines?

Discussion Questions

1. In what ways is the model of an atom used in this chapter a scientific model? How can you use it when it is not a completely correct description of an atom?

2. Can you think of classification systems commonly used to simplify what would otherwise be very complex measurements? Consider foods, movies, cars, grades, clothes, and so on.

Problems

1. Human body temperature is about 310 K (98.6°F). At what wavelength do humans radiate the most energy? What kind of radiation do we emit?

2. If a star has a surface temperature of 20,000 K, at what wavelength will it radiate the most energy?

3. Infrared observations of a star show that it is most intense at a wavelength of 2000 nm. What is the temperature of the star's surface?

4. If you double the temperature of a black body, by what factor will the total energy radiated per second per square meter increase?

5. If one star has a temperature of 6000 K and another star has a temperature of 7000 K, how much more energy per second will the hotter star radiate from each square meter of its surface?

6. Transition A produces light with a wavelength of 500 nm. Transition B involves twice as much energy as A. What wavelength light does it produce?

7. Determine the temperatures of the following stars based on their spectra. Use Figure 7-8.
   a. medium-strength Balmer lines, strong helium lines
   b. medium-strength Balmer lines, weak ionized calcium lines
   c. TiO bands very strong
   d. very weak Balmer lines, strong ionized calcium lines

8. To which spectral classes do the stars in Problem 7 belong?

9. In a laboratory, the Balmer beta line has a wavelength of 486.1 nm. If the line appears in a star's spectrum at 486.3 nm, what is the star's radial velocity? Is it approaching or receding?

10. The highest-velocity stars an astronomer might observe have velocities of about 400 km/s. What change in wavelength would this cause in the Balmer gamma line? (*Hint:* Wavelengths are given on page 147).

Media Cluster

ACTIVE FIGURES

Ace◐Astronomy™ To access the resources in the Media Cluster, log into AceAstronomy at **http://ace.brookscole .com/sf9** and select Chapter 7.

Kirchhoff's Laws
Kirchhoff studied and identified three different spectra in his experiments. Use this animation to study the conditions for producing each of the types of spectra.

ASTRONOMY EXERCISES

Black Body
In this animation, you can change the temperature of an object and see how its black body curve changes.

Stefan–Boltzmann Law
In this introduction to the Stefan–Boltzmann law, watch the luminosity of a star change as you adjust its temperature.

Emission and Absorption Spectra
This exercise will give you an inside look at atoms as they interact with photons. The photons are absorbed by the atom, sending the electron to a higher energy level.

Doppler Shift
Recall that the Doppler effect is the change in wavelength of radiation due to relative radial motion of source and observer. You can compare the Doppler shift for a source of sound with that for a source of light in this animation.

Stellar Rotation
This animation lets you increase or decrease the rotation speed of a star and see the effect on the star's spectrum.

Stellar Atomic Absorption Lines
In this exercise, you can build your own star. Select from the type and ratio of element mixes in your new star.

VIRTUAL ASTRONOMY LABS

Lab 2: Properties of Light and Its Interaction with Matter
This lab examines the wave properties of electromagnetic radiation and how different regions of the electromagnetic spectrum are related. It ends by looking at the interaction between matter and radiation on the atomic scale.

Lab 11: The Spectral Sequence and the H–R Diagram
This lab introduces the tools astronomers use to identify the stars of the main sequence and how that information is used to estimate the distance to the stars and the age of a star cluster.

Lab 3: The Doppler Effect
This lab provides a brief review of some basic properties of waves and investigates the shift in observed wavelengths of sound and light waves caused by the motion of an emitting source with respect to an observer.

Critical Inquiries for the Web

1. The name for the element helium has astronomical roots. Search the Internet for information on the discovery of helium. How and when was it discovered, and how did it get its name? Why do you suppose it took so long for helium to be recognized?

2. This chapter contains a model for an atom that may seem familiar to you. How was that model developed? What was the plum pudding model? Search the Web for information on historical models of the atom and compile a time line of important developments leading to our current understanding.

3. What can you find out about Anne J. Cannon and Cecilia Payne-Gaposchkin?

Exploring *TheSky*

1. Locate the following stars, click on them, and determine their spectral types: Antares in Scorpius, Betelgeuse in Orion, Aldebaran in Taurus, Sirius in Canis Major, Rigel in Orion.

2. How are spectral types correlated with the colors of stars? (*Hint:* Locate Orion and choose **Spectral Colors** under the **View** menu.)

Go to the Brooks/Cole Astronomy Resource Center **(http://astronomy.brookscole.com)** for critical thinking exercises, articles, and additional readings from InfoTrac College Edition, Brooks/Cole's online student library.

8 | The Sun

UV image

A WIT ONCE REMARKED that solar astronomers would know a lot more about the sun if it were farther away, and that contains a grain of truth; the sun is only a humdrum star, and although there are billions like it in the sky, the sun is the only one close enough to show surface detail. Solar astronomers can see so much detail in the swirling currents of gas and arching bridges of magnetic force that present theories seem inadequate to describe it. Yet the sun is not a complicated object. It is just a star. ▍ In their general properties, stars are very simple. They are great balls of hot gas held together by their own gravity. Their gravity would make them collapse into small, dense bodies were they not so hot. The tremendously hot gas inside stars has such a high pressure that the stars would surely explode were it not for their own confining gravity.

▍ Continued on page 162 ▍

160

Guidepost

Looking Back

The interaction of light and matter, which you studied in Chapter 7, is a powerful tool. It can reveal the secrets of the stars. The next step is to apply that tool to the sun.

This Chapter

In this chapter, you will discover how little astronomers could know about the sun were it not for the analysis of its spectrum. Just a bit of spectrographic ingenuity reveals that the brilliance of the sun hides a complex atmosphere of hot gases, powered by nuclear reactions and churned by powerful storms.

This chapter will help you answer four essential questions:

What do you see when you look at the sun?

How does the sun make its energy?

What are the dark sunspots and other forms of solar activity?

Why does the sun go through a cycle of activity?

Perhaps the most important question answered here is more general: How can astronomers learn about the sun and stars?

Looking Ahead

This is the first chapter that applies the methods of science to understand a celestial body. The tools developed in this chapter will help you explore stars, galaxies, and planets through the rest of this book.

Ace Astronomy™ The AceAstronomy icon throughout the text indicates an opportunity for you to test yourself on key concepts and to explore animations and interactions on the AceAstronomy website at: http://ace .brookscole.com/sf9

Like soap bubbles, stars are simple structures balanced between opposing forces that individually would destroy them. You can study our sun as a close-up example of a star.

Another reason to study the sun is that life on Earth depends critically on the sun. Very small changes in the sun's luminosity can alter Earth's climate, and a slightly larger change might make Earth uninhabitable. Nearly all of our energy comes from the sun—oil and coal are merely stored sunlight. Furthermore, the sun's atmosphere of very thin gas reaches out past Earth's orbit, and any change in the sun, such as an eruption or a magnetic storm, can have a direct effect on Earth.

Finally, you should study the sun because it is beautiful. Your analysis of sunlight will reveal that the sun is both powerful and delicate. So you should study the sun not only because it is *a star,* not only because it is *our star,* but because it is the sun.

8-1 The Solar Atmosphere

WHEN YOU WATCH the sun set in the west, you see a glowing disk. What are you seeing? Here you will see the sun sketched in its general properties and describe the surface you see. Later you can plunge inside to see how it makes its energy and then return to the surface to study the storms and eruptions not generally visible to the unaided eye. What do you see when you look at the sun? You see a star.

The sun is 109 times Earth's diameter and 333,000 times Earth's mass. This seems dramatic, but look at **Celestial Profile 1** and notice the sun's density. It is only a little bit denser than water. So, although the sun is very large and very massive, it must be a gas from its surface to its center. When you look at the sun you see only the outer layers of this vast sphere of gas. In fact, these outer layers, the solar atmosphere, extend high above the visible surface of the sun.

Heat Flow in the Sun

When you look at the sun you see a hot, glowing surface, and simple logic tells you that energy in the form of heat is flowing outward from the sun's interior. The solar spectrum reveals that the sun is a G2 star with a temperature of about 5800 K. At that temperature, every square millimeter of the sun's surface must be radiating more energy than a 60-watt lightbulb. With all that energy radiating into space, the sun's surface would cool rapidly if energy did not flow up from the interior to keep the surface hot.

Not until the 1930s did astronomers understand how the sun makes its energy. Nuclear reactions occur in the core of the sun and generate energy, which flows upward as heat and keeps the surface hot. These nuclear reactions are discussed in detail later in this chapter, but here you should notice the importance of the energy flowing outward through the sun's surface.

As you study the atmosphere of the sun, you will find many phenomena that are driven by the energy flowing outward. Like a pot of boiling soup on a hot stove, the surface of the sun is in constant activity as the heat flows up from below.

For now, you can consider the sun in its quiescent, unchanging state, and explore the layers in its atmosphere.

The Photosphere

The visible surface of the sun looks like a smooth layer of gas marked by a few dark **sunspots.** Although the photosphere seems to be a distinct surface, it is not solid. In fact, the sun is gaseous from its outer atmosphere right down to its center. The photosphere is the thin layer of gas from which Earth receives most of the sun's light. It is less than 500 km deep and has an average temperature of about 5800 K. If the sun magically shrank to the size of a bowling ball, the photosphere would be no thicker than a layer of tissue paper wrapped around the ball. For comparison, the chromosphere lies above the photosphere and is only a few times thicker in extent, but the corona, beginning above the chromosphere, extends far above the visible surface (■ Figure 8-1). Recall that you first met the terms *photosphere, chromosphere,* and *corona* in connection with solar eclipses in Chapter 3.

Below the photosphere, the gas is denser and hotter and therefore radiates plenty of light, but that light cannot escape from the sun because of the outer layers of gas. So you cannot detect light from these deeper layers. Above the photosphere, the gas is less dense and so is unable to radiate much light. The photosphere is the layer in the sun's atmosphere that is dense enough to emit plenty of light but not so dense that the light can't escape.

One reason the photosphere is so shallow is related to the hydrogen atom. Because the temperature of the photosphere is sufficient to ionize some atoms, there are a large number of free electrons in the gas. Neutral hydrogen atoms can add an extra electron and become an H⁻ (H-minus) ion, but this extra electron is held so loosely that almost any photon has energy enough to free it. In the process, of course, the photon is absorbed. That makes the H⁻ ions very good absorbers of photons and makes the gas of the photosphere opaque. Light from below cannot escape easily, and you see a well-defined surface glowing like a layer of hot fog—the photosphere.

Although the photosphere appears to be substantial, it is really a very-low-density gas. Even in the deepest and densest layers visible, the photosphere is 3400 times less dense than the air you breathe. To find gases as dense as the air you breathe, you would have to descend about 70,000 km below the photosphere, about 10 percent of the way to the sun's center. With fantastically efficient insulation, you could fly a spaceship right through the photosphere.

The spectrum of the sun is an absorption spectrum, and that can tell you a great deal about the photosphere. You know from Kirchhoff's third law that an absorption spectrum is produced when a source of a continuous spectrum is viewed through a gas. In the case of the photosphere, the deeper layers are dense enough to produce a continuous spectrum, but atoms in the photosphere

Chromosphere

Photosphere

a

Corona

b Visual-wavelength image

■ **Figure 8-1**

(a) A cross section at the edge of the sun shows the relative thickness of the photosphere and chromosphere. Earth is shown for scale. On this scale, the disk of the sun would be more than 1.5 m (5 feet) in diameter. (b) The corona extends from the top of the chromosphere to great height above the photosphere. (b) This photograph, made during a total solar eclipse, shows only the inner part of the corona. (Daniel Good)

absorb photons of specific wavelengths, producing the absorption lines you see.

In good photographs, the photosphere has a mottled appearance because it is made up of dark-edged regions called granules, and the visual pattern is called **granulation** (■ Figure 8-2a). Each granule is about the size of Texas and lasts for only 10 to 20 minutes before fading away. Faded granules are continuously replaced by new granules. Spectra of these granules show that the centers are a few hundred degrees hotter than the edges, and Doppler shifts reveal that the centers are rising and the edges are sinking at speeds of about 0.4 km/second.

From this evidence, astronomers recognize granulation as the surface effects of convection just below the photosphere.

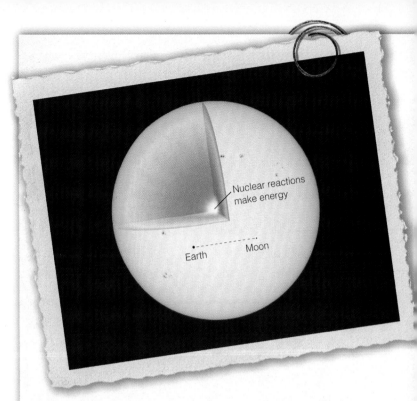

Nuclear reactions make energy

Earth   Moon

This visual-wavelength image of the sun shows a few sunspots and is cut away to show the location of energy generation at the sun's center. The Earth–moon system is shown for scale. (Dan Good)

Celestial Profile 1: The Sun

From Earth:

Average distance from Earth	1.00 AU (1.495979 × 10^8 km)
Maximum distance from Earth	1.0167 AU (1.5210 × 10^8 km)
Minimum distance from Earth	0.9833 AU (1.4710 × 10^8 km)
Average angular diameter	0.53° (32 minutes of arc)
Period of rotation	25.38 days at equator
Apparent visual magnitude	−26.74

Characteristics:

Radius	6.9599 × 10^5 km
Mass	1.989 × 10^{30} kg
Average density	1.409 g/cm^3
Escape velocity at surface	617.7 km/s
Luminosity	3.826 × 10^{26} J/s
Surface temperature	5800 K
Central temperature	15 × 10^6 K
Spectral type	G2 V
Absolute visual magnitude	4.83

Personality Profile:

In Greek mythology, the sun was carried across the sky in a golden chariot pulled by powerful horses and guided by the sun-god Helios. When Phaeton, son of Helios, drove the chariot one day, he lost control of the horses, and Earth was nearly set ablaze before Zeus smote Phaeton from the sky. Even in classical times, people understood that life on Earth depends critically on the sun.

a Visual-wavelength image

b

Granule

Sinking gas

Rising gas

■ **Figure 8-2**

(a) This ultra-high-resolution image of the photosphere shows granulation. The largest granules here are about the size of Texas. (P. N. Brandt, G. Scharmer, G. W. Simon, Swedish Vacuum Solar Telescope) (b) This model explains granulation as the tops of rising convection currents just below the photosphere. Heat flows upward as rising currents of hot gas and downward as sinking currents of cool gas. The rising currents heat the solar surface in small regions seen from Earth as granules.

Convection occurs when hot fluid rises and cool fluid sinks, as when, for example, a convection current of hot gas rises above a candle flame. You can create convection in a liquid by adding a bit of cool nondairy creamer to an unstirred cup of hot coffee. The cool creamer sinks, warms, rises, cools, sinks again, and so on, creating small regions on the surface of the coffee that mark the tops of convection currents. Viewed from above, these regions look much like solar granules.

In the sun, rising currents of hot gas heat small regions of the photosphere, which, being slightly hotter, emit more black body radiation and look brighter. The cool sinking gas of the edges emits less light and thus looks darker (Figure 8-2b). The presence of granulation is clear evidence that energy is flowing upward through the photosphere.

Spectroscopic studies of the solar surface have revealed another kind of granulation. **Supergranules** are regions about 30,000 km in diameter (about 2.3 times Earth's diameter) and include about 300 granules. These supergranules are regions of very slowly rising currents that last a day or two. They may be the surface traces of larger currents of rising gas deeper under the photosphere.

The edge, or **limb,** of the solar disk is dimmer than the center (see the figure in Celestial Profile 1). This **limb darkening** is caused by the absorption of light in the photosphere. When you look at the center of the solar disk, you are looking directly down into the sun, and you see deep, hot, bright layers in the photosphere. But when you look near the limb of the solar disk, you are looking at a steep angle and cannot see as deeply. The photons you see come from shallower, cooler, dimmer layers in the photosphere. Limb darkening proves that the temperature in the photosphere increases with depth, as you would expect if energy is flowing up from below.

The Chromosphere

Above the photosphere lies the chromosphere. Solar astronomers define the lower edge of the chromosphere as lying just above the visible surface of the sun with its upper regions blending gradually with the corona. You can think of the chromosphere as being an irregular layer with a depth on average less than Earth's diameter (see Figure 8-1). Because the chromosphere is roughly 1000 times fainter than the photosphere, you can see it with your unaided eyes only during a total solar eclipse when the moon covers the brilliant photosphere. Then, the chromosphere flashes into view as a thin line of pink just above the photosphere. The word *chromosphere* comes from the Greek word *chroma,* meaning "color." The pink color is produced by the combined light of three bright emission lines—the red, blue, and violet Balmer lines of hydrogen.

Astronomers know a great deal about the chromosphere from its spectrum. The chromosphere produces an emission spectrum, and Kirchhoff's second law tells you the chromosphere must be an excited, low-density gas. The density is about 10^8 times less dense than the air you breathe.

Atoms in the lower chromosphere are ionized, and atoms in the higher layers of the chromosphere are even more highly ionized. That is, they have lost more electrons. From this, astronomers can find the temperature in different parts of the chromosphere. Just above the photosphere the temperature falls to a minimum of about 4500 K and then rises rapidly (■ Figure 8-3). The region where the temperature increases fastest is called the **transition region** because it makes the transition from the lower temperatures of the photosphere and chromosphere to the extremely high temperatures of the corona.

Solar astronomers can take advantage of some elegant physics to study the chromosphere. The gases of the chromosphere are transparent to nearly all visible light, but atoms in the gas are

■ Figure 8-3

The chromosphere. If you could place thermometers in the sun's atmosphere, you would discover that the temperature increases from 5800 K at the photosphere to 10^6 K at the top of the chromosphere.

■ Figure 8-4 shows a filtergram made at the wavelength of the H_α Balmer line. This image shows complex structure in the chromosphere including long, dark **filaments** silhouetted against the brighter surface. **Spicules** are flamelike jets of gas extending upward into the chromosphere and lasting 5 to 15 minutes. Seen at the limb of the sun's disk, these spicules blend together and look like flames covering a burning prairie (Figure 8-1a), but they are not flames at all. Spectra show that spicules are cooler gas from the lower chromosphere extending upward into hotter regions. Images at the center of the solar disk show that spicules spring up around the edge of supergranules like weeds around flagstones (Figure 8-4). Although spicules are not yet well understood, they are clearly driven by the outward flow of energy in the sun.

Spectroscopic analysis of the chromosphere alerts you that it is a low-density gas in constant motion where the temperature increases rapidly with height. Just above the chromosphere lies even hotter gas.

The Solar Corona

The outermost part of the sun's atmosphere is called the corona, after the Greek word for crown. The corona is so dim that it is not visible in Earth's daytime sky because of the glare of scattered light from the brilliant photosphere. During a total solar eclipse, however, when the moon covers the photosphere, you can see the innermost parts of the corona, as shown in Figure 8-1b. Observations made with specialized telescopes called **coronagraphs** on Earth or in space can block the light of the photosphere and image the corona out beyond 20 solar radii, almost 10 percent of the

very good at absorbing photons of specific wavelengths. This produces certain dark absorption lines in the spectrum of the photosphere. A photon at one of those wavelengths is very unlikely to escape from deep layers. A **filtergram** is a photograph made using light in one of those dark absorption lines. Those photons can only have escaped from higher in the atmosphere. In this way filtergrams reveal detail in the upper layers of the chromosphere. In a similar way, an image recorded in the far-ultraviolet or in the X-ray part of the spectrum reveals other structures in the solar atmosphere.

■ Figure 8-4

H_α filtergrams reveal complex structure in the chromosphere, including long, dark filaments and spicules springing from the edges of supergranules twice the diameter of Earth. (NOAA/SEL/USAF; NOAO/NSO)

Two nearly simultaneous images show sunspots in the photosphere and excited regions in the chromosphere above the sunspots.

Visual-wavelength image

Ultraviolet

Twisted streamers in the corona suggest magnetic fields.

The corona extends far from the disk.

Background stars

Sun hidden behind mask

Visual image

Visual image

Sun hidden behind mask

■ Figure 8-5

Images of the photosphere, chromosphere, and corona show the relationships among the layers of the sun's atmosphere. The visual-wavelength image shows the sun in white light—that is, as you would see it with your eyes. (SOHO/ESA/NASA)

way to Earth. Such images reveal that magnetic fields link the sunspots with features in the chromosphere and corona (■ Figure 8-5).

The spectrum of the corona can tell you a great deal about the coronal gases and simultaneously illustrate how astronomers can analyze a spectrum. Some of the light from the outer corona produces a spectrum with absorption lines the same as the sun's spectrum. This light is just sunlight reflected from dust particles in the corona. In contrast, some of the light from the corona produces a continuous spectrum that lacks absorption lines, and that happens when sunlight from the photosphere is scattered off free electrons in the ionized coronal gas. Because the coronal gas has a temperature over 1 million K and the electrons travel very fast, the reflected photons suffer large, random Doppler shifts that smear out solar absorption lines to produce a continuous spectrum.

Superimposed on the corona's continuous spectrum are emission lines of highly ionized gases. In the lower corona, the atoms are not as highly ionized as they are at higher altitudes, and this tells you that the temperature of the corona rises with altitude. Just above the chromosphere, the temperature is about 500,000 K; but in the outer corona the temperature can be as high as 2 million K or more.

The spectrum of the corona reveals that it is exceedingly hot gas, but it is not very bright. Its density is very low, only 10^6 atoms/cm^3 in its lower regions. That is about a trillion times less dense than the air you breath. In its outer layers the corona contains only 1 to 10 atoms/cm^3, better than the best vacuum on Earth. Because of this low density, the hot gas does not emit much radiation.

Astronomers have wondered for years how the corona and chromosphere can be so hot. Heat flows from hot regions to cool regions, never from cool to hot. So how can the heat from the photosphere, with a temperature of only 5800 K, flow out into the much hotter chromosphere and corona? Observations made by the SOHO satellite have mapped a **magnetic carpet** of looped magnetic fields extending up through the photosphere (■ Figure 8-6). Turbulence below the surface may be whipping these fields about and heating the gases of the chromosphere and corona. Remember that the gas of the chromosphere and corona has very low density, so it can't resist the moving magnetic fields. The gas gets whipped about as the magnetic fields flick back and forth, and that heats the gas. In this instance, energy appears to flow outward as the agitation of the magnetic fields.

Gas from the solar atmosphere follows the magnetic fields pointing outward and flows away from the sun in a breeze called the **solar wind.** Like an extension of the corona, the low-density gases of the solar wind blow past Earth at 300 to 800 km/s with gusts as high as 1000 km/s. Earth is bathed in the corona's hot breath.

■ Figure 8-6

■ Figure 8-6

This extreme-ultraviolet image of a section of the sun's lower corona has been given a green color. White and black areas are regions of opposite magnetic polarity. Computer models show the location of the magnetic fields that connect these regions. The largest loops shown here could encircle Earth. The entire surface of the sun is covered by this magnetic carpet. (Stanford-Lockheed Institute for Space Research, Palo Alto, CA, and NASA GSFC)

Because of the solar wind, the sun is slowly losing mass, but this is only a minor loss for an object as massive as the sun. The sun loses about 10^7 tons per year, but that is only 10^{-14} of a solar mass per year. Later in life, the sun, like many other stars, will lose mass rapidly. You will see in future chapters how this affects stars.

Do other stars have chromospheres, coronae, and stellar winds like the sun? Ultraviolet and X-ray observations suggest that the answer is yes. The spectra of many stars contain emission lines in the far-ultraviolet that could only have formed in the low-density, high-temperature gases of a chromosphere and corona. Also, many stars are sources of X rays, which appear to have been produced by coronae. This observational evidence gives astronomers good reason to believe that the sun, for all its complexity, is a typical star.

Helioseismology

Almost no light emerges from below the photosphere, so you can't see into the solar interior. However, solar astronomers can use the vibrations in the sun to explore its depths in a process called **helioseismology.** Random motions in the sun constantly produce vibrations—rumbles too low to hear with human ears. Some of these vibrations resonate in the sun like sound waves in organ pipes. A vibration with a period of 5 minutes is strongest, but the periods range from 3 to 20 minutes. These are very, very low-pitched rumbles!

Astronomers can detect these vibrations by observing Doppler shifts in the solar surface. As a vibrational wave travels down into the sun, the increasing density and temperature curve its path and it returns to the surface, where it makes the photosphere heave up and down by small amounts—roughly plus or minus 15 km (■ Figure 8-7a). Short-wavelength waves penetrate less deeply and travel shorter distances than longer-wavelength waves. This covers the surface of the sun with a pattern of rising and falling regions that can be mapped using the Doppler effect (Figure 8-7b). By observing these motions, astronomers can determine which vibrations resonate the strongest. Just as

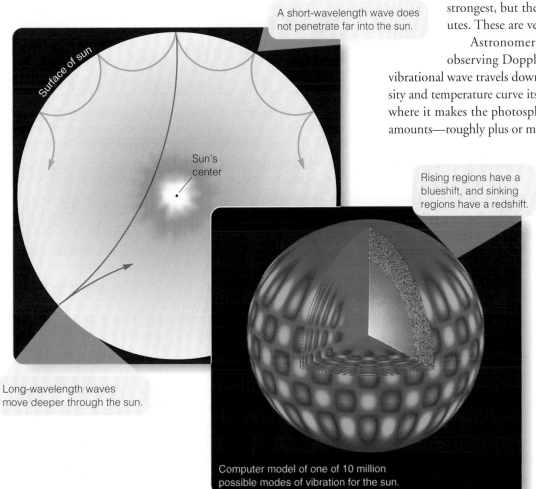

A short-wavelength wave does not penetrate far into the sun.

Surface of sun

Sun's center

Rising regions have a blueshift, and sinking regions have a redshift.

Long-wavelength waves move deeper through the sun.

Computer model of one of 10 million possible modes of vibration for the sun.

■ Figure 8-7

Helioseismology: The sun can vibrate in millions of different patterns or modes, and each mode corresponds to a different wavelength vibration penetrating to a different level. By measuring Doppler shifts as the surface moves gently up and down, astronomers can map the inside of the sun. (AURA/NOAO/NSF) geologists can study Earth's interior by analyzing vibrations from earthquakes, so solar astronomers can use helioseismology to explore the sun's interior.

Helioseismology sounds almost magical, but you can understand it better if you think of a duck pond. If you stood at the shore of a duck pond and looked down at the water, you would see ripples arriving from all parts of the pond. Because every duck on the pond contributes to the ripples, you could, in principle, study the ripples near the shore and draw a map showing the position and velocity of every duck on the pond. Of course, it would be difficult to untangle the different ripples, so you would need lots of data and a big computer. Nevertheless, all of the information would be there, lapping at the rocks at your feet.

The sun can oscillate in about 10 million different ways, and each mode of oscillation has its own characteristic wavelength and its own unique pattern on the solar surface. The waves producing different modes penetrate to different depths where conditions can weaken or strengthen a wave. By discovering which wavelength waves are actually present, solar astronomers can determine the temperature, density, pressure, composition, and motion at different depths inside the sun.

Of course, with 10 million possible wavelengths, the observations and analysis are difficult. A single wave produces a complicated pattern of motion on the solar surface. Large amounts of data are necessary, so helioseismologists have used a network of telescopes around the world operated by the Global Oscillation Network Group (GONG). The network can observe the sun continuously for weeks at a time as Earth rotates. The sun never sets on GONG. The SOHO satellite in space can observe solar oscillations continuously and can detect motions as slow as 1 mm/s (0.002 mph). Solar astronomers can then use high-speed computers to separate the different patterns on the solar surface and measure the strength of the waves at many different wavelengths.

Helioseismology has allowed astronomers to map the temperature, density, and rate of rotation inside the sun. They have been able to detect great currents of gas flowing below the photosphere and the emergence of sunspots before they appear in the photosphere. Helioseismology can even locate sunspots on the back side of the sun, sunspots that are not yet visible from Earth.

Building Scientific Arguments

How deeply into the sun can you see?

Scientific arguments usually involve observations, and it is always important to know how observations are made. When you look into the layers of the sun, your sight does not really penetrate into the sun. Rather, your eyes record photons that have escaped from the sun and traveled outward through the layers of the sun's atmosphere. If you observe at a wavelength at the center of a dark absorption line, then the photosphere and lower chromosphere are opaque, photons can't escape to your eyes, and the only photons you can see come from the upper chromosphere. What you see are the details of the upper chromosphere. On the other hand, if you observe at a wavelength that is not easily absorbed (a wavelength between spectral lines), the atmosphere is more transparent, and photons from deep inside the photosphere can escape to your eyes. There is a limit, however, set by the H^- ion, a hydrogen atom with an extra electron. At a certain depth, there is so much of this ion that the sun's atmosphere is opaque for almost all wavelengths, few photons can escape, and you can't see deeper.

By choosing the proper wavelength, solar astronomers can observe to different depths. But the corona is so thin and the gas below the photosphere so dense that this method doesn't work in these regions. Now it is time to build a new argument. **How can you observe the corona and the deeper layers of the sun?**

■   ■   ■

Connections: So far you have thought of the sun as a static, unchanging ball of gas with energy flowing outward from the interior and through the atmospheric layers. Now you are ready to plunge into the sun and find out how that energy is made.

(8-2) Nuclear Fusion in the Sun

ASTRONOMERS OFTEN USE the wrong words to describe energy generation in the sun and stars. Astronomers will say, "The star ignites hydrogen burning." In English, the word *ignite* means *catch on fire,* and *burn* means *on fire.* What goes on inside stars isn't really burning in the usual sense.

The sun is a star, and it is not burning. It is powered by nuclear reactions that occur near its center. The energy produced keeps the interior hot, and the gas is totally ionized. That is, the electrons are not attached to the atomic nuclei, and the gas is an atomic soup of rapidly moving particles colliding with each other at high velocity. When you discuss nuclear reactions inside the sun and stars, you should be careful to refer to atomic nuclei and not to atoms.

How exactly can the nucleus of an atom yield energy? The answer lies in the forces that hold the nuclei together.

Nuclear Binding Energy

The sun generates its energy by breaking and reconnecting the bonds between the particles *inside* atomic nuclei. This is quite different from the way you would generate energy by burning wood in a fireplace. The process of burning wood extracts energy by breaking and reconnecting chemical bonds between atoms in the wood. Chemical bonds are formed by the electrons in atoms, and you saw in Chapter 7 that the electrons are bound to the atoms by the electromagnetic force. So chemical energy originates in the electromagnetic force.

There are only four forces in nature: the force of gravity, the electromagnetic force, the **weak force,** and the **strong force.** The weak force is involved in the radioactive decay of certain kinds of nuclear particles, and the strong force binds together atomic nuclei. That means nuclear energy must come from the strong force.

Nuclear power plants on Earth generate energy through **nuclear fission** reactions that split uranium nuclei into less massive fragments. A uranium nucleus contains a total of 235 protons and neutrons, and it splits into a range of fragments containing roughly half as many particles. Because the fragments produced are more tightly bound than the uranium nuclei, binding energy is released during uranium fission.

Stars don't use nuclear fission. They make energy in **nuclear fusion** reactions that combine light nuclei into heavier nuclei. The most common reaction, including that in the sun, fuses hydrogen nuclei (single protons) into helium nuclei (two protons and two neutrons). Because the nuclei produced are more tightly bound than the original nuclei, energy is released. Notice in ▪ Figure 8-8

■ **Figure 8-8**

The red line in this graph shows the binding energy per particle, the energy that holds particles inside an atomic nucleus. The horizontal axis shows the atomic mass number of each element, the number of protons and neutrons in the nucleus. Both fission and fusion nuclear reactions move downward in the diagram (arrows) toward more tightly bound nuclei. Iron has the most tightly bound nucleus, so no nuclear reactions can begin with iron and release energy.

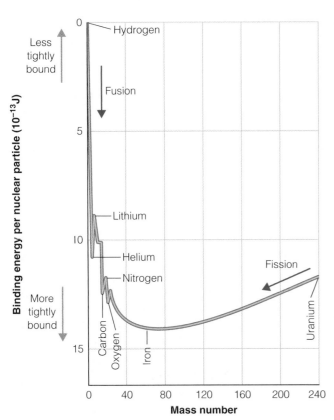

that both fusion and fission reactions move downward in the diagram toward more tightly bound nuclei. They both produce energy by releasing the binding energy of atomic nuclei.

Hydrogen Fusion

Not until the 1930s did astronomers realize how the sun generates energy. The sun fuses together four hydrogen nuclei to make one helium nucleus. Because one helium nucleus has 0.7 percent less mass than four hydrogen nuclei, it seems that some mass vanishes in the process. To see this, subtract the mass of a helium nucleus from the mass of four hydrogen nuclei.

$$\begin{aligned} 4 \text{ hydrogen nuclei} &= 6.693 \times 10^{-27} \text{ kg} \\ - 1 \text{ helium nucleus} &= 6.645 \times 10^{-27} \text{ kg} \\ \hline \text{Difference in mass} &= 0.048 \times 10^{-27} \text{ kg} \end{aligned}$$

It seems from this subtraction that a small amount of mass disappears, but it doesn't really vanish; it merely changes form. The equation $E = m_0 c^2$ reminds you that mass and energy are related, and under certain circumstances mass may become energy and vice versa. So the 0.048×10^{-27} kg does not vanish but merely becomes energy. To see how much, you use Einstein's equation:

$$\begin{aligned} E &= m_0 c^2 \\ &= (0.048 \times 10^{-27} \text{ kg})(3 \times 10^8 \text{ m/s})^2 \\ &= 0.43 \times 10^{-11} \text{ J} \end{aligned}$$

This is a very small amount of energy, hardly enough to raise a housefly one-thousandth of an inch. Because one reaction produces such a small amount of energy, it is obvious that many reactions are necessary to supply the energy needs of a star. The sun, for example, needs 10^{38} reactions per second, transforming 5 million tons of mass into energy every second, just to stay hot enough to resist its own gravity.

It seems from this that nuclear fusion is very powerful, especially if you calculate that the fusion of a milligram of hydrogen (roughly the mass of a match head) produces as much energy as burning 30 gallons of gasoline. However, the nuclear reactions in the sun are spread through a large volume in its core, and any single gram of matter produces only a little energy. A person of normal mass eating a normal diet produces about 4000 times more heat per gram than the matter in the core of the sun. The sun produces a lot of energy because it contains a lot of *grams* of matter in its core.

Fusion reactions can occur only when the nuclei of two atoms get very close to each other. Because atomic nuclei carry positive charges, they repel each other with an electrostatic force called the Coulomb force. Physicists commonly refer to this repulsion between nuclei as the **Coulomb barrier.** To overcome this barrier and get close together, atomic nuclei must collide violently. Violent collisions are rare unless the gas is very hot, in which case the nuclei move at high speeds and collide violently. (Remember, an object's temperature is related to the speed with which its particles move.)

So nuclear reactions in the sun take place only near the center, where the gas is hot and dense. A high temperature insures that some of the collisions between nuclei are violent enough to overcome the Coulomb barrier, and a high density ensures that there are enough collisions, and thus enough reactions, to meet the sun's energy needs.

You can symbolize the fusion reactions in the sun with a simple nuclear reaction:

$$4\,^1\text{H} \rightarrow\, ^4\text{He} + \text{energy}$$

In this equation, ^{1}H represents a proton, the nucleus of the hydrogen atom, and ^{4}He represents the nucleus of a helium atom. The superscripts indicate the approximate weight of the nuclei (the number of protons plus the number of neutrons). The actual steps in the process are more complicated than this convenient summary suggests. Instead of waiting for four hydrogen nuclei to collide simultaneously, a highly unlikely event, the process can proceed step by step in a chain of reactions—the proton–proton chain.

The **proton–proton chain** is a series of three nuclear reactions that builds a helium nucleus by adding together protons. This process is efficient at temperatures above 10,000,000 K. The sun, for example, manufactures over 90 percent of its energy in this way.

The three steps in the proton–proton chain entail these reactions:

$$^1\text{H} +\, ^1\text{H} \rightarrow\, ^2\text{H} + e^+ + \nu$$
$$^2\text{H} +\, ^1\text{H} \rightarrow\, ^3\text{He} + \gamma$$
$$^3\text{He} +\, ^3\text{He} \rightarrow\, ^4\text{He} +\, ^1\text{H} +\, ^1\text{H}$$

In the first reaction, two hydrogen nuclei (two protons) combine to form a heavy hydrogen nucleus called **deuterium,** emitting a particle called a **positron,** e^+ (a positively charged electron), and a **neutrino,** ν (a subatomic particle having an extremely low mass and a velocity nearly equal to the velocity of light). In the second reaction, the heavy hydrogen nucleus absorbs another proton and, with the emission of a gamma ray, γ, becomes a lightweight helium nucleus. Finally, two lightweight helium nuclei combine to form a common helium nucleus and two hydrogen nuclei. Because the last reaction needs two ^{3}He nuclei, the first and second reactions must occur twice (■ Figure 8-9). The net result of this chain reaction is the transformation of four hydrogen nuclei into one helium nucleus plus energy.

The energy appears in the form of gamma rays, positrons, the energy of motion of the particles, and neutrinos. The gamma rays are photons that are absorbed by the surrounding gas before they can travel more than a fraction of a millimeter. This heats the gas and helps maintain the pressure. The positrons produced in the first reaction combine with free electrons, and both particles

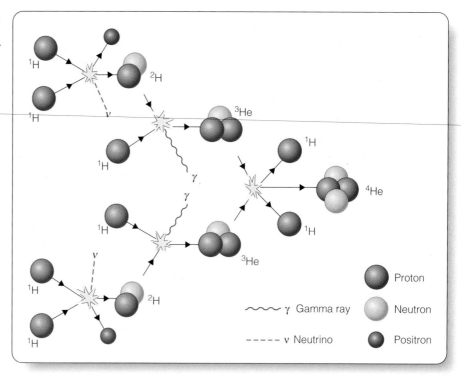

■ **Figure 8-9**

The proton–proton chain combines four protons (at far left) to produce one helium nucleus (at right). Energy is produced mostly as gamma rays and positrons, which combine with electrons and convert their mass into energy. Neutrinos escape, carrying away about 2 percent of the energy produced.

vanish, converting their mass into gamma rays, which are absorbed and also help keep the center of the star hot. In addition, when fusion produces new nuclei, they fly apart at high velocity. This energy of motion helps raise the temperature of the gas. The neutrinos resemble photons except that they almost never interact with other particles. The average neutrino could pass unhindered through a lead wall a light-year thick. Consequently, the neutrinos do not help heat the gas but race out of the star at nearly the speed of light, carrying away roughly 2 percent of the energy produced.

Energy Transport in the Sun

Now you are ready to follow the energy from the core of the sun to the surface. The surface is cool, only about 5800 K, and the center is over 10 million Kelvin, so energy must flow outward from the core.

Because the core is so hot, the photons being emitted are gamma rays. Each time a gamma ray encounters an electron, it is deflected or scattered in a random direction; and, as it bounces around, it slowly drifts outward toward the surface. That carries energy outward in the form of radiation, and astronomers refer to the inner parts of the sun as the **radiative zone.**

To examine this process, imagine picking a single gamma ray and following it to the surface. As your gamma ray is scattered over and over by the hot gas, it drifts outward into cooler layers, and the cooler gas tends to emit photons of longer wavelength. Your gamma ray will eventually be absorbed by the gas and reemitted as two X rays. Now you must follow those two X rays as they bounce around, and soon you will see them drifting outward into even cooler gas, where they will become a number of longer wavelength photons. The packet of energy that began as a single gamma ray gets broken down into a large number of lower-energy photons, and it eventually emerges from the sun's surface as about 1800 photons of visible light.

But something else happens along the way. The packet of energy that you began following from the core eventually reaches the outer layers of the sun where the gas is so cool it is not very transparent to radiation. The energy backs up like water behind a dam, and the gas begins to churn in convection. Hot blobs of gas rise, and cool blobs sink. In this region, known as the **convective zone,** the energy is carried outward as circulating gas. These two zones are shown in ■ Figure 8-10. The granulation visible on the photosphere is clear evidence that the sun has a convective zone just below the photosphere carrying energy upward to the surface.

Sunlight is nuclear energy produced in the core of the sun. The energy of a single gamma ray can take a million years to work its way outward, first as radiation and then as convection on its journey to the photosphere.

It is time to ask the critical question that lies at the heart of science. What is the evidence to support this theoretical explanation of how the sun makes its energy? The search for that evidence will introduce you to one of the great problems of modern astronomy.

Ace⊛Astronomy™ Log into AceAstronomy and select this chapter to see Astronomy Exercise "Nuclear Fusion" and take control of fusion in the sun's core.

The Solar Neutrino Problem

The center of a star seems forever hidden, but the sun is transparent to neutrinos because these subatomic particles almost never interact with normal matter. Nuclear reactions in the sun's core produce floods of neutrinos that rush out of the sun and off into space. If you could detect these neutrinos, you could probe the sun's interior.

Because neutrinos almost never interact with atoms, you never feel the flood of over 10^{12} solar neutrinos that flow through your body every second. Even at night, neutrinos from the sun rush through Earth as if it weren't there, up through your bed, through you, and onward into space. Obviously you are lucky to be transparent to neutrinos, but it means that neutrinos are extremely hard to detect. Certain nuclear reactions, however, can be triggered by a neutrino of the right energy; and, in the late 1960s, chemist Raymond Davis, Jr., began using such a reaction to detect solar neutrinos.

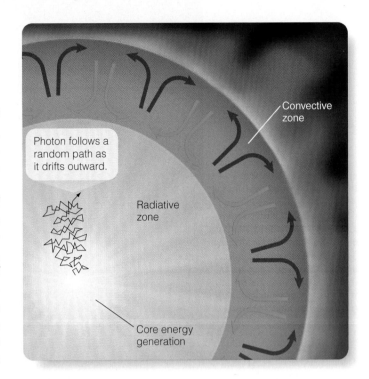

Photon follows a random path as it drifts outward.

Convective zone

Radiative zone

Core energy generation

Active Figure 8-10

A cross section of the sun. Near the center, nuclear fusion reactions generate high temperatures. Energy flows outward through the radiative zone as photons are randomly deflected over and over by electrons. In the cooler, more opaque outer layers, the energy is carried by rising convection currents of hot gas (red arrows) and sinking currents of cooler gas (blue arrows).

Ace⊛Astronomy™ Log into AceAstronomy and select this chapter to see Active Figure "The Sun." Watch energy flow outward from the core to the surface.

Davis filled a 100,000-gallon tank with the cleaning fluid perchloroethylene (C_2Cl_4). Theory predicts that about once a day, a solar neutrino will convert a chlorine atom in the tank into radioactive argon, which can be detected later by its radioactive decay. To protect the detector from cosmic rays from space, the tank was buried nearly a mile deep in a South Dakota gold mine (■ Figure 8-11a). Of course, the mile of rock overhead had no effect on the neutrinos.

The result of the Davis experiment startled astronomers. The cleaning fluid detected too few neutrinos—not one neutrino per day as predicted by models of the sun but about one every three days. The experiment was refined, tested, and calibrated for three decades; but it did not find the missing neutrinos. Other detectors were built, and they too counted too few neutrinos coming from the sun.

The missing solar neutrinos were one of the great mysteries of modern astronomy. Some scientists argued that astronomers didn't correctly understand how the sun and stars make their energy, but other scientists wondered if there was something about neutrinos that could explain the problem. Astronomers have great confidence in their theories of the sun's interior, and helio-

■ Figure 8-11

(a) The Davis solar neutrino experiment used cleaning fluid and could detect only one of the three flavors of neutrinos. (Brookhaven National Laboratory) (b) The Sudbury Neutrino Observatory is a 12-meter-diameter globe containing water rich in deuterium in place of hydrogen. Buried 6800 feet deep in an Ontario mine, it can detect all three flavors of neutrinos and confirms that neutrinos oscillate. (Photo courtesy of SNO)

seismology confirmed those theories, so astronomers did not abandon their theories immediately (**Window on Science 8-1**).

As the 21st century began, scientists were able to solve the mystery. Physicists know of three kinds of neutrinos, which they call flavors. The Davis experiment could detect (or taste) only one flavor, electron-neutrinos. Theory hinted that the electron-neutrinos produced in the core of the sun might oscillate among the three flavors as they rushed out through the sun and across space to Earth. Observations begun in 2000 confirm this theory (Figure 8-11b). Some of the electron-neutrinos produced in the sun transform into tau- and muon-neutrinos, which most detectors cannot detect.

This solution to the solar neutrino problem is exciting because neutrinos can't oscillate unless they have mass. Neutrinos were long thought to be massless, but if they have even a small mass, they are so common their gravity could affect the evolution of the universe as a whole—something you will read about in Chapter 18. The detection of neutrino oscillation excites astronomers for another reason. It confirms the theories that describe the interior of the sun and stars.

Building Scientific Arguments

Why does nuclear fusion require that the gas be very hot? This argument has to include some basic physics of atoms and thermal energy. Inside a star, the gas is so hot it is ionized, which means the electrons have been stripped off the atoms, and the nuclei are bare and have a positive charge. For hydrogen fusion, the nuclei are single protons. These atomic nuclei repel each other because of their positive charges, so they must collide with each other at high velocity to overcome that repulsion and get close enough together to fuse. If the atoms in a gas are moving rapidly, then it must have a

high temperature, and so nuclear fusion requires that the gas have a very high temperature. If the gas is cooler than about 10 million K, hydrogen can't fuse because the protons don't collide violently enough to overcome the repulsion of their positive charges.

It is easy to see why nuclear fusion in the sun requires high temperature, but now expand your argument. **Why does it require high density?**

■　■　■

Connections: You have been to the center of the sun and seen how it makes its energy, and you have followed that energy up to the surface. Now it is time for you to see how outrageous the sun can be as that energy bursts through the photosphere.

(8-3) Solar Activity

THE SUN IS UNQUIET. It is home to slowly changing spots larger than Earth and vast eruptions that dwarf human imagination. All of these seemingly different forms of solar activity have one thing in common—magnetic fields. The weather on the sun is magnetic.

Observing the Sun

Solar activity is often visible with even a small telescope, but you should be very careful about observing the sun. Sunlight is intense, and when it enters your eye it is absorbed and converted into thermal energy. Equally dangerous is the infrared radiation in sunlight. Your eyes can't detect the infrared, but it is converted to thermal energy in your eyes and can burn and scar the retina.

Avoiding Hasty Judgments: Scientific Confidence

Scientists like to claim that every scientific understanding is based on evidence, that every theory has been tested, and that the moment a theory fails a test, it is discarded or revised. The truth is much more complicated than that, and the solar neutrino problem is a good illustration. If the detection of too few solar neutrinos contradicted the theory of stellar structure, why wasn't the theory abandoned?

While scientists do indeed have tremendous respect for evidence, they also have confidence in theories that have been tested successfully many times. If a theory has been tested and confirmed over and over, they may even begin to call it a natural law, and that means scientists have great confidence that the theory or law is a good description of how nature works.

Nevertheless, it is not unusual for an experiment or an observation to contradict well-established theories. In many cases, the experiments and observations are simple mistakes,

or they have not been interpreted correctly. Scientists resist abandoning a well-tested theory even when an observation continues to contradict it. If confidence in the theory is stronger than the evidence, scientists begin by testing the evidence. Can it be right? Do we understand it correctly? Of course, if the evidence cannot be impeached and it continues to contradict the theory, scientists must eventually abandon or modify the theory no matter how many times it has previously been tested and confirmed. This ultimate reliance on evidence is the distinguishing characteristic of science.

It is only human nature to hang on to the principles you have come to trust, and that confidence in well-tested scientific principles helps scientists avoid rushing to faulty judgments. For example, claims for perpetual motion machines occasionally crop up in the news, but the world's scientists don't instantly abandon the laws of energy and motion pending an analysis of the latest claim. Of course, if such

a claim did prove true, the entire structure of scientific knowledge would come crashing down, but because the known laws of energy and motion have been well tested and no perpetual motion machine has ever been successful, scientists know which way to bet. Like the keel on a ship, confidence in well-tested theories and laws keeps the scientific boat from rocking before every little breeze.

Ultimately scientific confidence must be open to change. If, eventually, a single experiment or observation conclusively contradicts a cherished law of nature, scientists must abandon that law and find a new way to understand nature. You can see this scientific confidence at work in many controversies, from the origin of the human race, to the meaning of IQ measurements; but one of the best examples is the solar neutrino problem. While astronomers struggled to understand the origin of solar neutrinos, they continued to have confidence that they do indeed understand how stars make energy.

It is not safe to look directly at the sun, and it is even more dangerous to look at the sun through any optical instrument such as a telescope, binoculars, or even the viewfinder of a camera. The light-gathering power of such an optical system concentrates the sunlight and can cause severe injury. Never look at the sun with any optical instrument unless you are certain it is safe. ■ Figure 8-12 shows a safe way to observe the sun with a small telescope.

In the early 17th century, Galileo observed the sun and saw spots on its surface; day by day he saw the spots moving across the sun's disk. He rightly concluded that the sun was a sphere and was rotating. You could repeat his observations, and you would probably see something that looks like Figure 8-12b. You would see sunspots.

Sunspots

The dark sunspots that you see at visible wavelengths only hint at the complex processes that go on in the sun's atmosphere. To explore those processes, you must turn to the analysis of images and spectra at a wide range of wavelengths.

Study **Sunspots and the Sunspot Cycle** on pages 174–175 and notice five important points:

1 Sunspots are cool spots on the sun's surface caused by strong magnetic fields.

2 Also notice that sunspots follow an 11-year cycle not only in the number of spots visible but in their location on the sun.

3 The Zeeman effect gives astronomers a way to measure the strength of magnetic fields on the sun.

4 Notice that the sunspot cycle can vary over centuries and appears to affect Earth's climate.

5 Finally, notice the clear evidence that sunspots are part of a larger magnetic process that involves all layers of the sun's atmosphere.

The sunspot groups are merely the visible traces of magnetically active regions. But what causes this magnetic activity? The answer appears to be linked to the waxing and waning of the sun's overall magnetic field.

Ace ◉Astronomy™ Log into AceAstronomy and select this chapter to see Astronomy Exercise "Zeeman Effect" and see how astronomers measure magnetic fields.

Ace ◉Astronomy™ Log into AceAstronomy and select this chapter to see Astronomy Exercise "Sunspot Cycle I"; you can explore the link between sunspots and the sun's magnetic field.

Ace ◉Astronomy™ Log into AceAstronomy and select this chapter to see Astronomy Exercise "Sunspot Cycle II." You can observe sunspots over the centuries.

Sunspots and the Sunspot Cycle

1 The dark spots that appear on the sun are only the visible traces of complex regions of activity. Observations over many years and at a range of wavelengths tell you that sunspots are clearly linked to the sun's magnetic field.

Spectra show that sunspots are cooler than the photosphere with a temperature of about 4240 K. The photosphere has a temperature of about 5800 K. Because the total amount of energy radiated by a surface depends on its temperature raised to the fourth power, sunspots look dark in comparison. Actually, a sunspot emits quite a bit of radiation. If the sun were removed and only an average-size sunspot were left behind, it would be brighter than the full moon.

A typical sunspot is about twice the size of Earth, but there is a wide range of sizes. They appear, last a few weeks to as long as two months, and then shrink away. Usually, sunspots occur in pairs or complex groups.

Size of Earth

NASA

Royal Swedish Academy of Sciences

Umbra

Penumbra

Sunspots are not shadows, but astronomers refer to the dark core of a sunspot as its umbra and the outer, lighter region as the penumbra.

2 The number of spots visible on the sun varies in a cycle with a period of 11 years. At maximum, there are often over 100 spots visible. At minimum, there are very few.

2a Early in the cycle, spots appear at high latitudes north and south of the sun's equator. Later in the cycle, the spots appear closer to the sun's equator. If you plot the latitude of sunspots versus time, the graph looks like butterfly wings, as shown in this **Maunder butterfly diagram**, named after E. Walter Maunder of Greenwich Observatory.

3 Astronomers can measure magnetic fields on the sun using the **Zeeman effect** as shown below. When an atom is in a magnetic field, the electron orbits are altered, and the atom is able to absorb a number of different wavelength photons even though it was originally limited to a single wavelength. In the spectrum, you see single lines split into multiple components, with the separation between the components proportional to the strength of the magnetic field.

Sunspot groups

Magnetic fields around sunspot groups

Ultraviolet filtergram

Magnetic image

Simultaneous images

J. Harvey/NSO and HAO/NCAR

Slit allows light from sunspot to enter spectrograph.

Visual

AURA/NOAO/NSF

Spectral line split by Zeeman effect

3a Images of the sun above show that sunspots contain magnetic fields a few thousand times stronger than Earth's. The strong fields are believed to inhibit gas motion below the photosphere; consequently, convection is reduced below the sunspot, and the surface there is cooler. Heat prevented from emerging through the sunspot is deflected and emerges around the sunspot, which can be detected in infrared images.

Maunder minimum few spots colder winters

Winter severity in London and Paris

Warmer winters

↑ Warm
↓ Cold

Number of sunspots (y-axis: 0, 50, 100, 150, 200, 250, 300, 350)

Year (x-axis: 1650, 1700, 1750, 1800, 1850, 1900, 1950, 2000)

4 Historical records show that there were very few sunspots from about 1645 to 1715, a phenomenon known as the **Maunder minimum.** This coincides with a period called the "little ice age," a period of unusually cool weather in Europe and North America from about 1500 to about 1850, as shown in the graph at left. Other such periods of cooler climate are known. The evidence suggests that there is a link between solar activity and the amount of solar energy Earth receives. This link has been confirmed by measurements made by spacecraft above Earth's atmosphere.

M. Seeds

Magnetic fields can reveal themselves by their shape. For example, iron filings sprinkled over a bar magnet reveal an arched shape.

The complexity of an active region becomes visible at short wavelengths.

Far -UV image

5 Observations at nonvisible wavelengths reveal that the chromosphere and corona above sunspots are violently disturbed in what astronomers call **active regions.** Spectrographic observations show that active regions contain powerful magnetic fields.

Arched structures above an active region are evidence of gas trapped in magnetic fields.

Visual-wavelength image

Simultaneous images

Far-UV image

SOHO/EIT, ESA and NASA

NASA/TRACE

(a) Looking through a telescope at the sun is dangerous, but you can always view the sun safely with a small telescope by projecting its image on a white screen. (b) If you sketch the location and structure of sunspots on successive days, you will see the rotation of the sun and gradual changes in the size and structure of sunspots just as Galileo did in 1610.

The Sun's Magnetic Cycle

Sunspots are magnetic phenomena, so the 11-year cycle of sunspots must be caused by cyclical changes in the sun's magnetic field. To explore that idea, begin with the sun's rotation.

The sun does not rotate as a rigid body. It is a gas from its outermost layers down to its center, so some parts of the sun rotate faster than other parts. The equatorial region of the photosphere rotates faster than do regions at higher latitudes (■ Figure 8-13a). At the equator, the photosphere rotates once every 25 days, but at latitude 45° one rotation takes 27.8 days. Helioseismology can map the rotation throughout the interior (Figure 8-13b). This phenomenon is called **differential rotation,** and it is clearly linked with the magnetic cycle.

(a) In general, the photosphere of the sun rotates faster at the equator than at higher latitudes. If you started five sunspots in a row, they would not stay lined up as the sun rotates. (b) Detailed analysis of the sun's rotation from helioseismology reveals regions of slow rotation (blue) and rapid rotation (red). Such studies show that the interior of the sun rotates differentially and that currents similar to the trade winds in Earth's atmosphere flow through the sun. (NASA/SOI)

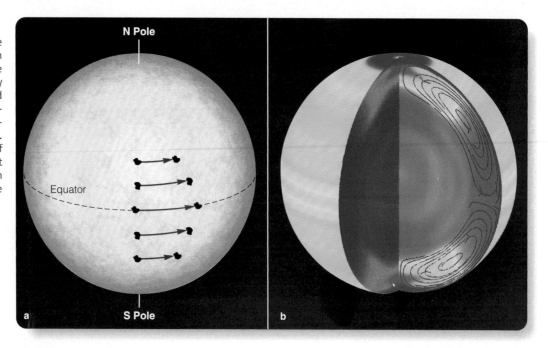

The sun's magnetic field appears to be powered by the energy flowing outward through the moving currents of gas. The gas is highly ionized, so it is a very good conductor of electricity. When an electrical conductor rotates rapidly and is stirred by convection, it can convert some of the energy flowing outward as convection into a magnetic field. This process is called the **dynamo effect,** and it is believed to produce Earth's magnetic field as well. Helioseismologists have found evidence that the sun's magnetic field is generated at the bottom of the convection zone deep under the photosphere. The details of this process are still poorly understood, but the sun's magnetic cycle is clearly related to the creation of its magnetic field.

The magnetic behavior of sunspots provides an insight into how the magnetic cycle works. Sunspots tend to occur in groups or pairs, and the magnetic field around the pair resembles that around a bar magnet with one end magnetic north and the other end magnetic south. At any one time, sunspot pairs south of the sun's equator have reversed polarity compared to those north of the sun's equator. ■ Figure 8-14 illustrates this by showing sunspot pairs south of the sun's equator with magnetic south poles leading and sunspots north of the sun's equator with magnetic north poles leading. At the end of an 11-year sunspot cycle, the new spots appear with reversed magnetic polarity.

This magnetic cycle is not fully understood, but the **Babcock model** (named for its inventor) explains the magnetic cycle as a progressive tangling of the solar magnetic field. Because the electrons in an ionized gas are free to move, the gas is a very good conductor of electricity, and any magnetic field in the gas is "frozen" into the gas. If the gas moves, the magnetic field must move with it. The sun's magnetic field is frozen into its gases, and the differential rotation wraps this field around the sun like a long string caught on a hubcap. Rising and sinking gas currents twist the field into ropelike tubes, which tend to float upward. Where these magnetic tubes burst through the sun's surface, sunspot pairs occur (■ Figure 8-15).

The Babcock model explains the reversal of the sun's magnetic field from cycle to cycle. As the magnetic field becomes tangled, adjacent regions of the sun are dominated by magnetic fields that point in different directions. After about 11 years of tangling, the field becomes so complex that adjacent regions of the sun begin changing their magnetic field to agree with neighboring regions. The entire field quickly rearranges itself into a simpler pattern, and differential rotation begins winding it up to start a new cycle. But the newly organized field is reversed, and the next sunspot cycle begins with magnetic north replaced by magnetic south. Evidently the complete magnetic cycle is 22 years long, and the sunspot cycle is 11 years long.

This magnetic cycle may explain the Maunder butterfly diagram. As a sunspot cycle begins, the twisted tubes of magnetic force first begin to float upward and produce sunspot pairs at higher latitude. Consequently the first sunspots in a cycle appear further north and south of the equator. Later in the cycle, when the field is more tightly wound, the tubes of magnetic force arch up through the surface closer to the equator. As a result, the later sunspot pairs in a cycle appear closer to the equator.

Notice the power of a scientific model. The Babcock model may in fact be incorrect in some or all details, but it provides a framework on which to organize all of the complex solar activity. Even though the models of the sky in Chapter 2 and the atom in Chapter 7 were only partially correct, they served as organizing themes to guide your thinking. Similarly, although the precise details of the solar magnetic cycle are not yet understood, the Babcock model gives you a general picture of the behavior of the sun's magnetic field (**Window on Science 8-2**).

Ace ⟲Astronomy™ Log into AceAstronomy and select this chapter to see Astronomy Exercise "Convection and Magnetic Fields." Discover the link between convection and magnetic fields.

Spots and Magnetic Cycles on Other Stars

The sun seems to be a representative star, so you should expect other stars to have similar cycles of starspots and magnetic fields. This is a difficult topic, because, except for the sun, the stars are so far away that no surface detail is visible. Some stars, however, vary in brightness in ways that suggest they are mottled by dark

■ Figure 8-14

In sunspot groups, here simplified into pairs of major spots, the leading spot and the trailing spot have opposite magnetic polarity. Spot pairs in the southern hemisphere have reversed polarity from those in the northern hemisphere.

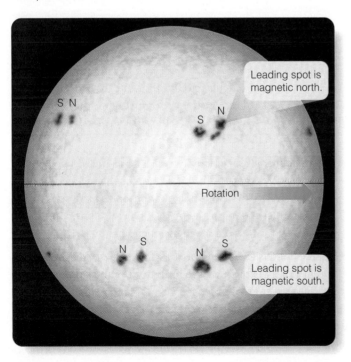

spots. As the star rotates, its total brightness changes slightly, depending on the number of spots facing Earth. High-precision spectroscopic analysis has allowed astronomers to map the loca-

■ Figure 8-15

■ Figure 8-15

The Babcock model of the solar magnetic cycle explains the sunspot cycle as a consequence of the sun's differential rotation gradually winding up the magnetic field.

The Solar Magnetic Cycle

For simplicity, a single line of the solar magnetic field is shown.

Differential rotation drags the equatorial part of the magnetic field ahead.

As the sun rotates, the magnetic field is eventually dragged all the way around.

Differential rotation wraps the sun in many turns of its magnetic field.

Where loops of tangled magnetic field rise through the surface, sunspots occur.

Bipolar sunspot pair

tion of spots on the surfaces of certain stars (■ Figure 8-16a). Such results tell you that the sunspots you see on our sun are not unusual.

Certain features in stellar spectra are associated with magnetic fields. Regions of strong magnetic fields on the solar surface emit strongly at the central wavelengths of the two strongest lines of ionized calcium. This calcium emission appears in the spectra of other sunlike stars and suggests that these stars, too, have strong magnetic fields on their surfaces. In some cases, the strength of this calcium emission varies over periods of days or weeks and suggests that the stars have active regions and are rotating with periods similar to that of the sun. These stars presumably have starspots as well.

In 1966, astronomers began a long-term project to monitor the strength of this calcium emission in the spectra of stars. The stars to be observed were selected to be similar to the sun. With temperatures ranging from 1000 K hotter than the sun to 3000 K cooler, these stars were considered most likely to have sunlike magnetic activity on their surfaces.

The observations show that the strength of the calcium emission varies over periods of years. The calcium emission averaged over the sun's disk varies with the sunspot cycle, and similar periodic variations can be seen in the spectra of some of the stars studied (Figure 8-16b). The star 107 Piscium, for instance, appears to have a starspot cycle lasting nine years. From this kind of evidence it appears that stars like the sun do have magnetic cycles.

These observations confirm that the sun is an average sort of star, that is, not peculiar. Most other stars like our sun have magnetic fields and starspots and go through magnetic cycles.

Ace✸Astronomy™ Log into AceAstronomy and select this chapter to see Astronomy Exercise "Convection and Magnetic Fields." You can adjust the sun's central temperature and change its magnetic field.

Chromospheric and Coronal Activity

The solar magnetic fields extend high into the chromosphere and corona, where they produce beautiful and powerful phenomena. Study **Magnetic Solar Phenomena** on pages 180–181 and notice three important points:

1 Notice that all solar activity is magnetic. You do not experience such events on Earth because Earth's magnetic field is weak, and Earth's atmosphere is not ionized and so is free to move independent of the magnetic field. On the sun, however, the weather is a magnetic phenomenon.

2 Tremendous energy can be stored in arches of magnetic field. These are visible near the limb of the sun as prominences, and, seen from above, as filaments (Figure 8-4). When that stored energy is released, it can trigger powerful eruptions, and, although these eruptions occur far from Earth, they can affect us in dramatic ways.

Building Confidence by Confirmation and Consolidation

While many textbooks describe science as the process of testing hypotheses by observation and experiment, you should not think that every astronomer approaches the telescope with the expectation of making an observation that will disprove long-held beliefs and trigger a revolution in science. Then what is the daily grind of science really about?

First, many observations and experiments merely confirm already tested hypotheses. The biologist knows that all worker bees in a hive are sisters, but a careful study of the DNA from different workers further confirms that hypothesis. By repeatedly confirming a hypothesis,

scientists build confidence in the hypothesis and may be able to extend it to a wider application. Of course, there is always the chance that a new observation or experiment will disprove the hypothesis, but that is usually very unlikely. Much of the daily grind of science is confirmation.

Another aspect of routine science is consolidation, the linking of a hypothesis to other well-studied phenomena. Chemists may understand certain kinds of carbon molecules shaped like rings, but by repeated study they find a carbon molecule shaped like a hollow sphere. To consolidate their findings, they must show that the chemical bonding in the two mol-

ecules follows the same rules and that the molecules have certain properties in common. No hypothesis is overthrown, but the chemists consolidate their knowledge and understand carbon molecules better.

The Babcock model of the solar magnetic cycle is an astronomical example of the scientific process. Solar astronomers know that the model explains some solar features but has shortcomings. Although most astronomers don't expect to discard the entire model, they work through confirmation and consolidation to better understand how the solar magnetic cycle works and how it is related to cycles in other stars.

Computer model of HD 12545

Dark spot
3500 K

Average temperature
4500 K

Sun

Bright spot
4800 K

■ Figure 8-16

Astronomers have found clear evidence that other stars have spots. Rotation broadens the shapes of absorption lines in the spectrum of the star HD12545, and variations in the line shapes allow astronomers to map the location of large spots. In addition, long-term studies of calcium emission show that some stars have active regions like those around sunspot groups on our sun. (Model: K. Strassmeier, Vienna, AURA/NOAO/NSF; CaII emission adapted from data by Baliunas and Saar)

Emission by ionized calcium is associated with sunspots and can be detected in spectra of other stars.

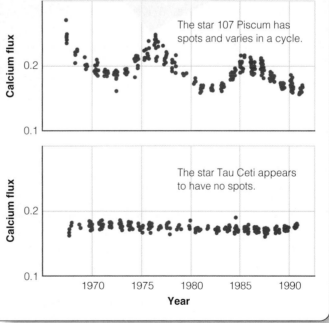

The star 107 Piscum has spots and varies in a cycle.

The star Tau Ceti appears to have no spots.

③ In some regions of the solar surface, the magnetic field does not loop back. High-energy gas from these regions flows outward and produces much of the solar wind.

Solar astronomers use every part of the electromagnetic spectrum to study the sun. Some things are easily visible to human eyes, but many phenomena on the sun can only be studied at nonvisible wavelengths. To understand our sun, astronomers need every piece of evidence they can find, and that means using the entire spectrum.

Magnetic Solar Phenomena

1 Magnetic phenomena in the chromosphere and corona, like magnetic weather, result as constantly changing magnetic fields on the sun trap ionized gas to produce beautiful arches and powerful outbursts. Some of this solar activity can affect Earth's magnetic field and atmosphere.

This ultraviolet image of the solar surface was made by the NASA TRACE spacecraft. It shows hot gas trapped in magnetic arches extending above active regions. At visual wavelengths, you would see sunspot groups in these active regions.

1a A **prominence** is composed of ionized gas trapped in a magnetic arch rising up through the photosphere and chromosphere into the lower corona. Seen during total solar eclipses at the edge of the solar disk, prominences look pink because of the three Balmer emission lines. The image below shows the arch shape suggestive of magnetic fields. Seen from above against the sun's bright surface, prominences form dark filaments.

Sacramento Peak Observatory

H-alpha filtergram

1b Quiescent prominences may hang in the lower corona for many days, whereas eruptive prominences burst upward in hours. The eruptive prominences below are many Earth diameters long.

Far-UV image

The gas in prominences may be 60,000 to 80,000 K, quite cold compared with the low-density gas in the corona, which may be as hot as a million Kelvin.

Trace/NASA

SOHO, EIT, ESA and NASA

An ultraviolet image shows an active region experiencing a flare.

Far-UV image

Helioseismology image

2   Solar **flares** rise to maximum in minutes and decay in an hour. They occur in active regions where oppositely directed magnetic fields meet and cancel each other out in what astronomers call **reconnections**. Energy stored in the magnetic fields is released as short-wavelength photons and as high-energy protons and electrons. X-ray and ultraviolet photons reach Earth in 8 minutes and increase ionization in our atmosphere, which can interfere with radio communications. Particles from flares reach Earth hours or days later as gusts in the solar wind, which can distort Earth's magnetic field and disrupt navigation systems. Solar flares can also cause surges in electrical power lines and damage to Earth satellites.

2b **Auroras** occur about 130 km above Earth's surface when energy in the solar wind guided by Earth's magnetic field excites gases in the upper atmosphere.

2a This sequence of three images of the photosphere taken 5 minutes apart shows sound waves rushing away from a solar flare at 50 km/s. This flare was equivalent to an earthquake 40,000 times stronger than the quake that destroyed San Francisco in 1906. The most powerful flares can be a billion times more powerful than a hydrogen bomb.

2c Magnetic reconnections can release enough energy to blow large amounts of ionized gas outward from the corona in **coronal mass ejections (CMEs).** These produce violent gusts in the solar wind, and if one blows past Earth it can create electrical currents up to a million megawatts, which flow down into Earth's magnetic poles and excite atoms in Earth's upper atmosphere to emit photons, which are seen as an aurora, as shown above.

Sun

CME

Visual-wavelength image

X-ray image

Coronal hole

3 Much of the solar wind comes from **coronal holes,** where the magnetic field does not loop back into the sun. These open magnetic fields allow ionized gas in the corona to flow away as the solar wind. The dark area in this X-ray image at right is a coronal hole.

Log into AceAstronomy and select this chapter to see Astronomy Exercise "Auroras." You can take control of Earth's magnetic field and the solar wind to create your own aurorae.

The Solar Constant

Even a small change in the sun's energy output could produce dramatic changes in Earth's climate. The continued existence of the human species depends on the constancy of the sun, but we humans know very little about the variation of the sun's energy output.

The energy production of the sun can be measured by adding up all of the energy falling on 1 square meter of Earth's surface during 1 second. Of course, some correction for the absorption of Earth's atmosphere is necessary, and you must count all wavelengths from X rays to radio waves. The result, which is called the **solar constant,** amounts to about 1360 joules per square meter per second. A change in the solar constant of only 1 percent could change Earth's average temperature by 1 to 2°C (about 1.8 to 3.6°F). For comparison, during the last ice age Earth's average temperature was about 5°C cooler than it is now.

Some of the best measurements of the solar constant were made by instruments aboard the Solar Maximum Mission satellite. These have shown variations in the energy received from the sun of about 0.1 percent that lasted for days or weeks. Superimposed on that random variation is a long-term decrease of about 0.018 percent per year that has been confirmed by observations made by sounding rockets, balloons, and satellites. This long-term decrease may be related to a cycle of activity on the sun with a period longer than the 22-year magnetic cycle.

Small, random fluctuations will not affect Earth's climate, but a long-term decrease over a decade or more could cause worldwide cooling. History contains some evidence that the solar constant may have varied in the past. As you saw on page 175, the "Little Ice Age" was a period of unusually cool weather in Europe and America that lasted from about 1500 to 1850.* The average temperature worldwide was about 1°C cooler than it is now. This period of cool weather corresponded to the Maunder minimum, a period of reduced solar activity—few sun spots, no auroral displays, and no solar coronae visible during solar eclipses.

In contrast, an earlier period called the Grand Maximum, lasting from about 1100 to about 1250 AD, saw a warming of Earth's

---

*Ironically, the Maunder minimum coincides with the reign of Louis XIV of France, the "Sun King."

climate. The Vikings were able to explore and colonize Greenland, and native communities in parts of North America were forced to abandon their settlements because of long droughts.

Other minima and maxima have been found in climate data taken from studies of the growth rings of trees. In good years, trees add a thicker growth ring than in poor years, so measuring tree rings can reveal the climate in the past. Evidently, solar activity can increase or decrease the solar constant very slightly and affect Earth's climate in dramatic ways. The future of our civilization on Earth may depend on our learning to understand the solar constant.

Building Scientific Arguments

What kind of activity would the sun have if it didn't rotate differentially?

This is a really difficult question because only one star is visible close up. Nevertheless, you can construct a scientific argument by thinking about the Babcock model. If the sun didn't rotate differentially, with its equator traveling faster than do the higher latitudes, then the magnetic field might not get twisted up, and there might not be a solar cycle. Twisted tubes of magnetic field might not form and rise through the photosphere to produce prominences and flares, although convection might tangle the magnetic field and produce some activity. Is the magnetic activity that heats the chromosphere and corona driven by differential rotation or by convection? It is hard to guess, but without differential rotation, the sun might not have a strong magnetic field and high-temperature gas above its photosphere.

This is very speculative, but sometimes in the critical analysis of ideas it helps to imagine a change in a single important factor and try to understand what might happen. For example, redo the argument above. **What do you think the sun would be like if it had no convection inside?**

■   ■   ■

Connections: The sun is beautiful and complex, with great eruptions, prominences, active regions, and dark spots sweeping across its surface like magnetic weather. All of this activity is driven by the outward flow of energy generated by hydrogen fusion in the core. Presumably other stars are like the sun, but other stars are so far away their surfaces are invisible. Through the largest telescopes, they look like nothing more than points of light. How can astronomers learn about the stars? You will begin to answer that question in the next chapter.

Study and Review Tools

Summary

8-1 | The Solar Atmosphere

What do you see when you look at the sun?

■ The solar atmosphere consists of three layers of hot, low-density gas: the photosphere, chromosphere, and corona. The photosphere, or visible surface, is the level in the sun from which visible photons most easily escape.

■ The granulation of the photosphere is produced by convection currents of gas rising from below.

■ The chromosphere is most easily visible during total solar eclipses, when it flashes into view for a few seconds. It is a thin, hot layer of gas just above the photosphere, and its pink color is caused by the Balmer emission lines in its spectrum.

■ Filtergrams of the chromosphere reveal spicules and filaments.

■ The corona is the sun's outermost atmospheric layer. It is composed of a very-low-density, very hot gas extending many solar radii from the visible sun. Its high temperature—up to three million Kelvin—is believed to be maintained by the magnetic field extending up through the photosphere—the magnetic carpet.

■ Parts of the corona give rise to the solar wind, a breeze of low-density ionized gas streaming away from the sun.

■ Solar astronomers can study the motion, density, and temperature of gases inside the sun by analyzing the way the solar surface oscillates. Known as helioseismology, this process requires large amounts of data and extensive computer analysis.

8-2 | Nuclear Fusion in the Sun

How does the sun make its energy?

■ The sun generates its energy near its center, where the temperature and density are high enough for nuclear fusion reactions to combine hydrogen nuclei to make helium nuclei. This is known as the proton–proton chain.

■ Observations of too few neutrinos coming from the sun's core are now explained by the oscillation of neutrinos among three different types. The neutrinos confirm that the sun makes its energy by hydrogen fusion.

8-3 | Solar Activity

What are the dark sunspots and other forms of solar activity?

■ Sunspots seem dark because they are slightly cooler than the rest of the photosphere. The average sunspot is about twice the size of Earth and contains magnetic fields a few thousand times stronger than Earth's. Sunspots are thought to form because the magnetic field inhibits rising currents of hot gas and allows the surface to cool.

■ Astronomers can use the Zeeman effect to measure magnetic fields on the sun.

Why does the sun go through a cycle of activity?

■ The sun is very bright, and its light and infrared radiation can burn your eyes, so you must take great care in observing the sun.

■ The average number of sunspots varies over a period of about 11 years and appears to be related to a magnetic cycle. The sunspot cycle does not repeat exactly each cycle, and the decades from 1645 to 1715, known as the Maunder minimum, seem to have been a time when solar activity was very low and Earth's climate was slightly colder.

■ The sun rotates differentially, with regions far from the equator rotating slower than equatorial regions.

■ Alternate sunspot cycles have reversed magnetic polarity, which has been explained by the Babcock model, in which the differential rotation of the sun winds up the magnetic field. Tangles in the field rise to the surface and cause active regions visible to your eyes as sunspot pairs. When the field becomes strongly tangled, it reorders itself into a simpler but reversed field, and the cycle starts over.

■ Other stars are too far away for starspots to be visible, but spectroscopic observations reveal that many other stars have spots and magnetic fields that follow long-term cycles like the sun's.

■ Prominences occur in the chromosphere; their arched shapes show that they are formed of ionized gas trapped in the magnetic field. You see prominences in filtergrams as dark filaments silhouetted against the bright chromosphere.

■ Flares are sudden eruptions of X-ray, ultraviolet, and visible radiation plus high-energy atomic particles produced when magnetic fields on the sun interact and reconnect. Flares are important because they can have dramatic effects on Earth such as communications blackouts and auroras.

■ Spacecraft images show long streamers extending from the corona out into space. Coronal mass ejections occur when magnetic fields on the surface of the sun eject bursts of ionized gas that flow outward in the solar wind. Such bursts can produce auroras and other phenomena if they strike Earth.

■ Small changes in the solar constant over decades can affect Earth's climate. The Maunder minimum, a time of low solar activity, coincides with the little ice age. Other climate fluctuations have occurred and appear to be linked to small changes in solar activity.

New Terms

sunspot (p. 162)	deuterium (p. 170)
granulation (p. 163)	positron (p. 170)
convection (p. 164)	neutrino (p. 170)
supergranule (p. 164)	radiative zone (p. 170)
limb (p. 164)	convective zone (p. 171)
limb darkening (p. 164)	Maunder butterfly diagram (p. 174)
transition region (p. 164)	Zeeman effect (p. 175)
filtergram (p. 165)	Maunder minimum (p. 175)
filament (p. 165)	active region (p. 175)
spicule (p. 165)	differential rotation (p. 176)
coronagraph (p. 165)	dynamo effect (p. 177)
magnetic carpet (p. 166)	Babcock model (p. 177)
solar wind (p. 166)	prominence (p. 180)
helioseismology (p. 167)	flare (p. 181)
weak force (p. 169)	reconnection (p. 181)
strong force (p. 169)	aurora (p. 181)
nuclear fission (p. 169)	coronal hole (p. 181)
nuclear fusion (p. 169)	coronal mass ejection (CME) (p. 181)
Coulomb barrier (p. 169)	
proton–proton chain (p. 170)	solar constant (p. 182)

Review Questions

Assess your understanding of this chapter's topics with additional quizzing and animations at http://ace .brookscole.com/sf9

1. Why can't you see deeper than the photosphere?

2. What evidence can you give that granulation is caused by convection?

3. How are granules and supergranules related? How do they differ?

4. How can astronomers detect structure in the chromosphere?

5. What evidence can you give that the corona has a very high temperature?

6. What heats the chromosphere and corona to high temperature?

7. How are astronomers able to explore the layers of the sun below the photosphere?

8. Why does nuclear fusion require high temperatures?

9. Why does nuclear fusion in the sun occur only near the center?

10. How can astronomers detect neutrinos from the sun?

11. How can neutrino oscillation explain the solar neutrino problem?

12. What evidence can you give that sunspots are magnetic?

13. How does the Babcock model explain the sunspot cycle?

14. What does the spectrum of a prominence reveal? What does its shape reveal?

15. How can solar flares affect Earth?

16. The upper two images here show two solar phenomena. What are they, and how are they related? How do they differ?

17. The lower image was recorded in the extreme ultraviolet by the SOHO spacecraft. Explain the features you see.

(Daniel Good and NOAO)

(NASA/SOHO)

Discussion Questions

1. What energy sources on Earth cannot be thought of as stored sunlight?

2. What would the spectrum of an auroral display look like? Why?

3. What observations would you make if you were ordered to set up a system that could warn astronauts in orbit of dangerous solar flares? Such a warning system exists.

Problems

1. The radius of the sun is 0.7 million km. What percentage of the radius is taken up by the chromosphere?

2. The smallest detail visible with ground-based solar telescopes is about 1 second of arc. How large a region does this represent on the sun? (*Hint:* Use the small-angle formula.)

3. What is the angular diameter of a star like the sun located 5 ly from Earth? Is the Hubble Space Telescope able to detect detail on the surface of such a star?

4. How much energy is produced when the sun converts 1 kg of mass into energy?

5. How much energy is produced when the sun converts 1 kg of hydrogen into helium? (*Hint:* How does this problem differ from Problem 4?)

6. A 1-megaton nuclear weapon produces about 4×10^{15} J of energy. How much mass must vanish when a 5-megaton weapon explodes?

7. Use the luminosity of the sun, the total amount of energy it emits each second, to calculate how much mass it converts to energy each second.

8. If a sunspot has a temperature of 4240 K and the solar surface has a temperature of 5800 K, how many times brighter is the surface compared to the sunspot? (*Hint:* Use the Stefan–Boltzmann law, Chapter 7.)

9. A solar flare can release 10^{25} J. How many megatons of TNT would be equivalent? (*Hint:* A 1-megaton bomb produces about 4×10^{15} J.)

10. The United States consumes about 2.5×10^{19} J of energy in all forms in a year. How many years could you run the United States on the energy released by the solar flare in Problem 9?

11. Neglecting energy absorbed or reflected by Earth's atmosphere, the solar energy hitting 1 square meter of Earth's surface is 1360 J/s (the solar constant). How long does it take a baseball diamond (90 ft on a side) to receive 1 megaton of solar energy?

Media Cluster

The Sun

Take a look inside the sun in this animation. You can click on various parts of the sun to see things like an animation of nuclear fusion and a movie of a solar flare.

Nuclear Fusion

Nuclear fusion is a reaction that joins the nuclei of atoms to form more massive nuclei. Vary the temperature of the star in this animation to see how this affects the rate of nuclear fusion and luminosity.

Zeeman Effect

The phenomenon known as the Zeeman effect allows astronomers to measure magnetic fields. In this animation, you can adjust the strength of the magnetic field around a star to study the effects on the star's spectrum.

Sunspot Cycle I

Sunspots are relatively dark spots on the sun that contain intense magnetic fields. This animation lets you study the number and location of sunspots occurring over an 11-year cycle.

Sunspot Cycle II

In this animation, you can track sunspot activity over the past 300 years.

Convection and Magnetic Fields

You can vary the temperature of the sun's core in this animation. As you do so, notice how the amount of convection changes, as well as what happens to the magnetic field around the sun.

Auroras

This animation lets you vary the strength of Earth's magnetic field as well as the speed and number of particles in the solar wind. See if you can determine what factors cause the most- and least-intensive auroras.

Lab 4: Solar Wind and Cosmic Rays

This lab begins with an overview of the properties of the sun's atmosphere and how energetic particles escape and travel through the solar system. It ends with a discussion of cosmic rays.

Lab 10: Helioseismology

This lab examines how helioseismology, the study of the sun's vibrations, allows you to obtain detailed information about its interior. Such information allows you to test your understanding of the sun very precisely.

Critical Inquiries for the Web

1. Do disturbances in one layer of the solar atmosphere produce effects in other layers? You have seen that filtergrams are useful in identifying the layers of the solar atmosphere and the structures within them. Visit a website that provides daily solar images. Then, choose today's date (or one near it) and examine the sun in several wavelengths to explore the relation between disturbances in various layers.

2. Explore the web to find out how auroral activity is affected as solar activity rises and falls through the solar cycle. What changes in auroral visibility occur during this cycle? In what other ways can the increased activity associated with a solar maximum affect Earth?

3. Explore the Web to find photos and observations of auroras. From what places on Earth are auroras most often seen?

4. What can you find on the Web about Earth-based efforts to generate energy through nuclear fusion? How do nuclear fusion power experiments attempt to trigger and control nuclear fusion? So-called cold fusion has been abandoned as a false trail. How did it resemble nuclear fusion?

Exploring *TheSky*

1. Locate the six photos of the sun provided in *TheSky* and attempt to draw in the sun's equator in each photo. (*Hint:* In the sun's information box, choose **More Information** and then **Multimedia.** What features are visible in these images that help us recognize the orientation of the sun's equator?)

 Go to the Brooks/Cole Astronomy Resource Center **(http:// astronomy.brookscole.com)** for critical thinking exercises, articles, and additional readings from InfoTrac College Edition, Brooks/Cole's online student library.

9 | The Family of Stars

Ice is the silent language

of the peak;

and fire the silent language

of the star.

CONRAD AIKEN,
AND IN THE HUMAN HEART

Visual-wavelength image

DOES YOUR FAMILY include some characters? The family of stars is amazingly diverse. In a photograph, the stars differ only slightly in color and brightness, but you are going to discover that some are huge and some are tiny, some are astonishingly hot and some are quite cool, some are ponderously massive and some are weeny little stars hardly massive enough to shine. If your family is as diverse as the family of stars, you must have some peculiar relatives. ▌ Unfortunately, finding out what a star is like is quite difficult. When you look at the stars, you look across vast distances and see only bright points of light. Just looking tells you almost nothing about the star's energy production, diameter, or mass. Rather than just looking at stars, you must analyze starlight with great care. This chapter concentrates on finding out three things

▌ Continued on page 188 ▌

The stars in this image of the Eagle Nebula look similar at first glance, but look carefully. The stars make up a diverse family of many different types.
(Mark McCaughrean and Morten Andersen of the Astrophysical Institute, Potsdam, and the European Southern Observatory)

Guidepost

Looking Back

If you want to study anything scientifically, the first thing you have to do is find a way to measure it. But measurement in astronomy is very difficult. In Chapter 6 you studied the sophisticated telescopes and instruments that astronomers use, and in Chapter 7 you read about the information that can be found in spectra. Chapter 8 showed you how astronomers use their understanding of the interaction of light and matter to study the nearest star, the sun. Now you are ready to expand your universe and study other stars.

This Chapter

Here you will discover how those same tools—telescopes, instruments, and basic physics—can tell you what stars are like. Here you will find answers to five essential questions:

How far away are the stars?

How much energy do stars make?

How big are stars?

How much matter do stars contain?

What is the typical star like?

With this chapter you leave our sun behind and begin your study of the billions of stars that dot the sky. In a sense, the star is the basic building block of the universe. If you hope to understand what the universe is and how it works, you must understand the stars.

Looking Ahead

Once you know how to find the basic properties of stars, you will be ready to trace the history of the stars from birth to death, a story that includes our sun, our Earth, and you and me. That story begins in the next chapter.

Ace Astronomy™ The AceAstronomy icon throughout the text indicates an opportunity for you to test yourself on key concepts and to explore animations and interactions on the AceAstronomy website at: http://ace .brookscole.com/sf9 about stars—how much energy they emit, how big they are, and how much mass they contain. These three properties, combined with stellar temperatures—already discussed in Chapter 7—will give you an insight into the family relations among stars.

Although you begin with three things to learn about stars, you immediately meet a detour. To find out how much energy a star emits, you must know how far away it is. If at night you see lights in the distance, you cannot tell whether you are looking at a powerful searchlight miles away or the fainter but nearby bulb on someone's front porch. Only when you know the distance to the lights can you judge their intrinsic brightness. In the same way, to find the intrinsic brightness of a star—or, more precisely, the amount of energy it emits—you must know its distance from you. A short detour will provide you with a method of measuring stellar distances.

After you reach these three goals, you will be able to put your data together to find out what the average star is like. Which types of stars are most common? Which are rare? Look at the photograph in ■ Figure 9-1. The stars are beautiful, but their light tells a cosmic story of great power and beauty, a story you can understand only if you understand the family of the stars.

■ **Figure 9-1**

The modern quest to understand the universe is symbolized in this photo of the Eta Carinae Nebula, roughly 7500 light-years from Earth. Only by the careful analysis of starlight have astronomers learned that the stars in the nebula are only 3 million years old and that the brightest star at lower left contains 100 times more matter than the sun. (2MASS Sky Survey and IPAC)

9-1 Measuring the Distances to Stars

DISTANCE IS THE MOST IMPORTANT and the most difficult measurement in astronomy, and astronomers have found many different ways to estimate the distance to stars. Yet each of those ways depends on a direct geometrical method that is much like the method surveyors would use to measure the distance across a river they cannot cross. You can begin by reviewing the surveyor's method and then applying it to stars.

The Surveyor's Method

To measure the distance across a river, a team of surveyors begins by driving two stakes into the ground. The distance between the stakes is the baseline of the measurement. The surveyors then choose a landmark on the opposite side of the river, a tree perhaps, and that establishes a large triangle marked by the two stakes and the tree. Using their surveyor's instruments, they sight the tree from the two ends of the baseline and measure the two angles on their side of the river.

Knowing two angles of this large triangle and the length of the side between them, the surveyors can then find the distance across the river by simple trigonometry. Another way to find the distance is to construct a scale drawing. For example, if the baseline is 50 m and the angles are 66° and 71°, you can draw a line 50 mm long to represent the baseline. Using a protractor, you can construct angles of 66° and 71° at each end of the baseline, and then, as shown in ■ Figure 9-2, extend the two sides until they

meet at *C*. Point *C* on your drawing is the location of the tree. Measuring the height of your triangle, you would find it to be 64 mm, and that would tell you that the distance across the river to the tree is 64 m.

Of course, modern surveyors don't make drawings. They make their trigonometric calculations in computers. The important

■ **Figure 9-2**

You can find the distance *d* across a river by measuring the baseline and the angles *A* and *B* and then constructing a scale drawing of the triangle.

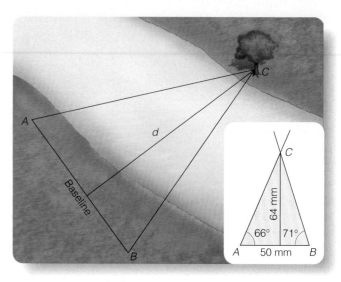

point is that if you measure the baseline and the two angles, you can figure out the distance across the river.

The more distant an object is, the longer the baseline you must use to measure the distance to the object. You could use a baseline 50 m long to find the distance across a river, but to measure the distance to a mountain on the horizon, you might need a baseline 1000 m long. Great distances require very long baselines.

The Astronomer's Method

To find the distance to a star, you must use an extremely long baseline, the diameter of Earth's orbit. If you take a photograph of a nearby star and then wait six months, Earth will have moved halfway around its orbit. You can then take another photograph of the star. This second photograph is taken at a point in space 2 AU (astronomical units) from the point where the first photograph was taken. So your baseline equals the diameter of Earth's orbit, or 2 AU.

You then have two photographs of the same part of the sky taken from slightly different locations in space. When you examine the photographs, you will discover that the nearby star is not in exactly the same place in the two photographs. This apparent shift in the position of the star is called *parallax.*

Parallax is the apparent change in the position of an object due to a change in the location of the observer. You saw in Chapter 4 that parallax is an everyday experience. Your thumb, held at arm's length, appears to shift position against a distant background when you look with first one eye and then with the other. In this case, the baseline is the distance between your eyes, and the parallax is the apparent movement of your thumb when you change eyes. The farther away you hold your thumb, the smaller the parallax.

Did you ever use a stereo viewer to see photographs in three dimensions? That's just parallax at work. Your brain uses the parallax of objects seen through your two eyes to determine distance and create a three-dimensional image.

Because the stars are so distant, their parallaxes are very small angles, usually expressed in seconds of arc. The quantity that astronomers call **stellar parallax (p)** is half the total shift of the star, as shown in ■ Figure 9-3. Astronomers measure the parallax, and surveyors measure the angles at the ends of the baseline, but both measurements tell you the same thing—the shape of the triangle and the distance to the object in question.

The distance to a star with parallax p is given by the simple formula

$$d = \frac{1}{p}$$

where the parallax, p, is measured in seconds of arc and the distance, d, is measured in a unit of distance invented by astronomers, the **parsec (pc),*** defined as the distance to an imaginary star with a parallax of 1 second of arc. One parsec turns out to equal 206,265 AU, or 3.26 ly. This makes it very easy to calculate the distance to a star given the parallax. For example, one of our Favorite Stars, Sirius, has a parallax of 0.375 second of arc. Then the distance to Sirius in parsecs is 1 divided by 0.375, which equals 2.7 pc. To convert to light-years, multiply 2.7 pc by 3.26 and discover that Sirius is 8.8 ly away.

Measuring the small angle p is very difficult. The nearest star to the sun is α Centauri, another of our Favorite Stars. It has a parallax of only 0.76 second of arc, and more distant stars have even smaller parallaxes. To see how small these angles are, hold a piece of paper edgewise at arm's length. The thickness of the paper covers an angle of about 30 seconds of arc. You can see that the parallax of a star, smaller than 1 second of arc, must be very difficult to measure accurately.

The blurring caused by Earth's atmosphere smears star images into tiny blobs of light about one second of arc in diameter, and that makes it difficult to measure parallax from Earth's sur-

---

*The parsec is used throughout astronomy because it simplifies the calculation of distance. However, there are instances when the light-year is also convenient. Consequently, the chapters that follow use either parsecs or light-years as convenience and custom dictate.

■ Figure 9-3

You can measure the parallax of a nearby star by photographing it from two points along Earth's orbit. For example, you might photograph it now and again 6 months from now. Half of the star's total change in position from one photograph to the other is its stellar parallax, p.

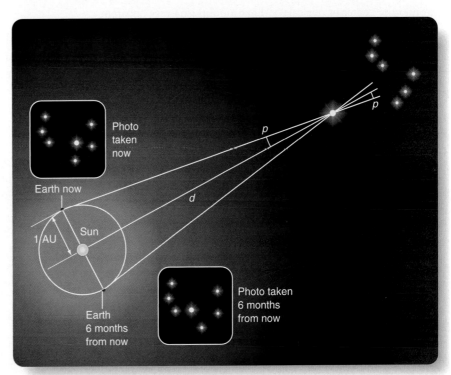

face. Even when astronomers average together many observations, they cannot measure parallax with an uncertainty smaller than about 0.002 second of arc. If you measure a parallax of 0.02 second of arc, the uncertainty is about 10 percent. That means that astronomers observing from Earth's surface can't measure accurate parallaxes smaller than about 0.02 second of arc, which corresponds to a distance of 50 pc; consequently, ground-based parallax measurements are limited to only the closest stars. Since the first stellar parallax was measured in 1838, ground-based astronomers have been able to measure accurate parallaxes for only about 10,000 stars.

In 1989, the European Space Agency launched the satellite Hipparcos to measure stellar parallaxes from orbit above the blurring effects of Earth's atmosphere. The little satellite observed for four years, and the data were reduced by the most sophisticated computers to produce two parallax catalogs in 1997. One catalog contains 120,000 stars with parallaxes 20 times more accurate than ground-based measurements. The other catalog contains over a million stars with parallaxes as accurate as ground-based parallaxes. The Hipparcos data have given astronomers new insights into the nature of stars.

Before dropping the subject of parallax measurement, you should learn about a related observation that reveals the motions of the stars through space.

Ace✺Astronomy™ Log into AceAstronomy and select this chapter to see Astronomy Exercises "Parallax I" and "Parallax II" to experiment with parallax.

Proper Motion

All the stars in the sky, including our sun, are moving along orbits around the center of our galaxy. That motion isn't very apparent over periods of years, but over the centuries it can significantly distort the shape of constellations (■ Figure 9-4). Astronomers use proper motion in a number of ways, but one thing it provides is an estimate of distance.

If you photograph a small area of the sky on two dates separated by 10 years or more, you can notice that some of the stars in the photograph have moved very slightly against the background stars. This motion, expressed in units of seconds of arc per year, is the **proper motion** of the stars. As examples, consider two stars from our list of Favorite Stars. Vega is a bright blue-white star in the summer sky, and it has a proper motion of 0.327 second of arc per year. Rigel, a bright blue-white star in the winter sky, has a proper motion of only 0.002 second of arc per year. The two stars are nearly the same brightness in the sky and nearly the same temperature, but the proper motion of Rigel is over a hundred times less than that of Vega. What does that mean?

A star might have a small proper motion if it is moving almost directly toward or away from you; then its position on the sky would change very slowly, so it would have a small proper motion. That is unlikely, but it does happen. Another reason a star

The Changing Shape of the Big Dipper

100,000 years ago the Big Dipper had a different shape.

Proper motion is moving the stars of the Big Dipper across the sky.

100,000 years in the future, the Big Dipper will have a distorted shape.

■ Figure 9-4

"Proper motion" refers to the slow movement of the stars across the sky.

might have a small proper motion is that it could be quite far away from you. Then even if the star were moving rapidly through space, it would not have a large proper motion. That explains why Rigel, at a distance of 237 pc, has a smaller proper motion, and Vega, at a distance of only 7.7 pc, has a larger proper motion. By the way, this alerts you to something interesting. Although Rigel is 31 times further away than Vega, they have nearly the same brightness in the sky. Rigel must be emitting a lot more light than Vega.

Astronomers can use proper motion to look for nearby stars. If you see a star with a small (or zero) proper motion, it is probably a distant star, but a star with a large proper motion is probably quite close. You may have seen this effect if you watch birds. Distant geese move slowly across the sky, but a nearby bird flits quickly across your field of view. In this way, proper motions can give astronomers a way to locate nearby stars.

Building Scientific Arguments

Why are parallax measurements made from space better than parallax measurements made from Earth?

At first you might suppose that a satellite in orbit can measure the parallax of the stars better because the satellite is closer to the stars, but that will lead your argument astray. When you recall the immense distances to the stars, you can see that being in space doesn't really put the satellite significantly closer to the stars. Rather, the reason for increased accuracy is that the satellite is above Earth's atmosphere. When astronomers try to measure parallax, the turbulence in Earth's atmosphere blurs the star images and smears them out into blobs roughly 1 second of arc in diameter. It isn't possible to measure the position of these fuzzy images accurately. Astronomers can't measure parallax smaller than about 0.02 second of arc. A parallax of 0.02 second of arc corresponds to a distance of 50 pc, so ground-based astronomers can't measure parallax accurately beyond that distance.

A satellite in orbit, however, is above Earth's atmosphere, so the only blurring in the star images is that produced by diffraction in the optics. In other words, the star images are very sharp, and a satellite in orbit can measure the positions of stars and thus their parallaxes to high accuracy.

Now extend your argument one more step. **If a satellite can measure parallaxes as small as 0.001 second of arc, then how far are the most distant stars it can measure?**

■　　■　　■

Connections: You started this chapter with three things you wanted to know about stars: energy production, size, and mass. Now you have found a way to locate nearby stars and measure their distances, so you are ready to discuss the first of your three goals. You are ready to find out how much energy stars emit.

9-2 Intrinsic Brightness

YOUR EYES TELL YOU that some stars look brighter than others, and in Chapter 2 you used the scale of apparent magnitudes to refer to stellar brightness. The faintest stars you can see with the naked eye are about sixth magnitude, with brighter stars having

magnitudes represented by smaller numbers. The brightest stars you see in the sky have negative magnitudes, such as Sirius, whose apparent magnitude is −1.47.

The scale of apparent magnitudes only tells you how bright stars look, however, and you need to know their true, or intrinsic, brightness. *Intrinsic* means "belonging to the thing." When astronomers refer to the intrinsic brightness of a star, they mean a measure of the total amount of light the star emits. Apparent magnitudes don't tell you the intrinsic brightness of the stars, only how bright they look. An intrinsically very bright star might appear faint if it were far away. To find the true brightness of stars, you must correct the apparent magnitudes for the influence of distance.

Brightness and Distance

If you see lights at night, it is difficult to determine which are less powerful but nearby and which are highly luminous but farther away (■ Figure 9-5). You face the same problem when you look at stars, and to resolve that problem, you must think carefully about how brightness depends on distance.

When you look at a bright light, your eyes respond to the visual-wavelength energy falling on the eye's retina, which tells you how bright the object looks. That means the brightness is related to the flux of energy entering your eye. **Flux** is the energy in joules (J) per second falling on one square meter. Recall that a joule is about as much energy as is released when an apple falls from a table onto the floor. Scientists commonly refer to 1 joule per second as 1 watt. The light flux entering your eye is directly related to the intensity of the light. The more flux entering your eye, the brighter the light looks.

If you placed a screen 1 meter square near a lightbulb, a certain amount of flux would fall on the screen. If you moved the screen twice as far from the bulb, the light that previously fell on the screen would be spread to cover an area four times larger, and the screen's surface would receive only one-fourth as much light per square meter. If you tripled the distance to the screen, its surface would receive only one-ninth as much light per square meter. In this way, the flux you receive from a light source is inversely proportional to the square of the distance to the source. This is known as the inverse square relation. (You first encoun-

Observer

■ Active Figure 9-5

To judge the true brightness of a light source, you need to know how far away it is. With no clues to distance, the distant headlight on a truck might look as bright as the nearby headlight on a bicycle.

Ace◐Astronomy™ Log into AceAstronomy and select this chapter to see Active Figure "Brightness and Distance." You can take control of this figure.

tered the inverse square relation in Chapter 5, where it was applied to the strength of gravity. See Figure 5-6.)

Now you understand how the brightness of a star depends on its distance. If you knew the apparent magnitude of a star and its distance from Earth, you could use the inverse square law to correct for the effect of distance. Astronomers do that using a special kind of magnitude scale as described in the next section.

Absolute Visual Magnitude

If all the stars were the same distance away, you could compare one with another and decide which was emitting more light and which less. Of course, the stars are scattered at different distances, and you can't shove them around to line them up for comparison. If, however, you knew the distance to a star, you could use the inverse square relation to calculate the brightness the star would have at some standard distance. Astronomers take 10 pc as the standard distance and refer to the intrinsic brightness of the star as its **absolute visual magnitude (M_V),** the apparent visual magnitude the star would have it if were 10 pc away.

The symbol for absolute visual magnitude is an uppercase M with a subscript V. Recall from Chapter 2 that the symbol for apparent visual magnitude is a lowercase m with a subscript V. The subscript tells you that the visual magnitude system is based only on the wavelengths of light human eyes can see. Other magnitude systems are based on other parts of the electromagnetic spectrum such as the infrared, ultraviolet, and so on.

The intrinsically brightest stars known have absolute visual magnitudes of about -8 and the faintest about $+19$. The sun has an absolute magnitude of 4.78. If the sun were only 10 pc away from Earth, it would look no brighter than the faintest star in the handle of the Little Dipper. Look at our list of Favorite Stars. The nearest star to the sun, alpha Centauri, is only 1.4 pc away, and its apparent magnitude is 0.0, indicating that it looks bright in the sky. However, its absolute magnitude is 4.39, about the same as the sun. Remember Vega and Rigel from the previous section? Vega has an absolute magnitude of 0.6, but Rigel has an absolute magnitude of -6.8. The two stars look the same in the sky, but Rigel is producing a lot more energy than Vega.

Calculating Absolute Visual Magnitude

How exactly do astronomers find the absolute visual magnitude of a star? This question leads to one of the most common formulas in astronomy, a formula that relates a star's magnitude and its distance.

The **magnitude–distance formula** relates the apparent magnitude m_V, the absolute magnitude M_V, and the distance d in parsecs:

$$m_V - M_V = -5 + 5\log_{10}(d)$$

If you know any two of the parameters in this formula, you can easily calculate the third. If you are interested in finding the ab-

solute magnitude of a star, then you should consider an example in which you know the distance and apparent magnitude of a star. Suppose a star has a distance of 50 pc and an apparent magnitude of 4.5. A pocket calculator tells you that the log of 50 is 1.70, and $-5 + 5 \times 1.70$ equals 3.5, so you know that the absolute magnitude is 3.5 magnitudes brighter than the apparent magnitude. That means the absolute magnitude is 1.0, since 4.5 minus 3.5 is 1.0. (Remember that smaller numbers mean brighter magnitudes.) If this star were 10 pc away, it would look bright in the sky, a first-magnitude star.

Astronomers also use the magnitude–distance formula to calculate the distance to a star if the apparent and absolute magnitudes are known. For that purpose, it is handy to rewrite the formula in the following form:

$$d = 10^{(m_V - M_V + 5)/5}$$

If you knew that a star had an apparent magnitude of 7 and an absolute magnitude of 2, then $m_V - M_V$ is 5 magnitudes, and the distance would be 10^2 or 100 parsecs.

The magnitude difference $m_V - M_V$ is known as the **distance modulus,** a measure of how far away the star is. The larger the distance modulus, the more distant the star. You could use the magnitude–distance formula to construct a table of distance and distance modulus (■ Table 9-1).

The magnitude–distance formula may seem awkward at first, but a pocket calculator makes it easy to use. It is important because it performs a critical function in astronomy: It allows astronomers to convert observations of distance and apparent magnitude into absolute magnitude, a measure of the true bright-

■ **Table 9-1 I Distance Moduli**

$m_V - M_V$	d (pc)
0	10
1	16
2	25
3	40
4	63
5	100
6	160
7	250
8	400
9	630
10	1000
⋮	⋮
15	10,000
⋮	⋮
20	100,000
⋮	⋮

ness of the star. Once you know the absolute magnitude, you can go one step further and figure out the total amount of energy a star is radiating into space.

Luminosity

The first of your three goals for this chapter was to find out how much energy stars emit. With the absolute magnitudes of the stars in hand, you can now compare other stars with the sun. That is easiest if you convert absolute magnitude into luminosity. The **luminosity (L)** of a star is the total amount of energy the star radiates in 1 second—not just visible light, but all wavelengths. To find a star's luminosity, you begin with its absolute visual magnitude, make a small correction, and compare the star with the sun.

The correction you must make adjusts for the radiation emitted at wavelengths humans cannot see. Absolute visual magnitude includes only visible light. The absolute visual magnitudes of hot stars and cool stars will underestimate their total luminosities because those stars radiate significant amounts of radiation in the ultraviolet or infrared parts of the spectrum. You can correct for the missing radiation because the amount of missing energy depends only on the star's temperature. For hot and cool stars, the correction can be large, but for medium-temperature stars like the sun, the correction is small. Adding the proper correction to the absolute visual magnitude changes it into the **absolute bolometric magnitude**—the absolute magnitude the star would have if you could see all wavelengths.

Once you know a star's absolute bolometric magnitude, you can find its luminosity by comparing it with the sun. The absolute bolometric magnitude of the sun is 4.7. For every magnitude a star is brighter than 4.7, it is 2.512 times more luminous than the sun. (Recall from Chapter 2 that a difference of 1 magnitude corresponds to an intensity ratio of 2.512.) That means that a star with an absolute bolometric magnitude of 2.7 is 2 magnitudes brighter than the sun, which means it is 6.3 times more luminous (6.3 is approximately 2.512 × 2.512).

Our Favorite Star Aldebaran makes a convenient example. It has an absolute bolometric magnitude of −0.39. That makes it just a bit over 5 magnitudes brighter than the sun. A difference of 5 magnitudes is defined to be a factor of 100 in brightness, so the luminosity of Aldebaran is 100 times the sun's luminosity, or 100 $L_\odot$. Aldebaran is the red eye of Taurus the Bull; next time you see it in the winter sky, nudge your friends and say, "See that star? It emits just over 100 times more energy than the sun."

Remember Favorite Stars Vega and Rigel? Earlier in this chapter you noted that they look the same in the sky, but Rigel is much farther away. If you analyze their brightnesses you will discover that Rigel is a hundred times more luminous than Vega.

The symbol $L_\odot$ represents the luminosity of the sun, a number astronomers can calculate in a direct way. Earth satellites can measure the total solar energy hitting 1 square meter in 1 second just above Earth's atmosphere (the solar constant defined in the previous chapter). The distance from Earth to the sun is known, so it is a simple matter to calculate how much energy the sun must radiate in all directions to provide Earth with the energy it receives per second (see Problem 9 at the end of this chapter). The measured luminosity of the sun is about 4×10^{26} joules/s. You can use that to convert luminosity in terms of the sun into actual joules per second. For example, if Aldebaran is 100 times the luminosity of the sun, it must be emitting a total luminosity of 4×10^{28} joules/s.

Review this point: If you can measure the parallax of a star, you can find its distance, calculate its absolute visual magnitude, correct for the light you can't see to find the absolute bolometric magnitude, and then find the luminosity in terms of the sun. Then you can multiply by the sun's luminosity to find the luminosity of the star in joules per second.

Some stars are a million times more luminous than the sun, and some are almost a million times less luminous. Clearly, the family of stars is filled with interesting characters.

Building Scientific Arguments

How can two stars look the same in the sky but have dramatically different luminosities?
You can answer this question by building a scientific argument that relates three factors: the appearance of a star, its true luminosity, and its distance. The farther away a star is, the fainter it looks, and that is just the inverse square law. If two stars such as Vega and Rigel have the same apparent visual magnitude, then your eyes must be receiving the same amount of light from them. But Rigel is much more luminous than Vega, so it must be farther away. Parallax observations from the Hipparcos satellite confirm that Rigel is 31 times farther away.

Distance is often the key to understanding the brightness of stars, but temperature can also be important. Build a scientific argument to answer the following: **Why must astronomers make a correction in converting the absolute visual magnitude of very hot or very cool stars into luminosities?**

■   ■   ■

Connections: Although you had to make a detour in order to find the distances to the stars, you have reached your first goal. You have found a way to discover the energy emitted by stars, and you have discovered that the range of stellar luminosities is very large. How can that be? The answer lies with your second goal—the diameters of the stars.

9-3 The Diameters of Stars

YOUR SECOND GOAL in this chapter is to find the diameters of stars. You know little about stars until you know their diameters. Are they all the same size as the sun, or are some larger and some smaller? You certainly can't see their diameters through a telescope; the images of stars are much too small for astronomers to resolve their disks and measure their diameters. But there is a way to find out how big stars really are. If you know their temperatures and luminosities, you can find their diameters. That relationship will introduce you to the most important diagram in astronomy, where you will discover more family relations among the stars.

Luminosity, Radius, and Temperature

The luminosity and temperature of a star can tell you its diameter if you understand the two factors that affect a star's luminosity: surface area and temperature. For example, you can eat dinner by candlelight because the candle flame has a small surface area, and, although it is very hot, it cannot radiate much heat; it has a low luminosity. However, if the candle flame were 12 ft tall, it would have a very large surface area from which to radiate, and although it might be no hotter than a normal candle flame, its luminosity would drive you from the table (■ Figure 9-6).

In a similar way, a hot star may not be very luminous if it has a small surface area. It could be highly luminous if it were larger, and even a cool star could be very luminous if it were very large and so had a large surface area from which to radiate. So you can see that both temperature and surface area help determine the luminosity of a star.

■ **Figure 9-6**

Molten lava pouring from a volcano is not as hot as a candle flame, but a lava flow has more surface area and radiates more energy than a candle flame. Approaching a lava flow without protective gear is dangerous. (Krafft/Photo Researchers, Inc.)

Stars are spheres, and the surface area of a sphere is $4\pi R^2$. If you express the radius in meters, the area is the number of square meters on the surface of the star. Each square meter radiates like a black body, and you will remember from Chapter 7 that the total energy given off each second from each square meter is σT^4. So the total luminosity of the star is the surface area multiplied by the energy radiated per square meter:

$$L = 4\pi R^2 \sigma T^4$$

If you divide by the same quantities for the sun, you can cancel out the constants and get a simple formula for the luminosity of a star in terms of its radius and temperature:

$$\frac{L}{L_\odot} = \left(\frac{R}{R_\odot}\right)^2 \left(\frac{T}{T_\odot}\right)^4$$

Here the symbol $\odot$ stands for the sun, and the formula says that the luminosity of a star in terms of the sun equals the radius of the star in terms of the sun squared times the temperature of the star in terms of the sun raised to the fourth power.

Suppose a star is 10 times the sun's radius but only half as hot. How luminous would it be?

$$\frac{L}{L_\odot} = \left(\frac{10}{1}\right)^2 \left(\frac{1}{2}\right)^4 = \left(\frac{100}{1}\right)\left(\frac{1}{16}\right) = 6.25$$

The formula says that the star would be 6.25 times more luminous than the sun.

How can you use this to find the diameters of the stars? If you see a cool star that is very luminous, you know it must be very large, and if you see a hot star that is not very luminous, you know it must be very small.

The formula allows you to calculate these sizes. For instance, suppose that a star is 40 times the luminosity of the sun and twice as hot. If you put these numbers into the formula, you get:

$$\frac{40}{1} = \left(\frac{R}{R_\odot}\right)^2 \left(\frac{2}{1}\right)^4$$

Solving for the radius, you get:

$$\left(\frac{R}{R_\odot}\right)^2 = \frac{40}{2^4} = \frac{40}{16} = 2.5$$

So the radius is:

$$\frac{R}{R_\odot} = \sqrt{2.5} = 1.58$$

The star is 1.58 times larger in radius than the sun.

Because a star's luminosity depends on its surface area and its temperature, you can use luminosity and temperature to sort the stars and discover which are large and which are small. Astronomers use a special diagram for that sorting.

The H–R Diagram

The **Hertzsprung–Russell (H–R) diagram,** named after its originators, Ejnar Hertzsprung and Henry Norris Russell, is a graph that separates the effects of temperature and surface area on stellar luminosities and sorts the stars according to their sizes. Before you study the details of the H–R diagram (as it is often called), try looking at a similar diagram you might use to sort automobiles.

You could plot a diagram such as ■ Figure 9-7 to show horsepower versus weight for various makes of cars. And in so doing, you would find that, in general, the more a car weighs, the more horsepower it has. Most cars fall somewhere along the sequence of cars running from heavy, high-powered cars to light, low-powered models. You might call this the main sequence of cars. But some cars have much more horsepower than normal for their weight—the sport or racing models—and the economy models have less power than normal for cars of the same weight. Just as this diagram sorts cars into family groups, the H–R diagram sorts the stars into groups based on size.

The H–R diagram is a graph with luminosity on the vertical axis and temperature on the horizontal axis. A star is represented by a point on the graph, which tells you the luminosity of the star and its temperature. The H–R diagram in ■ Figure 9-8 also contains a scale of spectral type across the top. Because a star's spectral type is determined by its temperature, you could use either spectral type or temperature on the horizontal axis. Notice that the vertical axis of the H–R diagram is an exponential scale.

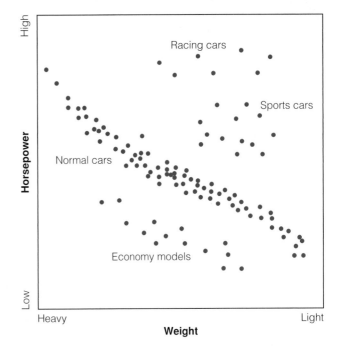

■ Figure 9-7

You could analyze automobiles by plotting their horsepower versus their weight and thus reveal relationships between various models. Most would lie somewhere along the main sequence of "normal" cars.

That's a convenient way to graph data, but you should make special note of what it means (**Window on Science 9-1**).

In an H–R diagram, the location of a point representing a star tells you a great deal about the star. Points near the top of the diagram represent very luminous stars, and points near the bottom represent very-low-luminosity stars. Also, points near the right edge of the diagram represent very cool stars, and points near the left edge of the diagram represent very hot stars. Notice in Figure 9-8 how the artist has used color to represent temperature. Cool stars are red, and hot stars are blue. You can see that dramatically in the photo of real stars in ■ Figure 9-9.

Astronomers use H–R diagrams so often that they usually skip the words "the point that represents the star." Rather they will say that a star is located in a certain place in the diagram. Of course, they mean the point that represents the luminosity and temperature of the star and not the star itself. The location of a star in the

■ Figure 9-8

In an H–R diagram, a star is represented by a dot that shows the luminosity and temperature of the star. The background color in this diagram indicates the temperature of the stars. The sun is a yellow-white G2 star. Most stars fall along the main sequence running from hot luminous stars at upper left to cool low luminosity stars at lower right.

Exponential Scales

Graphs are important in all branches of science, and some graphs have scales that are exponential (related to the power of some base, such as 10^2, 10^3, 10^4, and so on). Because you are accustomed to thinking about graphs with linear scales, exponential scales can be misleading. H–R diagrams are good examples of graphs with exponential scales, but such scales are common in all the sciences, including chemistry, biology, and geology. They also appear in economics, government, and other fields that deal with quantitative data.

In an exponential scale, each step (tick mark) along the axis corresponds to a constant factor. For example, suppose you plotted a diagram in which the vertical axis showed weight on an exponential scale. At the very bottom of the graph, you could plot the weight of a cat and, above that, your own weight. A little higher in the graph, you could plot the weight of a bull elephant and, just above that,

the weight of a whale. In fact, just above that, you could plot the weight of all the whales on Earth plus all the elephants, people, and cats. An exponential scale compresses the high numbers and spreads out the low numbers. That would make your weight on the graph seem much greater than that of a cat and surprisingly (perhaps distressingly) close to that of an elephant. Exponential scales are very helpful for compressing a wide range of numbers into a single graph, but they can be misleading.

For an astronomical example, notice that the luminosity scale on the vertical axis of the H–R diagram in Figure 9-10 has tick marks separated by a constant factor of 10 in stellar luminosity. If a star is just one tick mark higher than another star, it is *10 times* more luminous.

To see how misleading this can be, try drawing an H–R diagram with a linear scale on the vertical axis. Put zero luminosity at the bottom and make each tick mark equal to a step

of 100,000 $L_\odot$, with 1,000,000 $L_\odot$ at the top. You will discover that it is difficult to plot all of the different kinds of stars, but their relative luminosities will be much more clearly illustrated.

You are accustomed to judging the sizes of things from appearances, so exponential scales can be very confusing. (M. Seeds)

■ Figure 9-9

Notice the colors of the stars in the small star cluster M39. The brightest stars are either hot and blue or cool and red. Compare with the most luminous stars in Figure 9-8. (Heidi Schweiker/NAOA/AURA/NSF)

H–R diagram has nothing to do with the location of the star in space. Furthermore, a star may move in the H–R diagram as it ages and its luminosity and temperature change, but such motion in the diagram has nothing to do with the star's motion in space.

Giants, Supergiants, and Dwarfs

Look again at Figure 9-8 and notice the **main sequence,** a region of the H–R diagram running from upper left to lower right. It includes roughly 90 percent of all normal stars. As you might expect, the hot main-sequence stars are more luminous than the cool main-sequence stars. There are, however, stars that don't fall on the main sequence. That alerts you that temperature is not the only thing that determines the luminosity of a star. Size is important too.

The relation among luminosity, radius, and temperature is a precise mathematical relationship that can be used to draw lines of constant radius across an H–R diagram. ■ Figure 9-10 is an H–R diagram on which slanting dashed lines show the location of stars of certain radii. For example, locate the dashed line labeled 1 $R_\odot$. That line passes through the point marked "Sun" and represents the location of any star whose radius equals that of the sun. Of course, the line slants down to the right because cooler stars are always fainter than hotter stars of the same size.

The H–R diagram reveals relationships within the family of stars. The stars called **giants** lie at the right above the main sequence. Although these stars are cool, they are luminous because

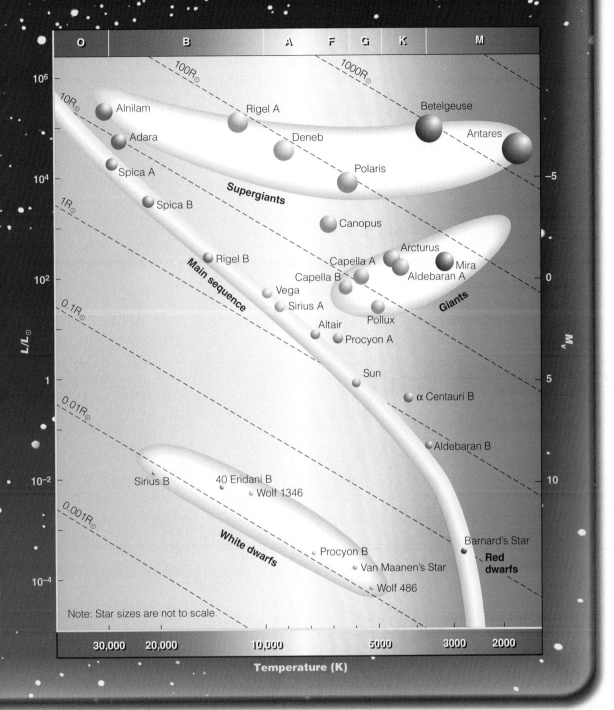

■ Figure 9-10

An H–R diagram showing the luminosity and temperature of many well-known stars. The dashed lines are lines of constant radius. The star sizes on this diagram are not to scale. (Individual stars that orbit each other are designated A and B, as in Spica A and Spica B.)

they are 10 to 100 times larger than the sun. The **supergiants** lie near the top of the H–R diagram and are 10 to 1000 times the size of the sun. Now you can understand why Rigel is so much more luminous than Vega. They have nearly the same temperature, but Rigel is a supergiant and has a much larger surface area from which to radiate. Another of our Favorite Stars is Betelgeuse in Orion, also a supergiant. If it magically replaced the sun at the center of our solar system, it would swallow up Mercury, Venus, Earth, and Mars. The largest stars known have radii of about 7 AU; if they replaced the sun, they would extend nearly to the orbit of Saturn.

At the bottom of the H–R diagram lie the economy models, stars that are very low in luminosity because they are very small. At the bottom end of the main sequence, the **red dwarfs** are not only small, but they are also cool, and that means they can't radiate much energy; they have very low luminosities. In contrast, the **white dwarfs** lie in the lower left of the H–R diagram, and, although some are very hot, they are so small they can't be very bright. They are all roughly the size of Earth and can't radiate much energy from their small surface areas.

The H–R diagram shows that there is a great range in the sizes of stars. The largest are 100,000 times larger than the smallest. Notice that the size of the dots in the H–R diagrams here are only symbolic of the true sizes of the stars. If those dots were plotted in true size, your book would need to be as big as a billboard (■ Figure 9-11).

The distribution of stars in the H–R diagram according to size is a clue to how stars are born and how they die. You will follow those clues in the chapters that follow, but first you need to gather more information about the diverse family of stars.

Ace✺Astronomy™ Log into AceAstronomy and select this chapter to see Astronomy Exercise "Stefan–Boltzmann Law II." You can control the size and temperature of a star and watch its luminosity change.

Luminosity Classification

You can tell from a star's spectrum what kind of star it is. Recall from Chapter 7 that collisional broadening can make spectral lines wider when the gas is dense and the atoms collide often.

Main-sequence stars are relatively small and have dense atmospheres in which the gas atoms collide often and distort their electron energy levels. That makes the lines in the spectra of main-sequence stars broad. On the other hand, giant stars are larger, their atmospheres are less dense, and the atoms disturb one another relatively little. Spectra of giant stars have narrower spectral lines, and spectra of supergiants have very narrow lines (■ Figure 9-12).

That means you can look at a star's spectrum and tell roughly how big it is. These are called **luminosity classes,** because the size of the star is the dominating factor in determining luminosity. Supergiants, for example, are very luminous because they are very large.

■ Figure 9-11

This H–R diagram shows the relative sizes of stars. Giant stars are 10 to 100 times larger than the sun, and white dwarfs are about the size of Earth. (The dots representing white dwarfs here are much too large.) The larger supergiants are too big for this diagram. To visualize the size of the largest stars, imagine that the sun is the size of one of your eyeballs. Then the largest supergiants would be the size of a hot-air balloon.

Luminosity effects on the widths of spectral lines

Supergiant

Giant

Main-sequence star

■ **Figure 9-12**

These schematic spectra show how the widths of spectral lines reveal a star's luminosity classification. Supergiants have very narrow spectral lines, and main-sequence stars have broad lines. In addition, certain spectral lines are more sensitive to this effect than others, so an experienced astronomer can inspect a star's spectrum and determine its luminosity classification.

The luminosity classes are represented by the Roman numerals I through V, with supergiants further subdivided into types Ia and Ib, as follows:

Luminosity Classes

Ia   Bright supergiant

Ib   Supergiant

II   Bright giant

III   Giant

IV   Subgiant

V   Main-sequence star

You can distinguish between the bright supergiants (Ia) such as Rigel and the regular supergiants (Ib) such as Polaris, the North Star. The star Adhara is a bright giant (II), Aldebaran is a giant (III), and Altair is a subgiant (IV). The sun is a main-sequence star (V). The luminosity class is written after the spectral type, as in G2 V for the sun. White dwarfs don't enter into this classification, because their spectra are peculiar. Some of our Favorite Stars are quite unusual. Next time you look at the North Star, remind yourself that it is a supergiant.

If you plot the positions of the luminosity classes on the H–R diagram you get a figure like ■ Figure 9-13. Remember that these are rather broad classifications and that the lines on the diagram are only approximate. A star of luminosity class III may lie slightly above or below the line labeled III. Luminosity classification is subtle and not too accurate, but it is an important tool in modern astronomy. As you will see in the next section, luminosity classification, combined with the H–R diagram, gives astronomers a way to find the distance to stars that are too far away to have measurable parallaxes.

Spectroscopic Parallax

Astronomers can measure the stellar parallax of nearby stars, but most stars are too distant to have measurable parallaxes. The distances to these stars can be estimated from their spectra and apparent magnitude in a process called **spectroscopic parallax.** Spectroscopic parallax is not an actual measure of parallax, but it does reveal the distance to the star.

The method of spectroscopic parallax depends on the H–R diagram. If you photographed the spectrum of a star, you could determine its spectral class, and that would tell you its horizontal location in the H–R diagram. You could also determine its luminosity class by looking at the widths of its spectral lines, and that would tell you the star's vertical location in the diagram. Once you plotted the point that represents the star in an H–R diagram such as Figure 9-13, you could read off its absolute magnitude. As you have seen earlier in this chapter, you can find the distance to a star by comparing its apparent and absolute magnitudes.

For example, our Favorite Star Betelgeuse is classified M2 Ia, and its apparent magnitude is about 0.05. You can plot this

■ **Figure 9-13**

The approximate location of the luminosity classes on the H–R diagram.

Bright supergiants are the most luminous stars.

Main-sequence stars, including the sun, are luminosity class V stars.

The luminosity classes are based on the appearance of absorption lines in the spectra of stars.

star in an H–R diagram such as that in Figure 9-13, where you would find that it should have an absolute magnitude of about −7.2. That means its distance modulus is 0.05 minus (−7.2), or about 7.7, and the distance (estimated from Table 9-1) is about 350 pc. Parallax from the Hipparcos satellite shows that the true distance to Betelgeuse is 520 pc, so the estimate from spectroscopic parallax is only approximate. An error of 1 magnitude changes the distance by a *factor* of 1.6. That's an error of 60 percent, so spectroscopic parallax isn't very accurate, but it can provide an estimate for stars so distant parallax can't be measured.

Building Scientific Arguments

What evidence can you give that giant stars really are bigger than the sun?

Scientific arguments are based on evidence, so you need to proceed step by step here. Stars exist that have the same spectral type as the sun but are clearly more luminous. Capella, for example, is a G star with an absolute magnitude of 0. Because it is a G star, it must have about the same temperature as the sun, but its absolute magnitude is almost 5 magnitudes brighter than the sun's. A magnitude difference of 5 magnitudes corresponds to an intensity ratio of 100, so Capella must be about 100 times more luminous than the sun. If it has the same surface temperature as the sun but is 100 times more luminous, then it must have a surface area 100 times greater than the sun's. Because the surface area of a sphere is proportional to the square of the radius, Capella must be 10 times larger in radius. That is clear observational evidence that Capella is a giant star.

In Figure 9-10, you can see that Procyon B is a white dwarf only slightly warmer than the sun but about 10,000 times less luminous than the sun. Build a scientific argument based on evidence to resolve this question. **Why do astronomers conclude that white dwarfs must be small stars?**

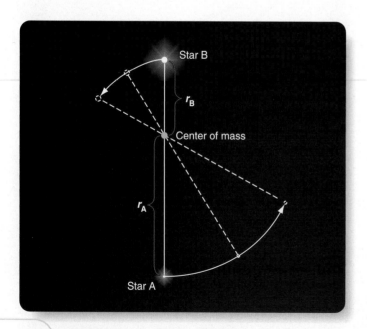

Star B

r_B

Center of mass

r_A

Star A

■   ■   ■

Connections: You have achieved the first two goals of this chapter and discovered the luminosities and radii of the stars. That has revealed surprising differences among stars that look similar in the sky. Now it is time to pursue the third goal. How massive are these stars?

(9-4) The Masses of Stars

YOUR THIRD GOAL is to find out how much matter stars contain, that is, to know their mass. Do they all contain about the same mass as our sun, or are some more massive and others less? Unfortunately, it's difficult to determine the mass of a star. Looking through a telescope at a star, you see only a point of light that tells you nothing about the star. Gravity is the key. Matter produces a gravitational field, and you can figure out how much matter a star contains if you watch an object move through the star's gravitational field. To find the masses of stars, you must study **binary stars,** pairs of stars that orbit each other.

Binary Stars in General

The key to finding the mass of a binary star is understanding orbital motion. Chapter 5 illustrated orbital motion by imagining a cannonball fired from a high mountain (see page 94). If Earth's gravity didn't act on the cannonball, it would follow a straight-line path and leave Earth forever. Because Earth's gravity pulls it away from its straight-line path, the cannonball follows a curved path around Earth—an orbit. When two stars orbit each other, their mutual gravitation pulls them away from straight-line paths and makes them follow closed orbits around a point between the stars.

Each star in a binary system moves in its own orbit around the system's center of mass, the balance point of the system. If the stars were connected by a massless rod and placed in a uniform gravitational field such as that near Earth's surface, the system would balance at its center of mass like a child's seesaw (see page 95). If one star were more massive than its companion, then the massive star would be closer to the center of mass and would travel in a smaller orbit, while the lower-mass star would whip around in a larger orbit (■ Figure 9-14). The ratio of the masses of the stars M_A/M_B equals r_B/r_A, the inverse of the ratio of the radii of the orbits. If one star has an orbit twice as large as the other star's orbit, then it must be half as massive. Getting the ratio of the masses is easy, but that doesn't tell you the individual masses

■ **Active Figure 9-14**

As stars in a binary star system revolve around each other, the line connecting them always passes through the center of mass, and the more massive star is always closer to the center of mass.

Ace✷Astronomy™ Log into AceAstronomy and select this chapter to see Active Figure "Center of Mass." Experiment to see how the ratio of the masses affects the relative size of the orbits.

Learning about Nature through Chains of Inference

Scientists can rarely observe the things they really want to know, so they must construct chains of inference. You can't observe the mass of stars directly, so you must find a way to use what you can observe, orbital period and angular separation, and step by step figure out the parameters you need to reach your goal. In this case, the goal is the masses of the two stars. Scientists follow chains of inference from the observable parameters to the unobservable quantities they want to know.

Chains of inference can be mathematical, as when geologists use earthquakes to calculate the temperature and density of Earth's interior. There is no way to drill a hole to Earth's center and lower a thermometer or recover a sample, so the internal temperature and density are not directly observable. Nevertheless, the speed of vibrations from distant earthquakes depends on the temperature and density of the rock they pass through. Geologists can't measure the speed of the vibrations deep inside Earth; but they can measure the delay in the arrival time at different locations on the surface, and that allows them to work their way back to the speed and, finally, the temperature and density.

Chains of inference can also be nonmathematical. Biologists studying the migration of whales can't follow individual whales for years at a time, but they can observe feeding and mating in different locations; take into consideration food sources, ocean currents, and water temperatures; and construct a chain of inference that leads back to the seasonal migration pattern for whales.

This chapter contains a number of chains of inference because it describes how astronomers know what stars are like. Almost all sci-

San Andreas fault: A chain of inference connects earthquakes to conditions inside Earth. (USGS)

ences use chains of inference. If you can link the observable parameters step by step to the final conclusions, you have gained a strong insight into the nature of science.

---

of the stars, which is what you really want to know. That takes further analysis.

To find the mass of a binary star system, you must know the size of the orbits and the orbital period—the length of time the stars take to complete one orbit. The smaller the orbits are and the shorter the orbital period is, the stronger the two stars' gravity must be to hold each other in orbit. For example, if two stars whirl rapidly around each other in small orbits, then their gravity must be very strong to prevent their flying apart. Such stars would have to be very massive. From the size of the orbits and the orbital period, astronomers can figure out the masses of the stars from Newton's laws.

Calculating the Masses of Binary Stars

According to Newton's laws of motion and gravity, the total mass of two stars orbiting each other is related to the average distance a between them and their orbital period P. If the masses are M_A and M_B, then

$$M_A + M_B = \frac{a^3}{P^2}$$

In this formula, a is expressed in AU, P in years, and the mass in solar masses.

Notice that this formula is related to Kepler's third law of planetary motion (see Table 4-1). Almost all of the mass of the solar system is in the sun. If you apply this formula to any planet in our solar system, the total mass is 1 solar mass. Then the formula becomes $P^2 = a^3$, which is Kepler's third law.

In other star systems, the total mass is not necessarily 1 solar mass, and this gives you a way to find the masses of binary stars. If you can find the average distance in AU between the two stars, a, and their orbital period in years, P, the sum of the masses of the two stars is just a^3/P^2.

Example 1: If you observe a binary system with a period of 32 years and an average separation of 16 AU, what is the total mass? *Solution:* The total mass equals $16^3/32^2$, which equals 4 solar masses.

Example 2: Let's call the two stars in the previous example A and B. Suppose star A is 12 AU away from the center of mass, and star B is 4 AU away. What are the individual masses? *Solution:* The ratio of the masses must be 12:4, which is the same as 3:1. What two numbers add up to 4 and have the ratio of 3:1? In this case, the answer is easy. Star B must be 3 solar masses, and Star A must be 1 solar mass.

Actually, figuring out the mass of a binary star system is not as easy as it might seem from this discussion. The orbits of the two stars may be elliptical, and, although the orbits lie in the same plane, that plane can be tipped at an unknown angle to your line of sight, further distorting the shapes of the orbits. Astronomers must find ways to correct for these distortions. In addition, astronomers analyzing binary systems must find the distances to the stars so they can estimate the true size of the orbits in astronomical units. Notice that finding the masses of binary stars requires a number of steps to get from what can be observed to what astronomers really want to know, the masses. Constructing such sequences of steps is an important part of science (**Window on Science 9-2**).

Although there are many different kinds of binary stars, three types are especially important for determining stellar masses. These are discussed separately in the next sections.

Visual Binary Systems

In a **visual binary system,** the two stars are separately visible in the telescope. Only a pair of stars with large orbits can be separated visually; if the orbits are small, the star images blend together in the telescope, and you see only a single point of light. Because visual binary systems must have large orbits, they also have long orbital periods. Some take hundreds or even thousands of years to complete a single orbit.

Astronomers study visual binary systems by measuring the position of the two stars directly at the telescope or in images. In either case, the astronomers need measurements over many years to map the orbits. The first frame of ■ Figure 9-15 shows a photograph of one of our Favorite Stars, Sirius, which is actually a visual binary system made up of the bright star Sirius A and its white dwarf companion Sirius B. The photo was taken in 1960. Successive frames in Figure 9-15 show the motion of the two stars as observed since 1960 and the orbits the stars follow. The orbital period is 50 years.

Binary systems are common; more than half of all stars are members of binary star systems. Our Favorite Star Polaris, for example, has a binary companion in an orbit with a period of 29. 59 years. Two smaller stars orbit even farther from Polaris, so it is a four-star binary system. Favorite Star Alpha Centauri, the nearest star to the sun, is a three-star system. Although binary stars are common, few can be analyzed completely. Many visual binaries are so far apart that their periods are much too long for practical mapping of their orbits. Other binary systems are so close together they are not visible as separate stars.

Spectroscopic Binary Systems

If the stars in a binary system are close together, the telescope, limited by diffraction and by seeing, reveals only a single point of light. Only by looking at a spectrum that is formed by light from both stars and contains spectral lines from both can astronomers tell that there are two stars present and not one. Such a system is a **spectroscopic binary system.**

■ Figure 9-16 shows a pair of stars orbiting each other in identical circular orbits. From the diagram you can guess that the two stars have equal masses, and you should notice that the circular orbit appears elliptical because you see it nearly edge-on. Of course, if this were a spectroscopic binary system, you would not see the separate stars. Nevertheless, the Doppler shift would alert you that there were two stars orbiting each other. In the first frame of the figure, star A is approaching Earth while star B recedes. In the spectrum, you see a spectral line from star A blueshifted while the same spectral line from star B is redshifted. As you watch the two stars revolve around their orbits, they alternately approach and recede from Earth, and you can see their spec-

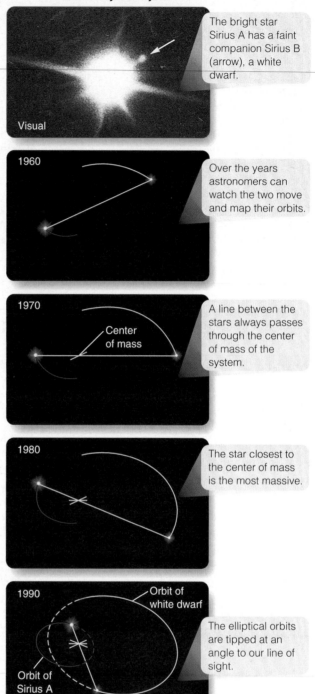

A Visual Binary Star System

The bright star Sirius A has a faint companion Sirius B (arrow), a white dwarf.

Over the years astronomers can watch the two move and map their orbits.

A line between the stars always passes through the center of mass of the system.

The star closest to the center of mass is the most massive.

The elliptical orbits are tipped at an angle to our line of sight.

■ Figure 9-15

The orbital motion of Sirius A and Sirius B can reveal their individual masses. (Photo © UC Regents/Lick Observatory)

tral lines Doppler shifted first toward the blue and then toward the red. In a real spectroscopic binary, you can't see the individual stars, but the sight of pairs of spectral lines moving back and forth across each other would alert you that you were observing a spectroscopic binary system (■ Figure 9-17).

A Spectroscopic Binary Star System

Approaching — **Receding**

A — B

Stars orbiting each other produce spectral lines with Doppler shifts.

← Blueshift   A   B   Redshift →

B

A

As the stars circle their orbits, the spectral lines move together.

← Blueshift   A B   Redshift →

B

A

When the stars move perpendicular to our line of sight, there are no Doppler shifts.

← Blueshift   A + B   Redshift →

B

A

Spectral lines shifting apart and then merging are a sign of a spectroscopic binary.

← Blueshift   B   A   Redshift →

B — A

The size of the Doppler shifts contains clues to the masses of the stars.

← Blueshift   B   A   Redshift →

▧ Figure 9-16

From Earth, a spectroscopic binary looks like a single point of light, but the Doppler shifts in its spectrum reveal the orbital motion of the two stars.

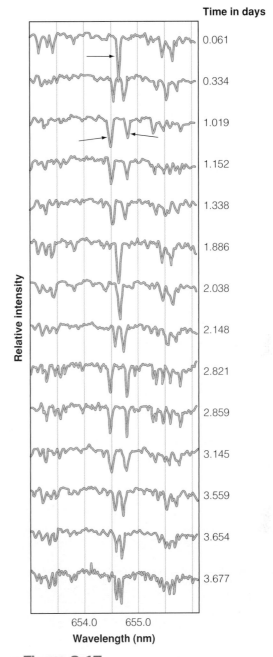

Time in days

0.061

0.334

1.019

1.152

1.338

1.886

2.038

2.148

2.821

2.859

3.145

3.559

3.654

3.677

Relative intensity

654.0   655.0

Wavelength (nm)

▧ Figure 9-17

Fourteen spectra of the star HD80715 are shown here as graphs of intensity versus wavelength. A single spectral line (arrow in top spectrum) splits into a pair of spectral lines (arrows in third spectrum), which then merge and split apart again. These changing Doppler shifts reveal that HD80715 is a spectroscopic binary. (Adapted from data courtesy of Samuel C. Barden and Harold L. Nations)

You can find the orbital period of a spectroscopic binary system by waiting to see how long it takes for the spectral lines to return to their starting positions. You can measure the size of the Doppler shifts to find the orbital velocities of the two stars. If you multiply velocity times orbital period, you can find the circum-ference of the orbit, and from that you can find the radius of the orbit. Of course, if you knew the orbital period and the size of the orbit, you could calculate the mass. One important detail is missing, however. You don't know how much the orbits are inclined to your line of sight.

You can find the inclination of a visual binary system because you can see the shape of the orbits. In a spectroscopic binary system, however, you cannot see the individual stars, so you can't find the inclination or untip the orbits. Recall that the Doppler effect only tells you the radial velocity, the part of the velocity directed toward or away from Earth. Because you cannot find the inclination, you cannot correct these radial velocities to find the true orbital velocities. Consequently, you cannot find the true masses. All you find from a spectroscopic binary system is a lower limit to the masses.

More than half of all stars are in binary systems, and most of those are spectroscopic binary systems. Many of the familiar stars in the sky are actually pairs of stars orbiting each other (■ Figure 9-18).

You might wonder what happens when the orbits of a spectroscopic binary system lie exactly edge-on to Earth. The result is the most useful kind of binary system.

Ace⊛Astronomy™ Log into AceAstronomy and select this chapter to see Astronomy Exercise "Spectroscopic Binaries." You can see how the motions of the two stars affect their shared spectrum.

■ **Figure 9-18**

(a) At the bend of the handle of the Big Dipper lies a pair of stars, Mizar and Alcor. Through a telescope you will discover that Mizar has a fainter companion and so is a member of a visual binary system. (b) Spectra of Mizar recorded at different times show that it is itself a spectroscopic binary system rather than a single star. In fact, both the faint companion to Mizar and the nearby star Alcor are also spectroscopic binary systems. (The Observatories of the Carnegie Institution of Washington)

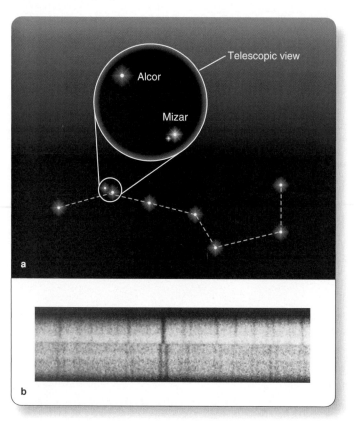

Eclipsing Binary Systems

As mentioned in the previous section, the orbits of the two stars in a binary system always lie in a single plane. If that plane is nearly edge-on to Earth, then the stars can cross in front of each other as seen from Earth. Imagine a model of a binary star system in which a cardboard disk represents the orbital plane and balls represent the stars, as in ■ Figure 9-19. If you see the model from the edge, then the balls that represent the stars can move in front of each other as they follow their orbits. The small star crosses in front of the large star, and then, half an orbit later, the large star crosses in front of the small star. When one star moves in front of the other, it blocks some of the light, and astronomers say the star is eclipsed. Such a system is called an **eclipsing binary system.**

Seen from Earth, the two stars are not visible separately. The system looks like a single point of light. But when one star moves in front of the other star, part of the light is blocked, and the total brightness of the point of light decreases. ■ Figure 9-20 shows a smaller star moving in an orbit around a larger star, first eclipsing the larger star and then being eclipsed as it moves behind. The resulting variation in the brightness of the system is shown as a graph of brightness versus time, a **light curve.**

■ **Figure 9-19**

Imagine a model of a binary system with balls for stars and a disk of cardboard for the plane of the orbits. Only if you view the system edge-on do you see the stars cross in front of each other.

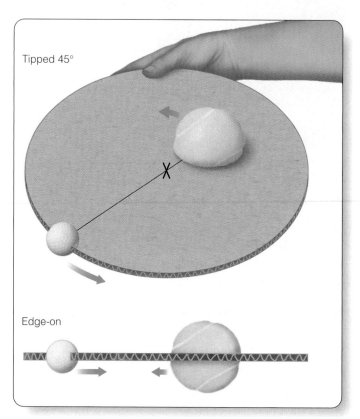

The light curves of eclipsing binary systems contain tremendous amounts of information about the stars, but the curves can be difficult to analyze. Figure 9-20 shows an idealized system.

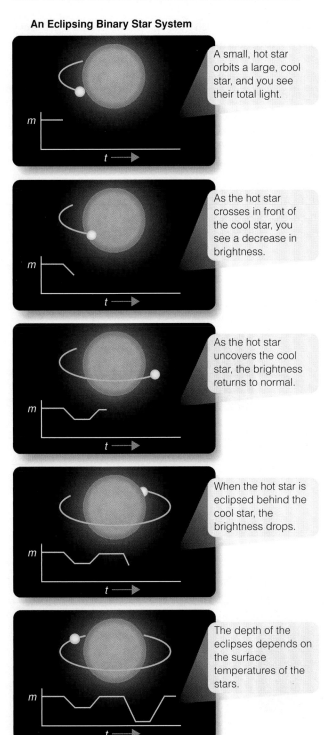

■ **Figure 9-20**

From Earth, an eclipsing binary looks like a single point of light, but changes in brightness reveal that two stars are eclipsing each other. Doppler shifts in the spectrum combined with the light curve, shown here as magnitude versus time, can reveal the size and mass of the individual stars.

An Eclipsing Binary Star System

A small, hot star orbits a large, cool star, and you see their total light.

As the hot star crosses in front of the cool star, you see a decrease in brightness.

As the hot star uncovers the cool star, the brightness returns to normal.

When the hot star is eclipsed behind the cool star, the brightness drops.

The depth of the eclipses depends on the surface temperatures of the stars.

■ Figure 9-21 shows the light curve of a real system in which the stars have dark spots on their surfaces and are so close to each other that their shapes are distorted.

Once the light curve of an eclipsing binary system has been accurately observed, you can construct a chain of inference that leads to the masses of the two stars. You could find the orbital period easily, and you could get spectra showing the Doppler shifts of the two stars. You could find the orbital velocity because you don't have to untip the orbits; you know they are nearly edge-on or there would not be eclipses. Then you can find the size of the orbits and the masses of the stars.

Earlier in this chapter, luminosity and temperature were used to calculate the radii of stars, but eclipsing binary systems provide a way to measure the sizes of stars directly. From the light curve you could tell how long it took for the small star to cross the large star. Multiplying this time interval by the orbital velocity of the small star would give you the diameter of the larger star. You could also determine the diameter of the small star by noting how long it took to disappear behind the edge of the large star.

■ **Figure 9-21**

The observed light curve of the binary star VW Cephei (lower curve) shows that the two stars are so close together their gravity distorts their shapes. Slight distortions in the light curve reveal the presence of dark spots at specific places on the star's surface. The upper curve shows what the light curve would look like if there were no spots. (Graphics created with Binary Maker 2.0)

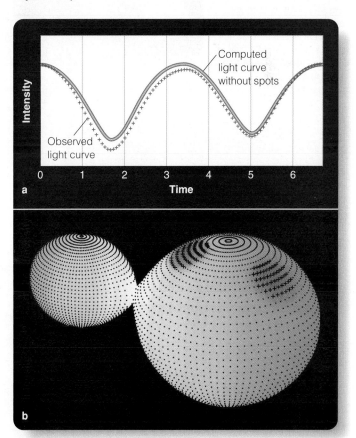

For example, if it took 300 seconds for the small star to disappear while traveling 500 km/s relative to the large star, then it would have to be 150,000 km in diameter.

Of course, there are complications due to the inclination and eccentricity of orbits, but often these effects can be taken into account, and the system can reveal not only the masses of its stars but also their diameters.

Algol (alpha Persei) is one of the best-known eclipsing binaries, because its eclipses are visible to the naked eye. Normally, its magnitude is about 2.15, but its brightness drops to 3.4 in eclipses that occur every 68.8 hours. Although the nature of the star was not recognized until 1783, its periodic dimming was probably known to the ancients. *Algol* comes from the Arabic for "the demon's head," and it is associated in constellation mythology with the severed head of Medusa, the sight of whose serpentine locks turned mortals to stone (■ Figure 9-22). Indeed, in some accounts, Algol is the winking eye of the demon.

From the study of binary stars, astronomers have found that the masses of stars range from 0.08 solar mass at the low end to as high as 150 solar masses at the high end. The most massive stars ever found in a binary system are a pair of stars with masses of 83 and 82 solar masses. A few other stars are believed to be more massive, but they do not lie in binary systems, so astronomers can only estimate their masses. Stars near the upper limit are very rare, and few are known, so this upper limit is uncertain.

Ace ➲ Astronomy™ Log into AceAstronomy and select this chapter to see Astronomy Exercise "Eclipsing Binaries." You can take control of a binary star system.

Building Scientific Arguments

When you look at the light curve for an eclipsing binary system with total eclipses, how can you tell which star is hotter?

This scientific argument brings together a number of things you have learned about light and binary stars. If you assume that the two stars in an eclipsing binary system are not the same size, then you can refer to them as the larger star and the smaller star. When the smaller star moves behind the larger star, you lose the light coming from the total area of the small star. And when the smaller star moves in front of the larger star, it blocks off light from the same amount of area on the larger star. In both cases, the same amount of area, the same number of square meters, is hidden from your sight. Then the amount of

■ **Figure 9-22**

The eclipsing binary Algol consists of a hot B star and a cooler G or K star. The eclipses are partial, meaning that neither star is completely hidden during eclipses. The orbit here is drawn as if the cooler star were stationary.

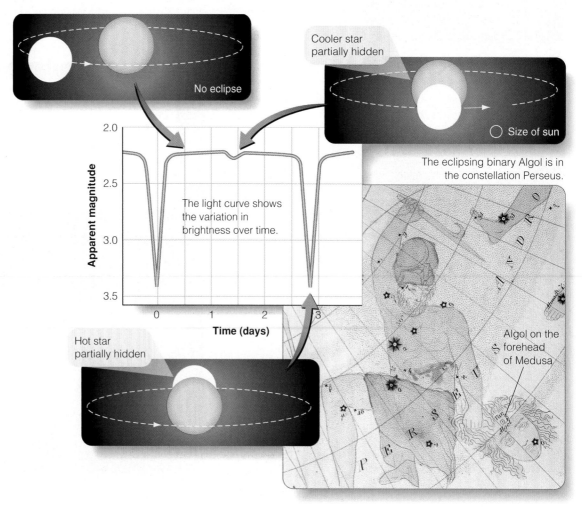

Information about Nature: Basic Scientific Data

In a simple sense, science is the process by which scientists look at data and search for relationships that show how nature works. But, as a result, science sometimes requires large amounts of data. For example, astronomers

Collecting mineral samples can be hard work, but it is also fun. Scientists sometimes collect large amounts of data because they enjoy the process. (M. Seeds)

need to know the masses and luminosities of many stars before they can begin looking for relationships.

Compiling basic data is one of the common forms of scientific study. This work may not seem very exciting to an outsider, but scientists often love their work not so much because they want to know nature's secrets but because they love the process of studying nature. Using a microscope or a telescope is fun. Gathering plants in the rain forest or geological samples from a cliff face can be tremendously exciting and satisfying. It is sometimes the process of science that is most rewarding, and enjoyment of the process can lead scientists to gather significant amounts of information.

Solving a single binary star system to find the masses of the stars does not tell an astronomer a great deal about nature, but solving a binary is like solving a puzzle. It is fun, and it is satisfying. Over the years, many astronomers have added their results to the growing data

file on stellar masses. Astronomers can now analyze that data to search for relationships between the masses of stars and other parameters such as diameter and luminosity.

The history of science is filled with stories of hardworking scientists who compiled large amounts of data that later scientists used to make important discoveries. For example, Tycho Brahe spent 20 years recording the positions of the stars and planets (Chapter 4). He must have loved his work to spend 20 years on his windy island, but he did not live to use his data. It was his successor, Johannes Kepler, who used Tycho Brahe's data to discover the laws of planetary motion.

Whatever science you study, you will encounter accumulations of measurements and observations that have been compiled over the years, including everything from the hardness of rocks to the attention span of infants. Determining these basic scientific data is as much a part of science as is testing a hypothesis.

---

light lost during an eclipse depends only on the temperature of the hidden surface because temperature is what determines how much each square meter can radiate per second. When the surface of the hotter star is hidden, the brightness will fall dramatically, but when the surface of the cooler star is hidden, the brightness will not fall as much. So you can look at the light curve and point to the deeper of the two eclipses and say, "That is where the hotter star is behind the cooler star."

Now change the argument to consider the diameters of the stars. **How could you look at the light curve of an eclipsing binary with total eclipses and find the ratio of the diameters?**

■　　■　　■

Connections: Perhaps the hardest thing to learn about a star is its mass, but stars orbiting around each other in binary systems can be analyzed to find their masses. Now you are ready to put the facts together.

9-5 A Survey of the Stars

YOU HAVE ACHIEVED the three goals set for you at the start of this chapter. You know how to find the luminosities, diameters, and masses of stars. Now you can put that data (**Window on**

Science 9-3) together to paint a family portrait of the stars; like most family portraits, both similarities and differences are important clues to the history of the family. As you begin trying to understand how stars are born and how they die, you can ask a simple question: What is the average star like? Answering that question is both challenging and illuminating.

Mass, Luminosity, and Density

With enough data plotted in an H–R diagram, you can see the patterns that hint as to how stars are born, how they age, and how they die.

If you label an H–R diagram with the masses of the plotted stars, as in ■ Figure 9-23, you will discover that the main-sequence stars are ordered by mass. The most massive main-sequence stars are the hot stars. As you run your eye down the main sequence, you find lower-mass stars; and the lowest-mass stars are the coolest, faintest main-sequence stars.

Stars that do not lie on the main sequence are not in order according to mass. Giant stars are a jumble of different masses, and supergiants, although they tend to be more massive than giants, are in no particular order in the H–R diagram. In contrast, all white dwarfs have about the same mass, somewhere in the narrow range of 0.5 to about 1 solar mass.

Because of the systematic ordering of mass along the main sequence, these main-sequence stars obey a **mass–luminosity** factor of 10^{12}. Clearly, a small difference in mass causes a large difference in luminosity.

Giants and supergiants do not follow the mass–luminosity relation very closely, and white dwarfs not at all. In the next chapters, the mass–luminosity relation will help you understand how stars generate their energy.

Though mass alone does not reveal any pattern among giants, supergiants, and white dwarfs, density does. Once you know a star's mass and diameter, you can calculate its average density by dividing its mass by its volume. Stars are not uniform in density but are most dense at their centers and least dense near their surface. The center of the sun, for instance, is about 100 times as dense as water; its density near the visible surface is about 3400 times less dense than Earth's atmosphere at sea level. A star's average density is intermediate between its central and surface densities. The sun's average density is approximately 1 g/cm³—about the density of water.

Main-sequence stars have average densities similar to the sun's, but giant stars, being large, have low average densities, ranging from 0.1 to 0.01 g/cm³. The enormous supergiants have still lower densities, ranging from 0.001 to 0.000001 g/cm³. These densities are thinner than the air you breathe; if you could insulate yourself from the heat, you could fly an airplane through these stars.

■ **Figure 9-23**

The masses of the plotted stars are labeled on this H–R diagram. Notice that the masses of main-sequence stars decrease from top to bottom but that masses of giants and supergiants are not arranged in any ordered pattern.

relation—the more massive a star is, the more luminous it is (■ Figure 9-24). In fact, the mass–luminosity relation can be expressed as a simple formula:

$$L = M^{3.5}$$

That is, a star's luminosity (in terms of the sun's luminosity) equals its mass (in solar masses) raised to the 3.5 power. For example, a star of 4 solar masses has a luminosity of approximately $4^{3.5}$, or $4 \times 4 \times 4 \times \sqrt{4}$. This equals 64×2, or 128. So a 4-solar-mass star will have a luminosity about 128 times the luminosity of the sun. This is only an approximate equation, as shown by the red line in Figure 9-24.

Notice how large the range in luminosity is. The observed range of masses extends from about 0.08 solar mass to about 83 solar masses—a factor of about 1000. But the range of luminosities extends from about 10^{-6} to about 10^{6} solar luminosities—a

■ **Figure 9-24**

The mass–luminosity relation shows that the more massive a main-sequence star is, the more luminous it is. The open circles represent white dwarfs, which do not obey the relation. The red line represents the equation $L = M^{3.5}$.

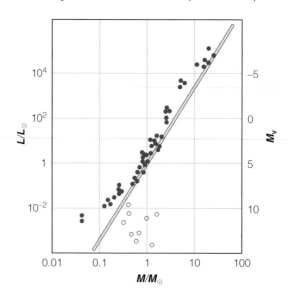

Only near the center would you be in any danger, for there the material is very dense—about 3,000,000 g/cm³.

The white dwarfs have masses about equal to the sun's but are very small, only about the size of Earth. That means the matter is compressed to densities of 3,000,000 g/cm³ or more. On Earth, a teaspoonful of this material would weigh about 15 tons.

Density divides stars into three groups. Most stars are main-sequence stars with densities like the sun's. Giants and supergiants are very-low-density stars, and white dwarfs are high-density stars. You will see in later chapters that these densities reflect different stages in the evolution of stars.

Ace ◎Astronomy™ Log into AceAstronomy and select this chapter to see Astronomy Exercise "Mass–Luminosity Relation." You can discover your own relation from real stars.

Ace ◎Astronomy™ Log into AceAstronomy and select this chapter to see Astronomy Exercise "H–R/Mass–Luminosity 3-D Graph." Look at the data in three dimentions and see how mass is related to position in the H–R diagram.

Surveying the Stars

If you want to know what the average person thinks about a certain subject, you take a survey. If you want to know what the average star is like, you must survey the stars. Such surveys reveal important relationships among the family of stars.

Not many decades ago, surveying large numbers of stars was an exhausting task, but modern computers have changed that. Specially designed telescopes controlled by computers can make millions of observations per night, and high-speed computers can compile and analyze these vast surveys and create easy-to-use databases. Chapter 6 mentioned the Sloan Digital Sky Survey and the 2Mass infrared survey. Those and other surveys produce mountains of data that astronomers can mine, searching for relationships within the family of stars.

What could you learn about stars from a survey of the stars near the sun? Because the sun is believed to be in a typical place in the universe, such a survey could reveal general characteristics of the stars and might reveal unexpected processes in the formation and evolution of stars. Study **The Family of Stars** on pages 210–211 and notice three important points:

1 First, taking a survey is difficult because you must be sure you get an honest sample. If you don't survey enough stars or if you don't notice some kinds of stars, you can get biased results.

2 The second important point is that most stars are faint, and the most luminous stars are rare. The most common kinds of stars are the lower-main-sequence red dwarfs and the white dwarfs.

3 Finally, notice that what you see in the sky is deceptive. Stars near the sun are quite faint, but luminous stars, although they are rare, are easily visible even at great distances. Many of the brighter stars in the sky are highly luminous stars that you see even though they lie far away.

The night sky is a beautiful carpet of stars, but they are not all the same. Some are giants and supergiants, and some are dwarfs. The family of the stars is rich in its diversity.

Building Scientific Arguments

What kind of stars do you see if you look at a few of the brightest stars in the sky?
This argument shows how careful you must be to interpret simple observations. When you look at the night sky, the brightest stars are mostly giants and supergiants. Most of the bright stars in Canis Major, for instance, are supergiants. Sirius, in Canis Major, is the brightest star in the sky, but it is a main-sequence star; it looks bright because it is very nearby, not because it is very luminous. In general, the supergiants and giants are so luminous that they stand out and look bright, even though they are not nearby. When you look at a bright star in the sky, you are probably looking at a highly luminous star—a supergiant or a giant. You can check the argument above by consulting the tables of the brightest and nearest stars in the Appendix.

Now revise your argument. **What kind of star do you see if you look at a few of the stars nearest to the sun?**

■　　■　　■

Connections: This chapter set out to find the basic properties of stars. Once you found the distance to the stars, you were able to find their luminosities, diameters, and masses—rather mundane data. But you have now discovered a puzzling situation. The largest and most luminous stars are so rare you might joke that they hardly exist, and the average stars are such dinky low-mass things they are hard to see even if they are near Earth in space. Why does nature make stars in this peculiar way? To answer that question, you must explore the birth, life, and death of stars. That quest begins in the next chapter.

1 What is the most common kind of star? Are some rare? Are some common? To answer those questions you must survey the stars. To do so you must know their spectral class, their luminosity class, and their distance. Your census of the family of stars produces some surprising demographic results.

1a You could survey the stars by observing every star within 62 pc of Earth. A sphere 62 pc in radius encloses a million cubic parsecs. Such a survey would tell you how many stars of each type are found within a volume of a million cubic parsecs.

62 pc

Earth

2 Your survey faces two problems.

1. The most luminous stars are so rare you find few in your survey region. There are no O stars at all within 62 pc of Earth.

2. Lower-main-sequence M stars, called red dwarfs, and white dwarfs are so faint they are hard to locate even when they are only a few parsecs from Earth. Finding every one of these stars in your survey sphere is a difficult task.

Spectral Class Color Key

- ● O and B
- ● A
- ● F
- ● G
- ● K
- ● M

The star chart in the background of these two pages shows most of the constellation Canis Major; stars are represented as dots with colors assigned according to spectral class. The brightest stars in the sky tend to be the rare, highly luminous stars, which look bright even though they are far away. Most stars are of very low luminosity, so nearby stars tend to be very faint red dwarfs.

Red dwarf
15 pc

o² Canis Majoris
B3Ia   790 pc

Red dwarf
17 pc

δ Canis Majoris
F8Ia   550 pc

σ Canis Majoris
M0Iab   370 pc

η Canis Majoris
B5Ia   980 pc

ε Canis Majoris
B2II   130 pc

Sirius A (α Canis Majoris) is the brightest star in the sky. With a spectral type of A1V, it is not a very luminous star. It looks bright because it is only 2.6 pc away.

Sirius B is a white dwarf that orbits Sirius A. Although Sirius B is not very far away, it is much too faint to see with the unaided eye.

In this histogram, bars rise from an H–R diagram to represent the frequency of stars in space.

Red dwarfs and white dwarfs are the most common kinds of stars.

2a O and B stars, super-giants, and giants are so rare their bars are not visible in this graph.

3 Luminous stars are rare but are easy to see. Most stars are very low luminosity. See H–R diagrams at right.

Brightest stars

Nearest stars

The nearest stars in space tend to be very faint stars — lower-main-sequence red dwarfs or white dwarfs. Nearly all of these stars are faint in the sky even though they are nearby. Only a few are visible to the unaided eye.

The brightest stars in the sky tend to be highly luminous stars — upper-main-sequence stars, giants, or supergiants. They look bright because they are luminous, not because they are nearby.

Although white dwarfs and red dwarfs are very common, not a single one is bright enough to be visible to the unaided eye.

Summary

9-1 | Measuring the Distances to Stars

How far away are the stars?

■ Distance is critical in astronomy. Your goal in this chapter was to characterize the stars by finding their luminosities, diameters, and masses. Before you could begin, you needed to find their distances. Only by first knowing the distance to a star could you find its other properties.

■ Astronomers can measure the distance to nearer stars by observing their parallaxes. The most distant stars are so far away that their parallaxes are unmeasurably small. To find the distances to these stars, astronomers must use spectroscopic parallax.

■ Stellar distances are commonly expressed in parsecs. One parsec is 206,265 AU—the distance to an imaginary star whose parallax is 1 second of arc.

9-2 | Intrinsic Brightness

How much energy do stars make?

■ Once you know the distance to a star, you can find its intrinsic brightness expressed as its absolute magnitude. The absolute magnitude of a star equals the apparent magnitude it would have if it were 10 pc away.

■ The luminosity of a star, found from its absolute magnitude and its distance, is a measure of the total energy radiated by the star in one second. Luminosity is often expressed in terms of the luminosity of the sun.

9-3 | The Diameters of Stars

How big are stars?

■ The H–R diagram is a plot of luminosity versus surface temperature. It is an important graph in astronomy because it sorts the stars into categories by size.

■ Roughly 90 percent of normal stars, including the sun, fall on the main sequence, with the more massive stars being hotter and more luminous. The giants and supergiants, however, are much larger and lie above the main sequence. Some of the white dwarfs are very hot stars, but they fall below the main sequence because they are so small.

■ The large size of the giants and supergiants means their atmospheres have low densities and their spectra have sharper spectral lines than the spectra of main-sequence stars. In fact, it is possible to assign stars to luminosity classes by the widths of their spectral lines. Class V stars are main-sequence stars. Giant stars, class III, have sharper lines; and supergiants, class I, have extremely sharp spectral lines.

■ Spectroscopic parallax allows astronomers to estimate the distance to stars too far away for direct parallax measurements. By classifying a star according to spectral type and luminosity class, an astronomer can estimate its absolute magnitude from the H–R diagram and then find its distance by comparing absolute and apparent magnitudes.

9-4 | The Masses of Stars

How much matter do stars contain?

■ The only direct way you can find the mass of a star is by studying binary stars. When two stars orbit a common center of mass, astronomers find their masses by observing the period and sizes of their orbits.

■ Given the mass and diameter of a star, you can find its average density. On the main sequence, the stars are about as dense as the sun, but

the giants and supergiants are very-low-density stars. Some are much thinner than air. The white dwarfs, lying below the main sequence, are tremendously dense.

■ The mass–luminosity relation says that the more massive a star is, the more luminous it is. Main-sequence stars follow this rule closely, the most massive being the upper-main-sequence stars and the least massive the lower-main-sequence stars. Giants and supergiants do not follow the relation precisely, and white dwarfs not at all.

9-5 | A Survey of the Stars

What is the typical star like?

■ A survey in the neighborhood of the sun shows that lower-main-sequence stars are the most common type. Giants and supergiants are rare, but white dwarfs are quite common, although they are faint and hard to find.

New Terms

stellar parallax (*p*) (p. 189)

parsec (pc) (p. 189)

proper motion (p. 190)

flux (p. 191)

absolute visual magnitude (M_V) (p. 192)

magnitude–distance formula (p. 192)

distance modulus ($m_V − M_V$) (p. 192)

luminosity (*L*) (p. 193)

absolute bolometric magnitude (p. 193)

H–R (Hertzsprung–Russell) diagram (p. 194)

main sequence (p. 196)

giant (p. 196)

supergiant (p. 197)

red dwarf (p. 197)

white dwarf (p. 197)

luminosity class (p. 198)

spectroscopic parallax (p. 199)

binary stars (p. 200)

visual binary system (p. 202)

spectroscopic binary system (p. 202)

eclipsing binary system (p. 204)

light curve (p. 204)

mass–luminosity relation (p. 207)

Review Questions

Ace⊛Astronomy™ Assess your understanding of this chapter's topics with additional quizzing and animations at **http://ace** **.brookscole.com/sf9**

1. Why are parallax measurements made from Earth-based telescopes limited to the nearest stars?

2. How would having an observatory on Mars help astronomers measure parallax more accurately?

3. How did having an observatory in orbit above Earth's atmosphere help astronomers measure parallax more accurately?

4. What do the words *absolute* and *visual* mean in the term *absolute visual magnitude*?

5. For which stars does absolute visual magnitude differ least from absolute bolometric magnitude? Why?

6. What does luminosity measure that absolute visual magnitude does not?

7. How can a cool star be more luminous than a hot star? Give some examples.

8. How can you be certain that the giant stars are actually larger than the sun?

9. How can astronomers be sure that white dwarfs really are small?

10. What observations would you make to classify a star according to its luminosity class?

11. Describe the steps in using the method of spectroscopic parallax. Do you really measure a parallax?

12. Give the approximate radii and intrinsic brightnesses of stars in the following classes: G2 V, G2 III, G2 Ia.

13. Why does the orbital period of a binary star depend on the mass of the stars?

14. Why can astronomers find the masses of the stars in visual binary systems and in eclipsing binary systems but not in spectroscopic binary systems?

15. How do the ordered masses along the main sequence illustrate the mass–luminosity relation?

16. Why is it difficult to find out how common the most luminous stars are? the least luminous stars?

17. If all of the stars in the photo here are members of the same star cluster, then they all have about the same distance. Then why are three of the brightest much redder than the rest? What kind of star are they?

(NASA)

Discussion Questions

1. If someone asked you to compile a list of the stars nearest the sun based on your own observations, what kind of observations would you make, and how would you analyze them to detect nearby stars?

2. The sun is sometimes described as an average star. What is the average star really like?

Problems

1. If a star has a parallax of 0.050 second of arc, what is its distance in parsecs? in light-years? in AU?

2. If a star has a parallax of 0.016 second of arc and an apparent magnitude of 6, how far away is it, and what is its absolute magnitude?

3. Complete the following table:

m_V	M_V	d (pc)	P (seconds of arc)
——	7	10	——
11	——	1000	——
——	−2	——	0.025
4	——	——	0.040

4. If a main-sequence star has a luminosity of 400 $L_\odot$, what is its spectral type? (*Hint:* See Figure 9-13.)

5. If a star has an apparent magnitude equal to its absolute magnitude, how far away is it in parsecs? in light-years?

6. If a star has an absolute bolometric magnitude that is eight magnitudes brighter than the sun, what is the star's luminosity?

7. If a star has an absolute bolometric magnitude that is one magnitude fainter than the sun, what is the star's luminosity?

8. An O8 V star has an apparent magnitude of +1. Use the method of spectroscopic parallax to find the distance to the star.

9. Find the luminosity of the sun, given the radius of Earth's orbit and the solar constant (Chapter 8). Make your calculation in two steps. First, use $4\pi R^2$ to calculate the surface area (in square meters) of a sphere surrounding the sun with a radius of 1 AU. Second, multiply by the solar constant to find the total solar energy passing through the sphere in 1 second. That is the luminosity of the sun. Compare your result with that in Celestial Profile 1.

10. In the following table, which star is brightest in apparent magnitude? most luminous in absolute magnitude? largest? farthest away?

Star	Spectral Type	m_V
a	G2 V	5
b	B1 V	8
c	G2 Ib	10
d	M5 III	19
e	White dwarf	15

11. What is the total mass of a visual binary system if its average separation is 8 AU and its period is 20 years?

12. Measure the wavelengths of the iron lines in Figure 9-17 and plot a graph showing the radial velocities of the two stars in this system. What is the period? Assuming that the orbit is circular and edge-on, what is the total mass? What are the individual masses?

13. If the eclipsing binary in Figure 9-20 has a period of 32 days, an orbital velocity of 153 km/s, and an orbit that is nearly edge-on, what is the circumference of the orbit? the radius of the orbit? the mass of the system?

14. If the orbital velocity of the eclipsing binary in Figure 9-20 is 153 km/s and the smaller star becomes completely eclipsed in 2.5 hours, what is its diameter?

15. What is the luminosity of a 4-solar-mass main-sequence star? of a 9-solar-mass main-sequence star? of a 7-solar-mass main-sequence star?

Media Cluster

Ace ✺ Astronomy™ To access the resources in the Media Cluster, log into AceAstronomy at **http://ace.brookscole .com/sf9** and select Chapter 9.

ACTIVE FIGURES

Brightness and Distance
In this animation, you can change the position of both the truck and the bicyclist in the figure and observe the effect on the intensity of light seen by the observer.

Center of Mass
You often think in terms of the moon orbiting Earth, but, more precisely, both bodies orbit each other around a common center of mass. Experiment with this animation to see how the ratio of the masses affects the relative size of the orbits.

ASTRONOMY EXERCISES

Parallax I and II
The apparent change in position of an object due to a change in the location of the observer is known as parallax. Use these two animations to study the concept of parallax in astronomy.

Apparent Brightness and Distance
The distance to stars affects how bright they look to you on Earth. You can change the distance to a star in this animation and see its apparent brightness change.

Stefan–Boltzmann Law II
In this exercise, you can vary the temperature of a star and its radius. As you change the temperature and its radius, notice what happens to the total energy output of the star as indicated by the luminosity meter.

Spectroscopic Binaries
This animation shows a close double star system where each star orbits a common center of gravity. As the stars change position, notice what happens to the absorption lines in the spectrum of the star system.

Eclipsing Binaries
Study an eclipsing binary system with this animation. As the stars move and periodically pass in front of one another, the total light output from the system is charted on a graph.

Mass–Luminosity Relation
This exercise provides a list of stars whose masses and luminosities have been measured. Click on each star to plot its point on the mass versus luminosity axes.

H–R/Mass–Luminosity 3-D Graph
Explore this 3-D graph and see how mass is related to position in the H–R diagram. You can select from main-sequence stars, red giants, white dwarfs, and variable stars.

Lab 11: The Spectral Sequence and the H–R Diagram

This lab introduces the tools astronomers use to identify the stars of the main sequence and demonstrates how that information is used to estimate the distance to the stars and the age of a star cluster.

Lab 12: Binary Stars

This lab investigates some things that can be learned from different types of binary stars.

Newton's form of Kepler's third law is used to determine stellar masses. You also see how the Doppler effect can be used in the study of binary systems.

Lab 16: Astronomical Distance Scales

In this lab, you explore some methods for determining distances in astronomy.

Critical Inquiries for the Web

1. The Hertzsprung–Russell diagram was named for two famous astronomers. Who were they? What did they do to earn such an honor?

2. Algol is a famous binary star. What can you find out about the mythology of Algol?

3. An entire class of binary stars is called the Algol binary stars. How would you characterize such stars?

Exploring *TheSky*

1. Locate the following stars and determine their apparent magnitude, parallax, distance in parsecs and in light-years, and spectral classification. (*Hint:* To center on an object, use **Find** under the **Edit** menu and type the object's name followed by a period.)

Sirius	Vega	Antares	Altair
Aldebaran	Deneb	Betelgeuse	

2. Use the spectral type and parallax of the stars above to estimate their distance. Compare with the distances given in *TheSky*.

3. Take a survey of the stars. Center on Orion and adjust the field until it is about 100° wide. Click on the 10 brightest stars and record their spectral types. Now zoom in until only a few dozen stars are in the frame. Click on the 10 faintest stars and record their spectral types. Is there a difference between the brightest and faintest stars? Is the re-+/sult what you expected? Are certain kinds of stars missing from this computer database?

4. Repeat Activity 3 for a region centered on the Big Dipper.

Go to the Brooks/Cole Astronomy Resource Center (**http://astronomy.brookscole.com**) for critical thinking exercises, articles, and additional readings from InfoTrac College Edition, Brooks/Cole's online student library.

10 | The Interstellar Medium

Visual-wavelength image

JULIET LOVED ROMEO so much she compared him to the beauty of the stars (see the quotation above). Had she known what was between the stars, she might have compared him to that instead. True, the gas and dust between the stars are mostly dark and cold, but where it is illuminated by stars, that matter creates beautiful nebulae, and where it is densest it gives birth to beautiful stars. If there is beauty in vast extent and sweeping power, then the **interstellar medium,** the gas and dust between the stars, could steal worship from the garish stars. One reason you need to know about the interstellar medium is that it gives birth to stars and is enriched by the deaths of stars. The interstellar medium is an important part of the story of stars. Another reason is that it illustrates once again how astronomers use the interaction of light

Continued on page 218

Guidepost

Looking Back

It is a thrill to photograph a faint galaxy a billion light-years from Earth; part of the fascination of astronomy is seeing the unseen. In previous chapters, you have studied the sun and stars in general, objects that are not too hard to detect. Now you are ready to study something that is nearly invisible.

This Chapter

Space is not empty. Great clouds of gas and dust drift between the stars, and this chapter will show you how astronomers use spectrographic analysis (Chapter 7) to understand that thinly scattered matter. In this chapter, you will answer three essential questions:

How do astronomers know that space isn't empty?

How do astronomers study the nearly invisible matter between the stars?

How does the matter between the stars interact with the stars?

Studying the matter between the stars is not an idle exercise. The gas and dust in space give birth to stars and are enriched by the deaths of stars.

Looking Ahead

As you begin studying the matter between the stars, you are beginning the story of stars. In the next few chapters you will see how stars are born, how they live, how they die, and what remains when they die. Are you surprised that stars die? That's just one of nature's little secrets.

and matter to understand the universe. But another reason to study the interstellar medium is that it is a beautiful part of the universe (■ Figure 10-1). Most people would use the phrase "the vacuum of space"; you are about to discover that space is not empty.

But be warned: In Chapter 1, you saw that we live in a disk-shaped galaxy, the Milky Way (Figure 1-11). The interstellar medium is confined mostly within a few hundred parsecs of the plane of the disk of our galaxy; if you journeyed out of the galaxy, out between the galaxies, you would find space a much better vacuum. In your study of the interstellar medium, you are learning about the disk of our galaxy in the neighborhood of the sun—probably a typical part of interstellar space.

Later you will need observations at radio, infrared, ultraviolet, and X-ray wavelengths to study the interstellar medium, but you can begin with one of your most sensitive instruments—eyeballs.

10-1) Visible-Wavelength Observations

A QUICK GLANCE at the night sky does not reveal any matter between the stars, but evidence of its existence is visible to the naked eye in a familiar constellation.

Nebulae

On a cold, clear winter night, Orion hangs high in the southern sky, a large constellation composed of brilliant stars. If you look carefully at Orion's sword, you will see that one of the stars is a hazy cloud (Figure 2-4). A small telescope reveals even more such clouds of gas and dust. Astronomers refer to these clouds as **nebulae** (singular, nebula), from the Latin word for mist or cloud.

Study **Three Kinds of Nebulae** on pages 220–221 and notice three important points:

1 First, very hot stars can excite clouds of gas and dust to emit light, and that reveals that the clouds contain mostly hydrogen gas at very low densities.

2 Second, where slightly cooler stars illuminate clouds you see evidence that the dust in the clouds is made up of very small particles.

3 The third thing to notice is that some dense clouds of gas and dust are detectable only where they are silhouetted against background regions filled with stars or bright nebulae.

A detailed analysis of the spectra of nebulae can tell astronomers even more. Certain lines in the spectra of emission nebulae are called **forbidden lines** because they are almost never seen in

■ **Figure 10-1**

A swirl of dusty gas forms the Horse Head Nebula. The background nebula glows with the pink color of ionized hydrogen, and a bright blue star at lower left is just able to shine through the gas and dust. (Daniel Good)

excited gas on Earth. Two good examples are the strong green lines at 495.9 nm and 500.7 nm produced by oxygen atoms that have lost two electrons. (Following the convention for naming ions, twice-ionized oxygen is OIII.) The oxygen ions can become excited by collision with a high-energy photon or a rapidly moving ion or electron, and the atom can emit various-wavelength photons as its electron cascades back down to lower energy levels. However, transitions between certain energy levels are so unlikely they are called "forbidden." If an electron enters such an energy level, it will remain there for a relatively long time before it can decay further and emit the appropriate photon. For a normal transition, the electron might wait 10^{-8} to 10^{-7} second. If it enters one of these **metastable levels,** the electron may be stuck there for as long as an hour before it becomes likely to fall to a lower level and emit the proper photon.

Now you can understand why these forbidden lines aren't visible in laboratories on Earth. The atoms in a dense gas collide with each other so often that there isn't time for an electron in a metastable level to decay to a lower level and emit a photon. In a jar of gas in a laboratory, the gas is dense and the atoms collide so often that such electrons get excited back up to higher levels before they can drop downward and emit a photon.

And that tells you something important about emission nebulae. In those nebulae, the gas has a very low density, and an atom could go for an hour or more between collisions, giving an electron stuck in a metastable level time to decay to a lower level and emit a photon at a so-called forbidden wavelength. The forbidden lines are clear evidence that nebulae have very low densities. This is a dramatic example of how astronomers can use knowledge of atomic physics to understand astronomical objects.

The three kinds of nebulae are a good introduction to the interstellar medium, but there is more to learn. The next step is to think carefully about light passing through this gas and dust.

Extinction and Reddening

The dust in space is called **interstellar dust.** Its presence is dramatically evident in dark nebulae, but simple observations tell astronomers that the interstellar dust is spread throughout space, making up roughly 1 percent of the mass of the interstellar medium.

One way astronomers know that dust is present in the interstellar medium is that it makes distant stars appear fainter than they would if space were perfectly transparent. This phenomenon is called **interstellar extinction,** and in the neighborhood of the

sun it amounts to about 2 magnitudes per thousand parsecs. That is, if a star is 1000 pc from Earth, it will look about 2 magnitudes fainter than it would if space were perfectly empty. If it were 2000 pc away, it would look about 4 magnitudes dimmer, and so on. This is a dramatic effect, and it shows that the interstellar medium is not confined to a few nebulae scattered here and there. The so-called vacuum of space is filled with a low-density, dusty gas. The spaces between the stars are far from empty.

Another way dust reveals its presence is through the effect it has on the colors of stars. An O star should be blue, but some stars with the spectrum of an O star look much redder than they should. Called **interstellar reddening,** this effect is produced by dust particles scattering light. As you saw in the case of the reflection nebulae, the dust particles are small, with diameters roughly equal to the wavelength of light, and they scatter shorter wavelengths better than longer wavelengths. That means they scatter blue photons more often than red photons. As light from a distant star travels toward Earth it loses some of its blue photons because of scattering, and consequently the star looks redder (■ Figure 10-2).

As was discussed earlier, this scattering of blue light is what makes the sky blue, but it is also what makes distant city lights look yellow. If you view the lights of a city at night from a high-flying aircraft or a distant mountaintop, the lights will look yellow. As you descend toward the city, the lights will look bluer.

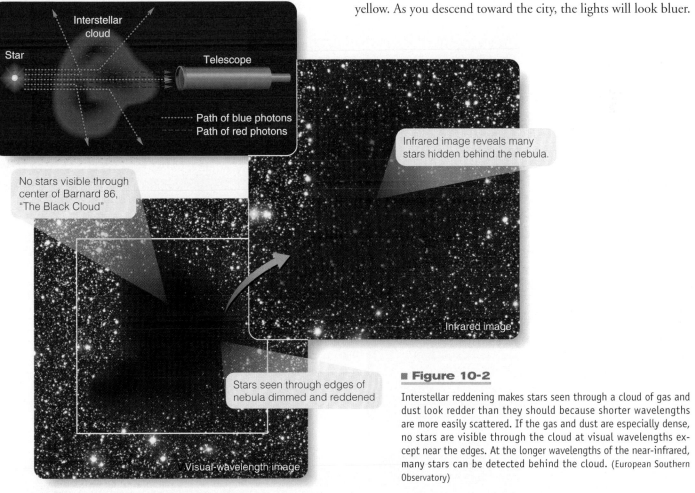

No stars visible through center of Barnard 86, "The Black Cloud"

Stars seen through edges of nebula dimmed and reddened

Visual-wavelength image

Infrared image reveals many stars hidden behind the nebula.

Infrared image

■ Figure 10-2

Interstellar reddening makes stars seen through a cloud of gas and dust look redder than they should because shorter wavelengths are more easily scattered. If the gas and dust are especially dense, no stars are visible through the cloud at visual wavelengths except near the edges. At the longer wavelengths of the near-infrared, many stars can be detected behind the cloud. (European Southern Observatory)

Three Kinds of Nebulae

1 Emission nebulae are produced when a hot star excites the gas near it to produce an emission spectrum. The star must be hotter than about B1 (25,000 K). Cooler stars do not emit enough ultraviolet radiation to ionize the gas. Emission nebulae have a distinctive pink color produced by the blending of the red, blue, and violet Balmer lines. Emission nebulae are also called HII regions, following the custom of naming gas with a roman numeral to show its state of ionization. HI is neutral hydrogen, and HII is ionized.

In an HII region, the ionized nuclei and free electrons are mixed. When a nucleus captures an electron, the electron falls down through the atomic energy levels, emitting photons at specific wavelengths. Spectra indicate that the nebulae have compositions much that like of the sun – mostly hydrogen. Emission nebulae have densities of 100 to 1000 atoms per cubic centimeter, better than the best vacuums produced in laboratories on Earth.

Visual-wavelength image

European Southern Observatory

Ace Astronomy™

Log into AceAstronomy and select this chapter to see Active Figure "Scattering." Watch photons scatter through Earth's atmosphere.

2 A reflection nebula is produced when starlight scatters from a dusty nebula. Consequently, the spectrum of a reflection nebula is just the reflected absorption spectrum of starlight. Gas is surely present in a reflection nebula, but it is not excited to emit photons. See image below.

Reflection nebulae NGC 1973, 1975, and 1977 lie just north of the Orion nebula. The pink tints are produced by ionized gases deep in the nebulae.

2a Reflection nebulae look blue for the same reason the sky looks blue. Short wavelengths scatter more easily than long wavelengths. See image below.

Sunlight enters Earth's atmosphere

Blue photons are scattered more easily than longer wavelengths and blue photons enter your eyes from all directions, making the sky look blue.

Visual-wavelength image

AATB

2b The blue color of reflection nebulae at left shows that the dust particles must be very small in order to preferentially scatter the blue photons. Interstellar dust grains must have diameters ranging from 0.001 mm down to 1 nm or so.

The hottest star in the Pleiades star cluster is Merope, a B3 star. It is not hot enough to ionize the gas so you see a reflection nebula rather than an emission nebula.

Merope

Visual-wavelength image

2c A dusty reflection nebula is located very close to the star Merope above. The glare from the star is caused by internal reflections in the telescope, but the wispy nature of the nebula is real. The intense light from the star is pushing the dust particles away and may destroy the little nebula over the next few thousand years.

Merope

Visual

NASA

Reflection

Emission

Trifid Nebula

The **Milky Way** in **Sagittarius** contains two **nebulae** that **dramatically demonstrate the difference** between **emission** and **reflection** nebulae.

Emission

Visual

Lagoon Nebula

Daniel Good

Dark Nebula Barnard 86

Star Cluster NGC6520

Hubble Heritage Team, NASA

Visual

Visual

AATB

3 **Dark nebulae** are dense clouds of gas and dust that obstruct the view of more distant stars. Some are generally round, but others are twisted and distorted, as shown at the left, suggesting that even when there are no nearby stars to ionize the gas or produce a reflection nebula, there are breezes and currents pushing through the interstellar medium.

Northern Coalsack

Cygnus

Milky Way

Great Rift

The Horsehead Nebula in Orion is a dark nebula silhouetted against a more distant emission nebula.

Large dark nebulae obstruct the view of more distant stars and form holes and rifts along the Milky Way. The Great Rift extends from Cygnus to Sagittarius.

The light from the city is reddened by microscopic particles in the air. If the particles are especially dense, people call them smog.

Astronomers can measure the amount of reddening by comparing two stars of the same spectral type, one of which is dimmed more than the other. The more obscured star will look redder. If you plot the difference in brightness between the two stars against wavelength, you get a curve that shows the reddening. That is, it shows how the starlight is dimmed at different wavelengths (■ Figure 10-3). In general, the light is dimmed in proportion to the reciprocal of the wavelength, a pattern typical of scattering from small dust particles. Laboratory measurements show that the high extinction at about 220 nm is caused by a form of carbon, evidence that some of the dust particles are carbon. Other evidence suggests that some grains contain silicates and metals and may have coatings of water ice, frozen ammonia, or carbon-based molecules.

Interstellar Absorption Lines

If you looked at the spectra of distant stars, you could see dramatic evidence of an interstellar medium. Of course, you would see spectral lines produced by the gas in the atmospheres of the stars, but you would also see sharp spectral lines produced by the gas in the interstellar medium. These **interstellar absorption lines** provide a new way to study the gas between the stars.

Astronomers can recognize interstellar absorption lines in three ways: by their ionization, by their widths, and by their multiple components. Some stellar spectra contain absorption lines that just don't belong because they represent the wrong ionization state. For example, if you looked at the spectrum of a very hot star such as an O star, you would expect to see no lines of once ionized calcium (CaII) because that ion cannot exist at the high temperatures in the atmosphere of such a hot star. But many O-star spectra contain lines of CaII. These lines must have been produced not in the star but in the interstellar medium.

In addition, the widths of the interstellar lines give away their identity. In the hot atmosphere of a star, the atoms move rapidly, and the random Doppler shifts broaden spectral lines. Also, the atoms collide with each other often enough to blur the electron energy levels, and that also broadens the spectral lines. These two effects were discussed as Doppler broadening and collisional broadening in Chapter 7. This blurring makes the lines in the spectrum of a main-sequence star quite broad, but even in the atmosphere of a giant or supergiant star, where the gas is less dense, the gas atoms move rapidly and collide often enough to broaden the spectral lines. Interstellar lines, on the other hand, are exceedingly sharp. This shows that the interstellar matter is extremely cold and has a low density. If it were hot, Doppler broadening would smear out the lines due to the motions of individual atoms. If the gas were dense, collisional broadening would produce wider lines. The exceedingly narrow widths of the interstellar absorption lines are typical of cold, low-density gas.

Another revealing characteristic of the interstellar lines is that they are often split into two or more components. The multiple components have slightly different wavelengths and appear to have been produced when the light from the star passed through different clouds of gas on its way to Earth. Because the clouds of gas have slightly different radial velocities, they produce absorption lines with slightly Doppler-shifted wavelengths. ■ Figure 10-4 illustrates

■ **Figure 10-4**

Interstellar absorption lines can be recognized in three ways. (a) The B0 supergiant ε Orionis is much too hot to show spectral lines of once-ionized calcium (CaII), yet this short segment of its spectrum reveals narrow, multiple lines of CaII (tick marks) that must have been produced in the interstellar medium. (The Observatories of the Carnegie Institution of Washington) (b) Spectral lines produced in the atmospheres of stars are much broader than the spectral lines produced in the interstellar medium. (Adapted from a diagram by Binnendijk) In both (a) and (b) the multiple interstellar lines are produced by separate interstellar clouds with slightly different radial velocities.

■ **Figure 10-3**

Interstellar extinction, the dimming of starlight by dust between the stars, depends strongly on wavelength. Infrared radiation is only slightly affected, but ultraviolet light is strongly scattered. The strong extinction at about 220 nm is caused by a certain form of carbon dust in the interstellar medium.

all three of these characteristics of interstellar absorption lines—the wrong ionization, narrow widths, and multiple components.

Clouds and What's in Between

Astronomers disagree as to the exact structure of the interstellar medium, and the boundaries and characteristics of clouds are ill defined. Nevertheless, you can get a general picture of the interstellar medium by classifying clouds into a few main varieties based on visual-wavelength observations.

Studies of interstellar absorption lines reveal clouds of neutral gas (and presumably dust) with densities of 10 to a few hundred atoms/cm^3. Because these clouds are not ionized, they are called **HI clouds.** These clouds must be 50 to 150 pc in diameter and have masses of a few solar masses. The gas temperature is only about 100 K. Starlight seems to pass through six to ten of these clouds for every 1000 pc near the plane of our galaxy. While it is easy to think of these clouds as more or less spherical blobs of gas, observations show that they are usually twisted into long filaments, flattened into thin sheets, or tangled into chaotic shapes (■ Figure 10-5)—further evidence that the interstellar medium is not static and motionless.

Between these HI clouds of neutral gas lies a hot **intercloud medium** with a temperature of a few thousand K and a density of only about 0.1 atom/cm^3. This intercloud medium consists of ionized hydrogen (HII). The gas is partially ionized by the ultraviolet photons in starlight. If it is to be in equilibrium with the HI clouds (if the HI clouds are not expanding or contracting), the pressure in both regions should be about the same.

The pressure in a gas depends on its density and its temperature (**Focus on Fundamentals 5**). In the interstellar medium, the HI clouds are cool but dense, and the HII gas of the intercloud medium is hot but low in density. That means the two regions could have similar pressures even though their temperatures are quite different. Where the pressures are equal, the HI clouds and the intercloud medium are stable—a bit like ice cubes floating in ice water. Nevertheless, observations reveal some regions where the pressures are not equal, and that is poorly understood.

By now you are probably wondering how the intercloud medium could be ionized when it is not close to hot stars. To understand why it is ionized, imagine that you are an atom floating in the intercloud medium. Ultraviolet photons from distant stars are not common, but they do whiz by now and then. Soon you become ionized by absorbing one of these photons and losing an electron. In a denser gas, you would quickly find another electron, capture it, and become a neutral atom again, but the interstellar medium has such a low density that the gas particles are spread far apart. You must wait a very long time before an electron comes close enough for you to capture. That is why the

■ **Figure 10-5**

Like a flag in a strong wind, this dark nebula is being distorted and twisted by intense radiation from very luminous stars out of the picture to the left. (Anglo-Australian Observatory)

gas of the intercloud medium is ionized; the atoms spend almost all their time waiting for an electron to happen past.

Studies of interstellar absorption lines can tell you more about the composition of the interstellar gas. It is much like the composition of the sun. Hydrogen is most abundant, with helium second. Light elements such as carbon, nitrogen, and oxygen are as common as in the sun, but elements such as iron, calcium, and titanium are less abundant than they are in the sun. Most likely, these elements are missing from the gas because they have condensed to form the dust.

Building Scientific Arguments

What evidence can you cite that there is an interstellar medium?

Everything in science is based on evidence, so your argument should discuss observations. First, there is the simple fact that certain parts of the interstellar medium are visible—the nebulae. Emission, reflection, and dark nebulae show that interstellar space is not totally empty. Further, astronomers can detect interstellar extinction and reddening; the more distant stars look fainter than they should, and they also look redder. This not only shows that all of interstellar space is filled by some thin material, it also shows that some of the material is in the form of tiny dust specks. Gas atoms would not be very effective at reddening starlight, so the interstellar medium must contain dust as well as gas.

Pressure

One of the most fundamental parameters in science is **pressure**, a measure of force per unit area. If you are going to understand astronomy, you must be sure you understand pressure. Doctors measure blood pressure, and astronomers measure gas pressure, but they are really measuring the same fundamental parameter.

Pressure is expressed in the units of force per unit area. If you inflate the tires on a car, you use the unit pounds per square inch. A typical pressure might be 34 lb/in². It is important to note that this is not the total force pushing out on the inside of the tire but only the force exerted on a single square inch. When you stand, your weight exerts a force on the floor, and the pressure under your shoes is your weight divided by the surface area of your shoes' soles. A typical pressure might be only 4 lb/in². If you step on someone's toe, that is the pressure you exert. Of course, if you were wearing ice skates, your weight would be spread over a much smaller area, the area of the bottom of the blade, and you might exert a pressure of 150 lb/in² or more. That's why skaters must be

Pressure pushes outward on the inside of a balloon.

careful not to step on someone's toe. The pressure would be dangerously high.

Astronomers are most commonly interested in the behavior of matter when it is a gas, and pressure in a gas arises when atoms or molecules collide. Consider, for example, how the

gas molecules colliding with the inside of a balloon exert an outward force on the rubber and keep the balloon inflated. If the gas is hot, the atoms or molecules move rapidly, and the resulting pressure is higher than for a cooler gas. If the gas is dense, there will be many gas particles colliding with the inside of the balloon, and the pressure will be higher than for a lower-density gas. In this way, pressure depends on both the temperature and the density of the gas.

Notice that pressure and density are related, but they are not at all the same thing. Density is a measure of the amount of matter in a given volume, and pressure is a measure of the force that matter exerts on its surroundings. A very-low-density gas and a very-high-density gas might have the same pressure if they had different temperatures.

In daily life, you think of pressure when you inflate an automobile tire, but pressure is common in nearly all of the sciences. Astronomers must consider pressure in thinking about the gas inside stars and the thin gas between the stars.

PRESSURE | MASS | ENERGY | TEMPERATURE AND HEAT | DENSITY

Perhaps this is enough evidence to convince someone that the spaces between the stars are not empty, but there is more. Expand your argument. **What do interstellar absorption lines reveal about the interstellar gas?**

■ ■ ■

Connections: Observations at visible wavelengths reveal important information about the interstellar medium, but to paint a complete picture of the matter between the stars, astronomers must observe at other wavelengths.

10-2 Long- and Short-Wavelength Observations

THE INTERSTELLAR MEDIUM is mostly invisible to human eyes, so astronomers must observe at the longer wavelengths of infrared and radio radiation and at the shorter wavelengths of ultraviolet and X-ray radiation to further explore the interstellar medium.

21-cm Observations

Cold, neutral hydrogen floating in space can emit electromagnetic radiation with a wavelength of 21 cm. This **21-cm radiation** allows radio astronomers to map the distribution of neutral hydrogen (HI) throughout our galaxy.

The existence of the 21-cm radiation was predicted theoretically in the 1940s by H. C. van de Hulst, but it was not detected until 1951. You can understand how a theoretical prediction could be made by thinking of the structure of a hydrogen atom. A hydrogen atom consists of a proton and an electron, and physicists know that both of these particles must spin like tiny tops. They can never stop. Because these particles have an electrostatic charge, their rapid spin is the same as the circulation of an electric current through a coil of wire; it creates a magnetic field. Because the charge on the proton is positive and the charge on the electron is negative, the magnetic fields are reversed when the particles spin in the same direction.

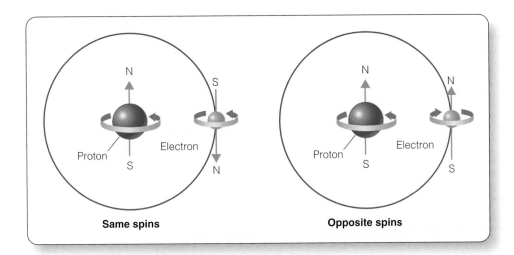

Both the proton and electron in a neutral hydrogen atom spin and consequently have small magnetic fields. When they spin in the same direction, their magnetic fields are reversed, and when they spin in opposite directions, their magnetic fields are aligned. As explained in the text, this allows cold, neutral hydrogen in space to emit radio photons with a wavelength of 21 cm.

Same spins　　　　**Opposite spins**

The magnetic fields of the spinning proton and electron affect the binding energy that holds the electron to the proton in a hydrogen atom. You have probably played with small magnets and noticed that they repel each other in one orientation and attract each other if you turn one around. In the same way, the small magnetic fields produced by the spinning proton and electron can repel or attract each other. In one orientation the electron is slightly less tightly bound, and in the other orientation it is slightly more tightly bound (■ Figure 10-6). The energy difference is very small, but it makes a big difference in astronomy.

Now you can follow van de Hulst's logic and predict the existence of 21-cm radiation. Because an electron in the ground state of a hydrogen atom could spin in either of two directions—the same way as the proton or the opposite way—the ground state must really be two energy levels separated by the tiny amount of energy produced by the spinning particles. If an electron is spinning such that it is in the higher of the two states, it can spontaneously flip over and spin in the other direction. Then the atom must drop to the lower of the two energy levels, and the excess energy is radiated away as a photon. The energy difference is small, so the photon has a long wavelength—21 cm (■ Figure 10-7).

The 21-cm radiation was predicted theoretically, but it could not be detected in laboratory experiments. Of the two closely spaced energy levels, the upper one is metastable. Once an electron gets caught in the upper energy level, it will, on average, stay there for 11 million years before spontaneously dropping to the lower energy level and emitting a photon of 21-cm wavelength. The atoms in a gas in a laboratory jar will collide with each other millions of times a second, so none of those atoms can remain undisturbed with its electron in the upper energy level long enough to produce a photon of 21-cm radiation. Atoms in space, however, collide much less often, so a few do manage to emit 21-cm radiation. That's why the existence of 21-cm radiation had to be confirmed observationally by radio astronomers who detected it coming from clouds of neutral hydrogen in space.

The 21-cm observations give astronomers a way to map the cold, neutral hydrogen that fills much of our galaxy. Astronomers can locate individual clouds and measure their motion by observing the Doppler shifts in the 21-cm radiation (■ Figure 10-8).

The 21-cm radiation can map only neutral hydrogen. Ionized hydrogen lacks an electron, so it can't emit 21-cm radiation. Further, hydrogen atoms locked in molecules are also unable to

■ **Figure 10-7**

The lowest energy level of the hydrogen atom, the ground state, is actually two closely spaced energy levels that differ because the proton and electron spin. In this diagram, you would need a magnifying glass to distinguish the energy levels that make up the ground state. When an atom decays from the upper energy level to the lower energy level, it emits a photon with a wavelength of 21 cm.

Ground state

21 cm

Powerful magnifier

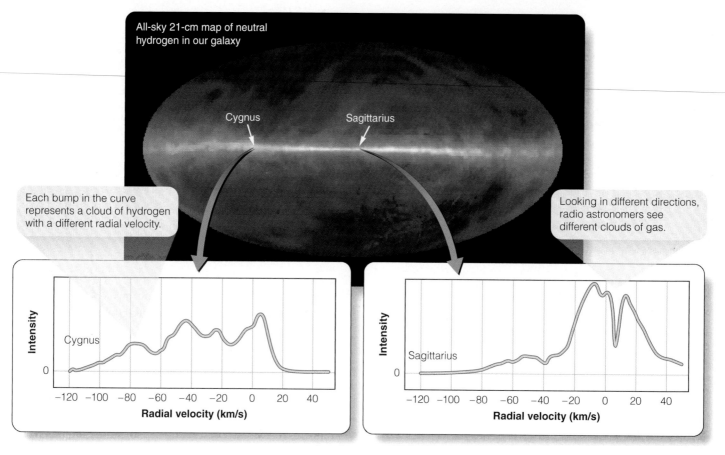

Figure 10-8

Most of the neutral hydrogen in the Milky Way Galaxy is in the plane of the disk-shaped galaxy—the bright band running from left to right. Notice how wispy and irregular the hydrogen is. (J. Dickey, UMn, F. Lockman, NRAO, SkyView) 21-cm radio observations reveal many clouds of neutral hydrogen orbiting the center of the galaxy. (Adapted from observations by Burton)

emit 21-cm wavelength photons. Astronomers must find other ways to study these parts of the interstellar medium.

Molecules in Space

Radio telescopes can also detect radiation from various molecules in the interstellar medium. A molecule can store energy in a number of different ways. For example, it can rotate at different rates, or the atoms in a molecule can vibrate as if they were linked together by small springs. If a molecule suffers a collision or absorbs a photon, it can be excited to vibrate and rotate in some higher energy state. Quickly, however, it will return to a lower energy state and radiate the excess energy as a photon. Because these energy levels are closely spaced, the emitted photons typically have low energies, and they fall in the radio or far-infrared part of the electromagnetic spectrum. Just as neutral hydrogen radiates at a specific wavelength of 21 cm, many natural molecules radiate at their own unique wavelengths.

Unfortunately for astronomers, molecules of hydrogen (H_2) are difficult to detect. In the far-ultraviolet, molecular hydrogen can be detected by the photons it absorbs, but these observations

must be made by specialized telescopes in space. Mapping the location of molecular hydrogen would be easier if the molecules emitted radio-wavelength photons, but a molecule containing two identical atoms does not radiate in the radio part of the spectrum. However, clouds of gas dense enough to form molecular hydrogen also form tiny amounts of other molecules, and many of them are good emitters of radio energy. Nearly 100 different molecules have been detected (■ Table 10-1). Some are quite complex, and it is not clear how they form. Most astronomers believe that the atoms meet and bond to form molecules on the surfaces of dust grains. Some of these molecules have not yet been synthesized on Earth, but others are common, such as N_2O (nitrous oxide), also known as laughing gas. Ethyl alcohol, which some humans drink, has also been detected. Although this is a very rare molecule compared to molecular hydrogen, a single interstellar cloud can contain ethyl alcohol in amounts equivalent to 10^{28} fifths of whisky (about 100 Earth masses).

One of the most important of the interstellar molecules may be carbon monoxide (CO). It is one of the poisonous gases that come out of the tailpipes of cars, but it is also a very good emitter

■ Table 10-1 │ Selected Molecules Detected in the Interstellar Medium

H_2	molecular hydrogen
H_2S	hydrogen sulfide
C_2	diatomic carbon
N_2O	nitrous oxide
CN	cyanogen
H_2CO	formaldehyde
CO	carbon monoxide
C_2H_2	acetylene
NO	nitric oxide
NH_3	ammonia
OH	hydroxyl
HCO_2H	formic acid
NaCl	common table salt
CH_4	methane
HCN	hydrogen cyanide
CH_3OH	methyl alcohol
H_2O	water
CH_3CH_2OH	ethyl alcohol

of radio energy at a wavelength of 2.6 mm. An interstellar cloud may contain only 1 CO molecule for every 10,000 molecules of hydrogen, but the CO can be detected while the molecular hydrogen cannot. Consequently, radio astronomers can map the interstellar clouds by searching for the CO radio emission; where they detect CO, they can be certain that molecular hydrogen is common.

Another important constituent of the interstellar medium is OH. Because this combination of an oxygen atom and a hydrogen atom is not electrically neutral, it is called a radical instead of a molecule. The OH radical forms in interstellar clouds and is a good emitter of radio energy.

These molecules are very fragile, and a high-energy photon such as an ultraviolet photon has enough energy to break the molecule into separate atoms. Consequently, the molecules cannot exist outside the dense clouds. Only deep inside the densest clouds, where dust absorbs and scatters the short-wavelength, high-energy photons, can the molecules survive. The very fact that these molecules are detected tells you that some of these **molecular clouds** must be very dense.

The molecules, especially CO, are such good radiators of energy that they cool the clouds to low temperatures. Thermal energy is present in the cloud as motion among the atoms and molecules. When a CO molecule collides with an atom or another molecule, some of the energy of motion can be stored in the rotation and vibration of the CO molecule. When the molecule emits that energy as a radio or infrared photon, the energy escapes from the cloud. In this way, molecular radiation can cool the interior of the cloud and keep it very cold.

The largest of these cool, dense clouds are called **giant molecular clouds.** They are 15 to 60 pc across and may contain 100 to 1,000,000 solar masses. The internal temperature is a frigid 10 K. Although these clouds are detected by their carbon monoxide molecular emission, remember that the gas is mostly hydrogen.

Giant molecular clouds are the nests of star birth. Deep inside these great clouds, gravity can pull the matter inward and create new stars. That is a story you will follow in detail in the next chapter. For now, you must meet the dirty part of the interstellar medium—the dust. To do this, you must trade your radio telescope for an infrared telescope.

Infrared Radiation from Dust

The dust in the interstellar medium makes up only about 1 percent of the mass, and at a temperature of 100 K ($-143°C$) or less it is very cold. Nevertheless, it is easy to detect at infrared wavelengths. To see how such cold dust can radiate a lot of energy, consider a simple experiment with the dust in an imaginary giant molecular cloud.

For the sake of quick calculation, assume that a giant molecular cloud has a mass of about 10^5 solar masses. Only 1 percent of that mass is dust, so all of the dust in the cloud amounts to a mass of 10^3 solar masses. Imagine that you could collect all of the dust into a single sphere. It would be only about 10 times the diameter of the sun, and its surface area would be about 100 times that of the sun. But suppose you left the dust as separate specks, each 0.0005 mm in diameter, a typical size. Then the cloud would contain about 10^{43} dust specks, and the total surface area of the dust would be about 10^{29} times that of the sun. When matter is finely divided into dust, it has a very large surface area, and even though it is much colder than the sun, its vast surface area can radiate tremendous amounts of infrared radiation.

In 1983, the Infrared Astronomy Satellite mapped the sky at far-infrared wavelengths and found the galaxy filled with infrared radiation from dust. Most of this dust is confined to the region near the plane of our galaxy, and, as you would expect, the dust is thickest where the gas is thickest. Much of the dust is distributed in wispy clouds that became known as the **infrared cirrus** because of their overall resemblance to cirrus clouds in Earth's atmosphere. The infrared cirrus consists of dusty clouds of interstellar matter with temperatures of about 30 K (■ Figure 10-9a). Studies of CO emission from molecular clouds show that at least some of the infrared cirrus is associated with molecular clouds within a few hundred parsecs of the sun. This suggests that the gas and dust in the molecular clouds is not uniform but rather very patchy.

As mentioned before, the interstellar dust appears to be composed of carbon, silicates, and iron, elements that are underabundant in the gas. Some grains appear to contain water ice contaminated with ammonia, methane, and other compounds.

About 450 ly in diameter, the cavity has been inflated by many supernova explosions.

The wispiness of the infrared cirrus reveals the turbulence of the interstellar medium.

Cygnus

Cygnus Superbubble

A nearby cavity only 110 ly in diameter formed by the explosion of a single star

Cygnus Loop

a Infrared image

b X-ray image

■ Figure 10-9

Infrared and X-ray observations reveal different components of the interstellar medium. (a) This far-infrared image shows the infrared cirrus, wispy clouds of very cold dusty gas spread all across the sky. (NASA/IPAC, Courtesy Deborah Levine) (b) This ROSAT X-ray image shows the Cygnus Superbubble, a cavity filled with gas from exploding stars. The gas is heated and emits X rays where the expanding bubble pushes into the surrounding gas. (Courtesy Steve Snowden and Max-Planck Institute for Extraterrestrial Physics, Germany)

Infrared observations can help you understand the interstellar medium, but X-ray and ultraviolet observations complete the picture by revealing a part of the gas between the stars that is otherwise invisible.

X Rays from the Interstellar Medium

The interstellar medium seems very cold, so you would hardly expect to detect X rays. These high-energy photons are commonly produced by high temperatures. Nevertheless, X-ray telescopes can help you understand the interstellar medium.

X-ray telescopes above Earth's atmosphere have detected X rays from a part of the interstellar medium with a very high temperature. This gas has been called the **coronal gas** because it has temperatures of 10^6 K or higher, as does the sun's corona. Of course, the coronal gas in the interstellar medium is not related in any way to the actual corona of the sun.

You might expect the coronal gas to have a very high pressure because it is so hot, but it has a very low density. It contains only 0.0004 to 0.003 particles/cm³. That is, you would have to search through a few hundred to a few thousand cubic centimeters of coronal gas to find a single particle—an ionized atom or an electron. Because of its very low density, the gas pressure in the coronal gas can be roughly the same as in the HI clouds and the intercloud medium. Nevertheless, exceptions are known. The cloud of interstellar matter near the sun has a pressure that is more than 20 times less than the pressure of neighboring coronal gas. Such differences in pressure are not well understood.

The coronal gas appears to originate when a massive star explodes violently in what astronomers call a supernova. Although rare, such explosions blast large amounts of very hot gas outward in expanding shells. Gas flowing away from very hot, young stars must add to the coronal gas. In some cases, neighboring shells of coronal gas may expand into each other and merge to form larger volumes, but an earlier theory that coronal gas fills a large network of tunnels and shells throughout interstellar space seems to be contradicted by evidence. Probably about 20 percent of the space between the stars is filled with isolated pockets of coronal gas.

The Cygnus superbubble appears to be related to the coronal gas. Located in Cygnus, it is a very large shell of hot gas about 450 pc in diameter (Figure 10-9b). The energy needed to create such a shell is equivalent to hundreds of supernova explosions. The bubble may have developed as a large cluster of stars was born and grew old and the most massive stars died in supernova explosions. Several of these large bubbles are known.

Ultraviolet Observations of the Interstellar Medium

You can divide the ultraviolet spectrum into the near-ultraviolet, with wavelengths only slightly shorter than visible light, and the far-ultraviolet, with much shorter wavelengths. Only a few decades ago, astronomers believed that far-ultraviolet photons could not travel far through the interstellar medium because they would be absorbed so easily by neutral hydrogen atoms. These atoms of neutral hydrogen are good absorbers of far-ultraviolet photons because the photons have enough energy to ionize the atoms. In fact, even one atom of HI per cubic centimeter would make the interstellar medium opaque to far-ultraviolet photons, and astronomers would be unable to see more than a light-year into space with a far-ultraviolet telescope.

When the Extreme Ultraviolet Explorer (EUVE) satellite was put into Earth orbit in 1992, it discovered that the interstellar medium was only partly cloudy. While some regions were filled with clouds of neutral hydrogen and so were opaque, other regions were filled with hot, ionized hydrogen that was transparent to far-ultraviolet photons. This confirms the description of the interstellar medium produced by X-ray observations.

The far-ultraviolet observations reveal that the sun is located just inside a large region of hot, ionized hydrogen, while only a few light-years from the sun lies a cool, neutral hydrogen cloud that is opaque to far-ultraviolet photons. This **local bubble or void** of gas in which the sun is located appears to be linked to other hot, transparent regions. In this way, ultraviolet observations, combined with X-ray observations, suggest that the apparently cold and empty regions between the stars are filled with a complex, evolving mixture of hot and cold gas.

Building Scientific Arguments

If hydrogen is the most common molecule, why do astronomers depend on the CO molecule to map molecular clouds?

This scientific argument links a bit of atomic physics with the chain of inference that leads from what can be observed to what astronomers need to know. Although hydrogen is the most common atom in the universe and molecular hydrogen the most common molecule, a molecule of hydrogen does not radiate in the radio part of the spectrum. Consequently, radio astronomers cannot detect it. But the much less common CO (carbon monoxide) molecule is a very efficient radiator of radio energy, so radio astronomers use it as a tracer of molecular clouds. Wherever radio telescopes reveal a great cloud of CO, you can be confident that most of the gas is molecular hydrogen.

You can also be confident that the molecular clouds contain dust, because it is the dust that protects the molecules in the cloud from the ultraviolet radiation that would otherwise break the molecules into atoms.

Dust doesn't radiate longer-wavelength radio energy, so you must study it in the infrared. Although the dust in a molecular cloud is very cold and makes up a small percentage of the total mass, it is a very good radiator of infrared radiation. Create a new argument using the physics of infrared emission from dust to answer a different question: **How can a small amount of cold dust radiate vast amounts of infrared radiation?**

■ ■ ■

Connections: From one end of the spectrum to the other, you have used every wavelength to study the interstellar medium. Now you can put your data in order and create a model of the gas and dust between the stars. That is the first step toward understanding how stars are born.

10-3 A Model of the Interstellar Medium

WHEN YOU LOOK at bright nebulae like the Great Nebula in Orion, you see a part of the interstellar medium, but what you see is only a very special region that happens to be close enough to hot, bright stars to become ionized and glow. Most of the interstellar medium is invisible to your eyes, so you must gather lots of evidence (**Window on Science 10-1**) and use it to develop a model of the interstellar medium.

For your model, you can divide the interstellar medium into four basic components (■ Table 10-2) and describe these components to explain how they interact and evolve. You will discover that their evolution is intimately connected to the process of star formation and star death.

Four Components of the Interstellar Medium

The interstellar medium is not at all uniform. Rather, it is lumpy, and the lumps differ dramatically in temperature and density.

HI clouds are cool, with temperatures of 50 to 150 K and densities of ten to a few hundreds of atoms per cubic centimeter. These clouds are only a few parsecs in diameter and contain a few solar masses.

Between the cool HI clouds lies the warm intercloud medium of HII, with temperatures of a few thousand Kelvin and densities of 0.01 atoms/cm^3. The intercloud medium is in approximate equilibrium with the HI clouds; they both have about the same pressure.

■ Table 10-2 | Four Components of the Interstellar Medium

Component	Temperature (K)	Density (atoms/cm^3)	Gas
HI clouds	50–150	1–1000	Neutral hydrogen; other atoms ionized
Intercloud medium	10^3–10^4	0.01	Partially ionized
Coronal gas	10^5–10^6	10^{-4}–10^{-3}	Highly ionized
Molecular clouds	20–50	10^3–10^5	Molecules

Understanding Science: Separating Facts from Theories

The fundamental work of science is testing theories by comparing them with facts. As you think about science, you need to distinguish clearly between facts and theories. The facts are the evidence against which scientists test theories.

Scientific facts are those observations or experimental results of which scientists are confident. An astronomer makes observations of stars, and a botanist collects samples of related plants. A fact could be a precise measurement, such as the mass of a star expressed as a specific number, or it could be a simple observation, such as that a certain butterfly no longer visits a certain mountain valley. In each case, the scientist is gathering facts.

A theory, however, is a conjecture as to how nature works. If you are uncertain of the theory, you might call it a hypothesis. In any case, these conjectures are not facts; they are attempts to explain how nature works. In a sense, a theory or hypothesis is a story that scientists have made up to explain how nature works in some specific case. These stories can

In science, evidence is made up of facts, which could range from precise numerical measurements to the observation of the shape of a flower. (M. Seeds)

be wonderfully detailed and ingenious, but without evidence they are nothing more than hunches.

When one of these stories is tested against the facts and confirmed, scientists have more confidence that the story is more or less right. The more a story is tested successfully, the more confidence scientists have in it. The facts rep-

resent reality, and every theory or hypothesis must be repeatedly tested against reality.

You can't test one theory against another theory. Theories are not evidence; they are conjectures. If you were allowed to test one theory against another theory, you might fall into the trap of circular reasoning. "Elves make the flowers bloom. I know that elves exist because the flowers bloom." That is using two theories to confirm each other, and it leads to nonsense. Only facts can be evidence.

Nor can you use a theory to deduce facts. You can use a theory to make predictions, which you can then test against facts, but the predictions themselves can never be certainties, so they can't be facts. The only way to arrive at facts is to consult nature and make direct measurements or observations.

As you study different problems in astronomy, you must carefully distinguish between facts and theories. Facts are the basic building blocks of all science, and they can come only from the careful study of nature. As Galileo would say, you must read the book of nature.

Molecular clouds are especially dense. Molecules cannot survive if they are exposed to ultraviolet photons in starlight, so they can form only in the densest clouds, where dust absorbs and scatters ultraviolet photons. The molecular clouds can be very large, with diameters up to 60 pc and masses up to a million solar masses, but they are also very cold. Strong evidence suggests that stars are born when giant molecular clouds contract under the influence of their own gravity, a subject explored in detail in the next chapter.

Pushing through this stew of interstellar clouds are regions of coronal gas. Most of this very hot gas is probably produced in supernova explosions, although some may be gas flowing away from very hot stars. With temperatures up to a million degrees and densities as low as 10^{-4} atoms/cm^3, the coronal gas occupies a significant part of the interstellar medium.

Astronomers believe that the HI clouds make up about 25 percent of the interstellar mass, and the intercloud medium about 50 percent. The coronal gas contributes only 5 percent of the mass, although it seems to occupy about one-fifth of the volume. The giant molecular clouds amount to about 25 percent of the mass. If you add up these percentages, you get slightly more than 100 percent, which illustrates the uncertainty inherent in this model of the interstellar medium.

As the stars move through the interstellar medium, they meet no resistance, but they do have a dramatic effect on the gas and dust. The Pleiades, for example, is a relatively young star cluster that is moving rapidly through space. As it moves through the interstellar medium, it is leaving behind a trail like that left by a boat in water (■ Figure 10-10). This wake has been detected in the infrared and is apparently produced by the ultraviolet radiation from the stars in the cluster. None of the stars is hot enough to ionize the gas, but there are a number of relatively hot stars; the ultraviolet radiation from those stars heats, but does not quite ionize, the gas. The heated gas then expands and forms a wake showing the path of the cluster through the interstellar medium.

The Pleiades clearly illustrates the close relationship between the stars and the interstellar medium. In fact, you are now ready to outline a cycle that links the stars to the gas and dust between them.

The Interstellar Cycle

The story of the interstellar medium is closely linked to star formation. You will see in the next chapter that stars form in giant molecular clouds. As soon as a group of stars forms, the hottest stars begin ionizing the gas to produce emission nebulae. The pressure of the starlight and the gas flowing away from the hot,

■ Figure 10-10

The Pleiades is a beautiful star cluster at visual wavelengths. (CalTech) An infrared image shows the brighter stars of the Pleiades as small crosses. The motion of the cluster from left to right has left a wake in the interstellar medium. The hottest star, Merope, is not quite hot enough to ionize the gas. (Courtesy Richard E. White; image rendered by Duncan Chesley of American Image, Inc.)

■ Figure 10-11

(a) The nebula N44C is part of a larger region where young, hot, massive stars have formed and where at least one supernova has blown a bubble filled with hot gas. (The Hubble Heritage Team, STScI/AURA/NASA) (b) This image of the entire N44 region combines a visual-wavelength image (white) with an X-ray image (red). X rays reveal regions of very-high-temperature gas that has been ejected from supernova explosions. (R. Chris Smith, CTIO and Y.-H. Chu, UICU)

young stars pushes the gas outward and may disrupt the cloud entirely.

The composition of the interstellar dust suggests that it is formed mostly in the atmospheres of cool stars. There the temperatures are low enough for some atoms to condense into specks of solid matter, much as soot can condense in a candle flame. The pressure of the starlight can push these dust specks out of the star and replenish the interstellar medium. Other stars that eject mass into space, such as supernovae, probably also add to the supply of interstellar dust.

The most massive stars die quickly in supernova explosions, and those tremendously violent events blast high-temperature gas outward that further disrupts the interstellar medium (■ Figure 10-11). Much of the motion in interstellar clouds and their twisted shapes are probably produced by these supernova explosions and the hot coronal gas they eject. It seems that our galaxy produces about two supernova explosions per century (although most are not visible from Earth because of obscuration by dust) and that these explosions keep the interstellar medium stirred and create the vast regions filled with coronal gas. In fact, the sun lies inside such a region of high-temperature, low-density gas called the local bubble or void. With a typical diameter of a few hundred parsecs, the local bubble may be a cavity in the interstellar medium inflated by a supernova explosion that occurred within the last million years or so.

While supernovae and hot stars keep the interstellar medium in motion, the natural gravitation of gas clouds and collisions between clouds may gradually build more massive clouds. In the most massive clouds, the dust protects the interior from ultraviolet photons, and molecules can form. These giant molecular clouds eventually give birth to new stars, and the cycle begins all over again.

The Trifid Nebula (■ Figure 10-12) is a dramatic illustration of this cycle. Measuring over 12 pc in diameter, the nebula is illuminated by a number of hot, young stars that have apparently formed recently from the gas. Near the stars, the gas is ionized and glows as a pink-red HII region; but farther from the stars, where the ultraviolet radiation is weaker, the gas is not ionized. Nevertheless, dust in the nebula scatters blue light, so this unionized part of the nebula is visible as a blue reflection nebula. Dark lanes of obscuring dust cross the face of the nebula as if to remind Earth's astronomers again of the importance of dust in the interstellar cycle.

Visual-wavelength image

■ Figure 10-12

The Trifid Nebula embodies an important part of the interstellar cycle. A dense cloud of gas has given birth to stars, and the hottest are ionizing the gas to produce the pink emission nebula, while dust scatters light to produce the blue reflection nebula. As the nebula is disrupted by the newborn stars, the gas and dust merge with the interstellar medium. (AURA/NOAO/NSF and Nigel Sharp)

Building Scientific Arguments

How can the coronal gas make up only a few percent of the mass but 20 percent of the volume of the interstellar medium?

Once again your argument needs to focus on some simple physics. The solution to this puzzle is the extremely low density of the coronal gas, roughly one particle for every thousand cubic centimeters. The coronal gas is the least dense part of the interstellar medium. It is a much better vacuum than any laboratory vacuum on Earth. It doesn't take much mass in coronal gas to occupy a large volume. In contrast, molecular clouds are the densest part, and they contain roughly a quarter of the mass, although they occupy only a small part of the volume.

The density and temperature of the different components of the interstellar medium determine how you could observe them. Create a new argument to consider observations. **What techniques would you use to observe the coronal gas and the molecular clouds?**

■   ■   ■

Connections: Your study of the interstellar medium is incomplete for two reasons. First, astronomers don't yet understand all of its components or how those components interact. Second, you can't fully understand the interstellar medium until you understand how stars are born and how they die. You begin that story in the next chapter.

Summary

10-1 | Visible-Wavelength Observations

How do astronomers know that space isn't empty?

- The interstellar medium, the gas and dust between the stars, is mostly concentrated near the plane of our Milky Way Galaxy.

- Emission nebulae are clear evidence of an interstellar medium. The gas is ionized by ultraviolet radiation from nearby hot stars, and that makes the nebula glow like a giant neon sign. The red, blue, and violet Balmer lines blend together to produce the characteristic pink-red glow of ionized hydrogen.

- A reflection nebula is produced by gas and dust illuminated by a star that is not hot enough to ionize the gas. Rather, the dust scatters the starlight to produce a reflection of the stellar absorption spectrum. Because shorter-wavelength photons scatter more than longer-wavelength photons, reflection nebulae look blue. The daytime sky looks blue for the same reason.

- A dark nebula is a cloud of gas and dust that is visible because it blocks the light of distant stars. The irregular shapes of these dark nebulae reveal the turbulence in the interstellar medium.

- Further evidence of an interstellar medium is the extinction, or dimming, of the light of distant stars and interstellar reddening. Light from distant stars suffers scattering by dust particles in the interstellar medium, and blue light is scattered more than red light. This makes distant stars look redder than their spectral types suggest. The dependence of this extinction on wavelength shows that the scattering dust particles are very small. The dust is made of carbon, silicates, iron, and ice.

- Interstellar absorption lines in the spectra of distant stars are very narrow. The interstellar gas is cold and has a very low density, and this makes the interstellar lines much narrower than the spectral lines produced in stars. Multiple interstellar lines reveal that the light has passed through more than one interstellar cloud on its way to Earth.

10-2 | Long- and Short-Wavelength Observations

How do astronomers study the nearly invisible matter between the stars?

- Radio observations at a wavelength of 21 cm reveal multiple clouds of neutral hydrogen. These neutral clouds drift through a warmer but lower-density intercloud medium.

- Radio telescopes tuned to other wavelengths have detected nearly 100 different molecules in the interstellar medium, most of them found in giant molecular clouds.

- Infrared observations made from orbit have detected dust in the interstellar medium in the form of the infrared cirrus.

- X-ray and far-ultraviolet observations have detected very hot coronal gas produced by supernova explosions.

10-3 | A Model of the Interstellar Medium

How does the matter between the stars interact with the stars?

- The four main components of the interstellar medium are the small neutral HI clouds, the warm intercloud medium, coronal gas, and molecular clouds.

- Stars are born in the dense molecular clouds, and the energy from hot stars and supernova explosions causes currents in the interstellar medium and creates the coronal gas. The dust in the interstellar medium is formed in the atmospheres of cool stars and from gas ejected by supernova explosions.

New Terms

interstellar medium (p. 216)	interstellar absorption lines (p. 222)
nebula (p. 218)	HI clouds (p. 223)
forbidden line (p. 218)	intercloud medium (p. 223)
metastable level (p. 218)	pressure (p. 224)
interstellar dust (p. 219)	21-cm radiation (p. 224)
interstellar extinction (p. 219)	molecular cloud (p. 227)
interstellar reddening (p. 219)	giant molecular cloud (p. 227)
emission nebula (p. 220)	infrared cirrus (p. 227)
HII region (p. 220)	coronal gas (p. 228)
reflection nebula (p. 220)	local bubble or void (p. 229)
dark nebula (p. 221)	

Review Questions

Ace⊙Astronomy™ Assess your understanding of this chapter's topics with additional quizzing and animations at http://ace .brookscole.com/sf9

1. What evidence can you cite that the spaces between the stars are not totally empty?

2. What evidence can you cite that the interstellar medium contains both gas and dust?

3. How do the spectra of HII regions differ from the spectra of reflection nebulae? Why?

4. Why are interstellar lines so narrow? Why do some spectral lines forbidden in spectra on Earth appear in spectra of interstellar clouds and nebulae? What does that tell you?

5. How is the blue color of a reflection nebula related to the blue color of the daytime sky?

6. Why do distant stars look redder than their spectral types suggest?

7. If starlight on its way to Earth passed through a cloud of interstellar gas that was hot instead of very cold, would you expect the interstellar absorption lines to be broader or narrower than usual? Why?

8. How can the HI clouds and the intercloud medium have similar pressures when their temperatures are so different?

9. Why can the 21-cm radio emission line of neutral hydrogen be observed in the interstellar medium but not in the laboratory?

10. What does the shape of the 21-cm radio emission line of neutral hydrogen tell you about the interstellar medium?

11. What produces the coronal gas?

12. The image here shows two nebulae, one pink in the background and one black in the foreground. What kind of nebulae are these?

(NASA and the Hubble Heritage Team, STScI/AURA)

Discussion Questions

1. When you see distant streetlights through smog, they look dimmer and redder than they do normally. But when you see the same streetlights through fog or falling snow, they look dimmer but not redder. Use your knowledge of the interstellar medium to discuss the relative sizes of the particles in smog, fog, and snowstorms compared to the wavelength of light.

2. If you could see a few stars through a dark nebula, how would you expect their spectra and colors to differ from similar stars just in front of the dark nebula?

Problems

1. A small nebula has a diameter of 20 seconds of arc and a distance of 1000 pc from Earth. What is the diameter of the nebula in parsecs? in meters?

2. The dust in a molecular cloud has a temperature of about 50 K. At what wavelength does it emit the maximum energy? (*Hint:* Consider black body radiation, Chapter 7.)

3. Extinction dims starlight by about 1.9 magnitudes per 1000 pc. What fraction of photons survives a trip of 1000 pc? (*Hint:* Consider the definition of the magnitude scale in Chapter 2.)

4. If the total extinction through a dark nebula is 10 magnitudes, what fraction of photons makes it through the cloud? (*Hint:* See Problem 3.)

5. The density of air in a child's balloon 20 cm in diameter is roughly the same as the density of air at sea level, 10^{19} particles/cm^3. To how large a diameter would you have to expand the balloon to make the gas inside the same density as the interstellar medium, about 1 particle/cm^3? (*Hint:* The volume of a sphere is $\frac{4}{3}\pi R^3$.)

6. If a giant molecular cloud has a diameter of 30 pc and drifts relative to neighboring clouds at 20 km/s, how long will it take to travel its own diameter?

7. An HI cloud is 4 pc in diameter and has a density of 100 hydrogen atoms/cm^3. What is its total mass in kilograms? (*Hints:* The volume of a sphere is $\frac{4}{3}\pi R^3$, and the mass of a hydrogen atom is 1.67×10^{-27} kg.)

8. Find the mass in kilograms of a giant molecular cloud that is 30 pc in diameter and has a density of 300 hydrogen molecules/cm^3. (*Hint:* See Problem 7.)

9. At what wavelength does the coronal gas radiate most strongly? (*Hint:* Consider black body radiation, Chapter 7.)

Media Cluster

Ace ✆ Astronomy™ To access the resources in the Media Cluster, log into AceAstronomy at **http://ace.brookscole.com/sf9** and select Chapter 10.

ACTIVE FIGURES

Scattering in Earth's Atmosphere
By absorption, reflection, and refraction, wavelengths of light are selected and filtered to produce the colorful world around you. In this animation, learn more about why you see blue skies and red sunsets and how the sun's position in the sky plays a part.

ASTRONOMY EXERCISES

Why Is the Sky Blue?
The sky is blue and the setting sun is red for the same reason that reflection nebulae are blue and distant stars are reddened. You can explore that process here in the topic begun in the Active Figure above.

Critical Inquiries for the Web

1. What causes the varied colors in images of gaseous nebulae that grace textbooks and websites? Search the Web for a color image of a nebula in the Messier catalog. Describe the structure and colors you see in terms of the concepts discussed in this chapter. (Be careful to distinguish real color from false color when answering this question. See Window on Science 6-2 for more information about false-color images.)

2. An interesting highlight in the history of astronomy is the discovery of "Nebulium." Search the Internet for information on this once-mysterious source of nebular spectral lines. How can astronomers today explain these lines in terms of known elements?

Exploring *TheSky*

1. The following nebulae are all star-formation regions. What kind of nebulae are they? (*Hint:* To center on an object, use **Find** under the **Edit** menu. Choose **Messier Objects** and pick from the list.)

   M42, M20, M8, M17

2. Locate M8 in *TheSky,* zoom in, and identify other nebulae in the region. Study the photo of NGC6559. What kind of nebula is it?

3. Locate M42, M20, and M8. Zoom in and identify the spectral type of the central star(s). What common characteristic do these stars share? Why?

Go to the Brooks/Cole Astronomy Resource Center (**http://astronomy.brookscole.com**) for critical thinking exercises, articles, and additional readings from InfoTrac College Edition, Brooks/Cole's online student library.

11 | The Formation of Stars

Jim he allowed [the stars]
was made, but I allowed
they happened. Jim said the
moon could'a laid them;
well, that looked kind of
reasonable, so I didn't say
nothing against it, because
I've seen a frog lay most
as many, so of course
it could be done.

MARK TWAIN, *THE ADVENTURES*
OF HUCKLEBERRY FINN

Visual + infrared image

THE STARS ARE NOT ETERNAL.
When you look at the sky, you see hundreds of
points of light, and each is an object like the sun held together
by its gravity and generating tremendous energy in its core through nuclear reactions. The stars you see tonight are the
same stars your parents, grandparents, and great grandparents saw. Stars change hardly at all in a human lifetime, but
they are not eternal. Stars are born and stars die. This chapter begins that story. ▌ In this chapter, you will see how
gravity creates stars from the thin gas of space and how nuclear reactions inside stars generate energy. You will see how
the flow of that energy outward toward the surface of the star balances gravity and makes the stars stable. In the next
two chapters, you will follow the life story of stars from the beginning of their stable lives to their ultimate deaths.

In this image of the Orion Nebula, every point of light is a star that was born recently from the gas and dust. At infrared wavelengths, you can see the warm glow of more stars taking form in the dark clouds behind the glowing nebula. (ESO)

Guidepost

Looking Back

Throughout the preceding chapters you have been building a general understanding of the sky and a facility with the tools of astronomy. The last chapter introduced you to the gas and dust between the stars. Now you are ready to make some stars.

This Chapter

This is the place where you will really begin to see how the universe works. Here you will begin putting together observations and theories to understand how nature makes stars. That will answer four essential questions:

How are stars born?

How do stars make energy?

How do stars maintain their stability?

What evidence do astronomers have that theories of star formation are correct?

Looking Ahead

The most important of the questions above may be the last. Testing theories against evidence is the basic skill required of all scientists, and you will use it over and over in the 15 chapters to come.

Ace⊙Astronomy™ The AceAstronomy icon throughout the text indicates an opportunity for you to test yourself on key concepts and to explore animations and interactions on the AceAstronomy website at: http://ace .brookscole.com/sf9

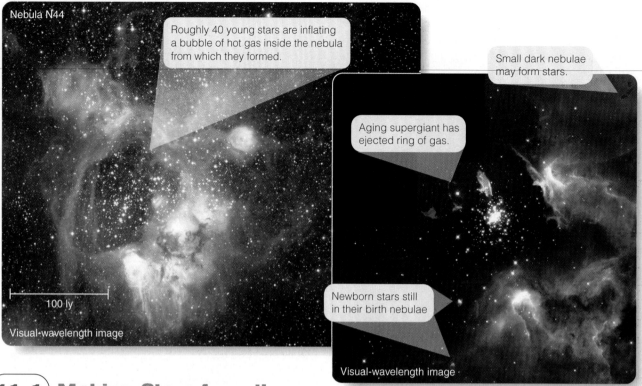

Nebula N44

Roughly 40 young stars are inflating a bubble of hot gas inside the nebula from which they formed.

Small dark nebulae may form stars.

Aging supergiant has ejected ring of gas.

Newborn stars still in their birth nebulae

100 ly

Visual-wavelength image

Visual-wavelength image

11-1 Making Stars from the Interstellar Medium

STARS HAVE BEEN FORMING continuously since our galaxy took shape over 10 billion years ago. You can depend on this for two reasons. First, the sun is only about 5 billion years old, a relative newcomer compared to the older stars in our galaxy. Second, you can see hot, blue stars such as one of our Favorite Stars Spica (α Virginis), a B1 main-sequence star. As you will see in the next chapter, such massive stars have very short lives. In fact, a star like Spica can last only 10 million years and must have formed recently.

The key to understanding star formation is the correlation between young stars and clouds of gas. Where you find the youngest groups of stars, you also find large clouds of gas illuminated by the hottest and brightest of the new stars (■ Figure 11-1). This should make you suspect that stars form from such clouds, much as raindrops condense from the water vapor in a thundercloud. Indeed, the giant molecular clouds discussed in the preceding chapter can give birth to entire clusters of new stars.

The central problem for any discussion of star formation is how these large, low-density, cold clouds of gas become comparatively small, high-density, hot stars. Gravity is the key.

Star Birth in Giant Molecular Clouds

The giant molecular clouds are the sites of active star formation, yet they are very unlike stars. With a typical diameter of 50 pc and a typical mass exceeding 10^5 solar masses, a giant molecular

■ **Figure 11-1**

Nebula N44 has given birth to a large cluster of stars. (ESO) Various stages of star formation are evident in NGC3603, including a massive star ejecting gas as it approaches its end. (W. Brandner, JPL/IPAC, E. K. Grebel, Univ. of Washington, Y. Chu, Univ. of Illinois, and NASA)

cloud is vastly larger than a star. Also, the gas in a giant molecular cloud is about 10^{20} times less dense than a star and has temperatures of only a few degrees Kelvin. These clouds can form stars because gravity can force some small regions of the clouds to contract to high density and high temperature.

Radio observations show that at least some giant molecular clouds develop dense cores that are only 0.1 pc in radius and that contain roughly 1 solar mass. A single giant molecular cloud may contain many of these dense cores and can give birth to star clusters containing hundreds of stars. However, both theory and observations suggest that many giant molecular clouds cannot begin the formation of dense cores spontaneously. At least four factors resist the contraction of a gas cloud, and gravity must overcome those four factors before star formation can begin.

First, thermal energy in the gas is present as motion among the atoms and molecules. Even at the very low temperature of 10 K, the average hydrogen molecule moves at about 0.35 km/s (almost 800 mph). This thermal motion would make the cloud drift apart if gravity were too weak to hold it together.

The interstellar magnetic field is the second factor that gravity must overcome to make the cloud contract. Neutral atoms and molecules are unaffected by a magnetic field, but ions, having an electric charge, cannot move freely through a magnetic field. Although the gas in a molecular cloud is mostly neutral, there are some ions, and that means a magnetic field can exert a force on the gas. The magnetic field present all through our galaxy averages only about 10^{-4} times as strong as that on Earth, but it can act like an internal spring to resist the contraction of the gas cloud.

The third factor is rotation. Everything in the universe rotates to some extent. As the gas cloud begins to contract, it spins more and more rapidly as it conserves angular momentum, just as ice skaters spin faster as they pull in their arms (Figure 5-7). This rotation can become so rapid that it resists further contraction of the cloud.

The turbulence in the interstellar medium is the fourth thing that could prevent a cloud from contracting. In the previous chapter, you learned that nebulae are often twisted and distorted by moving currents in the interstellar medium. This turbulence could make it difficult for a large molecular cloud to contract.

Given these four resistive factors, it seems surprising that any giant molecular clouds can contract, form dense cores, and eventually form stars. In some cases, the gas in giant molecular clouds can gradually recombine with free electrons and become less ionized. Neutral gas is free to "slip past" the magnetic field and contract. This gradual process has been observed to form dense cores inside isolated molecular clouds.

In contrast, both theory and observation suggest that many giant molecular clouds are triggered to form stars by a passing **shock wave,** the astronomical equivalent of a sonic boom (■ Figure 11-2). During such a triggering event, a few regions of the large cloud can be compressed to such high densities that the resistive factors can no longer oppose gravity, and star formation begins.

At least four different processes can produce shock waves that trigger star formation. Supernova explosions (Chapter 13) can produce powerful shock waves that rush through the interstellar medium. Also, the ignition of very hot stars can ionize nearby gas and drive it away to produce a shock wave where it pushes into the colder, denser interstellar matter. A third trigger is the collision of molecular clouds. Because the clouds are large, they are likely to run into each other occasionally; and, because they contain magnetic fields, they cannot pass through each other. A collision between such clouds can compress parts of the clouds and trigger star formation. The fourth trigger is the spiral pattern of our Milky Way Galaxy (see Figure 1-11). One theory suggests that the spiral arms are shock waves that travel around the galaxy like the moving hands of a clock (Chapter 15). As a cloud passes through a spiral arm, the cloud could be compressed, and star formation could begin. Astronomers have found regions of star formation where these processes can be identified (■ Figure 11-3).

Of course, a single giant molecular cloud containing a million solar masses does not contract to form a single humongous

Shock Wave Triggers Star Formation

A shock wave (red) approaches an interstellar gas cloud.

The shock wave passes through and compresses the cloud.

Motions in the cloud continue after the shock wave passes.

The densest parts of the cloud become gravitationally unstable.

Contracting regions of gas give birth to stars.

■ Figure 11-2

In this summary of a computer model, an interstellar gas cloud is triggered into star formation by a passing shock wave. The events summarized here might span about 6 million years.

star. The cloud fragments and the densest parts form a number of dense cores. Exactly why a cloud fragments isn't fully understood, but the rotation, magnetic field, and turbulence probably

■ Figure 11-3

A few million years ago, a massive star exploded as a supernova in the upper left quarter of this image. The expanding shock wave compressed a nearby cloud of gas and triggered the birth of new stars. In this infrared image, you can see the arc of compressed gas left behind as the shock wave rushed through the gas clouds. Most of the bright, young stars in this nebula are hidden deep in dust clouds and are not yet visible at visual wavelengths. (NASA/JPL-Caltech/ V. Gorjian, NOAO)

play important roles. In any case, a giant cloud of gas typically contracts to form a number of newborn stars.

Heating by Contraction

You can understand how low-density clouds of interstellar gas can become dense enough to make stars, but how can the horribly cold gas become hot enough to become a star? The answer, once again, is gravity.

To consider the formation of a star, shift your attention to a single dense core of gas destined to become a single star. Once a small cloud of gas begins to contract, gravity draws the atoms toward the center. That means the atoms are falling, and, like all things that fall, they gather speed as they fall. In fact, astronomers refer to this early stage in the formation of a star as **free-fall contraction.** Whereas the atoms may have had low velocities to start with, by the time they have fallen most of the way to the center of the cloud, they are traveling at high velocities. Thermal energy is the agitation of the particles in a gas, so this increase in velocity is a step toward heating the gas. But you can't say that the gas is hot simply because all of the atoms are moving rapidly. The air in the cabin of a jet airplane is traveling rapidly, but it isn't hot because all of the atoms are moving in generally the same direction with the plane. To convert the high velocity of

the infalling atoms into thermal energy, the motion must become randomized, and that happens when the atoms begin to collide with one another as they fall into the central region of the cloud. The jumbled, random motion of the atoms represents thermal energy, and the temperature of the gas increases.

This is an important principle in astronomy. Whenever a cloud of gas contracts, the atoms move downward in the gravitational field and pick up speed, and the gas grows hotter. Astronomers express this by saying that gravitational energy is converted into thermal energy. Whenever a gas cloud expands, gas atoms move upward against gravity and lose speed, and the gas becomes cooler. Astronomers say that in this case thermal energy is converted into gravitational energy. This principle applies not only to clouds of interstellar gas but also to contracting and expanding stars, as you will see in the following chapters.

Your study of gas clouds has shown how nature can begin the contraction of dense cores in giant molecular clouds and how contraction can heat the gas. Now you are ready to construct a detailed story of the transformation from gas cloud to star.

Protostars

To follow the story of star formation farther, you must concentrate on a single fragment of a collapsing cloud as it contracts, heats up, and begins to behave like a star. Although the term *protostar* is used rather loosely by astronomers, it will be defined here to be a prestellar object that is hot enough to radiate infrared radiation but not hot enough to generate energy by nuclear fusion.

A protostar begins life as a contracting cloud of gas and dust falling inward under the influence of gravity (■ Figure 11-4). As the cloud contracts, it develops a higher-density region at the center and a low-density envelope. Mass continues to flow inward from the outer parts of the cloud. That is, the cloud contracts from the inside out, with the protostar taking shape deep inside an enveloping cloud of cold, dusty gas. These clouds have been called **cocoons** because they hide the forming protostar from view as it takes shape.

The contraction of the cloud cannot lead directly to the formation of a small protostar because the cloud has some net rotation. Everything rotates one way or another. The rotation of the cloud may be undetectable at first, but its rotation must grow more and more pronounced as it contracts and conserves its angular momentum. The rapidly spinning core of the cloud must flatten into a spinning disk like a blob of pizza dough spun into the air. Gas that has lost its angular momentum through collisions can sink directly to the center of the cloud, where the protostar begins to grow, surrounded by the disk. As more gas falls inward, it passes through the disk, giving up much of its angular momentum before it sinks into the protostar (■ Figure 11-5).

Basic physics predicts that protostars should form at the centers of spinning disks of gas, which are called **protostellar disks.** That is important because astronomers believe that planets form within these disks. Earth formed in a disk around the protosun

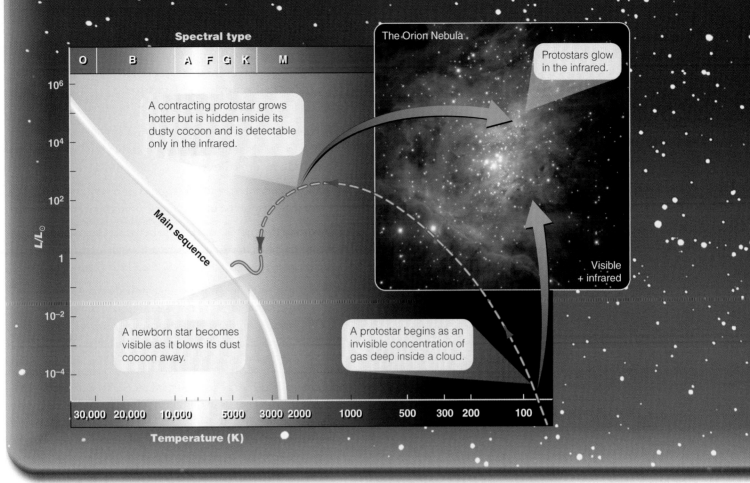

Spectral type

A contracting protostar grows hotter but is hidden inside its dusty cocoon and is detectable only in the infrared.

The Orion Nebula

Protostars glow in the infrared.

Main sequence

Visible + infrared

A newborn star becomes visible as it blows its dust cocoon away.

A protostar begins as an invisible concentration of gas deep inside a cloud.

Temperature (K)

■ **Figure 11-4**

This H–R diagram has been extended to very low temperatures to show schematically the contraction of a dim, cool protostar. At visual wavelengths, protostars are invisible because they are deep inside dusty clouds of gas, but they are detectable at infrared wavelengths. The Orion Nebula contains both protostars and newborn stars that are just blowing their dust cocoons away. (ESO)

4.6 billion years ago. As you will see in Chapter 19, the evidence is very strong that planetary systems form in these protostellar disks.

If you could see a protostar without its cocoon, it would be a very luminous and cool red star in the upper right part of the H–R diagram. The dust cocoon, however, absorbs almost all of the visible radiation and, growing warm, reradiates the energy as infrared radiation. So you should not expect to see protostars at visible wavelengths, but they should be emitting infrared radiation.

When the protostar becomes hot enough, it can drive away the surrounding gas and dust and become visible. The location in the H–R diagram where protostars first emerge from their cocoons and become visible is called the **birth line.** Once a star crosses the birth line, it continues to contract toward the main sequence with a speed that depends on its mass. More massive stars have

Formation of a Protostellar Disk

A slowly rotating cloud of gas begins to contract.

Conservation of angular momemtum spins the cloud faster and it flattens...

into a growing protostar at the center of a rotating disk of gas and dust.

■ **Figure 11-5**

The rotation of a contracting gas cloud forces it to flatten into a disk, and the protostar grows at the center.

stronger gravity and contract more rapidly (■ Figure 11-6). The sun took about 30 million years to reach the main sequence, but a 30-solar-mass star takes only 30,000 years. A 0.2-solar-mass star needs about 1 billion years to contract from a gas cloud to the main sequence.

The theory of star formation takes you into an unearthly realm filled with unfamiliar processes and objects. You might think it was nothing more than a fairy tale if observations did not support the theory.

Evidence of Star Formation

In astronomy, evidence means observations. Consequently, astronomers must ask what observations confirm their theories of star formation. Unfortunately, a protostar is not easy to observe. The protostar stage is less than 0.1 percent of a star's lifetime, so although that is a long time in human terms, you cannot expect to find many stars in the protostar stage. Furthermore, protostars

■ Figure 11-6

The more massive a protostar is, the faster it contracts. A 1-$M_\odot$ star requires 30 million years to reach the main sequence. (Recall that $M_\odot$ means "solar mass.") The dashed line is the birth line, where contracting protostars first become visible as they dissipate their surrounding clouds of gas and dust. Compare with Figure 11-4, which shows the evolution of a protostar of about 1 $M_\odot$ as a dashed line up to the birth line and as a solid line from the birth line to the main sequence. (Illustration design by author)

form deep inside clouds of dusty gas that absorb any light the protostar might emit. Astronomers must depend on observations at infrared wavelengths to search for hidden protostars.

Study **Observational Evidence of Star Formation** on pages 244–245 and notice four important points:

1 You can be sure that star formation is going on right now because you can find regions containing stars so young they must have formed recently.

2 Visual and infrared observations can reveal dusty clouds of gas that seem to be in the process of forming stars.

3 Notice that newborn stars lie between the main sequence and the birth line—just where you would expect to find stars that have recently blown away their dust cocoons.

4 Finally, notice how observations provide clues to the process by which stars form. Observations of jets coming from hidden protostars show that protostars form surrounded by disks of gas and dust. Also, observations of small globules of gas and dust within larger nebulae show how star formation begins.

The origin of jets from spinning protostellar disks is not understood. Certainly the contracting, spinning disk contains tremendous energy, and theorists believe that magnetic fields become twisted tightly around the disk. Exactly how those fields squeeze hot gas out above and below the disk along the axis of rotation is not yet clear. But the detection of these jets was one of the first pieces of evidence that protostars are born at the centers of spinning disks.

In some cases, these dark disks of gas and dust are clearly visible around newborn stars (■ Figure 11-7). Such disks are exciting not only because they are evidence of star formation but also because they are the regions where planets form. You will meet such disks later in this chapter and again when you learn about the origin of our solar system in Chapter 19.

Some evidence of star formation is not at first obvious. An **association** is a widely distributed star cluster that is not held together by its own gravity—its stars wander away as the association ages. These associations consist of young stars, because the stars wander apart so quickly. The constellation Orion, a known region of star formation, is filled with T Tauri stars in a **T association.** T Tauri stars are relatively low-mass objects ranging from 0.75 to 3 solar masses, but the stars in **O associations,** extended groups of O stars, are more massive. It is not clear why some gas clouds give birth to compact star clusters held together by their own gravity and others give birth to larger associations not bound together by gravity. Associations are dramatic evidence of recent star formation because they drift apart so easily. They must have formed recently.

Not only can astronomers locate evidence of star formation, but they have found evidence that star formation can stimulate more star formation. If a gas cloud produces massive stars, those massive stars can ionize the gas nearby and drive it away. Where

Jet

Forming star hidden at center of dusty disk

Disk

■ **Figure 11-7**

In the object HH30, a newly formed star lies at the center of a dense disk of dusty gas that is narrow near the star and thicker farther away. Although the star is hidden from you by the edge-on dusty disk, the star illuminates the inner surface of the disk. Interactions between the infalling material in the disk and the spinning star eject jets of gas along the axis of rotation. (C. Burrows, STScI & ESA, WFPC 2 Investigation Definition Team, NASA)

mation is through the explosion of a massive star in a supernova. You read earlier in this chapter how such an explosion can drive a shock wave through surrounding gas and trigger more star formation (Figure 11-3).

Like a grass fire spreading through the interstellar medium, star formation can reignite itself if it creates massive stars, and astronomers can locate the remains of such episodes (■ Figure 11-9). Of course, lower-mass stars also form from such star formation, but they are not hot enough, nor do they explode as supernovae, so they alone cannot trigger further star formation.

the intense radiation and hot gas pushes into surrounding gas, it can compress the gas and trigger more star formation. One sign of this process is the presence of **star-formation pillars,** columns of gas that point back at the source of the hot gas. Such pillars are produced by denser regions that protect the gas behind them as the blast of intense radiation and hot gas flows past (■ Figure 11-8). You can recognize this process in the overall shape of the Eagle Nebula on page 245 as well as in the smaller pillars within the nebula. Another way massive stars can trigger star for-

■ **Figure 11-8**

The hot, massive stars in the star cluster 30 Doradus are pouring out intense ultraviolet radiation and powerful winds of hot gas that are pushing back and compressing the surrounding nebula. Slightly denser regions in the nebula protect the gas behind them to form star-formation pillars a few light-years long that point back at the star cluster. The dense blobs are being compressed and may form more stars within the next few million years. (N. Walborn and J. Maiz- Apellániz, STScI, R. Barbá, La Plata Observatory, and NASA)

Star-formation pillar

Star-formation pillar

Origin of a star-formation pillar

Dense blob protects nebula behind it.

Radiation and high-speed gas

Star-formation pillar

UV + visual + IR image

Observational Evidence of Star Formation

1 The nebula around the star S Monocerotis is bright with hot stars. Such stars live short lives of only a few million years, so they must have formed recently. Such regions of young stars are common. The entire constellation of Orion is filled with young stars and clouds of gas and dust.

Visual-wavelength image

1a Nebulae containing young stars usually contain T Tauri stars. These stars fluctuate irregularly in brightness, and many are bright in the infrared, suggesting they are surrounded by dust clouds and in some cases by dust disks. Doppler shifts show that gas is flowing away from many T Tauri stars. The T Tauri stars appear to be newborn stars just blowing away their dust cocoons. T Tauri stars appear to have ages ranging from 100,000 years to 100,000,000 years. Spectra of T Tauri stars show signs of an active chromosphere as we might expect from young, rapidly rotating stars with powerful dynamos and strong magnetic fields.

3 The star cluster NGC2264, imbedded in the nebula on this page, is only a few million years old. Lower-mass stars have not yet reached the main sequence, and the cluster contains many T Tauri stars (open circles), which are found above and to the right of the main sequence, near the birth line. The faintest stars in the cluster were too faint to be observed in this study.

2 Bok globules, named after astronomer Bart Bok, are dense, dusty clouds seen silhouetted against glowing nebulae in the visual image below. Only a light-year or so in diameter, they contain from 10 to 1000 solar masses, and infrared observations show that most are very cold at their centers. Only a few have warm interiors and appear to be contracting to form protostars.

Visual

NASA

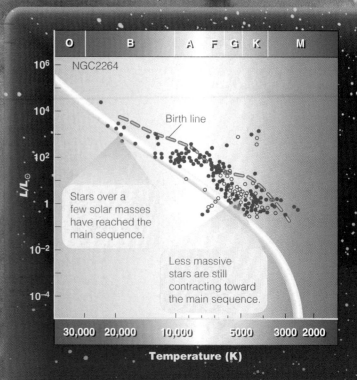

NGC2264

Birth line

Stars over a few solar masses have reached the main sequence.

Less massive stars are still contracting toward the main sequence.

O   B   A F G K   M

$L/L_\odot$

10^6
10^4
10^2
1
10^{-2}
10^{-4}

Temperature (K)

30,000 20,000   10,000   5000   3000 2000

4 At the center of this image, a newborn star is emitting powerful jets to left and right. Where the jets strike the interstellar medium, they produce **Herbig–Haro objects**. The inset shows how irregular the jet is. Such jets can be over a light-year long and contain gas traveling at 100 km/s or more.

HH34S

Gas jet

Young star

Visual

NASA

Visual

HH34N

Herbig–Haro object

4a Herbig–Haro objects, named after the two astronomers who first described them, are small nebulae that fluctuate in brightness. They appear to be produced by flickering jets from newborn stars exciting the interstellar medium.

Dusty disk

Jet

Jet

Herbig–Haro object

4b Matter flowing into a protostar swirls through a thick disk and, by a process believed to involve magnetic fields, ejects high-energy jets in opposite directions. Observation of these **bipolar flows** is evidence that protostars are surrounded by disks because only disks could focus the flows into jets.

NASA

EGGs

EGGs

Visual

4c The Eagle Nebula, shown at the right in a ground-based photograph, is a region of known star formation. Turn the page a quarter turn to the right to see the eagle flying with a salmon in its talons.

Visual

AURA/NOAO/NSF

4d This Hubble Space Telescope image at left of part of the Eagle Nebula shows that radiation from bright stars out of the field to the upper right is evaporating the dust and driving away the gas to expose small globules of denser gas and dust. Infrared observations show that about 15 percent of these objects have formed protostars, and at least one seems to contain a newborn star (arrow). In part because these objects were first found in the Eagle Nebula, astronomers have enjoyed calling them EGGs — evaporating gaseous globules. They are evidently stars exposed in the act of forming.

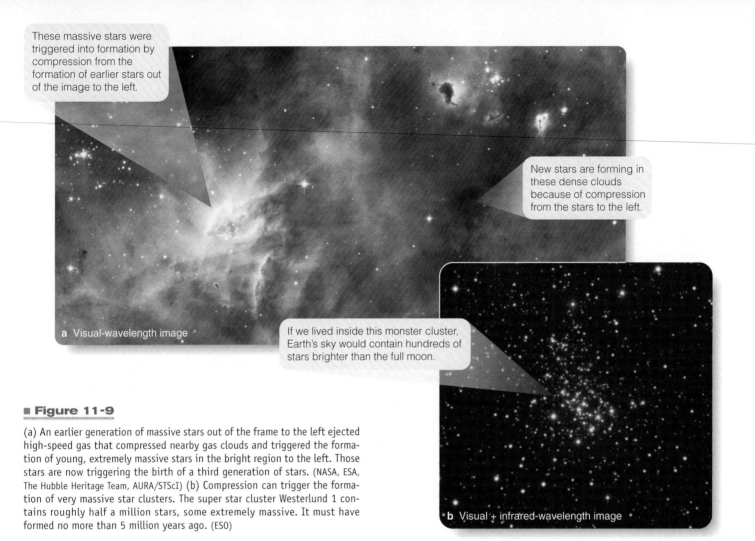

These massive stars were triggered into formation by compression from the formation of earlier stars out of the image to the left.

New stars are forming in these dense clouds because of compression from the stars to the left.

a Visual-wavelength image

If we lived inside this monster cluster, Earth's sky would contain hundreds of stars brighter than the full moon.

b Visual + infrared-wavelength image

■ **Figure 11-9**

(a) An earlier generation of massive stars out of the frame to the left ejected high-speed gas that compressed nearby gas clouds and triggered the formation of young, extremely massive stars in the bright region to the left. Those stars are now triggering the birth of a third generation of stars. (NASA, ESA, The Hubble Heritage Team, AURA/STScI) (b) Compression can trigger the formation of very massive star clusters. The super star cluster Westerlund 1 contains roughly half a million stars, some extremely massive. It must have formed no more than 5 million years ago. (ESO)

Observations provide plenty of evidence that star formation is a continuous process; you can be sure that stars are being born right now. The study of star formation is an exciting part of modern astronomy.

Building Scientific Arguments

What evidence can you cite that stars are forming right now? This is a very good example of building a scientific argument because you must cite different forms of evidence to confirm a theory. First, you should note that some extremely luminous stars can't live very long, so when you see such stars, such as the hot, blue stars in Orion, you know they must have formed in the last few million years. Star formation must be a continuous process, or you would not see any of these short-lived stars. But you have more direct evidence when you look at T Tauri stars, which lie above the main sequence and are often associated with gas and dust. In fact, you see entire associations of T Tauri stars as well as other associations of short-lived O stars. Many regions of gas and dust contain bright, hot stars that are blowing the nebulae apart; those stars must have formed recently.

There seems to be no doubt that star formation is an ongoing process, and you can revise your argument to discuss how it happens. **What evidence can you cite that protostars are often surrounded by disks of gas and dust?**

■   ■   ■

Connections: Gravity pulls the gas and dust together to make new stars, but this raises a natural question. If all this matter is falling inward to make a star, why does the contracting star stop? How do contracting protostars turn themselves into stable main-sequence stars? The answer is that the contracting stars begin to generate their own energy.

11-2 The Source of Stellar Energy

STARS ARE BORN when gravity pulls matter together, but what stops the contraction? Contracting stars somehow begin to make large amounts of energy, and somehow that energy stops the contraction and stabilizes the stars. In this section, you will see how stars make energy; in the next section, you will see how contracting stars reach stability. This story will lead your imagination into a region where your body can never go—the heart of a star.

A Review of the Proton–Proton Chain

In Chapter 8 you visited the center of the sun and discovered that it manufactured energy through hydrogen fusion using a series of nuclear reactions called the proton–proton chain. The reaction must begin with the fusion of two protons. Protons are the nuclei of hydrogen atoms and have positive charges and repel each other with a force called the Coulomb barrier. Consequently, the proton–proton chain cannot occur if the gas temperature is lower than about 10 million Kelvin. High-velocity collisions are required to overcome the Coulomb barrier, and high velocity means high temperature.

You also discovered that the gas must be dense if the proton–proton chain is to produce significant energy. The fusion of two protons is unlikely, so many collisions are necessary to produce a few fusion reactions. Furthermore, a single cycle through the proton–proton chain produces only a tiny amount of energy, so a vast number of fusion reactions are needed to supply the energy the sun needs. There will be a large number of fusion reactions only if the gas density is high, so high density is required.

Your study of the sun revealed that the proton–proton chain can produce energy only near the sun's center, where the temperature and density are high. In fact, only about 30 percent of the sun's mass is actually hot enough to fuse. The rest of the mass is too far from the center and isn't hot enough.

You might expect other stars to fuse hydrogen the same way the sun does, and you would be right for most stars. Some stars, however, can fuse hydrogen using a different recipe, and that makes a big difference to the structure of those stars.

Ace✲Astronomy™ Log into AceAstronomy and select this chapter to see Astronomy Exercise "Nuclear Fusion." Take control of fusion in the sun's core.

The CNO Cycle

Main-sequence stars more massive than 1.1 solar masses fuse hydrogen into helium using the **CNO cycle,** a hydrogen fusion process that uses carbon, nitrogen, and oxygen as stepping-stones.

Look carefully at ■ Figure 11-10 and notice the steps in the CNO cycle. The cycle begins with a carbon-12 nucleus absorbing a proton and becoming nitrogen-13, which decays to become carbon-13. The carbon-13 nucleus absorbs a second proton and becomes nitrogen-14, which absorbs a third proton and becomes oxygen-15. The oxygen-15 decays to become nitrogen-15, which absorbs a fourth proton, ejects a helium nucleus, and becomes carbon-12. Notice that carbon-12 begins the cycle and ends the cycle, so the carbon-12 nucleus can be reused over and over. Notice also that, along the way, four protons combine to make a

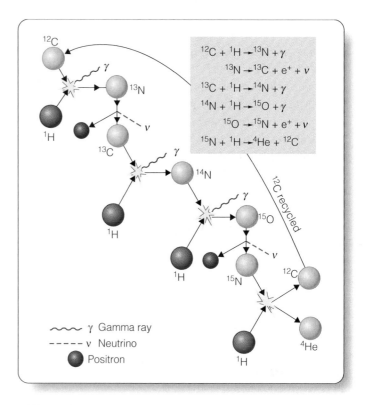

■ Figure 11-10

The CNO cycle uses ^{12}C as a catalyst to combine four hydrogen nuclei (^{1}H) to make one helium nucleus (^{4}He) plus energy. The carbon nucleus reappears at the end of the process, ready to start the cycle over.

helium nucleus. This CNO cycle has the same outcome as the proton–proton chain, but it is different in an important way.

The CNO cycle begins with a carbon nucleus combining with a proton, a bare hydrogen nucleus. Because a carbon nucleus has a positive charge six times higher than a proton, the Coulomb barrier is high, and temperatures higher than 16 million Kelvin are required to make the CNO cycle work. The CNO cycle needs much hotter gas than does the proton–proton chain.

The difference between the proton–proton chain and the CNO cycle is their dependence on temperature. The CNO cycle is so critically sensitive to temperature that it produces almost no energy at all at temperatures below 16 million Kelvin. The proton–proton chain can make energy at these lower temperatures, so the sun makes nearly all of its energy using the proton–proton chain. At slightly higher temperatures, the proton–proton chain makes slightly more energy, but the CNO cycle produces a flood of energy. That means that stars with higher-temperature cores make nearly all of their energy using the CNO cycle, while stars with cooler cores make their energy using the proton–proton chain.

The CNO cycle and the proton–proton chain both fuse hydrogen to make helium, but the CNO cycle's sensitivity to temperature is dramatic. You will see later in this chapter how this affects the internal structure of the stars.

Why does the CNO cycle require higher temperatures than the proton–proton cycle?

This scientific argument requires that you discuss a bit of nuclear physics. Nuclear fusion happens when atomic nuclei collide and fuse to form a new nucleus. Inside a star, the gas is ionized, so the nuclei have no orbiting electrons, and thus the nuclei have positive charges. Two objects with positive charges will repel each other in what is called the Coulomb barrier. To start fusion, the nuclei must collide at high enough velocity to overcome that barrier. Heavier atomic nuclei such as carbon have higher positive charges, so the Coulomb barrier is higher. That requires that the gas be hotter so the collisions will be violent enough to overcome the barrier.

Now redo your argument. **How are the CNO cycle and the proton–proton chain similar in some ways but different in other ways?**

■ ■ ■

Connections: Because the law of gravity and the rules of nuclear fusion determine how stars work, you can understand what the inside of a star is like. That is, you can describe the internal structure of a star.

(11-3) Stellar Structure

GRAVITY MAKES STARS CONTRACT, and, when the density and temperature at the centers of the stars are high enough, nuclear fusion begins making energy. Why do the stars stop contracting? Somehow, the energy generated at the center of the star flows outward to the surface, and that flow of energy stops the contraction. To understand the stable structure of a star, you must first understand how energy flows through the star and how that energy flow affects the entire star.

Energy Transport

The sun and other stars generate nuclear energy in their deep interiors, and that energy must flow outward to their surfaces to replace the energy radiated into space as light and heat. In Chapter 8, you studied the sun and discovered that energy moves through its deep interior as radiation and through its outer layers as convection. Other stars are similar to the sun, but there can be differences. Let's review how energy flows through the inside of a star and apply the results to stars in general.

Energy always flows from hot regions to cool regions, and the centers of stars are much hotter than their surfaces, so energy must flow outward. In the material of which stars are made, energy can move in three ways: by conduction, by radiation, or by convection.

Conduction is the most familiar form of heat flow. If you hold the bowl of a spoon in a candle flame, the handle of the spoon grows warmer. Heat, in the form of the motion of the particles in the metal, is conducted from particle to particle up the handle, until the particles under your fingers begin to move faster and you sense heat (■ Figure 11-11). That means that conduction requires close contact between the particles. Because matter in most stars is gaseous and the particles are not in close contact, conduction is an unimportant means of energy flow. Conduction is significant only in peculiar stars that have tremendous internal densities.

The transport of energy by radiation is another familiar experience. Put your hand beside a candle flame, and you can feel the heat. What you actually feel are infrared photons radiated by the flame (Figure 11-11). Because photons are packets of energy, your hand grows warm as it absorbs them. As you saw in Chapter 8, radiation is the principal means of energy transport in the sun's interior. Photons begin near the sun's center as high-energy gamma rays and are scattered over and over as they work their way outward. The high-energy photons common near the sun's center emerge from the sun's surface as large numbers of low-energy photons of visible light.

The flow of energy by radiation depends on how difficult it is for the photons to move through the gas. If the gas is cool and dense, the photons are more likely to be absorbed or scattered, and the radiation does not penetrate the gas very well. Physicists say such a gas is opaque. In a hotter, lower-density gas, the photons can penetrate more easily; such a gas is less opaque. The **opacity** of a gas—its resistance to the flow of radiation—depends strongly on its temperature.

■ Active Figure 11-11

The three modes by which energy may be transported from the flame of a candle, as shown here, are the three modes of energy transport within a star.

Ace ⑨ Astronomy™ Log into AceAstronomy and select this chapter to see Active Figure "Conduction, Convection, and Radiation." Experiment with the candle flame.

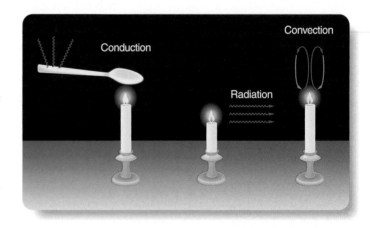

If the opacity of the gas in a star is low, energy flows outward as radiation and, as in the sun's deep interior, the region is called a radiative zone. If the opacity of the gas is high, radiation cannot flow through it easily. Like water behind a dam, energy builds up, raising the temperature until the gas begins to churn. Hot gas, being less dense, rises; and cool gas, being denser, sinks in convection, the third way energy can move in a star (Figure 11-11). As you learned when you studied the sun, a region inside a star where energy moves as convection is called a convective zone.

Convection is important in stars because it both carries energy and mixes the gas. Convection currents flowing through the layers of a star tend to homogenize the gas, giving it a uniform composition throughout the convective zone. As you might expect, this mixing affects the fuel supply of the nuclear reactions, just as the stirring of a campfire makes it burn more efficiently.

In the sun, the proton–proton chain converts hydrogen into helium, but the sun's interior is a radiative zone. There is no convection to mix the fuel and carry away the ashes. The hydrogen is gradually being used up, and the helium gradually accumulates. The sun is like a big pot of soup burning at the bottom and stirred only slightly by convection near its surface.

Now you are ready to understand the stability of the stars. They are elegantly simple. You can begin with our sun.

What Supports the Sun?

From its surface to its interior, the sun is gaseous. On Earth, a puff of gas, vapor from a smokestack perhaps, dissipates rapidly, driven by the motion of the air and by the random motions of the atoms in the gas. However, the sun differs from a puff of gas in a very important characteristic—its mass. The sun is over 300,000 times more massive than Earth, and that mass produces a tremendous gravitational field that draws the sun's gases into a sphere.

With such a strong gravitational field, it might seem as if the gases of the sun should compress into a tiny ball, but a second force balances gravity and prevents the sun from shrinking. The gas of which the sun is made is quite hot; and, because it is hot, it has a high pressure. The pressure pushes outward; indeed, without the restraint of the sun's gravity, it would blow the sun apart. In this way, the sun is balanced between two forces—gravity trying to squeeze it tighter and gas pressure trying to make it expand.

To discuss the forces inside the sun, you can imagine that the sun's interior is divided into concentric shells like those in an onion (■ Figure 11-12). You can then discuss the temperature, density, pressure, and so on in each shell. Keep in mind, however, that these helpful shells do not really exist. The sun and stars are not composed of separable layers. The layers are only a convenience for discussion.

The gravity–pressure balance that supports the sun is a fundamental part of stellar structure known as the law of **hydrostatic equilibrium.** It says that, in a stable star like the sun, the weight of the material pressing downward on a layer must be balanced

■ **Figure 11-12**

To discuss the structure of a star, it is helpful to divide its interior into concentric shells much like the layers in an onion. This model is, of course, only an aid to your imagination. Stars are not really divided into separable layers.

by the pressure of the gas in that layer. *Hydro* implies that you are discussing a fluid—the gases of the star. *Static* implies that the fluid is stable—neither expanding nor contracting.

The law of hydrostatic equilibrium can prove that the interior of the sun must be very hot. Near the sun's surface, there is little weight pressing down on the gas, so the pressure must be low, implying a low temperature. But as you go deeper into the sun, the weight becomes larger, so the pressure, and therefore the temperature, must also increase (■ Figure 11-13). Near the sun's surface, the pressure is only a few times Earth's atmospheric pressure, and the temperature is only about 5800 K. At the center of the sun, the weight pressing down equals a pressure of about 3×10^{14} times Earth's atmospheric pressure. To support that huge weight, the gas temperature has to be almost 15 million K.

Of course, the interior of the sun is kept hot by the nuclear reactions occuring at the core, and the outward flow of energy keeps each layer in the sun hot enough to support the weight pressing down from above. In a sense, the sun is supported by the flow of energy from its center to its surface. Turn off that energy, and gravity would gradually force the sun to collapse into its center.

Ace ✪ Astronomy ™   Log into AceAstronomy and select this chapter to see Astronomy Exercise "Hydrostatic Equilibrium." You can try, but you can't make this star violate hydrostatic equilibrium.

Inside Stars

In Chapter 9, you discovered that the stars on the main sequence are ordered according to mass, with the upper-main-sequence stars being most massive and the lower-main-sequence stars being

Figure 11-13

The law of hydrostatic equilibrium says the pressure in each layer must balance the weight on that layer. Consequently, as the weight increases from the surface of a star to its center, the pressure must also increase.

least massive. Combining this with what you know about hydrogen fusion reveals that there are two kinds of main-sequence stars: upper-main-sequence stars, which fuse hydrogen on the CNO cycle; and lower-main-sequence stars, which fuse hydrogen on the proton–proton chain. Viewed from the outside, these stars differ only in size, temperature, and luminosity, but inside they are quite different.

The upper-main-sequence stars are more massive and thus must have higher central temperatures to withstand their own gravity. These high central temperatures permit the star to fuse hydrogen on the CNO cycle, and that affects the internal structure of the star.

The CNO cycle is very temperature-sensitive. If the central temperature of the sun rose by 10 percent, energy production by the proton–proton chain would rise by about 46 percent, but energy production by the CNO cycle would shoot up 350 percent. This means that the more massive stars generate almost all of their energy in a tiny region at their very centers where the temperature is highest. A 10-solar-mass star, for instance, generates 50 percent of its energy in its central 2 percent of mass.

This concentration of energy production at the very center of the star causes a "traffic jam" as the energy tries to flow away from the center. Transport of energy by radiation can't drain away the energy fast enough, and the central core of the star churns in convection as hot gas rises upward and cooler gas sinks downward. Farther from the center, the traffic jam is less severe, and the energy can flow outward as radiation. That means that massive stars have convective cores at their centers and radiative envelopes extending from their cores to their surfaces (■ Figure 11-14).

Main-sequence stars less massive than about 1.1 solar masses cannot get hot enough to fuse much hydrogen on the CNO cycle. They generate nearly all of their energy by the proton–proton chain, which is not as sensitive to temperature, and thus the energy generation occurs in a larger region in the star's core. The sun, for example, generates 50 percent of its energy in a region that contains 11 percent of its mass. Because the energy generation is not concentrated at the very center of the star, no traffic jam develops, and the energy flows outward as radiation. Only near the surface, where the gas is cooler and therefore more opaque, does convection stir the gas. Consequently, the less massive stars on the main sequence have radiative cores and convective envelopes.

The lowest-mass stars have a slightly different kind of structure. For stars less than about 0.4 solar mass, the gas is relatively cool compared to the inside of more massive stars, and the radiation cannot flow outward easily. As a result, the entire bulk of these low-mass stars is stirred by convection.

The stars in the evening sky look much the same, but you have discovered that they are a diverse group. Different kinds of stars make their energy in different ways and have different internal structures. Now you are ready to answer the essential question: How do stars maintain their stability?

The Pressure–Temperature Thermostat

You can now understand what stops a newborn star from contracting into a tiny ball. Once nuclear fusion begins, energy flows outward, heats the layers of gas, raises the pressure, and stops the contraction. This leads to an interesting question: How does the star manage to make just the right amount of energy? The key is

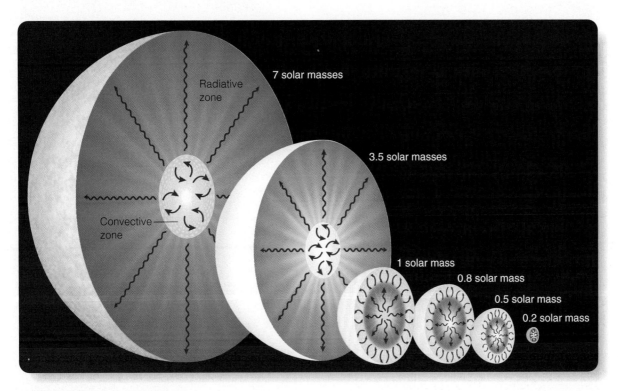

■ Active Figure 11-14

Inside stars. The more massive stars have small convective cores and radiative envelopes. Lower-mass stars, including the sun, have radiative cores and convective envelopes. The lowest-mass stars are convective throughout. The "cores" of the stars where nuclear fusion occurs (not shown) are smaller than the interiors. (Illustration design by author)

Ace ◐Astronomy™ Log into AceAstronomy and select this chapter to see Active Figure "Inside Stars." Explore the interiors of stars and compare them with the sun.

the relationship between pressure and temperature; it acts like a thermostat to keep the star burning steady.

In a star, the nuclear reactions generate just enough energy to balance the inward pull of gravity. Consider what would happen if the reactions began to produce too much energy. The star balances gravity by generating energy, so the extra energy flowing out of the star would force it to expand. The expansion would lower the central temperature and density and slow the nuclear reactions until the star regained stability. In this way the star has a built-in regulator that keeps the nuclear reactions from occurring too rapidly.

The same thermostat keeps the reactions from dying down. Suppose the nuclear reactions began making too little energy. Then the star would contract slightly, increasing the central temperature and density and increasing the nuclear energy generation until the star regained stability.

The stability of a star depends on this relation between pressure and temperature. If an increase or decrease in temperature produces a corresponding change in pressure, the thermostat functions correctly, and the star is stable. You will see in the next chapter how the thermostat accounts for the mass–luminosity relation. In Chapter 13, you will see what happens to a star when

the thermostat breaks down completely and the nuclear fires rage unregulated.

Building Scientific Arguments

What would happen if the sun stopped generating energy?
Sometimes one of the best ways to test your understanding is to build an argument based on an altered situation. Stars are supported by the outward flow of energy generated by nuclear fusion in their interiors. That energy keeps each layer of the star just hot enough for the gas pressure to support the weight of the layers above. Each layer in the star must be in hydrostatic equilibrium; that is, the inward weight must be balanced by outward pressure. If the sun stopped making energy in its interior, nothing would happen at first, but over many thousands of years the loss of energy from its surface would reduce the sun's ability to withstand its own gravity, and it would begin to contract. You wouldn't notice much for 100,000 years or so, but eventually the sun would lose its battle with gravity.

Stars are elegant in their simplicity—nothing more than a cloud of gas held together by gravity and warmed by nuclear

Star Formation in the Orion Nebula

1 The visible Orion Nebula shown below is a pocket of ionized gas on the near side of a vast, dusty molecular cloud that fills much of the southern part of the constellation Orion. The molecular cloud can be mapped by radio telescopes. To scale, the cloud would be many times larger than this page. As the stars of the Trapezium were born in the cloud, their radiation has ionized the gas and pushed it away. Where the expanding nebula pushes into the larger molecular cloud, it is compressing the gas (see diagram at right) and may be triggering the formation of the protostars that can be detected at infrared wavelengths within the molecular cloud.

Hundreds of stars lie within the nebula, but only the four brightest, those in the Trapezium, are easy to see with a small telescope. A fifth star, at the narrow end of the Trapezium, may be visible on nights of good seeing.

The cluster of stars in the nebula is less than 2 million years old. This must mean the nebula is similarly young.

Side view of Orion Nebula

Hot Trapezium stars   Protostars

To Earth

Expanding ionized hydrogen

Molecular cloud

Trapezium

Visual-wavelength image
Daniel Good

Infrared

NASA

The near-infrared image above reveals about 50 low-mass, very cool stars that must have formed recently.

X-ray

NASA/CXC/SAO

Roughly 1000 young stars with hot chromospheres appear in this X-ray image of the Orion Nebula.

Photons with enough energy to ionize H

Energy radiated by O6 star

Energy radiated by B1 star

Energy

Wavelength (nanometers)

0   100   200   300

2 Of all the stars in the Orion Nebula, only one is hot enough to ionize the gas. Only photons with wavelengths shorter than 91.2 nm can ionize hydrogen. The second-hottest stars in the nebula are B1 stars, and they emit little of this ionizing radiation. The hottest star, however, is an O6 star 30 times the mass of the sun. At a temperature of 40,000 K, it emits plenty of photons with wavelengths short enough to ionize hydrogen. Remove that one star, and the nebula would turn off its emission.

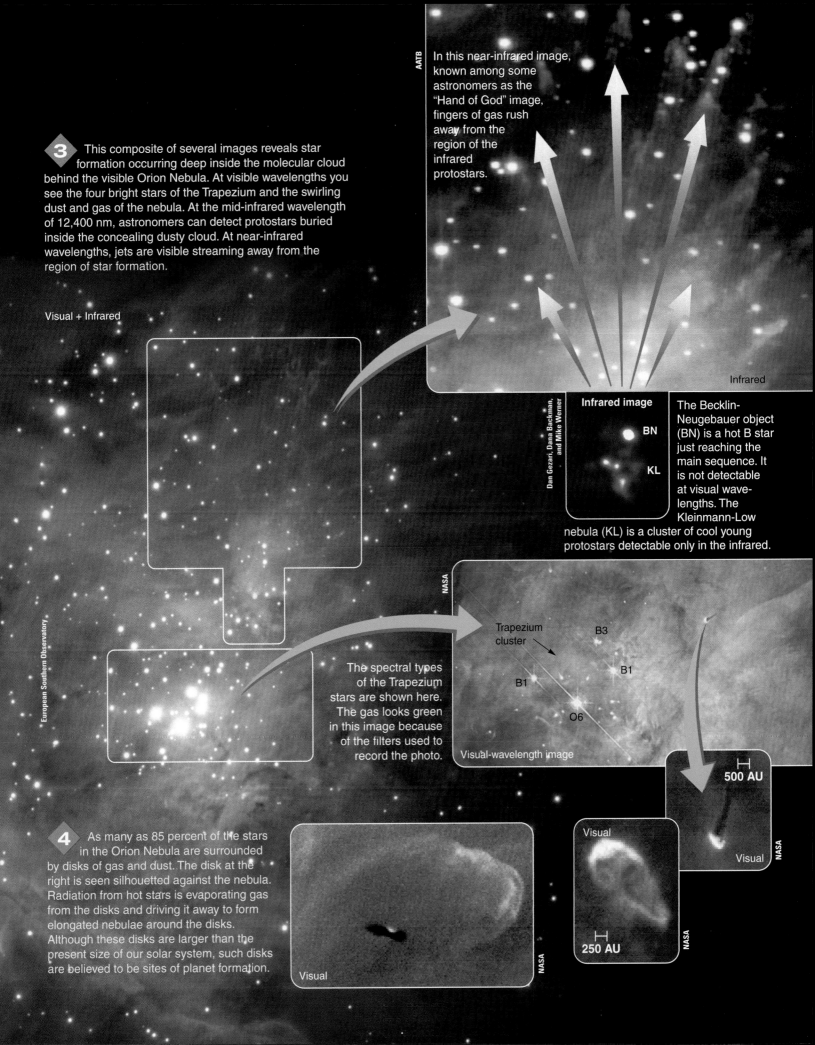

3   This composite of several images reveals star formation occurring deep inside the molecular cloud behind the visible Orion Nebula. At visible wavelengths you see the four bright stars of the Trapezium and the swirling dust and gas of the nebula. At the mid-infrared wavelength of 12,400 nm, astronomers can detect protostars buried inside the concealing dusty cloud. At near-infrared wavelengths, jets are visible streaming away from the region of star formation.

Visual + Infrared

European Southern Observatory

In this near-infrared image, known among some astronomers as the "Hand of God" image, fingers of gas rush away from the region of the infrared protostars.

AATB

Infrared

Dan Gezari, Dana Backman, and Mike Werner

Infrared image

BN

KL

The Becklin-Neugebauer object (BN) is a hot B star just reaching the main sequence. It is not detectable at visual wavelengths. The Kleinmann-Low nebula (KL) is a cluster of cool young protostars detectable only in the infrared.

NASA

The spectral types of the Trapezium stars are shown here. The gas looks green in this image because of the filters used to record the photo.

Trapezium cluster

B3

B1

B1

O6

Visual-wavelength image

500 AU

Visual

NASA

4   As many as 85 percent of the stars in the Orion Nebula are surrounded by disks of gas and dust. The disk at the right is seen silhouetted against the nebula. Radiation from hot stars is evaporating gas from the disks and driving it away to form elongated nebulae around the disks. Although these disks are larger than the present size of our solar system, such disks are believed to be sites of planet formation.

Visual

NASA

Visual

250 AU

NASA

fusion. Now build a different argument. **How does the star manage to make exactly the right amount of energy to support its weight?**

■   ■   ■

Connections: You have traced the birth of stars from the first instability in the interstellar medium to the final equilibrium of nuclear fusion and gravity. You have constructed a complete story of star formation, but you expect evidence to support theories. One of the best places to search for evidence of star formation is in the Great Nebula in Orion. It is just part of a great storm of star birth sweeping through Orion.

(11-4) The Orion Nebula

ON A CLEAR WINTER NIGHT, you can see with your naked eye the Great Nebula of Orion as a fuzzy wisp in Orion's sword. With binoculars or a small telescope it is striking, and through a large telescope it is breathtaking. At the center lie four brilliant blue-white stars known as the Trapezium, the brightest of a cluster of a few hundred stars. Surrounding the stars are the glowing filaments of a nebula more than 8 pc across. Like a great thundercloud illuminated from within, the churning currents of gas and dust suggest immense power. The significance of the Orion Nebula lies hidden, figuratively and literally, beyond the visible nebula. The region is ripe with star formation.

Evidence of Young Stars

You should not be surprised to find star formation in Orion. The constellation is a brilliant landmark in the winter sky because it is marked by hot, blue stars. These stars are bright in the sky, not because they are nearby, but because they are tremendously luminous. These O and B stars cannot live more than a few million years, so you can conclude that they must have been born recently. Furthermore, the constellation contains large numbers of T Tauri stars, which are known to be young. Orion is rich with young stars.

The history of star formation in the constellation of Orion is written in its stars. The stars at Orion's west shoulder are about 12 million years old, while the stars of Orion's belt are about 8 million years old. The stars of the Trapezium at the center of the Great Nebula are no older than 2 million years. Apparently, star formation began near the west shoulder, and the massive stars that formed there triggered the formation of the stars you see in Orion's belt. That star formation may have triggered the formation of the stars you see in the Great Nebula. Like a grass fire, star formation has swept across Orion from northwest to southeast.

Study **Star Formation in the Orion Nebula** on pages 252–253 and notice four points:

❶ The nebula you see is only a small part of a vast, dusty molecular cloud. You see the nebula because the stars born within it have ionized the gas and driven it outward, breaking out of the molecular cloud.

❷ Also notice that a single very hot star is almost entirely responsible for ionizing the gas.

❸ Notice how infrared observations reveal clear evidence of active star formation deeper in the molecular cloud just to the northwest of the Trapezium.

❹ Finally, notice that many stars visible in the Orion Nebula are surrounded by disks of gas and dust. Such disks do not last long and are clear evidence that the stars are very young.

In the next million years, the familiar outline of the Great Nebula will change, and a new nebula may begin to form as the protostars in the molecular cloud ionize the gas, drive it away, and become visible. Centers of star formation may develop and then dissipate as massive stars are born and force the gas to expand. If enough massive stars are born, they can blow the entire molecular cloud apart and bring the successive generations of star formation to a conclusion. The Great Nebula in Orion and its invisible molecular cloud are a beautiful and dramatic example of the continuing cycles of star formation.

Building Scientific Arguments

What did Orion look like to the ancient Egyptians, to the first humans, and to the dinosaurs?
Scientific arguments can do more than support a theory; they can change the way you think of the world around you. The Egyptian civilization had its beginning only a few thousand years ago, and that is not very long in terms of the history of Orion. The stars you see in the constellation are hot and young, but they are a few million years old, so the Egyptians saw the same constellation you see. (They called it Osiris.) Even the Orion Nebula hasn't changed very much in a few thousand years, and Egyptians may have admired it in the dark skies along the Nile.

Our oldest human ancestors lived about 3 million years ago, and that was about the time that the youngest stars in Orion were forming. Your earliest ancestors may have looked up and seen some of the stars you see, but some stars have formed since that time. Also, the Great Nebula is excited by the Trapezium stars, and they are not more than a few million years old, so your early ancestors probably didn't see the Great Nebula.

The dinosaurs saw something quite different. The last of the dinosaurs died about 65 million years ago, long before the birth of the brightest stars in the constellation you see. The dinosaurs, had they the brains to appreciate the view, might have seen bright stars along the Milky Way, but they didn't see Orion. All of the stars in the sky are moving through space, and the sun is orbiting the center of our galaxy. Over many millions of years, the stars move appreciable distances across the sky. The night sky above the dinosaurs contained totally different star patterns.

The Orion Nebula is the product of a giant molecular cloud, but such a cloud can't continue spawning new stars forever. Focus your argument to answer the following: **What processes limit star formation in a molecular cloud?**

■   ■   ■

Connections: The ancient Aztecs of central Mexico told the story of how the stars, known as the Four Hundred Southerners, were scattered across the sky when they lost a cosmic battle with their brother, the great war god Huitzilopochtli. Modern astronomy tells a less colorful story of the birth of stars, but the modern story is supported by evidence and leads you to further questions. If stars are born, then how do they die? You will begin that story in the next chapter.

Study and Review Tools

Summary

11-1 | Making Stars from the Interstellar Medium

How are stars born?
■ Stars are born from the gas and dust of the interstellar medium.
■ The existence of massive, hot stars such as Spica that cannot live very long and short-lived associations is strong evidence that star formation is a continuous process.
■ The gravity of giant molecular clouds makes them contract, but that is resisted by thermal energy in the gas, magnetic fields, rotation, and turbulence. In at least some cases, clouds are compressed by passing shock waves, and star formation is triggered. The birth of massive stars can produce shock waves that trigger further star formation.
■ The cold gas of interstellar space heats up as it contracts because the atoms fall inward and pick up speed. Astronomers say the gas is converting gravitational energy into thermal energy.
■ Protostars form deep inside dusty cocoons and are not directly visible at visual wavelengths until they cross the birth line in the H–R diagram.
■ Many, perhaps most, protostars form at the center of dusty disks, and jets of gas can be emitted along the axis of the spinning disk.

11-2 | The Source of Stellar Energy

How do stars make energy?
■ Many stars make their energy the same way the sun does, using the proton–proton cycle.
■ The CNO cycle is more efficient, but it requires a higher temperature than the proton–proton cycle. Both processes combine four hydrogen nuclei to make one helium nucleus plus energy.
■ Stars more massive than 1.1 solar masses can make energy with the CNO cycle. Less massive stars can only use the proton–proton cycle.

11-3 | Stellar Structure

How do stars maintain their stability?
■ Energy flows from the hot core to the cooler surface as radiation or as convection. Conduction is not important in most stars. Convection is important in stars because it stirs the material.

■ The law of hydrostatic equilibrium says that the weight pressing down on a layer of gas in a star must be balanced by the pressure in the gas. That shows that the inner layers of stars must be hotter because they must support more weight.
■ Upper-main-sequence stars, being more massive, must be hotter inside, and that allows them to use the CNO cycle. Because that cycle is so sensitive to temperature, the energy production occurs in a very small region near the center and the cores of the stars are convective zones. In their outer layers, these stars are radiative.
■ Lower-main-sequence stars, being less massive, are not as hot inside and can fuse hydrogen using the proton–proton chain. Because that chain is not very sensitive to temperature, the energy generation is more widely spread through the star's core, and the deep interior is radiative. The outer layers of these stars are convective.
■ The lowest-mass stars are so cool that their gases are rather opaque and they are convective throughout.

11-4 | The Orion Nebula

What evidence do astronomers have that theories of star formation are correct?
■ The visible Orion Nebula is only a small part of a much larger dusty molecular cloud. Ionization by ultraviolet photons from the hottest star is lighting up the nebula and making it glow brightly.
■ Infrared observations reveal clear evidence of active star formation deeper in the molecular cloud just to the northwest of the Trapezium.
■ Many stars visible in the Orion Nebula are surrounded by disks of gas and dust. Such disks do not last long and are clear evidence that the stars are very young.

New Terms

shock wave (p. 239)

free-fall contraction (p. 240)

protostar (p. 240)

cocoon (p. 240)

protostellar disk (p. 240)

birth line (p. 241)

association (p. 242)

T association (p. 242)

O association (p. 242)

star-formation pillar (p. 243)

T Tauri star (p. 244)

Bok globule (p. 244)

Herbig–Haro object (p. 245)   opacity (p. 248)

bipolar flow (p. 245)   hydrostatic equilibrium (p. 249)

CNO (carbon–nitrogen–oxygen)
  cycle (p. 247)

Review Questions

Ace⟲Astronomy™ Assess your understanding of this chapter's topics with additional quizzing and animations at **http://ace .brookscole.com/sf9**

1. What factors resist the contraction of a cloud of interstellar matter?

2. Explain four different ways a giant molecular cloud can be triggered to contract.

3. What evidence is there that (a) star formation is a continuing process? (b) protostars really exist? (c) the Orion region is actively forming stars?

4. How does a contracting protostar convert gravitational energy into thermal energy?

5. How does the geometry of bipolar flows and Herbig–Haro objects support the hypothesis that protostars are surrounded by rotating disks?

6. How does the CNO cycle differ from the proton–proton chain? How is it similar?

7. How does the extreme temperature sensitivity of the CNO cycle affect the structure of stars?

8. How does energy get from the core of a star, where it is generated, to the surface, where it is radiated into space?

9. Describe the principle of hydrostatic equilibrium as it relates to the internal structure of a star.

10. How does the pressure–temperature thermostat control the nuclear reactions inside stars?

11. The star at right appears to be ejecting a jet of gas. What is happening to this star?

(STScI and NASA)

Discussion Questions

1. Ancient astronomers, philosophers, and poets assumed that the stars were eternal and unchanging. Is there any observation they could have made or any line of reasoning that could have led them to conclude that stars don't live forever?

2. How does hydrostatic equilibrium relate to hot-air ballooning?

Problems

1. The ring ejected by the supergiant star in Figure 11-1 has a radius of about 7 seconds of arc. If the cluster is 20,000 ly from Earth, what is the radius of the ring in light-years?

2. If a giant molecular cloud is 50 pc in diameter and a shock wave can sweep through it in 2 million years, how fast is the shock wave going in kilometers per second?

3. If a giant molecular cloud has a mass of 10^{35} kg, and it converts 1 percent of its mass into stars during a single encounter with a shock wave, how many stars can it make? Assume the stars each contain 1 solar mass.

4. If a protostellar disk is 200 AU in radius, and the disk plus the forming star contain 2 solar masses, what is the orbital velocity at the outer edge of the disk in kilometers per second?

5. If a contracting protostar is five times the radius of the sun and has a temperature of only 2000 K, how luminous will it be? (*Hint:* See Chapter 9.)

6. The gas in a bipolar flow can travel as fast as 100 km/s. If the length of the jet is 1 ly, how long does it take for a blob of gas to travel from the protostar to the end of the jet?

7. If a T Tauri star is the same temperature as the sun but is ten times more luminous, what is its radius? (*Hint:* See Chapter 9.)

8. Circle all of the ^{1}H and ^{4}He nuclei in Figure 11-10 and explain how the CNO cycle can be summarized by $4\,^1H \rightarrow {}^4He + energy$.

9. How much energy is produced when the CNO cycle converts 1 kg of mass into energy? Is your answer different if the mass is fused by the proton–proton chain?

10. If the Orion Nebula is 8 pc in diameter and has a density of about 600 hydrogen atoms/cm^3, what is its total mass? (*Hint:* The volume of a sphere is $\frac{4}{3}\pi R^3$.)

11. The hottest star in the Orion Nebula has a surface temperature of 40,000 K. At what wavelength does it radiate the most energy? (*Hint:* See Chapter 7.)

Media Cluster

Ace Astronomy™ To access the resources in the Media Cluster, log into AceAstronomy at **http://ace.brookscole .com/sf9** and select Chapter 11.

ACTIVE FIGURES

Conduction, Convection, and Radiation
Whether inside a star or on the surface of a planet, the methods of heat transfer available are the same. Make sure you understand the difference between conduction, convection, and radiation with this animation.

Inside Stars
You can use the laws of physics to study the interior of the sun and compare it to other stars in the universe. This cutaway animation of the sun lets you explore the various structures of stars.

ASTRONOMY EXERCISES

Nuclear Fusion
Nuclear fusion is a reaction that joins the nuclei of atoms to form more massive nuclei. Vary the temperature of the star in this animation to see how this affects the rate of nuclear fusion and luminosity.

Hydrostatic Equilibrium
This animation allows you to adjust the rate of nuclear fusion and the diameter of a star in hydrostatic equilibrium.

Critical Inquiries for the Web

1. Use the Web to supply additional details concerning the evolution of protostars and T Tauri stars.

2. Astronomers continue to study the Orion Nebula and star formation in the molecular cloud behind the nebula. What is the latest news from Orion?

3. What is BM Orionis? Where is it located, and what does it do? Why might it be interesting to observe with even a small telescope?

Exploring *TheSky*

1. The following nebulae are all star-formation regions. What kind of nebulae are they? (*Hint:* To center on an object, use **Find** under **Edit.** Choose **Messier Objects** and pick from the list.)

   M42, M20, M8, M17

2. Locate M8 in *TheSky,* zoom in, and identify other nebulae in the region. Study the photo of NGC6559. What kind of nebula is it?

   ● Go to the Brooks/Cole Astronomy Resource Center (**http:// astronomy.brookscole.com**) for critical thinking exercises, articles, and additional readings from InfoTrac College Edition, Brooks/Cole's online student library.

12 | Stellar Evolution

We should be unwise to trust scientific inference very far when it becomes divorced from opportunity for observational test.

SIR ARTHUR EDDINGTON,
THE INTERNAL CONSTITUTION OF THE STARS

Visual-wavelength image

E VERY STAR HAD A beginning, and every star must have an ending, but in between they produce the light and energy that make our universe beautiful. The stars above you seem eternal, but the light and warmth they emit is produced by the fusion of nuclear fuels in their cores; and, even as you look at them, they are using up their fuels and drawing closer to their endings. The stars are, ever so slowly, going out. ▌ Although stars enjoy long, stable lives on the main sequence, they suffer the first pangs of stellar age as they gradually consume their hydrogen fuel. Once hydrogen is exhausted at their centers, they swell rapidly into giant stars 10 to 1000 times the size of the sun. For a short time, a few million to a billion years, they are majestic beacons visible across thousands of parsecs. ▌ Although astronomers say a star swells

▌ Continued on page 260 ▌

Guidepost

Looking Back

The life story of stars begins in the interstellar medium and continues through their birth in glowing nebulae. You have followed that story through the last two chapters, and now you are ready for the next step.

This Chapter

This chapter is about the long, stable middle age of stars on the main sequence and later as they swell to become giants. Here you will answer four essential questions:

How do astronomers know how stars evolve?

What are main-sequence stars like?

How do stars evolve after they exhaust hydrogen?

What evidence do astronomers have that stars really do evolve?

This discussion is a good example of how scientists use theory and evidence to understand how nature works.

Looking Ahead

This chapter is about how stars live. The next two chapters are about how stars die and the strange objects they leave behind.

rapidly and spends a short time as a giant, these changes happen slowly compared with a human life. How can anyone be sure stars really do age? Stellar evolution and the formation of giant stars might be just an astronomer's daydream.

In this chapter, you will see the full interplay of theory and evidence. You will use the basic laws of physics, combined with your knowledge of the nature of stars, to create a theory to describe how main-sequence stars maintain their equilibrium and how such stars change when they exhaust their nuclear fuels. As you are warned in the quotation that opens this chapter, theory alone is never enough. At each step, astronomers must compare their theories with the evidence. In this chapter, you will discover that clusters of stars can make the slow evolution of stars visible even during our short lifetime, and you will find that some evolving stars can become unstable and pulsate like beating hearts. Those pulsating stars reveal more about how stars evolve.

From beginning to end, this chapter tells the story of the evolution of stars from main sequence to giant. You will be able to trace the passage of stars through this phase of their existence and see how they generate their energy and why they swell so large. The place to begin is with the main-sequence stars.

12-1 Main-Sequence Stars

IF SHAKESPEARE WERE ALIVE today, he would probably have something sarcastic to say about modern astronomers. "How do these stargazers pretend to know the hearts of the eternal stars?" he might ask. In fact, one of the greatest triumphs of modern astronomy is the discovery that the stars are not eternal and that mere humans can indeed understand their hearts. Astronomers can know the conditions at the center of a star because stars are fundamentally very simple objects. You can begin by considering how astronomers use stellar models to understand the interior of main-sequence stars. Later you can use these same models to understand what happens to stars when they grow old.

Stellar Models

Every star is balanced between gravity trying to make it contract and internal pressure trying to make it expand (■ Figure 12-1). Astronomers describe the insides of stars by imagining that the stars are divided into concentric shells similar to the layers in an onion (Figure 11-12). Of course, stars are not made like onions, but defining these shells mathematically helps astronomers discuss the conditions at different levels inside the star—what astronomers call the "structure" of the star.

The structure of a star is described by four simple laws of physics, two of which you have already met. In Chapter 11, you met the law of hydrostatic equilibrium, which says that the weight pressing down on a shell in a star must be balanced by the pressure pushing outward in that shell. That is, at every level in a star,

■ **Figure 12-1**

Stars shine because energy flows out of their cores, and they are held together by their own gravity. Here Merope, the brightest star in the Pleiades, is imbedded in wisps of gas through which the star cluster is drifting. The circular rings and spikes are a diffraction pattern produced by light inside the telescope. The actual star is an unresolved point of light. (NOAO/AURA/NSF)

the weight of the overlying matter must be supported by the pressure. Of course, that means that there can be no empty shells; an empty gap could exert no pressure and would be unable to support the weight of the material above. Chapter 11 also introduced you to energy transport; the law of energy transport describes how energy flows from hot to cool regions by radiation, convection, or conduction.

To these two laws you can add two basic laws of nature. The **conservation of mass law** says that the total mass of the star must equal the sum of the masses in its shells. Of course, no shell can be empty, and there is no such thing as negative mass. The **conservation of energy law** says that the amount of energy flowing out the top of a shell in the star must be equal to the amount of energy coming in at the bottom of the shell plus whatever energy is generated within the shell. Of course, that's just logical. It means that the energy leaving the surface of the star, its luminosity, must equal the sum of the energy generated in all of the shells inside the star. This is like saying that the total number of new cars driving out of a factory must equal the sum of cars manufactured on each assembly line. No car can vanish into nothing or appear from nothing. Energy in a star may not vanish without a trace or appear out of nowhere.

The four laws of stellar structure, described in general terms in ■ Table 12-1, can be written as mathematical equations. By solving those equations in a special way, astronomers can build a mathematical model of the inside of a star.

If you wanted to build a model of a star, you would have to divide the star into about 100 concentric shells and then write down the four equations of stellar structure for each shell. You would then have 400 equations that would have 400 unknowns, namely, the temperature, density, mass, and energy flow in each shell. Solving 400 equations simultaneously is not easy, and the first such solutions, done by hand before the invention of electronic computers, took months of work. Now a properly programmed computer can solve the equations in a few seconds and print a

table of numbers that represent the conditions in each shell of the star. Such a table is a **stellar model.**

The table shown in ■ Figure 12-2 is a model of the sun. The bottom line, for radius equal to 0.00, represents the center of the sun, and the top line, for radius equal to 1.00, represents the surface. The other lines in the table show the temperature and density in each shell, the mass inside each shell, and the fraction of the sun's luminosity flowing outward through the shell. You can use the table to understand the sun. The bottom line tells you the temperature at the center of the sun is about 15 million Kelvin. At such a high temperature, the gas is highly transparent, and energy flows as radiation. Nearer the surface, the temperature is lower, the gas is more opaque, and the energy is carried by convection.

Notice that stellar models are quantitative; that is, properties have specific numerical values. Earlier in this book, you used models that were qualitative—the Babcock model of the sun's magnetic cycle, for instance. Both kinds of models are useful, but the quantitative model reveals deeper insights into how nature works, reflecting the power of mathematics as a precise way of thinking (**Window on Science 12-1**).

Stellar models let astronomers look into a star's past and future. In fact, they can use models as time machines to follow the evolution of stars over billions of years. To look into a star's future, for instance, you could use a stellar model to determine how

■ **Figure 12-2**

A stellar model is a table of numbers that represent conditions inside a star. Such tables can be computed using the four laws of stellar structure, shown here in mathematical form. The table in this figure describes the sun. (Illustration design by author)

$R/R_\odot$	T $(10^6$ K$)$	Density $(g/cm3)$	$M/M_\odot$	$L/L_\odot$
1.00	0.006	0.00	1.00	1.00
0.90	0.60	0.009	0.999	1.00
0.80	1.2	0.035	0.996	1.00
0.70	2.3	0.12	0.990	1.00
0.60	3.1	0.40	0.97	1.00
0.50	4.9	1.3	0.92	1.00
0.40	5.1	4.1	0.82	1.00
0.30	6.9	13.	0.63	0.99
0.20	9.3	36.	0.34	0.91
0.10	13.1	89.	0.073	0.40
0.00	15.7	150.	0.000	0.00

$$\frac{dM}{dr} = 4\pi r^2 \rho$$

$$\frac{dL}{dr} = 4\pi r^2 \rho e$$

$$\frac{dP}{dr} = -\frac{GM}{r^2}\rho$$

$$\frac{dT}{dr} = \frac{-3}{16\pi ac}\frac{\kappa\rho}{T^3}\frac{L}{r^2}$$

Quantitative Thinking with Mathematical Models

One of the most powerful tools in science is the mathematical model, a group of equations carefully designed to mimic the behavior of the object scientists want to study. Astronomers build mathematical models of stars using only four equations, but other systems are much more complicated and may require many more equations.

Many sciences use mathematical models. Medical scientists have built mathematical models of the nerves that control the human heart, and physicists have built mathematical models of the inside of an atomic nucleus. Economists have built mathematical models of certain aspects of economic systems, such as the municipal bond market, and Earth scientists have built mathematical models of Earth's atmosphere. In each case, the mathematical model allows the scientists to study something that is too difficult to study in the real world. The model can reveal regions that cannot be observed, speed up a slow process, slow down a fast process, and allow scientists to perform experiments that would be impossible in reality. Astronomers, for example, can change the

abundance of certain chemical elements in a model star to see how its structure depends on its composition.

Many of these mathematical models require very large and fast computers, in some cases supercomputers. A modern computer takes only seconds to compute a stellar model, but models such as the motions of the planets in our solar system millions of years into the future require the largest and fastest computers in the world and highly sophisticated mathematical methods. Most astronomers are apt computer programmers, and some are experts. A few astronomers have built their own specialized computers to calculate specific kinds of models. The mathematical model is one of the most important tools in astronomy.

As is true for any scientific model, a mathematical model is only as reliable as the assumptions that go into its creation. In Window on Science 2-1, you saw that the celestial sphere was an adequate model of the sky for some purposes but breaks down if extended too far. So, too, with mathematical models. You can think of a mathematical model as a numerical

Before any new airplane flies, engineers build mathematical models to test its stability. (The Boeing Company)

expression of one or more theories. While such models can be very helpful, they are always based on theory and so must be compared with the real world at every opportunity. Models must always be tested against experiment and observation. Otherwise, they might lead you astray.

---

fast the star uses its fuel in each shell. As the fuel is consumed, the chemical composition of the gas changes, and the amount of energy generated declines. By calculating the rate of these changes, you could predict what the star will look like at any point in the future.

Although this sounds simple, it is actually a highly challenging problem involving nuclear and atomic physics, thermodynamics, and sophisticated computational methods. Only since the 1950s have electronic computers made the rapid calculation of stellar models possible, and the advance of astronomy since then has been heavily influenced by the use of such models to study the structure and evolution of stars. The summary of star formation in this chapter is based on thousands of stellar models. You will continue to rely on theoretical models as you study main-sequence stars in the next section and the deaths of stars in the next chapter.

Why Is There a Main Sequence?

You can now understand why there is a main sequence. The clue is the law of hydrostatic equilibrium, and you can use models of stars to analyze this clue.

There is a main sequence because the centers of contracting protostars eventually grow hot enough to begin nuclear fusion.

Deuterium, the heavy isotope of hydrogen, is the first nuclear fuel to fuse, but it is so rare and produces so little energy it has no real effect on a contracting star. Hydrogen fusion is the big powerhouse; when it begins, it stops the contraction. Stars reach equilibrium somewhere along a line in the H–R diagram that astronomers call the main sequence. Hot stars are more luminous, and cool stars are less luminous, as you would expect. There is, however, a mystery about the main sequence that you can now solve. Why does the luminosity of a star depend on its mass?

In Chapter 9, you used binary stars to find the masses of stars, and you discovered that the masses of main-sequence stars are ordered along the main sequence. The least massive stars are at the bottom, and the most massive stars are at the top. Further, you discovered a direct relationship between the mass of a star and its luminosity—the mass–luminosity relation. This is one of the most fundamental observations in astronomy, and stellar models can tell you why there must be a mass–luminosity relation, and that tells you why there has to be a main sequence.

The keys to the mass–luminosity relation are the law of hydrostatic equilibrium, which says that pressure must balance weight, and the pressure–temperature thermostat, which regulates energy production. You have seen that a star's internal pressure stays high because the generation of thermonuclear energy keeps its interior hot. Because more massive stars have more weight pressing down on the inner layers, their interiors must have high pressures and thus must be hot. For example, the temperature at the center of a 15-solar-mass star is about 34,000,000 K, more than twice the central temperature of the sun.

Because massive stars have hotter cores, their nuclear reactions burn more fiercely. That is, their pressure–temperature thermostat is set higher. The nuclear fuel at the center of a 15-solar-mass star fuses over 3000 times more rapidly than the fuel at the center of the sun. The rapid reactions in the cores of massive stars produce more energy, but that energy cannot remain in the core. The energy must flow outward toward the cooler surface, and, in doing so, the flowing energy heats each level in the star and enables it to support the weight pressing inward. When all that energy reaches the surface, it radiates into space and makes the star highly luminous. That means there must be a mass–luminosity relation because each star must support its weight by generating nuclear energy, and more massive stars have more weight to support.

The main sequence is elegant in its simplicity. It exists because stars balance their weight by fusing hydrogen in their cores. To understand the main-sequence stars even better, you should next look at its top and bottom ends.

The Upper End of the Main Sequence

Stellar structure gives astronomers a way to think about the extreme ends of the main sequence, the most massive and least massive stars. The first are rare, but the latter are common. Nevertheless, both are very difficult to study.

There are two reasons why there is an upper limit to the mass of stars. First, it appears that gas clouds contracting to form a star can fragment and form two or more stars. If a gas cloud contains more mass, it is more likely to fragment, so there aren't many very massive stars because those gas clouds broke into smaller fragments and formed multiple-star systems.

But, second, stellar models reveal that stars of roughly 100 solar masses are unstable at formation. To support the tremendous weight in such stars, the internal gas must be very hot, and that means it must emit floods of radiation that flow outward through the star and blow gas away from the star's surface in powerful stellar winds. This mass loss from very massive stars could reduce a 60-solar-mass star to less than 30 solar masses in only a million years.

This process sets an upper limit on the mass of stars, but it is difficult to test the theory because it is hard to find truly massive stars. Most of the O and B stars in the sky have masses of 10 to 25 solar masses. The survey of stars at the end of Chapter 9 revealed that the stars at the upper end of the main sequence are very rare, so astronomers must search to great distances to find just a few. Nevertheless, a few stars are known that are thought to be very massive, and their spectra contain blueshifted emission lines. Kirchhoff's laws tell you that emission lines come from excited low-density gas, and the blueshift must be caused by a Doppler shift in gas coming toward Earth. These stars are losing mass.

■ Figure 12-3 shows a famous star, Eta Carinae, that is believed to be a massive binary in which the stars contain 60 solar masses and 70 solar masses. They may have formed with about 100 solar masses each, but mass loss is reducing their mass. An eruption 150 years ago made Eta Carinae the second brightest star in the sky and ejected the two expanding lobes of dusty gas. Although the star has faded, it is still very active, and more recent eruptions have ejected jets and an equatorial disk of gas and dust. Whatever their status, these massive stars are clearly unstable.

The Lower End of the Main Sequence

The lower end of the main sequence is difficult to study, not because the stars are rare but because they are dim. If a red dwarf from the lower end of the main sequence replaced the sun, it would shine only a few times brighter than the full moon. Such stars are difficult to find even when they are only a few light-years away.

Stellar models predict that stars less massive than 0.08 solar mass cannot get hot enough to ignite hydrogen fusion. These **brown dwarfs** should cool as they convert their gravitational energy into heat and radiate it away. A star of 0.08 solar mass fusing hydrogen has a surface temperature of about 2500 K, but brown dwarfs should have temperatures of 1000 K or so, giving them a color even ruddier than red dwarfs—thus the term "brown dwarf." Brown dwarfs were difficult to find because they are so faint, but large surveys and infrared studies have turned up lots. Some are located in binary systems with normal stars (■ Figure 12-4a), but large numbers are free-floating objects without stellar companions (Figure 12-4b).

Brown dwarfs are clearly different from normal stars. Some brown dwarfs have methane bands in their spectra, and that means they must be quite cool. Methane molecules would be broken up at the temperatures of true stars. Color variations suggest that some brown dwarfs may be cool enough to have weather patterns.

The discovery of brown dwarfs has created a controversy. Are they failed stars, or are they planets? To discuss this, astronomers use a unit of mass related to Jupiter, the giant planet in our solar system. Jupiter is one thousand times less massive than the sun, and astronomers refer to the mass of brown dwarfs in Jupiter masses. A brown dwarf must be less than about 80 Jupiter masses. Generally, astronomers think of stars as bodies that generate energy by nuclear fusion, and a star less than 80 Jupiter masses can't heat its center to the 2.7 million Kelvin temperature needed to start hydrogen fusion. Then brown dwarfs aren't stars. But astronomers tend to think of a planet as a nonluminous body that orbits a star, so can a brown dwarf floating free in space be called a planet?

Brown dwarfs more massive than 13 Jupiter masses can fuse deuterium, a heavy isotope of hydrogen, and some astronomers draw the line between stars and planets at 13 Jupiter masses.

X-ray bright gas

Far lobe

Disk

Near lobe

Visual-wavelength image

X-ray image

Visual

Gas expanding away
at 1.5 million miles
per hour

15 $M_\odot$ torus

Infrared image

■ **Figure 12-3**

The star Eta Carinae is a binary containing two stars that are so massive they are rapidly losing mass. At visual wavelengths, two inflating lobes are visible with a disk of ejected material between like a plate pressed between two basketballs. Each lobe is about half a light-year in diameter. At X-ray wavelengths, very hot gas is excited by collision with high-speed gas ejected from the stars. An infrared image reveals a 15-solar-mass torus (doughnut shape) of gas and dust squeezing the outflowing gas into the two lobes. (NASA/CXC/ SAO/HST; Jon Morse and J. Hester; IR image: ISO, Courtesy ESA)

However, deuterium fusion does not last very long, does not generate much energy, and does not stop the star from cooling. Not everyone agrees that this is a good dividing line.

Another approach to the controversy is to note how these objects form. A star forms from the contraction of a gas cloud, but astronomers believe planets form from the accumulation of solid bits of matter in disk-shaped nebulae around stars (a subject discussed in Chapter 19). In that case, free-floating brown dwarfs can't be planets.

The controversy over brown dwarfs is more an argument over the meaning of the words *star* and *planet*. In any case, you can understand the lower end of the main sequence. Stars less massive than 80 Jupiter masses cannot ignite hydrogen fusion; and, even if they are more massive than 13 Jupiter masses and deuterium fusion generates a little energy, they must continue contracting until the internal gas becomes so dense it cannot contract further. At that point, the object radiates its thermal energy away and slowly cools.

The Life of a Main-Sequence Star

A normal main-sequence star supports its weight by fusing hydrogen into helium, but its supply of hydrogen is limited. As it consumes its hydrogen, the chemical composition in its core changes, and the star evolves. Mathematical models of stars allow astronomers to follow that evolution.

Hydrogen fusion combines four nuclei into one. Consequently, as a main-sequence star consumes its hydrogen, the total number of nuclei in its interior decreases. Each newly made helium nucleus can exert the same pressure as a hydrogen nucleus; but, because the gas has fewer nuclei, its total pressure is less. This unbalances the gravity–pressure stability, and gravity squeezes the core of the star more tightly. As the core contracts, its temperature and density increase, and the nuclear reactions burn faster, releasing more energy and making the star more luminous. This additional energy flowing outward through the envelope

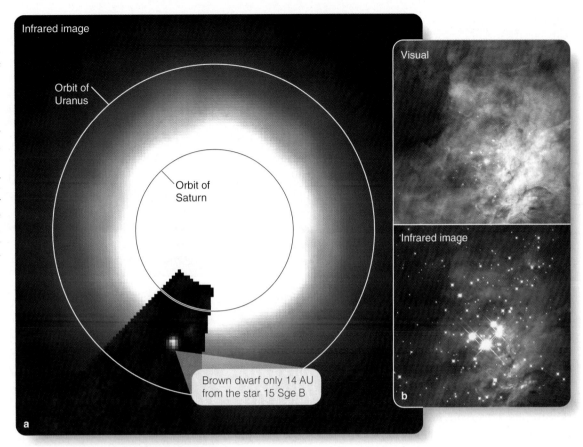

(a) Computer processing to remove glare from the star 15 Sge B reveals a brown dwarf only 14 AU from its star. It is about as far from its companion star as the giant planets Saturn and Uranus are from the sun. (Gemini Observatory/ University of Hawaii Institute for Astronomy/Michael Liu/NSF) (b) Too dim to see at visual wavelengths, about 50 free-floating brown dwarfs appear in an infrared image of the center of the Orion nebula. The bright stars of the Trapezium are located at the center. (STScI and NASA)

Infrared image

Orbit of Uranus

Orbit of Saturn

Brown dwarf only 14 AU from the star 15 Sge B

a

Visual

Infrared image

b

O B A F G K M

The Aging of Main-Sequence Stars

10 million years

−15M$_\odot$

Stars exhaust the last of the hydrogen in their cores as they leave the main sequence.

640 million years

3M$_\odot$

10 billion years

Present sun
Initial sun

Newly formed stars begin life at the lower edge of the main sequence.

ZAMS

L/L$_\odot$

30,000 20,000   10,000   5000   3000 2000

Temperature (K)

forces the outer layers to expand and cool, so the star becomes slightly larger, brighter, and cooler.

As a result of these gradual changes in main-sequence stars, the main sequence is not a sharp line across the H–R diagram but rather a band (■ Figure 12-5). A star begins its stable life fusing hydrogen and settles on the lower edge of this band, the **zero-age main sequence (ZAMS).** As it combines hydrogen nuclei to make helium nuclei, the star slowly changes. In the H–R diagram, the point that represents the star's luminosity and surface temperature moves upward and to the right, eventually reaching the upper edge of the main sequence, just as the star exhausts nearly all of the hydrogen in its center. Astronomers find main-sequence

■ **Active Figure 12-5**

Contracting protostars reach stability at the lower edge of the main sequence, the zero-age main sequence (ZAMS). As a star converts hydrogen in its core into helium, it moves slowly across the main sequence, becoming slightly more luminous and slightly cooler. Once a star consumes all of the hydrogen in its core, it can no longer remain a stable main-sequence star. More massive stars age rapidly, but less-massive stars use up the hydrogen in their cores more slowly and live longer main-sequence lives.

Ace◗Astronomy™ Log into AceAstronomy and select this chapter to see Active Figure "Future of the Sun." See how the future evolution of the sun will affect Earth.

stars plotted throughout this band at various stages of their main-sequence lives.

These gradual changes in the sun will spell trouble for Earth. When the sun began its main-sequence life about 5 billion years ago, it was only about 75 percent of its present luminosity. This, by the way, makes it difficult to explain how Earth has remained at roughly its present temperature for at least 3 billion years. Some experts suggest that Earth's atmosphere has gradually changed and compensated for the increasing luminosity of the sun.

By the time the sun leaves the main sequence in a few billion years, it will have twice its present luminosity. This will raise the average temperature on Earth by at least 19°C (34°F). As this happens over the next few billion years, the polar caps will melt, the oceans will evaporate, and much of the atmosphere will vanish into space. Clearly, the future of Earth as the home of life is limited by the evolution of the sun.

Once a star leaves the main sequence, it evolves rapidly and dies. The average star spends 90 percent of its life fusing hydrogen on the main sequence. This explains why 90 percent of all normal stars are main-sequence stars. You are most likely to see a star during that long, stable period when it is on the main sequence.

The number of years a star spends on the main sequence depends on its mass (■ Table 12-2). Massive stars use fuel rapidly and live short lives, but low-mass stars conserve their fuel and shine for billions of years. For example, a 25-solar-mass star will exhaust its hydrogen and die in only about 7 million years. This means, for one thing, that life is very unlikely to develop on planets orbiting massive stars. These stars do not live long enough for life to get started and evolve into complex creatures. You will learn about this problem in detail in Chapter 26.

Very-low-mass stars, the red dwarfs, use their fuel so slowly they should survive for 200 to 300 billion years. Because the universe seems to be only about 14 billion years old, red dwarfs must still be in their infancy. None of them should have exhausted their hydrogen fuel yet.

Nature makes more low-mass stars than massive stars, but this fact is not sufficient to explain the vast numbers of low-mass stars that fill the sky. An additional factor is the stellar lifetimes. Because low-mass stars live long lives, there are more of them in the sky than massive stars. Look at page 211 and notice how much more common the lower-main-sequence stars are than the massive O and B stars. The main-sequence K and M stars are so faint they are difficult to locate, but they are very common. The O and B stars are luminous and easy to locate; but, because of their fleeting lives, there are never more than a few on the main sequence at any one time.

The Life Expectancies of Stars

To understand how nature makes stars and how stars evolve, you must be able to estimate how long they can survive, and that turns out to be easy. In general, massive stars live short lives and lower-mass stars live long lives, but you can make more accurate

■ Table 12-2 I Main-Sequence Stars

Spectral Type	Mass (sun = 1)	Luminosity (sun = 1)	Approximate Years on Main Sequence
O5	40	405,000	1×10^6
B0	15	13,000	11×10^6
A0	3.5	80	440×10^6
F0	1.7	6.4	3×10^9
G0	1.1	1.4	8×10^9
K0	0.8	0.46	17×10^9
M0	0.5	0.08	56×10^9

estimates by using simple stellar models. In fact, you can calculate the approximate life expectancy of a star from its mass.

Because main-sequence stars consume their fuel at an approximately constant rate, you can estimate the amount of time a star spends on the main sequence—its life expectancy T—by dividing the amount of fuel by the rate of fuel consumption. This is a common calculation. If you drive a truck that carries 20 gallons of fuel and uses 5 gallons of fuel per hour, you know the truck can run for 4 hours.

The amount of fuel a star has is proportional to its mass, and the rate at which it burns its fuel is proportional to its luminosity; that means you could make a first estimate of the star's life expectancy by dividing mass by luminosity. A 2-solar-mass star is about 11 times more luminous than the sun and should live about 2/11, or 18 percent, as long as the sun. You can, however, make the calculation even easier if you remember that the mass–luminosity relation says that the luminosity of a star equals $M^{3.5}$. The life expectancy then is:

$$T = \frac{M}{L} = \frac{M}{M^{3.5}} \quad \text{or} \quad T = \frac{1}{M^{2.5}}$$

This means that you can estimate the life expectancy of a star by dividing 1 by the star's mass raised to the 2.5 power. If you express the mass in solar masses, the life expectancy will be in solar lifetimes.

For example, how long can a 4-solar-mass star live?

$$T = \frac{1}{4^{2.5}} = \frac{1}{4 \cdot 4 \cdot \sqrt{4}} = \frac{1}{32} \text{ solar lifetimes}$$

Detailed studies of models of the sun show that the sun, presently 5 billion years old, can last another 5 billion years. So, a solar lifetime is approximately 10 billion years, and a 4-solar-mass star will last about (10 billion years)/32, or about 310 million years.

This estimation of stellar life expectancies is very approximate. For example, the model ignores mass loss, which may affect the life expectancies of very luminous and very faint stars.

Nevertheless, it serves to illustrate an important point. Stars that are only slightly more massive than the sun have dramatically shorter lifetimes on the main sequence.

Why is there a main sequence?
Most scientific arguments are simple chains of ideas that begin from a well-understood idea and lead to a less obvious conclusion. You can begin with the simple observation that weight must be balanced by pressure. That is just the principle of hydrostatic equilibrium. A contracting protostar is not quite in equilibrium, and gravity squeezes it tighter and tighter. As it contracts, its interior heats up; and, when it gets hot enough to fuse hydrogen into helium, the pressure–temperature thermostat takes over and regulates energy production so that the star makes just enough energy to support its own weight. That means that massive stars, having more weight to support, must have higher internal pressures and therefore must make more energy. Energy must flow from hot to cool, so that energy flows from the hot core of the star out to the cooler surface and is radiated into space. Massive stars, having more weight to support, must have their pressure–temperature thermostats set higher, and that makes them more luminous. They reach stability along the upper main sequence. Less-massive stars support less weight and reach stability along the lower main sequence. There is a main sequence because stars fuse hydrogen to support their own weight.

Now revise your argument. **Why do massive stars have such short life expectancies?**

■ ■ ■

Connections: When a star finally exhausts its hydrogen fuel, it can no longer resist the pull of its own gravity. The contraction that began when it was a protostar resumes and begins the process that leads to the star's death. It can delay its end by fusing other fuels, but, as you will discover in Chapter 13, nothing can steal gravity's final victory.

(12-2) Post-Main-Sequence Evolution

IN EARLIER CHAPTERS, you probably had questions about giant stars: Why are giant stars so large? Why are they so uncommon? And why do they have such low densities? Now you are ready to answer those questions by discussing the evolution of stars after they leave the main sequence.

Expansion into a Giant

To understand how stars evolve, you must remember that they are not well mixed; that is, their interiors are not stirred. The centers of lower-mass stars like the sun are radiative, meaning the energy moves as radiation and not as circulating currents of heated gas. The gas does not move deep inside such stars, and that means they are not mixed at all. More massive stars have convective cores that mix the central regions (Figure 11-14), but these regions are not very large, and so, for the most part, these stars, too, are not mixed. (The lowest-mass stars are an exception that you will examine in the next chapter.)

In this respect, stars are like a campfire that is not stirred; the ashes accumulate at the center, and the fuel in the outer parts never gets used. Nuclear fusion consumes hydrogen nuclei and produces helium nuclei, the "ashes," at the star's center. Nothing mixes the interior of the star, so the helium nuclei remain where they are in the center of the star, and the hydrogen in the outer parts of the star is not mixed down to the center where it can be fused.

The helium ashes that accumulate in the star's core cannot fuse into heavier elements because the temperature is too low. As a result, the helium accumulates, the hydrogen is used up, and the core becomes an inert ball of helium. As this happens, the energy production in the core falls, and the weight of the outer layers forces the core to contract.

Although the contracting helium core cannot generate nuclear energy, it does grow hotter because it converts gravitational energy into thermal energy (see Chapter 11). The rising temperature heats the unprocessed hydrogen just outside the core, hydrogen that was never before hot enough to fuse. When the surrounding hydrogen becomes hot enough, it ignites in a shell of fusing hydrogen. Like a grass fire burning outward from an exhausted campfire, the hydrogen-fusing shell burns outward, leaving helium ash behind and increasing the mass of the helium core.

At this stage in its evolution, the star overproduces energy; that is, it produces more energy than it needs to balance its own gravity. The helium core, having no nuclear energy sources, must contract, and that contraction converts gravitational energy into thermal energy—the contraction heats the helium core. Some of that heat leaks outward through the star. At the same time, the hydrogen-fusing shell produces energy as the contracting core brings fresh hydrogen closer to the center of the star and heats it to high temperature. The result is a flood of energy flowing outward through the outer layers of the star, forcing them to puff up and swelling the star into a giant (■ Figure 12-6).

The expansion of the envelope dramatically changes the star's location in the H–R diagram. As the outer layers of gas expand, energy is absorbed in lifting and expanding the gas. The loss of that energy lowers the temperature of the gas. Consequently, the point that represents the star in the H–R diagram moves quickly to the right (in less than a million years for a star of 5 solar masses). A medium-mass star like the sun expands and cools to become a red giant (■ Figure 12-7). As the radius of a giant star continues to increase, the enlarging surface area makes the star more luminous, moving its point upward in the H–R diagram. One of our

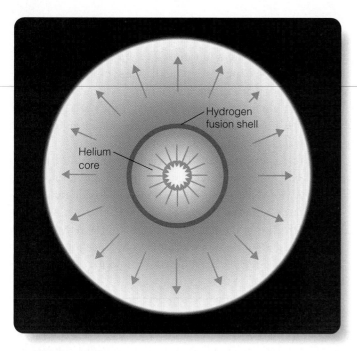

■ Figure 12-6

When a star runs out of hydrogen at its center, it ignites a hydrogen-fusing shell. The helium core contracts and heats while the envelope expands and cools. (For a scale drawing, see Figure 12-9.)

Favorite Stars, Aldebaran, the glowing red eye of Taurus the bull, is such a red giant, with a diameter 25 times that of the sun but with a surface temperature only half that of the sun.

More massive stars cross higher in the H–R diagram and become supergiants, stars even larger than giants. Consider two more Favorite Stars. Betelgeuse (α Orionis) is a very cool, red supergiant over 800 times the diameter of the sun. Rigel (β Orionis) is a supergiant 50 times larger than the sun. It may seem odd to say that Rigel, a blue star, has expanded and cooled. At a temperature of 12,000 K, Rigel looks quite blue to your eyes. When it was on the main sequence, Rigel had a much hotter surface, but it would not have looked much bluer because you cannot see the ultraviolet radiation that the hottest stars emit. Any star as hot or hotter than Rigel will look blue to your eyes, as shown in Figure 7-2. So, blue Rigel is indeed a star that has expanded and cooled.

■ Active Figure 12-7

The evolution of a massive star moves the point that represents it in the H–R diagram to the right of the main sequence into the region of the supergiants such as Rigel and Betelgeuse. The evolution of a medium-mass star moves its point in the H–R diagram into the region of the giants such as those shown here.

Ace Astronomy™ Log into AceAstronomy and select this chapter to see Active Figure "Evolution of Stars" and compare the evolution of a massive star to that of a low-mass star.

Now you can understand a few things you noticed in earlier chapters. Giants and supergiants are large because they have expanded, and they have very low densities for the same reason. Also, the giant and supergiant region of the H–R diagram contains a jumble of different mass stars because all main-sequence stars funnel through this part of the diagram as they die. (See Figure 9-23.)

Degenerate Matter

Although the hydrogen-fusing shell can force the envelope of the star to expand, it cannot stop the contraction of the helium core. Because the core has no energy source, gravity squeezes it tighter, and it becomes very small. If you represent the helium core of a 5-solar-mass star with a quarter, the outer envelope of the star would be about the size of a baseball diamond. Yet the core would contain about 12 percent of the star's mass compressed to very high density. When gas is compressed to such extreme densities, it begins to behave in astonishing ways that can alter the evolution of the star. To follow the story of stellar evolution, you must consider the behavior of gas at extremely high densities.

Normally, the pressure in a gas depends on its temperature. The hotter the gas is, the faster its particles move, and the more pressure it exerts. But the gas inside a star is ionized, so there are

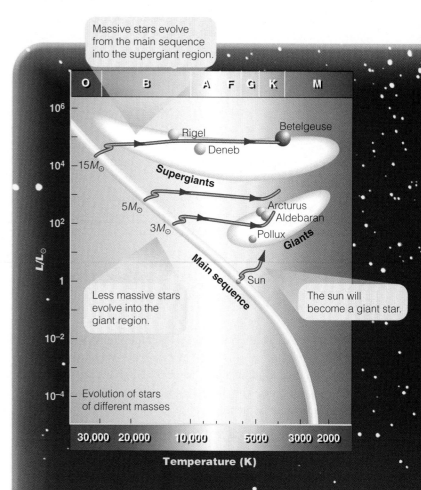

two kinds of particles, atomic nuclei and free electrons. If the gas is compressed to very high densities, in the core of a giant star, for example, the difference between these two kinds of particles is important.

If the density is very high, the particles of the gas are forced close together, and two laws of quantum mechanics become important. First, quantum mechanics says that the moving electrons confined in the star's core can have only certain amounts of energy, just as the electron in an atom can occupy only certain energy levels (see Chapter 7). You can think of these permitted energies as the rungs of an energy ladder. An electron can occupy any rung but not the spaces between.

The second quantum-mechanical law (called the Pauli Exclusion Principle) says that two identical electrons cannot occupy the same energy level. Because electrons spin in one direction or the other, two electrons can occupy an energy level if they spin in opposite directions. That level is then completely filled, and a third electron cannot enter because, whichever way it spins, it will be identical to one or the other of the two electrons already in the level. Thus, no more than two electrons can occupy the same energy level.

A low-density gas has few electrons per cubic centimeter, so there are plenty of energy levels available. If a gas becomes very dense, however, nearly all of the lower energy levels may be occupied (■ Figure 12-8). In such a gas, a moving electron cannot slow down, because slowing down would decrease its energy, and

■ Figure 12-8

Electron energy levels are arranged like rungs on a ladder. In a low-density gas, many levels are open, but in a degenerate gas, all lower energy levels are filled.

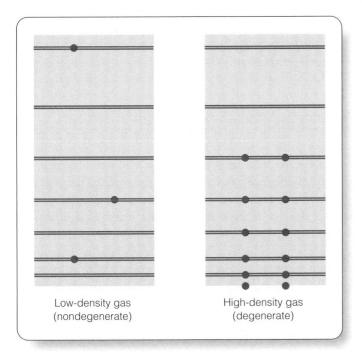

Low-density gas
(nondegenerate)

High-density gas
(degenerate)

there are no open energy levels for it to drop down to. It can speed up only if it can absorb enough energy to leap to the top of the energy ladder, where there are empty energy levels.

When a gas is so dense that the electrons are not free to change their energy, astronomers call it **degenerate matter.** Although it is a gas, it has two peculiar properties that can affect the star. First, the degenerate gas resists compression. To compress the gas, you would have to push against the moving electrons, and changing their motion means changing their energy. That requires tremendous effort, because you would have to boost them to the top of the energy ladder. That makes degenerate matter, though still a gas, harder to compress than the hardest steel.

Second, the pressure of degenerate gas does not depend on temperature but rather on the speed of the electrons, which cannot be changed without tremendous effort. The temperature, however, depends on the motion of all the particles in the gas, both electrons and nuclei. If you were to add thermal energy to the gas, most of that energy would go to speed up the motions of the nuclei, and only a few electrons would be able to absorb enough energy to reach the empty energy levels at the top of the energy ladder. This means that changing the temperature of the gas has almost no effect on the pressure.

These two properties of degenerate matter become important when stars end their main-sequence lives and approach their final collapse (**Window on Science 12-2**). Eventually, many stars collapse into white dwarfs, and you will discover that these tiny stars are made of degenerate matter. But long before that, the cores of many giant stars become so dense that they are degenerate, a situation that can produce a cosmic bomb when helium begins to fuse.

Helium Fusion

Hydrogen fusion leaves behind helium ash, which cannot fuse at the relatively low temperatures of hydrogen fusion. Helium nuclei have a positive charge twice that of a hydrogen nucleus (a proton), which means the Coulomb barrier is higher. At the temperature of hydrogen fusion, the particles move too slowly and collide too gently to produce helium fusion. Consequently, the helium that accumulates is an ash that cannot fuse until the temperature rises significantly.

As the star becomes a giant star, fusing hydrogen in a shell, the inner core of helium ash contracts and grows hotter. It may even become degenerate. Finally, as the temperature approaches 100,000,000 K, helium nuclei begin to fuse together to make carbon.

You can summarize the helium-fusing process in two steps:

$$^4\text{He} + {}^4\text{He} \rightarrow {}^8\text{Be} + \gamma$$
$$^8\text{Be} + {}^4\text{He} \rightarrow {}^{12}\text{C} + \gamma$$

This process is complicated by the fact that beryllium-8 is very unstable and may break up into two helium nuclei before it can

Causes in Nature:
The Very Small and the Very Large

One of the most interesting lessons of science is that the behavior of very small things often determines the structure and behavior of very large things. In degenerate matter, the quantum mechanical behavior of electrons helps determine the evolution of giant stars. Such links between the very small and the very large are common in astronomy, and you will see in later chapters how the nature of certain subatomic particles may determine the fate of the entire universe.

This is a second reason why astronomers study atoms. Recall that the first reason is that atoms interact with light, and astronomers must understand that interaction in order to analyze the light. The second reason is the link between the very small and the very large.

It would be impossible to understand the evolution of the largest stars if astronomers failed to understand how degenerate electrons behave. Similarly, astronomers studying galaxies must understand how microscopic dust in space reddens starlight, and astronomers studying planets must know how molecules form crystals.

As you study any science, look for the way very small things affect very large things. Sociologists and psychologists know that the mass behavior of very large groups of people can depend on the behavior of a few key individuals. Biologists study the visible consequences of atomic bonds in molecules called genes, and meteorologists know that tiny changes in temperature in one part of the

The properties of atoms and molecules determine the structure of the largest objects in the universe.

world can affect worldwide climate weeks or months later.

Nature is the sum of many tiny parts, and science is the study of nature. One way to do science is to search out the tiny causes that determine the way nature behaves on the grandest scales.

---

absorb another helium nucleus. Three helium nuclei can also form carbon directly, but such a triple collision is unlikely. Because a helium nucleus is called an alpha particle, astronomers often refer to helium fusion as the **triple alpha process.**

Some stars begin helium fusion gradually, but stars in a certain mass range begin helium fusion with an explosion called the **helium flash.** This explosion is caused by the density of the helium, which can reach 1,000,000 g/cm^3. On Earth, a teaspoon of this material would weigh as much as a large truck. At these densities, the gas is degenerate, and its pressure no longer depends on temperature. That means the pressure–temperature thermostat that controls the nuclear fusion reactions no longer works.

When the helium ignites, it generates energy, which raises the temperature. Because the pressure–temperature thermostat is not working in the degenerate gas, the core does not respond to the higher temperature by expanding. Rather, the higher temperature forces the reactions to go faster, which makes more energy, which raises the temperature, which makes the reactions go faster, and so on. The ignition of helium fusion in a degenerate gas results in a runaway explosion so violent that for a few minutes the helium core generates more energy than an entire galaxy. At its peak, the core generates 10^{14} times more energy per second than the sun.

Although the helium flash is sudden and powerful, it does not destroy the star. In fact, if you were observing a giant star as it experienced a helium flash, you would see no outward evidence of an eruption. The helium core is quite small (■ Figure 12-9), and all of the energy of the explosion is absorbed by the distended envelope. Also, the helium flash is a very short-lived event. In a

matter of minutes, the core of the star becomes so hot it is no longer degenerate, the pressure–temperature thermostat brings the helium fusion under control, and the star proceeds to fuse helium steadily in its core. You will learn about this post-helium-flash evolution later.

Not all stars experience a helium flash. Stars less massive than about 0.4 solar mass can never get hot enough to ignite helium, and stars more massive than about 3 solar masses ignite helium before their cores become degenerate (■ Figure 12-10). In such stars, pressure depends on temperature, so the pressure–temperature thermostat keeps the helium fusion under control.

If the helium flash occurs only in some stars and is a very short-lived event that is not visible from outside the star, why should you worry about it? The answer is that it limits the reliability of the mathematical models astronomers use to study stellar evolution. The medium-mass stars, which experience the helium flash, are those that astronomers must consider most carefully in studying stellar evolution. Massive stars are not very common; low-mass stars evolve so slowly that it is hard to find evidence of low-mass stellar evolution. For those reasons, studies of stellar evolution must concentrate on medium-mass stars, which do experience the helium flash. But the helium flash occurs so rapidly and so violently that computer programs cannot follow the changes in the star's internal structure in detail. To follow the evolution of medium-mass stars like the sun past the helium flash, astronomers must make assumptions about the way the helium flash affects stars' internal structures.

There is another reason why you should think about the helium flash. By seeing what happens when the pressure–temperature

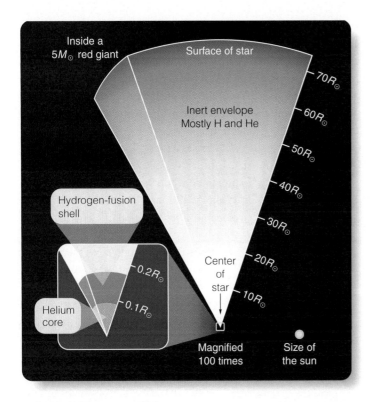

■ Active Figure 12-9

When a star runs out of hydrogen at its center, the core of helium contracts to a small size, becomes very hot, and begins nuclear fusion in a shell (blue). The outer layers of the star expand and cool. The red giant star shown here has an average density much lower than the air at Earth's surface. Here $M_\odot$ stands for the mass of the sun, and $R_\odot$ stands for the radius of the sun. (Illustration design by the author)

Ace⊙Astronomy™ Log into AceAstronomy and select this chapter to see Active Figure "Inside Stars." Compare the interior of stars with that of the sun.

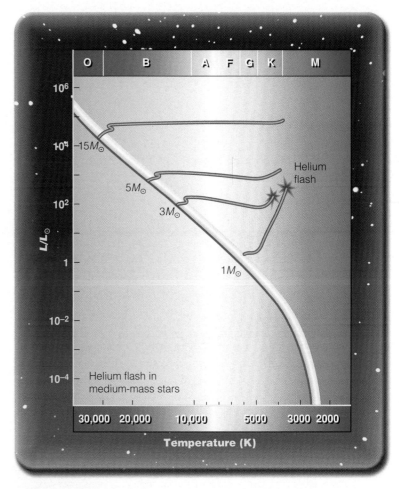

■ Figure 12-10

Stars like the sun suffer a helium flash, but more massive stars begin helium fusion without a helium flash. Stars less than 0.4 times the mass of the sun cannot get hot enough to ignite helium. The zero-age main sequence is shown here in red at the lower edge of the main sequence.

thermostat is turned off, you can appreciate its importance in stars like the sun.

Post-helium-flash evolution is generally understood. As the core temperature rises, the gas ceases to be degenerate, and helium fusion proceeds under the control of the pressure–temperature thermostat. Throughout these events, the hydrogen-fusion shell continues to produce energy, but the new helium-fusion energy produced in the core makes the core expand. That expansion absorbs energy previously used to support the outer layers of the star. As a result, the outer layers contract, and the surface of the star grows slightly hotter. In the H–R diagram, the point that represents the star initially moves down toward lower luminosity and to the left toward higher temperature. Later, as the star stabilizes fusing helium in its core, its luminosity recovers, but its surface grows hotter.

Helium fusion produces carbon and oxygen as "ash," and these accumulate at the center of the star. They do not fuse because the core is not hot enough. Thus, the helium in the core is gradually converted into carbon–oxygen ashes. As this happens, the core contracts, grows hotter, and ignites a helium-fusion shell. At this stage, the star has two shells producing energy. A hydrogen-fusion shell eats its way outward, leaving behind helium ash; a helium-fusion shell eats its way outward into the helium, leaving behind carbon–oxygen ash. Unable to generate energy in its carbon–oxygen core, the star expands its outer layers, and its point in the H–R diagram moves back toward the right, completing a loop. You can see that small loop in the giant region of the H–R diagram as shown in ■ Figure 12-11. Expanded drawings show how the interior of the star changes as it uses up its fuels.

Fusing Elements Heavier Than Helium

Modeling the evolution of stars after helium exhaustion requires great sophistication. The inert core of carbon–oxygen ash contracts and becomes hotter. Stars more massive than about 4 solar masses

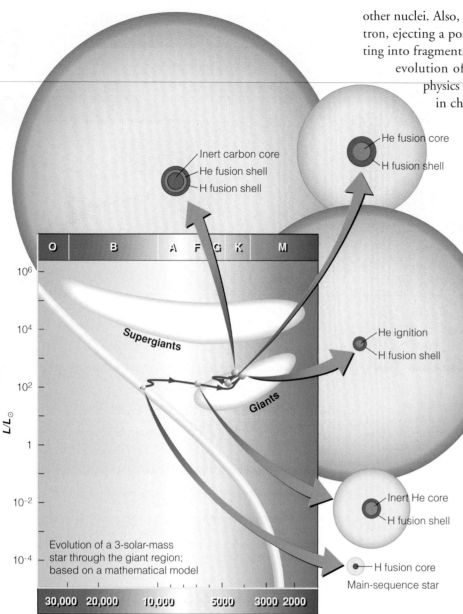

Inert carbon core
He fusion shell
H fusion shell

He fusion core
H fusion shell

He ignition
H fusion shell

Inert He core
H fusion shell

H fusion core
Main-sequence star

| O | B | A | F | G | K | M |

10^6

10^4

Supergiants

10^2

$L/L_\odot$

1

Giants

10^{-2}

Evolution of a 3-solar-mass
star through the giant region;
based on a mathematical model

10^{-4}

30,000 20,000   10,000   5000   3000 2000

Temperature (K)

■ **Figure 12-11**

When a main-sequence star exhausts the hydrogen in its core, it evolves rapidly to the right in the H–R diagram as it expands to become a cool giant. These changes are caused at first by the ignition of a hydrogen fusion shell. When helium ignites in the core, the star contracts and heats up slightly, but when helium begins to fuse in a shell, the star again expands and cools.

can heat their cores to temperatures as high as 600,000,000 K, hot enough to fuse carbon. Subsequently, the stars can fuse oxygen, silicon, and other heavy elements.

Carbon fusion in a star begins a complex network of reactions illustrated in ■ Figure 12-12, where each circle represents a possible nucleus and each arrow represents a different nuclear reaction. Nuclei can react by capturing a proton, capturing a neutron, capturing a helium nucleus, or by combining directly with

other nuclei. Also, unstable nuclei can decay by ejecting an electron, ejecting a positron, ejecting a helium nucleus or by splitting into fragments. Astronomers who model the structure and evolution of these stars must use sophisticated nuclear physics to compute the energy production and changes in chemical composition as the stars fuse heavier elements.

These stars develop complex internal structure as they fuse heavier and heavier elements in concentric shells. Like great spherical cakes with many layers, the stars may be fusing a number of different elements simultaneously in concentric shells.

Eventually, all stars face collapse. Less massive stars cannot get hot enough to ignite the heavier nuclei. The more massive stars can ignite heavy nuclei, but these stars consume their fuels at a tremendous rate, so they run through their available fuel quickly (■ Table 12-3). No matter what the star's mass, it will eventually run out of usable fuels, and gravity will win the struggle with pressure. You will explore the ultimate deaths of stars in the next chapter.

The time a star spends as a giant or supergiant is small compared with its life on the main sequence. The sun, for example, will spend about 10 billion years as a main-sequence star but only about a billion years as a giant. The more massive stars pass through the giant stage even more rapidly. Because of the short time a star spends as a giant, you are unlikely to see many such stars. This illustrates an important principle in astronomy: The shorter the time a given evolutionary stage takes, the less likely you are to see stars in that particular stage. That explains why you see a great many main-sequence stars but few giants (see page 211).

Building Scientific Arguments

What uncertainties limit the accuracy of our story of stellar evolution?

A scientific argument is not meant to convince others but to test your own understanding, and that means it should include a discussion of any uncertainties in the analysis. In the case of stellar evolution, there are two things that limit the accuracy of the models. First, some events in the history of a star happen so fast that computers can't predict the results in a reasonable amount of time. In fact, some astronomers estimate that a computer following the evolution of a model

■ **Figure 12-12**

Carbon fusion involves many possible reactions that build numerous heavy nuclei.

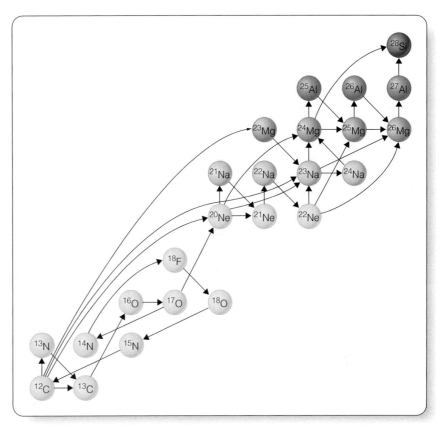

star undergoing the helium flash would have to recompute the model in such short time steps that the computer program would take millions of years to predict events that occur inside the star in just seconds. Truly rapid changes such as explosions in stars make the models uncertain.

Second, you have seen how the general properties of a giant star can depend on the subatomic properties of matter. Not all of those subatomic properties are well understood. For example, neutrinos oscillate, and scientists don't entirely understand how that happens. If some subatomic particle behaves differently than expected, then stellar models may be slightly different from the real stars they are supposed to represent.

Can you add to your argument? **What other physical processes have been omitted from stellar models? What features of the sun's surface suggest motions and forces that are ignored in most stellar models?**

■ ■ ■

Connections: Humans do not live long enough to see the stars evolving, so astronomers must test their theoretical story of stellar evolution against observation. To paraphrase the quotation at the beginning of this chapter, you shouldn't believe any theory until you compare it with reality—seeing is believing. Consequently, in the rest of this chapter, you will search for evidence to support the theory of stellar evolution. As you search, you will discover some interesting consequences of the fact that the stars are not eternal.

12-3 Evidence of Evolution: Star Clusters

THE THEORY OF STELLAR EVOLUTION is so complex and involves so many assumptions that astronomers would have little confidence in it were it not for H–R diagrams of clusters of stars. By observing the properties of star clusters of different ages, you can see clear evidence of the evolution of stars.

To grasp the difficulty of understanding stellar evolution, consider an analogy. Suppose a visitor to Earth who had never seen a tree wandered through a forest for an hour looking at mature trees, fallen seeds, young saplings, fallen logs, and rising

■ **Table 12-3 | Nuclear Reactions in Massive Stars**

Nuclear Fuel	Nuclear Products	Minimum Ignition Temperature	Main-Sequence Mass Needed to Ignite Fusion	Duration of Fusion in a 25-$M_\odot$ Star
H	He	4×10^6 K	0.1 $M_\odot$	7×10^6 yr
He	C, O	120×10^6 K	0.4 $M_\odot$	0.5×10^6 yr
C	Ne, Na, Mg, O	0.6×10^9 K	4 $M_\odot$	600 yr
Ne	O, Mg	1.2×10^9 K	~8 $M_\odot$	1 yr
O	Si, S, P	1.5×10^9 K	~8 $M_\odot$	~ 0.5 yr
Si	Ni to Fe	2.7×10^9 K	~8 $M_\odot$	~ 1 day

sprouts. Could such an observer understand the life cycle of the trees? Astronomers face the same problem when they try to understand the life story of the stars.

Humans do not live long enough to see stars evolve. You see only a momentary glimpse of the universe as it is during your lifetime, a snapshot in which all the stages in the life cycle of the stars are represented. Unscrambling these stages and putting them in order is a difficult task. Only by looking at selected groups of stars—star clusters—can you see the pattern.

Observing Star Clusters

Just as Sherlock Holmes studies peculiar dust on a lamp shade as a clue to a crime, astronomers look at star clusters and say, "Aha!" Each star cluster freezes a moment in the evolution of the cluster and makes the evolution of the stars visible.

The stars in a cluster all formed at about the same time and from the same cloud of gas, so they must be about the same age and composition. The differences you see among stars in a cluster must arise from differences in mass, and that makes stellar evolution visible. Study **Star Cluster H–R Diagrams** on pages 276–277 and notice three important points:

1 There are two kinds of star clusters, but they are similar in the way their stars evolve. You will learn more about these clusters in a later chapter.

2 Notice that you can estimate the age of a star cluster by observing the distribution of the points that represent its stars in the H–R diagram.

3 Finally, notice that the shape of a star cluster's H–R diagram is governed by the evolutionary path the stars take. By comparing clusters of different ages, you can visualize how stars evolve almost as if you were watching a film of a star cluster evolving over billions of years.

Were it not for star clusters, astronomers would have little confidence in the theories of stellar evolution. Star clusters make that evolution visible and assure astronomers that they really do understand how stars are born, live, and die.

The Evolution of Star Clusters

What you know about star formation and stellar evolution can help you understand star clusters and how they evolve.

A star cluster is formed when a cloud of gas contracts, fragments, and forms a group of stars. As you recall from Chapter 11, some groups of stars, called associations, are so widely scattered the group is not held together by its own gravity. The stars in an association wander away from each other rather quickly. Some groups of stars are more compact; even after the stars become luminous and the gas and dust are blown away, gravity holds the group together as a star cluster.

Open clusters are not as old or as crowded as the globular clusters, and that helps explain their appearance. Close encoun-

ters between stars are rare in an open cluster where the stars are further apart, and such clusters have an irregular appearance. The globular clusters appear to be nearly perfect globes because the stars are much closer together and encounters between stars are more common. The globular clusters have had time to evenly distribute the energy of motion among all of the stars so they have settled into a more uniform, spherical shape.

As a star cluster ages, some stars traveling a bit faster than others can escape. Globular clusters are compact and massive and have survived for 11 billion years or more. A star cluster with only a few stars widely distributed may evaporate completely as its stars escape one by one. Our sun probably formed in a star cluster 5 billion years ago, but there is no way to trace it back to its family home even if that star cluster still exists.

Building Scientific Arguments

Why do open clusters contain only a small number of giant stars but many main-sequence stars?
Sometimes the critical factor in a scientific argument is timing. When you look at a star cluster, you see a snapshot of the cluster that freezes it in time. In the H–R diagram of an open cluster, you see a main sequence containing many stars, but typically you see significantly fewer giant stars. Those giant stars were once main-sequence stars in the cluster, but they exhausted the hydrogen fuel in their cores and expanded to become giants. A star like the sun can fuse hydrogen on the main sequence for many billions of years but can remain a giant for only a billion years or less. More massive stars evolve even faster. Because of this, at any given moment, you see most of the stars in a star cluster on the main sequence and only a few, caught in the snapshot, as giant stars.

Now create a new argument. **How do H–R diagrams reveal the age of a star cluster?**

■   ■   ■

Connections: Star clusters provide dramatic confirmation that stars evolve and also reveal some of the details of stellar structure and evolution. Certain pulsating giant stars provide even further information about how stars work.

12-4 Evidence of Evolution: Variable Stars

FOR MILLENNIA, poets and astronomers believed that the stars were constant and unchanging. Modern theory shows that the stars are slowly evolving as they consume their fuels. But can you really be sure the stars are changing and evolving? Certain stars that pulsate and change their brightness provide evidence of stellar evolution and illustrate once more the importance of energy flowing outward through the layers of the stars.

A **variable star** is any star that changes its brightness in a periodic way. Variable stars include many different kinds of stars. Some variable stars are eclipsing binaries (Chapter 9), but many others are stars that expand and contract, heat and cool, and grow more luminous and less luminous because of internal characteristics. These stars are often called **intrinsic variables** to distinguish them from eclipsing binaries. Of these intrinsic variables, one particular kind is centrally important to modern astronomy even though the first star of that class was discovered centuries ago.

Cepheid and RR Lyrae Variable Stars

In 1784, the deaf and mute English astronomer John Goodricke, then 19 years old, discovered that the star δ Cephei is variable, changing its brightness by almost a magnitude over a period of 5.37 days (■ Figure 12-13a). Goodricke died at the age of 21, but his discovery ensures his place in the history of astronomy. Since 1784, hundreds of stars like δ Cephei have been found, and they are now known as **Cepheid variable stars.**

Cepheid variables are supergiant or bright giant stars of spectral type F or G. The fastest complete a cycle—bright to faint to bright—in about 2 days, whereas the slowest take as long as 60 days. A plot of a variable's magnitude versus time, a light curve, shows a typical rapid rise to maximum brightness and a slower decline (Figure 12-13b). Some Cepheids change their brightness

by only 0.5 magnitude (about 10 percent), while others change by more than a magnitude.

Supergiants and bright giants are rare, and not all are Cepheids. Nevertheless, some familiar stars are Cepheids. One of our Favorite Stars, Polaris, the North Star, is a Cepheid with a period of 3.9696 days and a very low amplitude.

The **RR Lyrae variable stars** are related to the Cepheids. The RR Lyrae stars pulsate with periods of less than a day and are fainter than the Cepheids.

Studies of Cepheids and RR Lyrae stars reveal two related facts of great importance. First, there is a **period–luminosity relation** that connects the period of pulsation of a Cepheid to its luminosity. The longer-period Cepheids are about 40,000 times more luminous than the sun, while the shortest-period Cepheids are only a few hundred times more luminous than the sun. Second, there are two types of Cepheids. Type I Cepheids, including δ Cephei itself, have chemical compositions like that of the sun, but type II Cepheids and the RR Lyrae stars are poor in elements heavier than helium. These two facts are clues that will help you understand variable stars as evidence of stellar evolution.

Pulsating Stars

Why should a star pulsate? Why should there be a period–luminosity relation? Why are there two types of Cepheids? The

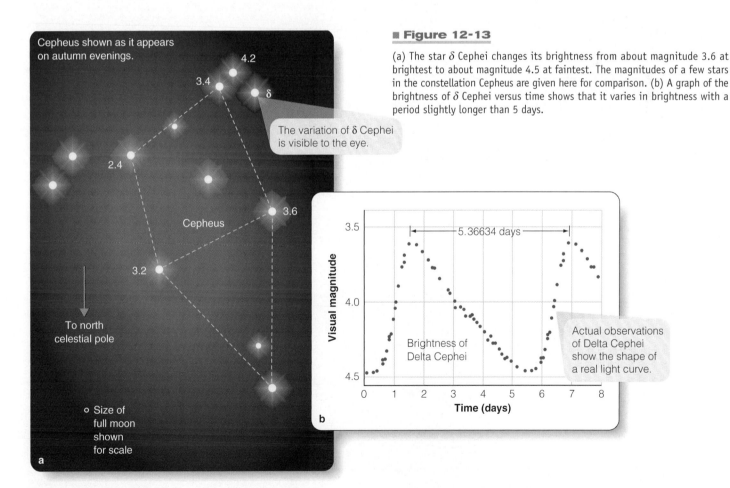

■ **Figure 12-13**

(a) The star δ Cephei changes its brightness from about magnitude 3.6 at brightest to about magnitude 4.5 at faintest. The magnitudes of a few stars in the constellation Cepheus are given here for comparison. (b) A graph of the brightness of δ Cephei versus time shows that it varies in brightness with a period slightly longer than 5 days.

1 An **open cluster** is a collection of 10 to 1000 stars in a region about 25 pc in diameter. Some open clusters are quite small and some are large, but they all have an open, transparent appearance because the stars are not crowded together.

In a star cluster each star follows its orbit around the center of mass of the cluster.

AURA/NOAO/NSF

Visual-wavelength image

**Open Cluster
The Jewel Box**

1a A **globular cluster** can contain 10^5 to 10^6 stars in a region only 10 to 30 pc in diameter. The term "globular cluster" comes from the word "globe," although globular cluster is pronounced like "glob of butter." These clusters are nearly spherical, and the stars are much closer together than the stars in an open cluster.

Astronomers can construct an H–R diagram for a star cluster by plotting a point to represent the luminosity and temperature of each star.

**Globular Cluster
47 Tucanae**

Visual-wavelength image

ATF8

The Hyades Star Cluster

| O | B | A F G K | M |

10^6

The most massive stars have died

Only a few stars are in the giant stage.

Main sequence

10^4

10^2

Giants

$L/L_\odot$

1

The lower-mass stars are still on the main sequence.

10^{-2}

The faintest stars were not observed in the study.

10^{-4}

30,000 20,000 10,000 5000 3000 2000

Temperature (K)

2 The H–R diagram of a star cluster can make the evolution of stars visible. The key is to remember that all of the stars in the star cluster have the same age but differ in mass. The H–R diagram of a star cluster provides a snapshot of the evolutionary state of the stars at the time you happen to be alive. The diagram here shows the 650- million-year-old star cluster called the Hyades. The upper main sequence is missing because the more massive stars have died, and our snapshot catches a few medium-mass stars leaving the main sequence to become giants.

As a star cluster ages, its main sequence grows shorter like a candle burning down. You can judge the age of a star cluster by looking at the turnoff point, the point on the main sequence where stars evolve to the right to become giants. Stars at the **turnoff point** have lived out their lives and are about to die. Consequently, the life expectancy of the stars at the turnoff point equals the age of the cluster.

Ace ☉ Astronomy™ Log into AceAstronomy and select this chapter to see Active Figure "Cluster Turnoff" and notice how the shape of a cluster's H–R diagram changes with time.

3 From theoretical models of stars, you could construct a film to show how the H–R diagram of a star cluster changes as it ages. You can then compare theory (left) with observation (right) to understand how stars evolve. Note that the time step for each frame in this film increases by a factor of 10.

Highest-mass stars evolving. Low-mass stars still contracting.

10^6 y

10^7 y

Upper main sequence stars have died.

10^8 y

10^9 y

Only the lower mass stars remain on the main sequence.

10^{10} y

NGC2264
Age 10^6 yr

Faintest stars not observed.

Pleiades
Age 10^8 yr

Turnoff point

Faintest stars not observed.

M67
Age 4×10^9 yr

Turnoff point

Faintest stars not observed.

$L/L_\odot$

Temperature (K)

NGC 2264 is a very young cluster still embedded in the nebula from which it formed. Its lower-mass stars are still contracting, and it is rich in T Tauri stars.

Visual — AATB

The nebula around the Pleiades is produced by gas and dust through which the cluster is passing. Its original nebula dissipated long ago.

Visual — Caltech

M67 is an old open cluster. In photographs, such clusters have a uniform appearance because they lack hot, bright stars. Compare with the Jewel Box on the opposite page.

Visual — Nigel Sharp, Mark Hanna/AURA/NOAO/NSF

Globular cluster H–R diagrams resemble the last frame in the film, which tells you that globular clusters are very old.

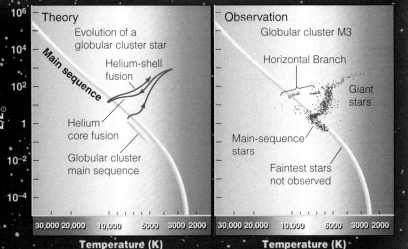

Theory

Evolution of a globular cluster star

Main sequence

Helium-shell fusion

Helium core fusion

Globular cluster main sequence

Observation

Globular cluster M3

Horizontal Branch

Giant stars

Main-sequence stars

Faintest stars not observed

$L/L_\odot$

Temperature (K)

3a The H–R diagrams of globular clusters have very faint turnoff points showing that they are very old clusters. The best analysis suggests these clusters are about 11 billion years old.

The **horizontal branch** stars are giants fusing helium in their cores and then in shells. The shape of the horizontal branch outlines the evolution of these stars.

The main-sequence stars in globular clusters are fainter and bluer than the zero-age main sequence. Spectra reveal that globular cluster stars are poor in elements heavier than helium, and that means their gases are less opaque. That means energy can flow outward more easily, which makes the stars slightly smaller and hotter. Again the shape of star cluster H–R diagrams illustrates principles of stellar evolution.

answers to these questions will reveal some of the deepest secrets of the stars. To find the answers, you should start with the evolution of stars in the H–R diagram.

After a star leaves the main sequence, it fuses hydrogen in a shell, and the point that represents it in the H–R diagram moves to the right as the star becomes a cool giant. When it ignites helium in its core, the point moves back to the left. Soon, however, the helium core is exhausted; helium fusion continues only in a shell, forcing the star to expand again; and the point in the H–R diagram moves back to the right, completing a loop. Because giant stars evolve through these stages, their sizes and temperatures change in complicated ways. Certain combinations of size and temperature make the star unstable, and it pulsates as an intrinsic variable star. In the H–R diagram, the region where size and temperature lead to pulsation is called the **instability strip,** as shown in the H-R diagram in ■ Figure 12-14.

Why do some stars pulsate? The answer has to do with the flow of energy out of the star's interior. You have seen that a star supports its own weight by generating energy and that the flow of that energy outward prevents the star from contracting. The star is balanced between the outward flow of energy and the inward force of gravity. You could make a stable star pulsate if you squeezed it and then released it. The star would rebound outward, and it might oscillate for a few cycles, but the oscillation would eventually run down because of friction. The energy of the moving gas would be converted to heat and radiated away, and the star would return to stability. To make a star pulsate continuously, some process must drive the oscillation, just as a spring in a wind-up clock is needed to keep it ticking.

Studies using stellar models show that variable stars in the instability strip pulsate like beating hearts because of an energy-absorbing layer in their outer envelopes. This layer is the region where helium is partially ionized. Above this layer, the temperature is too low to ionize helium, and below the layer it is hot enough to ionize all of the helium. Like a spring, the helium ionization zone can absorb energy when it is compressed and release it when the zone expands. That is enough to keep the star pulsating.

You can follow this pulsation in your imagination and see how the helium ionization zone makes a star pulsate. As the outer layers of a pulsating star expand, the ionization zone expands, and the ionized helium becomes less ionized and releases stored energy, which forces the expansion to go even faster. The surface of the star overshoots its equilibrium position—it expands too far—and eventually falls back. As the surface layers contract, the ionization zone is compressed and becomes more ionized, which absorbs energy. Robbed of some of their energy, the layers of the star can't support the weight, and they compress further. This allows the contraction to go even faster, and the in-falling layers overshoot the equilibrium point until the pressure inside the star slows them to a stop and makes them expand again. Thus, the surfaces of these stars are not in hydrostatic equilibrium but rather pulsate in and out.

Cepheids change their radius by 5 to 10 percent as they pulsate, and this motion can be observed as Doppler shifts in their spectra. When they expand, the surface layers approach, and astronomers see a blueshift. When the stars contract, the surface layers recede, and astronomers see a redshift. Although a 10 percent change in radius seems like a large change, it affects only the outer layers of the star. The center of the star is much too dense to be affected.

Now you can understand why there is an instability strip. Only stars in the instability strip have their helium ionization zone at the right depth to act as a spring and drive the pulsation. A star will pulsate if the zone lies at the right depth below the surface. In cool stars, the zone is too deep and cannot overcome the weight of the layers above it. That is, the spring is overloaded and can't expand. In hot stars, the helium ionization zone lies near the surface, and there is little weight above it to compress it. That is, the spring never gets squeezed. The stars in the instability strip have the right temperature and radius for the helium ionization zone to fall exactly where it is most effective, and that is why those stars pulsate.

You can also understand why there is a period–luminosity relation. The more massive stars are larger and more luminous. When they cross the instability strip, they cross higher in the H–R diagram. But the larger a star is, the more slowly it pulsates—just as large bells go "bong" and little bells go "ding." Pulsation period depends on size, which depends on mass, and mass also determines luminosity. So the most luminous Cepheids at the top of the instability strip are also the largest and pulsate the slowest. Less luminous Cepheids are smaller and pulsate faster. If you plot the average brightness of the stars against their period of pulsation, you can clearly see the relationship between luminosity and period. Look at the right half of Figure 12-14 to see this relationship. To sum up, there has to be a period–luminosity relation because period and luminosity both depend on mass.

You know enough about stars to understand why there are two types of Cepheid variable stars. Type I Cepheids have chemical abundances roughly like those of the sun, but type II Cepheids are poor in elements heavier than helium, and that means the gases in type II Cepheids are not as opaque as the gases in type I Cepheids. The all-important energy flowing outward can escape more easily in type II Cepheids, and they reach a slightly different equilibrium. For stars of the same period of pulsation, type II Cepheids are less luminous, and you can see that in the graph in Figure 12-14.

The RR Lyrae stars are related to the Cepheids. In the H–R diagram, they lie at the lower end of the instability strip. They are a bit smaller and hotter than Cepheids, but they pulsate for the same reason.

Period Changes in Variable Stars

Some Cepheids have been observed to change their periods of pulsation, and those changes are evidence that stars really do evolve.

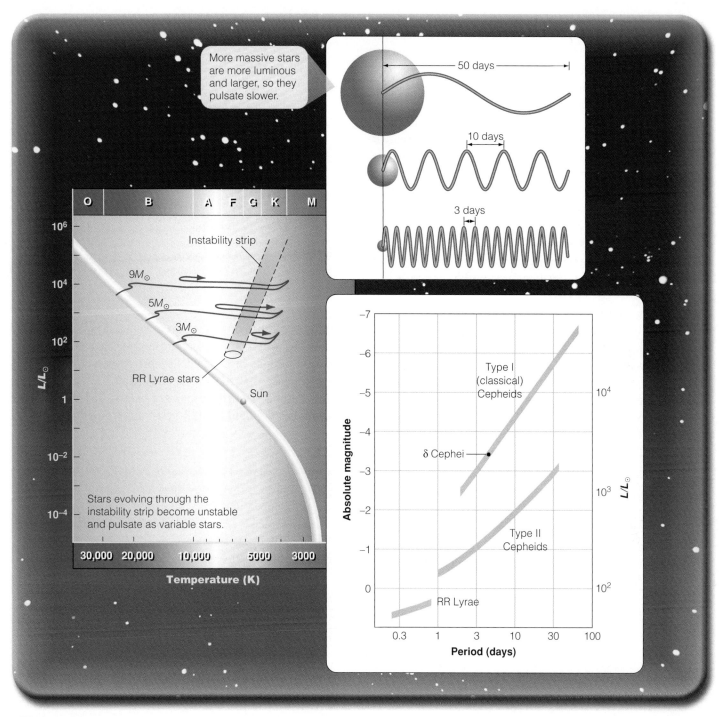

■ Figure 12-14

More massive stars are larger and more luminous and pulsate with longer periods than less massive stars. Consequently there is a period–luminosity relation: Type I and type II Cepheids have slightly different chemical compositions. The RR Lyrae stars have lower luminosities and shorter periods.

The evolution of a star may carry it through the instability strip a number of times, and each time it can become a variable star. If you could watch a star as it entered the instability strip, you could see it begin to pulsate. Similarly, if you could watch a star leave the instability strip, you could watch it stop pulsating. Such an observation would be dramatic evidence that the stars are evolving. Unfortunately, stars evolve so slowly that these events are surely rare.

One famous Cepheid, RU Camelopardalis, was observed to stop pulsating temporarily. Its pulsations died away in 1966, and many astronomers assumed that it had stopped pulsating because it was leaving the instability strip. But its pulsations resumed in 1967. Remember our Favorite Star, Polaris? It may be going through a similar change now, but it does not lie near the edge of the instability strip in the H–R diagram. Perhaps it is not an example of a star evolving out of the instability strip or perhaps there are other factors that astronomers have not discovered yet.

Even more dramatic evidence of stellar evolution can be found in the slowly changing periods of Cepheid variables. Even a tiny change in period can become easily observable because it accumulates, just as a clock that gains a second each day becomes obviously fast over a time as long as a year. Some Cepheids have periods that are growing shorter, and others have periods that are growing longer. These changes occur because the stars are evolving and changing their radii; contraction shortens the period, and expansion lengthens the period. The star X Cygni, for example, has a period that is gradually growing longer, and it has been "losing" since observations began in the late 19th century (■ Figure 12-15). These changes in the periods of pulsating stars provide dramatic evidence that the stars are evolving.

Some kinds of variable stars lie outside the instability strip because their pulsation is driven by different mechanisms, but you need not analyze them. It is sufficient here to discuss the Cepheids. They are common, they illustrate important ideas about stellar structure, and their changing periods make stellar evolution clearly evident. Also, you will use Cepheids in later chapters to study the distances to galaxies and the fate of the universe.

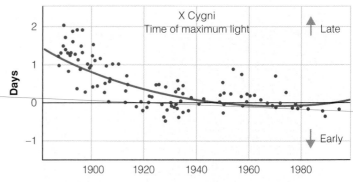

■ Figure 12-15

Like a clock running slower and slower, the Cepheid variable star X Cygni has been reaching maximum brightness later and later for most of this century. This is shown by the upward curve in this graph of its observed minus predicted times of maximum brightness. As it evolves to the right across the instability strip, it is slowly expanding, and its period is growing longer by about 1.46 seconds per year.

Building Scientific Arguments

How do Cepheid variable stars provide evidence that stars are evolving?

The key to scientific arguments is the relationship between theory and evidence; and, in astronomy, evidence means observations. If you observed Cepheids for a few years, you might notice that some pulsations are slowing down and others are speeding up. This is a very small change in the period of pulsation, but if you observe the stars for years, the change in their pulsation causes them to gain or lose just as a clock can gain or lose time over weeks. These changes in period are caused by the expansion or contraction of the stars as they evolve. The period of pulsation depends on the size of the star, so pulsations of a star that is expanding should slow down slightly, and those of a star that is contracting should speed up slightly. When you observed these small changes in the pulsation periods of Cepheid variable stars, you would have direct evidence that the stars are evolving.

Now build a scientific argument that is based on theory. **What makes Cepheid and RR Lyrae stars pulsate?**

■   ■   ■

Connections: This chapter has discussed the evolution of stars, a process that occurs so slowly you cannot expect to see much happen in a single human lifetime. Consequently, the study of stellar evolution depends heavily on theory and its verification by evidence from observations of real stars. You have seen that star clusters make the evolution of stars visible in the H–R diagram and that the pulsation of the intrinsic variable stars can give you insight into how giant stars evolve through the instability strip. You have not yet considered, however, the most interesting question of stellar evolution. What happens to a star when it uses up the last of its nuclear fuel? Clearly, stars must die—the question is how. You will explore that question in the next chapter.

Summary

12-1 | Main-Sequence Stars

How do astronomers know how stars evolve?

- Astronomers compute mathematical models of the interiors of stars based on four simple laws of stellar structure. Two laws say that mass and energy must be conserved and spread smoothly through the star. Another, the law of hydrostatic equilibrium, says that the star must balance the weight of its layers by its internal pressure. Yet another says that energy can flow outward only by conduction, convection, or radiation.

- Mathematical models of stars show how rapidly a star uses its fuel in each layer, and that allows astronomers to step forward in time and follow the evolution of the star as it ages.

What are main-sequence stars like?

- There is a main sequence because stars support their weight by hydrogen fusion. As energy flows outward, it heats the gas of the star, and pressure in the gas balances the inward pull of gravity.

- The mass–luminosity relation is explained by the requirement that a star support the weight of its layers by its internal pressure. The more massive a star is, the more weight it must support, and the higher its internal pressure must be. To keep its pressure high, it must be hot and generate large amounts of energy. Thus, the mass of a star determines its luminosity.

- The zero-age main sequence is the line in the H–R diagram where contracting stars begin fusing hydrogen and reach stability. As a star ages, it moves slightly upward and to the right, making the main sequence a band.

- Massive stars, having more weight to support, reach stability higher on the main sequence than lower-mass stars.

- The most massive stars are unstable and lose mass rapidly because of the pressure of radiation flowing outward. The least massive stars are brown dwarfs that are not massive enough to ignite hydrogen fusion.

- The life expectancy of a main-sequence star depends on its mass. The more massive stars use their fuels rapidly and remain on the main sequence only a few million years. The sun will last a total of about 10 billion years, and the least massive stars could survive for over 100 billion years.

12-2 | Post-Main-Sequence Evolution

How do stars evolve after they exhaust hydrogen?

- When a main-sequence star exhausts its hydrogen, its core contracts, and it begins to fuse hydrogen in a shell around its core. The outer parts of the star—its envelope—swell, and the star becomes a giant. Because of this expansion, the surface of the star cools, and it moves toward the right in the H–R diagram. The most massive stars move across the top of the diagram as supergiants.

- Because the core of a giant star contracts and the envelope expands, the nuclear fusion is confined to a very small volume at the center of the low-density star.

- When the core of the star becomes hot enough, helium fusion begins first in the core and then in a shell. This causes the star to describe a loop in the giant region of the H–R diagram.

- If the core becomes degenerate before helium ignites, the pressure of the gas does not depend on its temperature, and when helium ignites, the core explodes in the helium flash. Although the helium flash is violent, the star absorbs the extra energy and quickly brings the helium-fusing reactions under control.

- Helium fusion produces carbon. If a star is massive enough, it can ignite carbon and other fuels after helium fusion. Each fuel fuses more rapidly, producing heavier and heavier elements.

12-3 | Evidence of Evolution: Star Clusters

What evidence do astronomers have that stars really do evolve?

- Because all the stars in a cluster have about the same composition and age, you can see the effects of stellar evolution in the H–R diagram of a cluster. Massive stars evolve faster than low-mass stars, so in a given cluster the most massive stars leave the main sequence first.

- You can judge the age of a cluster by looking at the turnoff point, the location on the main sequence where the stars turn off to the right and become giants. The life expectancy of a star at the turnoff point equals the age of the cluster.

- There are two types of star clusters. Open clusters contain 10 to 1000 stars and have an open, transparent appearance. Globular clusters contain 10^5 to 10^6 stars densely packed into a spherical shape. The open clusters tend to be young to middle-aged, but globular clusters tend to be very old—11 billion years or more. Also, globular clusters tend to be poor in elements heavier than helium.

12-4 | Evidence of Evolution: Variable Stars

- The Cepheid and RR Lyrae variables provide evidence that stars evolve. These intrinsic variable stars lie in an instability strip in the H–R diagram; they contain a layer in their envelopes that stores and releases energy as they expand and contract. Stars outside the instability strip do not pulsate because the layer is too deep or too shallow to make the stars unstable.

- The Cepheids obey a period–luminosity relation because more massive stars, which are more luminous, are larger and pulsate more slowly.

- Some Cepheids have periods that are slowly changing, showing that the evolution of the star is changing the star's radius and thus its period of pulsation.

New Terms

conservation of mass law (p. 260)

conservation of energy law (p. 260)

stellar model (p. 261)

brown dwarf (p. 263)

zero-age main sequence (ZAMS) (p. 265)

degenerate matter (p. 269)

triple alpha process (p. 270)

helium flash (p. 270)

variable star (p. 275)

intrinsic variable (p. 275)

Cepheid variable star (p. 275)

RR Lyrae variable star (p. 275)

period–luminosity relation (p. 275)

open cluster (p. 276)

globular cluster (p. 276)

turnoff point (p. 276)

horizontal branch (p. 277)

instability strip (p. 278)

Review Questions

Assess your understanding of this chapter's topics with additional quizzing and animations at **http://ace .brookscole.com/sf9**

1. Why is there a main sequence?
2. Why is there a lower end to the main sequence? Why is there an upper end?
3. What is a brown dwarf?
4. Why is there a mass–luminosity relation?
5. Why does a star's life expectancy depend on mass?
6. Why do expanding stars become cooler and more luminous?
7. What causes the helium flash? Why does it make it difficult for astronomers to understand the later stages of stellar evolution?
8. How do some stars avoid the helium flash?
9. Why are giant stars so low in density?
10. Why are lower-mass stars unable to ignite more massive nuclear fuels such as carbon?
11. How can you estimate the age of a star cluster?
12. How do star clusters confirm that stars evolve?
13. How do some variable stars prove that stars are evolving?
14. The star cluster in the photo at the right contains many hot, blue, luminous stars. Sketch its H–R diagram and discuss its probable age.

(NASA/Walborn, Maiz-Apellaniz, and Barba)

Discussion Questions

1. How do you know that the helium flash occurs if it cannot be observed? Can you accept an event as real if you can never observe it?
2. Can you think of ways that chemical differences could arise in stars in a single star cluster? Consider the mechanism that triggered their formation.

Problems

1. In the model shown in Figure 12-2, how much of the sun's mass is hotter than 13,000,000 K?
2. What is the life expectancy of a 16-solar-mass star? of a 50-solar-mass star?
3. How massive could a star be and still survive for 5 billion years?
4. If the sun expanded to a radius 100 times its present radius, what would its density be? (*Hint:* The volume of a sphere is $\frac{4}{3}\pi R^3$.)
5. If a giant star 100 times the diameter of the sun were 1 pc from Earth, what would its angular diameter be? (*Hint:* Use the small-angle formula, in Chapter 3.)
6. What fraction of the volume of a 5-solar-mass giant star is occupied by its helium core? (*Hints:* See Figure 12-9. The volume of a sphere is $\frac{4}{3}\pi R^3$.)
7. If the stars at the turnoff point in a star cluster have masses of about 4 solar masses, how old is the cluster?
8. If an open cluster contains 500 stars and is 25 pc in diameter, what is the average distance between the stars? (*Hints:* What share of the volume of the cluster surrounds the average star? The volume of a sphere is $\frac{4}{3}\pi R^3$.)
9. Repeat Problem 8 for a typical globular cluster containing a million stars in a sphere 25 pc in diameter.
10. If a Cepheid variable star has a period of pulsation of 2 days and its period increases by 1 second, how late will it be in reaching maximum light after 1 year? after 10 years? (*Hint:* How many cycles will it complete in a year?)

Media Cluster

To access the resources in the Media Cluster, log into AceAstronomy at **http://ace.brookscole .com/sf9** and select Chapter 12.

ACTIVE FIGURES

Future of the Sun
The sun's evolution profoundly affects the planets and moons in our solar system. Here you can select the vantage point of Earth, Venus, Mars, or Callisto and get a quick view of the effects of the sun's evolution over 10 billion years.

Evolution of Stars
The evolution of a star depends on its mass. Here you can experiment with stars of different masses and see how they evolve.

Inside Stars

You can use the laws of physics to study the interior of the sun and compare it to other stars. This cutaway animation of the sun lets you explore the various structures of stars, understand their inner workings, and view photos and movies of the real thing.

Cluster Turnoff

The shape of a star cluster's H–R diagram reveals its age. Watch as a star cluster ages and its stars evolve off the main sequence.

Hydrostatic Equilibrium

This animation allows you to adjust the rate of nuclear fusion and the diameter of a star in hydrostatic equilibrium.

Mass–Luminosity Relation

This exercise provides a list of stars whose masses and luminosities have been measured. Click on each star to plot its point on the mass versus luminosity axes.

H–R/Mass–Luminosity 3-D Graph

Explore this 3-D graph and see how mass is related to position in the H–R diagram. You can select from main-sequence stars, red giants, white dwarfs, and variable stars.

Mass–Lifetime Relation

As stars age, they undergo a number of fundamental changes, such as in temperature and chemical composition. In this animation, experiment with different mass stars to see how fast they age and trace their evolution on the H–R diagram.

Cepheid Variable

Make your own observations and discover the relationship between period of pulsation and luminosity with this animation.

Critical Inquiries for the Web

1. Find Hertzsprung–Russell (or color-magnitude) diagrams for five star clusters. Based on the distributions of stars on the diagrams, rank the clusters in order from oldest to youngest. At what value (spectral type, temperature, or B-V color index) does each cluster have its turnoff?

2. What is the latest news about Eta Carinae? It is a complex system and is constantly under study as astronomers try to understand what is happening to its stars.

3. Both professional and amateur astronomers study variable stars. Search the Web for the American Association of Variable Star Observers and see what information you can find. Can you find recent data and light curves for known variables like δ Cephei and X Cyg?

4. What is the latest news about brown dwarfs?

Exploring *TheSky*

1. Locate the open clusters below and describe their location with respect to the Milky Way Galaxy.

   M7, M41, M44, M45

2. Locate the clusters in Activity 1 and zoom in until you can see individual stars. Plot an H–R diagram for each cluster using the spectral type and apparent magnitude of the stars you see. Discuss the ages of the clusters.

Go to the Brooks/Cole Astronomy Resource Center (**http:// astronomy.brookscole.com**) for critical thinking exercises, articles, and additional readings from InfoTrac College Edition, Brooks/Cole's online student library.

13 | The Deaths of Stars

Natural laws have no pity.

ROBERT HEINLEIN,
THE NOTEBOOKS OF LAZARUS LONG

Visual-wavelength image

GRAVITY IS PATIENT—so patient it can kill stars. All over the sky, astronomers find beautiful nebulae puffed gently into space by dying stars. In addition, astronomers occasionally see a new star appear in the sky, grow brighter, then fade away after a few weeks or a year. You will discover that a **nova,** what seems to be a new star in the sky, is in fact produced by the eruption of a very old dying star and that a **supernova,** a particularly luminous and long-lasting nova, is caused by the violent explosive death of a star. Modern astronomers find a few novae (plural of *nova*) each year, but supernovae are so rare that there are only one or two supernovae each century in our galaxy. Astronomers know that supernovae occur

Continued on page 286

Guidepost

Looking Back

Take a moment to be astonished and proud that the human race has figured out how stars die. You have seen in earlier chapters how hard it has been for astronomers to understand that Earth moves, how hard it is to find the distance to stars and their masses. You have seen how spectra contain clues to the nature of stars and reveal how stars form and age. Modern astronomers can tell the story of the stars right to the end.

This Chapter

Here you will learn how stars die, and at the same time you will see clear examples of how scientists think. You will have to proceed with care, testing theories against evidence to answer four essential questions:

How will the sun die?

Why are there so many white dwarfs?

What happens if an evolving star is in a binary system?

How do massive stars die?

Astronomy is exciting because it is about us. As you think about the deaths of stars, you are also thinking about the safety of Earth as a home for life and about the ultimate fate of our sun, our Earth, and the atoms of which you and I are made.

Looking Ahead

This is a chapter of endings, but it leads onward. Dying stars leave behind peculiar objects that will make your imagination ache. Furthermore, dying stars spread atoms back into the interstellar medium where they become part of new stars and new worlds. Stars die, but the story of astronomy continues.

because they occasionally flare in other galaxies and because telescopes reveal the shattered remains of these titanic explosions (■ Figure 13-1).

In the previous chapter, you saw how stars resist their own gravity by generating energy through nuclear fusion. The energy keeps their interiors hot, and the resulting high pressure balances gravity and prevents the star from collapsing. Stars, however, have only so much fuel, and gravity is patient. When the stars exhaust their fuel, gravity wins, and the stars die.

The mass of a star is critical in determining its fate. Massive stars use up their nuclear fuel at a furious rate and die after only a few million years, while the lowest-mass stars use their fuels sparingly and may be able to live hundreds of billions of years. Massive stars can die in violent supernova explosions, but low-mass stars die quiet deaths. To follow the evolution of stars to their graves, you can sort the stars into three categories according to their masses: low-mass red dwarfs, medium-mass sunlike stars, and massive upper-main-sequence stars. These stars lead dramatically different lives and die in different ways.

(13-1) Lower-Main-Sequence Stars

THE STARS OF THE LOWER MAIN SEQUENCE share a common characteristic: They have relatively low masses. That means that they face similar fates as they exhaust their nuclear fuels.

When a star exhausts one nuclear fuel, its interior contracts and grows hotter until the next nuclear fuel ignites. The contracting star heats up by converting gravitational energy into thermal energy; low-mass stars don't have a lot of gravitational energy, so they can't get very hot, and that limits the fuels they can ignite. The lowest-mass stars, for example, cannot get hot enough to ignite helium fusion.

Structural differences divide the low-mass stars into two subgroups—very-low-mass stars and medium-mass stars such as the sun. The critical difference between the two groups is the extent of interior convection (see Figure 11-14). If the star is convective, fuel is constantly mixed, and the resulting evolution of the star is drastically altered.

Red Dwarfs

Stars less massive than about 0.4 solar mass—the red dwarfs—have two advantages over more massive stars. First, they have very small masses, and consequently they have very little weight to support. Their pressure–temperature thermostats are set low, and they consume their hydrogen fuel very slowly. If you calculate the life expectancy of one of these stars (Chapter 12), you will discover a red dwarf could live over ten times longer than the sun.

The red dwarfs have a second advantage because they are totally convective. That is, they are stirred by circulating currents of hot gas rising from their interiors and cool gas sinking inward from their surfaces. This mixes the stars like pots of soup that are constantly stirred. Hydrogen is consumed, and helium accumulates

■ Figure 13-1

Supernova explosions are rare, but they leave behind expanding shells filled with hot, low-density gas. (IC443: 2MASS; Tycho's Supernova: NASA/CXC/SAO; N63A: NASA/CXC/Rutgers/J. Warren et al., STScI/U. Ill/Y. Chu; ATCA/U. Ill/J. Dickel et al.)

IC443
Age 30,000 yr
Infrared image

Tycho's Supernova
Supernova seen in 1572
X-ray image
Edge of detector

N63A
Age 2000 to 5000 yr
X-ray blue
Optical green
Radio red
Gas as hot as 10 million K fills the expanding shells.

uniformly throughout the stars; and that has two important consequences for red dwarfs.

First, because red dwarf stars are completely mixed and can fuse all of their hydrogen, their lives are prolonged even further. Models of a 0.1-solar-mass red dwarf predicted it would take 2 billion years to contract to the main sequence and 6 trillion years (6000 billion years) to use up its hydrogen fuel.

The second consequence is that they use up their hydrogen uniformly. That means they never develop a hydrogen fusion shell, and they never become giant stars. They can slowly convert their hydrogen into helium, a fuel they are not massive enough to ignite, and finally die slow, unremarkable deaths.

You will see in a later chapter that the universe is only about 14 billion years old, so none of these red dwarfs should have exhausted its fuel yet. Every red dwarf that has ever been born is still shining.

Sunlike Stars

Stars with masses between roughly 0.4 and 4 solar masses,* including the sun, ignite hydrogen and then helium and become giants, but they cannot get hot enough to ignite carbon, the next fuel in the sequence (see Table 12-3). When they reach that impasse, they can no longer maintain their stability, and their interiors contract while their envelopes expand.

To understand the fate of these stars, you need to consider two ideas, mixing and expansion. The interiors of these sunlike stars are not well mixed (see Figure 11-14). As you learned in Chapter 11, stars of 1.1 solar masses or less, including the sun, have no convection near their centers, so they are not mixed at all. Stars more massive than 1.1 solar masses have small zones of convection at their centers, but this mixes no more than about 12 percent of the star's mass. Consequently, medium-mass stars, whether they have convective cores or not, are not mixed, and the helium accumulates in an inert helium core surrounded by unprocessed hydrogen. When this core contracts, the unprocessed hydrogen just outside the core ignites in a shell and the outer layers of the star expand and cool, transforming the star into a giant.

As a giant, the star fuses helium in its core and then in a shell surrounding a core of carbon and oxygen. This core contracts and grows hotter, but it cannot become hot enough to ignite the carbon. The carbon–oxygen core is a dead end for these medium-mass stars.

Because no nuclear reactions can begin in the carbon–oxygen core, it cannot resist the weight pressing down on it. In addition, just outside the carbon–oxygen core, the helium-fusion shell converts helium into carbon and increases the mass of the carbon–oxygen core, so the core must contract. The energy released by the contracting core, plus the energy generated in the helium-

and hydrogen-fusing shells, flows outward and makes the envelope of the star expand and cool further.

This forces the star to become a very large giant. Its radius may become as large as the radius of Earth's orbit, and its surface becomes as cool as 2000 K. Such a star can lose large amounts of mass from its surface.

Mass Loss from Sunlike Stars

It's not hard to find evidence that stars like the sun lose mass. The sun itself is losing mass. The solar wind is a gentle breeze of gas blowing outward from the solar corona and carrying mass into space.

Other stars like the sun are losing mass, and the evidence lies in their spectra. Ultraviolet and X-ray spectra of many stars reveal strong emission lines, and you can use Kirchhoff's laws to conclude that the outer layers must be highly ionized; the stars must have hot chromospheres and coronas like the sun's. If these stars have atmospheres like the sun's, they presumably have similar winds of hot gas. You can find further evidence in the spectra of some giant stars, which contain blueshifted absorption lines. The blueshifts must be Doppler shifts produced as the gas flowing out of the star comes toward Earth.

The solar wind is not very strong; the sun loses about 0.001 solar mass per billion years. Even over the entire lifetime of the sun, that would not significantly alter the sun's mass. However, once the sun becomes a giant star, it may lose mass more rapidly. Because the spectra of some of these stars do not show the characteristic emission, you must assume that the stars do not have hot coronas; but other processes could drive mass loss. The stars are so large that gravity is weak at their surfaces, and convection in the cool gas can drive shock waves outward and power mass loss. In addition, some giants are so cool that specks of carbon dust condense in their atmospheres, just as soot can condense in a fireplace. The pressure of the star's radiation can push this dust and any atoms that collide with the dust completely out of the star.

Another factor that can cause mass loss is periodic eruptions in the helium-fusion shell. The triple-alpha process that fuses helium into carbon is extremely sensitive to temperature, and that can make the helium-fusion shell so unstable that it may experience eruptions called **thermal pulses.** Every 200,000 years or so, the shell can erupt and suddenly produce energy equivalent to a million times the luminosity of the sun. Almost all of that energy goes into lifting the outer layers of the star and driving gas away from the surface.

Whatever drives this mass loss from giants, it can affect the mass of the star appreciably in a short time. A star expanding as its carbon–oxygen core contracts could lose an entire solar mass in only 10^5 years, which is not a long time in the evolution of a star. That means that a star that began its existence on the main sequence with a mass of 8 solar masses might reduce its mass to only 3 solar masses in half a million years.

---

*This mass limit is uncertain, as are many of the masses stated here. The evolution of stars is highly complex, and such parameters are not precisely known.

Stellar mass loss confuses the story of stellar evolution. It would be nice to say that stars more massive than a certain limit will evolve one way, and stars less massive will evolve another way. But stars may lose enough mass to alter their own evolution. As a result, you need to consider both the initial mass a star has on the main sequence and the mass it retains after mass loss. Because astronomers don't know exactly how effective mass loss is, it is difficult to be exact about the mass limits in any discussion of stellar evolution.

The mass loss discussed so far has been a gradual process, but a sunlike star with a contracting carbon–oxygen core expands so far that it can expel all of its outer layers. This surface expulsion is helped by the growing pressure of the radiation inside the star and by instabilities in the helium-fusion shell. The expanding layers are visible as small nebulae.

Planetary Nebulae

When a medium-mass star like the sun becomes a distended giant, it can expel its outer atmosphere to form one of the most interesting objects in astronomy, a **planetary nebula,** so called because through a small telescope it looks like the greenish-blue disk of a planet such as Uranus or Neptune. In fact, a planetary nebula has nothing to do with a planet. It is composed of ionized gases expelled by a dying star.

Study **The Formation of Planetary Nebulae** on pages 290–291 and notice four things:

1 First, you can understand what planetary nebulae are like by using simple observational methods such as Kirchhoff's laws and the Doppler effect.

2 Second, notice the model astronomers have developed to explain planetary nebulae. The real nebulae are more complex than the simple model of a slow wind and a fast wind, but the model provides a way to organize the observed phenomena.

3 The third thing to notice is how oppositely directed jets (much like bipolar flows) produce many of the asymmetries seen in planetary nebulae.

4 Finally, notice the fate of the star itself; it must contract into a white dwarf.

As is always the case in science, the theories that describe the formation of planetary nebulae are open to revision. In 1996, astronomers saw the hot nucleus of a planetary nebula, a star that should have been contracting to form a white dwarf, flare back to life. Apparently, some stars flare one or more times as they push their surface layers away and collapse. Such contracting objects may temporarily re-ignite helium fusion a number of times before they finally sink into white-dwarf-hood. The flares may explain some of the complex structure visible in planetary nebulae.

Most astronomy books say that the sun will form a planetary nebula, but that may not happen. To light up a planetary nebula, a star must become a white dwarf with a temperature of at least 25,000 K. Some theorists think that slightly less massive stars, such as the sun, may not get hot enough when they collapse and consequently may not be able to ionize the gases they have ejected. Time will tell, of course, but you can't wait around. Astronomers continue to refine their theories and make new observations. They may eventually be able to say conclusively whether the sun will light up its own planetary nebula.

The medium-mass stars like the sun die by losing mass and contracting into white dwarfs. If they become hot enough white dwarfs, they light up planetary nebulae. This suggests another way to search for evidence of the deaths of medium-mass stars: study white dwarfs.

White Dwarfs

White dwarfs are among the most common stars. The survey in Chapter 9 revealed that white dwarfs and red dwarfs, although very faint, are the dominant kinds of stars (page 211). Now you can recognize white dwarfs as the remains of medium-mass stars that fused hydrogen and helium, failed to ignite carbon, drove away their outer layers to form planetary nebulae, and collapsed and cooled to form white dwarfs. The billions of white dwarfs in our galaxy must be the remains of medium-mass stars.

The first white dwarf discovered was the faint companion to one of our Favorite Stars, Sirius. In that visual binary system, the bright star is Sirius A. The white dwarf, Sirius B, is 10,000 times fainter than Sirius A. The orbital motions of the stars (shown in Figure 9-15) tell you that the white dwarf's mass is 0.98 solar mass, and its blue-white color tells you that its surface is hot, about 25,000 K. Although it is very hot, it has a very low luminosity, so it must have a small surface area—in fact, its diameter is about twice Earth's. Dividing its mass by its volume reveals that it is very dense—over 3×10^6 g/cm^3. On Earth, a teaspoonful of Sirius B material would weigh more than 15 tons (■ Figure 13-2). Thus, basic observations and simple physics lead to the conclusion that white dwarfs are astonishingly dense.

A normal star is supported by energy flowing outward from its core, but a white dwarf cannot generate energy by nuclear fusion. It has exhausted its hydrogen and helium fuel and produced carbon. As the star collapses into a white dwarf it converts gravitational energy into thermal energy, and its interior becomes very hot, but it cannot get hot enough to ignite carbon fusion. The star contracts until it becomes degenerate. Although a tremendous amount of energy flows out of the hot interior, it is not the energy flow that supports the star. The white dwarf is supported against its own gravity by the inability of its degenerate electrons to pack into a smaller volume.

The interior of a white dwarf is mostly carbon and oxygen ions floating among a whirling storm of degenerate electrons. It is the degenerate electrons that exert the pressure to support the star's weight, but most of its mass is represented by the carbon and oxygen ions. Theory predicts that, as the star cools, these ions will lock together to form a crystal lattice, so there may be some

■ Figure 13-2

The degenerate matter from inside a white dwarf is so dense that a lump the size of a beach ball would, transported to Earth, weigh as much as an ocean liner.

truth in thinking of aging white dwarfs as great crystals of carbon and oxygen. Near the surface, where the pressure is lower, a layer of ionized gases makes up a hot atmosphere.

The tremendous surface gravity of white dwarfs—100,000 times that of Earth—affects their atmospheres in strange ways. The heavier atoms in the atmosphere tend to sink, leaving the lightest gases at the surface. Astronomers see some white dwarfs with atmospheres of almost pure hydrogen, whereas others have atmospheres of nearly pure helium. Still others, for reasons not well understood, have atmospheres that contain traces of heavier atoms. In addition, the powerful surface gravity pulls the white dwarf's atmosphere down into a very shallow layer. If Earth's atmosphere were equally shallow, people on the top floors of skyscrapers would have to wear oxygen masks.

Clearly, a white dwarf is not a true star. It generates no nuclear energy, is almost totally degenerate, and, except for a thin layer at its surface, contains no gas. Instead of calling a white dwarf a "star," you can call it a **compact object.** In the next chapter, you will meet two other compact objects—neutron stars and black holes.

A white dwarf's future is bleak. As it radiates energy into space, its temperature gradually falls, but it cannot shrink any smaller because its degenerate electrons cannot get closer together. This degenerate matter is a very good thermal conductor, so heat flows to the surface and escapes into space, and the white dwarf gets fainter and cooler, moving downward and to the right in the H–R diagram. Because the white dwarf contains a tremendous amount of heat, it needs billions of years to radiate that heat

through its small surface area. Eventually, such objects may become cold and dark, so-called **black dwarfs.** Our galaxy is not old enough to contain black dwarfs. The coolest white dwarfs in our galaxy are about the temperature of the sun.

Perhaps the most interesting thing about white dwarfs has come from mathematical models. The equations predict that if you added mass to a white dwarf, its radius would *shrink,* because added mass would increase its gravity and squeeze it tighter. If you added enough to raise its total mass to about 1.4 solar masses, its radius would shrink to zero (■ Figure 13-3). This is called the **Chandrasekhar limit** after Subrahmanyan Chandrasekhar, the astronomer who discovered it. It seems to imply that a star more massive than 1.4 solar masses could not become a white dwarf unless it shed mass in some way.

Stars do lose mass. Young stars have strong stellar winds, and aging giants also lose mass (■ Figure 13-4). This suggests that stars more massive than the Chandrasekhar limit can eventually die as white dwarfs if they reduce their mass. Theoretical models show that an 8-solar-mass star should be able to reduce its mass to 1.4 solar masses before it collapses, and slightly more massive stars may also be able to get under the limit. Consequently, a wide range of medium-mass stars eventually die as white dwarfs.

Building Scientific Arguments

How will the sun die?

This question seems to call for an argument based mostly on theory, but remember that all theory is tested against observations. You have seen that lower-main-sequence stars like the sun die by producing planetary nebulae and then becoming white dwarfs. For the sun, this process will begin in a few billion years, when it exhausts the hydrogen in its core. It will then expand to become a giant star. Although the surface of the sun will become cooler, it will expand until it has such a large surface that it will be about 100 times more luminous than it is now. As a giant star, the sun will have a strong solar

■ Figure 13-3

The more massive a white dwarf, the smaller its radius. Stars more massive than the Chandrasekhar limit of 1.4 solar masses cannot be white dwarfs.

The Formation of Planetary Nebulae

1 Simple observations tell astronomers what planetary nebulae are like. Their angular size and their distance indicate that their radii range from 0.2 to 3 light-years. The presence of emission lines in their spectra assures astronomers they are excited, low-density gas. Doppler shifts show they are expanding at 10 to 20 km/s. If you divide radius by velocity, you find planetary nebulae are no more than 10,000 years old. Older planetary nebulae evidently become mixed into the interstellar medium.

1a Astronomers find about 1500 planetary nebulae in the sky. Because planetary nebulae are short-lived formations, you can conclude that they must be a common part of stellar evolution. Medium-mass stars up to a mass of about 8 solar masses are destined to die by forming planetary nebulae.

NGC6369   Visual

This nearly spherical planetary nebula has a low-luminosity outer envelope and a highly excited inner region.

The Ring Nebula in the constellation Lyra is visible even in small telescopes. Note the hot blue star at its center and the radial texture in the gas, suggesting outward motion.

Visual

Hubble Heritage Team, STScI/AURA/NASA

Hubble Heritage Team, AURA/STScI/NASA

2 The process that produces planetary nebulae involves two stellar winds. First, as an aging giant, the star gradually blows away its outer layers in a slow breeze of low-excitation gas that is not easily visible. Once the hot interior of the star is exposed, it ejects a high-speed wind that overtakes and compresses the gas of the slow wind like a snowplow, while ultraviolet radiation from the hot remains of the central star excites the gases to glow like a giant neon sign.

Slow stellar wind from a red giant

The gases of the slow wind are not easily detectable.

Fast wind from exposed interior

You see a planetary nebula where the fast wind compresses the slow wind.

Visual

NGC6826

Visual

NGC7009

Some planetary nebulae, such as M2-9, are highly elongated, and it has been suggested that the Ring Nebula (opposite page) is a tubular shape pointed roughly at Earth.

Infrared

M2-9   Visual

This infrared image of the Egg Nebula reveals an irregular, thick disk (red) from which beams of gas and dust emerge. Such beams may create many of the asymmetries in planetary nebulae.

NASA

The Egg Nebula

10^6 — Nuclei of planetary nebulae

10^4

Supergiants

10^2

Main sequence

Giants

$L/L_\odot$

1

10^{-2}

Mathematical model of an 0.8 solar mass stellar remnant contracting to become a white dwarf.

White dwarfs

10^{-4}

100,000 50,000 30,000   10,000   5000

Temperature (K)

4 Once an aging giant star blows its surface into space to form a planetary nebula, the remaining hot interior collapses into a small, intensely hot object containing a carbon and oxygen interior surrounded by hydrogen and helium fusion shells and a thin atmosphere of hydrogen. The fusion gradually dies out, and the core of the star evolves to the left of the conventional H–R diagram to become the intensely hot nucleus of a planetary nebulae. Mathematical models show that these nuclei cool slowly to become white dwarfs.

NGC3918   Visual

V838 Mon

Light from the eruption of a supergiant star illuminates gas and dust ejected previously.

Visual-wavelength image

Massive star WR124 is ejecting mass in a violent stellar wind.

Visual

Mass lost from λ Orionis

λ Orionis

Orion

Extremely hot stars such as those in Orion can drive gas away.

Orion Nebula

Infrared image

■ **Figure 13-4**

Stars can lose mass if they are very hot, very large, or both. The red supergiant V838 Mon has lost mass in the past, as revealed by light emitted when it erupted temporarily in January 2002. A young massive star such as WR124 and the hot, blue stars that make up the constellation Orion constantly lose mass into space. Warmed dust in these gas clouds can make them glow in the infrared. (V838 Mon: NASA and The Hubble Heritage Team AURA/STScI; WR124: NASA; Orion: NASA/IPAC courtesy Deborah Levine)

wind carrying gas into space and pushing back the interstellar medium. Eventually, the loss of mass will expose deeper layers in the sun. These extremely hot layers will heat the gas around the sun and propel it outward to scoop up previously expelled gases. If the sun becomes hot enough, it will ionize the gas and produce a planetary nebula. In any case, the last remains of the sun will contract and slowly cool to form a small, dense white dwarf. The surface will be very hot but so small that the sun will be about 100 times fainter than it is now.

The story of stellar evolution predicts that our sun will someday become a giant star and then a white dwarf. Focus your scientific argument on a slightly different part of the story. **As the sun becomes a white dwarf, what will have become of all the hydrogen the sun once contained?**

■   ■   ■

Connections: Lower-main-sequence stars like the sun die by producing beautiful planetary nebulae and shrinking to become white dwarfs. Now you are ready to see what happens when one of these dying stars has a stellar companion.

13-2 The Evolution of Binary Stars

SO FAR YOU HAVE BEEN THINKING about the deaths of stars as if they were all single objects that never interact. But more than half of all stars are members of binary star systems. Most such binaries are far apart, and one of the stars can swell into a giant and eventually collapse without affecting the companion star. In some binary systems, however, the two stars orbit close together. When the more massive star begins to expand, it interacts with its companion star in peculiar ways. These interacting binary stars are fascinating objects themselves, but they are also important because they help explain observed phenomena such as nova explosions. In the next chapter, you will see how they can help astronomers find black holes.

Mass Transfer

Binary stars can sometimes interact by transferring mass from one star to the other. To understand this process, you need to think about how the gravity of the two stars controls the matter in the binary system.

Each star in a binary system is held together by its own gravity: but, because the system is rotating, there are unexpected forces acting on loose matter between the stars. If you could gently release a pebble into such a system, it might fall into one star, fall into the other star, or be ejected from the system. What happened to your pebble would depend on where you dropped it. Each star controls a region of space near it, and if you could make these regions visible they would look like two teardrop-shaped lobes enclosing the stars and meeting tip to tip between the stars. These two volumes are called the **Roche lobes,** and the dumbbell-shaped surface of the two lobes is called the **Roche surface.** If you released your pebble inside one of the Roche lobes, it would belong gravitationally to the star that controls that lobe.

Because of the interaction of the gravity of the two stars and their rotation around their center of mass, there are five points of stability in the orbital plane called **Lagrangian points.*** ■ Figure 13-5 illustrates the arrangement of the Roche surface around the stars and the Lagrangian points.

The Lagrangian points are important because they are locations where matter is stable in the revolving binary system. If you could carefully release your pebble at a Lagrangian point, it would remain at that point and circle the center of mass along with the stars. Rather than a pebble, think of a bit of naturally occurring gas in the binary system. Gas at the L_2 or L_3 point is critically stable, meaning that any small disturbance will make it leak away. In contrast, gas at the L_4 or L_5 point can be trapped. (You will meet the L_4 and L_5 points again when you study planetary satellites and the orbits of asteroids in later chapters.) As you think about binary stars, the inner Lagrangian point L_1 located between the two stars is the most important. If matter from one star can reach the inner Lagrangian point, it can flow onto the other star. That is, the stars can transfer mass through the inner Lagrangian point.

In general, there are only two ways matter can escape from a star and reach the inner Lagrangian point. First, if a star has a strong stellar wind, some of the gas blowing away from the star can pass through the inner Lagrangian point and be captured by the other star. Second, if an evolving star expands so far that it fills its Roche surface, it will be forced to take on the teardrop shape of the Roche surface. You have seen that effect in your study of binary stars (Figure 9-21). If the star expands farther, matter will flow through the inner Lagrangian point (like water in a pond flowing over a dam) and will fall into the other star. Mass transfer driven by a stellar wind tends to be slow, but mass transfer driven by an expanding star can occur rapidly.

---

*The Lagrangian points are named after French mathematician Joseph Louis Lagrange, who solved this famous mathematical problem around the time of the French Revolution.

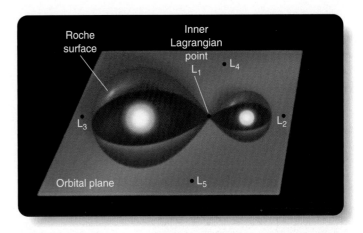

■ **Figure 13-5**

A pair of binary stars control the region of space located inside the Roche surface. The Lagrangian points are locations of stability, with the inner Lagrangian point making a connection through which the two stars can transfer matter.

Evolution with Mass Transfer

Mass transfer between stars can affect their evolution in surprising ways. In fact, it provides the explanation for a problem that puzzled astronomers for many years.

In some binary systems, the less massive star has become a giant, while the more massive star is still on the main sequence. That seems backward. If more massive stars evolve faster than lower-mass stars, how does the low-mass star in such binaries manage to leave the main sequence first? This is called the Algol paradox, after the binary system Algol (see Figure 9-22).

Mass transfer explains how this could happen. Imagine a binary system that contains a 5-solar-mass star and a 1-solar-mass companion (■ Figure 13-6). The two stars formed at the same time, so the more massive star must evolve faster and leave the main sequence first. When it expands into a giant, it can fill its Roche surface and transfer matter to the low-mass companion. In that way, the massive star will shrink into a lower-mass star, and the companion will gain mass and become a massive star still on the main sequence. If you observed such a system after the mass transfer ended, you might see a binary system such as Algol containing a 5-solar-mass main-sequence star and a 1-solar-mass giant.

The evolution of close binary stars could result in one of the stars having its outer layers "peeled" away. A massive star expanding to become a giant could lose its outer layers to its companion and then collapse to form a lower-mass peculiar star. A few such peculiar stars are known.

Another exotic result of the evolution of close binary systems is the merging of the stars. Astronomers see many binaries in which both stars have expanded to fill their Roche surfaces and spill mass out into space. If the stars are close enough together and expand rapidly enough, theorists believe, the two stars could merge into a single, rapidly rotating giant star. Most giants rotate slowly because they conserved angular momentum as they expanded and consequently slowed their rotation. Examples of rapidly rotating giant stars are known, however, and they are believed to be the results of merged binary stars. Inside the distended envelope of such a star, the cores of the two stars could continue to orbit each other until friction slowed them down and they sank to the center.

Mass transfer can lead to dramatic violence. The first four frames of Figure 13-6 show mass transfer producing a system like Algol. The last frame shows an additional stage in which the giant star has expelled its outer layers and has collapsed to form a white dwarf. The more massive companion has expanded, transferring matter back onto the white dwarf. Such systems can become the site of tremendous explosions. To see how this can happen, you need to consider in detail how mass falls into a star.

Accretion Disks

Matter flowing through the inner Lagrangian point cannot fall directly into the white dwarf because of the conservation of angular momentum (as described in Chapter 5). Instead, it falls into a whirlpool around the white dwarf. For a common example, consider a bathtub full of water. Gentle currents in the water give it some angular momentum, but you can't see its slow circulation until you pull the stopper. Then, as the water rushes toward the drain, conservation of angular momentum forces it to form a whirlpool. This same effect forces gas falling into a white dwarf to form a whirling disk of gas called an **accretion disk** (■ Figure 13-7).

Two important things happen in an accretion disk. First, the gas in the disk grows very hot due to friction and tidal forces. The disk also acts as a brake, ridding the gas of its angular momentum and allowing it to fall into the white dwarf. The temperature of the gas in the inner parts of an accretion disk can become very high, in some cases exceeding 1,000,000 K and emitting X rays.

Now you are ready to put the pieces together and explain one of the ways a star can explode. All you need is mass transfer onto a white dwarf.

Nova Explosions

At the beginning of this chapter you saw that the word *nova* refers to a new star that appears in the sky for a while and then fades

■ **Figure 13-6**

A pair of stars orbiting close to each other can exchange mass and modify their evolution.

The Evolution of a Binary System

Star B is more massive than Star A.

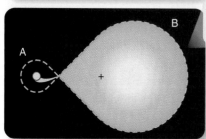

Star B becomes a giant and loses mass to Star A.

Star B loses mass, and Star A gains mass.

Star A is a massive main-sequence star with a lower-mass giant companion— an Algol system.

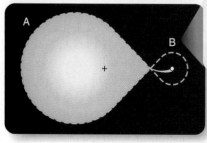

Star A has now become a giant and loses mass back to the white dwarf that remains of Star B.

■ **Figure 13-7**

Matter falling into a compact object forms a whirling accretion disk. Friction and tidal forces can make the disk very hot.

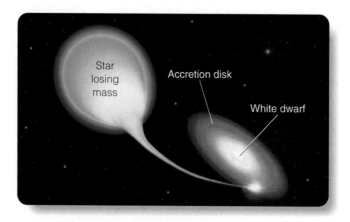

Star losing mass

Accretion disk

White dwarf

■ Figure 13-8

Nova Cygni 1975 near maximum at about second magnitude (a) and later when it had declined to about eleventh magnitude (b). (Photo © UC Regents)

tion lines, which tells you the gas is dense and coming toward Earth at a few thousand kilometers per second. After a few days, the spectral lines change to emission lines, which tells you the gas has thinned, but the blueshifts remain, showing that a cloud of debris has been ejected into space.

You can understand nova explosions as the result of mass transfer from a normal star through the inner Lagrangian point into an accretion disk around the white dwarf. As the matter loses its angular momentum in the accretion disk, it drifts inward and eventually settles onto the surface of the white dwarf. Because the matter came from the surface of a normal star, it is rich in unused fuel, mostly hydrogen, and when it accumulates on the surface of the white dwarf, it forms a layer of unprocessed fuel.

As the layer of fuel grows deeper, it becomes hotter and denser and eventually becomes degenerate, much like the core of a sun-like star approaching the helium flash. In such a gas, the pressure–temperature thermostat does not work, so the layer of unfused hydrogen is a thermonuclear bomb waiting to explode. By the time the white dwarf has accumulated about 100 Earth masses of hydrogen, the temperature at the base of the hydrogen layer reaches millions of degrees, and the density is 10,000 times the density of water. Suddenly, the hydrogen begins to fuse on the proton–proton chain, and the energy released drives the temperature so high that the CNO cycle begins. With no pressure–temperature thermostat to control the fusion, the temperature shoots to 100 million degrees in seconds. Quickly the temperature rises high enough to force the gas to expand, and it stops being degenerate. But it is too late. By the time the gas begins to expand, the amount of energy released is enough to blow the surface layers of the star into space as a violently expanding shell of hot gas, which is visible from Earth as a nova.

The explosion of its surface hardly disturbs the white dwarf or its companion star. Mass transfer quickly resumes, and a new layer of fuel begins to accumulate. How fast the fuel builds up depends on the rate of mass transfer. Accordingly, you can expect novae to repeat each time an explosive layer accumulates. Many novae take thousands of years to build an explosive layer, but some take only a few years (■ Figure 13-9).

away (■ Figure 13-8). The evidence shows that it is not a new star at all. After a nova fades, astronomers can photograph the spectrum of the object, and they invariably find a closely spaced binary star containing a normal star and a white dwarf. A nova is evidently an explosion involving a white dwarf in a binary system.

Observational evidence can tell you how nova explosions occur. As the explosion begins, spectra show blueshifted absorp-

■ Figure 13-9

Nova T Pyxidis erupts about every two decades, expelling shells of gas into space. The shells of gas are visible from ground-based telescopes, but the Hubble Space Telescope reveals much more detail. The shell consists of knots of excited gas that presumably form when a new shell collides with a previous shell. (M. Shara and R. Williams, STScI; R. Gilmozzi, ESO; and NASA)

The End of Earth

Astronomy is about us. Although you have been reading about the deaths of medium-mass stars, white dwarfs, and novae explosions, you have also been reading about the future of our planet. The sun is a medium-mass star and must eventually die by becoming a giant, possibly producing a planetary nebula, and collapsing into a white dwarf. That will spell the end of Earth.

Mathematical models of the sun suggest that it may survive for 5 billion years or so, but it is already growing more luminous as it fuses hydrogen into helium. In a few billion years, it will exhaust hydrogen in its core and swell into a giant star about 100 times its present radius. That giant sun will be about as large as the orbit of Earth, so that will mark the end of our world. Whether or not the expanding sun becomes large enough to totally engulf Earth, its growing luminosity will certainly evaporate our oceans, drive away our atmosphere, and even vaporize most of Earth's crust.

While it is a giant star, the sun will lose mass into space. This mass loss is a relatively gentle process, so any cinder that might remain of Earth would not be disturbed. Of course, when the sun finally collapses into a white dwarf, the solar system will become a much colder place, but not much of Earth will remain by then. If the collapsing sun becomes hot enough, it will ionize the expelled gas to form a planetary nebula. Some models suggest that the sun is not quite massive enough to light up a planetary nebula. It will collapse into a hot object that will cool to become a white dwarf, but it may not become hot enough to ionize the expelled gas and "turn on" a planetary nebula. An astronomer recently expressed disappointment that the dying sun might not leave behind a beautiful planetary nebula, but that embarrassment lies a few billion years in the future.

There is no danger that the sun will explode as a nova; it has no binary companion. Also, as you will see, the sun is not massive enough to die the violent death of the massive stars.

The most important lesson of astronomy is that we are part of the universe and not just observers. The atoms you are made of are destined to return to the interstellar medium in just a few billion years. That's a long time, and it is possible that the human race will migrate to other planetary systems. That might save the human race, but our planet is star dust.

Building Scientific Arguments

How does modern astronomy explain the Algol paradox?
Scientific arguments combine evidence with theory to explain nature, and some of the most interesting arguments explain what seem to be impossible situations. When you encounter a paradox in nature, it is usually a warning that you don't understand things as well as you thought you did. The Algol paradox is a good example. The binary star Algol contains a lower-mass giant star and a more massive main-sequence star. Because the two stars must have formed together, they must

be the same age. Then the more massive star should have evolved first and left the main sequence. But in binary systems such as Algol, the lower-mass star has left the main sequence first, or so it seems.

You can understand this paradox if you add mass transfer to your argument. The first star to leave the main sequence must be the more massive star; but if the stars are close together, the star that is initially more massive can fill its Roche surface and transfer mass back to its companion. The companion can grow more massive, and the giant can grow less massive. In this way, the lower-mass giant star may have originally been more massive and may have evolved away from the main sequence, leaving its companion behind to increase in mass as the giant decreased.

Mass transfer can explain the Algol paradox, and it can also explain the violent explosions called novae. Develop a new argument. **How does mass transfer explain why novae can explode over and over in the same binary system?**

■   ■   ■

Connections: Your study of the deaths of low- and medium-mass stars has led you to planetary nebulae, white dwarfs, and novae explosions. But there are more violent fish in the cosmic sea. What causes supernovae? To find out, you need to consider how massive stars die.

(13-3) The Deaths of Massive Stars

YOU HAVE SEEN that low- and medium-mass stars die relatively quietly as they exhaust their hydrogen and helium and then eject their surface layers to form planetary nebulae. In contrast, massive stars live spectacular lives and destroy themselves in violent explosions.

Nuclear Fusion in Massive Stars

Stars on the upper main sequence have too much mass to die as white dwarfs, but their evolution begins much like that of their lower-mass cousins. They consume the hydrogen in their cores and ignite hydrogen shells; as a result, they expand and become giants or, for the most massive stars, supergiants. Their cores contract and fuse helium first in the core and then in a shell, producing a carbon–oxygen core.

A massive star can lose significant mass as it ages; but, if it still has a mass over 4 solar masses when its carbon–oxygen core contracts, it can reach a temperature of 600 million Kelvin and ignite carbon fusion. The fusion of carbon produces heavier nuclei such as oxygen and neon. As soon as the carbon is exhausted in the core, the core contracts, and carbon ignites in a shell. This

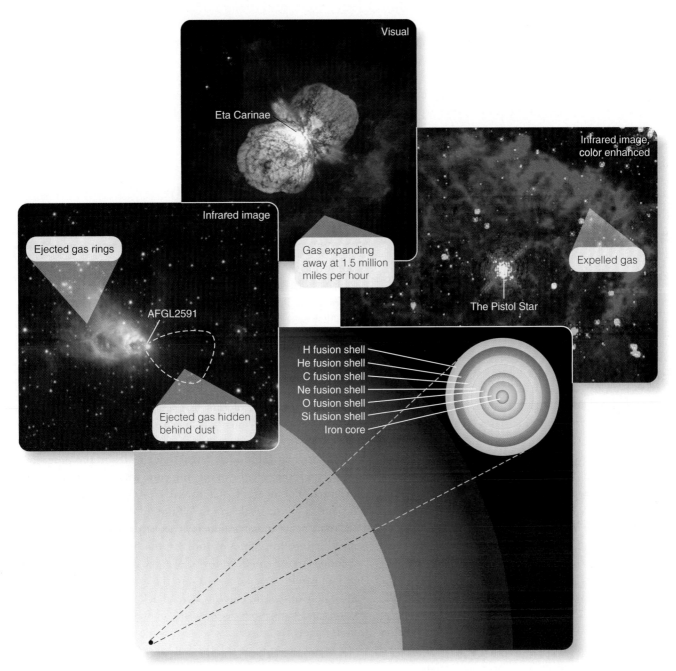

Visual

Eta Carinae

Infrared image, color enhanced

Gas expanding away at 1.5 million miles per hour

Expelled gas

The Pistol Star

Infrared image

Ejected gas rings

AFGL2591

Ejected gas hidden behind dust

H fusion shell
He fusion shell
C fusion shell
Ne fusion shell
O fusion shell
Si fusion shell
Iron core

■ Active Figure 13-10

Massive stars live fast and die young. The three shown here are among the most massive stars known, containing 100 solar masses or more. They are rapidly ejecting gas into space. The centers of these massive stars develop Earth-size cores (magnified 100,000 times in this figure) composed of concentric layers of gases undergoing nuclear fusion. The iron core at the center leads eventually to a star-destroying explosion. (AFGL2591: Gemini Observatory/NSF/C. Aspin; Eta Carinae and the Pistol Star: NASA)

Ace ✆ Astronomy™ Log into AceAstronomy and select this chapter to see the Active Figure "Inside Stars." Compare the interior of a sunlike star with that of a massive star.

pattern of core ignition followed by shell ignition continues with fuel after fuel, and the star develops a layered structure at its center (■ Figure 13-10), with a hydrogen-fusion shell above a helium-fusion shell above a carbon-fusion shell and so on. After

carbon fusion, oxygen, neon, magnesium, and heavier elements fuse, right up to iron.

The fusion of these nuclear fuels goes faster and faster as the massive star evolves rapidly. Recall that massive stars must consume their fuels rapidly to support their great weight, but other factors also cause the heavier fuels like carbon, oxygen, and silicon to fuse at increasing speed. For one thing, the amount of energy released per fusion reaction decreases as the mass of the fusing atom increases. To support its weight, a star must fuse oxygen much faster than it fused hydrogen. Also, there are fewer atoms in the core of the star by the time heavy atoms begin to fuse. Four hydrogens made a helium atom, and three heliums made a carbon, so there are 12 times fewer atoms of carbon available for fusion than there were of hydrogen. This means that heavy-element fusion goes very quickly in massive stars (■ Table 13-1). Hydrogen fusion can last 7 million years in a 25-solar-mass star, but that same star will fuse its oxygen in six months and its silicon in a day.

The Iron Core

Heavy-element fusion ends with iron, because nuclear reactions that use iron as a fuel cannot produce energy. Nuclear reactions can produce energy if they proceed from less tightly bound nuclei to more tightly bound nuclei. As shown in Figure 8-8, both nuclear fission and nuclear fusion produce nuclei that are more tightly bound than the starting fuel. Look again at Figure 8-8 and notice that iron is the most tightly bound nucleus of all. No nuclear reaction, fission or fusion, that starts with iron can produce a more tightly bound nucleus, and that means that iron is a dead end.

When a massive star develops an iron core, nuclear fusion cannot produce energy, and the core contracts and grows hotter. The shells around the core burn outward, fusing lighter elements and leaving behind more iron, which further increases the mass of the core. When the mass of the iron core exceeds 1.3 to 2 solar masses, the core must collapse.

As the core begins to collapse, two processes can make it contract even faster. Heavy nuclei in the core can capture high-energy electrons, thus removing thermal energy from the gas. This robs the gas of some of the pressure it needs to support the crushing weight of the outer layers. Also, in more massive stars, temperatures are so high that many photons have gamma-ray wavelengths and can break more massive nuclei into less massive nuclei. In the process, the gamma rays are absorbed. This reversal of nuclear fusion absorbs energy and allows the core to collapse even faster.

Although a massive star may live for millions of years, its iron core—about 500 km in diameter—collapses in only a few thousandths of a second. This collapse happens so rapidly that the most powerful computers are not capable of following the details. One thing is clear, however. The collapse of the iron core in a massive star triggers a star-destroying explosion—a supernova.

The Supernova Deaths of Massive Stars

Modern theory predicts that the collapse of a massive star can eject the outer layers of the star to produce a supernova explosion while the core of the star collapses to form a neutron star or a black hole. You will learn more about neutron stars and black holes in the next chapter; here you can concentrate on the process that triggers the supernova explosion.

A supernova explosion is rare, remote, rapid, and violent. It is an event in nature that is extremely difficult to study in person, which is why astronomers have used powerful mathematical modeling techniques and high-speed supercomputers to explore the inside of a star exploding as a supernova. Such models allow astronomers to experiment on an exploding star as if the star were in a laboratory beaker.

Mathematical models reveal that the key to the supernova explosion is the collapse of the iron core, which allows the rest of the interior of the star to fall inward, creating a tremendous "traffic jam" as all the nuclei try to fall toward the center. It is as if all the residents of Indiana suddenly tried to drive their cars as fast as possible into the center of Indianapolis. There would be a tremendous traffic jam downtown. As more cars rushed in, the traffic jam would spread outward into the suburbs. Similarly, as the inner core of the star falls inward, a shock wave (a traffic jam) develops and begins to move outward. Containing about 100 times more energy than necessary to destroy the star, such a shock wave was first thought to be the cause of the supernova explosion.

Computer models revealed, however, that the shock wave spreading outward through the collapsing star stalls within a few hundredths of a second. Matter flowing inward smothers the shock wave and pushes it back into the star. If this happened in a real star, it presumably would collapse without any visible explosion.

However, theory predicts that 99 percent of the energy released in the collapse will appear as neutrinos. In the sun, neutrinos zip outward unimpeded by the gas of the solar layers, but in a collapsing star the density is as high as 10^{12} g/cm^3, nearly as dense as an atomic nucleus. This gas is opaque to neutrinos, and they are partially absorbed in the gas. Not only does the tremendous burst of neutrinos remove energy from the core and allow the core to collapse even faster, but the neutrinos are absorbed outside the core and heat those layers. For some years, astronomers thought that these neutrinos could spread energy across the shock wave

■ **Table 13-1** | **Heavy-Element Fusion in a 25-$M_\odot$ Star**

Fuel	Time	Percentage of Lifetime
H	7,000,000 years	93.3
He	500,000 years	6.7
C	600 years	0.008
O	0.5 years	0.000007
Si	1 day	0.00000004

and, within a quarter of second or so, reaccelerate the stalled shock wave. The computer models, however, refused to explode.

The final ingredient that the models needed to make them explode was turbulent convection. When the collapse begins, the very center of the star forms a highly dense core, the beginnings

The Exploding Core of a Supernova

The core of a massive supergiant has begun to collapse at the lower left corner of this model.

Matter continues to fall inward (blue and green) as the core expands outward (yellow) creating a shock wave.

To show the entire star at this scale, this page would have to be 30 kilometers in diameter.

Only 0.4 s after beginning, violent convection in the expanding core (red) pushes outward.

The shock wave will blow the star apart as a neutron star forms at the extreme lower left corner.

of a neutron star, and the infalling material bounces off that core. With the temperature shooting up, the bouncing material produces highly turbulent convection currents that give the stalled shock wave an outward boost (■ Figure 13-11). Within a second or so, the shock wave begins to push outward, and after just a few hours it bursts out through the surface, blasting the star apart in a supernova explosion.

The supernova visible from Earth is the brightening of the star as its outer layers are blasted outward. As the months pass, the cloud of gas expands, thins, and begins to fade. But the way it fades in some cases tells astronomers about the death throes of the star. Essentially all of the iron in the core of the star is destroyed when the core collapses, but the violence in the outer layers can produce densities and temperatures high enough to trigger nuclear fusion reactions that produce as much as half a solar mass of radioactive nickel-56. The nickel gradually decays to form radioactive cobalt, which decays to form normal iron. The rate at which the supernova fades matches the rate at which these radioactive elements decay. Thus the destruction of iron in the core of the star is matched by the production of iron through nuclear fusion in the expanding outer layers.

The presence of nuclear fusion in the outer layers of the supernova testifies to the violence of the explosion. A typical supernova is equivalent to the explosion of 10^{28} megatons of TNT—about 3 million solar masses of high explosive.

Collapsing massive stars can trigger cosmic violence, but astronomers have observed more than one kind of supernova.

Types of Supernovae

Supernovae are rare, and only a few have been seen in our galaxy, but astronomers have been able to observe supernovae occurring in other galaxies (■ Figure 13-12). From these data accumulated over decades, astronomers have noticed that there are two main types. **Type I supernovae** have spectra that contain no hydrogen lines. They reach a maximum brightness about 4 billions times more luminous than the sun and decline rapidly at first and then more slowly. **Type II supernovae** have spectra containing hydrogen lines. They reach a maximum brightness up to about 0.6 billion times more luminous than the sun, decline to a standstill, and then fade rapidly. The light curves in ■ Figure 13-13 summarize the behavior of the two types of supernovae.

The evidence is clear that type II supernovae occur when a massive star develops an iron core and collapses. Such supernovae occur in regions of active star formation where there is plenty of

■ **Figure 13-11**

As the iron core of a massive star begins to collapse, intensely hot gas triggers violent convection. Even as the outer parts of the core continue to fall inward, the turbulence blasts outward and reaches the surface of the star within hours, creating a supernova eruption. This diagram is based on mathematical models and shows only the exploding core of the star. (Courtesy Adam Burrows, John Hayes, and Bruce Fryxell)

■ Figure 13-12

This supernova (arrow) was seen in the outskirts of the central disk of galaxy NGC4526. From its spectrum and from the way its brightness faded with time, astronomers recognize it as a type Ia supernova produced by the collapse and total destruction of a white dwarf. (High-Z Supernova Search Team, HST, NASA)

■ Figure 13-13

Type I supernovae decline rapidly at first and then more slowly, but type II supernovae pause for about 100 days before beginning a steep decline. Supernova 1987A was odd in that it did not rise directly to maximum brightness. These light curves have been adjusted to the same maximum brightness. Generally, type II supernovae are about two magnitudes fainter than type I.

gas and dust. These are the regions where you would expect to find massive stars. The spectra of type II supernovae contain hydrogen lines, as you would expect from the explosion of a massive star that contains large amounts of hydrogen in its outer layers.

Type I supernovae show no hydrogen in their spectra, but there are two kinds of type I supernovae that have dramatically different causes. Both involve binary stars. Type Ia supernovae are often found in regions where star formation ended long ago. That means they can't be caused by massive stars, which don't live very long. Also, type Ia supernovae lack hydrogen lines in their spectra, so they can't be caused by the deaths of massive stars. Rather, the evidence shows that a type Ia occurs when a white dwarf gaining mass in a binary star system exceeds the Chandrasekhar limit and collapses.

The collapse of a white dwarf is different from the collapse of a massive star because the core of the white dwarf contains usable fuel. As the collapse begins, the temperature shoots up, but the gas cannot halt the collapse because it is degenerate, and the pressure–temperature thermostat is turned off in the core. Even as carbon fusion begins, the increased temperature cannot increase the pressure and make the gas expand and slow the reactions. The core of the white dwarf is a bomb. The carbon–oxygen core fuses suddenly in violent nuclear reactions called **carbon deflagration.** The word *deflagration* means to be totally destroyed by fire, in this case by nuclear fusion. In a flicker of a stellar lifetime the entire star is consumed, and the outermost layers are blasted away in a violent explosion that at its brightest is three to six times more luminous than a type II supernova. The white dwarf is entirely destroyed. No neutron star or black hole is left behind. Of course, you would see no hydrogen lines in the spectrum of a type Ia supernova because white dwarfs contain very little hydrogen.

Type Ib supernovae are less common. Their spectra contain no hydrogen lines, but they do occur in regions where you would expect to find young stars. They are believed to occur when a massive star in a binary system loses its hydrogen-rich outer layers to its companion star. The remains of the massive star could develop an iron core and collapse, producing a supernova explosion that lacks hydrogen lines in the spectrum. Some astronomers have referred to these as "peeled" supernovae, meaning that the massive star has had its hydrogen-rich outer layers peeled away by its binary companion.

To summarize, a type II supernova is caused by the collapse of a massive star. A type Ia is caused by the collapse of a white dwarf. A type Ib is caused by the collapse of a massive star that has lost its outer envelope of hydrogen.

Much of what you have learned so far about supernovae has been based on theory, so it is time to compare theory with observations of real supernova explosions. These frequent reality checks are a distinguishing characteristic of science.

Observations of Supernovae

In 1054, Chinese astronomers saw a "guest star" appear in the constellation now known as Taurus, the bull. The star quickly became so bright it was visible in the daytime. After a month's time, it slowly faded, taking almost two years to vanish from sight. When modern astronomers turned their telescopes to the location of the guest star, they found a cloud of gas about 1.35 pc in radius, expanding at 1400 km/s. Projecting the expansion back in time, they concluded that the expansion must have begun about nine centuries ago, just when the guest star made its visit. From this and other evidence, astronomers conclude that the nebula, now called the Crab Nebula because of its shape (■ Figure 13-14), marks the site of the 1054 supernova.

The Crab Nebula contains a critical clue. The glowing filaments appear to be excited gas flung outward by the explosion, but the hazy glow in the inner nebula is something else. Radio observations show that the gas in the nebula is emitting **synchrotron radiation**—electromagnetic energy radiated by high-speed electrons spiraling through a magnetic field. Low-speed electrons radiate at longer wavelengths, and high-speed electrons radiate at shorter wavelengths, so synchrotron radiation is spread over a wide range of wavelengths. The foggy glow of light in the Crab Nebula is synchrotron radiation at the very short wavelengths of visible light, and that must mean the electrons are traveling at tremendous speeds. In the nine centuries since the Crab supernova explosion, the electrons should have radiated their energy away and slowed down. Evidently they have not, so there must be an energy source in the Crab Nebula that is producing very-high-speed electrons. You will follow this clue in the next chapter and discover a neutron star at the center of the Crab Nebula.

The Crab Nebula is important because it is linked to a supernova explosion that was actually seen to occur. Supernovae are rare. Only a few have been seen with the naked eye in recorded history. Arab astronomers saw one in 1006, and the Chinese saw the Crab supernova in 1054. European astronomers observed two—one in 1572 (Tycho's supernova) and one in 1604 (Kepler's supernova). Also, the guest stars of 185, 386, 393, and 1181 may have been supernovae. For 383 years following 1604, no naked-eye supernova appeared, and then in February 1987 a star in the southern sky exploded. You will meet the great supernova of 1987 later in this section, but for now let's continue to explore supernovae in general.

Most supernovae are discovered in distant galaxies, and searching for these supernovae was once a tedious job. Robotic telescopes now observe long lists of galaxies each night searching for the appearance of a new star. Detecting supernovae early is important because astronomers need to study the early stages of the explosion in order to understand the complex processes that occur in the rapidly expanding cloud of gas.

The supernova explosion fades to obscurity in a year or two, but an expanding shell of gas marks the explosion site. The gas, originally expelled at 10,000 to 20,000 km/s, may carry away one-fifth of the mass of the star. The collision of that expanding gas with the surrounding interstellar medium can sweep up even more gas and excite it to produce a **supernova remnant,** the nebulous remains of a supernova explosion.

Supernova remnants look quite delicate and do not survive very long—a few tens of thousands of years—before they gradually mix with the interstellar medium and vanish. The Crab Nebula is a young remnant, only about 950 years old, and it isn't very large, only a few parsecs in diameter. Older remnants can be larger. Some supernova remnants are detectable only at radio and X-ray wavelengths. They have become too tenuous to emit detectable light, but the collision of the expanding hot gas with the interstellar medium can generate radio and X-ray radiation and allows astronomers to create images of them at these nonvisible wavelengths. In general, supernova remnants are tenuous spheres of gas expanding into the interstellar medium (■ Figure 13-15). You saw in Chapter 11 that the compression of the interstellar medium by expanding supernova remnants can trigger star formation.

The Great Supernova of 1987

Until 1987, astronomers had never looked at a bright supernova through a telescope. Nearly all supernovae are in distant galaxies and thus are very faint. The last supernova visible to the naked eye, Kepler's supernova of 1604, predated the invention of the telescope. Then, in late February 1987, the news raced around the world: Astronomers in Chile had discovered a naked-eye supernova in the Large

The Crab Nebula

Filaments of gas rush away from the site of the supernova of 1054 AD

Glow produced by synchrotron radiation.

Visual-wavelength image

Photons

Magnetic line of force

Path of electron

■ **Figure 13-14**

The Crab Nebula is located in the constellation Taurus the bull, just where Chinese astronomers saw a brilliant guest star in AD 1054. Over tens of years, astronomers can measure the motions of the filaments as they expand away from the center. Doppler shifts confirm that the near side of the nebula is moving toward us. The foggy glow is synchrotron radiation produced by high-speed electrons moving through a magnetic field. (ESO)

The supernova remnant called the Cygnus Loop is 5000 to 10,000 years old and 80 ly in diameter.

Visible light produced by gas expanding into surrounding interstellar medium

Supernova remnant N132D is 3000 years old and 80 ly in diameter. It is 180,000 ly from Earth.

Shock waves in expanding gas heat it to millions of degrees, and it emits X rays.

Cassiopeia A (Cas A) is a supernova remnant about 300 years old and 10 ly in diameter.

X-ray image

Visual-wavelength image

Cas A is also bright at radio wavelengths.

Infrared radiation from dust condensing out of the gas

Infrared image

Radio image

■ **Figure 13-15**

A supernova remnant is an expanding bubble of hot gas created by a supernova explosion. As the remnant expands and pushes into neighboring gas, it can emit radiation at many wavelengths. (Cygnus Loop: Caltech; N132D: NASA/SAO/CXC; IR: 2MASS; Radio: NRAO/AUI/NSF)

Magellanic Cloud, a small galaxy very near our Milky Way Galaxy (■ Figure 13-16). Because the supernova was only 20 degrees from the south celestial pole, it could be studied only from southern latitudes. It was named SN1987A to denote the first supernova discovered in 1987.

The hydrogen-rich spectrum suggested that the supernova was a type II, caused by the collapse of the core of a massive star. As the months passed, however, the light curve proved to be odd (Figure 13-13) in that it paused for a few weeks before rising to its final maximum. From photographs of the area made some years before, astronomers were able to determine that the star that exploded, cataloged as Sanduleak −69°202, was not the ex-

pected red supergiant but rather a hot, blue supergiant of only 20 solar masses and 50 solar radii, not extreme for a supergiant. Theorists now believe that the star was chemically poor in elements heavier than helium and had consequently contracted and heated up after a phase as a cool, red supergiant, during which it lost mass into space. The relatively small size of the supergiant may explain the pause in the light curve. Much of the energy of the explosion went into blowing apart the smaller, denser-than-usual star and making it expand.

The brightening of the supernova after the first few weeks seems to have been caused by the decay of radioactive nickel into cobalt. Theory predicts the production of such nickel atoms in the

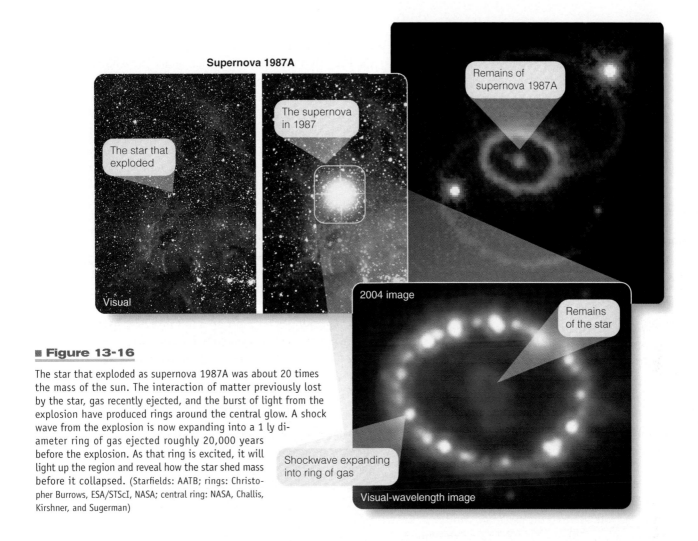

Supernova 1987A

The star that exploded

The supernova in 1987

Remains of supernova 1987A

Visual

2004 image

Remains of the star

Shockwave expanding into ring of gas

Visual-wavelength image

■ **Figure 13-16**

The star that exploded as supernova 1987A was about 20 times the mass of the sun. The interaction of matter previously lost by the star, gas recently ejected, and the burst of light from the explosion have produced rings around the central glow. A shock wave from the explosion is now expanding into a 1 ly diameter ring of gas ejected roughly 20,000 years before the explosion. As that ring is excited, it will light up the region and reveal how the star shed mass before it collapsed. (Starfields: AATB; rings: Christopher Burrows, ESA/STScI, NASA; central ring: NASA, Challis, Kirshner, and Sugerman)

explosion, and their decay into cobalt would release gamma rays that would heat the expanding shell of gas and make it brighter. About 0.07 solar mass of nickel was produced, about 20,000 times the mass of Earth.

The cobalt atoms are also unstable, but they decay more slowly, so it was not until sometime later, after much of the nickel had decayed, that the decay of cobalt into iron began providing energy to keep the expanding gas hot and luminous. Although these processes had been predicted, they were clearly observed in SN1987A. Gamma rays from the decay of cobalt to iron were detected, and cobalt and iron are clearly visible in the infrared spectra of the supernova.

As the supernova dimmed, photographs revealed bright rings of gas (Figure 13-16). Comparison with mathematical models shows that the rings were produced by two stellar winds. When the star was a red supergiant, it expelled a slow stellar wind. Later, when it became a blue supergiant, it expelled a fast stellar wind. The interaction of these two winds shaped by a magnetic field and illuminated by the supernova explosion produces the rings. Compare this with the rings seen in some of the planetary nebulae shown on page 291. In more recent years, a shock wave from

the supernova explosion has begun having an impact on the bright ring of gas and exciting it to glow brightly, as shown at the bottom of Figure 13-16.

Two independent observations confirm that SN1987A probably gave birth to a neutron star. Theory predicts that the collapse of a massive star's core should liberate a tremendous blast of neutrinos that leave the star hours before the shock wave from the interior blows the star apart. Two independent neutrino detectors, one in Ohio and one in Japan, recorded a burst of neutrinos passing through Earth at 2:35:41 AM EST on February 23, 1987, about 18 hours before the supernova was seen. The data show that the neutrinos rushed toward Earth from the direction of the supernova. The detectors caught only 19 neutrinos during a 12-second interval, but recall that neutrinos hardly ever react with normal matter. Only 19 were detected, but the full flood of neutrinos must have been immense. Within a few seconds of that time, roughly 20 trillion neutrinos passed harmlessly through each human body on Earth. The detection of the neutrino blast confirms that the collapsing core gave birth to a neutron star.

The expanding gas shell of the supernova will continue to thin and cool, and eventually astronomers on Earth will be able to peer inside and see what is left of Sanduleak −69°202. Will they see the expected neutron star, or will theories need further revision? For the first time in almost 400 years, astronomers can observe a bright supernova to test their theories.

Local Supernovae and Life on Earth

Although supernovae are rare events, they are very powerful and could affect life on planets orbiting nearby stars. In fact, supernovae explosions long ago may have affected Earth's climate and the evolution of life.

If a supernova occurred within about 50 ly of Earth, the human race would have to abandon the surface and live below ground for at least a few decades. The burst of gamma rays and high-energy particles from the supernova explosion could kill many life forms and cause serious genetic damage in others. The only way to avoid this radiation would be to move our population into tunnels below Earth's surface. Of course, if a supernova did occur, there would not be time to dig enough tunnels.

Even if you could survive in tunnels long enough for the radioactivity on Earth's surface to subside, you might not like Earth when you emerged. Genetic mutation induced by radioactivity might alter plant and animal life so seriously that the human population could not be supported. After all, humans depend almost totally on grass for food. The basic human foods—milk, butter, eggs, wheat, tomatoes, lettuce, and meat (Big Macs, in other words)—are grass and similar vegetation processed into different forms. Seafood is merely processed ocean plankton and plant life, which might also be altered or damaged by a local supernova explosion. Even if surface life survived the radiation, damage to the delicate upper layers of our atmosphere might alter the climate dramatically.

Local supernovae have been suggested as a possible cause of occasional climate changes and extinctions in Earth's past. It is quite possible that, as the sun moves through space, supernovae explode near Earth every few hundred million years. A really close call could trigger an extinction.

But do supernovae really affect Earth? In 1979, scientists studying Antarctic ice cores containing very old ice noticed that ice from the years 1181, 1320, 1572, and 1604 contains high abundances of nitrates. Bright supernovae were seen in the years 1181, 1572, and 1604. Ultraviolet radiation from those supernovae might have altered Earth's upper atmosphere and produced acid rain that was recorded in the ice as nitrates. But what of the ice from 1320? In 1999, X-ray astronomers noticed a supernova remnant in the southern sky. When they calculated its age, they found the supernova should have occurred about 1320. It isn't clear why there is no historical record of this supernova or why the nitrates appear in Antarctic ice cores but not in Greenland ice cores. In any case, this evidence may be a warning that even distant supernova explosions can affect Earth in subtle ways.

A study of cores from the seafloor has revealed a deposit of the isotope iron-60 in a layer that was laid down at the time of the Pliocene-Pleistocene extinction about two million years ago. Iron-60 is produced in supernova explosions and has a half-life of only 1.5 million years. To reach Earth before it decayed, this iron must have been produced in a supernova explosion no more than 100 ly away. Some scientists wonder if the Pliocene-Pleistocene extinction was caused by a nearby supernova explosion.

Earth seems to be fairly safe at the moment. No star within 50 ly is known to be a massive giant capable of exploding as a supernova. You need to worry only about nearby supernova explosions, and massive stars are so rare and so easy to locate that you can be sure that no type II supernova explosion will occur nearby for the time being.

Type Ia supernovae are caused by collapsing white dwarfs, and white dwarfs are both very common and very hard to locate. There is no way to be sure that a white dwarf teetering on the edge of the Chandrasekhar limit does not lurk near us in space. You can take comfort, however, in the extreme rarity of supernova explosions. Earth faces greater dangers from human ignorance than from exploding stars.

Building Scientific Arguments

Why does a type II supernova explode?

For this argument you need to review theory rather than evidence. A type II supernova occurs when a massive star reaches the end of its usable fuel and develops an iron core. The iron is the final ash produced by nuclear fusion, and it cannot fuse because iron is the most tightly bound nucleus. When energy generation begins to fall, the star contracts, but because iron can't ignite there is no new energy source to stop the contraction. In seconds, the core of the star falls inward, and a shock wave moves outward. Aided by a flood of neutrinos and sudden convective turbulence, the shock wave blasts the star apart; seen from Earth, it brightens as its surface gases expand into space.

Now revise your argument to consider evidence. **What evidence can you cite to support the theory that the core of a type II supernova can form a neutron star?**

■   ■   ■

Connections: The study of the deaths of stars has led you to discover astonishing objects of unbelievable density, temperature, and violence—all consequences of the victory of gravity over matter. But you have not considered the strangest circumstance of all. What happens when degenerate matter can't support the weight of the dying star—that is, when the mass of the compact object exceeds the Chandrasekhar limit? The answer is in the title of the next chapter.

Summary

13-1 | Lower-Main-Sequence Stars

How will the sun die?

■ Red dwarfs less massive than about 0.4 solar mass are completely mixed and will have very little hydrogen left when they die. They cannot ignite a hydrogen-fusion shell, so they cannot become giant stars. They will remain on the main sequence for many times the present age of the universe.

■ Medium-mass stars between about 0.4 and 4 solar masses, including the sun, become cool giants and fuse helium but cannot fuse carbon.

■ Medium-mass stars lose mass into space while they are giant stars and then collapse to become white dwarfs.

■ If the collapsed star becomes hot enough, it can ionize the gas it has ejected, first as a slow wind and then as a fast wind to turn on a planetary nebula.

Why are there so many white dwarfs?

■ White dwarfs are the cores of sunlike stars that have collapsed until their gas becomes degenerate. They have no nuclear fuels and slowly cool off; they will eventually become black dwarfs.

■ A white dwarf cannot support its own weight by degenerate pressure if its mass is greater than the Chandrasekhar limit of 1.4 solar masses.

■ Medium-mass stars up to about 8 solar masses can probably lose enough mass to die as white dwarfs. This explains the large number of white dwarfs found in space.

13-2 | The Evolution of Binary Stars

What happens if an evolving star is in a binary system?

■ Two stars orbiting each other control two regions of space called Roche lobes. The Lagrangian points mark places of stability in the system, and mass can flow from one star to the other through the inner Lagrangian point.

■ Close binary stars evolve in complex ways because they can transfer mass from one star to the other. This explains why some binary systems contain a main-sequence star more massive than its giant companion—the Algol paradox.

■ Mass is transferred from an evolving star through an accretion disk around the receiving star. Accretion disks can become hot enough to emit light and even X rays.

■ Mass transferred onto the surface of a white dwarf can build up a layer of fuel that erupts in a nova explosion. A white dwarf can erupt repeatedly so long as mass transfer continues to form new layers of fuel.

13-3 | The Deaths of Massive Stars

How do massive stars die?

■ Stars more massive than about 8 solar masses cannot lose mass fast enough to reduce their mass low enough to die by ejecting a planetary nebula and collapsing into a white dwarf. Such massive stars must die more violent deaths.

■ The massive stars on the upper main sequence fuse nuclear fuels up to iron but cannot generate further nuclear energy because iron is the most tightly bound of all atomic nuclei. When a massive star forms an iron core, the core collapses and triggers a supernova explosion known as a type II supernova.

■ A type Ia supernova can occur when mass transferred onto a white dwarf pushes it over the Chandrasekhar limit and it collapses suddenly, fusing all of its carbon at once.

■ A type Ib supernova occurs when a massive star loses its outer layers of hydrogen in a binary system. Later its iron core can collapse, and it can explode as a type Ib supernova.

■ The spectra of type II supernovae contain hydrogen lines, but the spectra of type Ia and type Ib supernovae do not.

■ Supernova remnants are the expanding shells of gas ejected by supernova explosions. They are filled with hot gas. The Crab Nebula is a supernova remnant formed by the supernova of AD 1054.

■ The supernova of 1987 is believed to have been produced by a peculiar type II supernova explosion. Neutrinos were detected and are evidence that the iron core collapsed and presumably formed a neutron star. The nebula the supernova produced is still developing.

New Terms

nova (p. 284)	Roche surface (p. 293)
supernova (p. 284)	Lagrangian point (p. 293)
thermal pulse (p. 287)	accretion disk (p. 294)
planetary nebula (p. 288)	supernova (type I) (p. 299)
compact object (p. 289)	supernova (type II) (p. 299)
black dwarf (p. 289)	carbon deflagration (p. 300)
Chandrasekhar limit (p. 289)	synchrotron radiation (p. 301)
Roche lobe (p. 293)	supernova remnant (p. 301)

Review Questions

Ace⊕Astronomy™ Assess your understanding of this chapter's topics with additional quizzing and animations at **http://ace** **.brookscole.com/sf9**

1. Why can't the lowest-mass stars become giants?

2. Presumably, all the white dwarfs in our galaxy were produced by sunlike stars of medium mass. Why couldn't any of these white dwarfs have been produced by the deaths of the lowest-mass stars?

3. What evidence can you cite that stars can lose mass?

4. What kind of spectrum does the gas in a planetary nebula produce? Where does it get the energy to radiate?

5. The coolest stars at the center of planetary nebulae are about 25,000 K. Why don't astronomers see planetary nebulae containing cooler central stars? (*Hint:* What kind of photons excite the gas in a planetary nebula?)

6. As white dwarfs cool, they move toward the lower right in the H–R diagram (Figure 9-10 and page 291), maintaining constant radius. Why don't they contract as they cool?

7. All white dwarfs are about the same mass. Why?

8. Why have no white dwarfs cooled to form black dwarfs in our galaxy?

9. What happens to a star in a close binary system when it becomes a giant?

10. Why do novae repeat while supernovae do not?

11. Why can't massive stars generate energy from iron fusion?

12. How could you use spectra to tell the difference between a type I supernova and a type II supernova? Why does this difference arise?

13. What processes produce type I and type II supernovae?

14. Why do supernova remnants emit X rays?

15. What processes caused a medium-mass star to produce the nebula at the right? The nebula is now about 0.1 ly in diameter and still expanding. What will happen to it?

(NASA/Hubble Heritage Team/STScI/AURA)

16. The image at right combines X-ray (blue), visible (green), and radio (red) images. Observations show the sphere is expanding at a high speed and is filled with very hot gas. What kind of object produced this nebula? Roughly how old do you think it must be?

(NASA/CAC/SAO/ CSIRO/ATNF/ATCA)

Discussion Questions

1. How certain can you be that the sun will never explode as a supernova? What does it mean when a scientist uses the word "certainty"?

2. Humans who lived for a billion years might notice a star that exploded occasionally as a nova and finally destroyed itself in a supernova explosion. What could account for this evolution?

Problems

1. Use the formula in Chapter 12 to compute the life expectancy of a 0.4-solar-mass star. Why might this be an underestimate if the star is fully mixed?

2. The Ring Nebula in Lyra is a planetary nebula with an angular diameter of 76 seconds of arc and a distance of 5000 ly. What is its linear diameter? (*Hint:* Use the small-angle formula.)

3. If the Ring Nebula is a light-year in diameter and is expanding at a velocity of 15 km/s, typical of planetary nebulae, how old is it? (*Hint:* 1 ly = 9×10^{12} km, and 1 y = 3.15×10^7 s.)

4. Suppose that a planetary nebula is 1 pc in diameter and the Doppler shifts in its spectrum show that it is expanding at 30 km/s. How old is it? (*Hint:* See Problem 3.)

5. A planetary nebula photographed 20 years ago and then photographed again today has increased its radius by 0.6 second of arc. If Doppler shifts in its spectrum show that it is expanding at a velocity of 20 km/s, how far away is it? (*Hint:* First figure out how many parsecs it has increased in radius in 20 years. Then use the small-angle formula.)

6. The Crab Nebula is now 1.35 pc in radius and is expanding at 1400 km/s. About when did the supernova occur?

7. The Cygnus Loop is now 2.6° in diameter and lies about 500 pc distant. If it is 10,000 years old, what was its average velocity of expansion? (*Hint:* Find its radius in parsecs first.)

8. The supernova remnant Cassiopeia A (Cas A) is expanding in radius at a rate of about 0.5 second of arc per year. Doppler shifts show that the velocity of expansion is about 5700 km/s. How far away is the nebula?

9. Cassiopeia A has a radius of about 2.5 minutes of arc. If it is expanding at 0.5 second of arc per year, when did the supernova explosion occur? (There is no record of a supernova being seen at that time.)

Media Cluster

Ace ☉ Astronomy™   To access the resources in the Media Cluster, log into AceAstronomy at **http://ace.brookscole.com/sf9** and select Chapter 13.

Inside Stars

You can use the laws of physics to study the interior of the sun and compare it to other stars in the universe. This cutaway animation of the sun lets you explore the various structures of stars, understand their inner workings, and view photos and movies of the real thing.

Lab 12: Binary Stars

This lab investigates some things that can be learned from different types of binary stars. Newton's form of Kepler's third law is used to determine stellar masses. You also see how the Doppler effect can be used in the study of binary systems.

Lab 13: Stellar Explosions, Novae, and Supernovae

This lab investigates two kinds of explosions that highly evolved stars can produce, novae and type Ia supernovae, and how the latter can be used to estimate the distance to faraway galaxies.

Critical Inquiries for the Web

1. As seen on pages 290 and 291, planetary nebulae show an incredible diversity in appearance. Browse the Web for images and information on these dying stars and discuss why astronomers see such a range of shapes of planetary nebulae.

2. Naked-eye supernovae in our galaxy are rare, but astronomers have noted supernovae in other galaxies for years. Search the Web for summaries of observations of recent supernovae. Are similar numbers of type I and type II supernovae being seen? Compare the number of supernovae seen during the last few years with that of two decades ago. Why are astronomers finding so many more supernovae in recent years than in the past?

3. The word *nova* is used in many ways, so you must be thoughtful to search out photos and observations of recently discovered novae. Limit your search to astronomy categories and also try pages about variable stars.

4. What is U Scorpii?

Exploring *TheSky*

1. Locate the planetary nebulae M57, M97, and M27. How does their shape distinguish them from the star-formation nebulae such as M42 and M8? (*Hint:* To find an object, use **Find** under **Edit.** Choose **Messier Objects** and pick from the list.)

2. The Crab Nebula is M1. Locate it, zoom in, measure its angular size in seconds of arc, and compute its diameter assuming it is about 6000 ly from Earth.

3. Locate the supernova remnant called the Cygnus Loop just south of ε Cygni. How big is this object in angular diameter compared to the full moon? (*Hint:* Under the **View** menu, choose **Labels** and **Setup.** Check **Bayer Designation,** go to Cygnus, and zoom in on ε Cygni until the Cygnus Loop appears.)

 Go to the Brooks/Cole Astronomy Resource Center **(http://astronomy.brookscole.com)** for critical thinking exercises, articles, and additional readings from InfoTrac College Edition, Brooks/Cole's online student library.

14 | Neutron Stars and Black Holes

Almost anything is easier to get into than out of.

AGNES ALLEN

X-ray image

G RAVITY ALWAYS WINS. In a star's struggle to withstand its own gravity, the star must eventually exhaust its fuel, and gravity must win. Gravity ensures that the star's last remains must eventually reach one of three final states—white dwarf, neutron star, or black hole. You studied the first of these compact objects in the previous chapter, and now you are ready to complete the story of the stars. ▌ Theory predicts that neutron stars and black holes exist, but science depends on evidence. Can astronomers find objects in the sky that are real neutron stars and black holes? Scientists always fall back on evidence—the final reality check on their theories.

▌ Continued on page 310 ▌

Yielding finally to its own gravity, a massive star collapsed and died, leaving behind this supernova remnant 40 ly in diameter. A neutron star located at the center is about the size of a mountain on Earth and emits 30,000 times more energy than the sun. (NASA/CXC/SAO)

Guidepost

Looking Back

Through the last three chapters you have traced the story of stars from their birth as clouds of gas in the interstellar medium to their final collapse. By now you are asking a simple question, "What is left?" The answer, of course, depends on the mass of the star. Stars like the sun leave behind white dwarfs, but more massive stars leave behind the strangest beasts in the cosmic zoo.

This Chapter

Now you are ready to meet neutron stars and black holes, and your exploration will answer five essential questions:

How did scientists predict the existence of neutron stars?

What is the evidence that neutron stars really exist?

How did scientists predict the existence of black holes?

What is the evidence that black holes really exist?

What happens when matter falls into a neutron star or black hole?

You can tell from the first four questions above that this chapter will show you clear examples of how astronomers combine theory and evidence to understand nature.

Looking Ahead

This chapter ends the story of individual stars, but it does not end the story of stars. In the next chapter, you will begin exploring the giant communities in which stars live—the galaxies.

A **NEUTRON STAR** is a star of a little over 1 solar mass compressed into a radius of about 10 km. Its density is so high that the matter is stable only as a fluid of neutrons. Theory predicts that such an object would spin a number of times a second, be nearly as hot at its surface as the inside of the sun, and have a magnetic field a trillion times stronger than Earth's (■ Figure 14-1). Two questions should occur to you immediately. First, how could any theory predict such a wondrously unbelievable star? And second, do such neutron stars really exist?

Theoretical Prediction of Neutron Stars

The neutron was discovered in the laboratory in 1932, and its properties suggested something fantastic. Neutrons spin in much the way that electrons do, which means that neutrons must obey the Pauli exclusion principle. In that case, if neutrons are packed together tightly enough, they can become degenerate just as electrons do. White dwarfs are supported by degenerate electrons; and, in 1932, the Russian physicist Lev Landau predicted that neutron stars might exist supported by degenerate neutrons. Of course, the inside of a neutron star would have to be much denser than the inside of a white dwarf.

Only two years later, in 1934, Walter Baade and Fritz Zwicky provided a pedigree for neutron stars. Not knowing of Landau's prediction, they suggested that some of the most luminous novae

■ Figure 14-1

A supernova explosion seen in AD 1181 left behind an expanding supernova remnant. The Chandra X-Ray Observatory has imaged the nebula in X rays and finds a tiny hot object within—a neutron star. (NASA/SAO/CXC/P. Slane et al.)

X-ray images

in the historical record were not true novae but were caused by the collapse of a massive star in an even larger explosion they called a supernova. What was left of the core of the star, they proposed, was a small, high-density neutron star.

Atomic physics provides an explanation of how the collapsing core of a massive star could form a neutron star. If the collapsing core is more massive than the Chandrasekhar limit of 1.4 solar masses, then it cannot reach stability as a white dwarf. The weight is too great to be supported by degenerate electrons. The collapse of the core continues, and the atomic nuclei are broken apart by gamma rays. Almost instantly, the increasing density forces the freed protons to combine with electrons and become neutrons by the emission of neutrinos:

$$e + p \rightarrow n + \nu$$

The burst of neutrinos, as you saw in the previous chapter, helps blow the star apart; and the core of the star, in a fraction of a second, is transformed into a dense ball of neutrons left behind by the supernova explosion as a neutron star.

As you saw in the previous chapter, a star of 8 solar masses or less could lose enough mass to die as a planetary nebula leaving behind a white dwarf. More massive stars, up to a limit of about 15 solar masses, will lose mass rapidly, but they cannot reduce their mass fast enough and will apparently face death as a supernova explosion leaving behind a neutron star (■ Figure 14-2). (You will see later in this chapter that the most massive stars probably leave behind black holes.)

Theoretical calculations predict that a neutron star will be only 10 or so kilometers in radius (■ Figure 14-3) and will have a density of about 10^{14} g/cm^3. On Earth, a sugar-cube-sized lump of this material would weigh 100 million tons. This is roughly the density of the atomic nucleus, and you can think of a neutron star as matter with all of the empty space squeezed out of it.

How massive can a neutron star be? That is a critical question, and it's a difficult one to answer because physicists don't know the strength of pure neutron material. Such matter can't be made in the laboratory, so its properties must be predicted theoretically. The most widely accepted calculations suggest that a neutron star cannot be more massive than 2 to 3 solar masses. If a neutron star were more massive than that, the degenerate neutrons would not be able to support the weight, and the object would collapse (presumably into a black hole).

Simple physics, the physics you have used in previous chapters to discuss normal stars, predicts that neutron stars should be hot, spin rapidly, and have strong magnetic fields. You have seen that contraction heats the gas in a star. As the gas particles fall inward, they pick up speed; and, when they collide, their high speeds become thermal energy. The sudden collapse of the core of a massive star to a radius of 10 km should heat it to millions of degrees. Furthermore, neutron stars should cool slowly because the heat can escape only from the surface, and neutron stars are so small they have little surface from which to radiate.

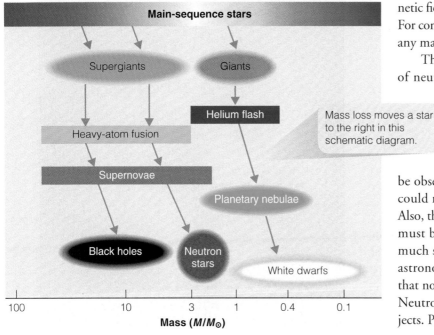

Mass loss moves a star to the right in this schematic diagram.

■ Figure 14-2

How a star evolves depends on its mass. Mass loss can change a star's fate by reducing its mass as it evolves. The mass limits of the categories shown here are not well known and are given only for purposes of illustration. (Figure design by author)

In this way, basic theory predicts that neutron stars should be very hot.

The conservation of angular momentum predicts that neutron stars should spin rapidly. All stars rotate to some extent because they form from swirling clouds of interstellar matter. As such a star collapses into a neutron star, it must rotate faster because it conserves angular momentum. Recall that you see this happen when ice skaters spin slowly with their arms extended and then speed up as they pull their arms closer to their bodies (see Figure 5-7). In the same way, a collapsing star must spin faster as it pulls its matter closer to its axis of rotation. If the sun collapsed to a radius of 10 km, its period of rotation would decrease from 25 days to about 0.001 second. You might expect the collapsed core of a massive star to rotate 10 or 100 times a second.

Basic theory also predicts that a neutron star should have a powerful magnetic field. Remember from your study of the sun's atmosphere in Chapter 8 that a magnetic field passing through an ionized gas is "frozen in." Whatever magnetic field a star has is frozen into the gas of the star. When the star collapses, the magnetic field is carried along and squeezed into a smaller area, which could make the field a billion times stronger. Some stars have magnetic fields over 1000 times stronger than the sun's, so you might expect a neutron star to have a mag-

netic field as much as a trillion times stronger than the sun's. For comparison, that is about 10 million times stronger than any magnetic field ever produced in the laboratory.

Theory allowed astronomers to predict the properties of neutron stars, but it also predicted that such objects should be difficult to observe. Neutron stars are very hot, so Wien's law of black body radiation (Chapter 7) told astronomers that neutron stars would radiate most of their energy in the X-ray part of the spectrum, radiation that could not be observed in the 1940s and 1950s because astronomers could not put their telescopes above Earth's atmosphere. Also, the small surface areas of neutron stars mean that they must be faint objects. They are hot, but they don't have much surface area from which to radiate. Consequently, astronomers of the mid-20th century were not surprised that none of the newly predicted neutron stars were found. Neutron stars were, at that point, entirely theoretical objects. Progress came not from theory but from observation.

The Discovery of Pulsars

In November 1967, Jocelyn Bell, a graduate student at Cambridge University in England, found a peculiar pattern on the paper chart from a radio telescope. Unlike other radio signals from celestial bodies, this was a series of regular pulses (■ Figure 14-4). At first she and the leader of the project, Anthony Hewish, thought the signal was interference, but they found it day after day in the same place in the sky. Clearly, it was celestial in origin.

■ Figure 14-3

A tennis ball and a road map illustrate the relative size of a neutron star. Such an object, containing slightly more than the mass of the sun, would fit with room to spare inside the beltway around Washington, DC. (M. Seeds)

■ **Figure 14-4**

The 1967 detection of regularly spaced pulses in the output of a radio telescope led to the discovery of pulsars. This record of the radio signal from the first pulsar, CP1919, contains regularly spaced pulses (marked by ticks). The period is 1.33730119 seconds.

Another possibility, that it was a radio signal from a distant civilization, led them to consider naming it LGM for Little Green Men. But within a few weeks, the team found three more objects in other parts of the sky, pulsing with different periods. The objects were clearly natural, and the team dropped the name LGM in favor of **pulsar**—a contraction of *pulsing star*. The pulsing radio source Bell had observed with her radio telescope was the first known pulsar.

As more pulsars were found, astronomers argued over their nature. Periods ranged from 0.033 to 3.75 seconds and were nearly as exact as an atomic clock. Months of observation showed that many of the periods were slowly growing longer by a few billionths of a second per day. Whatever produced the regular pulses had to be highly precise, nearly as exact as an atomic clock, but it had to gradually slow down.

It was easy to eliminate possibilities. Pulsars could not be stars. A normal star, even a small white dwarf, is too big to pulse that fast. Nor could a star with a hot spot on its surface spin fast enough to produce the pulses. Even a small white dwarf would fly apart if it spun 30 times a second.

The pulses themselves gave the astronomers a clue. The pulses last only about 0.001 second. This places an upper limit on the size of the object producing the pulse. If a white dwarf blinked on and then off in that interval, astronomers would not see a 0.001-second pulse. The near side of the white dwarf would be about 6000 km closer to Earth, and light from the near side would arrive 0.022 seconds before the light from the bulk of the white dwarf. In that way, its short blink would be smeared out into a longer pulse. This is an important principle in astronomy—an object cannot change its brightness appreciably in an interval shorter than the time light takes to cross its diameter. If pulses from pulsars are no longer than 0.001 second, then the objects cannot be larger than about 300 km (190 miles) in diameter and could be smaller.

Only a neutron star is small enough to be a pulsar. In fact, a neutron star is so small, it can't vibrate slowly enough, but it can spin as fast as 1000 times a second without flying apart. The missing link between pulsars and neutron stars was found in late 1968, when astronomers discovered a pulsar at the heart of the Crab Nebula (■ Figure 14-5). The Crab Nebula is a supernova rem-

■ **Figure 14-5**

The pulsar at the center of the Crab Nebula (arrow) is detectable in visual-wavelength photographs. The star to the right of the pulsar lies much closer to Earth and is not in the Crab Nebula. The white box outlines the area imaged on page 315. (Caltech)

nant, and theory predicts that some supernovae leave behind a neutron star.

The short pulses and the discovery of the pulsar in the Crab Nebula are strong evidence that pulsars are neutron stars. If you combine theory and observation, you can devise a model of a pulsar.

A Model Pulsar

Scientists often work by building a model of a natural phenomenon—not a physical model made of plastic and glue but an intellectual conception of how nature works in a specific instance. The model may be limited and incomplete, but it helps them organize their theories and observations. A model of a pulsar will help you draw together a lot of different ideas.

The modern model of a pulsar has been called the **lighthouse model** and is shown in **The Lighthouse Model of a Pulsar** on pages 314–315. Notice three important points:

1 A pulsar does not pulse but rather emits beams of radiation that sweep around the sky as the neutron star rotates. If the beams do not sweep over Earth, the pulses will not be detectable by Earth's radio telescopes.

2 Notice that the mechanism that produces the beams involves extremely high energies and is not fully understood.

3 Also notice how modern space telescopes observing at nonvisual wavelengths can help confirm and refine the model.

■ Figure 14-6

The effects of pulsar winds can be seen at X-ray wavelengths. The high-energy gas of the winds is sometimes detectable, as is the interaction of the winds with surrounding gas. Not all pulsars have detectable winds. (NASA/CXC/SAO/ U. Mass; F. Lu/McGill; V. Kaspi)

Neutron stars are not simple objects, and modern astronomers need both general relativity and quantum mechanics to try to understand them. Nevertheless, the life story of pulsars can be understood in terms of the lighthouse model.

The model of a pulsar as a spinning neutron star won the support of many astronomers because of two properties of pulsars. First, many pulsars are slowing down. Their periods are increasing by a few billionths of a second each day—a change radio astronomers can measure using atomic clocks. Evidently, the spinning neutron star is converting some of its energy of rotation into various kinds of electromagnetic energy and a powerful outflow of high-speed particles called a **pulsar wind** (■ Figure 14-6). About 99.9 percent of the energy released by the slowing of the neutron star is carried away by the pulsar wind, and only 0.1 percent goes into producing the radio beams. The energy that keeps the Crab Nebula glowing nearly 1000 years after the explosion is coming from the rotational energy of the neutron star and the pulsar wind that the neutron star produces.

The second property of pulsars that supports the neutron-star model of a pulsar is the **glitch**—a sudden increase in the pulse rate seen in some pulsars (■ Figure 14-7). Two theories have been

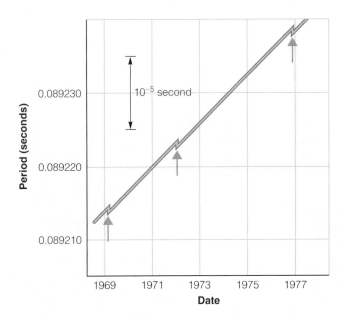

■ Figure 14-7

Soon after pulsars were discovered, radio astronomers accumulated enough data to show that pulsars were very gradually slowing down. That is, their pulses were growing longer. Some pulsars, such as the Vela pulsar whose data are shown here, experience glitches in which the pulsar suddenly speeds up only to resume its more leisurely decline.

The Lighthouse Model of a Pulsar

1 Astronomers think of pulsars not as pulsing objects, but rather as objects emitting beams. As they spin, the beams sweep around the sky; when a beam sweeps over Earth, observers detect a pulse of radiation. Understanding the details of this lighthouse model is a challenge, but the implications are clear. Although a neutron star is only a few kilometers in radius, it can produce powerful beams. Also, observers tend to notice only those pulsars whose beams happen to sweep over Earth.

In this artist's conception, gas trapped in the neutron star's magnetic field is excited to emit light and outline the otherwise invisible magnetic field.

Beams of electromagnetic radiation would probably be invisible unless they excited local gas to glow.

What color should an artist use to paint a neutron star? With a temperature of a million degrees, the surface emits most of its electromagnetic radiation at X-ray wavelengths. Nevertheless, it would probably look blue-white to your eyes.

2 How a neutron star can emit beams is one of the challenging problems of modern astronomy, but astronomers have a general idea. A neutron star contains a powerful magnetic field and spins very rapidly. The spinning magnetic field generates a tremendously powerful electric field, and the field causes the production of electron–positron pairs. As these charged particles are accelerated through the magnetic field, they emit photons in the direction of their motion, which produce powerful beams of electromagnetic radiation emerging from the magnetic poles.

Neutron Star Rotation with Beams

As in the case of Earth, the magnetic axis of a neutron star could be inclined to its rotational axis.

The rotation of the neutron star will sweep its beams around like beams from a lighthouse.

While a beam points roughly toward Earth, observers detect a pulse.

While neither beam is pointed toward Earth, observers detect no energy.

Beams may not be as exactly symmetric as in this model.

Ace Astronomy™

Log into AceAstronomy and select this chapter to see the Active Figure called "Neutron Star." Adjust the inclination of the neutron star's magnetic field to produce pulses.

3 Observations at many different wavelengths help astronomers understand pulsars. The hazy glow of the Crab Nebula is produced by synchrotron radiation. In the nearly 10 centuries since the supernova, the high-speed electrons should have radiated their energy away, and the synchrotron radiation should have faded. Evidently, the Crab pulsar powers the nebula. The Hubble Space Telescope image at right shows the region boxed in Figure 14-5. Circular wisps excited by the pulsar change and flicker from day to day.

Crab pulsar

Visual-wavelength image

NASA

3a Red and yellow in this image show the radio and visual extent of the Crab Nebula. Blue, an X-ray image, reveals that the central pulsar has flung off rings of highly ionized gas in a disk up to a light-year in radius. The pulsar is ejecting jets of high-energy particles perpendicular to the disk.

Disk

Pulsar

Jet

Composite image
Radio + visual + X-ray

NASA

X-ray image of Puppis supernova remnant

Neutron star

NASA

Hubble Space Telescope visual-wavelength image

Neutron star

NASA

3b If a pulsar's beams do not sweep over Earth, observers detect no pulses, and the neutron star is difficult to find. A few such objects are known, however. The Puppis A supernova remnant is about 4000 years old and contains a point source of X rays believed to be a neutron star. The isolated neutron star in the right-hand image has a temperature of 700,000 K.

proposed to explain these changes, and both depend on the internal structure of spinning neutron stars.

One theory suggests that "starquakes" occur on the surface of a neutron star and produce glitches. To understand this theory you need to note that theoretical models of neutron stars suggest that their interiors are fluids composed mostly of neutrons. Near the surface, where the pressure is lower, matter can exist as a rigid crust composed mostly of iron nuclei about 10^{16} times stronger than steel. This crust might be a few hundred meters thick. The rapidly spinning neutron star would be slightly flattened, but as it slowed, its gravity would squeeze it and try to make it more spherical until the crust broke in a "starquake"—a neutron-star earthquake. When the crust breaks, the neutron star could become slightly less flattened, and the conservation of angular momentum would cause the neutron star to spin slightly faster. From Earth, astronomers would see a sudden increase in the pulse rate.

The starquake theory has a drawback; starquakes should occur every few thousand years on any given neutron star, but glitches are more common. The Vela pulsar has glitched over a dozen times since it was discovered. A different theory may explain most glitches. Because the liquid in the core of a neutron star can circulate with almost no friction, vortexes should develop in the fluid. Like swarms of frictionless tornadoes, they should store large amounts of angular momentum. Calculations applying quantum mechanics to these vortices show that, as the neutron star slows, swarms of vortices should suddenly transfer their angular momentum to the crust. That would make the crust spin faster and increase the pulse rate. That is, it would produce a glitch. This theory has been confirmed in laboratory experiments with rotating fluids.

A rare glitch may be caused by a starquake, but most are probably caused by changes in internal vortices. Notice, in any case, that both explanations depend on the pulsar being a spinning neutron star. Consequently, glitches give astronomers further confidence that pulsars are spinning neutron stars.

Since the discovery of the first pulsar, over 1000 have been found, and there are probably about 100 million in our galaxy. Of course, Earthbound astronomers see only those pulsars whose beams sweep over Earth. Most neutron stars are invisible from Earth. Nevertheless, astronomers can describe how neutron stars are born, how they age, and how they die.

The Evolution of Pulsars

When a pulsar first forms, it is spinning quickly, perhaps nearly 100 times a second. The energy it radiates into space comes from its energy of rotation, so as it blows away its pulsar wind and blasts beams of radiation outward, its rotation slows. The average pulsar is apparently only a few million years old, and the oldest is about 10 million years old. Presumably, older neutron stars rotate too slowly to generate detectable radio beams.

From this you can expect that a young neutron star should emit powerful beams of radiation. The Crab Nebula provides an example of such a system. Only about 950 years old, the Crab pulsar is so powerful it emits photons of radio, infrared, visible, X-ray, and gamma-ray wavelengths (■ Figure 14-8). Careful measurements of its brightness with high-speed instruments show that it blinks twice for every rotation. One beam sweeps almost

■ **Figure 14-8**

High-speed images of the Crab Nebula pulsar show it pulsing at visual wavelengths and at X-ray wavelengths. The period of pulsation is 33 milliseconds, and each cycle includes two pulses as its two beams of unequal intensity sweep over Earth. (Visual: © AURA, Inc., NOAO, KPNO; X-ray: Figure 2 from F. R. Harnden, Jr., in *The Astrophysical Journal*, 283 (August 1, 1984): 279–285. Copyright 1984 by The American Astronomical Society. Reprinted with permission of F. R. Harnden, Jr., and the AAS.)

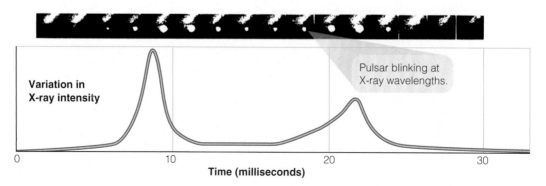

directly over Earth, and telescopes detect a strong pulse. Half a rotation later, the edge of the other beam sweeps over Earth, and telescopes detect a weaker pulse.

You would expect only the most energetic pulsars to produce short-wavelength photons and pulse at visible wavelengths. The Crab Nebula pulsar is young, fast, and powerful, and it produces visible pulses; so does a pulsar called the Vela pulsar (located in the Southern Hemisphere constellation Vela). The Vela pulsar is fast, pulsing about 11 times a second, and, like the Crab Nebula pulsar, is located inside a supernova remnant. Its age is estimated at about 20,000 to 30,000 years, young in terms of the average pulsar. This suggests that pulsars are capable of producing optical pulses only when they are young.

You might expect to find all pulsars inside supernova remnants, but the statistics must be examined with care. Not every supernova remnant contains a pulsar, and not every pulsar is located inside a supernova remnant. Many supernova remnants probably contain pulsars whose beams never sweep over Earth, and it is difficult to detect such pulsars. Also, some pulsars move through space at high velocity, which suggests that supernova explosions can occur asymmetrically, perhaps because of the violent turbulence in the exploding core. Such a violent, off-center explosion could give the neutron star a high velocity through space (■ Figure 14-9). Also, some supernovae probably occur in binary systems and fling the two stars apart at high velocity. In any case, pulsars are known to have such high velocities that many probably escape the disk of our galaxy. You should not be surprised that many neutron stars quickly leave their supernova remnant behind. And finally, you need to remember that a pulsar can remain detectable for 10 million years or so, but a supernova remnant cannot survive more than about 50,000 years before it is mixed into the interstellar medium. Consequently, you might expect

that most pulsars are not located in supernova remnants, and that most supernova remnants do not contain pulsars.

Careful analysis, however, shows that there are more empty supernova remnants than you should expect. Where are these missing pulsars? Searches suggest that many neutron stars may not be emitting powerful radio beacons. If they are born with very strong magnetic fields, they may have slowed down rapidly and may be rotating so slowly they do not emit beams of radio energy. These objects would have magnetic fields 1000 times stronger than those of a typical pulsar and have been dubbed **magnetars.** You will see more evidence that such magnetars do exist. This line of research is ongoing, but it suggests that there is much more to learn about the evolution of pulsars.

The explosion of Supernova 1987A in February 1987 apparently formed a neutron star. You can draw this conclusion because a burst of neutrinos was detected passing through Earth, and theory predicts that the collapse of a massive star's core into a neutron star would produce such a burst of neutrinos. At first the neutron star would be hidden at the center of the expanding shells of gas ejected into space; but, as the gas expands and thins, astronomers might be able to see it. Or, if its beams don't sweep over Earth, astronomers might be able to detect it from its X-ray and gamma-ray emission. Although, years after the explosion, no neutron star has been detected, astronomers continue to watch the site, hoping to see the newborn neutron star.

One reason pulsars are so fascinating is the extreme conditions found in spinning neutron stars. To see natural processes of even greater violence, you have only to look at pulsars in binary systems.

Binary Pulsars

Over a thousand pulsars are now known, and some are located in binary systems. These pulsars are of special interest because astronomers can learn more about the neutron star by studying the orbital motions of the binary. Also, in some cases, mass can flow from the companion star onto the neutron star, and that produces high temperatures and X rays.

The first binary pulsar was discovered in 1974 when astronomers Joseph Taylor and Russell Hulse noticed that the pulse period of the pulsar PSR 1913+16 was changing. The period first grew longer and then grew shorter in a cycle that took 7.75 hours. Thinking of the Doppler shifts seen in spectroscopic binaries, the radio astronomers realized that the pulsar had to be in a binary system with an orbital period of 7.75 hours. When the orbital motion of the pulsar carries it away from Earth, the pulse period is slightly lengthened, just as the wavelength of light emitted by a receding source is lengthened. That is a redshift. Then, when the pulsar rounds its orbit and approaches Earth, the pulse period is slightly shortened—a blueshift. From these changing Doppler shifts, the astronomers could calculate the radial velocity of the pulsar around its orbit just as if it were a spectroscopic binary star (Chapter 9). The resulting graph of radial velocity versus time

■ **Figure 14-9**

Many neutron stars have high velocities through space. Here the neutron star known as RX J185635−3754 was photographed on three different dates as it rushed past background stars. (NASA and F. M. Walter)

■ Figure 14-10

The radial velocity of pulsar PSR 1913+16 can be found from the Doppler shifts in its pulsation. Analysis of the radial velocity curve allows astronomers to determine the pulsar's orbit. Here, the center of mass does not appear to be at a focus of the elliptical orbit because the orbit is inclined. (Adapted from data by Joseph Taylor and Russell Hulse)

could be analyzed to find the shape of the pulsar's orbit (■ Figure 14-10). When Taylor and Hulse analyzed PSR 1913+16, they discovered that the binary system consisted of two neutron stars separated by a distance roughly equal to the radius of our sun.

Yet another surprise was hidden in the motion of PSR 1913+16. In 1916, Einstein's general theory of relativity described gravity as a curvature of space-time. Einstein realized that any rapid change in a gravitational field should spread outward at the speed of light as **gravitational radiation.** Gravity waves have not been detected, but Taylor and Hulse were able to show that the orbital period of the binary pulsar was slowly growing shorter because the stars are gradually spiraling toward each other. They are radiating orbital energy away as gravitational radiation. Taylor and Hulse won the Nobel prize in 1993 for their work with binary pulsars.

Dozens of binary pulsars have been found; by analyzing the Doppler shifts in their pulse periods, astronomers can estimate the mass of the neutron stars. Typical masses are about 1.35 solar masses, in good agreement with models of neutron stars.

In 2004, radio astronomers announced the discovery of a double pulsar. The two pulsars orbit each other in only 2.4 hours, and their spinning beams sweep over Earth (■ Figure 14-11). One spins with a period of 0.023 seconds (44 times a second), and the other spins in 2.8 seconds. This system is a pulsar jackpot because the orbits are nearly edge-on to Earth and the powerful magnetic fields eclipse each other, giving astronomers a chance to study their size and structure. Not only that, but the theory of general relativity predicts that they are emitting gravitational radiation and that their separation is decreasing by 7 mm per year. The two neutron stars will merge in 85 million years, presumably to trigger a supernova explosion. In the meantime, the steady decrease in orbital period can be measured and gives astronomers a further test of general relativity and gravitational radiation.

Binary pulsars can emit strong gravitational waves because the neutron stars contain large amounts of mass in a small volume. This also means that binary pulsars can be sites of tremendous violence because of the strength of gravity at the surface of a neutron star. An astronaut stepping onto the surface of a neutron star would be instantly smushed into a layer of matter only 1 atom thick. Matter falling onto a neutron star can release titanic amounts of energy. If you dropped a single marshmallow onto the surface of a neutron star from a distance of 1 AU, it would hit with an impact equivalent to a 3-megaton nuclear warhead. In general, a particle falling from a large distance to the surface of a neutron star will release energy equivalent to 0.2 mc^2, where m is the par-

■ Figure 14-11

Artist's impression of the double pulsar. One star must have exploded to form a pulsar, and later the other star did the same. Gravitational radiation causes the neutron stars to drift toward each other, and they will merge in 85 million years, presumably to trigger another supernova explosion. (John Rowe Animations)

ticle's mass at rest. Even a small amount of matter flowing from a companion star to a neutron star can generate high temperatures and release X rays and gamma rays.

As an example of such an active system, consider Hercules X-1. It emits pulses of X rays with a period of about 1.2 seconds, but every 1.7 days the pulses vanish for a few hours (■ Figure 14-12). You can understand this system by comparing it to an eclipsing binary star. Hercules X-1 seems to contain a 2-solar-mass star with a temperature of 7000 K and a neutron star that orbit each other with a period of 1.7 days. Matter flowing from the normal star into an accretion disk around the neutron star can reach temperatures of millions of degrees and emit a powerful X-ray glow. Interactions with the neutron star's magnetic field can produce beams of X rays that sweep around with the rotating neuron star. Earth receives a pulse of X rays every time a beam points our way. The X rays shut off every 1.7 days when the neutron star is eclipsed behind the normal star. The X rays from the neutron star and its accretion disk heat the near side of the normal star to about 20,000 K. As the system rotates, astronomers on Earth alternately see the hot side of the star and then the cool side, and its brightness at visible wavelengths varies. Hercules X-1 is a complex system and is still not well understood, but this quick analysis will show you how complex and powerful such binary systems are during mass transfer.

The Fastest Pulsars

This discussion of pulsars suggests that newborn pulsars should blink rapidly and old pulsars should blink slowly, but the handful that blink the fastest may be quite old. One of the fastest known pulsars is cataloged as PSR 1937+21 in the constellation Vulpecula. It pulses 642 times a second and is slowing down only slightly. The energy stored in the rotation of a neutron star at this rate is equal to the total energy of a supernova explosion, so it seemed difficult at first to explain this pulsar. It now appears that PSR 1937+21 is an old neutron star that has gained mass and rotational energy from its companion in the binary system. Like water hitting a mill wheel, the matter falling on the neutron star has spun it up to 642 rotations per second. With its old, weak magnetic field, it slows down very slowly and will continue to spin for a very long time.

A number of other very fast pulsars have been found. They are known generally as **millisecond pulsars** because their pulse periods are almost as short as a millisecond (0.001 s). This produces some fascinating physics because the pulse period of a pulsar equals the rotation period of the neutron star. If a neutron star 10 km in radius spins 642 times a second, as does PSR 1937+21, then the period is 0.0016 second, and the equator of the neutron star must be traveling about 40,000 km/s. That is fast enough to flatten the neutron star into an ellipsoidal shape and is nearly fast enough to break it up.

All scientists should be made honorary citizens of Missouri, the "Show Me" state, because scientists demand evidence. The hy-

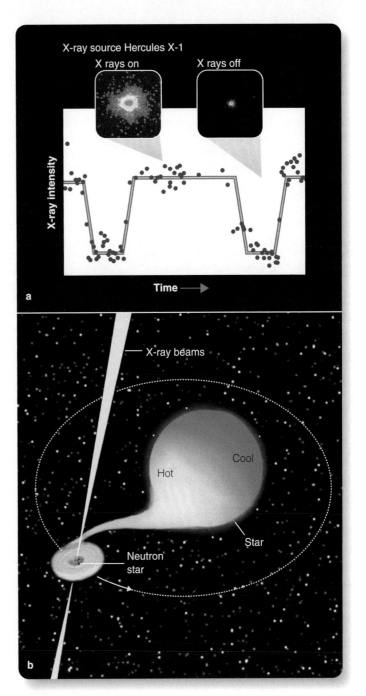

■ **Figure 14-12**

Sometimes the X-ray pulses from Hercules X-1 are on, and sometimes they are off. A graph of X-ray intensity versus time looks like the light curve of an eclipsing binary. (Insets: J. Trümper, Max-Planck Institute) (b) In Hercules X-1, matter flows from a star into an accretion disk around a neutron star producing X rays, which heat the near side of the star to 20,000 K compared with only 7000 K on the far side. X rays turn off when the neutron star is eclipsed behind the star.

pothesis that the millisecond pulsars are spun up by mass transfer from a companion star is quite reasonable, but astronomers demand reality checks, and evidence has been found. For example, the pulsar PSR J1740−5340 has a period of 42 milliseconds and

The Impossibility of Proof in Science

No scientific theory or hypothesis can be proved correct. You can test a theory over and over by performing experiments or making observations, but you can never prove that the theory is absolutely true. It is always possible that you have misunderstood the theory or the evidence, and the next observation you make might disprove the theory.

For example, you might propose the theory that the sun is mostly iron. You could test the theory by looking at the iron lines in the solar spectrum, and the strength of the iron lines would suggest that your theory is right. Although your observation has confirmed your theory, it has not proven the theory is right. You might confirm the theory many times before you realized that iron absorbs photons much more efficiently than hydrogen. Although the hydrogen lines are weak in the sun's spectrum, they show that most of the atoms in the sun are hydrogen and not iron.

The nature of scientific thinking can lead to two kinds of misconceptions. Sometimes nonscientists will say, "You scientists just want to tear everything down—you don't believe in anything." Scientists test a theory over and over to test its worth. If a theory survives many tests, scientists begin to have confidence it is true, but they continue to test their understanding of nature at every opportunity.

The second misconception nonscientists have is revealed when they say, "You scientists are never sure of anything." Again, the scientist knows that no theory can be proven correct. That the sun will rise tomorrow is very likely, and scientists have great confidence in that theory. But in the end it is still a theory.

People will say of an idea they dislike, "That is only a theory," as if a theory were simply a random guess. In fact, a theory can be a well-tested truth in which all scientists have great confidence. Yet you can never prove that any theory is absolutely true.

It is only a theory, but astronomers have tremendous confidence that the sun is made almost entirely of hydrogen and helium. (SOHO/MDI)

---

is orbiting with a bloated red star that is losing mass to the neutron star. This appears to be a pulsar in the act of being spun up to high speed. For another example, consider the X-ray source XTE J1751−305, a pulsar with a period of only 2.3 milliseconds. X-ray observations show that it is in the act of gaining mass from a companion star. The orbital period is only 42 minutes, and the mass of the companion star is only 0.014 solar masses. The evidence suggests this neutron star has devoured all but the last morsel of its binary partner.

Although some millisecond pulsars have binary companions, some are solitary neutron stars. How did they get spun up if they don't have a companion star? A pulsar known as the Black Widow may explain. The Black Widow has a period of 1.6 milliseconds, meaning it is spinning 622 times per second, and it orbits with a low-mass companion. Presumably the neutron star was spun up by mass flowing from the companion, but spectra show that the blast of radiation and high-energy particles from the neutron star is now boiling away the surface of the companion. The Black Widow has eaten its fill and is now evaporating the remains of its companion. It will soon be a solitary millisecond pulsar (■ Figure 14-13).

"Show me," say scientists; and, in the case of neutron stars, the evidence seems very strong. Of course, a theory can never be proven absolutely true (**Window on Science 14-1**), but the evidence for neutron stars is so strong that astronomers have great confidence that such objects really do exist. Other theories

■ **Figure 14-13**

The Black Widow pulsar and its companion star are moving rapidly through space, creating a shock wave like the bow wave of a speedboat. The shock wave confines high-energy particles shed by the pulsar into an elongated cocoon (red). (X-ray: NASA/CXC/ASTRON/B. Stappers et al.; Optical: AAO/J. Bland-Hawthorn & H. Jones)

that describe how they emit beams of radiation and how they form and evolve are less certain, but continuing observations at many wavelengths are revealing more about these last embers of massive stars. In fact, observations have turned up objects no one predicted.

Pulsar Planets

Finding planets orbiting stars other than the sun is very difficult, and hardly two hundred are known. Oddly, the first such planets were found orbiting a neutron star.

Because a pulsar's period is so precise, astronomers can detect tiny variations by comparison with atomic clocks. When astronomers checked pulsar PSR 1257+12, they found variations in the period of pulsation much like those caused by the orbital motion of the binary pulsar (■ Figure 14-14a). However, in the case of PSR 1257+12, the variations were much smaller; and, when they were interpreted as Doppler shifts, it became evident that the pulsar was being orbited by at least two objects with planetlike masses of 4.3 and 3.9 Earth masses. The gravitational tugs of the planets make the pulsar wobble about the center of mass

■ **Figure 14-14**

(a) The dots in this graph are observations showing that the period of pulsar PSR 1257+12 varies from its average value by a fraction of a billionth of a second. The blue line shows the variation that would be produced by planets orbiting the pulsar. (b) As the planets orbit the pulsar, they cause it to wobble by less than 800 km, a distance that is invisibly small in this diagram. (Adapted from data by Alexander Wolszczan)

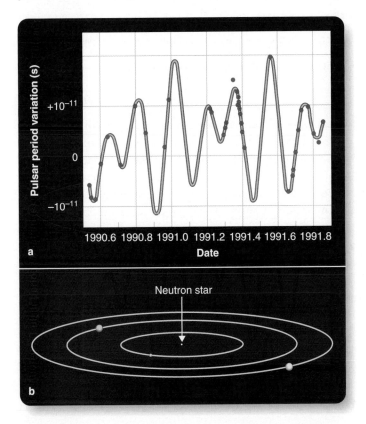

of the system by no more than 800 km, and that produces the tiny changes in period.

Astronomers greeted this discovery with both enthusiasm and skepticism. As usual, they looked for ways to test the hypothesis. Simple gravitational theory predicts that the planets should interact and slightly modify each other's orbit. When the data were analyzed, that interaction was found, further confirming the hypothesis that the variations in the period of the pulsar are caused by planets. In fact, further data revealed the presence of a third planet of about the mass of Earth's moon, and a fourth planet with a mass of about 100 Earth masses is now believed to follow a much larger orbit. This illustrates the astonishing precision of studies based on pulsar timing.

Astronomers wonder how a neutron star can have planets. The planets that orbit PSR 1257+12 are very close to the pulsar; the inner three orbit at 0.19 AU, 0.36 AU, and 0.47 AU— closer to the pulsar than Venus is to the sun. Any planets that orbit a star so closely would be lost or vaporized when the star exploded. Furthermore, a star about to explode as a supernova would be a large giant or a supergiant, and planets only a few astronomical units distant would be inside such a large star and could not survive. It seems more likely that these planets are the remains of a stellar companion that was devoured by the neutron star. In fact, the pulsar is very fast (162 pulses per second), suggesting that it was spun up in a binary system.

Another pulsar planet has been found orbiting in a binary system containing a neutron star and a white dwarf. Because this system is located in a very old star cluster and contains a white dwarf, astronomers suspect that the planet may be very old. Planets probably orbit other neutron stars, and small shifts in the timing of the pulses may eventually reveal their presence.

You can imagine what these worlds might be like. Planets formed from the remains of dying stars might be rich in heavy elements, but truly ancient planets might be poor in such elements. You can imagine visiting these worlds, landing on their surfaces, and hiking across their valleys and mountains. Above you, the neutron star would glitter in the sky, a tiny point of light.

Building Scientific Arguments

Why are neutron stars detectable at X-ray wavelengths?
This argument draws together a number of ideas you know from previous chapters. First, you should remember that a neutron star is very hot because of the heat released when it contracts to a radius of 10 km. It could easily have a surface temperature of 1,000,000 K, and Wien's law (Chapter 7) tells you that such an object will radiate most intensely at a very short wavelength, typical of X rays. However, you know that the total luminosity of a star depends on its surface temperature and its surface area, and a neutron star is so small it can't radiate much energy. X-ray telescopes have found such neutron stars, but they are not easy to locate.

There is, however, a second way a neutron star can radiate X rays. If a normal star in a binary system loses mass to a neutron star companion, the inflowing matter will form a very hot accretion disk that can radiate intense X rays easily detectable by X-ray telescopes orbiting above Earth's atmosphere. Further, any matter that hits the surface of the neutron star will impact with so much energy that it will be heated to very high temperatures and will radiate X rays.

Now build a new argument as if you were seeking funds for a research project. **What observations would you make to determine whether a newly discovered pulsar was young or old, single or a member of a binary system, alone or accompanied by planets?**

■　　■　　■

Connections: Perhaps the strangest planets in the universe are those orbiting pulsars. But however strange a pulsar planet may be, you can imagine going to one. The next topic, in contrast, seems beyond the reach of even your imagination.

14-2 Black Holes

YOU HAVE NOW STUDIED white dwarfs and neutron stars, two of the three end states of dying stars. Now you are ready to turn to the third—black holes.

Although the physics of black holes is difficult to discuss without sophisticated mathematics, simple logic is sufficient to predict that they should exist. The problem is to consider their predicted properties and try to find objects in the heavens that could be real black holes. More difficult than the search for neutron stars, the quest for black holes has nevertheless met with success.

To begin your study of black holes, consider a simple question. How fast must an object travel to escape from the surface of a celestial body?

Escape Velocity

Suppose you threw a baseball straight up. How fast must you throw it if it is not to come down? Of course, gravity would pull back on the ball, slowing it, but if the ball were traveling fast enough to start with, it would never come to a stop and fall back. Such a ball would escape from Earth. The escape velocity is the initial velocity an object needs to escape from a celestial body (■ Figure 14-15). (See Chapter 5.)

Whether you are discussing a baseball leaving Earth or a photon leaving a collapsing star, the escape velocity depends on two things, the mass of the celestial body and the distance from the center of mass to the escaping object. If the celestial body has a large mass, its gravity is strong, and you need a high velocity

■ **Figure 14-15**

Escape velocity, the velocity needed to escape from a celestial body, depends on mass. The escape velocity at the surface of a very small, low-mass body would be so low you could jump into space. Earth's escape velocity is much larger, about 11 km/s (25,000 mph).

to escape; but if you begin your journey farther from the center of mass, the velocity needed is less. For example, to escape from Earth, a spaceship would have to leave Earth's surface at 11 km/s (25,000 mph), but if you could launch spaceships from the top of a tower 1000 miles high, the escape velocity would be only 8.8 km/s (20,000 mph). If you could make an object massive enough or small enough, its escape velocity could be greater than the speed of light. Relativity shows that nothing can travel faster than the speed of light, so even photons, which have no mass, would be unable to escape. Such a small, massive object could never be seen because light could not leave it.

Long before Einstein and relativity, the Rev. John Mitchell, a British gentleman astronomer, realized that Newton's laws of gravity and motion had peculiar consequences. In 1783, Mitchell pointed out that an object 500 times the radius of the sun but of the same density would have an escape velocity greater than the speed of light. Then, "all light emitted from such a body would be made to return towards it." Mitchell didn't know it, but he was talking about a black hole.

Ace ⑤ Astronomy™ Log into AceAstronomy and select this chapter to see Astronomy Exercise "Escape Velocity." Try to escape from the planet.

Schwarzschild Black Holes

If the core of a star collapses and contains more than about three solar masses, no known force can stop it. The object cannot stop collapsing when it reaches the size of a white dwarf because degenerate electrons cannot support the weight. It cannot stop when it reaches the size of a neutron star because degenerate neutrons cannot support the weight. No force remains to stop the object from collapsing to zero radius.

As an object collapses, its density and the strength of its surface gravity increase; and, if an object collapses to zero radius, its density and gravity become infinite. Mathematicians call such a point a **singularity;** but in physical terms it is difficult to imagine an object of zero radius. Some theorists believe that a singularity is impossible and that when the laws of physics are better understood, they will show that the collapse halts at some subatomic radius and does not go to zero. Astronomically, it makes little difference.

If the contracting core of a star becomes small enough, the escape velocity in the region around it is so large that no light can escape. You can receive no information about the object or about the region of space near it, and such a region is called a **black hole.** If the core of an exploding star collapsed into a black hole, the core would vanish without a trace. Consequently, a supernova remnant that lacks a central pulsar may harbor a black hole, but you must be cautious. There are other reasons a supernova might not leave behind a compact object (■ Figure 14-16).

The boundary of the black hole is called the **event horizon,** because any event that takes place inside the event horizon is invisible to an outside observer. To see how a black hole can exist, you need to consider general relativity.

In 1916, Albert Einstein published a mathematical theory of space and time that became known as the general theory of relativity. Einstein treated space and time as a single entity—space-time. His equations showed that gravity could be described as a curvature of space-time, and almost immediately the astronomer Karl Schwarzschild found a way to solve the equations to describe the gravitational field around a single, nonrotating, electrically neutral lump of matter. That solution contained the first general relativistic description of a black hole; nonrotating, electrically neutral black holes are now known as Schwarzschild black holes. In most cases in astronomy, astronomers can use the Schwarzschild solution to think about black holes. You will see later in this chapter what difference rotation makes.

Schwarzschild's solution shows that if matter is packed into a small enough volume, then space-time curves back on itself. Objects can still follow paths that lead into the black hole, but no path leads out, so nothing can escape, not even light. For that reason, the inside of the black hole is totally beyond the view of an outside observer.

The event horizon is the boundary between the isolated volume of space-time and the rest of the universe, and the radius of the event horizon is called the **Schwarzschild radius, R_S**—the radius within which an object must shrink to become a black hole (■ Figure 14-17).

Although Schwarzschild's work was highly mathematical, his conclusion is quite simple. The Schwarzschild radius (in meters) depends only on the mass of the object (in kilograms):

$$R_S = \frac{2GM}{c^2}$$

In this simple formula, G is the gravitational constant, M is the mass, and c is the speed of light. A bit of arithmetic shows that a 1-solar-mass black hole will have a Schwarzschild radius of 3 km, a 10-solar-mass black hole will have a Schwarzschild radius of 30 km, and so on (■ Table 14-1). Even a very massive black hole would not be very large.

■ Figure 14-16

Some supernova remnants contain no neutron star, perhaps because the supernova formed a black hole instead. However, this supernova remnant, formed by Tycho Brahe's supernova of 1572, has the chemical composition typical of the remains of a type Ia supernova. Such supernovae are believed to destroy the white dwarf entirely and do not leave behind a neutron star or a black hole. (John P. Hughes, Rutgers University)

Tycho's Supernova

Edge of detector

X-ray image

■ Active Figure 14-17

A black hole forms when an object collapses to a small size (perhaps to a singularity) and the escape velocity in its neighborhood is so great that light cannot escape. The boundary of this region is called the event horizon because any event that occurs inside is invisible to outside observers. The radius of the region is R_S, the Schwarzschild radius.

Ace ◯ Astronomy™ Log into AceAstronomy and select this chapter to see the Active Figure "Schwarzschild Radius." Take control of this diagram.

Every object with mass has a Schwarzschild radius, but not every object is a black hole. For example, Earth has a Schwarzschild radius of about 1 cm, but it could become a black hole only if you squeezed it inside that radius. Fortunately, Earth will not collapse spontaneously into a black hole because it has a small mass and is supported by the mechanical strength of the iron and rock in its interior. Only exhausted stellar cores more massive than three solar masses can form black holes under the sole influence of their own gravity. In this chapter, you are interested in

■ Table 14-1 | The Schwarzschild Radius

Object	Mass ($M_\odot$)	Radius
Star	10	30 km
Star	3	9 km
Star	2	6 km
Sun	1	3 km
Earth	0.000003	0.9 cm

black holes that might originate from the deaths of stars. These black holes would have masses slightly larger than 3 solar masses. In later chapters, you will encounter black holes whose masses might exceed a million solar masses.

Do not think of black holes as giant vacuum cleaners that will pull in everything in the universe. A black hole is just a gravitational field, and at a reasonably large distance its gravity is no greater than that of a normal object of similar mass. If the sun were replaced by a 1-solar-mass black hole, the orbits of the planets would not change at all. The gravity of a black hole becomes extreme only if you approach close to it (■ Figure 14-18). The universe contains many black holes. So long as you and other objects stay at a safe distance from the black holes, they have no catastrophic effects.

Ace ◯ Astronomy™ Log into AceAstronomy and select this chapter to see Astronomy Exercise "Black Hole." You can change the mass of an object and watch it become a black hole.

Black Holes Have No Hair

Theorists who study black holes are fond of saying, "Black holes have no hair." By that they mean that once matter forms a black hole, it loses almost all of its normal properties. A black hole made of a collapsed star will be indistinguishable from a black hole of the same mass made from peanut butter and fake-fur mittens. Once the matter is inside the event horizon, it retains only three properties—mass, angular momentum, and electrical charge.

The Schwarzschild black hole is represented by a solution to Einstein's equations for the special case where the object has only mass. Schwarzschild black holes do not rotate or have charge. The solutions for rotating or charged black holes (or for rotating, charged black holes) are more difficult and have been found in only the last few decades. Generally, rotating, charged black holes are similar to Schwarzschild black holes.

It seems that astronomers need not worry about charged black holes because stars, whose collapse presumably forms black holes, cannot have large electrostatic charges. Suppose that you could give the sun a large positive charge. It would begin to repel protons in its corona and attract electrons and would soon return to neutral charge. For this reason, you can expect stars and black holes to be electrically neutral.

But everything in the universe seems to rotate, and collapsing stars spin rapidly as they conserve angular momentum. Consequently, you should probably expect black holes to have angular momentum. In 1963, New Zealand mathematician Roy P. Kerr found a solution to Einstein's equations that describes a rotating black hole. This is now known as the **Kerr black hole.**

The mass of a black hole curves neighboring space-time, and the Kerr solution predicts that the rotation of a black hole drags space-time around with it. The **ergosphere** is a region outside the

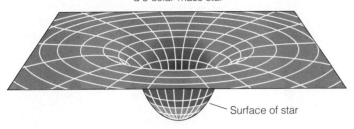

Gravitational field around
a 5-solar-mass star

Surface of star

Gravitational field around
a 5-solar-mass black hole

To the event horizon

■ **Figure 14-18**

If you fell into the gravitational field of a star, you would hit the star's surface before you fell very far. Because a black hole is so small, you could fall
much deeper into its gravitational field and eventually cross the event horizon. At a distance, the two gravitational fields are the same.

event horizon in which space-time rotates with the rotating black
hole so powerfully that nothing could avoid being dragged along.
No one has ever approached a rotating black hole, so no one has
any idea what it might feel like to enter a region where space-
time was whirling past.

The word *ergosphere* comes from the Greek word *ergo,* meaning "work," because the rotating space-time in the ergosphere
can do work on a particle; that is, the particle can gain energy. In
particular, the Kerr solution shows that a particle that enters the
ergosphere can break into two pieces, one falling into the black
hole and the other escaping with more energy than it had when
it entered. In this way, energy can be extracted from a rotating
black hole, and, as a result, the black hole slows its rotation very
slightly.

The Kerr solution is a fascinating bit of theoretical physics,
but it has an important application in astronomy. Almost certainly, black holes rotate, and matter falling into black holes must
pass through the ergosphere. This suggests you should expect to
find situations where energy is extracted from rotating black holes.

A Leap into a Black Hole

Before you can search for real black holes, you need to understand
what theory predicts about the appearance of a black hole. To
explore that idea, imagine that you leap, feet first, into a Schwarzschild black hole.

If you were to leap into a black hole of a few solar masses
from a distance of an astronomical unit, the gravitational pull
would not be very large, and you would fall slowly at first. Of
course, the longer you fell and the closer you came to the center,
the faster you would travel. Your wristwatch would tell you that
you fell for about 65 days before reaching the event horizon.

Your friends who stayed behind would see something different. They would see you falling more slowly as you came closer to
the event horizon because, as explained by general relativity, clocks
slow down in curved space-time. This is known as **time dilation.**

In fact, your friends would never actually see you cross the event horizon. To them you would fall more and more slowly until you seemed hardly to move. Generations later, your descendants could focus their telescopes on you and see you still inching closer to the event horizon. You, however, would have sensed no slowdown and would conclude that you had crossed the event horizon after only about 65 days.

Other relativistic effects would make it difficult for your descendents to see you. As light travels out of a gravitational field, it loses energy, and its wavelength grows longer. This is known as a **gravitational redshift** (■ Table 14-2). Light leaving you and traveling away from the black hole would suffer a larger and larger gravitational redshift. In addition, as your inward velocity grew higher and higher, a relativistic effect would cause more and more of the light leaving you to be emitted in the forward direction into the black hole. This would reduce the amount of light leaving you and traveling outward, making you even more difficult to see. Although you would notice none of these effects as you fell toward the black hole, your friends would need to observe at longer wavelengths and with larger telescopes to detect you.

While these relativistic effects seem merely peculiar, other effects would be quite unpleasant. Imagine again that you are falling feet first toward the event horizon of a black hole. You would feel your feet, which would be closer to the black hole, being pulled in more strongly than your head. This is a tidal force, and at first it would be minor. But as you fell closer, the tidal force would become very large. Another tidal force would compress you as your left side and your right side both fell toward the center of the black hole. For any black hole with a mass like that of a star, the tidal forces would crush you laterally and stretch you longitudinally long before you reached the event horizon (■ Figure 14-19). The friction from such severe distortions of your body would heat you to millions of degrees, and you would emit X rays and gamma rays. (Needless to say, this would render you inoperative as a thoughtful observer.)

Some years ago a popular book suggested that you could travel through the universe by jumping into a black hole in one place and popping out of another somewhere far across space. That might make for good science fiction, but tidal forces would make it an unpopular form of transportation even if it worked. You would certainly lose your luggage.

■ Figure 14-19

Leaping feet first into a black hole, a person of normal proportions (left) would be distorted by tidal forces (right) long before reaching the event horizon around a typical black hole of stellar mass. Tidal forces would stretch the body lengthwise while compressing it laterally. Friction from this distortion would heat the body to high temperatures.

Your imaginary leap into a black hole is not frivolous. You now know how to find a black hole: Look for a strong source of X rays. It may be a black hole into which matter is falling.

The Search for Black Holes

Do black holes really exist? Beginning in the 1970s, astronomers searched for observational evidence that their theories were correct. They tried to find one or more objects that were obviously black holes. That very difficult search is a good illustration of how the unwritten rules of science help you understand nature (**Window on Science 14-2**).

A black hole alone is totally invisible because nothing can escape from the event horizon. But a black hole into which matter is flowing would be a source of X rays. Of course, X rays can't escape from inside the event horizon, but X rays emitted by the heated matter flowing into the black hole could escape if the

■ Table 14-2 I The Gravitational Redshift	
Object	**Redshift (percent)**
Sun	0.0002
White dwarf	0.01
Neutron star	20
Black hole event horizon	Infinite

Natural Checks on Fraud in Science

Fraud is actually quite rare in science. The nature of science makes fraud difficult, and the way scientists publish their research makes it almost impossible. In fact, you can think of science as a set of unwritten rules of behavior that have evolved to prevent scientists from lying to one another or to themselves, even by accident.

Suppose for a moment that you wanted to commit scientific fraud. You would have to invent data supposedly obtained from experiment or observation. You might invent X-ray data supposedly obtained by observing an X-ray binary star. Or, if you were interested in theory, you might invent a fraudulent mathematical calculation of the physics going on in an X-ray binary. You might get away with it for a short time, but one of the most important rules in science is that good results must be reproducible.

Other people must be able to repeat your observations, experiments, and calculations.

In fact, most scientists routinely repeat other scientists' work as a way of getting started on a new research topic. As soon as someone tries to repeat your fraudulent research, you will be caught. The more important a scientific result is, the sooner other scientists will repeat it, so you don't have much of a chance of getting away with scientific fraud. In this way, science is self-correcting.

Even if you could invent some convincing scientific research, you would probably have difficulty publishing it. When a scientist submits an article to a scientific journal, it is subject to peer review. That is, the editor of the journal sends the article to one or two other experts in the field for comment and suggestions. These reviewers often make helpful suggestions, but they may also point out errors that have to be fixed before the journal can publish the article. In some cases, an article may be so flawed the editor will refuse to publish it at all. If you submitted your fraudulent

research on X-ray binaries, the reviewers would almost certainly notice things wrong with it, and it would never get into print.

Scientists know the rules, and they use them. If someone makes a big discovery and is interviewed by the press, scientists will begin asking each other, "Has this work been published in a peer-reviewed journal yet?" That is, they want to know if other experts have checked the work. Until research is peer reviewed and published, it isn't quite official, and most scientists would treat the results with care.

Fraud isn't impossible in science. Some cases have even been in the national news. Big grant money is a terrible temptation. But, in science, "the truth will out." Because of the way scientists reproduce research and because of the way research is published, fraud is quite rare among scientists.

X rays were emitted before the matter crossed the event horizon. An isolated black hole will not have much matter flowing into it, but a black hole in a binary system might receive a steady flow of matter transferred from the companion star. This suggests that you can search for black holes by searching among X-ray binaries.

Some X-ray binaries, such as Hercules X-1, contain a neutron star, and they will emit X rays much as would a binary containing a black hole. You can tell the difference between a neutron star and a black hole in an X-ray binary in two ways. If the compact object emits pulses, you know it is a neutron star. Otherwise, you must depend on the mass of the object. If the compact object has a mass greater than 3 solar masses, the object can't be a neutron star, and you can conclude that it must be a black hole.

The first X-ray binary suspected of harboring a black hole was Cygnus X-1, the first X-ray object discovered in Cygnus. It contains a supergiant B0 star and a compact object orbiting each other with a period of 5.6 days. Matter flows from the B0 star as a strong stellar wind, and some of that matter enters a hot accretion disk around the compact object (■ Figure 14-20). The accretion disk is about five times larger in diameter than the orbit of Earth's moon, and the inner few hundred kilometers of the disk have a temperature of about 2 million Kelvin—hot enough to radiate X rays. The compact object is invisible, but Doppler shifts in the spectrum reveal the motion of the B0 star around the center of mass of the binary. From the geometry of the orbit, astronomers can calculate the mass of the compact

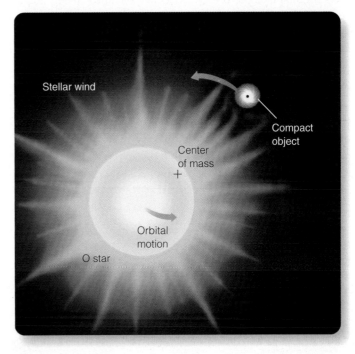

Labels: Stellar wind; Compact object; Center of mass; Orbital motion; O star

■ Figure 14-20

The X-ray source Cygnus X-1 is a supergiant B0 star and a compact object orbiting each other. Gas from the B0 star's stellar wind flows into the hot accretion disk, and the X rays detected come from the disk.

object—at least 3.8 solar masses, well above the maximum for a neutron star.

To confirm that black holes exist, astronomers needed to find a conclusive example, an object that couldn't be anything else. Cygnus X-1 didn't quite pass that test when it was first discovered. Perhaps the B0 star was not a normal star, said some astronomers, and that would make the mass of the compact object uncertain. Also, some astronomers suspected there might be a third star in the system, and that would distort the analysis. At the time, astronomers could not conclusively show that Cygnus X-1 contained a compact object with a mass greater than 3 solar masses. More recent research has given astronomers more confidence that the compact object is indeed a black hole with a mass slightly less than 10 solar masses.

Astronomers needed a more conclusive example than Cygnus X-1, and, as X-ray telescopes found more X-ray objects, the list of black hole candidates grew to a few dozen. A few of these objects, such as the first two in ■ Table 14-3, contain massive stellar companions, either giants, supergiants, or massive main-sequence stars. This makes such systems difficult to analyze, because such massive companions dominate the system. A good example of this sort is LMC X-3 (LMC refers to the Large Magellanic Cloud, a small galaxy near our own). The compact object in LMC X-3 has a mass of about 10 solar masses. One reason astronomers think the compact object is so massive is that it distorts the shape of the B main-sequence star into an egg shape; and, as the system rotates, the light from the B star varies because telescopes on Earth detect light from the side of the egg and then from the end. By analyzing the light change, astronomers can determine the shape of the B star and from that find the mass of the compact object.

Many of the black hole candidates are binary systems in which the normal star is a lower-mass main-sequence star. Such systems do not remain X-ray sources continuously but suffer X-ray nova outbursts as matter flows rapidly into the accretion disk. A year or so after an outburst, the flow has stopped, the accretion disk has dimmed, and astronomers can detect the spectrum of the main-sequence star. The spectral type and Doppler motions of the main-sequence companion reveal the mass of the compact object with only limited uncertainty. The lower-mass stellar companion makes these X-ray nova systems easier to analyze than are systems with massive companions.

A number of examples of these X-ray nova systems are shown in Table 14-3. A0620-00 is an old nova that erupted again in 1975. It contains an ordinary main-sequence K star and a compact object that orbit each other with a period of 7.75 hours. From the orbital motion and the distortion of the K star, astronomers conclude the compact object must have a mass between 5 and 15 solar masses. Three of the best understood examples are V404 Cygni; J1655-40, also known as Nova Scorpii 1994; and QZ Vul. The compact objects in these systems seem much too massive to be anything but black holes.

The growing list of X-ray binaries with compact objects exceeding 3 solar masses has convinced astronomers that black holes really do exist. The problem now is to understand how these objects interact with matter flowing into them through accretion disks to produce X rays, gamma rays, and jets of matter.

Building Scientific Arguments

How can a black hole emit X rays?
A common misconception about black holes is that they can emit nothing. By constructing a careful argument, you can show that a black hole can be a powerful source of energy. Of course, once a bit of matter falling into a black hole crosses the event horizon, no light or other electromagnetic radiation it emits can escape from the black hole. The matter becomes lost to view. But, if it emitted radiation before it crossed the event horizon, that radiation could escape, and you could detect it. Furthermore, the powerful gravitational field near a black hole stretches and distorts infalling matter, and internal friction heats the matter to millions of degrees. Wien's law

■ Table 14-3 | Nine Black-Hole Candidates

Object	Location	Companion Star	Orbital Period	Mass of Compact Object
Cygnus X-1	Cygnus	B0 supergiant	5.6 days	>3.8 $M_\odot$
LMC X-3	Dorado	B3 main-sequence	1.7 days	~10 $M_\odot$
A0620-00	Monocerotis	K main-sequence	7.75 hours	10 ± 5 $M_\odot$
V404 Cygni	Cygnus	K main-sequence	6.47 days	12 ± 2 $M_\odot$
J1655-40	Scorpius	F main-sequence	2.61 days	6.9 ± 1 $M_\odot$
QZ Vul	Vulpecula	K main-sequence	8 hours	10 ± 4 $M_\odot$
4U 1543-47	Lupus	A main-sequence	1.123 days	2.7–7.5 $M_\odot$
V4641 Sgr	Sagittarius	B supergiant	2.81678 days	8.7–11.7 $M_\odot$
XTE J1118+480	Ursa Major	K main-sequence	0.170113 days	>6 $M_\odot$

(Chapter 7) says that matter at such a high temperature should emit X rays. Any X rays emitted before the matter crosses the event horizon will escape, and you can look for black holes by looking for X-ray sources. Of course, an isolated black hole will probably not have much matter falling in, but black holes in binary systems may have large amounts of matter flowing in from the companion star. Consequently, the best place to search for black holes is in X-ray binaries.

A good scientific argument includes both theory and evidence. Expand your argument to include observations. **What observations would you make of an X-ray binary system to distinguish between a black hole and a neutron star?**

■   ■   ■

Connections: Today's astronomers are confident that black holes really exist. Modern research is revealing that both neutron stars and black holes can produce powerful phenomena as matter plunges in.

(14-3) Compact Objects with Disks and Jets

NEUTRON STARS AND BLACK HOLES seem to be exotic objects, and they generate equally exotic phenomena. By studying those phenomena, you can learn more about the strange objects.

X-Ray Bursters

Beginning in the 1970s, X-ray telescopes revealed that some objects emit irregularly spaced bursts of X rays. Typically, bursts that follow a long quiet period are especially large (■ Figure 14-21), and this suggests that some mechanism is accumulating energy that is released by the bursts. The longer the quiet phase, the more energy accumulates.

These **X-ray bursters** are thought to be binaries containing neutron stars. The X-ray burster 4U 1820−30 in the globular cluster NGC6624, for example, appears to be a neutron star pulling mass away from a white dwarf (■ Figure 14-22). Matter from the white dwarf first flows into an accretion disk and then falls to the surface of the neutron star where the impact heats the gas. Hydrogen fuses on the surface of the star, leaving helium ash to accumulate. When the helium reaches a depth of about a meter, it fuses explosively into carbon and produces an X-ray burst. The rapid increase in brightness (in a few seconds) and the total amount of energy produced (up to 100,000 times the

luminosity of the sun) fit well with theoretical models of an explosion on the surface of an object as small as a neutron star. Of course, mass continues to accumulate, producing burst after burst. Notice the similarity with the mechanism that produces nova explosions on the surfaces of white dwarfs.

Dozens of X-ray bursters are known, and astronomers suspect that they occur in binary systems where matter flows through accretion disks and falls into neutron stars.

Accretion Disk Observations

When material falls into a neutron star or black hole it forms an accretion disk, and high-speed observations have been able to reveal some of the processes that go on in such disks.

When a blob of matter gets caught in an accretion disk, it can orbit so fast its orbital period is measured in thousandths of a second. Because the blob of matter looses energy by friction, it spirals inward, and its orbital period grows shorter. This process can be detected as a quick flicker as the disk emits a short series of pulses of electromagnetic radiation. Some accretion disks produce pulses with a period as short as 0.00075 second, as blobs of material orbit only a dozen kilometers from the center of the compact object. Because the pulses last only a few cycles and because the period of the pulses grows rapidly shorter, they are called **quasi-periodic oscillations (QPOs).** Do not confuse these rapid flickering pulses with the regular click of pulses emitted by a pulsar.

QPOs can be observed in J1550−564, a star shedding mass into an accretion disk around a black hole. The flow of mass is irregular and causes X-ray flares. Short flickering strings of pulses are seen with periods as short as 0.003 second, and the period decreases rapidly. This suggests the energy is being emitted by

■ **Figure 14-21**

X-ray bursters emit bursts of X rays that rise to full intensity suddenly and then fade in seconds. Because of the way the size of a burst depends on the length of the pause that precedes it, astronomers conclude that the bursts release energy that accumulates during the pauses.

X-ray bursts last only seconds.

Time (seconds)

After a long pause, bursts are larger.

After a short pause, bursts are smaller.

Intensity

Time (hours)

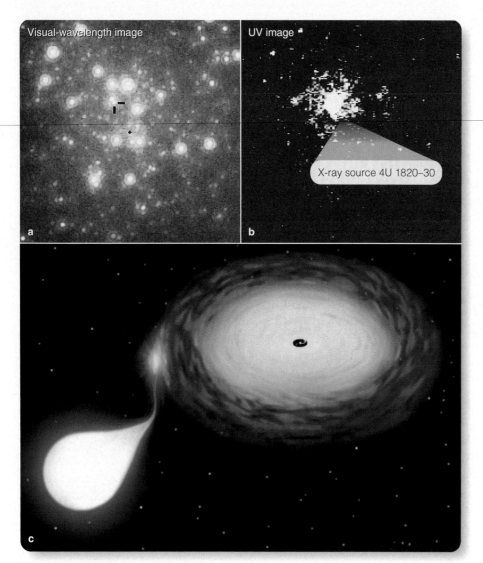

Visual-wavelength image

UV image

X-ray source 4U 1820–30

a

b

c

■ **Figure 14-22**

(a) At visible wavelengths, the center of star cluster NGC6624 is crowded with stars. (b) In the ultraviolet, one object stands out, the X-ray burster 4U 1820–30, consisting of a neutron star orbiting a white dwarf. (c) An artist's conception shows matter flowing from the white dwarf into an accretion disk around the neutron star. (a and b, Ivan King and NASA/ESA; c, Dana Berry, STScI)

material in the accretion disk moving rapidly inward toward orbits of shorter and shorter period.

Cyg X-1 appears to be a star shedding mass into a black hole between 3.8 and 10 solar masses. Pulse trains have been detected with periods of 0.018 second. The pulses grow dimmer and faster, which seems to mean the material is spiraling into the black hole. But no impact is seen, as you would expect of material falling into a neutron star. Evidently, the pulses fade away because the material approaches the event horizon and a powerful gravitational red shift stretches the photons to such long wavelengths they cannot be detected.

Further evidence of an event horizon was found in a detailed study of 12 X-ray novae. Six of the novae consisted of binary sys-

tems containing neutron stars, and bursts of energy were detected, produced by infalling material striking the surfaces of the neutron stars. In contrast, six of the novae were systems containing black holes, and no bursts were seen from impacts on a surface. Apparently, in these systems, the material spiraled inward and disappeared across event horizons (■ Figure 14-23).

Because space-time is so strongly curved near a compact object, very small orbits are unstable. Once material spirals inward to that smallest possible orbit, it must quickly fall into the object. Yet the Rossi X-ray Timing Explorer satellite found an X-ray binary containing a black hole in which a QPO had a period of 0.002 second. The smallest possible orbit around the black hole in that system has a period of a little over 0.003 second. This appears to be evidence that the spinning black hole is dragging space-time with it as theory predicts and making smaller, shorter-period orbits stable. Astronomers are searching for more of these very-short-period QPOs, because they confirm the Kerr solution's prediction that spinning black holes drag space-time.

High-speed observations at many wavelengths are allowing astronomers to follow the rapid processes that occur inside accretion disks. Those same accretion disks can eject high-energy jets extending out into space.

Jets of Energy from Compact Objects

As you can see, it is a misconception to think that it is impossible to get any energy out of a black hole. In later chapters, you will meet galaxies that extract vast amounts of energy from massive black holes, so you should pause here to see how a compact object and an accretion disk can eject powerful jets.

Whether a compact object is a black hole or a neutron star, it has a strong gravitational field. Any matter flowing into that field is accelerated inward; and, because it must conserve angular momentum, it flows into an accretion disk made so hot by friction that the inner regions can emit X rays and gamma rays. Somehow the spinning accretion disk can emit powerful beams of gas and radiation along its axis of rotation. The process isn't well understood, but it seems to involve magnetic fields that get caught in the accretion disk and are twisted into tightly wound tubes that squirt gas and radiation out of the disk and confine it in narrow beams (■ Figure 14-24).

■ Figure 14-23

Gas spiraling into an accretion disk grows hot; and, as it nears the central object, a strong gravitational redshift makes it appear redder and dimmer. Systems containing a neutron star emit bursts of energy when the gas hits the surface of the neutron star, but such bursts are not seen for systems containing black holes. In those systems, the matter vanishes across the event horizon. This is direct observational evidence of an event horizon around black holes. (NASA/CXC/SAO)

This process is similar to the bipolar outflows ejected by protostars, but it is much more powerful. You have seen in the X-ray image on page 315 that the Crab Nebula pulsar is ejecting jets of highly excited gas. The Vela pulsar does the same (Figure 14-6). Systems containing black holes can also eject jets. The black hole candidate J1655−40 is observed at radio wavelengths to be sporadically ejecting oppositely directed jets at 92 percent the speed of light.

One of the most powerful examples of this process is an X-ray binary called SS433. Its optical spectrum shows sets of spectral lines that

■ Active Figure 14-24

In this artist's impression, matter from a normal star at left flows into an accretion disk around a compact object at right. Processes in the spinning disk eject gas and radiation in jets perpendicular to the disk. The jet inclined slightly toward you has a blueshift, and the jet inclined slightly away from you has a redshift. (NASA)

Ace⊛Astronomy™ Log into AceAstronomy and select this chapter to see the Active Figure "End States of Stars." Compare the final states of high-, medium-, and low-mass stars.

are Doppler-shifted by about one-fourth the speed of light, with one set shifted to the red and one set shifted to the blue. Furthermore, the two sets of lines shift back and forth across each other with a period of 164 days. News media reported that astronomers had discovered an object that was both approaching and receding at a fantastic speed, but astronomers recognized the Doppler shifts as evidence of oppositely directed jets.

Apparently, SS433 is a binary system in which a compact object (probably a black hole) pulls matter from its companion star and forms an extremely hot accretion disk. Jets of high-temperature gas blast away in beams aimed in opposite directions. As the disk precesses, it sweeps these beams around the sky once every 164 days, and telescopes on Earth detect light from gas trapped in both beams. One beam produces a redshift, and the other produces a blueshift (■ Figure 14-25). SS433 is a prototype that illustrates how the gravitational field around a compact object can produce powerful beams of radiation and matter. You will meet this phenomenon again when you study peculiar galaxies in a later chapter.

Gamma-Ray Bursts

The Cold War has an odd connection to neutron stars and black holes. In 1963, a nuclear test ban treaty was signed; and, by 1968, the United States was able to put a series of Vela satellites in orbit to watch for nuclear tests that were violations of the

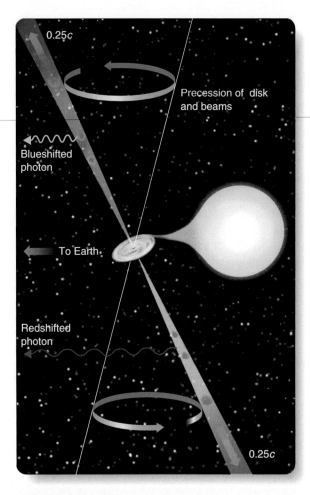

0.25*c*

Precession of disk and beams

Blueshifted photon

To Earth

Redshifted photon

0.25*c*

■ Figure 14-25

The generally accepted model of SS433 includes a compact object with a very hot accretion disk producing beams of radiation with embedded blobs of gas traveling at one-fourth the velocity of light. The precession of the disk swings the beams around a conical path every 164 days. (NASA/CXC/M. Weiss. Found at: http://www.chandra.harvard.edu/press/04_releases/press_010504.html)

treaty. A nuclear detonation emits gamma rays, so the Vela satellites were designed to watch for bursts of gamma rays coming from Earth. The experts were startled when the satellites began detecting about one gamma-ray burst a day coming from space. When those data were finally declassified, astronomers realized that the bursts might be coming from neutron stars or black holes. The sources of these eruptions are now known as **gamma-ray bursters.**

The Compton Gamma Ray Observatory reached orbit in 1991 and immediately began reporting gamma-ray bursts at the rate of a few a day. The intensity of the gamma rays rises to a maximum in seconds and then fades away quickly; a burst is usually over in seconds or minutes. Furthermore, the Compton Observatory discovered that the gamma-ray bursts were coming from all over the sky and not from any particular region. This helped astronomers sort out the different theories.

Some theories proposed that the gamma-ray bursts were being produced among the stars in our galaxy, but the Compton

Observatory observation that the bursts were coming from all over the sky eliminated that possibility. If the gamma-ray bursts were produced among stars in our galaxy, you would expect to see them most often from along the Milky Way, where many stars are located. The observation of bursts occurring all over the sky meant that bursts were probably coming from distant galaxies.

Gamma-ray bursts are hard to study because they occur without warning and fade so quickly; but, in 1997, the Italian-Dutch satellite BeppoSAX began giving Earth's astronomers immediate notice when a gamma-ray burst occurred, along with an accurate position in the sky. When astronomers quickly looked at those positions at visual wavelengths, they could see the fading glow of the explosion (■ Figure 14-26). The spectra of these glowing clouds of gas reveal that most gamma-ray bursts are in very distant galaxies.

Given the intensity of a gamma-ray burst and its distance, astronomers can calculate its luminosity. If the gamma-ray bursts occur at very great distances, out among the galaxies, then they must be very powerful—at least as powerful as the most violent supernova explosions. Or perhaps the eruption blasts gamma rays out in beams. In that case, the explosion doesn't have to be quite as violent and could still be detected at very great distance if a beam were pointed at Earth. But that would mean that Earth's telescopes don't see most of these events because their beams do not happen to be pointed toward Earth. In that case, the eruptions would have to be a few hundred times more common than they appear.

Some bursts of lower-energy (soft) gamma rays repeat and are therefore known as **soft gamma-ray repeaters (SGRs).** Only a few are known, and they lie along the Milky Way. Astronomers suspect that they occur on young neutron stars that have magnetic fields as much as 1000 times stronger than the average neutron stars. You have met these magnetars earlier in this chapter. Evidently, such a highly magnetic neutron star can produce bursts of gamma rays when shifts in the magnetic field break the crust of the neutron star. Such a neutron starquake can release tremendous energy. One of these objects produced a burst of gamma rays that reached Earth on August 27, 1998, and temporarily ionized Earth's upper atmosphere, disrupting radio communication worldwide.

Bursts of higher-energy gamma rays are spread uniformly across the sky and do not appear to repeat, so they can't be explained by magnetars in our galaxy. These gamma-ray bursts must be produced by violent explosions occurring in distant galaxies. Two theories have been proposed. One suggestion is that gamma-ray bursts occur when two neutron stars orbiting each other lose energy by gravitational radiation and fall together. Theoretical models show that the merger of two neutron stars would produce a tremendous eruption lasting only seconds as the neutron stars ripped each other apart with tides and vanished into a newborn black hole. Such an event could produce a burst of gamma rays, possibly emitted in beams.

A Hypernova Explosion

The collapsing core of a massive star drives its energy along the axis of rotation because. . .

the rotation of the star slows the collapse of the equatorial regions.

Within seconds, the remaining parts of the star fall in.

Beams of gas and radiation strike surrounding gas and generate beams of gamma rays.

The gamma-ray burst fades in seconds, and a hot accretion disk is left around the black hole.

Host galaxy

Visual-wavelength image

Visual-wavelength image

Visual-wavelength image

■ **Figure 14-26**

Alerted by gamma-ray detectors on satellites, observers used one of the VLT 8.2-meter telescopes on a mountaintop in Chile to image the location of a gamma-ray burst only hours after the burst. The image at top left shows that fading glow of the eruption. The image at top right, recorded 13 years before, reveals no trace of an object at the location of the gamma-ray burst. The Hubble Space Telescope image at bottom was recorded a year later and reveals a very faint galaxy at the location of the gamma-ray burst. (ESO and NASA)

The second theory proposes that gamma-ray bursts are produced when an extremely massive star exhausts its nuclear fuels. For the most massive stars, the resulting supernova explosion can be smothered by the massive outer layers of the star, and the star may collapse directly into a black hole. These have been called **hypernovae** or **collapsars.** Although a hypernova may produce no supernova explosion, it can emit jets of gamma rays because its rotation slows the collapse of its equatorial regions. The rotating star would collapse first along its poles, and moments later the equatorial regions would fall in. Models suggest this could produce powerful bursts of gamma rays shooting out of the poles of rotation (■ Figure 14-27). These hypernovae are thought to be much rarer than normal supernovae because they are produced by only the most massive stars.

■ **Active Figure 14-27**

The collapse of the cores of extremely massive stars can produce hypernova explosions, which are believed to be the source of at least some gamma-ray bursts. (NASA/Skyworks Digital)

Ace Astronomy™ Log into AceAstronomy and select this chapter to see the Active Figure "Hypernovae." Notice how the rotation of the star focuses the jets.

As satellites detect gamma-ray bursts, telescopes on Earth swivel to image the position of the burst. The results have shown that at least some gamma-ray bursts leave behind a fading glow like that from a supernova explosion in a distant galaxy. The turning point came on March 29, 2003, when astronomers rushed to study a powerful gamma-ray burst in the constellation Leo. The fading glow visible in ground-based telescopes had a complex spectrum that gradually organized itself into the spectrum of a hypernova.

The evidence seems conclusive that at least the longer gamma-ray bursts, those lasting more than 2 seconds, are produced by hypernovae. That does not rule out merging neutron stars. At least some short gamma-ray bursts are produced by magnetars, but it is certainly possible that some can be produced by merging neutron stars. In fact, two short bursts seen in 2005 appear to have been produced by merging neutron stars.

Incidentally, if a gamma-ray burst occurred only 1600 ly from Earth, the distance to the nearest known binary pulsar, the gamma-ray burst would shower the Earth with radiation equivalent to a 10,000-megaton nuclear blast. (The largest bombs ever made were a few megatons.) The gamma rays could create enough nitric oxide in the atmosphere to produce intense acid rain and could destroy the ozone layer and expose life on Earth to deadly levels of solar ultraviolet radiation. Gamma-ray bursts may occur relatively near the Earth as often as every few 100 million years and could be one of the causes of the mass extinctions that show up in the fossil record.

It may seem odd that such rare events as merging neutron stars and hypernovae are blamed for something so common that gamma-ray telescopes observe one or more every day. But remember that these events are so powerful they can be detected over very great distances. There may be 30,000 neutron star binaries in each galaxy, so mergers must occur now and then in any galaxy. Massive stars explode as hypernovae about once in 10 million years in any one galaxy, but gamma-ray bursts are so powerful they can be detected over huge distances, and that includes billions of galaxies. The evidence shows that the 1000 or so gamma-ray bursts that the Compton Observatory detected each year appear to be coming from the final deaths of stars far across space.

Building Scientific Arguments

Why do fluctuations from accretion disks have such short periods?

"Show me the evidence," say scientists, so scientific arguments always fall back on observations. When astronomers see an X-ray flicker coming from an accretion disk, they are seeing a blob of hot gas orbiting the neutron star or black hole at a very small radius. Earth takes a year to orbit the sun once, but it is very far from the sun. A blob of gas in an accretion disk around a neutron star is orbiting an object with a mass of about a solar mass, but it may have an orbit only a few dozen kilometers in radius. The orbital period must be very short, as you would expect from Kepler's third law.

The flickers coming from accretion disks tell of fantastic processes going on there. Extend your argument. **Why do these trains of pulses become more rapid as they fade away?**

■ ■ ■

Connections: Neutron stars and black holes give rise to violent processes and emit high-energy radiation, but they are only single stars. As you turn your attention to galaxies in the next few chapters, you will be studying communities containing billions of stars.

Study and Review Tools

Summary

14-1 | Neutron Stars

How did scientists predict the existence of neutron stars?
■ When a supernova explodes, the core collapses to very small size. Theory predicts that protons and electrons will combine to form a degenerate neutron gas.

■ The collapsing core cannot support itself as a white dwarf if its mass is greater than 1.4 solar masses, the Chandrasekhar limit. If its mass lies between 1.4 solar masses and about 3 solar masses, it can halt its contraction and form a neutron star.

■ A neutron star is supported by the pressure of the degenerate gas of neutrons. Theory predicts that a neutron star should be about 10 km in radius, spin very fast because it conserves angular momentum as it contracts, have a high temperature, and have a powerful magnetic field.

What is the evidence that neutron stars really exist?
■ Pulsars, rapidly pulsing radio sources, were discovered in 1967, and were eventually understood to be spinning neutron stars. The discovery of a pulsar in the supernova remnant called the Crab Nebula was a key link in the story.

■ Pulsars are evidently spinning neutron stars that emit beams of radiation from their magnetic poles. As they spin, they sweep the beams around the sky; if the beams sweep over Earth, pulses can be detected.

- A spinning neutron star slows as it radiates its energy into space. Most of the energy emitted by a pulsar is carried away as a pulsar wind.

- Dozens of pulsars have been found in binary systems, which allows astronomers to estimate the masses of the neutron stars. Such masses are consistent with the predicted masses of neutron stars.

- In some binary systems, mass flows into a hot accretion disk around the neutron star and causes the emission of X rays.

- The fastest pulsars, the millisecond pulsars, appear to be old pulsars that have been spun up to high speed by mass flowing from binary companions.

- Planets have been found orbiting at least one neutron star. They may be the remains of a companion star that was mostly devoured by the neutron star.

14-2 | Black Holes

How did scientists predict the existence of black holes?

- If the collapsing core of a supernova has a mass greater than 3 solar masses, then degenerate neutrons cannot stop the contraction, and it must contract to a very small size—perhaps to a singularity, an object of zero radius. Near such an object, gravity is so strong that not even light can escape, and the region is called a black hole.

- The outer boundary of a black hole is the event horizon; no event inside is detectable. The radius of the event horizon is the Schwarzschild radius, amounting to only a few kilometers for a black hole of a few solar masses.

- Once matter falls into a black hole, it loses all of its properties except for mass, electrical charge, and angular momentum. Rotating black holes are described by a model called the Kerr solution.

What is the evidence that black holes really exist?

- If you were to leap into a black hole, your friends who stayed behind would see two relativistic effects. They would see your clock slow relative to their own clock because of time dilation. Also, they would see your light redshifted to longer wavelengths.

- You would not notice these effects, but you would feel powerful tidal forces that would deform and heat your mass until you grew hot enough to emit X rays. Any X rays your mass emitted before crossing the event horizon could escape.

- To search for black holes, you must look for binary star systems in which mass flows into a compact object and emits X rays. If the mass of the compact object is greater than about 3 solar masses, then the object cannot be a neutron star and is presumably a black hole. A number of such objects have been located.

14-3 | Compact Objects with Disks and Jets

What happens when matter falls into a neutron star or black hole?

- X-ray bursters appear to be binary systems in which mass transfer deposits matter on the surface of a neutron star. When helium fusion ignites, the surface explodes and produces a burst of X-rays.

- Astronomers can observe rapid flickering, called quasi-periodic oscillations, produced by blobs of gas orbiting very rapidly and spiraling inward in accretion disks. Bursts of energy are seen when such matter hits the surface of neutron stars, but no bursts are seen if the matter crosses the event horizon around a black hole.

- Rapidly spinning accretion disks around neutron stars or black holes can twist magnetic fields into tubes and eject narrow, powerful jets of radiation and matter in a process that is not yet well understood.

- Gamma-ray bursters appear to be related to violent events involving neutron stars and black holes. The soft gamma-ray repeaters are thought to be related to disruptions in the crusts of magnetars. Many bursts appear to arise during hypernovae, the collapse of the most massive stars to form black holes and eject powerful beams of gamma rays. Some bursts appear to be caused by the merger of two neutron stars.

New Terms

neutron star (p. 310)	Kerr black hole (p. 324)
pulsar (p. 312)	ergosphere (p. 324)
lighthouse model (p. 312)	time dilation (p. 325)
pulsar wind (p. 313)	gravitational redshift (p. 326)
glitch (p. 313)	X-ray burster (p. 329)
magnetar (p. 317)	quasi-periodic oscillation (QPO) (p. 329)
gravitational radiation (p. 318)	
millisecond pulsar (p. 319)	gamma-ray burster (p. 332)
singularity (p. 323)	soft gamma-ray repeater (SGR) (p. 332)
black hole (p. 323)	
event horizon (p. 323)	hypernova (p. 333)
Schwarzschild radius (R_S) (p. 323)	collapsar (p. 333)

Review Questions

Ace⊙Astronomy™ Assess your understanding of this chapter's topics with additional quizzing and animations at **http://ace .brookscole.com/sf9**

1. Why is there an upper limit to the mass of neutron stars? Why is that upper limit not well known?

2. Explain in detail why you would expect neutron stars to be hot, spin fast, and have strong magnetic fields.

3. Why can't astronomers use visual-wavelength telescopes to locate neutron stars?

4. Why does the short length of pulsar pulses eliminate normal stars as possible pulsars?

5. What do you mean when you say, "Every pulsar is a neutron star, but not every neutron star is a pulsar"?

6. According to the modern model of a pulsar, if a neutron star formed with no magnetic field at all, could it be a pulsar? Why or why not?

7. Why did astronomers first assume that the millisecond pulsar was very young?

8. Why would you suspect that only very fast pulsars can emit visible pulses?

9. If the sun were replaced by a 1-solar-mass black hole, how would Earth's orbit change?

10. If the sun has a Schwarzschild radius, why isn't it a black hole?

11. How can a black hole emit X rays?

12. What do theorists mean when they say, "Black holes have no hair"?

13. What evidence can you cite that black holes really exist?

14. How can mass transfer into a compact object produce jets of high-speed gas?

15. How does an X-ray burster resemble a nova?

16. What observational evidence can you cite to show that black holes do have event horizons?

17. Discuss the possible causes of gamma-ray bursters.

18. The X-ray image at the right shows the supernova remnant G11.2−0.3 and its central pulsar in X rays. The blue nebula near the pulsar is caused by the pulsar wind. How old do you think this system is? Discuss the appearance of this system a million years from now.

(NASA/McGill, V. Kaspi et al.)

19. What is happening in the artist's impression at the right? How would you distinguish between a neutron star and a black hole in such a system?

(CXC/M. Weiss)

Discussion Questions

1. Has the existence of neutron stars been sufficiently tested to be called a theory, or should it be called a hypothesis? What about the existence of black holes?

2. Why would you expect an accretion disk around a star the size of the sun to be cooler than an accretion disk around a compact object?

3. In this chapter, you imagined what would happen if you jumped into a Schwarzschild black hole. From what you have read, what do you think would happen to you if you jumped into a Kerr black hole?

Problems

1. If a neutron star has a radius of 10 km and rotates 642 times a second, what is the speed of the surface at the neutron star's equator in terms of the speed of light?

2. Suppose that a neutron star has a radius of 10 km and a temperature of 1,000,000 K. How luminous is it? (*Hint:* See Chapter 9.)

3. A neutron star and a white dwarf have been found orbiting each other with a period of 11 minutes. If their masses are typical, what is their average separation? Compare the separation with the radius of the sun, 7×10^5 km. (*Hint:* See Chapter 9.)

4. If the accretion disk around a neutron star has a radius of 2×10^5 km, what is the orbital velocity of a particle at its outer edge? (*Hint:* Use circular velocity, Chapter 5.)

5. What is the escape velocity from the surface of a typical neutron star? How does that compare with the speed of light? (*Hint:* See Chapter 5.)

6. If Earth's moon were replaced by a typical neutron star, what would the angular diameter of the neutron star be as seen from Earth? (*Hint:* Use the small-angle formula, Chapter 3.)

7. If the inner accretion disk around a black hole has a temperature of 1,000,000 K, at what wavelength will it radiate the most energy? What part of the spectrum is this in? (*Hint:* Use Wien's law, Chapter 7.)

8. What is the orbital period of a bit of matter in an accretion disk 2×10^5 km from a 10-solar-mass black hole? (*Hint:* Use circular velocity, Chapter 5.)

9. If SS433 consists of a 20-solar-mass star and a neutron star orbiting each other every 13.1 days, what is their average separation? (*Hint:* See Chapter 9.)

Media Cluster

ACTIVE FIGURES

Neutron Star
Some neutron stars are pulsars; some are not. This animation shows you how the alignment of the star's axis of rotation and its polar axis creates the two effects. You can adjust the inclination of the neutron star's magnetic field to produce pulses.

Schwarzschild Radius
Nothing can approach closer than the event horizon of a black hole and still come back out. Its size—the Schwarzschild radius—depends only on the black hole's mass. In this animation, you take control of your own black hole to explore this relationship.

End States of Stars
In this animation, you can contrast the life spans and stages of nuclear fusion in high- and low-mass stars.

ASTRONOMY EXERCISES

Black Hole
You can change the mass and radius of the object in this animation. See if you can discover what it takes to make a black hole.

VIRTUAL ASTRONOMY LABS

Lab 14: Neutron Stars and Pulsars
This lab looks at the extraordinary properties of neutron stars, the dense balls of neutrons that may remain after some stars have exploded. Later the lab examines pulsars, neutron stars that appear to emit radiation in rapid pulses.

Lab 15: General Relativity and Black Holes
This lab explores the properties of black holes. It includes an exercise illustrating Einstein's general theory of relativity and an exercise on binary quasars. The lab concludes with a discussion and an exercise on black hole detection.

Critical Inquiries for the Web

1. Imagine that you are on a mission to explore one of the pulsar planets noted in this chapter. What would you find there? Look on the Web for information about pulsars and the known pulsar planets and describe what you might encounter on such a mission.

2. What would you experience if you were to pilot a spacecraft near a black hole? Visit related Internet sites to determine what the gravitational effects and general environment would be like. Also, use the Internet to find the limits of human tolerance to strong gravitational forces. (*Hint:* Look for information about astronaut training and find out how many Gs a human can withstand.) Using these sources, give a brief account of what your voyage would be like.

3. Search for information about gamma-ray bursters. What is the latest news in this developing story?

Exploring *TheSky*

1. The Crab Nebula pulsar is located in the Crab Nebula, also known as M1. Locate the nebula, zoom in, and compare its shape and size with Figure 14-5. (*TheSky* does not show the pulsar.)

Go to the Brooks/Cole Astronomy Resource Center **(http:// astronomy.brookscole.com)** for critical thinking exercises, articles, and additional readings from InfoTrac College Edition, Brooks/Cole's online student library.

15 | The Milky Way Galaxy

The Stars Are Yours.

JAMES S. PICKERING

Visual-wavelength image

THE STARS ARE YOURS was the title
of a popular astronomy book written by James
S. Pickering in 1948. You would have liked that book back in
1948, although it has gotten a bit out of date over the last half-century. The point of the title is that the stars belong to
everyone equally, and you can enjoy the wonder of the night sky as if you owned it.   In fact, the stars are indeed
yours. You live inside one of the largest star systems in the universe. Our Milky Way Galaxy is over 75,000 ly in diameter
and contains over 100 billion stars. If Earth is the only inhabited planet in the galaxy, then the galaxy belongs to all of us;
and, sharing equally, each person on Earth owns 50 stars plus any associated planets, moons, comets, and so on. Even if

Continued on page 340

338

Guidepost

Looking Back

In previous chapters, you learned how stars are born from clouds of gas and dust, how they evolve as they consume their nuclear fuels, and how they die. You discovered that most stars live long, quiet lives but that the most massive stars die in violent supernova explosions.

This Chapter

Now you are ready to see stars in their vast communities called galaxies. This chapter discusses our home galaxy, the Milky Way Galaxy, and attempts to answer four essential questions:

How do astronomers know we live in a galaxy?

How did our galaxy form and evolve?

What are the spiral arms?

What lies at the very center?

In this chapter you will see more examples of how scientists use evidence and theory to understand nature. If in some cases the evidence seems contradictory and the theories incomplete, do not be disappointed. The adventure of discovery is not yet over.

Looking Ahead

In the chapters that follow, you will meet some of the billions of galaxies that fill the depths of the universe. Understanding the Milky Way Galaxy is only a step in understanding the universe as a whole.

■ Figure 15-1

(a) Nearby stars look bright, and peoples around the world group them into constellations. Nevertheless, the vast majority of the stars in our galaxy merge into a faintly luminous path that circles the sky, the Milky Way. This artwork shows the location of a portion of the Milky Way near a few bright winter constellations. (See Figure 15-4 and the star charts at the end of this book to further locate the Milky Way in your sky.) (b) This photograph of the Andromeda Galaxy, a spiral galaxy about 2.3 million ly from Earth, shows what our own galaxy would look like if you could view it from a distance. (AURA/NOAO/NSF)

we must share with a few other inhabited planets, you are rich beyond any dream. You lack only transportation to visit your dominions.

As you read this chapter, you will be learning about our home galaxy, but you will also be learning how the stars of the galaxy have cooked up the atoms heavier than helium. You know already how the cores of massive stars make the heavier atoms, up to iron, and how supernovae blast those atoms back into space and add even heavier atoms made only during the supernova explosion. In this chapter you will see how the stars in the Milky Way Galaxy have, generation after generation, made the atoms in your body. As you begin this chapter it may seem that the stars belong to you; but, by the end of this chapter, you may decide that you belong to the stars.

15-1 The Nature of the Milky Way Galaxy

THE SECRET TO STUDYING any science lies in the simple question, "How do we know?" Because all scientific knowledge is based on evidence and logical arguments about natural processes, you need to know facts to get started. But you also need to understand how you know those facts. The next few pages set the

scene with a few facts, but the more interesting question is how astronomers know those facts. How do they know what our galaxy is like?

The Milky Way across the Sky

It isn't obvious that we live in a galaxy. We are inside, and we see nearby stars scattered all over the sky. From a dark site far from city lights, you can see great clouds of stars in our galaxy making a faint band of light circling the sky (■ Figure 15-1a). The ancient Greeks named that band *galaxies kuklos,* the "milky circle." The Romans changed the name to *via lactia,* "milky road" or "milky way." You will see in this chapter how the evidence reveals that we live inside a great wheel of stars, a galaxy.

Almost every object you see in the sky is part of our Milky Way Galaxy. Exceptions include the **Magellanic Clouds,** small irregular galaxies located near our galaxy in space. They are visible as faint patches of haze in the southern sky. Another exception is the Andromeda Galaxy, just visible to your unaided eyes as a faint patch of light in the constellation Andromeda. (Consult the star charts at the end of this book to locate the Androm-eda Galaxy.) You will see evidence that our galaxy looks much like the disk-shaped Andromeda Galaxy (Figure 15-1b).

The largest telescopes can detect an estimated 100 billion extremely faint, hazy objects, which astronomers, drawing on the Greek word for milk, call galaxies. They appear to be other systems of stars similar to our own. Our home galaxy appears to be special only in that it is our home.

The Structure of Our Galaxy

We live in a disk-shaped galaxy with beautiful spiral arms reaching from its center to its edge. Astronomers commonly give the diameter of the galaxy as 25,000 pc or 75,000 ly, with the sun located about two-thirds of the way from the center to the edge (■ Figure 15-2). The diameter of our galaxy is not known very accurately, as you can see from the two numbers given above. Convert 25,000 pc into light-years, and you get significantly more than 75,000 ly. That these two numbers don't quite match is a warning that the size of our galaxy is not known to better than about 10 percent.

The disk of the galaxy contains most of the stars and nearly all of the gas and dust. Although most of the stars in the disk are middle- to lower-main-sequence stars like the sun, a few are giants, and fewer still are brilliant O and B stars. These hot, luminous stars light up the disk and make it bright. It is difficult to judge just how big the disk is, because it is filled with gas and dust that dim starlight and make it difficult to see distant stars in the disk. Also, large dusty clouds block the view in many directions within the disk. In most directions in the plane of the disk, astronomers can't see farther than a few **kiloparsecs.** (A kiloparsec is 1000 pc.)

If you look up or down out of the disk, you look away from the dust and gas, and you can see out into the **halo** of our galaxy, a spherical cloud of stars and star clusters that contains almost no gas and dust. The halo completely encloses the disk of the galaxy. The size of the halo is uncertain, but it may be ten times the diameter of the disk.

Around the center of our galaxy lies the **nuclear bulge,** a flattened cloud of stars about 3 kpc in radius. It contains little gas and dust but billions of stars similar to those found in the halo.

This quick portrait of our galaxy is just a collection of bare facts. It raises two important questions that will guide you through the rest of this chapter. How do astronomers know what our galaxy is like? How did our galaxy get this way?

First Studies of the Galaxy

How do astronomers know what our galaxy is like? Humanity has known of the Milky Way, the hazy path of light around the sky, since antiquity, but not until Galileo looked at the sky with his telescope in 1610 did anyone know that the Milky Way was made of stars. Little more was known for two centuries after Galileo lived.

By the mid-18th century, astronomers generally understood that the stars were other suns, but they had little understanding of

■ Figure 15-2

An artist's conception of our Milky Way Galaxy, seen face-on and edge-on. Note the position of the sun and the distribution of globular clusters in the halo. Hot blue stars light up the spiral arms. Only the inner halo is shown here. At this scale, the entire halo would be larger than a dinner plate.

how the stars were distributed in space. One of the first people to study this problem was the English astronomer Sir William Herschel (1738–1822). He and his sister Caroline (1750–1848) mapped the stars in three dimensions by assuming that the sun was located inside a great cloud of stars and that they could see to the edges of the cloud in any direction. They believed that by counting the number of stars visible in different directions, they could gauge the relative distance to the edge of the cloud. If their telescope revealed few stars in one direction, they assumed that the edge was not far away; and, if they saw many stars in another direction, they assumed that the edge was very distant. Calling their method "star gauges," they counted stars in 683 directions in the sky and outlined a model of the star cloud. Their data showed that the cloud was a great disk with the sun near the center; and, using the technology of their day as an analogy, they called it the "grindstone model" (■ Figure 15-3).

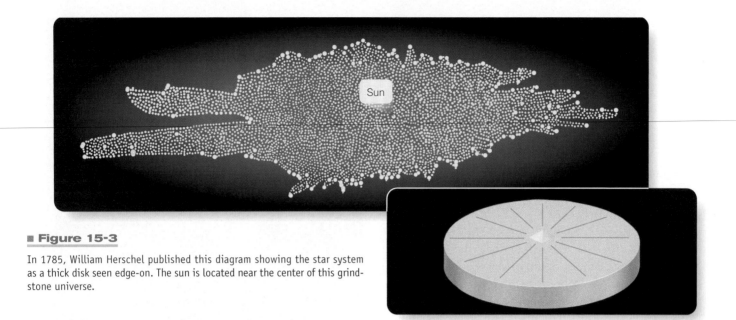

■ Figure 15-3

In 1785, William Herschel published this diagram showing the star system as a thick disk seen edge-on. The sun is located near the center of this grind-stone universe.

Open cluster M52 in Cassiopeia

Open clusters lie along the Milky Way all around the sky.

Visual-wavelength image

■ Figure 15-4

Nearly half of cataloged globular clusters (red dots) are located in or near Sagittarius and Scorpius. A few of the brighter globular clusters, labeled with their catalog designations, are visible in binoculars or small telescopes. Constellations are shown as they appear above the southern horizon on a summer night as seen from latitude 40° N, typical for most of the United States. (M52: NOAO/AURA/NSF; M19: Doug Williams, N.A. Sharp/ NOAO/AURA/NSF)

Globular clusters

M50

M22

M19

Center of
+
galaxy

M4

M54

M62

Sagittarius

Scorpius

M55

M70

Globular cluster M19

Globular clusters are scattered over the entire sky but are strongly concentrated toward Sagittarius.

Visual-wavelength image

The Calibration of One Parameter to Find Another

Astronomers often say that Shapley "calibrated" the Cepheids for the determination of distance, meaning that he did all the detailed background work so that the Cepheids could be used to find distances. Other astronomers could use the Cepheids without repeating the detailed calibration.

Calibration is actually very common in science. Chemists, for instance, have carefully calibrated the colors of certain compounds against acidity. They can quickly measure the acidity of a solution by dipping into it a slip of paper containing the indicator compound and looking at the color. They don't have to repeat the careful calibration every time they measure acidity.

Engineers in steel mills have calibrated the color of molten steel against its temperature. They can use a handheld device that measures the color of a ladle of molten steel and then looks up the temperature from a calibrated table. They don't have to repeat the calibration every time. Astronomers have made the same kind of color–temperature calibration for stars.

As you read about any science, notice how calibrations are used to simplify common measurements. But notice, too, how important it is to get the calibration right. An error in calibration can throw off every measurement made with that calibration. Some of the biggest errors in science have been errors of calibration.

This model of the star system explains the most obvious feature of the Milky Way—it is a glowing path that circles the sky. The grindstone model reproduces this as a disk of stars with the sun near the center. Unfortunately, there was no way for the Herschels to find the diameter of the disk.

Generations of astronomers studied the size of our star system, culminating with Jacobus C. Kapteyn (1851–1922). In the early 20th century, he analyzed the brightness, number, and motion of stars and concluded that our star system is a disk about 10 kpc in diameter and about 2 kpc thick, with the sun near the center.

As you saw in the previous section, modern astronomers believe our galaxy is much larger and that the sun is not at the center. Why did these first studies of the galaxy underestimate its size and mistakenly place the sun near the center? How astronomers discovered the true size of our galaxy and came to understand the nature of a galaxy is one of the great adventures of human knowledge. It began about a century ago when a man studying star clusters and a woman studying variable stars unlocked one of nature's biggest secrets—that our star system is a galaxy.

Discovering the Galaxy

It seems odd to say that astronomers discovered something that is all around us, but until early in the 20th century, no one knew that the star system we live in is a galaxy. That began to change in the second decade of the 20th century, when a young astronomer named Harlow Shapley (1885–1972) discovered how big our star system really is. Besides being one of the turning points of modern astronomy, Shapley's study illustrates one of the most common techniques in astronomy. If you want to know how astronomers know things about the universe, then Shapley's story is well worth tracing in detail.

Shapley began by noticing that open star clusters lie along the Milky Way but that more than half of all globular clusters lie in or near the constellation Sagittarius (■ Figure 15-4). The globular clusters seemed to be distributed in a great cloud whose center was located in the direction of Sagittarius. Shapley assumed that the orbital motion of these clusters was controlled by the gravitation of the entire star system, and, for that reason, the center of the star system could not be near the sun but must lie somewhere toward Sagittarius.

To find the distance to the center of the star system and thus the size of the star system as a whole, Shapley needed to find the distance to the star clusters, but that was difficult to do. The clusters are much too far away to have measurable parallaxes. They do, however, contain variable stars (Chapter 12), and those were the lampposts Shapley needed to find the distances to the clusters.

Shapley knew of the work of Henrietta S. Leavitt (1868–1921), who in 1912 had shown that a Cepheid variable star's period of pulsation was related to its brightness. Leavitt worked on stars whose distances were unknown, so she was unable to find their absolute magnitudes or luminosities, but astronomers realized that the Cepheids could be a powerful tool in astronomy if their true luminosities could be discovered.

Cepheids are giant and supergiant stars and so are relatively rare. None lies close enough to Earth to have a measurable parallax. Their proper motions—slow movement across the sky—can be measured, and the more distant a star is, the smaller its proper motion tends to be. In that way, proper motions contain clues to distance. Ejnar Hertzsprung (codiscoverer of the H–R diagram) made an early attempt to determine the absolute magnitudes of the Cepheids, but Shapley went further. He found 11 Cepheids with measured proper motions and used a statistical process to find their average distance and thus their average absolute magnitude. That meant he could erase Leavitt's apparent magnitudes from the period–luminosity diagram (see Figure 12-14) and write in absolute magnitudes. All of the stars in the diagram were thus calibrated as lampposts that astronomers could use to find distances (**Window on Science 15-1**), including the RR Lyrae stars that Shapley saw in the globular clusters.

Finding distance using Cepheid variable stars is so important in astronomy that you should pause to examine the process. Suppose you studied an open star cluster and discovered that it contained a type I Cepheid with a period of 10 days and an average apparent magnitude of 10.5. How far away is the cluster? From the period–luminosity diagram (Figure 12-14), you can see that the absolute magnitude of the star must be about −3. Then the distance modulus (Chapter 9) is:

$$m - M_v = 10.5 - (-3) = 13.5$$

You could use this distance modulus and a table such as Table 9-1 to estimate the distance to the cluster as somewhere between 1000 and 16,000 pc. You can be more accurate if you solve the magnitude–distance formula (Chapter 9) for distance:

$$d = 10^{(m_v - M_v + 5)/5}$$

Now you can substitute the apparent magnitude and absolute magnitude and discover that the distance is equal to $10^{3.7}$. A pocket calculator reveals that the distance is about 5000 pc. This is one of the most common calculations in astronomy.

Once Shapley had calibrated the period–luminosity relation, he could use it to find the distance to any cluster in which he could identify variable stars. By taking a series of photographic plates over a number of nights, Shapley could pick out the variable stars in a cluster, measure their average apparent magnitude, and find their periods of pulsation. Because the periods were very short, he knew the variable stars he saw in the clusters were the lowest-luminosity variable stars (known today as RR Lyrae stars). That allowed him to read their absolute magnitude off the period–luminosity diagram. Once he knew the apparent magnitude and the absolute magnitude, he could calculate the distance to the star cluster.

This worked well for the nearer globular clusters, but the variable stars in the more distant clusters are too faint to detect. Shapley estimated the distances to these more distant clusters by calibrating the diameters of the clusters. For the clusters whose distance he knew, he could use their angular diameters and the small-angle formula to calculate linear diameters in parsecs. He found that the nearby clusters are about 25 pc in diameter, which he assumed is the average diameter of all globular clusters. He then used the angular diameters of the more distant clusters to find their distances. This is another illustration of calibration in astronomy.

Shapley later wrote that it was late at night when he finally plotted the direction and distance to the globular clusters on graph paper and found that, just as he had supposed, they formed a great swarm whose center lay many thousands of light-years in the direction of Sagittarius. Evidently the center of the star system was not near the sun but rather was far away in Sagittarius (■ Figure 15-5). He found the only other person in the building, a cleaning lady, and the two stood looking at his graph as he explained that they were the only two people on Earth who under-

stood that humanity lives, not at the center of a small star system, but in the suburbs of a vast wheel of stars.

Why did astronomers before Shapley think we lived near the center of a small star system? Space is filled with gas and dust that dim the view of distant stars. When astronomers look toward the band of the Milky Way, they can see only the neighborhood near the sun. Most of the star system is invisible, and, like travelers in a fog, we seem to be at the center of a small region. Shapley was able to see the globular clusters at greater distances because they lie outside the plane of the star system and are not dimmed very much by the gas and dust.

Building on Shapley's work, other astronomers began to suspect that some of the faint patches of light visible through telescopes were other star systems like our own. The next chapter will continue this story and discuss galaxies in general. It is sufficient here to note that within a few years astronomers had found evidence that the faint patches of light were other galaxies much like our own Milky Way Galaxy.

Shapley's study of star clusters led to the discovery that we live in a galaxy and that the universe is filled with similar galaxies. Like Copernicus, Shapley moved us from the center to the suburbs. Nevertheless, our home galaxy is dramatically beautiful. To get a true impression of our home galaxy, you must conduct a careful analysis.

An Analysis of the Galaxy

The key to science is knowing how we know, and you are now ready to explore the modern understanding of our galaxy. You can survey the galaxy's components, understand how astronomers have learned about them, and begin to understand what those components reveal about the nature of our home galaxy.

You can divide our galaxy into two main components whose characteristics will help you understand how our galaxy formed. The **disk component** of the galaxy contains most of the stars, gas, and dust that orbit within the disk. In contrast, the **spherical component** includes the nuclear bulge and the halo.

In addition to stars, the disk of the galaxy contains vast amounts of gas and dust—the interstellar medium. You saw in Chapter 10 how astronomers can learn about this material. It contains mostly hydrogen gas and helium gas and is irregularly distributed as dense clouds separated by less dense regions. The densest of these clouds are the giant molecular clouds within which stars form (Chapter 11), and consequently nearly all star formation in our galaxy takes place in the disk. The most massive of these newborn stars are the most luminous, the O and B stars. They light up the disk and make it glow brightly. Some of these stars are hot enough to ionize the nearby gas to produce HII regions. The disk contains all of the emission nebulae in our galaxy.

Although astronomers can survey the location and distance of the stars (Chapter 9), they cannot cite a single number for the thickness of the disk because it lacks sharp boundaries. Stars become less crowded farther from the central plane of the galaxy.

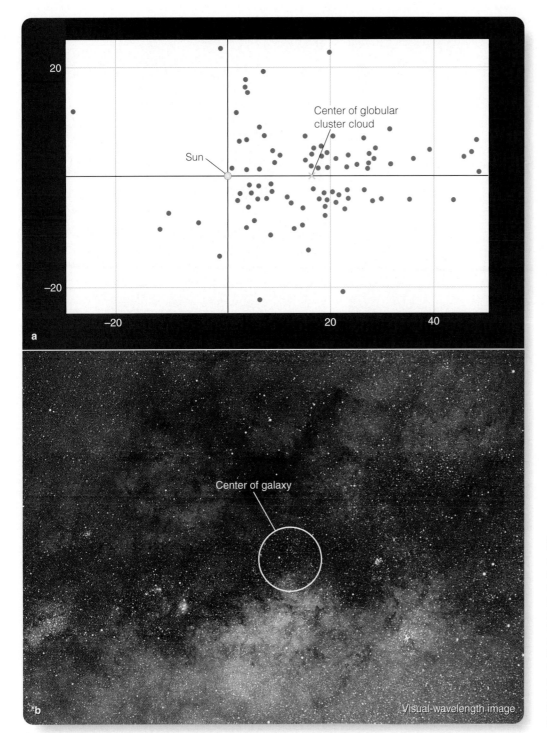

■ Figure 15-5

(a) Shapley's study of globular clusters showed that they were not centered on the sun, at the origin of this graph, but rather formed a great cloud centered far away in the direction of Sagittarius. Distances on this graph are given in thousands of parsecs. (b) Looking toward Sagittarius, you see nothing to suggest that this is the center of the galaxy. Gas and dust block your view. Only the distribution of globular clusters told Shapely the sun lay far from the center of the star system. (Daniel Good)

Also, the thickness of the disk depends on the kind of object studied. Stars like the sun with ages of a few billion years lie within about 500 pc above and below the central plane. But the youngest stars, including the O and B stars, and the gas and dust from which these young stars are forming are confined to a thin disk only about 100 pc thick (■ Figure 15-6). With a diameter of 25 kpc, the disk is, in proportion, thinner than a thin pizza crust.

From the shape and size of our galaxy, astronomers conclude that we live in a spiral galaxy (■ Figure 15-7). The most dramatic

features of these disk galaxies are the spiral arms—long spiral patterns of bright stars, HII regions, star clusters, and clouds of gas and dust. The sun is located on the inside edge of one of these spiral arms. You will see later in this chapter how astronomers can observe the spiral arms in our own galaxy.

There are two kinds of star groupings in the disk of our galaxy. Associations are groups of 10 to a few hundred stars so widely scattered in space that their mutual gravity cannot hold the association together. From the turnoff points in their H–R diagrams

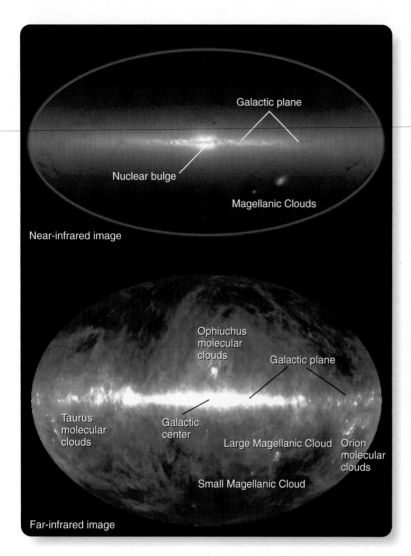

Galactic plane

Nuclear bulge

Magellanic Clouds

Near-infrared image

Ophiuchus
molecular
clouds

Galactic plane

Taurus
molecular
clouds

Galactic
center

Large Magellanic Cloud

Orion
molecular
clouds

Small Magellanic Cloud

Far-infrared image

The Sombrero Galaxy

a   Visual-wavelength image

NGC 2997

b   Visual-wavelength image

■ Active Figure 15-6

In these infrared images, the entire sky has been projected onto ovals with the center of the galaxy at the center. The Milky Way extends from left to right. In the near-infrared, the nuclear bulge is prominent, and dust clouds block your view. At longer wavelengths, the dust emits black body radiation and glows brightly. (Near-IR: 2Mass; Far-IR: DIRBE Image courtesy Henry Freudenreich)

Ace⊘Astronomy™   Log into AceAstronomy and select this chapter to see the Active Figure "Explorable Milky Way." Explore the Milky Way Galaxy at different wavelengths.

■ Figure 15-7

Our disk galaxy resembles the spiral galaxies visible through large telescopes. (a) Seen edge-on, spiral galaxies have nuclear bulges and dust-filled disks. (Todd Boroson/NOAO/AURA/NSF) (b) Seen face-on, the spiral arms are dramatically outlined by hot O and B stars, emission nebulae, and dust. (ESO)

(Chapter 12), you can tell they are very young groups of stars. The stars are moving together through space (■ Figure 15-8) because they formed from a single gas cloud and haven't had time to wander apart. Two kinds of association—O and B associations and T Tauri associations (Chapter 11)—are located along the spiral arms.

The second kind of cluster in the disk is the open cluster (Figure 15-4), a group of 100 to a few thousand stars in a region about 25 pc in diameter. Because they have more stars in less space than associations, open clusters are more firmly bound by gravity. Although they lose stars occasionally, they can survive for

a long time, and the turnoff points in their H–R diagrams give ages from a few million to a few billion years.

In contrast to the disk, the halo of our galaxy contains almost no gas and dust and only a thin scattering of stars and globular clusters. Because the halo contains no dense gas clouds, it cannot make new stars. Halo stars are old, cool, lower-main-sequence stars, red giants, and white dwarfs. It is difficult to judge the extent of the halo, but it could extend far beyond the edge of the visible disk and may be slightly flattened like a thick hamburger bun.

The halo contains roughly 200 globular clusters, each of which contains 50,000 to a million stars in a sphere about 25 pc in diameter (Figure 15-4). Because they contain so many stars in

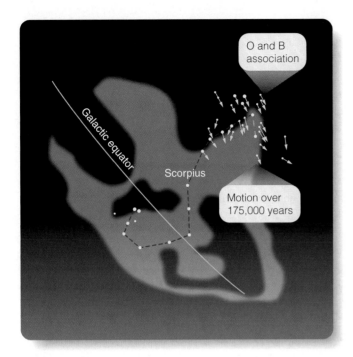

■ Figure 15-8

Many of the stars in the constellation Scorpius are members of an O and B association that has formed recently from a single cloud of gas. As the stars orbit the center of our galaxy, they are moving together southwest along the Milky Way.

such a small region, the clusters are very stable and have survived for billions of years. From the turnoff points in their H–R diagrams, you can tell they average about 11 billion years old.

The nuclear bulge at the center of the galaxy is the most crowded part of the spherical component. Visible above and below the plane of the galaxy and through gaps in the obscuring dust, the nuclear bulge contains stars that are old and cool like the stars in the halo.

This analysis of the components of our galaxy may leave you wondering about one critical question: How much matter does our galaxy contain? To answer that question, you must watch our galaxy rotate.

The Mass of the Galaxy

To find the mass of an object, astronomers must observe it in orbital motion as, for example, in a binary star system. Humans don't live long enough to see our galaxy rotate significantly, but astronomers can observe the radial velocities, proper motions, and distances of stars and then calculate their orbits. The results can reveal the mass of the galaxy and provide hints to its origin.

Stars in the disk of the galaxy follow nearly circular orbits in the plane of the disk (■ Figure 15-9a). By observing the radial velocities of other galaxies in various directions, astronomers can conclude that the sun is moving at about 220 km/s in the direction of Cygnus as it orbits the center of our galaxy in a nearly circular path. Given the distance to the center of our galaxy, 8.5 kpc,

you can find the circumference of the sun's orbit by multiplying by 2π. If you divide the circumference of its orbit by its orbital velocity, you will discover that the sun has an orbital period of about 240 million years.

Halo stars and globular clusters, however, follow highly elongated orbits tipped steeply to the plane of the disk (Figure 15-9b). Although the diagram shows these orbits as elliptical, they are more like rosettes because of the influence of the thick bulge. Where halo stars pass through the disk, they are cutting across the regular traffic of disk stars and seem to have very high velocities. For that reason, they are called **high-velocity stars.** The dramatic difference between the motions of halo stars and disk stars will be important evidence when you consider the formation of the galaxy in the next section.

The orbital motion of the sun can reveal the mass of the galaxy. When you studied binary stars in Chapter 9, you saw that the total mass (in solar masses) of a binary star system equals the cube of the separation of the stars a (in AU) divided by the square of the period P (in years):

$$M = \frac{a^3}{P^2}$$

The radius of the sun's orbit around the galaxy is about 8500 pc, and each parsec contains 206,265 AU. Multiplying, you find that the radius of the sun's orbit is 1.75×10^9 AU. The orbital period is 240 million years, so the mass is:

$$M = \frac{(1.75 \times 10^9)^3}{(240 \times 10^6)^2} = 0.93 \times 10^{11}\ M_\odot$$

■ Figure 15-9

(a) Stars in the galactic disk have nearly circular orbits that lie in the plane of the galaxy. (b) Stars in the halo have randomly oriented, highly elongated orbits.

■ Figure 15-10

The differential rotation of the galaxy means that stars at different distances from the center have different orbital periods. In this example, the star just inside the sun's orbit has a shorter period and pulls ahead of the sun, while the star outside falls behind.

This is only a rough estimate because it does not include any mass outside the orbit of the sun. Adding an estimate for the mass outside the sun's orbit yields a total mass for our galaxy of about 4×10^{11} solar masses.

The rotation of our galaxy is actually much more interesting than the description above. Stars at different distances from the center revolve around the center of the galaxy with different periods, so three stars near each other will draw apart as time passes (■ Figure 15-10). This motion is called differential rotation. (Recall that differential rotation was defined in Chapter 8.)

To fully describe the rotation of our galaxy, astronomers graph orbital velocity versus radius—a **rotation curve** (■ Figure 15-11). If all of the mass of the galaxy were concentrated near its center, then you would expect to see orbital velocities fall as you moved away from the center. This is what you see in our solar system, where nearly all of the mass is concentrated in the sun, and it is called **Keplerian motion** (a reference to Kepler's laws). The best observations of the rotation curve of the Milky Way show orbital velocities constant or rising in the outer disk rather than falling. These higher orbital velocities indicate that the larger orbits enclose more mass and seem to say that our galaxy is more massive than it appears to be.

The evidence tells astronomers that the extra mass lies in an extended halo sometimes called a **galactic corona.** It may extend up to 10 times farther than the edge of the visible disk and could contain a trillion solar masses. Much of this mass is invisible, so astronomers conclude it is not emitting or absorbing light and refer to it as **dark matter.** At least some of the mass in the galactic corona is made up of low-luminosity stars and white dwarfs, but much of the mass must be some other form of matter. You will learn more about the problem of the dark matter in the following two chapters. It is one of the fundamental problems of modern astronomy.

■ Figure 15-11

The rotation curve of our galaxy is plotted here as orbital velocity versus radius. Data points show measurements made by radio telescopes. Observations outside the orbit of the sun are much more uncertain, and the data points scatter widely. Orbital velocities do not decline outside the orbit of the sun, as you would expect if most of the mass of the galaxy were concentrated toward the center (Keplerian motion). Rather, the curve is approximately flat at great distances, suggesting that the galaxy contains significant mass outside the orbit of the sun. (Adapted from a diagram by Françoise Combes)

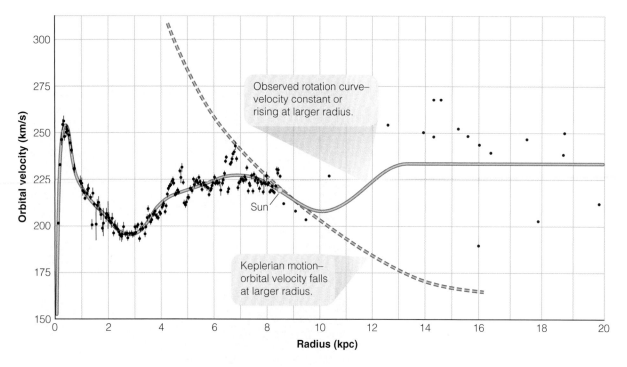

If gas and dust block the view, how do astronomers know how big our galaxy is?

Because scientific arguments depend ultimately on evidence, they must explain how scientists know what they know. The gas and dust in our galaxy block the view only in the plane of the galaxy. When telescopes look away from the plane of the galaxy, they look out of the gas and dust and can see to great distances. The globular clusters are scattered through the halo with a strong concentration in the direction of Sagittarius. When astronomers look above or below the plane of the galaxy, they can see those clusters, and careful observations reveal variable stars in the clusters. By using a modern calibration of the period–luminosity diagram, they can find the distance to those clusters. If they assume that the distribution of the clusters is controlled by the gravitation of the galaxy as a whole, then they can find the distance to the center of the galaxy by finding the center of the distribution of globular clusters.

In any logical argument, it is important to ask "How do we know?" and that is especially true in science. Create a new argument: **What measurements and assumptions reveal the total mass of our galaxy?**

■   ■   ■

Connections: The differences between the disk and spherical components extend even to their chemical composition. As you will see in the next section, those differences hold clues to the origin of our galaxy.

15-2 The Origin of the Milky Way

JUST AS DINOSAURS LEFT BEHIND fossilized footprints, our galaxy has left behind a fossil footprint of its youth. The stars of the spherical component are old and must have formed long ago when the galaxy was very young.

Stellar Populations

In the 1940s, astronomers realized that there are two families of stars in the galaxy. They form and evolve in similar ways, but they differ in the abundance of atoms that are heavier than helium—atoms that astronomers call **metals.** (Note that this is not the way the word *metal* is used by nonastronomers.) **Population I stars** are metal rich, containing 2 to 3 percent metals, whereas **population II stars** are metal poor, containing only about 0.1 percent metals or less. The difference may seem small, but it is dramatically evident in spectra (■ Figure 15-12).

Many metal lines

H_γ H_β H_α

Pop I

Pop II

Few metal lines

a

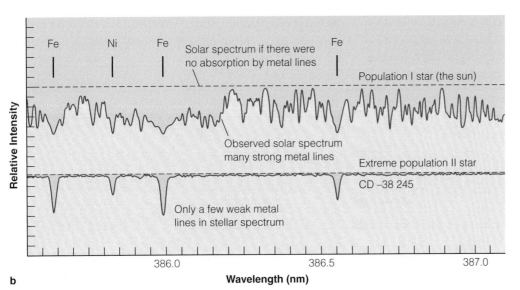

Fe   Ni   Fe   Solar spectrum if there were no absorption by metal lines   Fe

Population I star (the sun)

Observed solar spectrum many strong metal lines

Extreme population II star

CD –38 245

Only a few weak metal lines in stellar spectrum

Relative Intensity

386.0   386.5   387.0

Wavelength (nm)

b

■ **Figure 15-12**

(a) The difference between population I stars and population II stars is dramatic. Examine the upper spectrum here and notice the hundreds of faint spectral lines. The lower spectrum has fewer and weaker lines. (AURA/ NOAO/NSF) (b) A graph of such spectra reveals overlapping absorption lines of metals completely blanketing the population I spectrum. The lower spectrum is that of an extremely metal-poor star with only a few weak metal lines of iron (Fe) and nickel (Ni). This population II star contains about 10,000 times less metals than the sun. (Adapted from an ESO illustration)

Population I stars belong to the disk component of the galaxy and are sometimes called disk population stars. They have nearly circular orbits in the plane of the galaxy and are relatively young stars that formed within the last few billion years. The sun is a population I star, as are the type I Cepheid variables discussed in Chapter 12.

Population II stars belong to the spherical component of the galaxy and are sometimes called the halo population stars. These stars have randomly tipped orbits ranging from circular to highly elliptical. They are old stars that formed when the galaxy was young. The metal-poor globular clusters are part of the halo population, as are the RR Lyrae and type II Cepheids.

Since the discovery of stellar populations, astronomers have realized that there is a gradation between populations. Extreme population I stars are found only in the spiral arms. Slightly less metal-rich population I stars, called intermediate population I stars, are located throughout the disk. The sun is such a star. Stars even less metal rich, such as stars in the nuclear bulge, belong to the intermediate population II. The most metal-poor stars are those in the halo, including those in globular clusters. These are extreme population II stars.

Why do the disk and halo stars have different metal abundances? The answer to this question is the key to understanding of the history of our galaxy.

The different populations of stars differ in other ways, such as type of orbit and age (■ Table 15-1). These are clues to the formation of our galaxy. To follow these clues you need to consider the origin of the chemical elements.

The Element-Building Process

You can use what you have learned about stellar structure and stellar evolution (Chapters 11 to 13) to understand the origin of the metals. Remember that in astronomy, the term *metals* refers to all of the chemical elements heavier than helium. When the universe began, it contained only hydrogen and helium atoms. (You will see the evidence that supports this statement in Chapter 18.) All of the remaining chemical elements have been produced by **nucleosynthesis,** the process that fuses hydrogen and helium to make the heavier elements (**Window on Science 15-2**).

When our galaxy formed sometime after the beginning of the universe, the gas contained about 90 percent hydrogen atoms, 10 percent helium atoms, and almost no heavier atoms. So the first stars to form had to be metal poor. Succeeding generations of stars manufactured heavier atoms, and the metal abundance of the galaxy gradually increased.

The most massive stars fuse helium into carbon, nitrogen, oxygen, and heavier elements up to iron (Chapter 11), and when those stars die in supernova explosions, traces of those elements are spread back into the interstellar medium. Furthermore, the supernova explosion itself can fuse nuclei into atoms heavier than iron. These are rare atoms, such as gold, silver, and platinum, but they, too, get spread back into space during supernova explosions, and there they may become part of newly forming stars.

Elements heavier than helium, metals, are actually quite rare in the universe. Traditionally, astronomers graph the abundance of the elements using an exponential scale, but if you replot the data using a linear scale you can see how rare these atoms are (■ Figure 15-13).

When you look at population II stars, such as those in the halo, you are looking at the survivors of the early generations of stars in our galaxy. Those stars formed from gas that was metal poor, and their spectra show only weak metal lines. The survivors of these early generations are low-mass, long-lived stars that don't make heavy elements. Furthermore, any elements they might have made are locked in their cores. That means that their spectra reveal the composition of the gas from which they formed and that the gas was metal poor. Population I stars, such as the sun, formed more recently, after the interstellar medium had been enriched in metals, and their spectra show stronger metal lines. Stars forming now show the strongest metal lines.

Galactic Fountains

Supernova remnants are rich in metals and eventually mix back into the interstellar medium. In fact, supernovae continuously stir the interstellar medium. But a larger-scale process may be even more efficient at spreading newly created metals throughout the disk of the galaxy where they can be incorporated into the newly forming stars.

■ Table 15-1 I Stellar Populations

	Population I		Population II	
	Extreme	*Intermediate*	*Intermediate*	*Extreme*
Location	Spiral arms	Disk	Nuclear bulge	Halo
Metals (%)	3	1.6	0.8	Less than 0.8
Shape of orbit	Circular	Slightly elliptical	Moderately elliptical	Highly elliptical
Average age (yr)	100 million and younger	0.2–10 billion	2–10 billion	10–13 billion

Understanding Nature as Processes

One of the most common organizing themes in science is the concept of a process—a sequence of events that leads to some result or condition. Scientists try to discover the natural processes that shape the universe. Telling the stories of these processes explains why the universe is the way it is. At the same time, recognizing these processes and learning to tell their stories simplify much of science.

For example, astronomers try to assemble the sequence of events that led to the formation of the chemical elements. They know that the first stars formed from nearly pure hydrogen and helium and that they made certain heavier elements in their cores. Supernova explosions made other elements and mixed the elements back into the interstellar medium. New stars formed from this gas, and generation after generation of stars added to the heavier elements that we are made of. Telling that story explains much of the science of the stars.

As you study any science, be alert for processes as organizing themes. Science, at first glance, seems to be nothing but facts, but one of the ways to organize the facts is to fit them into the story of a process. If you study biology and learn the process by which plant cells know whether to become roots, stems, or leaves, you will have organized a great deal of factual information into a story that is not only easy to remember but also makes sense.

When you see a process in science, ask yourself a few basic questions. What conditions prevailed at the beginning of the process? What sequence of steps occurred? Can some steps occur simultaneously, or must one step occur before another can occur? What is the final state that this process produces? Most important, what evidence is there that nature really works this way?

Not everything in science is a process, but identifying a process will help you remember and understand a great many details and principles.

X-ray image

Supernova remnant N49B contains knots rich in magnesium (blue-green), which was created deep inside the star by the process that built the chemical elements. (NASA/CXC/PSU/S. Park et al.)

In Chapter 10, you saw how neighboring supernova remnants can merge to form a superbubble of hot gas (Figure 10-9b). A supernova remnant may be only a few tens of parsecs in diameter, but a superbubble can be more than ten times bigger. The denser gas of the galactic disk tends to confine an expanding superbubble, but the disk is only a few hundred parsecs thick. If a superbubble breaks out of the denser gas of the galactic disk, it can spew hot gas high above the plane of the galaxy in a **galactic fountain** (■ Figure 15-14). As this gas cools and falls back into the disk, it can spread metals through the galaxy.

a

b

■ Figure 15-13

The abundance of the elements in the universe. (a) When the elements are plotted on an exponential scale, you see that elements heavier than iron are about a million times less common than iron and that all elements heavier than helium (the metals) are quite rare. (b) The same data plotted on a linear scale provide a more realistic impression of how rare the metals are. Carbon, nitrogen, and oxygen make small peaks near atomic mass 15, and iron is just visible in the graph.

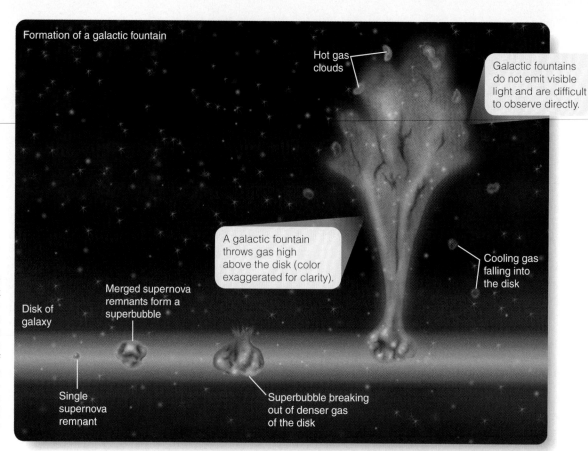

Formation of a galactic fountain

Hot gas clouds

Galactic fountains do not emit visible light and are difficult to observe directly.

A galactic fountain throws gas high above the disk (color exaggerated for clarity).

Cooling gas falling into the disk

Disk of galaxy

Merged supernova remnants form a superbubble

Single supernova remnant

Superbubble breaking out of denser gas of the disk

■ Figure 15-14

In this edge-on view of the galactic disk, you can see that superbubbles, formed of multiple supernova remnants, can be roughly as large as the thickness of the disk. Once a superbubble breaks out of the denser gas of the disk, it is believed to produce a galactic fountain. The metal-enriched gas could flow out into the halo where it would cool and fall back to enrich the interstellar medium in metals.

Galactic fountains are difficult to observe directly, but isolated clouds of gas have been found high in the halo, and cool clouds have been found falling toward the disk. Some of these clouds may be gas from outside the galaxy falling inward for the first time, but the lower-velocity clouds, only a few thousand parsecs above the disk, may be cooling gas spewed out of the disk by fountains. It seems very likely that this process contributes to spreading metals through the disk. Even without direct observational confirmation of galactic fountains, you can be sure that supernovae create metals and mix them back into the interstellar medium where they can be incorporated into newly forming stars.

Because the metal abundance of newborn stars increases as the galaxy ages, astronomers can use metal abundance as an index to the ages of the components of our galaxy. The halo is old, and the disk is younger. Such ages can help you understand the formation of the galaxy.

The Age of the Milky Way

Because astronomers know how to find the age of star clusters, they can estimate the age of our galaxy. It sounds straightforward, but uncertainties make the easy answer hard to interpret.

The oldest open clusters are 9 to 10 billion years old. These ages come from the turnoff point in the H–R diagram (see Chapter 12), but finding the age of an old cluster is difficult because old

clusters change so slowly. Also, the exact location of the turnoff point depends on chemical composition, which differs slightly among clusters. Finally, open clusters are not strongly bound by their gravity, so even older open clusters may have dissipated as their stars wandered away. So the galaxy could be older than the oldest remaining open clusters. All things considered, the open clusters suggest that the galactic disk is at least 9 billion years old.

Globular clusters have faint turnoff points in their H–R diagrams and are clearly old, but finding these ages is difficult. Clusters differ slightly in chemical composition, which must be accounted for in calculating the stellar models from which ages are determined. Also, to find the age of a cluster, astronomers must know the distance to the cluster. Precise parallaxes from the Hipparcos satellite have allowed astronomers to more precisely calibrate the Cepheid and RR Lyrae variable stars, and careful studies with the newest large telescopes have refined the chemical compositions and better defined the H–R diagrams of globular clusters. The best analysis of all of the data suggests that the average globular clusters are about 11 billion years old. Some globular clusters are younger than that, and some are older. Studies of the oldest globular clusters suggest that the halo of our galaxy is at least 13 billion years old.

Both populations and clusters show that the disk is younger than the halo. You can combine these ages with the process of nucleosynthesis to tell the story of our galaxy.

The History of the Milky Way Galaxy

In the 1950s, astronomers began to develop a hypothesis to explain the formation of our galaxy. Recent observations, however, are forcing a reevaluation of that traditional hypothesis.

The traditional hypothesis begins with the big bang beginning of the universe almost 14 billion years ago. (You will learn more about the big bang in Chapter 18). The big bang filled the universe with clouds of hydrogen and helium, and the traditional hypothesis proposes that one of these clouds contracted to form our galaxy. As gravity pulled the gas inward, the cloud began to fragment into smaller clouds; and, because the gas was turbulent, the smaller clouds had random velocities. Stars and star clusters that formed from these fragments went into orbits that had a wide range of shapes; a few were circular, but most were elliptical, and some were extremely elliptical. The orbits were also inclined at different angles, so the orbits produced a spherical cloud of stars. Of course, these first stars were metal poor because no stars had existed earlier to enrich the gas with metals. This is the way the traditional hypothesis explains the formation of the spherical component of the galaxy.

The second stage of this hypothesis accounts for the disk component. The contracting gas cloud was roughly spherical at first, but the turbulent motions canceled out, as do eddies in recently stirred coffee, leaving the cloud with uniform rotation. A rotating, low-density cloud of gas cannot remain spherical. A star is spherical because its high internal pressure balances its gravity; but, in a low-density cloud, the pressure cannot support the weight. Like a blob of pizza dough spun in the air, the cloud must flatten into a disk (■ Figure 15-15).

This contraction into a disk took billions of years, with the metal abundance gradually increasing as generations of stars were born from the gradually flattening gas cloud. The stars and globular clusters of the halo were left behind by the cloud when it was spherical, and subsequent generations of stars formed in flatter distributions. The gas distribution in the galaxy now is so flat that the youngest stars are confined to a disk only about 100 parsecs thick. These stars are metal rich and have nearly circular orbits.

This traditional hypothesis accounts for many of the Milky Way's properties. Advances in technology, however, have improved astronomical observation, and, beginning in the 1980s, contradictions arose. For example, many halo stars have metal abundances similar to those of globular clusters and may have escaped from such clusters, but some stars are even more metal poor. One star contains 200,000 times less metals than the sun. These stars may be older than the globular clusters. Also, some of the oldest stars in our galaxy are in the central bulge, not in the halo. Furthermore, not all globular clusters have the same age, and the younger clusters seem to be in the outer halo. The traditional hypothesis says that the halo formed first and that the clusters within it should have a uniform age.

Origin of the Halo and Disk

A spherical cloud of turbulent gas gives birth to the first stars and star clusters.

The rotating cloud of gas begins to contract toward its equatorial plane.

Stars and clusters are left behind in the halo as the gas cloud flattens.

New generations of stars have flatter distributions.

The disk of the galaxy is now very thin.

■ **Figure 15-15**

The traditional theory for the origin of our galaxy begins with a spherical gas cloud that flattens into a disk.

Ages are a key problem for the traditional hypothesis. The globular clusters are 10 to 13 billion years old, but the disk seems much younger. The oldest open star clusters in the disk are only 9 to 10 billion years old. Also, white dwarfs tend to accumulate over the eons, and there aren't as many white dwarfs in the disk as there should be if the disk were 10 to 13 billion years old. A successful hypothesis should account for these ages.

Another problem is that the oldest stars are metal poor but not metal free. There must have been at least a few massive stars to create these metals before the formation of the oldest stars astronomers see in the halo.

Can the traditional hypothesis be modified to explain these observations? Perhaps the galaxy began with the contraction of a gas cloud to form the central bulge and the later accumulation of the halo from gas clouds that had been slightly enriched in metals by an early generation of massive stars. A thick but low-density disk of stars may have formed at this early stage. This would explain the age of the central bulge and the metals in the oldest stars in the halo and in those scattered above and below the disk.

The thin part of the disk could have formed later as more gas fell into the galaxy and settled into the thin disk gas clouds where stars are forming today. Perhaps entire galaxies were captured by the growing Milky Way Galaxy. Astronomers have found streams and rings of scattered stars surrounding our galaxy and suspect they were produced when smaller galaxies were captured and pulled apart by our home galaxy (■ Figure 15-16). (You will see dramatic evidence in the next chapter that such galaxy mergers do occur.)

If the young Milky Way Galaxy absorbed a few small but partially evolved galaxies, then some of the globular clusters astronomers see in the halo may be hitchhikers. This would explain the range of globular cluster ages. In fact, some theories hold that our galaxy formed by a rapid merger of a number of smaller pregalaxy clouds of gas and stars with later additions of infalling gas and small merging galaxies.

The problem of the formation of our galaxy is frustrating because the theories are incomplete. Almost everyone prefers certainty to confusion, but astronomers are still gathering observations and testing hypotheses. The older theory has proven to be inadequate to explain the observations, and astronomers are attempting to refine the observations and devise new theories. The metal abundances and ages of the stars in our galaxy seem to be important clues, but metal abundance and age do not tell the whole story. Astronomer Bernard Pagel was thinking of this when he said, "Cats and dogs may have the same age and metallicity, but they are still cats and dogs."

Building Scientific Arguments

Why do metal-poor stars have the most elongated orbits?
A good argument makes connections between ideas, and sometimes those connections are not obvious. Certainly, the metal abundance of a star cannot affect its orbit, so an analysis must not confuse cause and effect with the relationship between these two factors. Both chemical composition and orbital shape

■ **Figure 15-16**

The Large and Small Magellanic Clouds, shown here rising above Walter Baade Telescope dome at the Magellan Observatory in Chile, are small galaxies that orbit the Milky Way Galaxy. They are being pulled apart and the Milky Way Galaxy is absorbing their stars, gas, and dust. A long trail of stars and gas stretching halfway around the sky reveals that they are being ripped apart as they orbit our galaxy. (Kris Koenig/Coast Learning Systems)

depend on a third factor—age. The oldest stars are metal poor because they formed before there had been many supernova explosions to create and scatter metals into the interstellar medium. Those stars formed long ago when the galaxy was young and motions were not organized into a disk, and they tended to take up randomly shaped orbits, many of which are quite elongated. Consequently, today, the most metal-poor stars tend to follow the most elongated orbits.

Nevertheless, even the oldest stars known in our galaxy contain some metals. They are metal poor, not metal free. Adjust your argument. **Where did these metal-poor stars get their metals?**

■   ■   ■

Connections: The story of our galaxy is an astronomical work in progress, but that is not the only mystery hidden in the Milky Way. The next two sections discuss two special problems that astronomers are trying to understand—the nature of spiral arms and the secret of the galactic nucleus.

15-3 Spiral Arms

THE MOST STRIKING FEATURE of galaxies like the Milky Way is the system of spiral arms that wind outward through the disk. These arms contain swarms of hot, blue stars; clouds of dust and gas; and young star clusters. The young objects hint that the spiral arms involve star formation. As you try to understand the spiral arms, you need to consider two questions. First, how can anyone be sure our galaxy has spiral arms when gas and dust obstruct the view? Second, why doesn't the differential rotation of the galaxy destroy the arms? The solution to both problems involves star formation.

Tracing the Spiral Arms

Studies of other galaxies show that spiral arms contain hot, blue stars (■ Figure 15-17). Consequently, one way to study the spiral arms of our own galaxy is to locate these stars. Fortunately, this is not difficult, because O and B stars are often found in associations and, being very bright, are easy to detect across great distances. Unfortunately, at these great distances their parallax is too small to measure, so their distances must be found by other means, usually by spectroscopic parallax (see Chapter 9).

O and B associations in the sky are not located randomly but lie along parts of three spiral arms near the sun, which have been named for the prominent constel-lations through which they pass, as shown in the diagram in Figure 15-17. If astronomers could penetrate the gas and dust, they could locate other O and B associations and trace the spiral arms farther; but, like travelers in a fog, Earth's astronomers see only the region near the sun.

Objects used to map spiral arms are called **spiral tracers.** O and B associations are good spiral tracers because they are bright and easy to see at great distances. Other tracers include young open clusters, clouds of hydrogen ionized by hot stars (emission nebulae), and certain higher-mass variable stars.

Notice that all spiral tracers are young objects. O stars, for example, live for only a few million years. If their orbital velocity is about 250 km/s, they cannot have moved more than about 500 pc since they formed. This is less than the width of a spiral arm. Because they don't live long enough to move away from the spiral arms, they must have formed there.

The youth of spiral tracers provides an important clue about spiral arms. Somehow they are associated with star formation. Before you can follow this clue, however, you need to extend your map of spiral arms to show the entire galaxy.

Visual-wavelength image

Spiral galaxy NGC 3370 contains many spiral arms.

Enhanced visual image

Galaxy M51 contains two main spiral arms.

Emission from hydrogen ionized by hot, young stars

■ Figure 15-17

Many of the galaxies in the sky are disk shaped, and most of those galaxies have spiral arms. You can suspect that our own disk-shaped galaxy also has spiral arms. Images of other galaxies show that spiral arms are marked by hot, luminous stars that must be very young; and this should make you suspect that spiral arms are related to star formation. Gas and dust block your view of most of the disk of our galaxy, but nearby young O and B stars fall along bands that appear to be segments of spiral arms. (Images: NASA, Hubble Heritage Team)

■ Figure 15-18

This map of our galaxy was made using 21-cm radio observations, and it confirms that our galaxy has spiral arms. Branches and spurs in the spiral arms are visible. The conical region around and beyond the center of the galaxy is the region where 21-cm radiation from gas clouds at different distances cannot be unscrambled. (Adapted from a radio map by Gart Westerhout)

Radio Maps of Spiral Arms

The dust that blocks the view at visual wavelengths is transparent at radio wavelengths because radio waves are much longer than the diameter of the dust particles. If you pointed a radio telescope at a section of the Milky Way, you would receive 21-cm radio signals coming from cool hydrogen in spiral arms at various distances across the galaxy. Fortunately, the signals can be unscrambled by measuring the Doppler shifts of the 21-cm radiation. Only in the direction toward the nucleus is there a problem. There, the orbital motions of gas clouds are perpendicular to the line of sight, and all of the radial velocities are zero. That is why the radio map shown in ■ Figure 15-18 reveals spiral arms throughout the disk of the galaxy but not in the wedge-shaped region toward the center.

The analysis of the 21-cm radial velocity data requires that astronomers estimate the orbital velocity of the clouds at different distances from the center of the galaxy. Because these velocities are not precisely known and because turbulent motions in the gas distort the radial velocities, the map doesn't trace a perfect spiral pattern in Figure 15-18. You can be confident our galaxy has spiral arms, but it isn't possible to see the overall pattern.

Radio astronomers can also use the strong radio emission from carbon monoxide (CO) to map the location of giant molecular clouds in the plane of the galaxy. Recall from Chapter 11 that these clouds are sites of active star formation. Maps constructed from such observations reveal that the giant molecular clouds are located in large structures that resemble segments of spiral arms (■ Figure 15-19).

Radio maps combined with studies at infrared and visible wavelengths paint a picture of our galaxy. Clearly we live in a spiral galaxy, but the spiral pattern appears slightly irregular with branches and spurs and gaps. The stars you see in Orion, for example, appear to be in a detached segment of a spiral arm, a spur. Some astronomers argue that our galaxy has a four-armed spiral pattern, but most believe it has a classic two-armed pattern with branches and

■ Figure 15-19

(a) At the wavelength emitted by the CO molecule astronomers find many molecular clouds along the Milky Way, but the clouds overlap in a confusing jumble. By using a model of the rotation of the galaxy and the radial velocities of the clouds, radio astronomers can find the distances to each cloud and use them to map spiral arms. (b) This map represents the view from a point 2 kpc directly above the sun. The molecular clouds, shown as hemispheres extending above the plane of our galaxy, are located along spiral arms. Angles in this diagram are galactic longitudes. (Adapted from a diagram by T. M. Dame, B. G. Elmegreen, R. S. Cohen, and P. Thaddeus)

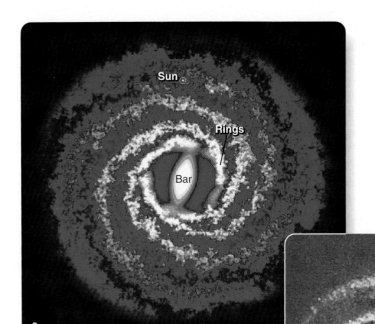

Astronomers believe that spiral arms are dynamically stable—they retain the same appearance even though the gas, dust, and stars in them are constantly changing. To see how this works, think of the traffic jam behind a slow-moving truck. Seen from an airplane, the traffic jam would be stable, moving slowly down the highway. But any given car would approach from behind, slow down, wait its turn, work its way to the front of the jam, pass the truck, and resume speed. The individual cars in the jam are constantly changing, but the traffic jam itself is dynamically stable.

In the **density wave theory,** the spiral arms are dynamically stable regions of compression that move slowly around the galaxy, just as the truck moves slowly down the highway. Gas clouds, moving at orbital velocity around the galaxy, overtake the slow-moving arms from behind and slam into the gas already in the arms. The sudden compression of the gas can trigger the collapse of the gas clouds and the formation of new stars (see Chapter 11). The newly formed stars and the re-

■ **Figure 15-20**

(a) This reconstruction of our galaxy is based on far-infrared observations of dust. The model includes a central bar surrounded by a ring from which spring four spiral arms. (Courtesy Henry Freudenreich) (b) This painting, based on radio and optical studies of the galaxy, includes a bar-shaped nucleus and a number of irregular arms with branches and spurs. (Painting by M. Seeds, based on a study by G. De Vaucouleurs and W. D. Pence)

spurs. The nuclear bulge of our galaxy does not seem to be a sphere but rather is an elongated bar pointing partially away from Earth. The spiral arms may spring from a ring that encloses this bar. In the next chapter, you will see that such structures are common in other galaxies. In this way, astronomers can construct a best guess at what our galaxy might look like for comparison with other galaxies (■ Figure 15-20).

The most important feature in the radio maps is easy to overlook—spiral arms are regions of higher gas density. Spiral tracers told you that the arms contain young objects, and you suspected active star formation. Now radio maps confirm your suspicion by telling you that the material needed to make stars is abundant in spiral arms.

The Density Wave Theory

Just what are spiral arms? You can be sure they are not physically connected structures like bands of magnetic field holding the gas in place. If they were, the strong differential rotation of the galaxy would destroy them within a billion years. They would get wound up and torn apart like paper streamers caught on the wheel of a speeding car. Yet spiral arms are common in galaxies (■ Figure 15-21) and must last billions of years.

■ **Figure 15-21**

Typical of spiral galaxies, M83 has a two-armed spiral pattern with prominent dust lanes and bright O and B stars along the arms. The galaxy also contains branches and spurs. (European Southern Observatory)

Visual-wavelength image

maining gas eventually move on through the arm and emerge from the front of the slow-moving arm to resume their travels around the galaxy (■ Figure 15-22). Whenever you look you see a spiral arm, although the stars, gas, and dust are constantly changing.

■ **Figure 15-22**

According to the density wave theory, star formation occurs as gas clouds pass through spiral arms.

The Spiral Density Wave

Orbiting gas clouds overtake the spiral arm from behind.

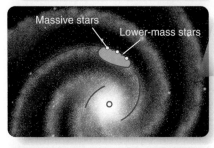

The compression of a gas cloud triggers star formation.

Massive stars are highly luminous and light up the spiral arm.

The most massive stars die quickly.

Low-mass stars live long lives but are not highly luminous.

Science depends on evidence, so you should ask what evidence astronomers have that spiral arms are caused by spiral density waves. For one thing, the stars in spiral arms fit the theory well. Stars of all masses are forming in spiral arms, but the O and B stars are the brightest. They live such short lifetimes that they die before they can move out of the spiral arm. These massive, high-luminosity stars, along with gas and dust, assure that spiral arms are sites of star formation. Lower-mass stars, such as the sun, also form in spiral arms. But because they live so long, they can leave the arms and circle the galaxy many times. The sun probably formed as part of a cluster in a spiral arm roughly 5 billion years ago, escaped from that star cluster, and has circled the galaxy about 20 times, passing through spiral arms many times.

Giant molecular clouds and thick dust in the arms (■ Figure 15-23) are evidence that material is compressed in the arms and spawns active star formation. These clouds can convert up to 30 percent of their mass into stars. The Orion complex and the Ophiuchus dark cloud are just two of these regions filled with infrared protostars. Such a cloud can form stars for 10 to 100 million years before it is disrupted by heat and light from the most massive new stars. The cloud needs about this long to pass through a spiral arm.

The evidence seems to fit the spiral density wave theory well, but the theory has two problems. First, how is the complicated spiral disturbance started and sustained? Spiral density waves should slowly die out in a billion years or so. Something must regenerate the spiral wave. Mathematical models show that the

■ **Figure 15-23**

This chance alignment of a small galaxy in front of a larger, more distant galaxy illustrates the nature of spiral arms. The small galaxy's spiral arms are bright with O and B stars where they are silhouetted against the darkness of space; but, seen against the glare of the distant galaxy, the arms are revealed to contain large amounts of dusty gas. (NASA and The Hubble Heritage Team)

galaxy is naturally unstable to certain disturbances, just as a guitar string is unstable to certain vibrations. Any sudden disturbance—the rumble of a passing truck, for example—can set the string vibrating at its natural frequencies. Similarly, minor fluctuations in the galaxy's disk, such as supernova explosions, might regenerate the waves.

Another suggestion is that gravitational disturbances can generate a spiral density wave. Observations show that the center of our galaxy is not a sphere but a bar. The rotation of such a bar could disturb the disk of the galaxy and stimulate the formation of a spiral density wave. A close encounter or collision with a nearby galaxy could also stimulate a spiral pattern.

The second problem for the density wave theory involves the spurs and branches in the arms of our own and other galaxies. Computer models of density waves produce regular, two-armed spiral patterns. Some galaxies, called grand-design galaxies, do indeed have symmetric two-armed patterns, but others do not. Some galaxies have a great many short spiral segments, giving them a fluffy appearance. These galaxies have been termed **flocculent,** meaning "woolly" (■ Figure 15-24). Our galaxy is probably an intermediate galaxy. How can astronomers explain these variations in the spiral pattern? Perhaps the answer lies in a process that sustains star formation once it begins.

Star Formation in Spiral Arms

Star formation is a critical process in the creation of the spiral patterns in disk galaxies like ours. It makes spiral arms visible and may shape the spiral pattern itself.

The spiral density wave creates spiral arms by the gravitational attraction of the stars and gas flowing through the arms. Even if there were no star formation at all, rotating disk galaxies could form spiral arms. But without star formation to make young, hot, luminous stars, the spiral arms would be difficult to see. It is the star formation that lights up the spiral arms and makes them so prominent. Star formation helps determine what you see when you look at a spiral galaxy.

But star formation can also control the shape of spiral patterns if the birth of stars in a cloud of gas can renew itself by triggering the formation of more new stars. Consider a newly formed star cluster with a single massive star. The intense radiation from that hot star can compress nearby parts of the gas cloud and trigger further star formation (■ Figure 15-25). Also, massive stars evolve so quickly that their lifetimes are only an instant in the history of a galaxy. When they explode as supernovae, the expanding gasses can compress neighboring clouds of gas and trigger more star formation. Examples of such a **self-sustaining star formation** have been found. The Orion complex, consisting of the Great Nebula in Orion and the star formation buried deep in the dark interstellar clouds behind the nebula, is such a region.

Self-sustaining star formation can produce growing clumps of new stars, and the differential rotation of the galaxy can drag the inner edge ahead and let the outer edge lag behind to pro-

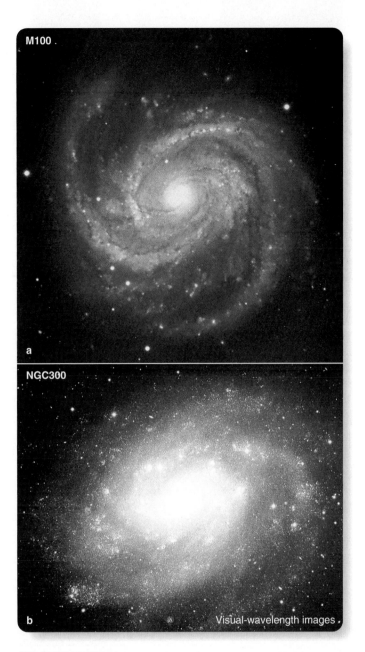

■ **Figure 15-24**

(a) Some galaxies are dominated by two spiral arms; but, even in these galaxies, minor spurs and branches are common. The spiral density wave can generate the two-armed, grand-design pattern, but self-sustained star formation may be responsible for the irregularities. (b) Many spiral galaxies do not appear to have two dominant spiral arms. Spurs and branches suggest that star formation is proceeding rapidly in such galaxies. (Anglo-Australian Telescope Board)

duce a cloud of star formation shaped like a segment of a spiral arm, a spur (■ Figure 15-26). Mathematical models of galaxies filled with such segments of spiral arms do have a spiral appearance, but they lack the bold, two-armed spiral that astronomers refer to as the grand-design spiral pattern (■ Figure 15-27). Rather, astronomers suspect that self-sustaining star formation can produce the branches and spurs so prominent in flocculent galaxies,

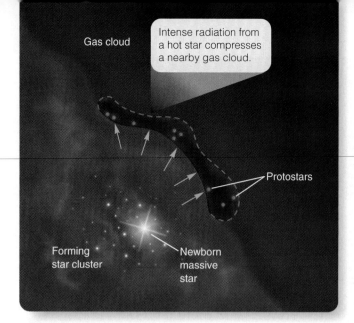

Gas cloud

Intense radiation from a hot star compresses a nearby gas cloud.

Protostars

Forming star cluster

Newborn massive star

■ Figure 15-25

Most stars in a star cluster are too small and too cool to affect nearby gas clouds; but, once a massive star forms, it becomes so hot and so luminous its radiation can push gas away and compress a nearby gas cloud. In the densest regions of the compressed cloud, new stars can begin forming.

but only the spiral density wave can generate the beautiful two-armed spiral patterns.

This discussion of star formation in spiral arms illustrates the importance of natural processes. The spiral density wave creates graceful arms, but it is the star formation in the arms that makes them stand out so prominently. Self-sustaining star formation can act in some galaxies to modify the spiral arms and produce branches and spurs. In some galaxies, it can make the spiral pattern flocculent. By searching out and understanding the details of such natural processes, astronomers can begin to understand the overall structure and evolution of the universe we live in.

Building Scientific Arguments

Why can't astronomers use solar-type stars as spiral tracers? Sometimes the timing of events is the critical factor in a scientific argument. In this case, you need to think about the evolution of stars and their orbital periods around the galaxy. Stars like the sun live about 10 billion years, but the sun's orbital period around the galaxy is 240 million years. The sun almost certainly formed when a gas cloud passed through a spiral arm, but since then the sun has circled the galaxy many times and has passed through spiral arms often. That means the sun's present location has nothing to do with the location of any spiral arms. An O star, however, lives only a few million years. It is born in a spiral arm and lives out its entire lifetime before it can leave the spiral arm. Short-lived stars such as O stars are found only in spiral arms, but G stars are found all around the galaxy.

Self-Sustaining Star Formation

Differential rotation drags the inner edge of a gas cloud ahead of its outer edge.

Center of galaxy

A cloud can become elongated by continuing differential rotation.

Star formation in a gas cloud can produce massive stars whose high luminosity...

and supernova explosions can compress surrounding gas and trigger more star formation.

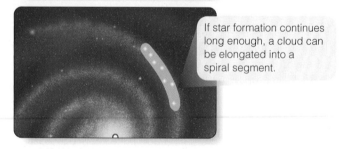

If star formation continues long enough, a cloud can be elongated into a spiral segment.

■ Figure 15-26

Continuing star formation may be able to produce long clouds of young stars that look like segments of spiral arms.

The spiral arms of our galaxy must make it beautiful in photographs taken from a distance, but we are trapped inside it. Create an argument based on evidence: **How do astrono-**

The beautiful symmetry of the grand-design spiral pattern is clear in this image of the barred spiral galaxy NGC1300. Young star clusters containing hot, bright stars and ionized hydrogen are located along the arms, while dark lanes of dust mark the inner edges of the arms. (NASA and The Hubble Heritage Team.)

mers know that the spiral arms mapped out near us by spiral tracers actually extend across the disk of our galaxy?

■   ■   ■

Connections: Spiral arms are a puzzle of great beauty, but the center of our galaxy hides a puzzle of great power.

15-4 The Nucleus

THE MOST MYSTERIOUS REGION of our galaxy is its very center, the nucleus. At visual wavelengths, this region is totally hidden by gas and dust that dim the light by 30 magnitudes. If a trillion (10^{12}) photons of light left the center of the galaxy on a journey to Earth, only one would make it through the gas and dust. Consequently, visual-wavelength photos reveal nothing about the nucleus. Observations at radio and infrared wavelengths can penetrate the dust and gas, and they paint a picture of tremendously crowded stars orbiting the center at high velocity. To understand what is happening at the center of our galaxy, you need to carefully compare observations and theories.

Observations

If you look up at the Milky Way on a dark night, you might notice a slight thickening in the direction of the constellation Sagit-

tarius, but nothing specifically identifies this as the direction of the heart of the galaxy. Even Shapley's study of globular clusters identified the center only approximately. When radio astronomers turned their telescopes toward Sagittarius, they found a complex collection of radio sources, with one, **Sagittarius A*** (abbreviated Sgr A* and usually pronounced "sadge A-star"), lying at the expected location of the galactic core.

Observations show that Sgr A* is less than a few astronomical units in diameter but is a strong source of radio energy. The tremendous amount of infrared radiation coming from the central area appears to be produced by crowded stars and by dust warmed by those stars. But what could be as small as Sgr A* and produce so much energy?

Study **Sagittarius A*** on pages 362–363 and notice three important points:

1 First, observations at radio wavelengths reveal complex structures near Sgr A* caused by magnetic fields and by rapid star formation. Supernova remnants show that massive stars have formed there recently and died supernova deaths.

2 A second thing to notice is how crowded the center is. Tremendous numbers of stars heat the dust and produce strong infrared radiation.

3 Third, notice the evidence that Sgr A* is a supermassive black hole into which gas is flowing.

Sagittarius A*

1 The constellation of Sagittarius is so filled with stars and with gas and dust you can see nothing at visual wavelengths of the center of our galaxy.

The image below is a wide-field radio image of the center of our galaxy. Many of the features are supernova remnants (SNR), and a few are clouds of star formation. Peculiar features such as threads, the Arc, and the Snake may be gas trapped in magnetic fields. At the center lies Sagittarius A, the center of our galaxy.

Arc

Radio image

NRAO/AUI/NSF

The radio map above shows Sgr A and the Arc filaments, 50 parsecs long. The image was made with the VLA radio telescope. The contents of the white box are shown on the opposite page.

Sgr D HII

Sgr D SNR

SNR 0.9 + 0.1

Sgr B2

Apparent angular size of the moon for comparison

Sgr B1

New SNR 0.3 + 0.0

Threads

Arc

The Cane

Background galaxy

Threads

Sgr A

2 Infrared photons with wavelengths longer than 4 microns (4000 nm) come almost entirely from warm interstellar dust. The radiation at these wavelengths coming from Sagittarius is intense, and that indicates that the region contains lots of dust and is crowded with stars that warm the dust.

Sgr C

The Pelican

Coherent structure?

Snake

Sgr E

SNR 359.1 − 00.5

Radio image

NRL

Infrared image

2MASS

1a This high-resolution radio image of Sgr A (the white boxed area on the opposite page) reveals a spiral swirl of gas around an intense radio source known as Sgr A*, the presumed central object in our galaxy. About 3 pc across, this spiral lies in a low-density cavity inside a larger disk of neutral gas. The arms of the spiral are thought to be streams of matter flowing into Sgr A* from the inner edge of the larger disk (drawing at right).

Sgr A*

Radio image

N. Killeen and Kwok-Yung Lo

Evidence of a Black Hole at the Center of Our Galaxy

3 Since the middle 1990s, astronomers have been able to use large infrared telescopes and active optics to follow the motions of stars orbiting around Sgr A*. A few of those orbits are shown here. The size and period of the orbit allows astronomers to calculate the mass of Sgr A* using Kepler's third law. The orbital period of the star SO-2, for example, is 15.2 years and the semimajor axis of its orbit is 950 AU. The combined motions of the observed stars suggest that Sgr A* has a mass of 2.6 million solar masses.

The Chandra X-ray Observatory has imaged Sgr A* and detected over 2000 other X-ray sources in the area.

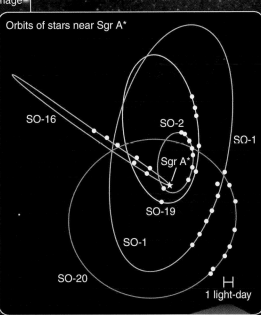

NASA/CXC/MIT/F.K. Baganoff et al.

Infrared Image

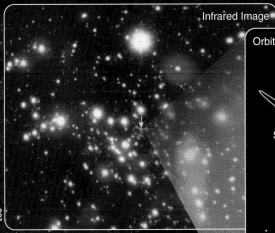

ESO

Orbits of stars near Sgr A*

SO-16
SO-2
SO-1
Sgr A*
SO-19
SO-1
SO-20

H
1 light-day

At its closest, SO-2 comes within 17 light-hours of Sgr A*. Alternative theories that Sgr A* is a cluster of stars, of neutron stars, or of stellar black holes are eliminated. Only a single black hole could contain so much mass in so small a region.

Our solar system is half a light-day in diameter.

3a A black hole with a mass of 2.6 million solar masses would have an event horizon smaller than the smallest dot in this diagram. A slow dribble of only 0.0002 solar masses of gas per year flowing into the black hole could produce the observed energy. A sudden increase as when a star falls in could produce a violent eruption.

The evidence of a massive black hole at the center of our galaxy seems conclusive. It is much too massive to be the remains of a dead star, however, and astronomers conclude that it probably formed as the galaxy first took shape.

A supermassive black hole is an exciting idea, but scientists must always be aware of the difference between adequacy and necessity. A supermassive black hole is adequate to explain the observations, but could there be other explanations? For example, astronomers have suggested that gas flowing inward could trigger tremendous bursts of star formation. Such theories have been considered and tested against the evidence, but none appears to be adequate to explain the observations. So far, the only theory that seems adequate is that our galaxy is home to a supermassive black hole. That is, a central black hole appears to be not only adequate but necessary to explain the observations. In later chapters you will see that such supermassive black holes are found at the centers of many galaxies.

Building Scientific Arguments

Why do astronomers think the center of our galaxy contains a large mass?

Because scientific arguments hinge so often on evidence, they can involve discussions of measurement—one of the keys to science. The best way to measure the mass of an astronomical object is to watch something orbit around it. You can then use Kepler's third law to find the mass inside the orbit. Because of gas and dust, astronomers can't see to the center of our galaxy at visual wavelengths, but infrared observations can detect individual stars orbiting Sgr A*. The star S2 has been particularly well observed, but a number of stars can be followed as they orbit the center. The sizes and periods of these orbits, interpreted using Kepler's laws, reveal that Sgr A* contains roughly 2.6 million solar masses.

Now build a new argument to analyze a different observation. **The strong infrared radiation at wavelengths longer than 4 microns (4000 nm) implies vast numbers of stars crowded into the center. Why?**

■   ■   ■

Connections: Observations of the Milky Way Galaxy suggest it is a place of great beauty and great power. Is it normal? Are all galaxies like ours? To answer that question you must survey other galaxies, which you will begin in the next chapter.

Summary

15-1 | The Nature of the Milky Way Galaxy

How do astronomers know we live in a galaxy?

- The hazy band of the Milky Way is our wheel-shaped galaxy seen from within, but its size and shape are not obvious. William and Caroline Herschel counted stars at many locations over the sky to show that our star system seemed to be shaped like a grindstone with the sun near the center. But they could not see very far into space because of gas and dust.

- In the early 20th century, Harlow Shapley calibrated Cepheid variable stars to find the distance to globular clusters and demonstrated that our galaxy is much larger than what we can see and that the sun is not at the center.

- Modern observations suggest that our galaxy contains a disk component about 75,000 ly in diameter and that the sun is two-thirds of the way from the center to the visible edge. The nuclear bulge around the center and an extensive halo containing old stars and little gas and dust make up the spherical component.

- The mass of the galaxy can be found from its rotation curve. Kepler's third law reveals that the galaxy contains over 100 billion solar masses, and the rotation curve at great distance from the center shows that the halo may contain much more mass. Because that mass is not emitting detectable electromagnetic radiation, astronomers call it dark matter.

15-2 | The Origin of the Milky Way

How did our galaxy form and evolve?

- The oldest star clusters reveal that the disk of our galaxy is younger than the halo, and the oldest globular clusters appear to be about 13 billion years old. So our galaxy must have formed about 13 billion years ago.

- Stellar populations are an important clue to the formation of our galaxy. The first stars to form, termed population II stars, were poor in elements heavier than helium—elements that astronomers call metals. As generations of stars manufactured metals in a process called nucleosynthesis and spread them back into the interstellar medium, the metal abundance of more recent generations increased. Population I stars, including the sun, are richer in metals.

- Galactic fountains produced by expanding supernova remnants may help spread metals throughout the disk.

- Because the halo is made up of population II stars and the disk is made up of population I stars, astronomers conclude that the halo formed first and the disk later. A theory that the galaxy formed from a single, roughly spherical cloud of gas and gradually flattened into a disk has been amended to include mergers with other galaxies and infalling gas contributing to the disk.

15-3 | Spiral Arms

What are the spiral arms?

- The most massive stars live such short lives they don't have time to move from their place of birth. Because they are found scattered along the spiral arms, astronomers conclude that the spiral arms are sites of star formation.

- You can trace the spiral arms through the sun's neighborhood by using spiral tracers such as O and B stars; but, to extend the map over the entire galaxy, astronomers must use radio telescopes to see through the gas and dust.

- The spiral density wave theory suggests that the spiral arms are regions of compression that move around the disk. When an orbiting gas cloud overtakes the compression wave, the gas cloud is compressed and forms stars. Another process, self-sustaining star formation, may act to modify the arms with branches and spurs as the birth of massive stars triggers the formation of more stars by compressing neighboring gas clouds.

15-4 | The Nucleus

What lies at the very center?

- The nucleus of the galaxy is invisible at visual wavelengths, but radio, infrared, and X-ray radiation can penetrate the gas and dust. These wavelengths reveal crowded central stars and warmed dust.

- The very center of the Milky Way Galaxy is marked by a radio source, Sagittarius A*. The core must be less than a few astronomical units in diameter, but the motion of stars around the center show that it must contain roughly 2.6 million solar masses. A supermassive black hole is the only object that could contain so much mass in such a small space.

New Terms

Magellanic Clouds (p. 340)	metals (p. 349)
kiloparsec (kpc) (p. 341)	population I star (p. 349)
halo (p. 341)	population II star (p. 349)
nuclear bulge (p. 341)	nucleosynthesis (p. 350)
disk component (p. 344)	galactic fountain (p. 351)
spherical component (p. 344)	spiral tracers (p. 355)
high-velocity star (p. 347)	density wave theory (p. 357)
rotation curve (p. 348)	flocculent (p. 359)
Keplerian motion (p. 348)	self-sustaining star formation (p. 359)
galactic corona (p. 348)	
dark matter (p. 348)	Sagittarius A* (p. 361)

Review Questions

Ace⊘Astronomy™ Assess your understanding of this chapter's topics with additional quizzing and animations at **http://ace .brookscole.com/sf9**

1. Why is it difficult to specify the dimensions of the disk and halo?
2. Why didn't astronomers before Shapley realize how large the galaxy is?
3. What evidence can you cite that our galaxy has a galactic corona?
4. Explain why some star clusters lose stars more slowly than others.
5. Contrast the motion of the disk stars and that of the halo stars. Why do their orbits differ?
6. Why do high-velocity stars have lower metal abundance than the sun?
7. Why are metals less abundant in older stars than in younger stars?
8. What evidence contradicts the traditional theory for the origin of our galaxy?
9. Why are all spiral tracers young?

10. Why couldn't spiral arms be physically connected structures? What would happen to them?

11. Why does self-sustaining star formation produce clouds of stars that look like segments of spiral arms?

12. Describe the kinds of observations you would make to study the galactic nucleus.

13. Why must astronomers use infrared telescopes to observe the motions of stars around Sgr A*?

14. What evidence can you cite that the nucleus of the galaxy contains a supermassive energy source that is very small in size?

15. Why does the galaxy shown at the right have so much dust in its disk? How big do you suppose the halo of this galaxy really is?

(NASA/Hubble Heritage Team/STScI/AURA)

16. Why are the spiral arms in the galaxy at the right blue? What color would the halo be if it were bright enough to see in this photo?

(NASA/Hubble Heritage Team and A. Riess, STScI)

Discussion Questions

1. How would this chapter be different if interstellar dust did not scatter light?

2. Why doesn't the Milky Way circle the sky along the celestial equator or the ecliptic?

Problems

1. Make a scale sketch of our galaxy in cross section. Include the disk, sun, nucleus, halo, and some globular clusters. Try to draw the globular clusters to scale.

2. Because of dust, astronomers can see only about 5 kpc into the disk of the galaxy. What percentage of the galactic disk does that include? (*Hint:* Consider the area of the entire disk and the area visible from Earth.)

3. If the fastest passenger aircraft can fly 0.45 km/s (1000 mph), how long would it take to reach the sun? the galactic center? (*Hint:* 1 pc = 3×10^{13} km.)

4. If a typical halo star has an orbital velocity of 250 km/s, how long does it take to pass through the disk of the galaxy? Assume that the disk is 1000 pc thick.

5. If the RR Lyrae stars in a globular cluster have apparent magnitudes of 14, how far away is the cluster? (*Hint:* See Figure 12-14.)

6. If interstellar dust makes an RR Lyrae variable star look 1 magnitude fainter than it should, by how much will you overestimate its distance? (*Hint:* Use the magnitude–distance formula or Table 9-1.)

7. If a globular cluster is 10 minutes of arc in diameter and 8.5 kpc away, what is its diameter? (*Hint:* Use the small-angle formula.)

8. If you assume that a globular cluster 4 minutes of arc in diameter is actually 25 pc in diameter, how far away is it? (*Hint:* Use the small-angle formula.)

9. If the sun is 5 billion years old, how many times has it orbited the galaxy?

10. If the true distance to the center of the galaxy is found to be 7 kpc and the orbital velocity of the sun is 220 km/s, what is the minimum mass of the galaxy? (*Hint:* Use Kepler's third law.)

11. What temperature would interstellar dust have to have to radiate most strongly at 100 μm? (*Hints:* 1 μm = 1000 nm. Use Wien's law, Chapter 7.)

12. Infrared radiation from the center of our galaxy with a wavelength of about 2 μm (2×10^{-6} m) comes mainly from cool stars. Use this wavelength as λ_{max} and find the temperature of the stars.

13. If an object at the center of our galaxy has a linear diameter of 10 AU, what will its angular diameter be as seen from Earth? (*Hint:* Use the small-angle formula, Chapter 3.)

Media Cluster

Ace ⓢ Astronomy™ To access the resources in the Media Cluster, log into AceAstronomy at **http://ace.brookscole .com/sf9** and select Chapter 15.

ACTIVE FIGURES

Explorable Milky Way
Viewing at different wavelengths has helped Earth astronomers to better see and understand the structures in surrounding galaxies and their own Milky Way. In this animation, explore the Milky Way Galaxy at different wavelengths.

ASTRONOMY EXERCISES

Cepheid Variable
Make your own observations and discover the relationship between period of pulsation and luminosity with this animation.

Milky Way Galaxy
What would your home galaxy look like if you could see it from the outside? Find out with this 3-D model of the Milky Way that you can spin around to view from any angle.

Critical Inquiries for the Web

1. Henrietta Leavitt discovered the period–luminosity relation for Cepheids while working on the staff at Harvard College Observatory under Edward Pickering. She was one of several women "computers" on staff there a century ago. Search the Web for information on Leavitt, her colleagues, and their work at Harvard. List three of the women employed by Pickering and note their contributions to astronomy. What was life like for a woman in astronomy at the beginning of the 20th century?

2. What if you lived near the center of the galaxy? Search the Web for research and information on the distribution of material near the center of our galaxy. Based on what you find, speculate as to how the sky would appear from a planet associated with a star near the galactic center.

3. Who was Wilhelm Heinrich Walter Baade, and why did he win the Bruce Medal? What other winners of the Bruce Medal do you recognize?

Exploring *TheSky*

1. Locate Sagittarius and examine the shape of the Milky Way there and the profusion of globular clusters. (*Hint:* To turn on Messier object labels, use **Labels** and **Setup** under the **View** menu.)

2. Locate the following globular clusters: M3, M4, M5, M10, M12, M13, M15, M22, M55, M92. Where are they located in the sky? (*Hint:* Use **Find** under the **Edit** menu.)

3. Compare the distribution of globular clusters with that of open clusters. (*Hint:* Use **Filters** under the **View** menu to turn off everything but globular clusters, the Milky Way, the Galactic Equator, and Constellation Boundaries. Use the thumbwheel at the bottom of the sky window to rotate the sky. Now repeat with globular clusters off and open clusters on.)

Go to the Brooks/Cole Astronomy Resource Center (**http:// astronomy.brookscole.com**) for critical thinking exercises, articles, and additional readings from InfoTrac College Edition, Brooks/Cole's online student library.

16 | Galaxies

A hypothesis or theory is clear,

decisive, and positive,

but it is believed by

no one but the man who

created it. Experimental

findings, on the other hand,

are messy, inexact things

which are believed by

everyone except the man

who did that work.

HARLOW SHAPLEY,
THROUGH RUGGED WAYS TO THE STARS

Visual-wavelength image

SCIENCE FICTION HEROES flit effortlessly between the stars, but almost none voyages between the galaxies. As you leave your home galaxy, the Milky Way, behind, you voyage out into the vast depths of the universe, out among the galaxies, into space so deep it is unexplored even in fiction. ▌ Less than a century ago, astronomers did not understand that there were galaxies. Nineteenth-century telescopes revealed faint nebulae scattered among the stars, and some were spiral. Astronomers argued about the nature of these nebulae, but it was not until the 1920s that astronomers understood that some were other galaxies much like our own; and it was not until recent decades that astronomical telescopes could reveal the tremendous beauty and intricacy of the galaxies (▪ Figure 16-1).

▌ Continued on page 370 ▌

The Sombrero Galaxy is 28 million light-years from Earth. Even if your spaceship could travel at the speed of light, it would take you 28 million years to reach this galaxy, which is one of the closer galaxies to our Milky Way Galaxy. Look for the more distant galaxies in the background of this image. (NASA/Hubble Heritage Team/STScI/AURA)

Guidepost

Looking Back

In the preceding chapter, you learned about our Milky Way Galaxy, an important object to Earthlings but only one of the many billions of galaxies visible in the sky. You can no more understand galaxies by understanding a single example, the Milky Way, than you could understand humanity by understanding a single person.

This Chapter

This chapter will expand your horizon to discuss the different kinds of galaxies and their complex histories. Here you can expect answers to five essential questions:

What do galaxies look like?

How do astronomers find the distances to galaxies?

How do galaxies differ in size, luminosity, and mass?

Do other galaxies contain supermassive black holes and dark matter, as does our own galaxy?

Why are there different kinds of galaxies?

These are perfectly reasonable questions, but the answers will lead you to an astonishing insight into how galaxies are born and how they evolve.

Looking Ahead

You can carry the lessons of this chapter into the next, which will discuss the violently active galaxies powered by their supermassive black holes, and on into Chapter 18, which will discuss the evolution of the universe as a whole.

Ace⊛Astronomy™ The AceAstronomy icon throughout the text indicates an opportunity for you to test yourself on key concepts and to explore animations and interactions on the AceAstronomy website at: http://ace .brookscole.com/sf9

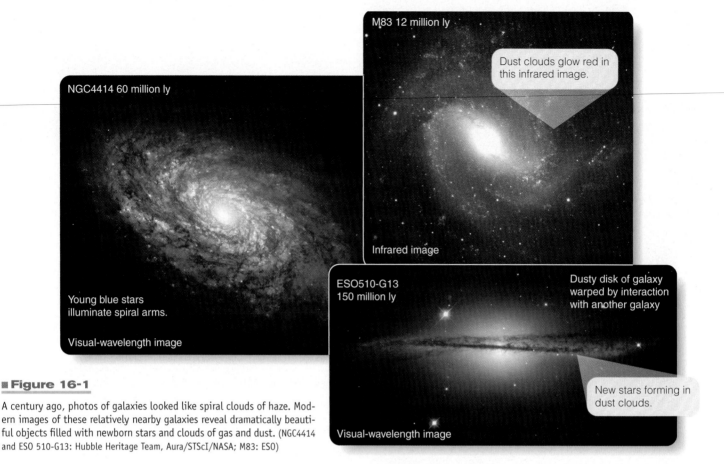

■ Figure 16-1

A century ago, photos of galaxies looked like spiral clouds of haze. Modern images of these relatively nearby galaxies reveal dramatically beautiful objects filled with newborn stars and clouds of gas and dust. (NGC4414 and ESO 510-G13: Hubble Heritage Team, Aura/STScI/NASA; M83: ESO)

In this chapter, you will try to understand how galaxies form and evolve, and you will discover that their complex shapes reflect long periods of peaceful star formation interrupted by episodes of cosmic violence.

Before you can build theories, however, you must gather some basic data concerning galaxies. How many kinds of galaxies are there? How big are they? How massive are they? That is, you need to characterize the family of galaxies just as you characterized the family of stars in Chapter 9.

16-1 The Family of Galaxies

LIKE LEAVES ON THE FOREST FLOOR, galaxies carpet the sky. Pick any spot on the sky away from the dust and gas of the Milky Way, and you are looking deep into space. Photons that have traveled for billions of years enter your eye, but they are too few to register on your retina. The largest telescopes can gather enough light to detect the most distant galaxies.

When astronomers picked a seemingly empty spot on the sky near the Big Dipper and used the Hubble Space Telescope to record an amazing time exposure of 10 days' duration, they found thousands of galaxies crowded into the image. That image, now known as a Hubble Deep Field, contains a few relatively nearby galaxies and others that are over 10 billion light-years away (■ Figure 16-2).

Since the first Hubble Deep Field was recorded, other deep fields have been observed in other parts of the sky and at other wavelengths. The GOODS program (Great Observatories Origins Deep Survey) has combined the Hubble Space Telescope with other space telescopes such as the Spitzer Space Telescope and the Chandra X-ray Observatory and with some of the largest telescopes on Earth's surface to image deep fields at many wavelengths. The GOODS images show that the first Hubble Deep Field was not unusual. The entire sky appears to be as thickly covered with galaxies as is the first Hubble Deep Field. Galaxies are very common, and they carpet the sky in every direction.

The GOODS study of distant galaxies is especially amazing when you think that, less than a century ago, humanity did not know that we live in a galaxy or that the universe was filled with other galaxies. The discovery of galaxies is one of the great discovery stories of astronomy.

The Discovery of Galaxies

Galaxies are faint objects, so they were not noticed until telescopes had grown large enough to gather significant amounts of light. In 1845, William Parsons, third Earl of Rosse in Ireland, built a telescope 72 in. in diameter. It was, for some years, the largest tele-

Foreground star in our galaxy

Visual-wavelength image

▪ Figure 16-2

An apparently empty spot on the sky only $\frac{1}{30}$ the diameter of the full moon contains over 1500 galaxies in this extremely long time exposure known as the Northern Hubble Deep Field. Only four stars are visible in this image; they are sharp points of light with diffraction spikes produced by the telescope optics. Presumably the entire sky is similarly filled with galaxies. (R. Williams and the Hubble Deep Field Team, STScI, NASA)

scope in the world. Before the invention of astronomical photography (in the late 1800s), Parsons had to view the faint nebulae directly at the eyepiece and sketch their shapes. He noticed that some have a spiral shape, and they became known as **spiral nebulae.** Parsons immediately concluded that the spiral nebulae were great spiral clouds of stars. The German philosopher Immanuel Kant in 1755 had proposed that the universe was filled with great wheels of stars that he called **island universes.** Parsons adopted that term for the spiral nebulae.

Not everyone agreed that the spiral nebulae were island universes. An alternative point of view was that we lived in a great disk-shaped star system in an otherwise empty universe. Sir William Herschel had counted stars in different directions and had shown that our star system was shaped approximately like a grindstone (Chapter 15). Beyond the edge of this grindstone star system, space was supposed to be a limitless void. The spiral nebulae,

in this view, were nothing more than whirls of gas or faint stars within our star system.

The nature of the spiral nebulae could not be resolved in the 19th century because the telescopes were not large enough and the photographic plates, when they became available, were not sensitive enough. The spiral nebulae remained foggy swirls on the best photographs, and astronomers continued to disagree on their true nature. In April 1920, two astronomers debated the issue at the National Academy of Science in Washington, D.C. Harlow Shapley of Mt. Wilson Observatory had recently shown that the Milky Way was a much larger star system than had been thought. He argued that the spiral nebulae were just nearby nebulae within the Milky Way star system. Heber D. Curtis of Lick Observatory argued that the spiral nebulae were island universes far outside the Milky Way star system. Historians of science mark this **Shapley–Curtis Debate** as a turning point in modern astronomy, but it was inconclusive. Shapley and Curtis cited the right evidence but drew incorrect conclusions. The disagreement was finally resolved, as is often the case in astronomy, by a bigger telescope.

On December 30, 1924, Mt. Wilson astronomer Edwin Hubble (namesake of the Hubble Space Telescope) announced that he had taken photographic plates of a few bright galaxies with

Searching for Clues through Classification in Science

Classification is one of the most basic and most powerful of scientific tools. It is often a way to begin studying a topic, and it often leads to dramatic insights. Charles Darwin, for example, sailed around the world with a scientific expedition aboard the ship HMS *Beagle*. Everywhere he went, he studied the living things he saw and tried to classify them. For example, he studied the tortoises, mockingbirds, and finches he saw on the Galapagos Islands. His classifications of these and other animals eventually led him to think about evolution and natural selection. Scientists in many fields depend on carefully designed systems of classification.

Classification is an everyday mode of thought. You use classifications when you order lunch, buy shoes, catch a bus, and so on. In science, classifications reveal the relationships between different kinds of objects and at the same time save the scientist valuable time by bringing order out of seeming disorder. An economist can classify different kinds of businesses and does not have to analyze each of the millions of businesses as if it were unique. Classifications of minerals, plants, psychological learning styles, modes of transportation, or sandwiches bring order to the world and help you deal with information.

Astronomers use classifications of galaxies, stars, moons, telescopes, and more. Whenever you encounter a scientific discussion, look for the classifications on which it is based. Classifications are the orderly framework on which much of science is built.

The careful classification of living things can reveal surprising relationships. The birds, including this flamingo, are descended from dinosaurs. (M. Seeds)

the new 100-in. telescope. Not only could he detect individual stars, but he could identify some of them as Cepheid variables with apparent magnitudes of about 18. For the brightest Cepheids, which are supergiants, to look that faint, they had to be very distant, and that meant the spiral nebulae had to be outside the Milky Way star system. They were galaxies.

Although it was spiral galaxies that first attracted the attention of astronomers, Hubble quickly realized that there were other kinds of galaxies. Not all galaxies had spiral shapes.

The Shapes of Galaxies

Look at the galaxies in the Hubble Deep Field, and you will see galaxies of different shapes. Some are spiral, some have elliptical shapes, and some are irregular. Astronomers classify galaxies according to their shape in photographs made at visual wavelengths using a system developed by Edwin Hubble in the 1920s. Such systems of classification are a fundamental technique in science (**Window on Science 16-1**).

Study **Galaxy Classification** on pages 374–375 and notice three important points:

1 Many galaxies have no disk, no spiral arms, and almost no gas and dust. These elliptical galaxies range from huge giants to small dwarfs.

2 Notice that disk-shaped galaxies usually have spiral arms and contain gas and dust. Many spiral galaxies have a barred structure, but a few disk galaxies contain little gas and dust.

3 Finally, notice that some galaxies are highly irregular in shape and tend to be rich in gas and dust.

The amount of gas and dust in a galaxy strongly influences its appearance. Galaxies rich in gas and dust usually have active star formation and contain hot, bright stars. Such galaxies tend to be bluer and contain emission nebulae. Galaxies that are poor in gas and dust contain few or none of these highly luminous stars, so those galaxies look redder and have a much more uniform appearance.

Spiral galaxies are clearly disk shaped, but the true, three-dimensional shape of elliptical galaxies isn't obvious from images. Some are spherical, but the more elongated elliptical galaxies could be shaped like flattened spheres (bun shaped) or like elongated spheres (football shaped). Some may even have three different diameters; they are longer than they are thick and thicker than they are wide, like a flattened loaf of bread. Statistical studies suggest that all of these shapes are represented among elliptical galaxies.

Traditionally astronomers have classified galaxies according to their appearance in visual-wavelength images, but infrared images reveal the location of the vast majority of stars—stars that are not as luminous as the hot stars that produce most of the light. These studies suggest that the structure of many galaxies is dominated by the gravitational fields of large masses of cool stars. For example, some spiral galaxies that appear at visible wavelengths to lack bars have massive bars evident in infrared images. As you think about galaxies you should remember that the classification of galaxies is based on where the light comes from in a galaxy and not necessarily where most of the mass is located.

Why do galaxies look so different? Later in this chapter you will begin to understand what determines the stellar content and shape of a galaxy. The basic classification system will be an important tool.

How Many Galaxies?

Looking at the Hubble Deep Field forces even the most sophisticated astronomer to ask what seems to be a silly question. Just how many galaxies are there? The answer is impressive.

The deep fields imaged by the Hubble Space Telescope and by the GOODS program are revealing tremendous numbers of distant galaxies in the small regions of the sky being studied. If the entire sky is similarly covered with galaxies, there must be over 100 billion galaxies visible with existing telescopes.

The faintest galaxies in the deep fields are about 500 million times too faint to be visible to your unaided eyes. Studies of those galaxies tell astronomers that they are not the most distant galaxies. There are even more galaxies too distant and too faint to see with current technology. You can conclude that there are at least a few hundred billion galaxies that will eventually be visible as astronomers build bigger telescopes with more sensitive detectors.

Looking at all these galaxies, you might wonder what proportion of the galaxies is elliptical, spiral, and irregular, but that is a surprisingly difficult question to answer. In catalogs of galaxies, about 70 percent are spiral, but that is because spiral galaxies are gaudy and easy to notice. Spiral galaxies contain hot, bright stars and clouds of ionized gas. Most ellipticals are fainter and harder to notice. Small galaxies such as dwarf ellipticals and dwarf irregulars are very common, but they are hard to detect. From careful statistical studies, astronomers estimate that ellipticals are more common than spirals and that irregulars make up only about 25 percent of all galaxies.

Building Scientific Arguments

What color are galaxies?
Scientific arguments must be based on evidence, and evidence means observations. But you must be careful to analyze even the simplest observations with care. Different kinds of galaxies have different colors, depending mostly on how much gas and dust they contain. If a galaxy contains large amounts of gas and dust, it probably contains lots of young stars, and a few of those young stars will be massive, hot, luminous O and B stars. They will produce most of the light and give the galaxy a distinct blue tint. In contrast, a galaxy that contains little gas and dust will probably contain few young stars. It will lack O and B stars, and the most luminous stars in such a galaxy will be red giants. They will give the galaxy a red tint. Because the light from a galaxy is a blend of the light from billions of stars, the colors are only tints. Nevertheless, the most luminous stars in a galaxy determine the color. From this you can conclude that elliptical galaxies tend to be red and the disks of spiral galaxies tend to be blue.

Now create your own scientific argument and analyze a different kind of observation. **Why are most galaxies in catalogs spiral in spite of the fact that the most common kind of galaxy is elliptical?**

■ ■ ■

Connections: If you want to create a theory to explain the shapes of galaxies, you need answers to some basic questions. How big are galaxies? How luminous are they? Are they all the same mass, or are some more massive than others? To answer those questions, you need to discover the basic properties of galaxies.

16-2 Measuring the Properties of Galaxies

LOOKING BEYOND THE EDGE of our Milky Way Galaxy, astronomers find billions of galaxies. Galaxies fill space as far as anyone can see. What are the properties of these star systems? What are the diameters, luminosities, and masses of galaxies? Just as in your study of stellar characteristics (Chapter 9), the first step in your study of galaxies is to find out how far away they are. Once you know a galaxy's distance, its size and luminosity are relatively easy to find. Later in this section, you will see that finding mass is more difficult.

Distance

The distances to galaxies are so large that it is not convenient to measure them in light-years, parsecs, or even kiloparsecs. Instead, astronomers use the unit **megaparsec (Mpc),** or 1 million pc. One Mpc equals 3.26 million ly, or approximately 2×10^{19} miles.

To find the distance to a galaxy, astronomers must search among its stars, nebulae, and star clusters for familiar objects whose luminosity or diameter they know. Such objects are called **distance indicators** because they can be used to find the distance to a galaxy. Most distance indicators are objects whose brightness is known, and astronomers often refer to them as **standard candles.** If you can find a standard candle in a galaxy, you can judge its distance.

Cepheid variable stars are reliable distance indicators because their period is related to their luminosity (see Figure 12-14). If you know the period of the star's variation, you can use the period–luminosity diagram to learn its absolute magnitude. Then, by comparing its absolute and apparent magnitudes, you can find its distance. ■ Figure 16-3 shows a galaxy in which the Hubble Space Telescope detected Cepheids.

Even with the Hubble Space Telescope, Cepheids are not visible in galaxies much beyond 80 million ly (25 Mpc), so astronomers must search for less common but brighter distance indicators and calibrate them using nearby galaxies containing visible

Galaxy Classification

1   **Elliptical galaxies** are round or elliptical, contain no visible gas and dust, and lack hot, bright stars. They are classified with a numerical index ranging from 1 to 7; E0s are round, and E7s are highly elliptical. The index is calculated from the largest and smallest diameter of the galaxy used in the following formula and rounded to the nearest integer.

$$\frac{10(a - b)}{a}$$

Outline of an E6 galaxy

Visual-wavelength image

Anglo-Australian Telescope Board

The Leo 1 dwarf elliptical galaxy is not many times bigger than a globular cluster.

Visual

M87 is a giant elliptical galaxy classified E1. It is a number of times larger in diameter than our own galaxy and is surrounded by a swarm of over 500 globular clusters.

---

2   **Spiral galaxies** contain a disk and spiral arms. Their halo stars are not visible, but presumably all spiral galaxies have halos. Spirals contain gas and dust and hot, bright O and B stars, as shown at right. The presence of short-lived O and B stars alerts us that star formation is occurring in these galaxies. Sa galaxies have larger nuclei, less gas and dust, and fewer hot, bright stars. Sc galaxies have small nuclei, lots of gas and dust, and many hot, bright stars. Sb galaxies are intermediate.

Anglo-Australian Telescope Board

Sa

Visual   NGC 3623

Sb

BAR

Anglo-Australian Telescope Board

NGC 3627   Visual

Ace ✺ **Astronomy**™

Log into AceAstronomy and select this chapter to see Active Figure "Galaxy Types" and review the classification of galaxies.

2a   Roughly 2/3 of all spiral galaxies are **barred spiral galaxies** classified SBa, SBb, and SBc. They have an elongated nucleus with spiral arms springing from the ends of the bar, as shown at left. Our own galaxy is a barred spiral.

NGC 1365   Visual

Sc

NGC 2997   Visual

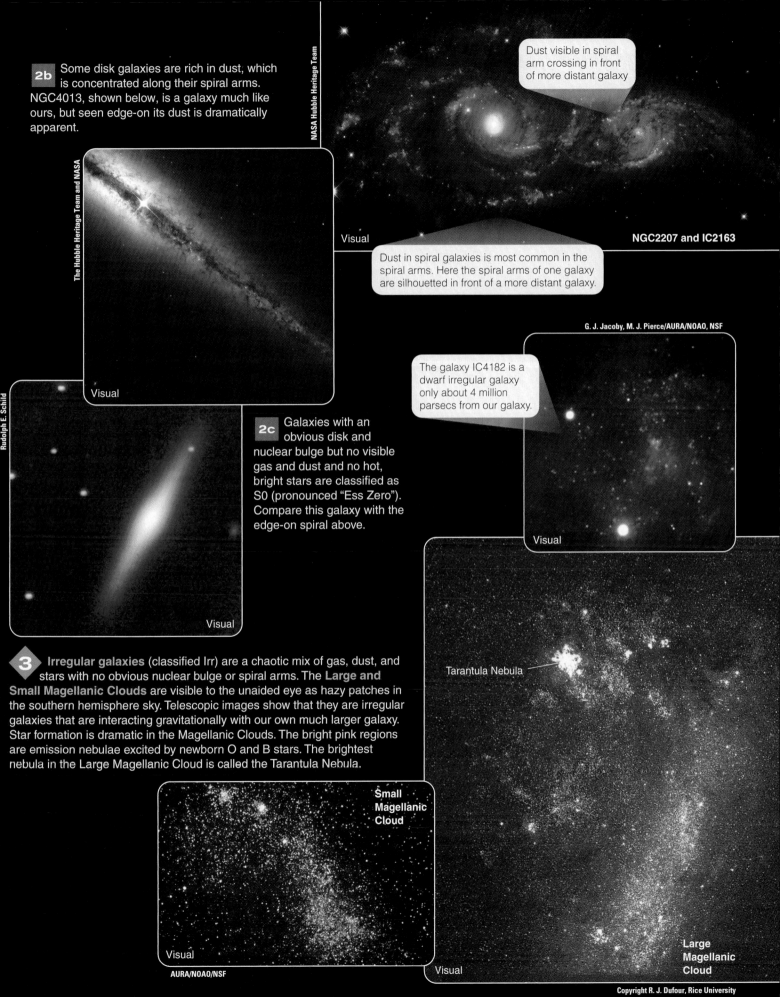

2b Some disk galaxies are rich in dust, which is concentrated along their spiral arms. NGC4013, shown below, is a galaxy much like ours, but seen edge-on its dust is dramatically apparent.

NASA Hubble Heritage Team

Dust visible in spiral arm crossing in front of more distant galaxy

Visual

NGC2207 and IC2163

The Hubble Heritage Team and NASA

Dust in spiral galaxies is most common in the spiral arms. Here the spiral arms of one galaxy are silhouetted in front of a more distant galaxy.

Visual

G. J. Jacoby, M. J. Pierce/AURA/NOAO, NSF

The galaxy IC4182 is a dwarf irregular galaxy only about 4 million parsecs from our galaxy.

Rudolph E. Schild

2c Galaxies with an obvious disk and nuclear bulge but no visible gas and dust and no hot, bright stars are classified as S0 (pronounced "Ess Zero"). Compare this galaxy with the edge-on spiral above.

Visual

Visual

3 Irregular galaxies (classified Irr) are a chaotic mix of gas, dust, and stars with no obvious nuclear bulge or spiral arms. The **Large and Small Magellanic Clouds** are visible to the unaided eye as hazy patches in the southern hemisphere sky. Telescopic images show that they are irregular galaxies that are interacting gravitationally with our own much larger galaxy. Star formation is dramatic in the Magellanic Clouds. The bright pink regions are emission nebulae excited by newborn O and B stars. The brightest nebula in the Large Magellanic Cloud is called the Tarantula Nebula.

Tarantula Nebula

Small Magellanic Cloud

Large Magellanic Cloud

Visual

Visual

AURA/NOAO/NSF

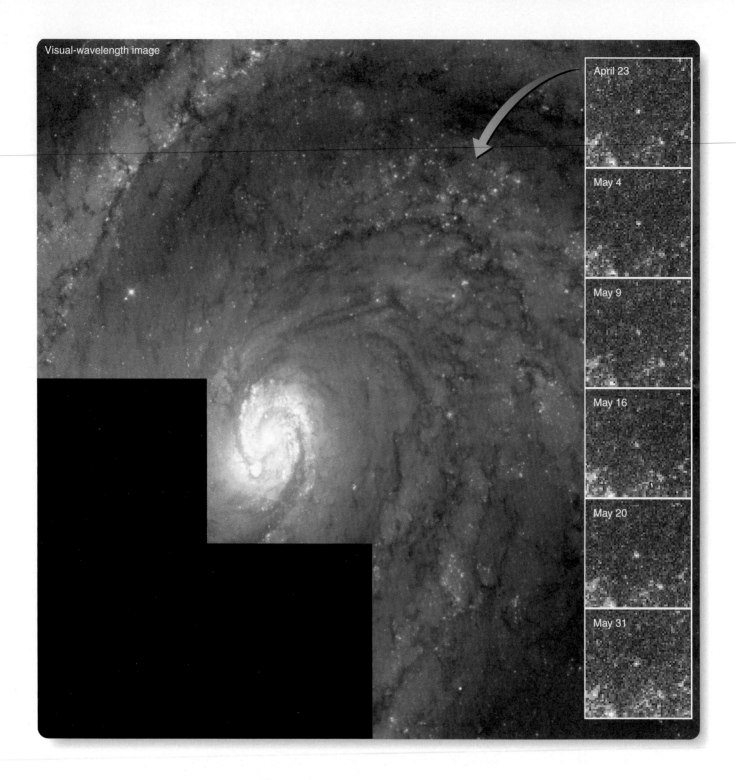

Visual-wavelength image

April 23

May 4

May 9

May 16

May 20

May 31

Cepheids. For example, by studying nearby galaxies with distances known from Cepheids, astronomers have found that the brightest globular clusters have absolute magnitudes of about −10. If you found globular clusters in a more distant galaxy, you could assume that the brightest of the globular clusters have absolute magnitudes of −10 and from that calculate the distance.

Planetary nebulae have proven to be very useful distance indicators, which is surprising considering how faint their central star is. Of course, the central stars of planetary nebulae are very

■ Figure 16-3

The vast majority of spiral galaxies are too distant for Earth-based telescopes to detect Cepheid variable stars. The Hubble Space Telescope, however, can locate Cepheids in some of these galaxies, as it has in the bright spiral galaxy M100. From a series of images taken on different dates, astronomers can locate Cepheids (inset), determine the period of pulsation, and measure the average apparent brightness. They can then deduce the distance to the galaxy—51 million ly for M100. (J. Trauger, JPL; Wendy Freedman, Observatories of the Carnegie Institution of Washington; and NASA)

Visual-wavelength image

small, and that limits the amount of energy they can radiate. But they are very hot, and most of the energy they radiate is in the ultraviolet. If you could see all of that energy, the central star of a planetary nebula would be almost as luminous as a giant star. The gas of the surrounding planetary nebula absorbs the ultraviolet radiation and reradiates it as visible emission lines, such as the oxygen emission line at a wavelength of 500.7 nm. Stars generally are not unusually bright at these wavelengths, so images of galaxies recorded at such wavelengths reveal planetary nebulae standing out like glowing paper lanterns illuminated from within. Astronomers have been able to calibrate the planetary nebulae for distance measurements by observing the absolute magnitude of planetary nebulae in nearby galaxies such as the Andromeda Galaxy, galaxies whose distances are known from other distance indicators. When planetary nebulae are detected in a more distant galaxy, a quick comparison of their apparent and absolute magnitudes reveals the distance to the galaxy. Notice once again the importance of calibration (Window on Science 15-1).

When a supernova explodes in a distant galaxy, astronomers rush to observe it. Type Ia supernovae, those produced by the collapse of a white dwarf, have been calibrated. Astronomers have observed type Ia supernovae occurring in galaxies whose distances are known from Cepheid variables and other distance indicators. For example, a type Ia supernova was observed in 1937, exploding in the small galaxy IC4182 shown on page 375. In such cases, astronomers can calculate the absolute magnitude of the supernova at maximum brightness. When astronomers see a type Ia supernova in a more distant galaxy, they observe the apparent magnitude of the supernova at maximum and compare that with the known absolute magnitude these supernovae reach at maximum to find the distance to the galaxy. As you will see in Chapter 18, this is a critical calibration in modern astronomy.

At the greatest distances, astronomers must calibrate the total luminosity of the galaxies themselves. For example, studies of nearby galaxies show that an average galaxy like our Milky Way Galaxy has a luminosity of about 16 billion times the sun's. If you see a similar galaxy far away, you can measure its apparent magnitude and calculate its distance. Of course, it is important to recognize the different types of galaxies, and that is difficult to do at great distances (■ Figure 16-4). Averaging the distances to the brightest galaxies in a cluster can reduce the uncertainty in this method.

Notice how astronomers use calibration (Window on Science 15-1) to build a **distance scale** reaching from the nearest galaxies to the most distant visible galaxies. Often astronomers refer to this as the distance pyramid or the distance ladder because each step depends on the steps below it. The bottom of the distance scale rests on the Cepheid variable stars, which rests on astronomers' understanding of the luminosities of the stars in the H–R diagram, and that rests on the very bottom step—measurements of the parallax of stars.

The most distant visible galaxies are over 10 billion ly (3000 Mpc) away, and at such distances you see an effect akin to time travel. When you look at a galaxy millions of light-years away, you do not see it as it is now but as it was millions of years ago when its light began the journey toward Earth. That means when you look at a distant galaxy, you look back into the past by an amount called the **look-back time,** a time in years equal to the distance to the galaxy in light-years.

You may have experienced look-back time if you have ever made a long-distance phone call carried by satellite or watched a TV newscaster interview someone on the other side of the world via satellite. A half-second delay occurs as a radio signal carries a question 23,000 miles out to a satellite then back to Earth and then carries the answer out to the satellite and again back to Earth. That half-second look-back delay can make people hesitate on long-distance phone calls and produces a seemingly awkward delay in intercontinental TV interviews.

The look-back time to nearby objects is usually not significant. The look-back time across a football field is a tiny fraction of a second. The look-back time to the moon is 1.3 seconds, to the sun only 8 minutes, and to the nearest star about 4 years. The Andromeda Galaxy has a look-back time of a bit over 2 million years, a mere eye blink in the lifetime of a galaxy. But if you look at more distant galaxies, the look-back time becomes an appreciable part of the age of the universe. You will see evidence in Chapter 18 that the universe began almost 14 billion years ago. That means when you look at the most distant visible galaxies, you are looking back over 10 billion years to a time when the universe was significantly different. The look-back time becomes an important factor as you begin to think about the origin and evolution of galaxies.

The Hubble Law

Although astronomers must work carefully to measure the distance to a galaxy, they often estimate such distances using a simple relationship that was first noticed at about the same time astronomers were beginning to understand the nature of galaxies.

In 1913, V. M. Slipher at Lowell Observatory reported on the spectra of faint, nebulous objects in the sky. Their spectra seemed to be composed of a mixture of stellar spectra: Some had Doppler shifts that suggested rotation; most had redshifts as if they were receding; and the faintest had the largest redshifts. Within two decades, astronomers concluded that the faint objects were galaxies similar to our own Milky Way and that the galaxies are indeed receding from us in a general expansion.

In the 1920s, astronomers Edwin Hubble and Milton Humason were able to measure the distances to a number of galaxies using Cepheid variable stars. In 1929, they published a graph that plotted the apparent velocity of recession versus distance for their galaxies. The points in the graph fell along a straight line (■ Figure 16-5). The straight-line relation from Hubble's diagram can be written as the simple equation

$$V = Hd$$

That is, the apparent velocity of recession V equals the distance d in millions of parsecs times the constant H. This relation between redshift and distance is known as the **Hubble law**, and the constant H is known as the **Hubble constant.**

Modern attempts to measure H have been difficult because of the uncertainty in the distances to galaxies. During the last decades of the 20th century, astronomers were caught up in a great controversy over the value of H. Different teams measuring distance differently got answers that ranged from 50 to 100 km/s/Mpc.*

---

*H has the units of a velocity divided by a distance. These are usually written as km/s/Mpc, meaning km/s per Mpc.

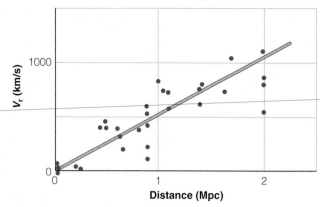

■ **Figure 16-5**

Edwin Hubble's first diagram of the apparent velocities of recession and distances of galaxies did not probe very deeply into space. It did show, however, that the galaxies are receding from one another.

The controversy was settled by better observations made with the Hubble Space Telescope and with new-generation giant telescopes on Earth. The value of H appears to be about 70 km/s/Mpc.

The Hubble law is important because it is commonly interpreted to show that the universe is expanding. In Chapter 18, you will study this expansion, but here you can use the Hubble law as a practical way to estimate the distance to a galaxy.

Simply stated, a galaxy's distance equals its apparent velocity of recession divided by H. For example, if a galaxy has an apparent radial velocity of 700 km/s, and H is 70 km/s/Mpc, then the distance to the galaxy is 700 divided by 70, or about 10 Mpc. This makes it relatively easy to estimate the distances to galaxies, because a large telescope can photograph the spectrum of a galaxy and determine its apparent velocity of recession even when it is too distant to have visible distance indicators.

Ace⊛Astronomy™ Log into AceAstronomy and select this chapter to see Astronomy Exercise "Hubble Relation." You can make your own observations to find the Hubble constant.

Diameter and Luminosity

The distance to a galaxy is the key to finding its diameter and its luminosity. With a good telescope and the right equipment, you could easily photograph a galaxy and measure its angular diameter in seconds of arc. If you knew the distance, you could use the small-angle formula (Chapter 3) to find its linear diameter. Also, if you measured the apparent magnitude of a galaxy, you could use the distance to find its absolute magnitude and from that its luminosity (Chapter 9).

The results of such observations show that galaxies differ dramatically in size and luminosity. Irregular galaxies tend to be small, 1 to 25 percent the size of our galaxy, and of low luminosity. Although they are common, they are easy to overlook. Our Milky Way Galaxy is large and luminous compared with most spiral galaxies, though astronomers know of a few spiral galaxies that are even larger and more luminous—nearly four times larger in diameter and about 10 times more luminous. Elliptical galaxies cover a wide range of diameters and luminosities. The largest, called giant ellipticals, are five times the size of our Milky Way. But many elliptical galaxies are very small, dwarf ellipticals, only 1 percent the diameter of our galaxy.

To put galaxies in perspective, you can use an analogy. If our galaxy were an 18-wheeler, the smallest dwarf galaxies would be the size of pocket-size toy cars, and the largest giant ellipticals would be the size of jumbo jets. Even among spiral galaxies that looked similar, the sizes could range from that of a go-cart to that of a large bus (■ Figure 16-6).

Clearly, the diameter and luminosity of a galaxy do not determine its type. Some small galaxies are irregular, and some are elliptical. Some large galaxies are spiral, and some are elliptical. You need more data before you can build a theory for the origin and evolution of galaxies.

Of the three basic parameters that describe a galaxy, you have found two—diameter and luminosity. The third, as was the case for stars, is more difficult to discover.

Mass

Although the mass of a galaxy is difficult to determine, it is an important quantity. It tells you how much matter the galaxy contains, which provides clues to the galaxy's origin and evolution. In this section, you will examine two fundamental ways to find the masses of galaxies.

One way to find the mass of a galaxy is to watch it rotate. You know the stars in the outer parts of the galaxy are in orbit around the galaxy, so you can use Newton's laws to find the mass. All you need to know is the radius of the galaxy (in AU) and the orbital period (in years) of the stars at the galaxy's outer edge. Then you can use Newton's version of Kepler's third law to find the mass just as you found the mass of binary stars in Chapter 9.

Recall from the previous section that it is easy to find the size of a galaxy if you know the distance. You just measure its angular size and use the small-angle formula (Chapter 3).

You could find the orbital velocity of the stars from the Doppler effect. If you focused the image of the galaxy on the slit of a spectrograph, you would see a bright spectrum formed by the bright nucleus of the galaxy, but you would also see fainter emission lines produced by ionized gas in the disk of the galaxy. Because the galaxy rotates, one side moves away from Earth and one side moves toward Earth, so the emission lines would be redshifted on one side of the galaxy and blueshifted on the other side. You could measure those changes in wavelength, use the Doppler formula to find the velocities, and plot a diagram showing the velocity of rotation at different distances from the center of the galaxy—a diagram called a **rotation curve.** The artwork in ■ Figure 16-7a shows the process of creating a rotation curve, and Figure 16-7b shows a real galaxy, its spectrum, and its rotation curve.

Visual-wavelength image

■ Figure 16-6

The small galaxy cluster known as Hickson Compact Group 87 appears to contain three spiral galaxies and one elliptical. It is not known whether the small spiral galaxy is in the distant background or is part of the group, but it could easily be a group member. Note the contrast between the sizes of the galaxies. (Hubble Heritage Team, STScI/AURA/NASA)

Of course, if you know the orbital velocity and the radius of a star's orbit around a galaxy, you can calculate its orbital period. The circumference of the orbit (2π times the radius) divided by the orbital velocity equals the orbital period. So knowing the velocity and size of an orbit easily tells you the orbital period, and from that you can find the mass.

The rotation curve of a galaxy contains all the information you need to find the mass of the galaxy, and the method is known as the **rotation curve method.** This is the most accurate way to find the mass of a galaxy, but it works only for the nearer galaxies, whose rotation curves can be observed. More distant galaxies appear so small astronomers cannot measure the radial velocity at different points along the galaxy.

You should note one warning about masses found from rotation curves. Recent studies of our own galaxy and others show that the outer parts of the rotation curve do not decline to lower velocities (Figure 15-11). This shows that the outermost visible parts of some galaxies do not travel more slowly and warns that the galaxies contain large amounts of mass outside this radius, perhaps in extended galactic coronas like the one that seems to surround our own galaxy (see Chapter 15). Because the rotation curve method can be applied only to nearby galaxies and because it cannot determine the masses of galactic coronas, astronomers have looked at another way to find the masses of galaxies.

The **cluster method** of finding galactic mass depends on the motions of galaxies within a cluster. If you measured the radial velocities of many galaxies in a cluster, you would find that some velocities are larger than others because of the orbital motions of the individual galaxies in the cluster. Given the range of velocities

■ Figure 16-7

(a) In the artwork in the upper half of this diagram, the astronomer has placed the image of the galaxy over a narrow slit so that light from the galaxy can enter the spectograph and produce a spectrum. A very short segment of the spectrum shows an emission line redshifted on the receding side of the rotating galaxy and blueshifted on the approaching side. Converting these Doppler shifts into velocities, the astronomer can plot the galaxy's rotation curve (right). (b) Real data are shown in the bottom half of this diagram. Galaxy NGC2998 is shown over the spectrograph slit, and the segment of the spectrum includes three emission lines. (Courtesy Vera Rubin)

and the size of the cluster, you could ask how massive a cluster of this size must be to hold itself together with this range of velocities. That is the same as asking how massive the cluster must be to have an escape velocity larger than the velocity of the fastest galaxy. That would tell you the mass of the entire cluster. Dividing the total mass of the cluster by the number of galaxies in the cluster yields the average mass of the galaxies. This method contains the built-in assumption that the cluster is not flying apart. If it is, your result is too large. Because it seems likely that most clusters are held together by their own gravity, the method is probably valid.

A related way of measuring a galaxy's mass is called the **velocity dispersion method.** It is really a version of the cluster method. Instead of observing the motions of galaxies in a cluster, you could observe the motions of matter within a galaxy. In the spectra of some galaxies, broad spectral lines indicate that stars and gas are moving at high velocities. If you assume the galaxy is bound by its own gravity, you can ask how massive it must be to hold this moving matter within the galaxy. This method, like the one before, assumes that the system is not coming apart.

Measuring the masses of galaxies reveals two things. First the range of masses is wide—from one-millionth that of the Milky Way Galaxy to 50 times more (■ Table 16-1). And second, there is more to a galaxy than meets the eye.

Supermassive Black Holes in Galaxies

Rotation curves show the motions of the outer parts of a galaxy, but it is possible to detect the Doppler shifts of stars orbiting close to the centers of galaxies. Although these motions are not usually shown on rotation curves, they reveal something astonishing.

Visual-wavelength image

■ **Figure 16-8**

The Pinwheel Galaxy is near our Milky Way Galaxy in space, and it can be studied in detail. The velocities at the center of the Pinwheel Galaxy are low, showing that it does not contain a supermassive black hole at its center. It also lacks a nuclear bulge, and that confirms the observation that the mass of a supermassive black hole is related to the size of a galaxy's nuclear bulge. (Bill Schoening/AURA/NOAO/NSF)

Measurements show that the stars near the centers of most galaxies are orbiting very rapidly. To hold stars in such small, short-period orbits, the centers of galaxies must contain masses of a million to a few billion solar masses in a very small region. The evidence shows that the nuclei of galaxies contain supermassive black holes. You saw in Chapter 15 that the Milky Way contains a supermassive black hole at its center. Evidently that is typical of galaxies.

Such a supermassive black hole cannot be the remains of a dead star. That would amount to only a few solar masses. Rather, these supermassive black holes must have formed as the galaxy formed, or they must have accumulated over billions of years as matter sank into the centers of the galaxies. Measurements show that the masses of these supermassive black holes are related to the mass of nuclear bulges. A galaxy with a large nuclear bulge has a supermassive black hole whose mass is greater than that in a galaxy with a small nuclear bulge. Rare galaxies lacking nuclear bulges also lack supermassive black holes (■ Figure 16-8). This suggests that the supermassive black holes formed long ago as

the galaxies formed. Of course, matter has continued to drain into the black holes, but they do not appear to have grown dramatically since they formed.

A billion-solar-mass black hole sounds like a lot of mass, but note that it is roughly 1 percent of the mass of a galaxy. The 2.6 million-solar-mass black hole at the center of the Milky Way Galaxy contains only a thousandth of a percent of the mass of our galaxy. In the next chapter you will discover that these supermassive black holes can produce titanic eruptions, but they represent only a small fraction of the mass of a galaxy.

Dark Matter in Galaxies

Given the size and luminosity of a galaxy, astronomers can make a rough guess as to the amount of matter it should contain. They know how much light stars produce, and they know about how much matter there is between the stars, so it is quite possible to estimate very roughly the mass of a galaxy from its luminosity. But when astronomers measure the masses of galaxies, they often find that the measured masses are much too large. You discovered this effect in Chapter 15 when you studied the rotation curve of our own galaxy and found that the rotation curve does not decline at large distances from the center. That must mean our galaxy contains dark matter. This seems to be true of most galaxies. Measured masses of galaxies amount to 10 to 100 times more mass than can be seen. Because this mass is not visible, it is called dark matter.

X-ray observations reveal more evidence of dark matter. X-ray images of galaxy clusters show that many of them are filled with very hot, low-density gas. The amount of gas present is much too small to account for the dark matter. Rather, the gas is impor-

■ **Table 16-1** | **The Properties of Galaxies***

	Elliptical	Spiral	Irregular
Mass	0.0001–50	0.005–2	<0.0005–0.15
Diameter	0.01–5	0.2–1.5	0.05–0.25
Luminosity	0.00005–5	0.005–10	0.00005–0.1

*In units of the mass, diameter, and luminosity of the Milky Way.

a Visual-wavelength image b X-ray image

■ **Figure 16-9**

(a) The Coma cluster of galaxies contains at least 1000 galaxies and is especially rich in E and S0 galaxies. Two giant galaxies lie near its center. Only the central area of the cluster is shown in this image. If the cluster were visible in the sky, it would span eight times the diameter of the full moon. (L. Thompson © NOAO) (b) In false colors, this X-ray image of the Coma cluster shows it filled and surrounded by hot gas. The box outlines the area of the cluster shown in part a. (Simon D. M. White, Ulrich G. Briel, and J. Patrick Henry)

tant because it is very hot and its rapidly moving atoms have not leaked away. Evidently the gas is held in the cluster by a strong gravitational field. To provide a high enough escape velocity to hold the hot gas, the cluster must contain much more matter than what astronomers see. The detectable galaxies in the Coma cluster, for instance, amount to only a small fraction of the total mass of the cluster (■ Figure 16-9).

Dark matter is not an insignificant issue. Observations of galaxies and clusters of galaxies show that approximately 95 percent of the matter in the universe is dark matter; the universe you see—the kind of matter that you and the stars are made of—has been compared to the foam on an invisible ocean. You will find further evidence of dark matter when you study cosmology in Chapter 18.

Gravitational Lensing and Dark Matter

When you solve a math problem and get an answer that doesn't seem right, you check your work for a mistake. Astronomers using Newton's laws (orbital velocity and escape velocity) to measure the mass of galaxies find evidence of large amounts of dark matter, so they look for ways to check their work. Fortunately, there is another way to detect dark matter that does not depend on Newton's laws. Astronomers can follow a light beam.

Albert Einstein described gravity as a curvature of space. The presence of mass actually distorts space, and that is what you feel as gravity. He predicted that a light beam traveling through a gravitational field would be deflected by the curvature of space much as a golf ball is deflected as it rolls over a curved putting green. That effect has been observed and is a strong confirmation of Einstein's theories.

Gravitational lensing occurs when light from a distant object passes a nearby massive object and is deflected by the gravitational field. The gravitational field of the nearby object is actually a region of curved space-time, and it acts as a lens to deflect the passing light. Astronomers have used gravitational lensing to detect dark matter. When light from very distant galaxies passes through a cluster of galaxies on its way to Earth, it can be deflected by the strong curvature. That distorts the images of the distant galaxies into arcs, and the amount of the distortion depends on the mass of the cluster of galaxies (■ Figure 16-10). Observations of gravitational lensing made with very large telescopes reveal that clusters of galaxies contain far more matter than what is seen. That is, they contain large amounts of dark matter. This confirmation of the existence of dark matter is independent of Newtons' laws and gives astronomers much greater confidence that dark matter is real.

Dark matter is difficult to detect, and it is even harder to explain. Some astronomers have suggested that dark matter consists of low-luminosity white dwarfs and brown dwarfs scattered through the halos of galaxies. Both observation and theory support the idea that galaxies have massive extended halos, and searches for white dwarfs and brown dwarfs in the halo of our galaxy have been successful. Nevertheless, the searches have not turned up enough of these low-luminosity objects to make up all of the dark matter.

The dark matter can't be hidden in vast numbers of black holes and neutron stars, because astronomers don't see the X rays these objects would emit. Remember, there must be roughly 10 times more dark matter than visible matter, and that many

■ Figure 16-10

The mass of galaxy cluster 0024+1654 curves space around it and bends the light of a much more distant galaxy to produce arcs that are actually distorted images of the distant galaxy. This gravitational lens effect reveals that many galaxy clusters contain much more mass than is visible as galaxies. Evidently galaxy clusters must contain large amounts of dark matter. (W. N. Colley and E. Turner, Princeton, and J. A. Tyson, Bell Labs, and NASA)

black holes and neutron stars would produce X rays that should be easy to detect.

Because observations seem to say that the dark matter can't be hidden in normal objects, some theorists have suggested that the dark matter is made up of unexpected forms of matter. Until recently neutrinos were thought to be massless, but studies now suggest they have a very small mass; neutrinos appear to represent part of the dark matter, but they are not massive enough to explain all of the dark matter that must exist.

Dark matter remains one of the fundamental unresolved problems of modern astronomy. You will learn more about this problem in Chapter 18 when you try to understand how dark matter affects the nature of the universe, its past, and its future.

Building Scientific Arguments

Why do you have to know the distance to a galaxy to find its mass?

A scientific argument must proceed step-by-step with no gaps. This is a good example. To find the mass of a galaxy, you need to know the size of the orbits and the orbital periods of stars at the galaxy's outer edge, and then use Kepler's third law to find the mass inside the orbits. Measuring the orbital velocity of the stars as the galaxy rotates is easy if you can obtain a rotation curve from a spectrum. But you must also know the radii of the stars' orbits in meters or astronomical units. That is where the distance comes in. To find the radii of the orbits, you must know the distance to the galaxy. Then

you can use the small-angle formula to convert the radius in seconds of arc into a radius in parsecs or AU. If your measurement of the distance to the galaxy isn't accurate, you will get inaccurate radii for the orbits and will compute an inaccurate mass for the galaxy.

Many different measurements in astronomy depend on the calibration of the distance scale. Build a step-by-step argument to analyze the following problem. **What would happen to your measurements of the diameters and luminosities of the galaxies if astronomers discovered that the Cepheid variable stars were slightly more luminous than had been believed?**

■　　　■　　　■

Connections: You have gathered the basic data to characterize galaxies. That is, you can describe their general characteristics. Now it is time to consider the more difficult and more interesting question. How did galaxies get to be the way they are?

  The Evolution of Galaxies

YOUR GOAL IN THIS CHAPTER has been to build a theory to explain the evolution of galaxies. In Chapter 15 you considered the origin of our own Milky Way Galaxy; presumably, other

NORTH

Visual-wavelength image

■ Figure 16-11

The Hercules galaxy cluster is named after the constellation in which it is found. It is a small cluster containing roughly a hundred galaxies both elliptical and spiral. It lies over 360 million light-years from our galaxy. (NOAO/AURA/NSF)

galaxies formed similarly. But why did some galaxies become spiral, some elliptical, and some irregular? An important clue to that mystery lies in the clustering of galaxies.

Clusters of Galaxies

Single galaxies are rare. Most occur in clusters containing a few to a few thousand galaxies in a volume 1 to 10 Mpc across (■ Figure 16-11). Our Milky Way Galaxy is a member of a cluster containing slightly over three dozen galaxies, and surveys have cataloged over 2700 other clusters within 4 billion ly.

For purposes of this study, you can sort clusters of galaxies into rich clusters and poor clusters. **Rich clusters** contain a thousand or more galaxies, many elliptical, scattered through a volume roughly 3 Mpc (10^7 ly) in diameter. Such a cluster is nearly always condensed; that is, the galaxies are concentrated toward the cluster center. And, at the center, such clusters often contain one or more giant elliptical galaxies.

The Virgo cluster is an example of a rich cluster. It contains over 2500 galaxies located about 17 Mpc (55 million ly) away. The Virgo cluster, like most rich clusters, is centrally condensed and contains the giant elliptical galaxy M87 at its center (page 374).

X-ray observations have found that many of these rich clusters are filled with a hot gas—an intracluster medium. Some of this gas has presumably been driven out of galaxies by supernovae explosions, but much of it appears to be left over from the earliest generations of galaxy formation.

Poor clusters contain fewer than a thousand (and often only a few) galaxies spread through a region that can be as large as a rich cluster. That means the galaxies are more widely separated.

Our Milky Way Galaxy is a member of a poor cluster known as the Local Group (■ Figure 16-12a). The total number of galaxies in the Local Group is uncertain, but it probably contains more than three dozen galaxies scattered irregularly through a volume roughly 1 Mpc in diameter. Of the brighter galaxies, 15 are elliptical, 4 are spiral, and 13 are irregular.

The total number of galaxies in the Local Group is uncertain because some lie in the plane of the Milky Way Galaxy and are difficult to detect. For example, a small dwarf galaxy, known as the Sagittarius Dwarf, has been found on the far side of our own galaxy where it is almost totally hidden behind the star clouds of Sagittarius (Figure 16-12b). Even closer to the center of the Milky Way Galaxy is the Canis Major Dwarf Galaxy (Figure 16-12c). The galaxy was found by mapping the distribution of red supergiants detected by the 2MASS infrared survey. There are, almost certainly, other small galaxies in our Local Group hidden behind the stars, gas, and dust of our Milky Way Galaxy.

The Local Group illustrates the subclustering found in poor galaxy clusters. The two largest galaxies, the Milky Way Galaxy and the Andromeda Galaxy, are the centers of two subclusters. The Milky Way Galaxy is accompanied by the Magellanic Clouds and seven other dwarf galaxies. The Andromeda Galaxy is attended by more dwarf elliptical galaxies (two of which are visible in Figure 15-1b) and a small spiral, M33 (Figure 16-8). The Andromeda Galaxy and our Milky Way Galaxy, with their retinues of smaller galaxies, are moving toward each other and will almost certainly collide in a few billion years.

From all of this, you can draw an important conclusion: Galaxies do not live in isolation. They are usually found in clusters, often interact with nearby companions, and may even collide with each other. That leads to a critical clue hidden in the clusters of galaxies—the relative abundance of the different types of galaxies.

In general, rich clusters tend to contain 80 to 90 percent E and S0 galaxies and a few spirals. Poor clusters contain a larger percentage of spirals; and, among isolated galaxies, those that are not in clusters, 80 to 90 percent are spirals. Galaxies that are crowded together tend to be E or S0 rather than spiral. Somehow the environment around a galaxy helps determine its type.

■ **Figure 16-12**

(a) The Local Group. Our galaxy is located at the center of this diagram. The vertical lines giving distances from the plane of the Milky Way are solid above the plane and dashed below. (b) The Sagittarius Dwarf Galaxy (Sgr Dwarf) lies on the other side of our galaxy. If you could see it in the sky, it would be 17 times larger than the full moon. (c) The Canis Major Dwarf Galaxy is even closer to the Milky Way Galaxy. It is the remains of a small galaxy that is being pulled apart by our galaxy and is hidden behind the stars of the constellation Canis Major.

Astronomers suspect that collisions between galaxies are an important process. In a rich cluster, the galaxies are crowded much closer together, and they must collide with each other more often than galaxies in a poor cluster. Could such collisions explain the excess of elliptical galaxies in the crowded rich clusters? Astronomers have discovered that galaxy smashups are both important and entertaining.

Colliding Galaxies

Galaxies should collide fairly often. The average separation between galaxies is only about 20 times their diameter. Like two blindfolded elephants blundering about at random under a circus tent, galaxies should bump into each other once in a while. Stars, on the other hand, almost never collide. In the region of the galaxy near the sun, the average separation between stars is about 10^7 times their diameter. Consequently, a collision between two stars is about as likely as collision between two blindfolded gnats flitting about at random in a football stadium.

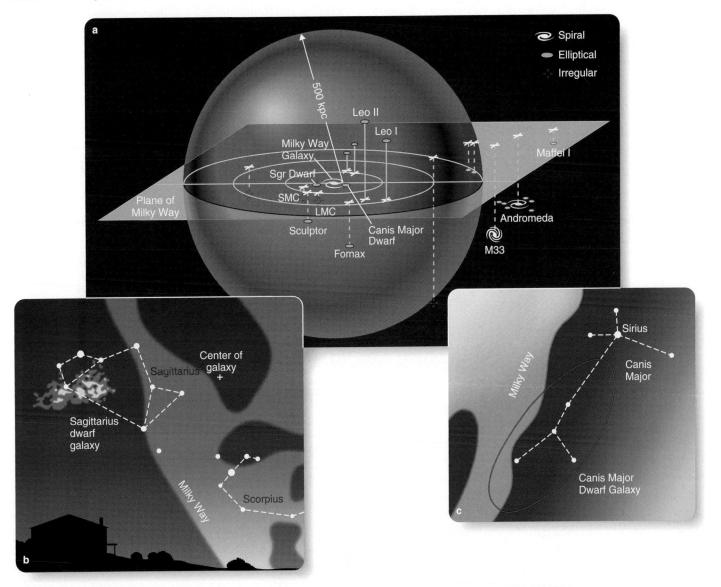

Study **Interacting Galaxies** on pages 388–389 and notice four important points:

1 Interacting galaxies can distort each other with tides producing tidal tails and shells of stars. They may even trigger the formation of spiral arms. In fact, large galaxies can even absorb smaller galaxies.

2 Notice how interactions can trigger star formation.

3 Also, notice how evidence left inside galaxies in the form of motions and multiple nuclei reveals that they have suffered past interactions and mergers.

4 Finally, notice the beautiful ring galaxies—bull's-eyes left behind by high-speed collisions.

Evidence of galaxy mergers is all around you. Our Milky Way Galaxy is a cannibal galaxy, snacking on the Magellanic Clouds as those two small galaxies orbit the Milky Way. Furthermore, our galaxy's tides are pulling the Sagittarius Dwarf Galaxy apart, and the Canis Major Dwarf Galaxy has been almost completely digested as tides have pulled stars away to form great streamers wrapped around the Milky Way galaxy (■ Figure 16-13). Almost certainly, our galaxy has dined on other small galaxies.

You have seen the evidence that collisions and mergers between galaxies are common and can produce dramatic changes in the structure of the galaxies; now you are ready to evaluate a hypothesis that may explain how galaxies form and evolve.

The Origin and Evolution of Galaxies

The test of any scientific understanding is whether you can put all the evidence and theory together to tell the history of the objects studied. Can you describe the origin and evolution of the galaxies? Just a few decades ago, it would have been impossible, but the evidence from space telescopes and new-generation telescopes on Earth, combined with advances in computer modeling and theory, allow astronomers to outline the story of the galaxies.

As you begin, you should eliminate a few older ideas immediately. It is easy to imagine that galaxies evolve from one type to another. But an elliptical galaxy cannot become a spiral galaxy or an irregular galaxy because ellipticals contain almost no gas and dust from which to make new stars. Elliptical galaxies can't be young galaxies. You can also argue that spiral and irregular galaxies cannot evolve into elliptical galaxies because spiral and irregular galaxies contain both young and old stars. The old stars

mean that spiral and irregular galaxies can't be young. The galaxy classes tell you something important, but a single galaxy does not change from one class to another any more than a cat can change into a dog.

Another old idea held that galaxies that form from rapidly rotating clouds of gas would have lots of angular momentum and would contract slowly to form disk-shaped spiral galaxies. Clouds of gas that rotated less rapidly would contract faster, form stars quickly, use up all the gas and dust, and become elliptical galaxies. The evidence clearly shows that galaxies didn't form and evolve in isolation. Collisions and mergers dominate the history of the galaxies.

The ellipticals appear to be the product of galaxy mergers, which triggered star formation that used up the gas and dust. In fact, astronomers see many **starburst galaxies** that are very luminous in the infrared because a collision has triggered a burst of star formation that is heating the dust (■ Figure 16-14). The warm dust reradiates the energy in the infrared. Some of these galaxies are a hundred times more luminous than our Milky Way Galaxy, but so deeply shrouded in dust that they are very dim at visible wavelengths. Such **ultraluminous infrared galaxies** show evidence of tidal tails and are probably the mergers of three or more galaxies triggering firestorms of star formation and gener-

■ Figure 16-13

As shown in this mathematical model, the Canis Major Dwarf Galaxy has orbited around the Milky Way Galaxy a number of times, and tides have ripped stars and gas away to form streamers. Seen from Earth, the dwarf galaxy is almost entirely hidden behind the dusty disk of the Milky Way. In this simulation, the thickness of the disk and its dust have been reduced to reveal the location of the Canis Major Galaxy. (Nicolas Martin and Rodrigo Ibata, Strasbourg Observatory) ating tremendous clouds of dust. The Antennae (page 389) contain over 15 billion solar masses of hydrogen gas and will become a starburst galaxy as the merger triggers rapid star formation. Such mergers can drive the newly formed galaxy to use up much of its gas and dust.

Another process can help clear the gas and dust out of elliptical galaxies. Supernovae in a starburst galaxy can blow away the remaining gas and dust, and a rapid burst of star formation would produce lots of supernova explosions.

In this way, a few collisions and mergers could leave a galaxy with no gas and dust from which to make new stars and could scramble the orbits of remaining stars to produce elliptical clouds of stars. Astronomers now suspect that most elliptical galaxies are formed by the merger of at least two or three galaxies of comparable size. The dwarf ellipticals, too small to be formed by mergers, may be fragments left over by the merger of larger galaxies.

In contrast, spirals seem never to have suffered major collisions. Their thin disks are delicate and would be destroyed by tidal forces in a collision with a massive galaxy. Also, they retain plenty of gas and dust and continue making stars. Of course, spiral galaxies can cannibalize smaller galaxies. You have seen plenty of evidence in our own galaxy, and some astronomers suspect that much of the halo consists of the remains of cannibalized galaxies. But our Milky Way Galaxy has never collided with a similarly large

galaxy. Such a collision would trigger star formation, using up gas and dust; scramble the orbits of stars; and destroy the thin disk. In a few billion years, our galaxy may merge with the approaching Andromeda Galaxy, and the final result may be an elliptical galaxy.

Barred spiral galaxies may be the product of tidal interactions. Mathematical models show that bars are not stable and eventually dissipate. It may take tidal interactions with other galaxies to regenerate the bars. Because well over half of all spiral galaxies have bars, you can suspect that these tidal interactions are common.

Other processes can alter galaxies. The S0 galaxies retain a disk shape but lack gas and dust. They may have lost gas and dust during a starburst, but another process involves their motion through the gas in a cluster of galaxies. A galaxy orbiting through such a gas would encounter a tremendous wind blowing its gas and dust away. A disk-shaped spiral galaxy could be reduced to an S0 galaxy. For example, X-ray observations show that the Coma cluster contains thin, hot gas between the galaxies (Figure 16-9), and astronomers have located, in a similar

■ **Figure 16-14**

Rapid star formation: (a) NGC1569 is a starburst galaxy filled with clouds of young stars and supernovae. At least some starbursts are triggered by interactions between galaxies. (ESA/NASA/P. Anders) (b) The dwarf irregular galaxy NGC1705 began a burst of star formation about 25 million years ago. (NASA/ESA/Hubble Heritage Team/AURA/STScI) (c) The inner parts of M64, known as the "black eye galaxy," are filled with dust produced by rapid star formation. Radio observations (page 389) show that the inner part of the galaxy rotates backward compared to the outer part of the galaxy, a product of a merger. Where the counter-rotating parts of the galaxy collide, star formation is stimulated. (NASA/Hubble Heritage Team/AURA/STScI)

NGC1569 — UV + visual image

NGC1705 — UV + visual + infrared image

M64 The Black Eye Galaxy — Visual + infrared image

Interacting Galaxies

Tidal Distortion

Small galaxy passing near a massive galaxy.

1 When two galaxies collide, they can pass through each other with the stars so small and so far apart that they never collide. Gas clouds and magnetic fields do collide, but the biggest effects may be tidal. Even when two galaxies just pass near each other, tides can cause dramatic effects, such as long streamers called **tidal tails**. In some cases, two galaxies can merge and form a single galaxy.

Gravity of a second galaxy represented as a single massive object

1a When a galaxy swings past a massive object such as another galaxy, tides are severe. Stars near the massive object try to move in smaller, faster orbits while stars further from the massive object follow larger, slower orbits. Such tides can distort a galaxy or even rip it apart.

Galaxy interactions can stimulate the formation of spiral arms

In this computer model, two uniform disk galaxies pass near each other.

The small galaxy passes behind the larger galaxy so they do not actually collide.

Visual false-color image

The Mice are a pair of galaxies whipping around each other and being distorted.

Allen Beechel

NOAO

Tidal forces deform the galaxies and trigger the formation of spiral arms.

Computer model of the Mice

Allen Beechel

The upper arm of the large galaxy passes in front of the small galaxy.

1b The merger of galaxies is called **galactic cannibalism**. Models show that merging galaxies spiral around their common center of mass while tides rip stars away and form shells.

Computer model

François Schweizer and Alan Toomre

Shells of stars

Allen Beechel

A photo of the well-known Whirlpool Galaxy resembles the computer model.

NOAO

Visual

Such shells have been found around elliptical galaxies such as NGC5128. It is peculiar in many ways and even has a belt of dusty gas. The shells revealed in this enhanced image are evidence that the giant galaxy has cannibalized at least one smaller galaxy. The giant galaxy itself may be the result of the merger of two large galaxies.

NOAO

Visual enhanced image

2 The collision of two galaxies can trigger firestorms of star formation as gas clouds are compressed. Galaxies NGC4038 and 4039 have been known for years as the Antennae because the long tails visible in Earth-based photos resemble the antennae of an insect. Hubble Space Telescope images reveal that the two galaxies are blazing with star formation. Roughly a thousand massive star clusters have been born.

Spectra show that the galaxy is 10 to 20 times richer in elements like magnesium and silicon. Such metals are produced by massive stars and spread by supernova explosions.

3 Evidence of past galaxy mergers shows up in the motions inside some galaxies. NCG7251 is a highly distorted galaxy with tidal tails in this ground-based image.

The Antennae

Giant star cluster

Brad Whitmore, STScI/NASA

Ground-based visual image

Hubble Space Telescope visual image

This Hubble Space Telescope image of the core of the galaxy reveals a small spiral spinning backward in the heart of the larger galaxy.

An X-ray image of the Antennae shows clouds of very hot gas heated by supernovae exploding 30 times more often than in our own galaxy.

X-ray image

NASA/CXC/SAO/ G. Fabbiano et al.

François Schweizer, Carnegie Inst. of Washington, Brad Whitmore, STScI

Visual

This counter rotation suggests that NCG7251 is the remains of two oppositely rotating galaxies that merged about a billion years ago.

3a Radio evidence of past mergers: Doppler shifts reveal the rotation of the spiral galaxy M64. The upper part of the galaxy has a redshift and is moving away from Earth, and the bottom part of the galaxy has a blueshift and is approaching. A radio map of the core of the galaxy reveals that it is rotating backward. This suggests a merger long ago between two galaxies that rotate in opposite directions.

Rotation of galaxy M64

Redshift

Blueshift

Robin Braun, NRAO/AUI/NSF

Evidence of galactic cannibalism: Giant elliptical galaxies in rich clusters sometimes have multiple nuclei, thought to be the densest parts of smaller galaxies that have been absorbed and only partly digested.

Multiple nuclei

Michael J. West

Visual false-color image

4 Galaxy AM 0644-741 below was once a normal spiral galaxy, but now it is a **ring galaxy**. Mathematical models show that such galaxies are produced by high-speed collisions in which a smaller galaxy passes through another galaxy almost perpendicular to the disk. Ring galaxies tend to have nearby companions, suggestive evidence of past encounters.

Visual-wavelength image

Hubble Heritage Team, NASA/ESA/AURA/STScI

Visible (white)
X-ray (purple)
Radio (red)

■ **Figure 16-15**

The distorted galaxy C153 is orbiting through the thin gas in its home cluster of galaxies at 4.5 million miles per hour. At that speed it feels a tremendous wind stripping gas out of the galaxy in a trail 200,000 ly long. Such galaxies could quickly loose almost all of their gas and dust. (NASA, W. Keel, F. Owen, M. Ledlow, and D. Wang)

Visual-wavelength image

■ **Figure 16-16**

Arrows point to three very distant red galaxies. The look-back time to these galaxies is so large they appear as they were when the universe was only a billion years old. Such highly redshifted galaxies are believed to be among the first to form stars and begin shining after the beginning of the universe about 14 billion years ago. (NASA, H.-J. Yan, R. Windhorst and S. Cohen, Arizona State University)

cluster, a galaxy plunging through such gas and being stripped of its gas and dust (■ Figure 16-15).

Small galaxies may arise in a number of ways. The dwarf ellipticals are too small to be produced by mergers, but they could be fragments of galaxies ripped free during interactions. Another possibility is that the dwarf ellipticals are small galaxies that plunged through the gas in a cluster of galaxies and lost their gas and dust. In contrast, the irregular galaxies may be small fragments ripped from larger galaxies but retaining enough gas and dust to continue forming stars.

Other factors must influence the evolution of galaxies. The hot gas in galaxy clusters warns that galaxies rarely form in isolation. Cooler gas clouds may fall into galaxies and add material for star formation. Most important, tidal interactions are complex and violent.

Only one in 10,000 galaxies is isolated from its neighbors Those galaxies tend to be disk galaxies, and some are flocculent, having many branches and spurs without a strong two-armed spiral pattern. Some isolated galaxies are rich in gas and dust but have little star formation. All of this suggests that collisions are necessary to stimulate spiral structure and star formation.

You may have found it surprising that the Virgo cluster is a rich, crowded cluster but contains lots of spiral galaxies. Astronomers understand that the cluster is so massive that the galaxies orbit the center of mass of the cluster at very high velocities; and, when they encounter each other, they rush past each other too

quickly to interact strongly. The galaxies are whittled down a bit, but many are still disk galaxies with spiral patterns.

A good theory helps you understand how nature works, and astronomers are just beginning to understand the exciting and complex story of the galaxies. Nevertheless, it is already clear that galaxy evolution is much like a pie-throwing contest and just about as neat.

The Farthest Galaxies

Observations with the largest and most sophisticated telescopes are taking astronomers back to the age of galaxy formation. At great distances the look-back time is so great that they see the universe as it was soon after the galaxies began to form. There were more spirals then and fewer ellipticals. On the whole, galaxies long ago were more compact and more irregular than they are now. The observations show that galaxies were closer together long ago; about a third of all distant galaxies are in close pairs, but only 7 percent of nearby galaxies are in pairs. The observational evidence clearly supports the hypothesis that galaxies have evolved by merger.

At great distances, astronomers see tremendous numbers of small, blue, irregularly shaped galaxies that have been called blue dwarfs. The blue color indicates they are rapidly forming stars, but the role of these blue dwarfs in the formation of galaxies is not clear. Blue dwarf galaxies are not found at small look-back times, so they no longer exist in the present universe. They may be clouds of gas and stars that were long ago absorbed in the formation of larger galaxies, or they may have used up their gas and dust and have faded to obscurity.

At the limits of the largest telescopes, astronomers see faint red galaxies (■ Figure 16-16). They look red from Earth because of their great redshifts. The floods of ultraviolet light emitted by these star-forming galaxies have been shifted into the far-red part of the spectrum. Their look-back times are so great that they appear as they were when the universe was only a billion years old. These galaxies seem to be among the first to begin shining after the beginning of the universe, a story told in Chapter 18.

Building Scientific Arguments

How did elliptical galaxies get that way?
Telling the story of a natural object lies at the heart of science, and many scientific arguments tell such tales. A growing body of evidence suggests that elliptical galaxies have been subject to collisions in their past and that spiral galaxies have not. During collisions, a galaxy can be driven to use up its gas and dust in a burst of star formation. The resulting supernova explosions can help drive gas and dust out of the galaxy. This explains why elliptical galaxies now contain little star-making material. The beautiful disk typical of spiral galaxies is very orderly, with all the stars following similar orbits. When galaxies collide, the stellar orbits get scrambled, and an orderly disk galaxy could be converted into a chaotic swarm of stars typical of elliptical galaxies. It seems likely that elliptical galaxies have had much more complex histories than spiral galaxies have had.

When scientists create a scientific argument to tell the story of a natural object, they use evidence to make the story as true as it can be. Expand your argument. **What evidence can you cite to support the story in the preceding paragraph?**

■   ■   ■

Connections: Before you can consider the universe as a whole, you need to examine the galaxies from a different perspective. Some galaxies are suffering tremendous eruptions as their cores blast radiation and matter outward. In the next chapter, you will discover that these peculiar galaxies are closely related to collisions and mergers between galaxies.

Study and Review Tools

Summary

16-1 | The Family of Galaxies

What do galaxies look like?
■ Through 19th-century telescopes, galaxies looked like hazy spirals, and astronomers debated the nature of these spiral nebulae. In the 1920s, with the construction of larger telescopes, astronomers could identify stars, including Cepheid variable stars, in the spiral nebulae. That showed that the spiral nebulae were galaxies.

■ Astronomers divide galaxies into three classes—elliptical, spiral, and irregular—with subclasses specifying the galaxy's shape.

■ Elliptical galaxies contain little gas and dust and cannot make new stars. Consequently, they lack hot, blue stars and have a reddish tint.

■ Spiral galaxies contain more gas and dust in their disks and support active star formation, especially along the spiral arms. Some of the newborn stars are massive, hot, and blue, and that gives the spiral arms a blue tint.

■ The halo and nuclear bulge of a spiral galaxy usually lack gas and dust and contain little star formation. The halos and nuclear bulges have a reddish tint because they lack hot, blue stars.

■ Irregular galaxies have no obvious shape but contain gas and dust and support star formation.

16-2 | Measuring the Properties of Galaxies

How do astronomers find the distances to galaxies?
■ Astronomers find the distance to galaxies using distance indicators sometimes called standard candles, objects of known luminosity. The most accurate distance indicators are the Cepheid variable stars. Others include globular clusters, planetary nebulae, and supernovae explosions.

■ Astronomers can estimate the distance to a galaxy by dividing its apparent rate of recession (its radial velocity) by the Hubble constant. The apparent rate of recession is related to distance by the Hubble Law.

How do galaxies differ in size, luminosity, and mass?
■ Some dwarf ellipticals and dwarf irregular galaxies are only a few percent the size and luminosity of our galaxy, but some giant elliptical galaxies are five times larger than the Milky Way galaxy.

■ Astronomers measure the masses of galaxies in two basic ways—the rotation curve method and the velocity dispersion method. The rotation curve method is more accurate but can be applied only to nearby galaxies.

Do other galaxies contain supermassive black holes and dark matter, as does our own galaxy?
■ Stars near the centers of galaxies are orbiting at high velocities, which suggests the presence of supermassive black holes in the centers of most galaxies.

■ The mass of a galaxy's supermassive black hole is proportional to the mass of its nuclear bulge. That shows that the supermassive black holes must have formed when the galaxy formed.

■ Observations of individual galaxies show that galaxies contain 10 to 100 times more dark matter than visible matter.

■ The hot gas held inside some clusters of galaxies and the gravitational lensing caused by the mass of galaxy clusters add further evidence of dark matter.

16-3 | The Evolution of Galaxies

Why are there different kinds of galaxies?
■ Rich clusters of galaxies contain fewer spirals than poor clusters of galaxies, and that is evidence that galaxies evolve by collisions and mergers.

- When galaxies collide, tides twist and distort their shapes, and the compression of gas clouds triggers star formation.

- Large galaxies can absorb smaller galaxies in what is called galactic cannibalism. You can see clear evidence that our own Milky Way Galaxy is devouring some of the small galaxies that orbit nearby and that our galaxy has consumed other small galaxies in the past.

- Shells of stars, counter-rotating parts of galaxies, streams of stars in the halos of galaxies, and multiple nuclei are evidence that galaxies can merge.

- The merger of two larger galaxies can scramble star orbits and drive bursts of star formation to use up gas and dust. Most larger ellipticals have evidently been produced by past mergers.

- Spiral galaxies have thin, delicate disks and appear not to have suffered mergers with large galaxies.

- A galaxy moving through the gas in a cluster of galaxies can be stripped of its own gas and dust and may become an S0 galaxy.

- At great distance and great look-back times, the largest telescopes reveal that galaxies were smaller, more irregular, and closer together. Also, there were more spirals and fewer ellipticals long ago.

- At the largest distances, astronomers find small irregular clouds of stars that may be the objects that fell together to begin forming galaxies when the universe was very young.

New Terms

spiral nebula (p. 371)

island universe (p. 371)

Shapley–Curtis Debate (p. 371)

megaparsec (Mpc) (p. 373)

distance indicator (p. 373)

standard candle (p. 373)

elliptical galaxy (p. 374)

spiral galaxy (p. 374)

barred spiral galaxy (p. 374)

irregular galaxy (p. 375)

distance scale (p. 377)

look-back time (p. 377)

Hubble law (p. 378)

Hubble constant (H) (p. 378)

rotation curve (p. 379)

rotation curve method (p. 379)

cluster method (p. 379)

velocity dispersion method (p. 380)

gravitational lensing (p. 382)

rich cluster (p. 384)

poor cluster (p. 384)

starburst galaxy (p. 386)

ultraluminous infrared galaxy (p. 386)

tidal tails (p. 388)

galactic cannibalism (p. 388)

ring galaxy (p. 389)

Review Questions

Ace ⊛ Astronomy™ Assess your understanding of this chapter's topics with additional quizzing and animations at **http://ace .brookscole.com/sf9**

1. If a civilization lived on a planet in an E0 galaxy, do you think they would have a Milky Way in their sky? Why or why not?

2. Why can't the evolution of galaxies go from elliptical to spiral? from spiral to elliptical?

3. If all elliptical galaxies had three different diameters, you would never see an elliptical galaxy with a circular outline on a photograph. True or false? Explain your answer. (*Hint:* Can a football ever cast a circular shadow?)

4. What is the difference between an Sa and an Sb galaxy? between an S0 and an Sa galaxy? between an Sb and an SBb galaxy? between an E7 and an S0 galaxy?

5. Why wouldn't white dwarfs make good distance indicators?

6. Why isn't the look-back time important among nearby galaxies?

7. Explain how the rotation curve method of finding a galaxy's mass is similar to the method used to find the masses of binary stars.

8. Explain how the Hubble law allows you to estimate the distances to galaxies.

9. How can collisions affect the shape of galaxies?

10. What evidence can you cite that galactic cannibalism really happens?

11. Describe the future evolution of a galaxy that astronomers now see as a starburst galaxy. What will happen to its interstellar medium?

12. Why does the intracluster medium help determine the nature of the galaxies in a cluster?

13. This image of M33, the Pinwheel Galaxy, has emission by ionized hydrogen enhanced as bright pink. Discuss the location of these clouds of gas and explain how that provides important evidence toward understanding spiral arms.

14. In the image at right you see two interacting galaxies; one is nearly face-on and the other is nearly edge-on. Discuss the shapes of these galaxies and describe what is happening.

(P. Massey, Lowell, N. King, STScI, S. Holes, Charleston, G. Jacoby, WIYN/AURA/NSF)

(WIYN/NOAO/NSF)

Discussion Questions

1. From what you know about star formation and the evolution of galaxies, do you think the Infrared Astronomy Satellite should have found irregular galaxies to be bright or faint in the infrared? Why or why not? What about starburst galaxies? What about elliptical galaxies?

2. Imagine that you could observe a gas cloud at such a high look-back time that it is just beginning to form one of the first galaxies. Further, suppose you discovered that the gas was metal rich. Would that support or contradict your understanding of galaxy formation?

Problems

1. If a galaxy contains a type I (classical) Cepheid with a period of 30 days and an apparent magnitude of 20, what is the distance to the galaxy?

2. If you find a galaxy that contains globular clusters that are 2 seconds of arc in diameter, how far away is the galaxy? (*Hints:* Assume that a globular cluster is 25 pc in diameter and use the small-angle formula.)

3. If a galaxy contains a supernova that at its brightest has an apparent magnitude of 17, how far away is the galaxy? (*Hint:* Assume that the absolute magnitude of the supernova is -19.)

4. If you find a galaxy that is the same size and mass as our Milky Way Galaxy, what orbital velocity would a small satellite galaxy have if it orbited 50 kpc from the center of the larger galaxy?

5. Find the orbital period of the satellite galaxy described in Problem 4.

6. If a galaxy has an apparent radial velocity of 2000 km/s and the Hubble constant is 70 km/s/Mpc, how far away is the galaxy? (*Hint:* Use the Hubble law.)

7. If you find a galaxy that is 20 minutes of arc in diameter and you measure its distance to be 1 Mpc, what is its diameter?

8. Suppose you found a galaxy in which the outer stars have orbital velocities of 150 km/s. If the radius of the galaxy is 4 kpc, what is the orbital period of the outer stars? (*Hints:* 1 pc = 3.08×10^{13} km, and 1 yr = 3.15×10^7 seconds.)

9. A galaxy has been found that is 5 kpc in radius and whose outer stars orbit the center with a period of 200 million years. What is the mass of the galaxy? On what assumptions does this result depend?

10. Among the globular clusters orbiting a distant galaxy, the fastest is traveling 420 km/s and is located 11 kpc from the center of the galaxy. Assuming the globular cluster is just barely gravitationally bound to the galaxy, what is the mass of the galaxy? (*Hint:* The galaxy had a slightly faster globular cluster, but it escaped some time ago. What is the escape velocity?)

Media Cluster

Ace Astronomy™ To access the resources in the Media Cluster, log into AceAstronomy at **http://ace.brookscole .com/sf9** and select Chapter 16.

ACTIVE FIGURES

Galaxy Types
Explore different kinds of galaxies and their structures and motions.

ASTRONOMY EXERCISES

Hubble Relation
This animation lets you plot the red-shifts of several galaxies and make your own observations to find the Hubble constant.

VIRTUAL ASTRONOMY LABS

Lab 16: Astronomical Distance Scales
In this lab, you explore some methods for determining distances in astronomy.

Lab 19: The Hubble Law
This lab discusses the Hubble law, shows how it can be used to estimate distances to remote objects, and illustrates that the law is a consequence of the expansion of space itself.

Critical Inquiries for the Web

1. How far out into the universe can astronomers see Cepheid variables? Research sources on the Internet to find other galaxies whose distances have been found through observation of Cepheids. List the more distant galaxies in which Cepheids have been identified and the distances determined from these data.

2. Who was Milton Humason, and how did he get started in astronomy?

3. In the early 1900s the nature of the "spiral nebulae" was not well understood. In 1920 a "great debate" was held between Harlow Shapley, who held that these objects were relatively nearby swirling clouds of gas, and Heber Curtis, who saw them as distant "island universes." Use the Internet to find information about the debate, outline the lines of evidence used by the two participants to present their views, and explain who was right and who was wrong.

4. Locate a Web page dedicated to the Messier Catalogue—a list of galaxies, clusters, and nebulae that is often used as a list of targets for small telescopes. Be sure that your destination includes images of the objects. For each of the galaxies in the Messier list, determine its Hubble classification. You may be given this information at the site, but examine the images to see if the features of these galaxies conform to a particular Hubble type.

Exploring *TheSky*

1. Locate the Andromeda Galaxy, also known as M31, and its companion galaxies. Zoom in on it and estimate its angular size compared to the full moon. (*Hint:* Use **Find** under the **Edit** menu.)

2. Take a survey of galaxies and see how many are spiral and how many are elliptical. Does your method produce a fair count, or are you biased to notice only the brighter galaxies? (*Hint:* Use **Filters** under the **View** menu to turn off everything but **Galaxies** and **Mixed Deep Sky** objects. Make sure the Messier labels are switched on using the **Labels** and **Setup** under the **View** menu.)

3. Locate the Sombrero Galaxy (M104). Study the photographs and discuss this galaxy's special properties. Zoom in on it and estimate its angular size compared to the moon.

4. Study the distribution of galaxies and notice how they cluster together. Can you find the Virgo cluster? Zoom in until more galaxies appear and then scroll north to find the Coma cluster. Zoom in on Leo to find the cluster there. What other clusters can you find?

5. Describe the galaxy located near the south celestial pole.

 Go to the Brooks/Cole Astronomy Resource Center **(http:// astronomy.brookscole.com)** for critical thinking exercises, articles, and additional readings from InfoTrac College Edition, Brooks/Cole's online student library.

17 Galaxies with Active Nuclei

Visual (blue) plus radio (red)

D O YOU LIKE FIREWORKS? A barrage of skyrockets in the dark and a few loud bangs make a great show; but, compared to astronomical eruptions, an earthly fireworks display is a whimper. You are about to meet some of the most energetic events in the universe. Nova and supernova explosions are tiny compared to the energy pouring out of the nuclei of certain galaxies. You will discover that the origin of these outbursts is related to the formation and history of galaxies. | The effort to understand these peculiar galaxies began in the 1950s when radio astronomers first noticed that some galaxies were sources of unusually strong radio waves. Those galaxies were called **radio galaxies.** By the 1970s, astronomers had put space telescopes in orbit that could observe in many other parts of the electromagnetic spectrum, and they discovered

| Continued on page 396 |

Guidepost

Looking Back

In the past few chapters, you have explored our own and other galaxies. In earlier chapters, you studied stars and discovered a process that ejects oppositely directed jets of gas and radiation from such diverse objects as protostars and X-ray binaries. Now you are ready to put those ideas together to study some of the most powerful objects in the universe.

This Chapter

Some galaxies have nuclei that are both small and powerful. Astronomers see some in the act of ejecting jets much more powerful than those from protostars or from X-ray binaries. As you study these active galaxies, you will be combining many of the ideas you have discovered so far to answer five essential questions:

What evidence shows that some galactic nuclei are active?

What is the energy source of this activity?

What triggers the nucleus of a galaxy into activity?

What are the most distant active galaxies?

What can active galaxies reveal about the history of the universe?

Looking Ahead

The active galaxies are the last background data you need before you try to understand the birth and evolution of the entire universe and the galaxies that fill it. You will start that journey in the next chapter.

Ace⊙Astronomy™ The AceAstronomy icon throughout the text indicates an opportunity for you to test yourself on key concepts and to explore animations and interactions on the AceAstronomy website at: http://ace.brookscole.com/sf9

that the radio galaxies were bright at other wavelengths. Consequently, these galaxies are now called **active galaxies.**

The floods of energy pouring out of active galaxies originate in the nuclei of the galaxies, which are referred to as **active galactic nuclei (AGN).** You are joining the research team at a time when the evidence has become overwhelming that these AGN are produced by matter plunging into supermassive black holes. But don't be disappointed that you are joining the team late; there are plenty of mysteries left to be solved.

17-1 Active Galactic Nuclei

ONLY A FEW PERCENT OF GALAXIES are active, so you must search carefully to find examples. The place to begin is with some galaxies that look normal at first glance but have violent secrets hidden in their cores.

Seyfert Galaxies

In 1943, Mount Wilson astronomer Carl K. Seyfert published his study of spiral galaxies. Observing at visual wavelengths, Seyfert found that some spiral galaxies have small, highly luminous nuclei (■ Figure 17-1). Today, these **Seyfert galaxies** are recognized by the peculiar spectra of these bright nuclei.

The light from a normal galaxy comes almost entirely from stars, and its spectrum is the combined spectra of many millions of stars. Consequently, the spectra of normal galaxies contain a few of the main absorption lines seen in the spectra of stars. But the spectrum of a Seyfert galaxy nucleus contains broad emission lines of highly ionized atoms. Emission lines suggest a hot, low-density gas, and ionized atoms suggest that the gas is very excited. The widths of the spectral lines suggest large Doppler shifts produced by high velocities in the nuclei; gas approaching Earth would produce blueshifted spectral lines, and gas going away from Earth would produce redshifted lines. The combined light would produce broad spectral lines. The velocities at the centers of Seyfert galaxies are roughly 10,000 km/s, about 30 times greater than velocities at the centers of normal galaxies.

About 2 percent of spiral galaxies appear to be Seyfert galaxies, and they are classified into two categories. Type 1 Seyfert galaxies are very luminous at X-ray and ultraviolet wavelengths and have the typical broad emission lines with sharp, narrow cores. Rapidly moving gas must be producing the broad part of the lines, but some lower-velocity gas must also be present to produce the narrow cores of the lines. Type 2 Seyfert galaxies lack strong X-ray emission and have emission lines that are narrower than those of type 1 Seyfert galaxies but still broader than spectral lines produced by a normal galaxy.

In addition, astronomers discovered that the brilliant nuclei of Seyfert galaxies fluctuate rapidly, especially at X-ray wavelengths. A Seyfert nucleus can change its X-ray brightness by a significant amount in a few minutes. As you saw in Chapter 14, an

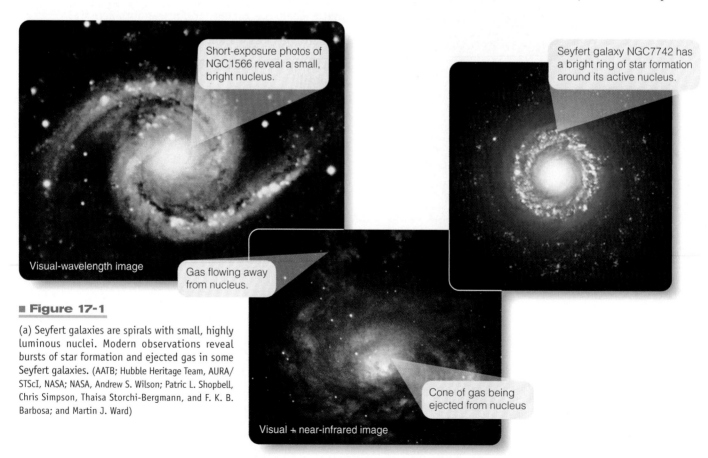

Short-exposure photos of NGC1566 reveal a small, bright nucleus.

Seyfert galaxy NGC7742 has a bright ring of star formation around its active nucleus.

Visual-wavelength image

Gas flowing away from nucleus.

Cone of gas being ejected from nucleus

Visual + near-infrared image

■ Figure 17-1

(a) Seyfert galaxies are spirals with small, highly luminous nuclei. Modern observations reveal bursts of star formation and ejected gas in some Seyfert galaxies. (AATB; Hubble Heritage Team, AURA/STScI, NASA; NASA, Andrew S. Wilson; Patric L. Shopbell, Chris Simpson, Thaisa Storchi-Bergmann, and F. K. B. Barbosa; and Martin J. Ward)

Figure 17-2

(a) Seyfert galaxy NGC7674 appears to be distorted by interacting with its companion galaxies. You can be sure they are a group because they all have similar apparent velocities of recession and must be at about the same distance. (John W. Mackenty, Institute for Astronomy, University of Hawaii) (b) Seyfert galaxy 3C219 is the brightest galaxy in a small, compact cluster of galaxies. Compare its radio image with that of the double-lobed radio galaxies discussed in the next section. (NRAO/AUI/NSF)

astronomical body cannot change its brightness in a time shorter than the time it takes light to cross its diameter. If the Seyfert nucleus can change in a few minutes, then it cannot be larger in diameter than a few light-minutes. In spite of their small size, the cores of Seyfert galaxies produce tremendous amounts of energy. The brightest emit a hundred times more energy than the entire Milky Way Galaxy. Something in the centers of these galaxies not much bigger than Earth's orbit produces a galaxy's worth of energy.

The shapes of Seyferts can give you clues to their energy source. They are three times more common in interacting pairs of galaxies than in isolated galaxies. Also, about 25 percent have peculiar shapes suggesting tidal interactions with other galaxies (■ Figure 17-2a). This statistical evidence (**Window on Science 17-1**) hints that Seyfert galaxies may have been triggered into activity by collisions or interactions with companions. Some Seyferts are expelling matter in oppositely directed flows (Figure 17-2b), a geometry you have seen on smaller scales when matter flows into accretion disks around neutron stars or black holes.

All of this evidence leads modern astronomers to suspect that the cores of Seyfert galaxies contain supermassive black holes—black holes with masses as high as a billion solar masses. The gas in the centers of Seyfert galaxies is traveling so fast it would escape from a normal galaxy. Only very large central masses could exert enough gravity to hold the gas inside the nuclei, and that suggests supermassive black holes. Encounters with other galaxies could throw matter into a black hole; and, as you have seen in Chapter 14, large amounts of energy can be liberated by matter flowing into a black hole. Later in this chapter, you will follow this theory further, but for now you need to search for more evidence of supermassive black holes by turning your attention to galaxies that emit powerful radio signals.

Double-Lobed Radio Sources

Beginning in the 1950s, radio astronomers found that some sources of radio energy in the sky consisted of pairs of radio-bright regions. When optical telescopes studied the locations of these **double-lobed radio sources,** they revealed galaxies located between the two radio lobes. Apparently, the central galaxies were producing the radio lobes.

Study **Cosmic Jets and Radio Lobes** on pages 398–399 and notice four important points:

1 First, the geometry suggests that radio lobes are inflated by jets of excited gas emerging from the central galaxy. This has been called the **double-exhaust model,** and the presence of hot spots and synchrotron radiation shows that the jets are very powerful.

2 Second, notice the evidence that jets and radio lobes are associated with interacting galaxies.

3 The third thing to notice is how the complex shapes of some jets and radio lobes can be explained by the motions of the active galactic nuclei. A good example of this is 3C31 (the 31st source in the *Third Cambridge Catalog of Radio Sources*) with its twisting radio lobes.

4 Finally, notice how these jets seem to be related to matter falling into a central black hole. You have seen similar jets produced by accretion disks around protostars, neutron stars, and stellar-mass black holes; although the details are not understood, the same process seems to be producing all of these jets.

Cosmic Jets and Radio Lobes

1 Many radio sources consist of two bright lobes — double-lobed radio sources — with a galaxy, often a peculiar or distorted galaxy, located between them. Evidence suggests these active galaxies are emitting jets of high-speed gas that inflate the lobes as cavities in the intergalactic medium. Where the jets impact the far side of the cavities, they create **hot spots**.

Size of Milky Way Galaxy

Hot spot

Radio image

Hot spots lie on the leading edge of a lobe where the jet pushes into the surrounding gas.

Jet

Jet

NRAO

Hot spot

Visible galaxy

1a Cygnus A, the brightest radio source in Cygnus, is a pair of lobes with jets leading from the nucleus of a highly disturbed galaxy. In this false-color image, the areas of strongest radio signals are shown in red and the weakest in blue. Because the radio energy detected is synchrotron radiation, astronomers conclude that the jets and lobes contain very-high-speed electrons, usually called relativistic electrons, spiraling through magnetic fields about 1000 times weaker than Earth's field. The total energy in a radio lobe is about 10^{53} J — what you would get if you turned the mass of a million suns directly into energy.

Used with permission, Fosbury R. A. E., Vernet, J., Villar-Martin, M., Cohen, M. H., Ogle, P. M., Tran, H. D. & Hook, R. N. 1998, Optical continuum structure of Cygnus A. "KNAW colloquium on: The most distant radio galaxies," Amsterdam, 15–17 October 1997, Roettgering H, Best P and Lehnert M eds, Reidel, astro-ph/9803310

2 Radio galaxy NGC5128 lies between two radio lobes and appears to be a giant elliptical galaxy experiencing a collision with a dusty spiral galaxy. It is known to radio astronomers as Centaurus A, an inner pair of radio lobes and a larger outer pair.

Infrared

NASA/CXC/NRAO/VLA

An infrared image of the center of the galaxy NGC5128 reveals a small, bright disk of hot gas surrounding the nucleus.

E. Schreier, STScl, NASA

Nucleus

Radio = Red
X-ray = Blue

The combined radio and X-ray image at the left shows a high-energy jet at the very center of the galaxy pointing to the upper left into the northern radio lobe.

If the outer radio lobes of Centaurus A were visible to your eyes, they would look 10 times larger than the full moon.

Radio (green)

X-ray (blue)

X-ray: NASA/CXC/M. Karovska et al., Radio: NRAO/VLA/Schiminovich, et al., Radio: NRAO/VLA/J. Condon et al., Optical: Digitized Sky Survey U.K. Schmidt Image/STScl.

Visual + radio + X-ray

Radio galaxy 3C31 NGC383

Visual (blue) + Radio (red)

NRAO/AUI/NSF

Ace Astronomy™

Log into AceAstronomy and select this chapter to see Active Figure "Jet Deflection" and take control of your own model of a moving radio galaxy.

3 The radio jets from NGC1265 are being left behind as the galaxy moves rapidly through the gas of the intergalactic medium. Twists in the tails are presumably caused by motions of the active nucleus.

Direction of motion

Radio image

NRAO/AUI/NSF and C. O'Dea & F. Owen

3a Radio galaxy 3C31 is one of a chain of galaxies. It has ejected jets from its core that twist, presumably because the active nucleus is orbiting another object such as the nucleus of a recently absorbed galaxy.

3b The radio source 3C75 is produced by two galaxies experiencing a close encounter. As the active nuclei whip around each other, their jets twist and turn. The size of the visible galaxies would be about the size of cherries at the scale of this image.

Nucleus

Nucleus

Radio image

NRAO/AUI/NSF and F.N. Owen, C.P. O'Dea, M. Inoue, & J. Eilek

4 High-energy jets appear to be caused by matter flowing into a supermassive black hole in the core of an active galaxy. Conservation of angular momentum forces the matter to form a whirling accretion disk around the black hole. How that produces a jet is not entirely understood, but it appears to involve magnetic fields that are drawn into the accretion disk and tightly wrapped to eject high-temperature gas. The twisted magnetic field confines the jets in a narrow beam and causes synchrotron radiation.

4a Jets from active galaxies may have velocities from thousands of kilometers per second up to a large fraction of the speed of light. Compare this with the jets in bipolar flows, where the velocities are only a few hundred kilometers per second. Active-galaxy jets can be millions of light-years long. Bipolar-flow jets are typically a few light-years long. The energy is different, but the geometry is the same.

Black hole

Magnetic field lines

Accretion disk

4b Excited matter traveling at very high speeds tends to emit photons in the direction of travel. Consequently, a jet pointed roughly toward Earth will look brighter than a jet pointed more or less away. This may explain why some radio galaxies have one jet brighter than the other, as in Cygnus A shown at the top of the opposite page. It may also explain why some radio galaxies appear to have only one jet. The other jet may point generally away from Earth and be too faint to detect, as in the case of NGC5128, also shown on the opposite page.

Adapted from a diagram by Ann Field, NASA, STScI.

It Wouldn't Stand Up in Court: Statistical Evidence

Notice that some scientific evidence is statistical. For example, you might note that Seyfert galaxies are three times more likely to have a nearby companion than a normal galaxy is. This is statistical evidence because you can't be certain that any single Seyfert galaxy will prove to have a companion. Yet the probability is higher than if it were a normal galaxy, and that leads you to suspect that interactions between galaxies are involved. Such statistical evidence can tell you something in general about the cause of Seyfert eruptions.

Statistical evidence is common in science, but it is inadmissible as evidence of guilt in a criminal trial. No judge would allow you to argue that the defendant must be guilty of robbing a store in the middle of the night because the defendant is an astronomer and astronomers are statistically more likely to be awake at night. The American legal system is based on the concept of reasonable doubt, so statistics can't be applied to prove guilt. But you could use statistics to learn something about the general sleep habits of astronomers.

You can use statistical evidence in science if you do not demand that the statistics demonstrate anything conclusively about any single example. To continue the example of Seyfert galaxies, you can't demand that the statistics predict that any specific galaxy has a companion. Rather, you can use the statistical evidence to gain a general insight into the cause of active galactic nuclei. That is, you are trying to understand galaxies in general, not to convict any single galaxy.

Of course, if you surveyed only a few galaxies, your statistics might not be very good, and a critic might be justified in making the common complaint, "Oh, that's only statistics." For example, if you surveyed only four galaxies and found that three had companions, your statistics would not be very reliable. But if you surveyed 1000 galaxies, your statistics could be very good indeed, and your conclusions could be highly significant.

Scientists can use statistical evidence if it passes two tests. It cannot be used to draw conclusions about specific cases, and it must be based on large enough samples so the sta-

Visual

Guilt by association is statistical evidence and can be applied to galaxies. (NASA)

tistics are significant. With these restrictions, statistical evidence can be a powerful scientific tool.

Study the image of NGC5128 on page 398 closely for a further clue. In visual-wavelength images, the galaxy looks like a round elliptical galaxy surrounded by a belt of dust and gas. Doppler shifts reveal that the spherical cloud of stars is rotating about an axis that lies in the plane of the dust disk. The dust disk is, of course, rotating about an axis that is perpendicular to the disk. Amazingly, the galaxy has two different axes of rotation. Furthermore, bright blue stars have been identified in the dust ring, and those stars must have formed recently. All of this is evidence of a recent collision, and that should add to your suspicion that collisions between galaxies can trigger eruptions in their cores.

All of the evidence suggests that supermassive black holes at the center of galaxies can eject powerful jets and inflate radio lobes Now it is time to try to understand what these supermassive energy machines are like.

Exploring Supermassive Black Holes

In the case of active galaxies, "the really difficult, central problem" is the source of the energy. The evidence seems conclusive that the energy is produced when matter flows into supermassive black holes. You have seen in earlier chapters how matter flowing into a stellar-mass black hole forms a hot accretion disk and can emit jets in opposite directions. It is not surprising that the hot accretion disk around a supermassive black hole can emit much more powerful jets.

Can you be sure the energy sources at the centers of active galaxies are really supermassive black holes? You can review the evidence and be the judge. The evidence takes the form of observations of size and observations of motion.

Earlier in this chapter, you saw evidence that the cores of some active galaxies are very small. The cores of Seyfert galaxies fluctuate in brightness over very short time periods, and that must mean they are very small—so small that astronomers know of nothing except a supermassive black hole that could produce so much energy in such a small region.

The motion of stars and gas near the centers of galaxies is important because it can tell you the amount of mass located there. Consider the giant elliptical galaxy M87 (page 374). It has a very small, bright nucleus and a visible jet of matter 1800 pc long racing out of its core (■ Figure 17-3). Radio observations show that the nucleus must be no more than a light-week in diameter. This evidence is suggestive, but observations of motions around the center are decisive. A high-resolution image shows that the core lies at the center of a spinning disk and the jet lies along the axis of the disk. Only 60 ly from the center, the gas in the disk is orbiting at 750 km/s. If you substitute radius and velocity into the

■ Figure 17-3

The jet at the center of giant elliptical galaxy M87 is only a few percent the diameter of the entire galaxy. Images reveal a small, rapidly spinning disk at the center of the galaxy with the jet emerging at nearly half the speed of light along the axis of the disk. X-ray images show rings and twisted plumes of hot gas produced in previous eruptions. (M87: AURA/NOAO/NSF; jet: NASA/STScI; radio image: NRAO and J. Biretta; X-ray: NASA/CXC/W. Forman et al.)

equation for circular velocity (Chapter 5), it will tell you that the central mass must be roughly 2.4 billion solar masses. Only a supermassive black hole could pack so much mass into such a small region.

For another example, look at ■ Figure 17-4. NGC4261 is a distorted galaxy ejecting jets into a pair of radio lobes. High-resolution images made by the Hubble Space Telescope reveal a

dusty disk surrounding the bright central core. The axis of the disk points along the jets into the radio lobes. In the case of NGC4261, you can see the energy source in the act of producing jets and radio lobes.

Although NGC7052 appears to be an innocuous elliptical galaxy at visual wavelengths, it has a powerful AGN at its center. High-resolution images made by the Hubble Space Telescope

■ Figure 17-4

(a) Galaxy NGC4261 looks like a distorted fuzzy blob in an Earth-based photograph at visible wavelengths (white), but radio telescopes reveal jets leading into radio lobes (orange). (b) A Hubble Space Telescope image of the core of the galaxy reveals a bright central spot surrounded by a disk with its axis pointing at the radio lobes. (L. Ferrarese, Johns Hopkins University, and NASA)

Visual-wavelength images

■ **Figure 17-5**

Near the center of active elliptical galaxy NGC7052, a dust disk circles a massive nucleus. Deep in a "haze" of billions of stars, the near side of the dust disk obscures more stars than the far side and looks darker. Scattering of shorter-wavelength photons in the dust gives it a red tint. Velocities near the center suggest a central black hole containing 300 million solar masses. (Roeland P. van der Marel, STScI, Frank C. van den Bosch, University of Washington, and NASA)

show that the center is filled with a haze composed of billions of stars. Surrounding the core, a dust disk rotates at high speed (■ Figure 17-5). The rotation of the disk reveals that the central object contains 300 million solar masses, evidently a supermassive black hole.

The appearance of active galaxies varies dramatically from galaxy to galaxy. The spiral galaxy NGC1068 has an elongated glow of X-rays emerging from its central region (■ Figure 17-6). High velocities around the center reveal that the galaxy must contain a 5-million-solar-mass black hole at its core. Radio wavelength observations show that a dusty doughnut of gas orbits from a few light-years out to about 300 ly from the central black hole.

In astronomy, evidence means observations, and the observations of small sizes and rapid orbital motions seem conclusive. Active galactic nuclei are powered by supermassive black holes. That probably leaves you with a few questions, and astronomers are struggling to answer those same questions. How do supermassive black holes and their accretion disks produce these eruptions? Furthermore, where did these supermassive black holes come from?

The Search for a Unified Model

When a field of research is young, scientists often find many seemingly different phenomena, such as double-lobed radio galaxies, Seyfert galaxies, cosmic jets, and so on. As the research matures,

the scientists begin seeing similarities and eventually are able to unify the different phenomena as different aspects of a single process. This is the real goal of science, to organize evidence and theory in logical arguments that explain how nature works (Window on Science 2-1). Astronomers studying active galaxies are now struggling to find a **unified model** of active galaxy cores.

It seems clear that the cores of active galaxies contain supermassive black holes surrounded by accretion disks that are extremely hot near the black hole but cooler further out. Theory predicts that the central cavity in the disk where the black hole lurks is very narrow but that the disk there is "puffed up" and thick.

This means the black hole may be hidden deep inside this narrow central well. The hot inner disk seems to be the source of the jets often seen coming out of active galaxy cores, but the process by which jets are generated is not understood. The outer part of the disk, according to calculations, is a fat, dense doughnut (a torus) of dusty gas (■ Figure 17-7).

What you see when you view the core of an active galaxy must depend on how this disk is tipped with respect to your line of sight. You should note, at this point, that the accretion disk may be tipped at a steep angle to the plane of a galaxy, so just because you see a galaxy face-on doesn't mean you are looking at the accretion disk face-on. The important factor is not the inclination of the galaxy but the inclination of the accretion disk.

For example, in 1968 astronomers realized that an object they thought was an irregular variable star in the constellation Lacerta was actually the core of a distant active galaxy. The visible spec-

Visual = red
X-ray = blue and green

Visual-wavelength image

■ **Figure 17-6**

At visual wavelengths, the galaxy NGC1068 looks like a normal spiral, but X-ray observations reveal hot gas blowing outward at high speed from a supermassive black hole hidden deep inside a thick doughnut of gas and dust. (X-ray: NASA/CXC/MIT/UCSB/P. Ogle et al.; Optical: NASA/STScI/A. Capetti et al.)

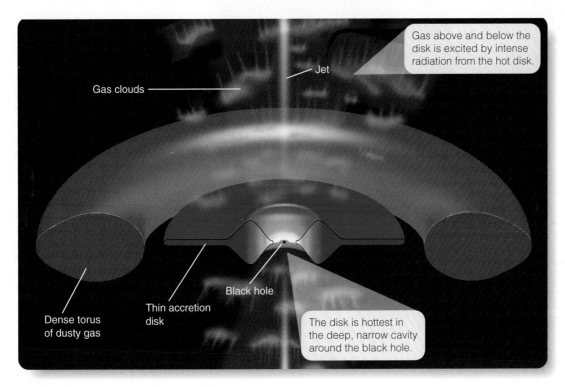

Figure 17-7

A sketch of the accretion disk around a supermassive black hole. Matter flowing inward passes first through a large, opaque torus; then into a thinner, hotter disk; and finally into a hot, narrow cavity around the black hole. This diagram is not to scale. The central cavity may be only 0.01 pc in radius, while the outer torus may be 1000 pc in radius.

trum is featureless, but the spectrum of surrounding nebulosity is that of a giant elliptical galaxy. Such objects, now known as **BL Lac objects** or **blazars,** are 10,000 times more luminous than the Milky Way Galaxy and fluctuate in only hours. They are believed to be the cores of active galaxies in which the brilliant jets happen to be pointed directly at Earth. In this case, the accretion disk is apparently face-on, and you are looking directly into the central cavity around the black hole—down the dragon's throat.

If the accretion disk is tipped slightly, you may not be looking directly into the jet, but you may be able to see some of the intensely hot gas in the central cavity. This region emits broad spectral lines because the gas is orbiting at high velocities and the high Doppler shifts smear out the lines. Seyfert 1 galaxies may be explained by this phenomenon.

If you view the accretion disk from the edge, you cannot see the central area at all because the thick dusty torus blocks your view. Instead you see radiation emitted by gas lying above and below the central disk. Because this gas is farther from the center, it orbits more slowly and has smaller Doppler shifts. For that reason you see narrower spectral lines. This might account for the Seyfert 2 galaxies.

This unified model is far from complete. The actual structure of accretion disks is poorly understood, as is the process by which the disks produce jets. Furthermore, the Seyfert galaxies are different from the giant elliptical galaxies that have double radio lobes. Also, your study of unification has ignored important factors such as differences in the rate at which mass flows inward and the influence of magnetic fields. A unified model does not explain all of the differences among active galaxies. Rather, it provides a framework for ideas as astronomers try to understand what is happening in the cores of active galaxies.

The Origin of Supermassive Black Holes

Naturally you are wondering where these supermassive black holes came from, and that question will lead you to understand why they erupt. Furthermore, it will give you an insight into the birth of galaxies.

In the previous chapter, you learned that most galaxies seem to contain a supermassive black hole at their center. Even our own Milky Way Galaxy contains a central black hole, as does the nearby Andromeda Galaxy (■ Figure 17-8). But only a few percent of galaxies have active galactic nuclei. That must mean that most of the supermassive black holes are dormant; presumably they are not being fed large amounts of matter. A slow trickle of matter flowing into the supermassive black hole at the center of our galaxy could explain the mild activity seen there. But it would take a larger meal to trigger an eruption such as those seen in active galactic nuclei.

What could trigger a supermassive black hole to erupt? The answer is something that you studied back in Chapter 5—tides. You have seen how tides twist interacting galaxies and rip matter away into tidal tails, but mathematical models show that those same interactions can throw matter inward. A sudden flood of matter flowing into the accretion disk around a supermassive black hole would trigger it into eruption. This explains why active galaxies are often distorted; they have been twisted by tidal forces as they interacted or merged with another galaxy. Some

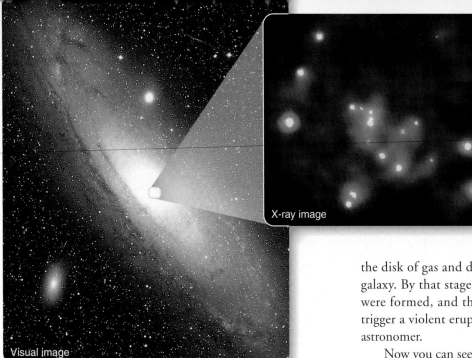

Visual image

X-ray image

active galaxies have nearby companions, and that makes you suspect that the companions are guilty of tidally distorting the other galaxy and triggering an eruption.

Tidal forces are implicated again when you imagine what happens to a lump of matter passing close to a supermassive black hole. ■ Figure 17-9 shows how a passing star would be ripped apart and, at least partially, consumed by the black hole. Inflowing gas, dust, and an occasional star would be an energy feast for a supermassive black hole.

A few dozen supermassive black holes have measured masses, and their masses are correlated with the masses of the host galaxies' nuclear bulges. In each case, the mass of the black hole is about 0.5 percent the mass of the surrounding nuclear bulge. But there is no relationship between the masses of the black holes and the masses of the disks of galaxies. This bit of statistics reveals an exciting insight into how galaxies form.

Apparently galaxies form their nuclear bulges and disks separately. As the nuclear bulge forms, a small fraction of the mass, lacking orbital momentum, sinks to the middle where it forms a supermassive black hole. All of that matter flowing together to form the black hole would release a tremendous amount of energy and trigger a violent eruption. Long ago, when galaxies were actively forming, the birth of the nuclear bulges must have triggered active galactic nuclei. The formation of a nuclear bulge was evidently a violent process.

Recall that when you studied the formation of the Milky Way Galaxy, you found evidence that the disk of gas and dust formed later as matter settled into the galaxy. By that stage, the nuclear bulge and central black hole were formed, and the gradual development of the disk didn't trigger a violent eruption. "Disk formation is wimpy," said one astronomer.

Now you can see how supermassive black holes were formed and how they erupt. Many are triggered into eruption by interactions or mergers with other galaxies. Some eruptions, however, may have been triggered by the formation of the nuclear bulge. That would have happened long ago, and astronomers would only see it in galaxies at great look-back times. Are these distant eruptions detectable? That introduces one of the biggest adventures of modern astronomy—the subject of the next section.

Building Scientific Arguments

What evidence can you cite to show that double-lobed radio galaxies and Seyfert galaxies have similar energy sources? Often a good argument, scientific or otherwise, compares two dissimilar objects and finds the connecting factors. In this case, observations show that double-lobed radio sources are produced by two jets flowing out of a galaxy's core and inflating the radio lobes. The evidence of small size and very-high-speed motions in the core shows that these galaxies must contain large amounts of mass in small regions. Seyfert galaxies also have small nuclei, as shown by their rapid fluctuations, and the high velocities in their cores imply that the cores must contain very large amounts of mass. Consequently, both kinds of objects are suspected to contain supermassive black holes at their centers.

This argument has cited evidence and theory to reach a conclusion. Construct a new argument: **Why do astronomers**

■ **Figure 17-9** ▶

Orbiting X-ray telescopes detected an X-ray flare in the galaxy RX J1242-11. Equaling the energy of a supernova explosion, the flare was evidently caused when a star wandered too close to the 100-million-solar-mass black hole at the center of the galaxy. When tidal forces ripped the star apart, some of the mass fell into the black hole, and the rest was flung away. That sudden meal for the black hole was enough to trigger an outburst. (ESA)

think that galaxy interactions are necessary to trigger active galactic nuclei?

■   ■   ■

Connections: Observational evidence and basic theory have led to the unified model of active galactic nuclei. Although the model seems successful, there are still difficulties, and astronomers continue to struggle to understand the details. Nevertheless, the unified model may help you understand the mysterious quasars.

Star Falling into a Black Hole

A star, perhaps disturbed by an encounter with another star, drifts toward a supermassive black hole.

As the near side of the star tries to orbit faster than the far side, the star is torn apart by tidal forces.

Most of the mass of the star is flung away from the black hole...

but roughly 1 percent falls into the black hole as an accretion disk forms.

THE LARGEST TELESCOPES detect multitudes of faint, starlike points of light with peculiar emission spectra, objects called **quasars** (also known as *quasi-stellar objects,* or *QSOs*). Although astronomers now recognize quasars as some of the most distant visible objects in the universe, they were a mystery when they were first identified. The discovery of these objects in the 1960s and the struggle to understand them comprise a classic example of how scientists explore hypotheses, gather evidence, and build confidence in new ways to understand nature. You can begin by reviewing the story of that discovery.

The Discovery of Quasars

In the early 1960s, radio interferometers (see Chapter 6) showed that a number of radio sources are much smaller than normal radio galaxies. Photographs of the location of these radio sources did not reveal a galaxy, not even a faint wisp, but rather a single starlike point of light. The first of these objects so identified was 3C48. Not long after that, the source 3C273 was added, and then more were found. Though these objects emitted radio signals like those from radio galaxies, they were obviously not normal radio galaxies. Even the most distant photographable galaxies look fuzzy, but these objects looked like stars. Their spectra, however, were totally unlike stellar spectra, so the objects were called quasistellar objects or quasars (■ Figure 17-10).

For a few years, the spectra of quasars were a mystery. A few unidentifiable emission lines were superimposed on a continuous spectrum. But, in 1963, Maarten Schmidt at Hale Observatories tried redshifting the hydrogen Balmer lines to see if they could be made to agree with the lines in 3C273's spectrum. At a redshift of 15.8 percent, three lines clicked into place (■ Figure 17-11). Other quasar spectra quickly yielded to this approach, revealing even larger redshifts.

The redshift z is the change in wavelength $\Delta\lambda$ divided by the unshifted wavelength λ_0:

$$\text{redshift} = z = \frac{\Delta\lambda}{\lambda_0}$$

Quasar spectra contain bright emission lines. The brightest are the Balmer lines of hydrogen, but other spectral lines also appear. The redshifts of quasars can be so large that spectral lines can be shifted completely out of the visible spectrum, and lines in the ultraviolet can be shifted into the visible spectrum. ■ Figure 17-12 shows the spectra of four quasars with relatively low redshifts in which the H_α Balmer line normally seen in the red part of the spectrum is shifted into the infrared. Quasars have been found with even larger redshifts.

To understand the significance of these large redshifts and the large velocities of recession they imply, you must draw on the

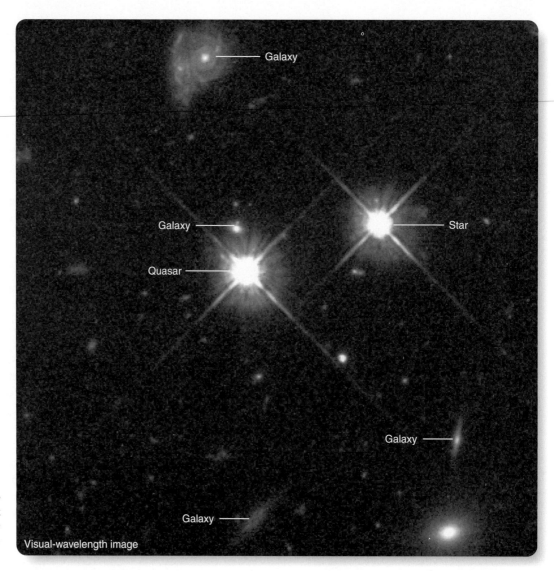

■ Figure 17-10

Quasars have starlike images clearly different from the images of distant galaxies. The spectra of quasars are unlike the spectra of stars or galaxies. (C. Steidel, Caltech, NASA)

Galaxy

Galaxy

Star

Quasar

Galaxy

Galaxy

Visual-wavelength image

■ Figure 17-11

(a) This image of 3C273 shows the bright quasar at the center surrounded by faint fuzz. Note the jet protruding to lower right. (b) The spectrum of 3C273 (top) contains three hydrogen Balmer lines redshifted by 15.8 percent. The drawing shows the unshifted positions of the lines. (Courtesy Maarten Schmidt)

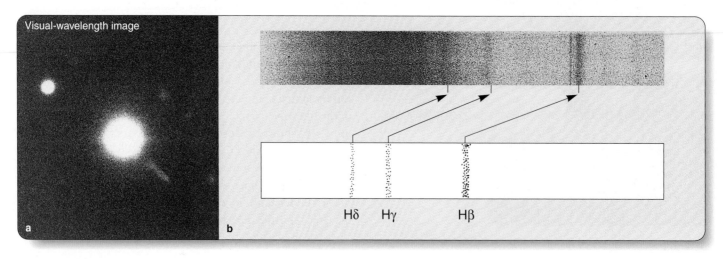

Visual-wavelength image

Hδ   Hγ   Hβ

a

b

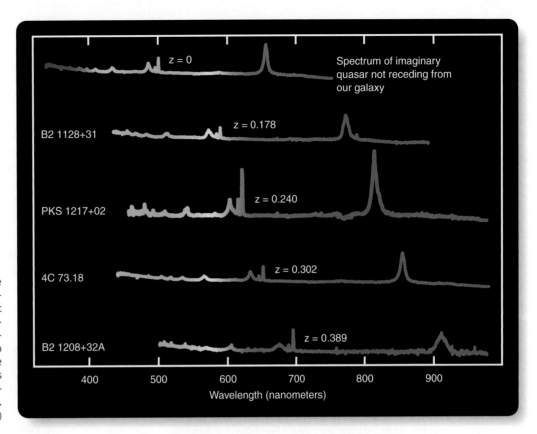

■ **Figure 17-12**

Spectra of four quasars are compared here with an idealized spectrum for an imaginary quasar that has no redshift. The first three Balmer lines of hydrogen are visible, plus lines of other atoms, but the redshifts of the quasars move these lines to longer wavelengths. Nevertheless, the relative spacing of the spectral lines is unchanged, and astronomers can recognize the lines even with a large redshift. (C. Pilachowski, M. Corbin, AURA/NOAO/NSF)

Hubble law. Recall that the Hubble law states that galaxies have apparent velocities of recession proportional to their distances, and thus the distance to a quasar is equal to its apparent velocity of recession divided by the Hubble constant. The large redshifts of the quasars imply that they must be at great distances.

The redshift of 3C273 is 0.158, and the redshift of 3C48 is 0.37. These are large redshifts, but not as large as the largest known for galaxies at the time of Schmidt's work, about 1.0. Soon, however, quasars were found with redshifts much larger than that of any known galaxy. Some quasars are evidently so far away that galaxies at those distances are very difficult to detect. Yet the quasars are easily photographed. This leads to the conclusion that quasars must be very far away and must have 10 to 1000 times the luminosity of a large galaxy. Quasars must be ultraluminous.

Soon after quasars were discovered, astronomers detected fluctuations in brightness over times as short as a few days. Recall from your study of pulsars (Chapter 14) that an object cannot change its brightness appreciably in less time than it takes light to cross its diameter. The rapid fluctuations in quasars showed that they are small objects, not more than a few light-weeks in diameter.

By the late 1960s, astronomers faced a problem: How could quasars be ultraluminous but also be very small? What could make 10 to 1000 times more energy than a galaxy in a region less than a light-year in diameter? Since that time, evidence has accumulated that quasars are the active cores of very distant galaxies, and the rest of this chapter will discuss that evidence. The key, as in

most astronomical arguments, is distance. The distance to quasars is so important and so much larger than the distances you have considered before that you need to pay special attention to how astronomers estimate the distances to these objects.

The Distance to Quasars

By now you may have detected a slipup in the logic of the previous section. Some quasars have redshifts greater than 1, and the largest known are over 6. If you substitute those redshifts into the Doppler formula you studied in Chapter 6, you get velocities greater than the velocity of light, and that is supposed to be impossible. You need to straighten this problem out before you can go on discussing quasars.

You should begin by eliminating a common but incorrect explanation. For over a generation, astronomy textbooks have listed a version of the Doppler formula called the **relativistic Doppler formula.** Einstein derived it to describe the Doppler effect for objects traveling at very high velocities. Tradition aside, that is the wrong formula because the redshifts of the galaxies and the quasars are not really Doppler shifts. In the next chapter you will discover a much more sophisticated way of understanding these redshifts.

Modern astronomers understand that the redshifts of the galaxies are caused by the expansion of the universe and that the Hubble Law allows them to find the distance to a galaxy from its apparent velocity of recession. For relatively nearby galaxies, astronomers commonly convert redshifts into velocities, but you should remember that they are not really velocities through space. That is why this book is careful to call them *apparent* velocities of recession.

Unfortunately the exact mathematical relationship between the redshift of a very distant galaxy or quasar and its apparent velocity of recession is not yet precisely known. That means astronomers can't calculate the precise distance to a quasar from its redshift, but they can make very good estimates. Certainly, you can be confident that the very large redshifts of the quasars assures that the quasars are very far away. Some quasars are over 10 billion light-years away; because of their large look-back times, you see those quasars as they were when the universe was only 10 percent of its present age.

Perhaps you have another question about quasar distances at this point: "How can you be sure quasars really are that far away?" Astronomers faced with explaining how a small object could produce so much energy asked themselves the same question. A discovery made in 1979 provides a dramatic illustration of the tremendous distances to quasars. The object cataloged as 0957+561 lies just a few degrees west of the bowl of the Big Dipper and consists of two quasars separated by only 6 seconds of arc. The spectra of these quasars are identical and even have the same redshift of 1.4136. Quasar spectra do resemble each other, but in detail they are as different as fingerprints. When two quasars so close together were discovered to have the same redshift and identical spectra, astronomers concluded that they were separate images of the same quasar.

The two images are formed by a gravitational lens. You will remember from Chapter 16 that the effect was first predicted by Einstein in 1936. The gravitational field of a galaxy between

Gravitational lensing

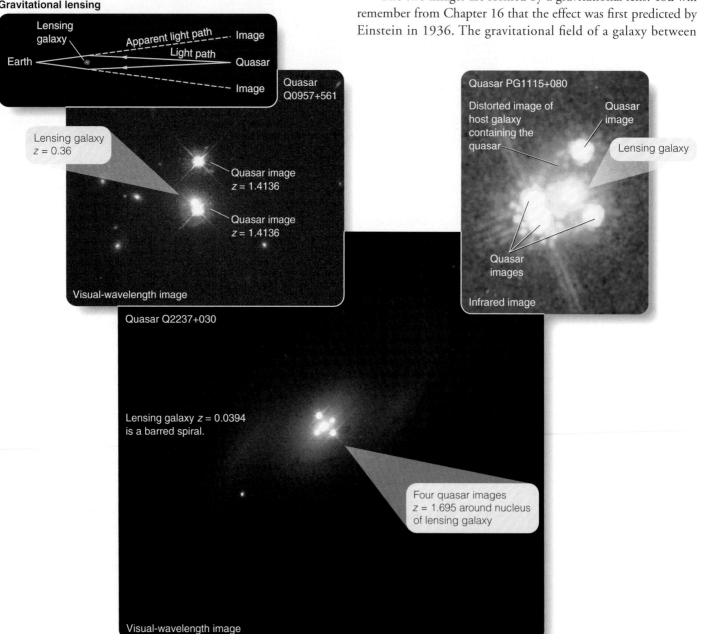

Earth and a distant quasar acts as a lens, bending the light from the quasar and focusing it into multiple images. A few dozen gravitationally lensed quasars are known (■ Figure 17-13); in most cases, the lensing galaxy is so far away it is difficult to detect. This provides powerful evidence that quasars really are distant objects as their large redshifts imply.

For more evidence that quasars are very distant, look again at Figure 17-10. Just above the image of the quasar lies the small, oval image of an elliptical galaxy, which, evidenced by its redshift, is about 7 billion light-years away. The spectrum of the quasar contains absorption lines with the same redshift as the elliptical galaxy. Evidently the lines are produced by the thin gas in the elliptical galaxy. From this you can conclude that the quasar, though very bright, is actually farther away than the distant elliptical galaxy.

About 10 percent of quasars have absorption lines as well as emission lines in their spectra. These absorption lines have a smaller redshift than the corresponding emission lines, and, in some cases, multiple sets of absorption lines are observed, each with its individual redshift. Some of the lines are hydrogen lines, but some are lines of metals, heavier elements such as carbon, silicon, and magnesium. Apparently, these absorption lines are formed when the light from the quasar passes through objects on its way to Earth. The hydrogen lines may be formed as the light from the quasar passes through low-density clouds of hydrogen between the galaxies. The metal lines must be formed when the light passes through galaxies on its way to Earth. But the galaxies themselves are so distant that they are not visible, further evidence that quasars are very distant objects.

The quasars are very distant and that means they must be ultraluminous. Other evidence shows they are quite small. How can anything produce that much power in such a small space? The key evidence is dramatic.

Evidence of Quasars in Distant Galaxies

Astronomers have good reason to suspect that quasars are not isolated objects in space; rather, they are located in galaxies. Thousands of quasars have been found, and the largest known redshifts are a bit over 6, meaning that the objects are exceedingly far away. The look-back time to these objects is so large that they appear as they were when the universe was only 10 per-

Visual image in false color

■ Figure 17-14

Quasar 0351+026 is located near a faint galaxy that has the same redshift and thus, presumably, the same distance from Earth as the quasar. The extended red region around the quasar (fuzz) is typical of lower-redshift quasars and reveals the spectrum of a normal galaxy. Thus, this image shows two interacting galaxies, one of which contains a quasar. (AURA/NOAO/NSF)

cent its present age. At such great distances, it is difficult to detect the host galaxy because of the glare of the quasar.

Quasars with lower redshifts are not quite so far away, however, and fuzzy wisps are visible around such objects. By blocking the glare of such a quasar, astronomers can often photograph the spectrum of the quasar fuzz, and they find that it is the same as the spectrum of a normal galaxy with the same redshift as the quasar. For example, 3C273 appears to lie in a giant elliptical galaxy.

Moreover, some quasars have faint objects near them in the sky. The spectra of these faint objects are the same as the spectra of galaxies, and they have the same redshift as that of the quasar they accompany. This is evidence that these quasars are located in galaxies that are part of larger clusters of galaxies (■ Figure 17-14).

Some quasars have characteristics related to active galaxies. For example, quasar 3C273 seems to be ejecting a jet much like the jets seen in active galaxies. Some quasars even have radio lobes, which further suggests that they are related to active galaxies (■ Figure 17-15).

Photographs made with the largest telescopes on Earth and with the Hubble Space Telescope provide dramatic evidence. The galaxies that host quasars are in many cases clearly visible with these large telescopes. Furthermore, the galaxies are distorted or have nearby companions (■ Figure 17-16). The host galaxies appear to be involved in collisions and interactions with other galaxies,

■ Figure 17-13 ◄

Gravitational lensing can occur when a relatively nearby lensing galaxy is aligned with a distant quasar. This can produce multiple images of the quasar. In the case of the quasar at upper right, the host galaxy, which contains the quasar, is visible as a faint, distorted ring around the lensing galaxy. Gravitational lenses provide direct evidence that quasars are at the great distances suggested by their redshifts. (Q0957+561: George Rhee, NASA, STScI; color composite courtesy Bill Keel; PG1115+080: Chris Impy, Univ. of Arizona and NASA; Q2237+030: Image by W. Keel from data in the NASA/ESA Hubble Space Telescope archive, originally obtained with J. Westphal as Principal Investigator)

■ **Figure 17-15**

This radio image of quasar 3C175 reveals that it is ejecting a jet and is flanked by radio lobes. Presumably you see only one jet because it is directed more or less toward Earth, and the other jet is invisible because it is directed more or less away from Earth. The presence of jets and radio lobes suggests that quasars are distant active galaxies. (NRAO/AUI/NSF)

Radio-wavelength image

■ **Figure 17-16**

The bright object at the center of each of these images is a quasar. Fainter objects near the quasar are galaxies distorted by collisions. Compare the upper left ring-shaped galaxy with that on page 389 and compare the tail in the lower left image with the Antennae Galaxies on the same page. (J. Bahcall, Institute for Advanced Study, Mike Disney, University of Wales, NASA)

Visual-wavelength image

Visual

Visual

Visual

or they have the distorted shapes that you recognize as the result of such collisions. You have seen that galaxy collisions can cause the core of a galaxy to erupt, so you can also suspect that quasars may have been triggered into existence in the cores of galaxies by interactions between galaxies.

Are you feeling confident that you know what quasars are? The evidence seems conclusive, but there is one piece of evidence that seems impossible. No good scientific argument can ignore evidence because it is inconvenient, so you need to figure out how quasars can appear to violate one of the most important laws of physics.

Superluminal Expansion

As astronomers gathered more and more evidence and grew ever more confident that quasars were located in distant galaxies, they hit a snag. A few quasars were found that seem to be expanding faster than the velocity of light, a behavior called **superluminal expansion** that is forbidden by the theory of relativity. Nothing can go faster than the speed of light.

Radio astronomers discovered superluminal expansion by using very long baseline interferometry (VLBI)—the interconnection of radio telescopes in different parts of the world. Such a network of antennas can resolve details about 0.002 second of arc in diameter. At the distance of quasar 3C273, for example, this corresponds to a few parsecs.

For a few quasars, these radio maps show small blobs nearby, and maps made over a few years show these blobs moving away from the quasar. Quasar 3C273 has a blob that is moving about 0.0008 second of arc farther from the quasar each year (■ Figure 17-17). If you accept that the quasar's redshift of 0.16 arises from the expansion of the universe, the quasar is about 960 Mpc distant, and the small-angle formula tells you

■ **Figure 17-17**

Very-high-resolution radio images of a few quasars show gas that seems to be moving faster than the speed of light.

Superluminal Expansion

1977.56
0.002″

High-resolution radio maps can resolve details as small as a thousandth of a second of arc.

1978.24

Images of quasar 3C273 recorded over a number of years show gas being ejected...

1978.92

in the direction of the much longer jet visible in Figure 17–11.

1979.44

The gas blobs are separating from the quasar at 0.0008 second of arc per year.

1980.52

At the distance of the quasar, the gas appears to be moving 12 times faster than light.

that the blob is moving away from the quasar at a speed of 3.8 pc per year. This is the same as 12 ly per year, which is impossible. Nothing can travel faster than light.

This superluminal expansion has been found in a number of quasars, and it is also seen in blazars. As startling as it is, it is only an illusion. If the quasar is ejecting a jet of matter at nearly the velocity of light and that jet is pointed nearly toward Earth, a blob of matter in the jet could appear to move away from the quasar faster than the velocity of light.

To see how this can happen, look at ■ Figure 17-18. A blob of matter in the jet emits radiation that will reach Earth in the year 2015. In the next 35 years, that radiation travels 35 ly toward Earth, but the blob, traveling at 98 percent the velocity of light, travels 34 ly closer to Earth. Radiation that it emits from this second position is only 1 ly behind the radiation it emitted 35 years before. The newly emitted radiation will arrive on Earth

■ **Figure 17-18**

When a jet from a quasar travels at nearly the speed of light and is pointed almost directly at Earth, it can produce what seems to be superluminal motion. It takes 35 years for a blob of matter to move a distance that seems to take only one year as seen from Earth. That makes the velocity appear to be 35 times greater than it really is, and the blob appears to move faster than the speed of light.

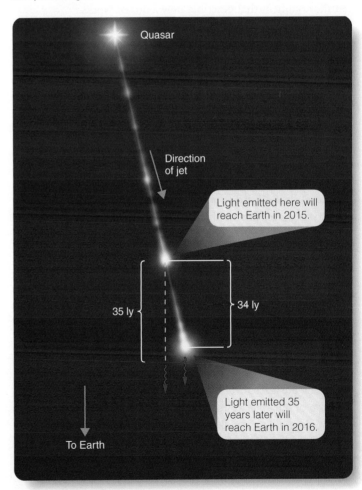

Quasar

Direction of jet

Light emitted here will reach Earth in 2015.

35 ly   34 ly

Light emitted 35 years later will reach Earth in 2016.

To Earth

in 2016, and Earth's astronomers will see the blob move in only 1 year a distance that actually took 35 years. This is a simplified explanation that omits certain relativistic effects, but it clearly shows that a blob of gas in a high-speed jet can appear to travel faster than light if the jet points nearly toward Earth.

In fact, this interesting bit of physics has been observed closer to home. The jet in M87 has been observed to have an apparent velocity 2.5 times the speed of light along a portion of its length. Evidently, the jet twists and turns as it leaves the nucleus of the galaxy, and one of those turns points the jet almost directly at Earth. Although the gas is moving at slightly over 0.9 times the speed of light, along that segment pointing toward Earth the gas appears to move 2.5 times faster than light. Even closer to home, within our own galaxy, a binary star containing a neutron star or a black hole has been observed ejecting material at speeds of about 1.25 times the speed of light. The gas is actually traveling at only 0.92 times the speed of light, but the material was ejected within 19° of the direction toward Earth, and that produces the apparent superluminal motion.

Superluminal expansion seems impossible at first glance, but once you understand how it happens, it ceases to be a stumbling block and becomes supporting evidence. By now you should feel comfortable thinking about quasars and be ready to build a scientific argument that describes how a quasar works.

A Model Quasar

The most satisfying part of science is combining evidence and theory to understand some aspect of nature. You have seen the evidence that quasars are very small but ultraluminous, and you are familiar with active galaxies, so now it is time to see how quasars fit into the picture. The best way to proceed is to build a model of a quasar.

Quasars seem similar to active galaxies, and some quasars even have jets and radio lobes. You can use the unified model of a hot accretion disk orbiting a supermassive black hole at the center of a galaxy. Matter flowing inward to feed the black hole can produce eruptions, jets, and radio lobes.

This model could explain the different kinds of radiation Earth receives from quasars (see Figure 17-7). A small percentage of quasars are strong radio sources, and the radio radiation may come from synchrotron radiation produced in the high-energy gas and magnetic fields in the jets. Much of the light from a quasar is spread in a continuous spectrum, and that is the light that fluctuates quickly. Consequently, it must come from a small region. Your model could produce this light in the innermost region surrounding the black hole. Because this is a very small region, roughly the size of our solar system, rapid fluctuations can occur, probably because of random fluctuations in the flow of matter into the accretion disk. The emission lines in quasar spectra don't fluctuate rapidly, which suggests that they are produced in a larger region.

In your model, emission lines would be emitted by clouds of gas surrounding the core in a region many light-years in diameter and excited by the intense synchrotron radiation streaming out of the central cavity.

Just as in the unified model of active galactic nuclei, the orientation of the accretion disk is important (see Figure 17-7). If the disk faces Earth, so that one of its jets points directly toward you, you may see one kind of quasar. But if the disk is tipped slightly so the jet is not pointing straight at Earth, you will see a slightly different kind of quasar. Astronomers are now using this unified model to sort out the different kinds of quasars and active galaxies so they can understand how such objects are related. For example, using infrared radiation to penetrate dust, astronomers observed the core of the double-lobed radio galaxy Cygnus A (see page 398) and found an object much like a quasar. Astronomers have begun to refer to such objects as buried quasars.

As in the case of active galaxies, many quasars are found in distorted galaxies that are interacting with nearby companions. Such a collision could throw matter into a central black hole and trigger a quasar outburst. That suggests you could use your model to explain how quasars evolve.

Quasars through Time

When you look at a photo of a quasar, the look-back time is large, and you are seeing the universe as it was long ago. The light journeying from quasars carries information about how the quasars formed, evolved, and died out. Now you can use that information to tell the story of quasars through time.

In the next chapter, you will see evidence that the universe began about 14 billion years ago. The first clouds of gas began forming stars and falling together to form galaxies, and astronomers suspect that some of that matter formed supermassive black holes at the centers of star clouds that became the nuclear bulges of galaxies. The abundance of matter flooding into these black holes could have triggered outbursts that are visible from Earth as quasars.

You should also note that galaxies were closer together when the universe was young and had not expanded very much. Because they were closer together, the forming galaxies collided more often, and you have seen how collisions between galaxies could throw matter into supermassive black holes and trigger eruptions. Recall that quasars are often located in host galaxies that are distorted as if they were interacting with other galaxies.

The evidence suggests that when the universe was young, forming galaxies created supermassive black holes; aided by interactions with other galaxies, large amounts of matter fell into those black holes, triggering quasar eruptions.

Quasars are most common with redshifts of about 2 and less common with redshifts above 2.7. The largest quasar redshifts are over 6, and such high-redshift quasars are quite rare. Evidently, if you looked at quasars with redshifts of 2 or so, you would be looking back to an age when galaxies were actively forming, colliding, and merging. In that age of quasars, they were about 1000 times more common than they are now. If you looked back to higher redshifts, you would see fewer quasars because you would be looking back to an age when the universe was so young it had not yet begun to form many galaxies. Nevertheless, even during the age of quasars, quasar eruptions must have been unusual. At any one time, only a few galaxies had quasars erupting in their cores.

Then where are all the dead quasars? An astronomer commented recently, "There is no way to get rid of supermassive black holes, so all of the galaxies that had short-lived quasars still have those black holes." Why don't astronomers see those dead quasars today? Actually, astronomers have found quite a few.

Many dead quasars are not truly dead—only sleeping. You have discovered that most galaxies contain supermassive black holes, and those black holes may have suffered quasar eruptions when the universe was younger, when galaxies were closer together, and when dust and gas were more plentiful. Quasar eruptions became less common as galaxies became more stable and as the abundance of gas and dust in the centers of galaxies was exhausted. The dormant black holes sleeping at the centers of the galaxies today can be triggered to become active galactic nuclei by collisions. Such AGN are much less energetic than quasars, evidently because there is now much less gas and dust available to feed the supermassive black holes.

Most galaxies contain sleeping supermassive black holes into which matter slowly trickles. Our own Milky Way Galaxy is a good example. It could have been a quasar a long time ago, but today its supermassive black hole is resting.

Building Scientific Arguments

Why are most quasars so far away?

Sometimes a scientific argument hinges on simple geometry, but in this case, you need to add more to your argument. To analyze this question you must combine two factors. First, quasars are the active cores of galaxies, and only a small percentage of galaxies contain active cores. In order to sample a large number of galaxies in your search, you must extend your search to great distances. Most of the galaxies lie far away from Earth because the amount of space you search increases rapidly with distance. Just as most seats in a baseball stadium are far from home plate, most galaxies are far from our Milky Way Galaxy. Consequently, most of the galaxies that might contain quasars lie at great distances.

But a second factor is much more important. The farther you look into space, the farther back in time you look. It seems that there was a time in the distant past when quasars were more common, and consequently you see most quasars at large look-back times, meaning at large distances. For these two reasons, most quasars lie at great distances.

Now expand your argument. **What observational evidence makes you think quasars must be triggered into eruption?**

■   ■   ■

Connections: Your study of active galaxies has led you far out in space and back in time to quasars. The light now arriving from the most distant quasars left them when the universe was only a fraction of its present age. The quasars stand at the very threshold of the study of the history of the universe itself, the subject of the next chapter.

Study and Review Tools

17-1 | Active Galactic Nuclei

What evidence shows that some galactic nuclei are active?
- Some galaxies have peculiar properties. Seyfert galaxies, for example, are spirals with small, highly luminous cores.
- Spectra of the nuclei of Seyfert galaxies show that they contain highly excited gas moving at very high velocities.
- Double-lobed radio galaxies emit radio energy from areas on either side of the galaxies. These lobes appear to be inflated by jets ejected from the nuclei of the galaxies.
- Some giant elliptical galaxies have small, energetic cores, with, in some cases, jets of matter rushing outward.

What is the energy source of this activity?
- Orbital motion around the nuclei of active galaxies reveals that the central objects have masses ranging from a few million to a few billion solar masses. These are presumably supermassive black holes.
- Matter flowing through hot accretion disks into supermassive black holes can release tremendous energy and eject jets in opposite directions. This process is not well understood, but it is also observed coming from accretion disks around protostars, around neutron stars, and around stellar-mass black holes.
- Relativistic jets of excited gas from the core of an active galaxy can push into the intergalactic medium and inflate radio lobes like balloons. Hot spots in the radio lobes are the points where the gas hits the intergalactic medium.

- Active galaxies moving through the intergalactic medium leave behind trails of hot gas from their jets. In other cases, the motions of the nucleus can produce twisted jets.
- Supermassive black holes cannot have been formed by dying stars but must have formed as the nuclear bulges of the galaxies began to form.

What triggers the nucleus of a galaxy into activity?

- Most galaxies appear to contain supermassive black holes at their centers, but they are dormant because large amounts of matter are not flowing inward. Only when a supermassive black hole is fed does it erupt.
- Interactions between galaxies can throw matter into the center, feed the black hole, and trigger eruptions. This explains why active galaxies are often distorted or have nearby companions.
- According to the unified model, what an observer sees depends on the tilt of the accretion disk. If you see into the core, you see broad spectral lines and rapid fluctuations. If you see the disk edge-on, you see only narrow spectral lines produced by slower moving gas above and below the disk.
- If the jet from the black hole points directly at Earth, you see a BL Lac object also known as a blazar.

17-2 | Quasars

What are the most distant active galaxies?

- The quasars appear to be very distant, highly luminous active galaxies.
- Astronomers know they are very far away because they have very high redshifts. To be visible at such great distances, they must be ultraluminous.
- Because quasars can change their brightness quickly, you can conclude they must be small—only a few times larger than our solar system.
- Observations of the spectra of hazy objects near quasars and the spectra of quasar fuzz show that quasars are the active cores of very distant galaxies.
- Gravitational lensing by very distant galaxies can form multiple images of quasars, and that shows that the quasars must be even farther away than the lensing galaxies.
- Superluminal expansion refers to blobs of material that appear to be rushing away from some quasars faster than the speed of light. This is an illusion caused when a jet points nearly at Earth and does not contradict the modern understanding of quasars.

What can active galaxies reveal about the history of the universe?

- Because quasars lie at great distances, they appear as they were long ago—over 10 billion years ago—when the universe was young and just forming galaxies.
- The best images show that the host galaxies of quasars are distorted, and that suggests that they have erupted because they have been involved in mergers or collisions. Such interactions were more common in the distant past before the universe had expanded very far and galaxies were closer together.
- It is also possible that at least some quasars are erupting while matter falls together to create a supermassive black hole as a galaxy begins to form. That is, some quasars may be caused by the formation of the first galaxies when the universe was young.
- Quasars are most common with redshifts of about 2, which shows that there was an age when galaxy formation, interaction, and mergers were more common. The so-called dead quasars today are the dormant black holes at the centers of galaxies in which little matter is flowing into the black hole.

New Terms

radio galaxy (p. 394)

active galaxy (p. 396)

active galactic nucleus (AGN) (p. 396)

Seyfert galaxy (p. 396)

double-lobed radio source (p. 397)

double-exhaust model (p. 397)

hot spot (p. 398)

unified model (p. 402)

BL Lac object (p. 403)

blazar (p. 403)

quasar (p. 405)

relativistic Doppler formula (p. 407)

superluminal expansion (p. 410)

Review Questions

Ace⊘Astronomy™ Assess your understanding of this chapter's topics with additional quizzing and animations at **http://ace .brookscole.com/sf9**

1. What is the difference between a type 1 Seyfert galaxy and a type 2 Seyfert galaxy? How might you explain the differences?
2. What evidence can you cite that radio lobes are inflated by jets from active galactic nuclei?
3. What is the significance of spiral jets from galaxies?
4. What evidence can you cite that AGNs contain supermassive black holes?
5. What properties of SS433 resemble the energy source in an AGN?
6. If you located a galaxy that had not interacted with another galaxy recently, would you expect it to be active or inactive? Why?
7. Seyfert galaxies are three times more common in close pairs of galaxies than in isolated galaxies. What does that suggest?
8. What evidence can you cite that quasars are small?
9. What evidence can you cite that quasars are very far away?
10. How do quasars resemble the AGN in Seyfert galaxies?
11. How does your model quasar account for the different components of a quasar's spectrum?
12. What does it mean that quasars are most common at a redshift of about 2?
13. Where are all the dead quasars?
14. The image at right combines visual (blue) with radio (red) to show the galaxy radio astronomers call Fornax A. Explain the features of this image. Is it significant that the object is a distorted elliptical galaxy or is that just a coincidence?

(NRAO/AUI)

15. Explain the features of this radio image of the galaxy IC708.

(NRAO/AUI)

16. A radio image of quasar 3C334 is shown at the right. Why do you see only one jet? What does it mean that this looks so much like a radio image of a double-lobed radio source?

(NRAO/AUI)

Discussion Questions

1. Why do quasars, active galaxies, SS433, and protostars have similar geometry?
2. By custom, astronomers refer to the unified *model* of AGN and not to the unified *hypothesis* or unified *theory*. In your opinion, which of the words seems best?
3. Do you think that our galaxy has ever been an active galaxy? Could it have hosted a quasar when it was young?
4. If a quasar is triggered in a galaxy's core, what would it look like to people living in the outer disk of the galaxy? Could life continue in that galaxy? (Begin by deciding how bright a quasar would be seen from the outer disk, considering both distance and dust.)

Problems

1. The total energy stored in a radio lobe is about 10^{53} J. How many solar masses would have to be converted into energy to produce this energy? (*Hints:* Use $E = m_0 c^2$. One solar mass equals 2×10^{30} kg.)

2. If the jet in NGC5128 is traveling 5000 km/s and is 40 kpc long, how long will it take for gas to flow from the core of the galaxy out to the end of the jet?

3. Cygnus A is 225 Mpc away, and its jet is about 50 seconds of arc long. What is the length of the jet in parsecs? (*Hint:* Use the small-angle formula.)

4. Use the small-angle formula to find the linear diameter of a radio source with an angular diameter of 0.0015 second of arc and a distance of 3.25 Mpc.

5. If the active core of a galaxy contains a black hole of 10^6 solar masses, what will the orbital period be for matter orbiting the black hole at a distance of 0.33 AU? (*Hint:* See circular velocity, Chapter 5.)

6. If a quasar is 1000 times more luminous than an entire galaxy, what is the absolute magnitude of such a quasar? (*Hint:* The absolute magnitude of a bright galaxy is about -21.)

7. If a quasar in Problem 6 were located at the center of our galaxy, what would its apparent magnitude be? (*Hint:* Use the magnitude–distance formula.)

8. Suppose you orbited a supermassive black hole with a mass of a million solar masses at a distance of one astronomical unit. What would your orbital velocity be?

9. If the Hubble constant is 70 km/s/Mpc, how far away is a quasar that has an apparent velocity of recession of 90,000 km/sec? (*Hint:* Use the Hubble law.)

10. The hydrogen Balmer line H_β has a wavelength of 486.1 nm. It is shifted to 563.9 nm in the spectrum of 3C273. What is the redshift of this quasar? (*Hint:* What is $\Delta\lambda$?)

Media Cluster

Ace ✦Astronomy™ To access the resources in the Media Cluster, log into AceAstronomy at **http://ace.brookscole .com/sf9** and select Chapter 17.

ACTIVE FIGURES

Jet Deflection
This animation lets you vary the velocity of a galaxy producing jets and observe how the jets bend in the "intergalactic wind." When the galaxy's velocity is higher, the bending is greater.

M31 at Many Wavelengths
In this animation, you can compare how five astronomical satellites, including Hubble and Chandra, view the Andromeda Galaxy.

VIRTUAL ASTRONOMY LABS

Lab 15: General Relativity and Black Holes
This lab explores the properties of black holes. It includes an exercise illustrating Einstein's general theory of relativity and an exercise on binary quasars. The lab concludes with a discussion and an exercise on black hole detection.

Lab 18: Active Galactic Nuclei
After surveying the types of active galaxies, this lab addresses evidence that active galaxies often have relativistic jets powered by supermassive black holes. It then looks at recent strong evidence that there is a supermassive black hole at the center of our own Milky Way Galaxy, thus confirming that such objects may exist at the centers of galaxies.

Critical Inquiries for the Web

1. What object currently holds the distinction as the farthest known galaxy? Search the Web for information on this distant object and find out its redshift and distance. What is the look-back time for this object?

2. The text discusses the superluminal expansion of material from 3C273, but how common is this faster-than-light effect? Search the Internet for information on superluminal motion related to quasars and blazars and list several examples—with the expansion speeds implied by their motion. What is the highest superluminal expansion rate you can find?

3. Gravitational lenses were first predicted by Einstein in 1936 but were not observed until recently. Search the Web for instances of gravitational lensing of galaxies and quasars. For a particular case, discuss how the lens effect allows astronomers to determine information about the lensing and/or lensed objects that might not have been available without the alignment.

 Go to the Brooks/Cole Astronomy Resource Center (**http:// astronomy.brookscole.com**) for critical thinking exercises, articles, and additional readings from InfoTrac College Edition, Brooks/Cole's online student library.

18 | Cosmology in the 21st Century

The Universe, as has been observed before, is an unsettlingly big place, a fact which for the sake of a quiet life most people tend to ignore.

DOUGLAS ADAMS, *THE RESTAURANT AT THE END OF THE UNIVERSE*

LOOK AT YOUR THUMB. The matter in your thumb was present in the fiery beginning of the universe. **Cosmology,** the study of the universe as a whole, can tell you where your matter came from, and it can tell you where your matter is going. One possibility suggested by the newest theories of cosmology is that the expansion of the universe will eventually rip the galaxies away from each other, rip stars away from galaxies, rip planets away from their stars, rip stars apart, rip molecules apart, and eventually rip every atom in the universe to pieces. Cosmologists like to call that the big rip, but you can relax; even if it happens, it won't be for 30 billion years or more. ▌ Cosmology is a mind-bendingly weird subject, and you can enjoy it for its strange ideas and theories. After all, it is fun to think about space stretching like a rubber sheet and

▌ Continued on page 418 ▌

All of the energy and matter in the universe, including the matter in your body, began in the big bang. In this computer model of the evolution of the universe soon after the big bang, matter is being drawn together to form great clouds of galaxies. (Courtesy MPA and Joerg Colberg)

Guidepost

Looking Back

Since Chapter 1 you have been on an outward journey through the universe. You have studied the appearance of the night sky seen from our planet, the birth and death of stars, and the interaction of galaxies. Now you have reached the limit of your journey in space and in time—the study of the universe as a whole.

This Chapter

The ideas in this chapter are the biggest and the most difficult in all of science. Can you imagine an edge to the universe or the first instant of time? Perhaps you have heard that the universe began in an event called the big bang. Now it is time to try to imagine what it was like.

As you explore cosmology, you will find answers to four essential questions:

Does the universe have an edge in space or time?

What evidence exists that the universe began with a big bang?

How can the universe expand if it has no edge?

Why is the universe the way it is?

These questions are deceptively simple, but they will lead you to a new understanding of what you are and where you are.

Looking Ahead

Once you have finished this chapter, you will have modern insight into the nature of the universe, and it will be time to focus on your place in that universe—the subject of the rest of this book.

Ace ☽Astronomy™ The AceAstronomy icon throughout the text indicates an opportunity for you to test yourself on key concepts and to explore animations and interactions on the AceAstronomy website at: http://ace .brookscole.com/sf9 invisible energy pushing the universe to expand faster and faster. As you wonder about the origin of vast walls of galaxy clusters, notice that it is all supported by evidence. Cosmology, however strange it may seem, is a serious and logical attempt to understand how the universe works.

The ideas of cosmology may seem weird at first glance, but this chapter will help you climb the cosmology pyramid one easy step at a time (■ Figure 18-1). You can begin with your common expectations; you have some ideas about what the universe is like. Start with those and test them against observations, compare them with theories, and step-by-step you can build a modern understanding of cosmology. Each step in the pyramid is small, but it leads to some astonishing insights into how the universe works and how you came to be a part of it.

18-1 Introduction to the Universe

EVERYONE KNOWS what a gold mine looks like, but few people have actually visited a gold mine. If you explored a real gold mine, you might be surprised at what you find. Similarly, everyone has

Visual + infrared image

■ **Figure 18-2**

The entire sky is filled with galaxies. Some lie in clusters of thousands, and others are isolated in nearly empty voids between the clusters. In this image of a typical spot on the sky, bright objects with spikes caused by diffraction in the telescope are nearby stars. All other objects are galaxies ranging from the nearby face-on spiral at upper right to the most distant galaxies visible only in the infrared, shown as red in this composite image. (R. Williams, STScI HDF-South Team, NASA)

■ **Figure 18-1**

Climbing the cosmology pyramid step-by-step isn't very difficult, and it leads to some fascinating ideas about the origin and evolution of the universe.

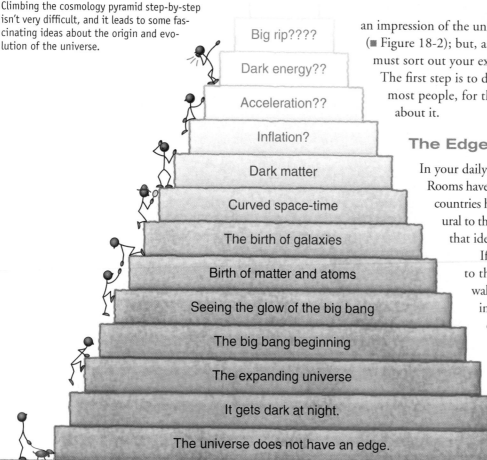

Big rip????

Dark energy??

Acceleration??

Inflation?

Dark matter

Curved space-time

The birth of galaxies

Birth of matter and atoms

Seeing the glow of the big bang

The big bang beginning

The expanding universe

It gets dark at night.

The universe does not have an edge.

an impression of the universe as a vast depth filled with galaxies (■ Figure 18-2); but, as you begin exploring the universe, you must sort out your expectations so they do not mislead you.

The first step is to deal with an expectation so obvious that most people, for the sake of a quiet life, don't even think about it.

The Edge–Center Problem

In your daily life, you are accustomed to boundaries. Rooms have walls, athletic fields have boundary lines, countries have borders, oceans have shores. It is natural to think of the universe as having an edge, but that idea can't be right.

If the universe had an edge, imagine going to that edge. What would you find there: a wall of cardboard? a great empty space? nothing? This is not an edge to the distribution of matter, but an edge to space itself, so presumably you could not reach beyond the edge and feel around.

An edge to the universe seems to violate common sense, and modern cosmologists assume that the universe cannot have an edge. That is, the universe must be unbounded.

If the universe has no edge, then it cannot have a center. You find the centers of things—galaxies, globular clusters, oceans, pizzas—by referring to their edges. If the universe cannot have an edge, then it cannot have a center.

You must think carefully about the universe and avoid using the ideas of an edge or a center. Those ideas are misleading. Of course, if the universe is infinite, if it extends without limit in all directions, then it has no edge or center. There is no edge–center problem if you think of the universe as infinite. Could the universe be finite—could it contain a limited volume? You will see later in this chapter how the universe could be finite but have no edge or center. But before you go further with that idea, you must deal with another common expectation, the beginning of the universe.

The Necessity of a Beginning

Of course you have noticed that the night sky is dark. However, reasonable assumptions about the geometry of the universe can lead to the conclusion that the night sky should glow as brightly as a star's surface. This conflict between observation and theory is called **Olbers's paradox** after Heinrich Olbers, a Viennese physician and astronomer, who discussed the problem in 1826.

However, Olbers's paradox is not Olbers's, and it isn't a paradox. The problem of the dark night sky was first discussed by Thomas Digges in 1576 and was further analyzed by astronomers such as Johannes Kepler in 1610 and Edmund Halley in 1721.

Olbers gets the credit through an accident of scholarship on the part of modern cosmologists who did not know of previous discussions. What's more, Olbers's paradox is not a paradox. You will be able to understand why the night sky is dark by revising your assumptions about the nature of the universe.

To begin, let's state the so-called paradox. Suppose you assume that the universe is infinite and uniformly filled with stars. (The aggregation of stars into galaxies makes no difference to your argument.) If you look in any direction, your line of sight must eventually reach the surface of a star. Look at ■ Figure 18-3, which uses the analogy of lines of sight in a forest (**Window on Science 18-1**). When you are deep in a forest, every line of sight terminates on a tree trunk, and you cannot see out of the forest. By analogy, every line of sight from Earth out into space should eventually terminate on the surface of a star, and the entire sky should be as bright as the surface of an average star. It should not get dark at night.

Of course, the more distant stars would be fainter than nearby stars because of the inverse square law. However, the farther you look into space, the larger the volume you see and the more stars you see; the two effects cancel out. Then given the assumptions, every spot on the sky must be occupied by the surface of a star, and the night sky should not be dark.

Can you imagine the entire sky glowing with the brightness of the surface of the sun? The glare would be overpowering. In fact, the radiation would rapidly heat Earth and all other celestial

■ **Figure 18-3**

(a) Every direction you look in a forest eventually reaches a tree trunk, and you cannot see out of the forest. (Photo courtesy Janet Seeds) (b) If the universe is infinite and uniformly filled with stars, then any line from Earth should eventually reach the surface of a star. This assumption predicts that the night sky should glow as brightly as the surface of the average star, a puzzle commonly referred to as Olbers's paradox.

Thinking about an Abstract Idea: Reasoning by Analogy

The economy is overheating and it may seize up," an economist might say. Economists like to talk in analogies because economics is often abstract, and one of the best ways to think about abstract problems is to find a more approachable analogy. Rather than discuss the details of the national economy, you might make conclusions about how the economy works by thinking about how a gasoline engine works. Of course, if your analogy is not a good one, it can mislead you.

Much of astronomy is abstract, and cosmology is the most abstract subject in astronomy. Furthermore, cosmology is highly mathematical, and unless you are prepared to learn some of the most difficult mathematics known, you must use analogies, such as lines of sight in a forest.

Reasoning by analogy is a powerful technique. An analogy can reveal unexpected insights and lead you to further discoveries. But carrying an analogy too far can be misleading. You might compare the human brain to a computer, and that would help you understand how data flow in and are processed and how new data flow out. But the analogy is flawed. Data in computers are stored in specific locations, but memories are stored in the brain in a distributed form. No single brain cell holds a specific memory. If you carry the analogy too far, it can mislead. Whenever you reason by analogy, you should be alert for potential problems.

As you study any science, be alert for analogies. They are tremendously helpful, but you must always take care not to carry them too far.

How is the economy like a gasoline engine? Thinking with analogies is a powerful tool when you respect the limits.

---

objects to the average temperature of the surface of the stars, a few thousand degrees. Consequently, you can pose Olbers's paradox in another way: The universe should not be so cold.

Olbers assumed that the sky was dark because clouds of matter in space absorb the radiation from distant stars. But this interstellar medium would gradually heat up to the average surface temperature of the stars, and the gas and dust clouds would glow as brightly as the stars.

It is interesting to note that Copernicus and other astronomers of his time saw no paradox in the dark night sky because they believed in a finite universe. Once they looked beyond the sphere of stars, there was nothing to see except, perhaps, the dark underside of the floor of heaven. Soon after the time of Galileo and Newton, Western astronomers began to think of an infinite universe, and that led to the paradox.

Today, cosmologists believe they understand why the sky is dark and the universe is cold. Olbers's paradox makes the incorrect prediction that the night sky should be bright because it is based on an incorrect assumption. The universe is not eternal. That is, it is not infinitely old.

This solution to Olbers's paradox was first suggested by Edgar Allan Poe in 1848. He proposed that the night sky was dark because the universe was not infinitely old but had been created at some time in the past. The more distant stars are so far away that light from them has not yet reached Earth. That is, if you look far enough, the look-back time is greater than the age of the universe, and you look back to a time before stars began to shine. The night sky is dark because the universe is not infinitely old.

This is a powerful idea because it clearly illustrates the difference between the universe and the observable universe. The universe is everything that exists, and it could be infinite. But the **observable universe** is the part that you can see. You will learn later that the universe is about 14 billion years old. In that case, the observable universe has a radius of about 14 billion light-years. Do not confuse the observable universe, which is finite, with the universe as a whole, which could be infinite.

When previous paragraphs described Olbers's paradox, they assumed without comment that the universe was infinitely old. This illustrates the importance of assumptions in cosmology and serves as a warning that commonsense expectations are not dependable. All of astronomy is reasonably unreasonable—that is, reasonable assumptions often lead to unreasonable results. That is especially true in cosmology, so you must proceed with care.

The common expectation that the universe is eternal and unchanging is clearly wrong. Olbers's paradox makes a beginning a necessity. In the next section you will discover that you should not expect the universe to be static; observations show that it is expanding.

Cosmic Expansion

In the early 20th century, astronomers began photographing the spectra of galaxies, and they noticed that the spectral lines in galaxy spectra had redshifts. In 1929, Edwin P. Hubble published his discovery that the size of the redshift was proportional to the distance to the galaxy. (You learned this as the Hubble law in Chapter 16, where you used it to estimate the distances to galaxies.)

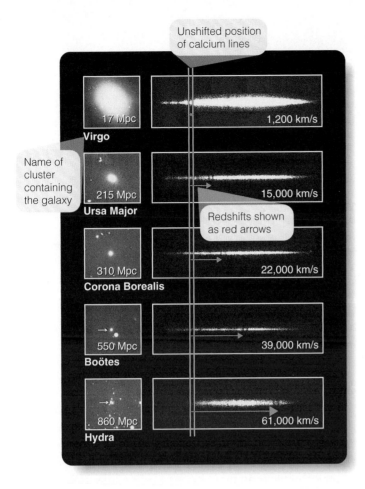

Unshifted position of calcium lines

Name of cluster containing the galaxy

Redshifts shown as red arrows

17 Mpc **Virgo**	1,200 km/s
215 Mpc **Ursa Major**	15,000 km/s
310 Mpc **Corona Borealis**	22,000 km/s
550 Mpc **Boötes**	39,000 km/s
860 Mpc **Hydra**	61,000 km/s

■ **Active Figure 18-4**

These galaxy spectra extend from the near-ultraviolet at left to the blue part of the visible spectrum at right. The two dark absorption lines of once-ionized calcium are prominent in the near-ultraviolet. The redshifts in galaxy spectra are expressed here as apparent velocities of recession. Note that the apparent velocity of recession is proportional to distance. (Caltech)

Ace ⬥ Astronomy™ Log into AceAstronomy and select this chapter to see the Active Figure "Cosmic Redshift." Graph your own observations of redshift and distance.

by once-ionized calcium are shifted from the near ultraviolet well into the visible part of the spectrum.

The important point about Hubble's discovery is that the redshifts are proportional to distance. That shows that the galaxies are receding from each other and the universe is expanding uniformly. But notice that the expansion does not imply that Earth is at the center of the universe. To see why, look at ■ Figure 18-5, which shows an analogy in which you imagine baking raisin bread. As the dough rises, the raisins are pushed away from each other uniformly at velocities that are proportional to distance. Two raisins that were originally close to each other are pushed apart slowly, but two raisins that were far apart, having more dough between them, are pushed apart faster. If bacterial astronomers lived on a raisin in your raisin bread, they could observe the redshifts of the other raisins and derive a bacterial Hubble law. They could conclude that their universe was expanding uniformly. But it does not matter which raisin the bacterial astronomers lived on. There is no center to the expansion, and they would get the same Hubble law no matter what raisin they lived on. Similarly, there is no center to the expansion of the universe and astronomers in any galaxy will see the same law of expansion.

When you look at Figure 18-4, you see the edge of the loaf of raisin bread, and you can identify a center to the loaf of bread. This happens because the raisin bread analogy breaks down when you look at the crust of the bread. Remember that our universe cannot have an edge or a center, so there can be no center to the expansion. You must think and speak very carefully as you discuss cosmology (**Window on Science 18-2**).

When interpreted using the Doppler effect, these redshifts imply that the galaxies are receding from each other and that the universe is expanding.

The redshifts of the galaxies are dramatic. Nearby galaxies have small redshifts, but when you look at more distant galaxies, the redshifts are quite large. ■ Figure 18-4 shows spectra of galaxies in galaxy clusters at various distances. The Virgo cluster is relatively nearby, and its redshift is small. The Hydra cluster is very distant, and its redshift is so large that the two dark lines formed

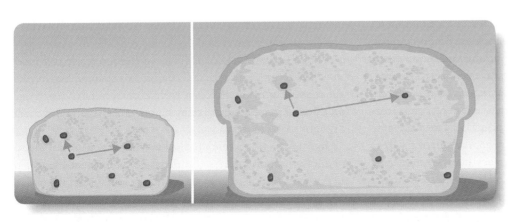

■ **Active Figure 18-5**

The raisin bread analogy for the expansion of the universe. As the dough rises, raisins are pushed apart with velocities proportional to distance. The expansion has no center. A colony of bacteria living on any raisin will find that the redshifts of the other raisins are proportional to their distances.

Ace ⬥ Astronomy™ Log into Ace Astronomy and select this chapter to see the Active Figure "Raisin Bread." You can watch raisin bread rise.

Why Scientists Speak Carefully: Words Lead Thoughts

There are certain words you should never say, not even as a joke. You should never call your friend "fool," for example, because you might begin to think of your friend as a fool. Words lead thoughts. It works in advertising and politics, and it can work in science, too. Using words carelessly can lead you to think carelessly.

In science, there are certain ways to say things, not because scientists are sticklers for good grammar, but because if they say things wrongly they may begin to think things wrongly. For example, a biologist would never let you say a beehive knows it must store food. "No, don't say it that way," the biologist would object. "The hive is just a collection of individual creatures, and it can't 'know' anything.

Even the individual bees don't really 'know' things. The instinctive behavior of individual bees causes food to accumulate in the hive. That's the way to say it."

All scientists are careful of language because careless words can mislead. You would never refer to the center of the universe, for example, but you must also be careful not to say things like "galaxies flying away from the big bang." Those words imply a center to the expansion of the universe, and you know the expansion must be centerless. Rather, you should say "galaxies flying away from each other."

Whatever science you study, notice the customary ways of using words. It is not just a matter of convention. It is a matter of careful thought.

You should not say that a bee "knows" how to make honey. How you speak guides how you think.

The Necessity of a Big Bang

The discovery of the expansion of the universe led astronomers to an astonishing conclusion. The universe must have begun with an event of cosmic fury.

Imagine that you have a videotape of the expanding universe, and you run it backward. You would see the galaxies moving toward each other. There is no center to the expansion of the universe, so you would not see galaxies approaching a single spot; but, rather, you would see the distances between all galaxies decreasing, and eventually the galaxies would begin to merge. If you ran your videotape far enough back, you would see the matter and energy of the universe compressed into a high-density, high-temperature state. The expanding universe must have begun from this moment of extreme conditions, which modern astronomers call the **big bang.**

You must not allow yourself to think of the big bang as an explosion. You have concluded that the universe cannot have a center, so the big bang cannot have occurred at a single point. The entire universe must have been filled with the high-density, high-temperature state called the big bang. Your challenge in this chapter is to understand how that could have happened without a center or an edge. That problem will be resolved when you refine your understanding of space and time.

For now, you can solve an easier problem. How long ago did the universe begin? You can estimate the age of the universe with a simple calculation. If you must drive to a city 100 miles away and you can travel 50 miles per hour, you divide distance by rate of travel and learn the travel time—in this example, 2 hours. To find the age of the universe, you can divide the distance between two galaxies by the velocity with which they are separating and find out how long they have taken to reach their present separation.

Dividing distance by velocity tells you the age of the universe, and the Hubble constant (Chapter 16) simplifies your task further. The Hubble constant H has the units km/s per Mpc, which is a velocity divided by a distance. If you calculate $\frac{1}{H}$ you have a distance divided by velocity. To make the division yield an age, you need to convert megaparsecs to kilometers, and then the distances will cancel out and you will have an age in seconds. To convert your answer from seconds to years, you need to divide by the number of seconds in a year. If you make these simple changes in units, the age of the universe in years is approximately 10^{12} divided by H in its normal units, km/s/Mp:

$$T = \frac{10^{12}}{H}$$

This is known as the **Hubble time.** For example, if H is 70 km/s/Mpc, then the age of the universe can be no older than $\frac{10^{12}}{70}$, which equals 14 billion years.

The Hubble time is an estimate of the age of the universe. You will fine-tune your estimate later in this chapter, but for the moment you can conclude that basic observations of the recession of the galaxies require that the universe began with a big bang about 14 billion years ago.

Your instinct is to think of the big bang as a historical event, like the Gettysburg Address—something that happened long ago and happened in a particular place. But the look-back time makes it possible to observe the big bang directly. The look-back time

■ Figure 18-6

This faint galaxy (arrow) is believed to be so distant that its light has been traveling for most of the history of the universe. Here you see it as it was about 1 billion years after the big bang. The galaxy is bright in the infrared but it is not detectable at visual wavelengths. (Ken Lanzetta and Amos Yahil, Stony Brook and NASA)

to nearby galaxies is only a few million years, and the look-back time to more distant galaxies is a large fraction of the age of the universe (■ Figure 18-6). Suppose you look between the distant galaxies, back to the time of the big bang. You should be able to detect the hot gas that filled the universe long ago.

Although your imagination tries to visualize the big bang as a localized event, you must keep firmly in mind that the big bang did not occur at a single place but filled the entire volume of the universe. The matter of which you are made was part of that big bang, so you are inside the remains of that event, and the universe continues to expand around you. You cannot point to any particular place and say, "The big bang occurred over there." The big bang occurred everywhere, and, in whatever direction you look, at great distance you look back to the age when the universe was filled with hot gas (■ Figure 18-7).

The radiation that comes from this great distance has a tremendous redshift. The most distant visible objects are faint gal-

■ Figure 18-7

Three views of a small region of the universe centered on our galaxy. (a) During the big bang, the region is filled with hot gas and radiation. (b) Later, the gas has formed galaxies, but you can't see the universe this way because the look-back time distorts what you see. (c) Near you, you see galaxies, but farther way you see young galaxies (dots), and at a great distance you see radiation (arrows) coming from the hot gas of the big bang.

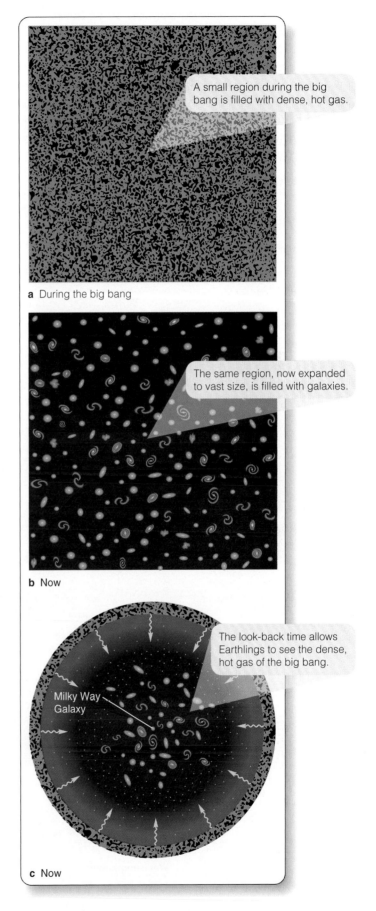

A small region during the big bang is filled with dense, hot gas.

a During the big bang

The same region, now expanded to vast size, is filled with galaxies.

b Now

The look-back time allows Earthlings to see the dense, hot gas of the big bang.

Milky Way Galaxy

c Now

axies and quasars, with redshifts a bit over 10. In contrast, the radiation from the hot gas of the big bang has a redshift of about 1100. That means the light emitted by the big bang gases arrives at Earth as infrared radiation and short radio waves. You can't see it with your eyes, but it should be detectable with infrared and radio telescopes. Unlike the Gettysburg Address, the big bang isn't over, and Earth's astronomers should be able to see it happening if they can detect the radiation it emitted. That amazing discovery is the subject of the next section.

The Cosmic Background Radiation

If radiation is now arriving from the big bang, then it should be detectable. The story of that discovery begins in the mid-1960s when two Bell Laboratories physicists, Arno Penzias and Robert Wilson, were using a horn antenna to measure the radio brightness of the sky (■ Figure 18-8). Their measurements showed a peculiar noise in the system, which they at first attributed to droppings from pigeons living inside the antenna. Perhaps they would have enjoyed cleaning the antenna more if they had known they would win the 1978 Nobel Prize for physics for the discovery they were about to make.

When the antenna was cleaned, they measured the brightness of the sky at radio wavelengths and again found the low-level radio noise. The pigeons were innocent, but what was causing the signal?

The explanation for the noise goes back to 1948, when George Gamow predicted that the early big bang would be very hot and

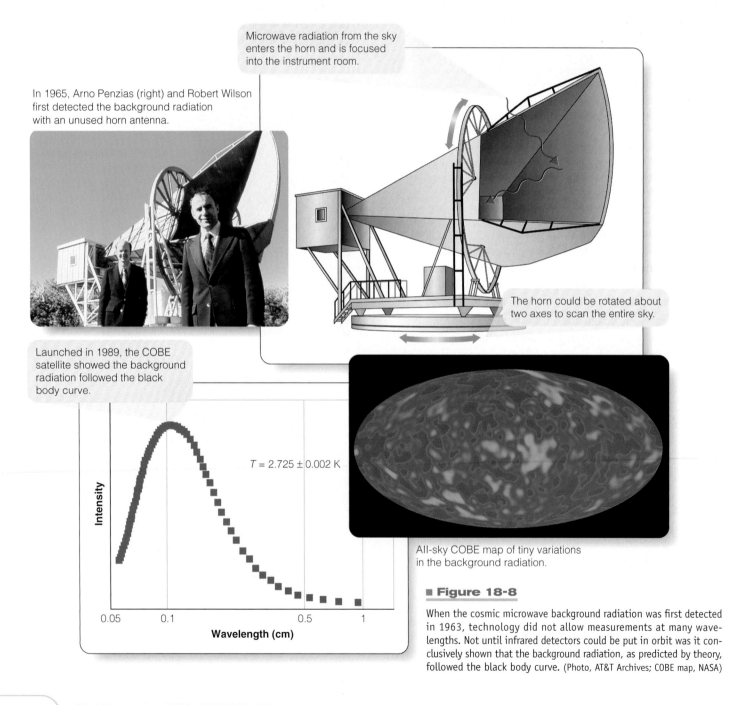

Microwave radiation from the sky enters the horn and is focused into the instrument room.

In 1965, Arno Penzias (right) and Robert Wilson first detected the background radiation with an unused horn antenna.

The horn could be rotated about two axes to scan the entire sky.

Launched in 1989, the COBE satellite showed the background radiation followed the black body curve.

$T = 2.725 \pm 0.002$ K

Intensity

Wavelength (cm)

0.05   0.1   0.5   1

All-sky COBE map of tiny variations in the background radiation.

■ Figure 18-8

When the cosmic microwave background radiation was first detected in 1963, technology did not allow measurements at many wavelengths. Not until infrared detectors could be put in orbit was it conclusively shown that the background radiation, as predicted by theory, followed the black body curve. (Photo, AT&T Archives; COBE map, NASA) would emit copious black body radiation. A year later, physicists Ralph Alpher and Robert Herman pointed out that the large redshift of the big bang would lengthen the wavelengths of the radiation into the far-infrared and radio part of the spectrum. There was no way to detect this radiation until the mid-1960s, when Robert Dicke at Princeton concluded the radiation should be just strong enough to detect with newly developed techniques. Dicke and his team began building a receiver.

When Penzias and Wilson heard of Dicke's work, they recognized the noise they had detected as radiation from the big bang, the **cosmic microwave background radiation.**

The detection of the background radiation was tremendously exciting, but astronomers wanted confirmation. Theory predicted that the radiation should look like black body radiation coming from a very cool source, but the critical observations in the far-infrared could not be made from the ground. It was not until January 1990 that satellite measurements confirmed that the background radiation was black body radiation with an apparent temperature of 2.725 ± 0.002 K—in good agreement with the theory.

It may seem strange that the hot gas of the big bang seems to have a temperature of only 2.7 K, but recall the tremendous redshift. Observers on Earth see light that has a redshift of about 1100—that is, the wavelengths of the photons are about 1100 times longer than when they were emitted. The gas clouds that emitted the photons had a temperature of about 3000 K, and they emitted black body radiation with a λ_{max} of about 1000 nm (Chapter 7). Although this is in the near infrared, the gas would also have emitted enough visible light to glow orange-red. But the redshift has made the wavelengths about 1100 times longer, so λ_{max} is about 1 million nm (which equals 1 mm). That is why the hot gas of the big bang seems to be 1100 times cooler, about 2.7 K.

The cosmic microwave background radiation is evidence that a big bang really did occur. In fact, the evidence is so strong that an alternative theory of the universe was abandoned. The **steady-state theory,** popular with some astronomers during the 1950s and 1960s, held that the universe was eternal and unchanging. If the universe had no beginning, then it could never have had a big bang. According to that theory, new matter continuously appeared from nothing to maintain the density of the expanding, steady-state universe. The discovery of the cosmic microwave background radiation in 1965 led astronomers to abandon the steady-state theory within a few years. You can be sure there was a big bang because, using radio and infrared telescopes, you can see it happening.

The Story of the Big Bang

The first few steps up the cosmology pyramid have not been very difficult. Simple observations of the darkness of the night sky and the redshifts of the galaxies tell you that the universe must have begun with a big bang. In fact, you have direct observational evidence of the big bang in the form of the cosmic microwave background radiation. Theorists can combine these observations with modern physics to tell the story of how the big bang occurred. As you review this story, remember that the big bang did not occur at a specific place. It filled the entire volume of the universe from that first moment.

Cosmologists cannot begin their history of the big bang at time zero, because no one understands the physics of matter and energy under such extreme conditions, but they can come close. If you could visit the universe when it was very young, only 10 millionths of a second old, for instance, you would find it filled with high-energy photons having a temperature well over 1 trillion (10^{12}) K and a density greater than 5×10^{13} g/cm^3, nearly the density of an atomic nucleus. When astronomers say the photons had a given temperature, they mean that the photons were the same as black body radiation emitted by an object of that temperature. Consequently, the photons in the early universe were gamma rays of very short wavelength and therefore very high energy. When astronomers say that the radiation had a certain density, they refer to Einstein's equation $E = mc^2$. You can express a given amount of energy in the form of radiation as if it were matter of a given density.

If photons have enough energy, two photons can combine and convert their energy into a pair of particles—a particle of normal matter and a particle of **antimatter.** When an antimatter particle meets its matching particle of normal matter—when an antiproton meets a normal proton, for example—the two particles annihilate each other and convert their mass into energy in the form of two gamma rays. In the early universe, the photons had enough energy to produce proton–antiproton pairs or neuron–antineutron pairs. But when the particles collided with their antiparticles, they converted their mass back into photons. Thus, the early universe was a dynamic soup of energy flickering from photons into particles and back again.

While all this went on, the expansion of the universe cooled the gamma rays to lower energy. By the time the universe was 0.0001 second old, its temperature had fallen to 10^{12} K. By this time, the average energy of the gamma rays had fallen below the energy equivalent to the mass of a proton or a neutron, so the gamma rays could no longer produce such heavy particles. The particles combined with their antiparticles and quickly converted most of the mass into photons.

It would seem from this that all of the protons and neutrons should have been annihilated with their antiparticles; but, for reasons that are poorly understood, a small excess of normal particles existed. For every billion protons annihilated by antiprotons, one survived with no antiparticles to destroy it. Consequently, you live in a world of normal matter, and antimatter is very rare.

Although the gamma rays did not have enough energy to produce protons and neutrons, they could produce electron–positron pairs, which are about 1800 times less massive than protons and neutrons. This continued until the universe was about four seconds old, at which time the expansion had stretched and cooled the gamma rays to the point where they could no longer create electron–positron pairs. The protons, neutrons, and electrons of which our universe is made were produced during the first four seconds of its history.

This soup of hot gas and radiation continued to cool, and eventually protons and neutrons began to join together to form atomic nuclei. High-energy gamma rays can break up a nucleus, so the formation of nuclei could not occur until the universe had cooled somewhat. By the time the universe was about two minutes old, protons and neutrons could link to form deuterium, the nucleus of a heavy hydrogen atom; by the end of the next minute, further reactions began converting deuterium into helium. But almost no heavier atoms could be built because there are no stable nuclei with atomic weights of 5 or 8 (in units of the hydrogen atom). Cosmic element building during the big bang had to proceed rapidly, step-by-step, like someone hopping up a flight of stairs (■ Figure 18-9). The lack of stable nuclei at atomic weights of 5 and 8 meant there were missing steps in the stairway, and the step-by-step reactions had great difficulty jumping over these gaps. A tiny amount of lithium was produced but nothing heavier.

By the time the universe was 30 minutes old, it had cooled sufficiently that all nuclear reactions had stopped. About 25 percent of the mass was helium nuclei, and the rest was in the form of protons—hydrogen nuclei. This is the cosmic abundance observed today in the oldest stars. (The heavier elements, remember, were built by nucleosynthesis inside generations of massive stars.) The cosmic abundance of helium was fixed during the first minutes of the universe.

At first, the universe was dominated by radiation. The gamma rays interacted continuously with the matter, and the radiation and matter cooled together as the universe expanded. The gas was ionized because it was too hot for the nuclei to capture electrons to form neutral atoms, and the free electrons made the gas very opaque. A photon could not travel very far before it collided with an electron and was deflected. Thus, radiation and matter were locked together and had to cool together.

As the young universe expanded, it went through three important changes. First, when the universe reached an age of roughly 50,000 years, the density of the energy present as photons became less than the density of the gas. Before this, matter could not clump together because the intense sea of photons smoothed the gas out. Once the density of the radiation fell below that of matter, the matter could begin to draw together and form the clouds that eventually became galaxies and stars.

The expansion of the universe spread the particles of the ionized gas farther and farther apart. As the universe reached the age of a few hundred thousand years, the second important change began. The free electrons were spread so far apart, the photons could travel for thousands of parsecs before getting scattered off an electron. That is, the universe began to grow more transparent. At about the same time, the third change happened. As the falling temperature of the universe reached 3000 K, protons were able

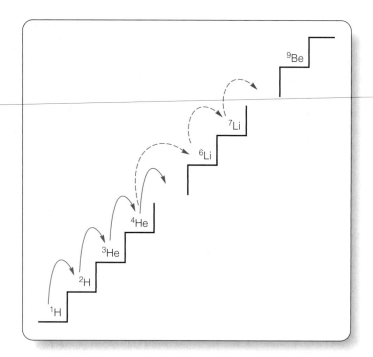

■ **Figure 18-9**

Cosmic element building. During the first few minutes of the big bang, temperatures and densities were high, and nuclear reactions built heavier elements. Because there are no stable nuclei with atomic weights of 5 or 8, the process built very few atoms heavier than helium.

to capture and hold free electrons to form neutral hydrogen, a process called **recombination.** As the free electrons were gobbled up, the gas finally became transparent, and the photons could travel through the gas without being absorbed (■ Figure 18-10). Although the gas continued to cool, the photons no longer interacted with the gas, and consequently the photons retained the black body temperature that the gas had at recombination. Those photons, with a black body temperature of 3000 K, are what is observed today as the cosmic microwave background radiation. Remember that the large redshift makes that gas appear to have a temperature of about 2.7 K.

Recombination left the gas of the big bang neutral and transparent. At first the universe was filled with the glow of the hot gas; but, as the universe expanded and cooled, the glow faded into the infrared, and the universe entered what cosmologists call the **dark age,** a period lasting hundreds of millions of years during which the universe expanded in darkness.

The dark age ended as the first stars began to form. The gas from which these first stars formed contained almost no metals and was consequently highly transparent. Mathematical models show that, because the first stars formed from this metal-poor gas, they were very massive, very luminous, and very short lived. That first violent burst of star formation produced enough ultraviolet light to begin ionizing the gas, and today's astronomers, looking back to the most distant visible quasars and galaxies, can see traces of that **reionization** of the universe (■ Figure 18-11). Reioniza-

| Dense ionized gas | Low-density ionized gas | Neutral gas after recombination |

a   b   c

■ Figure 18-10

Photons scatter from electrons (blue) easily but hardly at all from the much more massive protons (red). (a) When the universe was very dense and ionized, photons could not travel very far before they scattered off an electron. This made the gas opaque. (b) As the universe expanded, the electrons were spread further apart, and the photons could travel farther; this made the gas more transparent. (c) After recombination, most electrons were locked to protons to form neutral atoms, and the gas was highly transparent.

tion marks the end of the dark age and the beginning of the age of stars and galaxies in which you live today.

Look carefully at ■ Figure 18-12; it summarizes the story of the big bang, from the formation of helium in the first three minutes, through energy–matter equality, recombination, and finally reionization of the gas. It may seem amazing that mere humans trapped on Earth can draw such a diagram, but remember that it is based on evidence and on the best understanding of how matter and energy interact. This story of the big bang is far more than just an idle daydream (**Window on Science 18-3**).

■ Figure 18-11

In this artist's conception of reionization, the first stars produce floods of ultraviolet photons that ionize the gas in expanding bubbles. Such a storm of star formation ended an age when the universe had expanded in darkness. Spectra of the most distant quasars reveal that those first galaxies were surrounded by neutral gas that had not yet been fully ionized. Thus the look-back time allows modern astronomers to observe the age of reionization. (K. Lanzetta, SUNY, A. Schaller for STScI, and NASA)

Science: A System of Knowledge

If you hear a scientist asked a question such as, "Do you believe in the big bang?" you may hear a momentary hesitation. The question implies something that isn't true.

Science is a system of knowledge, but it is not a system of belief. Science is an attempt to logically understand how nature works, and it is based on observations and experiments used to confirm theories. A scientist does not really believe in a theory. Rather the scientist understands the theory and recognizes how the different pieces of evidence support or contradict the theory.

There are other ways to know things, and systems of belief are not unusual. Religion, for example, is a system of belief and is not based on observation and experiment. In some cases, politics is a system of belief; many people accept that democracy is the best form of government and do not ask for or expect to find evidence supporting their belief. A system of belief can be powerful and lead to deep insights, but it is different from science.

Scientists try to be careful with words, so thoughtful scientists would not say they *believe* in the big bang. They would say that the evidence is overwhelming that the big bang really did occur, and that they are forced by the logical analysis of the theory and observations to conclude that the theory is very likely correct.

A scientist once referred to "the terrible rule of evidence." Sometimes the evidence forces a scientist to an unpleasant conclusion, but science is not a system of belief, so the personal preferences of the scientist must take second place to the rule of evidence.

The Wilkinson Microwave Anisotropy Probe voyages far from Earth and the moon to gather observations that will help scientists test their theories of the universe. (NASA/WMAP Science Team)

Do you believe in the big bang? Or do you have some confidence the theory is right because of your analysis of the evidence? There is a big difference.

■ **Figure 18-12**

During the first few minutes of the big bang, some hydrogen was fused to produce helium, but the universe quickly became too cool for such fusion reactions to occur. The rate of cooling increased as matter began to dominate over radiation. Recombination freed the radiation from the influence of the gas, and reionization was caused by the birth of the first stars. Note how the exponential scale in time stretches early history and compresses recent history.

Building Scientific Arguments

How do you know there was a big bang?

A good scientific argument combines evidence and theory to describe how nature works, and this question calls for a detailed argument. The cosmic microwave background radiation consists of photons emitted by the hot gas of the big bang, so when astronomers detect those photons, they are "seeing" the big bang. Of course, all scientific evidence must be interpreted, so you must understand how the big bang could produce radiation all around the sky before you can accept the background radiation as evidence. First, you must remember that the big bang event filled all of the universe with hot, dense gas. The big bang didn't happen in a single place; it happened everywhere. At recombination, the expansion of the universe reached the stage where the matter became transparent and the radiation was free to travel through space. Today that radiation from the age of recombination arrives from all over the sky. It is all around you because you are part of the big bang event, and as you look out into space to great distance, you look back in time and see the hot gas in any direction you look. You can't see the radiation as light because of the large redshift that has lengthened the wavelengths by a factor of 1100 or so, but you could detect the radiation as photons with infrared and short radio wavelengths.

With this interpretation, the cosmic microwave background radiation is powerful evidence that there was a big bang. That tells you how the universe began, but your argument hinges on an important point. **Why do you think the universe cannot have a center or an edge?**

■   ■   ■

Connections: So far in this chapter you have climbed a number of steps up the cosmology pyramid. You began by comparing your expectations with basic observations, and you have discovered some fundamental properties of the universe. Now it is time to climb higher, time to stretch your imagination and refine your expectations further. The universe is much more interesting than you have imagined to this point.

(18-2) The Shape of Space and Time

BY NOW YOU ARE PROBABLY WONDERING how the universe can expand if it does not have an edge or a center. Solving that problem is the key to understanding modern cosmology, but your everyday expectations about how the world works are not a help. To solve the puzzle, you must put what seem to be reasonable expectations on hold and look carefully at how space and time behave on cosmic scales.

Looking at the Universe

Let's begin by carefully examining the general appearance of the universe. Not only do you see no edge to the universe and no center, there is no preferred direction.

The gas and dust in our own galaxy block the view of the distant universe along the Milky Way, but once you correct for that, the universe looks the same over the entire sky. Astronomers refer to this as **isotropy,** the property of being the same in all directions. No matter what direction you look, you will see the same kinds of galaxies and galaxy clusters. The universe seems isotropic.

Furthermore, the cosmic microwave background radiation is isotropic. Astronomers must correct for Doppler shifts caused by the motion of the Earth around the sun, the sun around our galaxy, and our galaxy around the center of mass of the local group, but those are only local variations. Once those local variations are taken into account, the background radiation is almost perfectly uniform over the entire sky.

It certainly appears that the universe is the same everywhere. Of course, there are local variations; some points in space are inside galaxies, and some are isolated far from any galaxy; but, overall, in its general properties, one place in the universe is the same as any other place. That is, there are no special places in the universe. You cannot point at any particular location in the universe and say, "That location is different from any other place in the universe." Astronomers refer to this as **homogeneity,** the property of being uniform. On the largest scales, the universe appears to be homogeneous.

Astronomers observe that the universe is isotropic and homogeneous on the largest scales, and that leads to an assumption that is critically important in cosmology. The **cosmological principle** says that any observer in any galaxy sees the same general features of the universe. Of course some observers live in spiral galaxies and some live in elliptical galaxies, but those are only local variations. According to the cosmological principle, when you look at the general features of the universe, you should see the same kind of universe no matter where your planet is located.

Evolutionary changes are not included in the cosmological principle. If the universe is expanding and the galaxies are evolving, observers living at different times may see galaxies at different stages. The cosmological principle says that once observers correct for evolutionary changes, they should see the same general features.

The cosmological principle is actually an extension of the Copernican principle. Copernicus said that Earth is not in a special place; it is just one of a number of planets orbiting the sun. The cosmological principle says that there are no special places in the universe. Local irregularities aside, one place is just like another. Your location in the universe is typical of all other locations.

If you accept the cosmological principle, then you may not imagine that the universe has an edge or a center. Such locations would not be the same as all other locations. Observations confirm this conclusion. No matter what direction you look in the sky, no matter where you look in space, the universe is generally the same. There is no edge and no center.

The Shape of the Expanding Universe

How can the universe expand if it does not have an edge or a center? You are now ready to resolve that central puzzle of cosmology. To do so you must understand the nature of space and time.

Remember that you are a small creature, and you live on a small world. The distances and times you experience each day are quite small. It should not surprise you that space and time are actually more complicated than your expectations suggest.

Study **The Nature of Space-Time** on pages 432–433 and notice three important ideas:

1 The redshifts of the galaxies are not produced by the Doppler effect. The galaxies do not move at high velocities through space. Space is expanding, carrying the galaxies away from each other, and stretching protons to longer and redder wavelengths.

2 Notice that the curvature of space-time allows the universe to have no edge or center even if it is finite.

3 Notice that you could make measurements in our three-dimensional universe to determine the curvature of space-time.

It seems to violate common sense that the universe can expand without having a center or an edge, but you must not depend on your limited imagination as a three-dimensional creature. You must speak carefully and thoughtfully when you discuss cosmology so as not to be misled (Window on Science 18-2). You must not try to visualize the expansion of the universe as an outer edge moving into previously unoccupied space. Open, flat, or closed, the universe has no edge, so it does not need additional room to expand. The universe contains all of the volume that exists, and the expansion is a gradual change in the nature of space-time that causes that volume to increase.

Model Universes

Modern cosmologists have been able to use general relativity to construct mathematical models of the universe under different assumptions. For example, they can describe the nature of the universe if it is open, closed, or flat. These models have had a strong influence on the development of cosmology.

The general curvature of the entire universe is determined by its density. If the average density of the universe is equal to the **critical density,** which is 9×10^{-30} g/cm^3, space-time will be flat. If the average density of the universe is less than the critical density, the universe is negatively curved and open. If the average density is greater than the critical density, the universe is positively curved and closed.

You learned earlier that an open universe and a flat universe are both infinite. If their behavior is ruled entirely by their matter density, both kinds of universes will expand forever. The gravitation of the matter in the universe, present as a curvature of space-time, will cause the expansion to slow; but, if the universe is open, it will never come to a stop.

The behavior of these cosmological models is shown in ■ Figure 18-13. The parameter R, which emerges from the mathematics of the models, is a measure of the extent to which the universe has expanded. As the universe expands, R gets larger. Of course R can't be the radius of the universe because there is no center or edge. It might be tempting to think that R is something like the average distance between galaxies, but remember that galaxies did not exist when the universe was young. For this discussion, you must be content to think of R as an indicator of expansion.

According to the models, if the universe is flat, it will barely slow to a stop after an infinite time. Thus the models predict that an open or flat universe will expand forever, and the galaxies will eventually become burnt-out, cold, solitary islands in a universe of darkness.

These models predict a different fate for a closed universe. In a closed universe, the gravitational field, present as curved space-time, is sufficient to slow the expansion to a stop and make the universe contract. Eventually, the contraction will compress all matter and energy back into the high-energy, high-density state from which the universe began. This end to the universe, a big bang in reverse, has been called the big crunch. Nothing in the universe could avoid being destroyed by this "crunch." Some theorists suggested that the big crunch will spring back to produce a new big bang and a new expanding universe. This theory, called the **oscillating universe theory,** predicted that the universe undergoes alternate stages of expansion and contraction. Theoretical work, however, suggests that successive bounces of an oscillating universe would be smaller until the oscillation ran down, and the theory is no longer taken seriously.

■ Figure 18-13

Open-universe models expand without end, and the corresponding curves fall in the region shaded orange. Closed models expand and then contract again (red curve). A flat universe (dotted line) marks the boundary between open and closed universes. The estimated age of the universe depends on the curvature of the universe.

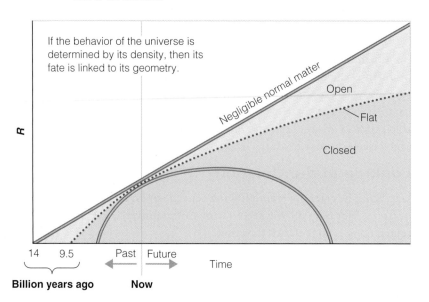

If the behavior of the universe is determined by its density, then its fate is linked to its geometry.

Notice in Figure 18-13 that you must know the geometry of the universe before you can determine its age. Earlier you calculated the Hubble time $\frac{1}{H}$ as an estimate of the age of the universe. Figure 18-13 shows that the Hubble time is the age of a model universe that contains so little matter its expansion is hardly slowed at all. If the density of a model universe is higher, the expansion will be slowed, and that means the true age will be less that $\frac{1}{H}$. If the universe is flat, its true age will be $\frac{2}{3}$ of $\frac{1}{H}$. If H equals 70 km/s/Mpc, then $\frac{1}{H}$ equals 14 billion years. But if the universe is flat and its evolution is controlled by its curvature, then its true age is $\frac{2}{3}$ of 14 billion years, which equals a bit over 9 billion years. This is less than the age of the globular clusters, which is impossible. You will solve this age problem later in this chapter.

Whether the universe is open, closed, or flat depends on its density, but it is quite difficult to measure the density of the universe. Astronomers could count galaxies in a given volume, multiply by the average mass of a galaxy, and divide by the volume; but the average mass of a galaxy is poorly known. Furthermore, many small galaxies are too faint to see even if they are nearby. Studies of light coming from distant galaxies suggest that large amounts of normal matter lie in great, nonluminous clouds between the galaxies, so any estimates must include that matter. Also, you must include the mass equivalent to all the energy in the universe because mass and energy are related. When all of the mass in the universe is added up, it provides a density of less than 5 percent of the critical density needed to close the universe.

Many astronomers will admit that, for theoretical reasons discussed later, they prefer to believe the universe is flat. Science, however, is not ruled by preferences or theory but rather by evidence. Observations of normal matter do not reveal enough density to make the universe flat, but an important factor is missing. Clear evidence shows that there is a large amount of dark matter in the universe. Could there be enough dark matter to make the universe flat?

Dark Matter in Cosmology

There is more to the universe than meets the eye. In Chapter 16, you discovered that galaxies have much stronger gravitational fields than you would expect based on the amount of visible matter. Even when you add in the nonluminous gas and dust that you expect to find in galaxies, their gravi-tational fields are much stronger than you would expect. You concluded that galaxies must contain dark matter.

In the last few chapters, you have seen further evidence for dark matter. Our galaxy rotates faster in its outer regions than expected. Galaxies in clusters move faster than expected, and the clusters hold onto very hot gas, so they must be much more massive than you would expect from what is visible. Look at ■ Figure 18-14. Gravitational lensing is dramatic evidence of dark matter. The galaxy cluster shown contains so much dark matter that it warps space-time and focuses the images of very distant galaxies into short arcs. Clearly the universe contains much more dark matter than normal matter.

■ **Figure 18-14**

Gravitational lensing shows that galaxy clusters contain much more mass than what is visible. The yellowish galaxies in this image are members of a relatively nearby cluster of galaxies. Most of the objects in this image are blue or red images of very distant galaxies focused by the gravitational field of the cluster. Some of these imaged galaxies may be over 13 billion light-years away. That they can be seen at all is evidence that the galaxy cluster contains large amounts of dark matter. (NASA, Benites, Broadhurst, Ford, Clampin, Hartig, Illingworth, ACS Science Team, and ESA)

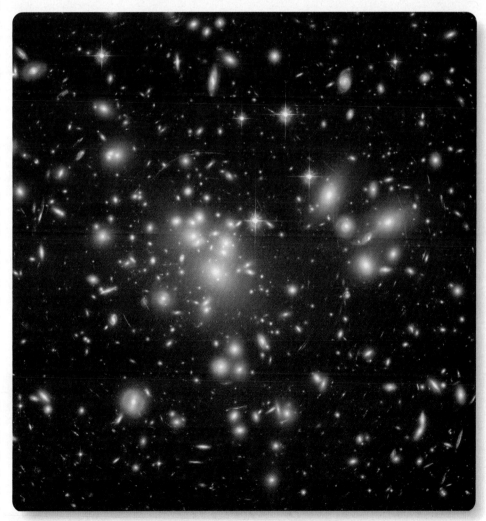

1 In 1929, Edwin Hubble discovered that the redshifts of the galaxies are proportional to distance — a relationship now known as the Hubble law. It was taken as dramatic evidence that the universe is expanding.

Distance is the separation between two points in space, and time is the separation between two events. Albert Einstein showed how to relate space to time and treat the whole as space-time. You can think of space-time as a canvas on which the universe is painted, a canvas that can stretch.

Hubble Heritage Team/STScI/AURA/NASA

1b The stretching of space-time not only moves the galaxies away from each other, but it lengthens the wavelength of photons traveling through space-time. See illustration below.

What are the cosmological redshifts?

1a Astronomers often express galaxy redshifts as apparent velocities of recession, but these redshifts are not Doppler shifts. They are caused by the expansion of space-time.

For decades textbooks have described the cosmological redshifts using Einstein's relativisitic Doppler formula, but that formula applies to motion through space and not to the behavior of space itself. The true relation between redshift and apparent velocity of recession is not accurately known.

Notice that this description of the expansion of the universe means the galaxies are not really moving any more than the raisins are swimming through the bread dough. Except for orbital motion among neighboring galaxies, the galaxies are motionless in space-time, and it is space-time that is stretching like a rubber sheet and carrying the galaxies away from each other.

A distant galaxy emits a short-wavelength photon toward our galaxy.

Grid shows expansion of space-time.

The expansion of space-time stretches the photon to longer wavelength as it travels.

The farther the photon has to travel, the more it is stretched.

When the photon arrives at our galaxy, you see it with a longer wavelength — a redshift that is proportional to distance.

2 To answer these questions you must use Einstein's theory of general relativity, which predicts that the presence of mass can curve space-time, a curvature you experience as gravity. Furthermore, the theory predicts that even empty space-time can have a curvature, so what you see in the universe depends on the overall curvature of space-time. Finally, the theory shows that space-time could have the property of expansion. That is, the expansion of space-time is not caused by external influences, but is part of space-time itself.

Could the universe be finite and still have no edge?

How can it expand if it has no edge?

How can the universe expand if there is no space outside for it to expand into?

2a To think about space-time curvature, you can use an analogy of a two-dimensional ant on an orange. If he is truly two-dimensional, he will not be able to understand up and down. He can only travel forward and back, right and left. Then as he walks over the surface of his spherical universe, he will eventually realize he has been everywhere because his universe is covered by his footprints. He will conclude that he lives in a finite universe that has no edge and no center.

Similarly, our universe could be curved back on itself so that it could be finite but you would never find an edge. To find out how our universe is really curved, astronomers must study very distant objects.

Notice that the center of the orange cannot be the center of the ant's two-dimensional universe because the center of the orange does not lie on the surface. That is, the center of the orange is not part of the ant's universe.

Analogy Ahead

Don't forget: These analogies are only two-dimensional. When you think about our universe, you must think of a three-dimensional universe. It is as difficult for you to imagine curvature in our three-dimensional universe as it is for ants to imagine curvature in their two-dimensional universes.

2b There are three possible ways our universe might be curved. It could have positive curvature, analogous to the surface of a sphere. Models of such universes are called **closed universes** because they are finite. Of course, ours could be a **flat universe** with zero curvature. Another possibility is that our universe has negative curvature, analogous to a two-dimensional universe shaped like a saddle or a potato chip. Models of these universes are called **open universes**.

r

Aha!

A positively curved universe, also called a closed universe, is finite and has no edge. In the two-dimensional analogy, ants on an orange would notice that the areas of large circles were less than πr^2.

A flat universe must be infinite or it will have an edge. In a two-dimensional analogy, ants on a flat sheet of paper would find that all circles have areas of πr^2.

Hey! C'mere!

r

3 Although our ant might be unable to sense a third dimension, it could still measure the curvature of its universe by drawing circles. On a flat universe, the area of a circle would always be πr^2 no matter how big the circle was. But on a spherical surface, large circles would contain less than πr^2. On a saddle-shaped surface, large circles would contain more than πr^2. In the same way, you could detect the curvature of our three-dimensional universe, but you would have to make measurements over very large distances.

Whether our universe is closed, flat, or open, it cannot have an edge or a center.

What?

r

A negatively curved universe, also called an open universe, must be infinite or it will have an edge. In a two-dimensional analogy, ants on a saddle shape will discover that large circles have areas greater than πr^2.

This is not a small issue. Judging by their gravitational fields, galaxies and clusters of galaxies contain as much as 10 times more matter than you would expect from what you see. To correct for this dark matter, you would not just add a small percentage. You have to multiply by a *factor* as large as 10.

Looking at the universe of visible galaxies is like looking at a ham sandwich and seeing only the mayonnaise. Most of the universe is invisible, and most astronomers now believe that the invisible matter is not the normal matter of which you and the stars are made. To follow that line of evidence, you must think again about the atoms made during the big bang.

During the first few minutes of the big bang, nuclear reactions converted some protons into helium and a small amount into other elements. How much of these elements was created depends critically on the density of the material. Deuterium, for example, is an isotope of hydrogen in which the nucleus contains a proton and a neutron. The amount of deuterium produced in the big bang depends strongly on the density of normal matter. If there were a lot of normal particles such as protons and neutrons, then they would have collided with the deuterium nuclei and converted them into helium. If the density of normal particles was less, more deuterium would have survived.

Lithium is another nucleus that could have been made in small amounts during the big bang. Figure 18-9 shows that there is a gap between helium and lithium; there is no stable nucleus with atomic mass 5, so regular nuclear reactions during the big bang could not convert helium into lithium. If, however, the density of normal matter such as protons and neutrons was high enough, a few nuclear reactions could have leaped the gap and produced a few atoms of the isotope lithium-7.

Deuterium is so easily converted into helium that none can be made in stars. In fact, stars destroy what deuterium they have by converting it into helium. Lithium too is destroyed in stars. Using the largest telescopes, astronomers have been able to measure the chemical abundance of gas clouds near quasars. The look-back times to these gas clouds is so great that they appear as they were before stars could have altered the abundance of the elements. As shown in ■ Figure 18-15, the observed amount of deuterium sets a lower limit on the density of the universe, and the observed abundance of lithium-7 sets an upper limit. The normal matter that you and the stars are made of cannot make up much more than 4 percent of the critical density. Yet observations show that galaxies and galaxy clusters contain large amounts of dark matter. The protons and neutrons that make up normal matter belong to a family of subatomic particles called *baryons,* so cosmologists believe that the dark matter cannot be baryons. Only a small amount of the mass in the universe can be baryonic; the dark matter must be **nonbaryonic matter.**

Theorists have suggested that the dark matter is made of weakly interacting massive particles (WIMPS), which could be hundreds of times more massive than a proton. If WIMPs almost never interact with other particles, they would be almost un-

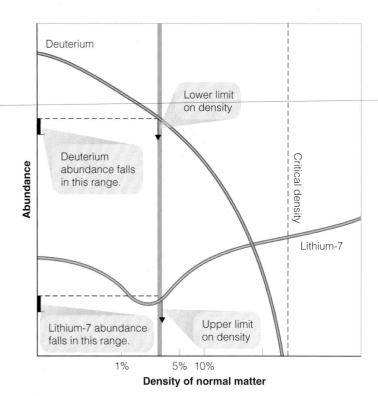

■ **Figure 18-15**

This diagram compares observation with theory. Theory predicts how much deuterium and lithium-7 you would observe for different densities of normal matter (red and blue curves). The observed density of deuterium falls in a narrow range shown at upper left and sets a lower limit on the possible density of normal matter. The observed density of lithium-7 sets an upper limit. This means the true density of normal matter must fall in a narrow range represented by the green column. Certainly, the density of normal matter is much less than the critical density.

detectable except for their gravitational influence. There is, however, no conclusive evidence that WIMPs actually exist. A controversial experiment has hinted that a WIMP called a neutralino may have been detected in a laboratory. If physicists can establish that WIMPs actually exist and measure their density in the universe, it may help cosmologists explain the dark matter.

For some years, astronomers thought that neutrinos could be an important part of the dark matter. You saw in Chapter 8 how observations made by underground detectors suggested that neutrinos coming from the sun oscillate, and that is important because, according to quantum mechanics, if neutrinos oscillate, they cannot have zero mass. Although the mass of neutrinos has not been measured, it is clearly very small. Even a small mass might be important because there are so many neutrinos in the universe—10^8 neutrinos for every normal particle.

Theoretical models of galaxy formation in the early universe show that the dark matter can't be mostly neutrinos. Dark matter composed of neutrinos and similar particles that travel at or near the speed of light is called **hot dark matter.** Such fast-moving particles do not clump together easily and could not have stimulated the formation of objects as small as galaxies and clusters of galaxies. The most successful models of galaxy formation require that the dark matter be made up of **cold dark matter,** meaning the particles move slowly and can clump into smaller structures. WIMPs, for example, are massive and slow moving and, if they exist, could clump together in the early universe and help pull matter into galaxy-size clouds.

Nonbaryonic dark matter does not interact significantly with normal matter or with photons, which is why you can't see it. But that means that dark matter was not affected by the intense radiation that dominated the universe when it was very young. That radiation prevented normal matter from contracting to begin forming galaxies. But the dark matter was immune, and it could contract to form clouds. Once the density of radiation fell low enough, normal matter could begin falling into the clouds of dark matter to form the first galaxies. Dark matter could have given galaxy formation a head start soon after the big bang. Models with cold, nonbaryonic dark matter are most successful at forming galaxies and clusters of galaxies early in the history of the universe.

Now that you understand more about dark matter, you can evaluate the possibility that the halos of galaxies contain large numbers of very-low-mass stars, brown dwarfs, or planets too faint to see. These **MACHOs** (for Massive Compact Halo Objects) can be detected if they pass between telescopes on Earth and a distant star. The gravitational field of the MACHO acts as a small gravitational lens, focusing the light of the star and making it grow brighter for a period of a few weeks as the MACHO passes between the star and Earth. Extensive searches have detected such events (■ Figure 18-16) but not in large numbers. Furthermore, other searches have turned up lots of white dwarfs in the halo of our galaxy. However common these halo objects are, they are made of baryonic matter, and the abundance of deuterium and lithium shows that the dark matter must be nonbaryonic. It does not appear that MACHOs and white dwarfs can make up a significant part of the dark matter.

Although there is good evidence that dark matter exists, it has proven difficult to identify. No form of dark matter has been found that is abundant enough to provide the critical density needed to make the universe flat. In fact, all forms of dark matter appear to add up to less than 30 percent of the critical density. As you will see later in this chapter, there is more to the universe than meets the eye, and more even than the dark matter.

Building Scientific Arguments

Why do astronomers think that dark matter can't be baryonic?
Good scientific arguments always fall back on evidence. In this case the evidence is very strong. Small amounts of isotopes like deuterium and lithium-7 were produced in the first minutes of the big bang, and the abundance of those elements depends strongly on the density of protons and neutrons. Because these particles belong to the family of particles called baryons,

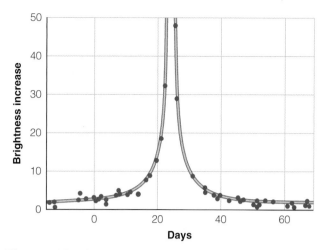

■ **Figure 18-16**

Measurements of the brightness of a distant star (dots) increased and then decreased exactly as predicted by general relativity (curve) for a star's passage behind a gravitational lens caused by a MACHO. In this example, the star brightened by a factor of over 40.

astronomers refer to normal matter as baryonic. Measurements of the abundance of deuterium and lithium-7 show that the universe cannot contain more baryons than about 4 percent of the critical density. Yet observations of galaxies and galaxy clusters show that dark matter must make up almost 30 percent of the critical density. Consequently, astronomers conclude that the dark matter must be made up of nonbaryonic particles.

Finding the dark matter is important because the density of matter in the universe determines its curvature. Expand your argument to discuss theory. **How does curvature allow you to avoid an edge or center in a finite model of the universe?**

■   ■   ■

Connections: By adding the curvature of space-time and effects of dark matter to your cosmology, you have made your theories much more sophisticated. You are now ready to explore the most recent and most exciting advances in cosmology.

18-3 21st-Century Cosmology

IF YOU ARE A LITTLE DIZZY from the weirdness of curved space-time and dark matter, make sure you are sitting down before you read much further. As the 21st century began, cosmologists made a startling discovery, and all around the world astronomers looked at each other and said, "What? What!" The most amazing thing about these amazing discoveries is that they fit so well with some of the things you have been learning in this chapter. To get a running start on these new discoveries, you'll have to go back a couple of decades.

Inflation

In 1980, astronomers faced a problem. The big bang models could not explain two important features of the universe. To solve those two problems, a new theory of the big bang was created, and that theory has been startlingly successful. To introduce the new theory, you can begin with the two problems.

One of the problems is called the **flatness problem.** The universe seems to be balanced near the boundary between an open and a closed universe. That is, it seems nearly flat. Given the vast range of possibilities, from zero to infinite, it seems peculiar that the density of the universe is within a factor of 10 of the critical density that would make it flat. If dark matter is as common as it seems, the density may be even closer than a factor of three to being perfectly flat.

Even a small departure from critical density when the universe was young would be magnified by subsequent expansion. To be so near critical density now, the density of the universe during its first moments must have been within 1 part in 10^{49} of the critical density. So the flatness problem is: Why is the universe so nearly flat?

The second problem with the big bang theory is the isotropy of the primordial microwave background radiation. When astronomers correct for the motion of Earth, they see the same background radiation in all directions to at least 1 part in 1000. Yet when you look at background radiation coming from two points in the sky separated by more than a degree, you look at two parts of the big bang that were not causally connected when the radiation was emitted. That is, when recombination occurred and the gas of the big bang became transparent to the radiation, the universe was not old enough for any signal to have traveled from one of these regions to the other. Thus, the two spots you look at did not have time to exchange heat and even out their temperatures. Then how did every part of the entire big bang universe get to be so nearly the same temperature by the time of recombination? This is called the **horizon problem** because the two spots are said to lie beyond their respective light-travel horizons.

The key to these two problems and to others involving subatomic physics may lie with the theory called the **inflationary universe** because it predicts a sudden inflation when the universe was very young, a violent expansion even more extreme than that predicted by the big bang theory.

To understand the inflationary universe, you need to recall that physicists know of only four forces—gravity, the electromagnetic force, the strong force, and the weak force (Chapter 8). You are familiar with gravity, and the electromagnetic force is responsible for making magnets stick to refrigerator doors and cat hair stick to wool sweaters charged with static electricity. The strong force holds atomic nuclei together, and the weak force is involved in certain kinds of radioactive decay.

For many years, theorists have tried to unify these forces; that is, they have tried to describe the forces with a single mathematical law. A century ago, James Clerk Maxwell showed that the electric force and the magnetic force were really the same effect, and physicists now count them as a single electromagnetic force. In the 1960s, theorists succeeded in unifying the electromagnetic force and the weak force in what they called the electroweak force, effective only for processes at very high energy. At lower energies, the electromagnetic force and the weak force behave differently. Now theorists have found ways of unifying the electroweak force and the strong force at even higher energies. These new theories are called **grand unified theories,** or **GUTs.**

According to the inflationary universe, the universe expanded and cooled until about 10^{-35} second after the big bang, when it became so cool that the electroweak force and the strong force began to disconnect from each other and behave in different ways. This released tremendous energy, which suddenly inflated the universe by a factor between 10^{20} and 10^{30} (■ Figure 18-17). At that time the part of the universe that is now visible from Earth, the entire observable universe, was no larger than the volume of an atom, but it suddenly inflated to the volume of a cherry pit and then continued its slower expansion to its present extent.

That sudden inflation can solve the flatness problem and the horizon problem. The sudden inflation of the universe would have forced whatever curvature it had toward zero, just as inflating a balloon makes a small spot on its surface flatter. Consequently you now live in a universe that is almost perfectly flat because of

■ **Figure 18-17**

When the universe was very young and hot (top), the four forces of nature were indistinguishable. As the universe began to expand and cool, the forces separated and triggered a sudden inflation in the size of the universe.

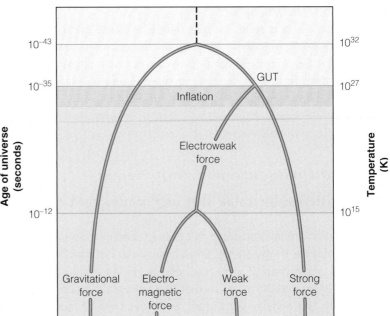

that sudden inflation long ago. In addition, because the observable part of the universe was once no larger in volume than an atom, it had plenty of time to equalize its temperature before inflation occurred. Now you live in a universe with the same temperature for the background radiation in all directions.

The inflationary universe is based, in part, on quantum mechanics, and a slightly different aspect of quantum mechanics may explain why there was a big bang at all. Theorists believe that a universe totally empty of matter could be unstable and decay spontaneously by creating pairs of particles until it was filled with the hot, dense state called the big bang. This theoretical discovery has led some cosmologists to believe that the universe could have been created by a chance fluctuation in space-time. In the words of physicist Frank Wilczyk, "The reason there is something instead of nothing is that 'nothing' is unstable."

The inflationary theory predicts that the universe is almost perfectly flat. That is, the true density must equal the critical density. A theory can never be used as evidence, but the beauty of the inflationary theory has given many cosmologists confidence that the universe must be flat. Observations, however, seem to say that the universe does not contain enough matter (baryonic plus dark) to be flat. Can there be more to the universe than baryonic matter and dark matter? What could be weirder than dark matter? Read on.

The Acceleration of the Universe

Ever since Hubble discovered the expansion of the universe, astronomers have known what to expect. They have also known that expectations are not reliable; and, as the 21st century approached, astronomers made a mind-boggling discovery that contradicted everyone's expectations.

Both common sense and mathematical models suggest that as the galaxies recede from each other, the expansion should be slowed by gravity trying to pull the galaxies toward each other. How much the expansion is slowed should depend on the amount of matter in the universe. If the density of matter is less than the critical density, the expansion should be slowed only slightly, and the universe should expand forever. If the density of matter in the universe is greater than the critical density, the expansion should be slowing down dramatically, and the universe should eventually stop expanding and begin contracting. Notice that this is the same as saying a low-density universe should be open and a high-enough-density universe should be closed.

For decades, astronomers struggled to measure the Hubble constant and detect the slowing of the expansion. A direct measurement of the rate of slowing would reveal the true curvature of the universe. This was one of the key projects for the Hubble Space Telescope, and two teams of astronomers spent years making the measurements.

Detecting a change in the rate of expansion is a difficult project because it requires accurate measurements of the distances to very remote galaxies, and both teams used the same technique.

They calibrated type Ia supernovae as distance indicators. A type Ia supernova occurs when a white dwarf gains matter from a companion star, exceeds the Chandrasekhar limit, and collapses in a supernova explosion. Because all such white dwarfs should collapse at the same mass limit, they should all produce explosions of the same luminosity, and that makes them good distance indicators.

In Chapter 16, you read that astronomers refer to distance indicators such as Cepheid variable stars as *standard candles,* objects of known luminosity. Type Ia supernovae have been described as *standard bombs.* They are titanic explosions, but they all reach the same peak brightness, and that makes them good distance indicators.

The two teams calibrated type Ia supernovae by locating such supernovae occurring in nearby galaxies whose distance was known from Cepheid variables and other reliable distance indicators. Once the peak luminosity of type Ia supernovae had been determined, they could be used to find the distance to much more distant galaxies.

Both teams announced their results in 1998. The expansion of the universe is not slowing down. It is speeding up! That is, the expansion of the universe is accelerating (■ Figure 18-18).

The announcement that the expansion of the universe is accelerating made astronomers stop and stare at each other. It was totally unexpected. It was simultaneously an exciting, puzzling, and revealing discovery, but astronomers immediately began testing it. It depends critically on the calibration of type Ia supernovae as distance indicators, and some astronomers suggested that the calibration might be wrong (Window on Science 15-1). Some unknown kind of dust might make very distant supernovae too faint without making them too red. No reddening had been detected, so normal dust seemed to be ruled out. Or perhaps the very distant supernovae were too faint because astronomers see them at great look-back times, and some unknown process made those early supernovae a bit less luminous than more recent supernovae.

Those problems with the calibration are ruled out by more recently discovered supernovae. For example, SN1997ff, recognized in 2001 as a type Ia supernova, is too faint for its redshift of 1.7 and confirms acceleration (■ Figure 18-19). More type Ia supernovae discovered by the Hubble Space Telescope rule out problems with the calibration. The universe really does seem to be expanding faster and faster.

Dark Energy and Acceleration

If the expansion of the universe is accelerating, then there must be a force of repulsion in the universe, and astronomers are struggling to understand what it could be. One possibility leads back to 1916.

When Albert Einstein published his theory of general relativity in 1916, he recognized that his equations describing space and time implied that space had to contract or expand. The galaxies could not float unmoving in space because their gravity would pull them toward each other. The only solutions seemed to be a

From the way a supernova fades over time, astronomers could determine which were type Ia. Once calibrated, those supernovae could be compared with their redshift, revealing that distant type Ia supernovae were about 25 percent fainter than expected. That must mean they are farther away than expected given their redshifts. This is strong evidence that the universe is accelerating. (ESO)

11 March 1997

Host galaxy

Supernova

16 March 1997

29 March 1997

Fading supernova

Visual-wavelength images

Flat decelerating universe

Flat accelerating universe

Observations of type Ia supernova fit an accelerating universe but not a decelerating universe.

Calibrated apparent magnitude

Redshift (z)

Visual-wavelength images.

Difference: 1997-1995

universe that was contracting under the influence of gravity or a universe in which the galaxies were rushing away from each other so rapidly that gravity could not pull them together. In 1916, astronomers did not yet know that the universe was expanding, so Einstein made what he later said was a mistake.

To balance the attractive force of gravity, Einstein added a constant to his equations called the **cosmological constant,** represented by an upper-case lambda (Λ). The constant plays the mathematical role of a force of repulsion that balances the gravitation in the universe so it does not have to contract or expand. Thirteen years later, in 1929, Edwin Hubble announced that the universe was expanding, and Einstein said introducing the cosmological constant was his biggest blunder. Modern astronomers aren't so sure.

■ Figure 18-19

In a 1997 follow-up image of the Hubble Deep Field made in 1995, astronomers noticed a faint galaxy that was brighter in the second image (lower left). Subtracting the earlier image from the later image revealed a supernova cataloged as SN1997ff (lower right). The unexpected faintness of the supernova confirms that the universe is accelerating. (Adam Riess, STScI/NASA)

One explanation for the acceleration of the universe is that the cosmological constant is not zero but rather represents a force that drives a continuing acceleration in the expansion of the universe. Because the cosmological constant remains constant with time, the universe would have to have experienced this acceleration throughout its history.

Another solution is to suppose that totally empty space, the vacuum, contains energy, which drives the acceleration. This is an interesting possibility because for years theoretical physicists have discussed an energy inherent in empty space. Astronomers have begun referring to this energy of the vacuum as **quintessence.** Unlike the cosmological constant, quintessence need not remain constant over time, so it may be more successful in explaining the observed distribution of galaxies and galaxy clusters.

Whichever explanation is right, the cosmological constant or quintessence, the acceleration is evidence that some form of energy is spread throughout space. Astronomers refer to this energy as **dark energy,** the energy that drives the acceleration of the universe but does not contribute to the formation of starlight or the cosmic microwave background radiation.

You will recall that acceleration and dark energy were first discovered when astronomers found that supernovae just a few billion light-years away were slightly fainter than expected. The acceleration of the expansion made those supernovae a bit farther away than expected, and so they looked fainter. Since then, astronomers have continued to find even more distant type Ia supernovae, some as distant as 12 billion light-years. The more distant of those supernovae are not too faint; they are too bright! That reveals even more about dark energy.

The very distant supernovae are a bit too bright because they are not as far away as expected, and that confirms a theoretical prediction based on dark energy. When the universe was young, galaxies were closer together, and their gravitational pull on each other could overpower dark energy and slow the expansion. That makes the very distant supernovae a bit too bright. As the universe expanded, galaxies moved farther apart, their gravitational pull on each other became weaker than dark energy, and acceleration began. That makes the less distant supernovae a bit too faint. Sometime about 6 billion years ago, the universe shifted gears from deceleration to acceleration, and supernovae are making that change visible.

Furthermore, dark energy can help you understand the curvature of the universe. The theory of inflation makes the specific prediction that the universe is flat. Dark energy seems to confirm that prediction. According to Einstein's most famous equation, $E = mc^2$, you know that energy and matter are equivalent. Then the dark energy is equivalent to a mass spread through space. Baryonic matter plus dark matter makes up about a third of the critical density, and dark energy appears to make up two-thirds. That is, the total density of the universe seems to equal the critical density, which means that the geometry of the universe is flat.

Step-by-step you have been climbing the cosmological pyramid. Each step has been small and logical, but look where it has led you. You know some of the deepest secrets of the universe, but there are more steps above and more secrets to explore.

The Age and Fate of the Universe

Acceleration helps with another problem. The Hubble constant equals 70 km/s/Mpc, and earlier in this chapter you calculated the Hubble time and found it was about 14 billion years. Further, you calculated the age of the universe on the assumption that it was flat as two-thirds of 14 billion years and got an answer of about 9 billion years. That was a problem because the globular clusters are older than that. Now you are ready to solve that age problem.

If the expansion of the universe has been accelerating, then it must have been expanding slower in the past, and that means it can be older than two-thirds of the Hubble time. The latest estimates suggest that acceleration increases the age of a flat universe from 9 billion years back up to about 14 billion years, clearly older than the oldest known star clusters, and that solves the age problem.

For many years cosmologists have enjoyed saying, "Geometry is destiny." By that they meant that the destiny of the universe is determined entirely by its geometry. An open universe must expand forever, and a closed universe must fall back. But that is true only if the universe is ruled by gravity. If acceleration dominates gravity, then geometry is not destiny, and even a closed universe might expand forever.

The ultimate fate of the universe depends on the nature of dark energy. If dark energy is described by the cosmological constant, then the force driving acceleration does not change with time, and our flat universe will expand forever with the galaxies getting farther and farther apart and using up their gas and dust making stars and the stars dying until each galaxy is isolated, burnt out, dark and alone. If, however, dark energy is described by quintessence, then it may be increasing with time, and the universe may accelerate faster and faster as space pulls the galaxies away from each other and eventually pulls the galaxies apart and then pulls the stars apart and finally rips individual atoms apart. This has been called the **big rip.** Don't worry. Even if a big rip is in our future, nothing will be happening for at least 30 billion years.

Maybe there will be no big rip. Critically important observations made by the Chandra X-Ray Observatory have been used to measure the amount of hot gas and dark matter in over two dozen galaxy clusters. Because distance is involved in the calculations, the X-ray astronomers could compare the amount of hot gas and dark matter and solve for distance. The most distant of these clusters was 8 billion light-years away. The observations are critical for two reasons. First, the redshifts and distance of these galaxies confirm that the universe was slowing down early on but that it shifted gears about 6 billion years ago and is now accelerating. Notice that this method does not depend on type Ia supernovae at all, so it is independent confirmation that acceleration is real. The results are also critical because they are almost good enough to rule out quintessence. If dark energy is described by the cosmological constant and not by quintessence, then there will be no big rip (■ Figure 18-20).

Do you believe this stuff? Or, rather, does the evidence give you confidence that these theories correctly describe the universe? Science is not a system of belief, but rather it is a system of knowledge based on evidence. Scientists work by building confidence in theories, and astronomers like to say, "Extraordinary claims require extraordinary evidence." Faced with the amazing discovery of acceleration, astronomers began searching for other ways to confirm that the universe is flat and accelerating.

One way to measure the curvature of space-time is to draw circles on it and measure the areas of the circles. But that is only practical in a simple cartoon analogy. In reality, astronomers must measure the size of things at great look-back times, and that measurement is yielding yet more exciting results, which are described in the next section.

The Origin of Structure and the Curvature of the Universe

For a number of years, astronomers have studied the distribution of galaxies in space, what they refer to as **large-scale structure.** Those studies have confirmed the modern theories of cosmology and led to astonishing insights that are changing cosmology.

When you look at galaxies in the sky you see them in clusters ranging from a few galaxies to thousands, and those clusters appear to be grouped into **superclusters** (■ Figure 18-21). The Local Supercluster, in which we live, is a roughly disk-shaped swarm of galaxy clusters 50 to 75 Mpc in diameter. By measuring the redshifts and positions of over 100,000 galaxies in great slices across the sky, astronomers have been able to create maps revealing that the superclusters are not scattered at random. They are distributed in long, narrow filaments and thin walls that outline great voids nearly empty of galaxies (■ Figure 18-22). These filaments and walls of superclusters stretching half a billion light-years are the largest structures in the universe. No larger objects exist.

These vast structures are a problem for cosmologists because the cosmic microwave background radiation is very uniform, and that means the gas of the big bang must have been extremely uniform at the time of recombination. Yet the look-back time to the furthest galaxies is about 93 percent of the way back to the big bang. How did the uniform gas at the time of recombination coagulate so quickly to form galaxies? Worse, how did that highly uniform gas form such large structures as the filaments, voids, and walls?

Baryonic matter is so rare in the universe it does not have enough gravity to pull itself together rapidly. Astronomers believe the nonbaryonic dark matter must provide the gravity, and models have attempted to describe how dark matter can form structure. As you have seen earlier in this chapter, hot dark matter does not clump into small enough structures, so the dark matter must be cold dark matter (■ Figure 18-23).

■ Figure 18-20

X-ray observations of hot gas in galaxy clusters confirm that in its early history the universe was decelerating because gravity was stronger than the dark energy. As expansion weakened the influence of gravity, dark energy began to force an acceleration. The evidence is not conclusive but it most directly supports the cosmological constant and weighs against quintessence, which means the universe may not face a big rip. This diagram is only schematic and the two curves are drawn separated for clarity; at the present the two curves have not diverged from each other. (NASA/CXC/IoA/S. Allen et al.)

The distribution of brighter galaxies in the sky reveals the great Virgo cluster (center), containing over 1000 galaxies only about 17 Mpc away. Other clusters fill the sky, such as the more distant Coma cluster just above the Virgo cluster in this diagram. The Virgo cluster is linked with others to form the Local Supercluster.

But what started the clumping of matter? Theorists say that subatomic quantum fluctuations in space would have been stretched at the moment of inflation to very large but very subtle variations in gravitational fields that could have stimulated the formation of clusters, filaments, and walls. The structure you see in Figure 18-22 may be the ghostly traces of quantum fluctuations in the infant universe.

The Two-Degree-Field (2df) Redshift Survey mapped the position and redshift of 250,000 galaxies and 30,000 quasars. As expected, the galaxies are spread in filaments and walls, but a careful statistical analysis of the clumpiness of the distribution confirms that the universe is accelerating. This is an important result because it is yet another confirmation that is independent of the brightness of type Ia supernovae or X-ray emitting gas in galaxy

■ **Figure 18-22**

Nearly 70,000 galaxies are plotted in this double slice of the universe extending outward in the plane of Earth's equator. The nearest galaxies are shown in red and the more distant in green and blue. The galaxies form filaments and walls enclosing empty voids. The Sloan Great Wall is almost 1.4 billion light-years long and is the largest known structure in the universe. The most distant galaxies in this diagram are roughly 3 billion light-years from Earth. (Sloan Digital Sky Survey)

Growth of Structure in the Universe

Soon after the big bang, radiation and hot gas are almost uniformly spread through the universe.

Cold dark matter, immune to the influence of light, can contract to form clouds…

which pull in normal gas to form superclusters of galaxies. Gravity continues to pull clusters together.

Statistical tests show the distribution in this model universe resembles the observed distribution of galaxies.

■ **Figure 18-23**

This computer model traces the formation of structure in the universe from soon after the big bang to the present. (Adapted from a model by Kauffmann, Colberg, Diaferio, & White: Max-Planck Institut für Astronomie)

clusters. When a theory is confirmed by observations of many different types, scientists have much more confidence that it is a true description of nature.

The inflationary theory of the universe makes very specific predictions about the sizes of the fluctuations an observer on Earth should see in the cosmic microwave background radiation. Observations made by the COBE satellite in 1992 detected the largest variations, but detecting the smaller variations is critical for testing the theory. A half-dozen teams of astronomers have built specialized telescopes to make these observations. Some have flown under

balloons high in the atmosphere, and others have observed from the ice of Antarctica. The most extensive observations have been made by the Wilkinson Microwave Anisotropy Probe (WMAP), a robotic infrared telescope that observed from space.

The background radiation is very isotropic—it looks almost exactly the same in all directions. However, when the average intensity is subtracted from each spot on the sky, small irregularities are evident. That is, some spots on the sky look a tiny bit hotter and brighter than other spots (■ Figure 18-24). Those irregularities contain lots of information.

If the irregularities in the cosmic microwave background radiation were produced by inflation, then their size must depend on the velocity of the pressure waves (sound) that could travel through the gas at the time of recombination. Theory predicts most of the irregularities should be about one degree in diameter if the universe is flat. If the universe were open, the most common irregularities would be smaller. Careful measurements of the size of the irregularities in the cosmic background radiation show that the observations fit the theory very well for a flat universe (■ Figure 18-25). Not only is the theory of inflation confirmed, an exciting result itself, but these data confirm that the universe is flat, which indirectly confirms the existence of dark energy and the acceleration of the universe.

Cosmologists can analyze the irregularities using sophisticated mathematics to find out how commonly the spots of different sizes recur. The mathematics confirm that spots about 1 degree in diameter are the most common, but spots of other sizes recur as well, and it is possible to plot a graph such as ■ Figure 18-26 to show how common different size irregularities are.

The results from the WMAP observations tell cosmologists a great deal about the universe. The universe is flat, accelerating, and will expand forever. The age of the universe derived from the data is 13.7 billion years. Furthermore, the smaller peaks in the curve reveal that the universe contains 4 percent baryonic (normal) matter, 23 percent dark matter, and 73 percent dark energy. The Hubble constant is confirmed to be 71 km/s/Mpc. The inflationary theory is confirmed, and the data support the cosmological constant version of dark energy, although quintessence is not ruled out. Hot dark matter is ruled out. The dark matter needs to be cold dark matter to clump together so rapidly after the big bang. In fact, the first stars began to produce light when the universe was only about 200 million years old. This is much earlier than most astronomers had expected.

Please reread the preceding paragraph. It is mind-blastingly amazing. WMAP and other studies of the cosmic microwave background radiation have revolutionized cosmology. At last, astronomers have accurate observations against which to test theories. The basic constants are known to a percent or so.

On reviewing these results, one cosmologist announced that "Cosmology is solved!" but that may be premature. Scientists don't understand dark matter or the dark energy that drives the

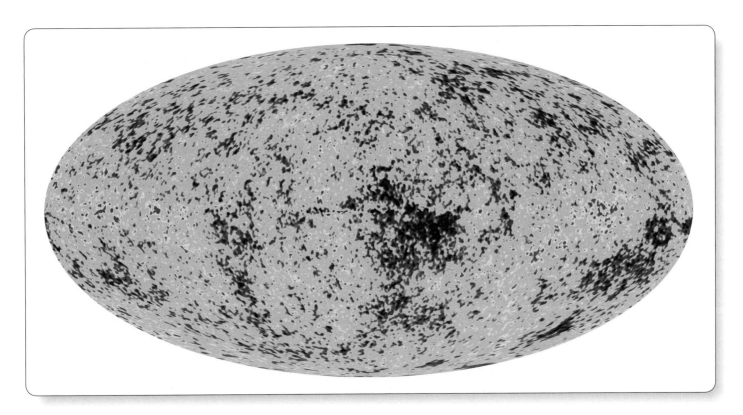

■ Figure 18-24

Data from the WMAP spacecraft were used to make this far-infrared map of the entire sky. The infrared glow from dust in our solar system and in the Milky Way Galaxy has been removed to reveal tiny irregularities in the cosmic microwave background radiation. Red and yellow spots are slightly warmer than the green and blue spots. (NASA/WMAP Science Team)

Simulated data
Closed model universe
Larger spots

Simulated data
Flat model universe
Spots about 1 degree

Simulated data
Open model universe
Smaller spots

Observational data

■ Figure 18-25

You can see the difference yourself. Compare the observations of the irregularities in the background radiation at right with the three simulations at left. The observed size of the irregularities fits best with cosmological models having flat geometry. Detailed mathematical analysis confirms your visual impression. The universe is flat. (Courtesy of the BOOMERANG Collaboration.)

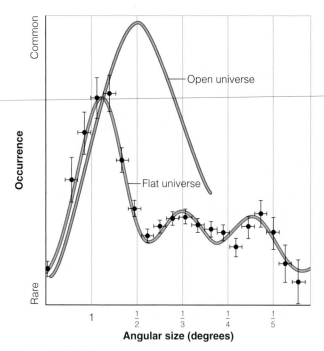

■ Figure 18-26

This graph shows how commonly irregularities of different sizes occur in the cosmic microwave background radiation. Irregularities of about 1 degree in diameter are most common. Models of the universe that are open or closed are ruled out. The data fit a flat model of the universe very well. Crosses on data points show the uncertainty in the measurements.

acceleration, so over 95 percent of the universe is not understood. Hearing this, another astronomer suggested a better phrase was "cosmology in crisis." Certainly there are further mysteries to be explored, but cosmologists are growing more confident that they can describe the overall properties of the universe.

How does inflation theory solve the flatness problem?
This question requires a carefully constructed argument. The flatness problem can be stated as a question: Why is the universe so nearly flat? After all, the density of matter in the universe could be anything from zero to infinite, but the observed density of baryonic and dark matter adds up to about 27 percent of the critical density that would produce a flat universe. Furthermore, the density must have been astonishingly close to the critical density when the universe was very young. Otherwise it would not be so close now. The inflationary theory solves this problem by proposing that the universe underwent a moment of rapid inflation when it was a tiny fraction of a second old. That inflation drove the universe toward flatness just as blowing up a balloon makes a spot on the balloon nearly flat.

Understanding theory in cosmology is critically important, but science depends on evidence. Now build an argument based on evidence. **What evidence can you cite that the expansion of the universe is accelerating?**

■　　■　　■

Connections: As you climbed the cosmology pyramid, you negotiated a lot of steps. Most were easy and all were logical. They have carried you up to some sweeping insights, and you have traced the origin of the universe, the origin of the elements, the birth of galaxies, and the birth and death of stars. Now it is time for you to add one final class of objects—planets. How do Earth and other planets fit into this grand scheme of origins? That is the subject of the following chapters.

18-1 | Introduction to the Universe

Does the universe have an edge in space or time?

- Astronomers conclude that it is impossible for the universe to have an edge because it introduces logical inconsistencies. That is, an edge to the universe does not make sense.

- If the universe has no edge, then it cannot have a center.

- The darkness of the night sky leads to the conclusion that the universe is not infinitely old. If it were infinite in extent and age, then every spot on the sky would glow as brightly as the surface of a star. This problem, known incorrectly as Olbers's paradox, leads to the conclusion that the universe had a beginning.

What evidence exists that the universe began with a big bang?

- Edwin Hubble's 1929 discovery that the redshift of a galaxy is proportional to its distance is known as the Hubble law. Tracing this expansion backward in time brings you to an initial high-density, high-temperature state commonly called the big bang.

- Although the expanding universe began from this big bang, it has no center. The galaxies do not recede from a single point. They recede from each other.

- The cosmic microwave background radiation is black body radiation with a temperature of about 2.73 K uniformly spread over the entire sky. It is the light from the big bang freed from the gas at the moment of recombination and redshifted by a factor of 1100.

- The background radiation is clear evidence that the universe began with a big bang.

- During the first three minutes of the big bang, nuclear fusion converted some of the hydrogen into helium but was unable to make many other heavy atoms because no stable nuclei exist with weights of 5 or 8. Today hydrogen and helium are common in the universe, but heavier atoms are rare.

18-2 | The Shape of Space and Time

How can the universe expand if it has no edge?

- The geometry of space and time is critical to understanding cosmology. The universe appears to be isotropic and homogeneous. That is, in its major features, the universe looks the same in all directions and in all locations.

- Isotropy and homogeneity lead to the cosmological principle, the assumption that there are no special places in the universe. Except for local differences, every place is the same.

- General relativity explains that the cosmic redshift is caused by the stretching of photons as they travel through expanding space-time.

- If the universe is infinite (open), then it has no edge and no center, and space-time can expand and carry the galaxies away from each other.

- If the universe is finite (closed) but cannot have an edge, then it must be curved back on itself. Then space-time can expand and carry the galaxies away from each other.

- Whether the universe is open, closed, or flat depends on its density. If the density of the universe is less than the critical density, it is open. If the density is more than the critical density, it is closed. If the density equals the critical density, the universe is flat—balanced between open and closed.

- The amount of deuterium and lithium-7 shows that normal baryonic matter can make up only about 4 percent of the critical density. Dark matter appears to make up less than 30 percent.

18-3 | 21st-Century Cosmology

Why is the universe the way it is?

- The inflationary theory proposes that the universe expanded dramatically when it was a tiny fraction of a second old.

- The flatness problem is explained by inflation because the sudden inflation forced the universe to become flat, just as a spot on an inflating balloon becomes flatter as the balloon inflates.

- The horizon problem is explained by inflation because the observable part of the universe was so small before inflation that energy could move and equalize the temperature everywhere.

- Observations of type Ia supernovae reveal that the expansion of the universe is speeding up because of energy present in empty space as dark energy.

- Dark energy added to dark matter and baryonic matter makes the observed density of the universe equal to the critical density and confirms the prediction made by inflation that the universe is flat.

- The nature of dark energy is unknown. It may be described by Einstein's cosmological constant, or it may change with time and be better called quintessence.

- Corrected for inflation, the observed value of the Hubble constant implies that the universe is about 14 billion years old. The future fate of the universe depends on the nature of dark energy.

- The sudden inflation of the universe is believed to have magnified tiny quantum mechanical fluctuations in space-time. These very large but very weak differences in gravity worked with dark matter to draw together baryonic matter and create the clusters of galaxies, filaments, and voids that astronomers call large-scale structure.

- Statistical observations of the large-scale structure of the universe confirm that it is flat and contains 4 percent baryonic matter, 23 percent dark matter, and 73 percent dark energy.

New Terms

cosmology (p. 416)

Olbers's paradox (p. 419)

observable universe (p. 420)

big bang (p. 422)

Hubble time (p. 422)

cosmic microwave background radiation (p. 425)

steady-state theory (p. 425)

antimatter (p. 425)

recombination (p. 426)

dark age (p. 426)

reionization (p. 426)

isotropy (p. 429)

homogeneity (p. 429)

cosmological principle (p. 429)

critical density (p. 430)

oscillating universe theory (p. 430)

closed universe (p. 433)

flat universe (p. 433)

open universe (p. 433)

nonbaryonic matter (p. 434)

hot dark matter (p. 434)

cold dark matter (p. 435)

MACHO (p. 435)

flatness problem (p. 436)

horizon problem (p. 436)

inflationary universe (p. 436)

grand unified theory (GUT) (p. 436)

cosmological constant (Λ) (p. 438)

quintessence (p. 439)

dark energy (p. 439)

big rip (p. 439)

large-scale structure (p. 440)

supercluster (p. 440)

Review Questions

1. How does the darkness of the night sky tell you something important about the universe?

2. How can Earth be located at the center of the observable universe if you accept the Copernican principle?

3. Why can't an open universe have a center? Why can't a closed universe have a center?

4. What evidence shows that the universe is expanding? that it began with a big bang?

5. Why couldn't atomic nuclei exist when the universe was younger than 2 minutes?

6. Why is it difficult to determine the present density of the universe?

7. How does the inflationary universe theory resolve the flatness problem? the horizon problem?

8. If the Hubble constant were really 100 km/s/Mpc, much of what astronomers understand about the evolution of stars and star clusters would have to be wrong. Explain why. (*Hint:* What would the age of the universe be?)

9. Why must the universe have been very uniform during its first million years?

10. What is the difference between hot dark matter and cold dark matter? What difference does it make to cosmology?

11. What evidence shows that the expansion of the universe is accelerating?

12. What evidence can you cite that the universe is flat?

13. Explain why some of the galaxies in this photo have elongated, slightly curved images. What do such observations tell you about the universe?

NASA/ESA

14. The image at the right shows irregularities in the background radiation. Why isn't the background radiation perfectly uniform? What does the size of these irregularities tell you?

Courtesy of BOOMER-ANG Collaboration

Discussion Questions

1. Do you think Copernicus would have accepted the cosmological principle? Why or why not?

2. If you reject any model of the universe that has an edge in space because you can't comprehend such a thing, shouldn't you also reject any model of the universe that has a beginning or an ending? Are those just edges in time, or is there a difference?

Problems

1. Use the data in Figure 18-4 to plot a velocity–distance diagram, find H, and estimate the Hubble time.

2. If a galaxy is 8 Mpc away from Earth and recedes at 456 km/s, what is H? What is the Hubble time? How old would the universe be if it were flat and there were no acceleration?

3. At what wavelength of maximum intensity did the gas of the big bang radiate at the time of recombination? By what factor is that different from the wavelength of maximum of the cosmic microwave background radiation?

4. If the average distance between galaxies is 2 Mpc, and the average mass of a galaxy is 10^{11} solar masses, what is the average density of matter in the universe? (*Hints:* The volume of a sphere is $\frac{4}{3}\pi R^3$, and the mass of the sun is 2×10^{33} g.)

5. Figure 18-13 is based on an assumed Hubble constant of 70 km/s/Mpc. How would you change the diagram to fit a Hubble constant of 50 km/s/Mpc?

6. Hubble's first estimate of the Hubble constant was 530 km/s/Mpc. If his distances were too small by a factor of 7, what answer should he have obtained?

7. What was the maximum age of the universe predicted by Hubble's first estimate of the Hubble constant? (See Problem 6.)

8. If the value of the Hubble constant were found to be 60 km/s/Mpc, what would the Hubble time be? How old would the universe be if it were flat and there were no acceleration? How would acceleration change your answer?

Media Cluster

ACTIVE FIGURES

Cosmic Redshift
Graph your own observations of redshift and distance.

Raisin Bread
You can make observations from any raisin as the dough rises.

ASTRONOMY EXERCISES

Age of the Universe
Take control of the value of the Hubble constant and the age of the oldest quasars to experiment with the age of the universe.

VIRTUAL ASTRONOMY LABS

Lab 17: Evidence for Dark Matter
This lab investigates two phenomena that provide evidence for the existence of dark matter—galactic rotation curves and gravitational lensing by unseen objects.

Lab 20: Fate of the Universe
In this lab, you investigate what modern cosmology has learned about the history and the ultimate fate of the universe.

Lab 18: The Hubble Law
This lab discusses the Hubble law, shows how it can be used to estimate distances to remote objects, and illustrates that the law is a consequence of the expansion of space itself.

Critical Inquiries for the Web

1. What is the latest news concerning acceleration and the flatness of the universe? Search the Web for the most recent observations.

2. The steady-state theory was once a rival cosmology of the big bang. Search for websites that provide information on steady-state cosmology. (Be careful to locate legitimate sites that discuss the theory, rather than sites where individuals use steady-state ideas as part of nonscientific arguments on cosmology.) What were the key predictions of steady-state cosmology? How did observational evidence lead to its decline?

3. Search for the Web pages of large surveys of galaxies, such as the 2dF Deep Field Survey and the Sloan Digital Sky Survey. What are they discovering about the largest and most distant objects in the universe?

4. What is the Wilkinson Microwave Anisotropy Probe? Where is it? What are the latest results?

Exploring *TheSky*

1 Search for galaxy clusters. (*Hint:* Use the **View** menu to turn off everything but galaxies, deep sky objects, constellations, and labels. Zoom in until the field of view is 40° or smaller and the fainter galaxies appear. Search for clusters of galaxies in Ursa Major, Canes Venatici, Lynx, Virgo, and Coma Bereneces.)

2. Locate small clusters of galaxies and compare the brightness of the galaxies with those in large clusters. Can you tell that the smaller clusters tend to be further away, or is the range of galaxy luminosities too great? (*Hint:* Click on a galaxy to find its magnitude.)

3. Can you find galaxies and clusters of galaxies along the Milky Way? Turn on the Milky Way and search for galaxy clusters within its outline.

Go to the Brooks/Cole Astronomy Resource Center (**http:// astronomy.brookscole.com**) for critical thinking exercises, articles, and additional readings from InfoTrac College Edition, Brooks/Cole's online student library.

19 | The Origin of the Solar System

What place is this?

Where are we now?

CARL SANDBURG, *GRASS*

MICROSCOPIC CREATURES
LIVE in the roots of your eyelashes. Don't worry.
Everyone has them, and they are harmless.* They hatch, fight
for survival, mate, lay eggs, and die in the tiny spaces around the roots of your eyelashes without doing any harm. Some
live in renowned places—the eyelashes of a glamorous movie star—but the tiny beasts are not self-aware; they never stop
to say, "Where are we?" Humans are more intelligent; humans have the ability to wonder where we are in the universe
and how we came here. ▮ You should study the solar system for many reasons. You need to understand Earth as a
planet because six billion of us are living on Earth and changing it in ways we don't understand. You should also study

Demodex folliculorum has been found in 97 percent of individuals and is a characteristic of healthy skin.

▮ Continued on page 450 ▮

The human race lives on a planet in a planetary system that appears to have formed in a nebula around the protostar that became the sun. This artist's impression shows the formation of the giant planet Jupiter. (NASA)

Guidepost

Looking Back

You have become an expert on the universe. You have studied the appearance, origin, structure, and evolution of stars, galaxies, and the universe itself. But so far, your studies have left out one important class of objects—planets. Now it is time for you to correct that omission.

This Chapter

In this chapter, you will look back on what you have learned and find your place. You are a planetwalker. What does that mean? Where do you fit into the cosmos? Most of all, you need to know how the solar system formed. That will tell you how your home planet was produced by the processes you have been studying. As you explore our solar system in space and time, you will find answers to four essential questions:

What theories account for the origin of the solar system?

What properties must these theories explain?

How do planets form?

Is our solar system unique?

Looking Ahead

One reason you should learn about the origin of the solar system is that you live here, but there is another reason. In the next three chapters you will explore in more detail each of the planets, plus the comets and asteroids. By studying the origin of the solar system first, you give yourself a framework for understanding these fascinating worlds.

Ace ◖Astronomy™ The AceAstronomy icon throughout the text indicates an opportunity for you to test yourself on key concepts and to explore animations and interactions on the AceAstronomy website at: http://ace .brookscole.com/sf9 the solar system because, as you are about to discover, there are more planets in the universe than stars. Above all, you should study the solar system because it is your home in the universe. Because humans are an intelligent species, we have the right and the responsibility to wonder what we are. Our kind have inhabited this solar system for at least a million years, but only within the last hundred years have we begun to understand what a solar system is. Like sleeping passengers on a train, we waken, look out at the passing scenery, and mutter, "What place is this? Where are we now?"

19-1 Theories of Earth's Origin

YOU ARE LINKED through a great chain of origins that leads backward through time to the first instant when the universe began 13.7 billion years ago. The gradual discovery of the links in that chain is one of the most exciting adventures of the human intellect. In earlier chapters, you studied some of that story: the origin of the universe in the big bang, the formation of galaxies, the origin of stars, and the growth of the chemical elements. Here you will explore further to consider the formation of planets.

Early Hypotheses

The earliest theories for Earth's origin are myths and folktales that go back beyond the beginning of recorded history. In addition, almost all religions contain an account of the origin of the world. It was not until about the time of Galileo that philosophers began searching for rational explanations for natural phenomena. While people like Copernicus, Kepler, and Galileo tried to find logical explanations for the motions of Earth and the other planets, other philosophers began thinking about the origin of the planets.

The first rational theory for Earth's origin was proposed by the French philosopher and mathematician René Descartes (1596–1650). Because he lived and wrote before the time of Newton, Descartes did not have the concept of gravitation as the dominant force in the universe. Rather, he believed that force was communicated by contact between bodies and that the entire universe was filled with vortices of whirling invisible particles. In 1644, he proposed that the sun and planets formed when a large vortex contracted and condensed. In this way, his hypothesis explained the general properties of the solar system known at the time.

A century later, in 1745, the French naturalist Georges-Louis de Buffon (1707–1788) proposed an alternative hypothesis that the planets were formed when a passing star collided with or passed close to the sun and pulled matter out of the sun and the star. The matter condensed to form the planets, and they fell into orbit around the sun (■ Figure 19-1a). This **passing star hypothesis** was popular off and on for two centuries, but it con-

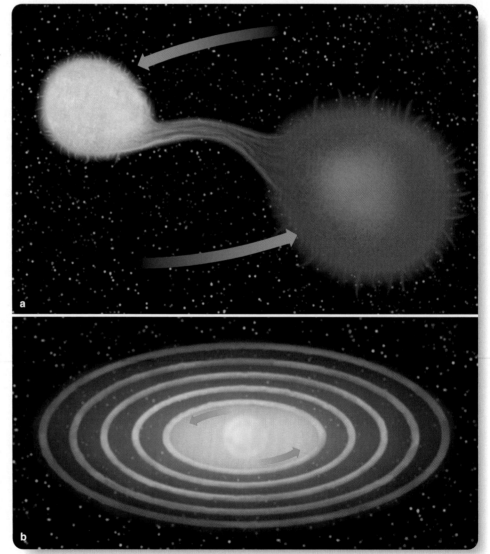

■ **Figure 19-1**

(a) The passing star hypothesis was catastrophic. It proposed that the sun was hit by or had a very close encounter with a passing star and that matter torn from the sun and the star formed planets orbiting the sun. The theory is no longer accepted.
(b) Laplace's nebular hypothesis was evolutionary. It suggested that a contracting disk of matter conserved angular momentum, spun faster, and shed rings of matter that formed planets.

Two Kinds of Theories: Evolution and Catastrophe

Many theories in science can be classified as either evolutionary, in that they involve gradual processes, or catastrophic, in that they depend on specific, unlikely events. Scientists have generally preferred evolutionary theories. Nevertheless, catastrophic events do occur.

Something in people prefers catastrophic theories, perhaps because everyone likes to see spectacular violence from a safe distance, which may explain the success of movies that include lots of car crashes and explosions. Also, cataclysmic theories resonate with Old Testament accounts of catastrophic events and special acts of creation. Thus, people have an understandable interest in catastrophic theories.

Nevertheless, most scientific theories are evolutionary. Such theories do not depend on unlikely events or special acts of biblical creation. For example, geologists study theories of mountain building that are evolutionary, with the mountains being pushed up slowly as centuries pass. All the evidence of erosion and the folding of rock layers shows that the process is gradual. Because most such natural processes are evolutionary, scientists sometimes find it difficult to accept any theory that depends on catastrophic events.

You will see in this and later chapters that catastrophes do occur. The planets, for example, are bombarded by debris from space, and some of those impacts are very large. As you study astronomy or any other natural science, notice that most theories are evolutionary but that you must allow for the possibility of unpredictable catastrophic events.

Mountains evolve to great heights by rising slowly, not catastrophically. (Janet Seeds)

---

tains serious flaws. First, stars are very small compared to the distances between them, and thus they collide very infrequently. In the entire history of our galaxy, stars have probably collided only a few times. More important, the gas pulled from the sun and the star would be much too hot to condense to make planets. Furthermore, even if planets formed, they would not go into stable orbits.

The hypotheses of Descartes and Buffon fall into two broad categories. Descartes's hypothesis is **evolutionary** in that it calls upon common, gradual events to produce the sun and planets. If it is correct, stars with planets are very common. Buffon's hypothesis, on the other hand, is **catastrophic.** It calls on unlikely, sudden events to produce the solar system, and thus it implies that solar systems are very rare. While your imagination may be tempted by colliding stars, modern hypotheses for the origins of the planets are evolutionary, with, as you will see, a few astonishing catastrophes thrown in (**Window on Science 19-1**).

The modern theory of the origin of the solar system had its true beginning with Pierre-Simon de Laplace (1749–1827), the brilliant French astronomer and mathematician. In 1796, he combined Descartes's vortex with Newton's gravity to produce a model of a rotating cloud of matter contracting under its own gravitation and flattening into a disk—the **nebular hypothesis.** As the disk grew smaller, it had to conserve angular momentum and spin faster and faster. Laplace reasoned that, when it was spinning as fast as it could, the disk would shed its outer edge to leave behind a ring of matter. Then the disk could contract further, speed up again, and leave another ring. In this way, he imagined, the contracting disk would leave behind a series of rings, each to become a planet circling the newborn sun at the center of the disk (Figure 19-1b).

According to the nebular hypothesis, the sun should be spinning very rapidly, or, to put it another way, the sun should have most of the angular momentum of the solar system. (Recall from Chapter 5 that angular momentum is the tendency of a rotating object to continue rotating.) As astronomers studied the planets and the sun, however, they found that the sun rotated slowly and that the planets moving in their orbits had most of the angular momentum in the solar system. In fact, the rotation of the sun contains only about 0.3 percent of the angular momentum of the solar system. Because the nebular hypothesis could not explain this **angular momentum problem,** it was never fully successful, and astronomers toyed with various versions of the passing star hypothesis for over a century.

In contrast to the astronomy of earlier centuries, 20th-century astronomy applied modern physics to the stars and galaxies, and that illuminated the origins of the stars and the elements. You can now trace that great chain of origins that began with the big bang and led to the matter of which you are made. First you should review the story of the origin of matter, and then you can add the story of how that matter formed your home world.

A Review of the Origin of Matter

The matter in your body came into existence within minutes of the beginning of the universe. Astronomers have strong evidence that the universe began in an event called the big bang

(Chapter 18); and, by the time the universe was three minutes old, the protons, neutrons, and electrons in your body had come into existence. You are made of very old matter.

Although those particles formed quickly, they were not linked together to form the atoms that are common today. Most of the matter was hydrogen, and about 25 percent was helium. Very few heavier atoms were made in the big bang. Although your body does not contain helium, it does contain many of those ancient hydrogen atoms unchanged since the universe began.

Within a few hundred million years after the big bang, matter began to collect to form galaxies containing billions of stars. You have learned how nuclear reactions inside stars combine low-mass atoms such as hydrogen to make heavier atoms (Chapter 8). Generation after generation of stars cooked the original particles, linking them together to build atoms such as carbon, nitrogen, and oxygen (Chapter 15). Those are common atoms in your body. Even the calcium atoms in your bones were assembled inside stars.

Massive stars produce iron in their cores, but much of that iron core is destroyed when the star collapses and explodes as a supernova. Most of the iron in your body was produced by carbon fusion in type Ia supernovae and by the expanding matter ejected by type II supernovae. Atoms heavier than iron were created by rapid nuclear reactions that can occur only during supernova explosions (Chapter 13). Gold and silver are rare in your body, but iodine is critical in your thyroid gland, and those heavy atoms were produced during the violent deaths of massive stars.

Our galaxy contains at least 100 billion stars, of which the sun is one. It formed from a cloud of gas and dust about 5 billion years ago, and the atoms in your body were part of that cloud. How the sun took shape, how the cloud gave birth to the planets, how the atoms in your body found their way onto Earth and into you is the story of this chapter. As you explore the origin of the solar system, keep in mind the great chain of origins that created the atoms. As the geologist Preston Cloud remarked, "Stars have died that we might live."

The Solar Nebula Hypothesis

The stars gave birth to the heavy atoms of which our world is made, and the modern theory of the origin of the planets is based on star formation. The **solar nebula hypothesis** proposes that the planets were formed from the disk of gas and dust that surrounded the sun as it formed (■ Figure 19-2).

As stars form in contracting clouds, they remain surrounded by cocoons of dust and gas, and the rotation of the cloud causes that dust and gas to form a spinning disk around the protostar. When the center of the star grows hot enough to ignite nuclear reactions, its surface quickly heats up, becomes more luminous, and blows away the gas and dust cocoon.

The solar nebula hypothesis supposes that planets form in the rotating disks of gas and dust around young stars. You have seen clear evidence that disks of gas and dust are common around young stars. Infrared observations of T Tauri stars, for instance, show that some are surrounded by gas clouds rich in dust, and spectra show that these stars are blowing away their nebulae at speeds up to 200 km/s. Bipolar flows from protostars (Chapter 11) were the first evidence of such disks, but modern techniques can image the disks directly (see Figure 11-7 and page 253).

Our own planetary system probably formed in such a disk-shaped cloud around the sun. When the sun became luminous enough, the remaining gas and dust were blown away into space, leaving the planets orbiting the sun. Notice that the solar nebula hypothesis also suffers from the angular momentum problem. Before modern astronomers could take the theory seriously, they had to understand why the sun now rotates so slowly, a puzzle you will be able to solve later in this chapter.

The Solar Nebula Hypothesis

A rotating cloud of gas contracts and flattens...

to form a thin disk of gas and dust around the forming sun at the center.

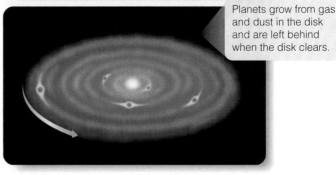

Planets grow from gas and dust in the disk and are left behind when the disk clears.

■ **Figure 19-2**

The solar nebula hypothesis proposes that the planets formed along with the sun.

If the solar nebula hypothesis is correct, then our Earth and the other planets of the solar system formed billions of years ago as the sun condensed from the interstellar medium. If that is true, then planets form as a by-product of star formation, and most stars should have planets.

Building Scientific Arguments

Why does the solar nebula theory imply planets are common? Often, the implications of a theory are more important in building a scientific argument than the theory's own conjecture about nature. The solar nebula theory is an evolutionary theory; and, if it is correct, the planets of our solar system formed from the disk of gas and dust that surrounded the sun as it condensed from the interstellar medium. That suggests it is a common process. Most stars form with disks of gas and dust around them, and planets should form in such disks. Planets should be very common in the universe.

Now build a new scientific argument to consider the old catastrophic theory. **Why did the passing star hypothesis suggest that planets are very rare?**

■   ■   ■

Connections: The solar nebula hypothesis seems to fit with what you know about star formation, so it is time to construct a detailed theory to describe planet formation. The first step is to survey the solar system to discover the distinguishing characteristics of the planetary system. Those are the characteristics that a successful theory must explain.

19-2 A Survey of the Solar System

TO TEST THEIR THEORIES, astronomers must search the present solar system for evidence of its past. In this section, you will survey the solar system and compile a list of its most significant characteristics, potential clues to how it formed.

You should begin with the most general view of the solar system. It is, in fact, almost entirely empty space (Figure 1-7). Imagine that you reduce the solar system until Earth is the size of a grain of table salt, about 0.3 mm (0.01 in.) in diameter. The moon is a speck of pepper about 1 cm (0.4 in.) away, and the sun is the size of a small plum 4 m (13 ft) from Earth. Mercury, Venus, and Mars are grains of salt. Jupiter is an apple seed 20 m (66 ft) from the sun, and Saturn is a smaller seed over 36 m (120 ft) away. Uranus and Neptune are slightly larger than average salt grains, and Pluto, at the edge of the solar system, is a speck of pepper over 150 m (500 ft) from the central plum. Although your model solar system would be larger than a football field, you would need a powerful microscope to detect the asteroids.

The planets are tiny specks of matter scattered around the sun—the last remains of the solar nebula.

Revolution and Rotation

The planets revolve* around the sun in orbits that lie close to a common plane. The orbit of Mercury, the closest planet to the sun, is tipped 7° to Earth's orbit, and Pluto's orbit is tipped 17.2°. The rest of the planets' orbital planes are inclined by no more than 3.4°. As you can see, the solar system is basically disk shaped.

The rotation of the sun and planets on their axes also seems related to this disk shape. The sun rotates with its equator inclined only 7.25° to Earth's orbit, and most of the other planets' equators are tipped less than 30°. The rotations of Venus, Uranus, and Pluto are peculiar, however. Venus rotates backward compared with the other planets, and both Uranus and Pluto rotate on their sides (with their equators almost perpendicular to their orbits). You will explore these planets in detail in Chapters 22 and 24, but later in this chapter you will be able to understand how they could have acquired their peculiar rotations.

Apparently, the preferred direction of motion in the solar system—counterclockwise as seen from the north—is also related to its disk shape. All the planets revolve counterclockwise around the sun; and, with the exception of Venus, Uranus, and Pluto, they rotate counterclockwise on their axes. Furthermore, nearly all of the moons in the solar system, including Earth's moon, orbit around their planets counterclockwise. With only a few exceptions, most of which are understood, revolution and rotation in the solar system follow a disk theme.

Two Kinds of Planets

Perhaps the most striking clue to the origin of the solar system comes from the division of the planets into two categories: the small Earthlike worlds and the giant Jupiterlike worlds. The difference is so dramatic that it is hard to keep from shouting, "Aha, this must mean something!" Study **Terrestrial and Jovian Planets** on pages 454–455 and notice three important points:

1 Notice how the two kinds of planets are distinguished by their location. The four inner planets are quite different from the next four outward.

2 Also notice how common craters are. Almost every solid surface in the solar system is covered with craters.

3 Finally, notice how the planets are distinguished by individual properties such as rings, clouds, and moons. Any theory of the origin of the planets needs to explain these properties.

The division of the planets into two families is a clue to how our solar system formed, but you can learn little from the

---

*Recall from Chapter 2 that the words *revolve* and *rotate* refer to different motions. A planet revolves around the sun but rotates on its axis. Cowboys in the old west didn't carry revolvers. They carried rotators.

1 The distinction between the terrestrial planets and the Jovian planets is dramatic. The inner four planets, Mercury, Venus, Earth, and Mars, are **terrestrial planets,** meaning they are small, dense, rocky worlds with little or no atmosphere. The outer four planets, Jupiter, Saturn, Uranus, and Neptune, are **Jovian planets,** meaning they are large, low-density worlds with thick atmospheres and liquid interiors. Pluto does not fit this scheme, being small but low density; you will see in a later chapter that it is a very special world and might not be considered a planet at all.

The planets and the sun to scale. Saturn's rings would just reach from Earth to the moon.

Planetary orbits to scale. The terrestrial planets lie quite close to the sun, whereas the Jovian planets are spread far from the sun outside the asteroid belt. The elliptical shape of Pluto's orbit is visible here.

Mercury
Venus
Earth
Mars
Jupiter
Saturn
Asteroids
Uranus
Neptune
Pluto

1a Of the terrestrial planets, Earth is most massive, but the Jovian planets are much more massive. Jupiter is over 300 Earth masses, and Saturn is nearly 100 Earth masses. Uranus and Neptune are 15 and 17 Earth masses.

Mercury is only 40 percent larger than Earth's moon, and its weak gravity cannot retain a permanent atmosphere. Like the moon, it is covered with craters from meteorite impacts.

Mercury

NASA

UCO/Lick Observatory

Earth's moon

2 Craters are common on all of the surfaces in the solar system that are strong enough to retain them. Earth has about 150 impact craters, but many more have been erased by erosion. Besides the planets, the asteroids and nearly all of the moons in the solar system are scarred by craters. Ranging from microscopic to hundreds of kilometers in diameter, these craters have been produced over the ages by meteorite impacts. When astronomers see a rocky or icy surface that contains few craters, they know that the surface is young.

Mercury is so close to the sun it is difficult to study from Earth. The Mariner 10 spacecraft flew past Mercury in 1974, but it was not able to take detailed photos of all of the planet's surface.

These five worlds are shown in proper relative size.

Mercury

Moon

Earth

The surface of Venus is not visible through its cloudy atmosphere, but radar maps reveal a dry desert world of craters and volcanoes.

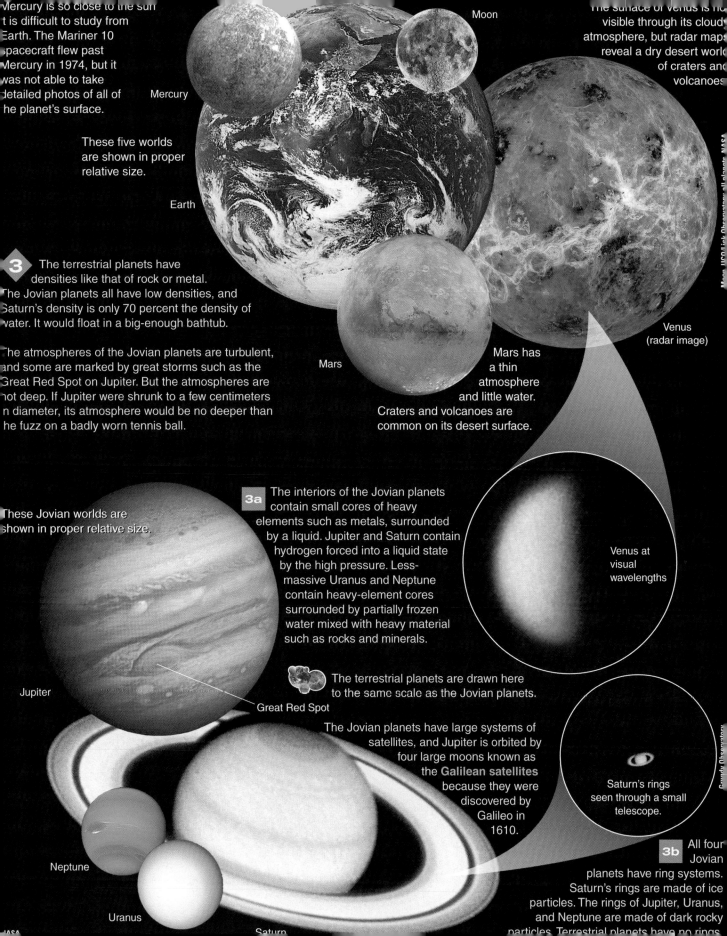

3 The terrestrial planets have densities like that of rock or metal. The Jovian planets all have low densities, and Saturn's density is only 70 percent the density of water. It would float in a big-enough bathtub.

The atmospheres of the Jovian planets are turbulent, and some are marked by great storms such as the Great Red Spot on Jupiter. But the atmospheres are not deep. If Jupiter were shrunk to a few centimeters in diameter, its atmosphere would be no deeper than the fuzz on a badly worn tennis ball.

Mars

Mars has a thin atmosphere and little water. Craters and volcanoes are common on its desert surface.

Venus (radar image)

These Jovian worlds are shown in proper relative size.

3a The interiors of the Jovian planets contain small cores of heavy elements such as metals, surrounded by a liquid. Jupiter and Saturn contain hydrogen forced into a liquid state by the high pressure. Less-massive Uranus and Neptune contain heavy-element cores surrounded by partially frozen water mixed with heavy material such as rocks and minerals.

Venus at visual wavelengths

The terrestrial planets are drawn here to the same scale as the Jovian planets.

Jupiter

Great Red Spot

The Jovian planets have large systems of satellites, and Jupiter is orbited by four large moons known as the **Galilean satellites** because they were discovered by Galileo in 1610.

Saturn's rings seen through a small telescope.

Neptune

Uranus

Saturn

3b All four Jovian planets have ring systems. Saturn's rings are made of ice particles. The rings of Jupiter, Uranus, and Neptune are made of dark rocky particles. Terrestrial planets have no rings.

individual planets because they have evolved since they formed. For further clues you must look at smaller objects that have remained largely unchanged since the birth of the solar system.

Space Debris

The sun and planets are not the only remains of the solar nebula. The solar system is littered with three kinds of space debris: asteroids, comets, and meteoroids. Although these objects represent a tiny fraction of the mass of the system, they are a rich source of information about the origin of the planets.

The **asteroids,** sometimes called minor planets, are small rocky worlds, most of which orbit the sun in a belt between the orbits of Mars and Jupiter. Roughly 20,000 asteroids have been charted, of which about 2000 follow orbits that bring them into the inner solar system, where they can occasionally collide with a planet. Earth has been struck many times in its history. Some asteroids are located in Jupiter's orbit, while some have been found beyond the orbit of Saturn.

About 200 asteroids are more than 100 km (60 mi) in diameter, and tens of thousands are estimated to be more than 10 km (6 mi) in diameter. There are probably a million or more that are larger than 1 km (0.6 mi) and billions that are smaller. Because even the largest are only a few hundred kilometers in diameter, Earth-based telescopes can detect no details on their surfaces, and the Hubble Space Telescope can image only the largest features.

Astronomers do, however, have clear evidence that asteroids are irregularly shaped cratered worlds. A number of spacecraft have visited asteroids and sent back photos. For instance, the NEAR spacecraft rendezvoused with the asteroid Eros, went into orbit, and studied the asteroid in detail. Like most asteroids, Eros is an irregular, rocky body pocked by craters (■ Figure 19-3). These observations will be discussed in detail in Chapter 25, but in this quick survey of the solar system you can note that all the evidence suggests the asteroids have suffered many impacts from collisions with other asteroids.

Older theories proposed that the asteroids are the remains of a planet that broke up, but modern astronomers recognize the asteroids as the debris left over by a planet that failed to form at a distance of 2.8 AU from the sun. When you study the formation of planets later in this chapter, you will discover why material in the solar nebula failed to form a planet at a distance of 2.8 AU.

Since 1992, astronomers have discovered roughly a thousand small, dark, icy bodies orbiting in the outer fringes of the solar system beyond Neptune and Pluto. This collection of objects is called the **Kuiper belt** after the Dutch-American astronomer Gerard Kuiper (KI-per), who predicted their existence in the 1950s. There are probably 70,000 of these small, asteroidlike bodies in the Kuiper belt, and any successful theory should explain how they came to orbit so far from the sun.

In contrast to the rocky asteroids and dark Kuiper belt objects, the brightest **comets** are impressively beautiful objects (■ Figure 19-4). Most comets are faint, however, and are difficult to locate even at their brightest. A comet may take months to sweep through the inner solar system, during which time it appears as a glowing head with an extended tail of gas and dust.

The beautiful tail of a comet can be longer than an astronomical unit, but it is produced by an icy nucleus only a few tens of kilometers in diameter. In its long, elliptical orbit, the nucleus remains frozen and inactive while it is far from the sun. As its orbit carries the nucleus into the inner solar system, the sun's heat begins to vaporize the ices, releasing gas and dust. The pressure of sunlight and the solar wind pushes the gas and dust away, forming a long tail. The motion of the nucleus along its orbit, the pressure of sunlight, and the outward flow of the solar wind can create comet tails that are long and straight or gently curved, but in

Visual-wavelength images

a

b

■ Figure 19-3

(a) Over a period of three weeks, the NEAR spacecraft approached the asteroid Eros and recorded a series of images arranged here in an entertaining pattern showing the irregular shape and 5-hour rotation of the asteroid. Eros is 34 km (21 mi) long. (b) This close-up of the surface of Eros shows an area about 11 km (7 mi) from top to bottom. (Johns Hopkins University, Applied Physics Laboratory, NASA)

(a) A comet may remain visible in the evening or morning sky for weeks as it moves through the inner solar system. Comet West was in the sky during March 1976. (b) A comet in a long, elliptical orbit becomes visible when the sun's heat vaporizes its ices and pushes the gas and dust away in a tail. (Celestron International)

either case the tails of comets always point approximately away from the sun (Figure 19-4b).

For decades astronomers described comet nuclei as dirty snowballs, meaning that they were icy bodies with a little bit of embedded rock and dust. Starting with the passage of Comet Halley in 1986, astronomers have found growing evidence that comet nuclei are not made of dirty ice but rather of icy dirt. That is, the nuclei are at least 50 percent rock and dust. One astronomer has suggested replacing the dirty snowball model with the icy mudball model.

The nuclei of comets are ice-rich bodies left over from the origin of the planets. From this you can conclude that at least some parts of the solar nebula were rich in ices. You will see later in this chapter how important ices were in the formation of the Jovian planets, and you will have a chance to study comets in more detail in Chapter 25.

Unlike the stately comets, **meteors** flash across the sky in momentary streaks of light (■ Figure 19-5). They are commonly called "shooting stars." Of course, they are not stars but small bits of rock and metal falling into Earth's atmosphere and bursting into incandescent vapor about 80 km (50 mi) above the ground because of friction with the air. This vapor condenses to form dust, which settles slowly to Earth, adding about 40,000 tons per year to the planet's mass.

Technically, the word *meteor* refers to the streak of light in the sky. In space, before its fiery plunge, the object is called a **meteoroid,** and any part of it that survives its fiery passage to Earth's surface is called a **meteorite.** Most meteoroids are specks of dust, grains of sand, or tiny pebbles. Almost all the meteors you see in the sky are produced by meteoroids that weigh less than 1 g. Only rarely is one massive enough and strong enough to survive its plunge and reach Earth's surface.

Thousands of meteorites have been found, and you will learn more about their particular forms in Chapter 25. Meteorites are mentioned here for one specific clue they can give you concerning the solar nebula: Meteorites can tell you the age of the solar system.

The Age of the Solar System

If the solar nebula theory is correct, the planets should be about the same age as the sun. The most accurate way to find the age of a celestial body is to bring a sample into the laboratory and determine its age by analyzing the radioactive elements it contains.

When a rock solidifies, it incorporates known percentages of the chemical elements. A few of these elements are radioactive and can decay into another element, called the daughter element. For example, the isotope uranium-238 can decay into an isotope of lead, lead-238. The **half-life** of a radioactive element is the time

■ Figure 19-5

A meteor is the streak of glowing gases produced by a bit of material falling into Earth's atmosphere. Friction with the air vaporizes the material about 80 km (50 mi) above Earth's surface. (Daniel Good)

it takes for half of the atoms to decay. The half-life of uranium-238 is 4.5 billion years. The abundance of a radioactive element gradually decreases as it decays, and the abundance of the daughter element gradually increases (■ Figure 19-6). If you know the abundance of the elements in the original rock, you can measure the present abundance and find the age of the rock. For example, if you studied a rock and found that only 50 percent of the uranium-238 remained and the rest had become lead-238, you could conclude that one half-life must have passed and that the rock was 4.5 billion years old.

Uranium isn't the only radioactive element used in radioactive dating. Potassium-40 decays with a half-life of 1.3 billion years to form calcium-40 and argon-40. Rubidium-87 decays to strontium-87, with a half-life of 47 billion years. Any of these elements can be used as a radioactive clock to find the age of mineral samples.

Of course, to find a radioactive age, you need a sample in the laboratory, and the only celestial bodies from which scientists have samples are Earth, the moon, Mars, and meteorites.

The oldest Earth rocks so far discovered and dated are tiny zircon crystals from Australia 4.3 billion years old. That does not mean that Earth formed 4.3 billion years ago. The surface of Earth is active, and the crust is continually destroyed and reformed from material welling up from beneath the crust (see Chapter 20). Consequently, the age of these oldest rocks tells you only that Earth is *at least* 4.3 billion years old.

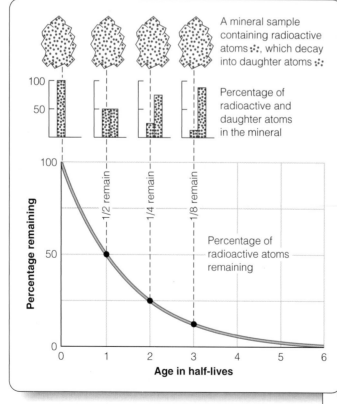

■ Active Figure 19-6

The radioactive atoms in a mineral sample (red) decay into daughter atoms (blue). Half the radioactive atoms are left after one half-life, a fourth after two half-lives, an eighth after three half-lives, and so on. Radioactive dating shows that this fragment of the Allende meteorite is 4.56 billion years old. It contains interstellar grains that formed long before our solar system. (R. Kempton, New England Meteoritical Services)

Ace✶Astronomy™ Log into AceAstronomy and select this chapter to see the Active Figure "Radioactive Decay" and graph the decay of different radioactive isotopes.

One of the most exciting goals of the Apollo lunar landings was bringing lunar rocks back to Earth's laboratories, where they could be dated. Because the moon's surface is not being recycled like Earth's, some parts of it might have survived unaltered since early in the history of the solar system. Dating the rocks showed the oldest to be 4.48 billion years old. That means the solar system must be *at least* 4.48 billion years old.

Although no one has yet been to Mars, over a dozen meteorites found on Earth have been identified by their chemical composition as having come from Mars. Most of these have ages of only 1.3 billion years, but one has an age of approximately 4.6 billion years. Mars must be at least that old.

Another important source for determining the age of the solar system is meteorites. Radioactive dating of meteorites yields a range of ages, with the oldest about 4.56 billion years old. This figure is widely accepted as the age of the solar system.

Meteorites also contain tiny grains containing mostly silicon and carbon, and the composition of those grains shows that they are from the interstellar medium. They cannot be dated radioactively, but they must have formed long before our solar system and were incorporated in the nebula that contracted to form our sun and our planetary system. These tiny particles are messengers from beyond the beginning of our solar system.

One last celestial body deserves mention: the sun. Astronomers estimate the age of the sun to be about 5 billion years, but this is not a radioactive date because they cannot obtain a sample of solar material. Instead, they estimate the age of the sun from mathematical models of the sun's interior. This yields an age of about 5 billion years plus or minus 1.5 billion years, a number that is in agreement with the age of the solar system derived from the age of meteorites.

Apparently, all the bodies of the solar system formed at about the same time some 4.6 billion years ago. You can add this as the final item to your list of characteristic properties of the solar system (■ Table 19-1).

■ **Table 19-1 | Characteristic Properties of the Solar System**

1. Disk shape of the solar system
   Orbits in nearly the same plane
   Common direction of rotation and revolution
2. Two planetary types
   Terrestrial—inner planets; high density
   Jovian—outer planets; low density
3. Planetary ring systems and large satellite systems for Jupiter, Saturn, Uranus, and Neptune
4. Space debris—asteroids, comets, and meteors
   Composition
   Orbits
5. Common ages of about 4.6 billion years for Earth, the moon, Mars, meteorites, and the sun

---

Building Scientific Arguments

In what ways is the solar system a disk?

Notice that this argument is really a summary of pieces of evidence. First, the general shape of the solar system is that of a disk. The planets follow orbits that lie in nearly the same plane. The orbit of Mercury is inclined 7° to the plane of Earth's orbit, and the orbit of Pluto is inclined a bit more than 17°. In this way, the planets follow orbits confined to a thin disk with the sun at its center.

Second, the motions of the sun and planets also follow this disk theme. The sun and most of the planets rotate in the same direction, counterclockwise as seen from the north, with their equators near the plane of the solar system. Also, all of the planets revolve around the sun in that same direction. Our solar system seems to prefer motion in the same direction, which further reflects a disk theme.

One of the basic characteristics of our solar system is its disk shape, but another dramatic characteristic is the division of the planets into two groups. Build an argument to detail that evidence. **What are the distinguishing differences between the terrestrial and Jovian planets?**

■   ■   ■

Connections: You have completed your survey of the solar system and gathered the preliminary evidence you can use to evaluate the solar nebula hypothesis. Now it is time to ask how the planets formed from the solar nebula and took on the characteristics they have today.

19-3) The Story of Planet Building

THE CHALLENGE FOR MODERN PLANETARY ASTRONOMERS is to compare the characteristics of the solar system with the solar nebula theory and tell the story of how the planets formed (**Window on Science 19-2**).

The Chemical Composition of the Solar Nebula

Everything astronomers know about the solar system and star formation suggests that the solar nebula was a fragment of an interstellar gas cloud. Such a cloud would have been mostly hydrogen with some helium and tiny traces of the heavier elements.

This is precisely what you see in the composition of the sun (see Table 7-2). Analysis of the solar spectrum shows that the sun is mostly hydrogen, with a quarter of its mass being helium and only about 2 percent being heavier elements. Of course, nuclear reactions have fused some hydrogen into helium, but this happens in the sun's core and has not affected its surface composition.

Reconstructing the Past from Evidence and Hypothesis

Scientists often face problems in which they must reconstruct the past. Some of these reconstructions are obvious, such as an archaeologist excavating the ruins of a burial tomb, but others are less obvious. In each case, success requires the interplay of hypotheses and evidence to re-create a past that no longer exists.

The reconstruction of the past is obvious when you use the chemical abundance of stars to reconstruct the story of the formation of our galaxy, but a biologist studying a centipede is also reconstructing the past. How did this creature come to have a segmented body with so many legs? How did it develop the metabolism that allows it to move quickly and hunt

prey? Although the problem might at first seem to be one of mere anatomy, the scientist must reconstruct an environment that no longer exists.

The astronomer's problem is not just to understand what the planets are like but to understand how they got that way. That means planetary astronomers must look at the evidence they can see today and reconstruct a history of the solar system, a past that is quite different from the present. If you had a time machine, it would be a fantastic adventure to go back and watch the planets form. Time machines are impossible, but scientists can use the grand interplay of evidence and theory, the distinguishing characteristic of

One way science can enrich and inform our lives is by recreating a world that no longer exists.

science, to journey back billions of years and reconstruct a past that no longer exists.

That means the composition revealed in the sun's spectrum is essentially the composition of the gases from which it formed.

This must have been the composition of the solar nebula, and you can see that composition reflected in the chemical compositions of the planets. The composition of a planet can be altered, however, depending on how it forms. The most common process of planet formation is believed to begin with the sticking together of solid bits of matter. Only when a planet has grown to a mass of about 15 Earth masses does it have enough gravitation to begin capturing gas directly from the solar nebula in a process called **gravitational collapse.** The Jovian planets began forming by the aggregation of bits of rock and ice. Once they accumulated enough mass, they began growing by gravitational collapse. That is, they began to capture large amounts of gas, mostly hydrogen and helium, directly from the solar nebula. Jupiter and Saturn grew rapidly and captured the most hydrogen and helium, whereas Uranus and Neptune apparently grew more slowly and didn't capture gas as rapidly as Jupiter and Saturn. That is how the Jovian worlds grew to be low-density, hydrogen-rich planets, with Jupiter and Saturn the largest and least dense.

The terrestrial planets, in contrast, contain very little hydrogen and helium. These small planets have such low masses they were unable to keep the hydrogen and helium from leaking into space. Indeed, because of their small masses, they were probably unable to capture very much of these lightweight gases from the solar nebula. The terrestrial planets are dense worlds because they are composed of the heavier elements from the solar nebula.

This account of planet formation is generally correct for most planets, but later in this chapter you will see how astronomers are refining the story to include other processes. Certainly, this general description of how planets form explains how the chemi-

cal composition of the solar nebula is reflected in the present composition of the sun and planets. Nevertheless, a very important question remains: How did gas and dust in the solar nebula come together to form the solid matter of the planets? You must answer that question in two stages. First, you must understand how the gas formed billions of small solid particles, and then you must explain how these particles built the planets.

The Condensation of Solids

The key to understanding the process that converted the nebular gas into solid matter is the variation in density among solar system objects. You have already noted that the four inner planets are high-density, terrestrial bodies, whereas the outermost planets are low-density, giant planets (except for Pluto, which is low in density but not a giant). This division is due to the different ways gases condensed into solids in the inner and outer regions of the solar nebula.

Even among the four terrestrial planets, you will find a pattern of subtle differences in density. Merely listing the observed densities of the terrestrial planets does not reveal the pattern because Earth and Venus, being more massive, have stronger gravity and have squeezed their interiors to higher densities. You must look at the **uncompressed densities**—the densities the planets would have if their gravity did not compress them. These densities (■ Table 19-2) show that, in general, the closer a planet is to the sun, the higher its uncompressed density.

This density variation originated when the solar system first formed solid grains. The kind of matter that condensed in a particular region would depend on the temperature of the gas there. In the inner regions, the temperature may have been 1500 K or so. The only materials that can form grains at this temperature are

Table 19-2 | Observed and Uncompressed Densities

Planet	Observed Density (g/cm³)	Uncompressed Density (g/cm³)
Mercury	5.44	5.30
Venus	5.24	3.96
Earth	5.50	4.07
Mars	3.94	3.73
(Moon)	3.36	3.40

compounds with high melting points, such as metal oxides and pure metals, which are very dense. Farther out in the nebula it was cooler, and silicates (rocky material) could condense. These are less dense than metal oxides and metals. In the cold outer regions, ices of water, methane, and ammonia could condense. These are low-density materials.

The sequence in which the different materials condense from the gas as you move away from the sun is called the **condensation sequence** (■ Table 19-3). It suggests that the planets, forming at different distances from the sun, accumulated from different kinds of materials. The inner planets formed from high-density metal oxides and metals, and the outer planets formed from low-density ices.

You must also remember that the solar nebula did not remain the same temperature throughout the formation of the planets but may have grown progressively cooler. That means a particular region of the nebula may have begun by producing solid particles of metals and metal oxides but, after cooling, began producing particles of silicates. Allowing for the cooling of the nebula makes the theory much more complex, but it also makes the processes by which the planets formed much more understandable.

The Formation of Planetesimals

In the development of a planet, three groups of processes operate. First, grains of solid matter grow larger, eventually reaching diameters ranging from a few centimeters to kilometers. The larger of these objects, called **planetcsimals,** are believed to be the bodies that the second group of processes collects into planets. Finally, a third set of processes clears away the solar nebula. The study of planet building is the study of these three groups of processes.

According to the solar nebula theory, planetary development in the solar nebula began with the growth of dust grains. These specks of matter, whatever their composition, grew from microscopic size by two processes, condensation and accretion.

A particle grows by **condensation** when it adds matter one atom at a time from a surrounding gas. Snowflakes, for example, grow by condensation in Earth's atmosphere. In the solar nebula, dust grains were continuously bombarded by atoms of gas, and

some of these stuck to the grains. A microscopic grain capturing a single gas atom increases its mass by a much larger fraction than a gigantic boulder capturing a single atom. That is why condensation can increase the mass of a small grain rapidly; but, as the grain grows larger, condensation becomes less effective.

The second process is **accretion,** the sticking together of solid particles. You may have seen accretion in action if you have walked through a snowstorm with big, fluffy flakes. If you caught one of those "flakes" on your mitten and looked closely, you saw that it was actually made up of many tiny, individual flakes that had collided as they fell and accreted to form larger particles. In the solar nebula the dust grains were, on the average, no more than a few centimeters apart, so they collided frequently. Their mutual gravitation was too small to hold them to each other, but other effects may have helped. Static electricity generated by their passage through the gas could have held them together, as could compounds of carbon that might have formed a sticky surface on the grains. Ice grains might have stuck together better than some other types. Of course, some collisions might have broken up clumps of grains; on the whole, however, accretion must have increased grain size. If it had not, the planets would not have formed.

There is no clear distinction between a very large grain and a very small planetesimal, but you can consider an object a planetesimal when its diameter becomes a kilometer or so (■ Figure 19-7). Objects larger than a centimeter were subject to new processes that tended to concentrate them. One important effect was that the growing planetesimals collapsed into the plane of the solar nebula. Dust grains could not fall into the plane because the turbulent motions of the gas kept them stirred up, but the larger objects had more mass, and the gas motions could not have prevented them from settling into the plane of the spinning nebula. This would have concentrated the solid particles into a thin plane about 0.01 AU thick and would have made further planetary growth more rapid.

Table 19-3 | The Condensation Sequence

Temperature (K)	Condensate	Planet (Estimated Temperature of Formation; K)
1500	Metal oxides	Mercury (1400)
1300	Metallic iron and nickel	
1200	Silicates	
1000	Feldspars	Venus (900)
680	Troilite (FeS)	Earth (600)
		Mars (450)
175	H_2O ice	Jovian (175)
150	Ammonia–water ice	
120	Methane–water ice	
65	Argon–neon ice	Pluto (65)

Visual-wavelength image

■ Figure 19-7

What did the planetesimals look like? You can get a clue from this photo of the 5-km-wide nucleus of Comet Wild2. Whether rocky or icy, the planetesimals must have been small, irregular bodies, pocked by craters from collisions with other planetesimals. (NASA)

This collapse of the planetesimals into the plane is analogous to the flattening of a forming galaxy. However, an entirely new process may have become important once the plane of planetesimals formed. Computer models show that the rotating disk of particles should have been gravitationally unstable and would have broken up into small clouds (■ Figure 19-8). This would fur-

■ Figure 19-8

Gravitational instabilities in the rotating disk of planetesimals may have forced them to collect in clumps, accelerating their growth.

ther concentrate the planetesimals and help them coalesce into objects up to 100 km (60 mi) in diameter.

Through these processes, the theory proposes, the nebula became filled with trillions of solid particles ranging in size from pebbles to tiny planets. As the largest began to exceed 100 km in diameter, new processes began to alter them, and a new stage in planet building began, the growth of protoplanets.

The Growth of Protoplanets

The coalescing of planetesimals eventually formed **protoplanets,** massive objects destined to become planets. As these larger bodies grew, new processes began making them grow faster and altered their physical structure.

If planetesimals collided at orbital velocities, it is unlikely that they would have stuck together. The average orbital velocity in the solar system is about 10 km/s (22,000 mph). Head-on collisions at this velocity would have vaporized the material. However, the planetesimals were moving in the same direction in the nebular plane and didn't collide head on. Instead, they merely rubbed shoulders at low relative velocities. Such gentle collisions would have been more likely to fuse them than to shatter them.

In addition, some adhesive effects probably helped. Sticky coatings and electrostatic charges on the surfaces of the smaller planetesimals probably aided formation of larger bodies. Collisions would have fragmented some of the surface rock; but, if the planetesimals were large enough, their gravity would have held on to some fragments, forming a layer of soil composed entirely of crushed rock. Such a layer on the larger planetesimals may have been effective in trapping smaller bodies.

The largest planetesimals would grow the fastest because they had the strongest gravitational field. Not only could they hold on to a cushioning layer to trap fragments, but their stronger gravity could also attract additional material. These planetesimals probably grew quickly to protoplanetary dimensions, sweeping up more and more material. When massive enough, they trapped some of the original nebular gas to form primitive atmospheres. At some point, they crossed the boundary between planetesimals and protoplanets.

To trace the growth of a protoplanet, you can think of the formation of Earth. In its simplest form, the theory of protoplanet growth supposes that all the planetesimals had about the same chemical composition. The planetesimals accumulated gradually to form a planet-size ball of material that was of homogeneous composition throughout. Once the planet formed, heat began to accumulate in its interior from the decay of short-lived radioactive elements, and this heat eventually melted the planet and allowed it to differentiate. **Differentiation** is the separation of material according to density. When the planet melted, the heavy metals such as iron and nickel settled to the core, while the lighter silicates floated to the surface to form a low-density crust. The story of planet formation from planetesimals of similar composition is shown in the left half of ■ Figure 19-9.

This process depends on the presence of short-lived radioactive elements in the solar nebula. Astronomers know such elements were present because very old rock found in meteorites contains daughter isotopes such as magnesium-26. That isotope is produced by the decay of aluminum-26 in a reaction that has a half-life of only 0.74 million years. The aluminum-26 and similar short-lived radioactive isotopes are gone now, but they must have been produced in a supernova explosion that occurred shortly before the formation of the solar nebula. In fact, many astronomers suspect that the supernova explosion compressed nearby gas and triggered the formation of stars, one of which became the sun. Thus our solar system may exist because of a supernova explosion that occurred about 4.6 billion years ago.

If planets formed in this gradual way and were later melted by radioactive decay, then Earth's present atmosphere was not its first. The first atmosphere consisted of small amounts of gases trapped from the solar nebula—mostly hydrogen and helium. Those gases were later driven off by the heat, aided perhaps by outbursts from the infant sun, and new gases released from the planet's interior formed a secondary atmosphere. This creation of a planetary atmosphere from a planet's interior is called **outgassing.**

This simple theory of planet formation can be improved in two ways. First, it seems likely that the solar nebula cooled during the formation of the planets, so they did not accumulate from planetesimals of common composition. As planet building began, the first particles to condense in the inner solar system were metals and metal oxides, so the protoplanets may have begun by accreting metallic cores. Later, as the nebula cooled, more silicates could form, and the protoplanets added silicate mantles. A second improvement in the theory proposes that the planets grew so rapidly that the heat released by the violent impacts of infalling particles, the **heat of formation,** did not have time to escape. This heat rapidly accumulated and melted the protoplanets as they formed. If this is true, then the planets would have differentiated as they formed. This improved story of planet formation is shown in the right half of Figure 19-9.

This improved theory suggests that Earth's first atmosphere was not captured from the solar nebula. If Earth formed in a molten state, then there was never a time when it had a primitive atmosphere of hydrogen and helium accumulated from the solar nebula. Rather, the gases of the atmosphere were released by the molten rock as the protoplanet grew. Those gases would not have included much water, however, so some astronomers now think that Earth's water

Two Models of Planet Building

Planetesimals contain both rock and metal.

The first planetesimals contain mostly metals.

A planet grows slowly from the uniform particles.

Later the planetesimals contain mostly rock.

The resulting planet is of uniform composition.

A rock mantle forms around the iron core.

Heat from radioactive decay causes differentiation.

Heat from rapid formation can melt the planet.

The resulting planet has a metal core and low-density crust.

The resulting planet has a metal core and low-density crust.

■ **Figure 19-9**

If the temperature of the solar nebula changed during planet building, the composition of the planetesimals may have changed. The simple model at left assumes no change occurred, but the model at the right incorporates a change from metallic to rocky planetesimals.

and much of its present atmosphere accumulated late in the formation of the planets as Earth swept up volatile-rich planetesimals forming in the cooling solar nebula. Such icy planetesimals may have formed in the outer parts of the solar nebula and been scattered by encounters with the Jovian planets in a bombardment of comets.

According to the solar nebula theory, the Jovian planets began growing by the same processes that built the terrestrial planets. But the Jovian planets grew faster. In the inner solar nebula, only metals and silicates could form solids, so the terrestrial planets grew slowly. The outer solar nebula not only contained solid bits of metals and silicates, but it was a blizzard of ices. The Jovian planets grew rapidly and quickly became massive enough to grow by gravitational collapse as they drew in large amounts of gas from the solar nebula. The terrestrial planets could not grow from ices, so they developed slowly and never became massive enough to grow by gravitational collapse.

If the planets formed in this way, the Jovian planets must have grown to their present size in about 10 million years. How do astronomers know this? Almost all T Tauri stars that have gas disks are younger than 10 million years; older T Tauri stars have blown their gas disks away. So the Jovian planets must have been complete within 10 million years or so. The terrestrial planets grew from solids and not from the gas, so they could have continued to grow by accretion from solid debris left behind when the gas was blown away. Apparently, the terrestrial planets were nearly complete within 10 million years but could have continued to grow for another 20 million years or so. All planet growth must have ended with the last of the solar nebula being blown away by 100 million years.

The solar nebula theory has been very successful in explaining the formation of the solar system. But there are some problems, and the Jovian planets are the troublemakers.

The Jovian Problem

Two observations are making it hard to explain the formation of the Jovian parents, and that is forcing astronomers to expand and revise their theories of planet formation (■ Figure 19-10).

In the final section of this chapter, you will see evidence that astronomers have found planets orbiting other stars, and those planets all have the mass of Jovian planets. There may be many terrestrial planets orbiting these stars that are too low in mass for astronomers to detect at present, but the important point is that there are lots of Jovian planets around. That is the first of the troublesome observations.

The second observation is the discovery that the gas and dust disks around newborn stars don't last long. In Chapter 11 you saw images of dusty gas disks around the young stars in the Orion Nebula (page 253). Those disks are being evaporated by the intense ultraviolet radiation from hot stars with the nebula. It seems that nearly all stars form in clusters containing massive stars, so this must happen to most disks. Even if a disk did not evaporate quickly, the gravitational influence of the crowded stars in a cluster should quickly strip away the outer parts of the disk. This is a troublesome observation because it seems to mean that disks don't last longer than about 7 million years at most, and many must evaporate within 100,000 years or so. That's not long enough to grow a Jovian planet. Yet Jovian worlds are common.

Mathematical models of the solar nebula have been computed using specially built computers running programs that take weeks to finish a calculation. The results show that the rotating gas and dust of the solar nebula may have become unstable and formed outer planets by direct gravitational collapse. That is, massive planets may have been able to form from the gas without first forming a dense core by accretion. Jupiters and Saturns form in these calculated models with a few hundred years.

If the Jovian planets formed in this way, they could have formed quickly before the solar nebula disappeared. Terrestrial planets grow from solid bits of material in the inner nebula, so they could have continued to grow by accretion even after the disk was cleared of gas and smaller particles of dust.

This new insight into the formation of the outer planets may help explain the formation of Uranus and Neptune. They are so far from the sun that accretion could not have built them rapidly. The gas and dust of the solar nebula must have been sparse out there, and Uranus and Neptune orbit so slowly they would not have swept up material very rapidly. The conventional view is that they grew by accretion so slowly that they never became quite mas-

■ **Figure 19-10**

Saturn is a beautiful planet, but the Jovian worlds are a problem for modern astronomers. The evidence suggests that accretion followed by gravitational collapse is too slow to build Jovian planets before the solar nebula was blown away. This has forced astronomers to revise their theories of how Jovian planets form. (JPL/NASA)

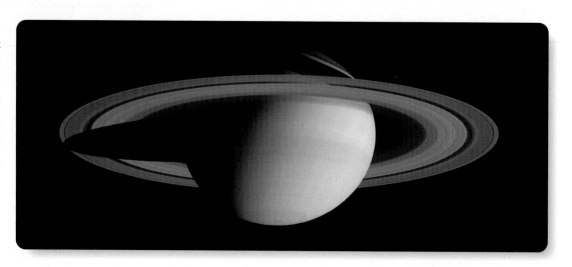

sive enough to begin growing by gravitational collapse. In fact, it is hard to understand how they could have reached their present mass if they started growing by accretion so far from the sun. Theoretical calculations show that they might have formed closer to the sun in the region of Jupiter and Saturn and then could have been shifted outward by gravitational interactions. In any case, the formation of Uranus and Neptune is part of the Jovian problem.

The traditional solar nebula theory proposes that the planets formed by accreting a core and then, if they became massive enough, by gravitational collapse. The new theories suggest that the outer planets may have skipped the core accretion phase.

Whatever the details of the story are, there must have been a solar nebula from which the planets formed. That is enough to allow you to explain the distinguishing characteristics of the solar system.

Explaining the Characteristics of the Solar System

Table 19-1 contains the list of distinguishing characteristics of the solar system. Any theory of the origin of the solar system should explain these characteristics.

The disk shape of the solar system is inherited from the solar nebula. The sun and planets revolve and rotate in the same direction because they formed from the same rotating gas cloud. The orbits of the planets lie in the same plane because the rotating solar nebula collapsed into a disk, and the planets formed in that disk.

The solar nebula hypothesis is evolutionary in that it calls on continuing processes to gradually build the planets. To explain the rotation of Venus, Uranus, and Pluto, however, you may need to consider catastrophic events. Uranus rotates on its side. This might have been caused by an off-center collision with a massive planetesimal when the planet was nearly formed. In Chapter 24 you will see evidence that Pluto's highly inclined rotation may be the result of a similar collision. Two theories have been proposed to explain the backward rotation of Venus. Theoretical models suggest that the sun can produce tides in the thick atmosphere of Venus and eventually reverse the planet's rotation—an evolutionary theory. It is also possible that the rotation of Venus was altered by an off-center impact late in the planet's formation, and that is a catastrophic theory.

The second item in Table 19-1, the division of the planets into terrestrial and Jovian worlds, can be understood through the condensation sequence. The terrestrial planets formed in the inner part of the solar nebula, where the temperature was high and most compounds remained gaseous. Only compounds such as the metals and silicates could condense to form solid particles. Terrestrial planets must have formed by the accumulation of solid particles because they never had enough gravitation to capture the hot gas of the inner solar nebula. That means the planets that began growing in the inner solar system had to form mainly from metals and silicates. Consequently, they became the small, dense terrestrial planets.

In contrast, the Jovian planets formed in the outer solar nebula, where the lower temperature allowed the gas to form large amounts of ices, perhaps three times more ices than silicates. That allowed the Jovian planets to grow rapidly and became massive.

The heat of formation (the energy released by infalling matter) was tremendous for these massive planets. Jupiter must have grown hot enough to glow with a luminosity of about 1 percent that of the present sun. However, because it never got hot enough to generate nuclear energy as a star would, it never generated its own energy. Jupiter is still hot inside. In fact, both Jupiter and Saturn radiate more heat than they absorb from the sun, so they are evidently still cooling.

This is an important point, so you'll want to summarize. The Jovian planets are low-density worlds because they grew in the outer solar nebula where there were large amounts of low-density ices in addition to bits of metal and silicates. Also, Jupiter and Saturn are so massive they have been able to grow by drawing in the cool gas directly from the solar nebula. The terrestrial planets could not do this because they never became massive enough and because the gas in the inner nebula was hotter and more difficult to capture.

A glance at the solar system suggests that you should expect to find a planet between Mars and Jupiter at the present location of the asteroid belt. Mathematical models show that there are asteroids there, and not a planet, because Jupiter grew into such a massive planet it was able to gravitationally disturb the motion of nearby planetesimals. The bodies that should have formed a planet just inward in the solar system from Jupiter were broken up, thrown into the sun, or ejected from the solar system. The asteroids seen today are the last remains of those rocky planetesimals.

The comets, in contrast, are evidently the last of the icy planetesimals. Some may have formed in the outer solar nebula beyond Neptune and Pluto, but many probably formed among the Jovian planets where ices could condense easily. Mathematical models show that the massive Jovian planets could have ejected some of these icy planetesimals into the far outer solar system. In a later chapter, you will see evidence that a comet appears in the sky when one of these icy bodies falls into the inner solar system.

The icy Kuiper belt objects appear to be ancient planetesimals that formed in the outer solar system but were never incorporated into a planet. They orbit slowly far from the light and warmth of the sun and, except for occasional collisions, have not changed much since the solar system was young.

The large satellite systems of the Jovian worlds may contain two kinds of moons. Some moons may have formed in orbit around the forming planet in a miniature of the solar nebula. But some of the smaller moons may be captured planetesimals and asteroids. The large masses of the Jovian planets would have made it easier for them to capture satellites.

In Table 19-1, you noted that all four Jovian worlds have ring systems, and you can understand this by considering the large mass of these worlds and their remote location in the solar system. A large mass makes it easier for a planet to hold onto orbiting ring particles; and, being farther from the sun, the ring particles are not as easily swept away by the pressure of sunlight and the solar wind. It is hardly surprising, then, that the terrestrial planets, low-mass worlds located near the sun, have no planetary rings.

The last entry in Table 19-1 is the common ages of solar system bodies, and the solar nebula hypothesis has no difficulty explaining that characteristic. If the hypothesis is correct, then the planets formed at the same time as the sun and should have roughly the same age. Although scientists have mineral samples for radioactive dating only from Earth, the moon, Mars, and meteorites, so far the ages seem to agree.

The solar nebula hypothesis can account for the distinguishing characteristics of the solar system, but there is yet another test you should apply to the hypothesis. What about the problem that troubled Laplace and his nebular hypothesis? What about the angular momentum problem? If the sun and planets formed from a contracting nebula, the sun should have been left spinning very rapidly. That is, it should have most of the angular momentum in the solar system, and instead it has very little. The solar nebula hypothesis can explain the angular momentum problem if you think about the relationship between the origin of planetary systems and star formation. What you know about star formation tells you that the sun must have had a strong stellar wind when it was young. Also, a rapidly spinning star is likely to generate a strong magnetic field. Like a great paddle wheel, the sun's magnetic field must have stuck out into space, and gas flowing away from the sun would have been dragged around with the rotating magnetic field. This would have slowed the sun and left it rotating slowly as it is today. Astronomers now believe that the angular momentum problem is no longer an objection to the solar nebula hypothesis.

Your general understanding of the origin of the solar system gives you a new way of thinking about asteroids, meteors, and comets. They are the last of the debris left behind by the solar nebula. These objects are such important sources of information about the history of our solar system that they will be discussed in detail in Chapter 25. But for now you are probably wondering: What happened to the solar nebula?

Clearing the Nebula

The sun probably formed along with many other stars in a swirling nebula. Observations of young stars in Orion suggest that radiation from nearby hot stars would have evaporated the disk of gas and dust around the sun and that the gravitational influence of nearby stars would have pulled gas away. Even without the external effects, four internal processes would have gradually destroyed the solar nebula.

The most important of these internal processes was **radiation pressure.** When the sun became a luminous object, light streaming from its surface pushed against the particles of the solar neb-

ula. Large bits of matter like planetesimals and planets were not affected, but low-mass specks of dust and individual gas atoms were pushed outward and eventually driven from the system.

The second effect that helped clear the nebula was the solar wind, the flow of ionized hydrogen and other atoms away from the sun's upper atmosphere. This flow is a steady breeze that rushes past Earth at about 400 km/s (250 mi/s). When the sun was young, it may have had an even stronger solar wind, and irregular fluctuations in its luminosity, like those observed in young stars such as T Tauri stars, may have produced surges in the wind that helped push dust and gas out of the nebula.

The third effect for clearing the nebula was the sweeping up of space debris by the planets. All of the old, solid surfaces in the solar system are heavily cratered by meteorite impacts (■ Figure 19-11). Earth's moon, Mercury, Venus, Mars, and most of the moons in the solar system are covered with craters. A few of these craters have been formed recently by the steady rain of meteorites that falls on all the planets in the solar system, but most of the craters appear to have been formed roughly 4 billion years ago in what is called the **heavy bombardment,** as the last of the debris in the solar nebula was swept up by the planets. You will find more evidence of this bombardment in later chapters. The craters show that many of the last remaining solid objects in the solar nebula were swept up by the planets soon after they formed.

The fourth effect was the ejection of material from the solar system by close encounters with planets. If a small object such as a planetesimal passes close to a planet, it can gain energy from the planet's gravitational field and be thrown out of the solar system. Ejection is most probable in encounters with massive planets, so the Jovian planets were probably very efficient at ejecting the icy planetesimals that formed in their region of the nebula.

Attacked by the radiation and gravity of nearby stars and racked by internal processes, the solar nebula could not survive very long. Once the gas and dust were gone and the last of the planetesimals were swept up, the planets could no longer gain mass, and planet building ended.

Building Scientific Arguments

Why are there two kinds of planets in our solar system?
This is an opportunity for you to build an argument that closely analyzes the solar nebula theory. Planets begin forming from solid bits of matter, not from gas. Consequently, the kind of planet that forms at a given distance from the sun depends on the kind of compounds that can condense out of the gas to form solid particles. In the inner parts of the solar nebula, the temperature was so high that most of the gas could not condense to form solids. Only metals and silicates could form solid grains, and the innermost planets grew from this dense material. Much of the mass of the solar nebula consisted of hydrogen, helium, water vapor, and other gases, and they were present in the inner solar nebula but couldn't form solid grains.

■ Figure 19-11

Every old, solid surface in the solar system is scarred by craters. (a) Earth's moon is scarred by craters ranging from basins hundreds of kilometers in diameter down to microscopic pits. (b) The surface of Mercury, as photographed by a passing spacecraft, shows vast numbers of overlapping craters. (NASA)

The small terrestrial planets couldn't grow from these gases, so the terrestrial planets are small and dense.

In the outer solar nebula, the composition of the gas was the same, but it was cold enough for water vapor to condense to form ice grains. Because hydrogen and oxygen are so abundant, there was lots of ice available. The outer planets grew from solid bits of metal and silicate combined with large amounts of ice. The outer planets grew so rapidly that they became massive enough to capture gas directly, and they became the hydrogen- and helium-rich Jovian worlds.

The condensation sequence combined with the solar nebula hypothesis gives you a way to understand the difference between the terrestrial and Jovian planets. Now expand your argument: **Why do some astronomers argue that the formation of the Jovian planets is a problem that needs further explanation?**

■ ■ ■

Connections: The great chain of origins leads from the first instant of the big bang through the birth of galaxies, the formation of stars, and the origin of the chemical elements to the formation of our solar system, and finally to us. You have traced the theory and compared it to evidence from our own solar system, but you must be wondering by now: Does this happen around other stars?

(19-4) Planets Orbiting Other Stars

ARE THERE OTHER EARTHS? You have read earlier in this chapter that astronomers have, indeed, detected planets orbiting other stars, so now it is time to get down to the details. Are those planets like Earth? How did they form? Can astronomers see planets forming around other stars?

Your questions can be boiled down to two main unknowns. Can astronomers see planet-forming disks around other stars? And, can astronomers see actual planets orbiting other stars? The answer in both cases is an amazing yes.

Planet-Forming Disks around Other Suns

Both visible- and radio-wavelength observations detect dense disks of gas orbiting young stars. For example, at least 50 percent of the stars in the Orion nebula are encircled by dense disks of gas and dust (■ Figure 19-12 and page 253). A young star is visible at the center of each disk, and astronomers estimate that the disks contain at least a few times Earth's mass in a region a few times larger in diameter than our solar system. The Orion star-forming region is only a few million years old, so it does not seem likely that planets could have formed in these disks yet.

Many of the young stars in the Orion Nebula are surrounded by disks of gas and dust, but intense light from the brightest star in the neighborhood is evaporating the disks to form expanding clouds of gas. These disks may evaporate before they can form planets, but the large number of such disks shows that disks around young stars are common. (C. R. O'Dell, Rice, NASA; Dark disk: M. McCaughrean, Max Plank Inst. for Astronomy, C. R. O'Dell, NASA; Lower left inset: J. Bally, H. Throop, C. R. O'Dell, NASA)

Gaseous cloud evaporated from dust disk

Dust disk

Light from central star scattered by dust

Dust disk seen edge-on

Dust disk

Gaseous cloud evaporated from dust disk

Visual-wavelength image

Furthermore, the intense radiation from the hot stars in the area is evaporating the disks so fast that planets may never have a chance to grow large. The important point for astronomers is that so many of these young stars have disks. Evidently, disks of gas and dust are a common feature of star formation. These disks appear to resemble the solar nebula from which our planetary system formed.

The Hubble Space Telescope can detect dense disks of gas and dust around young stars in a slightly different way. The disks show up by the shadows they cast in the nebulae that surround the newborn stars (■ Figure 19-13). These disks are related to the formation of bipolar flows (Figure 11-7) in that they focus the gas flowing away from a young star into two jets shooting in opposite directions. It seems likely that the sun and the solar nebula formed a bipolar flow when they were young.

Infrared astronomers have found very cold, low-density dust disks around stars such as Beta Pictoris. These stars are believed to have completed their formation, so they are clearly older than the newborn stars in Orion. The dust disk around Beta Pictoris is about 20 times the diameter of our solar system and, like the other known low-density disks, has an inner zone with even lower density. These inner regions may be places where planets have formed (■ Figure 19-14). Such tenuous dust disks are sometimes called **debris disks** because they are understood to be debris released in collisions among small bodies such as comets, asteroids,

and Kuiper belt objects. Our own solar system contains such dust, and astronomers believe it has an extensive debris disk of cold dust extending far beyond the orbits of the planets.

Any planets orbiting stars with debris disks are not detectable, but the presence of the cold dust disks assures you that small bodies like asteroids and comets are present. If those small objects are there, then you can expect that there are also planets orbiting those stars.

Remember Vega, one of our Favorite Stars? It is a very bright star high in the summer sky. Infrared observations reveal that it is surrounded by a disk of dust, and detailed studies show that some of the dust is tiny. The pressure of starlight from Vega would blow that dust way quickly, so astronomers conclude that the dust must have been produced by the collision of two large planetesimals within the last million years. Fragments from that collision are still smashing into each other now and then and producing more dust. This effect has also been found in the disk around Beta Pictoris and the faint star known as HIP8920. Such smashups happen rarely in a dust disk, but they happen often enough to keep the disk supplied with dust.

Notice the difference between the two kinds of disks that astronomers have found. The low-density dust disks such as the one around Beta Pictoris are produced by dust from collisions among comets, asteroids, and Kuiper belt objects. Such disks are evidence that planetary systems have already formed. The dense

■ **Figure 19-13**

Disks around young stars are evident in these Hubble Space Telescope infrared images. The stars are so young that material is still falling inward and being illuminated by the light from the star. Dark bands across the nebulae (arrows) are caused by dense disks of gas and dust that orbit the stars. (D. Padgett, IPAC/Caltech, W. Brandner, IPAC, K. Stapelfeldt, JPL, and NASA)

The first planet detected this way was discovered in 1995. It orbits the star 51 Pegasi. As the planet circles the star, the star wobbles slightly, and this very small motion of the star is detectable as Doppler shifts in the star's spectrum (■ Figure 19-15a). From the motion of the star and estimates of the star's mass, astronomers can deduce that the planet has half the mass of Jupiter and orbits only 0.05 AU from the star. Half the mass of Jupiter amounts to 160 Earth masses, so this is a large planet. Note also that it orbits very close to its star.

Astronomers were not surprised by the announcement that a planet orbited 51 Pegasi; for years astronomers had assumed that many stars had planets. Nevertheless, they greeted the discovery with typical skepticism (**Window on Science 19-3**). That skepticism led to careful tests of the data and further observations that confirmed the discovery. In fact, over 150 planets have been discovered in this way, including at least three planets orbiting the star Upsilon Andromedae—a true planetary system (Figure 19-15b). Over a dozen such planetary systems have been found.

Another way to search for planets is to look for changes in the brightness of the star when the orbiting planet crosses in front of or behind the star. The decrease in light is very small, but it is detectable, and astronomers have used this technique to detect a few planets as they crossed in front of their stars. All of these planets are roughly the size of Jupiter and even more massive. The Spitzer Infrared Space Telescope has detected two planets when they passed behind their stars. The systems are young, and the planets are hot and emit significant infrared radiation. When they passed behind the stars they orbit, the total infrared brightness of the system decreased. These observations further confirm the existence of extrasolar planets.

Notice how the techniques used to detect these planets resemble techniques used to study binary stars. Most of the planets were discovered using the same observational methods used to study spectroscopic binaries, but a few were found by observing the stars as if they were eclipsing binaries (Chapter 9).

disks of gas and dust such as those seen round the stars in Orion are sites where planets could be forming right now.

The observational evidence gives astronomers confidence that planets orbit many stars. Of course, you are wondering if there isn't some way to see these planets directly. That's not easy, but astronomers are making progress.

Planets Orbiting Other Stars

A planet orbiting another star is called an **extrasolar planet.** Such a planet would be quite faint and difficult to detect so close to the glare of its star. But there are ways to find these planets. All you have to do is imagine walking a dog.

You will remember that Earth and its moon orbit around their common center of mass, and two stars in a binary system orbit around their center of mass. When a planet orbits a star, the star moves very slightly as it orbits the center of mass of the planet–star system. Think of someone walking a poorly trained dog on a leash; the dog runs around pulling on the leash, and even if it were an invisible dog, you could plot its path by watching how its owner is jerked back and forth. Astronomers can detect a planet orbiting another star by watching how the star moves as the planet tugs on it.

Glare of star Beta Pictoris hidden behind central mask

Warp and rings in disk may be caused by one or more planets.

Infrared image

Visual-wavelength image

Size of Pluto's orbit

Dust grains contain ices and silicates.

Star

Narrow ring of dust may be confined by one or more planets.

Glare of star HR4796A hidden behind mask

Diameter of Neptune's orbit

Visual

The K2V star Epsilon Eridani is not very bright in the far-infrared.

Clumps in ring of dust may be related to planets.

Size of Pluto's orbit

Star HD107146 hidden behind mask

Visual

Neptune's orbit

Edge of detector

Visual

The star Fomalhaut is off-center, indicating the presence of a planet.

■ Active Figure 19-14

Dust disks have been detected orbiting a number of stars; but, in the visible part of the spectrum, the dust is at least 1000 times fainter than the stars, which must be hidden behind masks. In the far-infrared, the stars are not as bright as the dust. Warps, rings, clumps, and off-center rings in these disks suggest the gravitational influence of planets. (Beta Pic: NASA, Burrows and Krist; HR4769A: NASA, Weinberger, Becklin, G. Schneider; Eps Eri: Joint Astronomy Center; HD107146: NASA; Fomalhaut: NASA, ESA, Kalas, Graham, and Clampin)

Ace⊙Astronomy™ Log into AceAstronomy and select this chapter to see the Active Figure "Circumstellar Disk" and observe a model disk at different wavelengths.

The planets discovered so far tend to be massive and have short periods because lower-mass planets or longer-period planets are harder to detect. Low-mass planets don't tug on their stars very

■ Figure 19-15 ▶

(a) Just as someone walking a lively dog is tugged around, the star 51 Pegasi is tugged around by the planet that orbits it every 4.2 days. The wobble is detectable in precision observations of its Doppler shift. (b) Someone walking three dogs is pulled about in a more complicated pattern, and you can see something similar in the Doppler shifts of Upsilon Andromedae, which is orbited by three planets.

much, and present-day spectrographs can't detect the very small velocity changes that these gentle tugs produce. Planets with longer periods are harder to detect because Earth's astronomers have not been making high-precision observations for many years. Jupiter takes 11 years to circle the sun once, so it will take years for astronomers to see the longer-period wobbles produced by planets lying farther from their stars. So you should not be surprised that the first planets discovered are massive and have short periods.

Scientists: Courteous Skeptics

"Scientists are just a bunch of skeptics who don't believe in anything." That is a common complaint about scientists, but it misinterprets the fundamental characteristic of the scientist. Yes, scientists are skeptical about new ideas and discoveries, but they also hold strong beliefs about how nature works. Ultimately, scientists try to tell stories about how nature works, and skepticism is the scientists' guide.

Some people think that "telling a story" is telling a fib. The stories that scientists tell are exactly the opposite; perhaps you could call them antifibs, because they are as true as scientists can make them. The stories are explanations that have been tested over and over to make them accurate descriptions of nature. Skepticism is the tool scientists use to test every aspect of an explanation.

When a planet was discovered orbiting 51 Pegasi, astronomers were skeptical—not because they thought the observations were wrong but because that is how science works. Every observation is tested, and every discovery is confirmed. Only an idea that survives many tests begins to be accepted as a scientific truth.

To nonscientists, this makes scientists seem like irritable skeptics, but among scientists it is not bad manners to say, "Really, how do you know that?" or "Why do you think that?" or "Show me the evidence!"

Scientists try to tell stories about how nature works, stories that are sometimes called theories. A story by John Steinbeck can be a brilliant work of art invented to help you understand some aspect of your life, and another story, such as a TV script, can be little more

than chewing gum for the mind. But scientists' stories are different in a critical way. They are true. To create such stories, scientists cannot create facts to fit their story. Rather, every link in a scientific story must be based on evidence and logic. Of course, evidence can be misunderstood, and logical errors happen. To test every aspect of their story, scientists must be continuously skeptical. That is the only way to discover those scientific truths that help you understand how nature works.

Skepticism is not a refusal to hold beliefs. Scientists often believe sincerely in their theories, once those theories have been tested over and over. Rather, skepticism is the tool scientists use to find those natural principles worthy of belief.

The new planets may seem odd for another reason. In our own solar system, the large planets formed farther from the sun where the solar nebula was colder and ices could condense. How could big planets form so near their stars? Theorists find that planets that form in an especially dense disk of matter could spiral inward as they sweep up planetesimals. That means it is possible for a few planets to become the massive, short-period planets that are detected most easily.

A few of the newly discovered extrasolar planets have elliptical orbits, and that seems odd compared with our solar system in which the planetary orbits are nearly circular. Theorists point out, however, that planets may interact in some young planetary systems and can be thrown into elliptical orbits. This is probably rare among planetary systems, but astronomers find these extreme systems more easily because they tend to produce big wobbles.

The preceding paragraphs should reassure you that massive planets in small orbits or in elliptical orbits are not outrageous. They do not contradict the solar nebula hypothesis.

As astronomers refine their instruments to detect smaller velocity shifts in stars, they are finding lower-mass planets. A super-Earth was found in 2005. It is about 7.5 times the mass of Earth and twice Earth's diameter. It orbits its star in only a bit over two Earth days, so it must be very hot. It does not seem that it could be a Jovian world, so astronomers suspect it is made of rock and metal, as is Earth. Evidently the extrasolar planets are not all Jovian.

Actually photographing a planet orbiting another star is about as easy as photographing a bug crawling on the bulb of a searchlight miles away. Planets are small and dim and get lost in the glare of the stars they orbit. Nevertheless, a few objects have been detected that appear to be planets (■ Figure 19-16). Searches for more are being conducted, and space telescopes are being developed that will eventually be able to image Earth-size planets orbiting nearby stars.

The discovery of extrasolar planets gives astronomers added confidence in the solar nebula theory. The theory predicts that planets are common, and astronomers are finding them orbiting many stars.

Building Scientific Arguments

Why are dust disks evidence that planets have already formed?

Sometimes a good scientific argument combines evidence, theory, and an astronomer's past experience, a kind of scien-

Planet 2M1207b orbits 77AU from its brown dwarf "sun."

Infrared image

Brown dwarf

Artist's conception

■ Figure 19-16

Infrared observations reveal a planet of about 5 Jupiter masses orbiting a brown dwarf in an orbit roughly twice as large as that of Neptune around the sun. Spectra showing water vapor and the planet's infrared colors suggest it is relatively cool. In an artist's impression, the planet orbits round a brown dwarf still surrounded by a dusty disk. (ESO)

tific common sense. Certainly the cold dust disks seen around stars like Vega are not places where planets are forming. They are not hot enough or dense enough to be young disks. Rather, the dust disks must be older, and the dust is being produced by collisions among comets, asteroids, and Kuiper belt objects. Dust would be blown away quickly, so these collisions must be a continuing process. Astronomers know from experience that where you find comets, asteroids, and Kuiper belt objects, you should also find planets, so the dust disks seem to be evidence that planets have already formed in such systems.

Now build a new argument. **What direct evidence can you cite that planets orbit other stars?**

■   ■   ■

Connections: You have watched our solar system form, and you have seen clear evidence that planets form around other stars. Now it is time to voyage out among the planets that orbit the sun and search for more clues as to how they formed and evolved. Your journey begins in the next chapter.

Summary

19-1 | Theories of Earth's Origin

What theories account for the origin of the solar system?

■ René Descartes proposed that the solar system formed from a contracting vortex of matter—an evolutionary theory. Buffon later suggested that a passing star pulled matter out of the sun to form the planets— a catastrophic theory.

■ Laplace's nebular hypothesis required a contracting nebula to leave behind rings that formed each planet, but it could not explain the low angular momentum of the sun.

■ Modern astronomy reveals that the matter in our solar system was formed in the big bang, and the atoms heavier than helium were cooked up in a few generations of stars. The sun and planets evidently formed from a cloud of gas in the interstellar medium.

■ The solar nebula hypothesis proposes that the planets formed in a disk of gas and dust around the protostar that became the sun. Observations show that these disks are common.

19-2 | A Survey of the Solar System

What properties must these theories explain?

■ The solar system is disk shaped in the orbital revolution of the planets and their moons and in the rotation of the planets on their axes.

■ The planets are divided into two types. The inner four planets are terrestrial—small, rocky, dense Earthlike worlds. The next four outward are Jovian planets that are large and low density.

■ All four of the Jovian worlds have ring systems and large families of moons. The terrestrial planets have no rings and few moons.

■ Most of the asteroids, small, irregular rocky bodies, are located between the orbits of Mars and Jupiter.

■ Comets are icy bodies that fall into the inner solar system along long elliptical orbits. As the ices vaporize and release dust, the comet develops a tail that points approximately away from the sun.

■ Meteoroids that fall into Earth's atmosphere are vaporized by friction and are visible as meteors. Larger and stronger meteoroids may survive to reach the ground, where they are called meteorites.

■ The Kuiper belt is composed of small, icy bodies that orbit the sun beyond the orbit of Neptune.

■ The age of an object can be found by radioactive dating. The oldest rocks from Earth, the moon, and Mars have ages approaching 4.6 billion years. The oldest objects in our solar system are the meteorites, which have ages of 4.56 billion years. This is taken to be the age of the solar system.

19-3 | The Story of Planet Building

How do planets form?

■ Condensation in the solar nebula converted some of the gas into solid bits of matter, which accreted to form billions of planetesimals.

■ Planets begin growing by accreting solid material. But once a planet approaches about 15 Earth masses, it can begin growing by gravitational collapse as it pulls in gas from the solar nebula.

■ According to the condensation sequence, the inner part of the solar nebula was so hot that only metals and rocky minerals could form solid grains. The dense terrestrial planets grew from those solid particles and did not include many ices or low-density gases.

■ The outer solar nebula was cold enough for metals, rocky minerals, and large amounts of ices to form solid particles. The Jovian planets grew rapidly and incorporated large amounts of low-density ices and gases.

■ Evidence that the condensation sequence was important in the solar nebula can be found in the densities of the terrestrial planets compared to the Jovian planets. Even the uncompressed densities among the terrestrial planets show that the inner planets are the most dense.

■ The terrestrial planets may have formed slowly from the accretion of planetesimals of similar composition and then differentiated later when radioactive decay heated the planet's interiors.

■ It is also possible that the planets formed so rapidly that the heat of formation melted the planets and they differentiated as they formed. In that case, Earth's first atmosphere was not captured from the nebula but was baked out of Earth's interior.

■ Disks of gas and dust may not last long enough to form Jovian planets by accretion and gravitational collapse. Some models suggest the Jovian planets could have formed directly and rapidly by gravitational collapse.

■ In addition to intense light from hot nearby stars and the gravitational influence of passing stars, the solar nebula was eventually cleared away by radiation pressure, the solar wind, the sweeping up of debris by the planets, and ejection.

■ All of the old surfaces in the solar system were heavily cratered by an early bombardment of the debris that filled the solar system when it was young.

19-4 | Planets Orbiting Other Stars

Is our solar system unique?

■ Hot disks of gas and dust have been detected in early stages of star formation and are believed to be the kind of disk in which planets could form.

■ Cold dust disks, also known as debris disks, appear to be produced by dust released by collisions among comets, asteroids, and Kuiper belt objects. Such disks may be home to planets that have already formed.

■ Planets orbiting other stars have been detected by the way they tug their stars about, creating small Doppler shifts in the stars' spectra. Planets have also been detected as they cross in front of their star and dim the star's light. A few planets have been detected when they crossed behind their star and their infrared radiation was cut off.

■ Nearly all extrasolar planets found so far are massive, Jovian worlds. Lower-mass, terrestrial planets are harder to detect.

New Terms

passing star hypothesis (p. 450)	comet (p. 456)
evolutionary hypothesis (p. 451)	meteor (p. 457)
catastrophic hypothesis (p. 451)	meteoroid (p. 457)
nebular hypothesis (p. 451)	meteorite (p. 457)
angular momentum problem (p. 451)	half-life (p. 457)
solar nebula hypothesis (p. 452)	gravitational collapse (p. 460)
terrestrial planet (p. 454)	uncompressed density (p. 460)
Jovian planet (p. 454)	condensation sequence (p. 461)
Galilean satellites (p. 455)	planetesimal (p. 461)
asteroid (p. 456)	condensation (p. 461)
Kuiper belt (p. 456)	accretion (p. 461)

protoplanet (p. 462)

differentiation (p. 462)

outgassing (p. 463)

heat of formation (p. 463)

radiation pressure (p. 466)

heavy bombardment (p. 466)

debris disk (p. 468)

extrasolar planet (p. 469)

20. Why do astronomers conclude that the surface of Mercury, shown at right, is old? When did the majority of these craters form?

(NASA)

Review Questions

Ace Astronomy™ Assess your understanding of this chapter's topics with additional quizzing and animations at **http://ace .brookscole.com/sf9**

1. What were the objections to the passing star hypothesis? to the nebular hypothesis?

2. What produced the helium now present in the sun's atmosphere? in Jupiter's atmosphere? in the sun's core?

3. What produced the iron in Earth's core and the heavier elements like gold and silver in Earth's crust?

4. What evidence can you cite that disks of gas and dust are common around young stars?

5. According to the solar nebula theory, why is the sun's equator nearly in the plane of Earth's orbit?

6. Why does the solar nebula theory predict that planetary systems are common?

7. Why do astronomers think the solar system formed about 4.6 billion years ago?

8. If you visited another planetary system, would you be surprised to find planets older than Earth? Why or why not?

9. Why is almost every solid surface in our solar system scarred by craters?

10. What is the difference between condensation and accretion?

11. Why don't terrestrial planets have rings like the Jovian planets?

12. How does the solar nebula theory help you understand the location of asteroids?

13. How does the solar nebula theory explain the dramatic density difference between the terrestrial and Jovian planets?

14. If you visited some other planetary system in the act of building planets, would you expect to see the condensation sequence at work, or was it unique to our solar system?

15. Why would you expect to find that planets are differentiated?

16. What processes cleared the nebula away and ended planet building?

17. What evidence can you cite that planets orbit other stars?

18. What is the difference between the hot disks and cold disks seen around stars?

19. What do you see in the image at the right that indicates that the planet formed far from the sun?

(NASA)

Discussion Questions

1. In your opinion, should all solar systems have asteroid belts? Should all solar systems show evidence of an age of heavy bombardment?

2. If the solar nebula hypothesis is correct, then there are probably more planets in the universe than stars. Do you agree? Why or why not?

3. The human race has intelligence and consequently has both the right and the responsibility to wonder about its origins. Do you agree?

Problems

1. If you observed the solar system from the nearest star (1.3 pc), what would the maximum angular separation be between Jupiter and the sun? (*Hint:* Use the small-angle formula.)

2. The brightest planet in our sky is Venus, which is sometimes as bright as apparent magnitude −4 when it is at a distance of about 1 AU. How many times fainter would it look from a distance of 1 parsec (206,265 AU)? What would its apparent magnitude be? (*Hints:* Remember the inverse square law, from Chapter 5; also see Chapter 2.)

3. What is the smallest-diameter crater you can identify in the photo of Mercury on page 454? (*Hint:* See Appendix A, Properties of the Planets, to find the diameter of Mercury in kilometers.)

4. A sample of a meteorite has been analyzed, and the result shows that out of every 1000 nuclei of potassium-40 originally in the meteorite, only 100 have not decayed. How old is the meteorite? (*Hint:* See Figure 19-6.)

5. In Table 19-2, which object's observed density differs least from its uncompressed density? Why?

6. What composition might you expect for a planet that formed in a region of the solar nebula where the temperature was about 100 K?

7. Suppose that Earth grew to its present size in 1 million years through the accretion of particles averaging 100 g each. On the average, how many particles did Earth capture per second? (*Hint:* See Appendix A to find Earth's mass.)

8. If you stood on Earth during its formation as described in Problem 7 and watched a region covering 100 m², how many impacts would you expect to see in an hour? (*Hints:* Assume that Earth had its present radius. The surface area of a sphere is $4\pi r^2$.)

9. The velocity of the solar wind is roughly 400 km/s. How long does it take to travel from the sun to Pluto?

Ace Astronomy™ To access the resources in the Media Cluster, log into AceAstronomy at **http://ace.brookscole .com/sf9** and select Chapter 19.

Radioactive Decay
The characteristic decay patterns of different radioactive elements provide an invaluable dating tool. In this animation, see how radioactive atoms decay into daughter atoms.

Circumstellar Disk
In this animation, you can observe the disk surrounding a star in different kinds of light. Notice that what you observe depends on wavelength.

Extrasolar Planets
This simulation presents a hypothetical star system with one planet orbiting the star. Study what variables help astronomers to detect extrasolar planets.

Lab 8: Extrasolar Planets
This lab examines how indirect methods such as Doppler shift techniques are used to discover and study extrasolar planets. At the end of the lab, you will build your own solar system.

Critical Inquiries for the Web

1. How does our solar system compare with the others that have been found? Search the Internet for sites that give information about planetary systems around other stars. What kinds of planets have been detected by these searches so far? Discuss the selection effects (see Window on Science 25-2) that must be considered when interpreting these data.

2. The process of protoplanetary accretion is still not well understood. Search the Web for current research in this field. From the results of your search, outline the basic steps in the formation of a protoplanet through accretion. What specific factors are important in these models of planet building? Do these models produce planetary systems similar to the ones known to exist?

3. How is radioactive dating carried out on meteorites and rocks from surfaces of various bodies in the solar system? Look for websites on the details of radioactive dating and summarize the methods used to uncover the abundances of radioactive elements in a particular sample. (*Hint:* Try looking for information on how a particular meteorite—for example, the Martian meteorite ALH84001—was studied. What age range was determined, and what radioactive elements were used to arrive at the age?)

Exploring *TheSky*

1. Look at the solar system from space. Notice how thin the disk of the solar system is and how inclined the orbits of Jupiter and Mercury are. (*Hint:* Under the **View** menu, choose **3D Solar System Mode** and then zoom in or out. Tip the solar system up and down to see it edge-on.)

2. Look at the solar system from space and notice how small the orbits of the inner planets are compared to the orbits of the outer planets. They make two distinct groups. (*Hint:* Use **3D Solar System Mode.**)

3. Watch the comets orbiting around the sun. Can you locate the comet C/198 M5 (Linear)? What is its orbit like? (*Hint:* Use **3D Solar System Mode** and set the time step to 30 days [**30d**].)

 Go to the Brooks/Cole Astronomy Resource Center (**http:// astronomy.brookscole.com**) for critical thinking exercises, articles, and additional readings from InfoTrac College Edition, Brooks/Cole's online student library.

20 | Earth: The Standard of Comparative Planetology

Nature evolves. The world

was different yesterday.

PRESTON CLOUD,
COSMOS, EARTH AND MAN

S TARS DIE, BUT PLANETS
DON'T. That should give you a feeling of secu-
rity; the beautiful planet you live on will not explode or col-
lapse into a black hole. But Earth is not eternal. You know that the survival of your home world is limited by the life
expectancy of your star. Furthermore, you know that Earth has not always been as you see it today. Planets evolve, and
our planet has changed dramatically since it formed. Mountains rise and weather away, the climate warms and cools,
and the continents themselves drift and grind against each other. To fully know Earth, you need to know how planets in
general evolve. Why is Venus hot and Mars cold? Why does Mercury have many craters and Earth only a few? ▌ The
best way, perhaps the only way, to answer such questions is to use **comparative planetology,** the study of planets

▌ Continued on page 478 ▌

Guidepost

Looking Back

In the previous chapter, you learned how our solar system formed as a by-product of the formation of the sun. You saw how distance from the sun determined the general character of each planet. Now you are ready to visit the planets and get to know them as individuals.

This Chapter

As you study the individual planets, you need to avoid the pitfall of learning facts instead of understanding principles. One way to focus on principles is to compare the planets with each other and search for similarities and contrasts. Like people, the terrestrial planets are more alike than they are different, but it is the differences that are most memorable.

As you explore the terrestrial planets, you can use Earth as your basis of comparison, but that means you must think of Earth as a planet. In this chapter, you will see Earth in a new way and answer four essential questions:

How can comparison help you understand the terrestrial planets?

How has Earth changed since it formed?

What is the inside of Earth like?

How has Earth's atmosphere formed and evolved?

Looking Ahead

Like a mountain climber establishing a base camp before attempting the summit, you have established your basis of comparison on Earth. In the following chapters you will visit worlds that are un-Earthly but, in many ways, familiar.

through comparison and contrast, and Earth is the basis of comparison for your study. One reason to begin with Earth is that it is the planet you know best. You can study it carefully and then draw comparisons to other worlds. But another reason is that your home is a planet of extremes. The land above the oceans is quite different from the land under the oceans (**Celestial Profile 2**). Earth's interior is molten and generates a magnetic field. Its crust is active, with moving sections that push against each other and trigger earthquakes, volcanoes, and mountain building. Even Earth's atmosphere is extreme. Processes have altered Earth's air from its original composition to a highly unusual, oxygen-rich sea of gas. The complex properties of extreme Earth will help you understand the remaining planets in our solar system.

It may seem strange to begin a journey among the planets by looking at the dirt under your feet, so you should put Earth in perspective by taking a lightning-fast tour just to see how Earth compares with the worlds on your itinerary.

20-1 A Travel Guide to the Terrestrial Planets

IF YOU VISIT the city of Granada in Spain, you will probably consult a travel guide; and, if it is a good guide, it will do more than tell you where to find museums and restrooms. It will tell you to look at the palace called the Alhambra and compare it with

■ Figure 20-1

Planets in comparison. Earth and Venus are similar in size, but their atmospheres and surfaces are very different. The moon and Mercury are much smaller, and Mars is intermediate in size. (Moon: UCO/Lick Observatory; All planets: NASA)

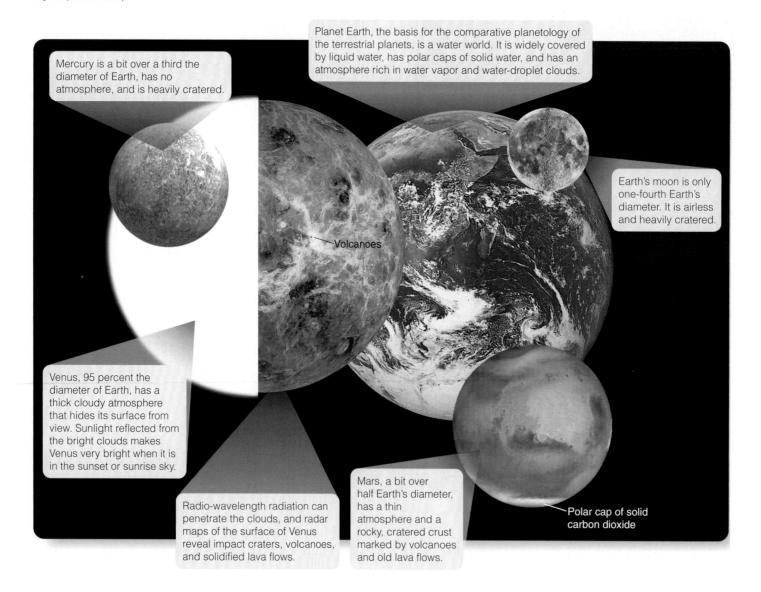

Mercury is a bit over a third the diameter of Earth, has no atmosphere, and is heavily cratered.

Planet Earth, the basis for the comparative planetology of the terrestrial planets, is a water world. It is widely covered by liquid water, has polar caps of solid water, and has an atmosphere rich in water vapor and water-droplet clouds.

Earth's moon is only one-fourth Earth's diameter. It is airless and heavily cratered.

Volcanoes

Venus, 95 percent the diameter of Earth, has a thick cloudy atmosphere that hides its surface from view. Sunlight reflected from the bright clouds makes Venus very bright when it is in the sunset or sunrise sky.

Radio-wavelength radiation can penetrate the clouds, and radar maps of the surface of Venus reveal impact craters, volcanoes, and solidified lava flows.

Mars, a bit over half Earth's diameter, has a thin atmosphere and a rocky, cratered crust marked by volcanoes and old lava flows.

Polar cap of solid carbon dioxide

buildings in Morocco. You are beginning a journey in which you will visit five Earthlike worlds, so you should consult a travel guide and see what is in store.

Five Worlds

The terrestrial planets include Mercury, Venus, Earth, Earth's moon, and Mars. It may surprise you that the moon is on your itinerary. It is, after all, just a natural satellite orbiting Earth and isn't one of the planets. But the moon is a fascinating world of its own, and it makes a striking comparison with the other worlds on your list. Just as you shouldn't tour Spain without seeing Granada, you should not tour the planets without seeing the moon.

■ Figure 20-1 compares the five worlds you will study. The first feature you should notice is diameter. The moon is small, and Mercury is hardly much bigger. Earth and Venus are large and quite similar in size, but Mars is a medium-sized world. You will discover that size is a critical factor in determining a world's personality. Small worlds tend to be cold and dead, but larger worlds can be hot and active.

Core, Mantle, and Crust

Notice that these terrestrial worlds are made up of rock and metal. They all have rocky, low-density crusts, while much of the metal in these worlds has sunk to their centers to form dense cores. You will discover that Earth has a molten iron core but that the moon lacks much iron at all. Between the crust and the core of a planet lies a deep layer of dense rock.

When the planets formed, their surfaces were cratered by the heavy bombardment of debris in the young solar system. You will see lots of craters on these worlds, especially on Mercury and the moon, but notice that cratered surfaces are old. If a lava flow covered up the craters after the end of the heavy bombardment, few new craters could be formed because most of the debris in the solar system was gone. When you see smooth plains on a planet such as those on the moon or on Mars, you can guess that the plains are younger than the cratered areas.

On your travels among the Earthlike worlds, look for signs of heat flowing up from the interior. In the preceding chapter you saw evidence that the planets probably form hot, and radioactive elements trapped in their interiors decay and generate more heat. That heat, flowing upward through the cooler crust, can make a world active with volcanoes and lava flows. Earth is a dramatic example of an active planet, but you will see evidence that Venus too is active. But watch for the small worlds that cooled fast and have little heat flowing outward now. You will find that Mercury and the moon, being small, are inactive, dead worlds.

Atmospheres

When you look at Mercury and the moon in Figure 20-1, you can see their craters and plains and mountains. But the surface of Venus is completely hidden by a cloudy atmosphere even thicker

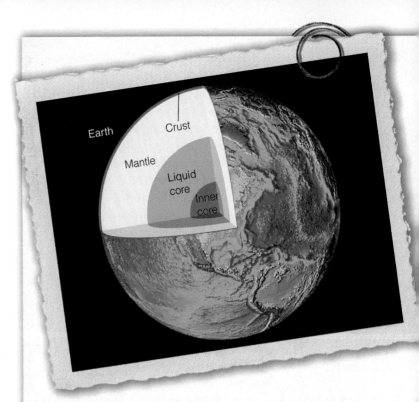

Earth's surface is marked by high continents and low sea floors, but the crust is only 10 to 60 km thick. Below that lie a deep mantle and an iron core. (NGDC)

Celestial Profile 2: Earth

Motion:

Average distance from the sun	1.00 AU (1.495979×10^8 km)
Eccentricity of orbit	0.0167
Maximum distance from the sun	1.0167 AU (1.5210×10^8 km)
Minimum distance from the sun	0.9833 AU (1.4710×10^8 km)
Inclination of orbit to ecliptic	0°
Average orbital velocity	29.79 km/s
Orbital period	1.00 y (365.25 days)
Period of rotation	$24^h00^m00^s$ (with respect to the sun)
Inclination of equator to orbit	23°27'

Characteristics:

Equatorial diameter	12,756 km
Mass	5.976×10^{24} kg
Average density	5.497 g/cm³ (4.07 g/cm³ uncompressed)
Surface gravity	1.0 Earth gravity
Escape velocity	11.2 km/s
Surface temperature	−50° to 50°C (−60° to 120°F)
Average albedo	0.39
Oblateness	0.0034

Personality Point:

Earth comes, through Old English *eorthe* and Greek *eraze,* from the Hebrew *erez,* which means ground. *Terra* comes from the Roman goddess of fertility and growth, *Terra Mater,* Mother Earth.

than Earth's. Mars, the medium planet, has a medium atmosphere of thin gases.

As you explore these worlds, you can ponder two questions. First, why do some worlds have atmospheres while others do not? You will discover that both size and temperature are important. The second question is more complex. Where did these atmospheres come from? To answer that question, you will have to study the geological history of these worlds.

The terrestrial planets are Earthlike worlds. Studying Earth will help you understand other planets, and studying the terrestrial planets will give you a new way to see Earth—as a planet.

Building Scientific Arguments

Why do you expect the inner planets to be high-density worlds?

A good scientific argument gives you a way to see nature—a way to understand why the universe behaves as it does. The best questions in science always begin with "why." In Chapter 19, you saw how the inner planets formed from hot inner parts of the solar nebula. No ice could form there, so the inner planets had to grow from the solid bits of rock and metal that condensed from the hot gas. So you expect the inner planets to be made mostly of rock and metal, which are dense materials.

As you begin studying planets one by one, keep thinking in scientific arguments. They will help you organize all of the information you will meet. Now build an argument to review an important point. **What made all of the craters that are spread through the solar system?**

■ ■ ■

Connections: Some people are satisfied just reading travel guides, but most people want to visit Granada rather than read about it. Of course, you can't really visit Venus or Mars, but one of the wonders of science is that it can take your imagination where your body cannot go. So grab your carry-on luggage. They are calling your flight, and the first world you will visit may look a little unfamiliar even though you have been there.

(20-2) The Early History of Earth

LIKE ALL THE TERRESTRIAL PLANETS, Earth formed from the inner solar nebula about 4.6 billion years ago. As it took form, it began to change, passing through four developmental stages (■ Figure 20-2). Processes that occurred during these four stages determined the present condition of Earth's interior, crust, and atmosphere.

Differentiation produces a dense core, thick mantle, and low-density crust.

The young Earth was heavily bombarded in the debris-filled early solar system.

Flooding by molten rock and later by water can fill lowlands.

Slow surface evolution continues due to geological processes, including erosion.

■ **Figure 20-2**

The four stages of planetary development are illustrated for Earth.

Four Stages of Planetary Development

The first stage of planetary evolution is *differentiation,* the separation of material according to density. Earth now has a dense core and a lower-density crust, and that structure must have originated very early.

Differentiation would have occurred easily if Earth had been molten when it was young. Two sources of heat could have heated Earth. First, heat of formation would be created by infalling material. A meteorite hitting Earth at high velocity converts most of its energy of motion into heat, and the impacts of a large number of meteorites could release tremendous heat. If Earth formed rapidly, this heat would have accumulated much more rapidly than it could leak away, and Earth was probably molten when it formed. A second source of heat requires more time to develop. The decay of radioactive elements trapped in the Earth releases heat gradually; but, as soon as Earth formed, that heat would have begun to accumulate and could have helped melt Earth to facilitate differentiation. Most of Earth's radioactive elements are now concentrated in the crust, where they continue to warm and soften the rock layers.

Earth was formed in a molten state by material falling together, but meteorites could leave no trace until a crust solidified. Once Earth had a hard surface, the meteorites could form craters. This second stage in planetary evolution, *cratering,* was violent. The heavy bombardment was intense because the solar nebula was filled with rocky and icy debris, and the young Earth was battered by meteorites that pulverized the newly forming crust. The largest meteorites blasted out crater basins hundreds of kilometers in diameter. As the solar nebula cleared, the amount of debris decreased, and the level of cratering fell to its present low level. Although meteorites still occasionally strike Earth and dig craters, cratering is no longer the dominant influence on Earth's geology. As you compare other worlds with Earth, you will discover traces of this intense period of cratering, the heavy bombardment, on every old surface in the solar system.

The third stage, *flooding,* no doubt began while cratering was still intense. The fracturing of the crust and the heating caused by radioactive decay allowed molten rock just below the crust to well up through fissures and flood the deeper basins. You will find such basins filled with solidified lava flows on other worlds, such as the moon, but all traces of this early lava flooding on Earth have been destroyed by later geological activity. On Earth, flooding continued as the atmosphere cooled and water fell as rain, filling the deepest basins to produce the first oceans. Notice that on Earth flooding involves both lava and water, a circumstance that you will not find on most worlds.

The fourth stage, *slow surface evolution,* has continued for the last 3.5 billion years or more. Earth's surface is constantly changing as sections of crust slide over each other, push up mountains, and shift continents. Also, moving air and water erode the surface and wear away geological features. Almost all traces of the first billion years of Earth's geology have been destroyed by the active crust and erosion.

Earth as a Planet

You are studying Earth in this chapter in order to use it as a basis for comparison with other worlds. Of course, four of the worlds in our solar system, the Jovian planets, are dramatically different from Earth. But the four terrestrial planets are rocky worlds much like Earth, and many of the moons in the solar system have geology that you can compare with Earth and its four stages of development.

All terrestrial planets pass through these four stages, so in that respect Earth is a good basic reference planet for comparative planetology. Some planets have emphasized one stage over another, and some planets have failed to progress fully through the four stages. Nevertheless, Earth is a good standard of comparison. Every major process on any rocky world in our solar system is represented in some form on Earth. Clearly, Earth is a highly complex world; for that reason, it makes a good standard of comparison.

On the other hand, Earth is peculiar in two ways. First, it has large amounts of liquid water on its surface. Fully 75 percent of its surface is covered by this liquid, and no other planet in our solar system is known to have such extensive liquid water on its surface. Not only does water fill the oceans, but it evaporates into the atmosphere, forms clouds, and then falls as rain. Water falling on the continents flows downhill to form rivers that flow back to the sea, and, in so doing, the water produces intense erosion. Entire mountain ranges can literally dissolve and wash away in only a few tens of millions of years, less than 1 percent of Earth's total age. You will not see such intense erosion on most worlds. Liquid water is, in fact, a rare material on most planets.

Your home planet is special in a second way. Some of the matter on the surface of this world is alive, and a small part of that living matter is aware. No one is sure how the presence of living matter has affected the evolution of Earth, but this process seems to be totally missing from other worlds in our solar system. Furthermore, the thinking part of the life on Earth, humankind, is actively altering our planet. It does not seem likely that humans have made major changes in Earth yet, but your study of Earth as a standard of comparison with other worlds will also give you new insight into your own planet and how modern civilization may be altering it.

Building Scientific Arguments

Why should you think Earth went through an early stage of cratering?

When you build a scientific argument, take great care to distinguish between theory and evidence. Recall from the previous chapter that the planets formed by the accretion of planetesimals from the solar nebula. The protoearth may have been molten as it formed, but as soon as it grew cool enough to form a solid crust, the remaining planetesimal impacts would have formed craters. So you can reason from the solar nebula hypothesis that Earth should have been cratered. But you can't use a theory as evidence to support some other theory. To find real observational evidence, you need only look at the moon. The moon has craters, and so does every other old surface in our solar system. There must have been a time, when the solar system was young, when there were large numbers of objects striking all the planets and moons and blasting out

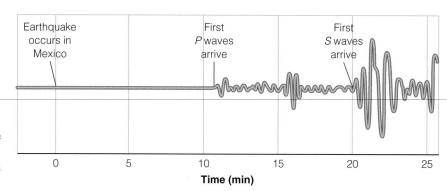

A seismograph in northern Canada made this record of seismic waves from an earthquake in Mexico. The first vibrations, *P* waves, arrived 11 minutes after the quake, but the slower *S* waves took 20 minutes to make the journey.

craters. If it happened to other worlds in our solar system, it must have happened to Earth, too.

The best evidence to support your argument would be lots of craters on Earth, but, of course, there are few craters on Earth. Extend your argument. **Why don't you see lots of craters on Earth today?**

■   ■   ■

Connections: As you explore our solar system in the chapters that follow, Earth will be your standard of comparison. But now you have to look under the hood and study a part of Earth most people comfortably ignore.

20-3 The Solid Earth

ALTHOUGH YOU MIGHT THINK OF EARTH as made of solid rock, it is in fact neither entirely solid nor entirely rock. The thin crust seems solid, but it floats and shifts on a semiliquid layer of molten rock just below the crust. Below that lies a deep, rocky mantle surrounding a core of liquid metal. Much of what you see on the surface of Earth is determined by its interior.

Earth's Interior

The theory of the origin of planets from the solar nebula predicts that Earth should have melted and differentiated into a dense metallic core, a dense mantle, and a low-density silicate crust. But did it differentiate? Where's the evidence? Clearly, Earth's surface is made of lower-density silicates, but what of the interior?

High temperature and tremendous pressure in Earth's interior make any direct exploration impossible. Even the deepest oil wells extend only a few kilometers down and don't reach through the crust. It is quite impossible to drill far enough to sample Earth's core. Yet Earth scientists have studied the interior and found clear evidence that Earth did differentiate.

This analysis of Earth's interior is possible because earthquakes produce vibrations called **seismic waves,** which travel through the interior and eventually register on sensitive detectors, known as **seismographs,** all over the world (■ Figure 20-3). Two kinds

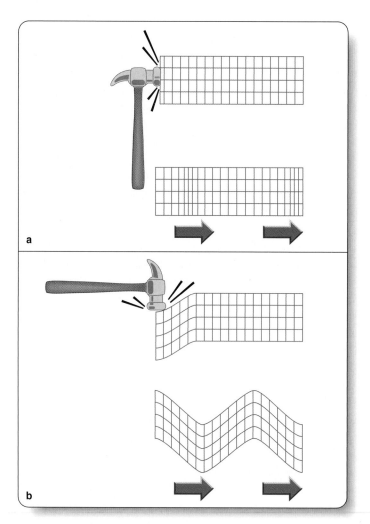

(a) *P,* or pressure, waves, like sound waves in air, travel as a region of compression. (b) *S,* or shear, waves, like vibrations in a bowl of jelly, travel as displacements perpendicular to the direction of travel. *S* waves tend to travel more slowly than *P* waves and cannot travel through liquids.

of seismic waves are important to us here. The **P,** or **pressure, waves** are much like sound waves in that they travel as a region of compression. As a *P* wave passes, particles of matter vibrate back and forth parallel to the direction of wave travel (■ Figure 20-4a). In contrast, the **S,** or **shear, waves** move as displacements of particles perpendicular to the waves' direction of travel (Figure 20-4b). Thus, *S* waves distort the material but do not compress it. Normal sound waves are pressure waves, whereas the vibrations in a bowl of jelly are shear waves. Because *P* waves are compression waves, they can move through a liquid, but *S* waves cannot. A glass of water can't shimmy like jelly because a liquid does not have the rigidity required to transmit *S* waves.

The *P* and *S* waves caused by an earthquake do not travel in straight lines or at constant speed within Earth. The waves may reflect off boundaries between layers of different density, or they may be refracted as they pass through such a boundary. In addition, the gradual increase in temperature and density toward Earth's center means the speed of sound is greater toward the center. Sound waves approaching such a region tend to curve away (refract), so seismic waves do not travel in straight lines through Earth's interior but are refracted to follow curved paths. Earth scientists can use the arrival times of reflected and refracted seismic waves from distant earthquakes to construct a picture of Earth's interior.

Such studies show that the interior consists of three parts: a central core, a thick mantle, and a thin crust. *S* waves provide an important clue to the nature of the core, because no such waves are observed to travel through the core. When an earthquake occurs, no direct *S* waves register on seismographs on the opposite side of Earth, as if the core were casting a shadow (■ Figure 20-5). This shows that the core is mostly liquid, and the size of the shadow fixes the size of the core at about 55 percent of Earth's radius. Mathematical models predict that the core is hot (about 6000 K), dense (about 14 g/cm³), and composed of iron and nickel.

Earth's core is as hot as the surface of the sun, but the tremendous pressure inside Earth keeps the material from vaporizing. The core is instead a liquid. Nearer the center the material is under even higher pressure, which in turn raises the melting point so high that the material cannot melt at the existing temperature (■ Figure 20-6). That is why there is an inner core of solid iron and nickel. Estimates suggest the inner core's radius is about 22 percent that of Earth.

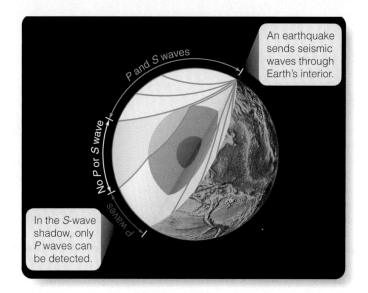

An earthquake sends seismic waves through Earth's interior.

In the *S*-wave shadow, only *P* waves can be detected.

■ Active Figure 20-5

P and *S* waves give you clues to Earth's interior. That no direct *S* waves reach the far side shows that Earth's core is liquid. The size of the *S* wave shadow tells you the size of the outer core.

Ace◐Astronomy™ Log into AceAstronomy and select this chapter to see the Active Figure "Seismic Waves" and follow *P* and *S* waves as they travel through Earth's interior.

The **mantle** is the layer of dense rock that lies between the molten core and the crust. The paths of seismic waves in the mantle show that it is not molten, but it is not precisely solid either. Mantle material behaves like a **plastic,** a material with the properties of a solid but capable of flowing under pressure. The asphalt used in paving roads is a common example of a plastic. It shatters if struck with a sledgehammer, but it bends under the steady weight of a heavy truck. Just below Earth's crust, where the pressure is less than at greater depths, the mantle is most plastic.

Earth's rocky crust is made up of low-density rocks and floats on the denser mantle. The crust is thickest under the continents,

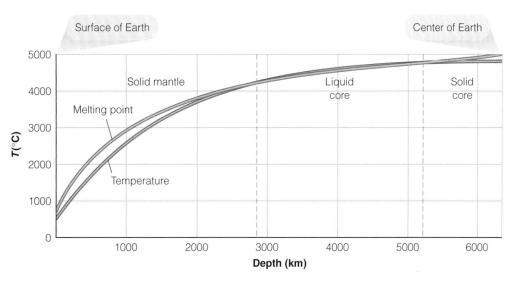

■ Figure 20-6

Theoretical models combined with observations of the velocity of seismic waves reveal the temperature inside Earth (blue line). The melting point of the material (red line) is determined by its composition and by the pressure. In the mantle and in the inner core, the melting point is higher than the existing temperature, and the material is not molten.

Scientific Imagination: Seeing the Unseen

One of the most fascinating aspects of science is its power to reveal the unseen. That is, it gives you an opportunity to learn about regions you can never visit. You saw this in earlier chapters when you studied the inside of the sun and stars, the surface of neutron stars, the event horizon around black holes, the cores of active galaxies, and more. In this chapter, you have seen Earth's core.

Not only does science take you to places you can never visit, but it takes you to scales you can never explore. For example, physicists can explore the inside of an atom, and biologists can study the structure of a virus. Not only can you stretch the scale of space, but you can also stretch the scale of time. Geologists can watch mountain ranges rising slowly from the plains and eroding away to grit. Astronomers can watch the 200-million-year-long whirl of two galaxies colliding.

In this respect, astronomy, geology, and psychology are particularly similar. All three sciences study realms that cannot be explored in person. The insides of stars, Earth's ancient history, and the workings of the human mind lie beyond immediate experience. Only by careful observation and careful thought can scientists learn to understand these aspects of the natural world.

Science can reveal the unseen because it can combine human imagination with the predictive power of the laws of nature and the precision of mathematics. This calls on scientists to be as creative as artists and as precise as surgeons to produce reliable theories and models, but the reward is one of the great thrills of science—exploring beyond the limits of normal human experience.

The electron microscope allows biologists to study the elegant structure of the virus that causes the common infection called pinkeye. (From *Virus Ultrastructure: Electron Micrograph Images.* Copyright 1995 by Linda M. Stannard. Reprinted with permission from Linda M. Stannard. Found at: http://www.web.uct.ac.za/depts/mmi/stannard/adeno.html)

---

up to 60 km thick, and thinnest under the oceans, where it is only about 10 km thick. Unlike the mantle, the crust is brittle and breaks much more easily than the plastic mantle.

Perhaps no region is more immediate yet more inaccessible than Earth's core. Only a few thousand kilometers from where you sit, Earth's interior is beyond your immediate reach. Yet you can send your imagination where your body can never go (**Window on Science 20-1**). Earth's seismic activity reveals some of Earth's innermost secrets. But there is another source of evidence about Earth's interior—its magnetic field.

The Magnetic Field

Apparently, Earth's magnetic field is a direct result of its rapid rotation and its molten metallic core. The internal heat forces the liquid core to churn with convection while Earth's rotation turns the core about an axis. This core is a highly conductive iron–nickel alloy, a better electrical conductor than copper, the material commonly used for electrical wiring. The rotation of this convecting, conducting liquid generates Earth's magnetic field in a process called the dynamo effect (■ Figure 20-7). This is believed to be the same process that generates the solar magnetic field in the convective layers of the sun, and you will meet it again when you explore other planets.

Earth's magnetic field protects it from the solar wind. Blowing outward from the sun at about 400 km/s, the solar wind consists of ionized gases carrying a small part of the sun's magnetic field. When the solar wind encounters Earth's magnetic field, it is deflected like water flowing around a boulder in a stream. The surface where the solar wind is first deflected is called the **bow shock,** and the cavity dominated by Earth's magnetic field is called the **magnetosphere** (■ Figure 20-8a). High-energy particles from the solar wind leak into the magnetosphere and become trapped

■ Figure 20-7

The dynamo effect couples convection in the liquid core with Earth's rotation to produce electric currents that are believed to be responsible for Earth's magnetic field.

Magnetosphere: the region controlled by the planet's magnetic field

Solar wind

Bow shock

Radiation belts

Rays follow Earth's magnetic field.

Dark clouds silhouetted against the much higher aurora

a

b

■ Figure 20-8

Earth's magnetic field dominates space around Earth by deflecting the solar wind and trapping high-energy particles in radiation belts. Around the north and south magnetic poles, where the magnetic field enters Earth's atmosphere, powerful currents can flow down and excite gas atoms to emit photons, which produces auroras. Colors are produced as different atoms are excited. Note the meteor (shooting star). (Jimmy Westlake)

within Earth's magnetic field to produce the **Van Allen belts** of radiation. You will see in later chapters that all planets that have magnetic fields have bow shocks, magnetospheres, and radiation belts.

One dramatic result of Earth's magnetic field are the auroras, glowing rays and curtains of light in the upper atmosphere (Figure 20-8b). The solar wind carries charged particles past Earth's extended magnetic field, and this generates tremendous electrical currents that flow into Earth's atmosphere near the north and south magnetic poles. The currents ionize gas atoms in Earth's atmosphere, and when the ionized atoms capture electrons and recombine, they emit light as if they were part of a vast "neon" sign. That is why the spectrum of an aurora is an emission spectrum.

Although you can be confident that Earth's magnetic field is generated within its molten core, many mysteries remain. For example, rocks retain traces of the magnetic field in which they solidify, and some contain fields that point backward. That is, they imply that Earth's magnetic field was reversed at the time they solidified. Careful analysis of such rocks indicates that Earth's field has reversed itself every million years or so, with the north magnetic pole becoming the south magnetic pole and vice versa. These reversals are poorly understood, but they may be related to changes in the core convection.

Convection in Earth's core is important because it generates the magnetic field. As you will see in the next section, convection in the mantle constantly remakes Earth's surface.

Ace Astronomy™ Log into AceAstronomy and select this chapter to see Astronomy Exercise "Convection and Magnetic Fields." Take control of a planetary interior.

Earth's Active Crust

Earth's crust is composed of lower-density rock that floats on the mantle. The image of a rock floating may seem odd, but recall that the rock of the mantle is very dense. Also, just below the crust,

the mantle rock tends to be highly plastic, so great sections of low-density crust do indeed float on the semiliquid mantle like great lily pads floating on a pond.

The motion of the crust and the erosive action of water make Earth's crust highly active. Look at **The Active Earth** on pages 486–487 and notice three important points:

1 The motion of crustal plates produces much of the geological activity on Earth. Earthquakes, volcanism, and mountain building are all linked to motions in the crust and the location of plate boundaries.

2 Notice how the continents on Earth's surface have moved and changed over periods of hundreds of millions of years. A hundred million years is only 0.1 billion years, so sections of Earth's crust are in rapid motion.

3 Notice that most of the geological features you know—mountain ranges, the Grand Canyon, and even the outline of the continents—are recent products of Earth's active surface. Earth's surface is constantly renewed. The oldest rocks on Earth, small crystals called zircons from western Australia, are 4.3 billion years old. Most of the crust is much younger than that. Most of the mountains and valleys you see around you are no more than a few tens of millions of years old.

The Active Earth

1 Our world is an astonishingly active planet. Not only is it rich in water and therefore subject to rapid erosion, but its crust is divided into moving sections called plates. Where plates spread apart, lava wells up to form new crust; where plates push against each other, they crumple the crust to form mountains. Where one plate slides over another, you see volcanism. This process is called **plate tectonics**, referring to the Greek word for "builder." (An architect is literally an arch builder.)

A typical view of planet Earth

Mountains are common on Earth, but they erode away rapidly because of the abundant water.

William K. Hartmann

Janet Seeds

A **rift valley** forms where continental plates begin to pull apart. The Red Sea has formed where Africa has begun to pull away from the Arabian peninsula.

Midocean rise

Red Sea

Midocean rise

National Geophysical Data Center

1a Evidence of plate tectonics was first found in ocean floors, where plates spread apart and magma rises to form **midocean rises** made of rock called **basalt**, a rock typical of solidified lava. Radioactive dating shows that the basalt is younger near the midocean rise. Also, the ocean floor carries less sediment near the midocean rise. As Earth's magnetic field reverses back and forth, it is recorded in the magnetic fields frozen into the basalt. This produces a magnetic pattern in the basalt that shows that the seafloor is spreading away from the midocean rise.

1b A **subduction zone** is a deep trench where one plate slides under another. Melting releases low-density magma that rises to form volcanoes such as those along the northwest coast of North America, including Mt. St. Helens.

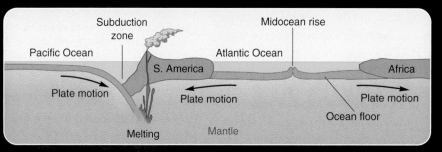

Subduction zone

Midocean rise

Pacific Ocean

Atlantic Ocean

S. America

Africa

Plate motion

Plate motion

Plate motion

Ocean floor

Melting

Mantle

Subduction zone

Appalachian Mountains

Ural Mountains

Himalaya Mountains

Hawaiian-Emperor chain

Midocean rise

Subduction zone

Hawaii

Red Sea

Midocean rise

Andes mountains

Midocean rise

Subduction zone

1c Hot spots caused by rising magma in the mantle can poke through a plate and cause volcanism such as that in Hawaii. As the Pacific plate has moved northwestward, the hot spot has punched through to form a chain of volcanic islands, now mostly worn below sea level. **Folded mountain ranges** can form where plates push against each other. For example, the Ural Mountains lie between Europe and Asia, and the Himalaya Mountains are formed by India pushing north into Asia. The Appalachian Mountains are the remains of a mountain range pushed up when North America was pushed against Africa.

Ace ◉ Astronomy™ Log into AceAstronomy and select this chapter to see Active Figure "Hot Spot Volcanoes." Notice how the moving plate can produce a chain of volcanic peaks, mostly under water in the case of Earth.

The floor of the Pacific Ocean is sliding into subduction zones in many places around its perimeter. This pushes up mountains such as the Andes and triggers earthquakes and active volcanism all round the Pacific in what is called the Ring of Fire. In places such as southern California, the plates slide past each other, causing frequent earthquakes.

Hawaii

National Geophysical Data Center

Yellow lines on this globe mark plate boundaries. Red dots mark earthquakes since 1980. Earthquakes within the plate, such as those at Hawaii, are related to volcanism over hot spots in the mantle.

Not long ago, Earth's continents came together to form one continent.

Pangaea broke into a northern and a southern continent.

Notice India moving north toward Asia.

The continents are still drifting on the highly plastic upper mantle.

Continental Drift

Pangaea

200 million years ago

Laurasia

Gondwanaland

135 million years ago

65 million years ago

Today

2 The floor of the Atlantic Ocean is not being subducted. It is locked to the continents and is pushing North and South America away from Europe and Africa at about 3 cm per year, a motion called *continental drift*. Radio astronomers can measure this motion by timing pulsars from European and from American radio telescopes. Roughly 200 million years ago, North and South America were joined to Europe and Africa. Evidence of that lies in similar fossils and similar rocks and minerals found in the matching parts of the continents. Notice how North and South America fit against Europe and Africa like a puzzle.

Mike Seeds

Formation of Earth
Heavy bombardment
?
Oldest fossil life
Formation of Grand Canyon
Age of dinosaurs
Breakup of Pangaea
First animals emerge on land

4.6   4   3   2   1   Now
Billions of years ago

3 Plate tectonics pushes up mountain ranges and causes bulges in the crust, and water erosion wears the rock away. The Colorado River began cutting the Grand Canyon only about 10 million years ago when the Colorado plateau warped upward under the pressure of moving plates. That sounds like a long time ago, but it is only 0.01 billion years. A mile down, at the bottom of the canyon, lie rocks 0.57 billion years old, the roots of an earlier mountain range that stood as high as the Himalayas. It was pushed up, worn away to nothing, and covered with sediment long ago. Many of the geological features we know on Earth have been produced by very recent events.

The continents move slowly on average, but plate margins can stick and accumulate stress and then release it in sudden motions. That's what happened on December 26, 2004, along a major subduction zone in the Indian Ocean. The total motion was as much as 15 m, and the resulting earthquake caused devastating tidal waves. Minor earthquakes occur every day, and stress builds in faults all around the world that will eventually be released in major earthquakes.

Earth's active crust explains why Earth contains so few impact craters. The moon is richly cratered, but Earth contains only about 150 impact craters. Plate tectonics and erosion have destroyed all but the most recent craters on Earth.

You can see that Earth's geology is dominated by two dramatic forces. Heat rising from the interior drives plate tectonics; just below the thin crust of solid rock lies a churning molten interior that rips the crust to fragments and pushes the pieces about like bits of algae on a pond. The second force modifying the crust is water. It falls as rain and snow and tears down the mountains, erodes the river valleys, and washes any raised ground into the sea. Like angry children, tectonics and erosion build mountains and continents and then rip them to nothing.

Ace✷Astronomy™ Log into AceAstronomy and select this chapter to see Astronomy Exercise "Convection and Plate Tectonics." Compare the interiors of the four terrestrial planets.

Building Scientific Arguments

What evidence can you cite that Earth has a molten metallic core?

In a scientific argument, the critical analysis of ideas must eventually return to evidence. In this case, the evidence is indirect because you can never visit Earth's core. Seismic waves from distant earthquakes pass through Earth, but a certain kind of wave, the S waves, cannot pass through the core. Because the S waves cannot propagate through a liquid, you can conclude that Earth's core is a liquid. Earth's magnetic field gives you further evidence of a metallic core. The theory for the generation of magnetic fields, the dynamo effect, requires a rotating liquid core composed of a conducting material and stirred by convection. If the core were not a liquid, it would not be able to generate a magnetic field. That gives you two different kinds of evidence that our planet has a liquid core.

Now build a new argument again focusing on evidence. **Why do scientists conclude that Earth's crust is broken into moving plates?**

■   ■   ■

Connections: For the astronomer, accustomed to the longest view of cosmic time, Earth's geology seems not so much the history of slow-moving rock as it does the churning surface of a pot of boiling soup. And just as steam rises from boil-ing soup, vapors rise from the crust. Earth's atmosphere not only acts to erode the rocky surface, but some of it came from the crust as a by-product of geological activity.

20-4 Earth's Atmosphere

YOU CAN'T TELL THE STORY OF EARTH without mentioning the atmosphere. Not only is it necessary for life, but it is also intimately related to the crust. It affects the surface by eroding geological features through wind and water, and in turn the chemistry of Earth's surface affects the composition of the atmosphere.

Origin of the Atmosphere

Until a few decades ago, Earth scientists believed that Earth's atmosphere had developed slowly. According to this picture, Earth formed by the slow accumulation of planetesimals into a large, cold ball of matter whose composition was uniform from surface to center (Figure 19-9). Early Earth might have attracted small amounts of gases such as hydrogen, helium, and methane from the solar nebula to form a **primeval atmosphere.** According to the old theory, the slow decay of radioactive elements heated Earth's interior, melted it, caused it to differentiate into core, mantle, and crust, and triggered widespread volcanism.

When a volcano erupts, 50 to 80 percent of the gas released is water vapor. The rest is mostly carbon dioxide, nitrogen, and smaller amounts of sulfur gases such as hydrogen sulfide—the rotten-egg gas that you smell if you visit geothermal pools and geysers such as those at Yellowstone National Park. These gases could have diluted the primeval atmosphere and eventually produced a **secondary atmosphere** rich in carbon dioxide, nitrogen, and water vapor. That's the old theory.

A modern understanding of how the solar system formed shows that Earth formed so rapidly that it was melted by the heat released by infalling material. If Earth's surface was molten as it formed, then outgassing would have been continuous, and the atmosphere would have been rich in carbon dioxide, nitrogen, and water vapor from the beginning. That means Earth never had a hydrogen-rich primeval atmosphere.

For some years, astronomers have suspected that the abundance of water on Earth arrived late in the formation process as a bombardment of volatile-rich planetesimals. These icy bodies, the theory goes, were scattered by the growing mass of the outer planets, and the inner solar system was bombarded by a storm of comets. This theory has faced a serious objection. Spectroscopic studies of a few comets revealed that the ratio of deuterium to hydrogen in comets does not match the ratio in the water on Earth. Some astronomers think this means that the water now on Earth could not have arrived in cometlike planetesimals. However, studies of Comet Linear, which broke up in 1999 as it passed the sun, show that the water in that comet had a composition of isotopes similar to water on Earth. There may be differences among comets. Icy planetesimals that formed far from the sun may be richer in deuterium, while comets that formed closer to the orbit of Jupiter than to the orbit of Neptune may contain water with isotopes more like those in Earth's water. This is an interesting controversy, and it shows that the origin of Earth's atmosphere and oceans is yet to be fully understood.

However the atmosphere originated, the mix of gases must have changed. The young atmosphere must have been rich in water vapor, carbon dioxide, and other gases. As the atmosphere cooled, the water condensed to form the first oceans; and, as the oceans grew bigger, carbon dioxide began to dissolve in the water. Carbon dioxide is easily soluble in water—which is why carbonated beverages are so easy to manufacture—but the first oceans could not have absorbed all of the atmospheric carbon dioxide if it had not reacted with dissolved substances in the seawater to form silicon dioxide, limestone, and other mineral sediments. Thanks to those chemical reactions in the oceans the carbon dioxide was transferred from the atmosphere to the seafloor.

Earth's lower atmosphere is now protected from ultraviolet radiation by an **ozone layer** about 25 km above the surface. However, the atmosphere of the young Earth did not contain free oxygen, so an ozone layer could not form, and the sun's ultraviolet radiation penetrated deep into the atmosphere. The energetic ultraviolet photons would have broken up weaker molecules such as water (H_2O). The hydrogen escaped to space, and the oxygen formed oxides in the crust. Earth's atmosphere could not reach its present composition (■ Table 20-1) until it was protected by an ozone layer, and that required oxygen.

The origin of the atmospheric oxygen is linked to the origin of life, the subject of Chapter 26, but it is sufficient here to note that life must have originated within a billion years of Earth's formation. This life did not significantly alter the atmosphere, however, until nature invented photosynthesis about 2.7 to 2.4 billion years ago. Photosynthesis absorbs carbon dioxide from the air or water and utilizes it for plant growth, releasing oxygen. Because oxygen is a very reactive element, it combines easily with other compounds, and the oxygen abundance could not grow rapidly

at first. Apparently, the development of large, shallow seas along the continental margins half a billion years ago allowed ocean plants to manufacture oxygen faster than chemical reactions could remove it. Atmospheric oxygen then increased rapidly, and it is still increasing at the rate of about 1 percent every 36 million years.

Earth's climate is critically sensitive to many different factors. For example, the **albedo** of a world is the fraction of the sunlight that hits it that gets reflected away. An albedo of 1 would be a perfectly white world, and an albedo of 0 would be perfectly black. Earth's overall albedo is 0.39, meaning it reflects back into space 39 percent of the sunlight that hits it. Much of this reflection is caused by clouds, and the formation of clouds depends critically on the presence of water vapor in the upper atmosphere, the temperature of the upper atmosphere, and the patterns of atmospheric circulation. Even a small change in any of these factors could change Earth's albedo and thus its climate. Your comfortable life on the surface of your world depends on Earth's delicate atmosphere.

(Ace Astronomy™) Log into AceAstronomy and select this chapter to see Astronomy Exercise "Primary Atmospheres." You can move your planet around to see how its atmosphere is affected.

Human Effects on Earth's Atmosphere

Earth is comfortable for human life because of its atmosphere. The temperature is moderate at sea level, but if you climb to the top of a high mountain, you find the temperature to be very low (■ Figure 20-9). Most clouds form at such altitudes. Higher still, you find the air much colder and so thin it cannot protect you

■ Figure 20-9

Thermometers placed in Earth's atmosphere at different levels would register the temperatures shown in the graph at the right. The lower few kilometers where you live are comfortable, but higher in the atmosphere the temperature is quite low. The ozone layer lies about 25 km above Earth's surface.

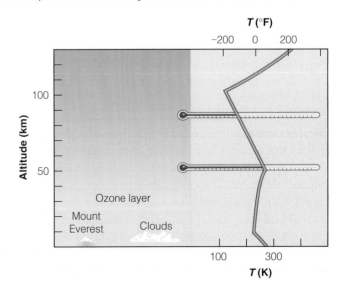

■ Table 20-1 | Earth's Atmosphere

Gas	Percent by Weight
N_2	75.5
O_2	23.1
Ar	1.29
CO_2	0.05
Ne	0.0013
He	0.00007
CH_4	0.0001
Kr	0.0003
H_2O (vapor)	1.7–0.06

from the intense ultraviolet radiation in sunlight. If you climb mountains or even hike in high mountain meadows, you should wear strong sunscreen.

You can live on Earth's surface in safety because Earth's atmosphere protects you, but modern civilization is altering Earth's atmosphere in at least two serious ways: It is adding carbon dioxide (CO_2) and destroying ozone.

The concentration of CO_2 in Earth's atmosphere is important because CO_2 can trap heat in a process called the **greenhouse effect** (■ Figure 20-10a). When sunlight shines through the glass roof of a greenhouse, it heats the benches and plants inside. The warmed interior radiates infrared radiation, but the infrared cannot get out through the glass. Heat is trapped within the greenhouse, and the temperature climbs until the glass itself grows warm enough to radiate heat away as fast as the sunlight enters. Of course, a greenhouse retains its heat because the walls prevent the warm air from mixing with the cooler air outside. This is the same process that heats a car when it is parked in the sun with the windows rolled up.

Earth's atmosphere is transparent to sunlight, and the absorbed sunshine makes the surface grow warm. However, the CO_2 makes the atmosphere less transparent to infrared radiation, so infrared radiation from the warm surface is absorbed and cannot escape back into space. That traps heat and makes Earth warmer (Figure 20-10b). Without the greenhouse effect, Earth would be 40 K cooler than it is. That amounts to 72°F colder than at present, so Earth would be uninhabitable without the greenhouse effect. The problem is that human civilization is increasing the amount of greenhouse gases in the atmosphere, and that is making Earth warmer.

CO_2 is not the only greenhouse gas. Water vapor, methane, and other gases also help warm Earth, but CO_2 is the most important because the human race is increasing its concentration in the atmosphere. For billions of years, natural processes on Earth have removed CO_2 from the atmosphere and buried the carbon in the form of coal, oil, and natural gas. Since the beginning of the industrial revolution in the mid-19th century, civilization has been digging up this carbon, burning it to get energy, and releasing CO_2 back into the atmosphere. At the same time, many nations are cutting down parts of the forests that absorb CO_2. Estimates are that the amount of CO_2 in Earth's atmosphere will double during the 21st century.

The increased concentration of CO_2 is increasing the greenhouse effect and warming Earth in what is known as **global warming.** The actual amount of warming in the future is difficult to predict. The best mathematical models predict a warming between 1.5 and 4.5°C (3 to 8°F). Predictions are uncertain because Earth's climate depends on so many factors. A slight warming should increase water vapor in the atmosphere, and water vapor is another greenhouse gas that would enhance the warming. But increased water vapor could increase cloud cover, increase Earth's albedo, and partially reduce the warming. High icy clouds,

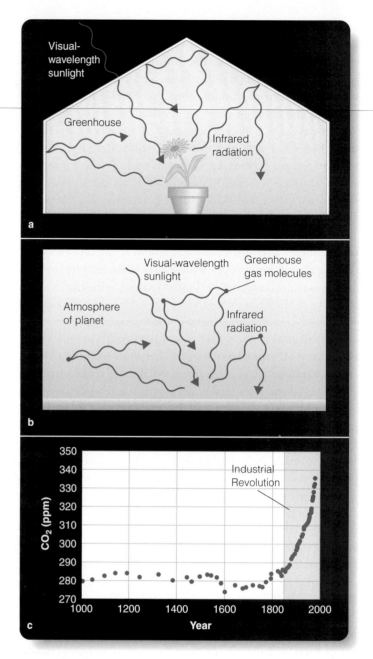

■ **Figure 20-10**

The greenhouse effect. (a) Visual-wavelength sunlight can enter a greenhouse and heat its contents, but the longer-wavelength infrared cannot get out. (b) The same process can heat a planet's surface if its atmosphere contains greenhouse gases such as CO_2. (c) The concentration of CO_2 in Earth's atmosphere as measured in Antarctic ice cores remained roughly constant until the beginning of the Industrial Revolution. (Ozone Processing Team/NASA/Goddard Space FLight Center)

however, tend to enhance the greenhouse effect. Even small changes in temperature can alter circulation patterns in the atmosphere and in the oceans, and the consequences of such changes are very difficult to predict.

The future is difficult to predict, but the evidence clearly shows that Earth is growing warmer. Studies of the growth rings in very old trees show that Earth's climate has been cooling for the last 1000 years, but the 20th century reversed that trend with a rise of 0.3 to 0.6°C (0.5 to 1.1°F). It is difficult to be sure how much of this change is due to the increased greenhouse effect and how much is natural variation in Earth's climate, but it is worrisome that general trends point toward warming. Mountain glaciers have melted back dramatically since the 19th century, and the warmest years of the last 100 years have all been recent. Measurements show that polar ice in the form of permafrost, ice shelves, and ice on the open Arctic Ocean is melting.

Although the temperature increase is small now, it is a serious issue for the future. Even a small rise in temperatures will alter climate and dramatically affect agriculture, not only through rising temperatures but also through changes in rainfall. Also, the melting of ice on polar landmasses such as Greenland can cause a rise in sea levels that will flood coastal regions and alter shore environments. A modest rise will cover huge low-lying areas such as the entire state of Florida.

There seems no doubt that civilization is warming Earth through an enhanced greenhouse effect, but a remedy is difficult to imagine. Reducing the amount of CO_2 and other greenhouse gases released to the atmosphere is difficult because modern society depends on burning fossil fuels for energy. Conserving forests is difficult because growing populations, especially in developing countries with large forest reserves, demand the wood and agricultural land produced when forests are cut. Political, business, and economic leaders argue that the issue is uncertain, but all around the world scientists of stature have reached an agreement: Global warming is real and will change Earth. What humanity can or will do about it is uncertain.

Human influences on Earth's atmosphere go beyond the greenhouse effect. Modern civilization is also reducing ozone in Earth's atmosphere. Ozone is an unstable molecule formed of three oxygen atoms (O_3). You may have heard of ozone because it is produced in auto emissions, and it is a pollutant in city air. But ozone is produced naturally in Earth's atmosphere at an altitude of about 25 km, and there it is beneficial. The ozone layer absorbs ultraviolet photons and prevents them from reaching the ground.

Certain chemicals called chlorofluorocarbons (CFCs), used in industrial processes and in refrigeration and air conditioning, can destroy ozone. As these CFCs escape into the atmosphere, they become mixed into the ozone layer and convert the ozone into normal oxygen molecules. Oxygen does not block ultraviolet radiation, so depleting the ozone layer causes an increase in ultraviolet radiation at Earth's surface. In small doses, ultraviolet radiation can produce a suntan, but in larger doses it can cause skin cancers.

The ozone layer over the Antarctic may be especially sensitive to CFCs. Since the late 1970s, the ozone concentration has been falling over the Antarctic, and a hole in the ozone layer now develops over the continent each October—spring for Antarctica (■ Figure 20-11). Satellite and ground-based measurements show that the same thing is happening at higher northern latitudes and that the amount of ultraviolet radiation reaching the ground is increasing. This is an early warning that human activity is modifying Earth's atmosphere in a potentially dangerous way.

The CO_2 problem and the ozone problem in Earth's atmosphere are paralleled on Venus and Mars. When you study Venus in Chapter 22, you will discover a runaway greenhouse effect that has made the surface of the planet hot enough to melt lead.

■ **Figure 20-11**

(a) Satellite observations of ozone concentrations over Antarctica are shown here as red for highest concentration and violet for lowest. Since the 1970s, a hole in the ozone layer has developed over the South Pole. (b) Although ozone depletion is most dramatic above the South Pole, ozone concentrations have declined at all latitudes. (Ozone Processing Team/NASA/Goddard Space Flight Center)

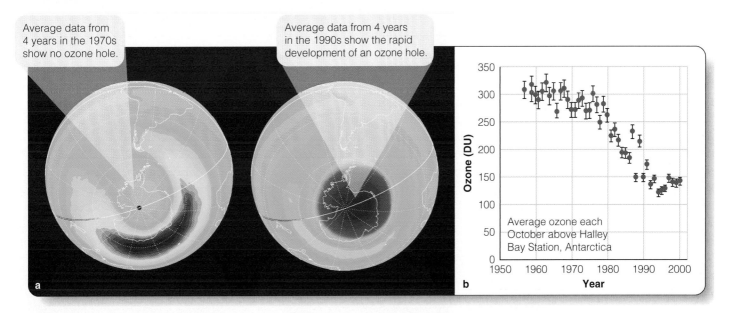

Average data from 4 years in the 1970s show no ozone hole.

Average data from 4 years in the 1990s show the rapid development of an ozone hole.

Average ozone each October above Halley Bay Station, Antarctica

On Mars you will discover an atmosphere without an ozone layer. A few minutes of sunbathing on Mars would kill you. Once again, you will learn more about your own planet by studying other planets.

Ace✸Astronomy™ Log into AceAstronomy and select this chapter to see Astronomy Exercise "Greenhouse Effect." Take control of the CO_2 in your experimental planet's atmosphere.

Building Scientific Arguments

Why does Earth's atmosphere contain little carbon dioxide and lots of oxygen?

Sometimes as you build a scientific argument, you must contradict what seems, at first glance, a simple truth. In this case, because outgassing releases mostly carbon dioxide, CO_2, and water vapor, you might expect Earth's atmosphere to be very rich in CO_2. Luckily for the human race, CO_2 is highly soluble in water, and Earth's surface temperature allows it to be covered with liquid water. The CO_2 dissolves in the oceans and combines with minerals in seawater to form deposits of silicon dioxide, limestone, and other mineral deposits. In this way, the CO_2 is removed from the atmosphere and buried in Earth's crust. Oxygen, in contrast, is highly reactive and forms oxides so easily you might expect it to be rare in the atmosphere. Happily it is continually replenished as green plants release oxygen into Earth's atmosphere faster than chemical reactions can remove it. Were it not for liquid-water oceans and plant life, Earth would be choking in a thick CO_2 atmosphere with no free oxygen.

Now follow up on your argument. **Why would an excess of CO_2 and a deficiency of free oxygen be harmful to all life on Earth in ways that go beyond mere respiration?**

■   ■   ■

Connections: Earth is a rich and beautiful world, and when you visit Venus and Mars, both of which have CO_2-rich atmospheres, you will appreciate Earth even more. But before you rush to compare Earth with planets that have atmospheres, you should examine two worlds that have no atmospheres at all—the moon and Mercury. As usual, the comparison of worlds will be revealing.

Study and Review Tools

Summary

20-1 | A Travel Guide to the Terrestrial Planets

How can comparison help you understand the terrestrial planets?

■ Earth is the standard of comparison in the study of the terrestrial planets.

■ The terrestrial planets include Earth, the moon, Mercury, Venus, and Mars. Earth's moon is included because it is a complex world and makes a striking comparison with Earth.

■ The terrestrial worlds differ mainly in size, but they all have low-density crusts, mantles of dense rock, and metallic cores.

■ Comparative planetology warns you to expect that cratered surfaces are old, that heat flowing out of a planet drives geological activity, and that the nature of a planet's atmosphere depends of the size of the planet and its temperature.

20-2 | The Early History of Earth

How has Earth changed since it formed?

■ Earth formed rapidly from the solar nebula and was hot enough to be molten.

■ Earth has passed through four stages as it has evolved: (1) differentiation; (2) cratering; (3) flooding by lava and water; and (4) slow surface evolution.

■ Earth is peculiar in that it has large amounts of liquid water on its surface, and that water drives strong erosion that alters the surface geology.

■ Earth is also peculiar in that it is the only known home for life.

20-3 | The Solid Earth

What is the inside of Earth like?

■ Seismology shows that *S*-waves cannot pass through Earth's core, and that is evidence that Earth has a liquid core. Heat flowing outward from the interior, combined with models, reveals that the core is very hot and composed of iron and nickel.

■ Earth's magnetic field is generated by the dynamo effect in the liquid, convecting, rotating, conducting core. The magnetic field shields Earth from the solar wind by producing a magnetosphere around the planet. Radiation belts and auroras are also produced by the field.

■ Earth is dominated by plate tectonics that breaks the crust into moving plates. Plate tectonics is driven by heat flowing upward from the interior.

■ Tectonic plates are made of low-density, brittle rock that floats on the hotter plastic upper layers of the mantle. Plates spread apart along midocean rises and sink into the mantle in subduction zones. Volcanism and earthquakes are common along the edge of the plates.

■ The continents are drifting slowly on the plastic mantle, and their arrangement changes with time.

■ Most geological features on Earth, such as mountain ranges and the Grand Canyon, have been formed recently. The first billion years of Earth's geology is almost entirely erased by plate tectonics and erosion.

20-4 | Earth's Atmosphere

How has Earth's atmosphere formed and evolved?

■ Because Earth formed hot, it never had a primeval atmosphere rich in hydrogen and helium. Earth's first atmosphere was probably mostly carbon dioxide, but that gas was dissolved in seawater, and plant life has created oxygen.

■ The greenhouse effect can warm a planet if gases in the atmosphere are transparent to light but opaque to infrared. The natural greenhouse effect warms Earth and makes it comfortable for life, but greenhouse gases added by industrial civilization are responsible for global warming.

■ The ozone layer high in Earth's atmosphere protects the surface from ultraviolet radiation. Certain chemicals called chlorofluorocarbons released in industrial processes attack the ozone layer and thin it. This is allowing more harmful ultraviolet radiation to reach Earth's surface.

New Terms

comparative planetology (p. 476)	midocean rise (p. 486)
seismic wave (p. 482)	subduction zone (p. 486)
seismograph (p. 482)	basalt (p. 486)
pressure *(P)* wave (p. 482)	folded mountain range (p. 486)
shear *(S)* wave (p. 482)	rift valley (p. 486)
mantle (p. 483)	primeval atmosphere (p. 488)
plastic (p. 483)	secondary atmosphere (p. 488)
bow shock (p. 484)	ozone layer (p. 489)
magnetosphere (p. 484)	albedo (p. 489)
Van Allen belts (p. 485)	greenhouse effect (p. 490)
plate tectonics (p. 486)	global warming (p. 490)

Review Questions

Ace ◗ Astronomy™ Assess your understanding of this chapter's topics with additional quizzing and animations at **http://ace .brookscole.com/sf9**

1. Why would you include the moon in a list of the terrestrial planets?

2. In what ways is Earth peculiar among the terrestrial planets?

3. What are the four stages of planetary development?

4. How do you know that Earth differentiated?

5. What evidence can you cite that Earth's metallic core is liquid?

6. How are earthquakes in Hawaii different from those in southern California?

7. What characteristics must Earth's core have in order to generate a magnetic field?

8. How does the Hawaiian-Emperor island chain help you understand plate tectonics?

9. What evidence can you cite that the Atlantic Ocean is growing wider?

10. How is your concept of Earth's first atmosphere related to the speed with which Earth formed from the solar nebula?

11. What has produced the oxygen in Earth's atmosphere?

12. How does the increasing abundance of CO_2 in Earth's atmosphere cause a rise in Earth's temperature?

13. Why would a decrease in the density of the ozone layer cause public health problems?

14. In what ways is the photo at the right a typical view of the surface of planet Earth? How is it unusual among planets in general?

William K. Hartmann

15. What do you see in this photo that suggests heat is flowing out of Earth's interior?

USGS

Discussion Questions

1. If you visited a planet in another solar system and discovered oxygen in its atmosphere, what might you guess about its surface?

2. If liquid water is rare on the surface of planets, then most terrestrial planets must have CO_2-rich atmospheres. Right? Why?

Problems

1. In Figure 20-3, the earthquake occurred 7440 km from the seismograph. How fast did the P waves travel in km/s? How fast did the S waves travel?

2. What percentage of Earth's volume is taken up by its metallic core?

3. If the Atlantic seafloor is spreading at 3 cm/year and is now 6400 km wide, how long ago were the continents in contact?

4. The Hawaiian-Emperor chain of undersea volcanoes is about 7500 km long. The Pacific plate is moving 9.2 cm a year; how old is the oldest detectable volcano in the chain? What has happened to older volcanoes in the chain?

5. From Hawaii to the bend in the Hawaiian-Emperor chain is about 4000 km. Use the speed of plate motion given in Problem 4 to estimate how long ago the direction of plate motion changed. (The San Andreas Fault in southern California became active at about the same time!)

6. Calculate the age of the Grand Canyon as a percent of Earth's age.

Media Cluster

ACTIVE FIGURES

Ace Astronomy™ To access the resources in the Media Cluster, log into AceAstronomy at **http://ace.brookscole.com/sf9** and select Chapter 20.

Seismic Waves
Use this animation to create an earthquake and see how tracking its seismic waves provides evidence about what lies beneath Earth's surface.

Hot Spot Volcanoes
If the crustal plate over a hot spot is moving, repeated eruptions can result in an island chain like the Hawaiian Islands. Set a series of eruptions in motion in this animation.

ASTRONOMY EXERCISES

Convection and Plate Tectonics
Study the convection currents of the four Earthlike planets in this animation and see if you can determine how this relates to the amount of plate tectonic activity on each planet.

Primary Atmospheres
This exercise lets you change the distance to the sun and mass of a hypothetical planet and study how these relate to the planet's retention of its atmosphere.

Convection and Magnetic Fields
A planet's core temperature and rotation speed are factors in the strength of its magnetic field. See the effect of these variables in this animation.

Greenhouse Effect
See how distance to the sun and amount of carbon dioxide in a planet's atmosphere affect its temperature in this animation.

VIRTUAL ASTRONOMY LABS

Lab 5: Planetary Geology
By thoroughly studying Earth's geology, you can leverage your knowledge to understand conditions on other planets. This lab includes an exercise in determining the interior characteristics of a planet by observing the travel times of its seismic waves.

Critical Inquiries for the Web

1. Search the Web to find the location of the most recent earthquakes. Earthquakes somewhere each day. Search for seismographs that will give you information about the most recent earthquakes.

2. Search for information about the arrangement of landmasses on Earth before Pangaea. Were there previous supercontinents? What was Rodinia? What is the evidence?

3. Earth's magnetic poles are not coincident with its axis of rotation, nor are their positions fixed. Search for information about the location of the magnetic poles and their motions. Why do the positions change?

4. What are the oldest known rocks on Earth? Where do they come from, and how were their ages determined?

5. What evidence can you find about global warming? Find examples of the effects of global warming. Business and political leaders do not agree with the consensus among scientists, so be prepared to analyze arguments with great care.

6. Can you locate a current image of the Antarctic ozone hole? Find data for the ozone concentration in the upper atmosphere over the place where you live. What is the latest news concerning ozone depletion?

 Go to the Brooks/Cole Astronomy Resource Center (**http://astronomy.brookscole.com**) for critical thinking exercises, articles, and additional readings from InfoTrac College Edition, Brooks/Cole's online student library.

21 | The Moon and Mercury: Comparing Airless Worlds

That's one small step

for a man . . . one giant leap

for mankind.

NEIL ARMSTRONG,
ON THE MOON

Beautiful, beautiful.

Magnificent desolation.

EDWIN E. ALDRIN, JR.,
ON THE MOON

I F YOU HAD BEEN the first person to step onto the surface of the moon, what would you have said? Neil Armstrong responded to the historic significance of the first human step onto the surface of another world. Buzz Aldrin was second, and he responded to the moon itself. It *is* desolate, and it *is* magnificent. But it is not unusual. Most planets in the universe probably look like Earth's moon, and astronauts may someday walk on such worlds and compare them with our moon. ▌ In this chapter, you begin your study of other worlds by applying comparative planetology to the study of the moon and Mercury. As you compare them with each other and with Earth, you will discover three important themes of planetary astronomy. First, you will see the importance of impact cratering. The surfaces of the moon and Mercury have been scarred by the

▌ Continued on page 498 ▌

Guidepost

Looking Back

Want to fly to the moon? You will need more than your lunch. There is no air and no water, and the sunlight is strong enough to kill you. Take shelter in the shade, and you will freeze to death in moments. You have never left Earth, so our planet seems normal to you and other worlds are, well, unearthly. In the previous chapter you studied your home world and found it a comfortable place. Now it is time to explore.

This Chapter

The moon and Mercury are small and airless, and that makes them good representatives of the smaller worlds in the universe. As you compare them with Earth, you will discover a number of important principles of comparative planetology, but you should also compare the moon and Mercury with each other. The differences are revealing. As you study these two airless worlds, you will find answers to four essential questions:

Why is the moon airless and cratered?

How did the moon form and evolve?

Why is Mercury different from the moon?

How did Mercury form and evolve?

Looking Ahead

Once you feel comfortable exploring airless worlds, you will be ready for bigger planets that have kept their heat and at least some of their atmospheres. They are not necessarily more interesting places, but they are just a tiny bit less unearthly.

impact of uncountable meteorites in a process that must have affected all the planets. Second, you will see the importance of internal heat in worlds smaller than Earth. Third, you will discover evidence of giant impacts so large they could shatter planets. By concentrating on these three themes, you will be able to organize the flood of details astronomers have learned about the moon and Mercury.

Your ultimate goal in the study of any world is to tell its story. How did the planet form, how has it evolved, and why is it the way it is? Each time you study a new planet, you should conclude your analysis by trying to combine what you have learned to tell the planet's story. For the moon and Mercury, you should use the three themes—cratering, internal heat, and giant impacts—to try to understand how small worlds can become such magnificently desolate places.

21-1 The Moon

ONLY 12 PEOPLE HAVE STOOD on the moon, but planetary astronomers know it well. The photographs, measurements, and samples brought back to Earth paint a picture of an airless, ancient, battered crust.

The View from Earth

A few billion years ago, the moon must have rotated faster, but Earth is four times larger in diameter than the moon (**Celestial Profile 3**), and its tidal forces on the moon are strong. Earth's gravity raised tidal bulges on the moon, and friction in the bulges slowed the moon until it now rotates once each orbit to keep the same side facing Earth in what is called **tidal coupling.** This means that those who live on Earth always see the same side of the moon; the back of the moon is never visible from Earth. The moon's familiar face has shone down on Earth since long before there were humans (■ Figure 21-1).

Even from Earth, you can tell that the moon is an airless world. How could you tell? First, your understanding of gravity tells you that a world as small as the moon must have a low escape velocity, and gas atoms near its surface escape easily into space. But that's theory and not evidence. What is the evidence? When you look at the moon you can see dramatic shadows near the **terminator,** the dividing line between daylight and darkness. There is no air on the moon to scatter light and soften shadows. Also, with even a small telescope you could watch stars disappear behind the **limb** of the moon—the edge of its disk—without dimming. Clearly, the moon is an airless (and soundless) world.

The most dramatic features on the moon, regions that are visible even to your unaided eyes, are **maria,** Latin for "seas." The dark maria (the singular is *mare*) that make up the face of the man in the moon were so named because early observers thought the

The dark, smooth areas of the moon are called seas (*mare* in Latin).

Much of the surface is covered with craters on top of craters.

Visual-wavelength image

■ Figure 21-1

The side of the moon that faces Earth is a familiar sight. Craters have been named for famous scientists and philosophers, and the so-called seas have been given romantic names. Mare Imbrium is the Sea of Rains and Mare Tranquillitatis is the Sea of Tranquillity. There is, in fact, no water on the moon. (Photo © UC Regents/Lick Observatory)

great smooth expanses were seas. Even a small telescope will show that the maria are marked by ridges, faults, smudges, and craters and therefore can't be water. The maria are actually great, smooth lowland plains covering 17 percent of the moon's surface.

Not many decades ago, astronomers argued about the nature of the maria. Some said that they were great solidified lava flows, but others argued that they were vast oceans of dust blasted from the lunar mountains by billions of years of meteorites and drawn to the lowlands by the lunar gravity. Science fiction writers wrote fascinating stories about the dust oceans. For example, *A Fall of Moon Dust* by Arthur C. Clarke tells of tourists in a lunar sightseeing boat trapped when their boat sinks in the dust to the bottom of a mare. Although it is a wonderful rescue story, it can never happen. Astronauts have walked on the maria and report that the maria are ancient, solidified lava flows with only a few centimeters of loose dust on their surfaces.

These lava flows suggest volcanism, but only small traces of past volcanic activity are visible from Earth. No major volcanic peaks are visible, and no active volcanism has ever been detected. The lava flows that created the maria happened long ago and were much too fluid to build peaks. With a good telescope and some diligent searching you can see a few small domes pushed up by lava below the surface, and you can see long, winding channels called **sinuous rilles** (■ Figure 21-2), often found near the edges of the maria. These channels were evidently cut by flowing lava. In some cases, such a channel may have had a roof of solid rock, forming a lava tube. When the lava drained away, meteorite impacts collapsed the roof to form a sinuous rille. The view from Earth provides only hints of ancient volcanic activity associated with the maria.

Even a small telescope reveals craters on the moon. The lunar surface is scarred with millions of craters formed by the impact of meteorites. Study **Impact Cratering** on pages 500–501 and notice three important points:

1 Impact craters have certain distinguishing characteristics, such as their shape and the way they eject material across the lunar surface.

2 Notice the range of sizes of impact craters from giant basins to microscopic pits.

3 Finally, notice that most of the craters on the moon are old; they were formed long ago when the solar system was young.

Meteorites strike the moon all the time, but large impacts are rare. Meteorites a few tens of meters in diameter probably strike the moon every 50 years or so, but no one has ever seen such an impact with certainty. Small flashes of light have been seen on the dark side of the moon during showers of meteors, but those impacts must have been made by very small objects. A flash of light photographed on the moon in 1953 has been tentatively linked to a small crater just visible in photographs made by spacecraft in orbit around the moon. No significant change has been seen on the moon since the invention of the telescope. Large impacts on the moon and Earth are quite rare, and nearly all of the lunar craters seen through telescopes date from the solar system's youth.

Craters are a key element in planetary study. They are one of the main themes of this chapter, and they will help you understand the moon, Mercury, and many other planets and moons throughout our solar system. You can begin here by using craters to unravel the history of the lunar maria.

Highlands and Lowlands

The surface of the moon is divided into two dramatically different kinds of terrain. The highlands are heavily cratered and are about 3 km higher than the lowlands, which are filled by the mostly uncratered, solidified lava flows known as maria. The number of craters on these surfaces will give you an important clue to understanding these two types of terrain.

There are no folded mountain ranges on the moon. No Rockies, Andes, or Urals stretch across the moon. What appears to be a mountainous region is really the highly cratered lunar highlands. The surface there is saturated with craters, meaning that it would be impossible to form a new crater without destroying the equivalent of one old crater (Figure 21-2).

The moon   Earth

Visual-wavelength images

Earth's moon is about one-fourth the diameter of Earth. Its low density indicates that it contains little iron, but the size of its iron core and the amount of remaining heat are unknown. (NASA)

Celestial Profile 3: The Moon

Motion:

Average distance from Earth	384,400 km (center to center)
Eccentricity of orbit	0.055
Maximum distance from Earth	405,500 km
Minimum distance from Earth	363,300 km
Inclination of orbit to ecliptic	5°9'
Average orbital velocity	1.022 km/s
Orbital period (sidereal)	27.321661 days
Orbital period (synodic)	29.5305882 days
Inclination of equator to orbit	6°41'

Characteristics:

Equatorial diameter	3476 km
Mass	7.35×10^{22} kg (0.0123 $M_{\oplus}$)
Average density	3.36 g/cm³ (3.35 g/cm³ uncompressed)
Surface gravity	0.167 Earth gravity
Escape velocity	2.38 km/s (0.21 $V_{\oplus}$)
Surface temperature	−170° to 130°C (−274° to 266°F)
Average albedo	0.07

Personality Point:

Lunar superstitions are very common. The words *lunatic* and *lunacy* come from *luna*, the moon. Someone who is *moonstruck* is supposed to be at least a bit nutty. Because the moon affects the ocean tides, many superstitions link the moon to water, to weather, and to women's cycle of fertility. According to legend, moonlight is supposed to be harmful to unborn children, but on the plus side, moonlight rituals are said to remove warts.

Impact Cratering

1 The craters that cover the moon and many other bodies in the solar system were produced by the high-speed impact of meteorites of all sizes. Meteorites striking the moon travel 10 to 60 km/s and can hit with the energy of many nuclear bombs.

A meteorite striking the moon's surface can deliver tremendous energy and can produce an impact crater 10 or more times larger in diameter than the meteorite. The vertical scale is exaggerated at right for clarity.

Impact Cratering

A meteorite approaches the lunar surface at high velocity.

On impact, the meteorite is deformed, heated, and vaporized.

The resulting explosion blasts out a round crater.

Slumping produces terraces in crater walls, and rebound can raise a central peak.

1a Lunar craters such as Euler, 27 km (17 mi) in diameter, look deep when you see them near the terminator where shadows are long, but a typical crater is only a fifth to a tenth as deep as its diameter, and large craters are even shallower.

Because craters are formed by shock waves rushing outward, by the rebound of the rock, and by the expansion of hot vapors, craters are almost always round, even when the meteorite strikes at a steep angle.

Debris blasted out of a crater is called ejecta, and it falls back to blanket the surface around the crater. Ejecta shot out along specific directions can form bright rays.

Euler

NASA

Visual-wavelength image

Ace Astronomy™

Log into AceAstronomy and select this chapter to see Active Figure "The Moon's Craters." Notice that the structure of the craters depends on their size.

1b Rock ejected from distant impacts can fall back to the surface and form smaller craters called secondary craters. The chain of craters here is a 45-km-long chain of secondary craters produced by ejecta from the large crater Copernicus 200 km out of the frame to the lower right.

Bright ejecta blankets and rays gradually darken as sunlight darkens minerals and small meteorites stir the dusty surface. Bright rays are signs of youth. Rays from the crater Tycho, perhaps only 100 million years old, extend halfway around the moon.

Visual

NASA

Rays

Tycho

Visual

NASA

2 Plum Crater, 40 m (130 ft) in diameter, was visited by Apollo 16 astronauts. Note the many smaller craters visible. Lunar craters range from giant impact basins to tiny pits in rocks struck by **micrometeorites**, meteorites of microscopic size.

Mare Orientale

Lunar rover

Sun glare in camera lens

NASA

Visual-wavelength images

Solidified lava

2b The energy of an impact can melt rock, some of which falls back into the crater and solidifies. When the moon was young, craters could also be flooded by lava welling up from below the crust.

A few meteorites found on Earth have been identified chemically as fragments of the moon's surface blasted into space by cratering impacts. The fragmented nature of these meteorites indicates that the moon's surface has been battered by impact craters.

2a In larger craters, the deformation of the rock can form one or more inner rings concentric with the outer rim. The largest of these craters are called **multiringed basins**. In Mare Orientale on the west edge of the visible moon, the outermost ring is almost 900 km (550 mi) in diameter.

NASA

3 Most of the craters on the moon were produced long ago when the solar system was filled with debris from planet building. As that debris was swept up, the cratering rate fell rapidly, as shown below.

Meteorite from moon

NASA

Rate of Crater Formation

Notice the exponential scale in this graph of the cratering rate. Only half a billion years after the origin of the solar system, the cratering rate had fallen by a factor of 10,000. Today, the rate is very low, and no crater visible from Earth is known to have appeared on the moon in historic times.

Cratering rate: 10^6, 10^5, 10^4, 10^3, 10^2, 10, 1

Time before present (billion years): 4, 3, 2, 1, Now

Hadly Rille is a sinuous rill near the edge of Mare Imbrium.

Faults where the crust has broken

Apollo 15 found Hadley Rille to be a 1200-foot-deep lava channel

The lunar highlands are saturated with craters on top of craters.

The smooth maria are generally circular with faults and wrinkles in the old lava surface.

Visual-wavelength images

Ancient craters partly flooded by lava

■ **Figure 21-2**

Details visible in photographs show that meteorite impacts long ago covered the moon with craters, but that lava flooded out and filled the largest basins covering the craters there with smooth plains. (Hadley: NASA; Moon disk, highlands, and mare: © UC Regents/Lick Observatory)

The maria fill the lunar lowlands, and they contain few craters, as you can see in Figure 21-2. Among the few craters that do scar the maria, many have bright rays, suggesting that they are younger craters. For examples, locate the craters Kepler and Copernicus in Figure 21-1.

You can conclude from this that the highlands are old regions where you see craters from early in the history of the solar system. Most of those craters were formed during the heavy bombardment over 4 billion years ago. The maria that fill the lowlands must have formed later, after the decline of the heavy bombardment. That means the maria are younger than the highlands. These are called **relative ages** because you can tell which features formed first, but you can't give ages in years.

The cratering curve shown on page 501 can reveal **absolute ages,** meaning ages in years. Studies of cratered surfaces in the solar system have allowed planetary scientists to calibrate the cratering curve. The lava flows that filled the lowlands and covered up the craters there could not have formed earlier than 3 to 4 billion years ago. Had they occurred earlier, the surfaces of the maria would show more craters.

Absolute ages derived from the cratering curve cannot be precise because planetary scientists suspect that the cratering rate may not have fallen as smoothly as the cratering curve suggests. Al-

though the debris in the early solar system was gradually removed, bursts of cratering may have occurred. A major impact between large planetesimals, for example, might shower the solar system with fragments and produce a storm of crater impacts. Mathematical models suggest that the final formation of Uranus and Neptune might have flung icy planetesimals through the solar system like giant shrapnel. The true cratering curve might have rises and dips superimposed on a gradual decline.

Whatever the exact timing of the heavy bombardment, it is clear that the craters in the highlands are old and that the later lava flows that formed the maria covered over any craters that had been formed in the lowlands. Using such clues, astronomers were able to begin telling the story of the moon's surface (**Window on Science 21-1**). But the view from Earth does not provide enough evidence. To really understand the lunar surface, humans had to go there.

The Apollo Missions

On May 25, 1961, President John Kennedy addressed Congress and committed the United States to landing a human being on the moon by 1970. The reasons for that decision are complicated

How Hypotheses and Theories Unify the Details

Like any technical subject, science includes a mass of details, facts, figures, measurements, and observations. It is easy to be overwhelmed by the flood of details, but one of the most important characteristics of science comes to your rescue. The goal of science is not to discover more details but to explain the details with a unifying hypothesis or theory. A good theory is like a basket that makes it easier for you to carry a large assortment of details.

This is true of all the sciences. When a psychologist begins studying the way the human eye and brain respond to moving points of light, the data are a sea of detailed measurements and observations. Once the psychologist forms a hypothesis about the way the eye and brain interact, the details fall into place as parts of a logical story. If you understand the hypothesis, the details all fit together and make sense, and you can remember the details without blindly memorizing tables of facts and figures. Remember that the goal of science is understanding nature, not memorizing details.

Scientists are in the storytelling business. The stories are often called hypotheses or theories, but they are really just stories to explain how nature works. The difference between scientific stories and works of fiction lies in the use of facts. Scientific stories are constructed to fit all known facts and are then tested over and over against new facts obtained by observation and experiment.

When you try to tell the story of each planet in our solar system, you pull together all the hypotheses, theories, and observations and try to make them into a logical history of how the planet got to be the way it is. Of course, your stories will be incomplete because astronomers don't understand all the factors in planetary astronomy. Nevertheless, your story of each planet will draw together the known facts and details and attempt to make them a logical whole.

When scientists create a hypothesis, it draws together a great many observations and measurements. (Phyllis Leber)

Memorizing a list of facts can give you a false feeling of security, just as when someone memorizes the names of things without understanding them (Window on Science 2-2). Rather than memorizing facts, you should search for the unifying hypothesis that pulls the details together into a single story. That is the goal of science.

and related more to economics, international politics, and the stimulation of technology than to science, but the Apollo program became a fantastic adventure in science, a flight to the moon that changed how we all think about Earth.

Flying to the moon is not particularly difficult; with powerful enough rockets and enough food, water, and air, it is a straightforward trip. Landing on the moon is more difficult but not impossible. The moon's gravity is only one-sixth that of Earth, and there is no atmosphere to disturb the trajectory of the spaceship. Getting to the moon is simple and landing is possible; the difficulty is doing both on one trip. The spaceship must carry food, water, and air for a number of days in space plus fuel and rockets for midcourse corrections and for a return to Earth. All of this adds up to a ship that is too massive to make a safe landing on the lunar surface. The solution was to take two spaceships to the moon, one to ride in and one to land in (■ Figure 21-3).

The command module was the long-term home and command center for the trip. Three astronauts had to live in it for a week, and it had to carry all the life-support equipment, navigation instruments, computers, power packs, and so on for a week's jaunt in space. The lunar landing module (LM for short) was tacked to the front of the command module like a bicycle strapped to the front of the family camper. It carried only enough fuel and supplies for the short trip to the lunar surface, and it was built to minimize weight and maximize maneuverability.

The weaker gravity of the moon made the design of the LM simpler. Landing on Earth requires reclining couches for the astronauts, but the trip to the lunar surface involved smaller accelerations. In an early version of the LM, the astronauts sat on what looked like bicycle seats, but these were later scrapped to save weight. The astronauts had no seats at all in the LM, and once they began their descent and acquired weight, they stood at the controls held by straps, riding the LM like two daredevils riding a rocket surfboard.

Lifting off from the lunar surface, the LM saved weight by leaving the larger descent rocket and support stage behind. Only the astronauts, their instruments, and their cargo of rocks returned to the command module orbiting above. Again, the astronauts in the LM blasted up from the lunar surface standing at the controls. The rocket engine that lifted them back into lunar orbit was not much bigger than a dishwasher.

The most complicated parts of the trip were the rendezvous and docking between the tiny remains of the LM and the command module, aided by radar systems and computers. The astronauts rejoined their colleague in the command module, transferred their moon rocks, and jettisoned the LM. Only the command module returned to Earth.

The first manned lunar landing was made on July 20, 1969. While Michael Collins waited in orbit around the moon, Neil Armstrong and Edwin Aldrin, Jr., took the LM down to the

Visual-wavelength image

■ **Figure 21-3**

The Apollo 12 lunar module. The two astronauts stood in a space hardly bigger than two telephone booths. The metal skin was so thin it was easily flexible, like metal foil, and the legs of the module, designed specifically for the moon's weak gravity, could not support the lander's weight on Earth. Only the upper half of the lander blasted off from the surface to return the astronauts to the command module in orbit around the moon. (NASA)

surface. Although much of the descent was controlled by computers, the astronauts had to override a number of computer alarms and take control of the LM to avoid a boulder-strewn crater bigger than a football field. Climbing down the ladder, Armstrong and then Aldrin stepped onto the lunar surface.

Between July 1969 and December 1972, 12 people reached the lunar surface and collected 380 kg (840 lb) of rocks and soil (■ Table 21-1). The flights were carefully planned to visit different regions and develop a comprehensive history of the lunar surface.

■ **Table 21-1 | Apollo Lunar Landings**

Apollo Mission*	Astronauts: Commander LM Pilot CM Pilot	Date	Mission Goals	Sample Weight (kg)	Typical Samples	Ages (10^9 y)
11	Armstrong Aldrin Collins	July 1969	First manned landing; Mare Tranquillitatis	21.7	Mare basalts	3.48–3.72
12	Conrad Bean Gordon	Nov. 1969	Visit Surveyor 3; sample Oceanus Procellarum (mare)	34.4	Mare basalts	3.15–3.37
14	Shepard Mitchell Roosa	Feb. 1971	Fra Mauro, Imbrium ejecta sheet	42.9	Breccia	3.85–3.96
15	Scott Irwin Worden	July 1971	Edge of Mare Imbrium and Apennine Mountains, Hadley Rille	76.8	Mare basalts; highland anorthosite	3.28–3.44 4.09
16	Young Duke Mattingley	April 1972	Sample highland crust; Cayley formation (ejecta); Descartes	94.7	Highland basalt; breccia	3.84 3.92
17	Cernan Schmitt Evans	Dec. 1972	Sample highland crust; dark halo craters; Taurus–Littrow	110.5	Mare basalt; highland breccia; fractured dunite	3.77 3.86 4.48

*The Apollo 13 mission suffered an explosion on the way to the moon and did not land.

Apollo 17, the last Apollo mission to the moon, landed in the highlands in December 1972.

Apollo 11 landed in the lunar lowlands in July 1969.

The first flights went into relatively safe landing sites (■ Figure 21-4)—Mare Tranquillitatis for Apollo 11 and Oceanus Procellarum for Apollo 12. Apollo 13 was aimed at a more complicated site, but an explosion in an oxygen tank on the way to the moon ended all chances of a landing and nearly cost the astronauts their lives. They succeeded in using the life support in the LM to survive the trip around the moon, and they eventually landed safely in the crippled command module.

The last four Apollo missions, 14 through 17, sampled geologically important places on the moon. Apollo 14 visited the Fra Mauro region, which is covered by ejecta from the impact that dug the basin now filled by Mare Imbrium. Apollo 15 visited the edge of Mare Imbrium at the foot of the Apennine Mountains and examined Hadley Rille (Figure 21-2). Apollo missions 16 and 17 visited highland regions to sample the older parts of the lunar crust (Figure 21-4). Almost all of the lunar samples from these six landings are now held at the Planetary Materials Laboratory at the Johnson Space Center in Houston. They are a national treasure containing clues to the beginnings of our solar system.

Moon Rocks

If you visited the moon, where the gravity is one-sixth that on Earth, you would discover that you weighed only one-sixth what you weigh on Earth. The Apollo astronauts found it easy to dance across the lunar surface conducting experiments and collecting geological specimens.

Of all of the rock samples that the Apollo astronauts carried back to Earth, every one is igneous. That is, they formed by the cooling and solidification of molten rock. No sedimentary rocks were found, which is consistent with the moon never having had liquid water on its surface. In addition, the rocks were extremely dry. Almost all Earth rocks contain 1 to 2 percent water, either as free water trapped in the rock or as water molecules chemically bonded with certain minerals. But moon rocks contain no water at all.

Rocks from the lunar maria are dark-colored, dense basalts much like the solidified lava produced by the Hawaiian volcanoes (■ Figure 21-5). These rocks are rich in heavy elements such as iron, manganese, and titanium, which give them their dark color. Some of the basalts are **vesicular,** meaning that they contain holes caused by bubbles of gas in the molten rock. Like bubbles in a carbonated beverage, these bubbles do not form while the magma is under pressure. Only when the molten rock flows out onto the surface, where the pressure is low, do bubbles appear. The presence of vesicular basalts assures that these rocks

The Apollo astronauts found that all moon rocks are igneous, meaning they solidified from molten rock.

Rocks exposed on the lunar surface become pitted by micrometeorites.

Vesicular basalt contains bubbles frozen into the rock when it was molten.

A breccia is formed by rock fragments bonded together by heat and pressure.

■ Figure 21-5

Rocks returned from the moon show that the moon formed in a molten state, that it was heavily fractured by cratering when it was young, and that it is now affected mainly by micrometeorites grinding away at surface rock. (NASA)

formed in lava flows that reached the surface and did not solidify underground.

Absolute ages of the mare basalts can be found from the radioactive atoms they contain (Chapter 19), and these ages range from about 3.1 to 3.8 billion years. That means the lava flows that formed these rocks must have happened some time after the end of the heavy bombardment.

The highlands are composed of low-density rock containing calcium-, aluminum- and oxygen-rich minerals that would be among the first to solidify and float to the top of molten rock. The crustal rocks range in age from 4.0 to 4.5 billion years old, significantly older than the mare basalts. The highlands contain **anorthosite,** a light-colored rock that contributes to the highlands' bright contrast with the dark, iron-rich basalts of the lowlands. The rocks of the highlands, although badly shattered by impacts, represent the moon's original low-density crust, whereas the mare basalts rose as molten rock from the deep crust and upper mantle.

A large fraction of the lunar rocks are **breccias,** rocks that are made up of fragments of earlier rocks cemented together by heat and pressure. Evidently, meteorite impacts have broken up the rocks and fused them together time after time.

If you went to the moon, you would get your spacesuit dirty. Both the highlands and the lowlands of the moon are covered by a layer of powdered rock and crushed fragments called the **regolith.** It is about 10 m deep on the maria but over 100 m deep in certain places in the highlands. About 1 percent of the regolith is meteoric fragments; the rest is the smashed remains of moon rocks that have been pulverized by the constant rain of meteorites. The smallest meteorites, the micrometeorites, do the most damage by their constant sandblasting of the lunar surface, which grinds the rock down to fine dust. The Apollo astronauts found that the dust coated their spacesuits and equipment and made a mess where it got tracked into the Lunar Landing Module. From all this you can see that impact cratering, a theme of this chapter, dominates the lunar surface and is even responsible for the lunar regolith.

Understanding Natural Processes: Follow the Energy

One of the best ways to think about a scientific problem is to follow the energy.

According to the principle of cause and effect, every effect must have a cause, and every cause must involve energy (Focus on Fundamentals 2). Energy moves from regions of high concentration to regions of low concentration and in doing so produces changes. For example, coal burns to make steam in a power plant, and the steam passes through a turbine and then escapes into the air. In flowing from the burning coal to the atmosphere, the heat spins the turbine and makes electricity.

It is surprising how commonly scientists use energy as a key to understanding nature. A biologist might ask where certain birds get the energy to fly thousands of miles, and an economist might ask where the economic energy to support the creation of new investments comes from. Energy is everywhere; and when it moves, whether it is birds, money, or molten magma, it causes change. Energy is the cause in cause and effect.

In earlier chapters, the flow of energy from the inside of a star to its surface helped you understand how stars, including the sun, work. You saw that the outward flow of energy supports the star against its own weight, drives convection currents that produce magnetic fields, and causes surface activity such as spots, prominences, and flares. You were able to understand stars because you could follow the flow of energy outward from their interiors.

You can think of a planet by following the energy. The heat in the interior of a planet may be left over from the formation of the planet, or it may be heat generated by radioactive decay, but it must flow outward toward the cooler surface, where it is radiated into space. In flowing outward, the heat can cause convection currents in the mantle, magnetic fields, plate motions, quakes, faults, volcanism, mountain building, and more.

When you think about any world, be it a small asteroid or a giant planet, think of it as a source of heat that flows outward through the planet's surface into space. If you can fol-

Heat flows out of Earth's interior and generates geological activity such as that at Yellowstone National Park. (M. Seeds)

low that energy flow, you can understand a great deal about the world. A planetary astronomer once said, "The most interesting thing about any planet is how its heat gets out."

The moon rocks are old, dry, igneous, and badly shattered by impacts. You can use these facts, combined with what you know about lunar features, to tell the story of the moon.

The History of the Moon

The four-stage history of the moon is dominated by a single fact—it is small, only one-fourth the diameter of Earth. The escape velocity is low, so it has been unable to hold any atmosphere, and it cooled rapidly as its internal heat flowed outward into space. As you will see when you visit other worlds, a planet's geology is largely driven by energy in the form of heat flowing outward from its interior (**Window on Science 21-2**). Small worlds have less heat and lose it more rapidly, so the moon's small size has been critical in its history.

The Apollo moon rocks show that the moon must have formed in a molten state. Planetary geologists now refer to the newborn moon as a sea of magma. Denser materials sank to form a small core; and, as the magma ocean cooled, low-density minerals floated to the top to form a low-density crust. In this way the moon must have differentiated into core, mantle, and crust. The radioactive ages of the moon rocks show that the surface solidified from 4.6 to 4.1 billion years ago. The moon has a low density and no magnetic field, so its dense core must be small. The core may still retain enough heat to be partially molten, but it

can't contain much iron or the dynamo effect would produce a magnetic field.

The second stage, cratering, began as soon as the crust solidified, and the older highlands show that the cratering was intense during the first 0.5 billion years—during the heavy bombardment at the end of planet building.

The moon's crust was shattered to a depth of 10 kilometers or so, and the largest impacts formed giant multiringed crater basins hundreds of kilometers in diameter, such as Mare Orientale. This led to the third stage—flooding. Although Earth's moon cooled rapidly after its formation, some process such as radioactive decay heated material deep in the crust, and part of it melted. Molten rock followed the cracks up to the surface and flooded the giant basins with successive lava flows of dark basalts from 3.8 to 3.2 billion years ago. This formed the maria (■ Figure 21-6).

Make a special note that the lava that floods out on the surfaces of planets does not come from the molten core. That is a common misconception. The lava comes from the lower crust and possibly from the top of the mantle. The pressure is lower there, and that lowers the melting point of the rock. Radioactive decay can generate enough heat to melt portions of the rock, and if there is enough heat and if there are faults and cracks, the magma can reach the surface and form volcanoes and lava flows. Whenever you see lava flows on a planet, you can be sure heat is flowing

a

|— 200 km —|

Mare Imbrium

Mare Serenitatis

Mare Crisium

Aitken Basin

b   Near side of moon   Far side of moon

■ **Figure 21-6**

(a) Major impacts broke the lunar crust and produced large basins. Lava flooded up through the fractures and flooded the basins to produce maria. (Adapted from a diagram by William Hartmann) (b) In these maps, color indicates elevation, with red the highest regions and purple the lowest. Maria on the near side are generally circular, low, lava-filled basins. Note the circular outline of the labeled maria. The crust on the far side is thicker, and there is less flooding. Even the Aitken Basin contains little lava flooding. (NASA/ Clementine)

out of the interior (one of the themes of this chapter), but the lava itself did not come all the way from the core.

Some maria on the moon, such as Mare Imbrium, Mare Serenitatis, Mare Humorum, and Mare Crisium, retain their round shapes, but others are irregular because the lava overflowed the edges of the basin or because the shape of the basin was modified by further cratering. The moving lava formed channels that are seen from Earth as sinuous rills. The weight of the maria pressed the crater basins downward, and the solidified lava was compressed and formed wrinkle ridges visible even in small telescopes. The tension at the edges of the maria broke the hard lava to produce straight fractures and faults. All of these features are visible in

Apollo 14 landed on rugged terrain suspected of being ejecta from the Imbrium impact. The large boulder here is ejecta that, at some time in the past, fell here from an impact beyond the horizon. (NASA)

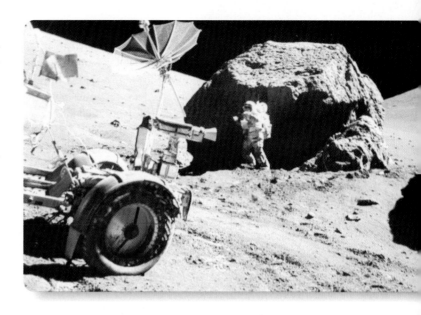

■ **Figure 21-8**

Lava flooding after the end of the heavy bombardment filled a giant multiringed basin and formed Mare Imbrium. (Don Davis)

Origin of Mare Imbrium

Four billion years ago, an impact forms a multiringed basin over 1000 km (700 mi) in diameter.

Continuing impacts crater the surface but do not erase the high walls of the multiringed basin.

Archimedes

Impacts form a few large craters, and, starting about 3.8 billion years ago, lava floods low regions.

Repeated lava flows cover most of the inner rings and overflow the basin to merge with other flows.

Impacts continue, including those that formed the relatively young craters Copernicus and Kepler.

Kepler   Copernicus

Figure 21-2. Later cratering and overlapping lava floods further modified the maria. Consequently you should think of the maria as accumulations of features reflecting the moon's complex history.

Mare Imbrium is a dramatic example of how the great basins became the maria. Near the end of the heavy bombardment, roughly 4 billion years ago, a planetesimal the size of Rhode Island struck the moon and blasted out a giant multiringed basin. The impact was so violent the ejecta blanketed 16 percent of the moon's surface. Apollo 14 landed at Fra Mauro to sample the ejecta (■ Figure 21-7). After the cratering rate fell at the end of the heavy bombardment, lava flows welled up time after time and flooded the Imbrium Basin, burying all but the highest parts of the giant multiringed basin. The Imbrium Basin is now a large, generally round mare marked by only a few craters that have formed since the last of the lava flows (■ Figure 21-8).

This story of the moon might suggest that it was a violent place during the cratering phase, but large impacts were in fact rare; and, except for a continuous rain of smaller meteorites, the moon was, for the most part, a peaceful place even during the heavy bombardment. Had you stood on the moon at that time you would have experienced a continuous rain of micrometeorites and much less common pebble-size impacts.

Centuries might pass between major impacts. Of course, a large impact far beyond the horizon might have threatened to bury you under ejecta or jolted you by seismic shocks. You could have felt the Imbrium impact anywhere on the moon, but had you been standing on the side of the moon opposite that impact, you would have been at the focus of seismic waves traveling around the moon from different directions. When the waves met under your feet, the surface would have jerked up and down by as much as 10 m. You would not have liked that. The place on the moon opposite the Imbrium Basin is a strangely disturbed landscape called **jumbled terrain.** You will see similar effects of large impacts on other worlds.

Studies of our moon show that its crust is thinner on the side toward Earth, perhaps due to tidal effects. Consequently, while lava flooded the basins on the Earthward side, it was unable to rise through the thicker crust to flood the lowlands on the far side. The largest impact basin in the solar system is the south pole Aitkin Basin on the moon (Figure 21-6b). It is about 2500 kilometers (1500 miles) in diameter and as deep as 13 kilometers (8 miles) in places, but flooding has never filled it with smooth lava flows.

The fourth stage, slow surface evolution, is limited because our moon lacks water and has cooled rapidly. Flooding on Earth included water, but the moon has never had an atmosphere and thus has never had liquid water. With no air and no water, erosion is limited to the constant bombardment of micrometeorites and rare larger impacts.

The moon is small, and small worlds cool rapidly because they have a large ratio of surface area to volume. You may have noticed that a small cupcake fresh from the oven cools more rapidly than a large cake. The moon has lost much of its internal heat, and it is the outward flow of heat that produces geological activity. That means the moon is mostly inactive. The crust of the moon rapidly grew thick and never divided into moving plates. There are no rift valleys or folded mountain chains on the moon. The last lava flows on the moon ended one or two billion years ago when the moon's temperature fell too low to maintain subsurface lava.

The overall terrain on the moon is almost fixed. On Earth a billion years from now, plate tectonics will have totally changed the shapes of the continents, and erosion will have long ago worn away the Rocky Mountains. On the moon, impacts will have formed only a few more large craters, and nearly all of the lunar scenery will be unchanged. Micrometeorites, however, will have blasted the soil, erasing the footprints left by the Apollo astronauts and reducing the equipment they left behind to peculiar chemical contamination in the regolith at the six Apollo landing sites.

You have studied the story of the moon's evolution in detail for later comparison with other planets and moons in our solar system, but the story has skipped one important question: Where did Earth get such a large satellite?

The Origin of Earth's Moon

Over the last two centuries, astronomers developed three different hypotheses for the origin of Earth's moon, but these traditional ideas have failed to survive comparison with the evidence. A the-

ory proposed in the 1980s may hold the answer. You can begin by testing the three unsuccessful theories against the evidence to see why they failed.

The first of the three traditional theories, the **fission hypothesis,** supposes that the moon formed by the fission of Earth. If the young Earth spun fast enough, tides raised by the sun might make it break into two parts (■ Figure 21-9a). If this separation occurred after Earth differentiated, the moon would have formed from crust material. In that way, the theory explains the moon's low density.

But the fission theory has problems. No one knows why the young Earth should have spun so fast, nearly ten times faster than today, nor where all that angular momentum went after the fission. In addition, the moon's orbit is not in the plane of Earth's equator, as it would be if it had formed by fission.

The second traditional theory is the **condensation** (or double-planet) **hypothesis.** It supposes that Earth and the moon condensed as a double planet from the same cloud of material (Figure 21-9b). However, if they formed from the same material, they should have the same chemical composition and density, which they don't. The moon is very poor in iron and related elements such as titanium and in volatiles such as water vapor and sodium. Yet the moon contains almost exactly the same ratios of oxygen isotopes as does Earth's mantle. The condensation theory cannot explain these compositional differences.

The third theory is the **capture hypothesis.** It supposes that the moon formed somewhere else and was later captured by Earth (Figure 21-9c). You might imagine the moon forming inside the orbit of Mercury, where the heat would prevent the condensation of solid metallic grains and only high-melting-point metal oxides could have solidified. A later encounter with Mercury could "kick" the moon out to Earth.

The capture theory was not popular because it requires highly unlikely events involving interactions with Mercury and Earth to move the moon from place to place. Scientists are always suspicious of explanations that require a chain of unlikely coincidences. Also, upon encountering Earth, the moon would have been moving so rapidly that Earth's gravity would have been unable to capture it without ripping the moon to fragments through tidal forces.

Until recently, astronomers were left with no acceptable theory to explain the origin of the moon, and they occasionally joked that the moon could not exist. But during the 1980s, plan-

■ **Figure 21-9**

Three traditional theories for the moon's origin. (a) Fission theories suppose that Earth and the moon were once one body and broke apart. (b) Condensation theories suppose that the moon formed at the same time and from the same material as Earth. (c) Capture theories suggest that the moon formed elsewhere and was captured by Earth. None of these theories explains all the facts.

etary astronomers developed a new theory that combines the best aspects of the fission hypothesis and the capture hypothesis.

The **large-impact hypothesis** supposes that the moon formed from debris ejected into a disk around Earth by the impact of a large body. The impacting body may have been twice as large as Mars. Instead of saying that Earth was hit by a large body, it may be more nearly correct to say that Earth and the moon resulted from the collision and merger of two very large planetesimals. The resulting large body became Earth, and the ejected debris formed the moon (■ Figure 21-10). Such an impact would have melted the proto-Earth, and the material falling together to form the moon would have been heated hot enough to melt. This fits the evidence from moon rocks that the moon formed as a sea of magma.

This hypothesis would explain a number of things. The collision had to occur at a steep angle to eject enough matter to make the moon. That is, the objects did not collide head-on. Such a glancing collision would have spun the resulting material rapidly enough to explain the observed angular momentum in the Earth–moon system. If the two colliding planetesimals had already differentiated, the ejected material would be mostly iron-poor mantle and crust. Calculations show that the iron core of the impacting body could have fallen into the larger body that became Earth. This would explain why the moon is so poor in iron and why the abundances of other elements are so similar to rocks from Earth's mantle. Also, the material that eventually became the moon would have remained in a disk long enough for volatile elements, which the moon lacks, to be lost to space.

The moon may be the remains of a giant impact. Until recently, astronomers have been reluctant to consider such catastrophic events, but a number of lines of evidence suggest that planets have been affected by giant impacts. Consequently, the third theme identified in the introduction to this chapter, giant impacts, has the potential to help you understand other worlds. Catastrophic events are rare, but they can occur.

Building Scientific Arguments

If the moon was intensely cratered by the heavy bombardment and then formed great lava plains, why didn't the same thing happen on Earth?

Is this argument obvious? It is still worth reviewing as a way to test your understanding. In fact, the same thing did happen on Earth. Although the moon has more craters than Earth, the moon and Earth are the same age, and both were battered by meteorites during the heavy bombardment. Some of those impacts on Earth must have been large and dug giant multiringed basins. Lava flows must have welled up through Earth's crust and flooded the lowlands to form great lava plains much like the lunar maria.

Earth, however, is a larger world and has more internal heat, which escapes more slowly than the moon's heat did. The moon

The Large-Impact Hypothesis

A protoplanet nearly the size of Earth differentiates to form an iron core.

Another body that has also formed an iron core strikes the larger body and merges, trapping most of the iron inside.

Iron-poor rock from the mantles of the two bodies forms a ring of debris.

Volatiles are lost to space as the particles in the ring begin to accrete larger bodies.

Eventually the moon forms from the iron-poor and volatile-poor matter in the disk.

■ Figure 21-10

Sometime before the solar system was 50 million years old, a collision produced Earth and the moon in its inclined orbit.

is now geologically dead, but Earth is very active, with heat flowing outward from the interior to drive plate tectonics. The moving plates long ago erased all evidence of the cratering and lava flows dating from Earth's youth.

Comparative planetology is a powerful tool in that it allows you to see similar processes occurring under different circumstances. For example, expand your argument to explain a different phenomenon. **Why doesn't the moon have a magnetic field?**

■　　■　　■

Connections: The moon's evolution has been restricted by its size. It is too small to have an atmosphere, and it lost its internal heat so rapidly that it never developed plate motion. Another world in our solar system has also suffered from its diminutive size. Mercury, like the moon, is a small, rocky world, cratered by impacts and flooded by ancient lava flows.

21-2　Mercury

EARTH'S MOON AND MERCURY are good subjects for comparative planetology. They are similar in a number of ways. Their rotation has been altered by tides, their surfaces are heavily cratered, their lowlands are flooded in places by ancient lava flows, and both are small, airless, and have ancient, inactive surfaces. Yet the differences between them are impressive and will help you understand the nature of these airless worlds.

Mercury is the innermost planet in the solar system (**Celestial Profile 4**), and thus its orbit keeps it near the sun in the sky. It is sometimes visible near the horizon in the evening sky after sunset or in the dawn sky just before sunrise. An Earth-based telescope shows Mercury passing through phases like the moon, but little detail is visible. Most of what is known about the surface of Mercury is based on photographs and measurements made by the Mariner 10 spacecraft when it looped through the inner solar system in 1974 and 1975 (■ Figure 21-11). Until then, Mercury was a mystery world.

Rotation and Revolution

During the 1880s, the Italian astronomer Giovanni Schiaparelli sketched the faint features he thought he saw on the disk of Mercury and concluded that the planet was tidally coupled to the sun just as the moon is tidally coupled to Earth. That is, he concluded that Mercury rotated on its axis to keep the same side facing the sun throughout its orbit. This was actually a very good guess because, as you will see, tidal coupling between rotation and revolution is common in the solar system. But the rotation of Mercury is more complex than Schiaparelli thought.

In 1962, radio astronomers detected radio emissions from the planet and concluded that the dark side was not as cold as it should have been if the planet kept one side perpetually in darkness. In 1965, radio astronomers made radar contact with Mercury. They used the 305-m Arecibo dish (Figure 6-19) to transmit a pulse of radio energy at Mercury and then waited for the re-

Visual-wavelength images

■ **Figure 21-11**

Photomosaic of Mercury made by the Mariner 10 spacecraft. Caloris Basin is in shadow at left. Almost no surface detail is visible from Earth (inset). (NASA; inset courtesy Lowell Observatory)

flected signal to return. Doppler shifts in the reflected radio pulse showed that the planet was rotating with a period of only about 59 days, much shorter than the orbital period of 87.969 days.

Apparently, Mercury is tidally coupled to the sun such that it rotates not once per orbit but 1.5 times per orbit. That is, its period of rotation is two-thirds its orbital period, or 58.65 days. This means that a mountain on Mercury directly below the sun at one place in its orbit will point away from the sun one orbit later and toward the sun after the next orbit (■ Figure 21-12).

If you flew to Mercury and landed your spaceship in the middle of the day side, the sun would be high overhead, and it would be noon. Your watch would show that almost 44 Earth days passed before the sun set in the west, and a total of 88 Earth days would have passed before the sun reached the midnight position. In those 88 Earth days, Mercury would have completed one orbit around the sun (Figure 21-12). It would require another entire orbit of Mercury for the sun to return to the noon position overhead. So a full day on Mercury is two Mercury years long!

The complex tidal coupling between the rotation and revolution of Mercury is an important illustration of the power of tides. Just as the tides in the Earth–moon system have slowed the moon's rotation and locked it to Earth, so have the sun–Mercury tides slowed the rotation of Mercury and coupled its rotation to its revolution. Astronomers refer to such a relationship as a **resonance.** You will see many such resonances as you explore the solar system.

Like its rotation, Mercury's orbital motion is also peculiar. Recall from Chapter 5 that Mercury's orbit precesses faster than

The Rotation of Mercury

Imagine a mountain on Mercury that points at the sun. It is noon at the mountain.

The planet orbits and rotates in the same direction, counterclockwise as seen from the north.

After half an orbit, Mercury has rotated 3/4 of a turn, and it is sunset at the mountain.

As the planet continues along its orbit, rotation carries the mountain into darkness.

After one orbit, Mercury has rotated 1.5 times, and it is midnight at the mountain.

■ **Figure 21-12**

Mercury's orbital motion and rotation are related. It orbits the sun in 88 days and rotates on its axis in two-thirds of that time. One full day on Mercury from noon to noon takes two full orbits.

can be explained by Newton's laws but at just the rate predicted by Einstein's theory of general relativity. The orbital motion of Mercury is taken as strong confirmation of the curvature of space-time as predicted by general relativity.

Mercury is a bit over one-third the diameter of Earth, but its high density must mean it has a large iron core. The amount of heat it retains is unknown. (NASA)

Celestial Profile 4: Mercury

Motion:

Average distance from the sun	0.387 AU (5.79×10^7 km)
Eccentricity of orbit	0.2056
Maximum distance from the sun	0.467 AU (6.97×10^7 km)
Minimum distance from the sun	0.306 AU (4.59×10^7 km)
Inclination of orbit to ecliptic	7°00'16"
Average orbital velocity	47.9 km/s
Orbital period	87.969 days (0.24085 y)
Period of rotation	58.646 days (direct)
Inclination of equator to orbit	0°

Characteristics:

Equatorial diameter	4878 km (0.382 $D_\oplus$)
Mass	3.31×10^{23} kg (0.0558 $M_\oplus$)
Average density	5.44 g/cm³ (5.4 g/cm³ uncompressed)
Surface gravity	0.38 Earth gravity
Escape velocity	4.3 km/s (0.38 $V_\oplus$)
Surface temperature	−173° to 430°C (−280° to 800°F)
Average albedo	0.1
Oblateness	0

Personality Point:

Mercury lies very close to the sun and completes an orbit in only 88 days. For this reason, the ancients named the planet after Mercury, the fleet-footed messenger of the gods. The name is also applied to the element mercury, which is also known as quicksilver because it is a heavy, quickly flowing silvery liquid at room temperatures.

The Surface of Mercury

Because the days are so long on Mercury and because it is close to the sun and too small to keep an atmosphere, the temperatures on Mercury are extreme. If you stood in direct sunlight on Mercury, you would hear your spacesuit's cooling system cranking up to top speed as it tried to keep you cool. Daytime temperatures can exceed 700 K (800°F), although about 500 K is a more usual high temperature. If you stepped into shadow on Mercury or took a walk at night, with no atmosphere to distribute heat, your spacesuit heaters would struggle to keep you warm. The surface can cool to 100 K (−280°F). Nights on Mercury are bitter cold. Don't go to Mercury in a cheap spacesuit.

In appearance, Mercury looks much like Earth's moon. It is heavily battered, with craters of all sizes, including some large basins. Some craters are obviously old and degraded; others seem quite young and have bright rays of ejecta. However, a quick glance at photos of Mercury shows no large, dark maria like the moon's flooded basins.

The largest basin on Mercury is called Caloris Basin after the Latin word for "heat," recognition of its location at one of the two "hot poles," which face the sun at alternate perihelions. At the times of the Mariner encounters, the Caloris Basin was half in shadow (■ Figure 21-13a). Although half cannot be seen, the low angle of illumination is ideal for the study of the lighted half because it produces dramatic shadows.

The Caloris Basin is a gigantic multiringed impact basin 1300 km in diameter with concentric mountain rings up to 3 km high. The impact threw ejecta 600 to 800 km across the planet, and the focusing of seismic waves on the far side produced peculiar terrain (called lineated terrain) that looks much like the jumbled surface of the moon that lies opposite the Imbrium basin (Figure 21-13b and c). The Caloris Basin is partially filled with lava flows. Some of this lava may be material melted by the energy of the impact, but some may be lava from below the crust that leaked up through cracks. The weight of this lava and the response of the crust have produced deep cracks in the central lava plains. The geophysics of such large, multiringed crater basins is not well understood at present, but Caloris Basin seems to be the same kind of structure as the Imbrium Basin on the moon, but it has not been as deeply flooded with lava.

When planetary scientists began looking at the Mariner photographs in detail, they discovered something not seen on the moon. Mercury is marked by great curved cliffs called **lobate scarps** (■ Figure 21-14). These seem to have formed when the planet cooled and shrank by a few kilometers, wrinkling its crust as a drying apple wrinkles its skin. Some of these scarps are as high as 3 km and reach hundreds of kilometers across the surface. Other faults in Mercury's crust are straight and may have been produced by tidal stresses generated when the sun slowed Mercury's rotation.

The Plains of Mercury

The most striking difference between Mercury and the moon is that Mercury lacks the great dark lava plains so obvious on the

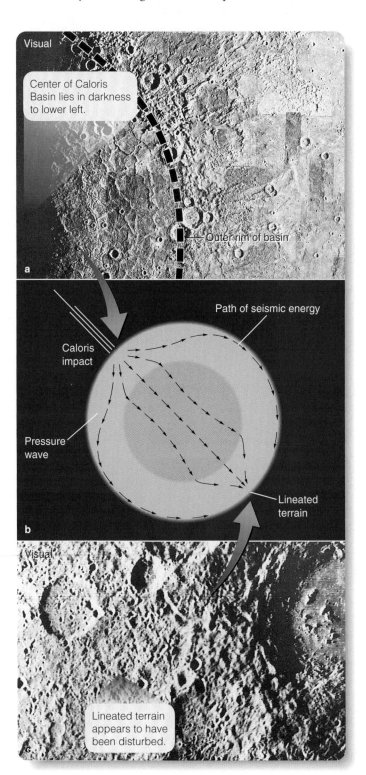

■ Figure 21-13

The huge impact that formed the Caloris Basin on Mercury sent seismic waves through the planet. Where the waves came together on the far side, they produced the lineated terrain, which resembles the jumbled terrain on Earth's moon opposite the Imbrium impact. (NASA)

Visual-wavelength image

■ **Figure 21-14**

A lobate scarp (arrow) crosses craters, indicating that Mercury cooled and shrank, wrinkling its crust, after many of its craters had formed. (NASA)

moon. Under careful examination, the Mariner 10 photographs show that Mercury has plains—two different kinds, in fact—but they are different from the moon's. Understanding the differences is the key to understanding the history of Mercury.

Much of Mercury's surface is old, cratered terrain (■ Figure 21-15), but other areas called **intercrater plains** are less heavily cratered. These plains are marked by meteorite craters less than 15 km in diameter and secondary craters produced by chunks of ejecta from larger impacts. Unlike the heavily cratered regions, the intercrater plains are not totally saturated with craters. As an expert of comparative planetology, you recognize that this must mean that the intercrater plains were produced by later lava flows that buried older terrain.

In contrast, smaller regions called **smooth plains** appear to be even younger than the intercrater plains. They have even fewer craters and appear to be ancient lava flows that occurred after most cratering had ended. Much of the region around the Caloris Basin is composed of these smooth plains (Figure 21-13), and they appear to have formed soon after the Caloris impact.

Given the available evidence, most planetary astronomers believe that the plains of Mercury are solidified lava flows, in which case they are much like the maria on the moon. Unlike the maria, Mercury's lava plains are not significantly darker than the rest of the crust. Except for a few bright crater rays, Mercury's surface is a uniform gray with an albedo of only about 0.1. That means Mercury's lava plains are not as dramatically obvious on photographs as the much darker maria on our own moon.

The Interior of Mercury

One of the most striking differences between Mercury and the moon is the difference between their interiors. You have seen that the moon is a low-density world that contains no more than a small core of metals. Yet evidence suggests that Mercury has a large metallic core.

Mercury is over 60 percent denser than the moon. Yet Mercury's surface appears to be normal rock, so you must conclude that its interior contains a large core of dense metals, probably iron. In proportion to its size, Mercury must have a larger metal-

■ **Figure 21-15**

Study this region of Mercury carefully and you will notice the plains between the craters. This surface is not saturated as are the lunar highlands, but it contains more craters than the surface the maria on Earth's moon. This shows that these plains formed after the heaviest of the cratering in the early solar system. (NASA)

lic core than Earth (see the diagram in Celestial Profile 4). If Mercury had a large metallic core that remained molten, then the dynamo effect would generate a magnetic field (Chapter 20). The Mariner 10 spacecraft found, however, a magnetic field only about 0.5 percent as strong as Earth's, and this weak field makes it difficult to understand the nature of the interior. Because Mercury is a relatively small world, about halfway between the size of Earth and the size of the moon, it should have lost most of its internal heat long ago. But if its metallic core is not molten, you would not expect to detect a magnetic field as strong as 0.5 percent of Earth's field. The verdict is still out on the nature of Mercury's large metallic core and weak magnetic field.

Similarly, it is difficult to explain the proportionally large size of the metallic core inside Mercury. In Chapter 19, you saw how planets that formed near the sun should incorporate more metals and be denser, but some models suggest that this cannot account for all of Mercury's large metallic core. One hypothesis that has been proposed involves a giant impact when Mercury was young, an impact much like the planet-shattering impact proposed to explain the origin of Earth's moon. If the forming planet had differentiated and was then struck by a large planetesimal, the impact could have shattered the crust and mantle and blasted much of the lower-density material into space. The denser core could have survived, re-formed, and attracted some of the lower-density debris to form a thin mantle and crust and so leave Mercury with a deficiency of low-density crust rock. Both Mercury and the moon may be products of giant impacts.

A History of Mercury

Can you combine evidence and theory to tell the story of Mercury? Presumably, it formed in the innermost part of the solar nebula, and, as you have seen, a giant impact may have robbed it of some of its lower-density rock and left it a small, dense world with a large metallic core.

Like the moon, Mercury suffered heavy cratering by debris in the young solar system. Planetary scientists don't know accurate ages for features on Mercury, but you can assume that cratering, the second stage of planetary evolution, occurred over the same period as did the cratering on the moon—during the heavy bombardment that lasted from formation until about 4 billion years ago. This intense cratering declined rapidly as the planets swept up the last of the debris left over from planet building.

The cratered surface of Mercury is not exactly like that of the moon. Because of Mercury's stronger gravity, the ejecta from an impact on Mercury would be thrown only about 65 percent as far as on the moon, and that means the ejecta from an impact does not blanket as much of the surface. Also, the intercrater plains appear to have formed when lava flows occurred during the heavy bombardment, burying the older surface, and then accumulated more craters. Sometime near the end of cratering, a planetesimal over 100 km in diameter smashed into the planet and blasted out

■ Figure 21-16

A radar map of Mercury reveals a bright region at its north pole (top). Observers suspect that ice deposits may survive just under the surface or in regions that never receive direct sunlight. (Martin Slade, JPL)

the great multiringed Caloris Basin. Only parts of that basin have been flooded by lava flows.

The smooth plains contain fewer craters and may date from the time of the Caloris impact. It may have been so big it fractured the crust and allowed lava flows to resurface some areas. Because this happened near the end of cratering, the smooth plains have few craters.

Finally, the cooling interior contracted, and the crust broke to form the lobate scarps. Lava flooding ended quickly, perhaps because the shrinkage of the planet squeezed off the lava channels to the surface. Mercury lacks a true atmosphere, so you would not expect flooding by water, but radar images show a bright spot at the planet's north pole (■ Figure 21-16) that may be caused by ice trapped in perpetually shaded crater floors where the temperature never exceeds 60 K (−350°F). This may be water from comets that occasionally collide with Mercury in a burst of water vapor. Similar water deposits have been found at the poles of Earth's moon.

The fourth stage in the story of Mercury, slow surface evolution, is now limited to micrometeorites, which grind the surface to dust; rare larger meteorites, which leave bright-rayed craters; and the slow but intense cycle of heat and cold, which weakens the rock at the surface. The planet's crust is now thick, and although its core may be partially molten, the heat flowing outward is unable to drive plate tectonics that would actively erase craters and build folded mountain ranges.

Building Scientific Arguments

Why don't Earth and the moon have lobate scarps?
Of course, this calls for a scientific argument based on comparative planetology. You might expect that any world with a large metallic interior should have lobate scarps. When the metallic core cools and contracts, the world should shrink, and the contraction should wrinkle and fracture the brittle crust to form lobate scarps. But there are other factors to consider. Earth has a fairly large metallic core; but it has not cooled very much, and the crust is thin, flexible, and active. If any lobate scarps ever did form on Earth, they would have been quickly destroyed by plate tectonics.

The moon does not have a large metal core. It may be that the rocky interior did contract slightly as the moon lost its internal heat, but that smaller contraction may not have produced major lobate scarps. Also, any surface features that formed early in the moon's history would have been destroyed by the heavy bombardment.

You can, in a general way, understand lobate scarps, but now add timing to your argument. **How do you know the lobate scarps on Mercury formed after most of the heavy bombardment was over?**

■   ■   ■

Connections: Mercury, like the moon, is now an inactive world. Impact cratering, internal heat, and possible giant impacts have left their marks on these worlds, but they are now nearly dead. To find planets where the heat flowing outward from the interior still drives geological activity, you must look at larger worlds such as Venus and Mars, the subjects of the next chapter.

Study and Review Tools

Summary

21-1 | The Moon

Why is the moon airless and cratered?
- The moon is small and has only one-sixth the gravity of Earth. It has such a low escape velocity that it is unable to retain an atmosphere.
- When a meteorite strikes the moon, it digs a large round impact crater and can eject debris that falls back and can form rays and secondary craters.
- The largest impacts on the moon dug huge multiringed basins.
- The highlands on the moon are saturated with craters. Rare meteorite impacts continue to form new craters, and micrometeorites constantly grind the surface down to dust, but most of the craters on the moon are very old.
- Lowlands contain ancient lava flows called maria, which contain features such as sinuous rilles, faults, and wrinkles that suggest they are solidified lava flows.
- Most of the craters on the moon were formed during the heavy bombardment at the end of planet building about 4 billion years ago. No crater is known with certainty to have been formed on the moon in historic times.
- Between 1969 and 1972, 12 Apollo astronauts landed on the moon and returned specimens to Earth.
- The moon rocks are all igneous, showing that they solidified from molten rock. Features such as vesicles show that some formed in lava flows, and the breccias show that much of the lunar crust was fractured by meteorite impacts.
- The surface of the moon is covered by rock crushed and powdered by meteorite impacts to form a soil called regolith.

How did the moon form and evolve?
- The fission, condensation, and capture hypotheses have all been abandoned. The commonly accepted theory is called the large-impact theory. The moon appears to have formed from a ring of debris ejected into space when a large planetesimal struck the protoearth after it had differentiated. This would explain, for example, the moon's low density and lack of volatiles.
- The moon rocks show that the moon formed in a molten state.
- The moon differentiated but contains little iron. Its low-density crust was heavily cratered and shattered to great depth.
- Lava, flowing up from below the crust, filled the lowlands to form the smooth plains known as maria. The maria formed after the end of the heavy bombardment and contain few craters.
- The near side of the moon, possibly due to tidal forces, has a thin crust, but the back side has a thicker crust and little lava flooding.
- Because the moon is small, it has lost its internal heat and is geologically dead. The only slow surface evolution taking place is the blasting of micrometeorites.

21-2 | Mercury

Why is Mercury different from the moon?
- Mercury is tidally locked to the sun but rotates 1.5 times per orbit.
- Mercury is a small world that has been unable to retain an atmosphere and has lost most of its internal heat. It is cratered and inactive.
- The Caloris Basin is a large multiringed basin on Mercury that has been partially flooded by lava flows.
- The intercrater plains appear to be later lava flows that covered older craters and then accumulated more craters. The smooth plains appear to be more recent lava flows that contain few craters. All of these plains are a similar shade of gray, so dark lava flows are not visible on Mercury as they are on the moon.

How did Mercury form and evolve?

- Mercury formed in the inner solar system and contains a larger proportion of dense metals. It is possible that a large impact shattered and drove off some of the planet's crust. This could explain why it has such a large metallic core.

- It is not clear how much heat remains in Mercury. It is not geologically active, and it does not have a strong magnetic field. Yet it has a weak magnetic field that suggests some internal heat.

- Mercury was heavily cratered during the heavy bombardment, but lava flows covered some of those craters and new craters formed the intercrater plains. Fractures produced by the Caloris impact may have triggered later lava flows that formed the smooth plains.

- As the metallic core cooled and contracted, the brittle crust broke to form lobate scarps like wrinkles in the skin of a drying apple.

New Terms

tidal coupling (p. 498)	anorthosite (p. 506)
terminator (p. 498)	breccia (p. 506)
limb (p. 498)	regolith (p. 506)
mare (p. 498)	jumbled terrain (p. 509)
sinuous rille (p. 499)	fission hypothesis (p. 510)
ejecta (p. 500)	condensation hypothesis (p. 510)
ray (p. 500)	capture hypothesis (p. 510)
secondary crater (p. 500)	large-impact hypothesis (p. 511)
micrometeorite (p. 501)	resonance (p. 512)
multiringed basin (p. 501)	lobate scarp (p. 514)
relative age (p. 502)	intercrater plain (p. 515)
absolute age (p. 502)	smooth plain (p. 515)
vesicular basalt (p. 505)	

Review Questions

Ace⟲Astronomy™ Assess your understanding of this chapter's topics with additional quizzing and animations at **http://ace .brookscole.com/sf9**

1. How could you find the relative ages of the moon's maria and highlands?
2. How can you tell that Copernicus is a young crater?
3. Why did the first Apollo missions land on the maria? Why were the other areas of more scientific interest?
4. Why do astronomers suppose that the moon formed with a molten surface?
5. Why are so many lunar samples breccias?
6. What do the vesicular basalts tell you about the evolution of the lunar surface?

7. What evidence would you expect to find on the moon if it had been subjected to plate tectonics? Do you find such evidence?
8. Cite objections to the fission, condensation, and capture hypotheses.
9. How does the large-impact hypothesis explain the moon's lack of iron? of volatiles?
10. How does the tidal coupling between Mercury and the sun differ from that between the moon and Earth?
11. What is the difference between the intercrater plains and the smooth plains in terms of time of formation?
12. What evidence can you cite that Mercury has a partially molten, metallic core?
13. In this photo, astronaut Alan Bean works at the Apollo 12 lander. Describe the surface you see. What kind of terrain did they land on for the second mission to the moon?

(NASA)

Discussion Questions

1. Old science-fiction paintings and drawings of colonies on the moon often showed very steep, jagged mountains. Why did the artists assume that the mountains would be more jagged than mountains on Earth? Why are lunar mountains actually less jagged than mountains on Earth?
2. From your knowledge of comparative planetology, propose a description of the view that astronauts would have if they landed on the surface of Mercury.

Problems

1. Calculate the escape velocity of the moon from its mass and diameter. (*Hint:* See Chapter 5.)
2. Why do small planets cool faster than large planets? Compare surface area to volume.
3. The smallest detail visible through Earth-based telescopes is about 1 second of arc in diameter. What size is this on the moon? (*Hint:* See Chapter 3.)
4. The trenches where Earth's seafloor slips downward are 1 km or less wide. Could Earth-based telescopes resolve such features on the moon? Why can you be sure that such features are not present on the moon?
5. The Apollo command module orbited the moon about 200 km above the surface. What was its orbital period? (*Hint:* See Chapter 5.)
6. From a distance of 200 km above the surface of the moon, what is the angular diameter of an astronaut in a space suit? Could someone have seen the astronauts from the command module? (*Hint:* See Chapter 3.)
7. If you transmitted radio signals to Mercury when it was closest to Earth and waited to hear the radar echo, how long would you wait?
8. Suppose you sent a spacecraft to land on Mercury, and it transmitted radio signals to Earth at a wavelength of 10 cm. If you saw Mercury at its greatest angular distance west of the sun, to what wavelength would you have to tune your radio telescope to detect the signals? (*Hints:* See Celestial Profile 4 and the section on the Doppler shift in Chapter 7.)
9. What would the wavelength of maximum be for infrared radiation from the surface of Mercury? How would that differ for the moon?
10. Calculate the escape velocity from Mercury. How does that compare with the escape velocity from the moon and from Earth?

Media Cluster

ACTIVE FIGURES

Ace⊗Astronomy™ To access the resources in the Media Cluster, log into AceAstronomy at **http://ace.brookscole.com/sf9** and select Chapter 21.

The Moon's Craters
The craters that cover the moon were produced by the high-speed impact of meteorites of all sizes. Explore three types of lunar craters in this animation.

ASTRONOMY EXERCISES

Cratering
Turn up your speakers for this one, where you can design collisions between a meteorite and a hypothetical planet.

VIRTUAL ASTRONOMY LABS

Lab 5: Planetary Geology
By thoroughly studying Earth's geology, you can leverage your knowledge to understand conditions on other planets. This lab includes an exercise in determining the interior characteristics of a planet by observing the travel times of its seismic waves.

Lab 7: Planetary Atmospheres and Their Retention
This lab investigates the retention of atmospheres. You will explore the factors that govern the loss of atmospheric gases, and you will see why certain bodies can retain some gases but not others.

Critical Inquiries for the Web

1. What did the Apollo astronauts do while they were on the moon? Search for websites that describe the experiments performed during the astronauts' extravehicular activities (EVAs). What types of experiments or activities were performed? What did planetary scientists learn from them?

2. Search for maps of the moon that show geological information such as chemical composition, elevation, age, and so on. Do the maps confirm your understanding of the lunar surface?

3. What would it be like to walk on the lunar surface? Apollo astronauts visited six different locations on the moon, exploring the variations in lunar terrain. Describe the horizons and general relief of the landing locations of the different missions by exploring websites that provide lunar surface photography from the missions. What differences do you see between images from landings in highlands and in maria?

4. What plans are being made to send a spacecraft to Mercury? What kind of observations will be made, and what theories will be tested?

Exploring *TheSky*

1. Locate the moon in the sky. What constellation is it in? Find the moon's phase as a percentage of full, its angular diameter, and its distance from Earth. Use **Find** to locate the moon.

2. Follow the moon through one complete orbit measuring its distance and angular diameter. How do these change? Under **Tools**, use the **Moon Phase Calendar.** Change dates by clicking on **Data** and then **Site Information.**

3. Locate Mercury and zoom in until you can see its phase. Follow it as it orbits the sun and describe how its phase changes with its orbital position. Use **Find** to locate Mercury.

Go to the Brooks/Cole Astronomy Resource Center **(http://astronomy.brookscole.com)** for critical thinking exercises, articles, and additional readings from InfoTrac College Edition, Brooks/Cole's online student library.

22 | Comparative Planetology of Venus and Mars

There wasn't a breath

in that land of death . . .

ROBERT SERVICE,
THE CREMATION OF SAM MCGEE

WITHOUT A SPACESUIT, you
could live only about 30 seconds on the surface
of Mars. The temperature might not be too bad. On a hot
summer day at noon, the temperature might reach 20°C (68°F). But the air is 96 percent carbon dioxide, is deadly dry,
and contains almost no oxygen. Also, the air pressure at the surface of Mars is only 0.01 that at the surface of Earth.
Stepping out onto the surface of Mars without a spacesuit would be about the same as opening the door of an airplane
flying at 30,000 m (100,000 ft) above Earth. (Commercial jets do not generally fly above 12,000 m, or 38,000 ft.)
No spacesuit would save you on Venus. The air pressure is 90 times Earth's, and the air is almost entirely carbon dioxide,
with traces of various acids. Worse yet, the surface is hot enough to melt lead.

Continued on page 522

Guidepost

Looking Back

The spacesuit you wore when you visited the moon and Mercury protected you and allowed you to explore those airless worlds. On your visits you discovered the fate that awaits small worlds. They can't keep atmospheres, become heavily cratered, lose their internal heat, and quickly become dead, inactive worlds of rock and dust. You will find that Venus and Mars are different.

This Chapter

Your spacesuit will serve you well on Mars, and you will find it an interesting world, but no spacesuit will protect you on the surface of Venus. It too is an interesting world, but it is also a very unpleasant place. You will need to observe it from a distance. As you explore you will discover answers to six essential questions:

Why is the atmosphere of Venus so thick?

What is the hidden surface of Venus really like?

How did Venus form and evolve?

Why is the atmosphere of Mars so thin?

Has Mars ever had water on its surface?

How did Mars form and evolve?

Of course, the key question of comparative planetology, the question always on your mind when you explore other worlds, is this: Why is this world so different from Earth? You will see that small initial differences can have big effects.

Looking Ahead

You are a planet walker, and you are becoming an expert on the kind of planets you can imagine walking on. But there are other worlds in our solar system so peculiar they have no surfaces to walk on, even in your imagination. Nevertheless, you will be able to use your experience with comparative planetology to explore those worlds in the next two chapters.

Why are Venus and Mars such rotten places for picnics? In many ways, they resemble Earth. They are made of silicates and metals, have atmospheres, and lie in the inner solar system. You might expect them to be similar to Earth. How did they get to be so different from each other and from Earth? Comparative planetology will give you some clues.

In this chapter, you will discover the important principle that the size of a world determines how fast it loses its internal heat. Venus, roughly the size of Earth, has lost internal heat slowly, but Mars, intermediate in size between Earth and the moon, is nearly as cold and dead as the moon. Furthermore, these larger worlds have enough gravity to hold onto an atmosphere, and that alters the evolution of surface features.

Venus and Mars are more than just two more planets to add to your gallery of weird worlds. They will give you an entirely new perspective on comparative planetology and a new perspective on Earth.

22-1 Venus

AN ASTRONOMER ONCE BECAME ANNOYED when someone referred to Venus as a planet gone wrong. "No," she argued, "Venus is probably a fairly normal planet. It is Earth that is peculiar. The universe probably contains more planets like Venus than like Earth." To understand how unusual, how special Earth is, you have only to compare it with Venus.

In many ways, Venus is a twin of Earth, and you might expect the two planets to be quite similar. Venus is nearly 95 percent

the diameter of Earth (**Celestial Profile 5**) and has a similar average density. It formed in the same general part of the solar nebula, so you might expect it to have a composition similar to Earth's. Planets of Earth's size cool slowly, so you could expect Venus to have a molten metallic interior and an active crust.

Unfortunately, the surface of Venus is perpetually hidden by thick clouds that completely envelop the planet. From the time of Galileo to the early 1960s, astronomers could only speculate about Earth's twin, and science fiction writers supposed that Venus was a steamy swamp planet inhabited by strange creatures.

Since the 1960s, astronomers have used radar to penetrate the clouds and image the surface. At least 23 spacecraft have flown past or orbited Venus, and over a dozen have landed on its surface. The resulting picture of Venus is dramatically different from the murky swamps of fiction. In fact, the surface of Venus is drier than any desert on Earth and twice as hot as a kitchen oven set to its highest temperature.

If Venus is not a planet gone wrong, it is certainly a planet gone down a different evolutionary path. How did Earth's twin become so different?

The Rotation of Venus

Nearly all of the planets in our solar system rotate counterclockwise as seen from the north. Uranus and Pluto are exceptions, and so is Venus.

In 1962, radio astronomers were able to transmit a radio pulse of precise wavelength toward Venus and detect the echo returning some minutes later. That is, they detected Venus by radar. The echo was not a precise wavelength, however. Part of the reflected signal had a longer wavelength, and part had a shorter wavelength. Evidently the planet was rotating; radio energy re-

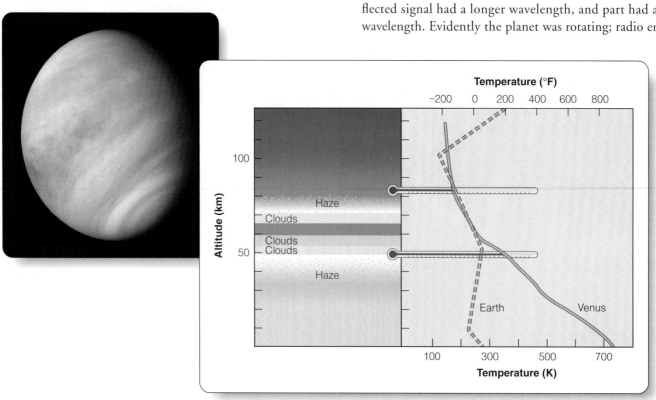

flected from the receding edge was redshifted, and radio energy reflected from the approaching edge was blueshifted. From this Doppler effect, the radio astronomers could tell that Venus was rotating once every 243.01 days. Because the west edge of Venus produced the blueshift, the planet had to be rotating in the backward direction.

This rotation period turns nearly the same face toward Earth each time the planets come closest together, and it is tempting to imagine that there is a tidal resonance locking the rotation of Venus to the orbital motion of Earth. However, a true resonance would require that Venus rotate 0.06 percent slower.

Why does Venus rotate backward? For decades, textbooks have suggested that proto-Venus was set spinning backward when it was struck off-center by a large planetesimal. That is a reasonable possibility; you have seen that a similar collision probably gave birth to Earth's moon. But there is an alternative. Sophisticated mathematical models suggest that a terrestrial planet with a molten core and a dense atmosphere can have its rotation gradually reversed by solar tides in its atmosphere. (Notice the contrast between the catastrophic theory of a giant impact and the evolutionary theory of atmospheric tides.)

Whichever theory explains the backward rotation of Venus, the near resonance with the motion of the Earth seems to be a coincidence. Earth is too small and too distant to raise tides on Venus and affect its rotation.

The Atmosphere of Venus

Although Venus is Earth's twin in size, the composition, temperature, and density of its atmosphere make it the most inhospitable of planets. About 96 percent of its atmosphere is carbon dioxide, and 3.5 percent is nitrogen. The remaining 0.5 percent is water vapor, sulfuric acid (H_2SO_4), hydrochloric acid (HCl), and hydrofluoric acid (HF). In fact, the thick clouds that hide the surface are composed of sulfuric acid droplets and microscopic sulfur crystals.

Soviet and American spacecraft have dropped probes into the atmosphere of Venus, and those probes have radioed data back to Earth as they fell toward the surface. These studies show that the cloud layers are much higher and much more stable than those on Earth. The highest layer of clouds, the layer visible from Earth, extends from 68 to 58 km above the surface (■ Figure 22-1). For comparison, the highest clouds on Earth do not

■ Active Figure 22-1 ◄

The four main cloud layers in the atmosphere of Venus are over 10 times higher above the surface than are Earth clouds. They completely hide the surface. If you could insert thermometers into the atmosphere at different levels, you would find that the lower atmosphere is much hotter than that of Earth, as indicated by the red line in the graph. (© Calvin J. Hamilton)

Ace ◆Astronomy™ Log into AceAstronomy and select this chapter to see the Active Figuree "Planetary Atmospheres." You can explore this figure and compare planets with each other.

Venus is only 5 percent smaller than Earth, but its atmosphere is perpetually cloudy and its surface is hot enough to melt lead. It is believed to have a hot core about the size of Earth's.

Celestial Profile 5: Venus

Motion:

Average distance from the sun	0.7233 AU (1.082×10^8 km)
Eccentricity of orbit	0.0068
Maximum distance from the sun	0.7282 AU (1.089×10^8 km)
Minimum distance from the sun	0.7184 AU (1.075×10^8 km)
Inclination of orbit to ecliptic	3°23'40"
Average orbital velocity	35.03 km/s
Orbital period	0.61515 y (224.68 days)
Period of rotation	243.01 days (retrograde)
Inclination of equator to orbit	177°

Characteristics:

Equatorial diameter	12,104 km (0.95 $D_\oplus$)
Mass	4.870×10^{24} kg (0.815 $M_\oplus$)
Average density	5.24 g/cm³ (4.2 g/cm³ uncompressed)
Surface gravity	0.903 Earth gravity
Escape velocity	10.3 km/s (0.92 $V_\oplus$)
Surface temperature	745 K (472°C, or 882°F)
Albedo (cloud tops)	0.76
Oblateness	0

Personality Point:

Venus is named for the Roman goddess of love, perhaps because the planet often shines so beautifully in the evening or dawn sky. In contrast, the ancient Maya identified Venus as their war god Kukulkan and sacrificed human victims to the planet when it rose in the dawn sky.

extend higher than about 16 km. Another cloud layer lies from 56 to 52 km, and a third, the densest, from 52 to 48 km. Thin haze extends down to about 33 km, below which the air is surprisingly clear.

These cloud layers are highly stable because the atmospheric circulation on Venus is much more regular than that on Earth. The heated atmosphere at the **subsolar point,** the point on the planet where the sun is directly overhead, rises and spreads out in the upper atmosphere. Convection circulates this gas toward the dark side of the planet and the poles, where it cools and sinks. This circulation produces 300-km/h jet streams in the upper atmosphere, which carry the atmosphere from east to west (the same direction the planet rotates) so rapidly that the entire atmosphere seems to rotate with a period of only four days.

The details of this atmospheric circulation are not well understood, but it seems that the slow rotation of the planet is an important factor. On Earth, large-scale circulation patterns are broken into smaller cyclonic disturbances by Earth's rapid rotation. Because Venus rotates more slowly, its atmospheric circulation is not broken up into small cyclonic storms but instead is organized as a planetwide wind pattern.

Although the upper atmosphere is cool, the lower atmosphere is quite hot (Figure 22-1b). Instrumented probes that have reached the surface report that the temperature is 745K (880°F), and the atmospheric pressure is 90 times that of Earth. Earth's atmosphere is 1000 times less dense than water, but on Venus the air is only 10 times less dense than water. If you could survive the unpleasant composition, intense heat, and high pressure, you could strap wings to your arms and fly.

The present atmosphere of Venus is extremely dry, but there is evidence that it once had significant amounts of water. As a Pioneer Venus spacecraft descended through the atmosphere of Venus in 1978, it discovered that deuterium is about 150 times more abundant compared to hydrogen atoms than it is on Earth. This abundance of deuterium, the heavy isotope of hydrogen, could develop if ultraviolet radiation in sunlight broke water molecules into hydrogen and oxygen. The hydrogen would escape into space, but the heavier deuterium atoms would escape more slowly than normal hydrogen. The excess of deuterium must mean that Venus once had more water than it has now. Calculations suggest that it may have once had enough water to make a planetwide ocean up to 25 meters deep. (For comparison, the water on Earth would make a uniform planetwide ocean 3000 meters deep.) Without an ozone layer to protect the atmosphere from ultraviolet radiation, the water was broken up and much of the hydrogen was lost to space. The oxygen presumably formed oxides in the soil. Consequently, Venus is now a deadly dry world with only enough water vapor in its atmosphere to make an ocean 0.3 meter deep.

Even if Venus once had lots of water, it was probably too warm for that water to fall from the atmosphere as rain and form oceans, and that lack of oceans is the biggest difference between Earth and Venus. Earth's oceans have made it a living paradise.

The lack of oceans on Venus, through the greenhouse effect, has made it a hell.

The Venusian Greenhouse

You saw in Chapter 20 how the greenhouse effect warms Earth. Carbon dioxide (CO_2) is transparent to light but opaque to infrared (heat) radiation. Consequently, energy can enter the atmosphere as light and warm the surface, but the surface cannot radiate an equal amount of energy back to space because the carbon dioxide is opaque to infrared radiation. Like Earth, Venus has a greenhouse effect, but on Venus the effect is fearsome. Whereas Earth's atmosphere contains only about 0.03 percent carbon dioxide, the atmosphere of Venus contains 96 percent carbon dioxide, and the surface of Venus is nearly twice as hot as a hot kitchen oven.

Planetary astronomers think they know how Venus got into such a jam. When Venus was young, it may have been cooler than it is now; but, because it formed 30 percent closer to the sun than did Earth, it was warmer than Earth, and that unleashed processes that made it even hotter. As you saw in the previous section, even if Venus once had water, it probably never had large liquid-water oceans. Carbon dioxide is highly soluble in water, but without large oceans to dissolve carbon dioxide and remove it from the atmosphere, Venus was trapped with a carbon dioxide–rich atmosphere, and the greenhouse effect made the planet warmer. As the planet warmed, any small oceans that did exist evaporated, and Venus lost the ability to cleanse its atmosphere of carbon dioxide. Volcanoes on Earth vent gases that are mostly carbon dioxide and water vapor, and presumably the volcanoes on Venus did the same. The water vapor was broken up by ultraviolet radiation, the hydrogen was lost to space, and the oxygen formed oxides in the soil; the carbon dioxide accumulated in the atmosphere. The high temperature baked even more carbon dioxide out of the surface, and the atmosphere became even less transparent to infrared radiation. Consequently, the temperature rose further. This runaway greenhouse effect has made the surface so hot that even sulfur, chlorine, and fluorine have baked out of the rock and formed sulfuric, hydrochloric, and hydrofluoric acid vapors.

Earth avoided this runaway greenhouse effect because it was farther from the sun and cooler. Consequently, it could form and preserve liquid-water oceans to absorb the carbon dioxide, which left a nitrogen atmosphere that was relatively transparent in some parts of the infrared. If all of the carbon dioxide in Earth's sediments was put back into the atmosphere, our air would be as dense as that of Venus, and Earth would suffer from a terrifying greenhouse effect. Recall from Chapter 20 how the use of fossil fuels and the destruction of forests is increasing the carbon dioxide concentration in our atmosphere and warming the planet. Venus warns us of what a greenhouse effect can do.

Ace◉Astronomy™   Log into AceAstronomy and select this chapter to see Astronomy Exercise "The Greenhouse Effect." Take control of the carbon dioxide in your experimental planet's atmosphere.

The Venera lander touched down on Venus in 1982 and carried a camera that swiveled from side to side to photograph the surface. The orange glow is produced by the thick atmosphere; when that is corrected, you can see that the rocks are dark gray. Isotopic analysis suggests they are basalts. (NASA)

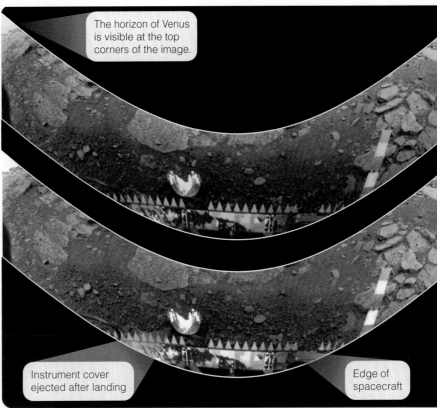

The horizon of Venus is visible at the top corners of the image.

Instrument cover ejected after landing

Edge of spacecraft

The Surface of Venus

Given that the surface of Venus is hidden perpetually by clouds, is hot enough to melt lead, and suffers under crushing atmospheric pressure, it is surprising how much planetary scientists know about the geology of Venus. Early radar maps made from Earth penetrated the clouds and showed that it had mountains, plains, and some craters. The Soviet Union launched a number of spacecraft that landed on Venus, and although the spacecraft failed within an hour or so of landing, they did analyze the rock and transmit a few images back to Earth. The rock seems to be basalt, typical of volcanism, and the images revealed dark-gray rocky plains bathed in a deep-orange glow caused by sunlight filtering down through the thick atmosphere (■ Figure 22-2).

Both the United States and the Soviet Union sent spacecraft to Venus. Some probed the atmosphere, and others landed on the surface and sent back data before succumbing, within an hour or so, to the intense heat and pressure. Some of the most fruitful missions orbited the planet and used radar to penetrate the clouds and map the surface. The NASA Pioneer Venus probe orbited Venus in 1978 and made radar maps showing features as small as 25 km in diameter. Later two Soviet Venera spacecraft mapped the north polar regions with a resolution of 2 km. From 1992 to 1994, the NASA Magellan spacecraft orbited Venus and created radar maps showing details as small as 100 m. These radar maps provide a unique look below the clouds.

The color of Venus radar maps is mostly arbitrary. Human eyes can't see radio waves, so the radar maps must be given some false colors to distinguish height or roughness or composition. Some maps use blues and greens for lowlands and yellows and reds for highlands. Coloring the lowlands blue like oceans on Earth maps might be deceptive, however, because there is no liquid water on Venus. Rather, Magellan scientists chose to use yellows and oranges for their radar maps in an effort to mimic the orange color of daylight caused by the thick atmosphere (■ Figure 22-3). When you look at these orange images, you need to remind yourself that the true color of the rock is dark gray.

■ **Figure 22-3**

Venus without its clouds: This mosaic of Magellan radar maps has been given an orange color to mimic the sunset coloration of daylight at the surface of the planet. The image shows scattered impact craters and volcanic regions such as Beta Regio and Atla Regio. (NASA)

Beta Regio

Atla Regio

Radar map

■ **Figure 22-4** ▲

Although it is nearly 1000 years old, this lava flow near Flagstaff, Arizona, is still such a rough jumble of sharp rock that it is dangerous to venture onto its surface. Rough surfaces are very good reflectors of radio waves and look bright in radar maps. Solidified lava flows on Venus show up as bright regions in the radar maps because they are rough. (M. Seeds)

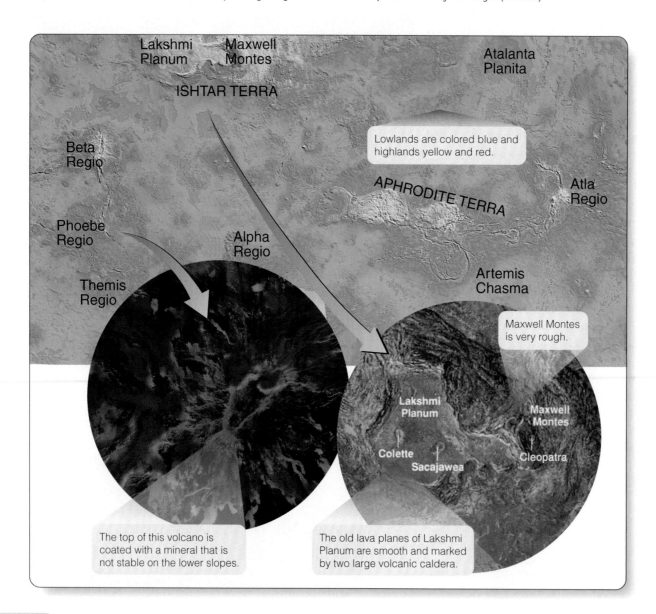

Radar maps of Venus can reveal a number of things about the surface. If you transmit a radio signal down through the clouds and measure the time until you hear the echo coming back up, you can measure the altitude of the surface. Part of the Magellan data is a detailed altitude map of the surface. You can also measure the amount of the signal that is reflected from each spot on the surface. Much of the surface of Venus is made up of old, smooth lava flows that do not look bright in radar maps, but faults and uneven terrain look brighter. Some young, rough lava flows are very rough and contain billions of tiny crevices (■ Figure 22-4) that bounce the radar signal around and shoot it back the way it came. These rough lava flows look bright in radar maps. Certain mineral deposits are also bright; consequently, radar maps do not show how the surface would look to human eyes but rather provide information about altitude, roughness, and, in some cases, chemical composition.

The big map in ■ Figure 22-5 is a map of all of Venus except the polar regions. (Note that this map has been color coded by altitude.) By international agreement, the names of celestial bodies and features on celestial bodies are assigned by the International Astronomical Union, which has decided that all names on Venus should be female. There are only a few exceptions, such as the mountain Maxwell Montes, 50 percent higher than Mt. Everest. It was discovered in early Earth-based radar maps and named for James Clark Maxwell, the 19th-century physicist who first described electromagnetic radiation. Alpha Regio and Beta Regio were also discovered by radar before the naming convention was adopted. Other names on Venus are female, such as the highland regions Ishtar Terra and Aphrodite Terra, named for the Babylonian and Greek goddesses of love. The insets in Figure 22-5 show roughness and composition of the surface, made by different satellites and color coded in different ways. All of these radar maps paint a picture of a violent, hot, desert world.

Radar maps show that the surface of Venus consists of low, rolling plains and highland regions. When comparing them with Earth, you might be tempted to think of these regions as ocean floors and continents, but there is no liquid water on Venus. Rather, the rolling plains appear to be large-scale smooth lava flows, and the highlands are regions of deformed crust.

Craters on Venus tell you that the surface is not old. With nearly 1000 impact craters on its surface, Venus has more craters than Earth but not nearly as many as the moon. The craters are uniformly scattered over the surface and look sharp and fresh

■ **Figure 22-6**

Impact crater Howe in the foreground of this Magellan radar image is 37 km in diameter. Craters in the background are 47 km and 63 km in diameter. This radar map has been digitally modified to represent the view from a spacecraft flying over the craters. (NASA)

(■ Figure 22-6). With no water and a thick, sluggish atmosphere, there is little erosion on Venus, and the thick atmosphere protects the surface from small meteorites. Consequently, there are no small craters. Putting all of this observational evidence together, planetary scientists conclude that the surface of Venus is older than Earth's surface but not nearly as old as the moon's. Unlike the moon, there are no old, cratered highlands. Lava flows have covered the entire surface within the last half-billion years.

Volcanism on Venus

Volcanism seems to dominate the surface of Venus. Much of Venus is covered by lava flows such as those photographed by the Venera landers (Figure 22-2). Also, volcanic peaks and other volcanic features are evident in the radar maps.

As usual, you can learn more about other worlds by comparing them with each other and with Earth. Study **Volcanoes** on pages 528–529 and notice three important ideas:

1 There are two main types of volcanoes found on Earth. One is associated with plate motion, and one is associated with hot spots.

2 Notice that you can recognize the volcanoes on Venus and Mars by their shapes, even when the images are manipulated in computer mapping (**Window on Science 22-1**). The shield volcanoes found on Venus and Mars are the kind produced by hot-spot volcanism and not by plate tectonics.

3 Also notice the large size of volcanoes on Venus and Mars. They have grown very large because of repeated eruptions at the same place in the crust. This is evidence that neither Venus nor Mars has been dominated by plate tectonics as on Earth.

■ **Figure 22-5** ◀

Notice how these three radar maps show different things. The main map here shows elevation over most of the surface of Venus. Only the polar areas are not shown. The inset map at left shows an electrical property of surface minerals related to chemical composition. The detailed map of Maxwell Montes and Lakshmi Planum at right is color coded to show roughness with purple smooth and orange rough. The rock on Venus is, in fact, dark gray. (Maxwell and Lakshmi Planum map: USGS; Other maps: NASA)

Volcanoes

1 Molten rock (magma) is less dense than its surroundings and tends to rise. Where it bursts through Earth's crust, you see volcanism. The two main types of volcanoes on Earth provide good examples for comparison with those on Venus and Mars.

On Earth, **composite volcanoes** form above subduction zones where the descending crust melts and the magma rises to the surface. This forms chains of volcanoes along the subduction zone, such as the Andes along the west coast of South America.

Magma rising above subduction zones is not very fluid, and it produces explosive volcanoes with sides as steep as 30°.

Oceanic plate · Subduction zone · Composite volcanoes · Continental crust · Upper mantle · Upper mantle

Chains of composite volcanoes are not found on Venus or Mars, which is evidence that subduction and plate motion does not occur on those worlds.

Based on *Physical Geology*, 4th edition, James S. Monroe and Reed Wicander, Wadsworth Publishing Company. Used with permission.

Mount St. Helens exploded northward on May 18, 1980, killing 63 people and destroying 600 km² (230 mi²) of forest with a blast of winds and suspended rock fragments that moved as fast as 480 km/hr (300 mph) and had temperatures as hot as 350°C (660°F). Note the steep slope of this composite volcano.

Shield volcano · Lava flow · Magma chamber · Oceanic crust

1a A shield volcano is formed by highly fluid lava (basalt) that flows easily and creates low-profile volcanic peaks with slopes of 3° to 10°. The volcanoes of Hawaii are shield volcanoes that occur over a hot spot in the middle of the Pacific plate.

Magma collects in a chamber in the crust and finds its way to the surface through cracks.

Magma forces its way upward through cracks in the upper mantle and causes small, deep earthquakes.

A hot spot is formed by a rising convection current of magma moving upward through the hot, deformable (plastic) rock of the mantle.

The Cascade Range composite volcanoes are produced by an oceanic plate being subducted below North America and partially melting.

Seattle · Washington · St Helens · Rainier · Pacific Ocean · Portland · Hood · Oregon · Shasta · Nevada · California · Lassen

USGS

Volcano Gula Mons

Volcano Sif Mons

Radar map

NASA

2 Volcanoes on Venus are shield volcanoes. They appear to be steep sided in some images created from Magellan radar maps, but that is because the vertical scale has been exaggerated to enhance detail. The volcanoes of Venus are actually shallow-sloped shield volcanoes.

3 Volcanism over a hot spot results in repeated eruptions that build up a shield volcano of many layers. Such volcanoes can grow very large.

Vertical scale exaggerated

Hot spot

2a This computer model of a mountain with the vertical scale magnified 10 times appears to have steep slopes such as those of a composite volcano.

A true profile of the computer model shows the mountain has very shallow slopes typical of shield volcanoes.

Mike Seeds

Old volcanic island eroded below sea level

Plate motion

Hot spot

If the crustal plate is moving, magma generated by the hot spot can repeatedly penetrate the crust to build a chain of volcanoes. Only the volcanoes over the hot spot are active. Older volcanoes slowly erode away. Such volcanoes cannot grow large because the moving plate carries them away from the hot spot.

Time since last eruption (million years)

5   3   1.5   1   0

Maui

Molokai   Hawaii   Active volcanoes

Oahu

Kauai

Plate motion

Lo'ihi—Newborn underwater volcano

3a The volcanoes that make up the Hawaiian Islands as shown at left have been produced by a hot spot poking upward through the middle of the moving Pacific plate.

NASA

3b The plate moves about 9 cm/yr and carries older volcanic islands northwest, away from the hot spot. The volcanoes cannot grow extremely large because they are carried away from the hot spot. New islands form to the southeast over the hot spot.

Olympus Mons at right is the largest volcano on Mars. It is a shield volcano 25 km (16 mi) high and 700 km (440 mi) in diameter at its base. Its vast size is evidence that the crustal plate must have remained stationary over the hot spot. This is evidence that Mars has not had plate tectonics.

Olympus Mons contains 95 times more volume than the largest volcano on Earth, Mauna Loa in Hawaii.

Caldera from repeated eruptions

Digital elevation map

Graphical Manipulation of Scientific Data

Because so much modern scientific data is stored inside computers, it is common to see data manipulated to exaggerate certain characteristics and suppress others. In some fields, this would be dishonest, but the goals of science make such manipulation valuable. By purposely distorting data, you may reveal things that would not have been seen otherwise.

Computers make it easy to manipulate the colors in an image and produce false-color images. Astronomers commonly change and exaggerate the colors in images and maps to reveal details invisible to our eyes. Radio maps of galaxies and X-ray maps of supernova remnants are only two examples. As long as you remember what real galaxies and real nebulae look like, you can learn a great deal from false-color images.

Geologists, too, manipulate data. Geological maps have been color coded for over a century, and now computers allow faster and more precise use of false color in maps and images. But geological data can also be manipulated to reveal other features. A relief map of the United States would look as smooth as glass if the map maker did not exaggerate the

height of mountain ranges, and having such data in a computer makes it easy to "stretch" the heights to make them more visible. Astronomers often do the same with radar maps of Venus and surface maps of Mars.

Such data distortion would be inadmissible in a court of law. In fact, photographic evidence is generally inadmissible in court because it is so easy to fake a photograph and show the defendant in almost any incriminating situation. Faking a photograph in journalism is at least disreputable, and in some cases it is libelous. In fact, modern computer graphics make it possible to put Humphrey Bogart into a modern TV commercial with no trace of fakery, so photographic manipulation is unlikely ever to become acceptable in court or in journalism.

In science, however, the quest is not for a first impression but for detailed analysis. Data can be manipulated and distorted in any way that is useful, as long as the form of the manipulation is specified. Then viewers are free to form their own opinions as to the meaning of the data. Of course, this assumes that the viewer is alert and takes into consideration the nature of any data manipulation. Thus, when

You are accustomed to seeing data manipulated and presented in convenient ways. (PhotoDisc/Getty Images)

you look at a relief map of the United States or a radar image of a mountain on Venus, it is your responsibility to take into account any data manipulation. If you don't, you might think that the Rocky Mountains are 100 miles high and that mountains on Venus are 20 times steeper than mountains on Earth.

The radar image of Sapas Mons in ■ Figure 22-7 shows a dramatic overhead view of this volcano, which is 400 km (250 mi) in diameter at its base and 1.5 km (0.9 mi) high. Many bright lava flows radiate outward, covering older, less bright flows. Remember that the colors in this image are artificial; if you could walk across these lava flows, you would find them solid stone of a dark-gray color. Sapas Mons and other nearby volcanoes are located along a system of faults where rising magma broke through the crust.

In addition to actual volcanoes, radar images reveal other volcanic features on the surface of Venus. Lava channels are common on Venus, and they appear similar to the sinuous rills visible on Earth's moon. The longest channel on Venus is also the longest known lava channel in the solar system. It stretches 6800 km (4200 miles), roughly twice the distance from Chicago to Los Angeles. These channels are 1 to 2 km wide and can sometimes be traced back to collapsed areas where lava appears to have drained from beneath the crust.

For further evidence of volcanism on Venus, you can look at circular features called **coronae,** circular bulges up to 2100 km in diameter containing volcanic peaks and lava flows and bordered by fractures. The coronae appear to be caused by rising currents

of molten magma below the crust that create an uplifted dome and then withdraw to allow the surface to subside. Coronae are sometimes accompanied by features called pancake domes, circular outpourings of viscous lava, and by domes and hills pushed up by molten rock below the surface. All of these volcanic features are shown in ■ Figure 22-8.

Volcanoes may be erupting on Venus right now, but the radar maps caught no evidence of actual eruptions in progress. Certainly there is no reason to suppose that all of these volcanoes are extinct.

If you want to learn the secrets of Venus, you might want to visit the huge landmass called Ishtar Terra. Western Ishtar consists of a high volcanic plateau called Lakshmi Planum. It rises 4 km above the rolling plains and appears to have formed from lava flows from volcanic vents such as Collette and Sacajawea (see Figure 22-5). Although folded mountain ranges do not occur on Venus as they do on Earth, the areas north and west of Lakshmi Planum appear to be ranges of wrinkled mountains where horizontal motion of the crust has pushed up against Ishtar Terra. Furthermore, faults and deep chasms are widespread over the surface of Venus, and that suggests stretching of the crust in those areas. So, although no direct evidence of plate motion is visible,

Radar map

■ **Figure 22-7**

(a) Volcano Sapas Mons, lying along a major fracture zone, is topped by two lava-filled calderas and flanked by rough lava flows. The orange color of this radar map mimics the orange light that filters through the thick atmosphere. (NASA) (b) Seen by light typical of Earth's surface, Sapas Mons might look more like this computer-generated landscape. Volcano Maat Mons rises in the background. The vertical scale has been exaggerated by a factor of 15 to reveal the shape of the volcanoes and lava flows. (Copyright © 1992, David P. Anderson, Southern Methodist University)

there has been compression and wrinkling of the crust in some areas and stretching and faulting in other areas, and that suggests some limited crustal motion.

The history of Venus has been dominated by volcanism. While Earth's crust has broken into rigid moving plates, the surface of Venus seems more pliable. In any case, to tell the story of Venus you will have to tell a fiery tale.

Ace ⟳ Astronomy ™   Log into AceAstronomy and select this chapter to see Astronomy Exercise "Convection and Plate Tectonics." Compare the interiors of the four terrestrial planets.

A History of Venus

Earth has passed through four major stages in its history (Chapter 20), and you have seen how the moon and Mercury were affected by the same stages. Venus, however, has had a peculiar passage through the four stages of planet development, and its history is difficult to understand. Planetary scientists are not sure how the planet formed and differentiated, how it was cratered and flooded, or how its surface has evolved.

Venus formed only slightly closer to the sun than did Earth, so you might expect it to be a similar planet differentiated into a silicate mantle and a molten iron core. The density and size of Venus require that it have a dense interior much like Earth's; but, if the core is molten, then you would expect the dynamo effect to generate a magnetic field. No spacecraft has ever detected such a field around Venus. The magnetic field must be at least 25,000

times weaker than Earth's. Some theorists wonder if the core of the planet is solid. If it is solid, then how did Venus get rid of its internal heat so much faster than Earth did?

Because the planet lacks a magnetic field, it is not protected from the solar wind. The solar wind slams into the uppermost layers of the planet's atmosphere, forming a bow shock where the wind is slowed and deflected (■ Figure 22-9). Planetary scientists know little about the differentiation of the planet into core and mantle, so the size of the core shown in Figure 22-9 is estimated by analogy with Earth's. The magnetic field carried by the solar wind drapes over Venus like seaweed over a fishhook, forming a long tail within which ions flow away from the planet. You will see in Chapter 25 that comets, which also lack magnetic fields, interact with the solar wind in the same way.

Presumably, Venus and Earth formed in molten states and never had primeval atmospheres rich in hydrogen. Rather, they formed carbon dioxide atmospheres as they formed. Calculations show that Venus and Earth have outgassed about the same amount of carbon dioxide, but Earth's oceans have dissolved it and converted it to sediments such as limestone. If all of Earth's carbon were dug up and converted back to carbon dioxide, Earth's atmosphere would be about as dense as the air on Venus. This suggests that the main difference between Earth and Venus is the lack of water on Venus. Venus may have had small oceans when it was young; but, being closer to the sun, it was warmer, and the carbon dioxide in the atmosphere created a greenhouse effect that made the planet even warmer. That process could have dried

Pancake domes

Lava flows

Volcanic domes

Corona

50km (30mi)

■ Figure 22-8

Volcanic features on Venus: (a) Arrows point to a 600-km segment of Baltis Vallis, the longest lava-flow channel in the solar system. It is at least 6800 km long. (b) Aine Corona, about 200 km in diameter, is marked by faults, lava flows, small volcanic domes, and pancake domes of solidified lava. (c) This perspective view shows a hill typical of those associated with a corona. Molten lava below the surface appears to have pushed up the hill, causing the network of radial faults. (d) Other regions of the crust appear to have collapsed as subsurface magma drained away. (NASA)

■ Figure 22-9

By analogy with Earth, the interior of Venus should contain a molten core (estimated here), but no spacecraft has detected a planetary magnetic field. Thus, Venus is unprotected from the solar wind, which strikes the planet's upper atmosphere and is deflected into an ion tail.

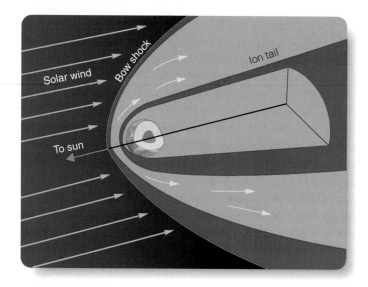

Solar wind

Bow shock

Ion tail

To sun

up any oceans that did exist and reduced the ability of the planet to clear its atmosphere of carbon dioxide. As more carbon dioxide was outgassed, the greenhouse effect grew even more severe. Venus seems to have been trapped in a runaway greenhouse effect.

Fully 70 percent of the heat from Earth's interior flows outward through midocean volcanism. But Venus lacks crustal rifts, and its numerous volcanoes cannot carry much of the heat out of the interior. Rather, Venus seems to get rid of its interior heat through large currents of hot magma that rise beneath the crust. Coronae, lava flows, and volcanism form above such currents. In fact, the surface rock is dark-gray basalt typical of basalts found in ocean crust on Earth.

True plate tectonics is not important on Venus. For one thing, the crust is very dry and is consequently about 12 percent less dense than Earth's crust. This low-density crust is more buoyant than Earth's crust and resists being pushed into the interior. Also, the crust is so hot it is halfway to the melting point of rock. Such hot rock is not very stiff, so it cannot form the rigid plates typical of plate tectonics on Earth.

There is no sign of plate tectonics on Venus, but rather evidence that convection currents below the crust are deforming the crust to create coronae and push up mountains such as Maxwell. Detailed measurements of the strength of gravity over the mountains on Venus show that some must be held up not by deep roots like mountains on Earth but by rising currents of magma. Other mountains, like those around Ishtar Terra, appear to be folded mountains caused by limited horizontal motions in the crust, driven perhaps by convection currents in the mantle.

The small number of craters on the surface of Venus hints that the entire crust has been replaced within the last half-billion years or so. This may have occurred in a planetwide overturning as the old crust broke up and sank and lava flows created a new crust. This could happen periodically on Venus, or the planet may have had a geology more like Earth's until a single resurfacing not too long ago. In either case, unearthly Venus may eventually reveal more about how our own world works.

Ace✩Astronomy™ Log into AceAstronomy and select this chapter to see Astronomy Exercise "Convection and Magnetic Fields." Take control of a planetary interior.

Building Scientific Arguments

What evidence can you point to that Venus does not have plate tectonics?
Sometimes a scientific argument can be helpful by eliminating a possibility. On Earth, plate tectonics is identifiable by the worldwide network of faults, subduction zones, volcanism, and folded mountain chains that outline the plates. Although some of these features are visible on Venus, they do not occur in a planetwide network that outlines plates. Volcanism is widespread, but folded mountain ranges occur in only a few places, such as near Lakshmi Planum and Maxwell Montes, and they do not make up long mountain chains as on Earth. Also, the large size of the shield volcanoes on Venus show that the crust is not moving over the hot spots as the Pacific seafloor is moving over the Hawaiian hot spot.

At first glance, you might think that Earth and Venus should be sister worlds, but comparative planetology reveals that they can be no more than distant cousins. You can blame the thick atmosphere of Venus for altering its geology, but that calls for a new scientific argument: **Why isn't Earth's atmosphere similar to that of Venus?**

Connections: Venus is certainly an unearthly world, but the familiar principles of comparative planetology can help you understand it. As you turn your attention to Mars, you will see the same principles in action, but you will discover that Mars is a world that lacks what Venus has in abundance—internal heat and air.

22-2 Mars

MERCURY AND THE MOON are small. Venus and Earth are, for terrestrial planets, large. But Mars occupies an intermediate position (**Celestial Profile 6** on page 535). It is twice the diameter of the moon but only 53 percent Earth's diameter. Its small size has allowed it to cool faster than Earth, and much of its atmosphere has leaked away. Its present carbon-dioxide atmosphere is only 1 percent as dense as Earth's.

In some ways, Mars is much like Earth. A day on Mars is nearly the same length as an Earth day—24 hours and 40 minutes—and its year lasts 1.88 Earth years. Also, just as Earth's axis is tipped 23.5°, that of Mars is tipped 25°. As the northern and southern hemispheres turn alternately toward the sun, seasonal changes are visible even through a small telescope. As spring comes to the southern hemisphere, the white polar cap shrinks, and the grayish surface markings grow darker and, according to some observers, greener. At one time, these seasonal changes led some to believe that plant life flourished on Mars when spring thawed the polar cap. You will see later that this is not so.

In other ways, Mars is very different from Earth. Its surface has never broken into moving plates like those that create the topography on Earth. Instead, Mars is a one-plate planet, with its crust frozen into a solid layer but bearing signs of geological activity that include some of the largest volcanoes in the solar system.

The Canals on Mars

Long before the space age, the planet Mars was a mysterious landscape in the public mind. In the century following Galileo's first astronomical use of the telescope, astronomers discovered dark markings on Mars as well as bright polar caps. Timing the motions of the markings, they concluded that a Martian day was about 24 hours long, and the similarity with Earth's day further supported the belief that Mars was another Earth.

The so-called canals of Mars began life innocently enough in 1858 when the Jesuit astronomer Angelo Secchi referred to a region on Mars known later as Syrtis Major as *Atlantic Canale*. This is the first use of the Italian word *canale* (channel) to refer to a feature on Mars. Then, in the late summer of 1877 the Italian astronomer Giovanni Virginio Schiaparelli, using a telescope only 8.75 in. in diameter, thought he glimpsed fine, straight lines on

(a) Early in the 20th century, Percival Lowell mapped canals over the face of Mars and concluded that intelligent life resided there. (b) Modern images recorded by spacecraft reveal a globe of Mars with no canals. Instead, the planet is marked by craters and, in some places, volcanoes. Both of these images are reproduced with south at the top, as they appear in telescopes. Lowell's globe is inclined more nearly vertically and is rotated slightly to the right compared with the modern globe. (a, Lowell Observatory; b, U.S. Geological Survey)

Mars. He too used the Italian word *canali* (plural) for these lines, and the word was translated into English not as "channel," a narrow body of water, but as "canal," an artificially dug channel. The "canals of Mars" were born. Many astronomers could not see the canals at all, but others drew maps showing hundreds (■ Figure 22-10).

In the decades that followed Schiaparelli's discovery, many people assumed that the canals were watercourses built by an intelligent race on Mars to carry water from the polar caps to the lower latitudes. Much of this excitement was generated by Percival Lowell, a wealthy Bostonian who founded Lowell Observatory in 1894, principally for the study of Mars. He not only mapped hundreds of canals but also popularized his results in books and lectures. Although some astronomers claimed the canals were merely illusions, by 1907 the general public was so sure that life existed on Mars that the *Wall Street Journal* could suggest that the most extraordinary event of the previous year had been "the proof by astronomical observations . . . that conscious, intelligent human life exists upon the planet Mars." Further sightings of bright clouds and flashes of light on Mars strengthened this belief, and some urged that gigantic geometrical diagrams be traced in the Sahara Desert to signal to the Martians that Earth, too, is inhabited. All seemed to agree that the Martians were older and wiser than humans.

This fascination with men from Mars was not a passing fancy. Beginning in 1912, Edgar Rice Burroughs wrote a series of 11 novels about the adventures of the Earthman John Carter, lost on Mars. Burroughs made the geography of Mars, named by Schiaparelli after Mediterranean lands both real and mythical, into household words. He also made his Martians small and gave them green skin.

By Halloween night of 1938, people were so familiar with life on Mars that they were ready to believe that Earth had been invaded. When a radio announcer repeatedly interrupted dance music to report the landing of a spaceship in New Jersey, the emergence of monstrous beings, and their destruction of whole cities, thousands of otherwise sensible people rushed to their cars and fled in panic, not knowing that Orson Welles and other actors were dramatizing H. G. Wells's book *The War of the Worlds*.

The fascination with Mars, its canals, and its little green men lasted right up to July 15, 1965, when Mariner 4, the first spacecraft to fly past Mars, radioed back photos of the surface and proved that there are no canals and no Martians. Apparently, the canals are optical illusions produced by the human brain's astounding ability to assemble a field of disconnected marks into a coherent image. If your brain could not do this, the photos on these pages would be nothing but swarms of dots, and the images on the screen of a television set would never make sense. The brain of an astronomer looking for something at the edge of visibility is capable of connecting faint, random markings on Mars into the straight lines of canals.

Even today, Mars holds a fascination for the general public. The grocery store tabloids regularly run stories about a giant face carved on Mars by an ancient race. Although planetary scientists recognize it as nothing more than chance shadows in a photograph and dismiss the issue as a silly hoax, the stories persist. A hundred years of speculation have raised high expectations for Mars. If there really was intelligent life on Mars and its representatives came to Earth, they would probably be a big disappointment to the readers of the tabloids.

The Atmosphere of Mars

If you visited Mars, your first concern, even before you opened the door of your spaceship, would be the atmosphere. Is it breathable? Even for the astronomer observing safely from Earth, the atmosphere of Mars is a major concern. The gases that cloak Mars are critical to understanding the history of the red planet.

The air on Mars is 95 percent carbon dioxide, with a few percent each of nitrogen and argon. It contains almost no water vapor or oxygen, and its density at the surface of the planet—only 1 percent that of Earth's atmosphere—does not provide enough pressure to prevent liquid water from boiling into vapor. If you stepped outdoors on Mars without a spacesuit, your own body heat would make your blood boil.

Although the air is thin, it is dense enough to be visible in photographs (■ Figure 22-11). Haze and clouds come and go, and occasional weather patterns are visible. Winds on Mars can be strong enough to produce dust storms that envelop the entire

■ Figure 22-11

The atmosphere of Mars is evident in this image made by the Hubble Space Telescope. The haze is made up of high, water-ice crystals in the thin CO_2 atmosphere. The spot at extreme left is the volcano Ascraeus Mons, 25 km (16 mi) high, poking up through the morning clouds. Note the north polar cap at the top. (Philip James, University of Toledo; Steven Lee, University of Colorado, Boulder; and NASA)

Visual-wavelength image

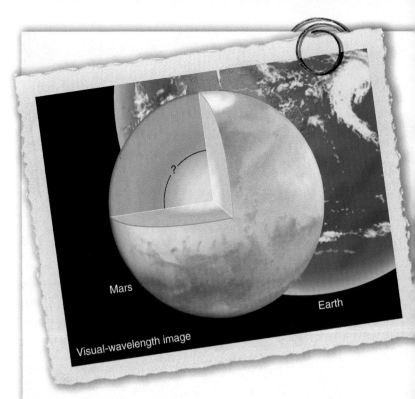

Mars

Earth

Visual-wavelength image

Mars is only half the diameter of Earth and probably retains some internal heat, but the size and composition of its core are not well known.

Celestial Profile 6: Mars

Motion:

Average distance from the sun	1.5237 AU (2.279×10^8 km)
Eccentricity of orbit	0.0934
Maximum distance from the sun	1.6660 AU (2.492×10^8 km)
Minimum distance from the sun	1.3814 AU (2.066×10^8 km)
Inclination of orbit to ecliptic	1°51'09"
Average orbital velocity	24.13 km/s
Orbital period	1.8808 y (686.95 days)
Period of rotation	$24^h37^m22.6^s$
Inclination of equator to orbit	25°11'

Characteristics:

Equatorial diameter	6792 km (0.53 $D_\oplus$)
Mass	0.6424 X 10^{24} kg (0.1075 $M_\oplus$)
Average density	3.94 g/cm³ (3.3 g/cm³ uncompressed)
Surface gravity	0.379 Earth gravity
Escape velocity	5.0 km/s (0.45 $V_\oplus$)
Surface temperature	−140° to 20°C (−220° to 68°F)
Average albedo	0.16
Oblateness	0.009

Personality Point:

Mars is named for the god of war. Minerva was the goddess of defensive war, but Bullfinch's *Mythology* refers to Mars's "savage love of violence and bloodshed." You can see how the planet glows blood red in the evening sky because of iron oxides in its soil.

Visual-wavelength image

■ **Figure 22-12**

Mars is a red desert planet, as shown in this true-color photo made by the Rover Spirit. Rock layers are fractured by meteorite impacts, and dust is blown around the planet by winds. Dust suspended in the atmosphere colors the sky red. (NASA)

planet. The reddish color of the soil is evidently caused by iron oxides (rusts), and that warns that the oxygen humans would prefer to find in the atmosphere is locked in chemical compounds in the soil. The polar caps visible in photos are also related to the Martian atmosphere. The ices in the polar caps are frozen carbon dioxide (dry ice) with unknown amounts of frozen water underneath.

If you could visit Mars you would find it a reddish, airless, bone-dry desert (■ Figure 22-12). To understand Mars, you should ask why its atmosphere is so thin and dry and why the surface is rich in oxides. The answers to these questions lie in the origin and evolution of the Martian atmosphere.

Presumably, the gases in the Martian atmosphere were mostly outgassed from its interior. Volcanism on terrestrial planets typically releases carbon dioxide and water vapor, plus other gases. Because Mars formed farther from the sun, you might expect that it incorporated more volatiles when it formed. But Mars is smaller than Earth, so it has had less internal heat to drive geological activity, and you might suspect that it has not outgassed as much as Earth. In any case, the outgassing occurred early in the planet's history, and Mars, being small, cooled rapidly and now releases little gas.

How much atmosphere a planet has depends on how rapidly it releases internal gas and how rapidly it loses gas from its atmosphere, and the rate at which a planet loses gas depends on its mass and temperature. The more massive the planet, the higher its escape velocity (Chapter 5), and the more difficult it is for gas atoms to leak into space. Mars has a mass less than 11 percent that of Earth, and its escape velocity is only 5 km/s, less than half Earth's. Consequently, gas atoms can escape from it much more easily than they can escape from Earth.

The temperature of a planet's atmosphere is also important. If the gas is hot, its molecules have a higher average velocity and are more likely to exceed escape velocity. That means a planet near the sun and very hot is less likely to retain an atmosphere

than a more distant, cooler planet. The velocity of a gas molecule, however, also depends on the mass of the molecule. On average, a low-mass molecule travels faster than a massive molecule. For that reason, a planet loses its lowest-mass gases more easily because those molecules travel fastest.

You can see this principle of comparative planetology if you plot a diagram such as that in ■ Figure 22-13. The points show the escape velocity versus temperature of the larger objects in our solar system. The temperature used in the diagram is the temperature of the gas that is in a position to escape. For the moon, which has essentially no atmosphere, this is the temperature of the sunlit surface. For Mars, the temperature that is important is that at the top of the atmosphere. The lines in Figure 22-13 show the typical velocity of the fastest-traveling examples of various molecules. At any given temperature, some water molecules, for example, travel faster than others, and it is the highest-velocity molecules that escape from a planet. The diagram shows that the Jovian planets have escape velocities so high that very few molecules can escape. Earth and Venus can't hold hydrogen, and Mars can hold only the more massive molecules. Earth's moon is too small to keep any of the common gases. Refer to this diagram again whenever you study the atmospheres of other worlds.

Over the 4.6 billion years since Mars formed, it has lost some of the lower-mass gases. Water molecules are massive enough for Mars to keep, but ultraviolet radiation can break up the water molecules. On Earth, the ozone layer protects water vapor from ultraviolet radiation, but Mars never had an oxygen-rich atmosphere, so it never had an ozone layer. Ultraviolet photons from the sun can penetrate deep into the atmosphere and break up molecules such as water. The hydrogen escapes, and the oxygen, a very reactive element, forms more oxides in the soil—the oxides that make Mars the red planet. In this way, molecules too massive to leak into space can be lost if they break into lower-mass fragments.

The argon in the Martian atmosphere is evidence of a denser blanket of air in the past. Argon atoms are massive, almost as massive as a carbon monoxide molecule, and would not be lost easily. In addition, argon is inert and would not form compounds in the soil. The 1.6 percent argon in the atmosphere of Mars is evidently left over from an ancient atmosphere that was 10 to 100 times denser than the present Martian air.

Finally, you should consider the interaction of the solar wind with the atmosphere of Mars. This is not an important process for Earth because Earth has a magnetic field that deflects the solar wind. Because Mars has no magnetic field, the solar wind interacts directly with the Martian atmosphere, and detailed calculations show that significant amounts of carbon dioxide could

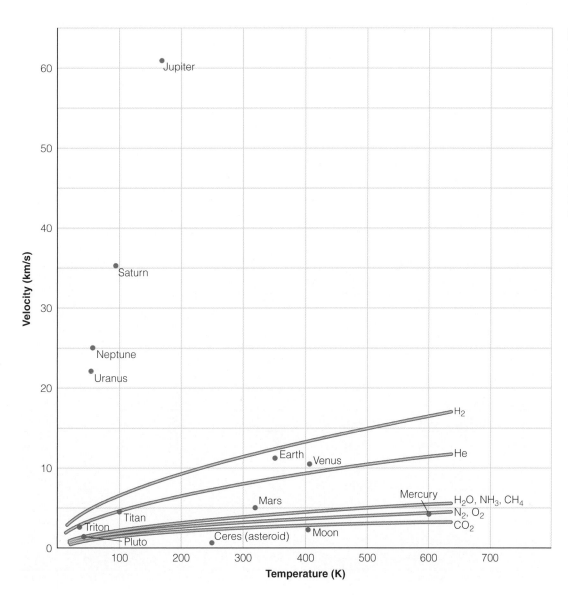

Figure 22-13

The loss of planetary gases. Dots represent the escape velocity and temperature of various solar-system bodies. The lines represent the typical highest velocities of molecules of various masses. The Jovian planets have high escape velocities and can hold on to even the lowest-mass molecules. Mars can hold only the more massive molecules, and the moon has such a low escape velocity that even the most massive molecules can escape.

have been carried away by the solar wind over the history of the planet. This process would have been most efficient long ago when the sun was more active and the solar wind was stronger. However, you should also consider that Mars probably had a magnetic field when it was younger and still retained significant internal heat. A magnetic field would have protected its atmosphere from the solar wind.

Although planetary scientists remain uncertain as to how much of an atmosphere Mars has had in its past and how much it has lost, it is a good example for your study of comparative planetology. When you look at Mars, you see what can happen to the atmosphere of a medium-size world. Like its atmosphere, the geology of Mars is typical of medium-size worlds.

Ace✷Astronomy™   Log into AceAstronomy and select this chapter to see Astronomy Exercise "Primary Atmospheres." You can move your planet around to see how its atmosphere is affected.

The Geology of Mars

If you ever decide to visit another world, Mars may be your best choice. You will need a heated, pressurized spacesuit with air, water, and food, but Mars is not as inhospitable as the moon. The nights on Mars are deadly cold, but a hot summer day would be comfortable (Celestial Profile 6). It is also more interesting than the moon, with weather, complex geology, and signs that water once flowed over its surface. You might even hope to find traces of ancient life hidden in the rocks.

Spacecraft have been visiting Mars for almost 40 years, but the pace has picked up recently. A small armada of spacecraft has gone into orbit around Mars to photograph and analyze its surface, and five spacecraft have landed. Two Viking landers touched down in 1976, and three rovers have landed in recent years. Rovers have an advantage because they are wheeled robots that can be directed from Earth to travel from feature to feature and make detailed measurements. Pathfinder and its rover, Sojourner, landed in 1997. Rovers Spirit and Opportunity landed in January 2004 and carried sophisticated instruments to explore the rocky surface (■ Figure 22-14).

Photographs made by rovers on the surface of Mars, such as Figure 22-12, show reddish deserts of broken rock. These appear to be lava plains fractured by meteorite impacts, but they don't look much like the surface of Earth's moon. The atmosphere of Mars, thin though it is, protects the surface from the blast of micrometeorites that grinds moon rocks to dust. Also, the Martian dust storms may sweep fine dust away from some areas and leave larger rocks exposed.

Spacecraft orbiting Mars have imaged the surface and measured elevations to reveal that all of Mars is divided into two parts. The southern highlands are heavily cratered, and the number of craters there shows that the highlands must be old. In contrast, northern lowlands are smooth (■ Figure 22-15) and so remarkably free of craters that the lowlands must have been resurfaced roughly a billion years ago. Some astronomers have suggested that volcanic floods filled the northern lowlands and buried the

■ Figure 22-14

If you had been standing beside the small crater now known as Endurance, you might have seen the Rover Opportunity roll up to the edge and pause to await further instructions from Earth. This true-color image of the layered rock in the crater wall was taken by Opportunity as it circled the crater looking for a safe way down to the bottom. NASA engineers added a digital image of the rover to recreate the scene. (NASA/JPL/Cornell)

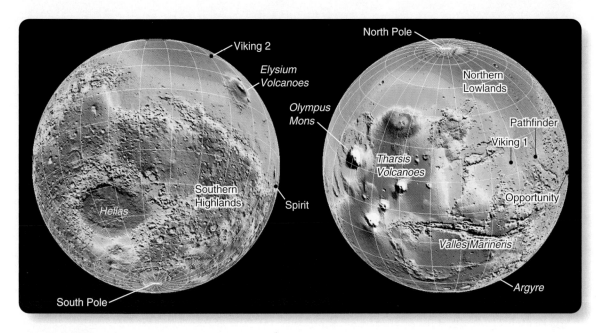

■ Figure 22-15

These globes of Mars are color coded to show elevation. The northern lowlands lie about 4 km below the southern highlands. Volcanoes are very high (white), and the giant impact basins, Hellas and Argyre, are low. Note the depth of the canyon Valles Marineris. The two Viking spacecraft landed on Mars in 1976. Pathfinder landed in 1997. Rovers Spirit and Opportunity landed in 2004. (NASA)

craters there. Growing evidence, however, suggests that the northern lowlands were once filled by an ocean of liquid water. This is a controversial theory, which will be discussed later, but it serves to illustrate how planetary scientists must work from the features seen today to reconstruct the past history of a world (**Window on Science 22-2**).

The cratering and volcanism on Mars will fit with what you know of comparative planetology. It is larger than Earth's moon, so it cooled more slowly, and its volcanism has continued longer than on the moon. But Mars is smaller than Earth and less geologically active, so some of its ancient cratered terrain has survived undamaged by volcanism and plate tectonics.

The Martian volcanoes are shield volcanoes with shallow slopes, showing that the lava flowed easily. Such volcanoes occur over hot spots of rising magma below the crust and are not related to plate tectonics. The largest volcano in the solar system is Olympus Mons on Mars (see page 529). The volcano Mauna Loa in Hawaii is so heavy it has sunk into Earth's crust to form an undersea moat. Olympus Mons is much larger than Mauna Loa but has not sunk into the crust of Mars, which is evidence that the crust of Mars is much thicker than the crust of Earth. You can see that in ■ Figure 22-16.

Other evidence shows that the Martian crust has been thinner and more active than the moon's. Valles Marineris is a network of canyons that stretches 4000 km (2500 mi) long and up to 600 km (400 mi) wide (Figure 22-16). At its deepest, it is four times deeper than the Grand Canyon on Earth. It is long enough to stretch from New York to Los Angeles. The canyon has been

produced by faults in the crust that allowed great blocks to sink. Landslides and some erosion have further modified the canyon. Although Valles Marineris is an old feature, it does show that the crust of Mars has been more active than the crusts of the moon or Mercury, worlds that lack such dramatic canyons.

The faults that created Valles Marineris seem linked at the west end to a great volcanic bulge in the crust of Mars called the Tharsis rise. Nearly as large as the United States, the Tharsis rise extends 10 km (6 mi) above the mean radius of Mars. Tharsis is home to many smaller volcanoes, but on its summit lie three giants, and just off of its northwest edge lies Olympus Mons (Figure 22-16). The origin of the Tharsis rise is not well understood, but it appears that magma rising from below the crust has broken through repeatedly to build a giant bulge of volcanic deposits. This bulge is large enough to have modified the climate and seasons on Mars and may be critical in understanding the history of the planet.

A similar uplifted volcanic bulge, the Elysium region visible in Figure 22-15, lies halfway around the planet. It appears to be similar to the Tharsis rise, but it is more heavily cratered and so must be older.

The vast sizes of the Tharsis rise and of Olympus Mons show that the crust of Mars has not been broken into moving plates. If a plate were moving over a hot spot, the rising magma would produce a long chain of shield volcanoes and not a single large peak (page 529). On Earth, the hot spot that creates the volcanic Hawaiian Islands has punched through the moving Pacific plate repeatedly to produce the Hawaiian-Emperor island chain ex-

The Present Is the Key to the Past

Geologists are fond of saying, "The present is the key to the past." By that they mean that you can learn about Earth's history by looking at the present condition of Earth's surface. The position and composition of various rock layers in the Grand Canyon, for example, tell you that the western United States was once at the floor of an ocean. This principle of geology was astonishing when it was first formulated in the late 1700s, and it continues to be relevant today as you try to understand other worlds, such as Venus and Mars.

In the late 18th century, naturalists recognized that the present gave them clues to Earth's history. This was astonishing because most people assumed that Earth had no history. That is, they assumed either that Earth had been created in its present state as described in the Old Testament or that Earth was eternal. In either case, people commonly assumed that the hills and mountains they saw around them had always existed more or less as they were. The 18th-century naturalists began to see evidence that the hills and mountains were not eternal but were the result of past processes and were slowly changing. That gave birth to the idea that Earth had a history.

As the naturalists of the 18th century made the first attempts to thoughtfully and logically explain Earth's nature by looking at the evidence, they were inventing modern geology as a way of understanding Earth. What Copernicus, Kepler, and Newton did for the heavens in the 1600s, the first geologists did for Earth beginning in the late 1700s. Of course, the invention of geology as the study of Earth led directly to modern attempts to understand the geology of other worlds.

Geologists and astronomers share a common goal: They are attempting to reconstruct the past (Window on Science 19-2). Whether you study Earth, Venus, or Mars, you are looking at the present evidence and trying to reconstruct the past history of the planet by drawing on observations and logic to test each step in the story. How did Venus get to be covered with lava, and how did Mars lose its atmosphere? The final goal of planetary astronomy is to draw together all of the available evidence (the present) to tell the story (the past) of how the planet got to be the way it is. Those first geologists of the late 1700s would be fascinated by the stories planetary astronomers tell today.

Layers of rock in the Martian crater Terby hint at a time when the crater was filled with a lake. (NASA/JPL/Malin Space Science Systems)

---

tending 7500 km (4700 mi) northwest across the Pacific seafloor (page 486). No such chains of volcanoes are evident on Mars, so you can conclude that the crust is not divided into moving plates.

No spacecraft has ever photographed an erupting volcano on Mars, but it is possible that some of the volcanoes are still active. Craters in the youngest lava flows in the Elysium region show that the volcanoes may have been active as recently as a few million years ago. Mars may still retain enough heat to trigger an eruption, but the interval between eruptions may be very long.

Hidden Water on Mars

You would not expect to find liquid water on Mars. The atmospheric pressure is too low to keep it from boiling away to vapor. Nevertheless, some geological features seem to have been formed by moving water, and sophisticated measurements from orbit and on-site observations by rovers suggest that significant amounts of water are hidden in the crust.

When the two Viking spacecraft reached orbit around Mars in 1976, they found exciting hints that water once flowed over the surface. More recent spacecraft such as Mars Global Surveyor,

which reached Mars in 1997, and Mars Odyssey, which orbited in 2001, have identified additional features that seem to be related to water. You must examine those features with care, however, because the water on Mars is not obvious at all.

■ Figure 22-17 summarizes water-related features. **Outflow channels** appear to have been cut by massive floods carrying as much as 10,000 times the water flowing down the Mississippi River. In a matter of hours or days, such floods swept away geological figures and left outflow channels. One way to account for these features is to suppose that the water was frozen in the soil as permafrost. Melted by volcanic heat and pressurized under the weight of the overlying rock, water could have burst out to scour the surface in sudden floods. The number of craters formed on top of the outflow channels shows they are billions of years old.

In contrast, the **valley networks** look like meandering riverbeds that may have formed over long periods. Some streambeds appear to have flowed repeatedly over millions of years. The valley networks are located in the old, cratered, southern hemisphere, and they have about the same age. Evidently the atmosphere was denser a few billion years ago.

Olympus Mons

Elevation
(meters)

27,000

-1,000

Ascraeus
Mons

Pavonis
Mons

Arsia
Mons

Green area: The Tharsis
rise is a 10-km-thick
dome of lava flows.

Olympus Mons is about 100 times
larger in volume than Mauna Loa,
the largest volcano on Earth.

10 km
100 km

Sea level   Moat

Olympus Mons

Mauna Loa

■ **Figure 22-16**

High volcanoes and deep canyons mark
the surface of Mars. Olympus Mons, a
shield volcano, is much larger than the
largest volcano on Earth. In this false-
color image, three other volcanoes are
visible. Those three volcanoes are also
visible in the photo, along with the
canon Valles Marineris, which stretches
as far as the distance from New York to
Los Angeles. (Four volcanoes: © Calvin J.
Hamilton, Columbia, Maryland; Photo: NASA/
USGS)

As shown in Figure 22-17, some features suggest water in
the crust of Mars, perhaps as ice. Jumbled valleys appear to be regions where water below the surface has drained away and allowed
the terrain to collapse. Some impact craters appear to be surrounded by a "splash" of material, as if the crust was rich in water
at the time of the impact. Slow seepage from underground water
or ice could gradually form a valley network over long periods.
The low atmospheric pressure at the surface of Mars today could
not prevent liquid water from boiling away to vapor, but water
trapped in the crust might be under enough pressure to remain a
liquid. Also, water frozen as ice could be trapped in the crust. This
gives planetary scientists confidence that water is trapped in the
crust of Mars.

More recent spacecraft have been able to photograph the surface of Mars at a much higher resolution than could the Viking
orbiters. Those new images show evidence of water. In many
places, the spacecraft found gullies leading down steep slopes,
and the lack of craters formed on the gullies shows that they are

young. Early interpretations suggested the gullies were cut by
subsurface water released in a gush when an ice plug melted.
Newer studies suggest the gullies could be debris flows caused by
rock and soil sliding downhill after being released by evaporating
ice. Another study suggests that the gullies formed beneath deep
layers of melting snow. Choosing among these hypotheses is not
easy, and you must keep in mind that more than one process may
have formed gullies. In any case, the observation of young gullies
is further evidence that there is still water on Mars.

Other images show the details of valley networks that contain
common features such as sandbars and islands much like rivers
on Earth (■ Figure 22-18a). Other features appear to show deltas
of material deposited by flowing water (Figure 22-18b). The similarity with features found on Earth is dramatic.

The atmosphere of Mars contains a tantalizing clue. Recall
that deuterium is an isotope of hydrogen that contains a proton
and a neutron as a nucleus. Because deuterium is twice as heavy
as normal hydrogen, it escapes from a planet's atmosphere more

■ Figure 22-17

These visual-wavelength images made by the Viking orbiters and Mars Global Surveyor show some of the features that suggest liquid water on Mars. Outflow channels and valley networks are old, but some gullies may be quite recent. (Malin Space Science Systems and NASA)

slowly. Earlier in this chapter you saw that the deuterium in the atmosphere of Venus was evidence that Venus once had more water. Observations of the Martian atmosphere show that deuterium is 5.5 times more abundant relative to hydrogen on Mars than it is on Earth. That suggests that Mars once had about 20 times more water than it does now.

The terrain at the edges of the northern lowlands has been compared to shorelines, and many planetary scientists suspect that the northern lowlands were at least once filled with an ocean. Study Figure 22-15, where the lowlands have been color-coded blue, and notice the major outflow channels leading from the highlands into the lowlands northwest of the Viking 1 landing site and southeast of the Pathfinder landing site. The evidence suggests that the northern lowlands were filled by a major ocean a

few billion years ago. Large, generally circular depressions such as Hellas and Argyre are believed to be ancient impact basins and were probably once filled with water.

The Mars Odyssey Spacecraft may have found the remains of that ancient water. Two instruments on Mars Odyssey are capable of detecting water within the top meter of the crust, and both agree that water ice makes up a significant fraction of the crust from latitude 60° north and from latitude −60° south. It isn't possible to determine how deep the ice goes, but the amount detected so far is enough to fill Lake Michigan twice.

Rovers Spirit and Opportunity were targeted to land in areas suspected of having had water on their surfaces. Images made from orbit showed flow features at the Spirit landing site, and hematite, a mineral that forms in water, was detected from orbit at the Opportunity landing site. Both rovers reported exciting discoveries, including evidence of past water. Using its analytic instruments, Opportunity found small spherical concretions (dubbed blue-

■ **Figure 22-18**

Details of past water flow are visible in these false-color images. At the left, an ancient flow channel meanders across the Apsus Vallis regions, complete with islands and bars. At the right, water flowing through a gap in the wall of the crater Moreux has deposited large amounts of sediment on the crater floor. (NASA/JPL/Arizona State University)

berries although they are, in fact, hematite) that must have formed in water. Later, Spirit found similar concretions in its area. In other rocks, Opportunity found layers of sediments with ripple marks and crossed layers showing they were deposited in moving water (■ Figure 22-19). Chemical analysis of the rocks at the Opportunity site showed the presence of sulfates much like Epsom salts, plus bromides and chlorides. On Earth, these compounds are left behind when bodies of water dry up. Halfway around Mars, Spirit found the mineral goethite, clear evidence that water was once present.

If liquid water once flowed on Mars, then its climate must have been different in the past. The atmosphere of Mars must have been thicker long ago before much of the gas had time to leak away. Also, volcanism may have been more active in the past, and outpourings of volcanic gases could have added to the atmosphere, increased the greenhouse effect, and vaporized more of the polar caps to further thicken the atmosphere.

Hints of climate change were found in the 1970s when the Viking orbiters photographed layered terrain near the polar caps. More modern Mars probes have imaged these layers in greater detail (■ Figure 22-20). These layers are laid down year after year as dust accumulates on the carbon dioxide polar caps and is then left behind when the polar caps vaporize in the spring. Over pe-

riods of thousands of years, deep layers can develop. What is significant is that newer layers are superimposed at angles over older layers, showing that the climate on Mars has changed repeatedly. Mars Global Surveyor has found similar layered features widespread on Mars. If these regions show traces of climate cycles, it supports studies that show that the climate on Mars may vary because of cyclic changes in the rotation and orbital revolution of the planet. Recall from Chapter 2 that Earth is affected by the Milankovitch cycles.

One astonishing bit of evidence of water on Mars is the analysis of rock samples from the planet. Of course, no astronaut has ever visited Mars and brought back a rock, but over the history of the solar system occasional impacts by asteroids have blasted giant craters on Mars and ejected bits of rock into space. A few of those bits of rock have fallen to Earth as meteorites, and over 30 have been found and identified (■ Figure 22-21). These meteorites include basalts, as you might expect from a planet so heavily covered by lava flows. They also contain small traces of water and minerals that are deposited by water. Chemical analysis shows that the magma from which the rocks solidified must have contained up to 1.8 percent water.

If all of the lava flows on Mars had contained that much water and it was all outgassed, it would have created a planetwide

■ Figure 22-19

(a) Rover Opportunity photographed these hematite concretions in a rock near its landing site. The spheres appear to have grown as minerals collected around small crystals in the presence of water. Similar concretions are found on Earth. (b) The layers in this rock were deposited as sand and silt in rapidly flowing water. From the way the layers curve and cross each other, geologists can estimate that the water was at least ten centimeters deep. A few "blueberries" and one small pebble are also visible in this image. (NASA/JPL/Cornell/USGS)

ocean 20 m deep. That wouldn't have been quite enough to explain all the flood features on Mars, but it would have provided much more water on the surface than the planet has today.

Mars has water, but it is hidden. When humans reach Mars, they will not need to dig far to find water, probably as ice. They can use solar power to break the water into hydrogen and oxygen. Hydrogen is fuel, and oxygen is the breath of life, so the water on Mars may prove to be buried treasure.

Even more exciting is the realization that Mars once had bodies of liquid water on its surface. It is a desert world now, but someday an astronaut may scramble down an ancient Martian streambed, turn over a rock, and find a fossil.

A History of Mars

The history of Mars is like a play where all the exciting stuff happens in the first act. After the first couple billion years on Mars, most of the activity was over, and it has been downhill ever since.

You can check off differentiation, the first stage of planetary development. Planetary scientists have good evidence that Mars differentiated when it formed and had a hot, molten core. Some of the evidence comes from exquisitely sensitive Doppler-shift measurements of radio signals coming from spacecraft orbiting Mars. The shifts allowed scientists to map the gravitational field and study the shape of Mars in such detail they could detect tides on Mars caused by the sun's gravity. Those tides are less than a centimeter high; but, by comparing them with models of the interior of Mars, the scientists can show that Mars has a very dense, liquid core, a dense mantle, and a low-density crust.

Observations made from orbit show that Mars has no magnetic field, but it does have traces of magnetism frozen into some sections of old crust. That shows that soon after Mars formed, it

(a) The north polar cap of Mars is made of many regions of ice separated by narrow valleys free of ice. (NASA) (b) Near the edge of the polar cap, a set of light-colored layers of dust overlies a set of darker layers. Sand eroding from the lower layers is forming dunes. (NASA/JPL/Malin Space Science Systems) (c) In places, different sets of layers are superimposed, suggesting periodic changes in the Martian climate. (Adapted from a diagram by J. A. Cutts, K. R. Balasius, G. A. Briggs, M. H. Carr, R. Greeley, and H. Masursky)

had a hot metallic core in which the dynamo effect generated a magnetic field. Because Mars is small, it lost its heat rapidly, and the central part of the core gradually froze solid; that may be why the dynamo finally shut down. Today Mars probably has a large solid core surrounded by a thin shell of liquid core material in which no dynamo can generate a magnetic field.

No one is sure what produced the southern highlands and northern lowlands. Powerful convection in the planet's mantle may have pushed crust together to form the southern highlands. The suggestion that a catastrophic impact produced the northern lowlands does not seem to fit with the structure of the highlands, but it is not impossible.

During the second stage of planetary evolution, cratering, the crust of Mars was battered

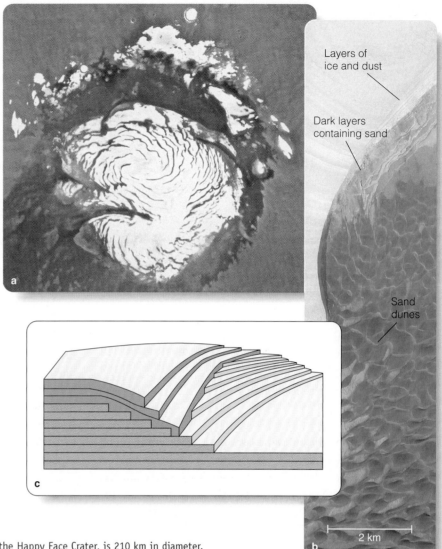

Layers of ice and dust

Dark layers containing sand

Sand dunes

2 km

■ Figure 22-21 ▼

(a) The impact crater Galle, known for obvious reasons as the Happy Face Crater, is 210 km in diameter. Such large impacts may occasionally eject fragments of the crust into space. A few fragments from Mars have fallen to Earth as meteorites. (Malin Space Science Systems/NASA) (b) Meteorite ALH84001, found in Antarctica, has been identified as an ancient rock from Mars. Minerals found in the meteorite were deposited in water and thus suggest that the Martian crust was once richer in water than it is now. (NASA)

ALH84001,

■ Figure 22-22

The Tharsis rise is a bulge on the side of Mars consisting of stacked lava flows extending up to 10 km high and over a third the diameter of Mars. It dominates this image of Mars with a few clouds clinging to the four largest volcanoes. This image was recorded in late winter for the southern hemisphere. Notice the size of the southern polar cap. (NASA/JPL/Malin Space Science Systems)

by the heavy bombardment as the last of the debris in the young solar system was swept up. The old southern hemisphere survives from this age, about 4 billion years ago. The largest impacts blasted out the great basins like Hellas and Argyre very late in the cratering, and there is no trace of magnetic field in those basins. Evidently the dynamo had shut down by then.

The third stage of planet formation, flooding, included flooding by great lava flows that smoothed some regions. Volcanism in the Tharsis and Elysium regions was very active, and the Tharsis rise grew into a huge bulge on the side of Mars (■ Figure 22-22). Some of the oldest lava flows on Mars are in the Tharsis rise, but it also contains some of the most recent lava flows. It has evidently been a major volcanic area for most of the planet's history.

Flooding included liquid water on Mars. You can check that off your list, but add a question mark because it is uncertain how much water was present and how long it lasted. Until nearly the end of cratering, Mars had a magnetic field, and that should have protected the atmosphere from the solar wind. Violent volcanism and the heat of formation should have released gases to keep the air pressure high enough to allow liquid water to exist on the surface. In fact, volcanism releases water vapor as well as gases, and impacts could have vaporized water trapped in the crust.

Most of the water features on Mars were formed in the first billion years or so. Valley networks drained water released by seeping groundwater or by rain and snow falling out of the atmosphere. Many of these valley networks flow away from the Tharsis rise, showing that they formed after the rise developed. Some very

old flood features are known, but most are a little younger than the valley networks and appear to have been formed by the sudden release huge amounts of groundwater. Water seems to have filled the northern lowlands and the deep basins to form oceans, but it's not known how long those bodies of water survived.

Because Mars is small, it lost its internal heat quickly, and atmospheric gases have escaped into space. It isn't clear how long liquid water remained stable on the surface, but it is clear that Mars dried up, chilled, and became the deadly cold desert world of today. There may have been episodes when the atmosphere grew denser and warmer and more water was available. Major volcanic eruptions could add gas and water vapor to the atmosphere. This may have produced short-lived episodes in which water flowed over the surface and collected in the lowlands to form new oceans. As the planet aged, the water became trapped in the crust as permafrost or at the poles where it froze out along with solid carbon dioxide.

The history of Mars may hinge on climate variations. Recently calculated models suggest that Mars may have once rotated at a much steeper angle to its orbit, as much as 45°. This could have supported a warmer climate and kept more of the carbon dioxide from freezing out at the poles. The rise of the Tharsis bulge could have tipped the axis to its present 25° and cooled the climate. Some calculations suggest that Mars goes through cycles as its rotational inclination fluctuates, and this may cause short-term variations in climate, much like the ice ages on Earth. That might explain the layered terrain near the polar caps.

The fourth stage of planet formation is unremarkable on Mars. The crust of Mars is now too thick to be active. The planet has lost much of its internal heat, and the core no longer generates a magnetic field. Although the crust was stressed by rising magma that produced the Tharsis and Elysium volcanic rises and the great canyon Valles Marineris, it is too thick for plate tectonics. There are no folded mountain ranges on Mars as there are on Earth. The huge size of the Martian volcanoes clearly shows that crustal plates have not moved on Mars, and repeated eruptions have built the volcanoes to vast size. Volcanism may still occur on Mars, but the crust has grown too thick for much geological activity beyond slow erosion by wind-borne dust and the occasional meteorite impact.

This history of Mars is far from complete, but it does illustrate how the size of Mars has influenced both its atmosphere and its geology. Neither small nor large, Mars is a medium world.

Building Scientific Arguments

Why doesn't Mars have coronae like those on Venus?
This argument is a good opportunity to apply the principles of comparative planetology. The coronae on Venus are caused by rising currents of molten magma in the mantle pushing upward under the crust and then withdrawing to leave the circular scars called coronae. Earth, Venus, and Mars have had signifi-

cant amounts of internal heat, and there is plenty of evidence that they have had rising convection currents of magma under their crusts. Of course, you wouldn't expect to see coronae on Earth; its surface is rapidly modified by erosion and plate tectonics. Furthermore, the mantle convection on Earth seems to produce plate tectonics rather than coronae. Mars, however, is a smaller world and must have cooled faster. There is no evidence of plate tectonics on Mars, and giant volcanoes suggest rising plumes of magma erupting up through the crust at the same point over and over. Perhaps there are no coronae on Mars because the crust of Mars rapidly grew too thick to deform easily over a rising plume. On the other hand, perhaps you could think of the entire Tharsis bulge as a single giant corona.

Planetary scientists haven't explored enough planets yet to see all the fascinating combinations nature has in store. But it does seem likely that the geology of Mars is typical of medium-size worlds. Of course, Mars is not medium in terms of its location. Of the terrestrial planets, Mars is the farthest from the sun. Build a scientific argument to analyze that factor. **How has the location of Mars affected the evolution of its atmosphere?**

■ ■ ■

Connections: If you want your history of Mars to be truly complete, it must include a phenomenon that was not part of your histories of Mercury and Venus. Unlike those worlds, Mars has moons. These moons, and the moons you will study in the outer solar system, contain important clues to the origin of the planets.

22-3 The Moons of Mars

IF YOU COULD CAMP OVERNIGHT on Mars, you might notice its two small moons, Phobos and Deimos. Phobos, shaped like a flattened loaf of bread 20 km × 23 km × 28 km, would appear less than half as large as Earth's full moon. Deimos, only 12 km in diameter and three times farther from Mars, would look only $\frac{1}{15}$ the diameter of Earth's moon.

Both moons are tidally locked to Mars, keeping the same side facing the planet as they orbit. Also, both moons revolve around Mars in the same direction that Mars rotates, but Phobos follows such a small orbit that it revolves faster than Mars rotates. That makes Phobos rise in the west and set in the east.

Origin and Evolution

Deimos and Phobos are typical of the small, rocky moons in our solar system (■ Figure 22-23). They are dark gray, with albedos of only about 0.06, making them as dark as coal. They have low densities, about 2 g/cm³.

Many of the properties of these moons hint that they are captured asteroids. In the outer parts of the asteroid belt, almost all asteroids are dark, low-density objects. Massive Jupiter, orbiting

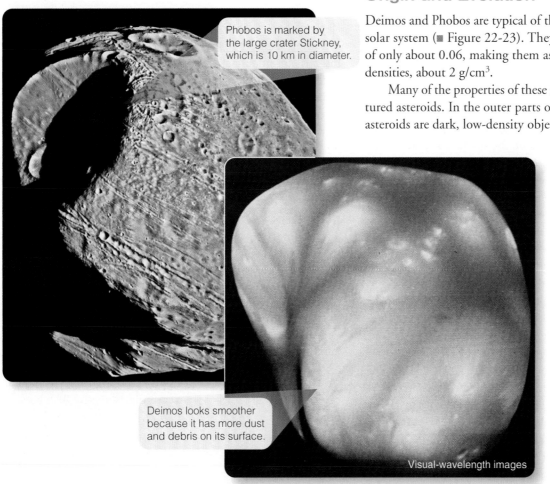

Phobos is marked by the large crater Stickney, which is 10 km in diameter.

Deimos looks smoother because it has more dust and debris on its surface.

Visual-wavelength images

■ **Figure 22-23**

The moons of Mars are too small to pull themselves into spherical shape. Deimos is about half the size of Phobos. The two moons were named for the mythical attendants of Mars, the god of war. Phobos was the god of panic, and Deimos was the god of fear. (Phobos: Damon Simonelli and Joseph Ververka, Cornell University/NASA; Deimos: NASA) just outside the asteroid belt, can scatter such bodies throughout the solar system, so you should not be surprised if Mars, the next planet inside the asteroid belt, has captured a few of these as satellites.

However, capturing a passing asteroid into a closed orbit is not so easy that it happens often. The asteroid approaches the planet along a hyperbolic (open) orbit and, if it is unimpeded, swings around the planet and disappears back into space. To convert the hyperbolic orbit into a closed orbit, the planet must slow the asteroid as it passes. Tidal forces might do this, but they would be rather weak. Interactions with other moons or grazing collisions with a thick atmosphere might also slow the asteroid.

Both satellites have been photographed by spacecraft, and those photos show that the satellites are heavily cratered. Such cratering could have occurred while the moons were either still in the asteroid belt or in orbit around Mars. In any case, the heavy battering has broken the satellites into irregular chunks of rock, and they cannot pull themselves into smooth spheres because their gravity is too weak to overcome the structural strength of the rock. You will discover that low-mass moons are typically irregular in shape, whereas more massive moons are more spherical.

Images of Phobos reveal a unique set of narrow, parallel grooves (see Figure 22-23). Averaging 150 m wide and 25 m deep, the grooves run from Stickney, the largest crater, to an oddly featureless region on the opposite side of the satellite. One theory suggests that the grooves are deep fractures produced by the impact that formed the crater. The featureless region opposite Stickney may be similar to the jumbled terrain on Earth's moon and on Mercury. All were produced by the focusing of seismic waves from a major impact on the far side of the body. High-resolution photographs show that the grooves are lines of pits, suggesting that the pulverized rock material on the surface has drained into the fractures or that gas, liberated by the heat of impact, escaped through the fractures and blew away the dusty soil.

The Mars Global Surveyor spacecraft revealed more about the dust. Observations made with the spacecraft's infrared spectrometer show that the moon's surface cools quickly from $-4°C$ to $-112°C$ (from 25°F to $-170°F$) as it passes from sunlight into the shadow of Mars. Solid rock would retain heat and cool more slowly, so the dust must be at least a meter deep and very fine. In photos made by the spacecraft camera, the dust blankets the terrain, but the photos also show boulders a few meters in diameter, thought to be ejecta from impacts.

Deimos not only has no grooves, it also looks smoother because of a thicker layer of dust on its surface (Figure 22-23). This material partially fills craters and covers minor surface irregularities. It seems certain that Deimos experienced collisions in its past, so fractures may be hidden below the debris.

The debris on the surfaces of the moons raises an interesting question. How can the weak gravity of small bodies hold any fragments from meteorite impacts? Escape velocity on Phobos is only 12 m/s. An athletic astronaut could almost jump into space. Certainly, most fragments from impacts should escape, but some could fall back and accumulate on the surface.

Deimos, smaller than Phobos, has a smaller escape velocity. But it has more debris on its surface because it is farther from Mars. Phobos is close enough to Mars that most ejecta from impacts on Phobos will be drawn into Mars. Deimos, being farther from Mars, is able to keep a larger fraction of its ejecta. Phobos is so close to Mars that tides are making its orbit shrink, and it will fall into Mars or be ripped apart by tidal forces within about 100 million years.

Deimos and Phobos illustrate three principles of comparative planetology that you will find helpful as you explore farther from the sun. First, some satellites are probably captured asteroids. Second, small satellites tend to be irregular in shape and heavily cratered. And third, tidal forces can affect small moons and gradually change their orbits. You will find even stronger tidal effects in Jupiter's satellite system (Chapter 23).

Building Scientific Arguments

Why would you be surprised if you found volcanism on Phobos or Deimos?

This is another obvious argument, isn't it? But remember, the purpose of a scientific argument is to test your own understanding, so it is a good way to review. In discussing Earth's moon, Mercury, Venus, Earth, and Mars, you have seen illustrations of the principle that the larger a world is, the more slowly it loses its internal heat. It is the flow of that heat from the interior through the surface into space that drives geological activity such as volcanism and plate motion. A small world, like Earth's moon, cools quickly and remains geologically active for a shorter time than a larger world like Earth. Phobos and Deimos are not just small, they are tiny. However they formed, any interior heat would have leaked away very quickly; with no energy flowing outward, there can be no volcanism.

Some futurists suggest that the first human missions to Mars will not land on the surface of the planet but will build a colony on Phobos or Deimos. These plans speculate that there may be water deep inside the moons that colonists could use. Build an argument based on what you know about water on Mars. **What would happen to water released in the sunlight on the surface of such small worlds?**

■　　■　　■

Connections: Venus is nearly the same size as Earth, but it is unlikely that humans will ever colonize its surface. The smaller and colder Mars, however, may someday be home to large colonies. In the next chapter, you will explore Jupiter and Saturn, giant worlds so totally unlike Earth that you will need entirely new principles of comparative planetology to describe them.

Summary

22-1 | Venus

Why is the atmosphere of Venus so thick?

■ Venus is Earth's twin in size but is slightly closer to the sun.

■ The atmosphere of Venus is 95 times thicker than Earth's and composed almost entirely of carbon dioxide. The temperature at the surface is about 745K (880°F).

■ Venus is heated by a runaway greenhouse effect.

■ The surface is so hot that compounds have cooked out of the crust to form traces of sulfuric, hydrochloric, and hydrofluoric acids in the atmosphere. The very high clouds on Venus are composed of small droplets of sulfuric acid and sulfur crystals.

What is the hidden surface of Venus really like?

■ Although it is perpetually hidden below thick clouds, the surface of Venus can be studied by radar mapping, which reveals higher uplands and low rolling plains. Volcanoes, lava flows, and impact craters are visible.

■ Volcanic features such as coronae appear to be produced by rising currents of molten rock pushing up under the crust. Where the magma breaks through it produces volcanoes, lava flows, and lava channels.

■ Landers have analyzed the surface rock and found it to be similar to basalts on Earth.

How did Venus form and evolve?

■ Because Venus formed closer to the sun, it was warmer, and large liquid-water oceans were not able to form and remove carbon dioxide from the atmosphere. The accumulating carbon dioxide produced an intense greenhouse effect, made the planet very hot, and evaporated any small bodies of water that did form.

■ Venus rotates slowly retrograde (backward). This may have been caused by an off-center impact by a very large planetesimal as the planet was forming or by solar tides raised in its very thick atmosphere.

■ Volcanism is common on Venus, and much of the surface is solidified lava flows. Volcanoes are probably still active on Venus.

■ The surface contains more impact craters than the maria on the moon, but fewer than does the Earth. The crust appears to be about half a billion years old. Planetary scientists suspect that the entire planet was resurfaced by an outpouring of lava about half a billion years ago.

■ Venus has no detectable magnetic field, so the state of its core is unknown.

22-2 | Mars

Why is the atmosphere of Mars so thin?

■ Mars is smaller than Earth but larger than the moon. It has lost much of its atmosphere because of its low escape velocity.

■ The atmosphere on Mars is very-low-density carbon dioxide, and the pressure at the surface is too low to prevent water from boiling away.

■ Mars has no magnetic field to protect it from the solar wind, and some of its atmosphere has been blasted away over its history by the pressure of the solar wind.

Has Mars ever had water on its surface?

■ Mars is now a dry, desert world on which liquid water cannot exist.

■ The southern hemisphere is old and heavily cratered, but the northern lowlands are smooth and mostly free of craters.

■ Images from orbit reveal outflow channels that appear to have been cut by massive floods and valley networks that resemble dry river beds. Crater counts show that these features are very old.

■ The smooth lowlands of the northern hemisphere appear to have once contained a liquid water ocean. Some outflow channels lead into the lowlands and features resembling shore lines have been found.

■ Evidence suggests that whatever water Mars retains is now frozen in the crust as permafrost. If it seeps out, it may cut gullies and form similar flow features, but rivers, lakes, and oceans cannot now exist on Mars because of the low atmospheric pressure.

How did Mars form and evolve?

■ Mars formed farther from the sun than Earth did and is much smaller.

■ Mars once had a molten core that generated a magnetic field, but it has no detectable magnetic field now. Evidently the inner core has solidified, and the molten-out layer of the core cannot generate a magnetic field.

■ Volcanism has been important in the history of Mars, and the Tharsis rise is a huge volcanic uplift. Volcanoes may still erupt on Mars, but, because of its small size, it has lost much of its internal heat.

■ Olympus Mons is a very large volcano, but it has not sunk into the crust, and that shows that the crust of Mars is now quite thick.

■ Mars must have had a thicker atmosphere in the past as evidenced by the liquid water flow features visible on it surface. But it has lost much of its water, and what remains is frozen in the crust or under the carbon dioxide polar caps.

■ The development of the Tharsis rise may have altered the rotation of Mars and changed its climate, but climate cycles driven by changes in rotation and revolution may explain the layered terrain near the polar caps.

22-3 | The Moons of Mars

■ Mars has captured two asteroids into orbit as moons. Phobos and Deimos are small, irregularly shaped, and cratered.

New Terms

subsolar point (p. 524)	corona (p. 530)
composite volcano (p. 528)	outflow channel (p. 540)
shield volcano (p. 528)	valley network (p. 540)

Review Questions

Ace Astronomy™ Assess your understanding of this chapter's topics with additional quizzing and animations at http://ace .brookscole.com/sf9

1. Why might you expect Venus and Earth to be similar?

2. What evidence can you cite that Venus and Mars once had more water than at present? Where did that water come from? Where did it go?

3. What features would you look for in high-resolution radar maps of Venus to search for plate tectonics?

4. Why doesn't Mars have mountain ranges like those on Earth? Why doesn't Earth have large volcanoes like those on Mars?

5. What were the canals on Mars? How do they differ from the dry streambeds on Mars?

6. Propose an explanation for the nearly pure carbon dioxide atmospheres of Venus and Mars. How did Earth avoid such a fate?

7. What evidence can you cite that the climate on Mars has changed?

8. How can planetary scientists estimate the ages of the streambeds on Mars?

9. Why are Phobos and Deimos nonspherical? Why is Earth's moon much more spherical?

10. Volcano Sif Mons on Venus is shown in this radar image. What kind of volcano is it, and why is it orange in this image? What color would the rock be if you could see it with your own eyes?

(NASA)

11. Olympus Mons on Mars is a very large volcano. In this image you can see multiple calderas at the top. What do those calderas and the immense size indicate about the geology of Mars?

(NASA)

Discussion Questions

1. From what you know of Earth, Venus, and Mars, do you expect the volcanoes on Venus and Mars to be active or extinct? Why?

2. If humans someday colonize Mars, the biggest problem may be finding water and oxygen. With plenty of solar energy beating down through the thin atmosphere, how might colonizers extract water and oxygen from the Martian environment?

Problems

1. How long would it take radio signals to travel from Earth to Venus and back if Venus were at its nearest point to Earth? at its farthest point from Earth?

2. The Pioneer Venus Orbiter circled Venus with a period of 24 hours. What was its average distance above the surface of Venus? (*Hint:* See Chapter 5.)

3. Calculate the velocity of Venus in its orbit around the sun.

4. What is the maximum angular diameter of Venus as seen from Earth? (*Hint:* Use the small-angle formula.)

5. If the Magellan spacecraft transmitted radio signals down through the clouds on Venus and heard an echo from a certain spot 0.000133 second before the main echo, how high is the spot above the average surface of Venus?

6. The smallest feature visible through an Earth-based telescope has an angular diameter of about 1 second of arc. If a canal on Mars was just visible when Mars was at its closest to Earth, how wide was the canal? (*Hint:* Use the small-angle formula.)

7. What is the maximum angular diameter of Phobos as seen from Earth? What surface features should you expect to see from Earth? from the surface of Mars? (*Hint:* Use the small-angle formula.)

8. What is the maximum angular diameter of Deimos as seen from the surface of Mars?

9. Deimos is about 12 km in diameter and has a density of 2 g/cm³. What is its mass? (*Hint:* The volume of a sphere is $\frac{4}{3}\pi R^3$.)

Media Cluster

Ace⊘Astronomy™ To access the resources in the Media Cluster, log into AceAstronomy at **http://ace.brookscole .com/sf9** and select Chapter 22.

ACTIVE FIGURES

Planetary Atmospheres
Experiment with the temperature throughout the atmospheres of several planets with this animation.

Hot Spot Volcanoes
If the crustal plate over a hot spot is moving, repeated eruptions can result in an island chain like the Hawaiian-Emperor island chain. Set a series of eruptions in motion in this animation.

ASTRONOMY EXERCISES

The Greenhouse Effect
See how distance to the sun and amount of carbon dioxide in a planet's atmosphere affect its temperature in this animation.

Convection and Plate Tectonics
Study the convection currents of the four Earthlike planets in this animation and see if you can determine how this relates to the amount of plate tectonic activity on each planet.

Convection and Magnetic Fields
A planet's core temperature and rotation speed are factors in the strength of its magnetic field. See the effect of these variables in this animation.

Primary Atmospheres
This animation lets you change the distance to the sun and mass of a hypothetical planet and study how these relate to the planet's retention of its atmosphere.

VIRTUAL ASTRONOMY LABS

Lab 5: Planetary Geology
By thoroughly studying Earth's geology, you can extend your knowledge to understand conditions on other planets. This lab includes an exercise in determining the interior characteristics of a planet by observing the travel times of its seismic waves.

Lab 7: Planetary Atmospheres and Their Retention
This lab investigates the retention of atmospheres. You will explore the factors that govern the loss of atmospheric gases, and you will see why certain bodies can retain some gases but not others.

Critical Inquiries for the Web

1. Have you ever wondered how the various features on the planets get their names? Surface features on Venus are (mostly) named for female figures from history and mythology. Who decides how planetary features are named? Look for information on planetary nomenclature and summarize the way different types of features on Venus are assigned names.

2. "Martians" have fascinated humans for the last century or more. Many online sources chronicle the representation of life on Mars by Earthlings throughout history and literature. Read about the Martians as represented by a particular literary work or nonfiction account and discuss to what extent they are (or are not) based on realistic views of the nature of Mars—in terms of your current understanding as well as the views of the period in which the work was written.

3. Spacecraft are studying Mars right now. Search for the latest discoveries and photographs.

Exploring *TheSky*

1. Venus and Mars exhibit phases similar to the moon's phases. (a) Use *TheSky* to sketch the appearance of Venus and Mars at the present time. Also give the phase as a percentage of full. How to proceed: Use **Find** to give you a highly magnified view of a particular planet. Click on the planet to obtain **Object Information.** (b) Explain the observed phase on the basis of geometrical relationship between the sun, Earth, and the planet as shown by the **3D Solar System Mode.**

Go to the Brooks/Cole Astronomy Resource Center **(http:// astronomy.brookscole.com)** for critical thinking exercises, articles, and additional readings from InfoTrac College Edition, Brooks/Cole's online student library.

23 | Comparative Planetology of Jupiter and Saturn

> *There is something fascinating about science. One gets such wholesale returns of conjecture out of such a trifling investment of fact.*
>
> MARK TWAIN,
> *LIFE ON THE MISSISSIPPI*

WHEN MARK TWAIN WROTE the sentences above, he was poking gentle fun at science, but he was right. The exciting thing about science isn't the so-called facts, the observations in which scientists have greatest confidence. Rather, the excitement lies in the understanding that scientists get by rubbing a few facts together. Science can take you to strange new worlds. Jupiter and Saturn are certainly strange, but you can combine the available observations with known principles of comparative planetology and understand these two giant planets. ❚ Jupiter and Saturn have been studied from Earth for centuries, but much of what is known has been radioed back by space probes. Pioneer 10 and Pioneer 11 explored Jupiter and Saturn in the mid-1970s, but their instruments were not very sensitive. The Voyager 1 and Voyager 2 spacecraft were

❚ Continued on page 554 ❚

Science can take you to other worlds. When the Cassini probe rounded Saturn, it fired its rockets to slow down and settle into orbit. Controlled from Earth, it studies the planet, the moons and the rings of the Saturn system. (NASA)

Guidepost

Looking Back

You have become an experienced planetwalker. You have been to Mercury, Venus, Earth, Mars, and Earth's moon. You know all about craters, volcanoes, lava flows, and plate tectonics. But your experience is limited to terrestrial planets. Now it is time to try something entirely different.

This Chapter

As you begin this chapter, you leave behind the psychological security of planetary surfaces. You can imagine standing on the moon, on Mars, or even on Venus, but Jupiter and Saturn have no surfaces. Here you face a new challenge—to use comparative planetology to study worlds so unearthly you cannot imagine being there. As you study these worlds you will find answers to five essential questions:

How do the outer planets compare with the inner planets?

How is Jupiter different from Earth?

How did Jupiter and its system of moons and rings form and evolve?

How is Saturn different from Jupiter?

How did Saturn and its system of moons and rings form and evolve?

There's no place to stand on Jupiter or Saturn, but be sure to bring your spacesuit. Both planets have big systems of moons, and you will be able to watch erupting volcanoes, stroll through a methane rain storm, and swim in Saturn's rings. It will be interesting, but it is no place like home.

Looking Ahead

Jupiter and its cousin Saturn make a good basis for comparison in your study of the rest of the outer solar system in the next chapter. Uranus and Neptune are small, peculiar relatives, and you will find that tiny Pluto may not even be a member of the family of planets.

launched in 1977 on a mission to visit all four Jovian worlds, concluding when Voyager 2 flew past Neptune in 1989. In December 1995, the Galileo spacecraft reached Jupiter. While an instrumented probe dropped into Jupiter's atmosphere, the main Galileo spacecraft went into orbit to begin a long-term study of the planet and its moons. The Cassini spacecraft went into orbit around Saturn in 2004 and dropped a probe into the atmosphere of Saturn's biggest moon, Titan. Throughout this discussion of the Jovian worlds, you will find images and data returned to Earth by these robotic explorers.

Jupiter and Saturn are unearthly; but, because they obey the same laws of nature, you can understand their peculiar personalities. They make a perfect introduction to the study of the outer solar system.

(23-1) A Travel Guide to the Outer Planets

IF YOU TRAVEL MUCH, you know that some cities make you feel at home, and some do not. In this and the next chapter, you will visit five worlds that are truly unearthly. You will not feel welcome, so this travel guide will warn you what to expect.

The Outer Planets

The outermost planets in our solar system are Jupiter, Saturn, Uranus, Neptune, and Pluto. The first four are often called the Jovian planets, meaning they are like Jupiter. In fact, they are all individuals with their own separate personalities.

■ Figure 23-1 compares the five outer worlds, and one striking feature is diameter. Jupiter is the largest of the Jovian worlds, over 11 times the diameter of Earth. Saturn is a bit smaller, but Uranus and Neptune are quite a bit smaller than Jupiter. Pluto, in contrast, is hardly visible in this diagram. It is a tiny world and doesn't seem to fit with its companions in the outer solar system. You will discover it is truly an oddball, and some astronomers argue that it is not a planet at all.

The other feature you will notice immediately when you look at Figure 23-1 is Saturn's rings. They are bright and beautiful and composed of billions of ice particles, each orbiting around the planet. Jupiter, Uranus, and Neptune also have rings, but they are not easily detected from Earth and are not visible in this figure. Nevertheless, as you visit these worlds you will be able to compare four different sets of planetary rings.

Atmospheres and Interiors

The four Jovian worlds have hydrogen-rich atmospheres filled with clouds. On Jupiter and Saturn, you can see that the clouds form belts and zones that circle the planets like the stripes on a child's ball. This form of atmospheric circulation is called **belt–zone circulation.** You will find traces of belts and zones on Uranus and Neptune, but they are not very distinct.

The atmospheres of the Jovian planets are not very deep. Jupiter's atmosphere makes up only about one percent of its radius. Below that, Jupiter and Saturn are composed of liquid hydrogen, so the older term for these planets, the *gas giants,* should probably be changed to the *liquid giants.* Only near their centers do these worlds have solid cores of dense material with the composition of rock and metal.

Uranus and Neptune are sometimes called the ice giants because they are rich in water, both as a liquid and as a solid. They too have denser material in their cores.

On your visits to these worlds, notice that they are low-density worlds that are rich in hydrogen. Jupiter and Saturn are mostly liquid hydrogen, and even Uranus and Neptune contain a much larger proportion of hydrogen than does Earth. Recall from Chapter 19 that these worlds are hydrogen-rich, low-density worlds because they formed in the outer solar nebula where water vapor could freeze to form tremendous amounts of tiny ice particles. Once these planets grew massive enough, they could draw in the hydrogen gas directly by gravitational collapse.

Satellite Systems

All of the Jovian worlds have large satellite systems, and even little Pluto has a moon. As you visit the moons of the Jovian worlds, look for two processes. In many cases, the moons interact with each other gravitationally and have adjusted their orbits. Their gravity can also influence planetary ring systems.

The second thing to notice is that a few moons are geologically active, and others show signs of past activity. You have learned that cratered surfaces are old, so when you see a section of a moon's surface that has few craters, you know that the moon must have been geologically active since the end of the heavy bombardment. Of course, geological activity depends on heat, so be alert for sources of internal heat as you visit these moons.

Building Scientific Arguments

Why do you expect the outer planets to be low-density worlds?
To build this scientific argument, you need to think about how the planets formed from the solar nebula. In Chapter 19, you discovered that the inner planets could not incorporate ice when they formed because it was too hot near the sun; but, in the outer solar nebula, the growing planets could accumulate lots of ice. Eventually they grew massive enough to grow by gravitational collapse, and that pulled in hydrogen and helium gas. That makes the outer planets low-density worlds.

The outer planets may be unearthly, but they are understandable. For example, extend your argument. **Why do you expect the outer planets to have rings and moons?**

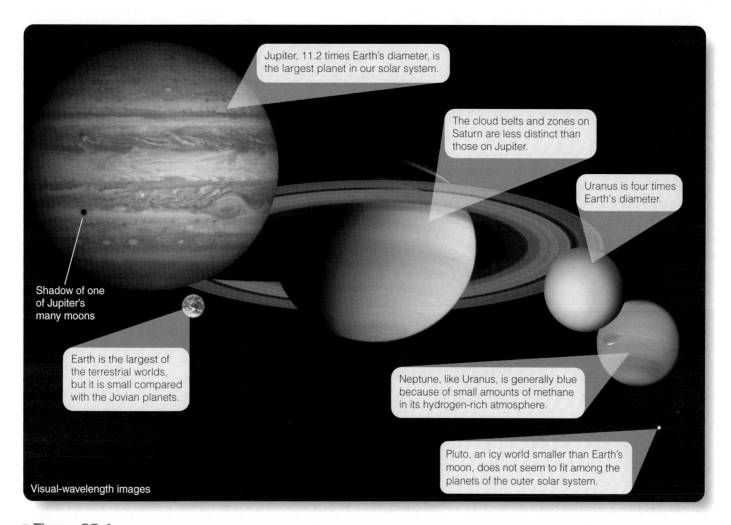

Jupiter, 11.2 times Earth's diameter, is the largest planet in our solar system.

The cloud belts and zones on Saturn are less distinct than those on Jupiter.

Uranus is four times Earth's diameter.

Shadow of one of Jupiter's many moons

Earth is the largest of the terrestrial worlds, but it is small compared with the Jovian planets.

Neptune, like Uranus, is generally blue because of small amounts of methane in its hydrogen-rich atmosphere.

Pluto, an icy world smaller than Earth's moon, does not seem to fit among the planets of the outer solar system.

Visual-wavelength images

■ Figure 23-1

The worlds of the outer solar system consist of the four Jovian planets, which are large, low-density worlds rich in hydrogen, and tiny, icy Pluto. (NASA/JPL/Space Science Institute/University of Arizona)

Connections: The outer solar system is so far from the sun that it is cold and dimly lit. The planets are far apart and orbit slowly. It is not a place that makes humans feel comfortable and welcome, but it is both mysterious and revealing. Exploring the outer solar system will give you a new appreciation for your home world.

23-2 Jupiter

JUPITER IS THE MOST MASSIVE of the Jovian planets. Over eleven times the diameter of Earth, it contains nearly three-fourths of all the planetary matter in our solar system. This high mass accentuates some processes that are less obvious or nearly absent on the other Jovian worlds. Just as you used Earth as the basis of comparison for your study of the terrestrial planets, you should examine Jupiter in detail so you can use it as a standard in your comparative study of the other Jovian planets.

Surveying Jupiter

Jupiter is interesting because it is big, massive, mostly liquid hydrogen, and very hot inside. The preceding facts are common knowledge among astronomers, but you should demand an explanation of how they know these facts. Often the most interesting thing about a fact isn't the fact itself but how it is known.

You can be sure that Jupiter is a big planet because it looks big. At its closest point to Earth, Jupiter is about eight times farther away than Mars, but even a small telescope will reveal that the disk of Jupiter is more than twice as big as the disk of Mars. If you use the small-angle formula, you can compute the diameter of Jupiter—a bit over 11 times Earth's diameter (**Celestial Profile 7**). That's a big planet!

You can be confident that Jupiter is massive because its moons race around it at high speed. Io is the innermost of the four Galilean moons, and its orbit is just a bit larger than the orbit of our moon around Earth. Io streaks around its orbit in less than two days, whereas Earth's moon takes a month. Jupiter has to be a very massive world to hold on to such a rapidly moving moon (■ Figure 23-2). In fact, you can put the radius of Io's orbit and its orbital period into Newton's version of Kepler's third law (Chapter 5) and calculate the mass of Jupiter—almost 318 times Earth's mass.

The size and mass of Jupiter are easy to find, but you might wonder how astronomers know that it is made of hydrogen. That fact is just a little bit harder to find, but the first step is to divide the mass by the volume and find Jupiter's average density, about 1.34 g/cm³. Of course, it is denser at the center and less dense near the surface, but this average density reveals that Jupiter can't be made totally of rock. Rock has a density of 2.5 to 4 g/cm³, so Jupiter must contain less dense material, such as hydrogen.

Spectra recorded from Earth and from spacecraft visiting Jupiter show that the composition of Jupiter is much like that of the sun—it is mostly hydrogen and helium. This fact was confirmed when a probe from the Galileo spacecraft fell into the atmosphere in 1995 and radioed its results back to Earth. Jupiter is mostly hydrogen and helium, with traces of heavier atoms forming molecules such as methane (CH₄), ammonia (NH₃), and water (■ Table 23-1).

Just as astronomers can build mathematical models of the interiors of stars, they can use the equations that describe gravity, energy, and how matter responds to pressure to build mathematical models of the interior of Jupiter. The models suggest that a planet as big and massive as Jupiter, having a low density and composed mostly of hydrogen and helium, must contain only a small core of elements heavier than helium. Astronomers often refer to this as a "rocky" core, but the pressure is so high there is nothing rocklike about the material. It would be better to call it a heavy-element core. Although the core is about the size of Earth, it must

Molecule	Jupiter (%)	Saturn (%)
H₂	78	88
He	19	11
H₂O	0.0001	—
CH₄	0.2?	0.6
NH₃	0.5?	0.2

■ **Table 23-1** I **Composition of Jupiter and Saturn (by mass)**

contain roughly 15 Earth masses. The models also reveal that the pressure below the atmosphere is so great that hydrogen cannot exist as a gas; most of Jupiter is liquid hydrogen.

Even the shape of the planet is revealing. By watching the cloud belts, astronomers can tell that Jupiter rotates rapidly, in about 10 hours. Simple measurements on a photograph show that it is slightly flattened; it is 6.37 percent larger in diameter through its equator than it is through its poles. Astronomers refer to this by saying that Jupiter's **oblateness,** its degree of flattening, is 0.0637. The amount of flattening depends on the speed of rotation and on the rigidity of the material. A rotating planet that is mostly liquid will be more oblate than a planet that is mostly rocky. In this way, astronomers can use the observed oblateness of Jupiter to check their models of the size of its heavy-element core.

It is easy to think of the heavy-element core as a ball of rock, but astronomers are quick to point out that it is very hot, about 30,000 K. That is five times hotter than the surface of the sun, but the pressure is so great that it keeps the material confined to the core of the planet. How do they know it is so hot? Because infrared observations show that Jupiter emits about 1.7 times as

■ **Figure 23-2**

It is obvious that Jupiter is a very massive planet when you compare Jupiter's moon Io with Earth's moon. Although Io is 10 percent farther from Jupiter, it travels 17 times faster in its orbit than does Earth's moon around Earth. Clearly, Jupiter's gravitational field is much stronger than Earth's, and that means Jupiter must be very massive.

much energy as it receives from the sun. It must be very hot inside to have so much heat flowing outward into space.

The models of Jupiter's interior reveal that ordinary liquid hydrogen cannot support the pressure deep inside the planet. It is compressed into a state in which the electrons are free to move easily from atom to atom, and, although the hydrogen can flow like a liquid, it can conduct electricity like a metal. This **liquid metallic hydrogen** does not occur naturally on Earth, but it has been created under tremendous pressure in laboratories.

Basic observations and the known laws of physics can tell you a great deal about Jupiter. Its vast magnetic field can tell you even more.

Jupiter's Magnetic Field

Astronomers detected a magnetic field around Jupiter as early as the 1950s, which is surprising considering that magnetic fields are invisible to the eye. In fact, Jupiter's giant magnetic field was first detected by radio.

In 1955, radio astronomers detected bursts of synchrotron radio waves coming from Jupiter. They called it **decameter radiation** because its wavelength was tens of meters. Synchrotron radiation is electromagnetic radiation from fast electrons spiraling in a magnetic field, so it was obvious that Jupiter had a magnetic field.

Because the radio signals fluctuated with a period of 9 hours 55.5 minutes, astronomers concluded that this period was the true period of rotation of Jupiter. This result is what you would expect from the dynamo theory for the generation of magnetic fields, which requires an electrically conducting fluid, convection driven by heat flowing outward, and rapid rotation. The liquid metallic hydrogen in Jupiter is a very good electrical conductor, and its infrared brightness is clear evidence that heat flows outward from inside Jupiter. That, coupled with Jupiter's rapid rotation, should generate a powerful dynamo effect. The fluctuation of the radio signals reveals the true rotation period of the planet's liquid interior.

By the early 1960s, radio astronomers had discovered short-wavelength **decimeter radiation,** with wavelengths a few tenths of a meter long. Although this type of radiation is also synchrotron radiation, radio telescopes revealed that it comes from a region at least three times larger than the disk of Jupiter. Detailed studies of the decimeter radiation showed that Jupiter's magnetic field was at least ten times stronger than Earth's and, like Earth's field, was inclined about 10° to the axis of rotation.

Radio studies were helpful, but nothing is as good as a visit. In 1973 and 1974, two Pioneer spacecraft flew past Jupiter, followed in 1979 by two Voyager spacecraft. The Galileo spacecraft entered orbit around Jupiter in late 1995, and the Cassini spacecraft flew past in December 2000. These spacecraft found Jupiter surrounded by a magnetosphere over 100 times larger than Earth's.

The magnetic field deflects the solar wind and traps high-energy particles in radiation belts much more intense than Earth's. The radiation is more intense because Jupiter's magnetic field is stronger and can trap and hold higher-energy particles, mostly

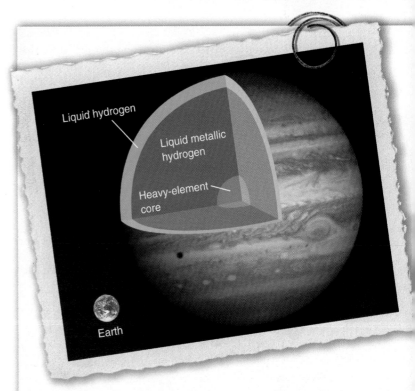

Jupiter is mostly a liquid planet with a small core of heavy elements that is not much bigger than Earth. (NASA/JPL/University of Arizona)

Celestial Profile 7: Jupiter

Motion:

Average distance from the sun	5.2028 AU (7.783 × 10⁸ km)
Eccentricity of orbit	0.0484
Maximum distance from the sun	5.455 AU (8.160 × 10⁸ km)
Minimum distance from the sun	4.951 AU (7.406 × 10⁸ km)
Inclination of orbit to ecliptic	1°18′29″
Average orbital velocity	13.06 km/s
Orbital period	11.867 y (4334.3 days)
Period of rotation	9ʰ55ᵐ30ˢ
Inclination of equator to orbit	3°5′

Characteristics:

Equatorial diameter	142,900 km (11.20 $D_\oplus$)
Mass	1.899 × 10²⁷ kg (317.83 $M_\oplus$)
Average density	1.34 g/cm³
Gravity at base of clouds	2.54 Earth gravities
Escape velocity	61 km/s (5.4 $V_\oplus$)
Temperature at cloud tops	−130°C (−200°F)
Albedo	0.51
Oblateness	0.0637

Personality Point:

Jupiter is named for the Roman king of the gods, and it is the largest planet in our solar system. It can be very bright in the night sky, and its cloud belts and four largest moons can be seen through even a small telescope. Its moons are even visible with a good pair of binoculars mounted on a tripod or braced against a wall.

electrons and protons. The spacecraft passing through the radiation belts received radiation doses equivalent to a billion chest X rays—at least 100 times a lethal dose for a human. Some of the electronics on the spacecraft were damaged by the radiation.

Jupiter's magnetic field is very strong, and it dominates a huge magnetosphere around the planet. Compare the size of Jupiter's field with that of Earth in ■ Figure 23-3.

You will recall that Earth's magnetosphere interacts with the solar wind to produce auroras, and the same is true on Jupiter. Charged particles in the magnetosphere leak downward along the magnetic field, and, where they enter the atmosphere, they produce auroras 1000 times more powerful than those on Earth. The auroras on Jupiter, like auroras on Earth, occur in rings around the magnetic poles (■ Figure 23-4).

Some of the heavier ions in the radiation belts appear to come from the inner Galilean moon, Io. As you will see later in this chapter, Io has active volcanoes that spew gas and ash. Because Io orbits with a period of 1.8 days, and Jupiter's magnetic field rotates in only 10 hours, the wobbling magnetic field rushes past Io at high speed, sweeping up stray particles, accelerating them to high energy, and spreading them around Io's orbit in a doughnut of ionized gas called the **Io plasma torus.** Jupiter's magnetic field interacts with Io to produce a powerful electric current (about a million amperes) that flows through a curving path called the **Io flux tube** from Jupiter out to Io and back to Jupiter. Small spots of bright auroras lie at the two points where the Io flux tube enters Jupiter's atmosphere (Figure 23-4).

This Io flux tube explains a mystery. Back in 1955, when radio astronomers discovered the decametric radio signals, they noticed that the signals were stronger when Io was in certain parts of its orbit. Now you can understand that. The electrical current flowing through the Io flux tube is at least partially responsible for generating the decametric radio signals and beams them out into space in certain directions.

Fluctuations in the auroras reveal that the solar wind buffets Jupiter's magnetosphere, but some of the fluctuations seem to be caused by changes in the magnetic dynamo deep inside the planet. In this way, studies of the auroras on Jupiter may help astronomers learn more about its liquid depths.

Much of the power and fascination of Jupiter is invisible to your eyes, but the swirling cloud belts are beautiful in their complexity.

Ace✸Astronomy™ Log into AceAstronomy and select this chapter to see Astronomy Exercise "Auroras." See how auroras are produced.

Ace✸Astronomy™ Log into AceAstronomy and select this chapter to see Astronomy Exercise "Convection and Magnetic Fields." How does convection in a planet generate a magnetic field?

Jupiter's Atmosphere

If you parachuted into Jupiter with a small rubber boat, expecting to go sailing on the liquid hydrogen ocean, you would be disappointed. Mathematical models of Jupiter show that there is no surface. Deep below the clouds, the temperature and pressure exceed the **critical point** for hydrogen, the temperature and pressure above which liquid hydrogen and gaseous hydrogen have the same density and are indistinguishable. As you parachuted down through Jupiter's atmosphere, the temperature, pressure, and density would rise, and gradually the gas would become a fluid. You would never splash down, and your rubber boat would be useless. Below the clouds of Jupiter lies the largest ocean in the solar system—and it has no surface and no waves.

When you look at Jupiter, you don't see a surface. You see those parts of the atmosphere that contain clouds. These cloud layers lie deep inside a nearly transparent atmosphere of hydrogen and helium. You can detect this nearly transparent hydrogen atmosphere by noticing that Jupiter has limb darkening just as the sun does (Chapter 8). When you look near the limb of Jupiter (the edge of its disk), the clouds are much dimmer (Figure 23-1) because it is nearly sunset or sunrise along the limb. If you were on Jupiter at that location, you would see the sun just above the horizon, and it would be dimmed by the atmosphere. In addition, light reflected from clouds must travel out at a steep angle through the atmosphere to reach Earth, dimming the light further. Jupiter

■ **Figure 23-3**

Jupiter's magnetic field is large and powerful. It traps particles from the solar wind to form powerful radiation belts. The rapid rotation of the planet forces the slightly inclined magnetic field to wobble up and down as the planet rotates. Earth's magnetosphere and radiation belts are shown to scale.

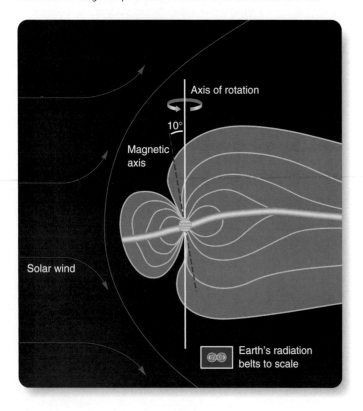

Axis of rotation

10°

Magnetic axis

Solar wind

Earth's radiation belts to scale

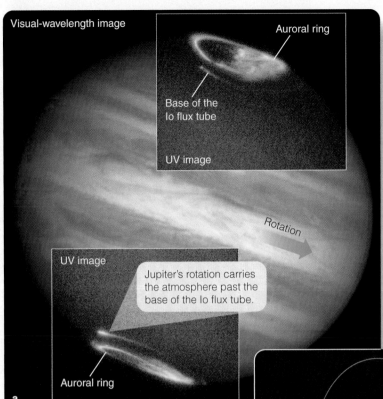

Visual-wavelength image

Auroral ring

Base of the
Io flux tube

UV image

Rotation

UV image

Jupiter's rotation carries
the atmosphere past the
base of the Io flux tube.

Auroral ring

a

■ **Active Figure 23-4**

Jupiter's huge magnetic field funnels energy from the solar wind down to form rings of auroras around its magnetic poles, which are tipped some distance from the rotational poles. The same thing happens on Earth. The Io flux tube connects the small moon Io to the planet and carries a powerful electric current that creates spots of auroras where it touches the planet's atmosphere. (Jupiter: John Clarke, University of Michigan, NASA; Flux tube: NASA)

Ace⊙Astronomy™ Log into AceAstronomy and select this chapter to see the Active Figure "Explorable Jupiter" and observe Jupiter at different wavelengths.

The Io flux tube is an
electric circuit connecting
Io to Jupiter's magnetic field.

Visual

Io

UV

b

is brighter near the center of the disk because the sunlight shines nearly straight down and the cloud layers look brighter.

Study **Jupiter's Atmosphere** on pages 560–561, and notice four important ideas:

1 The atmosphere is hydrogen rich, and the clouds are confined to a shallow layer.

2 Notice how the cloud layers lie at certain temperatures within the atmosphere where ammonia (NH_3), ammonium hydrosulfide (NH_4SH), and water (H_2O) can condense.

3 Also notice how the belt–zone circulation is related to the high- and low-pressure areas you find on Earth.

4 Finally, the major spots on Jupiter, although they are only circulating storms, can remain stable for decades or even centuries.

The circulation in Jupiter's atmosphere is not totally understood. Observations made by the Cassini spacecraft as it raced past Jupiter on its way to Saturn revealed that the dark belts contain small rising storm systems too small to see in previous images. Evidently, the general circulation usually attributed to the belts and zones is much more complex when it is observed in more detail. Further understanding of the small-scale motions in Jupiter's atmosphere may have to await future planetary probes to Jupiter.

The highly complex spacecraft that have visited Jupiter have produced a new understanding of the giant planet, but don't confuse the technology of spacecraft with the science of comparative planetology (**Window on Science 23-1**). Science is about understanding nature, and Jupiter is an entirely new kind of planet in your study. In fact, Jupiter has a feature that you did not find among the terrestrial planets. Jupiter has a ring.

Ace⊙Astronomy™ Log into AceAstronomy and select this chapter to see Astronomy Exercise "Convection and Turbulence." Study the influence that rotation and temperature have on this hypothetical planet.

Jupiter's Ring

Astronomers have known for centuries that Saturn has rings, but it was not until 1979, when the Voyager 1 spacecraft sent back photos, that Jupiter's ring was discovered. Less than 100 times as bright as Saturn's rings, the ghostly ring around Jupiter is a puzzle.

Jupiter's Atmosphere

1 You probably won't ever visit Jupiter's atmosphere. Its cloud layers are deathly cold, and the deeper layers that are warmer have a crushingly high pressure. There is no free oxygen to breathe; the gases are roughly three-quarters hydrogen and a quarter helium, plus traces of water vapor, methane, ammonia, and similar molecules. Traces of sulfur and molecules containing sulfur probably make it smell bad. Of course, Jupiter has no surface, so there isn't even a place to stand. Jupiter is a nice planet to look at, but it's not a place to visit.

Belts are dark bands of clouds.

Shadow of Europa

Zones are bright bands of clouds.

Jupiter's moon Europa

NASA/JPL/Univ. of AZ

NASA

1a The only spacecraft to enter Jupiter's atmosphere was the Galileo probe. Released from the Galileo spacecraft, the probe entered Jupiter's atmosphere in December 1995. It parachuted through the upper atmosphere of clear hydrogen, released its heat shield, and then fell through Jupiter's stormy atmosphere until it was crushed by the increasing pressure.

Jupiter's atmosphere is a very thin layer of turbulent gas above the liquid interior. It makes up only about 1 percent of the radius of the planet.

Lightning bolts are common in Jupiter's turbulent clouds.

Hughes Aircraft Co

The Great Red Spot at right is a giant circulating storm in one of the southern zones. It has lasted at least 300 years since astronomers first noticed it after the invention of the telescope. Smaller spots are also circulating storms.

NASA/JPL

2 The visible clouds on Jupiter are composed of ammonia crystals, but models predict that deeper layers of clouds contain ammonia hydrosulfide crystals, and deeper still lies a cloud layer of water droplets. These compounds are normally white, so planetary scientists think the colors arise from small amounts of other molecules formed by lightning or by sunlight.

If you could put thermometers in Jupiter's atmosphere at different levels, you would discover that the temperature rises below the uppermost clouds.

Far below the clouds, the temperature and pressure climb so high the gaseous atmosphere merges gradually with the liquid hydrogen interior and there is no surface.

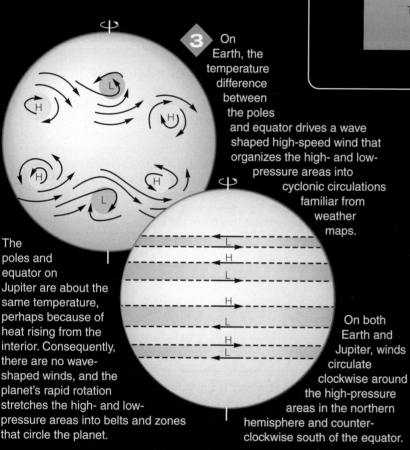

Clear hydrogen atmosphere

Ammonia
Ammonia hydrosulfide
Water

To liquid interior

Temperature (°F)
−300 −200 −100 0 100 212

Altitude (km)

Temperature (K)

Log into Ace Astronomy and select this chapter to see the **Ace ◒ Astronomy**™ Active Figure called "Planetary Atmospheres." Notice the temperatures at which the cloud layers form.

3 On Earth, the temperature difference between the poles and equator drives a wave shaped high-speed wind that organizes the high- and low-pressure areas into cyclonic circulations familiar from weather maps.

The poles and equator on Jupiter are about the same temperature, perhaps because of heat rising from the interior. Consequently, there are no wave-shaped winds, and the planet's rapid rotation stretches the high- and low-pressure areas into belts and zones that circle the planet.

On both Earth and Jupiter, winds circulate clockwise around the high-pressure areas in the northern hemisphere and counter-clockwise south of the equator.

Zones are brighter than **belts** because rising gas forms clouds high in the atmosphere, where sunlight is strong.

Belt   Zone

Altitude

North   Equator

4 The two white ovals here are counterclockwise, high-pressure weather systems that have been visible in Jupiter's southern hemisphere since they formed in the 1930s. These two huge storms merged into one spot in February 1998. The pear-shaped circulation visible between the two storms vanished during the merger.

Features in Jupiter's atmosphere may be stable for decades or centuries, but even the Giant Red Spot may someday vanish.

NASA/JPL/CalTech

Science, Technology, and Engineering

Sending spacecraft to another world is expensive, and it may seem pointless when that world seems totally hostile to human life. What practical value is there in sending a space probe to Jupiter? To resolve that question, you need to consider the distinction between science, technology, and engineering.

Science is nothing more than the logical study of nature, and the goal of science is understanding. Although much scientific knowledge proves to have tremendous practical value, the only goal of science is a better understanding of how nature works. Technology, in contrast, is the practical application of scientific knowledge to solve a specific problem. People trying to find a faster way to paint automobiles might use all the tools and techniques of science, but if their goal has some practical outcome, you should more properly call it technology rather than science. Engineering is the most practical form of technology. An engineer is likely to use well-understood technology to find a practical solution to a problem.

Of course, there are situations in which science and technology blur together. For example, humanity has a practical and tragic need to solve the AIDS problem; we need a cure and a preventative. Unfortunately, scientists don't understand the AIDS virus itself or viruses in general well enough to design a simple solution, and consequently much of the work involves going back to basic science and trying to better understand how viruses interact with the human body. Is this technology or science? It is hard to decide.

People often describe science that has no known practical value as basic science or basic research. The exploration of worlds such as Jupiter would be called basic science, and it is easy to argue that basic science is not worth the effort and expense because it has no known practical use. Of course, the problem is that no one has any way of knowing what knowledge will be of use until that knowledge is acquired. In the middle of the 19th century, Queen Victoria is supposed to have asked physicist Michael Faraday what good his experiments with electricity and magnetism were. He answered, "Madam, what good is a baby?" Of course, Faraday's experiments were the beginning of the electronic age. Many of the practical uses of scientific knowledge that fill your world—digital electronics, vaccines, plastics—began as basic research. Basic scientific research provides the raw materials that technology and engineering use to solve problems.

Exploring other worlds is of cultural value. It helps us as humans understand ourselves. (NASA/JPL/Space Science Institute)

Basic scientific research has yet one more important use that is so valuable it seems an insult to refer to it as merely practical. Science is the study of nature, and as you learn more about how nature works, you learn more about what your existence in this universe means. The seemingly impractical knowledge gained from space probes visiting other worlds tells you about your own planet and your own role in the scheme of nature. Science tells us where we are and what we are, and that knowledge is beyond value.

What is it made of? Why is it there? A few simple observations will help you solve some of these puzzles.

Saturn's rings are made of bright ice chunks, but the particles in Jupiter's ring are very dark and reddish. You can conclude that the ring is rocky rather than icy.

You can also conclude that the ring particles are mostly microscopic. Photos of the ring show that it is very bright when illuminated from behind (■ Figure 23-5)—it is scattering light forward. Efficient **forward scattering** occurs when particles have diameters roughly the same as the wavelength of light, a few millionths of a meter. Large particles do not scatter light forward, so a ring filled with basketball-size particles would look dark when illuminated from behind. Forward scattering tells you that the ring is made of particles about the size of those in cigarette smoke.

The location of these particles near the planet is understandable. They orbit inside the **Roche limit,** the distance from a planet within which a moon cannot hold itself together by its own gravity. If a moon stays far from its planet, then the moon's gravity will be much greater than the tidal forces caused by the planet, and the moon will be able to hold itself together. If, however, the planet's moon comes inside the Roche limit, the tidal forces overcome its gravity and pull the moon apart. The International Space Station can orbit inside Earth's Roche limit because it is held together by bolts and welds, but a moon held together by its own gravity cannot survive inside a planet's Roche limit. If a planet and its moon have similar densities, the Roche limit is 2.44 planetary radii. Jupiter's main ring has an outer radius of 130,000 km (1.8 planetary radii) and lies inside the Roche limit, as do the rings of Saturn, Uranus, and Neptune.

Now you can understand the dust near Jupiter. If a dust speck gets knocked loose from a small moon inside the Roche limit, the moon's gravity cannot hold the dust speck. And the billions of dust specks in the ring can't pull themselves together to make a new moon because of tidal forces inside the Roche limit.

You can also be sure that the ring particles are not old. The pressure of sunlight and the powerful magnetic field alter the orbits of the particles, and they gradually spiral into the planet. Images show faint ring material extending down toward the cloud tops, and this is evidently dust specs spiraling into the planet. Dust is also lost from the ring as electromagnetic effects force it out of the plane of the ring to form a low-density halo above and below the ring (Figure 23-5b). Another reason the

■ **Active Figure 23-5**

(a) The main ring of Jupiter, illuminated from behind, glows brightly in this visual-wavelength image made by the Galileo spacecraft from within Jupiter's shadow. (b) Digital enhancement and false color reveal the halo of ring particles that extends above and below the main ring. The halo is just visible in part a. (NASA)

Ace☯Astronomy™ Log into AceAstronomy and select this chapter to see the Active Figure "Roche Limit." Observe the powerful influence of tides on objects passing near planets.

Visual-wavelength images

ring particles can't be old is that the intense radiation around Jupiter will grind the dust specks down to nothing in a century or so. That means the rings seen today can't be material left over from the formation of Jupiter.

The rings of Jupiter must be continuously resupplied with new dust. Small moons that orbit near the outer edge of the rings may lose dust particles as they are hit by micrometeorite impacts. Observations made by the Galileo spacecraft show that the main ring is densest at its outer edge where the small moon Adrastea orbits and that another small moon, Metis, orbits inside the ring. Galileo images reveal much fainter rings, called the **gossamer rings,** extending twice as far from the planet as the main ring. These gossamer rings are most dense at the orbits of two small moons, Amalthea and Thebe, again providing evidence that the dust is being blasted into space by impacts on the moons.

Besides supplying the rings with particles, it is possible that the moons help confine the ring particles and keep them from spreading outward. You will find this an important process in planetary rings when you study the rings of Saturn later in this chapter.

Your exploration of Jupiter reveals that it is much more than just a big planet. It is the gravitational and magnetic center of an entire community of objects. Occasionally the community suffers an intruder.

Comet Impact on Jupiter

Comets are very common in the solar system, and Jupiter, because of its strong gravity, probably gets hit by comets more often than most planets. But no one had ever seen it happen until 1994, when fragments from a disrupted comet smashed into Jupiter.

The comet, named Shoemaker–Levy 9 after its discoverers, Eugene and Carolyn Shoemaker and David H. Levy, was captured into orbit around Jupiter sometime in the past. In 1992, it passed inside Jupiter's Roche limit and was pulled into at least 21 pieces that looped out away from Jupiter in long elliptical orbits. At this point, the objects were discovered on a photograph. Drawn out into a long strand of small comets, the pieces fell back and slammed into Jupiter over a period of six days in July 1994 (■ Figure 23-6).

The fragments were no bigger than a kilometer or so in diameter and contained a fluffy mixture of rock and ices. They hit Jupiter at speeds of about 60 km/s and released energy equivalent to a few million megatons of TNT. (At the height of the Cold War, all the nuclear weapons on Earth totaled about 80,000 megatons. The Hiroshima bomb was only 0.15 megaton.)

The impacts occurred just over Jupiter's horizon as seen from Earth, but they produced fireballs almost 3000 km high, and the rapid rotation of Jupiter brought the impact points within sight of Earth only 15 minutes later. Infrared telescopes easily spotted the glowing scars where the comets hit; and, as they cooled, the impact sites became dusty, dark spots at visible wavelengths as shown in Figure 23-6. Visible through even small telescopes, some of these dark smudges were larger than Earth.

The comet impact was an astonishing spectacle, but what does it reveal about Jupiter? In fact, the impact was revealing in two ways. First, astronomers used the impacts as probes of Jupiter's atmosphere. By making assumptions about the nature of Jupiter's atmosphere and by using the most powerful computers, astronomers created models of a high-velocity projectile penetrating into Jupiter's upper atmosphere. By comparing the observed impacts with the models, astronomers were able to fine-tune the models to better represent Jupiter's atmosphere. This method of comparing

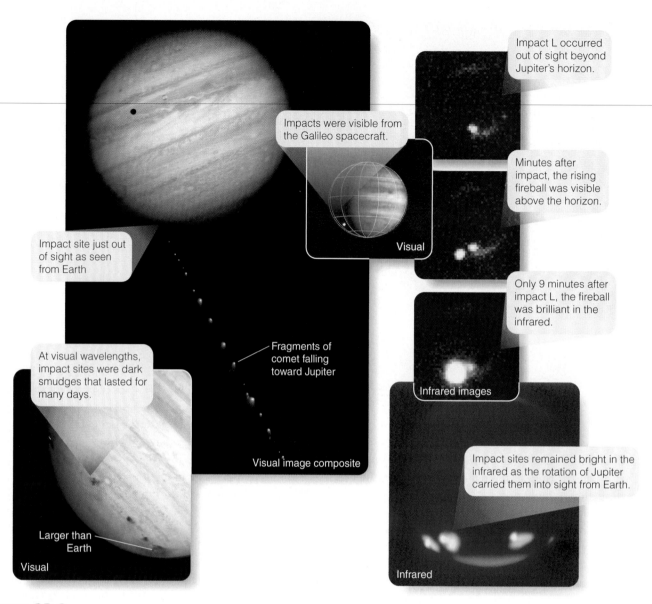

■ Figure 23-6

In 1994, fragments of Comet Shoemaker–Levy 9 struck Jupiter. Although these were the first comet impacts ever seen by Earth's inhabitants, such events are probably common occurrences in planetary systems. (Composite and visual images: NASA; IR sequence: Mike Skrutskie; IR image: University of Hawaii)

models with reality is a critical part of science; therefore, the impacts on Jupiter resulted in a better understanding of the nature of Jupiter's atmosphere.

Second, the spectacle is revealing in a more general way. It reminds you that planets are sometimes hit by large objects such as asteroids and the heads of comets. Jupiter probably is hit every century or so. In 1690, the Italian astronomer Cassini observed a dark spot that appeared on Jupiter and changed in a pattern over a period of days. Astronomers now recognize that pattern as the scar of a comet impact. You have seen evidence for large impacts on Earth, the moon, Venus, Mercury, and Mars, and you will see further evidence as you continue your exploration of the solar system.

Comets hit planets all the time. The impact in 1994 was just the first such event to occur since the rise of modern astronomy. You will see in Chapter 25 that a comet or asteroid impact on Earth may have changed the climate and killed the dinosaurs. A solar system is a dangerous place to put an inhabited planet.

The History of Jupiter

Your goal in studying any planet is to be able to tell its story—to describe how it got to be the way it is. While you can understand part of the story of Jupiter, there is still much to learn.

If the solar nebula theory for the origin of the solar system is correct, then Jupiter formed from the colder gases of the outer solar nebula, where ices were able to condense. Thus, Jupiter grew rapidly and eventually became massive enough to trap hydrogen and helium gas directly from the solar nebula. The denser material, such as rock and metal, sank to the center to form a core, and Jupiter became a low-density world of liquid hydrogen surrounding a hot core of heavier elements.

In the interior of Jupiter, the hydrogen takes the form of liquid metallic hydrogen, which is a very good electrical conductor. The planet's rapid rotation, coupled with the outward flow of heat from its hot interior, drives a dynamo effect that produces a powerful magnetic field. That vast magnetic field traps high-energy particles from the solar wind to form intense radiation belts and auroras.

The rapid rotation and large size of Jupiter cause the weather patterns to take the form of belt–zone circulation. Heat flowing upward from the interior causes rising currents in the bright zones, and cooler gas sinks in the dark belts. As on Earth, winds blow at the margins of these regions, and large spots appear to be cyclonic disturbances.

Although the age of planet building is long past, Jupiter continues to be hit by meteorites and occasional comets, as do all the planets. Any debris left over from the formation of Jupiter would have been blown away long ago by the solar wind, so the dust trapped in Jupiter's thin ring must be young. It probably comes from meteorites hitting the innermost moons.

Your study of Jupiter has been challenging because Jupiter lacks any surface—it is difficult to imagine being there. Most of the surface features and processes you found on the terrestrial planets are missing on Jupiter, but, as the prototype of the Jovian worlds, it earns its place as the ruler of the solar system.

Building Scientific Arguments

How do astronomers know Jupiter is hot inside?
A scientific argument is a way to test ideas, and sometimes it is helpful to test even the most basic ideas. You know that something is hot if you touch it and it burns your fingers, but you can't touch Jupiter. You also know something is hot if it is glowing bright red—it is red hot. But Jupiter is not glowing red hot. You can tell that something is hot if you can feel heat when you hold your hand near it. That is, you can detect infrared radiation with your skin. In the case of Jupiter, you would need greater sensitivity than the back of your hand, but infrared telescopes reveal that Jupiter is a source of infrared radiation. It really is glowing in the infrared. Sunlight would warm it a little bit, but it is emitting 70 percent more infrared than it should. That means it must be hot inside. From models of the interior, astronomers conclude that the center must be about 30,000 K in order to heat the surface of the planet and make it glow in the infrared.

Astronomical understanding is usually based on simple observations, so build an argument to answer the following simple question. **How do astronomers know that Jupiter has a low density?**

■   ■   ■

Connections: The study of Jupiter is more meteorology than geology, but when you examine its moons, you find all the excitement of impact cratering and active volcanism. The moons have surfaces and are places you could go. Bring your spacesuit.

(23-3) Jupiter's Family of Moons

HOW MANY MOONS does Jupiter have? You will have to check the World Wide Web to get the latest figure. Astronomers are finding more and more small moons, and the count is now well over two dozen. Most of these moons are small and rocky, and many are probably captured asteroids. Four of the moons, those discovered by Galileo and now called the Galilean moons, are large and have interesting geologies (■ Figure 23-7).

■ **Figure 23-7**

The Gallilean moons of Jupiter from left to right are Io, Europa, Ganymede, and Callisto. The circle shows the size of Earth's moon. (NASA)

Visual-wavelength images

Size of Earth's moon

Your study of the moons of Jupiter will illustrate three important principles in comparative planetology. First, a body's composition depends on the temperature of the material from which it formed. This is illustrated by the prevalence of ice as a building material in the outer solar system, where sunlight is weak. The second principle is that cratering can reveal the age of a surface. Finally, you will see that internal heat has a powerful influence over the geology of these larger moons.

Callisto: The Ancient Face

The outermost of Jupiter's four large moons, Callisto is half again as large in diameter as Earth's moon. Like all of Jupiter's larger satellites, Callisto is tidally locked to its planet, keeping the same side forever facing Jupiter. From its gravitational influence on passing spacecraft, astronomers can calculate Callisto's mass, and dividing by its volume shows that its density is 1.79 g/cm³. Ice has a density of about 1 and rock 2.5 to 4 g/cm³, so Callisto must be a mixture of rock and ice.

Images from the Voyager and Galileo spacecraft show that the surface of Callisto is dark, dirty ice heavily pocked with craters (■ Figure 23-8). Old, icy surfaces in the solar system become dark because of dust added by meteorites and because meteorite impacts vaporize water, leaving the rock in the ice behind to form a dirty crust. You may have seen the same thing happen to a city snowbank. As the snow evaporates over a few days, the crud in the snow is left behind to form a dirty rind. Break through that dirty surface, and the snow is much cleaner underneath.

Spectra of the surface show that it is mostly a 50/50 mix of ice and rock, but some areas are ice free. Nevertheless, the slumped

■ **Figure 23-8**

The dark surface of Callisto is dirty ice marked by craters in these visual-wavelength images. The youngest craters look bright, because they have dug down to cleaner ice. Valhalla is the 4000-km-diameter scar of a giant impact feature, one of the largest in the solar system. Valhalla is so large and old that the icy crust has flowed back to partially heal itself, and the outer rings of Valhalla are shallow troughs marking fractures in the crust. (NASA)

shape of craters suggests that the outer 10 km is mostly frozen water; ice isn't very strong, so big piles of it tend to slump under their own weight. The disagreement between the spectra and the shapes of craters can be understood when you recall that the spectra contain information about the outer 1 mm of the surface, which can be quite dirty, while the shapes of craters tell you about the outermost 10 km, which could be rich in ice.

Delicate measurements of the shape of Callisto's gravitational field were made by the Galileo spacecraft, and they show that Callisto has never fully differentiated to form a dense core and a lower-density mantle. Its interior is a mixture of rock and ice. This is consistent with the observation that it has only a weak magnetic field of its own. A strong magnetic field could be generated by the dynamo effect in a molten convecting core, and Callisto has no core. It does, however, interact with Jupiter's magnetic field in a way that suggests it has a layer of liquid water roughly 10 km thick about 100 km below its icy surface. Slow radioactive decay in its interior may provide enough heat to keep this layer of water from freezing.

Ganymede: A Hidden Past

The next Galilean moon inward is Ganymede, larger than Earth's moon, larger than Mercury, and over three-quarters the diameter of Mars. Its density is 1.9 g/cm³, and its influence on the Galileo spacecraft reveals that it has a rocky core, an ice-rich mantle, and a crust of ice 500 km thick. It may even have a small iron core at its center. It is large enough for radioactive decay to have melted its interior when it formed, allowing iron to drain to its center.

The surface hints at an active past. Although a third of the surface is old, dark, and cratered, the rest is marked by bright parallel grooves. Because this bright **grooved terrain** (■ Figure 23-9a) contains fewer craters, it must be younger.

Observations show that the bright terrain was produced when the icy crust broke and water flooded up from below. Some low-lying regions are smooth and appear to have been flooded by water. Spectra reveal concentrations of salts such as those that would be left behind by the evaporation of mineral-rich water. Also, some features in or near the bright terrain appear to be calderas formed when subsurface water drained away and the surface collapsed (Figure 23-9b).

The Galileo spacecraft found that Ganymede has a magnetic field about 10 percent as strong as Earth's. It even has its own magnetosphere inside the larger magnetosphere of Jupiter. Theoretical calculations find it difficult to produce a magnetic field in a water-rich mantle, and there does not appear to be enough heat in Ganymede for it to have a molten metallic core. Astronomers wonder if its magnetic field is left over from a time when it was hotter and more active.

Ganymede's magnetic field fluctuates with the 10-hour period of Jupiter's rotation. The rotation of the planet sweeps its tilted magnetic field past the moon, and the two fields interact. That interaction reveals that the moon has a layer of liquid water

■ Figure 23-9

(a) This color-enhanced image of Ganymede shows the frosty poles at top and bottom, the old dark terrain, and the brighter grooved terrain. (b) A band of bright terrain runs from lower left to upper right, and a collapsed area, a possible caldera, lies at the center in this visual-wavelength image. Calderas form where subsurface liquid has drained away, and the bright areas do contain other features associated with flooding by water. (NASA)

about 170 km (100 mi) below its surface. The water layer may be about 5 km (3 mi) thick. It is possible that the water layer was thicker and closer to the surface long ago when the interior of the moon was warmer. That might explain the flooding that appears to have helped form the bright grooved terrain.

■ Figure 23-10

Two effects on planetary satellites. (a) Tidal heating occurs when changing tides cause friction within a moon. (b) The focusing of meteoroids exposes satellites in small orbits to more impacts than satellites in larger orbits receive.

Ganymede orbits rather close to a massive planet, and that exposes it to two processes that many worlds never experience. **Tidal heating,** the frictional heating of a body by changing tides (■ Figure 23-10a) could have heated Ganymede's interior and added to the heat generated by radioactive decay. In a circular orbit, a moon experiences no tidal heating; but, at some point in the past, Ganymede could have interacted with the other moons and entered a slightly elliptical orbit. Jupiter's gravity would have deformed the moon. As Ganymede followed its elliptical orbit, tides would have flexed it, and friction would have heated it. Such an episode of tidal heating might have been enough to drive a dynamo to produce a magnetic field and break the crust to make the grooved terrain.

The second process that affects Ganymede exposes it to more than its share of meteorite impacts. Because massive planets like Jupiter draw debris inward, a moon orbiting near a massive planet should be struck by many meteorites (Figure 23-10b). Nevertheless, the bright terrain on Ganymede has few craters and appears to be only about 1 billion years old.

Europa: A Hidden Ocean

The next Galilean moon inward is Europa, a bit smaller than Earth's moon. Europa has a density of about 3 g/cm³, so it must be mostly rock and metal. Yet its surface is ice.

Europa lies closer to Jupiter than Ganymede, so it should be exposed to more meteorite impacts than Callisto or Ganymede, but the icy crust of Europa is almost free of craters. A recent crater such as Pwyll is bright, but most are hardly more than blemishes

Figure 23-11

(a) The icy surface of Europa is shown here in natural color. Many faults are visible on its surface, but very few craters. The bright crater is Pwyll, a young impact feature. (b) This circular bull's-eye is 140 km in diameter. It is the remains of an impact by an object about the size of a mountain. Notice the younger cracks and faults that cross the older impact feature. (c) Like icebergs on the Arctic Ocean, blocks of crust on Europa appear to have floated apart. Spectra show that the blue ice is stained by salts such as those that would be left behind by mineral-rich water welling up from below and evaporating. White areas are ejecta from the impact that formed Pwyll. (NASA)

in the ice (■ Figures 23-11a and b). Evidently the surface of Europa is active and erases craters almost as fast as they form. The number of impact scars on Europa suggests that the average age of its surface is only 10 million years, and that isn't very old at all. Other signs of activity include long cracks in the icy crust and regions where the crust has broken into sections that have moved apart as if they were icebergs floating on water (Figure 23-11c).

Even Europa's clean face tells you its surface is young. The albedo of the surface is 0.69, meaning that it reflects 69 percent of the light that hits it. This high albedo is produced by clean ice. You have discovered that old, icy surfaces tend to be very dark. Europa's high albedo must mean the surface is active with fresh ice covering older surfaces.

Europa is too small to have retained much heat from its formation or from radioactive decay, and the Galileo spacecraft found that Europa has no magnetic field of its own. It cannot have a molten conducting core. Tidal heating, however, is important for Europa and apparently provides enough heat to keep the little moon active. In fact, the curving cracks in its crust reveal the shape of the tidal forces that flex it as it orbits Jupiter.

Figure 23-12

The gravitational influence of Europa on the passing Galileo spacecraft shows that the little moon has differentiated into a dense core and rocky mantle. Magnetic interactions with Jupiter show that it has a liquid-water ocean below its icy crust. Heat produced by tidal heating could flow outward as convection in such an ocean and drive geological activity in the icy crust. (NASA)

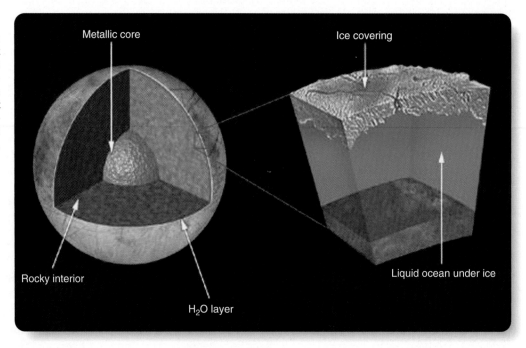

These enhanced-color images of volcanic features on Io were produced by combining visual and near-infrared images and digitally enhancing the color. To human eyes, most of Io would look pale yellow and light orange. (NASA)

Plume from volcano Pillan Patera rises 140 km.

Plume from volcano Prometheus

Shadow of plume

Visual + infrared images

Five months after the previous image, a new volcano has emerged.

Volcano Pele

Debris ejected from Pele

Volcano Pillan Patera

Hot lava at front of advancing lava flows

Volcanic caldera

Lava curtain erupting through a fault

50 km

Caldera Culann Patera has produced multiple lava flows.

If you hiked on Europa with a compass in your hand, you would be in big trouble. Europa has no magnetic field of its own, but Jupiter rotates rapidly and drags its field past the little moon. That induces a fluctuating magnetic field in Europa that would make your compass wander uselessly. Europa's interaction with Jupiter's magnetic field reveals the presence of a liquid-water ocean lying about 15 km (10 mi) below the icy surface. The ocean may be as deep as 100 miles (■ Figure 23-12) and could contain twice as much water as all the oceans on Earth. It would be rich in dissolved minerals, which would probably make it taste really bad. But those minerals would make the water a good electrical conductor and allow it to interact with Jupiter's magnetic field. No one knows what might be swimming through such an ocean, and many scientists hope for a future mission to Europa to drill through the ice crust and sample the ocean below for signs of life.

Tidal heating makes Europa geologically active. Apparently, rising currents of water can break through the icy crust or melt patches. Many of the cracks show evidence that they have spread apart and that fresh water has welled up and frozen between. In other regions, compression in the crust is revealed by networks of faults and low ridges. Compression on Earth pushes up mountain ranges, but no such ranges appear on Europa. The icy crust isn't strong enough to support ridges higher than a kilometer or so.

Orbiting deep inside Jupiter's radiation belts, Europa is bombarded by high-energy particles that damage the icy surface.

Water molecules are freed and broken up to be dispersed into a doughnut-shaped cloud spread round Jupiter and enclosing Europa's orbit. Flying past Jupiter in 2002, the Cassini spacecraft was able to image this cloud of excited gas. Europa's gas cloud serves as a warning that moons orbiting deep inside a massive planet's radiation belts are exposed to a form of erosion that is entirely lacking on Earth's moon.

Io: Bursting Energy

Geological activity is driven by heat flowing out of a planet's interior, and nothing could illustrate this principle better than Io, the innermost of Jupiter's Galilean moons. Photographs from the Voyager and Galileo spacecraft show no impact craters at all—surprising considering Jupiter's power to focus meteoroids inward (Figure 23-10b). No subtlety is needed to explain the missing craters. Over 150 active volcanoes are visible on Io's surface, blasting sulfurous gas and ash out over the surface to bury any newly formed craters (■ Figure 23-13). Unlike the dead or dormant outer Galilean moons, Io is bursting with energy.

Spectra reveal that Io has a tenuous atmosphere of gaseous sulfur and oxygen, but those gases can't be permanent. The erupting volcanoes pour out about one ton of gases per second; but, because of Io's low escape velocity, the gases leak into space easily. Also, any gas atoms that become ionized are swept away by Jupiter's rapidly rotating magnetic field. The ions produce a cloud of sulfur and sodium ions in a torus (a doughnut shape) enclosing Io's orbit (■ Figure 23-14).

You would need a very good spacesuit to visit the surface of Io. The temperature and atmospheric pressure would not be a big problem. Because of the continuous volcanism and the sulfurous gases, Io's thin atmosphere is dirty and probably smelly with sulfur. In fact, the reddish color of the small inner moon Amalthea may be caused by sulfur pollution escaping from Io. Your real problem walking across the surface of Io would be radiation. Io is deep inside Jupiter's magnetosphere and radiation belts; and, unless your spacesuit had astonishingly impressive shielding, the radiation would be lethal. Io, like Venus, may be a place that humans will never visit with any ease.

You can use basic observations to deduce the nature of Io's interior. From the density, 3.55 g/cm³, you can conclude that it is mostly rock. Spectra reveal no trace of water at all, so there is no ice on Io. It is the driest world in our solar system. The oblateness of Io caused by its rotation and by the slight distortion produced by Jupiter's gravity gives astronomers clues to its interior. Model calculations suggest it contains a modest core of metals and a deep rocky mantle.

The colors of Io have been compared to those of a badly made pizza. Certainly the reds, oranges, and browns of Io are caused by sulfur and sulfur compounds, but the crust is true rock. Infrared measurements show that volcanoes on Io erupt lava with a temperature over 1500°C (2700°F), about 300°C hotter than lavas on Earth. Sulfur on Io would boil at only 550°C. Also, a few isolated mountains exist that are as high as 18 km, twice the height of Mount Everest. Sulfur is not strong enough to support such high mountains. The lava flows and most of the crust must be silicate rock and not sulfur.

Volcanism is continuous on Io. Plumes come and go over periods of months, but some volcanic vents, such as Pele, have been active since the Voyager spacecraft first visited Io in 1979 (Figure 23-13). Earth's volcanoes eject lava and ash because of water dissolved in the lava. As rising lava reaches Earth's surface, the sudden decrease in pressure allows the water to come out of solution in the lava. It is like popping the cork on a bottle of champagne. The water flashes into vapor and blasts material out of the volcano, a process

■ **Figure 23-14**

Sulfur venting from Io is trapped in Jupiter's magnetic field to form a torus around the moon's orbit. Ions trapped in Jupiter's magnetosphere form radiation belts that enclose all four of the Galilean satellites. A fainter torus of water atoms is located around the orbit of Europa. (Sulfur cloud: NASA/JPL/Bruce A. Goldberg; Ion and neutral image: NASA/JPL/Johns Hopkins Univ. Applied Physics Lab; Radio: NASA/JPL)

Visible + ultraviolet

Jupiter   Io

Io sulfur cloud

Ion and neutral camera

Io plasma torus deep inside Jupiter's magnetic field

The most intense radiation belts are located near Jupiter.

Jupiter's radiation belts wobble as the planet rotates because the magnetic field is inclined to the axis of rotation.

Visual + radio

that was responsible for the Mount St. Helens explosion in 1980. But Io is dry. Its volcanoes appear to be powered by sulfur dioxide dissolved in the magma. When the pressure is released, the sulfur dioxide boils out of solution and blasts gas and ash high above the surface in plumes up to 500 km high. Ash falling back to the surface produces debris layers around the volcanoes, such as that around Pele in Figure 23-13. Whitish areas on the surface are thought to be frosts of sulfur dioxide.

Great lava flows are visible carrying molten material downhill, where it buries the surface under layer after layer. Sometimes lava bursts upward through faults to form long lava curtains, a form of eruption seen in Hawaii. Both of these processes are shown in Figure 23-13.

What powers Io? It is bursting with energy, but it is only five percent bigger than Earth's moon, which is cold and dead. Io is too small to have retained heat from its formation or to remain hot from radioactive decay. In fact, the energy blasting out of its volcanoes is about three times more energy than it could make by radioactive decay in its interior.

Io is heated by the same kind of tidal heating that is suspected in Ganymede's past and in Europa. Because Io is so close to Jupiter, tides are powerful and should have forced Io's orbit to become circular long ago. Io, however, has fallen in with a bad crowd. Io, Europa, and Ganymede are locked in an orbital resonance; in the time it takes Ganymede to orbit once, Europa orbits twice and Io four times. This gravitational interaction keeps the orbits slightly elliptical; and Io, being closest to Jupiter, suffers dramatic tides, with its surface rising and falling by about 100 m. Tides on Earth move the solid ground by only a few centimeters. The resulting friction in Io is enough to melt the interior and drive volcanism. In fact, the energy flowing outward is continually recycling Io's crust. Deep layers melt, are spewed out through the volcanoes, fall back to cover the surface, and are later covered themselves until they are buried so deeply that they are again melted.

Io is the most active world in our solar system because it lives so close to Jupiter. The other Galilean moons are more distant and have had less dramatic histories of tidal heating and geological activity.

The History of the Galilean Moons

Each time you have finished studying a world, you have tried to summarize its history. Now you have a system of four small worlds. Can you tell their story? To do that you need to draw on what you have learned about the moons, but also on what you have learned about Jupiter and the origin of the solar system (Chapter 19).

The minor moons of Jupiter are probably captured asteroids, but the Galilean moons seem to be primordial. That is, they formed with Jupiter. Also, they seem to be a set in that they share common features and their densities are related to their distance from Jupiter (■ Table 23-2). This suggests that the four moons

■ **Table 23-2 | The Galilean Satellites***

Name	Radius (km)	Density (g/cm³)	Orbital Period (days)
Io	1816	3.55	1.769
Europa	1563	3.04	3.551
Ganymede	2638	1.93	7.155
Callisto	2410	1.81	16.689

*For comparison, the radius of Earth's moon is 1738 km, and its density is 3.36 g/cm³.

formed in a disk-shaped nebula around Jupiter—a mini-solar nebula—in much the same way the planets formed from the solar nebula around the sun.

As Jupiter grew massive, it should have formed a hot, dense disk of matter around its equator. The moons could have condensed inside that disk with the innermost moons, Io and Europa, forming from rocky material and the outer moons, Ganymede and Callisto, incorporating more ices. This theory follows the condensation sequence that led to rocky planets forming near the sun and ice-rich worlds forming farther away.

There are objections to this theory. The disk around Jupiter would have been dense and hot, and moons would have formed rapidly, perhaps in only 1000 years. The heat of formation released as material fell into the moons would have heated them so hot they would have lost their water. Ganymede and Callisto are rich in water, and Callisto has never been hot enough to differentiate and form a core. Also, mathematical models show that moons orbiting in the dense disk would have swept up debris and lost orbital momentum; they would have spiraled into Jupiter within a century.

A newer theory proposes that Jupiter's early disk was indeed dense and hot and may have created moons, but those moons spiraled into the planet and were lost. Only later, as the disk grew thinner and cooler, did the present Galilean moons begin to form. Additional material may have dribbled slowly into the disk, but the moons must have formed slowly enough to retain their water and avoid spiraling into Jupiter.

You can understand the interiors of the moons by thinking about tidal heating. The moons formed slowly, over perhaps 100,000 years, and were not heated severely by infalling material. The inner moons, however, were cooked by tidal heating—possibly enhanced when an orbital resonance developed between Ganymede, Europa, and Io. The innermost moon, Io, was heated so much it lost all of its water, and Europa retained only a small amount. Ganymede was heated enough to differentiate but retained much of its water. Callisto, orbiting far from Jupiter and avoiding orbital resonances, was never heated enough to differentiate. The Galilean moons as they appear today seem to be the result of slow formation and tidal heating.

Io is bursting with energy, and the Galilean satellite system is bursting with history. As you learn to understand that history, you will learn more about how worlds form and evolve.

Building Scientific Arguments

What produces Io's internal heat?
Scientific arguments commonly draw on basic principles that are well understood. In this case, you understand that small worlds lose their internal heat quickly and become geologically inactive. Io is only slightly larger than Earth's moon, which is cold and dead, but Io is bursting with energy flowing outward.

Clearly, Io must have a powerful source of heat inside, and that heat source is tides. Io's orbit is slightly elliptical, so it is sometimes closer to Jupiter and sometimes farther away. That means that Jupiter's powerful gravity sometimes squeezes Io more than at other times, and the flexing of the little moon's interior produces heat through friction. Such tides would rapidly force Io's orbit to become circular, and then tidal heating would end and the planet would become inactive—except that the gravitational tugs of the other moons keep Io's orbit elliptical. Thus, it is the influence of its companions that keeps Io in such an active state.

Io has almost no impact craters, but Callisto has many. Build a new scientific argument drawing on a different principle. **What does the distribution of craters on the Galilean satellites tell you about their history?**

■   ■   ■

Connections: You have now applied the principles of comparative planetology not only to planets but also to moons. With the insight you have gained from Jupiter and its satellite system, you are ready to visit the most beautiful of the Jovian worlds.

(23-4) Saturn

SATURN HAS PLAYED SECOND FIDDLE to its own rings since Galileo first saw it in 1610. He didn't recognize the rings for what they are, but today astronomers know them as one of the great sights in the solar system. Nevertheless, Saturn itself, only slightly smaller than Jupiter (**Celestial Profile 8**), is a fascinating planet with a few mysteries of its own. Your exploration of Saturn can make use of the principles you have learned from Jupiter.

Planet Saturn

The basic characteristics of Saturn reveal its composition and interior. Only about a third of the mass of Jupiter and 16 percent smaller in radius, Saturn has an average density of 0.69 g/cm³. It

is less dense than water—it would float! Although Saturn rotates about as fast as Jupiter, it is almost twice as oblate, so you can conclude that it must not contain a large core of heavy elements. Spectra show that its atmosphere is rich in hydrogen and helium (see Table 23-1), and models predict that it is mostly liquid hydrogen.

Infrared observations show that Saturn is radiating 1.8 times as much energy as it receives from the sun, showing that heat is flowing out of its interior. It must be hot inside. In fact, it is too hot. It should have lost more heat since it formed. Astronomers suspect that helium in the liquid hydrogen interior is condensing into droplets and falling inward. The falling droplets, releasing energy as they pick up speed, heat the planet. This heating is similar to the heating produced when a star contracts.

You can learn more about Saturn's interior from its magnetic field. The Voyager spacecraft found that Saturn's magnetic field is about 20 times weaker than Jupiter's. It also has correspondingly weaker radiation belts. Models predict that the pressure inside Saturn, being weaker than that inside Jupiter, produces a smaller mass of liquid metallic hydrogen. Heat flowing outward must cause convection in this conducting core, and the rapid rotation drives a dynamo effect to produce the magnetic field. Unlike most magnetic fields, Saturn's field is not inclined to its axis of rotation, something you can see in ultraviolet images that show rings of auroras around Saturn's poles (■ Figure 23-15). This alignment

■ **Figure 23-15**

Auroras on Saturn occur in rings around the planet's magnetic poles. Because the magnetic field is not inclined very much to the axis of rotation, the rings occur nearly at the planet's geometrical poles. (Compare with Figure 23-4.) (NASA, ESA, J. Clarke, Boston Univ., and X. Levay, STScI)

Visual + ultraviolet

Auroras fluctuate day to day because of changes in the solar wind.

between the magnetic axis and the axis of rotation is peculiar and isn't understood.

The Cassini spacecraft arrived at Saturn 24 years after the Voyager spacecraft had flown past, and Cassini found that the rotation period of Saturn's magnetic field had slowed by 6 minutes. Surely the planet did not slow down, so planetary scientists suspect that the circulation in the core of the planet may have changed in some way.

Saturn's atmosphere is rich in hydrogen and contains belt–zone circulation, which appears to arise in the same way as that on Jupiter. Zones are higher clouds formed by rising gas, and belts are lower clouds formed by sinking gas.

But the clouds are not very distinct on Saturn (■ Figure 23-16a). Measurements from the Voyager and Cassini spacecraft indicate that Saturn's atmosphere is much colder than Jupiter's—something you would expect because Saturn is twice as far from the sun and receives only one-fourth as much solar energy per square meter. The clouds on Saturn form at about the same temperature as the clouds on Jupiter, but those temperatures are deeper in Saturn's cold atmosphere. Compare the cloud layers in Figure 23-16b with those on page 561. The atmospheres of Jupiter and Saturn are very similar once you remember that Saturn is colder.

One dramatic difference between Jupiter and Saturn concerns the winds. On Jupiter, winds bound each of the belts and zones, but on Saturn the pattern is not the same. Saturn has fewer such winds, but they are much stronger. The eastward wind at the equator of Saturn, for example, blows at 500 m/s (1100 mph), roughly five times faster than the eastward wind at Jupiter's equator. The reason for this difference is not clear.

Saturn's Rings

Looking at the beauty and complexity of Saturn's rings, an astronomer once said, "The rings are made of beautiful physics." You could add that the physics is actually rather simple, but the result is one of the most astonishing sights in our solar system.

In 1609, Galileo was the first to see the rings of Saturn; but, perhaps because of the poor optics in his telescopes, he did not recognize the rings as a disk. He drew Saturn as three objects—a central body and two smaller ones on either side. In 1659, Christian Huygens realized that the rings were a disk surrounding but not touching the planet.

Saturn's rings are governed by simple physics, but understanding what the rings are has demanded over a century of human ingenuity. In 1859, James Clerk Maxwell (for whom the large mountain on Venus is named) proved mathematically that solid rings would be unstable. Saturn's rings, he concluded, had to be made of particles. In 1867, Daniel Kirkwood demonstrated that gaps in the rings were caused by resonances with some of Saturn's moons. Spectra of the rings eventually showed that the particles were mostly water ice.

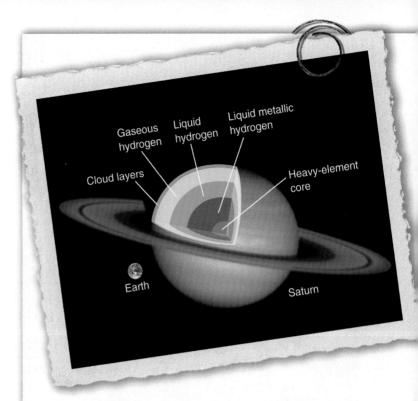

Density, oblatesness, and gravity measurements made by planetary probes allow planetary astronomers to model Saturn's interior. (NASA/STScI)

Celestial Profile 8: Saturn

Motion:

Average distance from the sun	9.5388 AU (14.27 × 10⁸ km)
Eccentricity of orbit	0.0560
Maximum distance from the sun	10.07 AU (15.07 × 10⁸ km)
Minimum distance from the sun	9.005 AU (13.47 × 10⁸ km)
Inclination of orbit to ecliptic	2°29'17"
Average orbital velocity	9.64 km/s
Orbital period	29.461 y (10,760 days)
Period of rotation	$10^h39^m25^s$
Inclination of equator to orbit	26°24'

Characteristics:

Equatorial diameter	120,660 km (9.42 $D_\oplus$)
Mass	5.69 × 10²⁶ kg (95.147 $M_\oplus$)
Average density	0.69 g/cm³
Gravity at base of clouds	1.16 Earth gravities
Escape velocity	35.6 km/s (3.2 $V_\oplus$)
Temperature at cloud tops	−180°C (−292°F)
Albedo	0.61
Oblateness	0.102

Personality Point:

The Greek god Cronus was forced to flee when his son Zeus took power. Cronus fled to Italy, where the Romans called him Saturn, protector of the sowing of seed. He was celebrated in a weeklong wild party called the Saturnalia at the time of the winter solstice in late December. Early Christians took over the holiday to celebrate Christmas.

(a) Saturn's belt–zone circulation is not very distinct at visible wavelengths. These images were recorded when Saturn's southern hemisphere was tipped toward Earth. (NASA and E. Karkoschka) (b) Because Saturn is colder than Jupiter, the clouds form deeper in the hazy atmosphere. Notice that the three cloud layers on Saturn form at about the same temperature as do the three cloud layers on Jupiter.

Ace◐Astronomy™ Log into AceAstronomy and select this chapter to see the Active Figure "Planetary Atmospheres" and compare the atmospheres of the planets.

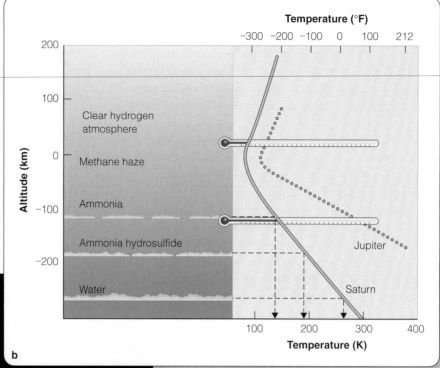

b

Ultraviolet

Visible

Infrared

a

Study **The Ice Rings of Saturn** on pages 576–577 and notice three things:

1 The rings are made up of billions of ice particles, each in its own orbit around the planet. But the rings can't be as old as Saturn. The rings must be replenished now and then by impacts on Saturn's icy moons or by the disruption of a small moon that wanders too close to the planet.

2 The gravitational effects of small moons can confine some rings in narrow strands or keep the edges of rings sharp. Moons can also produce waves in the rings that are visible as tightly wound ringlets.

3 The ring particles are confined in a thin plane spread among small moons and confined by gravitational interactions with larger moons. The rings of Saturn, and the rings of the other Jovian worlds, are created by and controlled by the planet's moons. Without the moons, there would be no rings.

Modern astronomers find simple gravitational interactions producing even more complex processes in the rings. Where particles orbit in resonance with a moon, the moon's gravity triggers spiral density waves—the same bit of physics that generates spiral arms in galaxies. The spiral density waves spread outward through the rings. If the moon follows an orbit that is inclined to the ring plane, the moon's gravity causes spiral bending waves—ripples extending above and below the ring plane—which spread inward. Both of these kinds of rings are shown in the inset ring image on page 577.

Many other processes occur in the rings. Specks of dust become electrically charged by sunlight, and Saturn's magnetic field lifts them out of the ring plane. Small moonlets imbedded in the rings produce gaps, waves, and scallops in the rings.

Particle size ranges from snowlike powder to large particles and aggregates of particles. The subtle colors of the rings must arise from contamination in the ice, and some areas have unusual compositions. Cassini's Division, for instance, is made up of particles richer in rock than is most of the ring. No one knows how these differences in composition arise, but they must be related to the way the rings formed and are replenished.

Like a beautiful flower, the rings of Saturn are controlled by many different natural processes. Observations from spacecraft

Who Pays for Science?

Science is an expensive enterprise, and that raises the question of payment. Some science has direct technological applications, but some basic science is of no immediate practical value. Who pays the bill?

In Window on Science 23-1, you saw that technology is of immediate practical use; consequently, business funds much of this kind of scientific research. Auto manufacturers need inexpensive, durable, quick-drying paint for their cars, and they find it worth the cost to hire chemists to study the way paint dries. Many industries have large research budgets; and some industries, such as pharmaceutical manufacturers, depend exclusively on scientific research to discover and develop new products.

If a field of research has immediate potential to help society, it is likely that the government will supply funds. Much of the research on public health is funded by government institutions such as the National Institutes of Health and the National Science Foundation. The practical benefit of finding new ways to prevent disease, for example, is well worth the tax dollars.

Basic science, however, has no immediate practical use. That doesn't mean it is useless, but it does mean that the practical-minded stockholders of a company will not approve major investments in such research. Digging up dinosaur bones, for instance, is very poorly funded because no industry can make a profit from the discovery of a new dinosaur. Astronomy is another field of science that has few direct applications, and consequently very little astronomical research is funded by industry.

The value of basic research is twofold. Discoveries that have no known practical use today may be critically important years from now. For that reason, society needs to continue basic research to protect its own future. But basic research, such as studying Saturn's rings or digging up dinosaur bones, is also of cultural value because it tells us what we are. Each person benefits in intangible ways from such research, and that is why society needs to fund basic research for the same reason it funds art galleries and national parks—to make everyone's life richer and more fulfilling.

Because there is no immediate financial return from this kind of research, it falls to government institutions and private foundations to pay the bill. The Keck Foundation has built two giant telescopes with no expectation of financial return, and the National Science Foundation has funded thousands of astronomy research projects for the benefit of society. Debates rage as to how much money is

Sending the Cassini spacecraft to Saturn cost each American 56¢ per year over the life of the project. (NASA)

enough and how much is too much, but, ultimately, funding basic scientific research is a public responsibility that society must balance against other needs. There isn't anyone else to pick up the tab.

---

such as Voyager and Cassini will reveal even more about the rings (**Window on Science 23-2**).

The History of Saturn

The farther you journey from the sun, the more difficult it is to understand the history of the planets. Any fully successful history of Saturn should explain its low density, its peculiar magnetic field, and its beautiful rings. Planetary scientists can't tell a fully successful story yet, but you can understand a few of the principles that led to the formation of Saturn and its rings.

Saturn formed in the outer solar nebula, where ice particles were stable. It grew rapidly and became massive enough to capture hydrogen and helium gas directly from the nebula. The denser elements sank to the middle to form a denser core, and the hydrogen formed a liquid mantle containing liquid metallic hydrogen. The outward flow of heat from the core is believed to drive convection currents in this mantle that, coupled with the rapid rotation of the planet, produce its magnetic field. Because Saturn is smaller than Jupiter, it has less liquid metallic hydrogen, and its magnetic field is weaker.

You might be tempted at first to suppose that the rings of Saturn are primordial, made of material left over from the formation of the planet. That seems unlikely, however. Saturn, like Jupiter, should have been very hot when it formed, and that heat would have vaporized and driven off any leftover material. Also, such a hot Saturn would have had a very distended atmosphere, which would have slowed ring particles by friction and caused the particles to fall into the planet.

Planetary rings do not seem to be stable over 4.6 billion years, so the ring material must be more recent. Saturn's beautiful rings may have been produced within the last 100,000 years. One suggestion is that a small moon or an icy planetesimal came within Saturn's Roche limit, and tides pulled it apart. At least some of the debris would settle into the ring plane. You will see in the next two chapters that icy planetesimals are common in the Kuiper Belt in the outer solar system. Another possibility is that a comet struck one of Saturn's moons. Because both comets and moons in the outer solar system are icy, such a collision would produce icy debris. Bright planetary rings such as Saturn's may be temporary phenomena, produced when fresh ice

The Ice Rings of Saturn

1 The brilliant rings of Saturn are made up of billions of ice particles ranging from microscopic specks to chunks bigger than a house. Each particle orbits Saturn in its own circular orbit. Much of what astronomers know about the rings was learned when the Voyager 1 spacecraft flew past Saturn in 1980, followed by the Voyager 2 spacecraft in 1981. The Cassini Spacecraft reached orbit around Saturn in 2004. From Earth, astronomers see three rings labeled A, B, and C. Voyager and Cassini images reveal over a thousand ringlets within the rings.

Saturn's rings can't be leftover material from the formation of Saturn. The rings are made of ice particles, and the planet would have been so hot when it formed that it would have vaporized and driven away any icy material. Rather, the rings must be debris from collisions between passing comets and Saturn's icy moons. Such impacts should occur every 10 million years or so, and they would scatter ice throughout Saturn's system of moons. The ice would quickly settle into the equatorial plane, and some would become trapped in rings. Although the ice may waste away due to meteorite impacts and damage from radiation in Saturn's magnetosphere, new impacts could replenish the rings with fresh ice. The bright, beautiful rings you see today may be only a temporary enhancement caused by an impact that occurred since the extinction of the dinosaurs.

Earth to scale

Visual-wavelength image

NASA

Encke's division

Cassini's division

A ring

B ring

C ring
The Crepe ring

As in the case of Jupiter's ring, Saturn's rings lie inside the planet's Roché limit where the ring particles cannot pull themselves together to form a moon.

Because it is so dark, the C ring has been called the Crepe ring, referring to the black, semitransparent cloth associated with funerals.

1a An astronaut could swim through the rings. Although the particles orbit Saturn at high velocity, all particles at the same distance from the planet orbit at about the same speed, so they collide gently at low velocities. If you could visit the rings, you could push your way from one icy particle to the next. This artwork is based on a model of particle sizes in the A ring.

The C ring contains boulder-size chunks of ice, whereas most particles in the A and B rings are more like golf balls, down to dust-size ice crystals. Further, C ring particles are less than half as bright as particles in the A and B rings. Cassini observations show that the C ring particles contain less ice and more minerals.

2 Because of collisions among ring particles, planetary rings should spread outward. The sharp outer edge of the A ring and the narrow F ring are confined by **shepherd satellites** that gravitationally usher straying particles back into the rings.

Some gaps in the rings, such as Cassini's Division, are caused by resonances with moons. A particle in Cassini's Division orbits Saturn twice for each orbit of the moon Mimas and three times for each orbit of Enceladus. On every other orbit, the particle feels a gravitational tug from Mimas and, on every third orbit, a tug from Enceladus. These tugs always occur at the same places in the orbit and force the orbit to become slightly elliptical. Such an orbit crosses the orbits of other particles, which results in collisions, and that removes the particle from the gap.

This image was recorded by the Cassini spacecraft looking up at the rings as they were illuminated by sunlight from above. Saturn's shadow falls across the upper side of the rings.

The F ring is clumpy and braided because of two shepherd satellites.

Pandora

Visual-wavelength image

F ring close up

Prometheus

F ring

Visual-wavelength images

Waves in the A ring

Encke's Division

Encke's Division is not empty. Note the ripples at the inner edge. A small moon orbits inside the division.

Saturn does not have enough moons to produce all of its ringlets by resonances. Many are produced by tightly wound waves, much like the spiral arms found in disk galaxies.

Cassini's Division

Encke's Division

A ring

This combination of UV images has been given false color to show the ratio of mineral material to pure ice. Blue regions such as the A ring are the purest ice, and red regions such Cassini's division are the dirtiest ice. How the particles become sorted by composition is unknown.

3 How do moons happen to be at just the right places to confine the rings? That puts the cosmic cart before the horse. The ring particles get caught in the most stable orbits among Saturn's innermost moons. The rings push against the inner moons, but those moons are locked in place by resonances with larger, outer moons. Without the moons, the rings would spread and dissipate.

Saturn's rings are a very thin layer of particles and nearly vanish when the rings turn edge-on to Earth. Although ripples in the rings caused by waves may be hundreds of meters high, the sheet of particles may be only a dozen meters thick.

Ultraviolet image

is scattered into the rings and then wasting away as the ice is gradually lost.

Why do the belts and zones on Saturn look so bland?
One of the most powerful tools of critical thought is simple comparison and contrast. You can make that the theme of this scientific argument by comparing and contrasting Saturn with Jupiter. In the atmosphere of Jupiter, the dark belts form in regions where gas sinks, and zones form where gas rises. The rising gas cools and condenses to form icy crystals of ammonia, which are visible as bright clouds. Clouds of ammonia hydrosulfide and water form deeper, below the ammonia clouds, and are not as visible. Saturn is twice as far from the sun as Jupiter, so sunlight is four times dimmer. (Remember the inverse square law from Chapter 5.) The atmosphere is colder, and gas currents do not have to rise as far to reach cold levels and form clouds. That means the clouds are deeper in Saturn's atmosphere than in Jupiter's atmosphere. Because the clouds are deeper, they are not as brightly illuminated by sunlight and look dimmer. Also, a layer of methane-ice-crystal haze high above the ammonia clouds makes the clouds even less distinct.

Now build a new argument comparing the ring systems. **How is Saturn's ring system similar to and different from Jupiter's ring system?**

■   ■   ■

Connections: In any contest, Saturn should win the ribbon for most beautiful planet. Its subtly marked disk and its delicate rings are both massive and graceful. Its extensive system of moons only adds to the mystery of this beautiful and curious world.

(23-5) Saturn's Moons

SATURN HAS OVER TWO DOZEN known satellites—far too many to examine individually—but these moons share characteristics common to icy worlds. Most of them are small and dead, but one is big enough to have an atmosphere and perhaps even oceans or lakes.

Titan

Saturn's largest satellite is a giant ice moon with a thick atmosphere and a mysterious surface. From Earth it is only a dot of light, and no detail is visible. Nevertheless, a few basic observations made from Earth can tell you a great deal about this strange world.

Titan's mass could be estimated from its influence on other moons, and its mass divided by its volume revealed that its density is 1.9 g/cm^3. Its uncompressed density (Chapter 19) is only

■ **Table 23-3 | Some Organic Compounds Detected on Titan**

C_2H_6	Ethane
C_2H_2	Acetylene
C_2H_4	Ethylene
C_3H_4	Methylacetylene
C_3H_8	Propane
C_4H_2	Diacetylene
HCN	Hydrogen cyanide
HC_3N	Cyanocetylene
C_2N_2	Cyanogen

1.2 g/cm^3. Although it must have a rocky core, it must also contain a large amount of ices. Titan is a bit larger than the planet Mercury and almost as large as Jupiter's moon Ganymede. Unlike these worlds, Titan has a thick atmosphere. Its escape velocity is low, but it is so far from the sun that it is very cold, and most gas atoms don't move fast enough to escape. (See Figure 22-13.) Methane was detected spectroscopically in 1944, as were various hydrocarbons beginning in about 1970 (■ Table 23-3), but most of Titan's air is nitrogen with only 2 to 3 percent methane.

When the Voyager 1 and Voyager 2 spacecraft few past Saturn in the early 1980s, their cameras could not penetrate Titan's hazy atmosphere (■ Figure 23-17). Measurements showed that the surface temperature was about 94 K ($-290°F$) and the surface atmospheric pressure was 50 percent greater than on Earth. Model calculations show that methane should condense from the atmosphere and fall as rain. Furthermore, sunlight converts methane (CH_4) into the gas ethane (C_2H_6) plus a collection of other organic molecules.* Some of these molecules produce the smoglike haze; and, as the smog particles gradually settle, they could deposit a layer of smelly, organic goo over the icy surface. This goo is exciting because such organic molecules may have been the precursors of life on Earth. You will examine this further in Chapter 26.

Before you imagine floundering through this gooey landscape, you should review the results of the Cassini spacecraft, which began exploring Saturn and its moons in 2005.

Infrared cameras and radar instruments on Cassini have been able to see through the hazy atmosphere. The surface is not a featureless ocean of methane, as some models had predicted. Rather, the surface consists of icy, irregular highlands and smoother dark areas (Figure 23-17). There are only a few craters, which suggests geological activity is erasing craters almost as fast as they are formed.

---

*Organic molecules are common in living things on Earth but do not have to be derived from living things. One chemist defined an organic molecule as "any molecule with a carbon backbone."

The Voyager spacecraft could not detect features on the surface of Titan because its atmosphere is filled with a haze of molecules and particles produced by sunlight breaking down methane. The Cassini spacecraft could image the surface at a number of wavelengths for which the atmosphere was transparent. (NASA/ESA/JPL/University of Arizona)

Visual-wavelength image

At visual wavelengths, Titan's hazy atmosphere hides its surface.

Suspected methane/ice volcano

Infrared images reveal detail on Titan's surface.

Infrared images

The Cassini spacecraft released a probe named Huygens, which parachuted down through the atmosphere of Titan and eventually landed on the surface. Huygens radioed back images of the surface as it descended under its parachute, and those images show dark drainage networks that lead into dark smooth areas (■ Figure 23-18). One astronomer said it looks like surf on a dark ocean rolling onto an icy beach, but there is no clear evidence of methane rain or liquid methane on the surface. Precipitation could wash the black goo off the highlands into the stream channels and lowlands, so that they would look smooth and dark even if all of the liquid had temporarily evaporated. If you visit Titan, you may get caught in a shower of methane rain, but it may not rain at your landing site for decades or even a century.

By the way, before you go to Titan, check your spacesuit for leaks. Nitrogen is not a very reactive gas, but methane is earthly cooking gas and is highly flammable. Of course, there is no free oxygen on Titan, so you are safe so long as your spacesuit does not leak oxygen. That might be exciting.

When the Huygens probe landed on the frigid surface, it radioed back measurements and images. The surface is mostly frozen water ice with some methane mixed in. The sunlight is orange because it has filtered down through the orange haze. Rocks littering the ground are actually steel-hard chunks of supercold ice rounded by erosion. Some rest in small depressions, suggesting that a liquid has flowed around them. You can see these depressions around the rocks in Figure 23-18.

The methane in the atmosphere is a puzzle. Sunlight can destroy methane by breaking the molecule up. The hydrogen escapes easily into space. In addition, sunlight can trigger chemical reactions that convert methane into organic compounds that settle out of the atmosphere. The 2 to 3 percent methane in Titan's atmosphere should be gone in only a million years. Obviously there is some source of fresh methane. There is no global methane ocean, so the methane must be produced by outgassing from Titan's icy crust. Detailed studies have identified at least one area where water ice and methane may be rising to the crust and forming a methane volcano (Figure 23-17). Such geological activity requires an internal heat source, and astronomers suspect that tidal heating may be helping radioactive decay keep the interior of Titan warm.

Titan is an exciting world to think about because it might harbor life in its organic deposits, but it is also exciting because it is so totally unearthly.

The Smaller Moons

In addition to Titan, Saturn has a large family of smaller moons. They are mixtures of rock and ice and are heavily cratered. Some of the smallest are probably captured objects. Some of the larger moons show traces of geological activity. You can compare the sizes of a few of these in ■ Figure 23-19.

The moon Phoebe orbits on the outer fringes of Saturn's satellite family, and it follows a retrograde orbit. That is, it orbits backward. It is quite small, only 214 km (133 mi) in diameter, and its surface is dark and cratered (■ Figure 23-20). Traces of ice are detected where impacts have excavated deeper layers or where landslides have exposed fresh ice. Nevertheless, the density

The Huygens probe descended by parachute through the thick atmosphere of Titan and photographed the surface. From an altitude of 8 km (26,000 ft), the surface showed clear signs that a liquid, almost certainly liquid methane, had drained over the surface and into the lowlands. Once Huygens landed on the surface, it radioed back photos showing a level plain and chunks of ice rounded by a moving liquid. (ESA/NASA/JPL/University of Arizona)

Visual-wavelength images

Drainage channels were cut by flowing liquid methane.

Possible methane fog

Icy grapefruit-size "rocks" on Titan are bathed in orange light from its hazy atmosphere.

of Phoebe is 1.6 g/cm³, which is high enough to show that it contains a significant amount of rock. It seems unlikely that it came from the asteroid belt, where ices are rare, but rather that it is a captured Kuiper Belt object that wandered in from an orbit beyond Neptune.

Larger moons such as Tethys, with a diameter over 1000 km (620 mi), are icy and cratered, but they show some signs of geological activity. Some smooth areas on Tethys appear to have been resurfaced by flowing water "lava," and long cracks and grooves may have formed when geological activity strained the icy crust (Figure 23-20).

With a diameter of 520 km (320 mi), the small moon Enceladus isn't much larger than Phoebe, but Enceladus shows dramatic signs of geological activity (■ Figure 23-21). For one thing, Enceladus has an albedo of 0.9. That is, it reflects 90 percent of

the sunlight that hits it, and that makes it the most reflective object in the solar system. You know that old icy surfaces become dark, so the surface of Enceladus must be quite young. Look closely at the surface, and you will see that some regions have few craters and that grooves and cracks are common. Observations made by the Cassini spacecraft show that Enceladus has a tenuous atmosphere. It is too small to keep such an atmosphere, so it must be releasing gas continuously, and Cassini detected a large cloud of water vapor over the moon's south pole where some cracks are actually warmer than its equator. Many planetary

A few of Saturn's moons compared with Earth's moon at the right. In general, larger moons are round and more likely to show signs of geological activity. Small moons such as Phoebe and Hyperion are cratered and do not have enough gravity to overcome the strength of their own ice and squeeze themselves into a spherical shape.

Enceladus   Dione   Iapetus   Phoebe

Mimas   Tethys   Rhea   Hyperion   Earth's moon for comparison

Titan

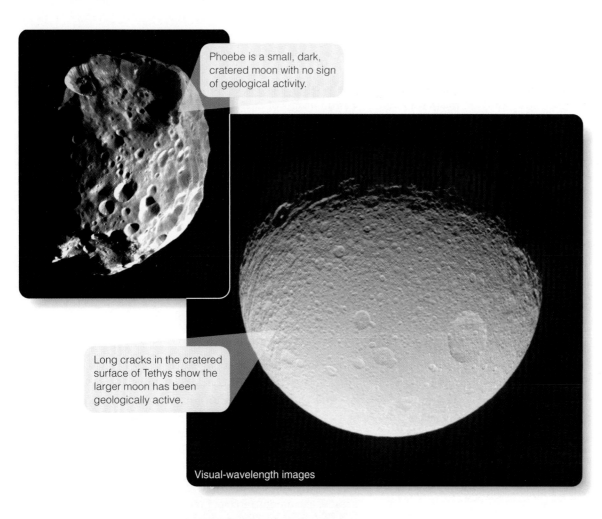

Phoebe is a small, dark, cratered moon with no sign of geological activity.

Long cracks in the cratered surface of Tethys show the larger moon has been geologically active.

Visual-wavelength images

■ **Figure 23-20**

Phoebe and Tethys have ancient cratered surfaces, but they differ in interesting ways. Phoebe is only 21 percent the diameter of Tethys and shows no sign of internal heat causing surface activity. Tethys has smooth areas and long cracks on its surface, showing that it has been active. In addition, Phoebe is probably a captured Kuiper Belt object, but Tethys might have formed with Saturn. (NASA/ JPL/Space Science Institute)

Visual-wavelength images

The terrain at the right appears to be older than the terrain at the left.

■ **Figure 23-21**

The bright, clean surface of Enceladus does not look old. Some areas have few craters, and the numerous cracks and lanes of grooved terrain resemble the surface of Jupiter's moon Ganymede. (NASA/JPL/Space Science Institute) scientists suspect that ice volcanism must be active on its surface. Ammonia dissolved in water can lower its melting point and might allow heat from inside Enceladus to produce eruptions. Cassini observations have detected tiny dust particles around Enceladus, and the E ring of Saturn is densest at the position of Enceladus. That suggests active eruptions.

Of course, you are wondering how a little moon like Enceladus can have heat flowing up from its interior. It is much too small to have remained active since it formed. A clue lies in the moon's orbit. Enceladus orbits Saturn in a resonance with the larger moon Dione. Each time Dione orbits Saturn once, Enceladus orbits twice. That means Dione's gravitational tugs on Enceladus always occur in the same places and make the orbit of the little moon into a slight ellipse. As Enceladus follows that elliptical orbit around Saturn, it feels tides flexing it, and tidal heating warms the interior. You saw how resonances can keep some of Jupiter's moons active, so you can add Enceladus to the list.

Saturn has too many moons to discuss in detail here, but you have to meet Iapetus. It is literally an odd ball. Iapetus is an asymmetric moon. Its trailing side, the side that always faces backward as it orbits Saturn, is old, cratered, icy, and about as bright as dingy snow. Its leading side, the side that always faces forward in its orbit, is also old and cratered, but it has an albedo of only 4 percent—about as dark as fresh asphalt on a highway

Figure 23-22

Like the windshield of a speeding car, the leading side of Saturn's moon Iapetus seems to have accumulated a coating of dark material. The poles and trailing side of the moon are much cleaner ice. The equatorial ridge is 20 km (12 mi) wide and up to 13 km (8 mi) high. It stretches roughly 1300 km (800 mi) along the moon's equator. (NASA/JPL/Space Science Institute)

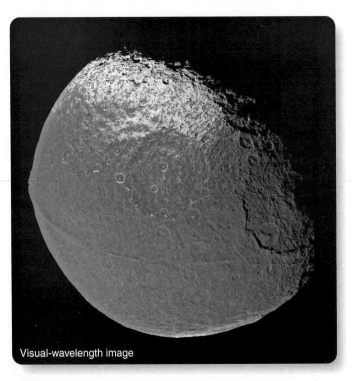

Visual-wavelength image

(■ Figure 23-22). The origin of this dark material is unknown, but theorists suspect that the little moon has swept up dark, silicon- and carbon-rich material on its leading side. This dust could be produced by meteorites striking the outermost moon, Phoebe. Cassini images of Iapetus reveal an equatorial ridge that stands as high as 13 km (8 mi) in some places. You can see the ridge clearly in Figure 23-22. The origin of this ridge is unknown, but it is not a minor feature. At 13 km high, it is over 50 percent higher than Mount Everest, and it extends for a long distance across the surface. That is a big pile of rock and ice.

Saturn's moons illustrate a number of principles of comparative planetology. Small moons are irregular in shape, and old surfaces are dark and cratered. Resonances can trigger tidal heating, and that can resurface moons and outgas atmospheres. Small moons can't keep atmospheres, but big, cold moons can. You are an expert in all of this, so you are ready to wonder where the moons came from.

The Origin of Saturn's Moons

Jupiter's four Galilean satellites seem clearly related to one another, and you can safely conclude that they formed with Jupiter. No such systematic relationship links Saturn's satellites. Planetary scientists suspect that they are captured icy planetesimals left over from the solar nebula or that the impacts of comets have so badly fractured them that they no longer show evidence of their common origin.

Understanding the origin of Saturn's moons is also difficult because the moons interact gravitationally in complicated ways, and the orbits they now occupy may differ from their earlier orbits. These interactions are dramatically illustrated by the two small moons that shepherd the F ring. An even more peculiar pair of moons is known as the coorbital satellites. These two irregularly shaped moonlets have orbits separated by only 100 km. Because one moon is about 200 km in diameter and the other about 100 km, they cannot pass in their orbits. Instead, the innermost moon gradually catches up with the outer moon. As the moons draw closer together, the gravity of the trailing moon slows the leading moon and makes it fall into a lower orbit. Simultaneously, the gravity of the leading moon pulls the trailing moon ahead, and it rises into a higher orbit (■ Figure 23-23). The higher orbit has a longer period, so the trailing moon begins to fall behind the leading moon, which is now in a smaller, faster orbit. Through this elegant dance, the moons exchange orbits and draw apart only to meet again and again. It seems very likely that these two moons are fragments of a larger moon destroyed by a major impact.

In addition to waltzing coorbital moons, the Voyager spacecraft discovered small moonlets trapped at the L4 and L5 Lagrangian points (see Figure 13-5) in the orbits of Dione and Tethys. These points of stability lie 60° ahead of and 60° behind the two moons, and small moonlets can become trapped in these regions. (You will see in Chapter 25 that asteroids are trapped in the Lagrangian points of Jupiter's orbit around the sun.) This

Coorbital Moons

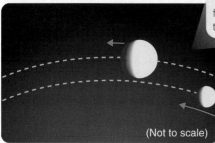

The inner moon orbits faster and overtakes the outer moon.

(Not to scale)

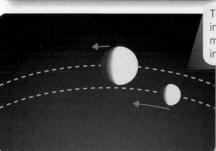

The gravitational interaction pulls the outer moon backward and the inner moon forward.

As they approach, the inner moon moves to a higher orbit, and the outer moon sinks to a lower orbit.

The moons have changed orbits, and the inner moon begins gaining on the outer moon.

The inner moon will eventually overtake the outer moon from behind once again.

■ **Active Figure 23-23**

Saturn's coorbital moons follow nearly identical orbits. The moon in the lower orbit travels faster and always overtakes the other moon from behind. The moons interact and change orbits over and over again.

Ace✪Astronomy™ Log into AceAstronomy and select this chapter to see the Active Figure "Coorbital Moons" and animate this diagram.

gravitational curiosity is surely not unique to the Saturn system, and it warns that interactions between moons can dramatically alter their orbits.

Some of Saturn's larger moons, especially Titan, may have formed with the planet, but the orbital relationships and intense cratering suggest that the moons have interacted and may have collided with large planetesimals or even the heads of comets in the past. Many of the moons may be icy bodies that Saturn has captured since the formation of the solar system. As you continue your exploration of the outer solar system, you must be alert for the presence of such small, icy worlds.

Building Scientific Arguments

What features on Enceladus suggest that it has been active? This argument is based on comparative planetology applied, in this case, to moons. The smaller moons of Saturn are icy worlds battered by impact craters, and you can suspect that they are cold and old. Small worlds lose their heat quickly; and, with no internal heat, there is no geological activity to erase impact craters. A small, icy world covered with craters is exactly what you would expect in the outer solar system. Enceladus, however, is peculiar. Although it is small and icy, its surface is highly reflective, and some areas contain fewer craters than you would expect. In fact, some regions seem almost free of craters. Grooves and faults mark some regions of the little moon and suggest motion in the crust. These features should have been destroyed long ago by impact cratering, so you must suppose that the moon has been geologically active at some time since the end of heavy cratering at the conclusion of planet building.

If you look at Titan, Saturn's largest moon, you see quite a different world. Build an argument based on a different principle of comparative planetology. **How can such a small world as Titan keep a thick atmosphere?**

■   ■   ■

Connections: Jupiter and Saturn reign over the solar system like their namesake ancient gods. Beyond them, out where the sunlight is dim and temperatures deadly cold, orbit three more planets that were unknown to ancient astronomers. In the next chapter, you will add these worlds to your study of comparative planetology.

Study and Review Tools

Summary

23-1 | A Travel Guide to the Outer Planets

How do the outer planets compare with the inner planets?

■ Of the outer planets, Jupiter, Saturn, Uranus, Neptune, and Pluto, the first four are much larger than Earth and lower in density. These Jovian planets are rich in hydrogen. Pluto is small and icy and does not seem to fit its role as a planet.

■ All four of the Jovian planets have rings, large satellite systems, and shallow atmospheres above liquid mantles. Jupiter and Saturn show strong belt–zone circulation, but this is harder to see on Uranus and Neptune.

■ Moons in the Jovian satellite systems interact gravitationally, and some have been heated by tides to produce geological activity. Most are old and cratered.

23-2 | Jupiter

How is Jupiter different from Earth?

■ Simple observations made from Earth show that Jupiter is 11 times Earth's diameter and 318 times Earth's mass. From that you can calculate its density, which is much lower than Earth's.

■ The low density combined with the oblateness and mathematical models reveal that Jupiter must have only a small core of heavy elements and a deep mantle of liquid hydrogen. The atmospheric composition is much like that of the sun—mostly hydrogen and helium.

■ Infrared observations show that Jupiter radiates more heat than it receives from the sun, so its interior must be five times hotter than the sun's surface and is prevented from flashing into vapor by the high pressure.

■ The pressure inside Jupiter converts much of its hydrogen into liquid metallic hydrogen, and the dynamo effect generates a powerful magnetic field that produces rings of auroras around the planet's magnetic poles, interacts with the small moon Io, and traps high-energy particles to form intense radiation belts.

■ Jupiter's magnetic field was first discovered by radio astronomers who detected radio signals coming from Jupiter.

■ Jupiter's shallow atmosphere is rich in hydrogen, with three layers of clouds at the temperatures where ammonia, ammonium hydrosulfide, and water condense out. High- and low-pressure areas form belt–zone circulation. Spots on Jupiter are long-lasting, circulating storm systems.

How did Jupiter and its system of moons and rings form and evolve?

■ Forward scattering shows that Jupiter's ring is composed of tiny dust specks orbiting inside Jupiter's Roche limit. The dust particles cannot have survived since the formation of the planet; rather, they are being produced by meteorite impacts on some of Jupiter's inner moons.

■ Jupiter was hit by the fragmented head of a comet in 1992. Such impacts had not been seen before but must be common in the history of the solar system.

■ Jupiter formed in the outer solar system where ice could condense from the solar nebula. It grew rapidly and captured large amounts of hydrogen and helium gas.

23-3 | Jupiter's Family of Moons

■ At least some of Jupiter's small moons are captured asteroids.

■ The four Galilean moons appear to have formed with Jupiter. Callisto, the outermost, is composed of ice and rock and has an old and cratered surface. Unlike the three inner moons, Calisto is not caught in an orbital resonance and has never been active.

■ Ganymede, Europa, and the innermost moon, Io, are locked in an orbital resonance, and that causes tidal heating. Ganymede's surface is old and cratered in some areas, but smooth, grooved areas must have been produced by a past episode of geological activity.

■ Europa is mostly rock with a thin icy crust that contains almost no craters. Cracks and lines show that the crust has broken repeatedly, and a subsurface ocean probably vents through the crust and deposits ice to cover craters as fast as they form. The subsurface ocean might be a place to look for life.

■ Io is strongly heated by tides and has no water at all. Over a hundred volcanoes erupt on its surface, and no impact craters are visible. Sulfur compounds color the surface yellow and orange and vent into space to be caught in Jupiter's magnetic field.

23-4 | Saturn

How is Saturn different from Jupiter?

■ Saturn is slightly smaller than Jupiter and less dense than water. It has a hot interior but contains less liquid metallic hydrogen, so its magnetic field is weaker than Jupiter's.

■ Saturn is twice as far from the sun as Jupiter and is much colder. The three cloud belts seen on Jupiter form deeper in Saturn's hazy atmosphere and are not as clearly visible.

How did Saturn and its system of moons and rings form and evolve?

■ Saturn must have formed much as Jupiter did, but it is smaller and less dense.

■ Saturn's rings are composed of ice particles and cannot have lasted since the formation of the planet. The rings must receive occasional additions of ice particles when a small moon wanders too close and is pulled apart by tides or when comets hit the planet's icy moons.

■ Icy particles can become trapped in stable places among the orbits of the innermost small moons, those within the Roche limit. Resonances with outer moons can produce gaps in the rings and generate waves that move like ripples through the rings. Small moons can shepherd sections of the ring to produce sharp edges, ripples, or narrow ringlets.

■ Without moons to confine them, the rings would have spread outward and dissipated long ago.

23-5 | Saturn's Moons

■ Many of Saturn's smaller moons, such as Phoebe, are probably captured asteroids or Kuiper Belt objects. All of the moons are mixtures of rock and ice.

■ Titan, the largest moon, is so cold it can retain a dense atmosphere of nitrogen with a few percent methane. The methane condenses from the atmosphere, falls as rain, and drains over the surface, washing dark, organic material into drainage channels and on into the lowlands.

■ The methane in Titan's atmosphere can be destroyed by sunlight, so it must be continuously replenished. It probably vents from the icy crust and may form methane volcanoes.

■ Some of the larger moons, such as Tethys, have old, dark, cratered surfaces with cracks and smoothed areas that suggest past activity. Enceladus, a rather small moon, is the most reflective object in the solar system and has large smooth areas, so it must have been very active. It orbits in a resonance with the moon Dione, and that may cause tidal heating inside Enceladus.

■ The moon Iapetus has a bright icy surface on its trailing side, but its leading side is coated with very dark material, possibly debris from

Phoebe. In addition, Iapetus has a long equatorial ridge higher than Mount Everest on Earth. The origin of the ridge is not known.

- The origin and evolution of Saturn's moons is not as clear as for Jupiter's Galilean moons. Orbital interactions and impacts have been important to these moons.

New Terms

belt–zone circulation (p. 554)

oblateness (p. 556)

liquid metallic hydrogen (p. 557)

decameter radiation (p. 557)

decimeter radiation (p. 557)

Io plasma torus (p. 558)

Io flux tube (p. 558)

critical point (p. 558)

belt (p. 561)

zone (p. 561)

forward scattering (p. 562)

Roche limit (p. 562)

gossamer ring (p. 563)

grooved terrain (p. 566)

tidal heating (p. 567)

shepherd satellite (p. 577)

Review Questions

Ace Astronomy™ Assess your understanding of this chapter's topics with additional quizzing and animations at **http://ace .brookscole.com/sf9**

1. Why is Jupiter more oblate than Earth? Do you expect all Jovian planets to be oblate? Why or why not?

2. How do the interiors of Jupiter and Saturn differ? How does this affect their magnetic fields?

3. What is the difference between a belt and a zone?

4. How can you be certain that Jupiter's ring does not date from the formation of the planet?

5. If Jupiter had a satellite the size of our own moon orbiting outside the orbit of Callisto, what would you predict for its density and surface features?

6. Why are there no craters on Io and few on Europa? Why should you expect Io to suffer more impacts per square kilometer than Callisto?

7. Why are the belts and zones on Saturn less distinct than those on Jupiter?

8. If Saturn had no moons, what do you suppose its rings would look like?

9. Where did the particles in Saturn's rings come from?

10. How can Titan keep an atmosphere when it is smaller than airless Ganymede?

11. If you piloted a spacecraft to visit Saturn's moons and wanted to land on a geologically old surface, what features would you look for? What features would you avoid?

12. Why does the leading side of some satellites differ from the trailing side?

13. Why would you expect small satellites to be irregular in shape? Give examples and exceptions.

14. This image shows a segment of the surface of Jupiter's moon Callisto. Why is the surface dark? Why are some craters dark and some white? What does this image tell you about the history of Callisto?

(NASA/JPL)

15. The Cassini spacecraft recorded this image of Saturn's A ring and Encke's division. What do you see in this photo that tells you about processes that confine and shape planetary rings?

(NASA/JPL/Space Science Institute)

Discussion Questions

1. Some astronomers argue that Jupiter and Saturn are unusual, while other astronomers argue that all solar systems should contain one or two such giant planets. What do you think? Support your argument with evidence.

2. Why don't the terrestrial planets have rings?

Problems

1. What is the maximum angular diameter of Jupiter as seen from Earth? What is the minimum? Repeat this calculation for Saturn and Titan. (*Hint:* Use the small-angle formula.)

2. The highest-speed winds on Jupiter are in the equatorial jet stream, which has a velocity of 150 m/s. How long does it take for these winds to circle Jupiter?

3. What are the orbital velocity and period of a ring particle at the outer edge of Jupiter's ring? at the outer edge of Saturn's A ring? (*Hints:* The radius of the edge of the A ring is 136,500 km. See Chapter 5.)

4. What is the angular diameter of Jupiter as seen from the surface of Callisto? (*Hint:* Use the small-angle formula.)

5. What is the escape velocity from the surface of Ganymede if its mass is 1.5×10^{26} g and its radius is 2628 km? (*Hint:* See Chapter 5.)

6. If you were to record the spectrum of Saturn and its rings, you would find light from one edge of the rings redshifted and light from the other edge blueshifted. If you observed at a wavelength of 500 nm, what difference in wavelength should you expect between the two edges of the rings? (*Hints:* See Problem 3 and Chapter 7.)

7. What is the difference in orbital velocity between particles at the outer edge of Saturn's B ring and particles at the inner edge of the B ring? (*Hint:* The outer edge of the B ring has a radius of 117,500 km, and the inner edge has a radius of 92,000 km.)

8. What is the difference in orbital velocity between the two coorbital satellites if the semimajor axes of their orbits are 151,400 km and 151,500 km? The mass of Saturn is 5.7×10^{26} kg. (*Hint:* See Chapter 5.)

Planetary Atmospheres
Experiment with the temperature throughout the atmospheres of several planets with this animation.

Ace ◯ Astronomy™ To access the resources in the Media Cluster, log into AceAstronomy at **http://ace.brookscole .com/sf9** and select Chapter 23.

ACTIVE FIGURES

Explorable Jupiter
Study Jupiter at visible, ultraviolet, radio, and X-ray wavelengths in this animation. What features can you see at the other wavelengths that are not visible to the human eye?

Coorbital Moons
Study the phenomenon of two of Saturn's moons repeatedly swapping orbits in this animation.

Roche Limit
Some moons remain intact in orbit, and some break apart to become rings like Saturn's rings. See if you can figure out why with this animation.

ASTRONOMY EXERCISES

Auroras
Use this simulation to study the formation of auroras. You can vary the strength of Earth's magnetic field and the speed and number of particles in the solar wind.

Convection and Turbulence
Core temperature and rotation speed both contribute to the amount of turbulence in a planet's atmosphere. Explore their relationship with this animation.

Convection and Magnetic Fields
A planet's core temperature and rotation speed are factors in the strength of its magnetic field. See the effect of these variables in this animation.

Lab 7: Planetary Atmospheres and Their Retention
This lab investigates the retention of atmospheres. You will explore the factors that govern the loss of atmospheric gases, and you will see why certain bodies can retain some gases but not others.

Critical Inquiries for the Web

1. Saturn has a great many satellites. Locate information on these moons, including discovery dates (make sure you use an updated source that includes discoveries from the past few years).

2. The collision of Comet Shoemaker–Levy 9 with Jupiter was one of the first events to be publicized directly on the Web. Search for websites dedicated to this comet and its spectacular demise. How long were the effects of the impacts visible in the atmosphere of Jupiter? How might they have appeared differently if this event had occurred at Saturn?

3. What is the status of scientists' understanding of the Jovian moon Europa? This moon was studied extensively by the Galileo mission. Visit Galileo-related websites to find the results of this exploration and discuss current ideas regarding the possibility of liquid water beneath Europa's crust. How might future missions to Europa probe such an underground ocean?

4. Cassini reached orbit around Saturn on July 1, 2004, and the Huygens probe parachuted through the atmosphere of Titan on January 14, 2005. Check the Cassini Web page to get the latest news from Saturn and Titan.

5. When will Saturn's rings next be edge-on to the Earth? When were they last edge-on? Find photographs of Saturn and its rings at different inclinations.

Exploring *TheSky*

1. Galileo recognized the significance of Jupiter's moons when he realized they were moving around Jupiter. Sketch the position of the four Galilean moons at intervals of 1 hour. Could you recognize their orbital motion in a long evening, or would you need observations over a number of evenings? (*Hint:* Use **Find** to **Center & Frame** Jupiter. Zoom in until you can see the moons. Use **Tracking Setup** to lock on Jupiter and then use the **Time Skip** buttons to make the moons move.)

2. Calculate the mass of Jupiter from your own observations of the orbital period and orbital radius of Jupiter's moons. (*Hint:* Use the equation for circular velocity in Chapter 5.)

3. Repeat Activity 2 for Saturn.

 Go to the Brooks/Cole Astronomy Resource Center (**http:// astronomy.brookscole.com**) for critical thinking exercises, articles, and additional readings from InfoTrac College Edition, Brooks/Cole's online student library.

24 | Comparative Planetology of Uranus, Neptune, and Pluto

A good many things go around in the dark besides Santa Claus.

HERBERT HOOVER

OUT IN THE DARKNESS be-
yond Saturn, out where sunlight is about 1000
times fainter than on Earth, there are things going around
that Aristotle, Galileo, and Newton never imagined. They knew about Mercury, Venus, Mars, Jupiter, and Saturn, but
our solar system includes three more planets, planets that were not discovered until after the invention of the telescope. ▌ Uranus and Neptune are Jovian planets and are similar in size, about four times Earth's diameter. That
makes them about half the size of Saturn. Through most Earth-based telescopes, you can see nothing but tiny disks, and
until recently what little scientists knew of these planets was deduced from a few observations compared with Jupiter
and Saturn. But in January 1986 the Voyager 2 spacecraft flew past Uranus and returned a wealth of photographs and

▌ Continued on page 590 ▌

Pluto and its moon Charon lie so far from the sun that no space-craft has ever visited them. No one knows what they are like, but someday a spacecraft will radio back images from the edge of the solar system. (NASA/Southwest Research Institute/ JHUAPL/SwRI)

Guidepost

Looking Back

Some people say they are seasoned travelers because they have been to Europe. You have visited six planets and a bunch of moons, braved the radiation belts of Jupiter, and swum through the rings of Saturn. Now you are ready to explore farther—out into the dark far from the sun.

This Chapter

Three worlds circle the sun in the twilight beyond Saturn. Uranus and Neptune are Jovian worlds, but you will find them strangely different from Jupiter and Saturn. Pluto is an oddity, and you will wonder if it is really a planet. As you explore you will answer five essential questions:

How is Uranus different from Jupiter?

How did Uranus, its rings, and its moons form and evolve?

How is Neptune different from Uranus?

How did Neptune, its rings, and its moons form and evolve?

Is Pluto a planet or something else?

Looking Ahead

As you finish this chapter, you will have visited all of the major worlds in our solar system. But there is more to see. Vast numbers of small rocky and icy bodies orbit among the planets, and you should visit them to complete your tour but also to search for clues to how the solar system formed.

measurements. Then, in August 1989, Voyager 2 flew past Neptune, revealing a blue world with complex weather, rings, and moons. In just a few years, Uranus and Neptune went from obscure points of light to known worlds. More recently, the Hubble Space Telescope and new-generation telescopes on Earth have been able to image Uranus and Neptune and monitor changing weather patterns.

Pluto, however, is a more difficult subject. Through an Earth-based telescope, it looks like nothing more than a faint point of light, and even through the Hubble Space Telescope it is a tiny disk. Voyagers 1 and 2 left the solar system without passing near Pluto, and, although a spacecraft is planned for a journey to Pluto, no spacecraft has been there yet. Consequently there are no photos of the little world's surface. Nevertheless, as an experienced traveler among planets, you may be able to guess Pluto's nature by combining the limited observations with the principles of comparative planetology.

Uranus, Neptune, and Pluto differ from the other planets in one special way. They were not known from antiquity. They had to be discovered. The stories of these discoveries highlight the nature of scientific discovery.

24-1 Uranus

IN MARCH 1781, BENJAMIN FRANKLIN was in France raising money, troops, and arms for the American Revolution. George Washington and his colonial army were only six months away from the defeat of Cornwallis at Yorktown and the end of the war. In England, King George III was beginning to show signs of madness. And a German-speaking music teacher in the English resort city of Bath was about to discover the planet Uranus.

The Discovery of Uranus

William Herschel (■ Figure 24-1) came from a musical family in Hanover, Germany, but immigrated to England as a young man and eventually obtained a prestigious job as the organist at the Octagon Chapel in Bath.

To compose exercises for his students and choral and organ works for the chapel, Herschel studied the mathematical prin-

■ Figure 24-1

When William Herschel discovered Uranus in 1781, he saw only a tiny blue dot. He never knew how interesting the planet is. Even the designer of this stamp marking the passage of Comet Halley in 1986 had to guess at the planet's appearance. Voyager 2 visited Uranus in 1986, and now the Hubble Space Telescope and Earth-based giant telescopes can image the planet's clouds and thin rings. (NASA) ciples of musical harmony from a book by Professor Robert Smith of Cambridge. The mathematics was so interesting that Herschel searched out other books by Smith, including a book on optics. Of course, an 18th-century book on optics written by a professor in England relied heavily on Isaac Newton's discoveries and described the principles of the reflecting telescope (see Chapter 6). Herschel and his brother Alexander began building telescopes, and William went on to study astronomy in his spare time.

Herschel's telescopes were similar to Newton's in that they had metal mirrors, but they were much larger. Newton's telescope had a mirror about 1 in. in diameter, but Herschel developed ways of making much larger mirrors. He soon had telescopes as long as 20 ft, although one of his favorite telescopes was 7 ft long with a mirror 6.2 in. in diameter. Using it, he began the research project that led to the discovery of Uranus.

Herschel did not set out to search for a planet; he was trying to detect stellar parallax produced by Earth's motion around the sun. No one had yet detected this effect, although by the 1700s all astronomers accepted Earth's motion. Galileo had pointed out that parallax might be detected if a nearby star and a very distant star lay so nearly in the same direction that they looked like a very close double star through a telescope. In such a case, Earth's orbital motion would produce a parallactic shift in the position of the nearby star with respect to the more distant star (■ Figure 24-2). Herschel began to examine all stars brighter than eighth magnitude to search for double stars that might show parallax. That project alone took over two years.

On the night of March 13, 1781, Herschel set up his 7-ft telescope in his back garden and continued his work. He later wrote, "In examining the small stars in the neighborhood of H Geminorum, I perceived one that appeared visibly larger than the rest." As seen from Earth, Uranus is never larger in angular diameter than 3.7 seconds of arc, so Herschel's detection of the disk illustrates the quality of his telescope and his eye. At first he suspected that the object was a comet, but other astronomers quickly showed that it was a planet orbiting the sun beyond Saturn.

The discovery of Uranus made Herschel world famous. Since antiquity, astronomers had known the five planets—Mercury, Venus, Mars, Jupiter, and Saturn—and they supposed that the universe was complete. Herschel's discovery extended the classical universe by adding a new planet. The English public accepted Herschel as their astronomer-hero, and, having named the new planet *Georgium Sidus* (George's Star) after King George III, Herschel received a royal pension. The former music teacher was welcomed into court society, where he later met and married a wealthy widow and took his place as one of the great English astronomers in history. With his new financial position, he built large telescopes on his estate and, with his sister Caroline, a talented astronomer herself, attempted to map the extent of the universe. You read about their research in Chapter 15.

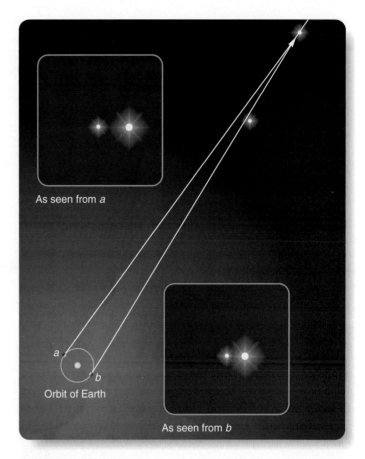

As seen from *a*

a
b
Orbit of Earth

As seen from *b*

■ **Figure 24-2**

A double star consisting of a nearby star and a more distant star could be used to detect stellar parallax. The separation between the stars would be different as seen from different places in Earth's orbit. The effect is actually much too small to detect by eye.

Continental astronomers were less than thrilled that an Englishman had made the great discovery, and even some professional English astronomers thought Herschel a mere amateur. They called his discovery a lucky accident. Herschel responded in his own defense by making three points. First, he had built some of the finest-quality telescopes then in existence. Second, he had been conducting a systematic research project and would have found Uranus eventually because he was inspecting all of the brighter stars visible with his telescope. And third, he had great experience seeing fine detail with his telescopes. As a musician, he knew the value of practice and applied it to the business of astronomical observing. In fact, records show that astronomers had seen Uranus more than 17 times before Herschel, but each time they failed to notice that it was not a star. They plotted Uranus on their charts as if it was just another faint star.

This illustrates one of the ways in which scientific discoveries are made. Often, discoveries seem accidental, but on closer examination you find that the scientist has earned the right to the discovery through many years of study and preparation. To quote a common saying, "Luck is what happens to people who work hard."

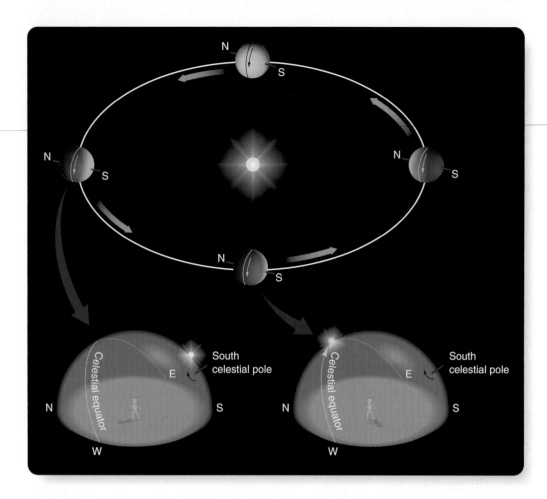

■ Figure 24-3

Uranus rotates on an axis that is tipped
97.9° from the perpendicular to its orbit,
so its seasons are extreme. When one
of its poles is pointed nearly at the sun
(a solstice), a citizen of Uranus would
see the sun near a celestial pole, and it
would never rise or set. As it orbits the
sun, the planet maintains the direction
of its axis in space, and thus the sun
moves from pole to pole. At the time of
an equinox on Uranus, the sun would
be on the celestial equator and would
rise and set with each rotation of the
planet. Compare with similar diagrams
for Earth on page 27.

observed position of the planet.
Tiny variations in the orbital motion of Uranus eventually led to
the discovery of Neptune, a controversial story you will hear later
in this chapter.

The Motion of Uranus

Uranus follows a slightly elliptical orbit with an average distance
from the sun of 19.18 AU and an
orbital period of 84.013 years
(**Celestial Profile 9**). The ancients thought of Saturn as the
slowest of the planets, but Saturn orbits in a bit over 29 years.
Uranus, being further from the sun, has a longer orbital period.

The rotation of Uranus is peculiar. Earth rotates approximately upright in its orbit. That is, its axis of rotation is inclined
only 23.5° from the perpendicular to its orbit. Uranus, in contrast, rotates on an axis that is inclined 97.9° from the perpendicular to its orbit. It rotates on its side (**■ Figure 24-3**).

Continental astronomers, especially the French, insisted that
the new planet not be named after an English king. The French
stubbornly called the planet Herschel, as did many other non-
English astronomers. Some years later, German astronomer
J. E. Bode suggested the name Uranus, one of the oldest of the
Greek gods.

Over the half-century following the discovery of Uranus,
astronomers noted that Newton's laws did not quite predict the

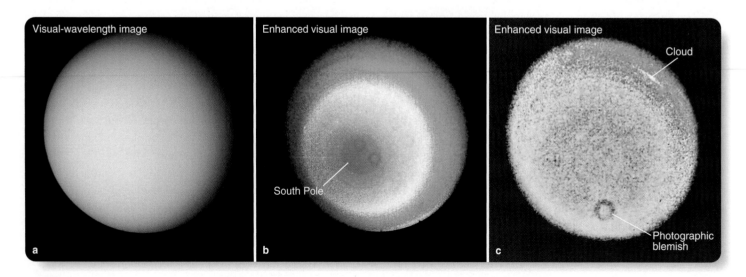

The seasons on Uranus are extreme. In 1986, when Voyager 2 flew past Uranus, the planet was in the segment of its orbit in which its south pole faced the sun. Consequently, its southern hemisphere was bathed in continuous sunlight, and a creature living on Uranus (an unlikely possibility, as you will discover later) would have seen the sun circling near the planet's south celestial pole. You can see this at lower left in Figure 24-3. Once Uranus moves about a quarter of the way around its orbit, the sun will be above its equator, and the citizen of Uranus would see the sun rise and set with the rotation of the planet. Look at the lower right in Figure 24-3 to see that geometry. As Uranus continues along its orbit, the sun will approach the planet's north celestial pole, and the southern hemisphere of the planet will experience a lightless winter lasting 21 Earth years. In other words, the ecliptic on Uranus passes very near the planet's celestial poles, and the resulting seasons are extreme.

The Atmosphere of Uranus

Like Jupiter and Saturn, Uranus has no surface. The gases of its atmosphere—mostly hydrogen, 15 percent helium, and a few percent methane, ammonia, and water vapor—blend gradually with a fluid interior.

Seen through Earth-based telescopes, Uranus is a small, featureless blue disk. The blue color arises because the atmosphere contains methane, which is a good absorber of longer-wavelength photons. As sunlight penetrates into the atmosphere and is scattered back out, the longer-wavelength (red) photons are more likely to be absorbed. Consequently, the light entering your eye is richer in blue photons and poorer in red photons, giving the planet a blue color.

As Voyager 2 drew closer to the planet in late 1985, astronomers studied the images radioed back to Earth. Uranus was a pale blue ball with no obvious clouds, and only when the images were carefully computer enhanced was any banded structure detected (■ Figure 24-4). A few very high clouds of methane ice particles were detected, and that allowed astronomers to measure the rotation period of the planet.

You can understand the appearance of the atmosphere by studying the temperature profile of Uranus shown in ■ Figure 24-5. The atmosphere of Uranus is much colder than that of Saturn or Jupiter. Consequently, the three cloud layers of ammonia, ammonia hydrosulfide, and water that occur in the atmosphere of Jupiter and Saturn would form very deep in the atmosphere of Uranus. These cloud layers are not visible because of the thick

■ **Figure 24-4** ◀

(a) This Voyager 2 image of Uranus was made in 1986 and shows no clouds. Only when the image is computer enhanced, as in (b), is a banded structure visible. At the time, the axis of rotation was pointed nearly at the sun. (c) Under extreme computer enhancement, small methane clouds were visible. The geometry of the banding and the clouds suggests belt–zone circulation. (NASA)

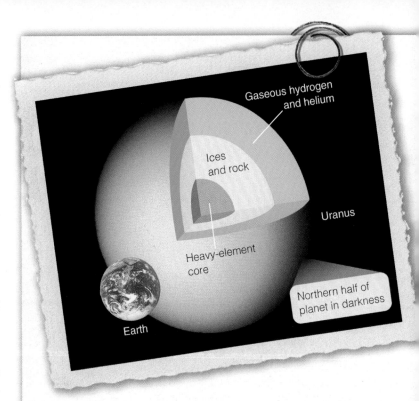

Uranus rotates on its side; when Voyager 2 flew past in 1986, the planet's south pole was pointed almost directly at the sun. (NASA)

Celestial Profile 9: Uranus

Motion:

Average distance from the sun	19.18 AU (28.69 $\times$ 10^8 km)
Eccentricity of orbit	0.0461
Maximum distance from the sun	20.1 AU (30.0 $\times$ 10^8 km)
Minimum distance from the sun	18.3 AU (27.4 $\times$ 10^8 km)
Inclination of orbit to ecliptic	0°46'23"
Average orbital velocity	6.81 km/s
Orbital period	84.013 y (30,685 days)
Period of rotation	17^h14^m
Inclination of equator to orbit	97°55'

Characteristics:

Equatorial diameter	51,118 km (4.01 $D_\oplus$)
Mass	8.69 $\times$ 10^{25} kg (14.54 $M_\oplus$)
Average density	1.29 g/cm^3
Gravity	0.919 Earth gravity
Escape velocity	22 km/s (1.96 $V_\oplus$)
Temperature above cloud tops	−220°C (−364°F)
Albedo	0.35

Personality Point:

Most creation stories begin with a separation of opposites, and Greek mythology is no different. Uranus (the sky) separated from Gaia (Earth), who was born from the void, Chaos. They gave birth to the giant Cyclops, Cronos (Saturn, father of Zeus), and his fellow Titans. Uranus is sometimes called the starry sky, but the sun (Helius), moon (Selene), and the stars were born later, so Uranus, one of the most ancient gods, began as the empty, dark sky.

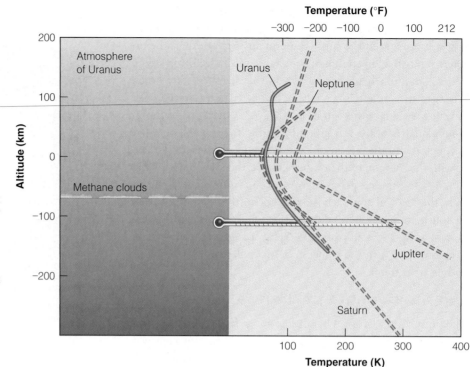

■ Active Figure 24-5

The atmosphere of Uranus is much colder than that of Jupiter or Saturn, and the only visible cloud layer is one formed of methane ice crystals deep in the hydrogen atmosphere. Other cloud layers would be much deeper in the atmosphere and are not visible. The temperature profile of Neptune is similar to that of Uranus, and it has methane clouds at about the same place in its atmosphere.

Ace ◑ Astronomy™ Log into AceAstronomy and select this chapter to see the Active Figure "Planetary Atmospheres" and compare the atmospheres of the planets.

atmosphere of hydrogen through which an observer has to look. The clouds that are visible on Uranus are clouds of methane ice crystals, which form at such a low temperature that they occur higher in the atmosphere of Uranus. Look again at Figure 24-5 and notice that there can be no methane clouds on Jupiter because that planet is too warm. The coldest part of the atmosphere of Saturn is just cold enough to form a thin methane haze high above its more visible cloud layers. (See Figure 23-16b.)

The clouds and atmospheric banding faintly visible on Uranus appear to be a form of belt–zone circulation, which is a bit surprising. Uranus rotates on its side, so solar energy strikes its surface in a geometry quite different than it does on Jupiter and Saturn. Evidently the belt–zone circulation is dominated by the rotation of the planet and not by the direction of sunlight.

Is Uranus always as featureless as it was then Voyager 2 flew past in early 1986? The Hubble Space Telescope and giant Earth-based telescopes have detected changing clouds on Uranus (■ Figure 24-6), and those clouds appear to follow belt–zone patterns. Furthermore, since 1986, Uranus has moved on along its orbit, and spring has been coming to its northern hemisphere. As sunlight has reached these latitudes, clouds have appeared there, probably generated by convection in the atmosphere. This may be part of a seasonal cycle on Uranus, but its year lasts 85 Earth years, so you will have to be patient.

The Interior of Uranus

Astronomers cannot describe the interiors of Uranus and Neptune as accurately as they can the interiors of Jupiter and Saturn.

Observational data are sparse, and the materials inside these planets are not as easy to model as simple liquid hydrogen.

The average density of Uranus, 1.29 g/cm³, tells you that the planet must contain a larger share of dense materials such as rock than does Jupiter or Saturn. Nearly all models of the interior of Uranus contain three layers. The atmosphere is rich in hydrogen and helium. A deep mantle must contain large amounts of ices of water, methane, and ammonia, mixed with hydrogen and with silicate matter. This mantle is sometimes described as "liquid" or as "ice," but it is probably quite unlike the materials these words suggest. The third layer in these three-layer models is a small core of rock. Because of the high pressure and high temperature, the material is probably not very rocklike. The term *rock* refers to its chemical composition and not its appearance. It would be better to call it a heavy-element core.

Planetary scientists make a subtle distinction among the Jovian planets. Jupiter and Saturn are sometimes called "gas giant planets," because they are rich in hydrogen and helium. A growing number of scientists are referring to Uranus and Neptune as "ice giant planets," in recognition of the larger proportion of ices in their interiors.

Because Uranus has a much lower mass than Jupiter, its internal pressure is not high enough to produce liquid metallic hydrogen. Consequently, you might expect it to lack a strong magnetic field, but the Voyager 2 spacecraft found that Uranus has a magnetic field about 75 percent as strong as Earth's. Furthermore, the field is tipped 60° to the axis of rotation and is offset from the center of the planet about 30 percent of the planet's radius

Infrared image

■ Figure 24-6

Images of Uranus recorded by the Keck near-infrared camera show both sides of Uranus. Banding and cloud features are visible, and some clouds are highly changeable. This suggests that seasonal changes are taking place on Uranus as spring comes to the northern hemisphere. The rings look red in this image because of image processing. (Lawrence Sromovsky, UW-Madison Space Science and Engineering Center)

(■ Figure 24-7). Theorists suggest that the magnetic field is produced by a dynamo effect, operating not at the center but nearer the surface in a layer of liquid water with dissolved ammonia and methane. Such a material would be a good conductor of electricity, and the rotation of the planet coupled with circulation in the fluid could generate the magnetic field.

The high inclination of the magnetic field and its offset from the center of Uranus give planetary scientists a good estimate of the true rotation period of the planet, otherwise difficult to estimate because of the lack of cloud features and high atmospheric winds. The magnetic field deflects the solar wind and traps some

charged particles to create weak radiation belts trapped in a magnetosphere. High-speed electrons spiraling along the magnetic field produce synchrotron radio emission just as on Jupiter, and the Voyager 2 spacecraft detected this radiation fluctuating with a period of 17.24 hours—the period of rotation of the magnetic field and, presumably, the planetary interior.

The magnetic field and the high inclination of the planet produce some peculiar effects. As is the case for all planets, the solar wind deforms the magnetosphere and draws it out into a long tail extending away from the planet in the direction opposite the sun. The rapid rotation of Uranus and its high inclination make

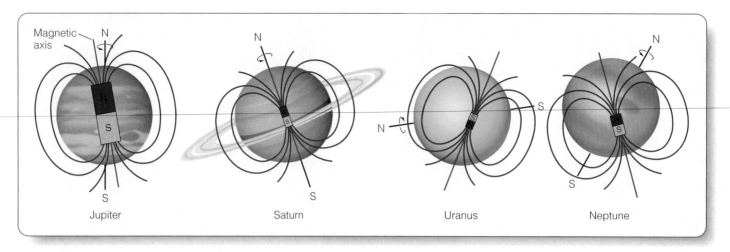

■ **Figure 24-7**

The magnetic fields of Uranus and Neptune are peculiar. While the magnetic axis of Jupiter is tipped only 10° from its axis of rotation and the magnetic axis of Saturn is not tipped at all, the magnetic axes of Uranus and Neptune are tipped at large angles. Furthermore, the magnetic fields of Uranus and Neptune are offset from the center of the planet. This suggests that the dynamo effect operates differently in Uranus and Neptune than in Jupiter and Saturn.

the magnetosphere and its long extension take on a corkscrew shape, which was partially mapped by Voyager 2. At the time Voyager 2 flew past in 1986, the south pole of Uranus was pointed nearly at the sun, and once each rotation the solar wind could pour directly down into the south magnetic pole. This interaction produced strong auroras at both magnetic poles, which were detected by the spacecraft in the ultraviolet (■ Figure 24-8). In the years since, Uranus has moved farther around its orbit, so the geometry

■ **Figure 24-8**

Auroras on Uranus were detected in the ultraviolet by the Voyager 2 spacecraft when it flew by in 1986. These maps of opposite sides of Uranus show the location of auroras near the magnetic poles. Recall that the magnetic field is highly inclined and offset from the planet's center, so the magnetic poles do not lie near the poles defined by rotation. In fact, the north magnetic pole (right) falls near the planet's equator. Also recall that Uranus is highly inclined and that in 1986 its south pole, located at the bottom of these diagrams, was pointed nearly at the sun. The white dashed line marks zero longitude. (Courtesy Floyd Herbert, LPL)

of its interaction with the solar wind is now different. Unfortunately, no spacecraft orbit Uranus at present, and there is no way to observe this interaction in detail.

The temperature of Uranus can reveal something about its interior. Jupiter and Saturn are warmer than you would expect, given the amount of energy they receive from the sun, and that must mean that heat is leaking out from their hot interiors. Uranus, in contrast, is about the temperature you would expect for a world at its distance from the sun. Apparently, it has lost much of its interior heat. Yet it must have some internal heat to cause convection in the fluid mantle and drive the dynamo effect. The decay of natural radioactive elements would generate heat, but some astronomers have suggested that the slow settling of heavier elements through the fluid mantle could release sufficient energy to warm the interior.

Laboratory studies of methane show that it can break down under the temperature and pressure found inside Uranus and form various compounds plus pure carbon in the form of diamonds. If this happens in Uranus, the diamond crystals would fall inward, warming the interior through friction. Whatever diamonds might exist in Uranus are forever beyond reach, but some of Uranus's internal heat may come from its diamond-making layers.

As you explored the atmosphere and interior of Uranus, you may have been a little disappointed. Uranus seems to be a small, modest, Jovian world. But now you are ready to visit one of its best features—its rings.

Ace⬦Astronomy™ Log into AceAstronomy and select this chapter to see Astronomy Exercise "Auroras." Review the factors that affect auroras on Earth.

Ace⬦Astronomy™ Log into AceAstronomy and select this chapter to see Astronomy Exercise "Convection and Magnetic Fields." How does convection in a planet produce a magnetic field?

The Rings of Uranus

Both Uranus and Neptune have rings that are more like those of Jupiter than those of Saturn. They are dark, faint, and confined by shepherd satellites. They are not easily visible from Earth, and the first hint that these planets had rings came from **occultations;** an occultation is the passage of a planet in front of a star. Nevertheless, most of what is known about these ring systems comes from observations by the Voyager 2 spacecraft.

Study **The Rings of Uranus and Neptune** on pages 598–599 and notice three important points:

❶ The rings of Uranus were discovered during an occultation when Uranus crossed in front of a star.

❷ The rings of Uranus are very dark, contain little dust, and are confined by small moons.

❸ The rings of Uranus cannot survive for long periods, so the rings need to be resupplied with material from impacts on

moons. When you read about Neptune's rings later in this chapter, you will return to this artwork and see how closely the two ring systems compare.

As you read about planetary rings, notice their close relationship with moons. Because of collisions among ring particles, planetary rings tend to spread outward almost like an expanding gas. If a planet had no moons, its rings would spread out into a more and more tenuous sheet until they were gone. The spreading rings can be anchored by small moons, which get pushed outward very slowly. Through orbital resonances, those small moons can be anchored by larger, more distant moons that are so massive they do not get pushed outward significantly. In this way, a system of moons can confine and preserve a system of planetary rings.

Planetary rings are beautiful blossoms created when debris falls into a system of small moons and becomes trapped in the most stable orbits among the moons. Without moons, there would be no rings.

The Moons of Uranus

Uranus has five large moons that were known from Earth-based observations. Those five moons (■ Figure 24-9), from the outermost inward, are Oberon, Titania, Umbriel, Ariel, and Miranda.

■ Figure 24-9

This infrared image shows Uranus, its disk, and its five major satellites. Because of photographic effects, the images of the satellites are much larger than the moons themselves. The largest, Titania, is nearly 32 times smaller in diameter than Uranus. Faint points of light to the left and right of the rings are the smaller moons Portia and Puck. Other faint points of light in this image are background stars. (ESO)

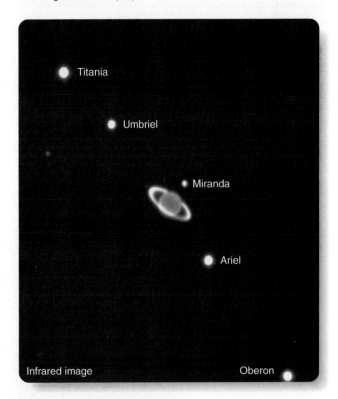

The Rings of Uranus and Neptune

1 The rings of Uranus were discovered in 1977, when Uranus crossed in front of a star. During this occultation, astronomers saw the star dim a number of times before and again after the planet crossed over the star. The dips in brightness were caused by rings circling Uranus.

More rings were discovered by Voyager 2. The rings are identified in different ways depending on when and how they were discovered.

Ace ☉ Astronomy™

Log into AceAstronomy and sele
this chapter to see the Activ
Figure "Uranus's Ring Detectior
and animate this diagran

Notice the eccentricity of the ε ring. It lies at different distances on opposite sides of the planet.

Minutes from midoccultation

Distance from center of Uranus (km)

2 The albedo of the ring particles is only about 0.015, darker than lumps of coal. If the ring particles are made of methane-rich ices, radiation from the planet's radiation belts could break the methane down to release carbon and darken the ices. The same process may darken the icy surface of Uranian moons.

The narrowness of the rings suggests they are shepherded by small moons. Voyager 2 found Ophelia and Cordelia shepherding the ε ring. Other small moons must be shepherding the other narrow rings. Such moons must be structurally strong to hold themselves together inside the planet's Roche limit.

The eccentricity of the ε ring is apparently caused by the eccentric orbits of Ophelia and Cordelia.

2a When the Voyager 2 spacecraft looked back at the rings illuminated from behind by the sun, the rings were not bright. That is, the rings are not bright in forward-scattered light. That means they must not contain much dust. The nine main rings contain particles no smaller than meter-sized boulders.

Uranus

3 Ring particles don't last forever as they collide with each other and are exposed to radiation. The rings of Uranus may need to be resupplied with fresh particles occasionally as impacts on icy moons scatter icy debris.

Collisions among ring particles produce dust, which is thinly scattered inward from the λ ring, but the high, tenuous atmosphere of Uranus is slowing the dust particles and making them fall into the planet. The rings of Uranus actually contain very little dust.

Disk of Neptune

Visual-wavelength image

The rings of Neptune are bright in forward-scattered light, as in the image above, and that indicates that the rings contain significant amounts of dust. The ring particles are as dark as those that circle Uranus, however, so they probably contain methane-rich ice darkened by radiation.

4 The brightness of Neptune is hidden behind the black bar in this Voyager 2 image. Two narrow rings are visible, and a wider, fainter ring lies closer to the planet. More ring material is visible between the two narrow rings.

Arc

Arc

5 When Neptune occulted stars, astronomers sometimes detected rings and sometimes did not. From that they concluded that Neptune might have ring arcs. Computer enhancement of this Voyager 2 visual-wavelength image shows arcs, regions of higher density, in the outer ring.

The ring arcs visible in the outer ring appear to be generated by the gravitational influence of the moon Galatea, but other moons must also be present to confine the rings.

Enhanced visual image

4a Neptune's rings lie in the plane of the planet's equator and inside the Roche limit. The narrowness of the rings suggests that shepherd moons must confine them, and a few such moons have been found among the rings. There must be more small moons to confine the rings.

Naiad

Galetea

Thalassa

Adams

LeVerrier

Galle

Despina

4c Like the rings of Uranus, the ring particles that orbit Neptune cannot have survived since the formation of the planet. Occasional impacts on Neptune's moons must scatter debris and resupply the rings with fresh particles.

4b Neptune's rings have been given names associated with the planet's history. English astronomer Adams and French astronomer LeVerrier predicted the existence of Neptune from the motion of Uranus. The German astronomer Galle discovered the planet in 1846 based on LeVerrier's prediction.

Neptune

Umbriel and Ariel are names from Alexander Pope's *The Rape of the Lock,* and the rest are from Shakespeare's *A Midsummer Night's Dream* and *The Tempest* (an Ariel also appears in *The Tempest*). Spectra show that they contain frozen water, and their surfaces are very dark, so planetary scientists assumed they were made of dirty ices. Little more was known of the moons before Voyager 2 flew through the system.

The Voyager 2 cameras discovered ten moons too small to have been seen from Earth. Also, with the construction of new-generation telescopes and new imaging techniques, astronomers have found even more small moons orbiting Uranus, and there are almost certainly more still to be found.

The smaller moons are all as dark as coal. They are probably icy worlds with surfaces that have been darkened by impacts. Also they orbit inside the radiation belts, and the radiation may convert methane trapped in their ices into dark carbon deposits and further darken their surfaces.

The five large moons are all tidally locked to Uranus, and that means their south poles were pointed toward the sun in 1986; it was these hemispheres that were photographed by Voyager 2. Their densities suggest that they contain relatively large rock cores surrounded by icy mantles (■ Figure 24-10).

Oberon, the outermost of the large moons, has a dark, cratered surface, but it was once an active moon (■ Figure 24-11). A large fault crosses the sunlit hemisphere, and dark material, perhaps dirty water "lava," appears to have flooded the floors of some craters.

Titania is the largest of the five moons and has a heavily cratered surface, but it lacks the largest craters (Figure 24-11). This suggests that soon after it formed, after most of the large debris had been swept up, Titania underwent an active phase in which internal melting flooded the surface with water and erased the early craters. Since then, more craters have formed, but few are as large as those that were erased. Another clue to Titania's past activity is the network of faults that crosses its surface.

Umbriel, the third moon inward, is a very dark, cratered world with no sign of faults or surface activity (Figure 24-11). It is the darkest of the moons, with an albedo of only 0.16 compared with 0.25 to 0.45 for the other moons. Planetary astronomers suspect that Umbriel's crust is a dark mixture of rock and ice. A bright crater floor in one region suggests that some clean ice may lie at shallow depths in some regions.

Ariel has the brightest surface of the five major satellites and shows clear signs of geological activity. It is crossed by faults over 10 km deep, and some regions appear to have been smoothed by resurfacing (Figure 24-11). Crater counts show that the smoothed regions are younger than the other regions. Apparently, past melting in the interior has caused geological activity on the moon's surface. Ariel may have been subject to tidal heating caused by an orbital resonance with Miranda and Umbriel. If you could visit the surface of Ariel, the boots of your spacesuit would probably kick up nothing more than dirty ices; the interesting features would be the broad valleys and faults.

Miranda is a mysterious moon. As you can see in Figure 24-11, it is the smallest of the five large moons, but it appears to have been the most active. In fact, its active past appears to have been quite unusual. One astronomer referred to it as a "moon designed by a committee." Miranda is marked by oval patterns of grooves known as **ovoids** (■ Figure 24-12). These features were originally thought to have been produced when Miranda was shattered by a major impact and the mutual gravity of the fragments pulled them back together to reassemble the little moon. More recent studies of the ovoids show that they are associated with faults, ice-lava flows, and rotated blocks of crust. These features suggest that a major impact was not involved. Rather, large convection currents in Miranda's icy mantle may have forced material upward to create the ovoids.

Certainly, Miranda has had a violent past. Near the equator, a huge cliff rises 20 km. If you stood in your spacesuit at the top of the cliff and dropped a rock over the edge, it would fall for 10 minutes before hitting the bottom. The cliff and the ovoids are old, however. Miranda is no longer geologically active, and you can read the history of its past on its disturbed surface.

The peculiar geology of Miranda illustrates the problem astronomers face in understanding these icy moons. Has Miranda

■ **Figure 24-10**

The densities of the five major satellites of Uranus suggest that they contain relatively large rocky cores with mantles of dirty ice. The size of Earth's moon is shown for comparison.

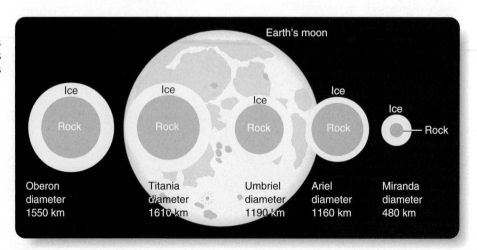

Earth's moon

Ice · Rock
Oberon diameter 1550 km

Ice · Rock
Titania diameter 1610 km

Ice · Rock
Umbriel diameter 1190 km

Ice · Rock
Ariel diameter 1160 km

Ice · Rock
Miranda diameter 480 km

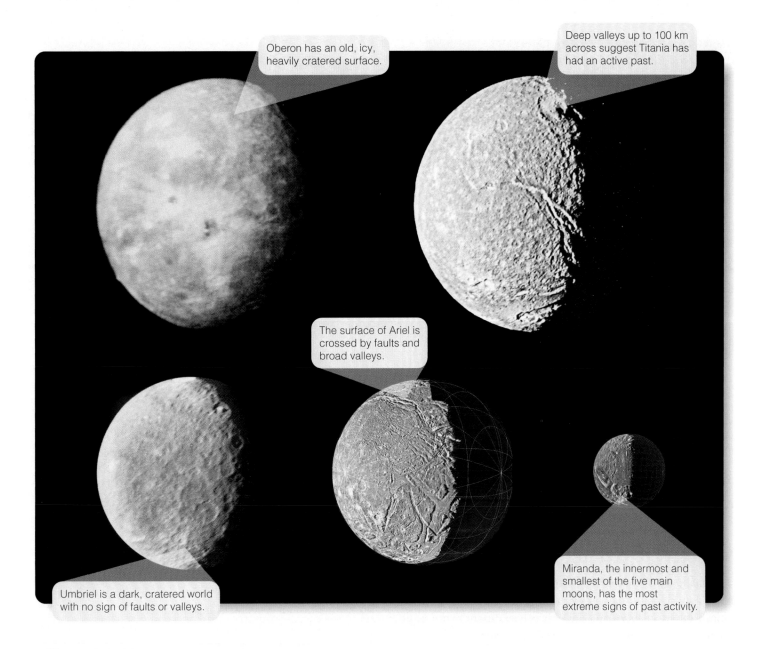

■ Figure 24-11

The five largest moons of Uranus range from the largest, Titania, 46 percent the diameter of Earth's moon, down to Miranda, only 14 percent the diameter of Earth's moon. For a better view of Miranda, see the next figure. The moons are shown to scale here. (Ariel and Miranda: U.S. Geological Survey, Flagstaff, Arizona; Other images: NASA)

been dominated by interior or exterior factors? If the ovoids were caused by convection in Miranda's icy mantle, then heat rising from its interior must be the dominant factor. Miranda is too small to have kept its heat of formation or to have been so severely heated by radioactive decay. Tidal heating could have heated Miranda if its orbit was made slightly elliptical by a resonance with other moons. Your experience with tidal heating in other moons can help you understand how internal factors could have driven slow convection currents rising through Miranda's icy mantle to create its ovoids.

But external factors are also a possibility. Comets sweep through the solar system, and they must occasionally strike the planets and moons. The Jovian planets have no surfaces and would retain no permanent scars of such impacts, as Earth's astronomers saw dramatically in the summer of 1994 when a comet struck Jupiter. The moons in the outer solar system, however, could be battered or even disrupted by such impacts, and a major impact could have broken Miranda up and allowed it to re-form.

Although planetary scientists now prefer internal heat to explain Miranda's geology, you should make a note that external factors such as major impacts do occur. Miranda is a reminder that modern astronomy recognizes the importance of catastrophic events in the history of the solar system. (See Window

■ Figure 24-12

Miranda, the smallest of the five major moons of Uranus, is only 480 km in diameter, but its surface shows signs of activity. This photomosaic of Voyager 2 images (a) reveals that it is marked by great oval systems of grooves. In the map (b), the smallest features detected are about 1.5 km in diameter. The cliff marked by an arrow near the bottom of the map is 20 km high. (U.S. Geological Survey, Flagstaff, Arizona; map by P. Bridges)

on Science 19-1.) You will see more hints of catastrophic events when you visit the moons of Neptune.

A History of Uranus

The challenge of comparative astronomy is to tell the story of a world, and Uranus may present the biggest challenge of all the objects in our solar system. Not only is it so distant that it is difficult to study, but it is also peculiar in many ways.

Presumably, Uranus formed from the solar nebula, as did the other Jovian planets, but it must have grown more slowly, perhaps because the solar nebula was less dense beyond Saturn. Also, objects orbit slowly in the outer solar system, so protoplanets don't sweep up material rapidly out there. By the time the proto-Uranus had grown massive enough to begin trapping hydrogen, the sun had turned on, and the solar nebula was being blown away. As one astronomer said, "By the time Uranus got all dressed up, the party was over." In fact, some theorists believe Uranus and Neptune could not have grown to their present size in the slow-moving orbits they now occupy. Sophisticated mathematical models of planet formation in the solar nebula suggest that Uranus and Neptune might have formed closer to the sun, in the neighborhood of Jupiter and Saturn. Gravitational interactions among the growing planets could have moved Uranus and Neptune outward to their present locations. This is an exciting hypothesis, and it illustrates how uncertain the histories of the outer planets really are.

The highly inclined axis of Uranus may have originated late in its formation when it was struck by a planetesimal as large as Earth. That impact may have disturbed the interior of the world and caused it to lose much of its heat. Uranus now radiates less than 10 percent more heat than it receives from the sun. There is now just enough heat flowing outward to drive circulation in its slushy mantle and generate its magnetic field.

Impacts have been important in the history of the moons and rings of Uranus. All of the moons are cratered, and some show signs of large impacts. Planetesimals and the nuclei of comets striking the moons may create debris that becomes trapped among the orbits of the smaller moons to produce the narrow rings. The rings could not have lasted since the formation of the planet, so they must be replenished with fresh material now and then. When the final history of Uranus is written, it will surely include some dramatic impact events.

Building Scientific Arguments

How do astronomers know what the interior of Uranus is like?

This argument must combine observation with theory. Obviously, you can't see inside the planet, so planetary scientists are limited to a few basic observations that they must connect with models in a chain of inference to describe the interior. First, the size of Uranus can be found from its angular diameter and distance, and you can find its mass by observing the orbital radii and periods of the orbits of its moons. Its mass divided by its volume equals its density, which implies that, for a Jovian planet, it must contain a higher proportion of dense material such as ice and rock. Add to that the oblateness, rotation period, and chemical composition obtained from spectra, and you have the data needed to build a mathematical model of the interior. Such models predict a core of heavy elements and a mantle of icy slush. The models cannot make the planet as flattened as is observed if the heavy elements are highly concentrated toward the core, so the mantle must contain denser material such as rock as well as ices.

Going from observable properties to unobservable properties is the heart of astronomical research. Now use this approach to build a new argument: **What observational properties of the rings of Uranus show that small moons must orbit among the rings?**

■　　■　　■

Connections: When Voyager 2 departed Uranus, Earth's astronomers were left grappling with new planetary puzzles. Over the following years, some of the puzzles were explained, but in August 1989 Voyager 2 flew past Neptune, and the game began all over again.

(24-2) Neptune

URANUS AND NEPTUNE are often discussed together. They are about the same size and density and are probably very similar planets in some ways. They do differ, however, in certain respects. Neptune has a peculiar ring and a peculiar satellite system. Even the discovery of Neptune was different from the discovery of Uranus.

The Discovery of Neptune

The discovery of Neptune triggered one of the greatest controversies in science. For over a century people told the story and took sides, but only in the last few years has the real story become known. You can follow the story and decide which side you favor.

In 1843, the young English astronomer John Couch Adams (1819–1892) completed his degree in astronomy and immediately began the analysis of one of the great problems of 19th-century astronomy. William Herschel had discovered Uranus in 1781, but earlier astronomers had seen the planet as early as 1690 and had plotted it on charts as a star. When 19th-century astronomers tried to combine this data, they found that no planet obeying Newton's laws of motion and gravity could follow such an orbit.

Some astronomers suggested that the gravitational attraction of an undiscovered planet was causing the discrepancies. Adams began with the variation in the positions of Uranus, never more

than 2 minutes of arc, and by October 1845 he had, through a laborious and difficult calculation, computed the orbit of the undiscovered planet. He sent his prediction to the Astronomer Royal, Sir George Airy, who passed it on to an observer who began a painstaking search of the area star by star.

Meanwhile, the French astronomer Urbain Jean Leverrier (1811–1877) made the same calculations and sent his predicted position of the planet to Johann Galle at the Berlin Observatory. Galle received Leverrier's prediction on the afternoon of September 23, 1846, and, after searching for 30 minutes that evening, found Neptune. It was only 1° from the position predicted by Leverrier.

The discovery of a new planet caused a sensation, but English astronomers didn't like a Frenchman getting all the credit. After all, they said, the planet was found only 2° from the position predicted by the orbit computed by their own astronomer, Adams. When the English announced Adams's work, the French suspected that he had plagiarized the calculations, and the controversy was bitter, as controversies between England and France often are. Historians of science have repeated the story for over a century of how the young English astronomer missed his chance because the astronomers in charge of the search were careless and slow.

Original papers related to Adam's calculations were lost for 30 years, but when they were found in 1998, they painted a different picture. Adams did the calculations correctly, but he only computed the orbit for the new planet. He didn't actually calculate its position in the sky; that was left to the English astronomers conducting the search. Once the new planet was discovered, the English astronomers, out of national pride, pressed Adam's case further than they should have.

For 150 years, astronomers told the story and took sides or gave both astronomers equal credit. In fact, it seems clear that Leverrier deserves credit for making an accurate, useful prediction of a position on the sky and pressing it aggressively on an astronomer who had the experience to make the search. But you should give both astronomers credit for solving a difficult problem that was one of the great challenges of the age.

Nevertheless, Leverrier and Adams could have been beaten to the discovery had Galileo paid less attention to Jupiter and more attention to what he saw in the background. Modern studies of Galileo's notebooks show that he saw Neptune on December 24, 1612, and again on January 28, 1613, but he plotted it as a star in the background of drawings of Jupiter. It is interesting to speculate about the response of the Inquisition had Galileo proposed that a planet existed beyond Saturn. Unfortunately for history (perhaps fortunately for Galileo), he did not recognize Neptune as a planet, and its discovery had to wait another 234 years.

The discovery of Neptune has been repeated by historians as a triumph for Newtonian physics: The three laws of motion and the law of gravity had proved sufficient to predict the position and orbit of an unseen planet. Thus, the discovery of Neptune was fundamentally different from the discovery of Uranus—the exis-

Neptune in 1989

Great Dark Spot

Visual-wavelength image
from Voyager 2

1996

2002

Visual-wavelength
images from
Hubble Space Telescope

1998

■ **Figure 24-13**

Because Neptune is inclined almost 29 degrees in its orbit, it experiences seasons that each last about 40 years. Since 1989, spring has come to the southern hemisphere, and the weather has clearly changed, which is surprising because sunlight on Neptune is 900 times dimmer than on Earth. (NASA, L. Sromovsky, and P. Fry, University of Wisconsin–Madison.)

tence of Neptune was predicted from basic laws. But a modern analysis shows that both Leverrier and Adams made unwarranted assumptions about Neptune's distance. By good fortune their assumptions made no difference in the 19th century, and the planet was close to the positions they predicted. Do you think they deserve less credit? They tried, and the other astronomers of the world didn't.

The Atmosphere and Interior of Neptune

Little was known about Neptune before the Voyager 2 spacecraft swept past it in August 1989. Seen from Earth, Neptune is a tiny, blue-green dot never more than 2.3 seconds of arc in diameter. Astronomers knew it was almost four times the diameter of Earth and about 4 percent smaller in diameter than Uranus (**Celestial Profile 10**). Spectra revealed that its blue-green color was caused by methane in its hydrogen-rich atmosphere. Methane absorbs red light, making Neptune look blue-green. Neptune's density showed that it was a Jovian planet rich in hydrogen, but almost no detail was visible from Earth, so even its period of rotation was poorly known.

Voyager 2 passed only 4905 km above Neptune's cloud tops, closer than any spacecraft had ever come to one of the Jovian planets. The Voyager images revealed that Neptune is marked by dramatic belt–zone circulation parallel to the planet's equator. Voyager 2 saw at least four cyclonic disturbances. The largest, dubbed the Great Dark Spot, looked similar to the Great Red Spot on Jupiter (■ Figure 24-13). The Great Dark Spot was located in the southern hemisphere and rotated counterclockwise, with a period of about 16 days. It appeared to be caused by gas rising from the planet's interior. When the Hubble Space Telescope began imaging Neptune in 1994, the Great Dark Spot was gone, and

new cloud features were seen appearing and disappearing in Neptune's atmosphere (Figure 24-13). Evidently, the cyclonic disturbances on Neptune are not nearly as long-lived as Jupiter's Great Red Spot.

The Voyager 2 images reveal other cloud features standing out against the deep blue of the methane-rich atmosphere. The white clouds are made up of crystals of frozen methane and range up to 50 km above the deeper layers, just where the temperature in Neptune's atmosphere is low enough for rising methane to freeze into crystals. (See Figure 24-5.) Presumably these features are related to rising convection currents that form clouds high in Neptune's atmosphere, where they catch sunlight and look bright. Special filters can reveal these bands in visual-wavelength images and at infrared wavelengths (■ Figure 24-14). Observations made by the Hubble Space Telescope suggest that atmospheric activity on Neptune may be related in some way to flares and other eruptions on the sun, but further observations are needed to explore this connection.

As on the other Jovian worlds, high winds circle Neptune parallel to its equator, but the winds blow at very high velocities and tend to blow backward—against the rotation of the planet. Why Neptune should have such high winds is not understood, and it is part of the larger problem of the belt–zone circulation. Now that you have seen at least traces of belts and zones on all four of the Jovian planets (assuming the faint clouds on Uranus are traces of belt–zone circulation), you can ask what drives this circulation. Because the belts and zones remain parallel to the planet's equator even when the planet rotates at a high inclination, as in the case of Uranus, it seems reasonable to believe that

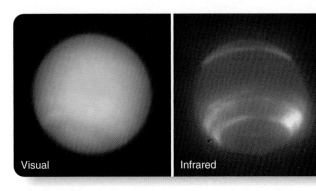

■ Figure 24-14

(a) This visual-wavelength image was made through filters that make belts of methane stand out. (NASA and Erich Karkoschka) (b) The color-corrected infrared image at right was recorded using one of the Keck 10-meter telescopes and adaptive objects. It shows the location of methane cloud belts around the planet. (U. C. Berkeley/W. M. Keck Observatory)

the atmospheric circulation is dominated by the rotation of the planet and not by solar heating.

The same observations that helped define the interior of Uranus can be applied to Neptune. The interior is believed to contain a heavy-element core, smaller than that in Jupiter, surrounded by a deep mantle of slushy ice mixed with heavier material such as rock. Neptune's magnetic field is a bit less than half as strong as Earth's and is tipped 47° from the axis of rotation. It is offset 55 percent of the way to the surface (Figure 24-7). As in the case of Uranus, the field is probably generated by the dynamo effect acting in the conducting fluid mantle.

Neptune has more internal heat than Uranus, and part of that may be generated by radioactive decay in the minerals in its interior. Some of the energy may be released by denser material falling inward, including, as in the case of Uranus, diamond crystals formed by the disruption of methane.

Neptune is a tantalizing world just big enough to be imaged by the Hubble Space Telescope but far enough away to make it difficult to study. The data from Voyager 2 revealed one detail invisible from Earth—rings.

The Rings of Neptune

Astronomers on Earth saw hints of rings when Neptune occulted stars, but the rings were not firmly recognized until Voyager 2 flew past the planet in 1989.

Look again at The Rings of Uranus and Neptune on pages 598–599 and compare the rings of Neptune with those of Uranus. Notice two additional points:

④ Neptune's rings, named after the astronomers involved in the discovery of the planet, are similar to those of Uranus but contain more dust.

⑤ Also notice a new way that rings can interact with moons; one of Neptune's moons is producing short arcs in the outermost ring.

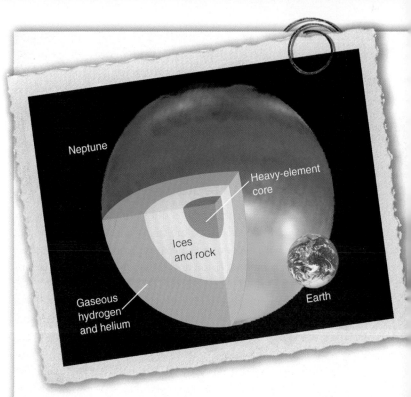

Neptune was tipped slightly away from the sun when the Hubble Space Telescope recorded this image. The interior is much like Uranus's, but there is more heat. (NASA)

Celestial Profile 10: Neptune

Motion:

Average distance from the sun	30.0611 AU (44.971 × 10⁸ km)
Eccentricity of orbit	0.0100
Maximum distance from the sun	30.4 AU (45.4 × 10⁸ km)
Minimum distance from the sun	29.8 AU (44.52 × 10⁸ km)
Inclination of orbit to ecliptic	1°46′27″
Average orbital velocity	5.43 km/s
Orbital period	164.793 y (60,189 days)
Period of rotation	16ʰ3ᵐ
Inclination of equator to orbit	28°48′

Characteristics:

Equatorial diameter	49,500 km (3.93 $D_\oplus$)
Mass	1.030 × 10²⁶ kg (17.23 $M_\oplus$)
Average density	1.66 g/cm³
Gravity	1.19 Earth gravities
Escape velocity	25 km/s (2.2 $V_\oplus$)
Temperature at cloud tops	−216°C (−357°F)
Albedo	0.35
Oblateness	0.027

Personality Point:

Because the planet Neptune looked so blue, astronomers named it after the Roman god of the sea, Neptune (Poseidon to the Greeks). His wife was Amphitrite, granddaughter of Ocean, one of the Titans. Neptune controlled the storms and waves and was a powerful god not to be trifled with. His three-pronged trident became the symbol for the planet Neptune.

The rings of Neptune resemble the rings of Uranus in one very important way. They can't be primordial. That is, they can't have lasted from the formation of Neptune. Evidently, impacts on moons occasionally scatter debris through the satellite system, and some of it falls into the rings in the most stable places among the orbits of the moons. Planetary rings are constantly being remade.

The Moons of Neptune

Before Voyager 2 visited Neptune, only two moons were known, but Voyager 2 discovered six more small moons, and a few more small moons have been found since using Earth-based telescopes. The two largest moons, Triton and Nereid, have been a puzzle for years because of their peculiar orbits.

Triton has a nearly circular orbit, but it travels backward—clockwise as seen from the north. This makes Triton the only large satellite in the solar system with a retrograde (backward) orbit. The other satellite visible from Earth, Nereid, goes in the right direction, but its orbit is highly elliptical and very large (■ Figure 24-15). Nereid takes 359.4 days to orbit Neptune once. Many astronomers have speculated that the orbits of the two moons are evidence of a violent event. An encounter with a massive planetesimal may have disturbed the moons, or Triton itself may have been captured into orbit during a close encounter with Neptune.

Voyager 2 found six more moons orbiting Neptune, but they all lie close to the planet among the rings. No new moons were found beyond the orbit of Triton, and some astronomers have suggested that Triton, in its retrograde orbit, would have consumed any moons near it.

The Voyager 2 photographs show that Triton is a highly complex moon. Although it is only 1360 km in radius (78 percent the radius of Earth's moon), it is so cold (34.5 K, or −398°F) that it can hold an atmosphere of nitrogen with some methane. Its air

is 10^5 times less dense than Earth's atmosphere. Although a few wisps of haze are visible in the photographs, the atmosphere is transparent, and the surface is easily visible (■ Figure 24-16).

The surface of Triton is evidently composed of ices. The surface ice is dominated by frozen nitrogen with some methane,

■ Figure 24-16

(a) Triton's south pole (bottom) had been in sunlight for 30 years when Voyager 2 flew past in 1989. The frozen nitrogen in the polar cap appears to be vaporizing, perhaps to refreeze in the darkness at the north pole. The dark smudges are produced when liquid nitrogen in the crust vaporizes and drives nitrogen geysers. Methane carried along in the nitrogen is turned to black compounds by sunlight and makes the black smudges. (b) Roughly round basins on Triton may be old impact basins flooded repeatedly by liquids from the interior. Notice the small number of craters on Triton, a clue that it is a partially active world. (NASA)

Visual-wavelength images

■ Figure 24-15

Nereid has a large, elliptical orbit around Neptune, but Triton follows a small, circular, retrograde orbit. At a distance of 30 AU from Earth, Triton is never more than 16 seconds of arc from the center of Neptune.

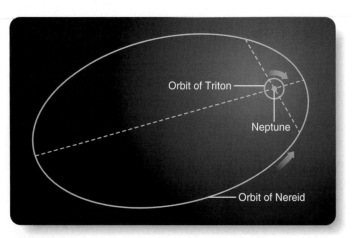

carbon monoxide, and carbon dioxide. This is consistent with the nitrogen-rich atmosphere. Some regions of the surface look dark and may be slightly older terrain that has grown dark as sunlight has converted methane into dark organic compounds. The south pole of Triton had been turned toward the sun for 30 years when Voyager 2 flew past, and deposits of nitrogen frost in a large polar cap appeared to be vaporizing there and refreezing in the darkness of the north pole (Figure 24-16a). The cycle of nitrogen on Triton resembles in some ways the cycle of carbon dioxide on Mars.

The icy surface of Triton suggests that it has been active recently and may still be active. Few craters mark the surface of the icy world, and comparison with the cratering rate you might expect for a moon located so close to the Kuiper Belt suggests the average age of the surface is no more than 100 million years and may be even less. Some process has erased older craters. Long linear features appear to be fractures in an icy crust, and some roughly round basins appear to have been flooded time after time by liquids from the interior (Figure 24-16b). Triton is much too cold for the liquid to be molten rock or even liquid water. Rather, the flooding may have been caused by water containing agents such as ammonia, which would lower the freezing point of the liquid. It isn't possible to say at present whether Triton is still active, but 100 million years isn't very long in the history of a world. Triton may still suffer periodic eruptions and flooding.

Another form of activity on Triton is the dark smudges visible in the bright nitrogen ices near the south pole. These appear to be caused by nitrogen ice beneath the crust. Warmed slightly by the sun, the nitrogen ice can change from one form of nitrogen ice to another and release heat that could vaporize some of the nitrogen. This nitrogen vapor vents through the crust to form nitrogen gas geysers. Sunlight converts some of the methane carried along with the nitrogen into dark organic material, which falls to the surface and accounts for the dark smudges on the ice.

If you want to visit Triton, you should schedule visits to sample the dirty crust contaminated by one of these nitrogen geysers and the material in one of the flooded basins. Mineral samples from those areas would help explain Triton's geological activity, but remember to refrigerate your sample boxes. The minerals there are frozen gases that are not stable at room temperature.

Active worlds must have a source of energy, so you are probably wondering where Triton gets its energy. Triton is big enough to retain some heat from low-level radioactivity in its interior. That may be enough to melt exotic ices and cause flooding on the surface. The nitrogen geysers appear to be powered not from the interior but from sunlight. In fact, Triton is very efficient at absorbing sunlight. Its thin atmosphere allows sunlight to fall on the crust, which is composed of ices that are partially transparent to light. As the light penetrates into the ice and is absorbed, it is converted to heat. The ices, however, are not very transparent to infrared radiation, and the heat is trapped in the crust. In this way, the crust of Triton appears to be heated, in part, by an icy form of the greenhouse effect. The heat could transform nitrogen ice

from one form into another form and release more heat to drive the geysers.

This low-level heating would not be enough to erase nearly all of Triton's craters, so some planetary astronomers wonder if Triton might have been captured by Neptune within the last billion years. Tidal forces from such a capture would cause enough tidal heating to melt Triton and totally resurface it. Triton may still be active with residual heat.

The History of Neptune

Can you tell the story of Neptune? In some ways it seems to be a simple world, but its magnetic field is peculiar, and its moons and rings demand careful attention.

Neptune is much like Uranus, and you can assume that it formed from the solar nebula much as did Uranus. It is a mostly liquid world with a dense core. Its hazy atmosphere, marked by changing cloud patterns, hides a mantle of partially frozen ices where astronomers suspect the dynamo effect generates its off-center magnetic field.

Neptune's satellite system suggests a peculiar history. Triton, the largest moon, revolves around Neptune in a retrograde orbit, while Nereid's long-period orbit is highly elliptical. These orbital oddities suggest that the satellite system may have been disturbed by an encounter with some other object, such as a large planetesimal or the capture of Triton. You have seen evidence in other satellite systems of impacts with large objects, so such an event is not unreasonable.

At least six smaller moons orbit Neptune, but they lie near its ring system. Because the rings are bright in forward scattering, you can conclude that they contain some dust, and the shepherding of small satellites must confine their width and produce the observed arcs. Such delicate rings could not have survived since the origin of the planets, so you must suppose that the rings are occasionally supplied with fresh particles generated by collisions between comets and Neptune's moons. Once again, the evidence of major impacts in the solar system's history assures you that such impacts do occur.

Building Scientific Arguments

Why is Neptune blue but its clouds white?

To solve this problem you must build a scientific argument that follows a process step by step. When you look at something, you really turn your eyes toward it and receive light from the object. When you look at Neptune, the light you receive is sunlight that is reflected from various layers of Neptune and journeys to your eyes. Because sunlight contains a distribution of photons of all visible wavelengths, it looks white to human eyes, but sunlight entering Neptune's atmosphere must pass through hydrogen gas that contains a small amount of methane. While hydrogen is transparent, methane is a good absorber of longer wavelengths, so red photons are more likely to be absorbed than blue photons. Once the light is scattered from deeper layers, it must run this methane gauntlet again to emerge from the atmosphere, and again red photons are more likely to be absorbed. The light that finally emerges from Neptune and eventually reaches your eyes is poor in longer wavelengths and thus looks blue.

The methane-ice-crystal clouds lie at high altitudes, so sunlight does not have to penetrate very far into Neptune's atmosphere to reflect off the clouds, and consequently it loses many fewer of its red photons. The clouds look white.

This discussion shows how a careful, step-by-step analysis of a natural process can help you better understand how nature works. For example, build a step-by-step argument to answer the following: **Where does the energy come from to power Triton's surface geysers?**

■   ■   ■

Connections: Neptune was the last planet visited by Voyager 2, which is now journeying out of our solar system. It will not pass near any other worlds unless, millions of years from now, it enters some distant planetary system. Nevertheless, you have one more planet to visit. Turn up the heater in your spacesuit. It is a cold place.

24-3 Pluto

PLUTO, THE FARTHEST OUTPOST of our solar system, is a small, icy world. Discovered through a combination of ingenuity, perseverance, and luck, it is now known to have a satellite. But some astronomers are wondering if it is really a planet.

The Discovery of Pluto

Percival Lowell (1855–1916) was fascinated with the idea that an intelligent race built the canals he thought he could see on Mars (Chapter 22). Lowell founded Lowell Observatory primarily for the study of Mars. Later, perhaps to improve the reputation of his observatory, he began to search for a planet beyond Neptune.

Lowell used the same method that Adams and Leverrier had used to predict the position of Neptune. Working from the observed irregularities in the motion of Neptune, Lowell predicted the location of an undiscovered planet of about 7 Earth masses, which, he concluded, would look like a 13th-magnitude stellar image in eastern Taurus. Lowell searched for the planet photographically until his death in 1916.

In the late 1920s, a 22-year-old amateur astronomer, Clyde Tombaugh, began using a homemade 9-in. telescope to sketch Jupiter and Mars from his family's wheat farm in western Kansas. He sent his drawings to Lowell Observatory, and the observatory director hired him without an interview. The young Tombaugh bought a one-way train ticket for Flagstaff not knowing what his new job would be like.

The observatory director set Tombaugh to work photographing the sky along the ecliptic around the predicted position of the planet. The search technique was a classic method in astronomy. Tombaugh obtained pairs of 14 × 17-inch glass plates exposed two or three days apart. To search a pair of plates, he mounted them in a blink comparator, a machine that allowed him to look through a microscope at a small spot on one plate and then at the flip of a lever see the same spot on the other plate. As he blinked back and forth, the star images did not move, but a planet would have moved along its orbit during the two or three

Visual-wavelength images

Pluto was here on the first plate.

a   b

days that elapsed before the second plate was exposed. So Tombaugh searched the giant plates, star image by star image, looking for an image that moved. A single pair of plates could contain 400,000 star images. He searched pair after pair and found nothing.

The observatory director turned to other projects, and Tombaugh, working alone, expanded his search to cover the entire ecliptic. For almost a year, Tombaugh exposed plates by night and blinked plates by day. Then on February 18, 1930, nearly a year after he had left Kansas, a quarter of the way through a pair of plates, he found a 15th magnitude image that moved (■ Figure 24-17). He later remembered, "'Oh,' I thought, 'I had better look at my watch; this could be a historic moment.' It was within about 2 minutes of 4 PM [MST]." The discovery was announced on March 13, the 149th anniversary of the discovery of Uranus and the 75th anniversary of the birth of Percival Lowell. The planet was named Pluto after the god of the underworld and, in a way, after Lowell, because the first two letters in Pluto are the initials of Percival Lowell.

The discovery of Pluto seemed a triumph of discovery by prediction, but Tombaugh sensed something was wrong from the first moment he saw the image. It was moving in the right direction by the right amount, but it was 2.5 magnitudes too faint. Clearly, Pluto was not the 7-Earth-mass planet that Lowell had predicted. The faint image implied that Pluto was a small planet with a mass too low to seriously alter the motion of Neptune.

Later analysis has shown that the variations in the motion of Neptune, which Lowell used to predict the location of Pluto, were random uncertainties of observation and could not have led to a trustworthy prediction. The discovery of the new planet only 6° from Lowell's predicted position was apparently an accident, which proves that if you search long enough and know what to expect, you are likely to find something (**Window on Science 24-1**).

Pluto as a World

Pluto is very difficult to observe from Earth. Only a bit larger than 0.1 second of arc in diameter, it is only 65 percent the diameter of Earth's moon and shows little surface detail. The best photos by the Hubble Space Telescope reveal large areas of light and dark terrain (**Celestial Profile 11**), promising that Pluto will reveal itself to be an interesting world when spacecraft finally visit it.

Most planetary orbits in our solar system are nearly circular, but Pluto's is quite elliptical (see Figure 1-7). On the average, it is the most distant planet, but it can come closer to the sun than

■ **Figure 24-17** ◄

Pluto is small and far away, so its image is indistinguishable from that of a star on most photographs. Clyde Tombaugh discovered the planet in 1930 by looking for an object that moved relative to the stars on a pair of photographs taken a few days apart. (Lowell Observatory photographs)

Pluto is a small, low-density world composed of ices and rock. No spacecraft has every visited it, so the best image available shows only bright and dark regions. (NASA)

Celestial Profile 11: Pluto

Motion:

Average distance from the sun	39.44 AU (59.00 × 10⁸ km)
Eccentricity of orbit	0.2484
Maximum distance from the sun	49.24 AU (73.66 × 10⁸ km)
Minimum distance from the sun	29.64 AU (44.34 × 10⁸ km)
Inclination of orbit to ecliptic	17°9′3″
Average orbital velocity	4.73 km/s
Orbital period	247.7 y (90,465 days)
Period of rotation	$6^d9^h21^m$
Inclination of equator to orbit	119.6°

Characteristics:

Equatorial diameter	2370 km (0.19 $D_\oplus$)
Mass	1.2×10^{22} kg (0.002 $M_\oplus$)
Average density	2.0 g/cm³
Surface gravity	0.06 Earth gravity
Escape velocity	1.2 km/s (0.11 $V_\oplus$)
Surface temperature	−230°C (−382°F)
Albedo	0.5

Personality Point:

Pluto (also known as Hades) was the ruler of the underworld. He kidnapped beautiful Persephone, maiden of spring, and made her his wife. But Zeus ruled that she need only spend a third of each year in the underworld; during that time, Earth grieves for her and experiences winter. Pluto possessed a helmet that made its wearer invisible—appropriate for a world so difficult to see.

Scientific Discoveries

In 1928, Alexander Fleming noticed that bacteria in a culture dish were avoiding a spot of mold. He went on to discover penicillin. In 1895, Conrad Roentgen noticed a fluorescent screen glowing in his laboratory when he experimented with other equipment. He discovered X rays. In 1896, Henri Becquerel happened to place a mineral containing uranium on a photographic plate safely wrapped in black paper. The plate was fogged, and Becquerel discovered natural radioactivity. Like many discoveries in science, these seem to be accidental; but, as you have seen in the case of the discovery of Uranus, Neptune, and Pluto, "accidental" doesn't quite describe what happened.

The most important discoveries in science are those that totally change the way people think about nature, and it is very unlikely that anyone would predict such discoveries. For the most part, scientists work within a paradigm (Window on Science 4-1), a set of models, theories, and expectations about nature, and it is very difficult to imagine natural events that lie beyond that paradigm. Ptolemy, for example, could not have imagined a universe filled with galaxies because they lie totally beyond his geocentric paradigm. That means that the most important discoveries in science are almost bound to arise unexpectedly.

An unexpected discovery, however, is not the same as an accidental discovery. Fleming discovered penicillin in his culture dish not because he was the first to see it, but because he had studied bacterial growth for many years; so, when he saw what many others must have seen, he recognized it as important. Roentgen recognized the importance of the glowing screen in his lab, and Becquerel didn't simply discard the fogged photographic plate. Long years of experience prepared them to recognize the significance of what they saw.

A study of discoveries made with new telescopes from Galileo to the present revealed that each time astronomers built a new telescope that significantly surpassed existing telescopes, they discovered new and wonderful things, such as the moons of Jupiter, binary stars, spiral galaxies, and quasars. But nearly all the discoveries made with these telescopes were unpredicted.

While it seemed possible to predict the presence of Neptune from the orbital motion of Uranus, this is not typical of scientific discoveries. Most important scientific discoveries lie

X-ray image

The discovery of pulsars, spinning neutron stars, was totally unexpected, as important scientific discoveries often are. (NASA/McGill/V. Kaspi et al.)

outside the current paradigm and so must be made by scientists who are searching for something else or studying some other phenomenon. That means that scientists pursuing true basic research are rarely able to explain the potential value of their work, just as astronomers can't predict what they will discover with a new telescope. That's why science is so exciting.

■ **Figure 24-18**

(a) A high-quality ground-based photo shows Pluto and its moon, Charon, badly blurred by seeing. (NASA)
(b) The Hubble Space Telescope image clearly separates the planet and its moon and allows more accurate measurements of the position of the moon. (R. Albrecht, ESA/ESO Space Telescope European Coordinating Facility, NASA)

Visual-wavelength image

a

Visual-wavelength image

b

Neptune. In fact, from January 21, 1979, to March 14, 1999, Pluto was closer to the sun than Neptune. The planets will never collide, however, because Pluto's orbit is inclined 17°.

If you land on the surface of Pluto, your spacesuit will have to work hard to keep you warm. Orbiting so far from the sun, Pluto is cold enough to freeze most compounds that you think of as gases, and spectroscopic observations have found evidence of solid nitrogen ice on its surface with traces of frozen methane and carbon monoxide. The daytime temperature of about 50 K (−370° F) is enough to vaporize some of the nitrogen and carbon monoxide and a little of the methane to form a thin atmosphere around Pluto. This atmosphere was detected in 1988 when Pluto passed in front of (occulted) a distant star, and the starlight faded gradually rather than winking out suddenly.

Pluto's largest moon was discovered on photographs in 1978. It is very faint and about half the diameter of Pluto (■ Figure 24-18). The moon was named Charon after the mythological ferryman who transports souls across the river Styx into the underworld. Two small moons were found in 2005.

The discovery of Charon was important for a number of reasons. Charon orbits Pluto in a nearly circular orbit in the plane of Pluto's equator. Observations show that the moon and Pluto are tidally locked to each other, and that Pluto's axis of rotation is highly inclined (■ Figure 24-19). Furthermore, the orbital motion of Charon allows the calculation of the mass of Pluto. Charon orbits 19,640 km from Pluto with an orbital period of 6.387 days. Kepler's third law reveals that the mass of the system is 6.5×10^{-9} solar masses or about 0.002 Earth mass. Most of this mass is Pluto, which seems to be about 12 times more massive than Charon.

The mass of a body is important in astronomy because mass divided by volume is density. The density of Pluto is about 2 g/cm³, and the density of Charon is just a bit less. Pluto and Charon must contain about 35 percent ice and 65 percent rock. Spectra of Charon show that the small moon has a surface that is mostly water ice. Perhaps it has lost its more volatile compounds because of its lower escape velocity. Water ice at Charon's surface temperature is no more volatile than is a piece of rock on Earth, so a water ice surface could last a long time.

Both Pluto and Charon go through dramatic seasons much like those on Uranus as they circle the sun in their highly inclined orbit. This may cause changes in Pluto's atmosphere, as it grows warmer when it is closest to the sun, as it was in the late 1980s, and then freezes as it draws away.

Although NASA is currently developing a mission to Pluto, no spacecraft has visited the little world. You must strain your imagination and try to visualize the surface of Pluto. Old surfaces should be cratered, so you might imagine an icy, cratered world. The tidal lock between Pluto and Charon suggests an age of tidal heating that may have driven geological activity and left traces on the surfaces of these worlds. However, the seasonal vaporization and refreezing of ices may have erased or covered much of the

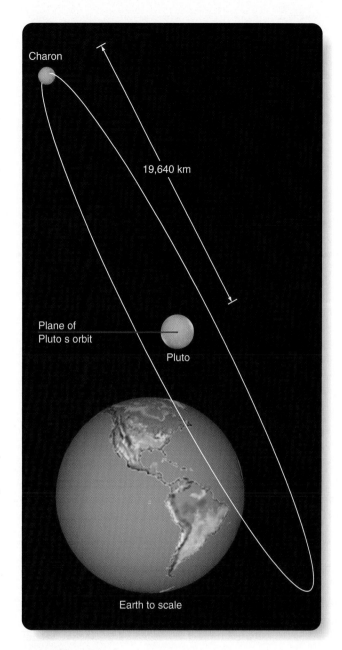

■ Figure 24-19

The nearly circular orbit of Charon is only a few times bigger than Earth. It is seen here nearly edge-on, and consequently it looks elliptical in this diagram. The orbit and the equator of Pluto are inclined 119.6° to the plane of Pluto's orbit around the sun.

older surface with ices. Thus Pluto and Charon may be worlds unlike any you have visited before. Yet the biggest puzzle may be not their geology, but their origin.

Pluto and the Plutinos

No, this section is not about a 1950s rock-and-roll band. It is about the history of Pluto, and it will take you back billions of years to watch the outer planets form. Pluto doesn't fit with the Jovian planets found in the outer solar system, and some astronomers have argued that Pluto is not a planet. As an expert on planets, you can vote; but, before you do, you should eliminate an old idea and review some new discoveries.

You should begin by eliminating that old idea. As soon as Pluto was discovered, astronomers realized that it was much smaller than its Jovian neighbors. Some suggested that it was a moon of Neptune that had escaped. After all, their orbits actually cross, although they will never collide because Pluto's orbit is so highly inclined. That escaped moon theory has been totally abandoned. It is hard to understand how Pluto could have escaped from Neptune and reached its present orbit. And how did it get a moon, if it was once a moon itself? Modern astronomers have a much better theory, and it links Pluto to the Kuiper Belt objects, the small icy bodies that orbit in the outer solar system beyond Neptune and Pluto.

Some of the Kuiper Belt objects are quite large. Two large objects named Sedna and 2004DW are about 63 percent the size of Pluto. Another object called Quaoarh (kwah-o-wahr) is 50 percent the diameter of Pluto. In 2005, an object roughly 50 percent bigger than Pluto was fund in a highly inclined, elliptical orbit with a period of 650 years. So at least one Kuiper Belt object is bigger than Pluto, and more large ones may be found. A few of these objects have moons of their own, or perhaps you should say that they are binaries with two icy bodies orbiting each other. In that way, they resemble Pluto and its big moon Charon.

Some Kuiper Belt objects have orbits similar to Pluto's. Over a dozen are known that are caught in a 3:2 resonance with Neptune. That is, they orbit the sun twice while Neptune orbits three times. You learned about orbital resonances when you studied Jupiter's Galilean moons. Because Pluto is also caught in the same 3:2 resonance, these Kuiper Belt objects have been named **plutinos.** Many other Kuiper Belt objects are caught in other resonances with Neptune.

If Pluto is related to the plutinos, then Pluto is related to the Kuiper Belt objects. They formed in the outer solar nebula, but how did they get caught in resonances with Neptune? Sophisticated models of the formation of the planets suggest that Uranus and Neptune may have formed closer to the sun, where the solar nebula was denser. Sometime later, gravitational interactions with Jupiter and Saturn could have gradually shifted the two ice giants outward, and, as Neptune migrated outward, its orbital resonances could have swept up icy bodies like strange nets pushed in front of a fishing boat.

Some icy Kuiper Belt objects may have been captured as moons, and astronomers have wondered if moons such as Triton and Charon could have been Kuiper Belt objects. Could Pluto have started life as a Kuiper Belt object and gotten swept up by a Neptune resonance?

Now you can vote: Is Pluto a planet? Is it the first object discovered in the Kuiper Belt? Whatever the final outcome, Pluto and the plutinos have rocked modern astronomers and forced them to reconsider their ideas on the origin of the outer worlds. As more Kuiper Belt objects are found beyond the orbit of Neptune, astronomers will see more clearly how the outer solar nebula formed planets.

Building Scientific Arguments

What evidence can you point to that cataclysmic impacts have occurred in our solar system?

To build this argument you must clearly distinguish between evidence and theory. You would expect objects that formed together from the solar nebula to rotate and revolve in the plane of the solar system, so the high inclination of Pluto and the orbit of Charon suggest a past collision or, at least, a close gravitational interaction with a passing body. And the Pluto–Charon system is not the only evidence for major impacts in the past. The peculiar orbits of Neptune's moons Triton and Nereid, the fractured and cratered state of the Jovian satellites, and the peculiar rotation of Uranus all hint that impacts and encounters with large planetesimals or the nuclei of comets have been important in the history of these worlds. Furthermore, the existence of planetary rings suggests that impacts have scattered particles into stable orbits among the small inner moons and replenished the ring systems. Even in the inner solar system, the backward rotation of Venus, the density of Mercury, and the formation of Earth's moon are seen as possible consequences of major impacts when the solar system was young.

Certainly, astronomers do not imagine that planets bounce around like billiard balls, but they do recognize that the planets are not totally isolated and may have been victims (or beneficiaries) of large impacts as they formed long ago. For example, build a new argument: **How do the size, orbits, and composition of Pluto and Charon suggest they originated in the Kuiper Belt?**

■   ■   ■

Connections: Like clues at the scene of a crime, the evidence in the outer solar system warns that all of the planets in our solar system have been and still are sitting ducks for impacts by the smaller objects that hurtle along orbits around the sun. What are these objects? What can they say about the origin and evolution of our solar system? You will explore these questions in the next chapter.

Summary

24-1 | Uranus

How is Uranus different from Jupiter?

- Uranus was discovered by William Herschel in 1781.

- Although Uranus is a Jovian planet, it is significantly smaller and less massive than Jupiter. Uranus is an ice giant planet about four times the diameter of Earth. It rotates on its side and has extreme seasons.

- The atmosphere of Uranus is mostly hydrogen and helium with some methane, which absorbs longer-wavelength photons and gives the planet a blue color.

- The atmosphere is so cold that only methane clouds are visible, and some traces of belt–zone circulation can be seen. As spring has come to the northern hemisphere, more cloud features have developed, suggesting that Uranus follows a seasonal cycle and is not always as bland as it was when Voyager 2 flew past in 1986.

- Uranus has a small core of dense matter and a deep slushy mantle of ice, water, and rock. Convection in the mantle may produce its highly inclined offset magnetic field.

- Uranus emits about the right amount of heat for a planet at its distance from the sun, which suggests that it is not extremely hot inside. The settling of heavy material, including diamonds, formed from the carbon in methane could generate some heat.

How did Uranus, its rings, and its moons form and evolve?

- The rings of Uranus are boulder-size objects darkened by radiation and trapped among the orbits of small moons. They make up a layer one particle thick. The ring material was probably produced by impacts on icy moons.

- The five large moons of Uranus appear to be icy and old, although some show signs of geological activity. Miranda appears to have suffered from dramatic activity. Tidal heating is a likely source of energy.

- Uranus appears to have formed slowly and never became massive enough to trap large amounts of hydrogen and helium from the solar nebula. It and Neptune may have formed closer to the sun and were moved outward by gravitational interactions with Jupiter and Saturn.

- An impact by a large planetesimal as Uranus was forming may have caused the planet to rotate around a highly inclined axis.

24-2 | Neptune

How is Neptune different from Uranus?

- Neptune was discovered in 1846 based on a position computed from irregularities in the orbital motion of Uranus.

- Neptune is an ice giant only slightly smaller than Uranus. It has a dense core; a mantle of ice, water, and rock; and an atmosphere of hydrogen and helium with traces of methane, which gives the planet a blue color.

- Methane clouds come and go in the cold atmosphere and appear to follow a belt–zone circulation.

- Circulation in the liquid mantle gives rise to the planet's magnetic field.

- Neptune has more internal heat than Uranus, and the sinking of dense material, including diamonds, may be adding to the internal heat.

How did Neptune, its rings, and its moons form and evolve?

- The rings of Neptune probably formed when impacts on moons scattered icy debris into stable places among the orbits of the small moons. The rings contain more dust than those of Uranus. Arcs in the rings appear to be caused by the gravitational influence of a small moon.

- Small moons are probably captured objects. Triton, the largest moon, follows a retrograde orbit, and Nereid follows a long elliptical orbit. These two moons may have been disturbed by the gravitational influence of a massive planetesimal or by the capture of Triton.

- Triton has an icy surface and a thin atmosphere of nitrogen. The lack of many craters and the presence of cracks and faults suggest that the moon may still be active. Sunlight appears to trigger nitrogen geysers in the crust. Radioactive decay and tidal heating during a close approach to Neptune when it was captured into orbit are other possible sources of heat.

- Neptune formed slowly, as did Uranus, and never accumulated a deep atmosphere of hydrogen and helium before the solar nebula was blown away.

- Neptune may have formed closer to the sun and was moved outward by interactions with Jupiter and Saturn.

24-3 | Pluto

Is Pluto a planet or something else?

- Pluto was discovered in 1930 during a search for a large planet orbiting beyond Neptune and disturbing Neptune's orbital motion. Pluto is much too small to affect Neptune.

- No spacecraft has visited Pluto, and it is too small and too far away to study in detail from Earth.

- Pluto has a frigid crust of solid nitrogen ice with traces of frozen methane and carbon monoxide. Its thin atmosphere is mostly nitrogen.

- Pluto's moon Charon orbits in a highly inclined orbit, and it and Pluto are tidally locked to face each other. That means Pluto rotates on its side much like Uranus.

- The density of Pluto shows that it and Charon must contain about two-thirds rock and one-third ices.

- Pluto and Charon resemble the plutinos, Kuiper Belt objects caught in a resonance with the orbital motion of Neptune. These plutinos could have been swept up as Neptune moved outward in the solar system, and some astronomers suggest that Pluto is not a planet but is a large Kuiper Belt object.

New Terms

occultation (p. 597)

ovoid (p. 600)

plutino (p. 612)

Review Questions

Ace⊛Astronomy™ Assess your understanding of this chapter's topics with additional quizzing and animations at **http://ace .brookscole.com/sf9**

1. Describe the location of the equinoxes and solstices in the Uranian sky. What are seasons like on Uranus?

2. Why is belt–zone circulation difficult to detect on Uranus?

3. Discuss the origin of the rings of Uranus and Neptune. Cite evidence to support hypotheses.

4. How do the magnetic fields of Uranus and Neptune suggest that the mantles inside those planets are fluid?

5. If Neptune had no satellites at all, would you expect it to have rings? Why or why not?

6. Why might the surface brightness of ring particles and small moons orbiting Uranus and Neptune depend on whether those planets have magnetic fields?

7. Both Uranus and Neptune have a blue-green tint when observed through a telescope. What does that tell you about their composition?

8. How can small worlds like Triton and Pluto have atmospheres when a larger world such as Ganymede has none?

9. Why do you suspect that Triton has had an active past? What source of energy could power such activity?

10. If you visited the surface of Pluto and found Charon a full moon directly overhead, where would Charon be in the sky when it was new? when it was first quarter?

11. What evidence can you cite that Pluto and Charon are made of mixtures of rock and ice?

12. Why might you suspect that Pluto is a Kuiper Belt object?

13. Two images of Uranus show it as it would look to the eye and through a red filter that enhances methane clouds in the northern hemisphere. Why didn't Voyager 2 photograph the northern hemisphere? What do the visible atmospheric features tell you about circulation on Uranus?

(NASA and Heidi Hammel)

Discussion Questions

1. Why might it be unfair to describe William Herschel's discovery of Uranus as accidental? Why might it be unfair to describe the discovery of the rings of Uranus as accidental?

2. Suggest a single phenomenon that could explain the inclination of the rotation axis of Uranus, the orbits of Neptune's satellites, and the existence of Pluto's moon.

Problems

1. What is the maximum angular diameter of Uranus as seen from Earth? of Neptune? of Pluto? (*Hint:* Use the small-angle formula.)

2. One way to recognize a distant planet is by its motion along its orbit. If Uranus circles the sun in 84 years, how many seconds of arc will it move in 24 hours? (This does not include the motion of Earth. Assume a circular orbit for Uranus.)

3. What is the orbital velocity of Miranda around Uranus?

4. What is the escape velocity from the surface of Miranda? (*Hints:* Radius = 242 km. Assume that the density is 2 g/cm^3. See Chapter 5.)

5. The magnetosphere of Uranus rotates with the planetary interior in 17.24 hours. What is the velocity of the outer portion of the magnetic field just beyond the orbit of Oberon? (*Hint:* The circumference of a circle is $2\pi r$.)

6. If the ε ring is 60 km wide and the orbital velocity of Uranus is 6.81 km/s, how long a blink should you expect to see when the ring crosses in front of a star? Is this consistent with the data on page 598?

7. What is the escape velocity from the surface of an icy moon with a diameter of 20 km? (*Hints:* The density of ice is 1 g/cm^3. The volume of a sphere is $\frac{4}{3}\pi R^3$. See Chapter 5.)

8. What is the difference in the orbital velocities of the two shepherd satellites Cordelia and Ophelia? (*Hints:* Orbital radii = 49,800 km and 53,800 km. See Chapter 5.)

9. Repeat Problem 2 for Pluto. That is, ignoring the motion of Earth, how far across the sky would Pluto move in 24 hours? (Assume a circular orbit for Pluto.)

10. Given the size of Triton's orbit (R = 355,000 km) and its orbital period (P = 5.877 days), calculate the mass of Neptune. (*Hint:* See Chapter 5.)

ACTIVE FIGURES

Ace ⊙Astronomy™ To access the resources in the Media Cluster, log into AceAstronomy at http://ace.brookscole.com/sf9 and select Chapter 24.

Planetary Atmospheres
Experiment with the temperature throughout the atmospheres of several planets with this animation.

Uranus's Ring Detection
See how the rings of Uranus were first discovered during an occultation.

ASTRONOMY EXERCISES

Auroras
Use this simulation to study the formation of auroras. You can vary the strength of Earth's magnetic field and the speed and number of particles in the solar wind.

Convection and Magnetic Fields
A planet's core temperature and rotation speed are factors in the strength of its magnetic field. See the effect of these variables in this animation.

VIRTUAL ASTRONOMY LABS

Lab 7: Planetary Atmospheres and Their Retention
This lab investigates the retention of atmospheres. You will explore the factors that govern the loss of atmospheric gases, and you will see why certain bodies can retain some gases but not others.

Critical Inquiries for the Web

1. Who was William Herschel? What can you find out about his life after he discovered Uranus?

2. What is weather on Uranus like now? Can you locate recent photos of cloud patterns there?

3. How many moons does Uranus have? More are discovered every year, it seems, so search for the latest discoveries.

4. What happened to John Couch Adams and Urbain Jean Leverrier after the discovery of Neptune? Did they make other important discoveries? What is the Adams Society? Can you find out what Leverrier did in the revolution of 1848?

5. Repeat Inquiries 2 and 3 for Neptune. Remember that finding new moons orbiting Neptune is difficult because it is farther away.

6. Who was Clyde Tombaugh? What can you find out about his life after he discovered Pluto?

Go to the Brooks/Cole Astronomy Resource Center (http://astronomy.brookscole.com) for critical thinking exercises, articles, and additional readings from InfoTrac College Edition, Brooks/Cole's online student library.

25 | Meteorites, Asteroids, and Comets

When they shall cry "PEACE, PEACE" then cometh sudden destruction!

COMET'S CHAOS?—
What Terrible events will the Comet bring?

FROM A RELIGIOUS PAMPHLET PREDICTING THE END OF THE WORLD BECAUSE OF THE APPEARANCE OF COMET KOHOUTEK, 1973

YOU ARE NOT AFRAID of comets, of course; but not long ago, people viewed them with terror. In 1910, Comet Halley was spectacular. On the night of May 19, Earth actually passed through the tail of the comet—and millions of people panicked. The spectrographic discovery of cyanide gas in the tails of comets led many to believe that life on Earth would end. Householders in Chicago stuffed rags around doors and windows to keep out the gas, and bottled oxygen was sold out. Con artists in Texas sold comet pills and inhalers to ward off the noxious fumes. An Oklahoma newspaper reported (in what was apparently a hoax) that a religious sect tried to sacrifice a virgin to the comet. ▌ Throughout history, bright comets have been seen as portents of doom. Even in our own time, the appearance of bright comets has generated predictions

▌ Continued on page 618 ▌

Comets can be terrifying to the superstitious, but they are dramatically beautiful and carry clues to the origin of the solar system. The Stardust spacecraft flew through the dust and gas spewing from comet Wild 2 in 2004. (NASA/JPL)

Guidepost

Looking Back

Everything has to have come from somewhere, and you began wondering where the solar system came from back in Chapter 19. You have now visited all the planets and many of the moons in the solar system, gathering evidence and making observations. But you found that the planets and moons did not offer many clues to the formation of the solar system because they have been heavily altered by cratering and by heat flowing out of their interiors. You will find the smaller objects in the solar system less altered.

This Chapter

Compared with planets, the comets and asteroids are unevolved objects. You will find them much as they were when they formed 4.6 billion years ago. The fragments of these objects that fall to Earth, the meteors and meteorites, will give you a close look at these ancient planetesimals. As you explore, you will find answers to four essential questions:

Where do meteors and meteorites come from?

What are the asteroids?

Where do comets come from?

What happens when an asteroid or comet hits Earth?

Looking Ahead

As you finish this chapter, you will have an astronomer's insight into your place in nature. You live on the surface of a planet. There are other planets. Are they inhabited too? That is the subject of the next chapter.

of the end of the world. Comet Kohoutek in 1973, Comet Halley in 1986, and Comet Hale–Bopp in 1997 caused deep concern among the superstitious. A bright comet moving slowly through the night sky is so out of the ordinary that you should not be surprised if it generates some instinctive alarm (■ Figure 25-1).

In fact, comets are graceful and beautiful visitors to our skies. Astronomers think of comets as messengers from the age of planet building. By studying comets, you can learn about the conditions in the solar nebula from which planets formed. But comets are only the icy remains of the solar nebula; the asteroids are the rocky debris left over from planet building. Because you cannot easily visit comets and asteroids, you need to begin your analysis with the fragments of those bodies that fall to Earth—the meteorites.

IN THE AFTERNOON of November 30, 1954, Mrs. E. Hulitt Hodges of Sylacauga, Alabama, lay napping on her living room couch. An explosion and a sharp pain jolted her awake, and she found that a meteorite had smashed through the ceiling and bruised her left leg. Mrs. Hodges is the only person known to have been injured by a meteorite. By the way, Mrs. Hodges lived right across the street from the Comet Drive-In Theater.

Meteorite impacts on homes are not common, but they do happen. Statistical calculations show that a meteorite should damage a building somewhere in the world about once every 16 months. About two meteorites large enough to produce visible impacts strike somewhere on Earth each day, but most meteors

■ **Figure 25-1**

Comet Hyakutake swept through the inner solar system in 1996 and was dramatic in the northern sky. Seen here from Kitt Peak National Observatory, the comet passed close to the north celestial pole (behind the observatory dome in this photo). Notice the Big Dipper below and to the left of the head of the comet. (Courtesy Tod Lauer)

Visual-wavelength image

are small particles ranging from a few centimeters down to microscopic dust. Earth gains about 40,000 tons of mass per year from meteorites of all sizes.

Recall from Chapter 19 that astronomers distinguish between the words *meteoroid, meteor,* and *meteorite*. A small body in space is a meteoroid, but once it begins to vaporize in Earth's atmosphere it is called a meteor. If it survives to strike the ground, it is called a meteorite.

You probably have two questions concerning meteorites: Where in the solar system do these objects come from? What can meteorites tell you about the origin of the solar system? To answer these questions, you must consider the orbits of meteoroids and the minerals found in meteorites.

Meteoroid Orbits

Meteoroids are much too small to be visible through even the largest telescope. They are visible only when they fall into Earth's atmosphere at 10 to 30 km/s, roughly 30 times faster than a rifle bullet, and are vaporized by friction with the air. The average meteoroid is about the mass of a paper clip and vaporizes at an altitude of about 80 km above Earth's surface, but it produces a bright streak of fire that you see as a meteor. The trail of a meteor points back along the path of the meteoroid, so if you could study the direction and speed of meteors, you could get clues to the orbits of the meteoroids.

One way to backtrack meteor trails is to observe meteor showers. On any clear night, you can see 3 to 15 meteors an hour, but on some nights you can see a shower of dozens of meteors an hour that are obviously related to each other. To confirm this, try observing a meteor shower. Pick a shower from ▪ Table 25-1 and on the appropriate night stretch out in a lawn chair and watch a large area of the sky. When you see a meteor, sketch its path on

▪ **Figure 25-2**

(a) Meteors in a meteor shower enter Earth's atmosphere along parallel paths, but perspective makes them appear to diverge from a radiant point. (b) Similarly, parallel railroad tracks appear to diverge from a point on the horizon.

the appropriate sky chart from the back of this book. In just an hour or so you will discover that most of the meteors you see seem to come from a single area of the sky, the **radiant** of the shower (▪ Figure 25-2a). In fact, meteor showers are named after the constellation from which they seem to radiate. The Perseid shower radiates from the constellation Perseus in mid-August. The Leonids were spectacular in November of 2001, when observers saw hundreds per hour. The Leonids are not often that dramatic.

Observing a meteor shower is relaxing, and the show is natural fireworks, but it is even more exciting when you understand what a meteor shower tells you (**Window on Science 25-1**). The significant fact is that the meteors in a shower appear to come from a single point in the sky, the radiant. That must mean the meteoroids were traveling through space along parallel paths. When they encounter Earth and are vaporized in the upper atmosphere, you see their fiery tracks in perspective; they appear to

▪ **Table 25-1 I Meteor Showers**

Shower	Dates	Hourly Rate	Radiant* R. A.	Radiant* Dec.	Associated Comet
Quadrantids	Jan. 2–4	30	15^h24^m	50°	
Lyrids	April 20–22	8	18^h4^m	33°	1861 I
η Aquarids	May 2–7	10	22^h24^m	0°	Halley?
δ Aquarids	July 26–31	15	22^h36^m	−10°	
Perseids	Aug. 10–14	40	3^h4^m	58°	1982 III
Orionids	Oct. 18–23	15	6^h20^m	15°	Halley?
Taurids	Nov. 1–7	8	3^h40^m	17°	Encke
Leonids	Nov. 14–19	6	10^h12^m	22°	1866 I Temp
Geminids	Dec. 10–13	50	7^h28^m	32°	

*R. A. and Dec. give the celestial coordinates (right ascension and declination) of the radiant of each shower.

Enjoying the Natural World

You can enjoy a meteor shower as Mother Nature's fireworks, but your enjoyment is much greater once you begin to understand what causes meteors and why meteors in a shower follow a pattern. When you know that meteor showers help you understand the origin of our world, the evening display of shooting stars is even more exciting. Science typically increases your enjoyment of the natural world by revealing the significance of things you might otherwise enjoy only in a casual way.

Everyone likes flowers, for example. You can enjoy them casually, admiring their colors and fragrances, their graceful petals and massed blooms. But botanists know that evolution has carefully designed flowers to welcome insects and spread pollen. The bright colors attract insects, and the shapes of the flowers provide little runways so the insect will find it easy to land. Some color patterns even guide the insect in like landing lights at an airport, and many flowers, such as orchids and snapdragons, force the insect to crawl inside in just the right way to exchange pollen and fertilize the flower. Nectar is bug bait. Once you begin to understand what flowers are for, a visit to a garden becomes not only an adventure of color and fragrance, but also an adventure in understanding the beauty of design and function, an adventure in meaning as well as an adventure of the senses.

And your understanding of a natural phenomenon can help you understand and enjoy related phenomena. For example, some flowers attract flies for pollination, and such blossoms smell like rotting meat. A few flowers depend on bats, and they open their blossoms at night. Many plants, such as pine trees, depend on the wind to spread their pollen, and those plants do not have colorful flowers at all; but flowers that depend on hummingbirds have long trumpet-shaped blossoms that just fit the hummingbirds' beaks.

The more science tells you about nature, the more enjoyable the natural world is. You can enjoy a meteor shower as nothing more than a visual treat, but the more you know about it, the more interesting and exciting it becomes. The natural world is filled with meaning, and science, as a way of discovering and understanding that meaning, gives you new ways to enjoy the world around you.

The beauty of flowers becomes more interesting when you know the reasons for their shapes and colors. (M. Seeds)

come from a single radiant point, just as railroad tracks seem to come from a single point on the horizon (Figure 25-2b).

Studies of meteor-shower radiants reveal that these meteors are produced by bits of matter orbiting the sun along the paths of comets. The vaporizing head of the comet releases bits of rock that become spread along the entire orbit (■ Figure 25-3). When Earth passes through this stream of material, you see a meteor shower. In some cases the comet has wasted away, has lost all of its ices, and is no longer visible, but in other cases the comet is still prominent though somewhere else along its orbit. For example, in May Earth comes near the orbit of Comet Halley, and Earthlings see the Eta Aquarids shower. Around October 20, Earth passes near the other side of the orbit of Comet Halley, and you can see the Orionids shower. These pleasant observations of meteor showers reveal that at least some meteoroids come from comets.

Even when there is no shower, you will still see meteors, which are called **sporadic meteors** because they are not part of specific showers. Many of these are produced by stray bits of matter that were released long ago by comets. Such comet debris gets spread throughout the inner solar system, and a few bits fall into Earth's atmosphere continuously even when there is no shower.

Another way to backtrack meteor trails is to photograph a meteor from two locations on Earth a few miles apart. Then astronomers can use triangulation to find the altitude, speed, and direction of the meteor as it moves through the atmosphere and work backward to find out what its orbit looked like before it entered Earth's atmosphere. These studies reveal that meteors belonging to showers and some sporadic meteors have orbits that are similar to the orbits of comets. In many cases, the comets can be identified, so the origin of these meteors is well known. But a few sporadic meteors have orbits that lead back to the asteroid belt between Mars and Jupiter. From this you can conclude that meteors have a dual source: Many come from comets, but some come from the asteroid belt. To learn more about meteors, you need to examine those that make it to Earth's surface, the meteorites. You can begin with their dramatic arrivals.

Meteorite Impacts on Earth

When a meteor is massive enough and strong enough to survive its plunge through Earth's atmosphere, it finally reaches Earth's surface, and it is called a meteorite. A large meteorite hitting Earth might dig an impact crater much like those on the moon (see page 500). Most meteorites are small and burn up in the atmosphere, but a few survive and hit Earth's surface with a bang. Big ones can dig craters.

Fresh material ejected from the comet forms a fan.

Orbit of comet

Orbit of comet

Earth

Sun

Infrared image

■ Figure 25-3

As a comet's ices evaporate, it releases rocky bits of material that spread along its orbit. If Earth passes through such material, it experiences a meteor shower. In this image of Comet Encke, the millimeter-size bits of rock along its orbit glow in the infrared as they are warmed by sunlight. The Taurid meteor shower occurs every October when Earth crosses the orbit. (NASA/JPL-Caltech/M. Kelley, Univ. of Minnesota)

Over 150 impact craters have been found on Earth. The Barringer Meteorite Crater near Flagstaff, Arizona, is a good example (■ Figure 25-4). It was created about 50,000 years ago by a meteorite about as large as a big building (40 m). Debris at the site shows that the meteorite was composed of iron, and astronomers can calculate from the size of the crater that the meteorite hit at a speed of about 11 km/s and released as much energy as a large thermonuclear bomb.

The Barringer crater is 1.2 km in diameter and 200 m deep. It seems large when you stand on the edge, and the hike around it, though beautiful, is long and dry. Nevertheless, the crater is actually small compared with some impact features on Earth. About 65 million years ago, an impact in the northern Yucatán of Central America produced a crater, now covered by sediment, that was between 180 and 300 km in diameter, almost as big as the state of Ohio. This impact has been blamed for changing Earth's climate and causing the extinction of the dinosaurs. Later in this chapter, you can come back to this Earth-shaking event. The 1994 comet impacts on Jupiter (shown in Figure 23-6) were even larger, but they formed no craters because Jupiter has no surface. During the Leonid meteor shower in recent years, astronomers watching the dark portion of the moon's disk have seen a number of bright flashes apparently caused by meteorite impacts. No new craters on the moon have been identified, but it is clear that meteorite impacts and the formation of impact craters are a continuing process in our solar system.

Large meteorites can produce craters, but most meteorites are small and don't create significant craters. Also, craters erode rapidly on Earth, so most meteorites are found without associated craters. The best place to look for meteorites turns out to be certain areas of Antarctica—not because more meteorites fall there but because the meteorites are easy to recognize on the icy terrain. Most meteorites look like Earth rocks, but on the ice of Antarctica there are no natural rocks. Also, the slow creep of the ice toward the sea concentrates the meteorites in certain areas where the ice runs into mountain ranges, slows down, and evaporates. Consequently, teams of scientists can recover meteorites from the ice (■ Figure 25-5). Such unbiased collections of meteorites are important in helping define the frequency of different meteorite types.

Meteorites that are seen to fall are called **falls.** A fall is known to have occurred at a given time and place, and thus the meteorite is well documented. A meteorite that is discovered but was not seen to fall is called a **find.** Such a meteorite could have fallen thousands of years ago. This distinction between falls and finds will be important as you analyze the different kinds of meteorites.

An Analysis of Meteorites

Meteorites can be divided into three broad categories: *Iron* meteorites are solid chunks of iron and nickel, *stony* meteorites are silicate masses that resemble Earth rocks, and *stony-iron* meteorites

■ **Active Figure 25-4**

(a) The Barringer Meteorite Crater (near Flagstaff, Arizona) is nearly a mile in diameter and was formed about 50,000 years ago by the impact of an iron meteorite roughly 40 m in diameter. It hit with energy equivalent to that of a 3-megaton hydrogen bomb. Notice the raised and deformed rock strata all around the crater. For scale, locate the brick building on the far rim at right. (M. Seeds) (b) Like all larger-impact features, the Barringer Meteorite Crater has a raised rim and scattered ejecta. (USGS)

Ace⊙Astronomy™ Log into AceAstronomy and select this chapter to see the Active Figuree "Impacts on Earth." Notice that large impacts are less common than smaller ones.

are mixtures of iron and stone. These types of meteorites are illustrated in ■ Figure 25-6.

Iron meteorites are easy to recognize because they are heavy, dense lumps of iron-nickel steel—a magnet will stick to them. That explains an important bit of statistics. Iron meteorites make up 66 percent of finds (■ Table 25-2) but only 6 percent of falls. Why? Because an iron meteorite does not look like a rock. If you trip over one on a hike, you are more likely to carry it home and show it to your local museum. Also, some stone meteorites deteriorate rapidly when exposed to weather; irons survive longer. That means there is a **selection effect** that makes it more likely that iron meteorites will be found (**Window on Science 25-2**). That only 6 percent of falls are irons shows that iron meteorites are fairly rare.

When iron meteorites are sliced open, polished, and etched with nitric acid, they reveal regular bands called **Widmanstätten patterns** (Figure 25-6). These patterns are caused by alloys of iron and nickel called kamacite and taenite that formed billions of years ago as the molten iron cooled and solidified. The size and shape of the bands indicate that the molten metal cooled very slowly, no faster than 20 K per million years.

A lump of molten metal floating in space would cool very quickly, so the presence of the Widmanstätten pattern tells you that the molten metal must have been well insulated to have cooled so slowly. Such slow cooling is typical of the interiors of bodies at least 30 to 50 km in diameter. On the other hand, the

iron meteorites show no evidence of the very high pressures that would develop in larger bodies. Evidently, the iron meteorites formed from the interiors of molten objects of planetesimal size.

In contrast to irons, **stony meteorites** are common. Among falls, 92 percent are stones. Although there are many different types of stony meteorites, you can classify them into two main categories depending on their physical and chemical content—chondrites and achondrites.

Roughly 80 percent of all meteorite falls are stony meteorites called **chondrites,** which look like dark gray, granular rocks. The chemical composition of chondrites is the same as a sample of matter from the sun with the volatile gases removed. The classification of meteorites has become quite sophisticated, and there are many types of chondrites, but in general they appear to be original samples of the material that condensed from the solar nebula. Some have slight mineral differences, which show that the solar nebula was not totally uniform when this material condensed.

Most types of chondrites contain **chondrules,** round bits of glassy rock no larger than 5 mm across (Figure 25-6). To be glassy rather than crystalline, the chondrules must have cooled from a molten state quickly, in a few hours, but their origin is not clear. One theory is that they are bits of matter that were suddenly melted by shock waves spreading through the solar nebula. Whatever their origin, they are very old.

All of the different types of chondrites contain some volatiles such as water, and this shows that the meteoroids were never

Each meteorite is assigned a number and photographed as it was found.

Figure 25-5

Braving bitter cold and high winds, teams of scientists riding snowmobiles search for meteorites that fell long ago in Antarctica and are exposed as the ice evaporates. Thousands of these meteorites have been collected, including rare meteorites from the moon and Mars. (Courtesy Dr. Monika Kress)

Large or small, meteorites are sealed airtight and refrigerated until they can be studied.

heated to high temperatures. Differences among the many types of chondrites arise from their condensation in different parts of the solar nebula and from processes that altered their composition after they formed. Some, for example, appear to have been altered by the presence of water released by the melting of ice. Among the chrondrites, the **carbonaceous chondrites** are rare, only about 5.7 percent of falls. These very dark gray,

rocky meteorites contain volatiles such as water and carbon compounds that would have been driven off if the meteoroid had been heated much above room temperature. It has been common for astronomers to think of the carbonaceous chondrites as

■ **Figure 25-6**

The three main types of meteorites, irons, stones, and stony-iron, have distinctive characteristics. (Lab photos courtesy of Russell Kempton, New England Meteoritical Services)

Iron meteorites are very heavy for their size and have a dark, irregular surface.

Stony meteorites tend to have a fusion crust caused by melting in Earth's atmosphere.

A stony-iron meteorite cut and polished reveals a mixture of iron and rock.

Chondrules are small, glassy spheres found in chondrites.

Cut, polished, and etched with acid, iron meteorites show a Widmanstätten pattern.

This carbonaceous chondrite contains chondrules and volatiles, including carbon, that make the rock very dark.

the least altered of the chondrites, but many types of chondrites are essentially unaltered samples of the solar nebula. A carbonaceous chondrite is shown in Figure 25-6.

One of the most important meteorites ever recovered was a carbonaceous chondrite that was seen falling on the night of February 8, 1969, near the Mexican village of Pueblito de Allende. The brilliant fireball was accompanied by tremendous sonic booms and showered an area about 50 km by 10 km with over 4 tons of fragments. About 2 tons were recovered.

Studies of the Allende meteorite disclosed that it contained—besides volatiles and chondrules—small, irregular inclusions rich in calcium, aluminum, and titanium. Now called **CAIs,** for calcium–aluminum-rich inclusions, these bits of matter are highly refractory (■ Figure 25-7); that is, they can survive very high temperatures. If you could scoop out a ton of the sun's surface matter and cool it, the CAIs would be the first particles to form. As the temperature fell, other materials would condense in accord with the condensation sequence described in Chapter 19. When the material finally reached room temperature, you would find that all of the hydrogen, helium, and a few other gases like argon and neon had escaped and that the remaining lump, weighing about 18 kg (40 lb), had almost the same composition as the Allende meteorite, including CAIs. The Allende meteorite seems to be a very old sample of the solar nebula.

Unlike the chondrites, stony meteorites called **achondrites** (7.1 percent of falls) are highly modified. They contain no chondrules and no volatiles. This suggests that they have been hot enough to melt chondrules and drive off volatiles, leaving behind rock with compositions similar to Earth's basalts.

Iron meteorites and stony meteorites make up most falls, but 2 percent of falls are meteorites that are made up of mixed iron and stone. These **stony-iron meteorites** appear to have solidified from a region of molten iron and rock—the kind of environment you might expect deep inside a planetesimal with a molten iron core and a rock mantle.

■ Table 25-2	Proportions of Meteorites	

Type	Falls (%)	Finds (%)
Stony	92	26
Iron	6	66
Stony-iron	2	8

Selection Effects in Science

Many different kinds of science depend on selecting objects to study. Scientists studying insects in the rain forest, for example, must choose which ones to catch. They can't catch every insect they see, so they might decide to catch and study any insect that is red. If they are not careful, a selection effect could bias their data and lead them to incorrect conclusions without their ever knowing it. The reason selection effects are dangerous in science is that they can have powerful influences without being evident. Only by carefully designing a research project can scientists avoid selection effects.

For example, suppose you decided to measure the speed of cars on a highway. There are too many cars to measure every one, so you reduce the workload and measure only red cars. It is quite possible that this selection criterion will mislead you because people who buy red cars may be more likely to be younger and drive faster. Should you measure only brown cars? No, because you might suspect that only older, more sedate people would buy a brown car. Should you measure the speed of any car in front of a truck? Perhaps you should pick any car following a truck? Again, you may be selecting cars that are traveling a bit faster or a bit slower than normal. Only by very carefully designing your experiment can you be certain that the cars you measure are traveling at representative speeds.

Public-opinion companies that survey voters in shopping malls must think carefully about selection effects. If they are seeking opinions about property taxes and decide to interview only men with gray hair, they will get a biased result. Interviewing only men wearing T-shirts would give them quite a different result. They must be very careful to avoid unnoticed selection effects.

Astronomers face the danger of selection effects quite often. Iron meteorites are easier to find than stone meteorites. Very luminous stars are easier to see at great distances than faint stars. Spiral galaxies are brighter, bluer, and more noticeable than elliptical galaxies. What you see through a telescope depends on what you notice, and that is powerfully influenced by selection effects.

Scientists engaged in observation must spend considerable time designing their experiments. They must be careful to observe an unbiased sample if they expect to make logical deductions from their results. The scientists in the rain forest, for example, should not catch

Visual-wavelength image

Things that are bright and beautiful, such as spiral galaxies, may attract a disproportionate amount of attention. Scientists must be aware of such selection effects. (Hubble Heritage Team/STScI/AURA/NASA)

and study only the red insects. Often, the most brightly colored insects are poisonous (or at least taste bad) to predators. Catching only brightly colored insects could produce a highly biased sample of the insect population. Careful scientists must plan their work with great care and avoid any possible selection effects.

The Origin of Meteorites

Where do meteorites come from? The properties of meteorites suggest that they formed in the solar nebula. Their radioactive ages are about 4.6 billion years (Chapter 19), and some appear not to have been modified since they formed. Others have been heated or melted sometime after formation.

Meteorites almost certainly do not come from comets. Most cometary particles are very small specks of low-density, almost fluffy, material. When these enter Earth's atmosphere, you see them incinerated as meteors, and most meteors are produced by this cometary debris. But such meteors are small and weak and do not survive to reach the ground. Although most meteors come from comets, most meteorites are stronger chunks of matter— more like fragments of asteroids.

Meteorites must come from asteroidlike planetesimals rich in metals and silicates, not the cometlike planetesimals that were rich in ices. Indeed, many of the properties of the meteorites suggest that they originated in rocky planetesimals, which evolved in complicated ways and were eventually broken up by collisions.

Chondrules

CAIs

■ Figure 25-7

A sliced portion of the Allende meteorite showing round chondrules and irregularly shaped white inclusions called CAIs. (NASA)

Of course, the solar nebula was full of rocky planetesimals, but you must be wondering about two things: How did these planetesimals evolve to produce both iron and stony meteorites? And when did these planetesimals break up?

The iron and achondritic meteorites show that at least some of the planetesimals melted and differentiated, but what produced this heat? Planets the size of Earth can accumulate a great deal of heat from the decay of radioactive atoms such as uranium, thorium, and the radioactive isotope of potassium. The heat is trapped deep underground by thousands of kilometers of insulating rock. But in a small planetesimal, the insulating layers are not as thick, and the heat leaks out into space as fast as the slowly decaying atoms can produce it. If a small planetesimal is to melt, it must have a more rapidly decaying heat source.

Modern studies have shown that some meteorites may have contained the radioactive element aluminum-26. These aluminum atoms are now gone, decayed to magnesium-26, which remains in the meteorites. If aluminum-26 was once present, its rapid decay could have melted the center of a planetesimal as small as 20 km in diameter. This is a sufficient heat source to melt the parent bodies of the meteorites.

Where did this aluminum-26 come from? Supernova explosions can manufacture aluminum-26, but it decays very rapidly, with a half-life of only 715,000 years. If the solar nebula was enriched by aluminum-26 from a nearby supernova explosion, the explosion must have occurred just before the formation of our solar system. In fact, some astronomers wonder if the shock waves from the supernova explosion compressed gas clouds and triggered the formation of the sun and planets.

If a planetesimal's interior melted, it would have differentiated as heavy metals sank to the center to form a molten metal core and the less-dense silicates floated upward to form a stony mantle. Once the aluminum-26 had decayed away, the planetesimals cooled and slowly solidified. In some cases, the outer surface may not have been melted. If such a planetesimal were broken up (■ Figure 25-8), fragments from the center would look much like iron meteorites with their Widmanstätten patterns.

The achondrites, which have been strongly heated or melted, appear to come from the mantles and surfaces of the larger planetesimals that melted and differentiated. The stony-iron meteorites probably come from the core–mantle boundaries in these bodies where iron and stone were mixed. But the chondrites are probably fragments of smaller bodies that never melted. Volatile-rich meteorites such as the carbonaceous chondrites may have formed in smaller bodies farther from the sun where temperatures were lower.

■ Figure 25-8 ▶

Planetesimals formed when the solar system was forming may have melted and separated into layers of different density and composition. The fragmentation of such a body could produce many types of meteorites. (Adapted from a diagram by C. R. Chapman)

The evidence seems strong that the meteorites are the result of the breaking up of larger bodies. For example, some meteorites are breccias—collections of fragments cemented together. Breccias are found on Earth and are very common on the moon, but studies of the meteoric breccias show that they were produced by impacts. A collision between planetesimals produces fragments, and the slower-moving particles fall back to the surface of the planetesimal to form a regolith, a soil of broken rock fragments.

The Origin of Meteorites

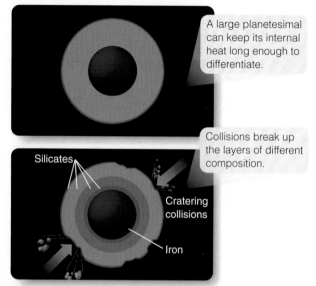

A large planetesimal can keep its internal heat long enough to differentiate.

Silicates

Cratering collisions

Iron

Collisions break up the layers of different composition.

Meteorites from deeper in the planetesimal were heated to higher temperatures.

Fragments from near the core might have been melted entirely.

Fragments of the iron core would fall to Earth as iron meteorites.

Later impacts may add to this regolith and stir it. Still later, an impact may be so violent that the fragments are pressed together and momentarily melt where they touch. Almost instantly, the material cools, and the fragments weld themselves together to form a breccia. Most of the breccias were formed as the planetesimals formed. Much later, as the planetesimals were broken up by collisions, the layers of breccia were exposed, shattered, and became brecciated meteoroids.

When did these bodies break up? Recall from Chapter 19 that radioactive dating shows that meteorites formed about 4.6 billion years ago. That is when the parent bodies formed, melted, differentiated, and cooled. However, the collisions that broke up the planetesimals could not have happened that long ago. Small meteoroids would have been swept up by the planets in a billion years or less. Also, cosmic rays striking meteoroids in space produce isotopes such as helium-3, neon-20, and argon-38. Studies of these atoms in meteorites show that most meteorites have not been exposed to cosmic rays for more than a few tens of millions of years. The meteorites must have been buried under protective layers until recently. That means that the thousands of meteorites now in museums around the world must have broken off planetesimals somewhere in our solar system within the last billion years or less. Where are those planetesimals?

To answer that question, you need only recall that many meteoroid orbits lead back to the asteroid belt. The asteroids are evidently the planetesimals from which meteorites are born.

Building Scientific Arguments

How can meteors come from comets, but meteorites come from asteroids?
This is a revealing argument because it contains a warning that seeing is not enough in science; thinking about seeing is critical. A selection effect can determine what you notice when you observe nature, and a very strong selection effect prevents people from finding meteorites that originated in comets. Cometary particles are physically weak, and they vaporize in Earth's atmosphere easily. Very few ever reach the ground, and people are unlikely to find them. Furthermore, even if a cometary particle reached the ground, it would be so fragile that it would weather away rapidly, and, again, people would be unlikely to find it. Asteroidal particles, however, are made from rock and metal and so are stronger. They are more likely to survive their plunge through the atmosphere and more likely to survive erosion on the ground. Meteors from the asteroid belt are rare. Almost all of the meteors you see come from comets, but not a single meteorite is known to be cometary.

The meteorites are valuable because they provide hints about the process of planet building in the solar nebula. Build a new argument but, as always, think carefully about what you see. **Why are most falls stony, but most finds are irons?**

■   ■   ■

Connections: The vast majority of meteorites must be coming from the asteroids. Your next goal is to try to understand the nature and origin of these tiny rocky worlds.

25-2 Asteroids

UNTIL RECENTLY, FEW ASTRONOMERS KNEW or cared much about asteroids. They were small chunks of rock drifting between the orbits of Mars and Jupiter that occasionally marred long-exposure photographs by drifting past more interesting objects. Asteroids were more irritation than fascination.

Now you know differently. The evidence from meteorites shows that the asteroids are the last remains of the rocky planetesimals that built the planets 4.6 billion years ago. The study of the asteroids gives you a way to explore the ancient past of our planetary system.

Properties of Asteroids

In Chapter 19, you learned that most of the asteroids orbit between Mars and Jupiter and that images recorded by passing spacecraft reveal them to be small, complex worlds with irregular shapes and heavily cratered surfaces (Figure 19-3).

Study **Observations of Asteroids** on pages 628–629 and notice four important points:

1 Most asteroids are irregular in shape and battered by impact cratering. In fact, some appear to be rubble piles of broken fragments.

2 Some asteroids are double objects or have small moons in orbit around them. This is further evidence of collisions among the asteroids.

3 A few larger asteroids show signs of geological activity on their surfaces that may have been caused by volcanic activity when the asteroid was young.

4 Asteroids can be classified by their albedo and color to reveal clues to their compositions.

Collisions among asteroids must have been occurring since the formation of the solar system, and astronomers have found evidence of catastrophic impacts powerful enough to break up an asteroid. Early in the 20th century, Japanese astronomer Kiyotsugu Hirayama discovered that some groups of asteroids share similar orbits. They have the same average distance from the sun, the same eccentricity, and the same inclination. Up to 20 of these **Hirayama families** are known, and modern observations show that the asteroids in a family typically share similar spectroscopic characteristics. Evidently, a family is produced by a catastrophic collision that breaks a single asteroid into a family of fragments that continue traveling along similar orbits around the sun. Evidence shows that one family was produced only 5.8 million years ago in a collision between asteroids 3 km and 16 km in diameter at

Observations of Asteroids

1 Seen from Earth, asteroids look like faint points of light moving in front of distant stars. Not many years ago they were known mostly for drifting slowly through the field of view and spoiling long time exposures. Some astronomers referred to them as "the vermin of the sky." Spacecraft have now visited asteroids, and the images radioed back to Earth show that the asteroids are mostly small, gray, irregular worlds heavily cratered by impacts.

The Near Earth Asteroid Rendezvous (NEAR) spacecraft visited the asteroid Eros in 2000 and found it to be heavily cratered by collisions and covered by a layer of crushed rock ranging from dust to large boulders. The NEAR spacecraft eventually landed on Eros.

Eros appears to be a solid fragment of rock.

Visual-wavelength image

Visual-wavelength image

5 meters

10 km

Most asteroids are too small for their gravity to pull them into a spherical shape. Impacts break them into irregularly shaped fragments.

Visual

The surface of Mathilde is very dark rock.

50 km

1a The mass of an asteroid can be found from its gravitational influence on passing spacecraft. Mathilde, at left, has a low mass, and that makes its density so low it cannot be solid rock. Like many asteroids, Mathilde may be a rubble pile of broken fragments with large empty spaces between fragments.

If you walked across the surface of an irregularly shaped asteroid such as Eros, you would find gravity very weak; and in many places, it would not be perpendicular to the surface.

Enhanced visual image

5 km

Like most asteroids, Gaspra would look gray to your eyes; but, in this enhanced image at left, color differences probably indicate difference in mineralogy.

Radar image

2 Asteroids that pass near Earth can be imaged by radar. The asteroid Toutatis is revealed to be a double object—two objects orbiting close to each other or actually in contact.

Ida

Double asteroids are more common than was once thought, reflecting a history of collisions and fragmentation. The asteroid Ida is orbited by a moon Dactyl only about 1.5 km in diameter.

Dactyl

30 km

Enhanced visual + infrared

Occasional collisions among the asteroids release fragments, and Jupiter's gravity scatters them into the inner solar system as a continuous supply of meteorites.

Visual-wavelength image   **Vesta**   Model

500 km

Elevation map

13-km-deep crater

Elevation

-12km   +12km

3 The large asteroid Vesta, as shown at right, provides evidence that some have suffered geological activity. No spacecraft has visited it, but its spectrum resembles that of solidified lava. Images made by the Hubble Space Telescope allow the creation of a model of its shape. It has a huge crater at its south pole. A family of small asteroids is evidently composed of fragments from Vesta, and a certain class of meteorites, spectroscopically identical to Vesta, are believed to be fragments from the asteroid. The meteorites appear to be solidified basalt.

Meteorite from Vesta

5 cm
2 in.

3a Vesta appears to have had internal heat at some point in its history, perhaps due to the decay of radioactive minerals. Lava flows have covered at least some of its surface.

4 Although asteroids would look gray to your eyes, they can be classified according to their albedos (reflected brightness) and spectroscopic colors. As shown at left, S-types are brighter and tend to be reddish. They are the most common kind of asteroid and appear to be the source of the most common chondrites.

M-type asteroids are not too dark but are also not very red. They may be mostly iron-nickel alloys.

C-type asteroids are as dark as lumps of sooty coal and appear to be carbonaceous.

Bright

Albedo (reflected brightness)

0.4
0.3
0.2
0.1
0.06
0.04

Dark

M   S

Common in the inner asteroid belt

Common in the outer asteroid belt

Grayer   Redder

C

0.8   1.0   1.2   1.4   1.6
Ultraviolet minus visual color index

about 5 km/s (11,000 mph), a typical speed for asteroid collisions. Evidently the fragmentation of asteroids is a continuing process.

In 1983, the Infrared Astronomy Satellite detected the infrared glow of sun-warmed dust scattered in bands throughout the asteroid belt. These dust bands appear to be the products of past collisions. The dust will eventually be blown away, but collisions occur constantly in the asteroid belt, so new dust bands will presumably be produced as the present bands dissipate.

What are the asteroids? You can be sure they have been modified and fragmented since they formed, but where did they come from? There are clues hidden among their orbits. You should begin that story at the beginning.

The Asteroid Belt

The first asteroid was discovered on January 1, 1801 (the first night of the 19th century), by the Sicilian monk Giuseppe Piazzi. It was later named Ceres after the Roman goddess of the harvest (thus our word *cereal*).

Astronomers were excited by Piazzi's discovery. There seemed to be a gap where a planet might exist between Mars and Jupiter at an average distance from the sun of 2.8 AU. Ceres fit right in. Its average distance from the sun is 2.766 AU. It was a bit small to be a planet, and because three more objects—Pallas, Juno, and Vesta—were discovered in the following years, astronomers realized that Ceres and the other asteroids were not true planets.

Today thousands of asteroids have accurately known orbits, and there are many more as yet undiscovered; but they are all small bodies. All of the larger asteroids in the asteroid belt have been found.

If you discovered an asteroid, you would be allowed to choose a name for it, and asteroids have been named for spouses, lovers, dogs, Greek gods, politicians, and others.* Once an orbit has been calculated, the asteroid is assigned a number listing its position in the catalog known as the *Ephemerides of Minor Planets*. Thus, Ceres is known as 1 Ceres, Pallas as 2 Pallas, and so on. Only three asteroids are larger than 400 km in diameter, and most are much smaller (■ Figure 25-9).

Most of these objects orbit the sun in the asteroid belt between Mars and Jupiter, and you might suspect that massive Jupiter was responsible for their origin. Certainly the distribution of asteroids in the belt shows the importance of Jupiter's gravitation. Certain regions of the belt, called **Kirkwood's gaps** after their discoverer, are almost free of asteroids (■ Figure 25-10). These gaps lie at certain distances from the sun where an aster-

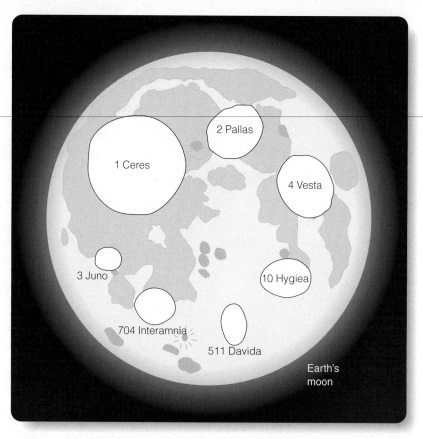

■ **Figure 25-9**

The relative size and approximate shape of the larger asteroids are shown here compared with the size of Earth's moon. Smaller asteroids can be highly irregular in shape.

oid would find itself in resonance with Jupiter. For example, if an asteroid lay 3.28 AU from the sun, it would revolve twice around its orbit in the time it took Jupiter to revolve once around the sun. On alternate orbits, the asteroid would find Jupiter at the same place in space tugging outward. The cumulative perturbations would rapidly change the asteroid's orbit until it was no longer in resonance with Jupiter. Our example is in a 2:1 resonance, but gaps occur in the asteroid belt at many resonances, including 3:1, 5:2, and 7:3. You will recognize that Kirkwood's gaps in the asteroid belt are produced in the same way as some gaps in Saturn's rings. Both were discovered by Daniel Kirkwood (Chapter 23).

Modern research shows that the motion of asteroids in Kirkwood's gaps is described by a theory in mathematics that deals with chaotic behavior. The smooth motion of water sliding over the edge of a waterfall decays rapidly into a chaotic jumble. The same mathematical theory of chaos that describes the motion of the water shows how the slowly changing orbit of an asteroid within one of Kirkwood's gaps can suddenly change into a long, elliptical orbit that carries the asteroid into the inner solar system, where it is likely to be removed by a collision with Mars, Earth, or Venus. This ejection process also explains how Jupiter's gravity can throw so many meteoroids from the asteroid belt into the inner solar system.

---

*Some sample asteroid names: Olga, Chicago, Vaticana, Noel, Ohio, Tea, Gaby, Fidelio, Hagar, Geisha, Dudu, Tata, Mimi, Dulu, Tito, Zulu, Beer, and Zappafrank (after the late musician Frank Zappa).

■ Figure 25-10

Here the red curve shows the number of asteroids at different distances from the sun. Purple bars mark Kirkwood's gaps, where there are few asteroids. Note that these gaps match resonances with the orbital motion of Jupiter.

Nonbelt Asteroids

You don't have to go all the way out to the asteroid belt if you want to visit an asteroid. Some of the most interesting follow orbits that cross the orbits of the terrestrial planets or wander among the Jovian worlds. In fact, some asteroids even share the orbits of the larger planets.

The **Apollo–Amor objects** are asteroids whose orbits carry them into the inner solar system. The Amor objects follow orbits that cross the orbit of Mars but don't reach the orbit of Earth. The Apollo objects have Earth-crossing orbits. These Apollo–Amor objects are dangerous. Jupiter's influence makes their orbits precess, so they are bound to interact with a planet eventually. About one-third will be thrown into the sun, and a few will be ejected from the solar system, but many of these objects are doomed to collide with a planet—perhaps ours. Earth is hit by an Apollo object once every 250,000 years, on average. With a diameter of up to 2 km, they hit with the power of a 100,000-megaton bomb and can dig craters 20 km in diameter.

Many Apollo objects have known orbits, and none of those will hit Earth in the foreseeable future. The bad news is that there are about 1000 of these near-Earth objects (NEOs) larger than 1 km in diameter, the minimum size of an impactor that could cause global effects on Earth. More than half a dozen teams are now searching for these NEOs. For example, LONEOS (Lowell Observatory Near Earth Object Search) is searching the entire sky visible from Lowell Observatory once a month. The LINEAR (Lincoln Near-Earth Asteroid Research) telescope in New Mexico has been very successful in finding NEOs and in finding new main-belt asteroids (■ Figure 25-11). The combined searches should be able to locate all of the largest NEOs by 2010.

This is a serious issue because even a small asteroid could do serious damage. For example, in late December 2004, an asteroid was discovered that had a 2.6 percent chance of striking Earth on April 13, 2029. The object is large enough to do significant damage over a wide area but would not be large enough to alter Earth's climate. Astronomers sprang into action and obtained further observations to refine the asteroid's orbit. The new orbit shows that the object will not hit Earth in 2029. Objects a few tens of meters or less fragment and explode in the atmosphere, but the shock waves from their explosions could cause serious damage on Earth's surface. Declassified data from military satellites show that Earth is hit about once a week by meter-size asteroids. Larger impacts produce more damage but are much less common.

It is easy to assume that the Apollo–Amor objects are rocky asteroids that have been thrown into their extreme orbits by events in the main asteroid belt. At least some of these objects, however, may be comets that have exhausted their ices and become trapped in short orbits that keep them in the inner solar system. You will see later in this chapter that the distinction between comets and asteroids is not totally clear.

While the Apollo–Amor objects rush through the inner solar system, there are also nonbelt asteroids in the outer solar system. These objects, being farther from the sun, move more slowly. The object Chiron, found in 1977, appears to be a body about 170 km in diameter. Its orbit carries it from just inside the orbit of Uranus to just inside the orbit of Saturn. Although it was first classified as an asteroid, its status is now less certain. Ten years after its discovery, Chiron surprised astronomers by suddenly brightening with a release of vapor and dust. Like a comet, Chiron was releasing a cloud of gas and dust in jets. Studies of older photographs showed that Chiron had been brighter even when it was farther from the sun, and astronomers now suspect that it may have a rocky crust covering deposits of ices such as solid nitrogen, methane, and carbon monoxide. Thus, Chiron may be more comet than asteroid, and it warns that the distinction is not clear-cut. Chiron's orbit is not stable. It cannot have remained in its present orbit for the entire history of the solar system but must have been perturbed into its present orbit within the last million years or so.

Jupiter ushers two groups of asteroids within its own orbit. These nonbelt asteroids have become trapped in the Lagrangian points along Jupiter's orbit. (See Figure 13-5.) These points lie 60° ahead of and 60° behind the planet and are regions where the gravitation of the sun and Jupiter combine to trap small bodies

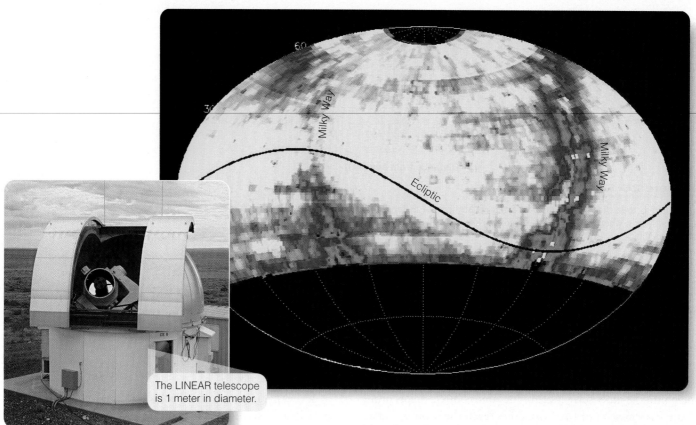

The LINEAR telescope
is 1 meter in diameter.

■ Figure 25-11

The LINEAR telescope searches for asteroids every clear night when the moon is not bright. The diagram shows the thoroughness of its search over the entire sky for one year. Asteroids are hard to discover in front of the starry Milky Way. (Reprinted with permission of MIT Lincoln Laboratory, Lexington, Massachusetts)

(■ Figure 25-12). Like cosmic sinkholes, the Lagrangian points have trapped chunks of debris now called **Trojan asteroids** and named after the heroes of the Trojan War (588 Achilles, 624 Hektor, 659 Nestor, and 1143 Odysseus, for example). Slightly over 1000 Trojan asteroids are known, but only the brightest have been given names.

Astronomers have also found a few objects in the Lagrangian points of the orbits of Mars and Neptune. Other planets may have Trojan asteroids trapped in their orbits. As astronomers search for smaller objects, they are finding that our solar system still contains large numbers of these small bodies. The challenge is to explain their origin.

The Origin of the Asteroids

You have concluded your study of each planet by trying to summarize its history. Can you tell the story of the asteroids? Begin with an idea that didn't work out.

An old theory held that the asteroids are the remains of a planet that broke up, but modern astronomers discount that idea. Once formed, a planet is very difficult to tear apart. Rather, astronomers believe the asteroids are the remains of planetesimals left over from the formation of the planets. Jupiter's gravity could have stirred the planetesimals just inside its orbit and caused collisions at unusually high velocities. These impacts would have tended to break up the planetesimals rather than assemble them into a planet. Resonances and chaos would have helped eject material, and most of the planetesimal objects have now been lost—swept up by planets, captured as satellites, or ejected from the solar system. The asteroid belt today contains hardly 4 percent the mass of Earth's moon.

The C-type asteroids have albedos smaller than 0.06 and would look very dark to your eyes. They are probably made of carbon-rich material similar to that in carbonaceous chondrites. The S-type asteroids have albedos of 0.1 to 0.2, so they would look brighter and spectroscopically redder. The M-type asteroids are bright but not as red. S-type asteroids are believed to be rocky, but M-type asteroids appear to be metal rich and may be the iron cores of fragmented asteroids.

Although S-type asteroids are very common in the inner asteroid belt, their spectroscopic colors are different from the chondrites—the most common kind of meteorite. New evidence from the analysis of moon rocks and from observations of Eros, an S-type asteroid, shows that bombardment by micrometeorites can redden and darken S-type asteroids. From this result, it seems that the most common kind of meteorites come from the most common kind of asteroid.

Main belt asteroids lie between the orbits of Mars and Jupiter.

Asteroids that could approach Earth are shown in red.

Trojan asteroids orbit in two clouds 60° ahead of Jupiter and 60° behind.

Jupiter

Mars

Earth

▮ Figure 25-12

This diagram plots the position of known asteroids inside or near the orbit of Jupiter on a specific day. Squares, filled or empty, show the location of known comets. Although asteroids and comets are small bodies and lie far apart, there are a great many of them in the inner solar system. (Gareth Williams/Minor Planet Center)

A few other types of asteroids are known, and a number of individual asteroids have been found that are unique, but these three classes contain a majority of the known asteroids.

How did these three types originate? A clue lies in their distribution in the asteroid belt. The S-type asteroids are much more common in the inner belt. In fact, there are almost none beyond a distance of about 3.45 AU. The C types are rather rare in the inner belt but are very common in the outer belt. Many astronomers believe that this distribution reflects the differences in temperature during the formation of the planetesimals. It was cooler in the outer belt, and the planetesimals there tended toward the volatile-rich composition of carbonaceous chondrites. In the warmer regions of the inner belt, the composition of the growing planetesimals was more like that of the chondrites.

Earlier you saw evidence that Vesta has at some time in the past been at least partly resurfaced by lava flowing up from its interior. How could a small asteroid be heated enough to produce

lava flows? One source of heat in newly formed asteroids would be the decay of short-lived radioactive elements such as aluminum-26. Such radioactive elements could have been produced in a supernova explosion occurring shortly before the formation of our solar system. The smallest asteroids lose their heat too fast to melt, but the decay of aluminum-26 occurs fast enough to melt the interiors of bodies larger than a few dozen kilometers in diameter. That makes it reasonable to suppose that Vesta and some other asteroids larger than about 100 km in diameter might have been strongly heated and modified by lava flows.

In contrast, the largest main-belt asteroid, Ceres, about 900 km in diameter, does not seem to have been modified by internal heating. The light reflected from its surface suggests a claylike surface related to carbonaceous chondrites. This is surprising, because clays form when minerals are exposed to water. In fact, spectroscopic observations reveal what seem to be traces of water ice on Ceres. Evidently some asteroids may have had significant amounts of water bound into their crusts when they were young.

All of the evidence suggests that the asteroids are the broken remains of planetesimals that formed in the solar nebula as planet building began. The largest remaining asteroids, such as Ceres and Vesta, may be largely unbroken planetesimals, and some of these may have evolved due to internal heat or the presence of water. Nevertheless, the vast majority of the asteroids are fragments, and many may consist of bodies that have been shattered and then reformed as gravity pulled the fragments back together. Some compositional differences seem to have originated when the asteroids formed in that the objects in the outer asteroid belt contain more carbon-rich compounds. The presence of massive Jupiter orbiting just beyond this region prevented the original planetesimals from accreting to build a planet. Instead, collisions fragmented the planetesimals, and nearly all of the material is now lost.

Some experts discuss the mining of the asteroid belt for its mineral riches, but the asteroid belt is nearly empty. The small, fragmented asteroids drift through space far from one another and only rarely collide with each other or with planets. A voyage to explore or exploit the asteroid belt would find the distance between bodies forbiddingly large.

Building Scientific Arguments

What evidence makes you think that the asteroids have been fragmented?

Perhaps the best scientific arguments test the interpretation of evidence. If you understand the evidence, you hold the key to the science. To begin, you might note that the solar nebula theory of the formation of the solar system predicts that planetesimals collided and either stuck together or fragmented. This is suggestive, but it is not evidence. A theory can never be used as evidence to support some other theory or hypothesis. Evidence means observations or the results of experiments, so you need to turn to observations of asteroids. Spacecraft photographs of asteroids such as Ida, Gaspra, and Eros show irregularly shaped little worlds heavily scarred by impact craters. In fact, observations of some asteroids show what may be pairs of bodies in contact, and the Galileo image of Ida reveals its small satellite, Dactyl. Furthermore, some meteorites appear to come from the asteroid belt, and a few have been linked to specific asteroids such as Vesta. There are even families of asteroids that seem to be fragments from a single collision. All of this evidence suggests that the asteroids have been broken up by violent impacts.

The impact fragmentation of asteroids has been important, but it has not erased all traces of the original planetesimals from which the asteroids formed. Build another argument based on evidence. **What evidence can you cite that reveals what those planetesimals were like?**

■ ■ ■

Connections: The asteroids are only the last crumbs left behind by failed planet building near Jupiter. Farther out in the solar system are other remains—ice shards from the age of planet formation.

(25-3) Comets

FEW THINGS IN ASTRONOMY are more beautiful than a bright comet hanging in the night sky (■ Figure 25-13). Whereas meteors shoot across the sky like demented fireflies, a comet moves with the stately grace of a great ship at sea, its motion hardly apparent. Night by night it shifts against the stars and may remain visible for months. Faint comets are common; a number are discovered every year. But a truly bright comet appears about once a decade. Comet Hyakutake in 1996 (Figure 25-1) and Comet Hale–Bopp in 1997 were both dramatic, but the later comet was so bright that you might class it with the great comets such as Comet Halley in 1910. A patient person might see half a dozen or more bright comets in a lifetime.

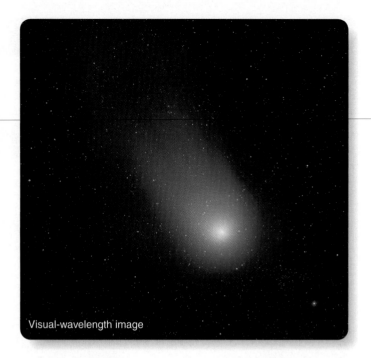

Visual-wavelength image

■ Figure 25-13

Comet C/2001 Q4 (NEAT) was discovered in 2001 by the Near Earth Asteroid Tracking (NEAT) system operated by NASA. The comet was brightest in 2004. A comet can remain bright in the sky for weeks or months as it sweeps along its orbit through the inner solar system. (T. Rector, Univ. of Alaska Anchorage, X. Levay and L. Frattare, STScI and WIYN/NOAO/AURA/NSF)

While everyone enjoys the beauty of comets, astronomers study them for their cargo of clues to the origin of our solar system.

Properties of Comets

As always, you should begin your study of a new kind of object by summarizing its observational properties. What do comets look like, and how do they behave? The observations are the evidence that reveals the secrets of the comets.

Study **Comet Observations** on pages 636–637 and notice three important properties of comets:

1 Comets have two kinds of tails shaped by the solar wind and solar radiation. The two kinds of tails show that the nucleus contains ices of water and other compounds plus rocky material most evident as dust.

2 Notice the importance of dust in comets. It not only produces dust tails but spreads throughout the solar system.

3 Notice the evidence that comet nuclei are fragile and can break into pieces.

Astronomers can put these and other observations together to discuss the structure of comet nuclei.

Ace✦Astronomy™ Log into AceAstronomy and select this chapter to see Astronomy Exercise "Comets." Design your own cometary nucleus and see what kind of tails it produces.

The Geology of Comet Nuclei

The nuclei of comets are quite small and cannot be studied in detail from Earth-based telescopes. Nevertheless, astronomers are beginning to understand the geology of these peculiar worlds.

Comet nuclei contain ices of water and other volatile compounds such as carbon dioxide, carbon monoxide, methane, ammonia, and so on. These ices are the kinds of compounds that should have condensed from the outer solar nebula, and that makes astronomers think that comets are ancient samples of the gases and dust from which the planets formed.

When the nuclei of comets approach the sun, the ices absorb energy from sunlight and sublime—change from a solid directly into a gas—to produce the observed tails. As the gases break down and combine chemically, they produce many compounds found in comet tails. Vast clouds of hydrogen gas observed around the heads of comets are derived from the breakup of molecules from the ices.

Five spacecraft flew past the nucleus of Comet Halley when it visited the inner solar system in 1985 and 1986. Spacecraft flew past the nuclei of Comet Borrelly in 2001 and Comet Wild 2 in 2004. The Deep Impact probe hit Comet Tempel 1 in 2005. Photos show that these comet nuclei are irregular in shape and very dark, with jets of gas and dust spewing from active regions on the nuclei (■ Figure 25-14). In general, these nuclei are darker than a lump of coal, which suggests the composition of the carbon-rich meteorites called carbonaceous chondrites.

From the gravitational influence of a nucleus on a passing spacecraft, astronomers can find the mass and density of the nucleus. Comet nuclei appear to have densities of 0.1 to 0.25 g/cm^3, much less than the density of ice. From these observations you can conclude that comet nuclei are not solid balls of ice but must be fluffy mixtures of ices and rocky dust with significant amounts of empty space.

Photographs of the comae (plural of coma) of comets often show jets springing from the nucleus and being swept back by the pressure of sunlight and by the solar wind to form the tail (Figure 25-14). Studies of the motions of these jets as the nucleus rotates reveal that the jets originate from active regions that may be faults or vents. As the rotation of a cometary nucleus carries an

■ Figure 25-14

Visual-wavelength images made by spacecraft and by the Hubble Space Telescope show how the nucleus of a comet produces jets of gases from regions where sunlight vaporizes ices. (Halley nucleus: © 1986 Max-Planck Institute; Halley coma: Steven Larson; Comets Borrelly, Hale–Bopp and Wild 2: NASA)

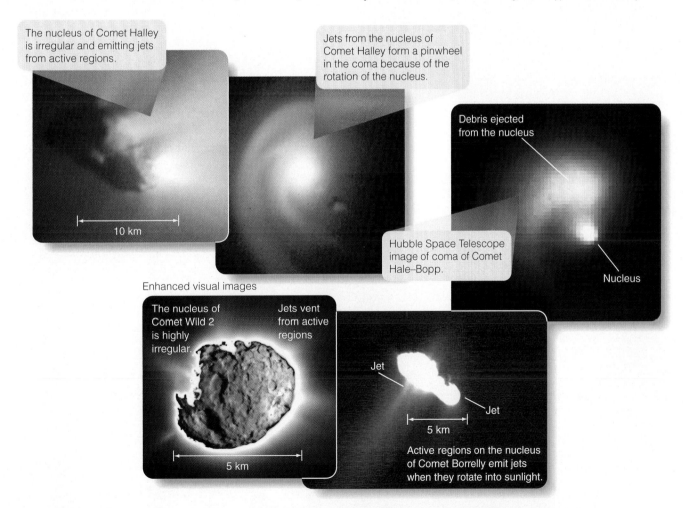

The nucleus of Comet Halley is irregular and emitting jets from active regions.

10 km

Jets from the nucleus of Comet Halley form a pinwheel in the coma because of the rotation of the nucleus.

Debris ejected from the nucleus

Hubble Space Telescope image of coma of Comet Hale–Bopp.

Nucleus

Enhanced visual images

The nucleus of Comet Wild 2 is highly irregular.

Jets vent from active regions

5 km

Jet

Jet

5 km

Active regions on the nucleus of Comet Borrelly emit jets when they rotate into sunlight.

Comet Observations

1 A **type I** or **gas tail** is produced by ionized gas carried away from the nucleus by the solar wind. The spectrum of a gas tail is an emission spectrum. The atoms are ionized by the ultraviolet light in sunlight. The wisps and kinks in gas tails are produced by the magnetic field embedded in the solar wind.

Spectra of gas tails reveal atoms and ions such as H_2O, CO_2, CO, H, OH, O, S, C, and so on. These are released by the vaporizing ices or produced by the breakdown of those molecules. Some gases, such as hydrogen cyanide (HCN), must be formed by chemical reactions.

Gas tail (Type I)

Dust tail (Type II)

1a A **type II** or **dust tail** is produced by dust from the vaporizing ices of the nucleus. The dust is pushed gently outward by the pressure of sunlight, and it reflects an absorption spectrum, the spectrum of sunlight. The dust is not affected by the magnetic field of the solar wind, so dust tails are more uniform than gas tails. Dust tails are often curved because the dust particles follow their individual orbits around the sun once they leave the nucleus.

When a spacecraft named ICE passed through the gas tail of a comet, it found a magnetic field from the solar wind draped over the nucleus like seaweed draped over a fishhook.

Nucleus

1b The nucleus of a comet (not visible here) is a small, fragile lump of porous rock containing ices of water, carbon dioxide, ammonia, and so on. Comet nuclei can be 10 to 100 km in diameter.

Coma

The **coma** of a comet is the cloud of gas and dust that surrounds the nucleus. It can be over 1,000,000 km in diameter, bigger than the sun.

1c Comet Mrkos in 1957 shows how the gas tail can change from night to night due to changes in the magnetic field in the solar wind.

Visual-wavelength images

Caltech

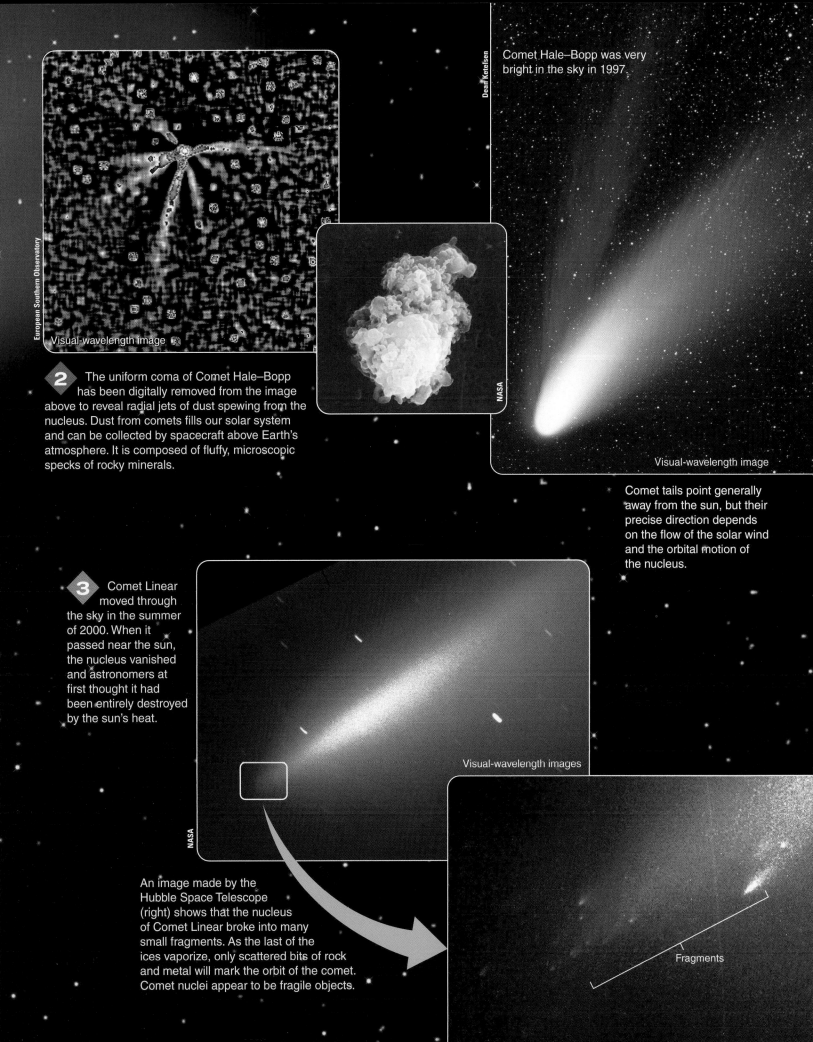

European Southern Observatory

Visual-wavelength image

2 The uniform coma of Comet Hale–Bopp has been digitally removed from the image above to reveal radial jets of dust spewing from the nucleus. Dust from comets fills our solar system and can be collected by spacecraft above Earth's atmosphere. It is composed of fluffy, microscopic specks of rocky minerals.

NASA

Dean Ketelsen

Comet Hale–Bopp was very bright in the sky in 1997.

Visual-wavelength image

Comet tails point generally away from the sun, but their precise direction depends on the flow of the solar wind and the orbital motion of the nucleus.

3 Comet Linear moved through the sky in the summer of 2000. When it passed near the sun, the nucleus vanished and astronomers at first thought it had been entirely destroyed by the sun's heat.

Visual-wavelength images

NASA

An image made by the Hubble Space Telescope (right) shows that the nucleus of Comet Linear broke into many small fragments. As the last of the ices vaporize, only scattered bits of rock and metal will mark the orbit of the comet. Comet nuclei appear to be fragile objects.

Fragments

active region into sunlight, it begins venting gas and dust, and as it rotates into darkness it shuts down. Evidently, the nuclei of comets have a crust of rocky dust that may be material left behind when the ices vaporize. Breaks in that crust can expose ices to sunlight, and vents can occur in those regions. It also seems that some comets have pockets of volatiles buried below the crust. When one of those pockets is exposed and begins to vaporize, the comet can suffer a dramatic outburst.

In 2005, the Deep Impact spacecraft released an instrumented impactor that weighed 816 lb on Earth into the path of comet Tempel 1. Exactly as planned, the nucleus of the comet ran into the impactor at almost 10 km/s (23,000 mph). The impact penetrated the crust of the nucleus and blasted material out into space where the mothership flying past could analyze it (■ Figure 25-15). The burst of vapor and dust was detected from the mothership and also from Earth-based telescopes.

The nuclei of comets are not strong. A nucleus can be ripped apart by the violence of gases bursting through the crust or by tidal forces produced if the comet passes near a massive body. A number of comets have been observed to break into two or more parts while near Jupiter or near the sun. On July 8, 1992, Comet Shoemaker–Levy 9 passed within 1.29 planetary radii of Jupiter's center, well within its Roche limit, and tidal forces ripped the nucleus into at least 21 pieces (■ Figure 25-16a). The fragmented pieces were as large as a few kilometers in diameter, and they released gas and dust and spread into a long string of objects that looped away from Jupiter and then fell back to strike the planet and produce massive impacts over a period of six days in July 1994 (Figure 23-6). The fragmentation of comet nuclei may explain long chains of craters found on the Earth's moon and on some other planetary surfaces in the solar system. A comet breaking into pieces could produce a chain of impact features such as the ones shown in Figure 25-16b and c. The Solar and Heliospheric Observatory (SOHO) was put into space to observe the sun, but it has discovered over 1000 comets called "sun grazers" as they zip in and around the sun. These comets follow long elliptical orbits that bring them 70 times closer to the sun's surface than the planet Mercury (■ Figure 25-17). Most of these comets belong to one of four groups, with the comets in each group having very similar orbits. Like the Hirayama families of asteroids, these comet groups

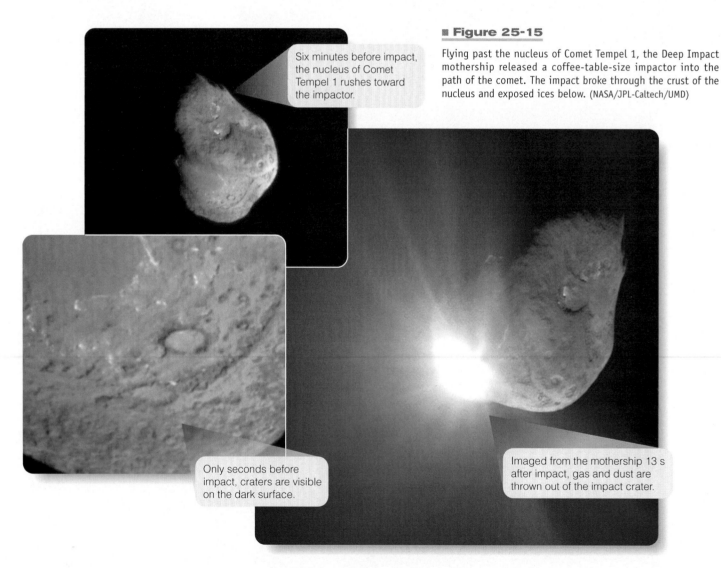

■ **Figure 25-15**

Flying past the nucleus of Comet Tempel 1, the Deep Impact mothership released a coffee-table-size impactor into the path of the comet. The impact broke through the crust of the nucleus and exposed ices below. (NASA/JPL-Caltech/UMD)

Six minutes before impact, the nucleus of Comet Tempel 1 rushes toward the impactor.

Only seconds before impact, craters are visible on the dark surface.

Imaged from the mothership 13 s after impact, gas and dust are thrown out of the impact crater.

Visual-wavelength image

■ Figure 25-16

(a) Tides from Jupiter pulled apart the nucleus of Comet Shoemaker–Levy 9 to form a long strand of icy bodies and dust that fell back to strike Jupiter two years later. (b) A 40-km-long crater chain on the moon and (c) another 140-km-long crater chain on Jupiter's moon Callisto were apparently formed by the impact of fragmented comet nuclei. The impacts that form such chains probably occur within a span of seconds. (NASA)

appear to be made up of fragments of larger bodies. You can jot this down as yet more evidence that comet nuclei are very fragile.

The nuclei of comets range from a few kilometers in diameter up to a few tens of kilometers. Each passage near the sun costs a nucleus many millions of tons of ices, so the nucleus slowly loses its ices until there is nothing left but dust and rock falling along an orbit around the sun. The fate of a comet is clear. The mystery is its origin.

The Origin of Comets

Family relationships among the comets can give you clues to their origin. Most comets have long, elliptical orbits with periods greater than 200 years. These are known as long-period comets. Their orbits are randomly inclined, with comets falling into the inner solar system from all directions. As many circle the sun clockwise as counterclockwise.

In contrast, about 100 or so of the 600 well-studied comets have orbits with periods less than 200 years. These short-period comets follow orbits that lie within 30° of the plane of the solar system, and most revolve around the sun counterclockwise—the same direction the planets orbit. Comet Halley, with a period of 76 years, is a short-period comet.

A comet cannot survive long in an orbit that brings it into the inner solar system. The heat of the sun vaporizes ices and reduces comets to inactive bodies of rock and dust. A comet may last only 100 to 1000 orbits around the sun. Also, encounters with planets, especially Jupiter, can fling a comet into the sun or out of the solar system. Of course, as you have seen, comets hit planets, and that's the end of any such unlucky comets. Even if it didn't vaporize in the sun's heat, a comet couldn't survive more than about half a million years before its career was ended by a planet. The comets visible in our skies can't have survived 4.6 billion years since the formation of the solar system, and that means there must be a continuous supply of new comets. Where do they come from?

In the 1950s, Dutch astronomer Jan Oort proposed that the long-period comets are objects that fall in from the **Oort cloud,** a spherical cloud of icy bodies believed to extend from about 10,000 to 100,000 AU from the sun (■ Figure 25-18). Astronomers estimate that the cloud contains several trillion icy bodies. Far from the sun, they are very cold, lack comae and tails, and are invisible. The gravitational influence of occasional passing stars could perturb a few of these objects to fall into the inner solar system, where the heat of the sun warms their ices and transforms them into comets. Because the Oort cloud is spherical, these long-period comets fall inward from random directions.

It is not outlandish that stars pass close enough to affect the Oort cloud. Data from the Hipparcos satellite show that the star Gliese 710 will pass within 1 ly (about 63,000 AU) of our solar system in about a million years. That may shower the inner solar

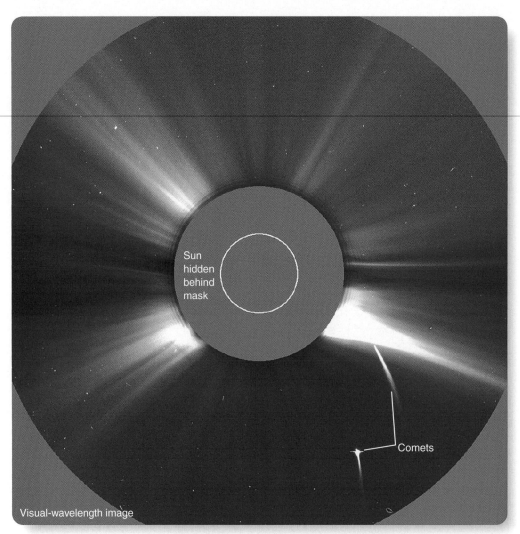

■ **Figure 25-17**

The SOHO observatory can see comets rounding the sun on very tight orbits. Some of these sun-grazing comets, like the two shown here, are destroyed by the heat and are not detected emerging on the other side of the sun. (SOHO)

system with Oort cloud objects, which, warmed by the sun, will become comets.

Saying that comets come from the Oort cloud only pushes the mystery back one step. Where did those icy bodies come from? Beginning in Chapter 19, you have studied the origin and evolution of our solar system so carefully that the answer leaps out at you. Those icy bodies are some of the icy planetesimals that formed in the outer solar nebula.

The bodies in the Oort cloud, however, could not have formed at their present location. The solar nebula would not have been dense enough so far from the sun. Also, had they formed from the solar nebula, you would expect them to be distributed in a disk and not in a sphere. Rather, astronomers think the future comets formed as icy planetesimals in the outer solar system near the present orbits of Uranus and Neptune. As those planets grew more massive, they swept up many of these planetesimals, but they

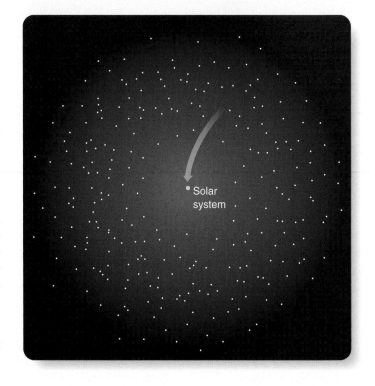

■ **Figure 25-18**

The long-period comets appear to originate in the Oort cloud. Objects that fall into the solar system from this cloud arrive from all directions.

would have ejected some out of the solar system. Most of those objects vanished into space, but perhaps 10 percent had their orbits modified by the gravity of stars passing nearby and became part of the Oort cloud. These are the objects that form the long-period comets.

Some of the short-period comets, including Comet Halley, appear to have originated in the Oort cloud and had their orbits altered by a close encounter with Jupiter. But many of the short-period comets cannot have begun in the Oort cloud. Interactions with a planet can't capture objects from the Oort cloud into the orbits that many short-period comets occupy. There must be another source of icy bodies in our solar system. To find the answer, you need only look out beyond the orbit of Neptune and study the icy Kuiper Belt objects.

The Kuiper Belt

If you want to have fun, get a bunch of astronomers together, order some pizza, and ask them to decide whether the Kuiper Belt objects are asteroids or comets. The argument will outlast the refreshments. Whatever they are, you first met the Kuiper Belt objects in Chapter 19 when you studied the origin of the solar system. It is important to study them carefully here because they seem to be one of the sources of comets.

In 1951, Dutch-American astronomer Gerard P. Kuiper proposed that the formation of the solar system should have left behind a belt of small, icy planetesimals beyond the Jovian planets and in the plane of the solar system. Such objects were first discovered in 1992 and are now known as Kuiper Belt objects. You should know, however, that in 1943 and 1949, astronomer Kenneth Edgeworth published papers that included a paragraph speculating about objects beyond Pluto. Consequently, you may see the Kuiper Belt referred to as the Edgeworth–Kuiper Belt. Actually, lots of astronomers have speculated about one or more bodies beyond Pluto, so the real controversy isn't in their name but in their origin.

The Kuiper Belt objects are small, icy bodies (■ Figure 25-19) that orbit in the plane of the solar system extending from the orbit of Neptune out to about 50 AU from the sun. Some objects loop out as far as 150 AU, but they seem to have been scattered by gravitational interactions with passing stars. The entire Kuiper Belt, containing as many as 70,000 objects as big as 100 km in diameter, would be hidden in Figure 25-18 behind the yellow dot symbolizing the solar system. Some Kuiper Belt objects are surprisingly large, and one is about 50 percent bigger than Pluto.

January 26, 2003
19:19 UT

06:50 UT

Visual-wavelength image

■ Figure 25-19

Kuiper Belt objects are small bodies with dark surfaces that are very hard to detect from Earth. In your imagination, you can see them as icy, cratered worlds orbiting far from the sun. The New Horizons spacecraft is planned to fly past at least one Kuiper Belt object after it passes Pluto. Dust, mostly from comets, produces the horizontal glow centered on the sun. (Image: NASA and G. Bernstein; Art: JHUAPL/SwRI)

Presumably these objects are icy planetesimals that formed in the outer solar system beyond Neptune. Planetesimals among the giant planets were ejected, thrown into the sun, or absorbed by the planets, but the KBOs orbiting beyond Neptune are safe from such interactions and could have survived since the formation of the solar system. In fact, they provide clues as to how our solar system formed. In Chapter 24, you learned that some KBOs orbit in resonances with Neptune, and the family called the plutinos provided evidence that Pluto may be one of the largest KBOs. The objects probably became caught in resonances when Neptune migrated outward during or soon after planet formation. The presence of a sharp outer edge to the Kuiper Belt may indicate that the sun formed as a member of a star cluster and that the crowded stars tidally limited the growth of the solar nebula beyond about 50 AU.

Can this belt of ancient, icy worlds generate short-period comets? Mathematical studies show that objects thrown inward from the Kuiper Belt could take up orbits resembling the short-period comets. Collisions and interactions among the KBOs must be a rare but continuing process, and small KBOs or fragments from a collision would resemble the nuclei of comets.

Some of these objects are quite large. One, called Quaoar (pronounced "KWAH-o-wahr"), is about 1270 km in diameter, a bit larger than Pluto's moon Charon. The KBO known as Sedna is even larger. Encounters with such a large body could throw a small object into an extreme orbit. Another object, known as 1998 W31, has its own moon. In fact, binary objects seem quite common in the Kuiper Belt, and that may be evidence that collisions are common.

You can find more evidence that our solar system has a Kuiper Belt by looking at other stars. A number of nearby stars such as Beta Pictoris (Chapter 19) are surrounded by disks of dust believed to be released by objects in Kuiper belts around these stars. Thus the detection of the dust implies the presence of Kuiper belts. It appears that our solar system would be peculiar if it didn't have a Kuiper Belt and plenty of dusty comets.

Building Scientific Arguments

How do comets help explain the formation of the planets?
This scientific argument pulls together a number of ideas. According to the solar nebula hypothesis, the planets formed from planetesimals that accreted in a disk-shaped nebula around the forming sun. In the outer solar nebula, it was cold, and the planetesimals would have contained large amounts of ices. Many of these planetesimals were destroyed when they fell together to make the planets, but some survived. The icy bodies of the Oort cloud and the Kuiper Belt may be the last surviving icy planetesimals in our solar system. When these bodies fall into the inner solar system, they become comets, and the gases they release reveal that they are rich in volatile

materials such as water, carbon dioxide, carbon monoxide, methane, and ammonia. These are the ices you would expect to find in the icy planetesimals. Furthermore, comets are rich in dust, and the planetesimals must have included large amounts of dust frozen into the ices when they formed. The nuclei of comets seem to be frozen samples of the ancient solar nebula.

Nearly all of the mass of a comet is in the nucleus, but the light you see comes from the coma and the tail. Build an argument pulling together ideas about cometary nuclei. **What kind of spectra do comets produce, and what does that tell you about their nuclei?**

■　　■　　■

Connections: Your study of comets has revealed that there is a continuous supply of these objects, that they fragment easily when they encounter planets, and that entire comets or fragments of comets can hit planets. Surely that makes you wonder what happens when a comet hits Earth.

25-4 Impacts on Earth

ASTEROIDS AND COMETS FALL through our solar system like runaway trucks on a busy highway. These objects must hit the planets. In fact, it is these collisions that built the planets, and most of the original planetesimals have now been either incorporated into the planets or ejected from the solar system. Nevertheless, planets still get hit, as was dramatically demonstrated in 1994 when Comet Shoemaker–Levy hit Jupiter. Like any other planet, Earth must get hit now and then; where is the evidence?

Impacts and Dinosaurs

Comet nuclei and asteroids may seem small compared to planets, but they hit Earth with tremendous energy. Calculations show that such impacts throw vast quantities of dust into the atmosphere and cause widespread changes in climate. The extinction of entire species, such as the dinosaurs, may have been caused by comet impacts. One of the most important clues is an element called iridium.

In 1980, Luis and Walter Alvarez announced the discovery of an excess of the element iridium in sediments laid down at the end of the Cretaceous period 65 million years ago—just when the dinosaurs and many other species became extinct. Iridium is rare in Earth's crust but common in meteorites, which led the Alvarezes to suggest that a major meteorite impact at the end of the Cretaceous threw vast amounts of iridium-rich dust into the atmosphere. This dust might have plunged Earth into a winter that lasted many years, killing off many species of plants and animals, including the dinosaurs.

■ **Figure 25-20**

The theory that the impact of one or more comets altered Earth's climate and drove dinosaurs to extinction has become so popular it appeared on this Hungarian stamp. The spacecraft shown (ICE) flew through the tail of Comet Giacobini–Zinner in 1985. Note the dead dinosaurs in the background. The giant impact scar buried in Earth's crust near the village of Chicxulub in the northern Yucatán was formed about 65 million years ago. This gravity map shows the extent of the crater hidden below limestone deposited since the impact. (Virgil L. Sharpton, Univ. of Alaska, Fairbanks)

This theory was met with skepticism at first, but soon scientists found the iridium anomaly in sediments of the same age all over the world. Others found related elements and mineral forms typical of meteorite impacts. At the same time, theorists studying nuclear weapons predicted that a nuclear war would throw so much dust into the atmosphere that our planet would be plunged into a "nuclear winter" that could last a number of years. A major impact would do the same thing. Within a few years, scientists generally agreed that the Cretaceous extinctions were caused by one or more major impacts.

Mathematical models and observations of the impact of the comet fragments on Jupiter in 1994 have combined to create a plausible scenario of the events following a major impact on Earth. Of course, creatures living near the site of the impact would probably die in the initial shock, and an impact at sea would create tsunamis (tidal waves) many hundreds of meters high that would devastate coastal regions for many kilometers inland even halfway around the world from the impact. But the worst effects would begin after the initial explosion. On land or sea, a major impact would excavate large amounts of pulverized rock and loft it high above the atmosphere. As this material fell back, Earth's atmosphere would be turned into a glowing oven of red-hot rock streaming through the air as a rain of meteors, and the heat would trigger massive forest fires around the world. Soot from such fires has been detected in the layers of clay laid down at the end of the Cretaceous. Once the firestorms cooled, the remaining dust in the atmosphere would block sunlight and produce deep darkness for a year or more. No matter where on Earth an impact occurred, it would almost certainly vaporize large amounts of limestone.

The carbon dioxide released from the limestone would produce intense acid rain. All of these consequences make it surprising that any life could have survived such an impact.

Geologists have located a crater at least 150 km in diameter centered near the village of Chicxulub in the northern Yucatán (■ Figure 25-20). Although the crater is totally covered by sediments, mineral samples show that it contains shocked quartz typical of impact sites and that it is the right age. A gigantic impact formed the crater about 65 million years ago, just when the dinosaurs and many other species died out, and many Earth scientists now believe that this is the scar of the impact that ended the Cretaceous.

At first, these climate-changing impacts were blamed on large meteorites, but the nucleus of a comet would do as much damage. In fact, you have seen in this chapter that the distinction between an asteroid and a comet is not clear-cut.

The theory that an impact caused the extinction at the end of the Cretaceous period was once highly controversial, but the accumulated evidence has made it a widely accepted idea. Now Earth scientists are extending the theory to other extinctions. Chemical evidence from Earth's crust suggests that the extinction at the end of the Permian 250 million years ago was caused by a giant impact. The Permian extinction killed off 95 percent of the species on Earth and set the stage for the rise of the dinosaurs. The dinosaurs may owe their origin as well as their extinction to giant impacts on Earth.

A solar system is a dangerous place to put a planet. Comets, asteroids, and meteoroids constantly rain down on the planets, and Earth gets hit quite often. The Chicxulub Crater isn't the only large impact scar on Earth. For example, a giant crater has been found buried under sediment in Iowa, and another giant impact crater may underlie most of Chesapeake Bay. What would it be like to live on Earth when such an impact occurred? Humanity got a hint in 1908 when something hit Siberia.

The Tunguska Event

On the morning of June 30, 1908, scattered reindeer herders and homesteaders in central Siberia were startled to see a brilliant blue-white fireball brighter than the sun streak across the sky. Still descending, it exploded with a blinding flash and an intense pulse of heat:

> The whole northern part of the sky appeared to be covered with fire. . . . I felt great heat as if my shirt had caught fire . . . there was a . . . mighty crash. . . . I was thrown on the ground about [7 m] from the porch. . . . A hot wind, as from a cannon, blew past the huts from the north.

The blast was heard up to 1000 km away, and the resulting pulse of air pressure circled Earth twice. For a number of nights following the blast, European astronomers, who knew nothing of the explosion, observed a glowing reddish haze high in the atmosphere.

Travel in the wilderness of Siberia was difficult early in the 20th century; moreover, World War I, the Bolshevik Revolution, and the Russian Civil War prevented expeditions from reaching the site before 1927. When at last an expedition arrived, it found that the blast had occurred above the Stony Tunguska River valley and had flattened trees in an irregular pattern extending out 30 km (■ Figure 25-21). The trees were knocked down radially away from the center of the blast, and limbs and leaves had been stripped away. The trunks of trees at the very center of the area were still standing, although they had lost all their limbs. No expedition has ever found a crater, so it seems that the explosion, equaling a 12-megaton nuclear weapon, occurred at least a few kilometers above the ground.

All kinds of speculative theories have been proposed in tabloid newspapers and occult books. One popular idea was that a flying saucer with engine trouble tried to make an emergency landing and exploded. More responsible ideas have proposed that Earth was hit by a piece of antimatter or a miniature black hole. The evidence does not necessarily rule out either of these last two ideas, but they violate an important principle in science. They assume more than needed. Science always prefers the simplest adequate explanation for any given phenomenon, and a small asteroid or comet would explain the event.

The comet hypothesis proposes that the head of a small comet struck Earth and exploded in the atmosphere. You might not see such a comet if it was coming from the direction of the sun. Witnesses reported that the fireball in the atmosphere came from the east, which in the morning means from the direction of the sun.

■ **Active Figure 25-21**

The 1908 Tunguska event in Siberia destroyed an area the size of a large city. Here the area of destruction is superimposed on a map of Washington, D.C., and its surrounding beltway. In the central area, trees were burned; and, in the outer area, trees were blown down in a pattern away from the path of the meteor.

Ace✪Astronomy™ Log into AceAstronomy and select this chapter to see the Active Figure "Impacts on Earth." Learn how often Earth is hit by meteorites of different sizes.

If the object was a comet, whatever tail it had would have been projected straight at Earth and would not have been visible. More likely, say some astronomers, the object was the exhausted core of a comet and had no appreciable tail.

A fragment of a partially exhausted comet should consist of silicate material resembling carbonaceous chondrites, mixed with some remaining ices. Modern studies of the area show that thousands of tons of powdered material resembling carbonaceous chondrites are scattered in the soil. A chunk of dirty ice 50 m in diameter would contain about the right mass and would be totally invisible approaching from the sun. Also, a comet fragment would be so fragile that it would explode in the atmosphere rather than dig a crater, as would a denser, stronger meteorite.

The falling-comet hypothesis has been very popular with astronomers. It does not make any fantastic assumptions, and it explains the observed phenomena, even the reddish glow over

Europe. Some astronomers claim the glow was the faint traces of the comet's scattered dust, trapped high in the atmosphere and illuminated by the summer sun shining over the North Pole.

As compelling as the comet hypothesis may seem, not all astronomers agree. In the early 1980s, a detailed analysis of all the evidence suggested that the object was not cometary. Its speed and direction were not consistent with the motion of a comet. The analysis suggested instead that the object was a small Apollo asteroid, 90 to 190 m in diameter.

In 1993, astronomers studied computer models of objects entering Earth's atmosphere at high speed and concluded that the fragile head of a comet would have exploded much too high in the atmosphere. An iron-rich asteroid, in contrast, would be so strong that it would survive to reach the ground and would form a large crater. The most likely candidate seems to be a stony asteroid about 30 m in diameter. It would explode at just about the right height to produce the Tunguska blast.

One natural question is: Why should a piece of rock explode at all? To understand, you need to recall that a meteor enters Earth's atmosphere at roughly 15 km/s. The air resistance is tremendous; and, as the meteor descends lower and meets denser air, the air resistance increases rapidly. The resistance that the meteor experiences depends on its surface area. When the meteor breaks into smaller pieces, its surface area suddenly increases, so the air resistance suddenly increases, and that causes even more fragmentation. Because air resistance depends on surface area, when the meteor begins to break up, it explodes and vaporizes almost instantly. That is most likely what created the blast that equaled a 12-megaton explosion at Tunguska.

Could such an impact happen again? Earth gets hit by small meteorites every day and by larger objects less often. Impacts by large asteroids may happen many millions of years apart, but they continue to happen. In mid-March 1998, newspaper headlines announced, "Mile-Wide Asteroid to Hit Earth in October 2028." The frightening story was true except that the news media did not mention the uncertainty in the orbit. Within days, astronomers found the asteroid on old photographs, refined the orbit with the new data, and concluded that the asteroid, known as 1997XF$_{11}$, will miss Earth by 600,000 miles. There will be no impact by this asteroid in 2028, but there are plenty more asteroids that haven't been discovered yet.

Some people have argued that the danger from asteroid and comet impacts is so great that governments should develop massive nuclear-tipped missiles, ready to blast a meteoroid to pieces long before it can reach Earth. Some astronomers point out that lots of small fragments slamming into Earth may be even worse than one big impact. Other astronomers argue that the big objects are so rare they can be ignored. The real danger, they say, lies in the smaller, more common meteoroids, which are too small to detect with current telescopes.

Some people may prefer to believe that the Tunguska event was an alien spaceship; that is a titillating idea. But the truth about the Tunguska event is far more exciting. Earth is struck by meteorites every day. Some impacts are large enough to alter the climate, and the future of our civilization on Earth may depend on how much time we have until the next big one.

Ace✸Astronomy™ Log into AceAstronomy and select this chapter to see Astronomy Exercise "Cratering." What factors determine the size of a crater?

Building Scientific Arguments

How can a large meteorite impact alter Earth's climate?
This argument is based on theory, but some evidence does exist. Even a meteorite no bigger than a large house strikes Earth with as much energy as a nuclear bomb and can produce a crater a mile in diameter. A large impact could produce a crater 100 km or more in diameter. Rock from Earth's crust would be pulverized and lifted high above the atmosphere. As it fell back in a rain of fiery meteors, it would be so hot it would trigger massive forest fires over much of Earth's surface. Scientists can detect traces of such fires at about the time of the Chicxulub impact, 65 million years ago. Even after the fires cooled, the dust in the atmosphere would prevent sunlight from reaching the ground, and plants dependent on sunlight would be starved of energy. This worldwide darkness might last a year or more before the dust settled out of the atmosphere. By that time, many species would have died out, and the climate could have changed dramatically.

As scary as this is, you need not ask if it will happen. The probability of such an event is 100 percent. Such impacts are common in the solar system. Now extend your argument: **Why does the presence of large amounts of limestone on Earth make a major impact even more devastating to the environment?**

■　　■　　■

Connections: Life on Earth seems fragile and exposed. Are humans just lucky to have survived so long? Does life exist on other worlds? If so, has it, too, survived long enough to develop intelligence? That is the story of the next chapter.

Summary

25-1 | Meteorites

Where do meteors and meteorites come from?

■ The vast majority of meteors, including those in meteor showers, appear to be low-density, fragile bits of debris from comets.

■ The evidence, including the orbits of meteorites seen to fall, suggests that the meteorites are fragments of asteroids.

■ Meteorites can be classified according to their mineral content.

■ Iron meteorites are mostly iron and nickel; when sliced open, polished, and etched, they show Widmanstätten patterns. That means the metal cooled from a molten state very slowly.

■ Stony meteorites include chondrites, which contain condrules—small, glassy particles that are believed to be very ancient droplets of molten material formed in the solar nebula.

■ Stony meteorites that are rich in volatiles and carbon are called carbonaceous chondrites. They are among the least-modified meteorites.

■ The achondrites are stony meteorites that contain no chondrules and no volatiles. They appear to have been melted after they formed and, in some cases, resemble solidified lavas.

■ Many meteorites appear to have formed as part of larger bodies that melted, differentiated, and cooled very slowly. Later they were broken up, and fragments from the core became iron meteorites, while fragments from the outer layers became stony meteorites.

■ Carbonaceous chondrites appear to have formed further from the sun.

25-2 | Asteroids

What are the asteroids?

■ Most asteroids lie in a belt between Mars and Jupiter, although some, the Apollo–Amor objects, follow orbits that cross into the inner solar system. If they pass near Earth, they are called Near Earth Objects (NEOs).

■ Kirkwood's gaps in the asteroid belt are caused by orbital resonances with Jupiter.

■ Two groups of asteroids called the Trojans are caught in the Lagrangian points along Jupiter's orbit 60° ahead and 60° behind the planet.

■ Asteroids are irregular in shape and heavily cratered from collisions. Their surfaces are covered by gray, pulverized rock, and some asteroids have such low densities they appear to be fragmented rubble piles.

■ C-type asteroids are more common in the outer asteroid belt, where it is cooler. They are darker and may be carbonaceous.

■ S-type asteroids are the most common and appear to be the source of the most common kind of meteorites, the chondrites. S-type asteroids are more common in the inner belt.

■ M-type asteroids appear to have nickel-iron compositions and may be the cores of broken asteroids.

■ A few asteroids, such as Vesta, show evidence of lava flows on their surfaces, but others, such as Ceres, show no evidence of geological activity.

■ The asteroids formed as rocky planetesimals between Mars and Jupiter, but Jupiter prevented them from forming a planet. Collisions have broken all but the largest of the asteroids.

25-3 | Comets

Where do comets come from?

■ A comet is produced by a lump of rock and ices 10 to 100 km in diameter. In long, elliptical orbits, the icy body stays frozen until it nears the sun. Then some of the ices vaporize and release dust and gas that are blown away to form a tail.

■ A type I, or gas, tail is ionized gas carried away by the solar wind.

■ A type II, or dust, tail is solid debris released from the nucleus and blown outward by the pressure of sunlight.

■ The coma of a comet can be up to a million kilometers in diameter, but the nucleus is rarely more than a few tens of kilometers across.

■ Spacecraft flying past comets have revealed that the nucleus of a comet has a very dark, rocky crust and that jets of vapor and dust issue from active regions on the sunlit side.

■ The low density of the nuclei show that they are irregular mixtures of ices and silicates, probably containing large voids.

■ Comet nuclei are not strong and have been seen to break apart when they pass near planets, especially Jupiter. A comet broke up and hit Jupiter in 1994, and such impacts are probably common in the history of the solar system.

■ Comets are believed to have formed as icy planetesimals in the outer solar system. Some were ejected to form the Oort cloud. Comets falling in from the Oort cloud become long-period comets.

■ A few short-period comets have been produced when comets passed near Jupiter and had their orbits changed.

■ Other icy bodies formed in the outer solar system and now make up the Kuiper Belt beyond Neptune. Objects from the Kuiper Belt that fall into the inner solar system can become short-period comets.

25-4 | Impacts on Earth

What happens when an asteroid or comet hits Earth?

■ A major impact on Earth can trigger extinctions because of changes to the climate.

■ Material falling back into the atmosphere after a major impact can trigger global forest fires.

■ Large amounts of carbon dioxide released into the atmosphere by a major impact can produce intense acid rain.

■ A major impact can fill the atmosphere with dust for years and plunge the planet into darkness.

■ An impact at Chicxulub in the Yucatán 65 million years ago appears to have triggered the extinction of 75 percent of the species then on Earth, including the dinosaurs. The Great Dying at the end of the Permian may also have been triggered by a major impact.

■ A small rocky object entered Earth's atmosphere and exploded over the Stony Tunguska River valley in Siberia in 1908. The explosion devastated forests in an area up to 60 km across.

New Terms

radiant (p. 619)

sporadic meteor (p. 620)

fall (p. 621)

find (p. 621)

iron meteorite (p. 622)

selection effect (p. 622)

Widmanstätten pattern (p. 622)

stony meteorite (p. 622)

chondrite (p. 622)

chondrule (p. 622)

carbonaceous chondrite (p. 623)

CAI (p. 624)

achondrite (p. 624)

stony-iron meteorite (p. 624)

Hirayama family (p. 627)

Kirkwood's gap (p. 630)

Apollo–Amor object (p. 631)

Trojan asteroid (p. 632)

gas (type I) tail (p. 636)

dust (type II) tail (p. 636)

coma (p. 636)

Oort cloud (p. 639)

Review Questions

Ace⊙Astronomy™ Assess your understanding of this chapter's topics with additional quizzing and animations at **http://ace .brookscole.com/sf9**

1. If most falls are stony meteorites, why are most finds iron?

2. How do observations of meteor showers reveal one of the sources of meteoroids?

3. How can most meteors be cometary if most, perhaps all, meteorites are asteroidal?

4. Why are meteorites easier to find in Antarctica? Why are stony meteorites better represented among these finds than among finds in the United States?

5. What evidence can you cite that some meteorites have originated inside large bodies?

6. Why couldn't uranium, thorium, and radioactive potassium have melted the planetesimals? What evidence can you cite that some asteroids have had molten interiors?

7. How do Kirkwood's gaps resemble Cassini's division?

8. The first asteroids discovered were much larger than the typical asteroid known today. How does this illustrate a selection effect?

9. Describe three types of asteroids (S, M, and C), and explain a theory to account for their differences.

10. How might the Apollo–Amor objects have originated?

11. What is the difference between a type I tail and a type II tail?

12. If comets are icy planetesimals left over from the formation of the solar system, why haven't they all vaporized by now?

13. What is the difference between the source of long-period comets and that of short-period comets?

14. How did the bodies in the Kuiper Belt and the Oort cloud form? Why are they in different locations in the solar system?

15. What do you see in this image that tells you how big planetesimals were when the solar system was forming?

(Russell Kempton, New England Meteoritical)

16. Discuss the surface of the asteroid Mathilde. What do you see that tells you something about the history of the asteroids?

Visual

(NASA)

17. What do you see in this image of the nucleus of Comet Borrelly that tells you how comets produce their comae and tails?

Visual

(NASA)

Discussion Questions

1. Futurists suggest that humans may someday mine the asteroids for materials to build and supply space colonies. What kinds of materials could they get from asteroids? (*Hint:* What are S-, M-, and C-type asteroids made of?)

2. If cometary nuclei were heated by internal radioactive decay rather than by solar heat, how would comets differ from what is observed?

3. From what you know now, do you think the government should spend money to locate near-Earth asteroids? How serious is the risk?

Problems

1. Large meteorites are hardly slowed by Earth's atmosphere. Assuming that the atmosphere is 100 km thick and that a large meteorite falls vertically toward the ground, how long does it spend in the atmosphere? (*Hint:* How fast do meteoroids travel?)

2. If a single asteroid 1 km in diameter were fragmented, how many meteoroids 1 m in diameter could it yield? (*Hint:* The volume of a sphere is $\frac{4}{3}\pi R^3$.)

3. What is the orbital velocity of a meteoroid orbiting the sun at the distance of Earth? (*Hint:* See Chapter 5.)

4. If a half-million asteroids, each 1 km in diameter, were assembled into one body, how large would it be? (*Hint:* The volume of a sphere is $\frac{4}{3}\pi R^3$.)

5. The asteroid Vesta has a mass of 2×10^{20} kg and a radius of about 250 km. What is its escape velocity? Could you jump off the asteroid? (*Hint:* See Chapter 5.)

6. The asteroid Pallas has a mass of 2.5×10^{20} kg. What is the orbital velocity of a small satellite orbiting 500 km from the center of Pallas? (*Hint:* See Chapter 5.)

7. What is the maximum angular diameter of Ceres as seen from Earth? Could Earth-based telescopes detect surface features? Could the Hubble Space Telescope (Chapter 6)? (*Hint:* Use the small-angle formula.)

8. At what distances from the sun would you expect to find Kirkwood's gaps where the orbital period of asteroids is one-half of and one-third of the orbital period of Jupiter? Compare your results with Figure 25-10. (*Hint:* Use Kepler's third law.)

9. If the velocity of the solar wind is about 400 km/s and the visible tail of a comet is 100 million km long, how long does it take an atom to travel from the nucleus to the end of the visible tail?

10. If you saw Comet Halley when it was 0.7 AU from Earth and it had a visible tail 5° long, how long was the tail in kilometers? Suppose that the tail was not perpendicular to the line of sight. Is your answer too large or too small? (*Hint:* Use the small-angle formula.)

11. Calculate the orbital velocity of a comet while it is in the Oort cloud. (*Hint:* See Chapter 5.)

12. The mass of a comet's nucleus is about 10^{14} kg. If the Oort cloud contains 2 trillion (2×10^{12}) cometary nuclei, what is the mass of the cloud in Earth masses? (*Hint:* Earth's mass = 6×10^{24} kg.)

Media Cluster

Ace Astronomy™ To access the resources in the Media Cluster, log into AceAstronomy at **http://ace.brookscole .com/sf9** and select Chapter 25.

ACTIVE FIGURES

Build a Comet
This animation shows how radiation from the sun creates the tails of comets. You can compare this to "heat from everywhere," which causes particles to be shed in every direction.

Impacts on Earth
Watch asteroids pummel Earth in this animation and learn about the probability of a "killer asteroid" hitting Earth.

ASTRONOMY EXERCISES

Comets
In this animation you can move a comet to different positions relative to the sun. You can also change the amount of gas and dust in the comet's nucleus and see how this affects the comet's tails.

Cratering
Turn up your speakers for this one, where you can design collisions between a meteorite and a hypothetical planet.

Lab 9: Asteroids and Kuiper Belt Objects
This lab presents an overview of asteroid properties, including an exercise on how to discover asteroids. You also investigate the class of objects known as Kuiper Belt objects.

Critical Inquiries for the Web

1. Some nights are better for looking for meteors than others (see Table 25-1). Meteor showers are associated with comets, but how are these associations made? Several sites on the Internet provide information on meteor showers, including historical data and information on parent comets. Pick a shower whose parent comet is known and summarize how astronomers came to know that the meteors and the comet are related.

2. Chances are very small that you will be killed by an asteroid impact, but if objects are out there that astronomers are not aware of whose orbits intersect Earth, humanity could be in for a surprise one day. Look for information on the LINEAR project and other searches for near-Earth asteroids. How many such objects have been discovered? What is the record for closest known passage of an asteroid to Earth?

3. Search for the IAU Minor Planet Center Web pages and find out what asteroids and what comets have come closest to Earth.

4. Search for information about comets in the sky right now. Are any comets bright enough for you to see? Are any bright comets expected soon?

Exploring *TheSky*

1. Of the five brightest asteroids, which has the most inclined orbit? (*Hint:* Use **Filters** in the **View** menu to turn off everything but the sun, stars, ecliptic, and minor planets. You can use **Tracking Setup** under **Time Skip** in the **Tools** menu to lock onto an object and follow it along the ecliptic as time passes.)

2. Do asteroids go through retrograde motion? (*Hint:* Use **Filters** in the **View** menu to turn off the stars and turn on the Equatorial Grid. See Activity 1 above.)

3. Of the five brightest asteroids, which has the most elliptical orbit? (*Hint:* Use **3D Solar System Mode** in the **View** menu and watch objects orbit the sun.)

4. Of the comets shown, which has the smallest orbit? (*Hint:* Use **3D Solar System Mode** in the **View** menu and watch objects orbit the sun.)

 Go to the Brooks/Cole Astronomy Resource Center (**http://astronomy.brookscole.com**) for critical thinking exercises, articles, and additional readings from InfoTrac College Edition, Brooks/Cole's online student library.

26 | Life on Other Worlds

Did I solicit thee from

Darkness to promote me?

JOHN MILTON, *PARADISE LOST*

As LIVING THINGS, we have been promoted from darkness. We are made of heavy atoms that could not have formed at the beginning of the universe. Successive generations of stars fusing light elements into heavier elements have built the atoms so important to our existence. When a dark cloud of interstellar gas enriched with these heavy atoms fell together to form our sun, a small part of the cloud gave birth to the planet you inhabit. Your atoms were in that cloud and have been part of Earth ever since it condensed from the solar nebula. ▌ If life can originate on other worlds, and if intelligence is a natural result of the evolution of life forms, then you might expect that other intelligent races inhabit other worlds. Visits between worlds seem impossible, but perhaps distant civilizations can be detected by radio. ▌ Alien life forms could

▌ Continued on page 652 ▌

Every life form we know of has evolved to live somewhere on Earth. The Wekiu bug lives with the astronomers at 13,600 feet atop Hawaiian volcano Mauna Kea. It lives in the icy cinders and eats insects carried up by ocean breezes. (Kris Koenig/Coast Learning Systems)

Guidepost

Looking Back

This chapter is either unnecessary or critical, depending on your point of view. If you believe that astronomy is the study of the physical universe above the clouds, then you are done; the last 25 chapters completed your study of astronomy. But if you believe that astronomy is the study of your position in the universe, not just your physical location but also your role as a living being in the evolution of the universe, then everything you have done so far was just preparation for this chapter.

This Chapter

You will begin by tackling a question that has troubled scientists and philosophers for centuries: What is life? You won't get an answer to the question, but just asking it will illuminate the problem and prepare you to search for life on other worlds. In fact, as you read this chapter, you will ask four essential questions:

What is life?

How did life originate on Earth?

Could life begin on other worlds?

Could Earthlings communicate with civilizations on other worlds?

You won't be able to answer these questions yet, but often in science asking a good question is more important than getting an answer.

Looking Ahead

You are different now. You have explored the universe from the phases of the moon to the big bang, from the origin of Earth to the death of the sun. Astronomy is important, not because it is about stars and galaxies but because it is about us. It tells us what we are, and once you know astronomy, you see yourself and your world in a different way. Astronomy changes us. You are different now.

Ace ✪ Astronomy™ The AceAstronomy icon throughout the text indicates an opportunity for you to test yourself on key concepts and to explore animations and interactions on the AceAstronomy website at: http://ace .brookscole.com/sf9

The Nature of Scientific Explanation

Science is a way of understanding the world around you, and the heart of that understanding is the explanations that science provides for natural phenomena. Whether you call these explanations stories, histories, theories, or hypotheses, they are attempts to describe how nature works based on fundamental rules of evidence and intellectual honesty. While you may take these explanations as factual truth, you should understand that they are not the only explanations that satisfy the rules of logic.

A separate class of explanations involves religion, and those explanations can be quite logical. The Old Testament description of the creation of the world, for instance, does not fit scientific observations, but if you accept the existence of an omnipotent being, then the biblical explanation is internally logical and acceptable. Of course, it is not a scientific explanation, but religion is a matter of faith and not subject to the rules of evidence. Religious explanations follow their own logic, and it is wrong to demand that they follow the rules of evidence that govern scientific explanations, just as it is wrong to demand that scientific explanations accept certain religious beliefs on

faith. Both kinds of explanations are logical, but the rules are different.

If scientific explanations are not the only logical explanations, then why do people give them such weight? First, you should notice the tremendous success of scientific explanations in producing technological advances in your daily lives. Diseases such as chickenpox are more childhood irritations than life-threatening illnesses thanks to the application of scientific explanations to modern medicine. The power of science to shape your world can lead you to think that its explanations are unique. But, second, the process called science depends on the use of evidence to test and perfect explanations, and the logical rigor of this process gives scientists great confidence in their conclusions.

Scientific explanations have given us tremendous insight into the workings of nature, and consequently both scientists and nonscientists tend to forget that there can be other logical explanations. The so-called conflict between science and religion has been symbolized for centuries by the trial of Galileo. That conflict is easier to understand when you consider the nature of scientific explanations

Galileo's telescope gave him a new way to know about the world.

and the role of evidence in testing scientific understanding.

---

be quite different from Earthlings, but if they are alive, then they must share certain characteristics. Your goal in this chapter is to use your knowledge of science to explain the greatest of mysteries—the origin and evolution of life on Earth and on other worlds (**Window on Science 26-1**).

26-1 The Nature of Life

WHAT IS LIFE? Philosophers have struggled with that question for thousands of years, so it is unlikely that you will find an answer here. But you must have a working model of life before you can speculate on its occurrence on other worlds. To that end, you can identify two important aspects in living things, a physical basis and a unit of controlling information.

The Physical Basis of Life

On Earth, the physical basis of life is the carbon atom (■ Figure 26-1). Because of the way this atom bonds to other atoms, it can form long, complex, stable chains that are capable of extracting, storing, and utilizing energy. Other chemical bases of life may

exist. Science fiction stories and movies abound with silicon creatures, living things whose body chemistry is based on silicon rather than carbon. However, silicon forms weaker bonds than carbon does, and it cannot form double bonds as easily. Consequently, it cannot form the long, complex, stable chains that carbon can. Silicon is 135 times more common on Earth than carbon is, yet there are no silicon creatures among us. All Earth life is carbon based. The likelihood that distant planets are inhabited by silicon people seems small, but you should not rule out life based on noncarbon chemistry.

In fact, nonchemical life might be possible. What is required is some mechanism capable of supporting the extraction and utilization of energy that has been identified as life. One could at least imagine life based on electromagnetic fields and ionized gas. No one has ever met such a creature, but science fiction writers conjure up all sorts.

Clearly, you could range far in space and time, theorizing about different bases for alien life, but to make progress you need to discuss what is known best—carbon-based life on Earth. How can a lump of carbon-rich matter live? The answer lies in the information that guides its life processes.

■ **Figure 26-1**

All living things on Earth are based on carbon chemistry. Even the long molecules that carry genetic information, DNA and RNA, have a framework defined by chains of carbon atoms. (a) Katie, a complex mammal, contains about 30 AU of DNA. (Michael Seeds) (b) Each rod of the tobacco mosaic virus contains a single spiral strand of RNA about 0.01 mm long. (L. D. Simon) All life on Earth stores its genetic information in such carbon-chain molecules.

Information Storage and Duplication

Living cells are tiny chemical factories. They must store all of the recipes for those chemicals in a safe place, use them to fulfill the cell's task, and hand down duplicates of the recipes to offspring. That information is encoded on long molecules.

Study **DNA: The Code of Life** on pages 654–655 and notice three important points:

❶ The chemical recipes of life are stored as templates on DNA molecules. The templates automatically guide specific chemical reactions within the cell.

❷ The instructions stored in DNA are the genetic information handed down to offspring. When people say, "You have your mother's eyes," they are talking about DNA codes.

❸ Notice how the DNA molecule reproduces itself when a cell divides so that each new cell contains a copy of the original information.

Although the DNA molecule must preserve its coded information from damage and make accurate copies, it must also be capable of making mistakes. To see why, you must consider how new DNA recipes are created.

Modifying the Information

You are probably a little surprised to hear that DNA needs to make mistakes, but it is true. If living things are to survive for many generations, then the information stored in their DNA must be able to change as the environment changes. Without change in DNA, a slight warming of the climate, for example, might kill a species of plant, in turn starving the rabbits, deer, and other plant eaters, and leaving the hawks, wolves, and mountain lions with no prey. If the information stored in DNA could never change, then environmental changes would quickly drive life forms to extinction. If life is to survive in a changing world, then the information in DNA must be changeable. Living things must evolve.

Species evolve by **natural selection.** Each time an organism reproduces, its offspring receives copies of the data stored in the DNA, but some variation is possible. For example, most of the rabbits in a litter may be normal, but it is possible for one to get a DNA recipe that gives it stronger teeth. If it has stronger teeth, it may be able to eat something other than the plant the others depend on; and, if that plant is becoming scarce, the rabbit with stronger teeth has a survival advantage. It can eat other plants and so will be healthier than its littermates and have more offspring. Some of these offspring will also have stronger teeth, as the altered DNA recipes are handed down to the new generation. In this way, nature selects and preserves those attributes that contribute to the survival of the species. Those creatures that are unfit die. Natural selection is merciless to the individual, but it gives the species the best possible chance to survive in a changing environment.

The only way nature can obtain new DNA patterns from which to select the best is from DNA molecules that have changed. This can happen through chance mismatching of base pairs—mistakes—in the reproduction of the DNA molecule. Another way

DNA: The Code of Life

1 The key to understanding life is information — the information that guides all of the processes in an organism. In most living things on Earth, that information is stored on a long spiral molecule called **DNA (deoxyribonucleic acid).**

1a The DNA molecule looks like a spiral ladder with rails made of phosphates and sugars. The rungs of the ladder are made of four chemical bases arranged in pairs. The bases always pair the same way. That is, base A always pairs with base T, and base G always pairs with base C.

1b Information is coded on the DNA molecule by the order in which the base pairs occur. To read that code, molecular biologists have to "sequence the DNA." That is, they must determine the order in which the base pairs occur along the DNA ladder.

The Four Bases

A — Adenine

C — Cytosine

G — Guanine

T — Thymine

2 DNA automatically combines raw materials to form important chemical compounds. The building blocks of these compounds are relatively simple **amino acids.** Segments of DNA act as templates that guide the amino acids to join together in the correct order to build specific **proteins,** chemical compounds important to the structure and function of organisms. Some proteins called **enzymes** regulate other processes. In this way, DNA recipes regulate the production of the compounds of life.

The traits you inherit from your parents, the chemical processes that animate you, and the structure of your body are all encoded in your DNA.

Nucleus
(information
storage)

Cell membrane
(transport of raw
materials and
finished product)

Material
storage

Manufacture
of proteins
and enzymes

Energy
production

2a A cell is a tiny factory that uses the DNA code to manufacture chemicals. Most of the DNA remains safe in the nucleus of a cell, and the code is copied to create a molecule of **RNA (ribonucleic acid).** Like a messenger carrying blueprints, the RNA carries the code out of the nucleus to the work site where the proteins and enzymes are made.

Original DNA

2b A single cell from a human being contains about 1.5 meters of DNA containing about 4.5 billion base pairs — enough to record the entire works of Shakespeare 200 times. A typical human contains a total of about 600 AU of DNA. Yet the DNA in each cell, only 1.5 meters in length, contains all of the information to create a new human. A clone is a new creature created from the DNA code found in a single cell.

Copy DNA

3 DNA, coiled into a tight spiral, makes up the **chromosomes** that are the genetic material in a cell. A **gene** is a segment of a chromosome that controls a certain function. When a cell divides, each of the new cells receives a copy of the chromosomes, as genetic information is handed down to new generations.

Copy DNA

3a To divide, a cell must duplicate its DNA. The DNA ladder splits, and new bases match to the exposed bases of the ladder to build two copies of the original DNA code. Because the base pairs almost always match correctly, errors in copying are rare. One set of the DNA code goes to each of the two new cells.

Cell Reproduction by Division

As a cell begins to divide, its DNA duplicates itself.

The duplicated chromosomes move to the middle.

The two sets of chromosomes separate, and . . .

the cell divides to produce . . .

two cells, each containing a full set of the DNA code.

this can occur is through damage to reproductive cells from exposure to radioactivity such as cosmic rays or natural radioactivity in the soil. In any case, an offspring born with altered DNA is called a **mutant.** Most mutations make no difference at all because they change segments of DNA that are not being used. Many mutations are fatal, and the individual dies long before it can have offspring of its own. But in rare cases, a mutation may give a species a new survival advantage. Then natural selection makes it likely that the new DNA message will survive and be handed down, making the species more capable of surviving.

Evolution is not random. Of course, the errors that occur in the DNA code are indeed random, but natural selection is not random. Those changes in the DNA that help a species survive are selected and preserved in future generations. With each passing generation, the species becomes more fit to survive in its environment.

Evolution is natural and automatic, but humans can control it. If you drink milk or eat cheese you are benefiting from the careful breeding of milk cows that has extended over centuries. Do you own a dog or a cat? Controlled evolution has created miniature poodles, huge Doberman pinschers, regal Siamese cats, and tailless Manx cats. But evolution works against us too. Some insecticides don't work anymore because the insects that were susceptible were killed off, and those that survived have a different gene that makes them immune. You have probably heard that some bacteria have built up a resistance to antibiotics. Again, the germs are evolving and handing on the genes—the bits of information—that help them survive.

Building Scientific Arguments

Why can't the information in DNA be permanent?
Sometimes the most valuable scientific arguments are those that challenge common misconceptions. It seems so obvious that DNA codes must not change, but that's a common misconception. The information stored in a creature's DNA provides all of the recipes that make the creature what it is. For example, the DNA in a starfish must contain all the recipes for making the various kinds of proteins needed to consume and digest food. That information must be passed on to offspring starfish, or they will be unable to survive. But the information must be changeable because the environment is changeable. Ice ages come and go, mountains rise, lakes dry up, and ocean currents shift. If the environment changes in some way, one or more of the recipes may no longer work. In this example, a change in the temperature of the ocean water may kill off the specific shellfish the starfish eat. If they can't digest other shellfish, the entire species will become extinct. Natural variation in DNA means that among all the infant starfish in any generation, a few of the recipes are different; if the environment changes, all of the old-style starfish may die, but a few—those with the different DNA—can carry on.

The survival of life depends on this delicate balance between reliable reproduction and the introduction of small variations in DNA information. Now go the next step in your argument. **What are some of the ways these small changes in DNA can arise?**

◼ ◼ ◼

Connections: Life is based not only on information but also on the duplication of information. Today, that process seems so complex that it is hard to imagine how it could have begun.

26-2 The Origin of Life

IF LIFE ON EARTH is based on the storage of information in these long, complex, carbon-chain molecules, how could it have ever gotten started? Obviously, 4.5 billion chemical bases didn't just happen to drift together to form the DNA formula for a human being. The key is evolution. Once a life form begins to reproduce itself, natural selection preserves the most advantageous traits. Over long periods of time spanning thousands, perhaps millions, of generations, the life form becomes more fit to survive. This often means the life form becomes more complex. That means life could have begun as a very simple process that gradually became more sophisticated as it was modified by evolution.

This is a critical point, so let's repeat it. Life is now very complex, but it didn't have to begin by some astonishingly unlikely combination of atoms to make a complicated creature. Life could have begun as a simple organism that automatically made copies of itself. Once reproduction began, natural selection made living things into the complex creatures in the world today. Of course, the key is evidence; what evidence exists to confirm this hypothesis?

The Origin of Life on Earth

The oldest fossils hint that life began in the sea. The first living things on Earth left behind a very poor fossil record. They were at first single-celled creatures; and, even when they became more complex, multicellular creatures, they contained no hard parts such as bones or shells that form good fossils. Yet a few traces of these earliest living things can be found, and they are all ocean creatures.

Although the oldest rocks contain no obvious fossils, microscopes reveal traces of ancient life. Rock from western Australia that is nearly 3.5 billion years old contains microscopic features that may be fossils of living things (◼ Figure 26-2), and spectroscopic analysis reveals the presence of organic matter. While the experts debate the details in this controversial field, you can conclude that life was present on Earth as simple organisms at least 3.5 billion years ago. That's roughly one billion years after Earth formed.

A little over a half-billion years ago something happened, perhaps a change in Earth's climate, and life exploded into a wide di-

Figure 26-2

Among the oldest fossils known, this microscopic filament resembles modern bacterial forms (artist's reconstruction at right). This fossil was found in the 3.5-billion-year-old chert of the Pribara Block in northwestern Australia. (Courtesy J. William Schopf)

Figure 26-3

Trilobites made their first appearance in the Cambrian oceans. The smallest were almost microscopic, and the largest were bigger than dinner plates. This example, about the size of a human hand, lived 400 million years ago in an ocean floor that is now a limestone deposit in Pennsylvania. (Grundy Observatory photograph)

versity of complex forms. This marks the beginning of the **Cambrian period** and is sometimes called the *Cambrian explosion.* The best known of the Cambrian creatures may be the trilobites (■ Figure 26-3). Although the Cambrian creatures were diverse, there are no Cambrian fossils of land plants or animals. Evidently, land surfaces were totally devoid of life until only 400 million years ago.

The fossil record shows that life began as simple organisms in the sea soon after Earth formed. Only recently, in geological terms, has life become complex. It isn't difficult to understand how simple creatures evolve to become complex. The problem is trying to understand how nonliving atoms became simple creatures. That's the hard bit right at the beginning.

The key to the origin of this life may lie in an experiment performed by Stanley Miller and Harold Urey in 1952. This **Miller experiment** sought to reproduce the conditions on Earth under which life began. In a closed glass container, the experimenters placed water (to represent the oceans); the gases hydrogen, ammonia, and methane (to represent the primitive atmosphere); and an electric arc (to represent lightning bolts). The apparatus was sterilized, sealed, and set in operation (■ Figure 26-4).

After a week, Miller and Urey stopped the experiment and analyzed the material in the flask. Among the many compounds the experiment produced, they found four amino acids (building blocks of protein), various fatty acids, and urea, a molecule common to many life processes. Evidently, the energy from the electric arc had molded the atmospheric gases into some of the basic components of living matter. Other energy sources, such as hot silica (to simulate hot lava spilling into the sea) and ultraviolet radiation (to simulate sunlight), produce similar results.

More recent studies of the composition of meteorites and models of planet formation suggest that Earth's first atmosphere did not resemble the gases used in the Miller experiment. Earth's first atmosphere was probably composed of carbon dioxide, nitrogen, and water vapor. This finding, however, does not invalidate the Miller experiment. When such gases are processed in a Miller apparatus, some organic molecules appear, although not as many as in the first trials.

The Miller experiment did not create life, of course, nor did it necessarily imitate the exact conditions on the young Earth. That is not the lesson to take from this famous experiment. Rather, it is important because it shows that complex organic molecules form naturally in a wide variety of circumstances. The chemical deck is stacked to deal nature a hand of complex molecules. If you could travel back in time, you would probably find Earth's first oceans filled with a rich mixture of organic compounds in what some have called the **primordial soup.**

The next step on the journey toward life is for the compounds dissolved in the oceans to link up and form larger molecules. Amino acids, for example, can link together to form proteins. This linkage occurs when amino acids join together end-to-end and release a water molecule (■ Figure 26-5). For many years, experts

■ Figure 26-4

(a) The Miller experiment circulated gases through water in the presence of an electric arc. This simulation of primitive conditions on Earth produced amino acids, the building blocks of proteins. (b) Stanley Miller with a Miller apparatus. (Courtesy Stanley Miller)

have assumed that this process must have happened in sun-warmed tidal pools where evaporation concentrated the broth. But more recent studies suggest that the young Earth was subject to extensive volcanism and large meteorite impacts that periodically modified the climate enough to destroy any life forms exposed on the surface. Rather it seems likely that the early growth of complex molecules took place among the hot springs along the midocean ridges. These complex molecules would not have been photosynthetic—taking energy from sunlight. Rather they would have taken energy from the heat and chemicals emerging

from the hot springs of the midocean ridge. That energy could have powered the growth of long protein chains. Deep in the oceans, they would have been safe from climate changes.

Although these proteins might have contained hundreds of amino acids, they would not have been alive. Not yet. Such molecules would not have reproduced but would have merely linked together and broken apart at random. Because some molecules are more stable than others, however, and because some molecules bond together more readily than others, this automatic **chemical evolution** would have led to the concentration of the varied

■ Figure 26-5

Amino acids can link together through the release of a water molecule to form long carbon-chain molecules. The amino acid in this hypothetical example is alanine, one of the simplest.

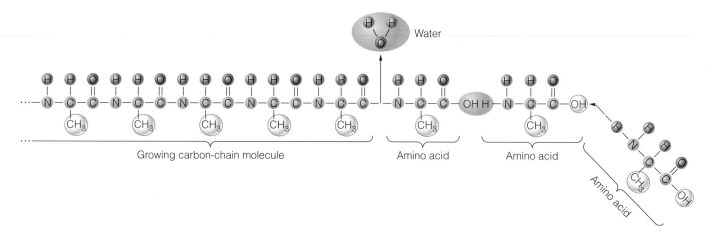

smaller molecules into the most stable larger forms. Eventually, somewhere in the oceans, a molecule took shape that automatically made copies of itself. At that point, the chemical evolution of molecules became the biological evolution of living things.

You might enjoy speculating about an alternative theory that proposes that primitive living things such as reproducing molecules did not originate on Earth but came here in meteorites or comets. Radio astronomers have found a wide variety of organic molecules in the interstellar medium, and some studies have found similar compounds inside meteorites (■ Figure 26-6). Such molecules form so readily that you should be surprised if they were not present in space. A few investigators, however, have speculated that living, reproducing molecules originated in space and came to Earth as a cosmic contamination. If this were true, every planet in the universe would be contaminated with the seeds of life. You may find this theory fun to think about, but it is presently untestable, and an untestable theory is of little use in science.

Whether life originated in the oceans or in space, you still face the problem of how life began—that hard bit at the beginning in which nonliving atoms become reproducing molecules. The exact details are not yet understood, but the important point is that natural processes could have done it. Just because you don't

understand something doesn't mean it is impossible. The evidence seems conclusive that life began as reproducing molecules in Earth's oceans.

Which came first, reproducing molecules or the cell? Because you probably think of the cell as the basic unit of life, this question seems to make no sense, but in fact the cell may have originated during chemical evolution. If a dry mixture of amino acids is heated, the acids form long, proteinlike molecules that, when poured into water, collect to form microscopic spheres that function in ways similar to cells (■ Figure 26-7). They have a thin membrane surface, they can absorb material from their surroundings, they grow in size, and they can divide and bud just as cells do. They contain no large molecule that copies itself, however. So it is possible that the structure of the cell originated first and the reproducing molecules later.

An alternative theory proposes that the replicating molecule developed first. Such a molecule would have been exposed to damage if it had been bare, so the first to manufacture or attract a protective coating of protein would have had a significant survival advantage. If this was the case, the protective cell membrane was a later development of biological evolution.

The first living things must have been single-celled organisms much like modern bacteria. Some of the oldest fossils known are **stromatolites,** structures produced by communities of photosynthesizing bacteria that grew in mats and, year by year, deposited layers of minerals that were later fossilized. One of the

■ Figure 26-6

A sample of the Murchison meteorite, a carbonaceous chondrite that fell in 1969 near Murchison, Australia. Analysis of the interior of the meteorite revealed evidence of amino acids. Whether the first building blocks of life originated in space is unknown, but the amino acids found in meteorites illustrate how commonly amino acids and other complex molecules occur even in the absence of living things. (Courtesy Chip Clark, National Museum of Natural History)

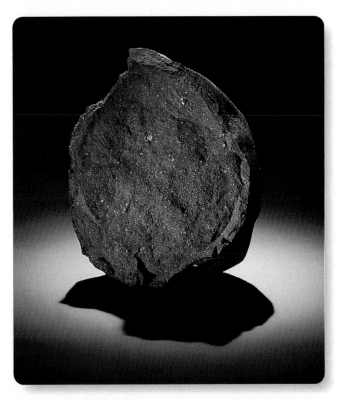

■ Figure 26-7

Single amino acids can be assembled into long proteinlike molecules. When such material cools in water, it can form microspheres, microscopic spheres with double-layered boundaries similar to cell membranes. Microspheres may have been an intermediate stage in the evolution of life between complex molecules and cells holding molecules reproducing genetic information. (Courtesy Sidney Fox and Randall Grubbs)

■ Active Figure 26-8

A 3.5-billion-year-old fossil stromatolite from western Australia is one of the oldest known fossils (inset). Stromatolites were formed, layer by layer, by mats of bacteria living in shallow water. Such life may have been common in shallow seas when Earth was young. Stromatolites are still being formed today in similar environments. (Mural by Peter Sawyer; photo courtesy Chip Clark, National Museum of Natural History)

Ace✺Astronomy™ Log into AceAstronomy and select this chapter to see the Active Figure "Future of the Sun." Watch Earth's temperature change as the sun ages.

oldest such fossils known is believed to be 3.5 billion years old (■ Figure 26-8). If such bacteria were common when Earth was young, the early atmosphere may have contained a small amount of oxygen produced by the photosynthesis. Atmospheric models suggest that an oxygen abundance of only 0.1 percent would have been sufficient to provide an ozone screen that would protect organisms from the sun's ultraviolet radiation.

How evolution shaped creatures to live in the ancient oceans, to photosynthesize and respire, to become multicellular, and to reproduce sexually is a fascinating story, but this chapter cannot explore those details. You can see that the hard bit in the story is how life began, and there appear to be natural chemical processes

that could have led to molecules automatically duplicating themselves. Once some DNA-like molecule formed, evolution was automatic; and, over billions of years, the genetic information stored in living things accumulated those qualities that favored survival, and unsuccessful variations did not get passed down to new generations. As Samuel Butler said, "The chicken is the egg's way of making another egg." In that sense, all living matter on Earth is the physical expression of DNA's automatic tendency to continue its existence.

Perhaps this seems harsh. Human experience goes far beyond mere reproduction. *Homo sapiens* has art, poetry, music, philosophy, religion, science. Perhaps all of the great accomplishments of our intelligence represent more than mere reproduction of DNA. Nevertheless, intelligence, the ability to analyze complex situations and respond with appropriate action, must have begun as a survival mechanism. For example, a fixed escape strategy stored in the DNA is a disadvantage for a creature that frequently moves from one environment to another. A rodent that always escapes from predators by automatically climbing the nearest tree would be in serious jeopardy if it met a hungry fox in a treeless clearing. Even a faint glimmer of intelligence might allow the rodent to analyze the situation and, finding no trees, to choose running over climbing. Intelligence, of which *Homo sapiens* is so proud, may have developed in ancient creatures as a way of making them more versatile without adding tremendous amounts of information to their DNA library.

Any discussion of the evolution of life seems to involve highly improbable coincidences until you consider how many years have passed in the history of Earth. You can read the words—4.6 billion years—so easily, but it is in truth hard to grasp the meaning of such a long period of time.

Geologic Time

Humanity is a very new experiment on planet Earth. You can fit the history of Earth into a single chart such as ■ Figure 26-9. You can take comfort in thinking that creatures like us have walked on Earth for roughly 3 million years, but when you add our history to the chart, you will discover that the entire history of humanity makes up no more than a thin line at the top. In fact, if you tried to represent the entire 4.6-billion-year history of Earth on the chart, the portion describing the rise of complex life after the Cambrian explosion would be an unreadably small segment. It is expanded in Figure 26-9 just to make it legible.

One way to represent the evolution of life is to compress the 4.6-billion-year history of Earth into a one-year-long video program. In such a program, Earth forms as the video begins on January 1; through all of January and February it cools and is cratered, and the first oceans form. Search as you might, you would find no trace of life in these oceans until sometime in March or early April, when the first living things develop. You would need a microscope to see them. The slow development of these simplest of living forms grinds on through the spring and summer of the video. The entire 4-billion-year history of Precambrian evolution lasts until the video reaches mid-November, when the primitive ocean life explodes into the more complex Cambrian organisms such as trilobites and many other specialized creatures (■ Figure 26-10).

While the yearlong video plays on and on, you might amuse yourself by looking at the land instead of the oceans, but you would be disappointed. The land is a lifeless waste with no plants or animals of any kind. Not until November 28 in the video does life appear on the land; but, once it does, it evolves rapidly into a wide range of plants and animals. Dinosaurs, for example, appear about December 12 and vanish by Christmas evening as mammals and birds flourish.

Throughout the one-year run of the video there have been no humans; and even during the last days of the year, as the mammals rise to dominate the landscape, there are no people. In the early evening of December 31, vaguely human forms move through the grasslands, and by late evening they begin making stone tools. The Stone Age lasts till about 11:45 PM, and the first signs of civilization, towns and cities, do not appear until 11:54 PM. The Christian era begins only 14 seconds before the New Year, and the Declaration of Independence is signed with but 1 second to spare.

By imagining the history of Earth as a yearlong video, you can place the rise of life in perspective. Tremendous amounts of time were needed for the first simple living things to evolve in the oceans, and even more time was needed for the evolution of complex creatures that could colonize the land. As life became more complex, it evolved and diversified faster and faster, as if evolution were drawing on a growing library of solutions that had been previously invented with great effort to solve earlier problems. The burst of diversity on land led eventually to the rise of intelligent creatures like us, a process that has taken 4.6 billion years.

If life could originate on Earth and develop into intelligent creatures, perhaps the same thing could have happened on other planets. This raises three questions. First, could life originate on another world if conditions were suitable? No one knows for sure, but you have found natural, chemical processes that could lead to living things, so your answer to this question should probably be yes. The second question is, will life always evolve toward intelligence? Perhaps intelligence arose on Earth under unusual conditions, but you should recall that intelligence appears to be a way to make a species more versatile and consequently more likely to survive. If that is true, then intelligence may develop under a wide range of conditions if enough time is available. But what of the third question: Are suitable conditions so rare that life almost never gets started? The only way to answer that is to search for life on other planets. You can begin in the next section by searching for life on the other planets in our solar system.

Life in Our Solar System

Although science fiction writers imagine life based on something other than carbon chemistry, no one knows of any real examples to illustrate how such life might arise and survive. To make progress in your search for life, you need to limit your discussion to life as it is known on Earth and the conditions it requires. The most important requirement is the presence of liquid water, not only as part of the chemical reactions of life but also as a medium to transport nutrients and wastes within the organism. Also, it seems that life on Earth began in the oceans and developed there for nearly 4 billion years before it was able to emerge onto the land. Certainly, any world where you hope to find life must have liquid water, and that means it must have moderate temperatures.

The liquid water requirement automatically eliminates many worlds in our solar system. The moon is airless, and although some data suggest ice frozen in the soil at its poles, it has never had liquid water on its surface. In the vacuum of the lunar surface, liquid water would boil away rapidly. Mercury too is airless and cannot have had liquid water on its surface for long periods of time. Venus has some traces of water vapor in its atmosphere, but it is much too hot for liquid water to survive. If there were any lakes or oceans of water on its surface when it was young, they must have evaporated quickly. Even if life began there, no traces would be left now.

The inner solar system seems too hot, and the outer solar system seems too cold. The Jovian planets have deep atmospheres; and, at a certain level, they have moderate temperatures where water might condense into liquid droplets. But it seems unlikely that life could begin there. The Jovian planets have no surfaces where oceans could nurture the beginning of life, and currents in

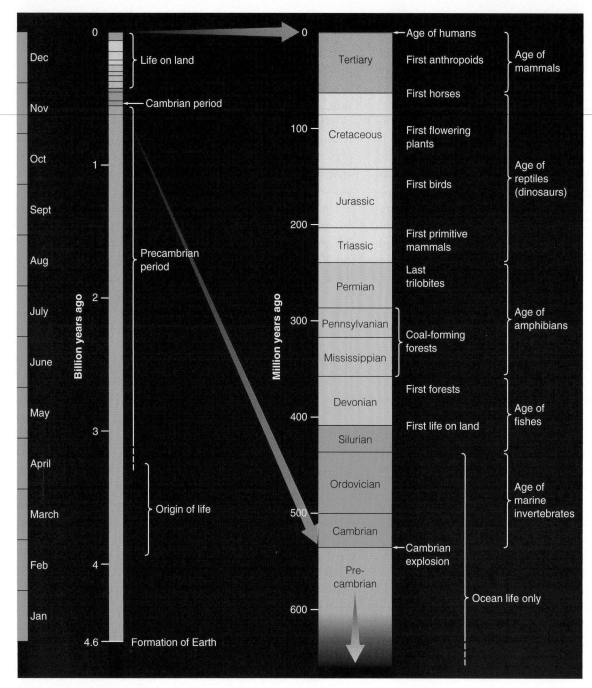

■ Active Figure 26-9

Complex life has developed on Earth only recently. If the entire history of Earth were represented in a time line (left), you would have to magnify the end of the line to see details such as life leaving the oceans and dinosaurs appearing. The age of humans would still be only a thin line at the top of your diagram. If the history of Earth were a yearlong videotape, humans would not appear until the last hours of December 31.

Ace Astronomy™ Log into AceAstronomy and select this chapter to see the Active Figure "Earth Calendar." Animate this diagram.

the atmosphere seem destined to circulate gas and water droplets from regions of moderate temperature to other levels that are much too hot or too cold for life to survive.

A few of the satellites of the Jovian planets might have suitable conditions for life. Jupiter's moon Europa seems to have a liquid-water ocean below its icy crust (see Figure 23-12), and minerals dissolved in that water would provide a rich broth of possibilities for chemical evolution. Nevertheless, Europa is not a promising site to search for life because conditions may not have remained stable for the billions of years needed for life to evolve beyond the microscopic stage. The subsurface ocean is kept from freezing by tidal heating. If Jupiter's moons interact gravitationally and modify their orbits, Europa may have been frozen solid at some points in history. Such periods of freezing would proba-

■ Figure 26-10

Before the Cambrian explosion, life consisted mostly of single-celled crea-tures living alone or in organized colonies. During the Cambrian period, life became complex. Anomalocaris (upper left) was about the size of your hand and had specialized organs including eyes, coordinated fins for active swim-ming, gripping mandibles capable of raking through the seafloor and captur-ing prey, and a powerful, toothed maw located below its head (upper right). Notice Opabinia at center with its long snout. (D. W. Miller)

Approximate true-color image

■ Figure 26-11

Mars rover Opportunity descended into the crater Endurance and then looked back up the crater wall. In places, Opportunity found layering in the rocks that showed they had formed as layers of sand and silt deposited by flow-ing water. The rover also found minerals that showed that water was once pres-ent at the site. Signs of past life may lie hidden in these rocks. (NASA/JPL/ Cornell)

bly prevent life from developing. Drilling through the icy crust of Europa to search for life in its ocean will be a wonderful adventure for future generations, but Europa does not seem to be a good bet to harbor any form of com-plex life.

Saturn's moon Titan has an atmosphere of nitrogen, argon, and methane and may have lakes or small oceans of liquid methane on its surface. You saw in Chapter 23 how sunlight can convert the methane in the atmosphere to organic smog particles that settle to the surface, and the images radioed back to Earth by the Huygens probe as it parachuted down to the surface of Titan showed that dark material filled drainage channels and lowlands. The chemistry of life that might crawl or swim on such a world is unknown, but life there may be unlikely because of the temperature. The surface of Titan is a deadly −179°C (−290°F). Chemical reactions occur slowly or not at all at such low temperatures, so the chemical evolution needed to begin life may never have occurred on Titan.

Mars is the most likely place for life in our solar system. The evidence, however, is not encouraging. In 1976, two robotic space-craft, Viking 1 and Viking 2, landed at two different places on the Martian surface. The spacecraft scooped up soil samples and subjected them to tests for the presence of living organisms. For example, they gave some soil samples a dose of nutrient-rich water and watched for signs the nutrients were taken up by bio-logical processes. The results seem negative. Although some pe-culiar chemical processes were detected, no evidence was found that clearly indicated the presence of any living things in the soil.

Rovers on Mars and orbiting spacecraft have found dramatic evidence that liquid water once flowed over the surface (■ Fig-ure 26-11), but that does not mean that life did originate there. Only detailed study of the rocky soil can answer that question, and that means Earth must send a geologist to Mars or bring rocks from Mars back to Earth. Nature has performed the latter task for us.

Meteorite ALH84001 (■ Figure 26-12a) was found on the Antarctic ice in 1984. Years later an analysis of its chemical com-position showed that it had originated on Mars. Being a good, skeptical scientist, you are probably wondering how anyone could know that the meteorite came from Mars. The evidence is solid.

■ Figure 26-12

(a) Meteorite ALH84001 is one of many meteorites known to have originated on Mars. It was claimed that the meteorite contained chemical and physical traces of ancient life on Mars, including what appear to be fossils of microscopic organisms (b). The evidence has not been confirmed, and the validity of the claim is highly questionable. (NASA)

The Viking landers measured the abundance of oxygen isotopes on Mars, and the pattern of isotopes was different from Earth and from meteorites. But the isotopes in ALH84001 are a precise match with the Viking results. The meteorite must have come from Mars.

ALH84001 was probably part of debris ejected into space by a large impact on Mars. Some of that debris fell to Earth, and a small number of these meteorites have been found, mostly in Antarctica because that's the easiest place to find meteorites. You might have heard of ALH84001; it was in the news because a team of scientists studied it and announced in 1996 that it contained chemical and physical traces of ancient life on Mars (Figure 26-12b).

The discovery was received with great excitement by the press and the public. It would, indeed, be dramatic evidence if true, because it would show that life could begin on another world. Scientists were excited too, but being professionally skeptical (see Window on Science 19-3), they began testing the results immediately. In many cases, the tests did not confirm the conclusion that life once existed on Mars. Some chemical contamination from water on Earth has occurred, and some chemicals in the meteorite may have originated without the presence of life. The physical features that look like fossil bacteria may be mineral formations in the rock. Although studies of ALH84001 continue, it does not provide unchallenged evidence that life once existed on Mars. Carl Sagan once said, "Extraordinary claims require extraordinary proof," and in this case the evidence is at best inconclusive.

Spacecraft now visiting Mars are revealing the past history of water there and painting a more detailed picture of present conditions. Spacecraft are planned that will eventually land on Mars, collect rocks, and return them to Earth. Nevertheless, conclusive evidence may have to wait until a geologist in a spacesuit can scramble down the dry streambeds of Mars cracking open rocks and searching for fossils.

Your search for life on the other worlds in our solar system has been inconclusive. You didn't find life, but you didn't find evidence to completely rule it out. So far as anyone knows at present, our solar system is bare of life except for Earth. Consequently the search for life in the universe takes you to other planetary systems.

Life in Other Planetary Systems

Might life exist in other solar systems? To consider this question, you need to consider how common planets are and what conditions a planet must fulfill for life to originate and evolve to intelligence. The first question is astronomical; the second is biological. The ability of scientists to discuss the problem of life outside our solar system is severely limited by lack of experience. They know only one planet well—Earth.

In Chapter 19 you learned that planets form as a natural byproduct of star formation and that a number of extrasolar planets have been found circling nearby stars. From this you can conclude that planetary systems are very common.

If a planet is to become a suitable home for life, it must have a stable orbit around its sun. This is simple in a solar system like our own, but in a binary system most planetary orbits are unstable. Most planets in such systems would not last long before they were swallowed up by one of the stars or ejected from the system. Consequently, single stars are the most likely to have planets suitable for life. Because our galaxy contains at least 10^{11} stars, half of which are single, there could be roughly 5×10^{10} planetary systems in which life might exist.

A few million years of suitable conditions does not seem to be enough time to originate life. On our planet it took about a billion years for the first living things to appear and 4.6 billion years for intelligence to evolve. Clearly, conditions on a planet must remain acceptable over a long time. This eliminates massive stars that remain stable on the main sequence for only a few million years. If it takes a few billion years for life to originate and evolve to intelligence, no star hotter than about F5 will do. This is not really a serious restriction, because upper-main-sequence stars are rare anyway.

In previous sections, you saw how life on Earth depends on water. In the past, astronomers have used that to define a **life zone** (or ecosphere) around a star, a region within which planets have temperatures that permit the existence of liquid water. A cool star has a small life zone, and a hot star has a large life zone. The life zone around the sun extends from about the orbit of Venus to the orbit of Mars.

For a number of decades astronomers have used the life zone to eliminate lower-main-sequence stars as candidates for life. The M dwarfs at the bottom of the main sequence are so faint that a planet would have to orbit very close to stay warm. Some astronomers have argued that such a planet would become tidally locked to its star, and one side of the planet would be in perpetual darkness. Water would tend to freeze out on the dark side and end chances for life to develop. Whether the atmosphere could distribute heat to the dark side is unknown. Also, M stars sometimes suffer violent flares that might make nearby planets unsuitable for life. Most astronomers thought the dinky red dwarfs were not good candidates for life. However, extrasolar planets have been discovered orbiting these M stars, and the large numbers of these extrasolar planets have lead some astronomers to suggest that at least some might be stable enough to develop life.

Recent discoveries, however, are making the whole idea of a life zone seem just a little bit of an oversimplification. For one thing, scientists on Earth are finding living things in astonishing places, such as the bottom of icy lakes in Antarctica and far underground in solid rock. Living things have been found in boiling springs where the water is highly acidic. It appears that life is much tougher than some scientists thought. Also, spacecraft visiting other worlds have found conditions that might support life outside the so-called life zone of the sun. The biggest water ocean in our solar system is under the ice on Jupiter's moon Europa. Liquid has flooded parts of Neptune's moon Triton, and Saturn's moon Titan has a complex surface chemistry of carbon-rich organic molecules. It isn't hard to imagine life in these environments, and they all lie outside the conventional life zone of the sun. Perhaps the life zone is most useful as a warning that life is proving to be more varied than many people expected.

One restriction on life is the slow evolution of a planet's star. You learned in Chapter 9 that main-sequence stars gradually grow more luminous as they convert their hydrogen to helium. The sun was only about 70 percent as luminous when it formed as it is today. As a star grows more luminous, it should slowly warm its planets. A planet might have liquid water on its surface long

enough for life to begin but then be sterilized as its star slowly grows more luminous and drives away the planet's atmosphere and oceans. Perhaps a planet must form at just the right distance from its star to remain hospitable to life for billions of years.

The fundamental question is simple. If conditions are right, will life begin? Early in this chapter you found natural chemical processes that seem capable of forming molecules that make copies of themselves. You saw that evolution would automatically shape these molecules to be most likely to survive. Perhaps the fundamental question should be, "What could prevent life from beginning?" Given what you know about life, it seems that it should arise whenever conditions permit, and our galaxy should be filled with planets that are inhabited with living creatures.

Building Scientific Arguments

What evidence can you cite that life is at least possible on other worlds?

A good scientific argument is based on evidence, but it also includes careful analysis. The evidence is limited almost entirely to Earth, but it is promising. Fossils show that life originated in Earth's oceans almost 4 billion years ago, and biologists have proposed relatively simple chemical processes that could have formed these first reproducing molecules. Fossils show that life developed from a very slow beginning into more and more complex creatures that filled the oceans. The pace of evolution quickened dramatically about half a billion years ago, the beginning of the Cambrian period, when life began taking on complex forms; later, when life emerged onto the surface of the land, it evolved rapidly to produce the tremendous diversity of today. Human intelligence has been a very recent development; it is only a few million years old.

If this process occurred naturally on Earth, then it seems reasonable that it could have occurred on other worlds as well. This conclusion hinges mostly on a reasonable belief that the laws of nature apply on other worlds as they do on Earth. Now expand your argument: **What conditions should you expect of other worlds that host life?**

■   ■   ■

Connections: It is both easy and fun to speculate about life on other worlds, but it leads to a simple question: If there really is life on other worlds, why can't we detect it? Why haven't we heard from alien civilizations?

(26-3) Communication with Distant Civilizations

IF OTHER CIVILIZATIONS EXIST, perhaps we Earthlings can communicate with them in some way. Sadly, travel between the stars is more difficult in real life than in science fiction—and may

UFOs and Space Aliens

If you discuss life on other worlds, then you might be tempted to use UFO sightings and supposed visits by aliens from outer space as evidence to test your hypotheses. Scientists don't do so for two reasons, both related to the reliability of these observations.

First, the reputation of the sources of UFO sightings and alien encounters does not inspire confidence that these data are reliable. Most people hear of such events via grocery store tabloids, daytime talk shows, or sensational "specials" on viewer-hungry cable networks. You must consider the low reputation of the media that report UFOs and space aliens. Most of these reports are simply made up for the sake of sensation, and you should not try to use them as reliable evidence.

Second, the remaining UFO sightings, those not simply made up, do not survive careful examination. Most are mistakes or unconscious misinterpretations of natural events made by honest people. A number of unbiased studies have found no grounds for believing in UFOs. In short, there is no conclusive evidence that Earth has been visited by aliens from space.

That's too bad. A confirmed visit by intelligent creatures from beyond our solar system would answer many questions. It would be exciting, enlightening, and, like any real adventure, a bit scary. But none of the UFO sightings is dependable, and we are left with no direct evidence of intelligent life on other worlds.

UFOs from space are fun to think about, but there is no evidence that they are real.

in fact be impossible. If physical visits are impossible, perhaps radio communication is the way to say hello. Again, nature places restrictions on such conversations, but the restrictions are not too severe. As you will see, the real problem lies with the life expectancy of civilizations.

Travel between the Stars

Practically speaking, roaming among the stars is tremendously difficult because of three limitations: distance, speed, and fuel. The distances between stars are almost beyond comprehension. It does little good to explain that if you use a golf ball in New York City to represent the sun, the nearest star would be another golf ball in Chicago. It is only slightly better to note that the fastest commercial jet would take about 4 million years to reach the nearest star.

The second limitation is a speed limit—you cannot travel faster than the speed of light. Though science fiction writers invent hyperspace drives so their heroes can zip from star to star, the speed of light is a natural and unavoidable limit that cannot be exceeded. This, combined with the large distances between stars, makes interstellar travel very time consuming.

The third limitation is that you can't even approach the speed of light without using a fantastic amount of fuel. Even if you ignore the problem of escaping from Earth's gravity, you must still use energy stored in fuel to accelerate to high speed and to decelerate to a stop when you reach our destination. To return to Earth, assuming you wish to, you have to repeat the process, and that takes more fuel.

These changes in velocity require a tremendous amount of fuel. If you flew a spaceship as big as a large yacht to a star 5 light-

years (1.5 pc) away and wanted to get there in only 10 years, you would use 40,000 times as much energy as the United States consumes in a year.

Travel for a few individuals might be possible if they accept very long travel times. That would require some form of suspended animation (currently unknown) or colony ships that carry a complete, though small, society in which people are born, live, and die generation after generation as the ship travels through space. Whether the occupants of such a ship would retain the social and emotional characteristics of humans over a long voyage is questionable.

These three limitations not only make it difficult for us to leave our solar system but would also make it difficult for aliens to visit Earth. Reputable scientists have studied "unidentified flying objects" (UFOs) and related phenomena and have never found any evidence that Earth is being visited or has ever been visited by aliens from other worlds (**Window on Science 26-2**). Consequently, humans are unlikely ever to meet an alien face-to-face. The only way to communicate with other civilizations is via radio.

Radio Communication

Nature places two restrictions on radio communication with distant societies. One has to do with simple physics, is well understood, and merely makes the communication difficult. The second has to do with the fate of technological civilizations, is still unresolved, and may severely limit the number of societies detectable by radio.

Radio signals are electromagnetic waves that travel at the speed of light. Because even the nearest civilizations must be a few light-years away, this limits conversations with distant be-

An anticoded message

1 0 1 0 0 1 1 1 1 1
0 0 1 0 1 0 0 1 0 0
0 1 0 1 0 0 1 0 1 0
0 1 0 1 0

5 rows of 7

1	0	1	0	0	1	1
1	1	1	0	0	1	0
1	0	0	1	0	0	0
1	0	1	0	0	1	0
1	0	0	1	0	1	0

7 rows of 5

1	0	1	0	0
1	1	1	1	1
0	0	1	0	1
0	0	1	0	0
0	1	0	1	0
0	1	0	1	0
0	1	0	1	0

■ Figure 26-13

An anticoded message is designed for easy decoding. Here a string of 35 radio pulses, represented as 1s and 0s, can be arranged in only two ways, as 5 rows of 7 or 7 rows of 5. The second way produces a friendly message. Any number of pulses can be used so long as it is the product of two prime numbers. Then the pulses can be arranged in only two ways.

ings. If you ask a question of a creature 10 light-years away, you will have to wait 20 years for a reply. Clearly, the give-and-take of normal conversation will be impossible.

Instead, you could simply broadcast a radio beacon of friendship to announce your presence as an intelligent being. In fact, the human race is already broadcasting a recognizable beacon. Short-wavelength radio signals, such as TV and FM, have been leaking into space for the last 50 years at least. Any civilization within 50 light-years might already have detected Earth.

If you intentionally broadcast a signal, you can anticode it. That is, you could arrange it to make it easy to decode. You could transmit pulses to represent 1s and gaps to represent 0s. A message counting through the first few prime numbers would distinguish your signal from natural sources of radio noise. You could even transmit a picture by sending a string of 1s and 0s that can be arranged in only two ways. One way produces nonsense, but the other way produces a meaningful picture (■ Figure 26-13).

In 1974, at the dedication of the 1000-ft radio telescope at Arecibo, radio astronomers transmitted such a signal toward the globular cluster M13, which is located 26,000 ly from Earth. When the signal finally arrives, any aliens who detect it will be able to arrange its 1679 pulses in only two ways, as 23 rows of 73 or 73 rows of 23. The second arrangement will form a picture that describes life on Earth (■ Figure 26-14).

It took only minutes to transmit the Arecibo message. If more time were taken, a more detailed picture could be sent; and, if you were sure your radio telescope was pointed at a listening civilization, you

could send a long series of pictures. With pictures you could teach aliens your language and tell them all about human life, its difficulties, and its accomplishments.

If you can think of sending such signals, aliens could think of it too. If you pointed your radio telescope in the right direction and listened at the right wavelength, you might hear other intelligent races calling out to one another. This raises two questions: Which stars are the best candidates, and what wavelengths are most likely? You have already answered the first question. Main-sequence G and K stars have the most favorable characteristics,

■ Active Figure 26-14

The Arecibo message to M13 begins by counting from 1 to 10 and goes on to describe our solar system and the biochemistry of life on Earth (color added for clarity). Binary numbers give the height of the human figure (1110) and the diameter of the telescope dish (100101111110) in units of the wavelength of the signal, 12.3 cm. (NASA)

Ace◉Astronomy™ Log into AceAstronomy and select this chapter to see the Active Figure "Interstellar Communication." Send messages and probes to nearby stars.

Start of number markers — Binary numbers 1 to 10

Atomic numbers of hydrogen, carbon, nitrogen, oxygen, and phosphorus

Formulas for sugars and bases in DNA

DNA double helix

Number of units in DNA

Start of number markers —

Human figure — Start of number marker

Population of Earth —

Height of human in wavelengths

Arecibo radio dish transmitting signal — Sun and planets with Earth offset

Diameter of dish in wavelengths

Start of number marker —

and maybe M stars are worth a look. But the second question is more complex.

Only certain wavelengths are useful for communication. Wavelengths longer than about 30 cm can't be used because the signal would be lost in the background radio noise from our galaxy. Nor can wavelengths much shorter than 1 cm be used because of absorption within our atmosphere. So only a certain range of wavelengths, a radio window, is open for communication.

This communications window is very wide, so a radio telescope would take a long time to tune over all the wavelengths in the window searching for intelligent signals. Nature may have provided a way to narrow the search, however. Within the communications window lie the 21-cm line of neutral hydrogen and the 18-cm line of OH. The interval between these two lines has been dubbed the **water hole** because the combination of H and OH yields water (H_2O). Water is the fundamental solvent in our life form, so it might seem natural for similar water creatures to call out to each other at wavelengths in the water hole (■ Figure 26-15). But even silicon creatures, if they were competent to build radio telescopes, would be familiar with the 21-cm line of hydrogen and the 18-cm line of OH.

This is not idle speculation. A number of searches for extraterrestrial radio signals have been made, and some major searches are now under way. The field has become known as **SETI,** Search for Extra-Terrestrial Intelligence, and it has generated heated debate. Some scientists and philosophers argue that life on other worlds can't possibly exist, so it is a waste of money to search. Others argue that life on other worlds is common and could be de-

tected with present technology. Congress funded a NASA search for a short time but then ended support in the early 1990s because political leaders feared public reaction. In fact, the annual cost of a major search is only about as much as a single Air Force attack helicopter. The controversy may spring in part from the theological and philosophical controversy that would result from the discovery of intelligent life on another world.

In spite of the controversy, major searches have been made, and others are under way. The NASA SETI project canceled by Congress was completed with private funds as Project Phoenix. The SETI Institute was founded in 1984 and has pursued a number of important searches using radio telescopes at major radio astronomy observatories to listen to the radio emissions of a long list of candidate stars. About two billion radio frequency bands must be examined for each star, so the project depends heavily on computers to search for signs of artificial radio transmissions in the recorded radio emissions of each star.

Astonishing amounts of computer power are needed to search for weak or unusual signals. Consequently, the Berkeley SETI team, with the support of the Planetary Society, has recruited owners of personal computers linked to the Internet to participate in a project called seti@home. Participants download a screen saver that searches data files from the Arecibo radio telescope for meaningful signals whenever the owner is not using the computer. About 4 million people have signed up and are providing 1000 years of computer time per day for the project. For information, locate the seti@home project at http://setiathome.ssl.berkeley.edu/.

META II is a major search for radio signals in the southern sky, and the ingenious search called SERENDIP (Search for Extraterrestrial Radio Emission from Nearby Developed Intelligent Populations) rides piggyback on the 305-m Arecibo Telescope searching for signals wherever the radio astronomers point the telescope. Another major search is scanning the sky for intelligent signals in the form of pulsating light flashes.

The SETI Institute, working with the University of California, Berkeley, is building the Allen Telescope Array, which will eventually include 350 dish antennas (■ Figure 26-16). It will be a powerful radio telescope, but it will also be used to search for radio transmissions from other worlds.

At radio wavelengths, noise is a problem. Poorly designed radio transmitters generate noise at unexpected frequencies, and other electronic devices, such as computers, emit radio frequency signals. Furthermore, society, industry, and governments press to use wider and wider sections of the electromagnetic spectrum. Radio astronomers struggle to hear through the radio babble, and SETI searches suffer because many of the noise signals mimic the patterns expected from other civilizations. It would be ironic if we fail to detect signals from another world because our own world has become too noisy.

Ultimately, the chances of success depend on the number of inhabited worlds in our galaxy, and that number is difficult to estimate.

■ Active Figure 26-15

Radio noise from various sources makes it difficult to detect distant signals at wavelengths longer than 30 cm or shorter than 1 cm. In this range, radio emission from H atoms and from OH molecules marks a small wavelength range dubbed the water hole, which may be a likely place for communication.

Ace ◐Astronomy™ Log into AceAstronomy and select this chapter to see the Active Figure "Drake Equation." How many civilizations might be ready for communication?

Figure 26-16

The Allen Telescope Array now being built in California will eventually grow to include 350 radio dishes, each 6 meters in diameter, in an arrangement precisely designed to maximize resolution. As radio astronomers aim the telescope at galaxies and nebulae of interest, state-of-the-art computer systems will analyze stars in the field of view searching for signals from distant civilizations. (Isaac Gary/SETI Institute)

How Many Inhabited Worlds?

The technology exists, and given enough time the searches will find other inhabited worlds, assuming there are at least a few out there. If intelligence is common, then the signals should be found soon—in the next few decades—but if intelligence is rare in the universe, it may be a very long time before SETI confirms that we are not alone.

Simple arithmetic can provide an estimate of the number of technological civilizations available for communication, N_c. The first proposed formula for N_c is now known as the **Drake equation,** named after radio astronomer Frank Drake, a pioneer in the search for extraterrestrial intelligence. It is better to use a version of the Drake equation modified slightly to make it a bit easier to understand. The formula gives N_c, the number of communicative civilizations in a galaxy, as

$$N_c = N^* \cdot f_P \cdot n_{LZ} \cdot f_L \cdot f_I \cdot F_S$$

N^* is the number of stars in a galaxy, and f_P represents the fraction of all stars that have planets. If all single stars have planets, f_P is about 0.5. The factor n_{LZ} is the average number of planets in a solar system suitably placed in the life zone, f_L is the fraction of suitable planets on which life begins, and f_I is the fraction of planets on which life forms evolve to intelligence. These factors can be roughly estimated, but the remaining factor is much more uncertain.

F_S is the fraction of a star's life during which the life form is communicative. Here you can assume that a star lives about 10 billion years. If a society survives at a technological level for only 100 years, the chances of communicating with it are small. But a society that stabilizes and remains technological for a long time is much more likely to be in the communicative phase at the proper time to signal to Earth. If you assume that technological societies destroy themselves in about 100 years, F_S is 100 divided by 10 billion, or 10^{-8}. But if societies can remain technological for a million years, then F_S is 10^{-4}. The influence of the factors in the formula is shown in ■ Table 26-1.

If the optimistic estimates are true, there may be a communicative civilization within a few dozen light-years of Earth, and it could be found by searching through only a few thousand stars. On the other hand, if the pessimistic estimates are correct, Earth may be the only planet in our galaxy capable of communication. It all depends on how technological societies function and survive.

Building Scientific Arguments

Why does the number of inhabited worlds that could be detected depend on how long civilizations survive at a technological level?

This scientific argument depends on timing and requires careful analysis. If you turned a radio telescope toward the sky and scanned millions of frequency bands for lots of stars, you

■ **Table 26-1** **|** **The Number of Technological Civilizations per Galaxy**

	Variables	Estimates Pessimistic	Estimates Optimistic
N^*	Number of stars per galaxy	2×10^{11}	2×10^{11}
f_P	Fraction of stars with planets	0.01	0.5
n_{LZ}	Planets per star in life zone for over 4 billion years	0.01	1
f_L	Fraction of suitable planets on which life begins	0.01	1
f_I	Fraction of life forms that evolve to intelligence	0.01	1
F_S	Fraction of star's life during which a technological society survives	10^{-8}	10^{-4}
N_C	Number of communicative civilizations per galaxy	2×10^{-5}	10×10^6

would be taking a snapshot of the universe at a particular time when you are alive and able to build a radio telescope. To detect other civilizations, they must be in a similar technological stage so they will be broadcasting either intentionally or accidentally. If you searched for decades and detected no signals, it could mean not that life is rare but rather that civilizations do not survive at a technological level for very long. Only a few decades ago, our civilization was threatened by nuclear war, and now we are threatened by the pollution of our environment. If nearly all civilizations in our galaxy are either on the long road up from primitive life forms in oceans or on the long road down from nuclear war or environmental collapse, there may be no one transmitting during the short interval when Earthlings are capable of building radio telescopes with which to listen.

Radio communication between inhabited worlds is limited because it requires very fast computers to search many frequency intervals. Build a new argument to explain: **Why must so many frequencies be searched when the water hole would be a good place to listen?**

■　　■　　■

Connections: Is the human race the only thinking species? If we are, we bear the sole responsibility to understand and admire the universe. Then we are the sole representatives of that state of matter called intelligence. The mere detection of signals from another civilization would demonstrate that we share the universe with others. Although humans may never leave our solar system, such communication would end the self-centered isolation of humanity and stimulate a reevaluation of the meaning of our existence. We may never realize our full potential as humans until we communicate with non-human intelligent life.

Study and Review Tools

Summary

26-1 | The Nature of Life

What is life?
- The process of life extracts energy from the surroundings, maintains the organism, and modifies the surroundings to promote the organism's survival.
- Living things have a physical basis—the arrangement of matter and energy that makes life possible. Life on Earth is based on carbon chemistry.
- Living things must also have a controlling unit of information, which is the genetic information passed to each new generation.
- Genetic information for life on Earth is stored in long carbon-chain molecules such as DNA.
- The DNA molecule stores information in the form of chemical bases linked together like the rungs of a ladder. Copied by the RNA molecule, the patterns of bases act as recipes for the manufacture of proteins and enzymes.
- When a cell divides, the DNA molecules split lengthwise and duplicate themselves so that each of the new cells can receive a copy of the genetic information.
- Errors in duplication or damage to the DNA molecule can produce mutants, organisms that contain new DNA information and have new properties.
- Natural selection determines which of these new organisms are best suited to survive, and the species evolves to fit its environment.

26-2 | The Origin of Life

How did life originate on Earth?
- The oldest fossils on Earth are as old as 3.5 billion years. The fossils show that life began in the oceans.
- Life began on Earth as simple organisms like bacteria, and it evolved into more complex creatures.
- The Miller experiment shows that the building blocks of life form naturally under a wide range of circumstances.
- Chemical evolution concentrated simple molecules into larger, stronger molecules, but those molecules did not reproduce copies of themselves.
- Biological evolution begins when molecules begin reproducing.
- Not until about half a billion years ago did life become complex in what is called the Cambrian explosion.
- Life emerged from the oceans only about 0.4 billion years ago, and human intelligence developed over the last 3 million years.

Could life begin on other worlds?
- Life on Earth requires liquid water and moderate temperatures, and that is probably true of life on other worlds.
- No planet in our solar system appears to harbor life at present. Most are too hot or too cold, but life might have begun on Mars before it became too cold and dry.
- Life in the oceans of Europa or Titan seems possible but unlikely.
- Because the origin of life and its evolution into intelligent creatures took so long on Earth, scientists eliminate short-lived stars such as middle- and upper-main-sequence stars as homes for life.

- Main-sequence G, K, and possibly M stars are thought to be likely candidates for searches for life.

- The life zone around a star may be larger than expected, given the wide variety of living things being found in extreme environments on Earth.

26-3 | Communication with Distant Civilizations

Could Earthlings communicate with civilizations on other worlds?

- Because of distance, speed, and fuel, travel between the stars seems almost impossible for humans or for aliens who might visit Earth.

- Radio communication may be possible, but a conversation would be impossible because of very long travel times for radio signals.

- Broadcasting a radio beacon of pulses in mathematical patterns would distinguish the signal from naturally occurring radio emissions and identify a civilization as technological.

- A signal could be anti-coded so it would be easy for another civilization to decode.

- The best place in the radio spectrum for communication is in the water hole, the wavelength range from the 21 cm line of hydrogen to the 18 cm emission line of OH. Even so, millions of radio wavelengths need to be tested to fully survey the water hole for a given target.

- Sophisticated searches are now under way to detect radio transmissions from civilizations on other worlds, but such SETI programs are hampered by limited computer power and radio noise pollution.

- The number of civilizations in our galaxy that are at a technological level and able to communicate may be limited by the lifetime of such civilizations.

New Terms

natural selection (p. 653)

DNA (deoxyribonucleic acid) (p. 654)

amino acid (p. 654)

protein (p. 654)

enzyme (p. 654)

RNA (ribonucleic acid) (p. 655)

chromosome (p. 655)

gene (p. 655)

mutant (p. 656)

Cambrian period (p. 657)

Miller experiment (p. 657)

primordial soup (p. 657)

chemical evolution (p. 658)

stromatolite (p. 659)

life zone (p. 665)

water hole (p. 668)

SETI (p. 668)

Drake equation (p. 669)

Review Questions

Ace Astronomy™ Assess your understanding of this chapter's topics with additional quizzing and animations at **http://ace .brookscole.com/sf9**

1. If life is based on information, what is that information?

2. What would happen to a life form if the information handed down to off-spring was always the same? How would that endanger the future of the life form?

3. How does the DNA molecule produce a copy of itself?

4. Give an example of natural selection acting on new DNA patterns to select the most advantageous characteristics.

5. What is the evidence that life on Earth began in the sea?

6. What would make you conclude that liquid water is necessary for the origin of life?

7. What is the difference between chemical evolution and biological evolution?

8. What was the significance of the Miller experiment?

9. How does intelligence make a creature more likely to survive?

10. Why are upper-main-sequence stars unlikely sites for intelligent civilizations?

11. What factors make travel between stars nearly impossible?

12. How does the stability of technological civilizations affect the probability that they can be detected from Earth?

13. What is the water hole, and why would it be a good place to look for other civilizations?

14. The star cluster NGC2264 shown here contains cool red giants plus main-sequence stars ranging from hot blue stars all the way down to red dwarfs. Discuss the likelihood that planets orbiting any of these stars might be home to intelligent life. Don't neglect to estimate the age of the cluster.

(T. A. Rector, B. A. Wolpa, NRAO/NOAO/ AURA/NSF)

15. If you could search for life in the galaxy shown, would you look among disk stars or halo stars? Discuss the factors that influence your decision.

(ESO)

Discussion Questions

1. What would you change in the Arecibo message if humanity lived on Mars instead of Earth?

2. What do you think it would mean if decades of careful searches for radio signals for extraterrestrial intelligence turned up nothing? What might that suggest about the future of our own civilization?

3. What philosophical or theological questions might arise if conclusive evidence of intelligent life on other worlds was found?

Problems

1. A single human cell encloses about 1.5 m of DNA, containing 4.5 billion base pairs. What is the spacing between these base pairs in nanometers? That is, how far apart are the rungs on the DNA ladder?

2. If you represented the history of the Earth by a line 1 m long, how long a segment would represent the 400 million years since life moved onto the land? How long a segment would represent the 3-million-year history of human life?

3. If a human generation, the time from birth to childbearing, is 20 years, how many generations have passed in the last million years?

4. If a star must remain on the main sequence for at least 5 billion years for life to evolve to intelligence, how massive could a star be and still harbor intelligent life on one of its planets? (*Hint:* See Chapter 12.)

5. If there are about 1.4×10^{-4} stars like the sun per cubic light-year, how many lie within 100 light-years of Earth? (*Hint:* The volume of a sphere is $\frac{4}{3}\pi R^3$.)

6. Mathematician Karl Gauss suggested planting forests and fields in a gigantic geometric proof to signal to possible Martians that intelligent life exists on Earth. If Martians had telescopes that could resolve details no smaller than 1 second of arc, how large would the smallest element of Gauss's proof have to be? (*Hint:* Use the small-angle formula.)

7. If you detected radio signals with an average wavelength of 20 cm and suspected that they came from a civilization on a distant planet, roughly how much of a change in wavelength should you expect to see because of the orbital motion of the distant planet? (*Hint:* See the Doppler shift in Chapter 7.)

8. Calculate the number of communicative civilizations per galaxy from your own estimates of the factors in Table 26-1.

Media Cluster

Ace ◑ Astronomy™ To access the resources in the Media Cluster, log into AceAstronomy at http://ace.brookscole.com/sf9 and select Chapter 26.

ACTIVE FIGURES

DNA
Explore DNA in detail and see how individual atoms are linked together to produce each of the four bases. Notice how the molecules "hitch" to the DNA rails and fit with their matching bases.

Future of the Sun
Over billions of years, the sun will gradually mature and die. This animation lets you observe the death of the sun from four different vantage points in our solar system.

Earth Calendar
This animation shows how the formation of Earth and its life forms would look if the entire history of Earth were condensed into one calendar year.

Interstellar Communication
In this animation, you can launch a space probe between distant stars or send radio signals and compare the two methods of communication.

Drake Equation
Explore how different values for factors in the Drake equation affect predictions about how plentiful intelligent life is in a galaxy.

ASTRONOMY EXERCISES

Extrasolar Planets
This simulation presents a hypothetical star system with one planet orbiting a star. Study what variables help astronomers detect extrasolar planets.

Lab 8: Extrasolar Planets
This lab examines how indirect methods such as Doppler shift techniques are used to discover and study extrasolar planets. At the end of the lab you will build your own solar system.

Critical Inquiries for the Web

1. The popular movie *Contact* focused interest on the SETI program by profiling the work of a radio astronomer dedicated to the search for extraterrestrial intelligence. Visit websites that give information about the movie, SETI programs, and radio astronomy, and discuss how realistic the movie was in capturing how such research is done.

2. Where outside the solar system would you look for habitable planets? NASA has increasingly focused its interest on this question. Look for information online about programs dedicated to detecting which planets might support life. What criteria are used to choose targets for the planned searches? What methods will be used to carry out the searches?

 Go to the Brooks/Cole Astronomy Resource Center **(http://astronomy.brookscole.com)** for critical thinking exercises, articles, and additional readings from InfoTrac College Edition, Brooks/Cole's online student library.

Afterword

The aggregate of all our joys and sufferings, thousands of confident religions, ideologies and economic doctrines, every hunter and forager, every hero and coward, every creator and destroyer of civilizations, every king and peasant, every young couple in love, every hopeful child, every mother and father, every inventor and explorer, every teacher of morals, every corrupt politician, every superstar, every supreme leader, every saint and sinner in the history of our species, lived there on a mote of dust, suspended in a sunbeam.

CARL SAGAN (1934–1996)

Earth photographed by Voyager 1 from the edge of the solar system. (NASA)

Our journey together is over, but before we part company, there is one last thing to discuss—the place of humanity in the universe. Astronomy gives us some comprehension of the workings of stars, galaxies, and planets, but its greatest value lies in what it teaches us about ourselves. Now that you have surveyed astronomical knowledge, you can better understand your own position in nature.

To some, the word *nature* conjures up visions of furry rabbits hopping about in a forest glade. To others, nature is the blue-green ocean depths, and still others think of nature as windswept mountaintops. As diverse as these images are, they are all Earth-bound. Having studied astronomy, you see nature as a beautiful mechanism composed of matter and energy interacting according to simple rules to form galaxies, stars, planets, mountaintops, ocean depths, and forest glades.

Perhaps the most important astronomical lesson is that humanity is a small but important part of the universe. Most of the universe is lifeless. The vast reaches between the galaxies appear to be empty of all but the thinnest gas, and the stars are much too hot to preserve the chemical bonds that seem necessary for life to survive and develop. Only on the surfaces of a few planets, where temperatures are moderate, could atoms link together to form living matter.

If life is special, then intelligence is precious. The universe must contain many planets devoid of life, planets where the wind has blown unfelt for billions of years. There may also exist planets where life has developed but has not become complex, planets on which the wind stirs wide plains of grass and rustles dark forests. On some planets, insects, fish, birds, and animals may watch the passing days unaware of their own existence. It is intelligence, human or alien, that gives meaning to the landscape.

Science is the process by which intelligence tries to understand the universe. Science is not the invention of new devices or processes. It does not create home computers, cure the mumps, or manufacture plastic spoons—that is engineering and technology, the adaptation of scientific understanding for practical purposes. Science is understanding nature, and astronomy is understanding on the grandest scale. Astronomy is the science by which the universe, through its intelligent lumps of matter, tries to understand its own existence.

As the primary intelligent species on this planet, we are the custodians of a priceless gift—a planet filled with living things. This is especially true if life is rare in the universe. In fact, if Earth is the only inhabited planet, our responsibility is overwhelming. We are the only creatures who can take action to preserve the existence of life on Earth, and, ironically, our own actions are the most serious hazards.

The future of humanity is not secure. We are trapped on a tiny planet with limited resources and a population growing faster than our ability to produce food. In our efforts to survive, we have already driven some creatures to extinction and now threaten others. If our civilization collapses because of starvation, or if our race destroys itself somehow, the only bright spot is that the rest of the creatures on Earth will be better off for our absence.

But even if we control our population and conserve and recycle our resources, life on Earth is doomed. In 5 billion years, the sun will leave the main sequence and swell into a red giant, incinerating Earth. However, Earth will be lifeless long before that. Within the next few billion years, the growing luminosity of the sun will first alter Earth's climate and then drive off its atmosphere and oceans. Earth, like everything else in the universe, is only temporary.

To survive, humanity must eventually leave Earth and search for other planets. Colonizing the moon and other planets of our solar system will not save us, because they will face the same fate as Earth when the sun dies. But travel to other stars is tremendously difficult and may be impossible with the limited resources we have in our small solar system. We and all of the living things on Earth may be trapped.

This is a depressing prospect, but a few factors are comforting. First, everything in the universe is temporary. Stars die, galaxies die, perhaps the entire universe will someday end. That our distant future is limited only assures us that we are a part of a much larger whole. Second, we have a few billion years to prepare, and a billion years is a very long time. Only a few million years ago, our ancestors were learning to walk erect and communicate.

A billion years ago, our ancestors were microscopic organisms living in the primeval oceans. To suppose that a billion years hence we humans will still be human, that we will still be the dominant species on Earth, or that we will still be the sole intelligence on Earth is the ultimate conceit.

Our responsibility is not to save our race for all eternity but to behave as dependable custodians of our planet, preserving it, admiring it, and trying to understand it. That calls for drastic changes in our behavior toward other living things and a revolution in our attitude toward our planet's resources. Whether we can change our ways is debatable—humanity is far from perfect in its understanding, abilities, or intentions. However, you must not imagine that we and our civilization are less than precious. We have the gift of intelligence, and that is the finest thing this planet has ever produced.

We shall not cease from exploration
And the end of all our exploring
Will be to arrive where we started
And know the place for the first time.

—*T. S. Eliot, "Little Gidding"*

Appendix A

Units and Astronomical Data

Introduction

The metric system is used worldwide as the system of units, not only in science but also in engineering, business, sports, and daily life. Developed in 18th-century France, the metric system has gained acceptance in almost every country in the world because it simplifies computations.

A system of units is based on the three fundamental units for length, mass, and time. Other quantities, such as density and force, are derived from these fundamental units. In the English (or British) system of units (commonly used only in the United States, Myanmar, and Liberia, but not in Great Britain) the fundamental unit of length is the foot, composed of 12 inches. The metric system is based on the decimal system of numbers, and the fundamental unit of length is the meter, composed of 100 centimeters.

Because the metric system is a decimal system, it is easy to express quantities in larger or smaller units as is convenient. You can express distances in centimeters, meters, kilometers, and so on. The prefixes specify the relation of the unit to the meter. Just as a cent is $\frac{1}{100}$ of a dollar, so a centimeter is $\frac{1}{100}$ of a meter. A kilometer is 1000 m, and a kilogram is 1000 g. The meanings of the commonly used prefixes are given in ■ Table A-1.

The SI Units

Any system of units based on the decimal system would be easy to use, but by international agreement, the preferred set of units, known as the *Système International d'Unités* (SI units) is based on the meter, kilogram, and second. These three fundamental units define the rest of the units, as given in ■ Table A-2.

The SI unit of force is the newton (N), named after Isaac Newton. It is the force needed to accelerate a 1 kg mass by 1 m/s², or the force roughly equivalent to the weight of an apple at Earth's surface. The SI unit of energy is the joule (J), the energy produced by a force of 1 N acting through a distance of 1 m. A joule is roughly the energy in the impact of an apple falling off a table.

■ Table A-1 | Metric Prefixes

Prefix	Symbol	Factor
Mega	M	10^6
Kilo	k	10^3
Centi	c	10^{-2}
Milli	m	10^{-3}
Micro	μ	10^{-6}
Nano	n	10^{-9}

■ Table A-2 | SI Metric Units

Quantity	SI Unit	English Unit
Length	Meter (m)	Foot
Mass	Kilogram (kg)	Slug (sl)
Time	Second (s)	Second (s)
Force	Newton (N)	Pound (lb)
Energy	Joule (J)	Foot-pound (fp)

Exceptions

Units can help you in two ways. They make it possible to make calculations, and they can help you to conceive of certain quantities. For calculations, the metric system is far superior, and it is used for calculations throughout this book.

But Americans commonly use the English system of units, so for conceptual purposes this book also expresses quantities in English units. Instead of saying the average person would weigh 133 N on the moon, it might be more helpful to express the weight as 30 lb. Consequently, this text commonly gives quantities in metric form followed by the English form in parentheses: the radius of the moon is 1738 km (1080 miles).

In SI units, density should be expressed as kilograms per cubic meter, but no human hand can enclose a cubic meter, so this unit does not help you grasp the significance of a given density. This book refers to density in grams per cubic centimeter. A gram is roughly the mass of a paperclip, and a cubic centimeter is the size of a small sugar cube, so you can conceive of a density of 1 g/cm^3, roughly the density of water. This is not a bothersome departure from SI units because you will not have to make complex calculations using density.

Conversions

To convert from one metric unit to another (from meters to kilometers, for example), you have only to look at the prefix. However, converting from metric to English or English to metric is more complicated. The conversion factors are given in ■ Table A-3.

Example: The radius of the moon is 1738 km. What is this in miles? Table A-3 indicates that 1 mile equals 1.609 km, so

$$1738 \text{ km} \times \frac{1 \text{ mile}}{1.609 \text{ km}} = 1080 \text{ miles}$$

Temperature Scales

In astronomy, as in most other sciences, temperatures are expressed on the Kelvin scale, although the centigrade (or Celsius) scale is also used. The Fahrenheit scale commonly used in the United States is not used in scientific work.

Temperatures on the Kelvin scale are measured from absolute zero, the temperature of an object that contains no extractable heat. In practice, no object can be as cold as absolute zero, although laboratory apparatuses have reached temperatures less than 10^{-6} K. The scale is named after the Scottish mathematical physicist William Thomson, Lord Kelvin (1824–1907).

The centigrade scale refers temperatures to the freezing point of water (0°C) and to the boiling point of water (100°C). One degree centigrade is 1/100th the temperature difference between the freezing and boiling points of water, thus the prefix *centi*. The centigrade scale is also called the Celsius scale after its inventor, the Swedish astronomer Anders Celsius (1701–1744).

The Fahrenheit scale fixes the freezing point of water at 32°F and the boiling point at 212°F. Named after the German physicist Gabriel Daniel Fahrenheit (1686–1736), who made the first successful mercury thermometer in 1720, the Fahrenheit scale is used only in the United States.

It is easy to convert temperatures from one scale to another using the information given in ■ Table A-4.

Powers of 10 Notation

Powers of 10 make writing very large numbers much simpler. For example, the nearest star is about 43,000,000,000,000 km from the sun. Writing this number as 4.3 × 10^{13} km is much easier.

Very small numbers can also be written with powers of 10. For example, the wavelength of visible light is about 0.0000005 m. In powers of 10 this becomes 5 × 10^{-7} m.

The powers of 10 used in this notation appear below. The exponent tells you how to move the decimal point. If the exponent is positive, move the decimal point to the right. If the exponent is negative, move the decimal point to the left. For example, 2 × 10^3 equals 2000.0, and 2 × 10^{-3} equals 0.002.

$$\vdots$$

10^5	=	100,000
10^4	=	10,000
10^3	=	1,000
10^2	=	100
10^1	=	10
10^0	=	1
10^{-1}	=	0.1
10^{-2}	=	0.01
10^{-3}	=	0.001
10^{-4}	=	0.0001

$$\vdots$$

■ Table A-3 I Conversion Factors

1 inch = 2.54 centimeters	1 centimeter = 0.394 inch
1 foot = 0.3048 meter	1 meter = 39.36 inches = 3.28 feet
1 mile = 1.6093 kilometers	1 kilometer = 0.6214 mile
1 slug = 14.594 kilograms	1 kilogram = 0.0685 slug
1 pound = 4.4482 newtons	1 newton = 0.2248 pound
1 foot-pound = 1.35582 joules	1 joule = 0.7376 foot-pound
1 horsepower = 745.7 joules/s	1 joule/s = 1 watt

■ Table A-4 I Temperature Scales

	Kelvin (K)	Centigrade (°C)	Fahrenheit (°F)
Absolute zero	0 K	−273°C	−459°F
Freezing point of water	273 K	0°C	32°F
Boiling point of water	373 K	100°C	212°F
Conversions:			
K = °C + 273			
°C = $\frac{5}{9}$(°F − 32)			
°F = $\frac{9}{5}$°C + 32			

If you use scientific notation in calculations, be sure you correctly enter numbers into your calculator. Not all calculators accept scientific notation, but those that can have a key labeled EXP, EEX, or perhaps EE that allows you to enter the exponent of ten. To enter a number such as 3×10^8, press the keys 3 EXP 8. To enter a number with a negative exponent, you must use the change-sign key, usually labeled $+/-$ or CHS. To enter the number 5.2×10^{-3}, press the keys 5.2 EXP $+/-$ 3. Try a few examples.

To read a number in scientific notation from a calculator you must read the exponent separately. The number 3.1×10^{25} may appear in a calculator display as 3.1 25 or on some calculators as 3.1 10^{25}. Examine your calculator to determine how such numbers are displayed.

* * *

Astronomy, and science in general, is a way of learning about nature and understanding the universe. To test hypotheses about how nature works, scientists use observations of nature. The tables that follow contain some of the basic observations that support our best understanding of the astronomical universe. Of course, these data are expressed in the form of numbers, not because science reduces all understanding to mere numbers, but because the struggle to understand nature is so demanding that science must use every tool available. Quantitative thinking—reasoning mathematically—is one of the most powerful tools ever invented by the human brain. Thus these tables are not nature reduced to mere numbers but numbers supporting humanity's growing understanding of the natural world around us.

■ Table A-5 | Constants

Astronomical unit (AU)	$= 1.495979 \times 10^{11}$ m
Parsec (pc)	$= 206{,}265$ AU
	$= 3.085678 \times 10^{16}$ m
	$= 3.261633$ ly
Light-year (ly)	$= 9.46053 \times 10^{15}$ m
Velocity of light (c)	$= 2.997925 \times 10^8$ m/s
Gravitational constant (G)	$= 6.67 \times 10^{-11}$ m^3/s^{2}kg
Mass of Earth ($M_\oplus$)	$= 5.976 \times 10^{24}$ kg
Earth equatorial radius ($R_\oplus$)	$= 6378.164$ km
Mass of sun ($M_\odot$)	$= 1.989 \times 10^{30}$ kg
Radius of sun ($R_\odot$)	$= 6.9599 \times 10^8$ m
Solar luminosity ($L_\odot$)	$= 3.826 \times 10^{26}$ J/s
Mass of moon	$= 7.350 \times 10^{22}$ kg
Radius of moon	$= 1738$ km
Mass of H atom	$= 1.67352 \times 10^{-27}$ kg

■ Table A-6 | Units Used in Astronomy

1 angstrom (Å)	$= 10^{-8}$ cm
	$= 10^{-10}$ m
1 astronomical unit (AU)	$= 1.495979 \times 10^{11}$ m
	$= 92.95582 \times 10^6$ miles
1 light-year (ly)	$= 6.3240 \times 10^4$ AU
	$= 9.46053 \times 10^{15}$ m
	$= 5.9 \times 10^{12}$ miles
1 parsec (pc)	$= 206{,}265$ AU
	$= 3.085678 \times 10^{16}$ m
	$= 3.261633$ ly
1 kiloparsec (kpc)	$= 1000$ pc
1 megaparsec (Mpc)	$= 1{,}000{,}000$ pc

Table A-7 | Properties of Main-Sequence Stars

Spectral Type	Absolute Visual Magnitude (M_v)	Luminosity*	Temp. (K)	λ_{max} (nm)	Mass*	Radius*	Average Density (g/cm³)
O5	−5.8	501,000	40,000	72.4	40	17.8	0.01
B0	−4.1	20,000	28,000	100	18	7.4	0.1
B5	−1.1	790	15,000	190	6.4	3.8	0.2
A0	+0.7	79	9900	290	3.2	2.5	0.3
A5	+2.0	20	8500	340	2.1	1.7	0.6
F0	+2.6	6.3	7400	390	1.7	1.4	1.0
F5	+3.4	2.5	6600	440	1.3	1.2	1.1
G0	+4.4	1.3	6000	480	1.1	1.0	1.4
G5	+5.1	0.8	5500	520	0.9	0.9	1.6
K0	+5.9	0.4	4900	590	0.8	0.8	1.8
K5	+7.3	0.2	4100	700	0.7	0.7	2.4
M0	+9.0	0.1	3500	830	0.5	0.6	2.5
M5	+11.8	0.01	2800	1000	0.2	0.3	10.0
M8	+16	0.001	2400	1200	0.1	0.1	63

*Luminosity, mass, and radius are given in terms of the sun's luminosity, mass, and radius.

Table A-8 | The Brightest Stars

Star	Name	Apparent Visual Magnitude (m_v)	Spectral Type	Absolute Visual Magnitude (M_v)	Distance (ly)
α CMa A	Sirius	−1.47	A1	1.4	8.7
α Car	Canopus	−0.72	F0	−3.1	98
α Cen	Rigil Kentaurus	−0.01	G2	4.4	4.3
α Boo	Arcturus	−0.06	K2	−0.3	36
α Lyr	Vega	0.04	A0	0.5	26.5
α Aur	Capella	0.05	G8	−0.6	45
β Ori A	Rigel	0.14	B8	−7.1	900
α CMi A	Procyon	0.37	F5	2.7	11.3
α Ori	Betelgeuse	0.41	M2	−5.6	520
α Eri	Achernar	0.51	B3	−2.3	118
β Cen AB	Hadar	0.63	B1	−5.2	490
α Aql	Altair	0.77	A7	2.2	16.5
α Tau A	Aldebaran	0.86	K5	−0.7	68
α Cru	Acrux	0.90	B2	−3.5	260
α Vir	Spica	0.91	B1	−3.3	220
α Sco A	Antares	0.92	M1	−5.1	520
α PsA	Fomalhaut	1.15	A3	2.0	22.6
β Gem	Pollux	1.16	K0	1.0	35
α Cyg	Deneb	1.26	A2	−7.1	1600
β Cru	Beta Crucis	1.28	B0.5	−4.6	490

Name	Absolute Magnitude (M_v)	Distance (ly)	Spectral Type	Apparent Visual Magnitude (m_v)
Sun	4.83		G2	−26.8
Proxima Cen	15.45	4.28	M5	11.05
α Cen A	4.38	4.3	G2	0.1
α Cen B	5.76	4.3	K5	1.5
Barnard's Star	13.21	5.9	M5	9.5
Wolf 359	16.80	7.6	M6	13.5
Lalande 21185	10.42	8.1	M2	7.5
Sirius A	1.41	8.6	A1	−1.5
Sirius B	11.54	8.6	white dwarf	7.2
Luyten 726-8A	15.27	8.9	M5	12.5
Luyten 726-8B (UV Cet)	15.8	8.9	M6	13.0
Ross 154	13.3	9.4	M5	10.6
Ross 248	14.8	10.3	M6	12.2
ε Eri	6.13	10.7	K2	3.7
Luyten 789-6	14.6	10.8	M7	12.2
Ross 128	13.5	10.8	M5	11.1
61 CYG A	7.58	11.2	K5	5.2
61 CYG B	8.39	11.2	K7	6.0
ε Ind	7.0	11.2	K5	4.7
Procyon A	2.64	11.4	F5	0.3
Procyon B	13.1	11.4	white dwarf	10.8
Σ 2398 A	11.15	11.5	M4	8.9
Σ 2398 B	11.94	11.5	M5	9.7
Groombridge 34 A	10.32	11.6	M1	8.1
Groombridge 34 B	13.29	11.6	M6	11.0
Lacaille 9352	9.59	11.7	M2	7.4
τ Ceti	5.72	11.9	G8	3.5
BD + 5° 1668	11.98	12.2	M5	9.8
L 725-32	15.27	12.4	M5	11.5
Lacaille 8760	8.75	12.5	M0	6.7
Kapteyn's Star	10.85	12.7	M0	8.8
Kruger 60 A	11.87	12.8	M3	9.7
Kruger 60 B	13.3	12.8	M4	11.2

■ **Table A-10** | **Properties of the Planets**

PHYSICAL PROPERTIES (EARTH = ⊕)

Planet	Equatorial Radius (km)	Equatorial Radius (⊕ = 1)	Mass (⊕ = 1)	Average Density (g/cm³)	Surface Gravity (⊕ = 1)	Escape Velocity (km/s)	Sidereal Period of Rotation	Inclination of Equator to Orbit
Mercury	2439	0.382	0.0558	5.44	0.378	4.3	58.646^d	0°
Venus	6052	0.95	0.815	5.24	0.903	10.3	244.3^d	177°
Earth	6378	1.00	1.00	5.497	1.00	11.2	23^{h}56^{m}04.1^s	23°27′
Mars	3396	0.53	0.1075	3.94	0.379	5.0	24^{h}37^{m}22.6^s	25°19′
Jupiter	71,494	11.20	317.83	1.34	2.54	61	9^{h}55^{m}30^s	3°5′
Saturn	60,330	9.42	95.147	0.69	1.16	35.6	10^{h}13^{m}59^s	26°24′
Uranus	25,559	4.01	14.54	1.19	0.919	22	17^{h}14^m	97°55′
Neptune	24,750	3.93	17.23	1.66	1.19	25	16^{h}3^m	28°48′
Pluto	1185	0.18	0.0022	2.0	0.06	1.2	6^{d}9^{h}21^m	122°

ORBITAL PROPERTIES

Planet	Semimajor Axis (AU)	Semimajor Axis (10⁶ km)	Orbital Period (y)	Orbital Period (days)	Average Orbital Velocity (km/s)	Orbital Eccentricity	Inclination to Ecliptic
Mercury	0.3871	57.9	0.24084	87.969	47.89	0.2056	7°0′16″
Venus	0.7233	108.2	0.61515	224.68	35.03	0.0068	3°23′40″
Earth	1	149.6	1	365.26	29.79	0.0167	0°
Mars	1.5237	227.9	1.8808	686.95	24.13	0.0934	1°51′9″
Jupiter	5.2028	778.3	11.867	4334.3	13.06	0.0484	1°18′29″
Saturn	9.5388	1427.0	29.461	10,760	9.64	0.0560	2°29′17″
Uranus	19.18	2869.0	84.013	30,685	6.81	0.0461	0°46′23″
Neptune	30.0611	4497.1	164.793	60,189	5.43	0.0100	1°46′27″
Pluto	39.44	5900	247.7	90,465	4.74	0.2484	17°9′3″

Planet	Satellite	Radius (km)	Distance from Planet (10^3 km)	Orbital Period (days)	Orbital Eccentricity	Orbital Inclination
Earth	Moon	1738	384.4	27.322	0.055	5°8′43″
Mars	Phobos	14 × 12 × 10	9.38	0.3189	0.018	1.0°
	Deimos	8 × 6 × 5	23.5	1.262	0.002	2.8°
Jupiter	Metis	20	126	0.29	0.0	0.0°
	Adrastea	12 × 8 × 10	128	0.294	0.0	0.0°
	Amalthea	135 × 100 × 78	182	0.4982	0.003	0.45°
	Thebe	50	223	0.674	0.0	1.3°
	Io	1820	422	1.769	0.000	0.3°
	Europa	1565	671	3.551	0.000	0.46°
	Ganymede	2640	1071	7.155	0.002	0.18°
	Callisto	2420	1884	16.689	0.008	0.25°
	Leda	~8	11,110	240	0.146	26.7°
	Himalia	~85	11,470	250.6	0.158	27.6°
	Lysithea	~20	11,710	260	0.12	29°
	Elara	~30	11,740	260.1	0.207	24.8°
	Ananke	15	21,200	631	0.169	147°
	Carme	22	22,350	692	0.207	163°
	Pasiphae	35	23,300	735	0.40	147°
	Sinope	20	23,700	758	0.275	156°
Saturn	Pan	10	133.570	0.574	0.000	0°
	Atlas	20 × 15 × 15	137.7	0.601	0.002	0.3°
	Prometheus	70 × 40 × 50	139.4	0.613	0.003	0.0°
	Pandora	55 × 35 × 50	141.7	0.629	0.004	0.05°
	Epimetheus	70 × 50 × 50	151.42	0.694	0.009	0.34°
	Janus	110 × 80 × 100	151.47	0.695	0.007	0.14°
	Mimas	196	185.54	0.942	0.020	1.5°
	Enceladus	250	238.04	1.370	0.004	0.0°
	Tethys	530	294.67	1.888	0.000	1.1°
	Calypso	17 × 11 × 12	294.67	1.888	0.0	~1°?
	Telesto	12	294.67	1.888	0.0	~1°?
	Dione	560	377	2.737	0.002	0.0°
	Helene	20 × 15 × 15	377	2.74	0.005	0.15°
	Rhea	765	527	4.518	0.001	0.4°
	Titan	2575	1222	15.94	0.029	0.3°
	Hyperion	205 × 130 × 110	1484	21.28	0.104	~0.5°
	Iapetus	720	3562	79.33	0.028	14.72°
	Phoebe	110	12,930	550.4	0.163	150°
Uranus	Cordelia	20	49.8	0.3333	~0	~0°
	Ophelia	15	53.8	0.375	~0	~0°
	Bianca	25	59.1	0.433	~0	~0°
	Cressida	30	61.8	0.462	~0	~0°
	Desdemona	30	62.7	0.475	~0	~0°

(continued)

Planet	Satellite	Radius (km)	Distance from Planet (10^3 km)	Orbital Period (days)	Orbital Eccentricity	Orbital Inclination
Uranus	Juliet	40	64.4	0.492	~0	~0°
(continued)	Portia	55	66.1	0.512	~0	~0°
	Rosalind	30	69.9	0.558	~0	~0°
	Belinda	30	75.2	0.621	~0	~0°
	Puck	85 ± 5	85.9	0.762	~0	~0°
	Miranda	242 ± 5	129.9	1.414	0.017	3.4°
	Ariel	580 ± 5	190.9	2.520	0.003	0°
	Umbriel	595 ± 10	266.0	4.144	0.003	0°
	Titania	805 ± 5	436.3	8.706	0.002	0°
	Oberon	775 ± 10	583.4	13.463	0.001	0°
	Caliban	40	7164	579	0.082	139.2°
	Stephano	~20	7900	676		
	Sycorax	80	12,174	1284	0.509	152.7°
	Prospero	~20	16,100	1950		
	Setebos	~20	17,600	2240	0.539	
Neptune	Naiad	30	48.2	0.296	~0	~0°
	Thalassa	40	50.0	0.312	~0	~0°
	Despina	90	52.5	0.333	~0	~0°
	Galatea	75	62.0	0.396	~0	~0°
	Larissa	95	73.6	0.554	~0	~0°
	Proteus	205	117.6	1.121	~0	~0°
	Triton	1352	354.59	5.875	0.00	160°
	Nereid	170	5588.6	360.125	0.76	27.7°
Pluto	Charon	626	19.7	6.38718	~0	122°

Table A-12 | **Meteor Showers**

Shower	Dates	Hourly Rate	Radiant		Associated Comet
			R.A.	*Dec.*	
Quadrantids	Jan. 2–4	30	15^h24^m	50°	
Lyrids	April 20–22	8	18^h4^m	33°	1861 I
η Aquarids	May 2–7	10	22^h24^m	0°	Halley?
δ Aquarids	July 26–31	15	22^h36^m	−10°	
Perseids	Aug. 10–14	40	3^h4^m	58°	1982 III
Orionids	Oct. 18–23	15	6^h20^m	15°	Halley?
Taurids	Nov. 1–7	8	3^h40^m	17°	Encke
Leonids	Nov. 14–19	6	10^h12^m	22°	1866 I Temp
Geminids	Dec. 10–13	50	7^h28^m	32°	

Table A-13 | Greatest Elongations of Mercury

Evening Sky	Morning Sky
Feb. 24, 2006*	April 8, 2006
June 20, 2006	Aug. 7, 2006
Oct. 16, 2006	Nov. 25, 2006*
Feb. 7, 2007	March 22, 2007
June 2, 2007	July 20, 2007
Sept. 29, 2007	Nov. 8, 2007*
Jan. 22, 2008	March 3, 2008
May 14, 2008	July 1, 2008
Sept. 11, 2008	Oct. 22, 2008*
Jan. 4, 2009	Feb. 13, 2009
April 26, 2009*	June 13, 2009
Aug. 24, 2009	Oct. 6, 2009*
Dec 18, 2009	Jan 27, 2010
Apr. 8, 2010*	May 26, 2010
Aug. 7, 2010	Sep 19, 2010*
Dec. 1, 2010	

Elongation is the angular distance from the sun to a planet.

*Most favorable elongations.

Table A-14 | Greatest Elongations of Venus

Evening Sky	Morning Sky
	March 25, 2006
June 9, 2007	Oct. 28, 2007
Jan. 14, 2009	June 5, 2009
Aug. 20, 2010	Jan. 8, 2011
March 27, 2012	Aug. 15, 2012
Nov. 1, 2013	March 22, 2014
June 6, 2015	Oct. 26, 2015
Jan. 12, 2017	June 3, 2017
Aug. 17, 2018	

Table A-15 | The Greek Alphabet

A, α	alpha	H, η	eta	N, ν	nu	T, τ	tau
B, β	beta	Θ, θ	theta	Ξ, ξ	xi	Y, υ	upsilon
Γ, γ	gamma	I, ι	iota	O, o	omicron	Φ, ϕ	phi
Δ, δ	delta	K, κ	kappa	Π, π	pi	X, χ	chi
E, ε	epsilon	Λ, λ	lambda	P, ρ	rho	Ψ, ψ	psi
Z, ζ	zeta	M, μ	mu	Σ, σ	sigma	Ω, ω	omega

Group

Noble Gases

IA(1)

(18)

Atomic number → 11
Symbol → Na
Atomic mass → 22.99

Atomic masses are based on carbon-12. Numbers in parentheses are mass numbers of most stable or best-known isotopes of radioactive elements.

Transition Elements

Period	IA(1)	IIA(2)	IIIB(3)	IVB(4)	VB(5)	VIB(6)	VIIB(7)	(8)	(9)	(10)	IB(11)	IIB(12)	IIIA(13)	IVA(14)	VA(15)	VIA(16)	VIIA(17)	(18)
1	1 H 1.008																	2 He 4.003
2	3 Li 6.941	4 Be 9.012											5 B 10.81	6 C 12.01	7 N 14.01	8 O 16.00	9 F 19.00	10 Ne 20.18
3	11 Na 22.99	12 Mg 24.31						VIII					13 Al 26.98	14 Si 28.09	15 P 30.97	16 S 32.06	17 Cl 35.45	18 Ar 39.95
4	19 K 39.10	20 Ca 40.08	21 Sc 44.96	22 Ti 47.90	23 V 50.94	24 Cr 52.00	25 Mn 54.94	26 Fe 55.85	27 Co 58.93	28 Ni 58.7	29 Cu 63.55	30 Zn 65.38	31 Ga 69.72	32 Ge 72.59	33 As 74.92	34 Se 78.96	35 Br 79.90	36 Kr 83.80
5	37 Rb 85.47	38 Sr 87.62	39 Y 88.91	40 Zr 91.22	41 Nb 92.91	42 Mo 95.94	43 Tc 98.91	44 Ru 101.1	45 Rh 102.9	46 Pd 106.4	47 Ag 107.9	48 Cd 112.4	49 In 114.8	50 Sn 118.7	51 Sb 121.8	52 Te 127.6	53 I 126.9	54 Xe 131.3
6	55 Cs 132.9	56 Ba 137.3	57* La 138.9	72 Hf 178.5	73 Ta 180.9	74 W 183.9	75 Re 186.2	76 Os 190.2	77 Ir 192.2	78 Pt 195.1	79 Au 197.0	80 Hg 200.6	81 Tl 204.4	82 Pb 207.2	83 Bi 209.0	84 Po (210)	85 At (210)	86 Rn (222)
7	87 Fr (223)	88 Ra 226.0	89** Ac (227)	104 Rf (261)	105 Db (262)	106 Sg (263)	107 Bh (262)	108 Hs (265)	109 Mt (266)	110 Ds (269)	111 Uuu (272)	112 Uub (277)	113 Uub (284)	114 Uuq (285)	115 Uub (288)	116 Uuh (289)		

Inner Transition Elements

	58	59	60	61	62	63	64	65	66	67	68	69	70	71
Lanthanide Series 6*	Ce 140.1	Pr 140.9	Nd 144.2	Pm (145)	Sm 150.4	Eu 152.0	Gd 157.3	Tb 158.9	Dy 162.5	Ho 164.9	Er 167.3	Tm 168.9	Yb 173.0	Lu 175.0
	90	91	92	93	94	95	96	97	98	99	100	101	102	103
Actinide Series 7**	Th 232.0	Pa 231.0	U 238.0	Np 237.0	Pu (244)	Am (243)	Cm (247)	Bk (247)	Cf (251)	Es (252)	Fm (257)	Md (258)	No (259)	Lr (260)

The Elements and Their Symbols

Element	Symbol	Element	Symbol	Element	Symbol	Element	Symbol	Element	Symbol	Element	Symbol
Actinium	Ac	Cesium	Cs	Hafnium	Hf	Mercury	Hg	Protactinium	Pa	Tellurium	Te
Aluminum	Al	Chlorine	Cl	Hassium	Hs	Molybdenum	Mo	Radium	Ra	Terbium	Tb
Americium	Am	Chromium	Cr	Helium	He	Neodymium	Nd	Radon	Rn	Thallium	Tl
Antimony	Sb	Cobalt	Co	Holmium	Ho	Neon	Ne	Rhenium	Re	Thorium	Th
Argon	Ar	Copper	Cu	Hydrogen	H	Neptunium	Np	Rhodium	Rh	Thulium	Tm
Arsenic	As	Curium	Cm	Indium	In	Nickel	Ni	Rubidium	Rb	Tin	Sn
Astatine	At	Darmstadtium	Ds	Iodine	I	Niobium	Nb	Ruthenium	Ru	Titanium	Ti
Barium	Ba	Dubnium	Db	Iridium	Ir	Nitrogen	N	Rutherfordium	Rf	Tungsten	W
Berkelium	Bk	Dysprosium	Dy	Iron	Fe	Nobelium	No	Samarium	Sm	Uranium	U
Beryllium	Be	Einsteinium	Es	Krypton	Kr	Osmium	Os	Scandium	Sc	Vanadium	V
Bismuth	Bi	Erbium	Er	Lanthanum	La	Oxygen	O	Seaborgium	Sg	Xenon	Xe
Bohrium	Bh	Europium	Eu	Lawrencium	Lr	Palladium	Pd	Selenium	Se	Ytterbium	Yb
Boron	B	Fermium	Fm	Lead	Pb	Phosphorous	P	Silicon	Si	Yttrium	Y
Bromine	Br	Fluorine	F	Lithium	Li	Platinum	Pt	Silver	Ag	Zinc	Zn
Cadmium	Cd	Francium	Fr	Lutetium	Lu	Plutonium	Pu	Sodium	Na	Zirconium	Zr
Calcium	Ca	Gadolinium	Gd	Magnesium	Mg	Polonium	Po	Strontium	Sr		
Californium	Cf	Gallium	Ga	Manganese	Mn	Potassium	K	Sulfur	S		
Carbon	C	Germanium	Ge	Meitnerium	Mt	Praseodymium	Pr	Tantalum	Ta		
Cerium	Ce	Gold	Au	Mendelevium	Md	Promethium	Pm	Technetium	Tc		

Appendix B

Observing the Sky

Observing the sky with the naked eye is of no more importance to modern astronomy than picking up pretty pebbles is to modern geology. But the sky is a natural wonder unimaginably bigger than the Grand Canyon, the Rocky Mountains, or any other natural wonder that tourists visit every year. To neglect the beauty of the sky is equivalent to geologists neglecting the beauty of the minerals they study. This supplement is meant to act as a tourist's guide to the sky. You analyzed the universe in the regular chapters, but here you will admire it.

The brighter stars in the sky are visible even from the centers of cities with their air and light pollution. But in the countryside, only a few miles beyond the cities, the night sky is a velvety blackness strewn with thousands of glittering stars. From a wilderness location, far from the city's glare, and especially from high mountains, the night sky is spectacular.

Using Star Charts

The constellations are a fascinating cultural heritage of our planet, but they are sometimes a bit difficult to learn because of Earth's motion. The constellations above the horizon change with the time of night and the seasons.

Because Earth rotates eastward, the sky appears to rotate westward around Earth. A constellation visible in the southern sky soon after sunset will appear to move westward, and in a few hours it will disappear below the horizon. Other constellations will rise in the east, so the sky changes gradually through the night.

In addition, Earth's orbital motion makes the sun appear to move eastward among the stars. Each day the sun moves about twice its own diameter, about one degree, eastward along the ecliptic, and consequently each night at sunset, the constellations are about one degree farther toward the west.

Orion, for instance, is visible in the evening sky in January; but, as the days pass, the sun moves closer to Orion. By March, Orion is difficult to see in the western sky soon after sunset. By June, the sun is so close to Orion it sets with the sun and is invisible. Not until late July is the sun far enough past Orion for the constellation to become visible rising in the eastern sky just before dawn.

Because of the rotation and orbital motion of Earth, you need more than one star chart to map the sky. Which chart you select depends on the month and the time of night. The charts given in this appendix show the evening sky for each month.

Two sets of charts are included for two typical locations on Earth. The northern hemisphere charts show the sky as seen from a northern latitude typical of the United States and central Europe. The southern hemisphere star charts are appropriate for readers in Earth's southern hemisphere, including Australia, southern South America, and southern Africa.

To use the charts, select the appropriate chart and hold it overhead as shown in ■ Figure B-1. If you face south, turn the chart until the words *southern horizon* are at the bottom of the chart. If you face other directions, turn the chart appropriately.

■ **Figure B-1**

To use the star charts in this book, select the appropriate chart for the date and time. Hold it overhead, and turn it until the direction at the bottom of the chart is the same as the direction you are facing.

Northern Hemisphere Sky

JANUARY

Early in Month	9 P.M.
Midmonth	8 P.M.
End of Month	7 P.M.

Months along the ecliptic show the location of the sun during the year.

Numbers along the celestial equator show right ascension.

Northern Hemisphere Sky

FEBRUARY

Early in Month	9 P.M.
Midmonth	8 P.M.
End of Month	7 P.M.

Months along the ecliptic show the location of the sun during the year.

Numbers along the celestial equator show right ascension.

Northern Hemisphere Sky

MARCH

Early in Month	9 P.M.
Midmonth	8 P.M.
End of Month	7 P.M.

Months along the ecliptic show the location of the sun during the year.

Numbers along the celestial equator show right ascension.

Northern Hemisphere Sky

APRIL

Early in Month	9 P.M.
Midmonth	8 P.M.
End of Month	7 P.M.

Months along the ecliptic show the location of the sun during the year.

Numbers along the celestial equator show right ascension.

Northern Hemisphere Sky

MAY

Early in Month	9 P.M.
Midmonth	8 P.M.
End of Month	7 P.M.

Months along the ecliptic show the location of the sun during the year.

Numbers along the celestial equator show right ascension.

Northern Hemisphere Sky

JUNE

Early in Month	9 P.M.
Midmonth	8 P.M.
End of Month	7 P.M.

Months along the ecliptic show the location of the sun during the year.

Numbers along the celestial equator show right ascension.

Northern Hemisphere Sky

JULY

Early in Month	9 P.M.
Midmonth	8 P.M.
End of Month	7 P.M.

Months along the ecliptic show the location of the sun during the year.

Numbers along the celestial equator show right ascension.

Northern Hemisphere Sky

AUGUST

Early in Month	9 P.M.
Midmonth	8 P.M.
End of Month	7 P.M.

Months along the ecliptic show the location of the sun during the year.

Numbers along the celestial equator show right ascension.

Northern Hemisphere Sky

SEPTEMBER

Early in Month	9 P.M.
Midmonth	8 P.M.
End of Month	7 P.M.

Months along the ecliptic show the location of the sun during the year.

Numbers along the celestial equator show right ascension.

Northern Hemisphere Sky

OCTOBER

Early in Month	9 P.M.
Midmonth	8 P.M.
End of Month	7 P.M.

Months along the ecliptic show the location of the sun during the year.

Numbers along the celestial equator show right ascension.

Northern Hemisphere Sky

NOVEMBER

Early in Month	9 P.M.
Midmonth	8 P.M.
End of Month	7 P.M.

Months along the ecliptic show the location of the sun during the year.

Numbers along the celestial equator show right ascension.

Northern Hemisphere Sky

DECEMBER

Early in Month	9 P.M.
Midmonth	8 P.M.
End of Month	7 P.M.

Months along the ecliptic show the location of the sun during the year.

Numbers along the celestial equator show right ascension.

Southern Hemisphere Sky

JANUARY

Early in Month	9 P.M.
Midmonth	8 P.M.
End of Month	7 P.M.

Months along the ecliptic show the location of the sun during the year.

Numbers along the celestial equator show right ascension.

Southern Hemisphere Sky

FEBRUARY

Early in Month	9 P.M.
Midmonth	8 P.M.
End of Month	7 P.M.

Months along the ecliptic show the location of the sun during the year.

Numbers along the celestial equator show right ascension.

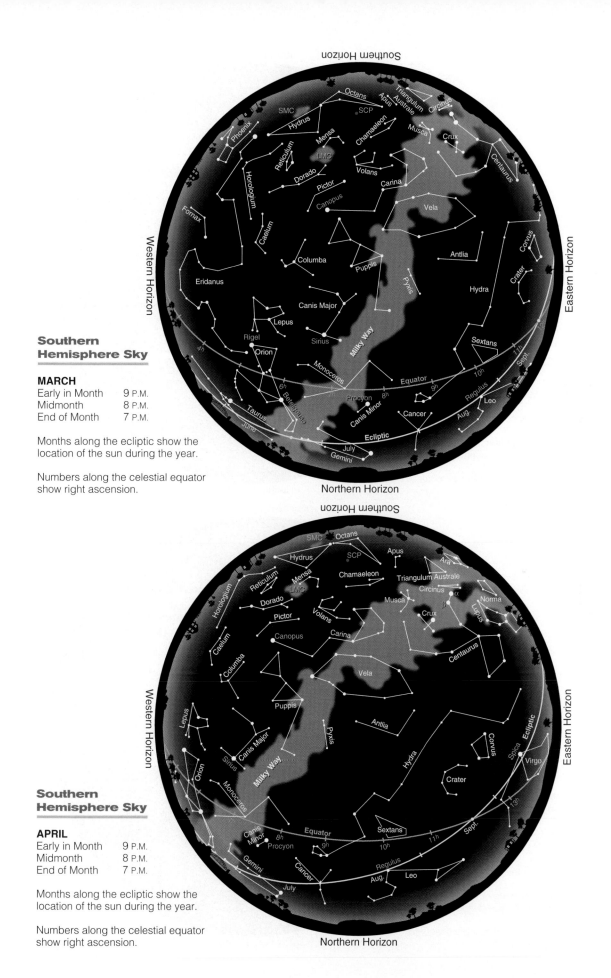

Southern Horizon

**Southern
Hemisphere Sky**

MARCH
Early in Month   9 P.M.
Midmonth   8 P.M.
End of Month   7 P.M.

Months along the ecliptic show the
location of the sun during the year.

Numbers along the celestial equator
show right ascension.

Western Horizon

Eastern Horizon

Northern Horizon

Southern Horizon

**Southern
Hemisphere Sky**

APRIL
Early in Month   9 P.M.
Midmonth   8 P.M.
End of Month   7 P.M.

Months along the ecliptic show the
location of the sun during the year.

Numbers along the celestial equator
show right ascension.

Western Horizon

Eastern Horizon

Northern Horizon

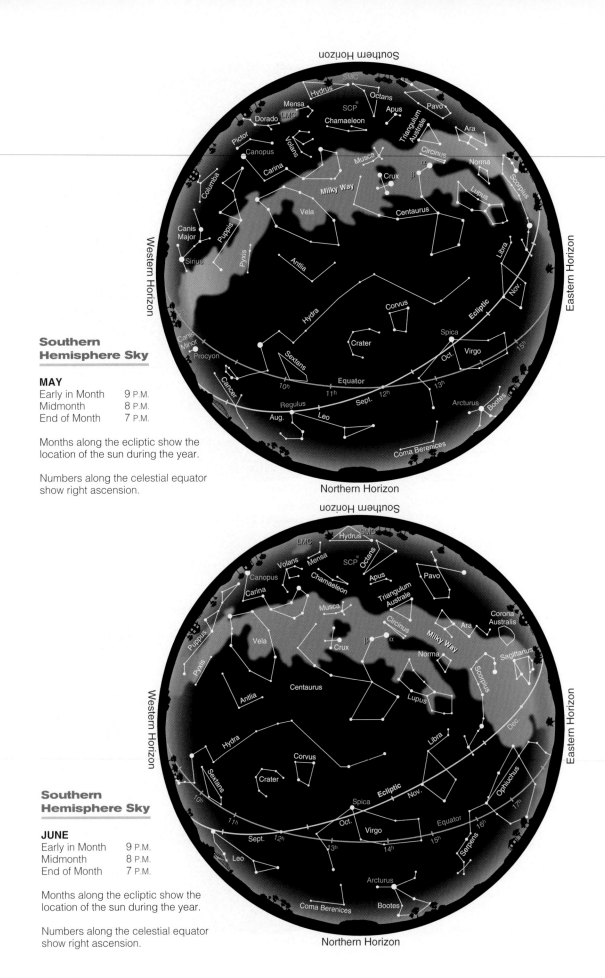

Southern Hemisphere Sky

MAY

Early in Month 9 P.M.
Midmonth 8 P.M.
End of Month 7 P.M.

Months along the ecliptic show the location of the sun during the year.

Numbers along the celestial equator show right ascension.

Southern Hemisphere Sky

JUNE

Early in Month 9 P.M.
Midmonth 8 P.M.
End of Month 7 P.M.

Months along the ecliptic show the location of the sun during the year.

Numbers along the celestial equator show right ascension.

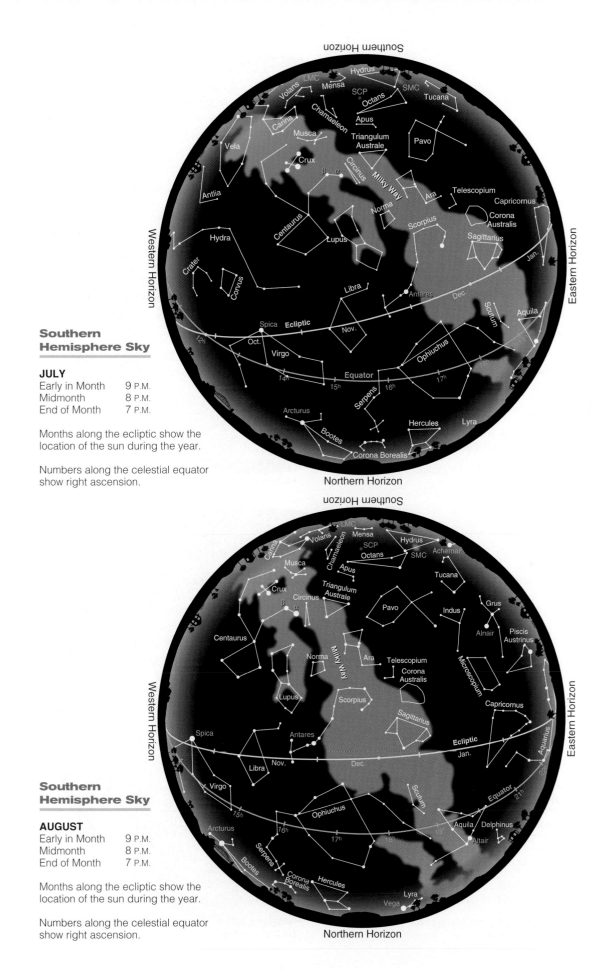

Southern Hemisphere Sky

JULY

Early in Month	9 P.M.
Midmonth	8 P.M.
End of Month	7 P.M.

Months along the ecliptic show the location of the sun during the year.

Numbers along the celestial equator show right ascension.

Southern Hemisphere Sky

AUGUST

Early in Month	9 P.M.
Midmonth	8 P.M.
End of Month	7 P.M.

Months along the ecliptic show the location of the sun during the year.

Numbers along the celestial equator show right ascension.

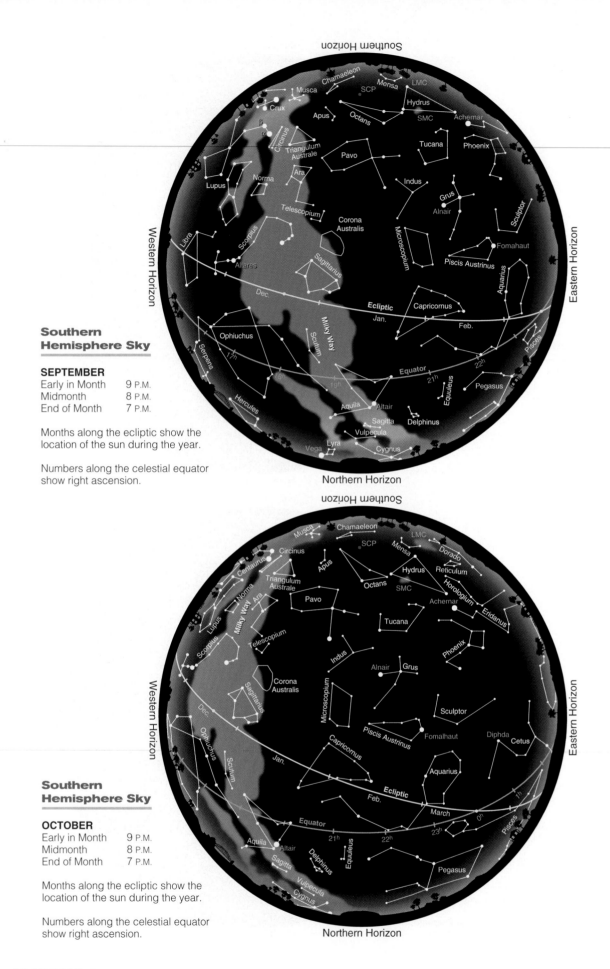

Southern Hemisphere Sky

SEPTEMBER

Early in Month	9 P.M.
Midmonth	8 P.M.
End of Month	7 P.M.

Months along the ecliptic show the location of the sun during the year.

Numbers along the celestial equator show right ascension.

Southern Hemisphere Sky

OCTOBER

Early in Month	9 P.M.
Midmonth	8 P.M.
End of Month	7 P.M.

Months along the ecliptic show the location of the sun during the year.

Numbers along the celestial equator show right ascension.

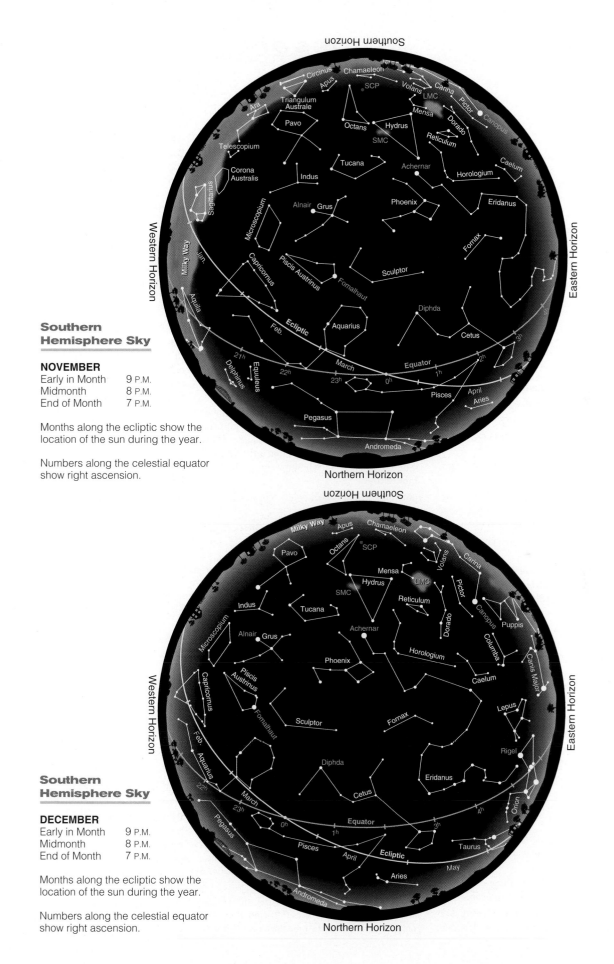

Southern Hemisphere Sky

NOVEMBER

Early in Month	9 P.M.
Midmonth	8 P.M.
End of Month	7 P.M.

Months along the ecliptic show the location of the sun during the year.

Numbers along the celestial equator show right ascension.

Southern Hemisphere Sky

DECEMBER

Early in Month	9 P.M.
Midmonth	8 P.M.
End of Month	7 P.M.

Months along the ecliptic show the location of the sun during the year.

Numbers along the celestial equator show right ascension.

Glossary

Numbers in parentheses refer to the page where the term is first discussed in the text.

absolute age (502) An age determined in years, as from radioactive dating (see also *relative age*).

absolute bolometric magnitude (193) The absolute magnitude you would observe if you could detect all wavelengths.

absolute visual magnitude (M_V) (192) Intrinsic brightness of a star; the apparent visual magnitude the star would have if it were 10 pc away.

absolute zero (138) The lowest possible temperature; the temperature at which the particles in a material, atoms or molecules, contain no energy of motion that can be extracted from the body.

absorption line (146) A dark line in a spectrum; produced by the absence of photons absorbed by atoms or molecules.

absorption spectrum (dark-line spectrum) (146) A spectrum that contains absorption lines.

acceleration (89) A change in a velocity; a change in either speed or direction. (See *velocity*.)

acceleration of gravity (87) A measure of the strength of gravity at a planet's surface.

accretion (461) The sticking together of solid particles to produce a larger particle.

accretion disk (294) The whirling disk of gas that forms around a compact object such as a white dwarf, neutron star, or black hole as matter is drawn in.

achondrite (624) Stony meteorite containing no chondrules or volatiles.

achromatic lens (114) A telescope lens composed of two lenses ground from different kinds of glass and designed to bring two selected colors to the same focus and correct for chromatic aberration.

active galactic nucleus (AGN) (396) The central energy source of an active galaxy.

active galaxy (396) A galaxy that is a source of excess radiation, usually radio waves, X rays, gamma rays, or some combination.

active optics (121) Optical elements whose position or shape is continuously controlled by computers.

active region (175) An area on the sun where sunspots, prominences, flares, and the like occur.

adaptive optics (121) Computer-controlled telescope mirrors that can at least partially compensate for seeing.

albedo (489) The fraction of the light hitting an object that is reflected.

alt-azimuth mounting (121) A telescope mounting capable of motion parallel to and perpendicular to the horizon.

amino acid (654) One of the carbon-chain molecules that are the building blocks of protein.

Angstrom (Å) (111) A unit of distance; 1 Å = 10^{-10} m; often used to measure the wavelength of light.

angular diameter (21) A measure of the size of an object in the sky; numerically equal to the angle in degrees between two lines extending from the observer's eye to opposite edges of the object.

angular distance (21) A measure of the separation between two objects in the sky; numerically equal to the angle in degrees between two lines extending from the observer's eye to the two objects.

angular momentum (93) The tendency of a rotating body to continue rotating; mathematically, the product of mass, velocity, and radius.

angular momentum problem (451) An objection to Laplace's nebular hypothesis that cited the slow rotation of the sun.

annular eclipse (48) A solar eclipse in which the solar photosphere appears around the edge of the moon in a bright ring, or annulus. The corona, chromosphere, and prominences cannot be seen.

anorthosite (506) Rock of aluminum and calcium silicates found in the lunar highlands.

antimatter (425) Matter composed of antiparticles, which upon colliding with a matching particle of normal matter annihilate and convert the mass of both particles into energy. The antiproton is the antiparticle of the proton, and the positron is the antiparticle of the electron.

aphelion (27) The orbital point of greatest distance from the sun.

apogee (48) The orbital point of greatest distance from Earth.

Apollo–Amor object (631) Asteroid whose orbit crosses that of Earth (Apollo) and Mars (Amor).

apparent visual magnitude (m_V) (16) The brightness of a star as seen by human eyes on Earth.

archaeoastronomy (58) The study of the astronomy of ancient cultures.

association (242) Group of widely scattered stars (10 to 1000) moving together through space; not gravitationally bound into a cluster.

asterism (14) A named group of stars not identified as a constellation, e.g., the Big Dipper.

asteroid (456) Small rocky world; most asteroids lie between Mars and Jupiter in the asteroid belt.

astronomical unit (AU) (6) Average distance from Earth to the sun; 1.5×10^8 km, or 93×10^6 miles.

atmospheric window (112) Wavelength regions in which Earth's atmosphere is transparent—at visual, infrared, and radio wavelengths.

aurora (181) The glowing light display that results when a planet's magnetic field guides charged particles toward the north and south magnetic poles, where they strike the upper atmosphere and excite atoms to emit photons.

autumnal equinox (26) The point on the celestial sphere where the sun crosses the celestial equator going southward. Also, the time when the sun reaches this point and autumn begins in the northern hemisphere—about September 22.

Babcock model (177) A model of the sun's magnetic cycle in which the differential rotation of the sun winds up and tangles the solar magnetic field in a 22-year cycle. This is thought to be responsible for the 11-year sunspot cycle.

Balmer series (147) Spectral lines in the visible and near-ultraviolet spectrum of hydrogen produced by transitions whose lowest orbit is the second.

barred spiral galaxy (374) A spiral galaxy with an elongated nucleus resembling a bar from which the arms originate.

basalt (486) Dark, igneous rock characteristic of solidified lava.

belt (561) One of the dark bands of clouds that circle Jupiter parallel to its equator; generally red, brown, or blue-green; believed to be regions of descending gas.

belt–zone circulation (585) The atmospheric circulation typical of Jovian planets. Dark belts and bright zones encircle the planet parallel to its equator.

big bang (422) The theory that the universe began with a violent explosion from which the expanding universe of galaxies eventually formed.

big rip (439) The possible fate of the universe if dark energy increases rapidly and the expansion of spacetime pulls galaxies, stars, and ultimately atoms apart.

binary star (200) One of a pair of stars that orbit around their common center of mass.

binding energy (143) The energy needed to pull an electron away from its atom.

bipolar flow (245) Oppositely directed jets of gas ejected by some protostellar objects.

birth line (241) In the H–R diagram, the line above the main sequence where protostars first become visible.

black body radiation (139) Radiation emitted by a hypothetical perfect radiator; the spectrum is continuous, and the wavelength of maximum emission depends only on the body's temperature.

black dwarf (289) The end state of a white dwarf that has cooled to low temperature.

black hole (323) A mass that has collapsed to such a small volume that its gravity prevents the escape of all radiation; also, the volume of space from which radiation may not escape.

blazar (403) See *BL Lac object*.

BL Lac object (403) Object that resembles a quasar; thought to be the highly luminous core of a distant galaxy emitting a jet almost directly toward Earth.

blueshift (153) The shortening of the wavelengths of light observed when the source and observer are approaching each other.

Bok globule (244) Small, dark cloud only about 1 ly in diameter that contains 10 to 1000 $M_\odot$ of gas and dust; believed related to star formation.

bow shock (484) The boundary between the undisturbed solar wind and the region being deflected around a planet or comet.

breccia (506) A rock composed of fragments of earlier rocks bonded together.

bright-line spectrum (146) See *emission spectrum*.

brown dwarf (263) A very cool, low-luminosity star whose mass is not sufficient to ignite nuclear fusion.

butterfly diagram (174) See *Maunder butterfly diagram*.

CAI (624) Calcium–aluminum-rich inclusions found in some meteorites; believed to be very old.

Cambrian period (657) A geological period 0.6 to 0.5 billion years ago during which life on Earth became diverse and complex. Cambrian rocks contain the oldest easily identifiable fossils.

capture hypothesis (510) The theory that the moon formed elsewhere in the solar system and was later captured by Earth.

carbonaceous chondrite (623) Stony meteorite that contains both chondrules and volatiles. These may be the least altered remains of the solar nebula still present in the solar system.

carbon deflagration (300) The process in which the carbon in a white dwarf is completely consumed by nuclear fusion, producing a type Ia supernova explosion.

carbon–nitrogen–oxygen (CNO) cycle (247) A series of nuclear reactions that use carbon as a catalyst to combine four hydrogen atoms to make one helium atom plus energy; effective in stars more massive than the sun.

Cassegrain focus (120) The optical design of a reflecting telescope in which the secondary mirror reflects light back down the tube through a hole in the center of the objective mirror.

catastrophic hypothesis (451) Explanation for natural processes that depends on dramatic and unlikely events, such as the collision of two stars to produce our solar system.

celestial equator (20) The imaginary line around the sky directly above Earth's equator.

celestial sphere (20) An imaginary sphere of very large radius surrounding Earth and to which the planets, stars, sun, and moon seem to be attached.

center of mass (95) The balance point of a body or system of bodies.

Cepheid variable star (275) Variable star with a period of 1 to 60 days; the period of variation is related to luminosity.

Chandrasekhar limit (289) The maximum mass of a white dwarf, about 1.4 $M_\odot$; a white dwarf of greater mass cannot support itself and will collapse.

charge-coupled device (CCD) (123) An electronic device consisting of a large array of light-sensitive elements used to record very faint images.

chemical evolution (658) The chemical process that led to the growth of complex molecules on the primitive Earth. This did not involve the reproduction of molecules.

chondrite (622) A stony meteorite that contains chondrules.

chondrule (622) Round, glassy body in some stony meteorites; believed to have solidified very quickly from molten drops of silicate material.

chromatic aberration (114) A distortion found in refracting telescopes because lenses focus different colors at slightly different distances. Images are consequently surrounded by color fringes.

chromosome (655) One of the bodies in a cell that contains the DNA carrying genetic information.

chromosphere (48) Bright gases just above the photosphere of the sun.

circular velocity (94) The velocity required to remain in a circular orbit about a body.

circumpolar constellation (21) Any of the constellations so close to the celestial pole that they never set (or never rise) as seen from a given latitude.

closed orbit (95) An orbit that returns to its starting point; a circular or elliptical orbit. (See *open orbit*.)

closed universe (433) A model universe in which the average density is great enough to stop the expansion and make the universe contract.

cluster method (379) The method of determining the masses of galaxies based on the motions of galaxies in a cluster.

CNO cycle (247) See *carbon–nitrogen–oxygen cycle*.

cocoon (240) The cloud of gas and dust around a contracting protostar that conceals it at visible wavelengths.

cold dark matter (435) Invisible matter in the universe composed by heavy, slow-moving particles such as WIMPs.

collapsar (333) See *hypernova*.

collisional broadening (155) The smearing out of a spectral line because of collisions among the atoms of the gas.

coma (636) The glowing head of a comet.

comet (456) One of the small, icy bodies that orbit the sun and produce tails of gas and dust when they near the sun.

compact object (289) A star that has collapsed to form a white dwarf, neutron star, or black hole.

comparative planetology (476) The study of planets by comparing the characteristics of different examples.

comparison spectrum (124) A spectrum of known spectral lines used to identify unknown wavelengths in an object's spectrum.

composite volcano (528) A volcano built up of layers of lava flows and ash falls. These are steep sided and typically associated with subduction zones.

condensation (461) The growth of a particle by addition of material from surrounding gas, one atom or molecule at a time.

condensation hypothesis (510) The hypothesis that the moon and Earth formed by condensing together from the solar nebula.

condensation sequence (461) The sequence in which different materials condense from the solar nebula at increasing distances from the sun.

conservation of energy law (266) One of the basic laws of stellar structure. The amount of energy flowing out of the top of a shell must equal the amount

coming in at the bottom plus whatever energy is generated within the shell.

conservation of mass law (266) One of the basic laws of stellar structure. The total mass of the star must equal the sum of the masses of the shells, and the mass must be distributed smoothly throughout the star.

constellation (14) One of the stellar patterns identified by name, usually of mythological gods, people, animals, or objects; also, the region of the sky containing that star pattern.

continuous spectrum (146) A spectrum in which there are no absorption or emission lines.

convection (164) Circulation in a fluid driven by heat; hot material rises, and cool material sinks.

convective zone (171) The region inside a star where energy is carried outward as rising hot gas and sinking cool gas.

corona (54, 530) The faint outer atmosphere of the sun; composed of low-density, very hot, ionized gas. On Venus, round networks of fractures and ridges up to 1000 km in diameter.

coronagraph (165) A telescope designed to photograph the inner corona of the sun.

coronal gas (228) Extremely high-temperature, low-density gas in the interstellar medium.

coronal hole (181) An area of the solar surface that is dark at X-ray wavelengths; thought to be associated with divergent magnetic fields and the source of the solar wind.

coronal mass ejection (CME) (181) Gas trapped in the sun's magnetic field.

cosmic microwave background radiation (425) Radiation from the hot clouds of the big bang explosion. Because of its large redshift, it appears to come from a body whose temperature is only 2.7 K.

cosmic ray (132) A subatomic particle traveling at tremendous velocity that strikes Earth's atmosphere from space.

cosmological constant (Λ) (438) Einstein's constant that represents a repulsion in space to oppose gravity.

cosmological principle (429) The assumption that any observer in any galaxy sees the same general features of the universe.

cosmology (416) The study of the nature, origin, and evolution of the universe.

Coulomb barrier (169) The electrostatic force of repulsion between bodies of like charge; commonly applied to atomic nuclei.

Coulomb force (142) The repulsive force between particles with like electrostatic charge.

critical density (430) The average density of the universe needed to make its curvature flat.

critical point (558) The temperature and pressure at which the vapor and liquid phases of a material have the same density.

dark age (426) The period of a few hundred million years during which the universe expanded in darkness. Extends from soon after the big bang glow faded into the infrared to the formation of the first stars.

dark energy (439) The energy of empty space that drives the acceleration of the expanding universe.

dark-line spectrum (146) See *absorption spectrum*.

dark matter (348) Nonluminous material that is detected only by its gravitational influence.

dark nebula (221) A nonluminous cloud of gas and dust visible because it blocks light from more distant stars and nebulae.

debris disk (468) A disk of dust found by infrared observations around some stars. The dust is debris from collisions among asteroids, comets, and Kuiper belt objects.

decameter radiation (557) Radio signals from Jupiter with wavelengths of about 10 m.

decimeter radiation (557) Radio signals from Jupiter with wavelengths of about 0.1 m.

deferent (65) In the Ptolemaic theory, the large circle around Earth along which the center of the epicycle moved.

degenerate matter (269) Extremely high-density matter in which pressure no longer depends on temperature, due to quantum mechanical effects.

density (156) The amount of matter per unit volume in a material; measured in grams per cubic centimeter, for example.

density wave theory (357) Theory proposed to account for spiral arms as compressions of the interstellar medium in the disk of the galaxy.

deuterium (170) An isotope of hydrogen in which the nucleus contains a proton and a neutron.

diamond ring effect (48) A momentary phenomenon seen during some total solar eclipses when the ring of the corona and a bright spot of photosphere resemble a large diamond set in a silvery ring.

differential rotation (176) The rotation of a body in which different parts of the body have different periods of rotation; this is true of the sun, the Jovian planets, and the disk of the galaxy.

differentiation (462) The separation of planetary material according to density.

diffraction fringe (115) Blurred fringe surrounding any image caused by the wave properties of light. Because of this, no image detail smaller than the fringe can be seen.

disk component (344) All material confined to the plane of the galaxy.

distance indicator (373) Object whose luminosity or diameter is known; used to find the distance to a star cluster or galaxy.

distance modulus $(m_V - M_V)$ (192) The difference between the apparent and absolute magnitude of a star; a measure of how far away the star is.

distance scale (377) The combined calibration of distance indicators used by astronomers to find the distances to remote galaxies.

DNA (deoxyribonucleic acid) (654) The long carbon-chain molecule that records information to govern the biological activity of the organism. DNA carries the genetic data passed to offspring.

Doppler broadening (155) The smearing of spectral lines because of the motion of the atoms in the gas.

Doppler effect (153) The change in the wavelength of radiation due to relative radial motion of source and observer.

double-exhaust model (397) The theory that double radio lobes are produced by pairs of jets emitted in opposite directions from the centers of active galaxies.

double-lobed radio source (397) A galaxy that emits radio energy from two regions (lobes) located on opposite sides of the galaxy.

Drake equation (669) A formula for the number of communicative civilizations in our galaxy.

dust (type II) tail (636) The tail of a comet formed of dust blown outward by the pressure of sunlight. (See *gas tail*.)

dynamo effect (177) The process by which a rotating, convecting body of conducting matter, such as Earth's core, can generate a magnetic field.

east point (20) The point on the eastern horizon exactly halfway between the north point and the south point; exactly east.

eccentric (62) In astronomy, an off-center circular path.

eccentricity, *e* (76) A measure of the flattening of an ellipse. An ellipse of $e = 0$ is circular. The closer to 1 *e* becomes, the more flattened the ellipse.

eclipse season (50) That period when the sun is near a node of the moon's orbit and eclipses are possible.

eclipse year (51) The time the sun takes to circle the sky and return to a node of the moon's orbit; 346.62 days.

eclipsing binary system (204) A binary star system in which the stars eclipse each other.

ecliptic (24) The apparent path of the sun around the sky.

ejecta (518) Pulverized rock scattered by meteorite impacts on a planetary surface.

electromagnetic radiation (110) Changing electric and magnetic fields that travel through space and transfer energy from one place to another—for example, light, radio waves, and the like.

electron (138) Low-mass atomic particle carrying a negative charge.

ellipse (76) A closed curve enclosing two points (foci) such that the total distance from one focus to any point on the curve back to the other focus equals a constant.

elliptical galaxy (371) A galaxy that is round or elliptical in outline; it contains little gas and dust, no disk or spiral arms, and few hot, bright stars.

emission line (146) A bright line in a spectrum caused by the emission of photons from atoms.

emission nebula (220) A cloud of glowing gas excited by ultraviolet radiation from hot stars.

emission spectrum (bright-line spectrum) (146) A spectrum containing emission lines.

energy (96) The capacity of a natural system to perform work—for example, thermal energy.

energy level (144) One of a number of states an electron may occupy in an atom, depending on its binding energy.

enzyme (654) Special protein that controls processes in an organism.

epicycle (65) The small circle followed by a planet in the Ptolemaic theory. The center of the epicycle follows a larger circle (deferent) around Earth.

equant (66) The point off-center in the deferent from which the center of the epicycle appears to move uniformly.

equatorial mounting (121) A telescope mounting that allows motion parallel to and perpendicular to the celestial equator.

ergosphere (324) The region surrounding a rotating black hole within which one could not resist being dragged around the black hole. It is possible for a particle to escape from the ergosphere and extract energy from the black hole.

escape velocity (95) The initial velocity an object needs to escape from the surface of a celestial body.

evening star (28) Any planet visible in the sky just after sunset.

event horizon (323) The boundary of the region of a black hole from which no radiation may escape. No event that occurs within the event horizon is visible to a distant observer.

evolutionary hypothesis (451) Explanation for natural events that involves gradual changes as opposed to sudden catastrophic changes—for example, the formation of the planets in the gas cloud around the forming sun.

excited atom (144) An atom in which an electron has moved from a lower to a higher orbit.

extrasolar planet (469) A planet orbiting a star other than the sun.

eyepiece (114) A short-focal-length lens used to enlarge the image in a telescope; the lens nearest the eye.

fall (621) A meteorite seen to fall. (See *find*.)

false-color image (123) A representation of graphical data in which the colors are altered or added to reveal details.

field (92) A way of explaining action at a distance; a particle produces a field of influence (gravitational, electric, or magnetic) to which another particle in the field responds.

filament (165) On the sun, a prominence seen silhouetted against the solar surface.

filtergram (165) An image (usually of the sun) taken in the light of a specific region of the spectrum—e.g., an H-alpha filtergram.

find (621) A meteorite that is found but was not seen to fall. (See *fall*.)

fission hypothesis (510) The theory that the moon formed by breaking away from Earth.

flare (181) A violent eruption on the sun's surface.

flatness problem (436) In cosmology, the circumstance that the early universe must have contained almost exactly the right amount of matter to close space-time (to make space-time flat).

flat universe (433) A model of the universe in which space-time is not curved.

flocculent (359) Woolly, fluffy; used to refer to certain galaxies that have a woolly appearance.

flux (191) A measure of the flow of energy onto or through a surface. Usually applied to light.

focal length (114) The distance from a lens to the point where it focuses parallel rays of light.

folded mountain range (486) A long range of mountains formed by the compression of a planet's crust—for example, the Andes on Earth.

forbidden line (218) A spectral line that does not occur in the laboratory because it depends on an atomic transition that is highly unlikely.

forward scattering (562) The optical property of finely divided particles to preferentially direct light in the original direction of the light's travel.

free-fall contraction (240) The early contraction of a gas cloud to form a star during which internal pressure is too low to resist contraction.

frequency (111) The number of times a given event occurs in a given time; for a wave, the number of cycles that pass the observer in 1 second.

galactic cannibalism (388) The theory that large galaxies absorb smaller galaxies.

galactic corona (348) The low-density extension of the halo of a galaxy; now suspected to extend many times the visible diameter of the galaxy.

galactic fountain (351) A region of the galaxy's disk in which gas heated by supernova explosions throws gas out of the disk where it can fall back and spread metals through the disk.

galaxy (8) A very large collection of gas, dust, and stars orbiting a common center of mass. The sun and Earth are located in the Milky Way Galaxy.

Galilean satellites (455) The four largest satellites of Jupiter, named after their discoverer, Galileo.

gamma-ray burster (332) An object that produces a sudden burst of gamma rays; thought to be associated with neutron stars and black holes.

gas (type I) tail (636) The tail of a comet produced by gas blown outward by the solar wind. (See *dust tail.*)

gene (655) A unit of DNA containing genetic information that influences a particular inherited trait.

general theory of relativity (102) Einstein's more sophisticated theory of space and time, which describes gravity as a curvature of space-time.

geocentric universe (61) A model universe with Earth at the center, such as the Ptolemaic universe.

geosynchronous satellite (94) An Earth satellite in an eastward orbit whose period is 24 hours. A satellite in such an orbit remains above the same spot on Earth's surface.

giant molecular cloud (227) Very large, cool cloud of dense gas in which stars form.

giants (196) Large, cool, highly luminous stars in the upper right of the H–R diagram; typically 10 to 100 times the diameter of the sun.

glitch (313) A sudden change in the period of a pulsar.

global warming (490) The gradual increase in the surface temperature of Earth caused by human modifications to Earth's atmosphere.

globular cluster (276) A star cluster containing 50,000 to 1 million stars in a sphere about 75 ly in diameter; generally old, metal-poor, and found in the spherical component of the galaxy.

gossamer ring (563) The dimmest part of Jupiter's ring produced by dust particles orbiting near small moons.

grand unified theory (GUT) (436) Theory that attempts to unify (describe in a similar way) the electromagnetic, weak, and strong forces of nature.

granulation (163) The fine structure visible on the solar surface caused by rising currents of hot gas and sinking currents of cool gas below the surface.

grating (124) A piece of material in which numerous microscopic parallel lines are scribed; light encountering a grating is dispersed to form a spectrum.

gravitational collapse (460) The stage in the formation of a massive planet when it grows massive enough to begin capturing gas directly from the nebula around it.

gravitational lensing (382) The effect of the focusing of light from a distant galaxy or quasar by an intervening galaxy to produce multiple images of the distant body.

gravitational radiation (318) As predicted by general relativity, expanding waves in a gravitational field that transport energy through space.

gravitational redshift (326) The lengthening of the wavelength of a photon due to its escape from a gravitational field.

greenhouse effect (490) The process by which a carbon dioxide atmosphere traps heat and raises the temperature of a planetary surface.

grooved terrain (566) Region of the surface of Ganymede consisting of parallel grooves; believed to have formed by repeated fracture of the icy crust.

ground state (145) The lowest permitted electron orbit in an atom.

half-life (457) The time required for half of the atoms in a radioactive sample to decay.

halo (341) The spherical region of a spiral galaxy containing a thin scattering of stars, star clusters, and small amounts of gas.

heat of formation (463) In planetology, the heat released by the infall of matter during the formation of a planetary body.

heavy bombardment (466) The period of intense meteorite impacts early in the formation of the planets when the solar system was filled with debris.

heliocentric universe (63) A model of the universe with the sun at the center, such as the Copernican universe.

helioseismology (167) The study of the interior of the sun by the analysis of its modes of vibration.

helium flash (270) The explosive ignition of helium burning that takes place in some giant stars.

Herbig–Haro object (245) Small nebula that varies irregularly in brightness; believed to be associated with star formation.

Hertzsprung–Russell diagram (194) A plot of the intrinsic brightness versus the surface temperature of stars; it separates the effects of temperature and surface area on stellar luminosity; commonly absolute magnitude versus spectral type, but also luminosity versus surface temperature or color.

HI cloud (233) An interstellar cloud of neutral hydrogen.

HII region (220) A region of ionized hydrogen around a hot star.

high-velocity star (347) A star with a large space velocity. Such stars are halo stars passing through the disk of the galaxy at steep angles.

Hirayama family (627) Family of asteroids with orbits of similar size, shape, and orientation; believed to be fragments of larger bodies.

homogeneity (429) The assumption that, on the large scale, matter is uniformly spread through the universe.

horizon (20) The line that marks the apparent intersection of Earth and the sky.

horizon problem (436) In cosmology, the circumstance that the primordial background radiation seems much more isotropic than could be explained by the standard big bang theory.

horizontal branch (277) In the H–R diagram of a globular cluster, the sequence of stars extending from the red giants toward the blue side of the diagram; includes RR Lyrae stars.

horoscope (28) A chart showing the positions of the sun, moon, planets, and constellations at the time of a person's birth; used in astrology to attempt to read character or foretell the future.

hot dark matter (434) Invisible matter in the universe composed of low-mass, high-velocity particles such as neutrinos.

hot spot (398) In radio astronomy, a bright spot in a radio lobe.

H–R diagram (194) (See *Hertzsprung–Russell diagram.*)

Hubble constant (*H*) (378) A measure of the rate of expansion of the universe; the average value of velocity of recession divided by distance; about 70 km/s/ megaparsec.

Hubble law (378) The linear relation between the distance to a galaxy and its apparent radial velocity.

Hubble time (422) An upper limit on the age of the universe derived from the Hubble constant.

hydrostatic equilibrium (249) The balance between the weight of the material pressing downward on a layer in a star and the pressure in that layer.

hypernova (333) The explosion produced as a very massive star collapses into a black hole; thought to be responsible for at least some gamma-ray bursts.

hypothesis (77) A conjecture, subject to further tests, that accounts for a set of facts.

inflationary universe (436) A version of the big bang theory that includes a rapid expansion when the universe was very young.

infrared cirrus (227) A fine network of filaments covering the sky detected in the far infrared by the IRAS satellite; believed to be associated with dust in the interstellar medium.

infrared radiation (111) Electromagnetic radiation with wavelengths intermediate between visible light and radio waves.

instability strip (278) The region of the H–R diagram in which stars are unstable to pulsation; a star passing through this strip becomes a variable star.

intercloud medium (223) The hot, low-density gas between cooler clouds in the interstellar medium.

intercrater plain (515) The relatively smooth terrain on Mercury.

interferometry (122) The observing technique in which separated telescopes combine to produce a virtual telescope with the resolution of a much-larger-diameter telescope.

interstellar absorption lines (222) Dark lines in some stellar spectra that are formed by interstellar gas.

interstellar dust (219) Microscopic solid grains in the interstellar medium.

interstellar extinction (219) The dimming of starlight by gas and dust in the interstellar medium.

interstellar medium (216) The gas and dust distributed between the stars.

interstellar reddening (219) The process in which dust scatters blue light out of starlight and makes the stars look redder.

intrinsic variable (275) A variable star driven to pulsate by processes in its interior.

inverse square law (90) The rule that the strength of an effect (such as gravity) decreases in proportion as the distance squared increases.

Io flux tube (558) A tube of magnetic lines and electric currents connecting Io and Jupiter.

ion (142) An atom that has lost or gained one or more electrons.

ionization (142) The process in which atoms lose or gain electrons.

Io plasma torus (558) The doughnut-shaped cloud of ionized gas that encloses the orbit of Jupiter's moon Io.

iron meteorite (622) A meteorite composed mainly of iron–nickel alloy.

irregular galaxy (375) A galaxy with a chaotic appearance, large clouds of gas and dust, and both population I and population II stars, but without spiral arms.

island universe (371) An older term for a galaxy.

isotopes (142) Atoms that have the same number of protons but a different number of neutrons.

isotropy (429) The assumption that in its general properties the universe looks the same in every direction.

joule (J) (96) A unit of energy equivalent to a force of 1 newton acting over a distance of 1 meter; 1 joule per second equals 1 watt of power.

Jovian planet (454) Jupiterlike planet with large diameter and low density.

jumbled terrain (509) Strangely disturbed regions of the moon opposite the locations of the Imbrium Basin and Mare Orientale.

Kelvin temperature scale (138) The temperature, in Celsius (centigrade) degrees, measured above absolute zero.

Keplerian motion (348) Orbital motion in accord with Kepler's laws of planetary motion.

Kerr black hole (324) A solution to the equations of general relativity that describes the properties of a rotating black hole.

kiloparsec (kpc) (341) A unit of distance equal to 1000 pc, or 3260 ly.

kinetic energy (96) Energy of motion; depends on mass and velocity of a moving body.

Kirchhoff's laws (146) A set of laws that describe the absorption and emission of light by matter.

Kirkwood's gap (630) Region in the asteroid belt in which there are very few asteroids; caused by resonances with Jupiter.

Kuiper belt (456) The collection of icy planetesimals believed to orbit in a region from just beyond Neptune out to about 50 AU.

Lagrangian point (293) Point of stability in the orbital plane of a binary star system, planet, or moon. One is located 60° ahead and one 60° behind the orbiting bodies; another is located between the orbiting bodies.

large-impact hypothesis (511) The theory that the moon formed from debris ejected during a collision between Earth and a large planetesimal.

large-scale structure (440) The distribution of galaxy clusters and superclusters in walls and filaments surrounding voids mostly empty of galaxies.

L dwarf (150) A type of star that is even cooler than the M stars.

life zone (665) A region around a star within which a planet can have temperatures that permit the existence of liquid water.

light curve (204) A graph of brightness versus time commonly used in analyzing variable stars and eclipsing binaries.

light-gathering power (115) The ability of a telescope to collect light; proportional to the area of the telescope objective lens or mirror.

lighthouse model (312) The explanation of a pulsar as a spinning neutron star sweeping beams of radio radiation around the sky.

light pollution (117) The illumination of the night sky by waste light from cities and outdoor lighting, which prevents the observation of faint objects.

light-year (ly) (7) The distance light travels in 1 year.

limb (164, 498) The edge of the apparent disk of a body, as in "the limb of the moon."

limb darkening (164) The decrease in brightness of the sun or other body from its center to its limb.

line of nodes (50) The line across an orbit connecting the nodes; commonly applied to the orbit of the moon.

liquid metallic hydrogen (557) A form of hydrogen under high pressure that is a good electrical conductor.

lobate scarp (514) A curved cliff such as those found on Mercury.

local bubble or void (229) A region of high-temperature, low-density gas in the interstellar medium in which the sun happens to be located.

look-back time (377) The amount by which you look into the past when you look at a distant galaxy; a time equal to the distance to the galaxy in light-years.

luminosity (L) (193) The total amount of energy a star radiates in 1 second.

luminosity class (198) A category of stars of similar luminosity; determined by the widths of lines in their spectra.

lunar eclipse (42) The darkening of the moon when it moves through Earth's shadow.

Lyman series (147) Spectral lines in the ultraviolet spectrum of hydrogen produced by transitions whose lowest orbit is the ground state.

MACHO (435) Massive Compact Halo Object; a low-luminosity object such as a planet or a brown dwarf that contributes to the mass of the halo.

Magellanic Clouds (340) Small, irregular galaxies that are companions to the Milky Way; visible in the southern sky.

magnetar (317) A class of neutron stars that have exceedingly strong magnetic fields; thought to be responsible for soft gamma-ray repeaters.

magnetic carpet (166) The widely distributed, low-level magnetic field extending up through the sun's visible surface.

magnetosphere (484) The volume of space around a planet within which the motion of charged particles is dominated by the planetary magnetic field rather than the solar wind.

magnifying power (116) The ability of a telescope to make an image larger.

magnitude–distance formula (192) The mathematical formula that relates the apparent magnitude and absolute magnitude of a star to its distance.

magnitude scale (16) The astronomical brightness scale; the larger the number, the fainter the star.

main sequence (196) The region of the H–R diagram running from upper left to lower right, which includes roughly 90 percent of all stars.

mantle (483) The layer of dense rock and metal oxides that lies between the molten core and Earth's surface; also, similar layers in other planets.

mare (498) One of the lunar lowlands filled by successive flows of dark lava; from the Latin word for *sea*.

mass (90) A measure of the amount of matter making up an object.

mass–luminosity relation (207) The more massive a star is, the more luminous it is.

Maunder butterfly diagram (174) A graph showing the latitude of sunspots versus time; first plotted by W. W. Maunder in 1904.

Maunder minimum (175) A period of less numerous sunspots and other solar activity from 1645 to 1715.

megaparsec (Mpc) (373) A unit of distance equal to 1 million pc.

metals (349) In astronomical usage, all atoms heavier than helium.

metastable level (218) An atomic energy level from which an electron takes a long time to decay; responsible for producing forbidden lines.

meteor (457) A small bit of matter heated by friction to incandescent vapor as it falls into Earth's atmosphere.

meteorite (457) A meteor that has survived its passage through the atmosphere and strikes the ground.

meteoroid (457) A meteor in space before it enters Earth's atmosphere.

micrometeorite (501) Meteorite of microscopic size.

midocean rise (486) One of the undersea mountain ranges that push up from the seafloor in the center of the oceans.

Milankovitch hypothesis (30) The hypothesis that small changes in Earth's orbital and rotational motions cause the ice ages.

Milky Way (9) The hazy band of light that circles the sky, produced by the combined light of billions of stars in our Milky Way Galaxy.

Milky Way Galaxy (9) The spiral galaxy containing the sun; visible at night as the Milky Way.

Miller experiment (657) An experiment that reproduced the conditions under which life began on Earth and manufactured amino acids and other organic compounds.

millisecond pulsar (319) A pulsar with a period of approximately a millisecond, a thousandth of a second.

minute of arc (21) An angular measure; each degree is divided into 60 minutes of arc.

molecular cloud (227) An interstellar gas cloud that is dense enough for the formation of molecules; discovered and studied through the radio emissions of such molecules.

molecule (142) Two or more atoms bonded together.

momentum (89) The tendency of a moving object to continue moving; mathematically, the product of mass and velocity.

morning star (28) Any planet visible in the sky just before sunrise.

multiringed basin (501) Very large impact basin in which there are concentric rings of mountains.

mutant (656) Offspring born with altered DNA.

nadir (20) The point on the bottom of the sky directly under your feet.

nanometer (nm) (111) A unit of length equal to 10^{-9} m.

natural law (77) A conjecture about how nature works in which scientists have overwhelming confidence.

natural motion (87) In Aristotelian physics, the motion of objects toward their natural places—fire and air upward and earth and water downward.

natural selection (653) The process by which the best traits are passed on, allowing the most able to survive.

neap tide (98) Ocean tide of low amplitude occurring at first- and third-quarter moon.

nebula (218) A cloud of gas and dust in space.

nebular hypothesis (451) The proposal that the solar system formed from a rotating cloud of gas.

neutrino (170) A neutral, massless atomic particle that travels at or nearly at the speed of light.

neutron (141) An atomic particle with no charge and about the same mass as a proton.

neutron star (310) A small, highly dense star composed almost entirely of tightly packed neutrons; radius about 10 km.

Newtonian focus (120) The focal arrangement of a reflecting telescope in which a diagonal mirror reflects light out the side of the telescope tube for easier access.

node (50) A point where an object's orbit passes through the plane of Earth's orbit.

nonbaryonic matter (434) In cosmology, a suspected component of the dark matter composed of matter that does not contain protons and neutrons.

north celestial pole (20) The point on the celestial sphere directly above Earth's North Pole.

north point (20) The point on the horizon directly below the north celestial pole; exactly north.

nova (284) From the Latin "new," a sudden brightening of a star, making it appear as a "new" star in the sky; believed associated with eruptions on white dwarfs in binary systems.

nuclear bulge (341) The spherical cloud of stars that lies at the center of spiral galaxies.

nuclear fission (169) Reaction that splits nuclei into less massive fragments.

nuclear fusion (169) Reaction that joins the nuclei of atoms to form more massive nuclei.

nucleosynthesis (350) The production of elements heavier than helium by the fusion of atomic nuclei in stars and during supernovae explosions.

nucleus (of an atom) (141) The central core of an atom containing protons and neutrons; carries a net positive charge.

O association (242) A large, loosely bound cluster of very young stars.

objective lens or mirror (114) The main optical element in an astronomical telescope. The large lens at the top of the telescope or large mirror at the bottom.

oblateness (556) The flattening of a spherical body; usually caused by rotation.

observable universe (420) The part of the universe that is visible from Earth's location in space and time.

occultation (597) The passage of a larger body in front of a smaller body.

Olbers's paradox (419) The conflict between observation and theory as to why the night sky should or should not be dark.

Oort cloud (639) The cloud of icy bodies—extending from the outer part of our solar system out to roughly 100,000 AU from the sun—that acts as the source of most comets.

opacity (248) The resistance of a gas to the passage of radiation.

open cluster (276) A cluster of 10 to 10,000 stars with an open, transparent appearance and stars not tightly grouped; usually relatively young and located in the disk of the galaxy.

open orbit (95) An orbit that does not return to its starting point; an escape orbit. (See *closed orbit.*)

open universe (433) A model universe in which the average density is less than the critical density needed to halt the expansion.

oscillating universe theory (430) The theory that the universe begins with a big bang, expands, is slowed by its own gravity, and then falls back to create another big bang.

outflow channel (540) Geological feature on Mars that appears to have been caused by sudden flooding.

outgassing (463) The release of gases from a planet's interior.

ovoid (600) Geological feature on Uranus's moon Miranda thought to be produced by circulation in the solid icy mantle and crust.

ozone layer (489) In Earth's atmosphere, a layer of oxygen ions (O_3) lying 15 to 30 km high that protects the surface by absorbing ultraviolet radiation.

paradigm (68) A commonly accepted set of scientific ideas and assumptions.

parallax (64) The apparent change in the position of an object due to a change in the location of the observer. Astronomical parallax is measured in seconds of arc.

parsec (pc) (189) The distance to a hypothetical star whose parallax is one second of arc; 1 pc = 206,265 AU = 3.26 ly.

partial eclipse (44, 45) A lunar eclipse in which the moon does not completely enter Earth's shadow; a solar eclipse in which the moon does not completely cover the sun.

Paschen series (147) Spectral lines in the infrared spectrum of hydrogen produced by transitions whose lowest orbit is the third.

passing star hypothesis (450) The proposal that our solar system formed when two stars passed near each other and material was pulled out of one to form the planets.

path of totality (46) The track of the moon's umbral shadow over Earth's surface. The sun is totally eclipsed as seen from within this path.

penumbra (42) The portion of a shadow that is only partially shaded.

penumbral eclipse (54) A lunar eclipse in which the moon enters the penumbra of Earth's shadow but does not reach the umbra.

perigee (48) The orbital point of closest approach to Earth.

perihelion (27) The orbital point of closest approach to the sun.

period–luminosity relation (275) The relation between period of pulsation and intrinsic brightness among Cepheid variable stars.

permitted orbit (143) One of the energy levels in an atom that an electron may occupy.

photon (111) A quantum of electromagnetic energy; carries an amount of energy that depends inversely on its wavelength.

photosphere (47) The bright visible surface of the sun.

planet (6) A nonluminous object, larger than a comet or asteroid, that orbits a star.

planetary nebula (288) An expanding shell of gas ejected from a star during the latter stages of its evolution.

planetesimal (461) One of the small bodies that formed from the solar nebula and eventually grew into protoplanets.

plastic (483) A material with the properties of a solid but capable of flowing under pressure.

plate tectonics (486) The constant destruction and renewal of Earth's surface by the motion of sections of crust.

plutino (612) One of the icy Kuiper belt objects that, like Pluto, are caught in a 3:2 orbital resonance with Neptune.

polar axis (121) The axis around which a celestial body rotates.

poor cluster (384) An irregularly shaped cluster that contains fewer than 1000 galaxies, many spiral, and no giant ellipticals.

population I star (349) Star rich in atoms heavier than helium; nearly always a relatively young star found in the disk of the galaxy.

population II star (349) Star poor in atoms heavier than helium; nearly always a relatively old star found in the halo, globular clusters, or the nuclear bulge.

positron (170) The anti-particle of the electron.

potential energy (96) The energy a body has by virtue of its position. A weight on a high shelf has more potential energy than a weight on a low shelf.

precession (22) The slow change in the direction of Earth's axis of rotation; one cycle takes nearly 26,000 years.

pressure (224) A force exerted over a surface. Expressed as force per unit area.

pressure *(P)* wave (482) In geophysics, a mechanical wave of compression and rarefaction that travels through Earth's interior.

primary lens or mirror (114) The main optical element in an astronomical telescope. The large lens at the top of the telescope tube or the large mirror at the bottom.

prime focus (120) The point at which the objective mirror forms an image in a reflecting telescope.

primeval atmosphere (488) Earth's first air, composed of gases from the solar nebula.

primordial soup (657) The rich solution of organic molecules in Earth's first oceans.

prominence (48, 180) Eruption on the solar surface; visible during total solar eclipses.

proper motion (190) The rate at which a star moves across the sky; measured in seconds of arc per year.

protein (654) Complex molecule composed of amino acid units.

proton (141) A positively charged atomic particle contained in the nucleus of atoms; the nucleus of a hydrogen atom.

proton–proton chain (170) A series of three nuclear reactions that build a helium atom by adding together protons; the main energy source in the sun.

protoplanet (462) Massive object resulting from the coalescence of planetesimals in the solar nebula and destined to become a planet.

protostar (240) A collapsing cloud of gas and dust destined to become a star.

protostellar disk (240) A gas cloud around a forming star flattened by its rotation.

pulsar (335) A source of short, precisely timed radio bursts; believed to be a spinning neutron star.

pulsar wind (313) The flow of high-energy particles that carries most of the energy away from a spinning neutron star.

quantum mechanics (143) The study of the behavior of atoms and atomic particles.

quasar (quasi-stellar object, or QSO) (405) Small, powerful source of energy believed to be the active core of a very distant galaxy.

quasi-periodic oscillation (QPO) (329) A high-speed flickering in the radiation from an accretion disk evidently caused by material spiraling inward.

quintessence (439) The proposed energy of empty space that causes the acceleration of the expanding universe.

radial velocity (V_r) (153) That component of an object's velocity directed away from or toward Earth.

radiant (619) The point in the sky from which meteors in a shower seem to come.

radiation pressure (466) The force exerted on the surface of a body by its absorption of light. Small particles floating in the solar system can be blown outward by the pressure of the sunlight.

radiative zone (170) The region inside a star where energy is carried outward as photons.

radio galaxy (394) A galaxy that is a strong source of radio signals.

radio interferometer (127) Two or more radio telescopes that combine their signals to achieve the resolving power of a larger telescope.

ray (500) Ejecta from a meteorite impact, forming white streamers radiating from some lunar craters.

recombination (426) The stage within a million years of the big bang when the gas became transparent to radiation.

reconnection (181) The process in the sun's atmosphere by which opposing magnetic fields combine and release energy to power solar flares.

red dwarf (197) Cool, low-mass star on the lower main sequence.

redshift (153) The lengthening of the wavelengths of light seen when the source and observer are receding from each other.

reflecting telescope (114) A telescope that uses a concave mirror to focus light into an image.

reflection nebula (220) A nebula produced by starlight reflecting off dust particles in the interstellar medium.

refracting telescope (114) A telescope that forms images by bending (refracting) light with a lens.

regolith (506) A soil made up of crushed rock fragments.

reionization (426) The stage in the early history of the universe when ultraviolet photons from the first stars ionized the gas filling space.

relative age (502) The age of a geological feature referred to other features. For example, relative ages reveal that the lunar maria are younger than the highlands.

relativistic Doppler formula (407) Einstein's description of the redshift or blueshift observed when an object moves through space at a very high velocity.

resolving power (115) The ability of a telescope to reveal fine detail; depends on the diameter of the telescope objective.

resonance (512) The coincidental agreement between two periodic phenomena; commonly applied to agreements between orbital periods, which can make orbits more or less stable.

retrograde motion (64) The apparent backward (westward) motion of planets as seen against the background of stars.

revolution (23) The motion of an object in a closed path about a point outside its volume; Earth revolves around the sun.

rich cluster (384) A cluster containing over 1000 galaxies, mostly elliptical, scattered over a volume about 3 Mpc in diameter.

rift valley (486) A long, straight, deep valley produced by the separation of crustal plates.

ring galaxy (389) A galaxy that resembles a ring around a bright nucleus; believed to be the result of a head-on collision of two galaxies.

RNA (ribonucleic acid) (655) A long carbon-chain molecule that uses the information stored in DNA to manufacture complex molecules necessary to the organism.

Roche limit (562) The minimum distance between a planet and a satellite that holds itself together by its own gravity. If a satellite's orbit brings it within its planet's Roche limit, tidal forces will pull the satellite apart.

Roche lobe (293) In a system with two bodies orbiting each other, the volume of space dominated by the gravitation of one of the bodies.

Roche surface (293) In a system with two bodies orbiting each other, the outer boundary of the volume of space dominated by the gravitation of one of the bodies.

rotation (23) The turning of a body about an axis that passes through its volume; Earth rotates on its axis.

rotation curve (348, 379) A graph of orbital velocity versus radius in the disk of a galaxy.

rotation curve method (379) The procedure for finding the mass of a galaxy from its rotation curve.

RR Lyrae variable star (275) Variable star with a period of 12 to 24 hours; common in some globular clusters.

Sagittarius A* (361) The powerful radio source located at the core of the Milky Way Galaxy.

saros cycle (52) An 18-year $11\frac{1}{3}$-day period after which the pattern of lunar and solar eclipses repeats.

Schmidt-Cassegrain focus (120) The optical design of a reflecting telescope in which a thin correcting lens is placed at the top of a Cassegrain telescope.

Schwarzschild radius (R_s) (323) The radius of the event horizon around a black hole.

scientific model (19) An intellectual concept designed to help you think about a natural process without necessarily being a conjecture of truth.

scientific notation (6) The system of recording very large or very small numbers by using powers of 10.

secondary atmosphere (488) The gases outgassed from a planet's interior; rich in carbon dioxide.

secondary crater (500) A crater formed by the impact of debris ejected from a larger crater.

secondary mirror (120) In a reflecting telescope, the mirror that reflects the light to a point of easy observation.

second of arc (21) An angular measure; each minute of arc is divided into 60 seconds of arc.

seeing (115) Atmospheric conditions on a given night. When the atmosphere is unsteady, producing blurred images, the seeing is said to be poor.

seismic wave (482) A mechanical vibration that travels through Earth; usually caused by an earthquake.

seismograph (482) An instrument that records seismic waves.

selection effect (622) An influence on the probability that certain phenomena will be detected or selected, which can alter the outcome of a survey.

self-sustaining star formation (359) The process by which the birth of stars compresses the surrounding gas clouds and triggers the formation of more stars; proposed to explain spiral arms.

semimajor axis, _a_ (76) Half of the longest axis of an ellipse.

SETI (668) Search for Extra-Terrestrial Intelligence.

Seyfert galaxy (396) An otherwise normal spiral galaxy with an unusually bright, small core that fluctuates in brightness; believed to indicate the core is erupting.

Shapley–Curtis Debate (371) The 1920 debate between Harlow Shapley and Heber Curtis over the nature of the spiral nebulae.

shear (_S_) wave (482) A mechanical wave that travels through Earth's interior by the vibration of particles perpendicular to the direction of wave travel.

shepherd satellite (577) A satellite that, by its gravitational field, confines particles to a planetary ring.

shield volcano (528) Wide, low-profile volcanic cone produced by highly liquid lava.

shock wave (239) A sudden change in pressure that travels as an intense sound wave.

sidereal drive (121) The motor and gears on a telescope that turn it westward to keep it pointed at a star.

sidereal period (41) The period of rotation or revolution of an astronomical body relative to the stars.

singularity (323) The object of zero radius into which the matter in a black hole is believed to fall.

sinuous rille (499) A narrow, winding valley on the moon caused by ancient lava flows along narrow channels.

small-angle formula (54) The mathematical formula that relates an object's linear diameter and distance to its angular diameter.

smooth plain (515) Apparently young plain on Mercury formed by lava flows at or soon after the formation of the Caloris Basin.

soft gamma-ray repeater (SGR) (332) An object that produces repeated bursts of lower-energy gamma rays; thought to be produced by magnetars.

solar constant (182) A measure of the energy output of the sun; the total solar energy striking 1 m² just above Earth's atmosphere in 1 second.

solar eclipse (45) The event that occurs when the moon passes directly between Earth and the sun, blocking your view of the sun.

solar nebula hypothesis (452) The theory that the planets formed from the same cloud of gas and dust that formed the sun.

solar system (6) The sun and the nonluminous objects that orbit it, including the planets, comets, and asteroids.

solar wind (166) Rapidly moving atoms and ions that escape from the solar corona and blow outward through the solar system.

south celestial pole (20) The point of the celestial sphere directly above Earth's South Pole.

south point (20) The point on the horizon directly above the south celestial pole; exactly south.

special relativity (101) The first of Einstein's theories of relativity, which dealt with uniform motion.

spectral class or type (149) A star's position in the temperature classification system O, B, A F, G, K, and M. Based on the appearance of the star's spectrum.

spectral sequence (149) The arrangement of spectral classes (O, B, A, F, G, K, M) ranging from hot to cool.

spectrograph (123) A device that separates light by wavelength to produce a spectrum.

spectroscopic binary system (202) A star system in which the stars are too close together to be visible separately. You see a single point of light, and only by taking a spectrum can you determine that there are two stars.

spectroscopic parallax (199) The method of determining a star's distance by comparing its apparent magnitude with its absolute magnitude, as estimated from its spectrum.

spherical component (344) The part of the galaxy including all matter in a spherical distribution around the center (the halo and nuclear bulge).

spicule (165) Small, flamelike projection in the chromosphere of the sun.

spiral arm (9) Long, spiral pattern of bright stars, star clusters, gas, and dust that extends from the center to the edge of the disk of spiral galaxies.

spiral galaxy (374) A galaxy with an obvious disk component containing gas; dust; hot, bright stars; and spiral arms.

spiral nebula (371) Nebulous object with a spiral appearance observed in early telescopes; later recognized as a spiral galaxy.

spiral tracer (355) Object used to map the spiral arms (e.g., O and B associations, open clusters, clouds of ionized hydrogen, and some types of variable stars).

sporadic meteor (620) A meteor not part of a meteor shower.

spring tide (98) Ocean tide of high amplitude that occurs at full and new moon.

standard candle (373) Object of known brightness which astronomers use to find distance—for example, Cepheid variable stars and supernovae.

star (6) A celestial object composed of gas held together by its own gravity and supported by nuclear fusion occurring in its interior.

starburst galaxy (386) A bright blue galaxy in which many new stars are forming, believed to be caused by collisions between galaxies.

star-formation pillar (243) The column of gas produced when a dense core of gas protects the nebula behind it from the energy of a nearby hot star that is evaporating and driving away a star-forming nebula.

steady state theory (425) The theory (now generally abandoned) that the universe does not evolve.

stellar model (261) A table of numbers representing the conditions in various layers within a star.

stellar parallax (_p_) (189) A measure of stellar distance. (See parallax.)

stony-iron meteorite (624) A meteorite that is a mixture of stone and iron.

stony meteorite (622) A meteorite composed of silicate (rocky) material.

stromatolite (659) A layered fossil formation caused by ancient mats of algae or bacteria that build up mineral deposits season after season.

strong force (169) One of the four forces of nature; the strong force binds protons and neutrons together in atomic nuclei.

subduction zone (486) A region of a planetary crust where a tectonic plate slides downward.

subsolar point (524) The point on a planet that is directly below the sun.

summer solstice (26) The point on the celestial sphere where the sun is at its most northerly point; also, the time when the sun passes this point, about June 22, and summer begins in the northern hemisphere.

sunspot (162) Relatively dark spot on the sun that contains intense magnetic fields.

supercluster (440) A cluster of galaxy clusters.

supergiant (197) Exceptionally luminous star 10 to 1000 times the sun's diameter.

supergranule (164) A large granule on the sun's surface including many smaller granules.

superluminal expansion (410) The apparent expansion of parts of a quasar at speeds greater than the speed of light.

supernova (284) A new star appearing in Earth's sky and lasting for a year or so before fading. Caused by the violent explosion of a star.

supernova remnant (301) The expanding shell of gas marking the site of a supernova explosion.

supernova (type I) (299) The explosion of a star, believed caused by the transfer of matter to a white dwarf.

supernova (type II) (299) The explosion of a star, believed caused by the collapse of a massive star.

synchrotron radiation (301) Radiation emitted when high-speed electrons move through a magnetic field.

synodic period (41) The period of rotation or revolution of a celestial body with respect to the sun.

T association (242) A large, loosely bound group of T Tauri stars.

T dwarf (151) A very-low-mass star at the bottom end of the main sequence with a cool surface and a low luminosity.

temperature (138) A measure of the velocity of random motions among the atoms or molecules in a material.

terminator (498) The dividing line between daylight and darkness on a planet or moon.

terrestrial planet (454) Earthlike planet—small, dense, rocky.

theory (77) A system of assumptions and principles applicable to a wide range of phenomena that have been repeatedly verified.

thermal energy (139) The energy stored in an object as agitation among its atoms and molecules.

thermal pulse (287) Periodic eruptions in the helium fusion shell in an aging giant star; thought to aid in ejecting the surface layers of the stars to form planetary nebulae.

tidal coupling (498) The locking of the rotation of a body to its revolution around another body.

tidal heating (567) The heating of a planet or satellite because of friction caused by tides.

tidal tail (388) A long strand of gas, dust, and stars drawn out of a galaxy interacting gravitationally with another galaxy.

time dilation (325) The slowing of moving clocks or clocks in strong gravitational fields.

total eclipse (42, 45) A solar eclipse in which the moon completely covers the bright surface of the sun; a lunar eclipse in which the moon completely enters Earth's dark shadow.

transition (147) The movement of an electron from one atomic orbit to another.

transition region (164) The layer in the solar atmosphere between the chromosphere and the corona.

triple alpha process (270) The nuclear fusion process that combines three helium nuclei (alpha particles) to make one carbon nucleus.

Trojan asteroid (632) Small, rocky body caught in Jupiter's orbit at the Lagrangian points, 60° ahead of and behind the planet.

T Tauri star (244) Young star surrounded by gas and dust, believed to be contracting toward the main sequence.

turnoff point (276) The point in an H–R diagram where a cluster's stars turn off the main sequence and move toward the red giant region, revealing the approximate age of the cluster.

21-cm radiation (224) Radio emission produced by cold, low-density hydrogen in interstellar space.

ultraluminous infrared galaxy (386) A highly luminous galaxy so filled with dust that most of its energy escapes as infrared photons emitted by warmed dust.

ultraviolet radiation (112) Electromagnetic radiation with wavelengths shorter than visible light but longer than X rays.

umbra (42) The region of a shadow that is totally shaded.

uncompressed density (460) The density a planet would have if its gravity did not compress it.

unified model (402) The attempt to explain the different kinds of active galaxies and quasars by a single model.

uniform circular motion (60) The classical belief that the perfect heavens could move only by the combination of constant motion along circular orbits.

valley networks (540) Dry drainage channels resembling streambeds found on Mars.

Van Allen belt (485) One of the radiation belts of high-energy particles trapped in Earth's magnetosphere.

variable star (275) A star whose brightness changes periodically.

velocity (89) A rate of travel that specifies both speed and direction.

velocity dispersion method (380) A method of finding a galaxy's mass by observing the range of velocities within the galaxy.

vernal equinox (26) The place on the celestial sphere where the sun crosses the celestial equator moving northward; also, the time of year when the sun crosses this point, about March 21, and spring begins in the northern hemisphere.

vesicular basalt (505) A porous rock formed by solidified lava with trapped bubbles.

violent motion (87) In Aristotelian physics, motion other than natural motion. (See *natural motion*.)

visual binary system (202) A binary star system in which the two stars are separately visible in the telescope.

water hole (668) The interval of the radio spectrum between the 21-cm hydrogen radiation and the 18-cm OH radiation; likely wavelengths to use in the search for extraterrestrial life.

wavelength (110) The distance between successive peaks or troughs of a wave; usually represented by λ.

wavelength of maximum intensity (λ_{max}) (139) The wavelength at which a perfect radiator emits the maximum amount of energy; depends only on the object's temperature.

weak force (169) One of the four forces of nature; the weak force is responsible for some forms of radioactive decay.

west point (20) The point on the western horizon exactly halfway between the north point and the south point; exactly west.

white dwarf (197) The remains of a dying star that has collapsed to the size of Earth and is slowly cooling off; at the lower left of the H–R diagram.

Widmanstätten pattern (622) Bands in iron meteorites due to large crystals of nickel–iron alloys.

winter solstice (26) The point on the celestial sphere where the sun is farthest south; also, the time of year when the sun passes this point, about December 22, and winter begins in the northern hemisphere.

X-ray burster (329) An object that produces occasional X-ray flares; believed to be caused by mass transfer in a close binary star system.

Zeeman effect (175) The splitting of spectral lines into multiple components when the atoms are in a magnetic field.

zenith (20) The point on the sky directly overhead.

zero-age main sequence (ZAMS) (265) The locus in the H–R diagram where stars first reach stability as hydrogen-burning stars.

zodiac (28) The band around the sky centered on the ecliptic within which the planets move.

zone (561) One of the yellow-white regions that circle Jupiter parallel to its equator; believed to be areas of rising gas.

Answers to
Even-Numbered Problems

Chapter 1:
2. 3475 km; **4.** 1.05×10^8 km; **6.** about 1.2 seconds; **8.** 75,000 years;
10. about 27

Chapter 2:
2. 4; **4.** 2800; **6.** A is brighter than B by a factor of 170; **8.** 66.5°;
113.5°

Chapter 3:
2. a) full; b) first quarter; c) waxing gibbous; d) waxing crescent;
4. 29.5 days later on about March 30; 27.3 days later on about March 24;
6. 6850 arc seconds or about 1.9°; **8.** a) The moon won't be full until
Oct. 17; b) The moon will no longer be near the node of its orbit;
10. August 12, 2026 [July 10, 1972 + 3 × ($6585\frac{1}{3}$ days)]. In order
to get Aug. 12 instead of Aug. 11, you must take into account the number
of leap days in the interval.

Chapter 4:
2. Retrograde motion: Jupiter, Saturn, Uranus, Neptune, and Pluto; Never
seen as crescents: Jupiter, Saturn, Uranus, Neptune, and Pluto; **4.** Mars,
about 18 seconds of arc; the maximum angular diameter of Jupiter is 50 seconds of arc; **6.** $\sqrt{27} = 5.2$ years

Chapter 5:
2. The force of gravity on the moon is about $\frac{1}{6}$ the force of gravity on
Earth; **4.** 7350 m/s; **6.** 5070 s (1 hr and 25 min); **8.** The cannonball
would move in an elliptical orbit with Earth's center at one focus of the
ellipse; **10.** 6320 s (1 hr and 45 min)

Chapter 6:
2. 3m; **4.** Either Keck telescope has a light-gathering power that is
1.56 million times greater than the human eye; **6.** No, his resolving power
should have been about 5.8 seconds of arc at best; **8.** 0.013 m (1.3 cm
or about 0.5 inches); **10.** about 50 cm (From 400 km above, a human is
about 0.25 seconds of arc from shoulder to shoulder.)

Chapter 7:
2. 150 nm; **4.** by a factor of 16; **6.** 250 nm; **8.** a) B; b) F; c) M; d) K;
10. about 0.58 nm

Chapter 8:
2. 730 km; **4.** 9×10^{16} J; **6.** 0.222 kg; **8.** about 3.5 times;
10. 400,000 years

Chapter 9:
2. 63 pc; absolute magnitude is 2; **4.** about B7; **6.** about 1580 solar
luminosities; **8.** 160 pc; **10.** a, c, c, d (use Figure 9-13 to determine the
absolute magnitudes); **12.** 3.69 days or 0.010 years; about 1.2 solar masses;
about 0.67 and 0.53 solar masses; **14.** 1.38×10^6 km, about the size of
the sun

Chapter 10:
2. 60,000 nm; **4.** 0.0001; **6.** 1.5×10^6 years; **8.** 4.2×10^{35} kg or
210,000 solar masses (*Note:* Each hydrogen molecule contains two H atoms.)

Chapter 11:
2. 24.5 km/s; **4.** 2.98 km/s; **6.** 9.46×10^{10} s (about 3000 years);
8. There are four ^{1}H nuclei in the figure. They are used in a series of
reactions to build up nuclei from ^{12}C to ^{15}N. In the last reaction (with
the ^{15}N) a ^{12}C and a ^{4}He are produced. Because the ^{12}C can be used
in the next cycle, the net is four ^{1}H nuclei in with one ^{4}He nucleus out;
10. 7.9×10^{33} kg

Chapter 12:
2. about 9.8×10^6 years for a 16-solar-mass star; about 5.7×10^5 years
for a 50-solar-mass star; **4.** about 1×10^6 times less than present or about
1.4×10^{-6} g/cm^3; **6.** 2.4×10^{-9} or 1/420,000,000; **8.** about 3 pc;
10. 3.04 minutes early after 1 year; 30.4 minutes early after 10 years

Chapter 13:
2. about 1.8 ly; **4.** about 16,000 years old; **6.** about 940 years ago
(approximately 1060 AD); **8.** 2400 pc

Chapter 14:
2. 7.1×10^{25} J/s or about 0.19 solar luminosity; **4.** 820 km/s (assuming
mass is one solar mass); **6.** about 11 seconds of arc; **8.** about 490 seconds

Chapter 15:
2. about 16 percent; **4.** 3.8×10^6 years; **6.** overestimate by a factor of
1.58; **8.** about 21 kpc; **10.** 7.8×10^{10} solar masses; **12.** 1500 K

Chapter 16:
2. 2.58 Mpc; **4.** 131 km/s; **6.** 28.6 Mpc; **8.** 1.64×10^8 yr;
10. 4.49×10^{41} kg

Chapter 17:
2. 7.8×10^6 years; **4.** 0.024 pc; **6.** −28.5; **8.** about 29,800 km/s;
10. 0.16

Chapter 18:
2. 57; 17.5 billion years; 11.7 billion years; **4.** 1.6×10^{-30} gm/cm^3;
6. 76 km/Mpc; **8.** 16.6 billion years; 11 billion years; the universe could
be older

Chapter 19:
2. It will look $206,265^2 = 4.3 \times 10^{10}$ times fainter, which is about
26.6 magnitudes fainter; 22.6 mag; **4.** about 3.3 times the half-life,
or 4.3 billion years; **6.** large amounts of methane and water ices;
8. about 1300

Chapter 20:
2. about 17 percent; **4.** 81×10^6 yr; they have been subducted;
6. 0.22 percent

Chapter 21:
2. The rate at which an object radiates energy is proportional to its surface
area, which is proportional to its radius squared (r^2). However, the energy
an object has stored as heat is proportional to its mass and hence to its volume, and that is proportional to its radius cubed (r^3). So the cooling time
will be proportional to the amount of heat divided by the rate of cooling,
which is the same as the radius cubed divided by the radius squared (r^3/r^2).
That shows that the cooling time is proportional to the radius (r); and
that means that the bigger an object is, the longer it takes to cool; **4.** No.
Their angular diameter would be only 0.5 second of arc. They would be
visible in photos taken from orbit around the moon; **6.** 0.5 second of arc
(assuming an astronaut seen from above is 0.5 meter in diameter); no;
8. 10.0016 cm; **10.** Mercury, $V_e = 4250$ m/s; moon, $V_e = 2380$ m/s;
Earth, $V_e = 11,200$ m/s

Chapter 22:
2. 33,400 km (39,500 km from the center of Venus); **4.** 61 seconds of
arc; **6.** 260 km; **8.** 82 seconds of arc

Chapter 23:
2. 35 Earth days; **4.** 4.4°; **6.** about 0.056 nm; **8.** 5.2 m/s

Chapter 24:
2. 42 seconds of arc; **4.** 256 m/s; **6.** 8.8 s; yes; **8.** 12.3 km/s;
10. 1.04×10^{26} kg (17.2 Earth masses)

Chapter 25:
2. one billion; **4.** 79 km; **6.** 0.18 km/s; **8.** 3.28 AU and 2.5 AU;
10. 9 million km (0.06 AU); too small; **12.** 33 Earth masses

Chapter 26:
2. 8.9 cm; 0.67 mm; **4.** about 1.3 solar masses; **6.** 380 km;
8. pessimistic, 2×10^{-5}; optimistic, 10^7

Credits

Chapter 1. 2–3: T. Rector, University of Alaska, and WIYN/NOAO/AURA/NSF 4, left: M. Seeds 4, right: USGS 5, left: NASA infrared photograph 5, right: NASA 6: NOAO 6: NASA 8: NOAO 9, bottom: Anglo-Australian Telescope Board 9, top: Detail of galaxy map from M. Seldner, B. L. Siebers, E. J. Groth, and P. J. E. Peebles, *Astronomical Journal 82* (1977) 10: Bill Schoening/NOAO/AURA/NSF

Chapter 2. 12–13: Courtesy of Kris Koenig 33: USPS 15, top: From Duncan Bradford, *Wonders of the Heavens*, Boston: John B. Russell, 1837 16: William Hartman 18: Roger Ressmeyer/Corbis 20: AURA/NOAO/NSF 23: Royalty-Free/Corbis 27: NASA

Chapter 3. 36–37: NSO/AURA/NSF 38, top: Yerkes Observatory 38, bottom: T. Scott Smith 39: UCO/Lick Observatory 43: © 1982 by Dr. Jack B. Marling 44: Celestron International 47: Daniel Good 48, left: Daniel Good 48, right: NOAO 49: Daniel Good 49, left: UCO/Lick Observatory 54, top: Laurence Marschall

Chapter 4. 59, top: Jamie Backman 59, bottom: Benelux Press/Index Stock Imagery 60: Courtesy NPS Chaco Culture National Historical Park 64: Yerkes Observatory 64: From *Cosmographica* by Peter Apian (1539) 68: NOAO and Nigel Sharp 69: Grundy Observatory 71: Yerkes Observatory 74: The Granger Collection, New York 77: From the collection of John Coolidge III

Chapter 5. 86: NASA/JSC 88: NASA 91: ESA/STScI and NASA 95: NASA 99: Nigel Sharp/NOAO/AURA/NSF 106: Larry Mulvehill/The Image Works 107: NASA/JSC

Chapter 6. 108: Robert Frost, "The Star-Splitter," *New Hampshire: A Poem with Notes and Grace Notes* (New York: Henry Holt and Co., 1923), pp. 27–30. 108–109: NOAO/AURA/NSF 110: NOAO/AURA/NSF 116: NASA 117, top: Courtesy of William Keel 117, center: NOAA 117, bottom: ESO 119, bottom: NOAO/AURA/NSF 119, left: NOAO/AURA/NSF 119, right: ESO 120: AURA/NOAO/NSF 121, left: W. M. Keck Observatory 121, right: Paul Kalas 123, left and bottom: C. Hawk, B. Savage, N. A. Sharp, NOAO/WIYN/NSF 123, right: NOAO/WIYN/NSF 124: NASA/CXC/Rutgers/J. Hughes 126, bottom: NRAO 126, top: Courtesy Seth Shostak/SETI In-

stitute 127: NRAO/AUI/NSF 127, bottom: NRAO/AUI 127, center: David Parker/SPL/Photo Researchers, Inc. 129, left: William Keel/IRAF 129, center: NASA 129, bottom: Kris Koenig/Coast Learning Systems 130, 131: NASA 134, left: ESO 134, right: NASA/CXC/PSU/S. Park

Chapter 7. 136, 137: ESO 138: NASA Hubble Heritage Team/STScI/AURA 147: AURA/NOAO/NSF 150: AURA/NOAO/NSF 152: Adapted from Thomas R. Geballe, Gemini Observatory, from a graph that originally appeared in *Sky and Telescope Magazine*, February 2005, p. 37. 158: T. Rector, University of Alaska and WIYN/NURO/AURA/NSF

Chapter 8. 160–161: NASA/SOHO 163: Daniel Good 164: P. N. Brant, G. Scharmer, G. W. Simpson, Swedish Vacuum Solar Telescope, La Palma 165, left: NOAA/SEL/USAF 165, right: NOAO/NSO 166: SOHO/ESA/NASA 167: Stanford-Lockheed Institute for Space Research, Palo Alto, CA and NASA GSFC 172, left: Brookhaven National Laboratory 172, right: Courtesy SNO 174, background: Royal Swedish Academy of Science 174, right, and 175, top left: NASA 175, top right: J. Harvey/NSO and HAO/NCAR 175, center: M. Seeds 175, bottom left: SOHO/EIT, ESA and NASA 175, bottom right: TRACE/NASA 179: K. Strassmeier, Vienna, AURA/NOAO /NSF 180, left: TRACE/NASA 180, right: Sacramento Peak Observatory 180, bottom: SOHO, EIT, ESA and NASA 181, top left: NASA 181, center left: SOHO/MDI, ESA and NASA 181, background: © Jan Curtis 181, bottom left: SOHO/LASCO, ESA, and NASA 181, bottom right: Yohkoh/ISAS/NASA 184, top : NOAO and Daniel Good 184, bottom: NASA/SOHO

Chapter 9. 186–187: Mark McCaughrean and Morten Andersen of the Astrophysical Institute, Potsdam, and the ESO 188: 2MASS Sky Survey and IPAC 194: Krafft/Photo Researchers, Inc. 196: M. Seeds 196: Heidi Schweiker/NAOA/AURA/NSF 201: USGS 202: UCO/Lick Observatory 204: The Observations of the Carnegie Institution of Washington 207: M. Seeds 211, bottom left, and 213: NASA

Chapter 10. 216–217: NASA, ESA, and the Hubble Heritage Team, AURA/STScI 218: Daniel Good 219 and 220, top: ESO 220, bottom: Anglo-Australian Telescope Board 221: Caltech 221, top center: NASA 221, top right: Daniel Good 221, center: Anglo-Australian Telescope Board 221, bottom left: Hubble Heritage Team, NASA 222: The Observations of the Carnegie Institution of Washington 223: Anglo-Australian Telescope Board 226: J. Dickey, UMn, F. Lockman, NRAO, Sky View 228: NASA/IPAC, Courtesy Deborah Levine 230: M. Seeds 231, top: Caltech 231, center:

Courtesy Richard E. White; image rendered by Duncan Chesley of American Image, Inc. 231, bottom left: NASA and The Hubble Heritage Team, AURA/STScI 231, bottom right: R. Chris Smith, CTIO and Y.-H. Chu, UICU 232: AURA/NOAO/NSF and Nigel Sharp 233: NASA and The Hubble Heritage Team, AURA/STScI

Chapter 11. 236–237: ESO 238, left: ESO 238, right: W. Brander, JPL/IPAC, E. K. Grebel, University of Washington, Y. Chu, University of Illinois, and NASA 240: NASA/JPL-Caltech/V. Gorjian and NOAO 241: ESO 243, top: C. Burrows, STScI & ESA, WFPC 2 Investigation Definition Team, NASA 243, bottom: N. Walborn and J. Maiz-Apellaniz, STScI, R. Barba, La Plata Observatory, and NASA 244, background, and 244, bottom right: NASA 245, top left: Richard Mundt, Calar Alto Observatory 245, top, and 245, bottom left: NASA 245, bottom right: AURA/NOAO/NSF 246, left: NASA, ESA, The Heritage Hubble Team 246, right: ESO 252, center: NASA 252, bottom: NASA/CXC/SAO 253, left: ESO 253, top right: Anglo-Australian Telescope Board 253, center: Dan Gezari, Dana Backman, and Mike Werner 253, bottom: NASA 256: STScI/NASA

Chapter 12. 258–259: T. A. Rector and B. A. Wolpa, NRAO/NOAO/AUI/NSF 260: NOAO/AURA/NSF 262: The Boeing Company 264, left: NASA/CXC/SAO/HST 264, bottom: ISO, and courtesy ESA 264, inset: Jon Morse and J. Hester 265, left: Gemini Observatory/University of Hawaii Institute for Astronomy/Michael Liu/NSF 265, right: STSci and NASA 276, right: Anglo-Australian Telescope Board 276, background: AURA/NOAO/NSF 277, top: Anglo-Australian Telescope Board 277, center: Caltech 277, bottom: Nigel Sharp, Mark Hanna/AURA/NOAO/NSF 282: NASA/Walborn, Maiz-Apellaniz, and Barba

Chapter 13. 284–285: NASA/ESA/HEIC/Hubble Heritage Team 286, left: 2MASS 286, center: NASA/CXC/SAO 286, right: NASA/CXC/Rutgers/J. Warren et al., STScI, University of Illinois 290: Hubble Heritage Team, STScI/AURA/NASA 291: NASA 292, left: Hubble Heritage Team, STScI/AURA/NASA 292, center and right: NASA/IPAC courtesy of Deborah Levine 295, top: © UC Regents 295, bottom: M. Shara and R. Williams, STScI 295, bottom: R. Gilmozzi, ESO, and NASA 297, left: Gemini Observatory/NSF/C. Aspin 297, right: NASA 299: Adam Burrows/University of Arizona, Department of Astronomy 300: High-Z Supernova Search Team, HST, NASA 301: ESO 302, top left: Caltech 302, top right: NASA/SAO/CXC 302, bottom left: 2MASS 302, bottom right: NRAO/AUI/NSF 303, left: Anglo-Australian Telescope Board 303, bottom: NASA, Challis, Kirshner and Sugerman 303, right: Christopher

Burrows, ESA/STcI, NASA 306, top: NASA/Hubble Heritage Team/STSci/AURA 306, bottom: NASA/CAC/SAO/CSIRO/ATNF/ATCA

Chapter 14. 308–309: NASA/CXC/SAO 310: NASA/SAO/CXC/P. Slane et al. 311: M. Seeds 312: Caltech 313, left: NASA/CXC/SAO/University of Massachusetts 313, right: F. Lu/McGill 313, bottom: V. Kaspi 315: NASA 316, top: AURA/NOAO/KPNO 316, bottom: Figure 2 from F. R. Harnden, Jr., in *The Astrophysical Journal,* 283 (August 1, 1984): 279–285. Copyright 1984 by The American Astronomical Society. Reprinted with permission of F. R. Harnden, Jr., and the AAS.

317: NASA and F. M. Walter 318: John Rowe Animations 319: J. Trumper, Max-Planck Institute 320, bottom: NASA/CXC/ASTRON/B. Stappers et al. 320, bottom: AAO/J. Bland-Hawthorn & H. Jones 320, top: SOHO/MDI 323: John P. Hughes, Rutgers University 330: Ivan King and NASA/ESA 331, bottom: NASA 331, top: NASA/CXC/SAO 332: NASA/CXC/M. Weiss. Found at: http://www.chandra.harvard.edu/press/04_releases/press_010504.html 333, left: ESO and NASA 333, right: NASA/Skyworks Digital 335: NASA/McGill, V. Kaspi et al. 336: CXC/M. Weiss

Chapter 15. 338–339: ESO and Nico Housen 340: AURA/NOAO/NSF 342, center: AURA/NOAO/NSF 342, bottom: Doug Williams, N. A. Sharp/NOAO/AURA/NSF 345: Daniel Good 346, top left: 2MASS 346, bottom left: DIRBE courtesy Henry Freudenreich 346, top right: Todd Boroson/NOAO/AURA/NSF 346, bottom right: NASA/CXC/PSU/S. Park et al. 354: Kris Koenig/Coast Learning Systems 355: NASA/Heritage Hubble Team 357, center: Painting by M. Seeds, based on a study by G. De Vaucouleurs and W. D. Pence 357, bottom: ESO 358: NASA/Hubble Heritage Team 359: Anglo-Australian Telescope Board 361: NASA/Hubble Heritage Team 362, top right: NRAO/AUI/NSF 362, bottom left: 2MASS 362, center: N. Killeen and Kwok-Yung Lo 363, center: NASA/CXC/MIT/F. K. Baganoff et al. 363, bottom left: ESO 366: NASA/Hubble Heritage Team/STScI/AURA

Chapter 16. 368–369: NASA/Hubble Heritage Team/STScI/AURA 370, left: Hubble Heritage Team/AURA/StScI/NASA 370, top right: ESO 370, bottom right: Hubble Heritage Team/Aura/StScI/NASA 371: R. Williams and the Hubble Deep Field Team/STScI/NASA 372: M. Seeds 374, bottom: Anglo-Australian Telescope Board 374, top right: AURA/NOAO/NSF 374, top right: Anglo-Australian Telescope Board 375, top: NASA/Hubble Heritage Team 375, center left: Rudolph E. Schild 375, center right: G. J. Jacoby, M. J. Pierce/AURA/NOAO/NSF 375, bottom left: AURA/NOAO/

NSF 375, **bottom right:** © R. J. Dufour, Rice University 376: J. Trauger, JPL; Wendy Freedman, Observatories of the Carnegie Institution of Washington 376: NASA 377: NASA/ESA/GOODS Team and M. Giavgalisco./STScI 379: Hubble Heritage Team, STScI/AURA/NASA 380: Courtesy Vera Rubin 381: Bill Schoening/AURA/NOAO/NSF 382, **left:** L. Thompson/NOAO 382, **right:** Simon D. M. White, Ulrich G. Briel, and J. Patrick Henry 383: W. N. Colley and E. Turner, Princeton, and J. A. Tyson Bell Labs, and NASA 384: NOAO/AURA/NSF 386: Courtesy Nicolas Martin and Rodrigo Ibata, Observatoire de Strasbourg-France 387, **left:** ESA/NASA, P. Anders 387, **center and right:** NASA/ESA/Hubble Heritage Team/AURA/STScI 388, **top and bottom right:** NOAO 388, **bottom left:** NOAO 389, **top right:** Brad Whitmore, STScI/NASA 389, **center right:** NASA/CXC/SAO/G. Fabbiano et al. 389, **center left:** Francois Schweizer, Carnegie Institution of Washington, Brad Whitmore, STScI 389, **center left:** NASA/Heritage Hubble Team 389, **bottom left:** Michael J. West 389, **right center:** Robin Braun, NRAO /AUI/NSF 389, **bottom right:** NASA/Hubble Heritage Team/ESA/AURA/STScI 390, **left:** NASA, W. Keel, F. Owen, M. Ledlow, and D. Wang 390, **right:** NASA, H.-J. Yan, R. Windhorst, and S. Cohen, Arizona State University 392, **top:** P. Massey, Lowell, N. King, STScI, S. Holmes, Charleston, G. Jacoby, WIYN/AURA/NSF 392, **bottom:** WIYN/NOAO/NSF

Chapter 17. 394–395: Courtesy NRAO/AUI 396, **left:** Anglo-Australian Telescope Board 396, **center:** Hubble Heritage Team, AURA/STScI, NASA 396, **right:** NASA, Andrew S. Wilson; Patrick L. Shopbell, Chris Simpson, Thaisa Storchi-Bergman, F. K. B. Barbosa and Martin J. Ward 397, **top:** John W. Mackenty, Institute for Astronomy, University of Hawaii 397, **center:** NRAO/AUI/NSF 398, **top:** NASA/CXC/NRAO 398, **right:** E. Schreier, STScI, NASA 398, **center right :** Used with permission, Fosbury R. A. E., Vernet, J., Vilar-Martin, M., Cohen, M. H., Ogle, P. M., Tran, H. D, & Hook, R. N. 1998, Optical continuum structure of Cygnus A. "KNAW colloquium on: The most distant radio galaxies," Amsterdam, 15–17 October 1997, Roettgering H, Best, P., and Lehnert, M., eds, Reidel, astro-ph/9803310 398, **bottom left :** X-ray: NASA/CXC/M. Karovska, et al., Radio: NRAO/VLA/Schiminovich, et al., Radio: NRAO/VLA/J. Condon et al., Optical: Digitized Sky Survey U. K. Schmidt Image/STScI 398, **center left:** NRAO/AUI/NSF 399, **top right:** NRAO/AUI/NSF and C. O'Dea and F. Owen 399, **center right:** NRAO/AUI/NSF and C. O'Dea, F. Owen, M. Inoue, and J. Eilek 399, **bottom:** Adapted from a diagram by Ann Field, NASA, STScI 400: NASA 401, **left:** AURA/NOAO/NSF 401, **top center:** NASA/STScI 401, **center:** NASA/CXC/W. Forman et al. 401, **right:**

NRAO and J. Biretta 401, **left:** L. Ferrarese, Johns Hopkins University, and NASA 401, **right:** The Hubble Heritage Team/NASA 402, **top:** Roeland, P. van der Marel, STScI, Frank C. van den Bosch, University of Washington, and NASA 402, **bottom left:** NASA/CXC/MIT/UCSB/P/Ogle et al. 402, **bottom right:** NASA/STSci/A. Capetti et al. 404, **left:** AURA/NOAO /NSFA; NASA; CXC/SAO 404, **right:** NASA/CXC/SAO 405: ESA 406, **top:** C. Steidel, Caltech, NASA 406, **bottom:** Courtesy Maarten Schmidt 408, **top left:** George Rhee, NASA, STScI 408, **top right:** Chris Impy, University of Arizona and NASA 408, **bottom:** Image by W. Keel from data in the NASA/ESA Hubble Space Telescope Archive, originally obtained with J. Westphal as Principal Investigator 409: AURA/NOAO/NSF 410, **top:** NRAO/AUI/NSF 410, **bottom:** J. Bahcall, Institute for Advanced Study, Mike Disney, University of Wales, NASA 414, **top, center, and bottom:** NRAO/AUI

Chapter 18. 416–417: Courtesy of MPA and Joerg Colberg 418: R. Williams, STScI HDF-South Team, NASA 419: Courtesy of Janet Seeds 423: Ken Lanzetta and Amos Yahil, Stony Brook and NASA 424, **left:** AT&T Archives 424, **right:** NASA 427: K. Lanzetta, SUNY, A. Schaller for STScI, and NASA 428: NASA/WMAP Science Team 431: NASA, Benites, Broadhurst, Ford, Clampin, Hartig, Illingworth, ACS Science Team and ESA 432: Hubble Heritage Team/STScI/AURA/NASA 438: ESO 438: Adam Riess, STScI/NASA 440: NASA/CXC/IoA/S. Allen et al. 441: Sloan Digital Sky Survey 442: Adapted from a model by Kauffmann, Colberg, Diaferio, & White: Max-Planck Institute für Astronomie 443, **top:** NASA/WMAP Science Team 443, **bottom:** Courtesy of the BOOMERANG Collaboration 445, **top:** NASA, Benites, Broadhurst, Ford, Clampin, Hartig, Illingworth, ACS Science Team and ESA 445, **bottom:** Courtesy of the BOOMERANG Collaboration

Chapter 19. 448–449: NASA 451: Janet Seeds 454: John Hopkins University, Applied Physics Laboratory, NASA 455: Celestron International 456, 457: UCO/Lick Observatory 457, **bottom right:** Grundy Observatory 458, **left:** Daniel Good 458, **right:** R. Kempton, New England Meteoritical Services 462: NASA 464: JPL/NASA 467: NASA 468, **center:** C. R. O'Dell, Rice, NASA 468, **top right:** M. McCaughrean, Max Plank Inst. für Astronomie, C. R. O'Dell, Rice, NASA 468, **bottom left:** J. Bally, H. Throop, C. R. O'Dell, NASA 469: D. Padgett, IPAC/Caltech, W. Brandner IPAC, K. Stapelfeldt, JPL/NASA 470, **top:** C. Burrows and J. Krist, STScI and NASA 470, **center left:** A. Weinberger, E. Becklin, UCLA, G. Schneider, U. of Arizona, and NASA 470 **center right:** Joint Astronomy Center 470, **bottom left:** NASA 470, **bottom right:** NASA, ESA, Kalas, Graham, and Clampin 473: ESO 475: NASA

Chapter 20. 476–477: Jean-Bernard Carillet/Getty Images 478: NASA 478, **top right:** UCO/Lick Observatory 484: From *Virus Ultrastructure: Electron Micrograph Images.* Copyright 1995 by Linda M. Stannard. Reprinted with permission from Linda M. Stannard. Found at: http://web.uct.ac .za/depts/mmi/stannard/adeno.html 485: Jimmy Westlake 486, **left:** William K. Hartmann 486 **top center:** Janet Seeds 486, **top right and top left:** National Geophysical Data Center 487, **bottom:** M. Seeds 491: Ozone Processing Team/NASA/Goddard Space Flight Center 493, **top:** William K. Hartmann 493, **bottom:** USGS

Chapter 21. 496–497: JSC/NASA 498: UCO/Lick Observatory 499: NASA 500, 501: NASA 502, **left:** NASA 502 **center and right:** UCO/Lick Observatory 503: Phyllis Leber 504: NASA 505, **left:** UCO/Lick Observatory 505, **right and bottom, and 506:** NASA 507 : M. Seeds 508: NASA/Clementine 509, **right:** NASA 509, **left:** Don Davis 512, 513, 514, 515: NASA 516: Martin Slade/JPL/NASA 518, 520, 521: NASA

Chapter 22. 522: © Calvin J. Hamilton 523 and 525, **top and bottom:** NASA 526, **top:** M. Seeds 526, **top:** USGS 527: NASA 528, **bottom left:** USGS 529: NASA 530: PhotoDisc/Getty Images 531, **left:** NASA 531, **right:** © 1992, David P. Anderson, Southern Methodist University 532: NASA 534, **left:** Courtesy of Lowell Observatory 534, **right:** USGS 535: NASA 535: Philip James, University of Toledo, Steven Lee, University of CO, Boulder, and NASA 536, 537: NASA 538: NASA/JPL/Cornell 539: NASA 540: NASA/JPL/Malin Space Science Systems 541, **left:** NASA/USGS 541, **left:** © Calvin J. Hamilton 542: Malin Space Science Systems/NASA 543: NASA/JPL/Arizona State University 544: NASA/JPL/Cornell/USGS 545, **top:** NASA 545, **bottom left:** Malin Space Science Systems/NASA 545, **bottom right:** NASA 546: NASA/JPL/Malin Space Science Systems 547, **left:** Damon Simonelli and Joseph Ververka, Cornell University/NASA 547, **right, and 550:** NASA

Chapter 23. 552–553: NASA/JPL 555: NASA/JPL/Space Science Institute/University of Arizona 557: NASA/JPL/University of Arizona 559, **top:** John P. Clarke, University of Michigan, NASA 559, **bottom and 560, top left:** NASA 560, **top right:** NASA/JPL/University of Arizona 560, **left:** Hughes Aircraft Company 560, **bottom:** NASA/JPL 561, **bottom:** NASA/JPL/Caltech 562: NASA/JPL/Space Science Institute 563, **bottom, and 564, left:** NASA 564, **center:** Mike Skrutskie 564, **right:** University of Hawaii 565-569: NASA 570, **top:** NASA/JPL/Bruce A. Goldberg 570, **center:** NASA/JPL/John Hopkins University Applied Physics Lab 570, **bottom:** NASA/JPL 572: NASA, ESA, J. Clarke, Boston University and X. Levay, STScI

573: NASA/STScI 574: NASA and E. Karkoschka 575, 576, 577: NASA 577: NASA/JPL/Science Institute 579, 580 NASA/ESA/JPL/University of Arizona 581, 582: NASA/JPL/Space Science Institute 585: NASA/JPL

Chapter 24. : 588–589: NASA/JHUAPL/SwRI 590, 592, 593: NASA 595: Lawrence Sromovsky, UW-Madison Space Science and Engineering Center 596: Courtesy Floyd Herbert, LPL 597: ESO 599, 601: NASA 601, **center and bottom right; 602, bottom left:** USGS 602, **bottom right:** P. Bridges 604: NASA, L. Sromovsky, and P. Fry, UW-Madison 605: NASA 605, **left:** NASA and Erich Karkoschka 605, **right:** U.C. Berkeley/W. M. Keck Observatory 608: Courtesy of Lowell Observatory 609: NASA 610, **top:** NASA/McGill/V. Kaspi et al. 610, **bottom left:** NASA 610, **bottom right:** R. Albrecht, ESA/ESO Space Telescope European Coordinating Facility, NASA 614: NASA and Heidi Hammel

Chapter 25. 616–617: NASA/JPL 618: Courtesy Tod Lauer 620: M. Seeds 621: NASA/JPL-Caltech/M. Kelley, University of Minnesota 622: USGS 622, 623: M. Seeds 623: Courtesy Dr. Monica Kress 624: Courtesy of Russell Kemton, New England Meteoritical 625, **top:** Hubble Heritage Team/STScI/AURA/NASA 625, **bottom; 628, 629:** NASA 632: Reprinted with permission of MIT Lincoln Laboratory, Lexington, Massachusetts 633: Courtesy of Gareth Williams, Minor Planet Center 634: T. Rector, University of Alaska Anchorage, Z. Levy, and L. Frattare, Space Telescope Science Institute and WIYN/NOAO /AURA/NSF 635, **top left:** © 1986 Max-Planck Institute 635, **center:** Steven Larson 635: NASA 636: Caltech 637, **top left:** ESO 637, **center and bottom:** NASA 637, **top right:** Dean Ketelsen 638, **top left:** NASA/JPL-Caltech/UMD 639: NASA 640: SOHO 641, **left:** NASA and G. Bernstein 641, **right:** John Hopkins University Applied Physics Lab./Southwest Research Institute, JHUAPL/SwRI 643, **center:** Virgil L. Sharpton, University of Alaska, Fairbanks 647: NASA 647, **bottom:** Courtesy Russell Kempton, New England Meteoritical

Chapter 26. 650–651: Kris Koenig/Coast Learning Systems 653, **left:** M. Seeds 653, **right:** L. D. Simon 654: M. Seeds 657, **left:** Courtesy J. William Schopf 657, **right:** Grundy Observatory 658: Courtesy Stanley Miller 659, **left:** Courtesy Chip Clark, National Museum of Natural History 659, **right:** Courtesy Sidney Fox and Randall Grubbs 660, **inset:** Courtesy Chip Clark, National Museum of Natural History 660: Peter Sawyer/National Museum of Natural History 663, **left:** D. W. Miller 663, **right:** NASA/JPL/Cornell 664: NASA 669: Isaac Gary/SETI Institute 671, **top:** T. A. Rector, B. A. Wolpa, NRAO /NOAO/AURA/NSF 671, **bottom:** ESO

Afterword. 675: NASA

Index

eclipse season, **50**, 51, 52

eclipse year, **51**

eclipsing binary systems, **204–207**
  variable stars as, 275

ecliptic, **24**

Eddington, Sir Arthur, 258

edge–center problem, 418–419

Edgeworth, Kenneth, 641

Egg Nebula, 291

Egyptian civilization, 58, 254

Einstein, Albert, 86, 100–105, 323, 382, 432, 437–438

ejecta, **500**

electromagnetic radiation, **110**–112, 128

electromagnetic spectrum, 111–112

electron, **138**

electron shells, 142–143

element-building process, 350, 351

Eliot, T. S., 676

ellipses, **76**

elliptical galaxies, 372, 373, **374**
  origin and evolution of, 386–387, 391
  size and luminosity of, 379

emission lines, **146**

emission nebulae, **220, 221**

emission spectrum, **146**

Enceladus, 580, 581, 582, 583

energy, **96**
  binding, **143**
  conservation law, **260**, 261
  dark, **439**
  flow of, 507
  stellar, 246–248
  thermal, **139**

energy levels, **144**

energy transport, 248–249

engineering, 562

enzymes, **654**

epicycle, **65**

equant, **65**

equatorial mounting, **121**

equinox, 26

equivalence principle, **102**, 105

Eratosthenes, 62

ergosphere, **324**–325

Eros asteroid, 456, 628

escape velocity, 93, **95**, 322–323

Eta Aquarids meteor shower, 620

Eta Carinae, 188, 263, 264

ethane, 578

Eudoxus of Cnidus, 61

Europa, 567–569, 662–663, 665

European Southern Observatory, 119

European Space Agency, 190

evening star, **28**

event horizon, **323**, 324, 325

evidence, scientific, 31, 400

evolution
  of binary stars, 292–296
  of galaxies, 386–387, 390
  of life, 653, 656, 660–661, 662
  of pulsars, 316–317
  of star clusters, 274
  of stars, 258–283
  of the sun, 266

evolutionary hypothesis, **451**

excited atom, **144**

explanations, 652

exponential scales, 196

extinction
  interstellar, **219**, 222
  meteorite impacts and, 642–643

extrasolar planets, **469**–472

extraterrestrial communication, 665–670

Extreme Ultraviolet Explorer (EUVE), 229

eyepiece, **114**

F

facts versus theories, 230

Fahrenheit temperature scale, 678

Fall of Moon Dust, A (Clarke), 498

falls, **621**

false-color images, **123**, 124

family of stars, 210–211

field, **92**

51 Pegasi, 469, 470, 471

filaments, **165**, 180

filtergram, **165**

finds, **621**

fission, nuclear, 169

fission hypothesis, **510**

FitzGerald, Edward, 136

flares, **181**

flatness problem, **436**, 444

flat universe, 430, 431, **433**

Fleming, Alexander, 610

flocculent galaxies, **359**

flooding stage, 481

floppy mirrors, 121

flux, **191**

focal length, **114**

foci, **76**

folded mountain ranges, **486**

forbidden lines, **218**

Forerunner of Dissertations on the Universe, Containing the

Mystery of the Universe, The (Kepler), 75

formation of stars. *See* star formation

Fornax A Galaxy, 395

forward scattering, **562**

fossils, 656, 657, 660, 665

Fox Fur Nebula, 259

Franklin, Benjamin, 590

fraud in science, 327

Fraunhofer, Joseph von, 136

free-fall contraction, **240**

frequency, **111**

Frost, Robert, 108

full moon, 40, 41

fusion. *See* nuclear fusion

G

G stars, 360, 667

galactic cannibalism, **388**, 389

galactic corona, **348**

galactic fountains, **351**–352

galaxies, 8, 368–393
  active, **396**–405
  barred spiral, **374**, 387
  blue dwarf, 390
  classification of, 372–373, 374–375
  clusters of, 8–9, 384–385, 440, 441
  colliding, 385–386, 388–389
  color of, 373
  dark matter in, 381–383
  diameter of, 378–379
  discovery of, 370–372
  distances to, 373, 376–378, 383
  elliptical, 372, 373, **374**, 379, 386–387, 391
  evolution of, 386–387, 390
  flocculent, **359**
  formation of, 390–391, 403–404
  Hubble law and, **378**
  intelligent life in, 669
  irregular, 373, **375**
  luminosity of, 378–379
  mass of, 379–380, 383
  merging of, 388, 389
  number of, 373
  properties of, 373, 376–383
  quasars in, 409–410
  radio, **394**, 396
  redshifts of, 390–391, 420–421, 432
  ring, **389**

rotation of, 379

Seyfert, **396**–397

shapes of, 372–373

spiral, 372, 373, **374**, 378–379, 387

starburst, **386**

supermassive black holes in, 364, 380–381, 400–402, 403–404

ultraluminous infrared, **386**–387

See also Milky Way Galaxy

Galilean satellites, **455**, 565–572
  Callisto, 566
  comparative data on, 571
  Europa, 567–569
  Ganymede, 566–567
  history of, 571–572
  Io, 555–556, 558, 559, 569–571, 572

Galileo Galilei, 56, 68–72, 87–89, 572, 573, 603

Galileo spacecraft, 554, 556, 557, 560

Galle, Johann, 603

gamma rays, 112, 331–334
  gamma-ray bursters, **332**–334
  mass converted into, 425
  soft gamma-ray repeaters, **332**
  solar energy and, 170–171

gamma-ray telescopes, 130

Gamow, George, 424

Ganymede, 566–567

Gaposchkin, Sergei, 152

gas giant planets, 594

gas (type I) tails, **636**

Gaspra asteroid, 628

Gemini telescopes, 110, 119

general theory of relativity, **102**–105, 323, 433, 437

genes, **655**

geocentric universe, **61**, 64

geologic time, 661

geosynchronous satellites, **94**

giant molecular clouds, **227**
  star birth in, 238–240, 358

giant stars, 196–197, 200, 267–273
  carbon fusion in, 272
  degenerate matter and, 268–**269**
  expansion of stars into, 267–268
  helium fusion in, 269–271
  nuclear reactions in, 273
  star clusters and, 274

gibbous moon, 40, 41

glitches, **313**, 316

Global Oscillation Network Group (GONG), 168

global warming, **490**–491

morning star, **28**

motion
    Einstein's theories of, 100–102
    Galileo's study of, 87–89
    Kepler's laws of, 76–78, 93, 95, 96–97, 348
    lunar, 38–39
    natural vs. violent, 87
    Newton's laws of, 89–90, 91, 94, 97, 99–100
    orbital, 92–100
    planetary, 6, 64, 72–79
    proper, **190**
    relativity of, 100–101
    retrograde, **64**, 67
    solar, 23–24
    space-time curvature and, 102–103, 104
    uniform circular, **60**, **64**, 67

mountain ranges, 486

Mount St. Helens, 528

multiringed basins, **501**

Murchison meteorite, 659

mutant, **656**

mutual gravitation, 90–92

Mysterium Cosmographicum (Kepler), 75, 76

N

nadir, **20**

nanometer (nm), **111**

National Institutes of Health, 575

National Radio Astronomy Observatory, 127

National Science Foundation, 575

natural law, **77**

natural motion, **87**

natural selection, **653**

nature, 676

neap tides, **98**

Near Earth Asteroid Tracking (NEAT) system, 634

near-Earth objects (NEOs), 631

NEAR spacecraft, 456, 628

nebula (nebulae), **218–219**
    dark, **221**
    emission, **220**, 221
    planetary, **288**, 290–291, 376–377
    reflection, **220**, 221
    star formation in, 244–245, 252–253, 254
    supernova remnants as, 301
    three kinds of, 220–221

nebular hypothesis, 451

Neptune, 603–608
    atmosphere of, 554, 604–605, 607–608
    discovery of, 603–604
    history of, 607
    interior of, 605
    magnetic field of, 596, 605
    moons of, 606–607, 684
    rings of, 599, 605–606
    rotation of, 604–605
    spacecraft exploration of, 590, 606
    statistical data about, 605

Nereid, 606

neutrinos, **170**, 171–172, 303, 434

neutrons, **141**, 310

neutron stars, **310–322**
    accretion disks around, 329–330
    binary pulsars as, 317–319
    discovery of, 311–312
    energy produced by, 330–331
    evolution of, 316–317
    gamma-ray bursts from, 332
    glitches produced by, **313**, 316
    lighthouse model of, **312**, 314–315
    magnetic field of, 311
    millisecond pulsars, **319–320**
    planets orbiting, 321
    pulse period of, 319–321
    rotation of, 319–320
    theoretical prediction of, 310–311
    X rays from, 319, 321–322, 329

Newgrange, 58, 59

New Horizons spacecraft, 641

new moon, 40, 41

newton (N), **677**

Newton, Isaac, 76, 84, 86–100, 123, 677

Newtonian focus, **120**

nodes, **50**, 51

nonbaryonic matter, **434**, 435

north celestial pole, **20**

northern hemisphere star charts, 688–693

north point, **20**

nova (novae), **284**, 294–295

Nova Cygni, 295

Nova T Pyxidis, 295

nuclear bulge, **341**, 346, 347

nuclear fission, **169**

nuclear fusion, **169**
    CNO cycle and, 247
    heavy-element, 271–272, 296–298

helium, 269–271

hydrogen, 169–170

solar, 168–172

stellar, 247, 250–251, 269–272, 296–298

temperature and, 172, 247, 248

nucleosynthesis, **350**

nucleus
    of atoms, **141**
    of comets, 457, 635, 636, 638–639
    of the Milky Way Galaxy, 361–364

O

O associations, **242**, 346, 347, 355

O stars, 242, 344, 345, 355, 358, 360, 373

Oberon, 600, 601

objective lens, **114**

objective mirror, **114**

oblateness, **556**

observable universe, **420**

observatories
    location of, 117–118, 122
    See also names of specific observatories

observed densities, 461

occultations, **597**

ocean tides, 97–99

Olbers, Heinrich, 419

Olbers's paradox, **419–420**

Olympus Mons volcano, 529, 539, 541

Oort, Jan, 639

Oort cloud, **639–641**

opacity, **248–249**

open orbit, **95**

open star clusters, **274**, **276**, 346, 352

open universe, 430, **433**

optical telescopes, 113–122, 128

orbital motion, 92–100

orbital velocity, 92–93

orbits
    asteroid, 630, 631–632
    closed, **95**
    comet, 456–457, 639
    Earth, 6, 24, 25, 30, 32
    galactic, 347–348, 354
    Kepler's laws of, 76–78
    lunar, 38–39
    Newton's laws of, 92, 97, 99–100
    open, **95**
    permitted, **143**

planetary, 25, 28, 72–79, 453, 454, 682

space-time curvature and, 102–103, 104

organic molecules, 578, 657, 659

Orion constellation, 14, 16, 22, 292, 687

Orionids meteor shower, 620

Orion Nebula, 237
    disks surrounding stars in, 467–468
    star formation in, 241, 252–253, 254, 359
    viewed from Earth, 254–255

oscillating universe theory, **430**

outflow channels, **540**, 542

outgassing, **463**

ovoids, **600**

oxygen, 489, 492

ozone layer, **489**, 491–492

P

P (pressure) waves, **482–483**

Pagel, Bernard, 354

paradigms, **68**

parallax, **64**, 73
    measuring distance using, 189–190, 191
    spectroscopic, **199–200**
    stellar, **189–190**, 591

parsec (pc), **189**, 679

Parsons, William, 370–371

partial lunar eclipse, **44**

partial solar eclipse, **45**, 46

Paschen series, **147**

passing star hypothesis, **450–451**

path of totality, **46**

Pauli Exclusion Principle, 269

Payne-Gaposchkin, Cecilia, 151–152

Pegasus, 14, 15

penumbra, **42**
    lunar eclipse, 42
    solar eclipse, 46
    sunspot, 174

penumbral eclipse, **44**

Penzias, Arno, 424, 425

perigee, **48**

perihelion, **27**

periodic table of the elements, 686

period–luminosity relation, **275**, 279, 344

permitted orbits, **143**

Perseids meteor shower, 619

phases of the moon, 39, 40–41